BIOLOGICAL AND MEDICAL PHYSICS
BIOMEDICAL ENGINEERING

BIOLOGICAL AND MEDICAL PHYSICS BIOMEDICAL ENGINEERING

The fields of biological and medical physics and biomedical engineering are broad, multidisciplinary and dynamic. They lie at the crossroads of frontier research in physics, biology, chemistry, and medicine. The Biological & Medical Physics/Biomedical Engineering Series is intended to be comprehensive, covering a broad range of topics important to the study of the physical, chemical and biological sciences. Its goal is to provide scientists and engineers with textbooks, monographs, and reference works to address the growing need for information.

Continued After Index

Conversion Factors

Some of the more useful conversion factors for converting from older units to SI units are listed in Table Q.1. They are taken from *Standard for Metric Practice*, ASTM E 380-76, Copyright 1976 by the American Society for Testing and Materials, Philadelphia.

TABLE Q.1. *Useful conversion factors.*

To convert from	To	Multiply by
angstrom	meter	$1.000\ 000\times10^{-10}$
atmosphere (standard)	pascal	$1.013\ 250\times10^{5}$
bar	pascal	$1.000\ 000\times10^{5}$
barn	meter^2	$1.000\ 000\times10^{-28}$
calorie (thermochemical)	joule	4.184 000
centimeter of mercury (0 °C)	pascal	$1.333\ 22\times10^{3}$
centimeter of water (4 °C)	pascal	$9.806\ 38\times10^{1}$
centipoise	pascal second	$1.000\ 000\times10^{-3}$
curie	becquerel	$3.700\ 000\times10^{10}$
dyne	newton	$1.000\ 000\times10^{-5}$
electron volt	joule	$1.602\ 18\times10^{-19}$
erg	joule	$1.000\ 000\times10^{-7}$
fermi (femtometer)	meter	$1.000\ 000\times10^{-15}$
gauss	tesla	$1.000\ 000\times10^{-4}$
liter	meter^3	$1.000\ 000\times10^{-3}$
mho	siemens	1.000 000
millimeter of mercury	pascal	$1.333\ 22\times10^{2}$
poise	pascal second	$1.000\ 000\times10^{-1}$
roentgen	coulomb per kilogram	2.58×10^{-4}
torr	pascal	$1.333\ 22\times10^{2}$

Russell K. Hobbie

Intermediate Physics for Medicine and Biology

Third Edition

Russell K. Hobbie
School of Physics and Astronomy
University of Minnesota
Minneapolis, MN 55455
USA

The cover illustration shows tracings of the ^{31}P chemical shift in different parts of the brain, superimposed on an MRI image of the brain (see Fig. 17.38). Courtesy of Professor Xiaoping Hu, Magnetic Resonance Imaging Center, University of Minnesota. (From X. Hu, W. Chen, M. Patel, and K. Ugurbil. Chemical Shift Imaging: An Introduction to its Theory and Practice. Chapter 65.4 in J.D. Bronzino, Ed. The Biomedical Engineering Handbook. Copyright—CRC Press, 1995.)

Library of Congress Cataloging-in-Publication Data
Hobbie, Russell K.
Intermediate physics for medicine and biology / Russell K. Hobbie.
– 3rd ed.
p. cm.
Includes bibliographical references and index.
ISBN 1-56396-458-9 (hardcover : alk. paper)
1. Medical physics. 2. Biophysics. 3. Physics. I. Title.
R895.H6 1997
610'.1'53–dc21 97-34134

ISBN-10: 1-56396-458-9
ISBN-13: 978-156396-458-9

Printed on acid-free paper

Printed in the United States of America. (MVY)

9 8 7 6 5 4 3

springeronline.com

Series Preface

The fields of biological and medical physics and biomedical engineering are broad, multidisciplinary, and dynamic. They lie at the crossroads of frontier research in physics, biology, chemistry, and medicine. The Biological and Medical Physics/Biomedical Engineering series is intended to be comprehensive, covering a broad range of topics important to the study of the physical, chemical, and biological sciences. Its goal is to provide scientists and engineers with textbooks, monographs, and reference works to address the growing need for information.

Books in the series emphasize established and emergent areas of science including molecular, membrane, and mathematical biophysics; photosynthetic energy harvesting and conversion; information processing; physical principles of genetics; sensory communications; and automata networks, neural networks, and cellular automata. Equally important will be coverage of applied aspects of biological and medical physics and biomedical engineering, such as molecular electronic components and devices, biosensors, medicine, imaging, physical principles of renewable energy production, advanced prostheses, and environmental control and engineering.

Oak Ridge, Tennessee

ELIAS GREENBAUM
Series Editor-in-Chief

Preface

Between 1971 and 1973 I audited all the courses medical students take in their first two years at the University of Minnesota. I was amazed at the amount of physics I found in these courses and how little of it is discussed in the general physics course.

I found a great discrepancy between the physics in some papers in the biological research literature and what I knew to be the level of understanding of most biology majors or pre–med students who have taken a year of physics. It was clear that an intermediate level physics course would help these students. It would provide the physics they need and would relate it directly to the biological problems where it is useful.

This book is the result of my having taught such a course since 1973. It is intended to serve as a text for an intermediate course taught in a physics department and taken by a variety of majors. Since its primary content is *physics*, I hope that physics faculty who might shy away from teaching a conventional biophysics course will consider teaching it. I also hope that research workers in biology and medicine will find it a useful reference to brush up on the physics they need or to find a few pointers to the current literature in a number of areas of biophysics. (The bibliography in each chapter is by no means exhaustive; however, the references should lead you quickly into a field.) The course offered at the University of Minnesota is taken by undergraduates in a number of majors who want to see more physics with biological applications and by graduate students in physics, biophysical sciences, biomedical engineering, physiology, and cell biology.

Because the book is intended primarily for students who have taken only one year of physics, I have tried to adhere to the following principles in writing it:

1. Calculus is used without apology. When an important idea in calculus is used for the first time, it is reviewed in detail. These reviews are found in the appendices.
2. The reader is assumed to have taken physics and know the basic vocabulary. However, I have tried to present a logical development from first principles, but shorter than what would be found in an introductory course. An exception is found in Chaps. 13–17, where some results from quantum mechanics are used without deriving them from first principles. (My students have often expressed surprise at this change of pace.)
3. I have not intentionally left out steps in most derivations. Some readers may feel that the pace could be faster, particularly after a few chapters. My students have objected strongly when I have suggested stepping up the pace in class.
4. Each subject is approached in as simple a fashion as possible. I feel that sophisticated mathematics, such as vector analysis or complex exponential notation, often hides physical reality from the student. I have seen electrical engineering students who could not tell me what is *happening* in an *RC* circuit but could solve the equations with Laplace transforms.

The third edition has added a significant amount of new material and dropped some topics. The most notable additions include the circulatory system and the Poiseuille–Bernoulli equation; the logistic equation; countercurrent transport; the bidomain model for anisotropy in myocardial conductivity; electric and magnetic stimulation of nerve and muscle; noise and its frequency spectrum; sensory transducers; nonlinear systems, difference equations, and chaotic behavior; radiometry and photometry; image formation and spatial frequencies in an image; refraction in the eye; and the diffusion approximation to photon transport in turbid media. New topics in radiological physics include charged-particle equilibrium, buildup, radiobiology, Auger electrons, brachytherapy, and radon. There is a brief discussion of the physical constraints on possible effects of weak electric and magnetic fields.

Chapter 1 reviews mechanics. Translational and rotational equilibrium are introduced, with the forces in the hip joint as a clinical example. Hydrostatics, incompressible viscous flow, and the Poiseuille–Bernoulli equation are discussed, with an example from the circulatory system.

Chapter 2 is essential to nearly every other chapter in the book. It discusses exponential growth and decay and gives examples from pharmacology and physiology (including clearance). The logistic equation has been added in this edition. Students are also shown how to use semilog and log–log graph paper. (I have found that most students have been exposed to these techniques but are still unsure about them.) A technique for determining power-law coefficients using a spreadsheet has been added. The chapter concludes with a brief discussion of scaling.

Chapter 3 is a condensed treatment of statistical physics: average quantities, probability, thermal equilibrium, entropy, and the first and second laws of thermodynamics. The Boltzmann factor and the principle of equipartition of energy are introduced here and used in later chapters. The chemical potential is discussed because its use is so central to the transport of particles. The Gibbs free energy is also described. You can either plow through this chapter if you are a slave to thoroughness, touch on the highlights, or use it as a reference as the topics are needed in later chapters.

Chapter 4 treats diffusion and transport of solute in an infinite medium. Fick's first and second laws of diffusion are developed. An important model is the spherical cell with pores providing transport through the cell membrane. It is shown that only a small number of pores are required to keep up with the rate of diffusion toward or away from the cell, so there is plenty of room on the cell surface for many different kinds of pores and receptor sites.

Chapter 5 discusses transport of fluid and neutral solutes through a membrane. This might be a cell membrane, the basement membrane in the glomerulus of the kidney, or a capillary wall. The phenomenological transport equations are introduced as the first (linear) approximation to describe these flows. A hydrodynamic model is developed for right-cylindrical pores. The third edition adds a section on countercurrent transport and deletes two sections found in earlier editions.

Chapter 6 describes the electrochemical changes that cause an impulse to travel along a nerve axon or along a muscle fiber before contraction. The electric field, electric potential, and circuits are reviewed, followed by a discussion of electrotonus (when the membrane obeys Ohm's law) and the Hodgkin–Huxley model (when the membrane is nonlinear).

Chapter 7 shows how an electric potential is generated in the medium surrounding a nerve or muscle cell. This leads to a discussion of the electrocardiogram. The model is then refined to account for the anisotropy of the electrical conductivity of the heart. Electrical stimulation, important for pacemakers and measurements of nerve conduction velocity, is described.

Chapter 8 shows how the currents in a conducting nerve or muscle cell generate a magnetic field, leading to the magnetocardiogram and the magnetoencephalogram. Some bacteria (and probably some higher organisms) contain magnetic particles used for determining spatial orientation in the earth's magnetic field. The mechanism by which these bacteria are oriented is described. The detection of weak magnetic fields and the use of changing magnetic fields to stimulate nerve or muscle cells are described.

Chapter 9 covers a number of topics at the cellular and membrane level. It begins with Donnan equilibrium, where the presence of an impermeant ion on only one side of a membrane leads to the buildup of a potential difference across the membrane, and the Gouy–Chapman model for how ions redistribute near the membrane to generate this potential difference. The Debye–Hückel model is a simple description of the neutralization of ions by surrounding counterions. The Nernst–Planck equation provides the basic model for describing combined diffusion and drift in an applied electric field. It also forms the basis for the Goldman–Hodgkin–Katz model for zero total current in a membrane with a constant electric field. Gated membrane channels and noise in membranes are then discussed. After showing how a properly adapted shark can detect very weak electric fields with a reasonable signal-to-noise ratio, the chapter concludes with a discussion of the basic physical principles that must be kept in mind when assessing the possibility of biological effects of weak electric and magnetic fields.

Chapter 10 discusses feedback systems in the body. It starts with the regulation of breathing rate to stabilize the carbon dioxide level in the blood, moves to linear feedback systems with one and two time constants, and then describes nonlinear models. It shows how nonlinear systems described by simple difference equations can exhibit chaotic behavior, and how chaotic behavior can arise in continuous systems as well. Examples include oscillating white-blood-cell counts, waves in excitable media, and period doubling and chaos in the heart.

Chapter 11 shows how to use the method of least squares to fit curves to data. The least-squares technique is then used to introduce discrete and continuous Fourier series, power spectra, correlation functions, and the Fourier transform. This leads to the frequency response of a linear system and allows describing noise by its frequency spectrum. There is a very brief discussion of testing data for chaotic behavior.

Chapter 12 describes images from the standpoint of linear system analysis and convolution. This leads to the use of Fourier analysis to describe the spatial frequencies in an image and the reconstruction of an image from its projections. Both Fourier techniques and filtered back-projection are discussed.

Chapter 13 discusses the visible, infrared, and ultraviolet regions of the electromagnetic spectrum. The scattering and absorption cross sections are introduced and will be used here and in the next three chapters. Thermal radiation and the emission of infrared radiation by the body are described. A new section in this edition discusses photon transport in turbid media, including the diffusion approximation, time-resolved absorption, and the integrating-sphere photometer. Thermal radiation from the sun includes ultraviolet light, which injures skin. Protection from ultraviolet light is both possible and prudent. A new section describes how to spectacle lenses are used to correct errors of refraction in the eye. The chapter closes with a description of the quantum limitations to dark-adapted vision.

Chapter 14, like Chap. 3, has few biological examples but sets the stage for later work. It describes how photons and ionizing charged particles such as electrons lose energy in traversing matter. These interaction mechanisms, both in the body and in the detector, are fundamental to the formation of a radiographic image and to the use of radiation to treat cancer. The discussion of charged-particle stopping power has been extensively revised in this edition.

Chapter 15 describes the use of x rays for both medical

diagnosis and for treatment. It moves from production to detection, to the diagnostic radiograph. A new section discusses image quality, including noise. The therapy section has expanded treatments of radiobiology and dose measurement. The chapter closes with a new section that provides a very brief overview of radiation risk and suggests that the linear-no-threshold model for assessing radiation risk—standard for half a century—may overestimate risk at low doses.

Chapter 16 introduces the reader to nuclear physics and nuclear medicine. The different kinds of radioactive decay are described. Dose calculations are made using the fractional absorbed dose method recommended by the Medical Internal Radiation Dose committee of the Society of Nuclear Medicine. Diagnostic imaging is described, followed by brachytherapy and internal radiotherapy. A brief section on the nuclear physics of radon is new in this edition.

Chapter 17 develops the physics of magnetic resonance imaging. The section on imaging has been updated and expanded in this edition.

Biophysics is a very diverse field, and many topics are left out. Molecular biology is an important field that occupies much of the *Biophysical Journal*. However, it is traditionally taught by biochemists, and it is outside my sphere of knowledge. I have less excuse for not talking more about hearing and vision, and for completely ignoring ultrasound as a diagnostic modality in medical physics. I can only plead exhaustion, and point out that the book is long enough as it is.

I have taught this material in three separate one-quarter courses that can be taken independently. Each course meets five days a week, with four days devoted to lecture and one to problems. Early in the quarter one day each week is devoted to reviews of the mathematics found in some of the appendices. In spite of meeting five days a week, I cannot cover all of the material in every chapter in one year. The first quarter of my course covers Chaps. 1–5 and Chap. 10; the second quarter covers Chaps. 6–9 and 11. The third quarter quickly reviews Chap. 11 and then covers Chaps. 12–17.

A two-semester course meeting five days a week could cover Chaps. 1–8 the first semester and Chaps. 9–17 the second. A three-semester course meeting three days a week should use the same schedule as the 5-credit three quarter course described above.

Many of the end-of-chapter problems extend the material in the text. A solutions manual is available to those teaching the course. Instructors can use it as a reference or provide solutions to students at appropriate times. The solutions manual makes it much easier for a student to take the course by independent study. Instructors should contact the author or the publisher for information.

I am very grateful to the University of Minnesota for support in the form of an administrative transitional leave during the 1995–1996 academic year. It was exciting to learn about the developments in the field during my twelve years as an administrator. I hope that the reader will find the effort useful.

There are many people who made this book possible. If the late W. Albert Sullivan, Jr., Assistant Dean of the University of Minnesota Medical School had not suggested that I attend class, this book never would have been written. Richard L. Reece introduced me to laboratory medicine and started me on this odyssey. I am grateful for help and discussions from the following people in the preparation of the third edition: Amy Alving, Gordon Beavers, John Cameron, Richard Geise, Bruce Gerbi, Clayton Giese, Leon Glass, Matthew A. Hall, Bruce Hasselquist, Sarah Hobbie, Xiaoping Hu, Daniel Kaplan, Faiz Khan, Tuong Huu Le, David Levitt, John Moulder, E. Russell Ritenour, Bradley Roth, Julia Stephen, Gerhard Stroink, Stephen Strother, John P. Wikswo, Jr., and James Ziegler. James Roberts converted the computer programs from Pascal to C. Shuling Li, a student in my course during fall 1996, found many inconsistencies and confusing statements in the first few chapters. Ernest Madsen and another reviewer who chose to remain anonymous made many valuable suggestions. I am also grateful to Jennifer VanCura of AIP Press for her design and attention to detail in the production of the book. My son, Erik Hobbie, read the entire manuscript and offered suggestions from his perspective as a polymer physicist. We had many enjoyable discussions, often while fishing (sometimes successfully). In spite of all this help, errors remain for three reasons. First, I followed most, but not all of the many suggestions I received. Second, I had to study a lot of material from many diverse areas of biophysics and may have missed something important. Finally, in spite of extensive proofreading, some typographical errors surely remain. I would appreciate receiving suggestions and corrections.

Every list of acknowledgments seems to close with thanks to a long-suffering family. I never knew what these words really mean, nor how deep the indebtedness, until I wrote this book.

Russell K. Hobbie
University of Minnesota

For Lynn, Erik, Sarah, and Ann

Contents

CHAPTER 1

Mechanics

This chapter introduces some concepts from mechanics that are of biological or medical interest. We begin with a discussion of the forces on an object that is in equilibrium and calculate the forces experienced by various bones and muscles. In Sec. 1.8 we introduce the concept of mechanical work, which will recur throughout the book. The next two sections describe how materials deform when certain forces act on them. Sections 1.11 through 1.13 discuss the forces in stationary and moving fluids. These concepts are then applied to laminar viscous flow in a pipe, which is a model for the flow of blood and the flow of fluid through pores in cell membranes. The chapter ends with a discussion of work applied to compressible fluids and the flow of blood.

1.1. FORCES AND TRANSLATIONAL EQUILIBRIUM

There are several ways that we could introduce the idea of force, depending on the problem at hand and our philosophical bent. For our present purposes it will suffice to say that a force is a *push or a pull*, that forces have both a magnitude and a direction, and that they give rise to accelerations through Newton's second law, $\mathbf{F}=m\mathbf{a}$. Experiments show that forces add like *displacements*, so they can be represented by *vectors*. (Some of the properties of vectors are reviewed in Appendix B; others are introduced as needed.) Vectors will be denoted by boldfaced characters.

One finds experimentally that an object is in *translational equilibrium* if the vector sum of all the forces acting on the body is zero. *Equilibrium* means that the object either remains at rest or continues to move with a constant velocity. That is, it has no acceleration. *Translational* means that only changes of position are being considered; changes of orientation of the object with respect to the axes are ignored.

We must consider all the forces that act *on the object*. If the object is a person standing on both feet, the forces are the upward force of the floor on each foot and the downward force of gravity on the person (more accurately, the vector sum of the gravitational force on every cell in the person). We do *not* consider the downward force that the person's feet exert *on the floor*. It is also possible to replace the sum of the gravitational force on each cell of the body with a single downward gravitational force acting at one point, the *center of gravity* of the body.

The forces that add to zero to give translational equilibrium need not all act at one point on the object. If the object is a person's leg and the leg is at rest, there are three forces acting on the leg by other objects (Fig. 1.1). Force $\mathbf{F}_1$ is the push of the floor up on the bottom of the foot. The various pushes and pulls of the rest of the body on the leg through the hip joint and surrounding muscles have been added together to give $\mathbf{F}_2$. The gravitational pull of the earth downward on the leg is $\mathbf{F}_3$. Force $\mathbf{F}_1$ acts on the bottom of the leg, $\mathbf{F}_2$ acts on the top, and $\mathbf{F}_3$ acts somewhere in between. If the leg is in equilibrium the sum of these forces is zero, as shown in Fig. 1.1(b). Although the points of application of the forces can be ignored in considering translational equilibrium, they are important in determining whether or not the object is in rotational equilibrium. This is discussed shortly.

The Greek letter Σ (capital sigma) is usually used to mean a sum of things. With this notation, the condition for translational equilibrium can be written

$$\sum_i \mathbf{F}_i=\mathbf{0}. \tag{1.1}$$

The subscript i is used to label the different forces acting on the body. A notation this compact has a lot hidden in it. This is a vector equation, standing for three equations:

$$\sum_i F_{ix}=0,$$
$$\sum_i F_{iy}=0, \tag{1.2}$$
$$\sum_i F_{iz}=0.$$

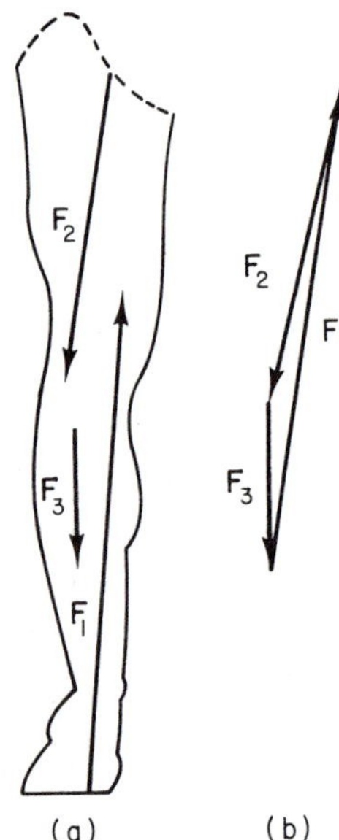

FIGURE 1.1. Forces on the leg in equilibrium. Each force is exerted by some other object. (a) The points of application are widely separated. (b) The sum of the forces is zero.

Often the subscript i is omitted and the equations are written as $\Sigma F_x=0$, $\Sigma F_y=0$, $\Sigma F_z=0$. In this notation, a component is positive if it points along the positive axis and negative if it points the other way.

Sometimes, as in the next example, we draw forces in particular directions and assume that these directions are positive. If the subsequent algebra happens to give a solution that is negative, the force points in the opposite direction.

As an example, consider the person standing on both feet as in Fig. 1.2. The earth pulls down at some point with force $\mathbf{W}$. The floor pushes up on the right foot with force $\mathbf{F}_1$ and on the left foot with force $\mathbf{F}_2$. To determine what the condition for translational equilibrium tells us about the forces, draw the force diagram or *free-body diagram* of Fig. 1.2(b). This diagram is an abstraction that ignores the points at which the forces are applied to the body. We can get away with this abstraction because we are considering only translation. When we consider rotational equilibrium, we will have to redraw the diagram showing the points at which the various forces act on the person. If all the forces are vertical, then there is only one component of each force to worry about, and the equilibrium condition gives $F_1+F_2-W=0$, or $F_1+F_2=W$. The total force of the floor pushing up on both feet is equal to the pull of the earth down.

If there is a sideways force on each foot, translational equilibrium provides two conditions:

$$F_{1x}+F_{2x}=0,$$

$$F_{1y}+F_{2y}-W=0.$$

This is all that can be learned from the condition for translational equilibrium. If the person stands on one foot, then $F_1=0$ and $F_2=W$. If the person stands with equal force on each foot, then $F_1=F_2=W/2$.

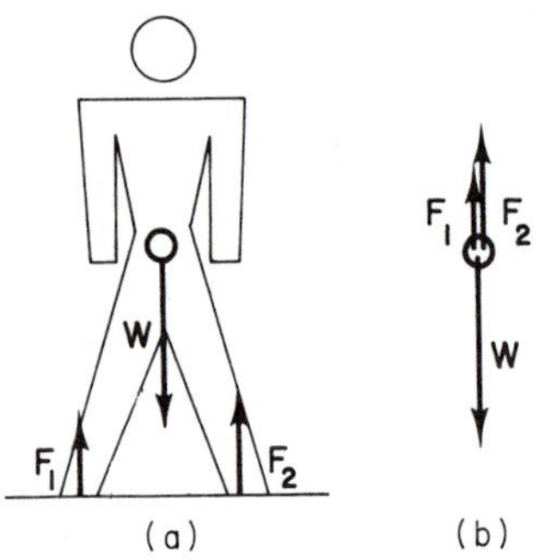

FIGURE 1.2. A person standing. (a) The forces on the person. (b) A free-body or force diagram.

1.2. ROTATIONAL EQUILIBRIUM

If the object is in rotational equilibrium, then another condition must be placed upon the forces. Rotational equilibrium means that the object either does not rotate or continues to rotate at a constant rate (with a constant number of rotations per second). Consider the object of Fig. 1.3, which is a rigid rod pivoted at point X so that it can rotate in the plane of the paper. Forces $\mathbf{F}_1$ and $\mathbf{F}_2$ are applied to the rod in the plane of the paper at distances r_1 and r_2 from the pivot and perpendicular to the rod. The pivot exerts the force $\mathbf{F}_3$ necessary to maintain translational equilibrium. If both $\mathbf{F}_1$ and $\mathbf{F}_2$ are perpendicular to the rod, they are parallel. They must also be parallel to $\mathbf{F}_3$, and translational equilibrium requires that $F_3=F_1+F_2$.

Experiment shows that there is no rotation of the rod if $F_1r_1=F_2r_2$. The condition for rotational equilibrium can be stated in a form analogous to that for translational equilibrium if we define the *torque*, τ, to be

$$\tau_i=r_iF_i. \tag{1.3}$$

With this definition goes an algebraic sign convention: the torque is positive if it tends to produce a counterclockwise rotation.

Experiment shows that the body is in rotational equilibrium if the algebraic sum of all the torques is zero:

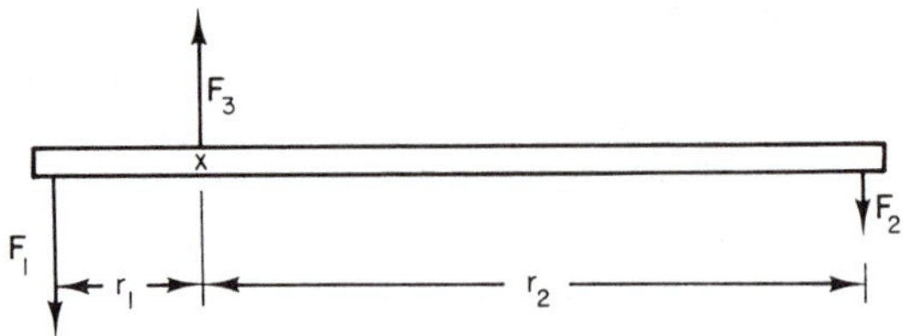

FIGURE 1.3. A rigid rod free to rotate about a pivot at point X.

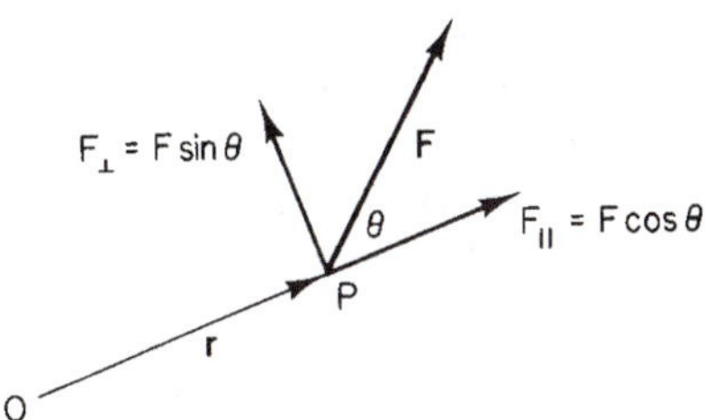

FIGURE 1.4. A force **F** is applied to an object at point *P*. The object can rotate about point *O*. Vectors **r** and **F** determine the plane of the paper.

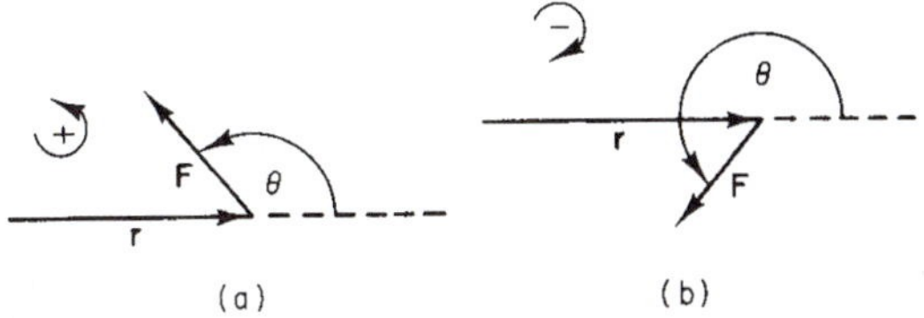

FIGURE 1.5. (a) When θ is between 0° and 180°, both sin θ and the torque are positive. (b) When θ is between 180° and 360°, both sin θ and the torque are negative.

$$\sum_i \tau_i = \sum_i r_i F_i = 0. \tag{1.4}$$

Note that F_3 contributes nothing to the torque because r_3 is zero.

The torque is defined about a certain point, X. It depends on the distance from the point of application of each force to X.[1] (We will see shortly that in equilibrium, the torque can be calculated about any point X.)

The torque can also be calculated if the force is not at right angles to the rod. Imagine an object free to rotate about point O in Fig. 1.4. Force **F** lies in the plane of the paper but is applied in some arbitrary direction at point P. The vectors **r** and **F** determine the plane of the paper if they are not parallel. Force **F** can be resolved into two components: one parallel to **r**, $F_{\parallel} = F \cos \theta$, and the other perpendicular to **r**, $F_{\perp} = F \sin \theta$. The component parallel to **r** will not cause any rotation about point O. (Pull on an open door parallel to the plane of the door; there is no rotation.) The torque is therefore

$$\tau = rF_{\perp} = rF \sin \theta. \tag{1.5}$$

The perpendicular distance from the line along which the force acts to point X is $r \sin \theta$. It is often called the *moment arm*, and the torque is the magnitude of the force multiplied by the moment arm.

The angle θ is the angle of rotation from the direction of **r** to the direction of **F**. It is called positive if the rotation is counterclockwise. For the angle shown in Fig. 1.4 sin θ has a positive value, and the torque is positive. Figure 1.5(a) shows an angle between 90° and 180° for which the torque and sin θ are still positive. Figure 1.5(b) shows an angle between 180° and 360°, for which both the torque and sin θ are negative. In all cases, Eq. (1.5) gives the correct sign for the torque.

To summarize: the torque due to force **F** applied to a body at point P must be calculated about some point O. If **r** is the vector from O to P, the magnitude of the torque is equal to the magnitude of **r** times the magnitude of **F**, times the sine of the angle between **r** and **F**. The angle is measured counterclockwise from **r** to **F**.

[1]The discussion associated with Fig. 1.3 suggests that the torque is taken about an axis, rather than a point. In a three-dimensional problem the torque is taken about a point in space.

1.3. VECTOR PRODUCT

Torque may be thought of as a vector, $\boldsymbol{\tau}$. Its magnitude is $Fr \sin \theta$. The only direction uniquely defined by vectors **r** and **F** is perpendicular to the plane in which they lie. This is also the direction of an axis about which the torque would cause a rotation. However, there is ambiguity about which direction along this line to assign to the torque. This ambiguity can be settled only by convention. The convention is to say that a positive torque points in the direction of the thumb of the right hand when the fingers curl in the direction of positive rotation from **r** to **F**. (This arbitrariness in assigning the sense of $\boldsymbol{\tau}$ means that it does not have quite all the properties that vectors usually have. It is called an axial vector or a pseudovector. It will not be necessary in this book to worry about the difference between a real vector and an axial vector. When **r** and **F** point in the same direction, so that no plane is defined, the magnitude of the torque is zero.)

The product of two vectors according to the foregoing rules is called the *cross product* or *vector product* of the two vectors. One can use a shorthand notation

$$\boldsymbol{\tau} = \mathbf{r} \times \mathbf{F}. \tag{1.6}$$

There is another way to write the cross product. If both **r** and **F** are resolved into components, as shown in Fig. 1.6, then the cross product can be calculated by applying the rules above to the components. Since $\mathbf{F}_y$ is perpendicular to $\mathbf{r}_x$ and parallel to $\mathbf{r}_y$, its only contribution is a counterclockwise torque $r_x F_y$. The only contribution from F_x is a clockwise torque, $-r_y F_x$. The magnitude of the cross product is therefore

$$\tau = r_x F_y - r_y F_x. \tag{1.7}$$

Note that this is the (signed) sum of each component of the force multiplied by its moment arm.

The equivalence of this result to Eq. (1.5) can be verified by writing Eq. (1.6) as

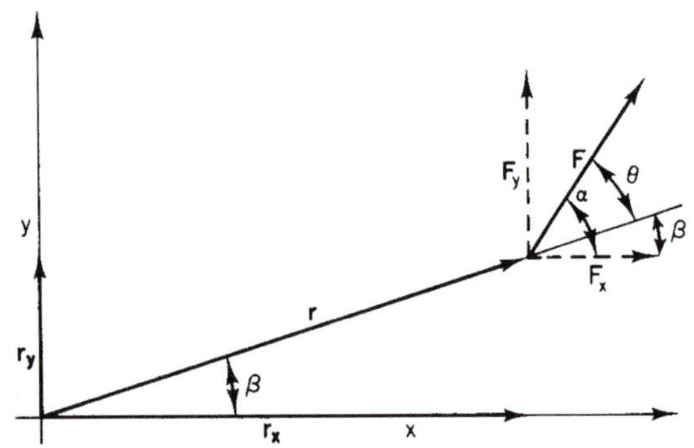

FIGURE 1.6. The cross product **r**×**F** is calculated by resolving **r** and **F** into components.

$$\tau=(r\cos\beta)(F\sin\alpha)-(r\sin\beta)(F\cos\alpha),$$

$$\tau=rF(\sin\alpha\cos\beta-\cos\alpha\sin\beta).$$

There is a trigonometric identity that

$$\sin(\alpha-\beta)=\sin\alpha\cos\beta-\cos\alpha\sin\beta.$$

Since $\theta=\alpha-\beta$ (from Fig. 1.6), this is equivalent to $\tau=rF\sin\theta$.

When vectors **r** and **F** lie in the xy plane, $\boldsymbol{\tau}$ points along the z axis. If **r** and **F** point in arbitrary directions, Eq. (1.7) gives the z component of $\boldsymbol{\tau}$. One can apply the same reasoning for other components and show that

$$\tau_x=r_yF_z-r_zF_y,$$

$$\tau_y=r_zF_x-r_xF_z, \tag{1.8}$$

$$\tau_z=r_xF_y-r_yF_x.$$

If you are familiar with the rules for evaluating determinants, you will see that this is equivalent to the notation

$$\tau=\begin{vmatrix}\hat{\mathbf{x}} & \hat{\mathbf{y}} & \hat{\mathbf{z}}\\ r_x & r_y & r_z\\ F_x & F_y & F_z\end{vmatrix}. \tag{1.9}$$

1.4. WHEN THE BODY IS IN TRANSLATIONAL EQUILIBRIUM, THE TORQUE CAN BE EVALUATED ABOUT ANY POINT

In the previous discussion, it was assumed that there was some point in space about which the object could rotate and about which the torques were to be evaluated. In fact, as long as the object is in translational equilibrium, the torque can be evaluated around any point. This theorem is easily proved; the result often allows calculations to be simplified, because taking torques about certain points can cause some forces not to contribute to the torque equation. You can skip this section (and use the result) if you are not interested in a proof of the theorem.

Figure 1.7 shows an object in equilibrium with forces $\mathbf{F}_1$, $\mathbf{F}_2$, and $\mathbf{F}_3$ applied at points displaced $\mathbf{r}_1$, $\mathbf{r}_2$, and $\mathbf{r}_3$ from some point X. The conditions for equilibrium are

$$F_{1x}+F_{2x}+F_{3x}=0,$$

$$F_{1y}+F_{2y}+F_{3y}=0,$$

and for the torque about point X,

$$r_{1x}F_{1y}-r_{1y}F_{1x}+r_{2x}F_{2y}-r_{2y}F_{2x}+r_{3x}F_{3y}-r_{3y}F_{3x}=0.$$

If torques are taken about point O instead of point X, the sum is

$$\sum\tau'=r'_{1x}F_{1y}-r'_{1y}F_{1x}+r'_{2x}F_{2y}-r'_{2y}F_{2x}+r'_{3x}F_{3y}-r'_{3y}F_{3x}.$$

Since $\mathbf{r}'_i=\mathbf{r}_i+\mathbf{R}$, this can be rewritten as

$$\sum\tau'=r_{1x}F_{1y}+R_xF_{1y}-r_{1y}F_{1x}-R_yF_{1x}+r_{2x}F_{2y}+R_xF_{2y}-r_{2y}F_{2x}-R_yF_{2x}+r_{3x}F_{3y}+R_xF_{3y}-r_{3y}F_{3x}-R_yF_{3x}.$$

These terms can be regrouped to give

$$\sum\tau'=\sum\tau+R_x(F_{1y}+F_{2y}+F_{3y})-R_y(F_{1x}+F_{2x}+F_{3x}).$$

Since the object is in translational equilibrium, the sums of the forces in each of the parentheses is zero, and

$$\sum\tau'=\sum\tau.$$

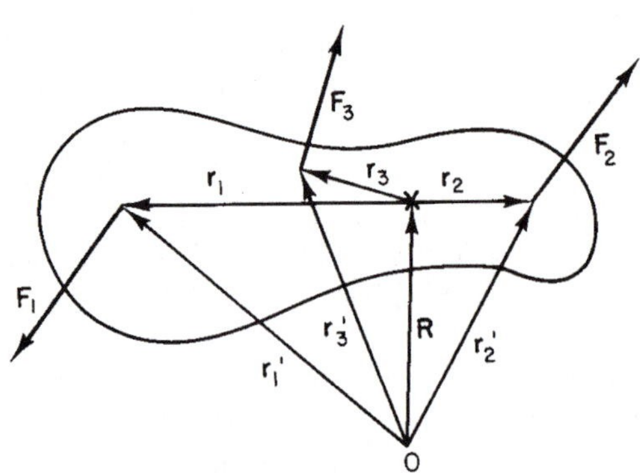

FIGURE 1.7. Three forces are applied to a body in equilibrium. Torques are taken about point X and then about point O.

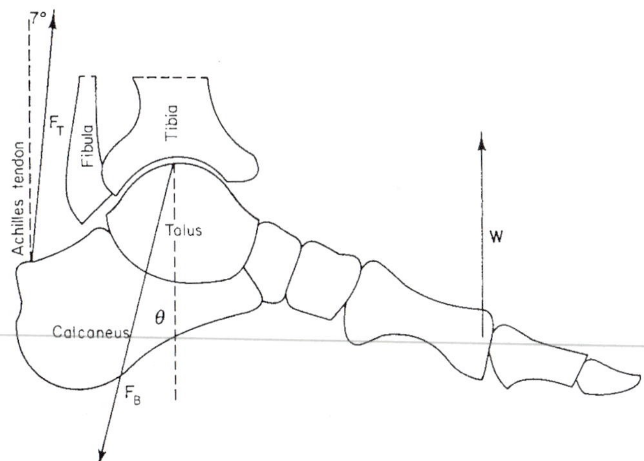

FIGURE 1.8. Simplified anatomy of the foot.

1.5. FORCE IN THE ACHILLES' TENDON

The equilibrium conditions can be used to understand many problems in clinical orthopedics. Two are discussed in this book: forces that sometimes cause the Achilles' tendon at the back of the heel to break, and forces in the hip joint.

The Achilles' tendon connects the calf muscles (the gastrocnemius and the soleus) to the calcaneus at the back of the heel (Fig. 1.8). To calculate the force exerted by this tendon on the calcaneus when a person is standing on the ball of one foot, assume that the entire foot can be regarded as a rigid body. This is our first example of creating a *model* of the actual situation. We try to simplify the real situation to make the calculation possible while keeping the features that are important to what is happening. In this model the internal forces within the foot are being ignored. Figure 1.9 shows the force exerted by the tendon on the foot ($\mathbf{F}_T$), the force of the leg bones (tibia and fibula) on the foot ($\mathbf{F}_B$), and the force of the floor upward, which is equal to the weight of the body (W). The weight of the foot is small compared to these forces and will be neglected. Measurements on a

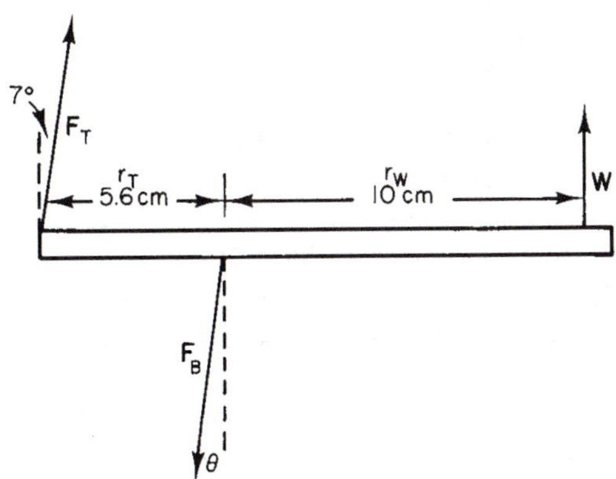

FIGURE 1.9. Forces on the foot, neglecting its own weight and internal forces.

few people suggest that the angle the Achilles' tendon makes with the vertical is about 7°.

Translational equilibrium requires that the following equations be satisfied:

$$F_T \cos(7°) + W - F_B \cos\theta = 0,$$
$$F_T \sin(7°) - F_B \sin\theta = 0. \tag{1.10}$$

To write the condition for rotational equilibrium, we need to know the lengths of the appropriate vectors $\mathbf{r}_T$ and $\mathbf{r}_W$, assuming that the torques are taken about the point where $\mathbf{F}_B$ is applied to the foot. The horizontal components of these vectors and the vertical components of the forces give the largest contributions to the torque, so in our simple model we will ignore the contributions of the horizontal components of any forces to the torque equation. This is not essential (if we are willing to make more detailed measurements), but it simplifies the equations and thereby makes the process clearer. The horizontal distances measured on the author are $r_T = 5.6$ cm and $r_W = 10$ cm, as shown in Fig. 1.9. Ignoring the horizontal components of the forces gives the torque equation, which is

$$10W - 5.6F_T \cos 7° = 0. \tag{1.11}$$

This equation can be solved to give the tension in the tendon:

$$F_T = \frac{10W}{5.6} = 1.8W. \tag{1.12}$$

This result can now be used in Eq. (1.10) to find $F_{By} = F_B \cos\theta$:

$$(1.8)(W)(0.993) + W = F_B \cos\theta,$$
$$2.8W = F_B \cos\theta. \tag{1.13}$$

From Eqs. (1.10) and (1.12), we get

$$(1.8)(W)(0.122) = F_B \sin\theta,$$
$$0.22W = F_B \sin\theta. \tag{1.14}$$

Equations (1.13) and (1.14) are squared and summed and the square root taken to give

$$2.8W = F_B,$$

while they can be divided to give

$$\tan\theta = \frac{0.22}{2.8} = 0.079,$$

$$\theta = 4.5°.$$

The tension in the Achilles' tendon is nearly twice the

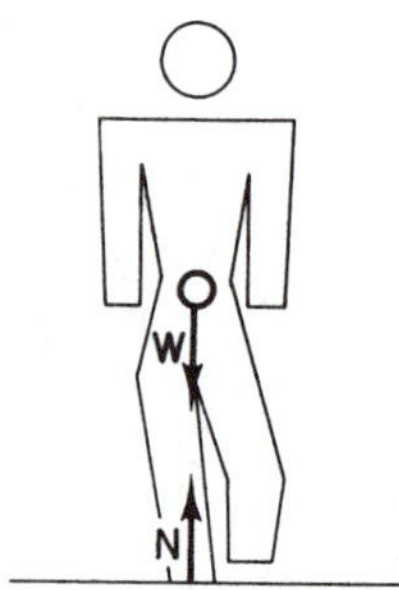

FIGURE 1.10. A person standing on one foot must place the foot under the center of gravity, which is on or near the midline.

person's weight, while the force exerted on the leg by the talus is nearly three times the body weight. One can understand why the tendon might rupture.

1.6. FORCES ON THE HIP

The forces in the hip joint can be several times the person's weight, and the use of a cane can be very effective in reducing them.

As a person walks, there are moments when only one foot is on the ground. There are then two forces acting on the body as a whole: the downward pull of the earth W and the upward push of the ground on the foot N. The pull of the earth may be regarded as acting at the center of gravity of the body [Halliday, Resnick, and Kane, (1992, Chap. 13); Williams and Lissner (1962), Chap. 5]. The center of gravity is located on the midline (if the limbs are placed symmetrically), usually in the lower abdomen [Williams and Lissner (1962), Chap. 5]. If torques are taken about the foot, then the center of gravity must be directly over the foot so that there will be no torque from either force. This situation is shown in Fig. 1.10. The condition for translational equilibrium requires that $N = W$. The anatomy of the pelvis, hips, and leg is shown schematically in Fig. 1.11. Fourteen muscles and several ligaments connect the pelvis to the femur. Extensive measurements of the forces exerted by the abductor[2] muscles in the hip have been made by Inman (1947). If the leg is considered an isolated system as in Fig. 1.11, the following forces act:

F: The net force of the abductor muscles, acting on the greater trochanter at an angle of about 70° with the horizontal. These muscles are primarily the gluteus medius and gluteus minimus, shown schematically in Fig. 1.11.

R: The force of the acetabulum (the socket of the pelvis) on the head of the femur.

N: The upward force of the floor on the bottom of the foot (in this case, equal to W).

[2]*To abduct* means to move away from the midline of the body.

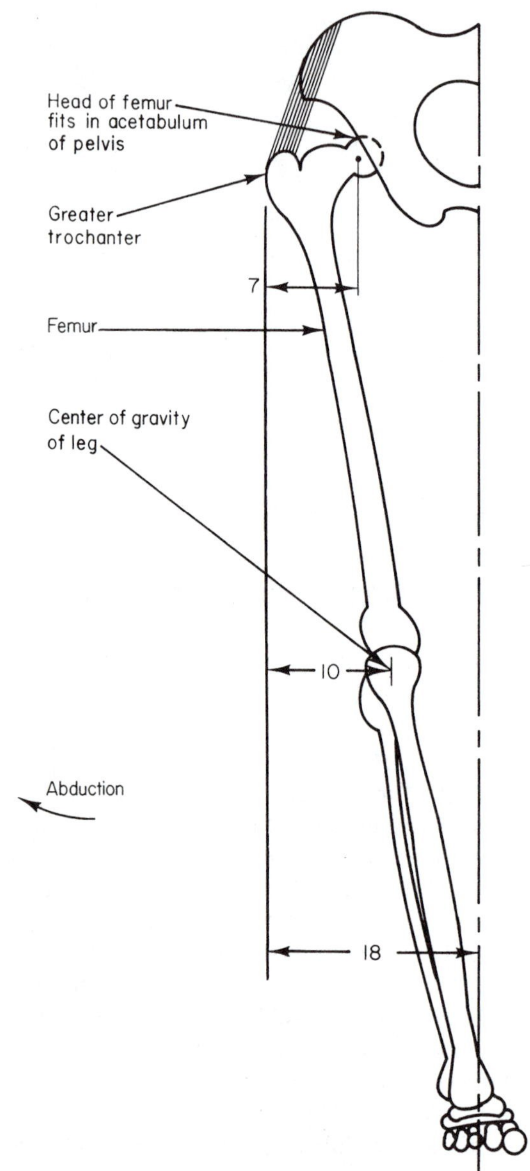

FIGURE 1.11. Pertinent features of the anatomy of the leg.

$\mathbf{W}_L$: The weight of the leg, acting vertically downward at the center of gravity of the leg. $W_L \approx W/7$ [Williams and Lissner (1962), Chap. 5].

Inman found that **F** acts at about a 70° angle to the horizontal. In a typical adult, the distance from the greater trochanter to the midline is about 18 cm, the horizontal distance from the greater trochanter to the center of gravity of the leg is about 10 cm, and the distance from the greater trochanter to the middle of the head of the femur is about 7 cm.

A free body diagram is shown in Fig. 1.12. The middle of the head of the femur will turn out to be very close to the intersection of the line along which **R** acts and a horizontal line drawn from the point where **F** acts. This means that if torques are taken about this intersection point (point O),

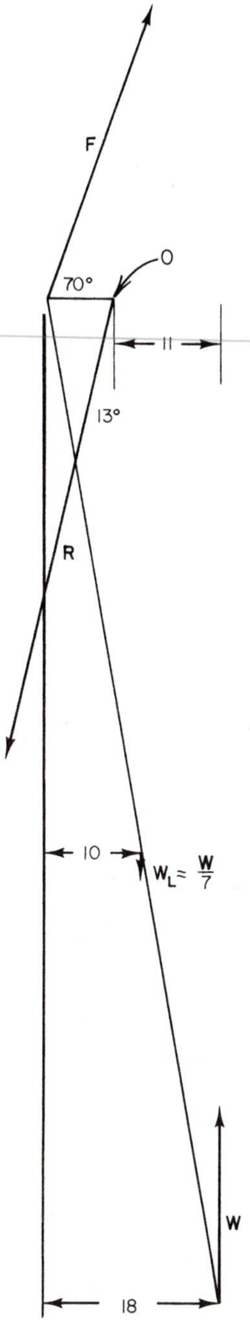

FIGURE 1.12. A free-body diagram for the forces acting on the leg. Torques are taken about a point that is the intersection of a line along which **R** acts and a horizontal line through the point at which **F** is applied. This point is 7 cm toward the midline (medially) from the greater trochanter.

there will be no contributions from **R** or from the horizontal component of **F**. The intersection is about 7 cm toward the midline from the point of application of **F**. Since $N=W$ and $W_L \approx W/7$, the equilibrium equations are

$$\sum F_y = F\ \sin(70°) - R_y - W/7 + W = 0, \qquad (1.15)$$

$$\sum F_x = F\ \cos(70°) - R_x = 0, \qquad (1.16)$$

$$\sum \tau = -F\ \sin(70°)(7) - (W/7)(10-7) + W(18-7) = 0.$$

The last of these equations can be written as

$$11W - \tfrac{3}{7}W - 6.6F = 0,$$

from which

$$F = 1.6W.$$

The magnitude of the force in the abductor muscles is about 1.6 times the body weight. (If the patient had not had to put the foot under the center of gravity of the body, the moment arm of the only positive torque, $11W$, could have been much less, and this would have been balanced by a smaller value of F. This can be done by having the patient use a cane on the *opposite* side, so that the foot need not be right under the center of gravity. Conversely, if the patient were carrying a suitcase in the opposite hand, the center of mass would be moved away from the midline, the foot would still have to be placed under the center of mass, and the moment arm, and hence F, would be even larger.)

Equations (1.15) and (1.16) can now be used to find R_x and R_y:

$$R_x = F\ \cos(70°) = (1.6)(W)(0.342) = 0.55W,$$

$$R_y = F\ \sin(70°) + \tfrac{6}{7}W = (1.6)(W)(0.94) + 0.86W = 2.36W.$$

The angle that **R** makes with the vertical is given by

$$\tan\phi = \frac{R_x}{R_y} = 0.23$$

$$\phi = 13°.$$

The magnitude of **R** is

$$R = (R_x^2 + R_y^2)^{1/2} = 2.4W.$$

One very interesting conclusion of Inman's study was that the force **R** always acts along the neck of the femur in such a direction that the femoral epiphysis has very little sideways force on it. The epiphysis is the growing portion of the bone (Fig. 1.13) and is not very well attached to the rest of the bone. If there were an appreciable sideways force, the epiphysis would slip sideways, and indeed it sometimes does (Fig. 1.14). This is a serious problem, since if the blood supply to the epiphysis is compromised, there will be no more bone growth.

Suppose that, for some reason, the gluteal muscles are severed. The patient can no longer apply force **F** to the greater trochanter; Eq. (1.16) shows that then $\mathbf{R}_x$ must be zero. This change in the direction of $\mathbf{R}_x$ causes a rotation of the epiphyseal plate and a gradual reshaping of the femur.

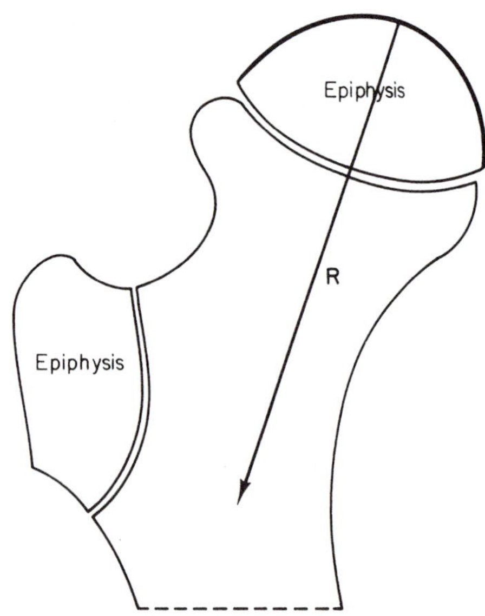

FIGURE 1.13. The femoral epiphysis and the direction of **R**.

1.7. THE USE OF A CANE

A cane is beneficial if used on the side opposite to the affected hip (Fig. 1.15). We ignore the fact that the arm holding the cane has moved, thereby shifting slightly the center of mass, and we assume that the force of the ground on the cane is vertical. If we assume that the tip of the cane is about 30 cm (12 in.) from the midline and supports one-sixth of the body weight, then we can apply the equilibrium conditions to learn that

$$N+\tfrac{1}{6}W-W=0,$$

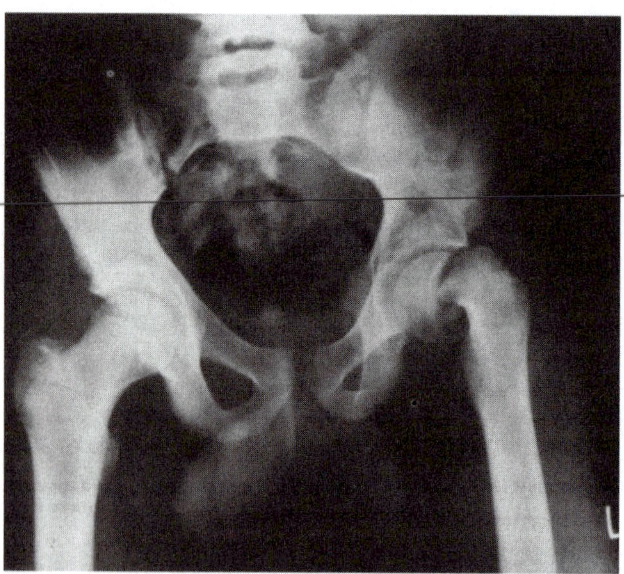

FIGURE 1.14. X ray of a slipped femoral epiphysis in an adolescent male. (X ray courtesy of the Department of Diagnostic Radiology, University of Minnesota.)

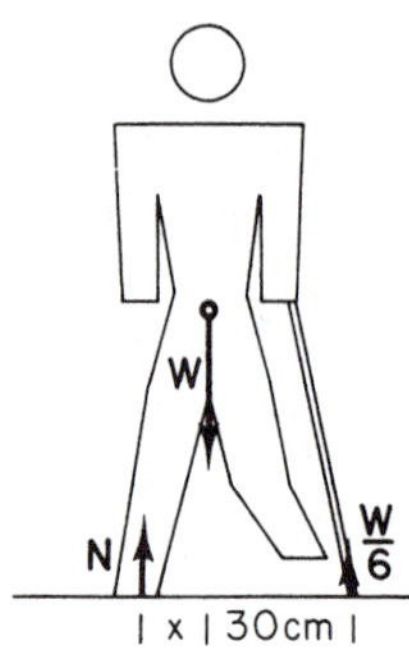

FIGURE 1.15. A person using a cane on the left side to favor the right hip.

$$N=\tfrac{5}{6}W.$$

Torques taken about the center of mass give

$$(30)\left(\frac{W}{6}\right)-x\left(\frac{5}{6}\right)W=0,$$

$$x=6 \text{ cm}.$$

(Figure 1.15 is not to scale.) Having the foot 6 cm from the midline reduces the force in the muscle and the joint. To find out how much, consider the force diagram in Fig. 1.16. The most difficult part of the problem is working out the various moment arms. Assume that the slight movement of the leg has not changed the point about which we take torques (the intersection of a horizontal line from the greater trochanter through the line of application of the force **R**). Again, **R** contributes no torque about this point. The horizontal distance of **F** from this point is still 7 cm. The force of the ground on the leg is now $5W/6$, and its moment arm is $18-6-7=5$ cm. The weight of the leg, $W/7$, acts at the center of mass of the leg, which is still $\frac{10}{18}$ of the distance from the greater trochanter to the foot. Its horizontal position is therefore $\frac{10}{18}$ of the horizontal distance from the greater trochanter to the foot:

$$\frac{(10)(12)}{18}=6.67 \text{ cm}.$$

The moment arm is $7-6.67$ cm$=0.33$ cm. The torque equation is

$$-F\ \sin(70°)(7)+\left(\frac{W}{7}\right)(0.33)+\left(\frac{5W}{6}\right)(5)=0.$$

It is solved by writing it as

$$-6.58F+0.047W+4.17W=0,$$

$$F=0.6W.$$

Even though the cane supports only one-sixth of the body weight, F has been reduced from $1.6W$ to $0.6W$ by the change in the moment arm.

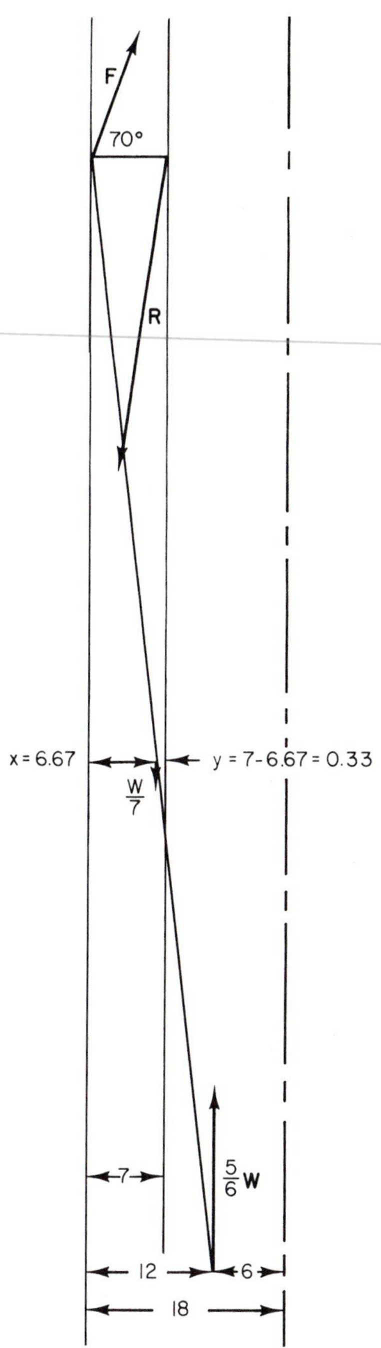

FIGURE 1.16. A force diagram for the leg when a cane is being used and the leg is 6 cm from the midline.

The force of the acetabulum on the head of the femur can be determined from the conditions for translational equilibrium:

$$F\cos(70°)-R_x=0,$$

$$R_x=0.2W,$$

$$F\sin(70°)-R_y-\frac{W}{7}+\tfrac{5}{6}W=0,$$

$$R_y=1.25W.$$

The resultant force **R** has magnitude $(R_x^2+R_y^2)^{1/2}=1.26W$. This compares to the value $2.4W$ without the cane. The force in the joint has been reduced by slightly more than the body weight. It is interesting to read what an orthopedic surgeon had to say about the use of a cane. The following is from the presidential address of W. P. Blount, M.D., to the Annual Meeting of the American Academy of Orthopedic Surgeons, January 30, 1956:

> The patient with a wise orthopedic surgeon walks with crutches for six months after a fracture of the neck of the femur. He uses a stick for a longer time—the wiser the doctor, the longer the time. If his medical adviser, his physical therapist, his friends, and his pride finally drive him to abandon the cane while he still needs one, he limps. He limps in a subconscious effort to reduce the strain on the weakened hip. If there is restricted motion, he cannot shift his body weight, but he hurries to remove the weight from the painful hip joint when his pride makes him reduce the limp to a minimum. The excessive force pressing on the aging hip takes its toll in producing degenerative changes. He should not have thrown away the stick.[3]

1.8. WORK

So far this chapter has considered only situations in which an object is in equilibrium. If the total force on the object is not zero, the object experiences an acceleration **a** given by Newton's second law:

$$\mathbf{F}=m\mathbf{a}.$$

The study of how forces produce accelerations is called *dynamics*. It is an extensive field that will be discussed only briefly here.

Suppose an object moves along the x axis with velocity v_x. If it is subject to a force in the x direction F_x, it will be accelerated, and the velocity will change according to

$$F_x=ma_x=m\left(\frac{dv_x}{dt}\right).$$

If F_x is known as a function of *time*, then this equation can be written as

$$dv_x=\frac{1}{m}F_x(t)dt,$$

and it can be integrated, at least numerically.

In this context it is useful to define the *kinetic energy*

$$E_k=\tfrac{1}{2}mv_x^2. \tag{1.17}$$

[3]Quoted from W. P. Blount (1956). Don't throw away the cane. *J. Bone Joint Surg.* **38A**: 695–708. Used with permission of *J. Bone Joint Surg.* This article was first quoted to the physics community by Benedek and Villars (1973), pp. 3–8.

As long as F_x acts, the object is accelerated and the kinetic energy changes. We can gain some understanding of how it changes by noting that

$$\frac{d}{dt}\left(\tfrac{1}{2}mv_x^2\right)=mv_x\frac{dv_x}{dt}=F_xv_x. \qquad (1.18)$$

Therefore F_xv_x is the rate at which the kinetic energy is changing with time. It is called the *power* due to force F_x. The units of kinetic energy are kg m^2 s^{-2} or joules (J); the units of power are J s^{-1} or watts (W).

If v_x and F_x are both positive, the acceleration increases the object's velocity, the kinetic energy increases, and the power is positive. If v_x and F_x are both negative, the object's velocity vector decreases—becomes more negative—but the magnitude of the velocity increases. The kinetic energy again increases with time, and the power is positive. If v_x and F_x point in opposite directions, then the effect of the acceleration is to reduce the magnitude of v_x, the kinetic energy decreases, and the power is negative.

Equation (1.18) can be written as

$$\frac{d}{dt}\left(\tfrac{1}{2}mv_x^2\right)=F_x\frac{dx}{dt}.$$

Both sides of this equation can be integrated with respect to t:

$$\int_{t_1}^{t_2}\frac{d}{dt}\left(\tfrac{1}{2}mv_x^2\right)dt=\int_{t_1}^{t_2}F_x(t)\frac{dx}{dt}\,dt.$$

The indefinite integral corresponding to the left-hand side is the integral with respect to time of the derivative of $\tfrac{1}{2}mv_x^2$ and is therefore $\tfrac{1}{2}mv_x^2$. If F_x is known not as a function of t but as a function of x, it is convenient to write the right-hand side as

$$\int_{x_1}^{x_2}F_x(x)dx=W.$$

This quantity is called the *work* done by force F_x on the object as it moves from x_1 to x_2. The complete equation is therefore

$$\left[\frac{1}{2}mv_x^2\right]_2-\left[\frac{1}{2}mv_x^2\right]_1=\int_{x_1}^{x_2}F_xdx=W. \qquad (1.19)$$

The increase in kinetic energy of the body as it moves from position 1 (at time 1) to position 2 (at time 2) is equal to the work done *on* the body *by* the force F_x. The work done on the body by force F_x is the area under the curve of F_x vs x, between points x_1 and x_2. This is shown in Fig. 1.17.

If several forces act on the body, then the acceleration is given by Newton's second law, where **F** is the total force on the body. The change in kinetic energy is therefore the work done by the *total* force or the sum of the work done by each individual force.

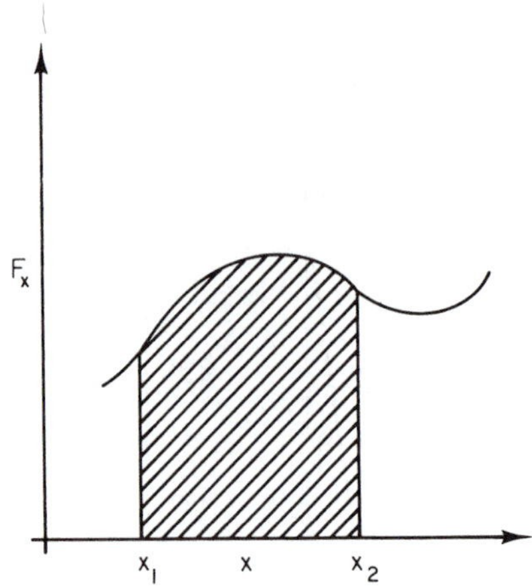

FIGURE 1.17. The work done by F_x is the shaded area under the curve between x_1 and x_2.

When the force and displacement vectors point in any direction, the kinetic energy is defined to be

$$E_k=\tfrac{1}{2}mv^2=\tfrac{1}{2}m(v_x^2+v_y^2+v_z^2). \qquad (1.20)$$

Differentiating this expression with respect to time shows that the power is given by an extension of Eq. (1.18):

$$\frac{dE_k}{dt}=F_xv_x+F_yv_y+F_zv_z.$$

This particular combination of vectors **F** and **v** is called the *scalar product*. It is written as $\mathbf{F}\cdot\mathbf{v}$. There is another way to write the scalar product. If **F** and **v** are not parallel, they define a plane. Align the x axis with **v** so that v_y and v_z are zero, and choose the direction of y so that **F** is in the xy plane (Fig. 1.18). Then it is easy to see that $\mathbf{F}\cdot\mathbf{v}=F_xv_x=Fv\cos\theta$, where θ is the angle between **F** and **v**. To summarize, the power is

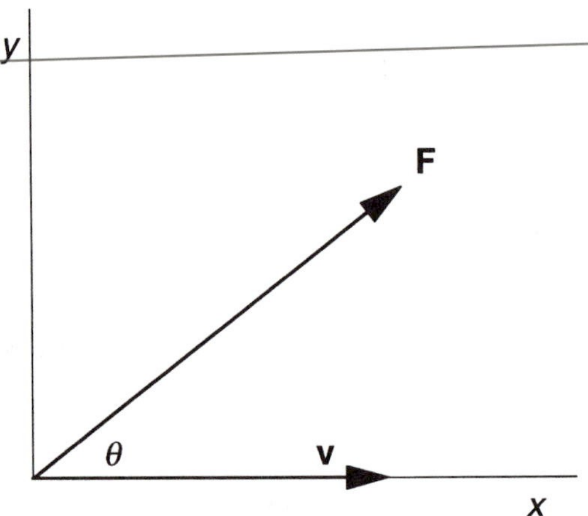

FIGURE 1.18. Aligning the axes so that **v** is along the x axis and **F** is in the xy plane shows that an alternative expression for $\mathbf{F}\cdot\mathbf{v}$ is $Fv\cos\theta$.

$$P=\frac{dE_k}{dt}=\mathbf{F}\cdot\mathbf{v}=Fv\cos\theta=F_xv_x+F_yv_y+F_zv_z. \tag{1.21}$$

Equation (1.21) can be integrated in the same manner as above to obtain

$$\Delta E_k=\int F_x dx+\int F_y dy+\int F_z dz=\int \mathbf{F}\cdot d\mathbf{s}. \tag{1.22}$$

This is the general expression for the work done by force **F** on a point mass that undergoes displacement **s**.

1.9. STRESS AND STRAIN

Whenever a force acts on an object, it undergoes a change of shape or deformation. Often these deformations can be ignored, as they were in the previous sections. In other cases, such as the contraction of a muscle, the expansion of the lungs, or the propagation of a sound wave, the deformation is central to the problem and must be considered. This book will not develop the properties of deformable bodies extensively; nevertheless deformable body mechanics is important in many areas of biology. We will develop the subject only enough to be able to consider viscous forces in fluids.

Consider a rod of cross-sectional area S. One end is anchored, and a force F is exerted on the other end parallel to the rod (Fig. 1.19). Effects of weight will be ignored. A *surface force* is transmitted across any surface defined by an imaginary cut perpendicular to the axis of the rod. A surface force is exerted by the substance to the right of the cut on the substance to the left (and vice versa, in accordance with Newton's third law: when object A exerts a force on object B, object B exerts an equal and opposite force on object A). The surface force per unit area is called the *stress*. In this case, when the surface is perpendicular to the axis of the rod and the force is along the axis of the rod, it is called a *normal stress*:

$$s_n=\frac{F}{S}. \tag{1.23}$$

In the general case there can also be a component of stress parallel to the surface.

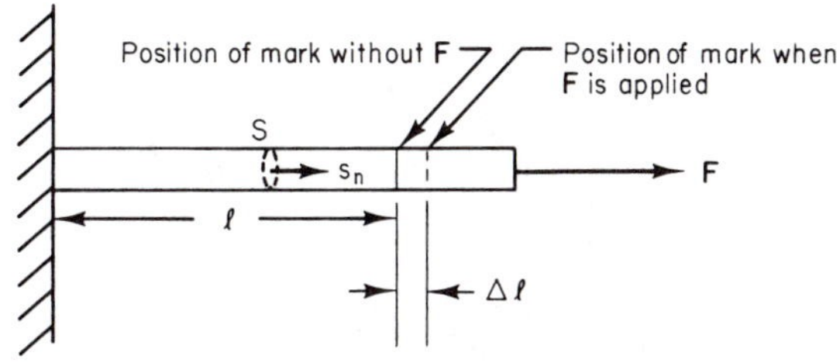

FIGURE 1.19. A rod subject to a force **F** along it.

The *strain* ϵ_n is the fractional change in the length of the rod:

$$\epsilon_n=\frac{\Delta l}{l}. \tag{1.24}$$

If increasing stress is applied to a typical substance, the strain increases linearly with the stress for small stresses. Then it increases even more rapidly and finally may increase even if the stress is reduced. Finally the sample breaks. This is plotted in Fig. 1.20. Because of the double-valuedness of the strain as a function of stress, the strain is usually plotted as the independent variable, as in Fig. 1.20(b).

In the linear region, the relationship between stress and strain is written as

$$s_n=E\epsilon_n. \tag{1.25}$$

The proportionality constant E is called *Young's modulus*. Since the strain is dimensionless, E has the dimensions of stress. Various units are N m^{-2} or pascal (Pa), dyn cm^{-2}, psi (pound per square inch), and bar (1 bar=14.5 psi=10^5 Pa=10^6 dyn cm^{-2}).

If the stress is increased enough, the bar breaks. The value of the stress when the bar breaks under tension is called the *tensile strength*. The material will also rupture under compressive stress; the rupture value is called the *compressive strength*. Table 1.1 gives values of Young's modulus, the tensile strength, and the compressive strength for steel, long bone (femur), and wood (walnut).

In some materials, the stress depends not only on the strain, but on the rate at which the strain is produced. It may take more stress to stretch the material rapidly than to stretch it slowly, and more stress to stretch it than to maintain a fixed strain. Such materials are called *viscoelastic*. They are often important biologically but will not be discussed here.

Still other materials exhibit *hysteresis*. The stress–strain relationship is different when the material is being stretched than when it is allowed to return to its unstretched state. This difference is observed even if the strain is changed so slowly that viscoelastic effects are unimportant. Figure 1.21

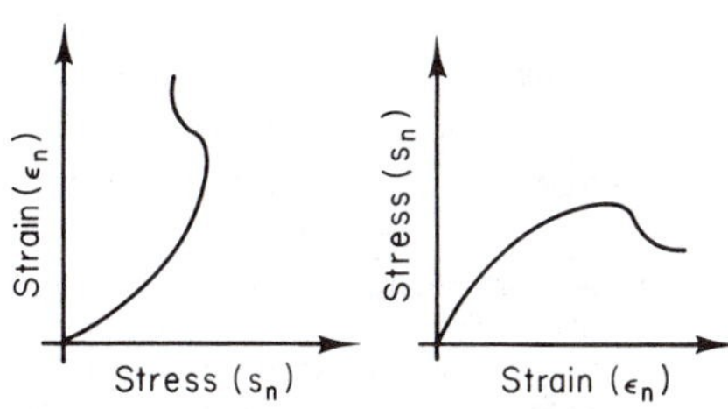

FIGURE 1.20. A typical stress–strain relationship. (a) Stress is the independent variable. (b) Strain is the independent variable. Plot (b) is usually used, since strain is often a double-valued function of the stress.

TABLE 1.1. *Young's modulus, tensile strength, and compressive strength of various materials in Pa.*

Material	E	Tensile strength	Compressive strength	Reference
Steel (approx.)	20×10^{10}	50×10^{7}		*a*
Femur (wet)	1.4×10^{10}	8.3×10^{7}	$18.\times10^{7}$	*b*
Walnut	0.8×10^{10}	4.1×10^{7}	5.2×10^{7}	*c*

[a]*American Institute of Physics Handbook* (1957). New York, McGraw-Hill, p. 2-70.
[b]B. K. F. Kummer (1972), Biomechanics of bone. In Y. C. Fung *et al.*, eds., *Biomechanics—Its Foundations and Objectives.* Englewood Cliffs, NJ, Prentice-Hall, p. 237.
[c]U.S. Department of Agriculture (1955). *Wood Handbook*, Handbook No. 72. Washington, D.C., U.S. Government Printing Office, p. 74.

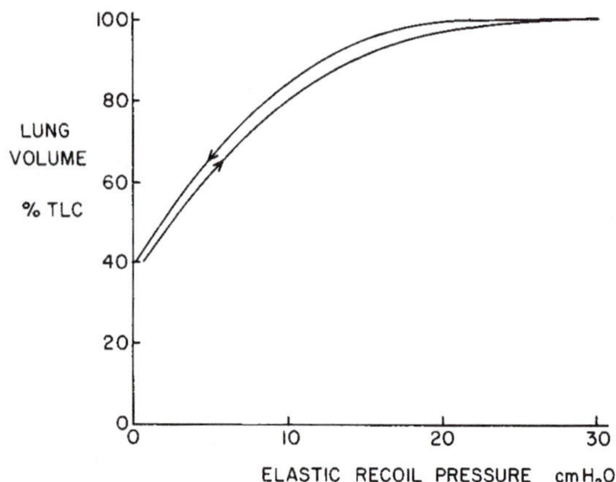

FIGURE 1.21. A pressure–volume curve for a normal lung, showing hysteresis. The elastic recoil pressure is the difference between the pressure in the alveoli (air sacs) of the lung and the thorax just outside the lung. From P. T. Macklem (1975). Tests of lung mechanics. *N. Engl. J. Med.* **293**: 339–342. Reprinted by permission. Drawing courtesy of Prof. Macklem.

shows a pressure–volume curve for the lung. It is related to the stress–strain relationship for the lung tissue and shows hysteresis.

1.10. SHEAR

In a shear stress, the force is parallel to the surface across which it is transmitted.[4] In a shear strain, the deformation increases as one moves in a direction perpendicular to the deformation. Examples of shear stress and strain are shown in Fig. 1.22. The shear stress is

$$s_s = \frac{F}{S}, \tag{1.26}$$

and the shear strain is

$$\epsilon_s = \frac{\delta}{h}. \tag{1.27}$$

It is possible to define a shear modulus G analogous to Young's modulus when the shear strain is small:

$$s_s = G\epsilon_s. \tag{1.28}$$

1.11. HYDROSTATICS

We now turn to some topics in the mechanics of fluids that will be useful to us in discussing fluid movement through membranes in Chap. 5. *Hydrostatics* is the description of fluids at rest. A fluid is a substance that will not support a shear when it is at rest. When the fluid is in motion, there can be a shear force called *viscosity*.

An immediate consequence of the definition of a fluid is that when the fluid is at rest, all the stress is normal. The normal stress is called the pressure. The pressure at any point in the fluid is the same in all directions. This can be demonstrated experimentally, and it can be derived from the conditions for equilibrium. Consider the small volume of fluid shown in Fig. 1.23. It has a length a perpendicular to the page. This volume is in equilibrium. Since the fluid at rest cannot support a shear, the pressure is perpendicular to each face, and there is no other force across each face. For the moment, assume that the pressures perpendicular to the three faces can be different, and call them p_1, p_2, and p_3. The total force exerted across face 1 is $p_1 ab \sin\theta$, acting downward. The force across face 2 is $p_2 ab \cos\theta$ acting to the right. Across face 3 it is $p_3 ab$, with vertical component $p_3 ab \sin\theta$ and horizontal component $p_3 ab \cos\theta$. The vertical components sum to zero only if $p_1 = p_3$, while the horizontal components sum to zero only if $p_3 = p_2$. Since this result is independent of the value of θ, the pressure must be the same in every direction.

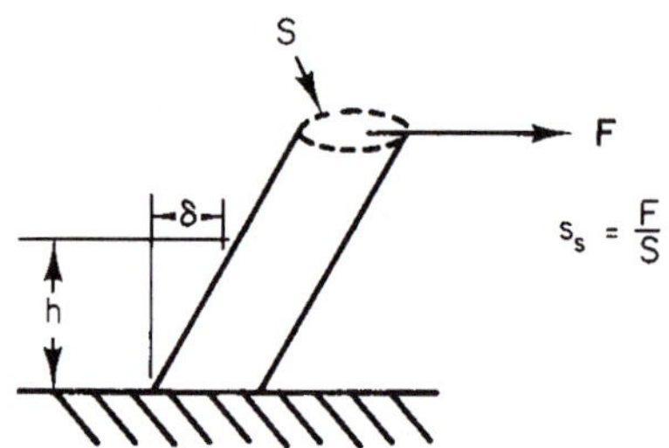

FIGURE 1.22. Shear stress and strain.

[4]This discussion of stress and strain has been made simpler than is often the case. In general, the force **F** across any surface is a vector. It can be resolved into a component perpendicular to the surface and two components parallel to the surface. One can speak of nine components of stress: $s_{xx}, s_{xy}, s_{xz}, s_{yx}, s_{yy}, s_{yz}, s_{zx}, s_{zy}, s_{zz}$. The first subscript denotes the direction of the force and the second denotes the normal to the surface across which the force acts. Components s_{xx}, s_{yy} and s_{zz} are normal stresses; the others are shear stresses. It can be shown that $s_{xy} = s_{yx}$, and so forth.

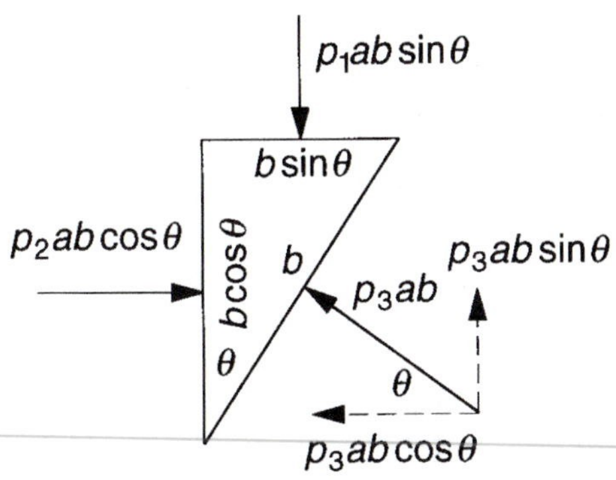

FIGURE 1.23. A volume element of fluid used to show that the pressure in a fluid at rest is the same in all directions.

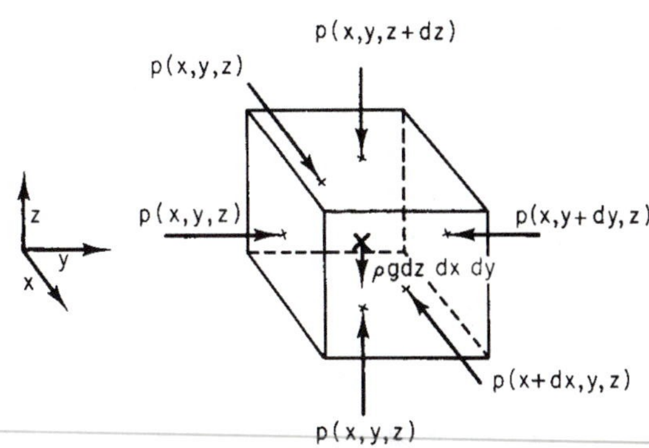

FIGURE 1.24. The fluid in volume *dx dy dz* is in equilibrium.

Next, consider how the pressure changes with position. Suppose that p depends on the coordinates $p=p(x,y,z)$ and that the density of the fluid is ρ kg m^{-3}. The only external force acting is gravity in the direction of the $-z$ axis. The fluid in the volume $dx\ dy\ dz$ of Fig. 1.24 is in equilibrium. In the y direction there is a force to the right across the left-hand face equal to $p(x,y,z)\ dx\ dz$ and to the left across the right-hand face equal to $-p(x,y+dy,z)\ dx\ dz$. These are the only forces in the y direction, and their magnitudes must be the same. Therefore p does not change in the y direction. A similar argument shows that p does not change in the x direction. In the z direction there are three terms: the upward force across the bottom face, the downward force across the top face, and the pull of gravity. The weight of the fluid is its mass ($\rho\ dx\ dy\ dz$) times the gravitational acceleration g ($g=9.8$ m s^{-2}). The three forces must add to zero:

$$p(x,y,z)\ dx\ dy-p(x,y,z+dz)\ dx\ dy-\rho g\ dx\ dy\ dz=0.$$

For small changes in height, dz, it is possible to approximate[5] $p(x,y,z+dz)$ by $p(x,y,z)+(dp/dz)\ dz$. With this approximation, the equilibrium equation is

$$dx\ dy\ dz\left(-\frac{dp}{dz}-\rho g\right)=0.$$

This equation can be satisfied only if

$$\frac{dp}{dz}=-\rho g. \tag{1.29}$$

This is a differential equation in $p(z)$. It is a particularly simple one, since the right-hand side is constant if ρ and g are constant:

$$dp=-\rho g\ dz.$$

Integrating this gives

$$\int dp=-\rho g\int dz,$$

$$p=-\rho gz+c.$$

The constant of integration is determined by knowing the value of p for some value of z. If $p=p_0$ when $z=0$, then $p_0=c$ and

$$p=p_0-\rho gz. \tag{1.30}$$

With a constant gravitational force per unit volume acting on the fluid, the pressure decreases with increasing height.

The SI unit of pressure is N m^{-2} (or pascal, Pa). The density is expressed in kg m^{-3}, so that ρg has units of N m^{-3} and ρgz is in N m^{-2}. Pressures are often given as values of z, for example, in millimeters of mercury (torr) or centimeters of water. In such cases, the value of z must be converted to an equivalent value of ρgz before calculations involving anything besides pressure are done. The density of water is 1 g cm^{-3} or 10^3 kg m^{-3}, and the density of mercury is 13.55×10^3 kg m^{-3}.

1.12. COMPRESSIBILITY

Increasing the pressure on a fluid causes a deformation and a decrease in volume. The compressibility κ is defined as

$$\frac{\Delta V}{V}=-\kappa\Delta p. \tag{1.31}$$

Since $\Delta V/V$ is dimensionless, κ has the units of inverse pressure, N^{-1} m^2 or Pa^{-1}. In many liquids the compressibility is quite small (e.g., 5×10^{-10} Pa^{-1} for water), and for many purposes, such as flow through pipes, compressibility can be ignored. Other effects, such as the transmission of sound through a fluid, depend on deformation and compressibility cannot be ignored.

[5]See Appendix D on Taylor's series for a more complete discussion of this approximation.

1.13. VISCOSITY

A fluid at rest does not support a shear. If the fluid is moving, a shear force can exist. At large velocities the flow of the fluid is turbulent and may be difficult or impossible to calculate. We will consider only cases in which the velocity is low enough so that the flow is "smooth." This means that particles of dye introduced into the fluid to monitor its motion flow along smooth lines called *streamlines*. There is no mixing of fluid across these lines. A streamline is tangent to the velocity vector of the fluid at every point along its path.

A fluid can support a viscous shear stress if the shear strain is changing. One way to create such a situation is to immerse two parallel plates in the fluid, and to move one parallel to the other as in Fig. 1.25. If the fluid in contact with each plate sticks to the plate, the fluid in contact with the lower plate is at rest and that in contact with the upper plate moves with the same velocity as the plate. Between the plates the fluid flows parallel to the plates, with a speed that depends on position as shown in Fig. 1.25. Such flow is called *laminar flow* because the fluid flows in layers. There is no mixing across layers as there would be in turbulent flow. (Laminar flow is often used in rooms where dirt or bacterial contamination is to be avoided, such as operating rooms or manufacturing clean rooms. Clean air enters and passes through the room without mixing. Any contaminants picked up are carried out in the air.) The variation of velocity between the plates gives rise to a velocity gradient dv_x/dy. Note that this is the rate of change of the shear strain, Eq. (1.27).

In order to keep the top plate moving and the bottom one stationary, it is necessary to exert a force of magnitude F on each plate: to the right on the upper plate and to the left on the lower plate. The resulting shear stress or force per unit area is in many cases proportional to the velocity gradient:

$$\frac{F}{S} = \eta \frac{dv_x}{dy}. \tag{1.32}$$

The constant η is called the *coefficient of viscosity*. Often this equation is written with a minus sign, in which case **F** is the force of the fluid on the plate rather than the plate on the fluid. The units of η are N s m^{-2} or kg m^{-1} s^{-1} or Pa s. Older units are the dyn s cm^{-2} or poise, the centipoise, and the micropoise. 1 poise=0.1 Pa s. Equation (1.32) gives the force exerted by fluid above the plane at height y on the fluid below the plane. In the case of the parallel plates, the force from above on fluid in the slab between y and $y+dy$ is the same in magnitude as (and opposite in direction to) the force exerted by the fluid below the slab. Therefore there is no net force on the fluid in the slab, and the fluid moves with constant velocity.

1.14. VISCOUS FLOW IN A TUBE

Biological fluid dynamics is a well-developed area of study [Lighthill (1975); Mazumdar (1992)]. External biological fluid dynamics is concerned with locomotion—from single-celled organisms to swimming fish and flying birds. Internal biological fluid dynamics deals with mass transport within the organism. Two obvious examples are flow in the airways and the flow of blood.

Consider laminar viscous flow of fluid through a pipe of constant radius R_p and length Δx. Ignore for now the gravitational force. The pressure at the left end of a segment of pipe is $p(x)$; at the right end it is $p(x+\Delta x)$. For now consider the special case in which none of the fluid is accelerated, so the total force on any volume element of the fluid is zero. The velocity profile must be as shown in Fig. 1.26: zero at the walls and a maximum at the center. Our problem is to determine $v(r)$.

We apply Newton's first law to the shaded cylinder of fluid of radius r shown in Fig. 1.26. Since gravity is ignored, there are only three forces acting on the volume. The fluid on the left exerts a force $\pi r^2 p(x)$ acting to the right in the direction of the positive x axis. The fluid on the right exerts a force $-\pi r^2 p(x+\Delta x)$ (the minus sign because it points to the left). The slower moving fluid outside the shaded region exerts a viscous drag force across the cylindrical surface at radius r. The area of the surface is $2\pi r\ \Delta x$. The force points to the left. Its magnitude is $2\pi r\ \Delta x\ \eta |dv/dr|$. Since dv/dr is negative, we obtain the correct sign by writing it as $2\pi r\ \Delta x\ \eta (dv/dr)$. Newton's first law then becomes

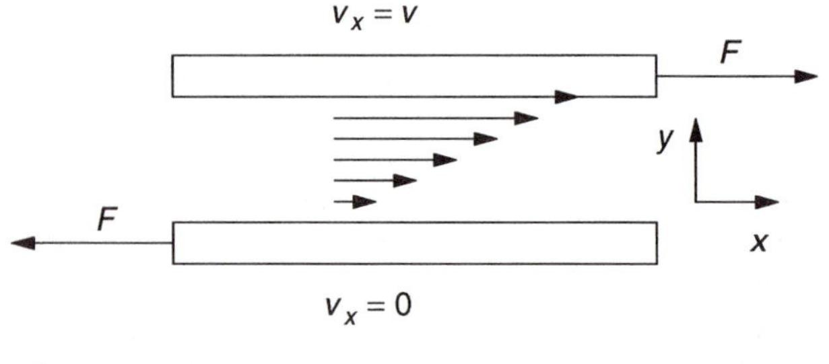

FIGURE 1.25. Forces **F** and −**F** are needed to make the top plate move in a viscous fluid while the bottom plate remains stationary. The velocity profile is also shown.

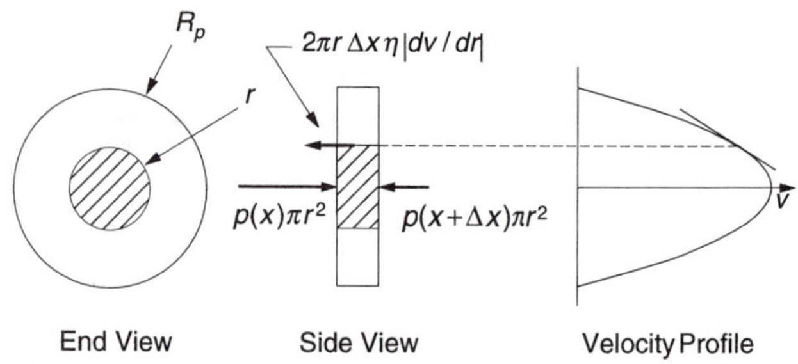

FIGURE 1.26. Longitudinal and transverse cross sections of the tube. Newton's first law is applied to the shaded volume.

$$\pi r^2[p(x)-p(x+\Delta x)]+2\pi r\ \Delta x\ \eta(dv/dr)=0, \tag{1.33}$$

which can be rearranged to give

$$\frac{dv}{dr}=\frac{r}{2\eta}\left(\frac{p(x+\Delta x)-p(x)}{\Delta x}\right)=\left(\frac{dp}{dx}\right)\frac{r}{2\eta}. \tag{1.34}$$

This can be integrated:

$$\int dv=\frac{1}{2\eta}\left(\frac{dp}{dx}\right)\int r\ dr,$$

$$v(r)=\frac{1}{4\eta}\left(\frac{dp}{dx}\right)r^2+A. \tag{1.35}$$

For flow to the right dp/dx is negative. Therefore it is convenient to write Δp as the pressure drop from x to $x+dx$: $\Delta p=p(x)-p(x+\Delta x)$. Then the first term in Eq. (1.35) is $-(1/4\eta)(\Delta p/\Delta x)r^2$. The constant of integration can be determined from the fact that the velocity is zero at the walls ($r=R_p$). The final result is

$$v(r)=\frac{1}{4\eta}\frac{\Delta p}{\Delta x}(R_p^2-r^2). \tag{1.36}$$

It is now necessary to calculate the total flow through the tube. The total flow rate or *volume flux* or *volume current i* is the volume of fluid per second moving through a cross section of the tube. Its units are $\mathrm{m^3\,s^{-1}}$. The *volume fluence rate* or *volume flux density*[6] or *current density* j_v is the volume per unit area per unit time across some small area in the tube. The units of j_v are $\mathrm{m^3\,s^{-1}\,m^{-2}}$ or $\mathrm{m\,s^{-1}}$. In fact, j_v is just the velocity of the fluid at that point. To see this, consider the flow of an incompressible fluid during time Δt. In Fig. 1.27 the fluid moves to the right with velocity v. At $t=0$, the fluid just to the left of plane B crosses the plane; at $t=\Delta t$, that fluid that was at A at $t=0$ crosses plane B. All the fluid between plane A and plane B crosses plane B during the time interval Δt. The volume fluence rate is

$$j_v=\frac{\text{(volume transported)}}{\text{(area)(time)}}=\frac{Sv\Delta t}{S\Delta t}=v. \tag{1.37}$$

It may seem unnecessarily confusing to call the fluence rate or flux density j_v instead of v; however, this notation corresponds to a more general notation in which j means the fluence rate or flux density of anything per unit area per unit time, and the subscript v, s, or q tells us whether it is the fluence rate of volume, solute particles, or electric charge.

To find the volume current i, j_v must be integrated over the cross-sectional area of the pipe. The volume of fluid crossing the washer-shaped area $2\pi r\ dr$ is $j_v 2\pi r\ dr=v2\pi r\ dr$. The total flux through the tube is therefore

[6] Some authors call j_v the flux. The nomenclature used here is consistent throughout the book.

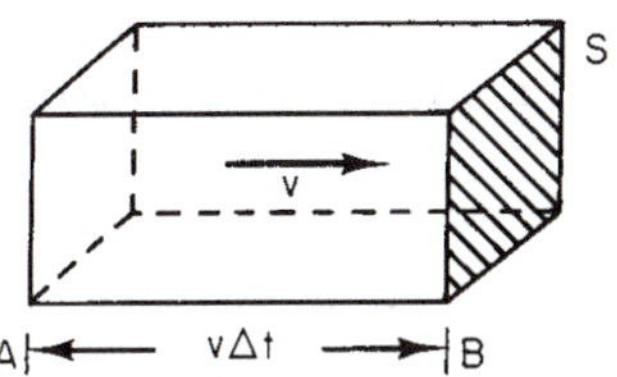

FIGURE 1.27. Flow of fluid across the plane at B.

$$i=\int_0^{R_p} j_v(r)2\pi r\ dr,$$

$$i=\frac{2\pi}{4\eta}\frac{\Delta p}{\Delta x}\int_0^{R_p}(R_p^2-r^2)r\ dr. \tag{1.38}$$

To integrate this, let $u=R_p^2-r^2$. Then $du=-2r\ dr$ and the integral is $R_p^4/4$. Inserting this in Eq. (1.38) gives

$$i=\frac{\pi R_p^4}{8\eta}\frac{\Delta p}{\Delta x} \tag{1.39}$$

as the flux of a viscous fluid through a pipe of radius R_p due to a pressure gradient $\Delta p/\Delta x$ along the pipe.

This relationship was determined experimentally in painstaking detail by a French physician, Jean Leonard Marie Poiseuille, in 1835. He wanted to understand the flow of blood through capillaries. His work and knowledge of blood circulation at that time have been described by Herrick (1942). The dependence of i on R_p^4 means that small changes in diameter cause large changes in flow.

As an example of the use of Eq. (1.39), consider a pore of the following size, which might be found in the basement membrane of the glomerulus of the kidney:

$$\begin{aligned} R_p&=5\times10^{-9}\ \mathrm{m},\\ \Delta p&=15.4\ \mathrm{torr},\\ \eta&=1.4\times10^{-3}\ \mathrm{kg\,m^{-1}\,s^{-1}},\\ \Delta x&=50\times10^{-9}\ \mathrm{m}. \end{aligned} \tag{1.40}$$

It is first necessary to convert 15.4 torr to Pa using Eq. (1.30) and the value of ρ for mercury, $13.55\times10^3\ \mathrm{kg\,m^{-3}}$:

$$\begin{aligned}\Delta p&=\rho g\ \Delta z=(13.55\times10^3)(9.8)(15.4\times10^{-3})\\ &=2.04\times10^3\ \mathrm{N\,m^{-2}}.\end{aligned}$$

Then Eq. (1.39) can be used:

$$i=\frac{(3.14)(5\times10^{-9})^4(2.04\times10^3)}{(8)(1.4\times10^{-3})(50\times10^{-9})}=7.2\times10^{-21}\ \mathrm{m^3\,s^{-1}}.$$

Now consider the general case in which we have not only viscosity, but the fluid may be accelerated and gravity is important. We continue to write Δp as the pressure drop and consider four contributions, each of which will be discussed:

$$\Delta p = -\int_{x_1}^{x_2} dp/dx = \Delta p_{\text{visc}} + \Delta p_{\text{grav}} + \Delta p_{\text{accel 1}} + \Delta p_{\text{accel 2}}. \tag{1.41}$$

For simplicity, we restrict the derivation to an incompressible fluid and a pipe of circular cross section where the radius can change. Distance along the pipe is x and the radius of the pipe is $R_p(x)$. Gravitational force acts on the fluid, and the height of the axis of the pipe above some reference plane is z, as shown in Fig. 1.28. Because the fluid is incompressible, the total current i is independent of x. If the pipe narrows, the velocity increases. Assume that changes in pipe radius occur slowly enough so that the velocity profile remains parabolic at every point in the pipe and we can treat x as though it were distance along the axis of the cylinder. If we define the average velocity as

$$\bar{v}(x) = \frac{i}{\pi R_p^2(x)}, \tag{1.42}$$

we can use Eq. (1.38) to rewrite the velocity profile as

$$v(r,x) = 2\bar{v}\,[1 - r^2/R_p^2(x)] = \frac{2i}{\pi R_p^2(x)}\left(1 - \frac{r^2}{R_p^2(x)}\right). \tag{1.43}$$

The first term in Eq. (1.41) is the pressure to overcome viscous drag. We can rewrite Eq. (1.34) as

$$\frac{dp_{\text{visc}}}{dx} = \frac{2\eta}{r}\frac{dv}{dr}.$$

Using Eq. (1.43) we can write

$$\frac{dp_{\text{visc}}}{dx} = -\frac{8\eta i}{\pi R_p^4(x)}. \tag{1.44}$$

We saw this earlier, solved for i in a pipe of constant radius, as Eq. (1.39). The pressure drop is obtained by integration:

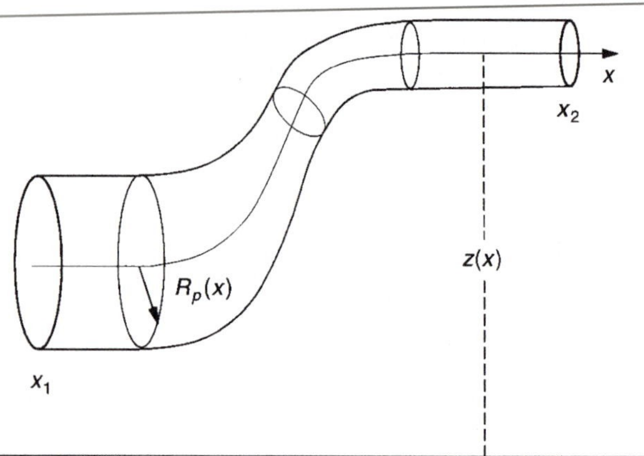

FIGURE 1.28. A pipe of circular cross section with radius and height varying along the pipe.

$$\Delta p_{\text{visc}} = -\int_{x_1}^{x_2} dp_{\text{visc}} = -\int_{x_1}^{x_2}\left(\frac{dp_{\text{visc}}}{dx}\right)dx$$
$$= +\frac{8\eta i}{\pi}\int_{x_1}^{x_2}\frac{dx}{R_p^4(x)}. \tag{1.45}$$

To go further requires knowing $R_p(x)$.

The next term Δp_{grav} is the hydrostatic pressure change that we saw in Eq. (1.30):

$$\Delta p_{\text{grav}} = -\int_{x_1}^{x_2} dp_{\text{grav}} = -\int \frac{dp_{\text{grav}}}{dz}\,dz = \rho g(z_2 - z_1), \tag{1.46}$$

The last two terms of Eq. (1.41) are pressure differences required to accelerate the fluid. When the flow is steady—that is, the velocity depends only on position, and the velocity at a fixed position does not change with time—there can still be an acceleration if the cross section of the pipe changes. The third term, $\Delta p_{\text{accel 1}}$, provides the force to accelerate the fluid if the cross section of the pipe changes. It can be derived as follows. Imagine a streamline in the fluid. (In some cases, such as turbulent flow, it may be hard to identify.) No fluid crosses the streamline. Consider a small length of streamline ds and a small area dA perpendicular to it. The edge of dA defines another set of streamlines that form a "tube of flow," and $dA\ ds$ defines a small volume of fluid. Make ds and dA small enough so that v is nearly the same at all points within the volume. The mass of fluid in the volume is $dm = \rho\ dA\ ds$. We ignore viscosity and gravity, so the only pressure difference is due to acceleration. The net force on the volume is

$$dF = -\left(\frac{dp}{ds}\right)ds\ dA. \tag{1.47}$$

The acceleration of the fluid in the element is

$$\frac{dv}{dt} = \frac{dF}{dm} = \frac{-\left(\frac{dp}{ds}\right)ds\ dA}{\rho\ ds\ dA} = -\frac{1}{\rho}\left(\frac{dp}{ds}\right). \tag{1.48}$$

Fluid that enters the volume is accelerated this much by the pressure difference by the time it leaves. We are considering only velocity changes that occur because the fluid moves along a streamline to a different position. We use the chain rule to write

$$\frac{dv}{dt} = \left(\frac{dv}{ds}\right)\left(\frac{ds}{dt}\right) = v\left(\frac{dv}{ds}\right).$$

Combining these gives

$$\frac{dp_{\text{accel 1}}}{ds} = -\rho v\left(\frac{dv}{ds}\right). \tag{1.49}$$

This can be integrated along the streamline to give

$$\Delta p_{\text{accel }1} = -\int_{s_1}^{s_2} \left(\frac{dp_{\text{accel }1}}{ds} \right) ds = +\rho \int_{x_1}^{x_2} v \left(\frac{dv}{ds} \right) ds$$

$$= \frac{\rho v_2^2}{2} - \frac{\rho v_1^2}{2}. \tag{1.50}$$

The final term $\Delta p_{\text{accel }2}$ is the pressure change required to accelerate the fluid between points 1 and 2 if the velocity of the fluid at a fixed position is changing with time. This happens, for example, to blood that is accelerated as it is ejected from the heart during systole, or to fluid that is sloshing back and forth in a U tube. To derive this term, again imagine a small length of streamline ds and a small area dA perpendicular to it. In addition to ignoring gravity and viscosity, we ignore changes in velocity because of changes in cross section. There is acceleration only if the velocity at a fixed location is changing. The acceleration is $\partial v/\partial t$. The derivative is written with ∂'s to signify the fact that we are considering only changes in the velocity with time that occur at a fixed position. The net force required to accelerate this mass is provided by the pressure difference (1.47):

$$dF = -dA\ dp_{\text{accel }2} = dm \left(\frac{\partial v}{\partial t} \right) = \rho \left(\frac{\partial v}{\partial t} \right) dA\ ds,$$

$$dp_{\text{accel }2} = -\rho \left(\frac{\partial v}{\partial t} \right) ds,$$

$$\Delta p_{\text{accel }2} = -\int_{s_1}^{s_2} dp_{\text{accel }2} = \rho \int_{s_1}^{s_2} \left(\frac{\partial v}{\partial t} \right) ds. \tag{1.51}$$

All of these effects can be summarized in the *generalized Bernoulli equation*:

$$p_1 - p_2 = \Delta p = \underbrace{\rho \int_{s_1}^{s_2} \frac{\partial v}{\partial t}\, ds}_{\Delta p_{\text{accel }2}} + \underbrace{\int_{s_1}^{s_2} \left(-\frac{dp_{\text{visc}}}{ds} \right) ds}_{\Delta p_{\text{visc}}}$$

$$+ \underbrace{\frac{\rho v_2^2}{2} - \frac{\rho v_1^2}{2}}_{\Delta p_{\text{accel }1}} + \underbrace{\rho g (z_2 - z_1)}_{\Delta p_{\text{grav}}}. \tag{1.52}$$

Equation (1.52) is valid for nonuniform viscous flow that may be laminar or turbulent if the integral is taken along a streamline [see, for example, Synolakis and Badeer (1989)].

1.15. PRESSURE–VOLUME WORK

An important example of work is that done in a biological system when the volume of a container (such as the lungs or the heart or a blood vessel) changes while the fluid within the container exerts a force on the walls.

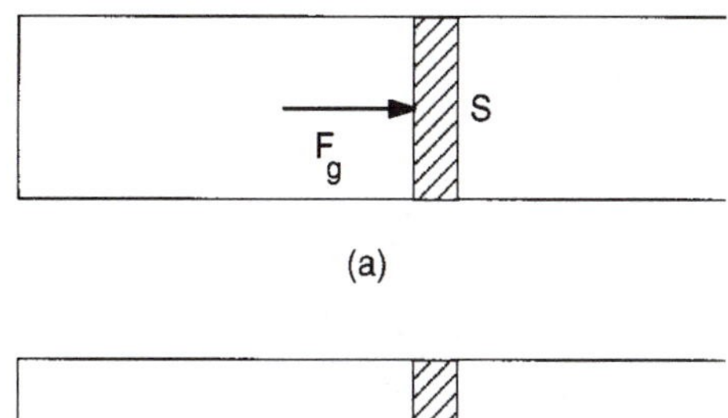

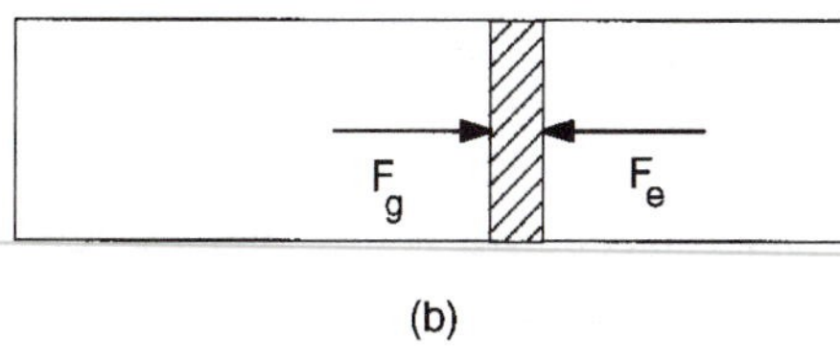

FIGURE 1.29. (a) A cylinder containing gas has a piston of area S at one end. (b) The force exerted on the piston by the gas is balanced by an external force of the piston is at rest.

To deduce an expression for pressure–volume work, consider a cylinder of gas fitted with a piston, Fig. 1.29(a). If the piston has area S, the gas exerts a force $F_g = pS$ on the piston. If no other force is exerted on the piston to restrain it, it will be accelerated to the right and gain kinetic energy as the gas does work on it:

$$(\text{work done by gas}) = F_g dx = pS\ dx = p\ dV. \tag{1.53}$$

If the piston is prevented from accelerating by an external force $\mathbf{F}_e$ equal and opposite to that exerted by the gas [Fig. 1.27(b)], then the external force does work on the piston:

$$(\text{work done by external force}) = -F_e dx = -pS\ dx$$
$$= -p\ dV, \tag{1.54}$$

which is the negative of the work done on the piston by the expanding gas. The result is that the kinetic energy of the piston does not change. The gas does work on the surroundings as it expands, increasing the energy of the surroundings; the surroundings, through the external force, do *negative* work on the gas; that is, they decrease the energy of the gas. (The meaning of "energy of the gas" and "energy of the surroundings" is discussed in Chap. 3.) If the gas is compressed, the situation is reversed: the surroundings do positive work on the gas and the gas does negative work on the surroundings.

For a large change in volume from V_1 to V_2, the pressure may change as the volume changes. In that case the work done by the gas on the surroundings is

$$W_{\text{by gas}} = \int_{V_1}^{V_2} p\ dV. \tag{1.55}$$

This work is the shaded area in Fig. 1.30. If the gas is compressed, the change in volume is negative and the work done *by* the gas is negative.

Suppose that the left ventricle of the heart contracts at constant pressure, so that it changes volume by $\Delta V = V_2 - V_1$. (Since $V_2 < V_1$ the quantity ΔV is negative.

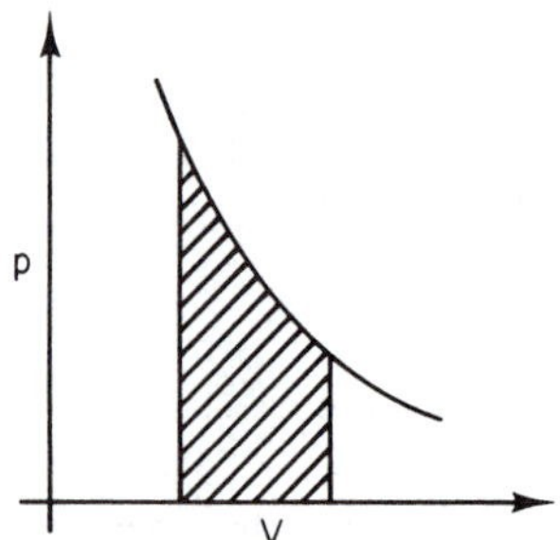

FIGURE 1.30. A plot of p vs V, showing the work done by the gas as it expands.

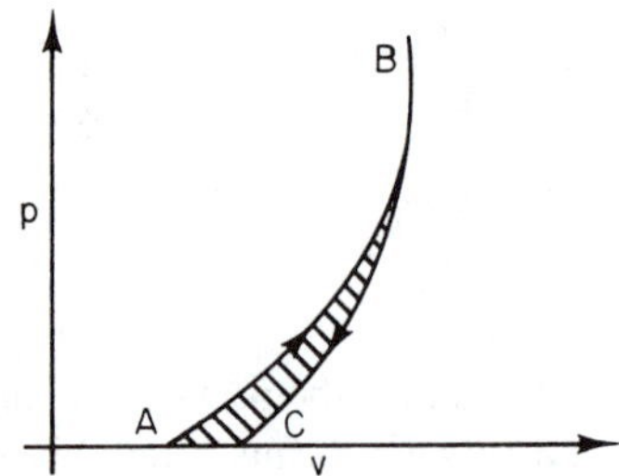

FIGURE 1.32. A hypothetical plot of the pressure–volume relationship for inhalation and exhalation.

A volume of blood $-\Delta V$ is ejected into the aorta.) The work done by the heart wall on the blood is $-p\ \Delta V$ and is positive, since ΔV is negative.

As another example of pressure–volume work, we can develop a model to estimate the work necessary to breathe. Consider the model of the lungs and airways shown in Fig. 1.31. The pressure at the nose is the atmospheric pressure p. In the alveoli (air sacs) the pressure is p_a. If there is no flow taking place, $p_a = p$. For air to flow in, p_a must be less than p; for it to flow out, p_a must be greater than atmospheric. The work done by the walls of the alveoli on the gas in them is $-\int p_a dV$. The net value of this integral for a respiratory cycle is positive. [Perhaps the easiest way to see this is to imagine an inspiration, in which the alveolar pressure is $p_a = p - \Delta p$ and the volume change is ΔV. The work done on the gas is $-(p-\Delta p)\Delta V$. This is followed by an expiration at pressure $p_a = p + \delta p$, for which the work is $-(p+\delta p)(-\Delta V)$. The net work done on the gas is $(\Delta p + \delta p)\Delta V$.] The energy imparted to the gas shows up as a mixture of heating because of frictional losses and kinetic energy of the exhaled air.

There is another mechanism by which work is done in breathing. Refer again to Fig. 1.31. The pressure in the chest cavity (thorax) it is p_t. (The pressure measured in midesophagus is a good estimate of p_t.) Because of

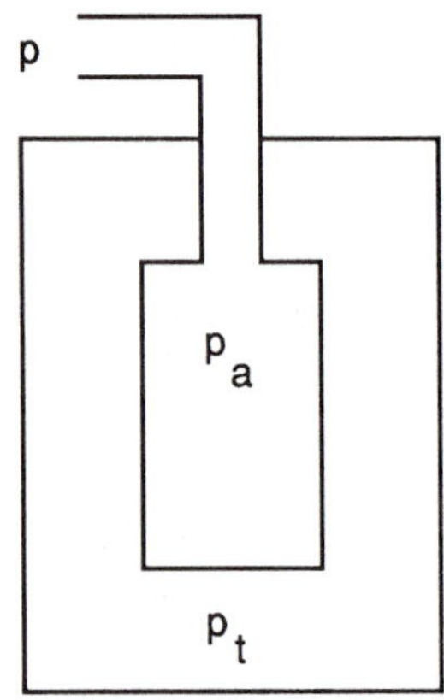

FIGURE 1.31. A model of the thorax, lungs, and airways that can be used to understand some features of breathing.

contractile forces in the lung tissue, $p_a > p_t$. The quantity $p_a - p_t$ is the "elastic recoil pressure" of Fig. 1.21 [Macklem (1975)]. We can think of the gas in the alveoli as doing work on the lung tissue, and the fluid in the thorax also doing work on the lung tissue. The latter has opposite sign, since a positive displacement dx of a portion of the alveolar wall is in the direction of the force exerted by the alveolar gas but is opposite to the direction of the force exerted by the thoracic fluid. The elastic recoil pressure, multiplied by dV, gives the net work done by both forces on the wall of the lung. Figure 1.21 shows elastic recoil pressure versus lung volume. It is redrawn in Fig. 1.32. During inspiration (curve AB), the elastic recoil pressure $p_a - p_t$ is greater than that during expiration (curve BC). The net work done on the lung wall during the respiratory cycle goes into frictional heating of the lung tissue.

1.16. BLOOD FLOW

The circulatory system has two parts: the *systemic circulation* and the *pulmonary circulation.* The left heart pumps blood into the systemic circulation: organs, muscles, etc. The right heart pumps blood through the lungs. As the heart beats, the pressure in the blood leaving the heart rises and falls. The maximum pressure during the cardiac cycle is the *systolic pressure.* The minimum is the *diastolic pressure.* (A blood pressure reading is in the form systolic/diastolic, measured in torr.) The blood flows from the aorta to several large arteries, to medium-sized arteries, to small arteries, to arterioles, and finally to the capillaries, where exchange with the tissues of oxygen, carbon dioxide, and nutrients takes place. The blood emerging from the capillaries is collected by venules, flows into increasingly larger veins, and finally returns to the heart through the vena cava.

At any given time, blood is flowing in only a fraction of the capillaries. The state of flow in the capillaries is continually changing to provide the amount of oxygen required by each organ. In skeletal muscle, terminal arterioles constrict and dilate to control distribution of blood to groups of capillaries. In smooth muscle and skin, a precapillary sphincter muscle controls the flow to each capillary [Patton *et al.* (1989), p. 860]. Since the blood is incom-

TABLE 1.2. *Typical values for the average pressure at the entrance to each generation of the major branches of the cardiovascular tree, the average blood volume in certain branches, and typical dimensions of the vessels.*

Location	Average pressure (torr)	Blood volume[a] (ml)	Diameter[b]	Length[b] (m)	Wall thickness[b] (m)	Avg. velocity[b] ($m\ s^{-1}$)	Reynolds number at maximum flow[c]
			Systemic circulation				
Left atrium	5						
Left ventricle	100						
Aorta	100	156	2.00×10^{-2}	0.5	2.00×10^{-3}	4.80×10^{-1}	9 400
Arteries	95	608	4.00×10^{-3}	0.5	1.00×10^{-3}	4.50×10^{-1}	1 300
Arterioles	86	94	5.00×10^{-5}	1.00×10^{-2}	2.00×10^{-4}	5.00×10^{-2}	
Capillaries	30	260	8.00×10^{-6}	1.00×10^{-3}	1.00×10^{-6}	1.00×10^{-3}	
Venules	10	470	2.00×10^{-5}	2.00×10^{-3}	2.00×10^{-6}	2.00×10^{-3}	
Veins	4	2682	5.00×10^{-3}	2.50×10^{-2}	5.00×10^{-4}	1.00×10^{-2}	
Vena cava	3	125	3.00×10^{-2}	5.00×10^{-1}	1.50×10^{-3}	3.80×10^{-1}	3 000
Right atrium	3						
			Pulmonary circulation				
Right atrium	5						
Right ventricle	25						
Pulmonary artery	25	52					
Arteries	20	91					7 800
Arterioles	15	6					
Capillaries	10	104					
Veins	5	215					2 200
Left atrium	5						

[a]From R. Plonsey (1995). Physiologic Systems. In J. R. Bronzino, ed. *The Biomedical Engineering Handbook*, Boca Raton, CRC Press, pp. 9–10.
[b]From J. N. Mazumdar (1992). *Biofluid Mechanics.* Singapore, World Scientific, p. 38.
[c]From W. R. Milnor (1989). *Hemodynamics*, 2nd ed. Baltimore, Williams & Wilkins, p. 148.

pressible and is conserved,[7] the total volume flow i remains the same at all generations of branching in the vascular tree. Table 1.2 shows average values for the pressure and vessel sizes at different generations of branching. Most of the pressure drop occurs in the arterioles.

We define the *vascular resistance* R in a pipe or a segment of the circulatory system as the ratio of pressure difference across the pipe or segment to the flow through it:

$$R=\frac{\Delta p}{i}. \tag{1.56}$$

The units are Pa m^{-3} s. Physiologists use the peripheral resistance unit (PRU), which is torr ml^{-1} min.

For Poiseuille flow the resistance can be calculated from Eq. (1.39):

$$R=\frac{8\,\eta\Delta x}{\pi R_p^4}. \tag{1.57}$$

The resistance decreases rapidly as the radius of the vessel increases.

[7]This is not strictly true. Some fluid leaves the capillaries and returns to the heart through the lymphatic system instead of the venous system. See Chap. 5.

If vessels of different diameters are connected in series so that the flow i is the same through each one and the total pressure drop is the sum of the drops across each vessel, then the total resistance is the sum of the resistances of each vessel:

$$R_{\mathrm{tot}}=R_1+R_2+R_3+\cdots . \tag{1.58}$$

If there is branching so that several vessels are in parallel with the same pressure drop across each one, the total flow through all the branches equals the flow in the vessel feeding them. The total resistance is then given by

$$\frac{1}{R_{\mathrm{tot}}}=\frac{1}{R_1}+\frac{1}{R_2}+\frac{1}{R_3}+\cdots . \tag{1.59}$$

The average flow from the heart is the stroke volume—the volume of blood ejected in each beat—multiplied by the number of beats per second. A typical value might be

$$i=(60\ \text{ml beat}^{-1})(80\ \text{beat min}^{-1})=4800\ \text{ml min}^{-1}=80\times10^{-6}\ \text{m}^3\,\text{s}^{-1}.$$

The total resistance would then be the average pressure divided by the flow:

$$R=\frac{(100\ \text{torr})(133\ \text{Pa torr}^{-1})}{80\times10^{-6}\ \text{m}^3\ \text{s}^{-1}}=1.66\times10^8\ \text{Pa m}^{-3}\ \text{s}.$$

The pressure in the left ventricle changes during the cardiac cycle. It can be plotted vs time. It can also be plotted vs ventricular volume, as in Fig. 1.33. The $p-V$ relationship moves counterclockwise around the curve during the cycle. Filling occurs at nearly zero pressure until the ventricle begins to distend when the volume exceeds 60 ml. There is then a period of contraction at nearly constant volume that causes the ventricular pressure to rise until it exceeds the (diastolic) pressure in the aorta, and the aortic valve opens. The contraction continues, and the pressure rises further, but the ventricular volume decreases as blood flows into the aorta. The ventricle then relaxes. The aortic valve closes when the ventricular pressure drops below that in the aorta. The work done in one cycle is the area enclosed by the curve. For the curve shown, it is 6600 torr ml=0.88 J. At 80 beats per minute the power is 1.2 W. In this drawing the stroke volume is 100−35=65 ml, and the cardiac output is

$$i=(65\ \text{ml beat}^{-1})[(80\ \text{beats/60 s})]=87\times10^{-6}\ \text{m}^3\ \text{s}^{-1}.$$

Many features of the circulation can be modeled by Poiseuille flow. However, at least four effects—in addition to those in Eq. (1.41)—cause departures from Poiseuille flow: (1) there may be turbulence; (2) there are departures from a parabolic velocity profile; (3) the vessel walls are elastic; and (4) the apparent viscosity depends on the both fraction of the blood volume occupied by red cells and on the size of the vessel.

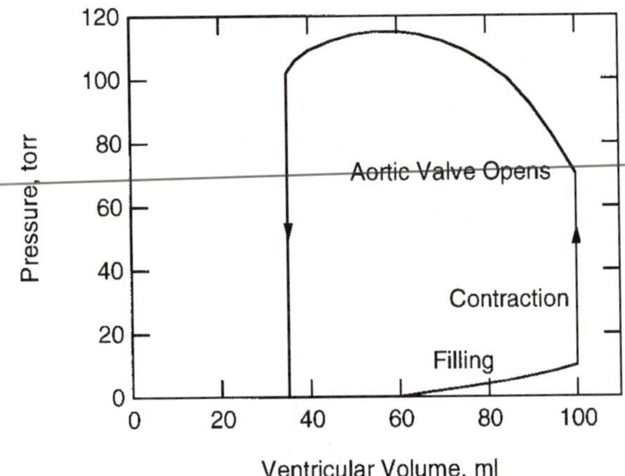

FIGURE 1.33. Pressure–volume relationship in the left ventricle. The curve is traversed counterclockwise with increasing time. The stroke volume is 100−35=65 ml. Systolic pressure is 118 torr, and diastolic pressure is 70 torr. The ventricular pressure drops below diastolic while the pressure in the arteries remains about 70 torr because the aortic valve has closed and prevents back flow.

The importance of turbulence (nonlaminar) flow is determined by a dimensionless number characteristic of the system called the *Reynolds number* N_R. It is defined by

$$N_R=\frac{LV\rho}{\eta}, \tag{1.60}$$

where L is a length characteristic of the problem, V a velocity characteristic of the problem, ρ the density, and η the viscosity of the fluid. When N_R is greater than a few thousand, turbulence usually occurs.

The Reynolds number arises in the following way. If we were to write Newton's second law for a fluid (which we have not done) in terms of dimensionless primed variables such as $\mathbf{r}'=\mathbf{r}/L$, $\mathbf{v}'=\mathbf{v}/V$, and $t'=t/(L/V)$, we would find that the equations depended on the properties of the fluid only through the combination N_R [Mazumdar (1992), p. 14]. With appropriate scaling of dimensions and times, flows with the same Reynolds number are identical.

There is ambiguity in defining the characteristic length and the characteristic velocity. Should one use the radius or the diameter of a tube? The maximum velocity or the average velocity? If one is solving the equations of motion, one knows what values of L and V were used to transform the equations. They are used to transform the solution back to "real world" coordinates. However, if one is making a statement such as "turbulence usually occurs for values of N_R greater than a few thousand," there is ambiguity. On the other hand, the statement is not very precise. Sometimes an additional subscript is used to specify how N_R was determined.

As the viscosity increases (for fixed L, V, and ρ) the Reynolds number decreases. When the Reynolds number is small, viscous effects are important. The system is not accelerated, and external forces that cause the flow are balanced by viscous forces. Work done on the system is transformed into thermal energy. When N_R is large, inertial effects are important. External forces accelerate the fluid. This happens when the mass is large and the viscosity is small. The low-Reynolds-number regime is so different from our everyday experience that the effects often seem counterintuitive. They are nicely described by Purcell (1977).

Here is an example of an estimate expressed in terms of the Reynolds number. A pressure difference Δp acts on a segment of fluid of length Δx undergoing Poiseuille flow. The difference between the force exerted on the segment of fluid by the fluid "upstream" and that exerted by the fluid "downstream" is $\pi R_p^2 \Delta p$. If the average speed of the fluid is $\bar{v}$, then the net work done on the segment by the fluid upstream and downstream in time Δt is $W_{\text{visc}}=\pi R_p^2 \Delta p \bar{v} \Delta t$. Since the fluid is not accelerated, this work is converted into thermal energy. We can solve Eq. (1.44) for Δp and write

$$W_{\text{visc}}=\pi R_p^2 \Delta p\ \bar{v}\ \Delta t=8\,\eta\pi\bar{v}^2 \Delta x\ \Delta t.$$

The kinetic energy of the moving fluid in a cylinder of length $\bar{v}\,\Delta t$ is

$$E_k=\frac{m\bar{v}^2}{2}=\frac{\rho\pi R_p^2(\bar{v}\Delta t)\bar{v}^2}{2}=\frac{\rho\pi R_p^2\bar{v}^3\Delta t}{2},$$

and the ratio of the kinetic energy to the work done is

$$\frac{E_k}{W_{\text{visc}}}=\frac{\rho\bar{v}R_p^2}{16\eta\Delta x}=\frac{1}{16\xi}\frac{\rho\bar{v}R_p}{\eta}=\frac{1}{16\xi}N_R.$$

(The last step was done by writing the Δx as ξR_p.) This result shows that the ratio of kinetic energy to viscous work is proportional to the Reynolds number. Another example is given in the Problem section.

A large range of values of N_R occurs in the circulatory system. Typical values corresponding to the peak flow are given in Table 1.2. Blood flow is laminar except in the ascending aorta and main pulmonary artery where turbulence may occur during peak flow.

There are two main causes of departures from the parabolic velocity profile. First, a red cell is about the same diameter as a capillary. Red cells in capillaries line up single file, each nearly blocking the capillary. The plasma flows in small cylinders between red cells, with a velocity profile that is nearly independent of radius. Second, the *entry region* causes deviations from Poiseuille flow in larger vessels. Suppose that blood flowing with a nearly flat velocity profile enters a vessel, as might happen when blood flowing in a large vessel enters the vessel of interest, which has a smaller radius. At the wall of the smaller vessel the flow is zero. Since the blood is incompressible, the *average* velocity is the same at all values of x, the distance along the vessel. (We assume the vessel has constant cross-sectional area.) However, the velocity profile $v(r)$ changes with distance x along the vessel. At the entrance to the vessel ($x=0$) there is a very abrupt velocity change near the walls. As x increases a parabolic velocity profile is attained. The transition or entry region is shown in Fig. 1.34. In the entry region the pressure gradient is different from the value for Poiseuille flow. The velocity profile cannot be calculated analytically in the entry region. Various numerical calculations have been made, and the results can be expressed in terms of scaled variables [see, for example, Cebeci and Bradshaw (1977)]. The Reynolds number used in these calculations was based on the diameter of the pipe, $D=2R_p$, and the average velocity. The length of the entry region is

$$L=0.05DN_{R,D}=0.1R_pN_{R,D}=0.2R_pN_{R,R_p}. \quad (1.61)$$

Blood pressure is, of course, pulsatile. This means that the average velocity and $v(r)$ are changing with time and also departing from the parabolic profile. Also, at the peak pressure during systole, the aorta and arteries expand, storing some of the blood and releasing it gradually during the rest of the cardiac cycle. Pulsatile flow and the elasticity of vessel walls are discussed extensively by Caro *et al.* (1978) and by Milnor (1989).

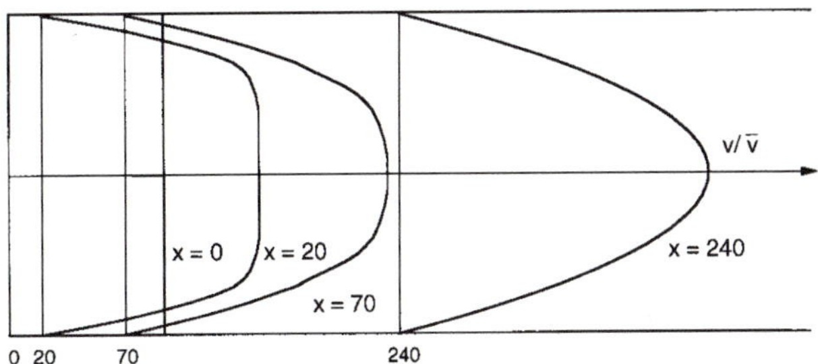

FIGURE 1.34. Velocity profiles in steady laminar flow at the entrance to a tube, showing the development of the parabolic velocity profile. The velocity is given as $v/\bar{v}$. At the entrance $v/\bar{v}=1$. When the Poiseuille flow is fully developed, $v/\bar{v}$ is 2 at the center of the tube. These curves are calculated from a graph by Cebeci and Bradshaw (1977) for laminar flow in a tube of radius 2 mm and a pressure gradient of 20 torr m^{-1}, carrying a fluid with a viscosity of 3×10^{-3} N s m^{-2} and a density of 10^3 kg m^{-3}. The scales are different along the axis and radius of the tube; the tube radius is 2 mm and the entrance region is 240 mm long.

Blood is not a Newtonian fluid. The viscosity depends strongly on the fraction of volume occupied by red cells (the hematocrit). In blood vessels of less than 100 μm radius, the apparent viscosity decreases with tube radius. Since a red cell barely fits in a capillary, the velocity profile in capillaries is not parabolic. Flow in arterioles and arteries is often modeled as individual particles surrounded by plasma and transported by laminar flow, each red cell staying at its own distance from the central axis. However, high-speed motion pictures show that the red cells often collide with other red cells and with the wall. [See the articles by Trowbridge (1982, 1983) and Trowbridge and Meadowcroft (1983), and also the Caro *et al.* and Milnor articles.]

SYMBOLS USED IN CHAPTER 1

Symbol	Use	Units	First used on page
a, a	Acceleration	m s^{-2}	9
c	Constant of integration		13
g	Acceleration due to gravity	m s^{-2}	13
h	Small distance	m	12
i	Total volume flux or flow rate or volume current	m^3 s^{-1}	15
j_v	Volume fluence rate or flux density (flow of volume per unit area per second)	m s^{-1}	15

Symbol	Use	Units	First used on page
l	Length of rod	m	11
m	Mass	kg	9
p	Pressure	Pa	12
p_t	Pressure in thorax	Pa	18
p_a	Pressure in alveoli	Pa	18
$\mathbf{r}$	Position	m	3
r	Distance from origin (radius) in polar or cylindrical coordinates	m	2
$\mathbf{s}$	Displacement	m	11
s_n	Normal stress	Pa	11
s_s	Shear stress	Pa	12
s	Distance along a streamline	m	16
t	Time	s	9
v, $\mathbf{v}$	Velocity	m s^{-1}	9
x,y,z	Coordinates	m	1
$\hat{\mathbf{x}}$, $\hat{\mathbf{y}}$, $\hat{\mathbf{z}}$	Unit vectors along the x, y, and z axes		4
dA	Small area perpendicular to a streamline	m^2	16
D	Pipe diameter	m	21
E	Young's modulus	Pa	11
E_k	Kinetic energy	J	9
F,$\mathbf{F}$	Force	N	1
G	Shear modulus	Pa	12
L	Characteristic length	m	20
N	Force	N	6
N_R	Reynolds number		20
$N_{R,D}$	Reynolds number based on diameter		21
N_{R,R_p}	Reynolds number based on pipe radius		21
P	Power	W	11
R	Displacement	m	4
R, $\mathbf{R}$	Force	N	6
R_p	Radius of pipe	m	14
R	Vascular resistance	Pa m^{-3} s	19
S	Cross-sectional area	m^2	11
V	Volume	m^3	13
V	Velocity	m s^{-1}	20
W	Weight	N	2
W	Work	J	10
δ	A small distance	m	12
ϵ_n	Normal strain		11
ϵ_s	Shear strain		12
η	Viscosity	Pa s	14
α, β, θ, ϕ	Angle		3
κ	Compressibility	Pa^{-1}	13
ρ	Mass density	kg m^{-3}	13
τ	Torque	N m	3

PROBLEMS

Section 1.2

1.1. A person with mass $m=70$ kg has a weight (mg) of about 700 N. If the person is doing push-ups as shown, what are the vertical components of the forces exerted by the floor on the hands and feet?

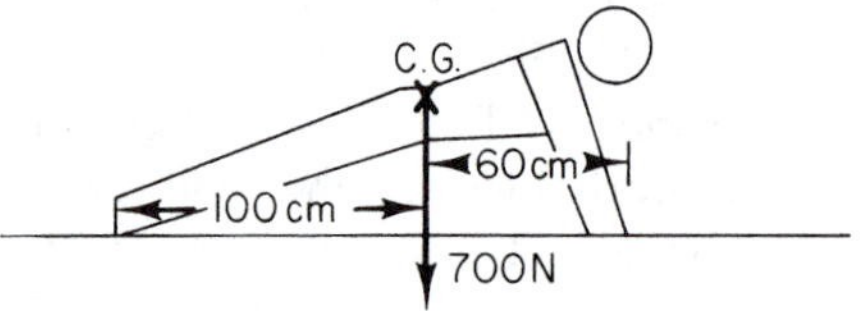

1.2. A person with upper arm vertical and forearm horizontal holds a mass of 4 kg. The mass of the forearm is 1.5 kg. Consider four forces acting on the forearm: **F** by the bones and ligaments of the upper arm at the elbow, **T** by the biceps, 40 N by the mass, and 15 N as the weight of the arm. The points of application are shown in the drawing. Calculate the vertical components of **F** and **T**.

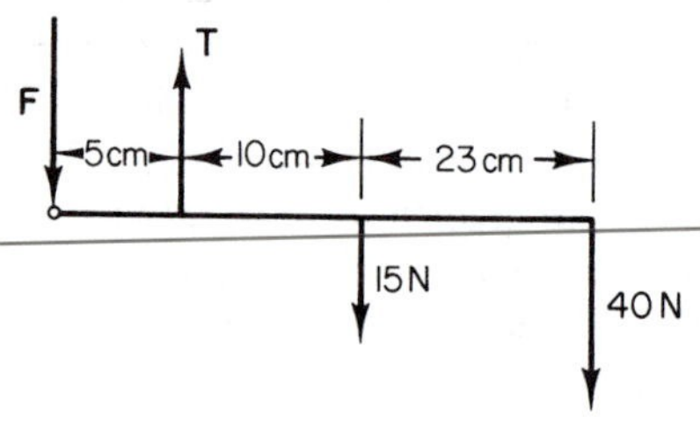

1.3. When the arm is stretched out horizontally, it is held by the deltoid muscle. The situation is shown schematically. Determine **T** and **F**.

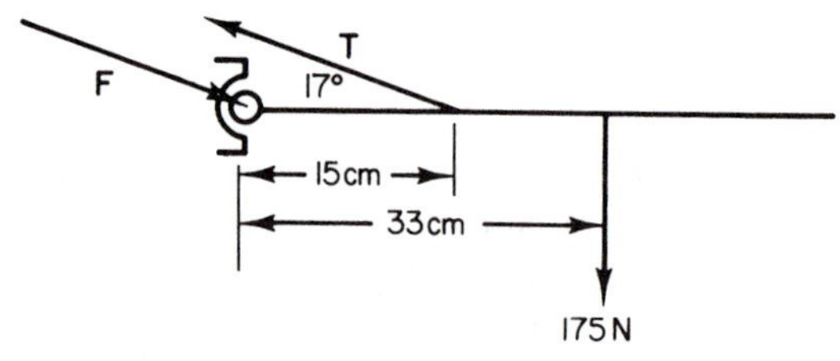

1.4. When a person crouches, the geometry of the heel is as shown. Determine **T** and **F**. Assume they act in the same horizontal plane.

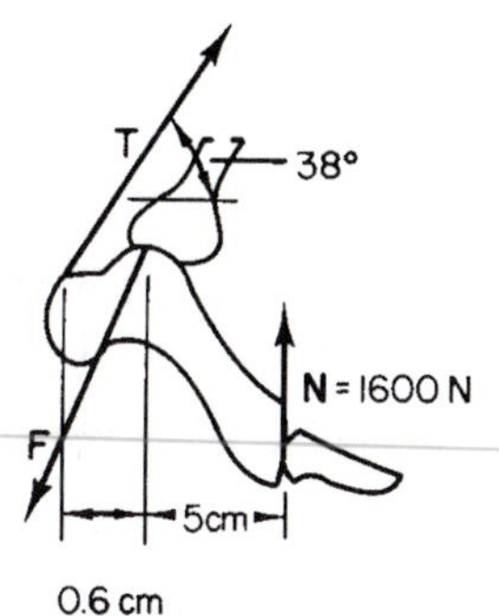

1.5. A person is suspended by both hands from a high bar as shown. The center of mass is directly below the bar.

(a) Find F_x and F_y, where **F** is the force exerted by the bar on each of the two hands.

(b) Given the additional information about the arm shown in the second drawing, calculate the components of **R**, the force exerted by the humerus on the forearm through the elbow, and the tension **T** in the biceps tendon. Neglect the weight of the arm, and assume that **T** and **R** are the only forces exerted on the forearm by the upper arm.

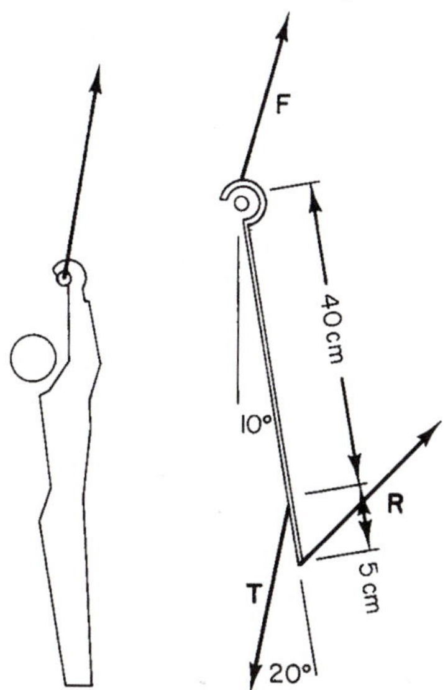

Section 1.3

1.6. Use the definition of the angular momentum,

$$\mathbf{L}=\mathbf{r}\times\mathbf{p}=\begin{vmatrix} \hat{\mathbf{x}} & \hat{\mathbf{y}} & \hat{\mathbf{z}} \\ r_x & r_y & r_z \\ p_x & p_y & p_z \end{vmatrix},$$

and the fact that $d\mathbf{r}/dt=\mathbf{p}/m$ to show that $d(\mathbf{r}\times\mathbf{p})/dt=\mathbf{r}\times d\mathbf{p}/dt$ and therefore the rotational form of Newton's second law is $\boldsymbol{\tau}=d\mathbf{L}/dt$.

Section 1.7

1.7. Suppose that instead of using a cane, a person holds a suitcase of weight $W/4$ in one hand, 0.4 m from the midline. Calculate the force exerted by the hip abductor muscles and by the acetabulum on the leg on the other side.

Section 1.9

1.8. A wire of length L, radius a, and Young's modulus E hangs vertically with mass M on the end. Neglect the mass of the wire. Find the spring constant k in the equation $F=-kx$. F is the force exerted by the wire on the mass; x is the departure of the mass from the equilibrium position. Also find the period of oscillation as the mass bounces up and down on this "spring." The period is given by $T=2\pi(M/k)^{1/2}$. Use values $L=300$ m, $M=100$ kg, $a=1$ mm, $E=2\times10^{11}$ N m^{-2}. Do you think the wire will break?

Section 1.11

1.9. The walls of a cylindrical pipe that has an excess pressure p inside are subject to a tension force per unit length T. (Consider only the force per unit length in the walls of the cylinder, not the force in any end caps of the pipe.) The force per unit length in the walls can be calculated by considering a different "pipe" made up of two parts as shown in the figure: a semicircular half-cylinder of radius R and length L attached to a flat plate of width $2R$ and length L. What is the force that the excess pressure exerts on the flat plate? What force per unit length must the semicircular piece exert on the flat plate where they are in contact? (Do not worry about any deformation.) See if you can obtain the same answer by direct integration of the horizontal and vertical components of the force due to the excess pressure. Sometimes a patient will have an aneurysm of the aorta, and a portion of the aortic wall will balloon out. Comment on this phenomenon in light of the R dependence of the force per unit length.

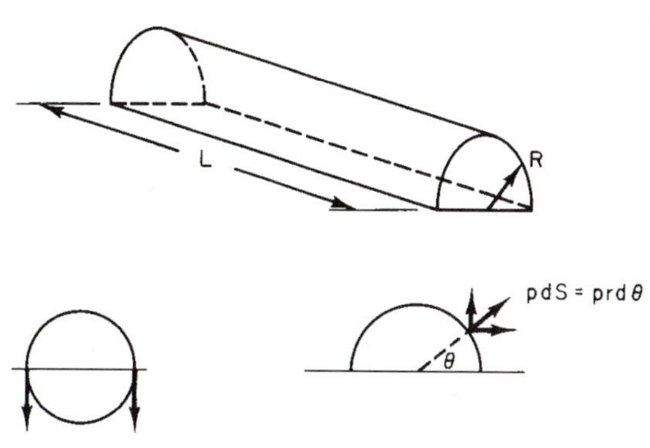

1.10. Find a relationship among the tension per unit length T across any element of the wall of a soap bubble, the excess pressure inside the bubble, Δp, and the radius of the bubble, R. (Hint: Use the same technique as for the previous problem.)

1.11. The inspirational pressure difference p_{in} that the lung can generate is about 86 torr. What would be the absolute maximum depth at which a person could breathe through a snorkel device? (A safe depth is only about half this maximum, since the lung ventilation becomes very small at the maximum depth. Assume the lungs are 30 cm below the mouth.)

1.12. This problem explores the physics of a centrifuge. A cylinder of fluid of density ρ and length L is rotated at an angular velocity ω (rad s^{-1}) in a horizontal plane about a vertical axis through one end of the tube. Neglect gravity. An object moving in a circle with constant angular velocity has an acceleration $a=-r\omega^2$ toward the center of the circle. Find the pressure in the fluid as a function of distance from the axis of rotation, assuming the pressure is p_0 at $r=0$.

Section 1.12

1.13. What is the compressibility of a gas for which pV=const? Make a numerical comparison of a gas at atmospheric pressure with a liquid.

Section 1.14

1.14. Consider laminar flow in a pipe of length Δx and radius R_p. Find the total viscous drag exerted by the pipe on the fluid.

1.15. The maximum flow rate from the heart is 500 ml s^{-1}. If the aorta has a diameter of 2.5 cm and the flow is Poiseuille, what are the average velocity, the maximum velocity at the center of the vessel, and the pressure gradient along the vessel? Plot the velocity vs distance from the center of the vessel. As an approximation to the viscosity of blood, use $\eta=10^{-3}$ kg m^{-1} s^{-1}.

1.16. The glomerular pore described in Eq. (1.40) has a flow $i=7.2\times10^{-21}$ m^3 s^{-1}. How many molecules of water per second flow through it? What is their average speed?

1.17. A sphere of radius R moving with speed v in laminar flow through a viscous fluid experiences a drag force

$$F_{visc}=6\pi\eta Rv.$$

At higher speeds inertial effects (the acceleration of the fluid displaced by the sphere) become important, and the drag becomes proportional to v^2 and to ρ, the density of the fluid. The force also depends on the radius of the sphere to some unknown power:

$$F_{drag}=K\rho v^2R^a,$$

where K is a dimensionless constant.

(a) Make a dimensional analysis to find the power a.

(b) Find the critical velocity at which $F_{visc}=F_{drag}$.

1.18. Consider fluid flowing between two slabs as shown in Fig. 1.25. Since the work done by the external force on the system in time dt is $dW=Fv\ dt$, the rate of doing work is $P=dW/dt=Fv$, where v is the speed of the moving plate. Find the power dissipated per unit volume of the fluid in terms of the velocity gradient.

Section 1.15

1.19. The accompanying figure shows the negative pressure (below atmospheric) that must be maintained in the thorax during the respiratory cycle by a patient with airway obstruction in order to breathe. Viscous effects are included. Estimate the work in joules done by the body during a breath.

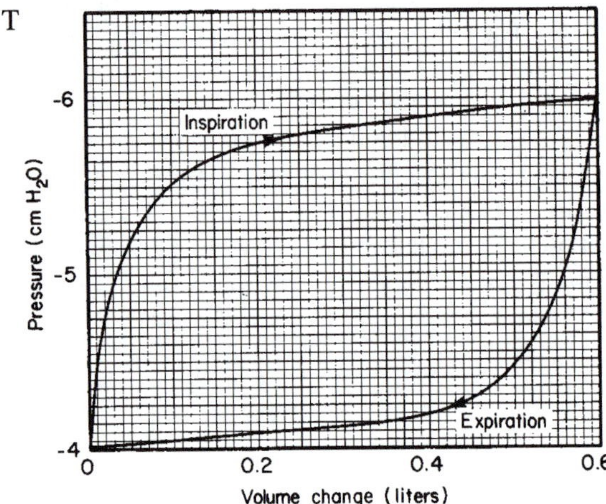

Section 1.16

1.20. Find the conversion factor between PRU and Pa m^{-3} s.

1.21. Derive the equations for resistance in a collection of vessels in series and in parallel. Remember that when several vessels are in series, the current is constant and the total pressure change is the sum of the pressure changes along the length of each vessel. When vessels are in parallel, each has the same pressure drop, but the current before the vessels branch is the sum of the currents in each branch.

1.22. For Poiseuille flow, find an expression for the maximum shear rate in each vessel from Eq. (1.43). Where in the vessel does it occur? Typical maximum shear rates are 50 s^{-1} in the aorta, 150 s^{-1} in the femoral artery, and 400 s^{-1} in an arteriole.

1.23. A sphere of radius a moving through a fluid is subject to a viscous drag $F_{drag}=6\pi\eta a\bar{v}$. Make an argument similar to that in the text to show that the ratio of kinetic energy of a sphere of fluid of the same size moving at the same speed to the viscous work done to displace the sphere by its own diameter is $N_R/18$.

1.24. Find an expression for the entry length in terms of the tube size, the pressure gradient, and the properties of the fluid. Estimate the length of the entry region in the aorta, in an artery, and in an arteriole of radius 20 μm. Use $\eta = 10^{-3}$ kg m^{-1} s^{-1}.

1.25. Estimate the tension per unit length and the stress in the walls of various blood vessels using the data in Table 1.2.

1.26. Consider laminar viscous flow in the following situation, which models flow in the bronchi or a network of branching blood vessels. A vessel of radius R connects to N smaller vessels, each of radius xR.

(a) What is the relationship between total cross-sectional area of the smaller vessels and that of the larger vessel if the pressure gradient is the same in both sets of vessels?

(b) How do the pressure gradients compare if the total cross-sectional area is the same in both sets of vessels? (Neither assumption is realistic.)

1.27. Compare the magnitude of the four terms in Eq. (1.41) in the following two cases. Ignore branching. Assume the vessels are vertical. Use $\rho = 10^3$ kg m^{-3} and $\eta = 10^{-3}$ Pa s.

(a) The descending aorta. Assume the length is 35 cm, the radius is 1 cm (independent of distance along the aorta), the peak acceleration of the blood is 1800 cm s^{-2}, and the peak velocity (during the cardiac cycle) is 70 cm s^{-1} at the entrance and 60 cm s^{-1} at the exit. (These velocities are different because some of the blood leaves the aorta in major arteries.)

(b) An arteriole of radius 50 μm, length 1 mm, and constant velocity of 5 mm s^{-1} at both entrance and exit.

REFERENCES

Benedek, G. B., and F. M. H. Villars (1973). *Physics with Illustrative Examples from Medicine and Biology*. Vol. 1. *Mechanics*. Reading, MA, Addison-Wesley.

Caro, C. G., T. J. Pedley, R. C. Schroter, and W. A. Seed (1978). *The Mechanics of the Circulation*. Oxford, Oxford University Press.

Cebeci, T., and P. Bradshaw (1977). *Momentum Transfer in Boundary Layers*. Washington, Hemisphere.

Halliday, D., R. Resnick, and K. S. Krane (1992). *Fundamentals of Physics*, 4th ed., Vol. 1. New York, Wiley.

Herrick, J. F. (1942). Poiseuille's observations on blood flow lead to a new law in hydrodynamics. *Am. J. Phys.* **10**: 33–39.

Inman, V. T. (1947). Functional aspects of the abductor muscles of the hip. *J. Bone Joint Surg.* **29**: 607–619.

Lighthill, J. (1975). *Mathematical Biofluiddynamics*. Philadelphia, Society for Industrial and Applied Mathematics.

Macklem, P. T. (1975). Tests of lung mechanics. *N. Engl. J. Med.* **293**: 339–342.

Mazumdar, J. N. (1992). *Biofluid Mechanics*. Singapore, World Scientific.

Milnor, William R. (1989). *Hemodynamics*, 2nd ed. Baltimore, Williams & Wilkins.

Patton, H., *et al.* (1989). *Textbook of Physiology*. 21st ed. Philadelphia, Saunders.

Purcell, E. M. (1977). Life at low Reynolds number. *Am. J. Phys.* **45**: 3–11.

Synolakis, C. E., and H. S. Badeer (1989). On combining the Bernoulli and Poiseuille equation—A plea to authors of college physics texts. *Am. J. Phys.* **57**(11): 1013–1019.

Trowbridge, E. A. (1982). The fluid mechanics of blood: equilibrium and sedimentation. *Clin. Phys. Physiol. Meas.* **3**(4): 249–265.

Trowbridge, E. A. (1983). The physics of arteriole blood flow. I. General Theory. *Clin. Phys. Physiol. Meas.* **4**(2): 151–175.

Trowbridge, E. A. and P. M. Meadowcroft (1983). The physics of arteriole blood flow. II. Comparison of theory with experiment. *Clin. Phys. Physiol. Meas.* **4**(2): 177–196.

Williams, M., and H. R. Lissner (1962). *Biomechanics of Human Motion*. Philadelphia, Saunders.

CHAPTER 2

Exponential Growth and Decay

The exponential function is one of the most important and widely occurring functions in physics and biology. In biology it may describe the growth of bacteria or animal populations, the decrease of the number of bacteria in response to a sterilization process, the growth of a tumor, or the absorption or excretion of a drug. (Exponential growth cannot continue forever because of limitations of nutrients, etc.) Knowledge of the exponential function makes it easier to understand birth and death rates, even when they are not constant. In physics, the exponential function describes the decay of radioactive nuclei, the emission of light by atoms, the absorption of light as it passes through matter, the change of voltage or current in some electrical circuits, the variation of temperature with time as an object comes to equilibrium with its surroundings, and the rate of some chemical reactions.

In this book, the exponential function will be needed to describe certain probability distributions, the concentration ratio of ions across a cell membrane, the flow of solute particles through membranes, the decay of a signal traveling along a nerve axon, and the return of some physiologic variables to their equilibrium values after they have been disturbed.

Because the exponential function is so important, and because I have seen many students who did not understand it even after having been exposed to it, the chapter starts with a gentle introduction to exponential growth (Sec. 2.1) and decay (Sec. 2.2). Section 2.3 shows how to analyze exponential data using semilogarithmic graph paper. The next section shows how to use semilogarithmic graph paper to find instantaneous growth or decay rates when the rate varies. Some would argue that the availability of computer programs that automatically produce logarithmic scales for plots makes these sections unnecessary. I feel that intelligent use of semilogarithmic and logarithmic (log–log) plots requires an understanding of the basic principles.

Clearance, discussed in Sec. 2.5, is an exponential decay process that is important in physiology. Sometimes there are competing paths for exponential removal of a substance; multiple decay paths are introduced in Sec. 2.6. A very basic and simple model for many processes is the combination of input at a fixed rate, accompanied by exponential decay. This is described in Sec. 2.7. Sometimes a substance exists in two forms, each with its own decay rate. One then must fit two or more exponentials to the set of data, as shown in Sec. 2.8.

Section 2.9 discusses the logistic equation, one possible model for a situation in which the growth rate decreases as the amount of substance increases. The chapter closes with a section on power-law relationships. While not exponential, they are included because data analysis can be done with log–log graph paper, a technique similar to that for semilog paper.

If you feel mathematically secure, you may wish to skim the first four sections, but you will probably find the rest of the chapter worth reading.

2.1. EXPONENTIAL GROWTH

An exponential growth process is one in which the rate of increase of a quantity is proportional to the present value of that quantity. The simplest example is a savings account. If the interest rate is 5% and if the interest is credited to the account once a year, the account increases in value by 5% of its present value each year. If the account starts out with $100, then at the end of the first year, $5 is credited to the account and the value becomes $105. At the end of the second year, 5 percent of $105 is credited to the account and the value grows by $5.25 to $110.25. The growth of such an account is shown in Table 2.1 and Fig. 2.1. These amounts can be calculated as follows. At the end of the first year, the original amount, y_0, has been augmented by $(0.05)y_0$:

TABLE 2.1. *Growth of a savings account earning 5% interest compounded annually.*

Year	Amount	Year	Amount	Year	Amount
0	\$100.00	10	\$162.88	100	\$13,150.13
1	105.00	20	265.33	200	1,729,258.09
2	110.25	30	432.19	300	2.27×10^{8}
3	115.76	40	704.00	400	2.99×10^{10}
4	121.55	50	1146.74	500	3.93×10^{12}
5	127.63	60	1867.92	600	5.17×10^{14}
6	134.01	70	3042.64	700	6.80×10^{16}
7	140.71	80	4956.14		
8	147.75	90	8073.04		
9	155.13				

$$y_1=y_0(1+0.05).$$

During the second year, the amount y_1 increases by 5%, so

$$y_2=y_1(1.05)=y_0(1.05)(1.05)=y_0(1.05)^2.$$

After x years, the amount in the account is

$$y_x=y_0(1.05)^x.$$

In general, if the growth rate is b per compounding period, the amount after x periods is

$$y_x=y_0(1+b)^x. \tag{2.1}$$

It is possible to keep the same annual growth (interest) rate, but to compound more often than once a year. Table 2.2 shows the effect of different compounding rates on the amount, when the interest rate is 5%. The last two columns,

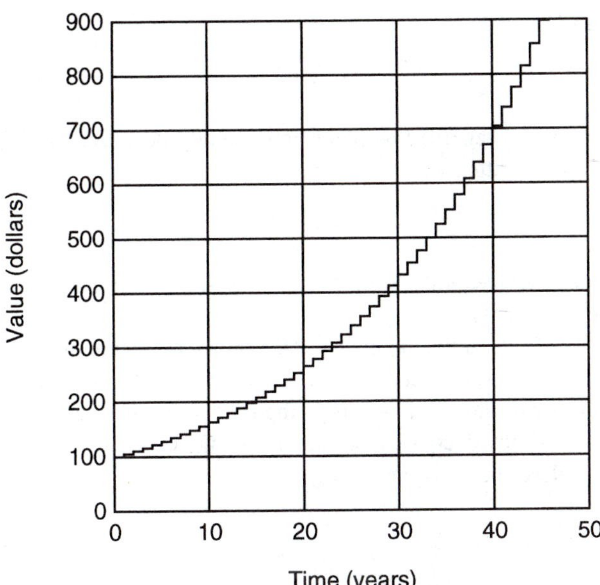

FIGURE 2.1. The amount in a savings account after x years, when the amount is compounded annually at 5% interest.

TABLE 2.2. *Amount on an initial investment of \$100, at 5% annual interest, with different methods of compounding.*

Month	Annual	Semiannual	Quarterly	Monthly	"Instant"
0	\$100.00	\$100.00	\$100.00	\$100.00	\$100.00
1	100.00	100.00	100.00	100.417	100.418
2	100.00	100.00	100.00	100.835	100.837
3	100.00	100.00	101.25	101.255	101.258
4	100.00	100.00	101.25	101.677	101.681
5	100.00	100.00	101.25	102.101	102.105
6	100.00	102.50	102.52	102.526	102.532
7	100.00	102.50	102.52	102.953	102.960
8	100.00	102.50	102.52	103.382	103.390
9	100.00	102.50	103.80	103.813	103.821
10	100.00	102.50	103.80	104.246	104.255
11	100.00	102.50	103.80	104.680	104.690
12	105.00	105.06	105.09	105.116	105.127

for monthly compounding and for "instant interest," are listed to the nearest tenth of a cent to show the slight difference between them. The table entries were calculated in the following way. Suppose that compounding is done N times a year. In x years, the number of compoundings is Nx. If the annual fractional rate of increase is b, the increase per compounding is b/N. [For six months at 5% ($b=0.05$), the increase is 2.5%, for three months, it is 1.25%, etc.] The amount after x units of time (years) is, in analogy with Eq. (2.1),

$$y=y_0\left[1+\left(\frac{b}{N}\right)\right]^{Nx}. \tag{2.2}$$

Recall (refer to Appendix C) that $(a)^{bc}=(a^b)^c$. The expression for y can be written as

$$y=y_0\left[\left(1+\frac{b}{N}\right)^N\right]^x. \tag{2.3}$$

Most calculus textbooks show that the quantity

$$\left(1+\frac{b}{N}\right)^N\rightarrow e^b$$

as N becomes very large. (Rather than prove this fact here, we give numerical examples in Table 2.3 for two different values of b.) Therefore, Eq. (2.3) can be rewritten as

$$y=y_0e^{bx}. \tag{2.4}$$

To calculate the amount for instant interest, it is necessary only to multiply the fractional growth rate per unit time b by the length of the time interval and then look up the exponential function of this amount in a table or evaluate it with a computer or calculator. The exponential function is plotted in Fig. 2.2. (The meaning of negative values of x will be considered in the next section.) This function increases more and more rapidly as x increases. This is

TABLE 2.3. *Numerical examples of the convergence of* $(1+b/N)^N$ to e^b as N becomes large.

N	$b=1$	$b=.05$
10	2.594	1.0511
100	2.705	1.0513
1000	2.717	1.0513
e^b	2.718	1.0513

expected, since the rate of growth is always proportional to the present amount. This is also reflected in the following property of the exponential function:

$$\frac{d}{dx}(e^{bx})=be^{bx}. \tag{2.5}$$

This means that the function $y=y_0e^{bx}$ has the property that

$$\frac{dy}{dx}=by. \tag{2.6}$$

Any constant multiple of the exponential function e^{bx} has the property that its rate of growth is b times the function itself. Whenever we see the exponential function, we know that an equation such as Eq. (2.6) is satisfied by it. Whenever we have a problem in which the growth rate of something is proportional to the present amount, we can expect to have an exponential solution. Notice that for time intervals Δx that are not too large, Eq. (2.6) implies that

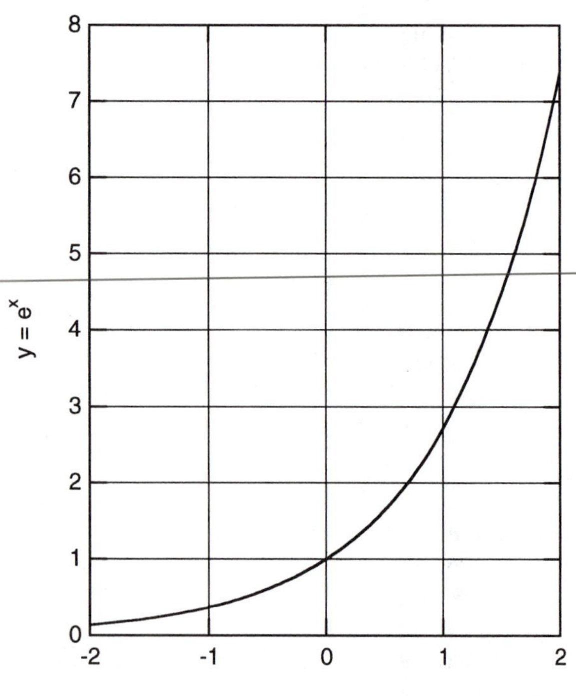

FIGURE 2.2. A graph of the exponential function $y=e^x$.

$$\Delta y=(b\Delta x)y.$$

This again says that the increase in y is proportional to y itself.

The independent variable in this discussion has been x. It can represent time, in which case b is the fractional growth rate per unit time; distance, in which case b is the fractional growth per unit distance; or something else. We could, of course, use another symbol such as t for the independent variable, in which case we would have

$$\frac{dy}{dt}=by,$$

$$y=y_0e^{bt}.$$

2.2. EXPONENTIAL DECAY

Figure 2.2 shows the exponential function for negative values of x as well as positive ones. (Remember that $e^{-x}=1/e^x$.) To see what this means, consider a bank account in which no interest is credited, but from which 5% of what remains is taken each year. If the initial balance is \$100, \$5 is removed the first year to leave \$95.00. In the second year 5% of \$95 or \$4.75 is removed. In the third year 5% of \$90.25 or \$4.51 is removed. The annual decrease in y becomes less and less as y becomes less and less. The mathematical equations developed in the preceding section will also handle this situation: it is necessary only to call b the fractional rate of decay and let it have a negative value, $-|b|$. Equation (2.1) then has the form

$$y=y_0(1-|b|)^x$$

and Eq. (2.4) is

$$y=y_0e^{-|b|x}. \tag{2.7}$$

Often b is regarded as being intrinsically positive, and Eq. (2.7) is written as

$$y=y_0e^{-bx}. \tag{2.8}$$

One could equally well write $y=y_0e^{bx}$ and regard b as being negative.

The radioactive isotope ^{99m}Tc (read as technetium-99) has a fractional decay rate $b=0.1155\ \text{h}^{-1}$. If the relative number of atoms at $t=0$ is 1, the relative number at subsequent times decreases as shown in Fig. 2.3. The equation that describes this curve is

$$f=\frac{y}{y_0}=e^{-bt}, \tag{2.9}$$

where t is the elapsed time in hours and $b=0.1155\ \text{h}^{-1}$. The product bt must be dimensionless, since it is in the exponent.

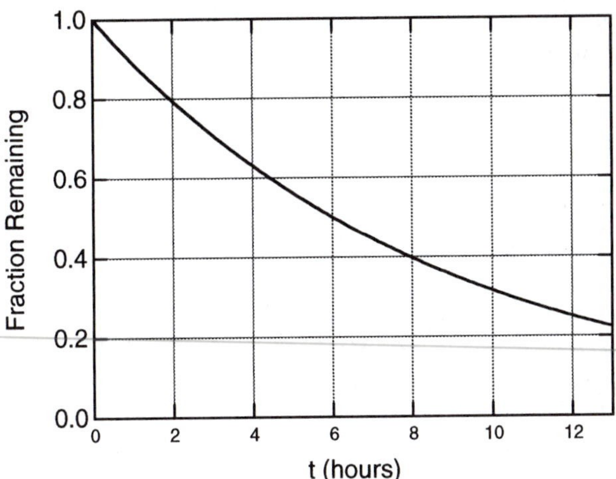

FIGURE 2.3. A plot of the fraction of nuclei of ^{99m}Tc surviving at time *t*.

People often talk about the *half-life* $T_{1/2}$, which is the length of time required for f to decrease to one-half. From inspection of Fig. 2.3, the half-life is 6 h. This can also be determined from Eq. (2.9):

$$0.5 = e^{-bT_{1/2}}.$$

From a table of exponentials, one finds that $e^{-x}=0.5$ when $x=0.693\ 15$. This leads to the very useful relationship

$$bT_{1/2}=0.693$$

or

$$T_{1/2}=\frac{0.693}{b}. \tag{2.10}$$

For the case of ^{99m}Tc, $T_{1/2}=0.693/0.1155=6$ h.

One can also speak of a doubling time if the exponential is positive. In that case, $2=e^{bT_2}$, from which

$$T_2=\frac{0.693}{b}. \tag{2.11}$$

2.3. SEMILOG PAPER

A special kind of graph paper makes the analysis of exponential growth and decay problems much simpler. If one takes logarithms (to any base) of Eq. (2.4) one has

$$\log y = \log y_0 + bx \log e. \tag{2.12}$$

If the dependent variable is considered to be $u=\log y$, and since $\log y_0$ and $\log e$ are constants, this equation is of the form

$$u=c_1+c_2x. \tag{2.13}$$

The graph of u vs x is a straight line with positive slope if b is positive and negative slope if b is negative.

Special graph paper is available on which the vertical axis is marked in a logarithmic fashion. The graph can be plotted without having to calculate any logarithms. Figure 2.4 shows a plot of the exponential function of Fig. 2.2, for both positive and negative values of x. First, note how to read the vertical axis. A given distance along the axis always corresponds to the same multiplicative factor. Each cycle represents a factor of 10. To use the paper, it is necessary first to mark off the decades with the desired values. In Fig. 2.4 the decades have been marked 0.1, 1.0, 10.0, 100. The "6" that lies between 0.1 and 1.0 is 0.6; the "6" between 1.0 and 10.0 is 6.0; the "6" between 10 and 100 represents 60 and so forth. The paper can be imagined to go vertically forever in either direction; one never reaches zero. Figure 2.4 has two examples marked on it with dashed lines. The first shows that for $x=-1.0$, $y=0.36$; the second shows that for $x=+1.5$, $y=4.5$.

The graph paper is most useful for plotting data that you suspect may have an exponential relationship. If the data plot as a straight line, your suspicions are confirmed. From the straight line, you can determine the value of b. Figure 2.5 is a plot of the intensity of light that passed through an absorber in a hypothetical example. The independent variable is absorber thickness. The decay is exponential, except for the last few points, which may be high because of

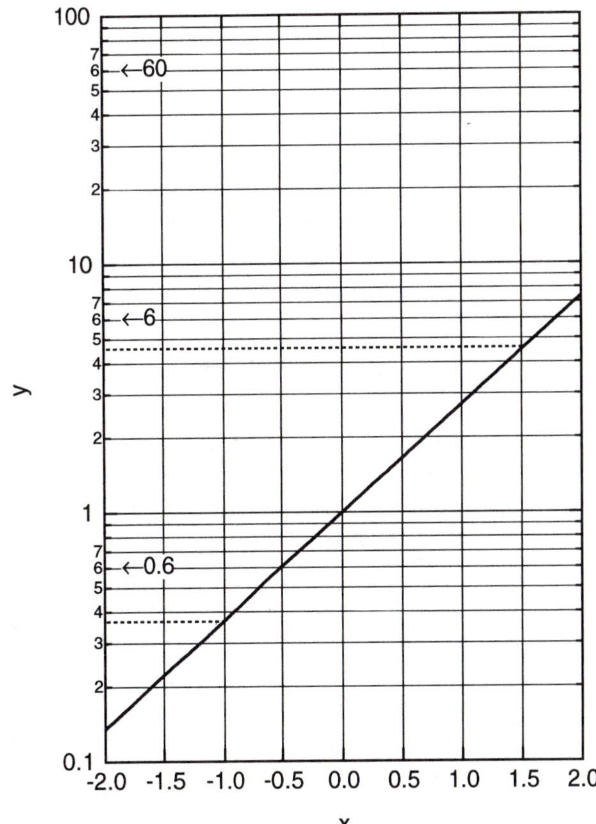

FIGURE 2.4. A plot of the exponential function on semilogarithmic graph paper.

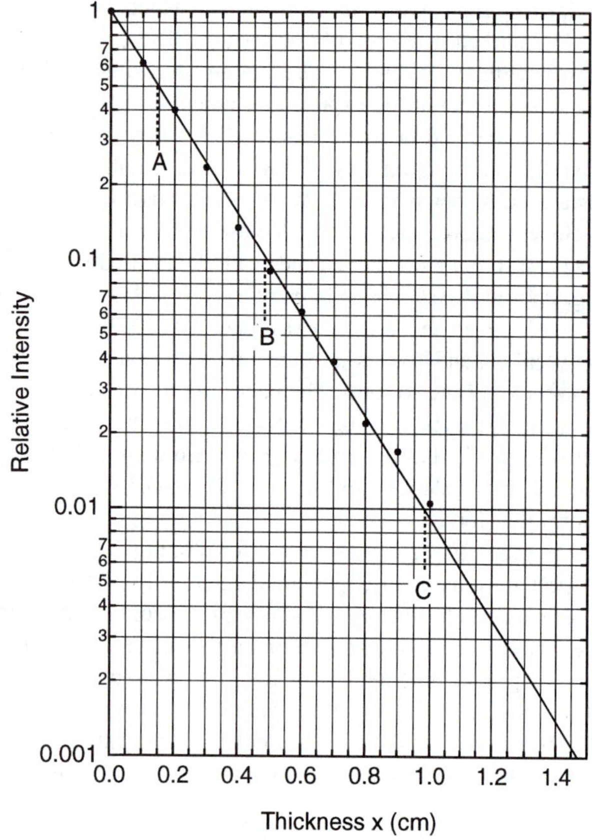

FIGURE 2.5. A semilogarithmic plot of the intensity of light after it has passed through an absorber of thickness *x*.

experimental error. (As the intensity of the light decreases, it becomes harder to measure accurately.) We wish to determine the decay constant in $y=y_0e^{-bx}$. One way to do it would be to note (dashed line *A* in Fig. 2.5) that the half-distance is 0.145 cm, so that, from Eq. (2.10), since b is negative,

$$b=-\frac{0.693}{0.145}=-4.8 \text{ cm}^{-1}.$$

This technique can be inaccurate because it is difficult to read the graph accurately. It is more accurate to use a portion of the curve for which y changes by a factor of 10 or 100. The general relationship is

$$y=y_0e^{bx},$$

where the value of b can be positive or negative. If two different values of x are selected, one can write

$$\frac{y_2}{y_1}=\frac{y_0e^{bx_2}}{y_0e^{bx_1}}=e^{b(x_2-x_1)}.$$

If $y_2/y_1=10$, then this equation has the form

$$10=e^{bX_{10}}$$

where $X_{10}=x_2-x_1$ when $y_2/y_1=10$. From a table of exponentials, $bX_{10}=2.303$, so that

$$b=\frac{2.303}{X_{10}}. \tag{2.14}$$

The same procedure can be used to find b using a factor of 100 change in y:

$$b=\frac{4.605}{X_{100}}. \tag{2.15}$$

If the curve represents a decaying exponential, then $y_2/y_1=10$ when $x_2<x_1$, so that $X_{10}=x_2-x_1$ is negative. Equation (2.14) then gives the negative value of b.

As an example, consider the exponential decay in Fig. 2.5. Using points B and C, we have $x_1=0.97$, $y_1=10^{-2}$, $x_2=0.48$, $y_2=10^{-1}$, $X_{10}=0.48-0.97=-0.49$. Therefore

$$b=\frac{2.303}{-0.49}=-4.7,$$

which is a more accurate determination than the one we made using the half-life.

2.4. VARIABLE RATES

The equation $dy/dx=by$ says that y grows or decays at a rate that is proportional to y. The constant b is the fractional rate of growth or decay. It is possible to define the fractional rate of growth or decay even if it is not constant but is a function of x:

$$b(x)=\frac{1}{y}\frac{dy}{dx}. \tag{2.16}$$

Semilogarithmic graph paper can be used to analyze the curve even if b is not constant. Since $d(\ln y)/dy=1/y$, the chain rule for evaluating derivatives gives

$$\frac{d}{dx}(\ln y)=\frac{1}{y}\frac{dy}{dx}=b.$$

This means that $b(x)$ is the slope of a plot of ln y vs x. A semilogarithmic plot of y vs x is shown in Fig. 2.6. The straight line is tangent to the curve and decays with a constant rate equal to $b(x)$ at the point of tangency. The value of b for the tangent line can be determined using the methods in the previous section. A second tangent line at a later time in Fig. 2.6 has a smaller value of the decay rate.

If finite changes Δx and Δy have been measured, they may be used to estimate $b(x)$ directly from Eq. (2.16). For example, suppose that $y=100\,000$ people and that in $\Delta x=1$ year there is a change $\Delta y=-37$. In this case Δy is very small compared to y, so we can say that $b=(1/y)(\Delta y/\Delta x)=-37\times10^{-5}$. If the only cause of change in this population is deaths, the absolute value of b is called the death rate.

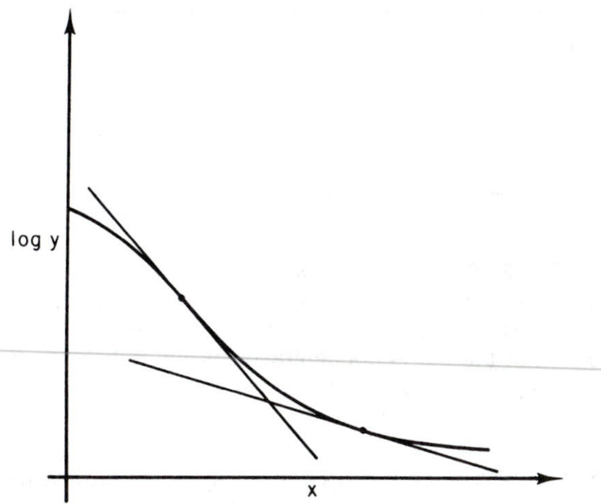

FIGURE 2.6. A semilogarithmic plot of y vs x when the decay rate is not constant. Each tangent line represents the instantaneous decay rate for that value of x.

A plot of the number of people surviving in a population, all of whom have the same disease, can provide information about the prognosis for that disease. The death rate is equivalent to the decay constant. An example of such a plot is shown in Fig. 2.7. Curve A shows a disease for which the death rate is constant. Curve B shows a disease with an initially high death rate which decreases with time; if the patient survives the initial period, the prognosis is much better. Curve C shows a disease for which the death rate increases with time.

Surprisingly, there are a few diseases that actually have death rates independent of the duration of the disease (Zumoff 1966). Any discussion of mortality should be made in terms of the surviving population, since any further deaths must come from that group. Nonetheless, one sometimes finds results in the literature reported in terms of the cumulative fraction of patients who have died. Figure 2.8 shows the survival of patients with congestive heart failure for a period of nine years. The data are taken from the Framingham study [McKee *et al.* (1971)]; the death rate is constant during this period. For a more detailed discussion of various possible survival distributions, see Clark (1975).

As long as b has a constant value, it makes no difference what time is selected to be $t=0$. To see this, consider two different time scales, shifted with respect to each other so that

$$t'=t_0+t.$$

The value of y decays exponentially with constant rate:

$$y=y_0e^{-bt}.$$

In terms of the shifted time t', the value of y is

$$y=y_0e^{-bt}=y_0e^{-b(t'-t_0)}=(y_0e^{bt_0})e^{-bt'}.$$

This has the same form as the original expression for $y(t)$. The value of y_0' is $y_0e^{bt_0}$, which reflects the fact that $t'=0$ occurs at an earlier time than $t=0$, so $y_0'>y_0$.

If the decay rate is not constant, then the origin of time becomes quite important. Usually there is something about the problem that allows $t=0$ to be determined. Figure 2.9 shows survival after a heart attack (myocardial infarct). The

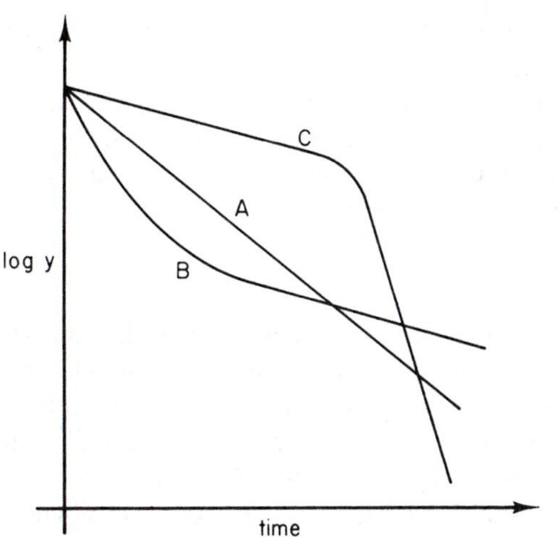

FIGURE 2.7. Semilogarithmic plots of the fraction of a population surviving in three different diseases. The death rates (decay constants) depend on the duration of the disease.

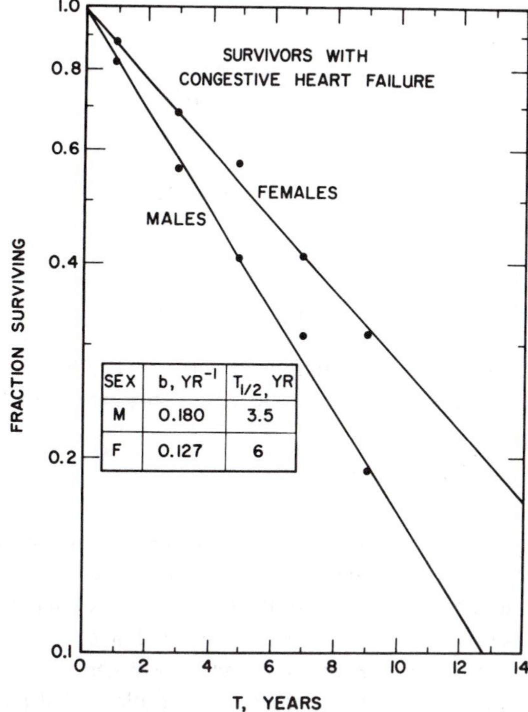

FIGURE 2.8. Survival of patients with congestive heart failure. From R. K. Hobbie (1973). Teaching exponential growth and decay: Examples from medicine. *Am. J. Phys.* **41**: 389–393. Reproduced by permission of the *American Journal of Physics.* Data are from McKee *et al.* (1971).

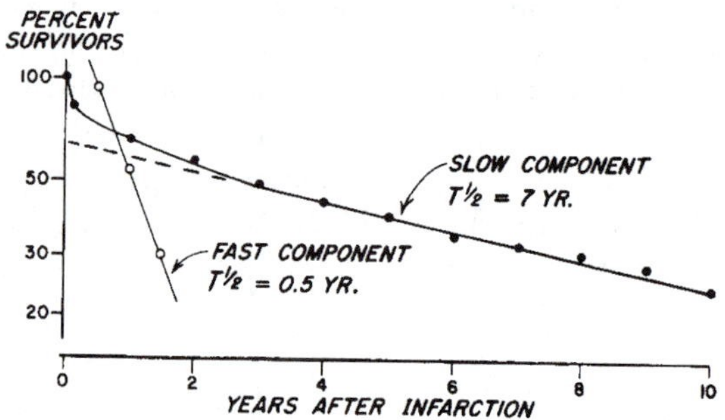

FIGURE 2.9. The fraction of patients surviving after a myocardial infarction (heart attack) at $t=0$. From B. Zumoff, H. Hart, and L. Hellman (1966). Considerations of mortality in certain chronic diseases. *Ann. Intern. Med.* **64**: 595–601. Reproduced by permission of *Annals of Internal Medicine.* Drawing courtesy of Prof. Zumoff.

time of the initial infarct defines $t=0$; if the origin had been started two or three years after the infarct, the large initial death rate would not have been seen.

As long as the rate of increase can be written as a function of the independent variable, Eq. (2.16) can be rewritten as

$$\frac{dy}{y}=b(x)dx.$$

This can be integrated:

$$\int_{y_1}^{y_2}\frac{dy}{y}=\int_{x_1}^{x_2}b(x)dx,$$

$$\ln(y_2/y_1)=\int_{x_1}^{x_2}b(x)dx,$$

$$\frac{y_2}{y_1}=\exp\left(\int_{x_1}^{x_2}b(x)dx\right). \tag{2.17}$$

If we can integrate the right-hand side analytically, numerically, or graphically, we can determine the ratio y_2/y_1.

2.5. CLEARANCE

In some cases in physiology, the amount of a substance may decay exponentially because the rate of removal is proportional to the concentration of the substance (amount per unit volume) instead of to the total amount. For example, the rate at which the kidneys excrete a substance may be proportional to the concentration in the blood that passes through the kidneys, while the total amount depends on the total blood volume. This is shown schematically in Fig. 2.10. The large box on the left represents the total blood volume V. It contains a total amount of some substance, y. If the blood is well mixed, the concentration is $C=y/V$.

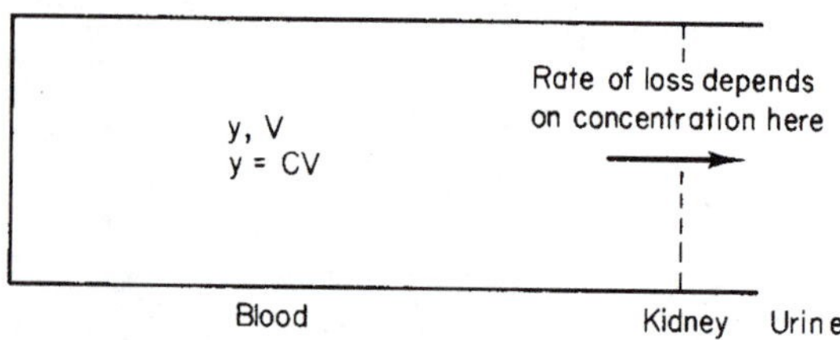

FIGURE 2.10. A case in which the rate of removal of a substance from the blood depends on the concentration, not on the total amount of substance in the blood. Adding more blood with the same concentration of the substance would not change the rate of removal.

The removal process takes place only at the dashed line, at a rate proportional to C. The equation describing the change of y is

$$\frac{dy}{dt}=-KC=-K\left(\frac{y}{V}\right). \tag{2.18}$$

The proportionality constant K is called the *clearance*. Its units are $m^3\,s^{-1}$. The equation is the same as Eq. (2.6) if K/V is substituted for b. The solution is

$$y=y_0e^{-(K/V)t}. \tag{2.19}$$

The basic concept of clearance is best remembered in terms of Fig. 2.10. Other definitions are found in the literature. It sometimes takes considerable thought to show that the definitions are equivalent. A common one is *clearance is the volume of plasma from which y is completely removed per unit time*. To see that this definition is equivalent, imagine that all removal of y is accomplished by removing a volume V of the blood. The rate of loss of y is the concentration times the rate of volume removal:

$$\frac{dy}{dt}=\frac{dV}{dt}C. \tag{2.20}$$

(dV/dt is negative for removal.) Comparison with Eq. (2.18) shows that $|dV/dt|=K$.

As long as the compartment containing the substance is well mixed, the concentration will decrease uniformly throughout the compartment as y is removed. The concentration also decreases exponentially:

$$C=C_0e^{-(K/V)t}. \tag{2.21}$$

An example may help to clarify the distinction between b and K. Suppose that the substance is distributed in a fluid volume $V=18$ l. The substance has an initial concentration $C_0=3$ mg l^{-1} and the clearance is $K=2$ l h^{-1}. The total amount is $y_0=C_0V=3\times 18=54$ mg. The fractional decay rate is $b=K/V=1/9$ h^{-1}. The equations for C and y are

$$C=(3\text{ mg l}^{-1})e^{-t/9},$$

$$y=(54\text{ mg})e^{-t/9}.$$

At $t=0$ the initial rate of removal is $-dy/dt=54/9=6$ mg h^{-1}.

Now double the fluid volume to $V=36$ l without adding any more of the substance. The concentration falls to 1.5 mg l^{-1} although y_0 is unchanged. The rate of removal is also cut in half, since it is proportional to K/V and the clearance is unchanged. The concentration and amount are now

$$C=1.5e^{-t/18},$$

$$y=54e^{-t/18}.$$

The initial rate of removal is $-dy/dt=54/18=3$ mg h^{-1}. It is half as large as above, because C is now half as large.

If more of the substance were added along with the additional blood, the initial concentration would be unchanged, but y_0 would be doubled. The fractional decay rate would still be $K/V=1/18$ h^{-1}:

$$C=3.0e^{-t/18},$$

$$y=108e^{-t/18}.$$

The initial rate of disappearance would be $-dy/dt=108/18=6$ mg h^{-1}. It is the same as in the first case, because the initial concentration is the same.

2.6. MULTIPLE DECAY PATHS

It is possible to have several independent paths by which y can disappear. For example, there may be several competing ways by which a radioactive nucleus can decay, a radioactive isotope given to a patient may decay radioactively and be excreted biologically at the same time, a substance in the body can be excreted in the urine and metabolized by the liver, or patients may die of several different diseases.

In such situations the total decay rate b is the sum of the individual rates for each process, as long as the processes act independently and the rate of each is proportional to the present amount (or concentration) of y:

$$\frac{dy}{dt}=-b_1y-b_2y-b_3y+\cdots=-(b_1+b_2+b_3+\cdots)y$$

$$=-by. \qquad (2.22)$$

The equation for the disappearance of y is the same as before, with the total decay rate being the sum of the individual rates. The rate of disappearance of y by the ith process *is not* dy/dt *but is* $-b_iy$.

Instead of decay rates, one can use half-lives. Since

$$b=b_1+b_2+b_3+\cdots,$$

the total half-life T is given by

$$\frac{0.693}{T}=\frac{0.693}{T_1}+\frac{0.693}{T_2}+\frac{0.693}{T_3}+\cdots$$

or

$$\frac{1}{T}=\frac{1}{T_1}+\frac{1}{T_2}+\frac{1}{T_3}+\cdots. \qquad (2.23)$$

2.7. DECAY PLUS INPUT AT A CONSTANT RATE

Suppose that in addition to the removal of y from the system at a rate $-by$, there is leakage of y *into* the system at a constant rate a, independent of y and t. The net rate of change of y is given by

$$\frac{dy}{dt}=a-by. \qquad (2.24)$$

It is often easier to write down a differential equation describing a problem than it is to solve it. In this case the solution to the equation and the techniques for solving it are well known. However, a good deal can be learned about the solution by examining the equation itself.

Suppose that $y(0)=0$. Then the equation at $t=0$ is

$$\frac{dy}{dt}=a,$$

and y initially grows at a constant rate a. As y builds up, the rate of growth decreases from this value because of the $(-by)$ term. Finally when

$$a-by=0,$$

dy/dt is zero and y stops growing. This is enough information to make the sketch in Fig. 2.11.

The equation is solved in Appendix F. The solution is

$$y=\frac{a}{b}(1-e^{-bt}). \qquad (2.25)$$

The derivative of y is

$$\frac{dy}{dt}=\frac{a}{b}(-1)(-b)e^{-bt}=ae^{-bt}.$$

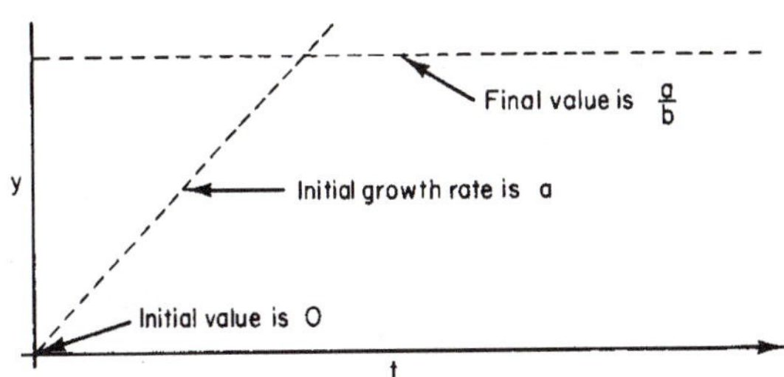

FIGURE 2.11. Sketch of the initial slope a and final value a/b of y when $y(0)=0$.

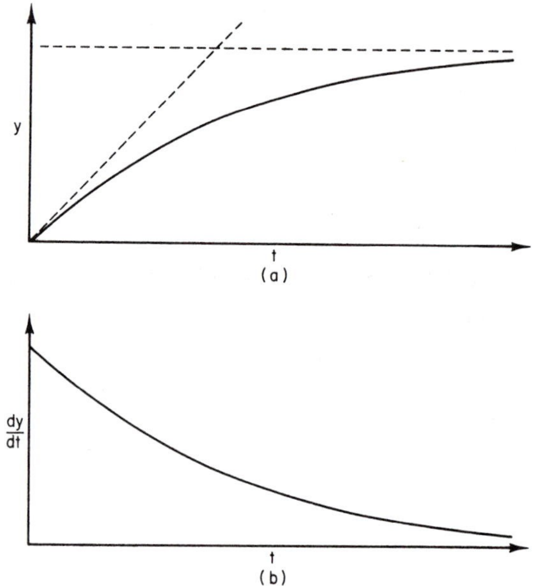

FIGURE 2.12. (a) Plot of $y(t)$. (b) Plot of dy/dt.

You can verify by substitution that Eq. (2.25) satisfies Eq. (2.24). The solution does have the properties sketched in Fig. 2.11, as you can see from Fig. 2.12. The initial value of dy/dt is a, and it decreases exponentially to zero. When t is large, the exponential term in y vanishes, leaving $y=a/b$.

2.8. DECAY WITH MULTIPLE HALF-LIVES AND FITTING EXPONENTIALS

Sometimes y is a mixture of two or more quantities, each decaying at a constant rate. It might represent a mixture of radioactive isotopes, each decaying at its own rate. A biological example is the survival of patients after a myocardial infarct (Fig. 2.9). The death rate is not constant, and many models can be proposed to explain why. One possible model is that there are two distinct classes of patients immediately after the infarct. Each class has an associated death rate which is constant. After three years, virtually none of the subgroup with the higher death rate remains. Another model is that the death rate is higher right after the infarct for all patients. This higher death rate is due to causes associated with the myocardial injury: irritability of the muscle, arrhythmias in the heartbeat, the weakening of the heart wall at the site of the infarct, and so forth. After many months, the heart has healed, scar tissue has replaced the necrotic muscle, and deaths from these causes no longer occur.

Whatever the cause, it is sometimes useful to fit a set of experimental data with a sum of exponentials. It should be clear from the discussion of survival after myocardial infarction that simply fitting with an exponential or a sum of exponentials does not prove anything about the decay mechanism.

If y consists of two quantities, y_1 and y_2, each with its own decay rate, then

$$y=y_1+y_2=A_1e^{-b_1t}+A_2e^{-b_2t}. \tag{2.26}$$

Suppose that $b_1>b_2$, so that y_1 decays more rapidly than y_2. After enough time has elapsed, y_1 will be much less than y_2, and its effect on a semilog plot will be negligible. A typical plot of y is curve A in Fig. 2.13. Line B can then be drawn through the data and used to determine A_2 and b_2. This line is extrapolated back to earlier times, so that y_2 can be subtracted from y to give an estimate for y_1. For example, at point C ($t=4$), $y=400$, $y_2=300$, and $y_1=100$. At $t=0$, $y_1=1500-500=1000$. For times greater than 5 s, the curves for y and y_2 are close together, and error in reading the graph produces considerable scatter in y_1. Once several values of y_1 have been determined, line D is drawn, and parameters A_1 and b_1 are estimated.

This technique can be extended to several exponentials. However it becomes increasingly difficult to extract meaningful parameters as more exponentials are used, because the estimated parameters for the short-lived terms are very

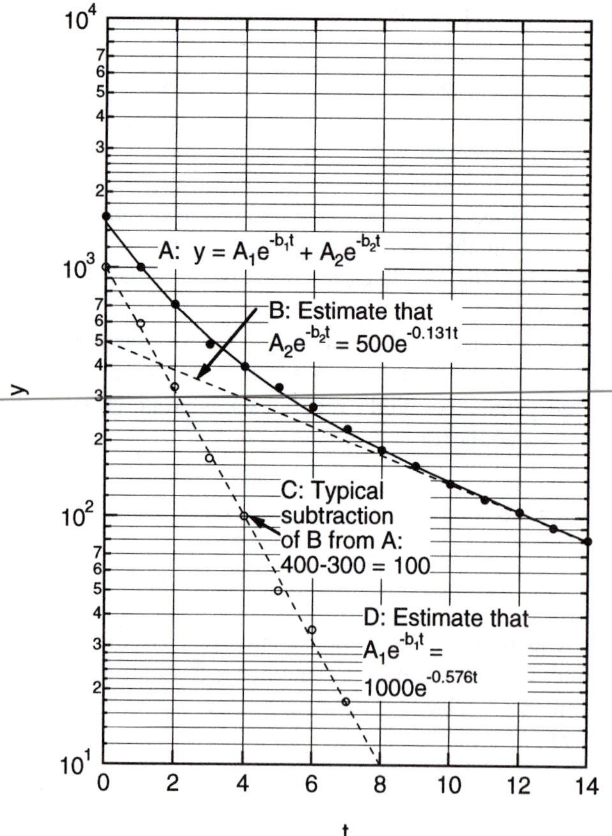

FIGURE 2.13. Fitting a curve with two exponentials.

sensitive to the initial guess for the parameters of the longest-lived term. For a discussion of this problem, see Riggs (1970), pp. 146–163.

2.9. THE LOGISTIC EQUATION

Exponential growth cannot go on forever. Sometimes a growing population will level off at some constant value. Other times the population will grow and then crash. One model that exhibits leveling off is the *logistic model*, described by the differential equation

$$\frac{dy}{dt}=b_0 y\left(1-\frac{y}{y_\infty}\right), \tag{2.27}$$

where b_0 is a constant. This equation has constant solutions $y=0$ and $y=y_\infty$. If $y \ll y_\infty$, then the equation is approximately $dy/dt=b_0 y$ and y grows exponentially. As y becomes larger, the term in parentheses reduces the rate of increase of y, until y reaches the saturation value y_∞. This might happen, for example, as the population begins to consume a significant fraction of the food supply, causing the birth rate to decrease or the mortality rate to increase.

If the initial value of y is y_0, the solution of Eq. (2.27) is

$$y(t)=\frac{1}{\dfrac{1}{y_\infty}+\left(\dfrac{1}{y_0}-\dfrac{1}{y_\infty}\right)e^{-b_0 t}}=\frac{y_0 y_\infty}{y_0+(y_\infty-y_0)e^{-b_0 t}}. \tag{2.28}$$

You can easily verify that $y(0)=y_0$ and $y(\infty)=y_\infty$. A plot of the solution is given in Fig. 2.14, along with exponential growth with the same value of b_0.

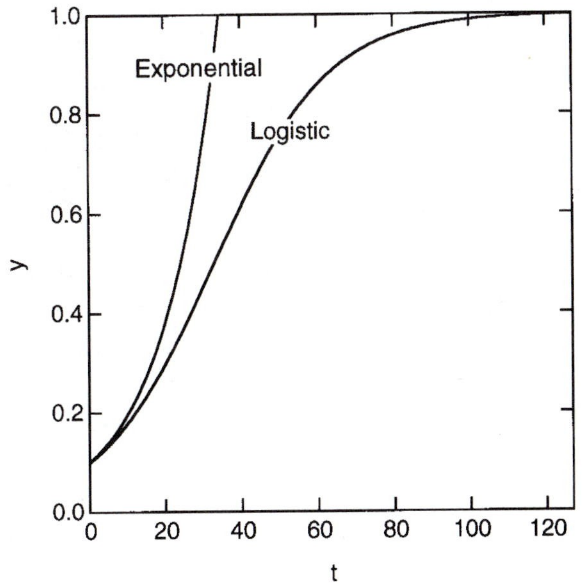

FIGURE 2.14. Plot of the solution of the logistic equation when y_0=0.1, y_∞=1.0, b_0=0.0667. Exponential growth with the same values of y_0 and b_0 is also shown.

Another way to think of Eq. (2.27) is that it has the form $dy/dt=b(y)y$, where $b(y)=b_0(1-y/y_\infty)$ is now a function of the dependent variable y instead of the independent variable t. As y grows toward the asymptotic value, the growth rate $b(y)$ decreases linearly to zero.

The logistic model was an early and very important model for population growth. It provides good fits in a few cases, but there are now many more sophisticated models in population biology.

2.10. LOG–LOG GRAPH PAPER, POWER LAWS, AND SCALING

This section considers the use of graph paper on which both scales are logarithmic. It is usually called log–log graph paper. It is useful when x and y are related by the function

$$y=Bx^n. \tag{2.29}$$

Notice the difference between this and the exponential function: here the independent variable x is *raised to a constant power*, while in the exponential case x is *in the exponent*. It also leads to a discussion of *scaling*, whereby simple physical arguments lead to important conclusions about the variations between species in size, shape, metabolic rate, and the like.

By taking logarithms of both sides of Eq. (2.29), we get

$$\log y=\log B+n \log x. \tag{2.30}$$

This is a linear relationship between $u=\log y$ and $v=\log x$:

$$u=\text{const}+nv. \tag{2.31}$$

Therefore a graph of u vs v is a straight line with slope n. The slope can be positive or negative and need not be an integer. Figure 2.15 shows plots of $y=x$, $y=x^2$, $y=x^{1/2}$, and $y=x^{-1}$. The slope can be determined from the graph by taking $\Delta u/\Delta v$. The value of B is determined either by substituting particular values of y and x in Eq. (2.29) after n is known, or by determining the value of y when $x=1$, in which case $x^n=1$ for any value of n, so n need not be known.

Figure 2.16 shows how the curves change when B is changed while $n=1$. The curves are all parallel to each other. Multiplying by B is equivalent to adding a constant to $\log y$.

If the expression is not of the form $y=Bx^n$ but has an added term, it will not plot as a straight line on log–log paper. Figure 2.16 also shows a plot of $y=1+x$, which is not a straight line. (Of course, for very large values of x, the logarithm of $1+x$ becomes nearly indistinguishable from $\log x$, and the line appears straight.)

When the slope is constant, n can be determined from the slope $\Delta v/\Delta u$ measured with a ruler on the log–log paper. When determining the slope in this way *one must be sure that the length of a cycle is the same in each direction on*

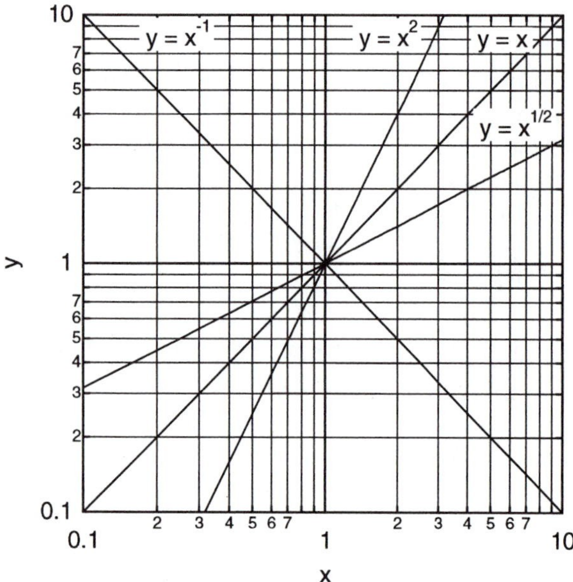

FIGURE 2.15. log–log plots of $y=x^n$ for different values of n. When $x=1$, $y=1$ in every case.

the graph paper. To repeat the warning: it is easy to get a rough idea of the exponent from inspection of the slope of the log–log plot in Fig. 2.15 because on commercial log–log graph paper the distance spanned by a decade or cycle is the same on both axes. Some magazines routinely show log–log plots in which the distance spanned by a decade is not the same on both axes. Moreover, commercial graphing software does not impose this constraint on log–log plots, so it is becoming less and less likely that you can determine the exponent by glancing at the plot. Be careful!

When using a spreadsheet or other graphing software, it is often useful to make an extra column that contains the calculated variable $y_{\text{calc}}=Ax^m$ with the values for A and m stored in two cells of the spreadsheet. If you plot this column as a line, and your real data as points without a line, then you can change the parameters while inspecting the graph to find the values that give the best fit.

An example of the use of log–log graph paper is a plot of Poiseuille flow of fluid through a tube vs tube radius when the pressure gradient along the tube is constant. It was shown in Chap. 1 that an r^4 dependence is expected.

As a final example, consider the relation of daily food consumption to body mass. This will introduce us to simple scaling arguments. As a first model, we might suppose that each kilogram of tissue has the same metabolic requirement, so that food consumption should be proportional to body mass. However, there is a problem with this argument. Most of the food that we consume is converted to heat. The various mechanisms to lose heat—radiation, convection and perspiration—are all roughly proportional to the surface area of the body rather than its mass. (This statement neglects the fact that considerable evaporation takes place through the lungs and that the body can control the rate of heat loss through sweating and shivering.) If all persons were the same shape, then the total surface area would be proportional to H^2, where H is the height. The total volume and mass would be proportional to H^3, so H would be proportional to $M^{1/3}$. Therefore the surface area would be proportional to $(M^{1/3})^2$ or $M^{2/3}$. Figure 2.17 plots H and M

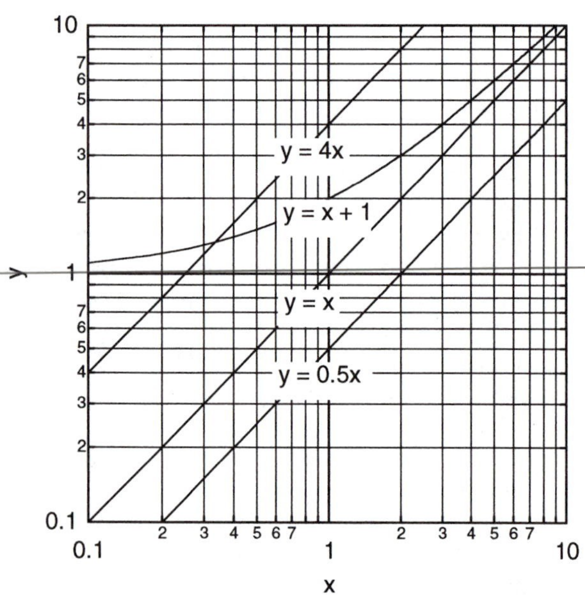

FIGURE 2.16. log–log plots of $y=Bx$, showing how the curves shift on the paper as B changes. Since $n=1$ for all the curves, they all have the same slope. There is also a plot of $y=x+1$, to show that a polynomial does not plot as a straight line.

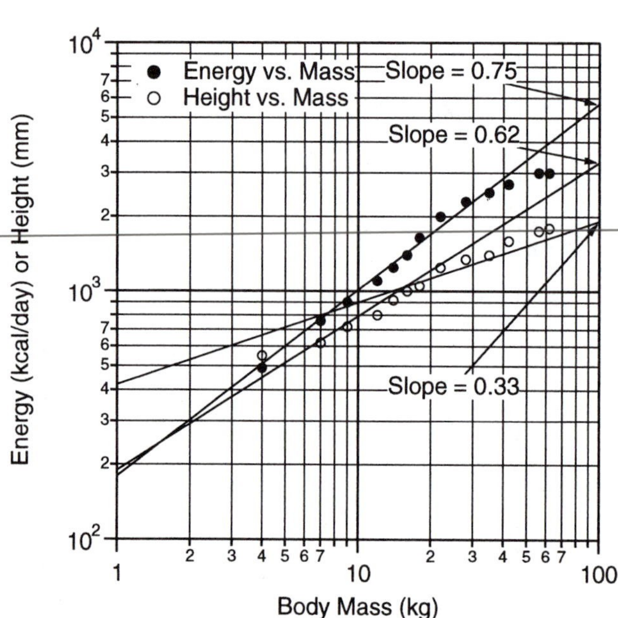

FIGURE 2.17. Plot of daily food requirement F and height H vs mass M for growing children. Data are from Kempe, Silver, and O'Brien (1970), p. 90.

and total daily food requirement F vs body mass M for growing children [Kempe, Silver, and O'Brien (1970), p. 90].

Neither of the models proposed above fits the data very well. At early ages H is more nearly proportional to $M^{0.62}$ than to $M^{1/3}$. For older children, when the shape of the body has stopped changing, an $M^{0.33}$ dependence does fit better. This better fit occurs for masses greater than 23 kg, which correspond to ages over 6 years. The slope of the $F(M)$ curve is 0.75. This is less than the 1.0 of the model that food consumption is proportional to the mass and greater than the 0.67 of the model that food consumption is proportional to surface area. This 3/4-power dependence is remarkable and is seen in nearly all species, from one-celled organisms to large mammals. Peters (1983) quotes work by Hemmingsen (1960) that shows the standard metabolic rates for many species can be fitted by the following. The standard metabolic rate is in watts and mass in kilograms. (*Standard* means as close to resting or basal as possible.) For unicellular organisms at 20 °C,

$$R_{\text{unicellular}} = 0.018M^{0.751}. \tag{2.32a}$$

The range of masses extended from 10^{-15} to 10^{-6} kg. For poikilotherms (organisms such as fish whose body temperature is the same as the surroundings) at 20 °C (masses from 10^{-8} to 10^{2} kg),

$$R_{\text{poikilotherm}} = 0.14M^{0.751}, \tag{2.32b}$$

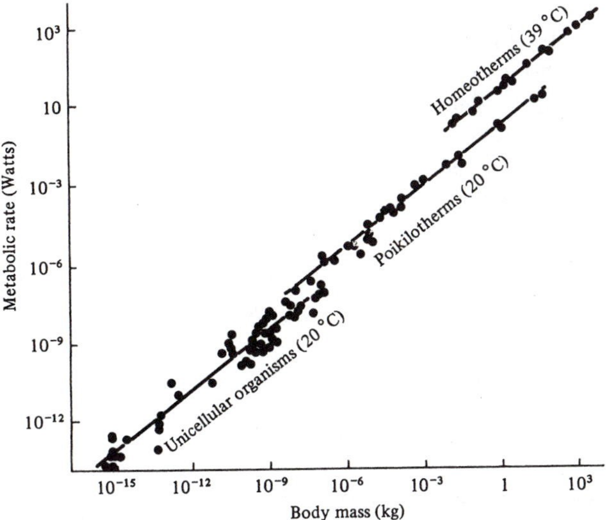

FIGURE 2.18. Plot of resting metabolic rate vs body mass for many different organisms. Graph is from R. H. Peters (1983). *The Ecological Implications of Body Size.* Cambridge, Cambridge University Press. Modified from A. M. Hemmingsen (1960). Energy metabolism as related to body size and respiratory surfaces, and its evolution. *Reports of the Steno Memorial Hospital and Nordisk Insulin Laboratorium.* **9** (Part II): 6–110. Used with permission.

and for homeotherms (animals that can maintain their body temperature independent of the surroundings) at 39 °C (masses from 10^{-2} to 10^{3} kg),

$$R_{\text{homeotherm}} = 4.1M^{0.751}. \tag{2.32c}$$

Peters's graph is shown in Fig. 2.18. Various models to explain this nearly 3/4-power dependence have been proposed [see McMahon (1973) or Peters (1983)].

SYMBOLS USED IN CHAPTER 2

Symbol	Use	Units	First used on page
a	Rate of input of a substance	s^{-1}	33
b, b_0	Rate of growth or decay	s^{-1}, h^{-1}	27
c_1, c_2	Constants		29
f	Fraction		28
m, n	Exponent in power law relationship		35
t	Time	s	28
u	Logarithm of dependent variable		29
v	Logarithm of independent variable		35
x	General independent variable, time		27
y	General dependent variable		27
y	Amount of substance in plasma	kg, mg	32
x_0, y_0	Initial value of x or y		26
y_∞	Saturation value of y		35
A	Constant		34
B	Constant		35
C	Concentration	kg m^{-3}, etc.	32
F	Food requirement	kcal	36
H	Body height	m	36
K	Clearance	$m^3\,s^{-1}$	32
M	Body mass	kg	36
N	Number of compoundings per year		27
R	Standard metabolic rate	W	37
$T_{1/2}$	Half-life	s, etc.	29
T_2	Doubling time	s	29
V	Volume	m^3	32
X_{10}	Change in x for a factor-of-10 change in y		30
X_{100}	Change in x for a factor-of-100 change in y		30

PROBLEMS

Section 2.1

2.1. Suppose that you are 20 years old and have an annual income of $20,000. You plan to work for 40 years. If inflation takes place at a rate of 10% per year, what income would you need at age 60 to have the same buying power you have now? Ignore taxes. Make the calculation assuming that (a) the inflation is 10% and occurs once a year and (b) the inflation is continuous but at a 10% annual rate.

2.2. The number e is defined by $\lim_{n\to\infty}[(1+1/n)^n]$.

(a) Calculate values of $(1+1/n)^n$ for $n=1$, 2, 4, 8, and 16.

(b) The binomial formula is

$$(1+a)^n=1+na+\frac{n(n-1)}{2!}a^2+\frac{n(n-1)(n-2)}{3!}a^3+\cdots.$$

Use the binomial formula to obtain a series for

$$e^x=\lim_{n\to\infty}\left(1+\frac{x}{n}\right)^n.$$

[See also Appendix D, Eq. (D2).]

2.3. A child with acute lymphocytic leukemia has approximately 10^{12} leukemic cells when the disease is clinically apparent.

(a) If a cell is about 8 μm in diameter, estimate the total mass of leukemic cells.

(b) Cure requires killing every single cell. The doubling time for the cells is about 5 days. If all cells were killed except for one, how long would it take for the disease to become apparent again?

(c) A standard regimen of chemotherapy reduces the number of cells to 10^9. How long a remission would you expect? What if the number were reduced to 10^6?

2.4. Suppose that tumor cells within the body reproduce at rate r, so that the number is given by $y=y_0e^{rt}$. Each time a chemotherapeutic agent is given it destroys a fraction f of the cells then existing. Make a semilog plot showing y as a function of time for several administrations of the drug, separated by time T. What different cases must you consider for the relation among f, T, and r?

2.5. An exponentially growing culture of bacteria increases from 10^6 to 5×10^8 cells in 6 h. What is the time between successive cell divisions if (a) there is no cell mortality? and (b) 10% of the cells die during each cell cycle, so that in the time between divisions the number of cells increases by $(2\times0.9=1.8)$ instead of a factor of 2?

2.6. The following data on railroad tracks were obtained from R. H. Romer [(1991). The mathematics of exponential growth—keep it simple, *Phys. Teach.* Sept: 344–345]:

Year	Miles of railroad track
1860	30 626
1870	52 922
1880	93 262
1890	166 703

(a) What is the doubling time?

(b) Estimate the surface area of the contiguous United States. Assume that a railroad track is 7 m wide. In what year would an extrapolation predict that the surface of the United States would be completely covered with railroad track?

Section 2.2

2.7. A dose D of drug is given that causes the plasma concentration to rise from 0 to C_0. The concentration then falls according to $C=C_0e^{-bt}$. At time T, what dose must be given to raise the concentration to C_0 again? What will happen if the original dose is administered over and over again at intervals of T?

2.8. Consider the atmosphere to be at constant temperature but to have a pressure p that varies with height y. A slab between y and $y+dy$ has a different pressure on the top than on the bottom because of the weight of the air in the slab. (The weight of the air is the number of molecules N times mg, where m is the mass of a molecule and g is the gravitational acceleration.) Use the ideal gas law, $pV=Nk_BT$ (where k_B is the Boltzmann constant and T is the absolute temperature), and the fact that the air is in equilibrium to write a differential equation for p as a function of y. The equation should be familiar. Show that $p(y)=Ce^{-mgy/k_BT}$.

2.9. The mean life of a radioactive substance is defined by the equation

$$\tau=\frac{-\int_0^\infty t(dy/dt)dt}{-\int_0^\infty(dy/dt)dt}.$$

Show that if $y=y_0e^{-bt}$, then $\tau=1/b$.

2.10. R. Guttman [(1996). *J. Gen. Physiol.* **49**: 1007] measured the temperature dependence of the current pulse necessary to excite the squid axon. He found that for pulses shorter than a certain length τ, a fixed amount of electric charge was necessary to make the nerve fire; for longer pulses the current was fixed. This suggests that the axon integrates the current for a time τ but no longer. The following data are for the integrating time τ vs temperature T (°C). Find an empirical exponential relationship between T and τ.

T (°C)	τ (ms)
5	4.1
10	3.4
15	1.9
20	1.4
25	0.7
30	0.6
35	0.4

2.11. A normal rabbit was injected with 1 cm^3 of staphylococcus aureus culture containing 10^8 organisms. At various later times, 0.2 cm^3 of blood was taken from the rabbit's ear. The number of organisms per cm^3 was calculated by diluting the material, smearing it on culture plates, and counting the number of colonies formed. The results were as follows:

t (min)	Bacteria (per cm^3)
0	5×10^5
3	2×10^5
6	5×10^4
10	7×10^3
20	3×10^2
30	1.7×10^2

Plot these data and see if they can be fitted by a single exponential. Can you also estimate the blood volume of the rabbit?

Section 2.4

2.12. The death rate in a certain population (deaths per unit population per unit time) is found to increase linearly with age:

$$(\text{death rate})=a+bt.$$

Find the population as a function of time if the initial population is y_0.

2.13. The accompanying table gives death rates as a function of age. Plot these data on linear graph paper and on semilog paper. Find a region over which the death rate rises approximately exponentially with age, and determine parameters to describe that region.

Age	Death rate	Age	Death rate
0	0.000 863	45	0.005 776
5	0.000 421	50	0.008 986
10	0.000 147	55	0.013 748
15	0.001 027	60	0.020 281
20	0.001 341	65	0.030 705
25	0.001 368	70	0.046 031
30	0.001 697	75	0.066 196
35	0.002 467	80	0.101 443
40	0.003 702	85	0.194 197

2.14. Suppose that the amount of a resource at time t is $y(t)$. At $t=0$ the amount is y_0. The rate at which it is consumed is $r=-dy/dt$. Let $r=r_0e^{bt}$, that is, the rate of use increases exponentially with time. (For example, the world use of crude oil has been increasing about 7% per year since 1890.)

(a) Show that the amount remaining at time t is

$$y(t)=y_0-\left(\frac{r_0}{b}\right)(e^{bt}-1).$$

(b) If the present supply of the resource were used up at constant rate r_0 it would last for a time T_c. Show that when the rate of consumption grows exponentially at rate b, the resource lasts a time

$$T_b=\left(\frac{1}{b}\right)\ln(1+bT_c).$$

(c) An advertisement in *Scientific American*, September 1978, p. 181, said, ''There's still twice as much gas underground as we've used in the past 50 years—at our present rate of use, that's enough to last about 60 years.'' Calculate how long the gas would last if it were used at a rate which increases 7% per year.

(d) If the supply of gas were doubled, how would the answer to part (c) change?

(e) Repeat parts (c) and (d) if the growth rate is 3% per year.

2.15. When we are dealing with death or component failure, we often write Eq. (2.17) in the form $y(t)=y_0\exp[-\int_0^t m(t)dt]$ and call $m(t)$ the *mortality function*. Various forms for the mortality function can represent failure of computer components, batteries in pacemakers, or the death of organisms. (This is not the most general possible mortality model. For example, it ignores any interaction between organisms, so it cannot account for effects such as overcrowding or a limited supply of nutrients.)

(a) For human populations the mortality function is often written as $m(t)=m_1e^{-b_1t}+m_2+m_3e^{+b_3t}$. What sort of processes does each of these terms represent?

(b) Assume that m_1 and m_2 are zero. Then $m(t)$ is called the Gompertz mortality function. Obtain an expression for $y(t)$ with the Gompertz mortality function. Time t_{max} is sometimes defined to be the time when $y(t)=1$. It depends on y_0. Obtain an expression for t_{max}.

2.16. The *incidence* of a disease is the number of new cases per unit time per unit population (or per 100 000). The *prevalence* of the disease is the number of cases per unit population. In each case below the size of the general population remains fixed at the constant value y, and the disease has been present for many years.

(a) The incidence of the disease is a constant, i cases per year. Each person has the disease for a fixed time of T years, after which the person is either cured or dies. What is the prevalence p? Hint: the number who are sick at time t is the total number who became sick between $t-T$ and t.

(b) The patients in part (a) who are sick die with a constant death rate b. What is the prevalence?

(c) A new epidemic begins at $t=0$, and the incidence increases exponentially with time: $i=i_0e^{kt}$. What is the prevalence if each person has the disease for T years?

Section 2.5

2.17. The creatinine clearance test measures a patient's kidney function. Creatinine is produced by muscle at a rate p g h^{-1}. The concentration in the blood is C g l^{-1}. The volume of urine collected in time T (usually 24 h) is V l. The creatinine concentration in the urine is U g l^{-1}. The clearance is K. The plasma volume is V_p. Assume that creatinine is stored only in the plasma.

(a) Draw a block diagram for the process and write a differential equation for C.

(b) Find an expression for the creatinine clearance K in terms of p and C when C is not changing with time.

(c) If C is constant all creatinine produced in time T appears in the urine. Find K in terms of C, V, U, and T.

(d) If p were somehow doubled, what would be the new steady-state value of C? What would be the time constant for change to the new value?

2.18. A liquid is injected in muscle and spreads throughout a spherical volume $V=4\pi r^3/3$. The volume is well supplied with blood, so that the liquid is removed at a rate proportional to the remaining mass per unit volume. Let the mass be m and assume that r remains fixed. Find a differential equation for $m(t)$ and show that m decays exponentially.

2.19. A liquid is injected as in Problem 2.18, but this time a cyst is formed. The rate of removal of mass is proportional to both the pressure of liquid within the cyst, and to the surface area of the cyst, which is $4\pi r^2$. Assume that the cyst shrinks so that the pressure of liquid within the cyst remains constant. Find a differential equation for the rate of mass removal and show that dm/dt is proportional to $m^{2/3}$.

2.20. The following data showing ethanol concentration in the blood vs time after ethanol ingestion are from L. J. Bennison and T. K. Li [(1976). *New Engl. J. Med.* **294**: 9–13]. Plot the data and discuss the process by which alcohol is metabolized.

t (min)	Ethanol concentration (mg/dl)
90	134
120	120
150	106
180	93
210	79
240	65
270	50

2.21. Consider the following two-compartment model. Compartment 1 is damaged myocardium (heart muscle). Compartment 2 is the blood of volume V. At $t=0$ the patient has a heart attack and compartment 1 is created. It contains q molecules of some chemical which was released by the dead cells. Over the next several days the chemical moves from compartment 1 to compartment 2 at a rate $i(t)$, such that

$$q=\int_0^\infty i(t)dt.$$

The amount of substance in compartment 2 is $y(t)$ and the concentration is $C(t)$. The only mode of removal from compartment 2 is clearance with clearance constant K.

(a) Write a differential equation for $C(t)$ that may also involve $i(t)$.

(b) Integrate the equation and show that q can be determined by numerical integration if $C(t)$ and K are known.

(c) Show that volume V need not be known if $C(0)=C(\infty)$.

Section 2.6

2.22. The radioactive nucleus ^{64}Cu decays independently by three different paths. The relative decay rates of these three modes are in the ratio 2:2:1. The half-life is 12.8 h. Calculate the total decay rate b, and the three partial decay rates b_1, b_2, and b_3.

2.23. The following data were taken from Berg *et al.* [(1982). *N. Engl. J. Med.* **307**: 642]. At $t=0$, a 70-kg subject was given an intravenous injection of 200 mg of phenobarbital. The initial concentration in the blood was 6 mg l^{-1}. The concentration decayed exponentially with a half-life of 110 h. The experiment was repeated, but this time the subject was fed 200 g of activated charcoal every 6 h. The concentration of phenobarbital again fell exponentially, but with a half-life of 45 h.

(a) What was the volume in which the phenobarbital is distributed?

(b) What was the clearance in the first experiment?

(c) What was the clearance due to charcoal?

Section 2.7

2.24. You are treating a severely ill patient with an intravenous antibiotic. You give a "loading dose" D mg, which distributes immediately through blood volume V to give a concentration C mg dl^{-1} (1 deciliter=0.1 liter). The half-life of this antibiotic in the blood is T h. If you are giving an intravenous glucose solution at a rate R ml h^{-1}, what concentration of antibiotic should be in the glucose solution to maintain the concentration in the blood at the desired value?

2.25. The solution to the differential equation $dy/dt=a-by$ for the initial condition $y(0)=0$ is $y=(a/b)(1-e^{-bt})$. Plot the solution for $a=5$ g min^{-1} and for $b=0.1$, 0.5, and 1.0 min^{-1}. Discuss why the final value and the time to reach the final value change as they do. Also make a plot for $b=0.1$ and $a=10$ to see how that changes the situation.

2.26. We can model the repayment of a mortgage with a differential equation. Suppose that $y(t)$ is the amount still owed on the mortgage at time t, the rate of repayment per unit time is a, b is the interest rate, and that the initial amount of the mortgage is y_0.

(a) Find the differential equation for $y(t)$.

(b) Try a solution of the form $y(t)=a/b+Ce^{bt}$, where C is a constant to be determined from the initial conditions. Find C, plot the solution, and determine the time required to pay off the mortgage.

2.27. A drug is infused into the body through an intravenous "drip" at a rate of 100 mg/h. The total amount of drug in the body is y. The drug distributes uniformly and instantaneously throughout the body in a compartment of volume $V=18$ l. It is cleared from the body by a single exponential process. In the steady state the total amount in the body is 200 mg.

(a) At noon ($t=0$) the intravenous line is removed. What is $y(t)$ for $t>0$?

(b) What is the clearance of the drug?

Section 2.8

2.28. You are given the following data:

x	y	x	y
0	1.000	5	0.444
1	0.800	6	0.400
2	0.667	7	0.364
3	0.571	8	0.333
4	0.500	9	0.308
		10	0.286

Plot these data on semilog graph paper. Is this a single exponential? Is it two exponentials? Plot $1/y$ vs x. Does this alter your answer?

Section 2.9

2.29. Suppose that the rate of growth of y is described by $dy/dt=b(y)y$. Expand $b(y)$ in a Taylor's series and relate the coefficients to the terms in the logistic equation.

Section 2.10

2.30. Below are the molecular weights and radii of some molecules. Use log–log graph paper to develop an empirical relationship between them.

Substance	M	R (m)
Water	18	1.5×10^{-10}
Oxygen	32	2.0
Glucose	180	3.9
Mannitol	180	3.6
Sucrose	390	4.8
Raffinose	580	5.6
Inulin	5 000	12.5
Ribonuclease	13 500	18
β-lactoglobulin	35 000	27
Hemoglobin	68 000	31
Albumin	68 000	37
Catalase	250 000	52

2.31. How well does (Eq. 2.32c) explain the data of Fig. 2.17? Discuss any differences.

2.32. Compare the mass and metabolic requirements (and hence waste output, including water vapor) of 180 people each weighing 70 kg with 12 600 chickens of average mass 1 kg.

2.33. Figure 2.17 shows that in young children, height is more nearly proportional to $M^{0.62}$ than to $M^{1/3}$. Find pictures of children and adults and compare ratios of height to width, to see what the differences are.

2.34. If food consumption is proportional to $M^{3/4}$ across species, how does the food consumption per unit mass scale with mass? Qualitatively compare the eating habits of hummingbirds and eagles and of mice and elephants.

REFERENCES

Clark, V. A. (1975). Survival distributions. *Ann. Rev. Biophys. Bioeng.* **4**: 431–438.

Hemmingsen, A. M. (1960). Energy metabolism as related to body size and respiratory surfaces, and its evolution. *Reports of the Steno Memorial Hospital and Nordinsk Insulin Laboratorium* **9**: 6–110.

Hobbie, R. K. (1973). Teaching exponential growth and decay: Examples from medicine. *Am. J. Phys.* **41**: 389–393.

Kempe, C. H., H. K. Silver, and D. O'Brien (1970). *Current Pediatric Diagnosis and Treatment*, 2nd ed. Los Altos, CA, Lange.

McKee, P. A., W. P. Castelli, P. M. McNamara, and W. B. Kannel (1971). The natural history of congestive heart failure: The Framingham study. *New Eng. J. Med.* **285**: 1441–1446.

McMahon, T. (1973). Size and shape in biology. *Science* **179**: 1201–1204.

Peters, R. H. (1983). *The Ecological Implications of Body Size.* Cambridge, Cambridge University Press.

Riggs, D. S. (1970). *The Mathematical Approach to Physiological Problems.* Cambridge, MA, MIT Press.

Zumoff, B., H. Hart, and L. Hellman (1966). Considerations of mortality in certain chronic diseases. *Ann. Intern. Med.* **64**: 595–601.

CHAPTER 3

Systems of Many Particles

It is possible to identify all the external forces acting on a simple system and use Newton's second law ($\mathbf{F}=m\mathbf{a}$) to calculate how the system moves. (Applying this technique in a complicated case such as the femur may require the development of a simplified model, because so many muscles, other bones, and ligaments apply forces at so many different points.) In an atomic-size system consisting of a single atom or molecule, it is possible to use the quantum-mechanical equivalent of $\mathbf{F}=m\mathbf{a}$, the Schrödinger equation, to do the same thing. (The Schrödinger equation takes into account the wave properties that are important in small systems.)

In systems of many particles, such calculations become impossible. Consider, for example, how many particles there are in a cubic millimeter of blood. Table 3.1 shows some of the constituents of such a sample. To calculate the translational motion in three dimensions, it would be necessary to write three equations for each particle using Newton's second law. Suppose that at time t the force on a molecule is $\mathbf{F}$. Between t and $t+\Delta t$, the velocity of the particle changes according to the equations

$$v_x(t+\Delta t)=v_x(t)+F_x\Delta t/m,$$

$$v_y(t+\Delta t)=v_y(t)+F_y\Delta t/m,$$

$$v_z(t+\Delta t)=v_z(t)+F_z\Delta t/m.$$

If Δt is short enough so that the force does not change appreciably, the three equations for the change of position of the particle are of the form

$$x(t+\Delta t)=x(t)+v_x(t)\Delta t+F_x(t)(\Delta t)^2/(2m).$$

If Δt is small enough the last term can be neglected, and the pairs of equations for x and v_x and so forth can be solved. To solve these equations requires at least six multiplications and six additions for each particle. For 10^{19} particles, this means about 10^{20} arithmetic operations per time interval. If a supercomputer can do 10^{12} operations per second, then the complete calculation for a single time interval will require 10^8 seconds or three years! Another limitation arises in the physics of the processes. It is now known that relatively simple systems can exhibit deterministic chaos: a collection of identical systems differing in their

TABLE 3.1. *Some constituents of 1 mm^3 of human blood.*

Constituent	Concentration in customary units	Number in 1 mm^3 ($=10^{-9}$ m^3$=10^{-3}$ cm^3)
Water	1 g cm^{-3}	3.3×10^{19}
Sodium	3.2 mg cm^{-3}	8.3×10^{16}
Albumin	4.5 g dl^{-1}	3.9×10^{14}
Cholesterol	200 mg dl^{-1}	3.1×10^{15}
Glucose	100 mg dl^{-1}	3.3×10^{15}
Hemoglobin	15 g dl^{-1}	1.4×10^{15}
Erythrocytes	5×10^6 mm^{-3}	5×10^6

initial conditions by an infinitesimally small amount can become completely different in their subsequent behavior in a surprisingly short period of time. It is impossible to trace the behavior of this many molecules on an individual basis.

Nor is it necessary. We do not care which water molecule is where. The properties of a system that are of interest are averages over many molecules: pressure, concentration, average speed, and so forth. These average macroscopic properties are studied in *statistical* or *thermal physics* or *statistical mechanics*.

Unfortunately, this chapter relies heavily on your ability to accept delayed gratification. It has only a few biological examples, but the material developed here is necessary for understanding some topics in most of the later chapters, especially Chaps. 4–9 and 13–17. In addition to developing a statistical understanding of pressure, temperature, and concentration, this chapter derives four quantities or concepts that are used later:

1. The *Boltzmann factor*, which tells how concentrations of particles vary with potential energy (Sec. 3.7).
2. The *principle of equipartition of energy*, which underlies the diffusion process that is so important in the body (Sec. 3.11).
3. The *chemical potential*, which describes the condition for equilibrium of two systems for the exchange of particles, and how the particles flow when the systems are not in equilibrium (Secs. 3.12, 3.13, and 3.18).
4. The *Gibbs free energy*, which tells the direction in which a chemical reaction proceeds and allows us to understand how the cells in the body use energy (Sec. 3.17).

The first six sections form the basis for the rest of the chapter, developing the concepts of microstates, heat flow, temperature, and entropy. Sections 3.7 and 3.8 develop the Boltzmann factor and its corollary, the Nernst equation. Section 3.9 applies the Boltzmann factor to the air molecules in the atmosphere. Section 3.10 derives the Maxwell–Boltzmann velocity distribution, which is then used to obtain the equipartition of energy theorem in Sec. 3.11.

The transport of particles between two systems is described most efficiently using the chemical potential. The chemical potential is introduced in Sec. 3.12, and an example of its use is shown in Sec. 3.13.

Section 3.14 considers systems that can exchange volume. An idealized example is two systems separated by a flexible membrane or a movable piston. Of greater interest is a contracting muscle, the volume of which is not constant. The next two sections extend the idea of systems that exchange energy, particles, or volume to the exchange of other variables such as electric charge.

The Gibbs free energy, introduced in Sec. 3.17, is used to describe chemical reactions that take place at constant temperature and pressure. It is closely related to the chemical potential. The chemical potential of an ideal solution is derived in Sec. 3.18 and is used extensively in Chap. 5.

3.1. GAS MOLECULES IN A BOX

Statistical physics or *statistical mechanics* deals with average quantities such as pressure, temperature, and particle concentration and with probability distributions of variables such as velocity. Some of the properties of these averages can be displayed by considering a simple example: the number of particles in each half of a box containing a fixed number of gas molecules. (This is a simple analog for the concentration.) We will not be concerned with the position and velocity of each molecule, since we have already decided not to use Newtonian mechanics. Nor will we ask for the velocity distribution at this time. This simplified example will describe only how many molecules are in the volume of interest. The number will fluctuate with time. We will deal with probabilities: if the number of particles in the volume is measured repeatedly, what values are obtained, and with what relative frequency?

If we were willing to use Newtonian mechanics, we could count periodically how many molecules are in the volume of interest. [This has actually been done for small numbers of particles. See Reif (1964), pp. 8–9]. For larger numbers of particles, it is easier to use statistical arguments to obtain the probabilities. The particles travel back and forth, colliding with the walls of the box and occasionally with one another. After some time has elapsed, all memory of the particles' original positions and velocities has been lost, because of collisions with the walls of the box, which have microscopic inhomogeneities. Therefore, the result can be obtained by imagining a whole succession of completely different boxes, in which the particles have been placed at random. We can count the number of molecules in the volume of interest in each box. Such a collection of similar boxes is called an *ensemble*. Ensembles of similar systems will be central to the ideas of this chapter.

Imagine an ensemble of boxes, each divided in half as in Fig. 3.1. We want to know how often a certain number of particles is found in the left half. If one particle is in a box

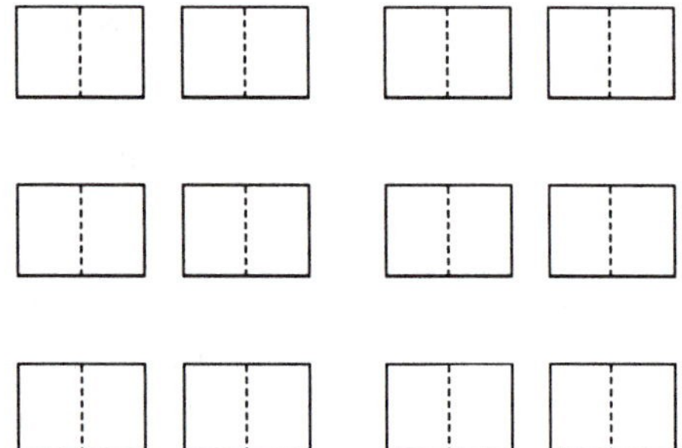

FIGURE 3.1. An ensemble of boxes, each divided in half by an imaginary partition.

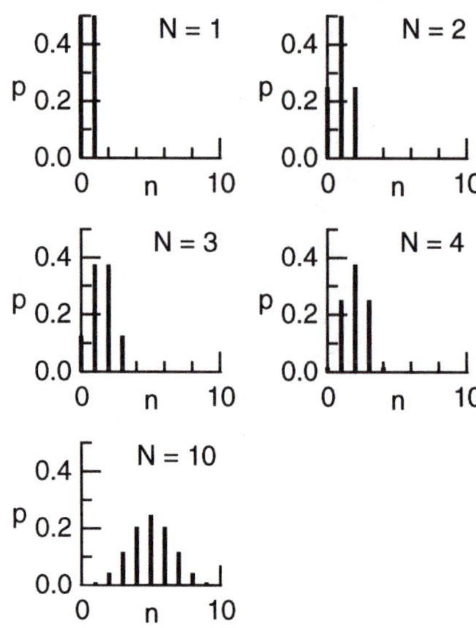

FIGURE 3.2. Histograms of $P(n;N)$ for different values of N.

($N=1$), two cases can be distinguished, depending on which half the particle is in. Call them L and R. Each case is equally likely to occur, since nothing distinguishes one half of a box from the other. If n is the number of particles in the left half, then case L corresponds to $n=1$ and case R corresponds to $n=0$.

The probability of having a particular value of n is defined to be

$$P(n)=\frac{(\text{number of systems in the ensemble in which } n \text{ is found})}{(\text{total number of systems})} \tag{3.1}$$

in the limit as the number of systems becomes very large.

Because there are only two possible values of n, $n=0$ or $n=1$, and because each corresponds to one of the equally likely configurations,

$$P(0)=0.5,$$

$$P(1)=0.5.$$

The sum of the probabilities is 1, since one of the possible values of n must always occur, and the sum over all n is the denominator of Eq. (3.1). A histogram of $P(n)$ for $N=1$ is given in Fig. 3.2(a). To recapitulate: n is the number of molecules in the left half of the box, and N is the total number of molecules in the entire box. Since N will be changed in the discussion below, we will call the probability $P(n;N)$. (The fixed parameters that determine the probability distribution are located after the semicolon.)

Now let $N=2$. Each molecule can be on the left or the right with equal probability. The possible outcomes are listed shortly, along with the corresponding values of n and $P(n;2)$. Each of the four outcomes is equally probable. To see this, note that L or R is equally likely for each molecule. In half of the boxes in the ensemble, the first molecule is found on the left. In half of these, the second molecule is also on the left. Therefore LL occurs in one-fourth of the systems in the ensemble. (This is not strictly true, because there can be fluctuations. If we throw a coin six times, we cannot say that heads will always occur three times. If we repeat the experiment many times, the average number of heads will be three.) The outcomes are listed in the following table:

Molecule 1	Molecule 2	n	$P(n,2)$
R	R	0	$\frac{1}{4}$
R	L	1	$\frac{1}{2}$
L	R	1	
L	L	2	$\frac{1}{4}$

If three molecules are placed in each box, there are two possible locations for the first particle, two for the second, and two for the third. If the three particles are all independent, then there are $2^3=8$ different ways to locate the particles in a box. If a box is divided in half, each of these ways has a probability of 1/8.

Molecule 1	Molecule 2	Molecule 3	n	$P(n;3)$
R	R	R	0	$\frac{1}{8}$
R	R	L	1	
R	L	R	1	$\frac{3}{8}$
L	R	R	1	
L	L	R	2	
L	R	L	2	$\frac{3}{8}$
R	L	L	2	
L	L	L	3	$\frac{1}{8}$

The cases of two and three molecules in the box are also plotted in Fig. 3.2.

In each case, $P(n;N)$ has been determined by listing all the ways that the N particles can go into a box. This can become tedious if the number of particles is large. Furthermore, it does not provide a way to calculate P if the two volumes of the box are not equal. We will now introduce a more general technique that can be used for any number of particles and for any fractional volume of the box.

Each box is divided into two volumes, v and v', with total volume $V=v+v'$. Call p the probability that a single particle is in volume v. The probability that the particle is in the remainder of the box, v', is q:

$$p+q=1. \tag{3.2}$$

As long as there is nothing to distinguish one part of a box from the other, p is the ratio of v to the total volume:

$$p=\frac{v}{V}. \tag{3.3}$$

By the same argument, $q=v'/V$. These values satisfy Eq. (3.2). If N particles are distributed between the two volumes of the box, the number in v is n and the number in v' is $n'=N-n$. The probability that n of the N particles are found in volume v is given by the binomial probability distribution (Appendix H):

$$P(n;N)=P(n;N,p)=\frac{N!}{n!(N-n)!}\,p^n(1-p)^{N-n}. \tag{3.4}$$

Table 3.2 shows the calculation of $P(n;10)$ using this equation. A histogram for this case is also plotted in Fig. 3.2. In each case there is a value of n for which P is a maximum. When N is even, this value is $N/2$; when N is odd, the values on either side of $N/2$ share the maximum value. The probability is significantly different from zero only for a few values of n on either side of the maximum.

A probability distribution, in the form of an expression, a table of values, or a histogram, usually gives all the information that is needed about the number of molecules in v; it is not necessary to ask which molecules are in v. The number of molecules in v is not fixed but fluctuates about the number for which P is a maximum. For example, if $N=10$, and we measure the number of molecules in the left half many times, we find $n=5$ only about 25% of the time. On the other hand, we find that $n=4$, 5, or 6 about 65% of the time, while $n=3$, 4, 5, 6, or 7 about 90% of the time.

TABLE 3.2. *Calculation of P(n;10) using the binomial probability distribution. Note that 0!=1.*

$$P(0;10)=\frac{10!}{0!\ \ 10!}\,(\tfrac{1}{2})^0(\tfrac{1}{2})^{10}=(\tfrac{1}{2})^{10}=0.001$$

$$P(1;10)=\frac{10!}{1!\ \ 9!}\,(\tfrac{1}{2})^1(\tfrac{1}{2})^{9}=10(\tfrac{1}{2})^{10}=0.010$$

$$P(2;10)=\frac{10!}{2!\ \ 8!}\,(\tfrac{1}{2})^2(\tfrac{1}{2})^{8}=45(\tfrac{1}{2})^{10}=0.044$$

$$P(3;10)=\frac{10!}{3!\ \ 7!}\,(\tfrac{1}{2})^3(\tfrac{1}{2})^{7}=120(\tfrac{1}{2})^{10}=0.117$$

$$P(4;10)=\frac{10!}{4!\ \ 6!}\,(\tfrac{1}{2})^4(\tfrac{1}{2})^{6}=210(\tfrac{1}{2})^{10}=0.205$$

$$P(5;10)=\frac{10!}{5!\ \ 5!}\,(\tfrac{1}{2})^5(\tfrac{1}{2})^{5}=252(\tfrac{1}{2})^{10}=0.246$$

$$P(6;10)=\frac{10!}{6!\ \ 4!}\,(\tfrac{1}{2})^6(\tfrac{1}{2})^{4}=210(\tfrac{1}{2})^{10}=0.205$$

$$P(7;10)=\frac{10!}{7!\ \ 3!}\,(\tfrac{1}{2})^7(\tfrac{1}{2})^{3}=120(\tfrac{1}{2})^{10}=0.117$$

$$P(8;10)=\frac{10!}{8!\ \ 2!}\,(\tfrac{1}{2})^8(\tfrac{1}{2})^{2}=45(\tfrac{1}{2})^{10}=0.044$$

$$P(9;10)=\frac{10!}{9!\ \ 1!}\,(\tfrac{1}{2})^9(\tfrac{1}{2})^{1}=10(\tfrac{1}{2})^{10}=0.010$$

$$P(10;10)=\frac{10!}{10!\ \ 0!}\,(\tfrac{1}{2})^{10}(\tfrac{1}{2})^{0}=(\tfrac{1}{2})^{10}=0.001$$

3.2. MICROSTATES AND MACROSTATES

If we know "enough" about the detailed properties (such as position and momentum) of every particle in a system,[1] then we say that the *microstate* of the system is specified. (The criterion for "enough" will be discussed shortly.) We may know less than this but know the *macrostate* of the system. (In an ideal gas, for example, the macrostate would be defined by knowing the number of molecules and volume, and the pressure, temperature, or total energy.) Usually there are many microstates corresponding to each macrostate. The large-scale average properties (such as pressure and number of particles per unit volume in the ideal gas) fluctuate slightly about well-defined mean values.

In the problem of how many molecules are in half of a box, the macrostate is specified if we know how many molecules there are, while a microstate would specify the position and momentum of every molecule. In other cases, internal motions of the molecule may be important, and it will be necessary to know more than just the position and momentum of each particle.

The relation between microstates and macrostates may be clarified by the following example, which contains the essential features, although it is oversimplified and somewhat artificial. A room is empty except for some toys on the floor. Specifying the location of each of the toys on the floor would specify the microstate of the system. If the toys are in the shaded corner in Fig. 3.3, the macrostate is "picked up." If the toys are any place else in the room, the macrostate is "mess." There are many more microstates corresponding to the macrostate "mess" than there are corresponding to the macrostate "picked up." We know from experience that children tend to regard any microstate as equally satisfactory; the chances of spontaneously finding the macrostate "picked up" are relatively small.

[1] A *system* is that part of the universe that we choose to examine. The surroundings are the rest of the universe. The system may or may not be isolated from the surroundings.

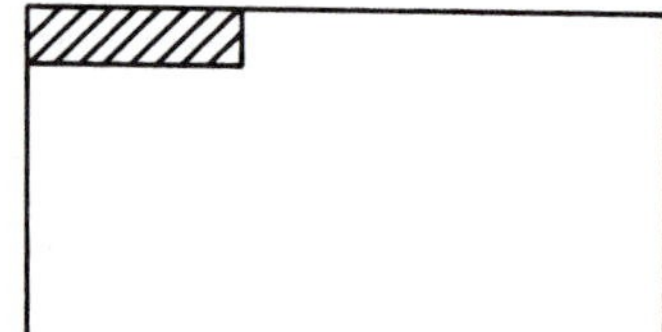

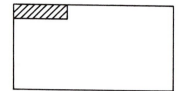

FIGURE 3.3. A room with toys. If all the toys are in the shaded area, the macrostate is "picked up." Otherwise, the macrostate is "mess."

A situation in which P is small is called *ordered* or *nonrandom*. A situation in which P is large is called *disordered* or *random*. Macrostate "mess" is more probable than macrostate "picked up" and is disordered or random.

The same idea can be applied to a box of gas molecules. Initially, the molecules are all kept in the left half of the box by a partition. If the partition is suddenly removed, a large number of additional microstates are suddenly available to the molecules. The macrostate in which they find themselves—all in the left half of the box, even though the partition has been removed—is very improbable or highly ordered. The molecules soon fill the entire box; it is quite unlikely that they will all be in the left half again if the number of molecules is very large. (Suppose that there are 80 molecules in the box. The probability that all are in the left half is $(\frac{1}{2})^{80}=10^{-24}$. If samples were taken 10^6 times per second, it would take 10^{18} seconds to sample 10^{24} boxes, one of which, on the average, would have all of the molecules in the left half. This is 10 times the age of the universe.)

Just after the partition in the box was removed, the situation was very ordered. The system spontaneously approached a much more random situation in which nearly half the molecules were in each half of the box. The actual number n fluctuates about $N/2$, but in such a way that the average $\langle n \rangle$ (taken, say, over several seconds) no longer changes with time. Typical fluctuations with a constant $\langle n \rangle$ are shown in Fig. 3.4(a). When the average[2] of the macroscopic parameters is not changing with time, we say that the system is in an *equilibrium* state. Figure 3.4(b) shows the system moving toward the equilibrium state after the partition is removed.

An equilibrium state is characterized by macroscopic parameters whose average values remain constant with time, although the parameters may fluctuate about the average value. It is also the most random (i.e., most probable) macrostate possible under the prescribed conditions. It

[2]There is a subtlety about the meaning of average that we are glossing over here. If we take a whole ensemble of identical systems, which were all prepared the same way, and measure n in each one, we have the *ensemble average* $\bar{n}$. This is calculated in the way described in Appendix F. If we watch one system over some long time interval, as in Fig. 3.4, we can take the *time average* $\langle n \rangle$. It is taken by recording values of n for a large number of discrete times in some interval. Strictly speaking, an equilibrium state is one in which the ensemble average is not changing with time.

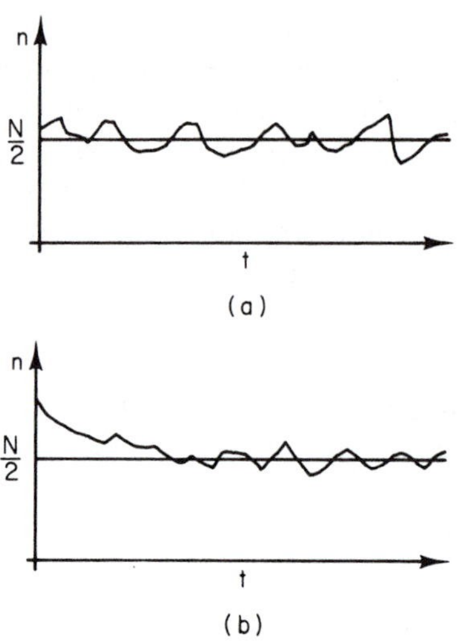

FIGURE 3.4. (a) Fluctuations of n about $N/2$. (b) The approach of the system to the equilibrium state after the partition is removed.

is independent of the past history of the system and is specified by a few macroscopic parameters.[3]

The definition of a microstate of a system has so far been rather vague; we have not said precisely what is required to specify it. It is actually easier to specify the microstate of a system when using quantum mechanics than when using classical mechanics. When the energy of an individual particle in a system (such as one of the molecules in the box) is measured with sufficient accuracy, it is found that only certain discrete values of the energy occur. This is because of the wave nature of the particles. The allowed values of the energy are called *energy* levels. You are probably familiar with the idea of energy levels from a previous physics or chemistry course; for example, the spectral lines of atoms are due to the emission of light when an atom changes from one energy level to another. Because the energy levels are well defined, the energy difference, and hence the frequency or color of the light, is also well defined (see Chap. 12).

A particle in a box has a whole set of energy levels at energies determined by the size and shape of the box. Compared to macroscopic measurements of energy, these levels are very close together. The particle can be in any one of these levels; which energy the particle has is specified by a set of *quantum numbers*. If the particle moves in three dimensions, three quantum numbers are needed to specify the energy level. If there are N particles, it will be necessary to specify three quantum numbers for each particle or $3N$ numbers in all. If the particle can rotate or have other internal motion, then more quantum numbers are needed. The total number of quantum numbers required to specify the state of all the particles in the system is called the number of *degrees of freedom* of the system.

[3]A more detailed discussion of equilibrium states is found in Reif (1964).

A microstate of a system is specified if all the quantum numbers for all the particles in the system are specified.

In most of this chapter, it will not be necessary to consider the energy levels in detail. The important fact is that each particle in a system has discrete energy levels, and a microstate is specified if the energy level occupied by each particle is known.

3.3. THE ENERGY OF A SYSTEM: THE FIRST LAW OF THERMODYNAMICS

Figure 3.5 shows some energy levels in a system occupied by a few particles. The total energy of the system U is the sum of the energy of each particle. In making this drawing, I have assumed that all the particles are the same and that they do not interact with one another very much. Then each particle has the same set of energy levels, and the presence of other particles does not change them. In that case, we can say that there is a certain set of energy levels in the system and that each level can be occupied by any number of particles. The energy of the ith level, occupied or not, will be called u_i. For the example of Fig. 3.5, the total energy is

$$U=2u_{23}+u_{25}+u_{26}+3u_{28}.$$

Suppose that the system is isolated so that it does not gain or lose energy. It is still possible for particles within the system to exchange energy and move to different energy levels, as long as the total energy does not change. (Classically, two particles could collide, so that one gained and one lost energy.) Therefore the number of particles occupying each energy level can change, as long as the total energy remains constant. For a system in equilibrium, the average number of particles in each level does not change with time.

There are two ways in which the total energy of a system can change. *Work* can be done on the system by the surroundings, or *heat* can flow from the surroundings to the system. The meaning of work and heat in terms of the energy levels of the system is quite specific and is discussed shortly. First, we define the sign conventions associated with them.

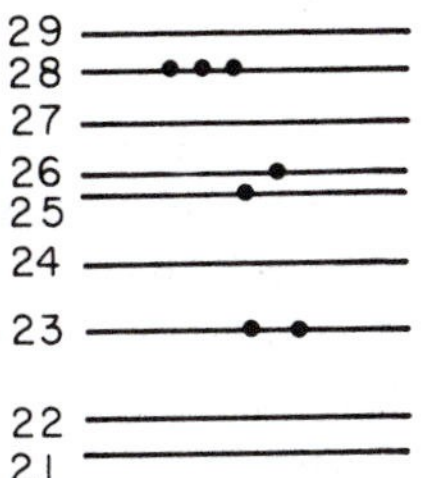

FIGURE 3.5. A few of the energy levels in a system. If a particle has a particular energy, a dot is drawn on the level. More than one particle in this system can have the same quantum numbers.

It is customary to define Q to be the heat flow *into* a system. If no work is done, the energy change in the system is

$$\Delta U=Q.$$

It is also customary to call W the work done *by* the system *on* the surroundings. When W is positive, energy flows from the system to the surroundings. If there is no accompanying heat flow, the energy change of the system is

$$\Delta U=-W.$$

The most general way the energy of a system can change is to have both work done by the system and heat flow into the system. The statement of the conservation of energy in that case is called the *first law of thermodynamics*:

$$\Delta U=Q-W. \tag{3.5}$$

The positions of the energy levels in a system are determined by some macroscopic properties of the system. For a gas of particles in a box, for example, the positions of the levels are determined by the size and shape of the box. For charged particles in an electric field, the positions of the levels are determined by the electric field. If the macroscopic parameters that determine the positions of the energy levels are not changed, the only way to change the total energy of a system is to change the average number of particles occupying each energy level, as in Fig. 3.6. This energy change is called *heat flow*.

Work is associated with the change in the macroscopic parameters (such as volume) that determine the positions of the energy levels. If the energy levels are shifted by doing work without an accompanying heat flow, the change is called *adiabatic*. An adiabatic change is shown in Fig. 3.7. In general, there is also a shift of the populations of the levels in an adiabatic change; the average occupancy of each level can be calculated using the Boltzmann factor, described in Sec. 3.10. There is no heat flow, but work is done on or by the system, and its energy changes.

To summarize: Pure heat flow involves a change in the average number of particles in each level without a change in the positions of the levels. Work involves a change in the macroscopic parameters, which changes the positions of at least some of the energy levels. In general, this means that

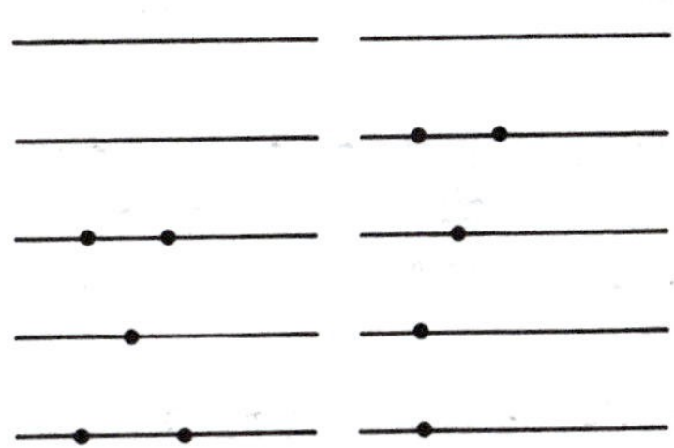

FIGURE 3.6. No work is done on the system, but heat is added. The positions of the levels do not change; their average population does change.

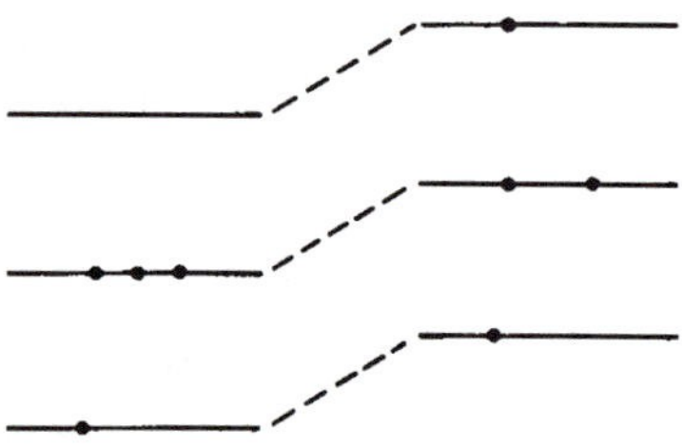

FIGURE 3.7. Work is done on the system, but no heat flows. Each level has been shifted to a higher energy.

there is also a shift in the average population of each level. The most general energy change of a system involves both work and heat flow. In that case the total energy change is the sum of the changes due to work and to heat flow.

It is customary in drawing systems to use the symbols in Fig. 3.8 to describe how the system can interact with the surroundings. A double-walled box means that no heat flows, and any processes that occur are adiabatic. This is shown in Fig. 3.8(a). If work can be done on the system, a piston is shown as in Fig. 3.8(b). If heat can flow to or from the system, a single wall is used as in Fig. 3.8(c).

3.4. ENSEMBLES AND THE BASIC POSTULATES

In the next few sections we will develop some quite remarkable results from statistical mechanics. Making the postulate that when a system is in equilibrium each microstate is equally probable, and arguing that as the energy, volume, or number of particles in the system is increased the number of microstates available to the system increases, we will obtain several well-known results from thermodynamics as follows. Heat flows from one system to another in thermal contact until their temperatures are the same; if their volumes can change they adjust themselves until the pressures are the same; and the systems exchange particles until their chemical potentials are the same. We will obtain the concept of entropy, the Boltzmann factor, the theorem of equipartition of energy, and the Gibbs free energy. The Gibbs free energy is useful in chemical reactions in living systems where the temperature and pressure are constant. The initial postulates are deceptively simple. Unfortunately, a fair amount of mathematics is required to get from them to the final results. We start with the basic postulates.

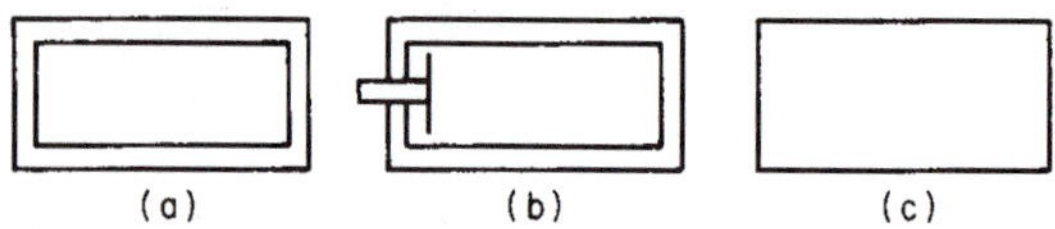

FIGURE 3.8. Symbols used to indicate various types of isolation in a system. (a) This system is completely isolated. (b) There is no heat flow through the double wall, but work can be done (symbolized by a piston). (c) No work can be done, but there can be heat flow through the single wall.

The microstate of a system is determined by specifying the quantum numbers of each particle in the system. The total number of quantum numbers is the number of degrees of freedom f. The macrostate of a system is determined by specifying two things:

1. All of the external parameters, such as the volume of a box of gas or any external electric or magnetic field, on which the positions of the energy levels depend. (Classically, all the external parameters that affect the motion of the particles in the system.)
2. The total energy of the system. The total energy of the system is not known precisely, but is known to be between U and $U+\Delta U$.

The external parameters determine a set of energy levels for the particles in the system; the total energy determines which energy levels are accessible to the system.

Another word about degrees of freedom is in order. Suppose that the system contains N particles in three-dimensional space. Then there are $f=3N$ degrees of freedom. However, if some of the N particles are connected together, as in a molecule, there are fewer degrees of freedom. If there are M molecules, each made up of a atoms, then $N=aM$. The number of quantum numbers is less than $3N$ because the atoms cannot all move independently. If the molecules were thought of as single particles, there would be $3M$ degrees of freedom. But the molecules can rotate and vibrate, so that the number of degrees of freedom is greater than $3M$ and less than $3N$.

Statistical physics deals with average quantities and probabilities. We imagine a whole set or ensemble of "identical" systems, as we did in Eq. (3.1). The systems are identical in that they all are in the same macrostate. Different systems within the ensemble will be in different microstates. Imagine that at some instant of time we "freeze" all the systems in the ensemble and examine which microstate each is in. From this we can determine the probability that a system in the ensemble is in microstate i:

$$P(\text{of being in microstate } i)$$

$$=\frac{(\text{number of systems in microstate } i)}{(\text{total number of systems in the ensemble})}.$$

Imagine that we now "unfreeze" all the systems in the ensemble and let the particles move however they want. At some later time we "freeze" them again and examine the probability that a system is in each microstate. These probabilities may have changed with time. For example, if the system is a group of particles in a box, and if the initial "freeze" was done just after a partition confining all the particles to the left half of the box had been removed, we would have found many systems in the ensemble in microstates for which most of the particles are on the left-hand side. Later, this would not be true. We would find microstates corresponding to particles in both halves of the box.

We will make two basic *postulates* about the systems in the ensemble.[4]

1. If an isolated system (really, an ensemble of isolated systems) is found with equal probability in each one of its accessible microstates, it is in equilibrium.[5] If it is in equilibrium, it is found with equal probability in each one of its accessible microstates.

2. If it is not in equilibrium, it tends to change with time until it is in equilibrium. Therefore the equilibrium state is the most random, most probable state.

For the rest of this chapter, we deal with equilibrium systems. According to our first postulate, each microstate that is accessible to the system (that is, consistent with the total energy that the system has) is equally probable. We will discover that this statement has some far-reaching consequences.

Statistical physics deals with probabilities. Suppose that we want to consider some variable x, which takes on various values. This variable might be the pressure of a gas, the number of gas molecules in some volume of the box, or the energy that one of the molecules has. For each value of x, there will be some number of microstates in which the system could be that are consistent with that value of x. There will also be some total number of microstates in which the system could be, consistent with its initial preparation. We will use the Greek letter Ω to denote the number of microstates. The total number of accessible microstates (for all possible values of x) is Ω; the number for which x has some particular value is Ω_x. It is consistent with the first assumption to say that the probability that the variable has a value x when the system is in equilibrium is

$$P_x = \frac{\Omega_x}{\Omega}. \tag{3.6}$$

We have been considering ensemble averages. For example, the variable of interest might be the pressure, and we could find the ensemble average by calculating $\bar{p} = \Sigma P_p p$, where P_p is the probability of having pressure p. In equilibrium P_p is given by Eq. (3.6), and $\bar{p}$ does not change with time. We could also consider a single system, measure $p(t)$ M times, and compute the time average, $\langle p(t) \rangle = \Sigma_i p(t_i)/M$. (The equivalence of the time average and the ensemble average for systems in equilibrium is called the ergodic hypothesis.)

[4]For a more detailed discussion of these assumptions, see Reif (1964), Ch. 3.

[5]In thermodynamics and statistical mechanics, *equilibrium* and *steady state* do not mean the same thing. Steady state means that some variable is not changing with time. The concentration of sodium in a salt solution flowing through a pipe could be in steady state as the solution flowed through, but the system would not be in equilibrium. Only a few microstates corresponding to bulk motion of the fluid are occupied. In other areas, such as feedback systems, the words *equilibrium* and *steady state* are used almost interchangeably.

3.5. THERMAL EQUILIBRIUM

A system that never interacts with its surroundings is an idealization. The adiabatic walls of Fig. 3.8(a) can never be completely realized. However, much can be learned by considering two systems that can exchange heat, work, or particles, but that, taken together, are isolated from the rest of the universe. Once we have learned how these two systems interact, the second system can be taken to be the rest of the universe. Eventually, we will allow all three exchanges—heat, work, and particles—to take place; for now, it will be convenient to consider only exchanges of heat energy. Figure 3.9 shows the two systems, A and A', isolated from the rest of the universe. The total system will be called A^*. The total number of particles is $N^* = N + N'$. For now N and N' are fixed. The total energy is $U^* = U + U'$. The two systems can exchange heat energy, so that U and U' may change, as long as their sum remains constant.

The number of microstates accessible to the total system is Ω^*. The system was originally given a total energy between U^* and $U^* + \Delta U^*$ before it was sealed off from the rest of the universe. The barrier between A and A' prevents exchange of particles or work. The total number of microstates depends on how much energy is in each system: when system A has energy between U and $U + \Delta U$, the total number of microstates is $\Omega^*(U)$. In the discussion ΔU, $\Delta U'$, and ΔU^* are fixed, and we will not carry them along in the notation. There are many microstates accessible to the system, with U and U' having different values, subject always to $U^* = U + U'$. Let the total number of microstates, including all possible values of U, be Ω^*_{tot}. Then, according to the postulate, the probability of finding system A with energy U is

$$P(U) = \frac{\Omega^*(U)}{\Omega^*_{\text{tot}}} = C\Omega^*(U). \tag{3.7}$$

C is a constant (independent of U) and is just $1/\Omega^*_{\text{tot}}$.

If the meaning of Eq. (3.7) is obscure, consider the following example. Systems A and A' each consist of two

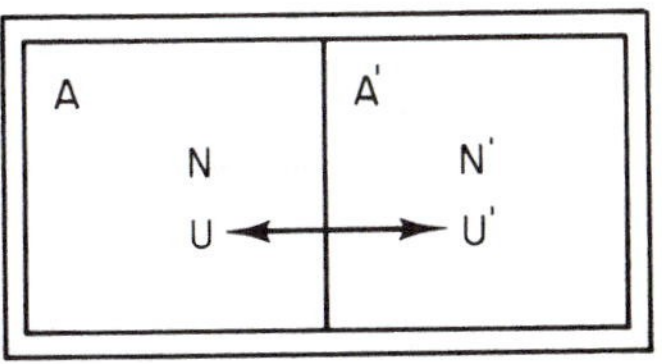

FIGURE 3.9. Two systems are in thermal contact with each other but are isolated from the rest of the universe. They can exchange energy only by heat flow.

particles, the energy levels for each particle being at u, $2u$, $3u$, and so forth. The total energy available to the combined system is $U^*=10u$. The smallest possible energy for system A is $U=2u$, both particles having energy u. If $U=3u$, there are two states: in one, the first particle has energy u and the second $2u$; in the second, the particles are reversed. Label these states $(u,2u)$ and $(2u,u)$. For $U=4u$, there are three possibilities: $(u,3u)$, $(2u,2u)$, and $(3u,u)$. In general, if $U=nu$, there are $n-1$ states, corresponding to the first particle having energy $u,2u,3u,\ldots,(n-1)u$. Table 3.3 shows values for U, U', Ω, and Ω'.

It is now necessary to consider Ω^* in more detail. If there are two microstates available to system A and 6 available to system A', there are $2\times6=12$ states available to the total system. System A can be in either of its two states, while A' can be in any of its six. $\Omega^*=\Omega\Omega'$ is also given in Table 3.3. In a more general case, *the number of microstates for the total system is the product of the number for each subsystem:*

$$\Omega^*(U)=\Omega(U)\Omega'(U'). \tag{3.8}$$

For the specific example, there are a total of 84 microstates accessible to the system when $U^*=10u$. Equation (3.7) says that since each microstate is postulated to be equally probable, the probability that the energy of system A is $3u$ is $12/84=0.14$. The most probable state of the combined system is that for which A has energy $5u$ and A' has energy $5u$.

The next question is how Ω and Ω' depend on energy in the general case. In the example, Ω is proportional to U. For three particles, one can show that Ω increases as U^2. In general, the more particles there are in a system, the more rapidly Ω increases with U. For a system with a large number of particles, *increasing the energy drastically increases the number of microstates accessible to the system.*

As more energy is given to system A and $\Omega(U)$ increases, there is less energy available for system A' and $\Omega'(U')$ decreases. The product $\Omega^*=\Omega\Omega'$ goes through a maximum at some value of U, and that value of U is therefore the most probable. These features are shown in Fig. 3.10, which assumes that U and Ω are continuous variables. The continuous approximation becomes excellent when we deal with a large number of particles and very closely spaced energy levels. The solid line in Fig. 3.10(a) represents $\Omega(U)$; $\Omega'(U')$ is the solid line in Fig. 3.10(b). The function Ω' is also plotted vs U, rather than U', as the

TABLE 3.3. *An example of two systems that can exchange heat energy.*[a]

System A		System A'		System A^*
U	Ω	U'	Ω'	Ω^*
$2u$	1	$8u$	7	7
$3u$	2	$7u$	6	12
$4u$	3	$6u$	5	15
$5u$	4	$5u$	4	16
$6u$	5	$4u$	3	15
$7u$	6	$3u$	2	12
$8u$	7	$2u$	1	7
				$\Omega^*_{tot}=84$

[a]The total energy is $U^*=10u$. Each system contains two particles for which the energy levels are u, $2u$, $3u$, etc.

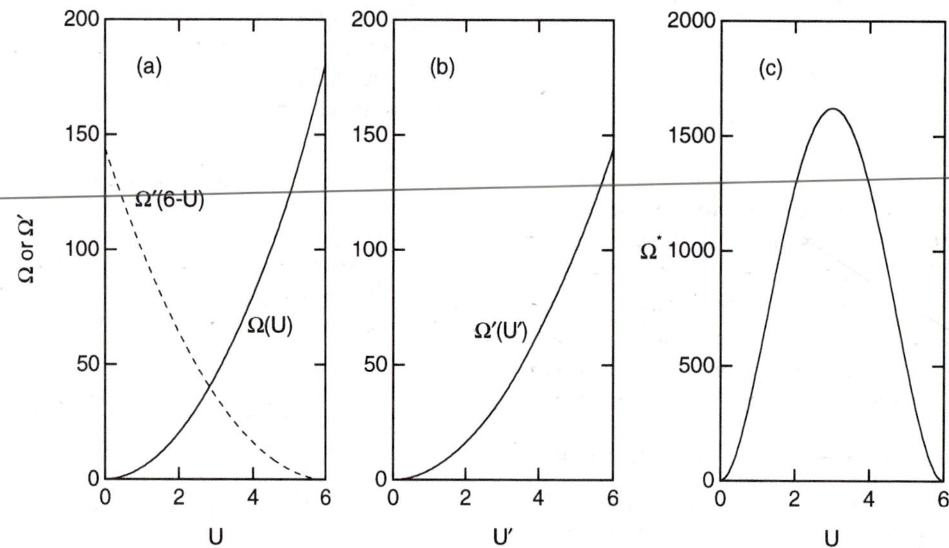

FIGURE 3.10. Example of the behavior of Ω, Ω', and Ω^*. In this case, the values used are $\Omega(U)=5U^2$ and $\Omega'(U')=4(U')^2$. The total energy is 6, so only values of U between 0 and 6 are allowed. (a) Plot of $\Omega(U)$. The dashed line is $\Omega'(6-U)$. (b) Plot of $\Omega'(U')$. (c) Plot of $\Omega^*=\Omega\Omega'$. These functions give $\Omega=0$ when $U=0$. This is not correct. But they are simple and behave properly at higher energies.

dashed line in Fig. 3.10(a). As more energy is given to A, Ω increases but Ω' decreases. The product, $\Omega^*=\Omega\Omega'$, shown in Fig. 3.10(c), reaches a maximum at $U=3$.

The most probable value of U is that for which $P(U)$ is a maximum. Since P is proportional to Ω^*, $\Omega^*(U)$ is also a maximum. Therefore,

$$\frac{d}{dU}[\Omega^*(U)]=0 \tag{3.9}$$

at the most probable value of U. This derivative can be evaluated using Eq. (3.8). Since $U+U'=U^*$, Eq. (3.8) can be rewritten as

$$\Omega^*(U)=\Omega(U)\Omega'(U^*-U). \tag{3.10}$$

The derivative is

$$\frac{d}{dU}(\Omega^*)=\frac{d\Omega}{dU}\Omega'+\Omega\frac{d\Omega'}{dU}.$$

By the chain rule for taking derivatives,

$$\frac{d\Omega'}{dU}=\left(\frac{d\Omega'}{dU'}\right)\left(\frac{dU'}{dU}\right).$$

Since $U'=U^*-U$, $dU'/dU=-1$. Therefore

$$\frac{d\Omega^*}{dU}=\Omega'\frac{d\Omega}{dU}-\Omega\frac{d\Omega'}{dU'}. \tag{3.11}$$

Factoring out the factor $\Omega\Omega'$ gives

$$\frac{d\Omega^*}{dU}=\Omega\Omega'\left(\frac{1}{\Omega}\frac{d\Omega}{dU}-\frac{1}{\Omega'}\frac{d\Omega'}{dU'}\right). \tag{3.12}$$

In equilibrium, this must be zero by Eq. (3.9). Since $\Omega^*=\Omega\Omega'$ cannot be zero, the most probable state or the equilibrium state exists when

$$\frac{1}{\Omega}\frac{d\Omega}{dU}=\frac{1}{\Omega'}\frac{d\Omega'}{dU'}. \tag{3.13}$$

It is convenient to define the quantity τ by

$$\frac{1}{\tau}\equiv\frac{1}{\Omega}\frac{d\Omega}{dU}$$

for any system. We must remember that this derivative was taken when the number of particles and the parameters that determine the energy levels were held fixed. These parameters are such things as volume and electric and magnetic fields. To remind ourselves that everything *but* U is being held fixed, it is customary to use the notation for a *partial derivative*: ∂ instead of d (Appendix N). Therefore, we write

$$\frac{1}{\tau}=\frac{1}{\Omega}\left(\frac{\partial\Omega}{\partial U}\right)_{N,V,\text{etc}}. \tag{3.14}$$

Often we will be careless and just write $\partial\Omega/\partial U$.

The quantity τ defined by Eq. (3.14) depends only on the variables of one system. It is therefore a property of that system. Thermal equilibrium occurs when $\tau=\tau'$. Since Ω is just a number, Eq. (3.14) shows that τ has the dimensions of energy.

Systems A and A', which are in thermal contact, will be in equilibrium (the state of greatest probability) when $\tau=\tau'$. This is reminiscent of something that is familiar to all of us: if a hot system is placed in contact with a cold one, the hotter one cools off and the cooler one gets warmer. The systems come to equilibrium when they are both at the same temperature. This suggests that τ is in some way related to temperature, even though it has the dimensions of energy. We will not prove it, but many things work out right if the absolute temperature T is defined by the relationship

$$\tau=k_BT. \tag{3.15}$$

The proportionality constant is called *Boltzmann's constant*. If T is measured in kelvin (K), k_B has the value

$$\begin{aligned}k_B&=1.380\,658\times10^{-23}\ \text{J K}^{-1}\\&=0.861\,739\times10^{-4}\ \text{eV K}^{-1}.\end{aligned} \tag{3.16}$$

The most convincing evidence in this book that Eq. (3.15) is reasonable is the derivation of the thermodynamic identity in Sec. 3.16.

The absolute temperature T is related to the temperature in degrees centigrade or Celsius by

$$T=(\text{temperature in °C})+273.15. \tag{3.17}$$

3.6. ENTROPY

The preceding section used the idea that the number of microstates accessible to a system increases as the energy of the system increases, to develop a condition for thermal equilibrium. There are two features of those arguments that suggest that there are advantages to working with the natural logarithm of the number of microstates. First, the total number of microstates is the product of the number in each subsystem: $\Omega^*=\Omega\Omega'$. Taking natural logarithms of this gives

$$\ln\Omega^*=\ln\Omega+\ln\Omega'. \tag{3.18}$$

The other feature is the appearance of $(1/\Omega)(\partial\Omega/\partial U)$ in the equilibrium condition. For any non negative, differentiable function $y(x)$,

$$\frac{d}{dx}(\ln y)=\frac{1}{y}\frac{dy}{dx}.$$

Therefore, Eq. (3.14) can be written as

$$\frac{1}{\tau}=\frac{\partial}{\partial U}(\ln\Omega). \tag{3.19}$$

The entropy S is defined by

$$S = k_B \ln \Omega, \quad \Omega = e^{S/k_B}. \tag{3.20}$$

If both sides of Eq. (3.19) are multiplied by k_B, it is seen that

$$\left(\frac{\partial S}{\partial U}\right)_{N,\text{parameters}} = \frac{k_B}{\tau} = \frac{1}{T}. \tag{3.21}$$

This is a fundamental property of entropy that may be familiar to you from other thermodynamics textbooks; if so, it forms a justification for defining temperature as we did.

Another important property of the entropy is that the entropy of system A^* is the sum of the entropy of A and the entropy of A':

$$S^* = S + S'. \tag{3.22}$$

This can be proved by multiplying Eq. (3.18) by k_B.

A third property of the entropy is that S^* is a maximum when systems A and A' are in thermal equilibrium. This result follows from the fact that Ω^* is a maximum at equilibrium, since $S^* = k_B \ln \Omega^*$ and the logarithm is a monotonic function.

Finally, the entropy change in the system can be related to the heat flow into it. Equation (3.21) shows that if there is an energy change in the system *when N and the parameters that govern the spacing of the energy levels are fixed*, then

$$dS = \left\{\frac{\partial S}{\partial U}\, dU\right\}_{N,\text{parameters fixed}} = \left\{\frac{dU}{T}\right\}_{N,\text{parameters fixed}}.$$

But the energy change when N and the parameters are fixed is the heat flow dQ:

$$dS = \frac{dQ}{T}. \tag{3.23}$$

3.7. THE BOLTZMANN FACTOR

Section 3.5 considered the equilibrium state of two systems that were in thermal contact. It is often useful to consider systems in thermal contact when one of the systems is a single particle. This leads to an expression for the total number of microstates as a function of the energy in the single-particle system, known as the *Boltzmann factor*. The Boltzmann factor is used in many situations, as is its alternate form, the *Nernst equation* (Sec. 3.8).

We saw that when two systems in thermal contact reach the same temperature, the entropy and the number of accessible microstates of the total system are a maximum. The total number of microstates, the product of the number in each system, goes through a maximum because increasing the energy of system A at the expense of system A' causes Ω to increase while Ω' decreases.

Let system A be a single particle in thermal contact with a large system or *reservoir* A'. Transferring energy from A' to A decreases the number of microstates in A'. The number of microstates in A may change by some factor G or remain the same. We will discuss G at the end of this section.

To make this argument quantitative, consider system A when it has two different energies, U_r and U_s. Reservoir A' is very large and has many energy levels almost continuously distributed. Let $\Omega'(U')$ be the number of microstates in A' when it has energy U'. As before, the relative probability that A has energy U_s compared to having energy U_r is given by the ratio of the total number of microstates accessible to the combined system:

$$\frac{P(U_s)}{P(U_r)} = \frac{\Omega^*(U=U_s)}{\Omega^*(U=U_r)} = \frac{\Omega(U_s)\Omega'(U^*-U_s)}{\Omega(U_r)\Omega'(U^*-U_r)}. \tag{3.24}$$

This probability is the product of two functions, one depending on system A and one on reservoir A':

$$\frac{P(U_s)}{P(U_r)} = GR,$$

where

$$G = \frac{\Omega(U_s)}{\Omega(U_r)}, \qquad R = \frac{\Omega'(U^*-U_s)}{\Omega'(U^*-U_r)}. \tag{3.25}$$

Ratio R is calculated most easily by using Eq. (3.14), remembering the definition $\tau = k_B T$. Since neither the volume nor number of particles is interchanged, we use an ordinary derivative. We write it in terms of the temperature of the reservoir:

$$\frac{1}{\Omega'}\left(\frac{d\Omega'}{dU'}\right) = \frac{1}{k_B T'}, \qquad \frac{d\Omega'}{dU'} = \left(\frac{1}{k_B T'}\right)\Omega'. \tag{3.26}$$

This is easily integrated:

$$\Omega'(U') = \text{const} \times e^{U'/k_B T'}.$$

Therefore the ratio is

$$R = \frac{\text{const} \times e^{(U^*-U_s)/k_B T'}}{\text{const} \times e^{(U^*-U_r)/k_B T'}} = e^{-(U_s-U_r)/k_B T'} = e^{-(U_s-U_r)/k_B T}. \tag{3.27}$$

Although the temperature T' is a property of the reservoir, we drop the prime. This ratio is called the *Boltzmann factor*. It gives the factor by which the number of microstates in the *reservoir* decreases when the reservoir gives up energy $U_s - U_r$ to the system A.

The relative probability of finding system A with energy U_r or U_s is then given by

$$\frac{P(U_s)}{P(U_r)}=Ge^{-(U_s-U_r)/k_BT}=\left[\frac{\Omega(U_s)}{\Omega(U_r)}\right]e^{-(U_s-U_r)/k_BT}. \tag{3.28}$$

The exponential Boltzmann factor is a property of the reservoir. The factor G is called the *density of states factor*. It is a property of the system. If system A is a single atom with discrete energy levels and we want to know the relative probability that the atom has a particular value of its allowed energy, G may be unity. In other cases, there may be two or more sets of quantum numbers corresponding to the same energy, a situation called *degeneracy*. In that case G may be a small number. We would have to know the details to calculate it.

In other cases, we may be interested in the ratio of the concentration of some ion in two different regions, which is proportional to the probability that an ion is found in a certain volume element. It is argued in Sec. 3.10 that in such a case the number of microstates is proportional to the size of the volume being considered. To relate concentration ratios to probabilities we would consider equal volumes, and G would be unity.

3.8. THE NERNST EQUATION

The Nernst equation is widely used in physiology to relate the concentration of ions on either side of a membrane. It is an example of the Boltzmann factor.

Suppose that certain ions can pass easily through a membrane. If the membrane has an electrical potential difference across it, the ions will have different energy on each side of the membrane. As a result, when equilibrium exists they will be at different concentrations. The ratio of the probability of finding an ion on either side of the membrane is the ratio of the concentrations on the two sides:

$$\frac{C_2}{C_1}=\frac{P(2)}{P(1)}.$$

The total energy of an ion is its kinetic energy plus its potential energy:

$$U=E_k+E_p.$$

Chapter 6 will show that when the electrical potential is v, the potential energy is

$$E_p=zev.$$

In this equation z is the valence of the ion (+1, −1, +2, etc.) and e is the electronic charge (1.6×10^{-19} C).

The concentration ratio is therefore

$$\frac{C_2}{C_1}=\left[\frac{\Omega(2)}{\Omega(1)}\right]e^{-(U_2-U_1)/k_BT}. \tag{3.29}$$

We must now evaluate the quantity in square brackets. It is the ratio of the number of microstates available to the ion on each side of the membrane. The concentration is the number of ions per unit volume and is proportional to the probability that an ion is in volume $\Delta x\Delta y\Delta z$. We will argue in Sec. 3.10 that for a particle that can undergo translational motion in three dimensions, $\Omega(U)$ is $\alpha\Delta x\Delta y\Delta z$, where α is a proportionality constant. Therefore

$$\frac{\Omega(2)}{\Omega(1)}=\frac{\alpha\Delta x\Delta y\Delta z}{\alpha\Delta x\Delta y\Delta z}=1.$$

The energy difference is

$$U_2-U_1=E_k(2)-E_k(1)+ze(v_2-v_1).$$

It will be shown in Sec. 3.12 that the average kinetic energy on both sides of the membrane is the same if the temperature is the same. Therefore,

$$\frac{C_2}{C_1}=e^{-ze(v_2-v_1)/k_BT}. \tag{3.30}$$

If the potential difference is v_2-v_1 then the ions will be in equilibrium if the concentration ratio is as given by Eq. (3.30). If the ratio is not as given, then the ions, since they are free to move through the membrane, will do so until equilibrium is attained or the potential changes.

If the ions are positively charged and $v_2>v_1$, then the exponent is negative and $C_2<C_1$. If the ions are negatively charged, then $C_2>C_1$.

The concentration difference is explained qualitatively by the electrical force within the membrane that causes the potential difference. If $v_2>v_1$, the force within the membrane on a positive ion acts from region 2 toward region 1. It slows ions moving from 1 to 2 and accelerates ions moving from 2 to 1. Thus it tends to increase C_1.

The Nernst equation is obtained by taking logarithms of both sides of Eq. (3.30):

$$\ln\left(\frac{C_2}{C_1}\right)=-\frac{ze}{k_BT}(v_2-v_1).$$

From this,

$$v_2-v_1=\frac{k_BT}{ze}\ln\left(\frac{C_1}{C_2}\right).$$

Multiplying both numerator and denominator of k_BT/ze by Avogadro's number N_A gives the quantities N_Ak_B and N_Ae. The former is the gas constant:

$$N_Ak_B=R=8.314\,51\ \text{J mol}^{-1}\ \text{K}^{-1}. \tag{3.31}$$

The latter is the Faraday constant:

$$N_Ae=F=96\,485.31\ \text{C mol}^{-1}. \tag{3.32}$$

The coefficient is, therefore,

$$\frac{k_B T}{ze} = \frac{RT}{zF}. \tag{3.33}$$

At body temperature, $T = 37\ °\mathrm{C} = 310$ K, the value of the coefficient is 0.0267 J $\mathrm{C}^{-1} = 0.0267$ V. In the form

$$v_2 - v_1 = \frac{RT}{zF} \ln\left(\frac{C_1}{C_2}\right), \tag{3.34}$$

the Boltzmann factor is called the Nernst equation. It is sometimes said that the concentration determines the potential difference. Although in a given situation this may be partially true, it is not what the Nernst equation by itself says. The derivation showed that the concentration ratio given by Eq. (3.34) is the one for which equilibrium will exist, given the potential difference $v_2 - v_1$.

3.9. THE PRESSURE VARIATION IN THE ATMOSPHERE

That atmospheric pressure decreases with altitude is well known. This truth has medical significance because of the effects of lower oxygen at high altitudes. We will derive the decrease using the Boltzmann factor and then do it again using hydrostatic equilibrium.

The gravitational potential energy of an air molecule at height y is mgy, where m is the mass of the molecule and g is the gravitational acceleration. If the atmosphere has a constant temperature, there will be no change of kinetic energy with altitude. For a molecule to increase its potential energy, and therefore its total energy, by mgy, the energy of all the other molecules (the reservoir) must decrease, with a corresponding decrease in the number of accessible microstates. The number of particles per unit volume is given by a Boltzmann factor:

$$C(y) = C(0) e^{-mgy/k_B T}. \tag{3.35}$$

Since for an ideal gas $p = Nk_B T/V = Ck_B T$, the pressure also decreases exponentially with height.

The same result can be obtained without using statistical physics, by considering a small volume of the atmosphere that is in static equilibrium. Let the volume have thickness dy and horizontal cross-sectional area S, as shown in Fig. 3.11. The force exerted upward across the bottom face of the element is $p(y)S$. The force down on the top face is $p(y+dy)S$. The N molecules in the volume each experience the downward force of gravity. The total gravitational force is Nmg. In terms of the concentration, $N = CS\,dy$. Therefore, the condition for equilibrium is

$$p(y)S - p(y+dy)S - CSmg\,dy = 0.$$

Since $p(y) - p(y+dy) = -(dp/dy)dy$, this can be written as

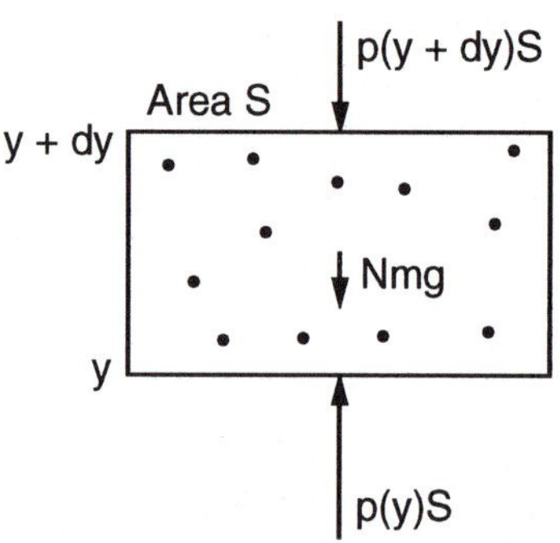

FIGURE 3.11. Forces on a small volume element of the atmosphere.

$$\left[-\left(\frac{dp}{dy}\right) - Cgm\right] S\,dy = 0.$$

The next step is to use the ideal gas law to write $p = Ck_B T$:

$$-k_B T \frac{dC}{dy} - Cgm = 0.$$

If this is written in the form

$$\frac{dC}{dy} = -\frac{mg}{k_B T} C \tag{3.36}$$

it will be recognized as the equation for exponential decay. The solution is that given by Eq. (3.35).

3.10. THE MAXWELL–BOLTZMANN DISTRIBUTION

This section derives the velocity distribution of particles of any kind that have translational kinetic energy because they are in thermal equilibrium. You (or your instructor) may wish to skip this section, since it has no direct biological application. It is included for completeness and to show an example of probability distributions in several variables, which are sometimes found in biological problems. The result obtained in this section is used in Sec. 3.11 to derive an important theorem about the equipartition of energy.

It was shown that the probability ratio for a particle to have energy U_r or U_s is given by Eq. (3.28):

$$\frac{P(U_s)}{P(U_r)} = \frac{\Omega(U_s)}{\Omega(U_r)} e^{-(U_s - U_r)/k_B T}.$$

For the discussion of the Nernst equation and the atmosphere, the factor $\Omega(U_s)/\Omega(U_r)$ was set equal to unity without any real justification. This section examines that factor in the classical limit—when the energy levels are so closely spaced that classical physics can be used.

To simplify the notation, suppose that the zero of the energy scale is picked so that $U_r = 0$. Then Eq. (3.28) becomes

$$P(U_s)=\frac{P(0)}{\Omega(0)}\,\Omega(U_s)e^{-U_s/k_BT}.$$

When the levels become very closely spaced, it is impossible to identify which level the particle is in. Instead, it is necessary to speak of a range of energy between U and $U+\Delta U$. In this case, $P(U)\Delta U$ is the probability that the particle has energy in the interval, and $\Omega(U,\Delta U)$ is the number of microstates in the interval:

$$P(U)\Delta U=\left[\frac{P(0)}{\Omega(0)}\right]\Omega(U,\Delta U)e^{-U/k_BT}. \quad (3.37)$$

The factor $\Omega(U,\Delta U)$ is the number of microstates in which the particle can be when its energy is between U and $U+\Delta U$. We state the result and then try to make it plausible. The factor is

$$\Omega(U,\Delta U)=\frac{\Delta p_x\Delta p_y\Delta p_z\Delta x\Delta y\Delta z}{\hbar^3}, \quad (3.38)$$

where $p_x=mv_x$ is the x component of the particle's momentum, etc., and x,y,z are its location in space. The factor $\hbar$ is Planck's constant h divided by 2π:

$$\hbar=1.054\,572\,7\times10^{-34}\ \text{J s}. \quad (3.39)$$

This result for Ω includes only energy resulting from translation, that is, for a particle moving in three dimensions in a box or in solution. Additional microstates corresponding to rotation and other internal motions may be important in extended bodies such as molecules.

Here are some justifications for Eq. (3.38). In classical mechanics the state of a particle undergoing translational motion in three dimensions is specified at any instant by its position (x,y,z) and velocity or momentum (p_x,p_y,p_z). Its subsequent motion can be calculated if its initial position and momentum are known along with the force on the particle at later times. The initial position and momentum are the constants of integration needed when integrating Newton's second law. The state of the particle at any instant, given by the six values x,y,z,p_x,p_y,p_z, corresponds to a point in six-dimensional *phase space.*

Imagine a collection of gas molecules in a box. Even in the classical case, we might not care about the motion of each individual molecule. We might simply want to know the average number in some small volume of the box. This average is given by the total number of molecules times the probability of any molecule being in the small volume. If the volume is small enough, then the probability is proportional to the volume and can be written as $P(x,y,z)\,dx\,dy\,dz$. This is a joint probability distribution (Appendix M). It is the probability that a molecule is in the volume element $dx\,dy\,dz$, that is, simultaneously between x and $x+dx$ *and* between y and $y+dy$ *and* between z and $z+dz$. In a small box, $P(x,y,z)$ would be the reciprocal of the volume of the box. However, P can depend on position. For example, in a box tall enough so that exponential variations in the atmosphere are important, $P(x,y,z)$ would depend on height.

This joint probability distribution is independent of the momentum of the molecules. One could extend the concept and ask for the probability that a molecule is in the volume element $dx\,dy\,dz$ while at the same time having momentum components between p_x and p_x+dp_x, p_y and p_y+dp_y, and p_z and p_z+dp_z. This probability is proportional to both $dx\,dy\,dz$ and $dp_xdp_ydp_z$, and therefore to the volume in phase space $dx\,dy\,dz\,dp_xdp_ydp_z$:

$$P(x,y,z,p_x,p_y,p_z)\,dx\,dy\,dz\,dp_xdp_ydp_z.$$

This is rather cumbersome to write out, so the abbreviation

$$P(\mathbf{r},\mathbf{p})\,d^3\mathbf{r}\,d^3\mathbf{p}$$

is often used.

If we have an ensemble of identical systems we can at any instant describe the ensemble as a collection of points in phase space. The density of points per unit volume of phase space is proportional to the probability function $P(\mathbf{r},\mathbf{p})$. Liouville's theorem of advanced classical mechanics (Reif, 1965, Appendix 13) shows that for an ensemble of systems conserving energy, the density of points in phase space does not change with time. This means that if the density of points in phase space is uniform it remains that way. This is consistent with the postulates of Sec. 3.4, with the equilibrium state corresponding to an ensemble with a uniform density of points in phase space.

Quantum mechanics does not allow precise determination of both the position and the momentum of a particle. The Heisenberg uncertainty principle gives the limitation in precision of the simultaneous measurement of both quantities:

$$(\Delta x)(\Delta p_x)\geq\hbar.$$

Similar relations hold for the y direction and z direction. This says that there is a fundamental limitation on how small one can make the products $dx\,dp_x$, $dy\,dp_y$, and $dz\,dp_z$. Because $\hbar$ is so small, we can conceive of intervals $dx\,dp_x$ that are quite small on a classical scale but that contain many multiples of $\hbar$. If each element of phase space of size $\hbar$ (in one dimension) is associated with a quantum-mechanical energy level, then the factor $dx\,dp_x/\hbar$ is the number of levels in interval $dx\,dp_x$. In three dimensions, the number of levels is

$$\frac{dx\,dy\,dz\,dp_xdp_ydp_z}{\hbar^3},$$

that is, the volume in phase space divided by $\hbar^3$.

For a single particle moving in three dimensions and in thermal equilibrium, therefore, Eq. (3.37) becomes

$$P(U)dU=C(e^{-U/k_BT})\,dx\,dy\,dz\,dp_xdp_ydp_z. \quad (3.40)$$

[In writing this equation, the factor $P(0)/[\hbar^3\Omega(0)]$ has been called the constant C.] This is the probability that the particle is between x and $x+dx$, between y and $y+dy$, and between z and $z+dz$, *and* has momentum components between p_x and p_x+dp_x, p_y and p_y+dp_y, and p_z and p_z+dp_z. This is called a joint probability distribution; see Appendix M. The total energy U in the exponent is related to position and momentum by

$$U=E_k+E_p,$$
$$U=\frac{p_x^2+p_y^2+p_z^2}{2m}+E_p(x,y,z). \tag{3.41}$$

This is a rather complex equation, and it will take several pages to show what it means. There are two different ways to proceed: we can either calculate the probability that the particle has energy between U and $U+dU$, or we can ask for the probability that the position and momentum are in the range just described. The latter is easier and will be done first. It is only necessary to write the left-hand side as

$$P(\mathbf{r},\mathbf{p})\ d^3\mathbf{r}\ d^3\mathbf{p}$$

and substitute Eq. (3.41) in the exponent.

We will work through a specific example: an oxygen molecule at 300 K. Since 1 mol of O_2 has a mass of 32×10^{-3} kg, the mass of a single molecule is the mass of 1 mol divided by Avogadros's number:

$$m=\frac{32\times10^{-3}}{6\times10^{23}}=5.314\times10^{-26}\text{ kg}.$$

The factor $2mk_BT$, which occurs in the expressions below, is

$$2mk_BT=(2)(5.31\times10^{-26})(1.38\times10^{-23})(300)$$
$$=4.40\times10^{-46}\text{ kg}^2\text{ m}^2\text{ s}^{-2}.$$

If the potential energy does not depend on position, then the exponent is

$$\frac{U}{k_BT}=\frac{p_x^2+p_y^2+p_z^2}{2mk_BT}.$$

The factor $P(\mathbf{r},\mathbf{p})$ is given by

$$Ce^{-U/k_BT}=C\exp\left[-\left(\frac{p_x^2+p_y^2+p_z^2}{2mk_BT}\right)\right].$$

Since $e^{a+b}=e^ae^b$, this can also be written as

$$(e^{-p_x^2/2mk_BT})(e^{-p_y^2/2mk_BT})(e^{-p_z^2/2mk_BT}).$$

Therefore, the probability can be written

$$P(\mathbf{r},\mathbf{p})\ d^3\mathbf{r}\ d^3\mathbf{p}=C(e^{-p_x^2/2mk_BT}dp_x)(e^{-p_y^2/2mk_BT}dp_y)$$
$$\times(e^{-p_z^2/2mk_BT}dp_z)\ dx\ dy\ dz. \tag{3.42}$$

Consider the dependence of this expression on p_x. The probability that p_x is in the interval (p_x,dp_x) is proportional to $e^{-p_x^2/2mk_BT}dp_x$. The exponential term is plotted in Fig. 3.12. It is the same term that appears in the Gaussian probability distribution (Appendix I). It is a maximum when the exponent is zero and is symmetric about the origin because the exponent contains p_x^2. The area under the curve is finite and is equal to $\sqrt{2\pi mk_BT}$ (see Appendix K). The fact that the curve is nearly zero for momenta greater than 5×10^{-23} kg m s^{-1} reflects the fact that very few oxygen molecules at room temperatures have momenta exceeding that.

If we want to know the probability that the molecule has a velocity between $+200$ and $+210$ m s^{-1} (the + sign meaning that it is traveling to the right), we can find it from Eq. (3.42). First, these velocities are multiplied by the mass of an oxygen molecule to get the range of p_x, which is from $+1.063\times10^{-23}$ to $+1.116\times10^{-23}$ kg m s^{-1}. These two values are represented by the vertical lines in Fig. 3.12. Suppose the molecules are confined to a box with dimensions L_x, L_y, and L_z. The probability that the molecule has a value of p_x between p_x and p_x+dp_x, while it can be anywhere within the box and have any value of p_y or p_z, is obtained by integrating over all the other variables:

$$P(p_x)dp_x=\int_{p_y=-\infty}^{p_y=\infty}\int_{p_z=-\infty}^{p_z=\infty}\int_{x=0}^{L_x}\int_{y=0}^{L_y}\int_{z=0}^{L_z}$$
$$\times P(\mathbf{r},\mathbf{p})dp_ydp_zdx\ dy\ dz\ dp_x.$$

It becomes a bit cumbersome to keep track of which integral sign has limits for which variable, so it is customary to rearrange the terms:

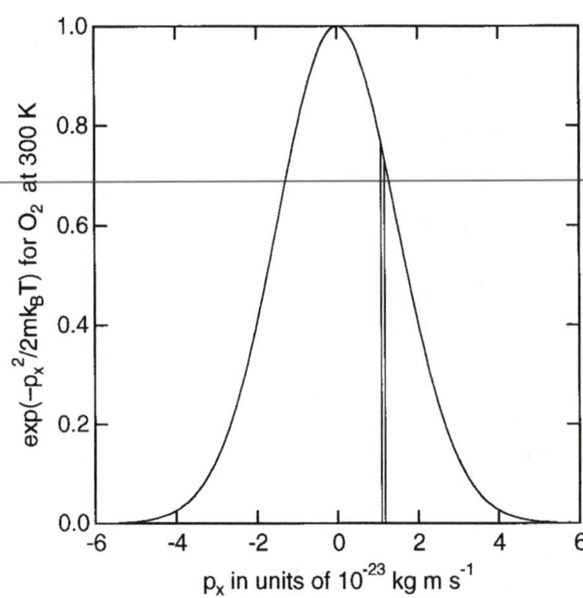

FIGURE 3.12. Plot of $e^{-p_x^2/2mk_BT}$ for an oxygen molecule at 300 K.

$$P(p_x)dp_x=\left[\int_{-\infty}^{\infty}dp_y\int_{-\infty}^{\infty}dp_z\int_0^{L_x}dx\int_0^{L_y}dy \times\int_0^{L_z}dz\, P(\mathbf{r},\mathbf{p})\right]dp_x.$$

If Eq. (3.42) is used for $P(\mathbf{r},\mathbf{p})$, this becomes

$$P(p_x)dp_x=C\int_{-\infty}^{\infty}dp_y e^{-p_y^2/2mk_BT}\int_{-\infty}^{\infty}dp_z e^{-p_z^2/2mk_BT} \times\int_0^{L_x}dx\int_0^{L_y}dy\int_0^{L_z}dz e^{-p_x^2/2mk_BT}dp_x.$$

Each of these integrals can be evaluated. They are

$$\int_{-\infty}^{\infty}dp_y e^{-p_y^2/2mk_BT}=(2\pi mk_BT)^{1/2},$$

$$\int_{-\infty}^{\infty}dp_z e^{-p_z^2/2mk_BT}=(2\pi mk_BT)^{1/2},$$

$$\int_0^{L_x}dx=L_x,$$

$$\int_0^{L_y}dy=L_y,$$

$$\int_0^{L_z}dz=L_z.$$

When these values are substituted, the expression becomes

$$P(p_x)dp_x=C(2\pi mk_BT)(L_xL_yL_z)e^{-p_x^2/2mk_BT}dp_x.$$

The next step is to determine the constant C. The easiest way is to use the fact that the particle must have some value of p_x, so that

$$\int_{-\infty}^{\infty}P(p_x)dp_x=1.$$

Therefore,

$$C(2\pi mk_BT)(L_xL_yL_z)\int_{-\infty}^{\infty}dp_x e^{-p_x^2/2mk_BT}=1.$$

This integral is just like the other momentum integrals above and has the value $(2\pi mk_BT)^{1/2}$:

$$C(2\pi mk_BT)^{3/2}(L_xL_yL_z)=1.$$

It is customary to call $L_xL_yL_z$ the volume of the box, V. Then

$$C=\frac{1}{V(2\pi mk_BT)^{3/2}}. \quad (3.43)$$

When this is put in the expression for $P(p_x)$, the result is

$$P(p_x)dp_x=\frac{1}{(2\pi mk_BT)^{1/2}}e^{-p_x^2/2mk_BT}dp_x. \quad (3.44)$$

The probability that p_x has a value between p_1 and p_2 is then

$$P(p_1,p_2)=(2\pi mk_BT)^{-1/2}\int_{p_1}^{p_2}dp_x e^{-p_x^2/2mk_BT}.$$

For the case we are considering, this integral is most easily evaluated by assuming that the integrand is nearly constant because the two values of p are so close together. (The integrand is plotted over this range in Fig. 3.13.) The integral is equal to the shaded area in Fig. 3.13. If that area is assumed to be a rectangle of height 0.774 (the value of the integrand at $p=1.063\times10^{-23}$ kg m s^{-1}), the area is 4.10×10^{-25}; if it is approximated by a trapezoid, the width of the base must be multiplied by the average height, and the area is 4.04×10^{-25}. The probability that the velocity or momentum is between these limits is therefore

$$P=(4.04\times10^{-25})(1.38\times10^{-45})^{-1/2}=0.011.$$

The probability that a particle has an x component of velocity between 200 and 210 m s^{-1} to the right is about 1%.

If we had asked for the probability that the particle had this velocity *and* at the same time was in the left half of the box (say, between $x=0$ and $x=L_x/2$), the answer would have been

$$\int_{p_1}^{p_2}dp_x\int_{-\infty}^{\infty}dp_y\int_{-\infty}^{\infty}dp_z\int_0^{L_x/2}dx\int_0^{L_y}dy\int_0^{L_z}dz\, P(\mathbf{r},\mathbf{p}).$$

This works out to be

$$\frac{(2\pi mk_BT)(L_x/2)(L_y)(L_z)}{(2\pi mk_BT)^{3/2}(L_xL_yL_z)}\int_{p_1}^{p_2}e^{-p_x^2/2mk_BT}dp_x.$$

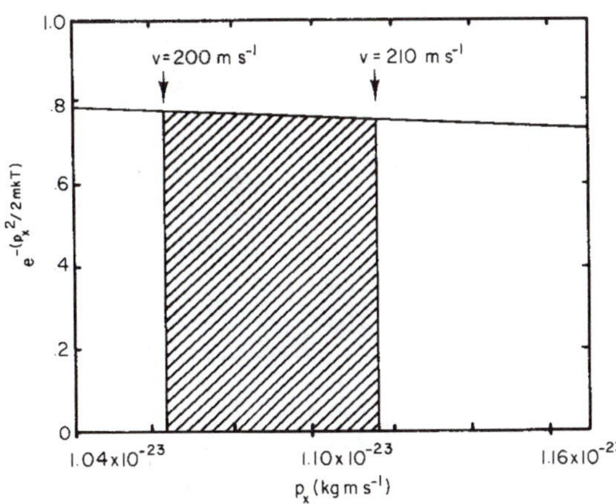

FIGURE 3.13. Expanded scale for a portion of the plot in Fig. 3.12, showing the exponential between 200 and 210 m s^{-1} (p_x=1.063×10^{-23} and 1.116×10^{-23} kg m s^{-1}).

The only difference between this and the previous case is the factor of 1/2 introduced because only half the volume is considered. The final result is 0.005.

Another thing we can find with the Maxwell–Boltzmann distribution is the probability that the particle has an energy between U and $U+dU$. To find this, it is necessary to return to Eq. (3.40) and write it in terms of U instead of the momentum. The exponential part, e^{-U/k_BT}, is correct as it stands. We will restrict ourselves to the case that there is no potential energy so that

$$U=E_k=\frac{p^2}{2m}=\frac{p_x^2+p_y^2+p_z^2}{2m}.$$

It is necessary to relate the fact that the energy may be between U and $U+dU$, to the volume element $d^3\mathbf{p}$.

Instead of working with rectangular coordinates, it is more convenient to use the spherical coordinates shown in Fig. 3.14. The three independent variables are p, the magnitude of the momentum, θ, the angle that $\mathbf{p}$ makes with the p_z axis, and ϕ, the angle that the projection of $\mathbf{p}$ in the (p_x,p_y) plane makes with the p_x axis. The components can be written in terms of these new variables by considering the projections in Fig. 3.14:

$$p_x=p\sin\theta\cos\phi,$$

$$p_y=p\sin\theta\sin\phi,$$

$$p_z=p\cos\theta. \tag{3.45}$$

With these variables, the total energy is $p^2/2m$ and does not depend on the angles.

In rectangular coordinates, the volume element $d^3\mathbf{p}$ was determined by lines along which only one of the three variables changed at a time. In spherical coordinates, the volume element is determined the same way. First, p is varied by an amount dp while θ and ϕ are fixed. Then θ is changed by $d\theta$ while p and ϕ are fixed; the edge of the "cube" traced out has length $p\,d\theta$. Finally, ϕ is changed, giving an edge $p\sin\theta\,d\phi$. The volume of the element is the product of these three edges, as long as the displacements are so small that the element is like a rectangular solid:

$$d^3\mathbf{p}=p^2\sin\theta\,d\theta\,d\phi\,dp.$$

The dimensions are cubic momentum. This replaces $dp_xdp_ydp_z$ in Eq. (3.42):

$$P(\mathbf{r},\mathbf{p})\,d^3\mathbf{r}\,d^3\mathbf{p}=Ce^{-U/k_BT}dx\,dy\,dz\,p^2\sin\theta\,d\theta\,d\phi\,dp. \tag{3.46}$$

The angular integrations can be carried out because U does not depend on θ and ϕ. The limits of the ϕ integration are 0 to 2π and give the shaded strip shown in Fig. 3.15. The θ integration is carried from 0 to π (north pole to south pole) to cover the entire surface of the sphere. The angular integrals are

$$\int_0^{2\pi}d\phi$$

and

$$\int_0^{\pi}\sin\theta\,d\theta. \tag{3.47}$$

The ϕ integral is 2π. The θ integral is

$$\int_0^{\pi}\sin\theta\,d\theta=[-\cos\theta]_0^{\pi}=-[(-1)-(1)]=2.$$

The angular integrals together are therefore 4π. Equation (3.46) now has become

$$P(\mathbf{r},\mathbf{p})\,d^3\mathbf{r}\,dp=4\pi Ce^{-U/k_BT}dx\,dy\,dz\,p^2dp. \tag{3.48}$$

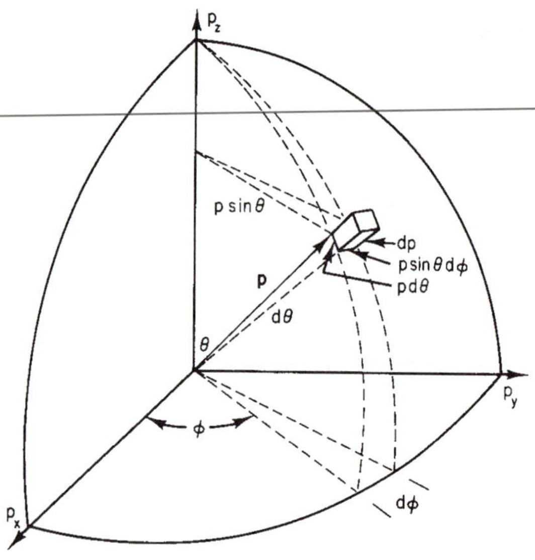

FIGURE 3.14. The momentum in spherical coordinates.

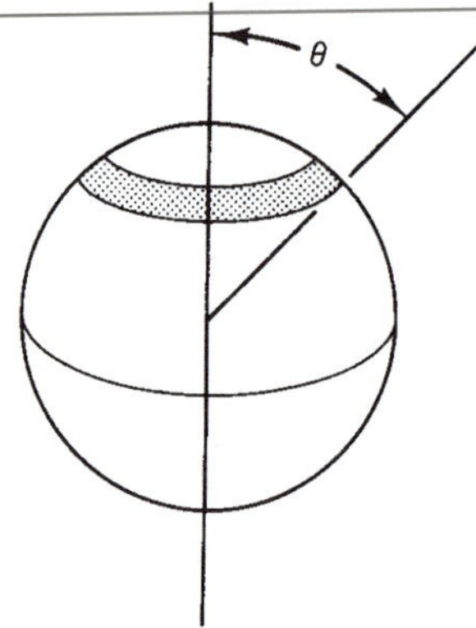

FIGURE 3.15. The shaded area on the surface of a sphere of radius p is $2\pi p^2\sin\theta\,d\theta$. The ϕ integration has already been carried out. The θ integration goes from 0 to π to cover the entire sphere.

Note that the differential is now a fourth-order differential instead of a sixth-order differential, because two of the integrations have already been made.

The next step is to write p^2dp in terms of $U=p^2/2m$. Since

$$dU=\frac{2p\ dp}{2m}=\frac{p\ dp}{m}.$$

p^2dp can be written as $pm\ dU$. Finally, the fact that $p=(2mU)^{1/2}$ is used to write

$$p^2dp=m^{3/2}2^{1/2}U^{1/2}dU.$$

The probability as a function of U is therefore

$$P(\mathbf{r},U)\ d^3\mathbf{r}\ dU=C(4\pi 2^{1/2}m^{3/2})\times e^{-U/k_BT}U^{1/2}dU\ dx\ dy\ dz. \tag{3.49}$$

This is the Maxwell–Boltzmann distribution as a function of energy for particles whose energy is $U=p^2/2m$. Figure 3.16 is a plot of $U^{1/2}e^{-U}$, which gives the general shape of this distribution. The probability of the particle having no energy is zero, while the probability that p_x was zero is a maximum. The difference occurs because $d^3\mathbf{p}$ is proportional to p^2dp, which vanishes for zero total momentum.

Finally, as an example of using these integrals, we will verify that the probability is still correctly normalized. Imagine that the particle is confined to a box with sides L_x, L_y, L_z and total volume V. The spatial integrals again give V. The constant was determined for this condition in Eq. (3.43). Therefore

$$P(U)\ dU=\int_0^{L_x}dx\int_0^{L_y}dy\int_0^{L_z}dz\ P(\mathbf{r},U)\ dU$$

$$=\frac{1}{(2\pi mk_BT)^{3/2}}(4\pi 2^{1/2}m^{3/2})e^{-U/k_BT}U^{1/2}dU.$$

The possible values of U range from 0 to infinity. Therefore we must evaluate

$$\int_0^\infty e^{-U/k_BT}U^{1/2}dU.$$

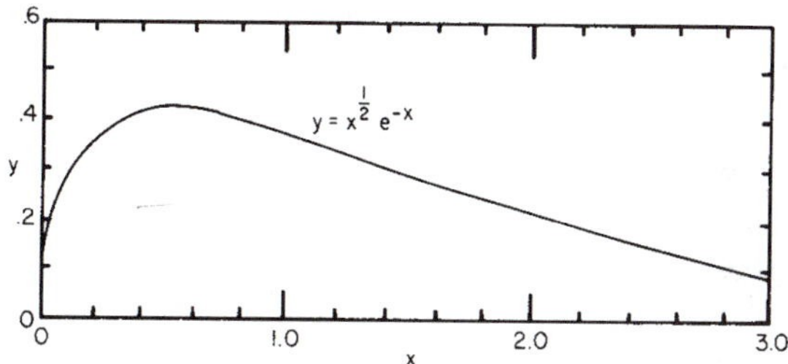

FIGURE 3.16. Plot of $y=x^{1/2}e^{-x}$ to show the general shape of the Maxwell–Boltzmann energy distribution.

Standard tables of integrals show that

$$\int_0^\infty x^n e^{-ax}dx=\frac{\Gamma(n+1)}{a^{n+1}}. \tag{3.50}$$

The quantity $\Gamma(n+1)$ is a tabulated function called the gamma function. In this case, $n=\frac{1}{2}$, so that

$$\int_0^\infty U^{1/2}e^{-U/k_BT}dU=\Gamma(\tfrac{3}{2})(k_BT)^{3/2}.$$

The Γ function is evaluated using the integral tables again

$$\Gamma\left(m+\frac{1}{2}\right)=\frac{1\times3\times5\cdots\times(2m-1)}{2^m}(\pi)^{1/2}.$$

Setting $m=1$ in this expression gives

$$\Gamma(\tfrac{3}{2})=\tfrac{1}{2}\sqrt{\pi}.$$

When this result is put in the expression for $P(U)$, we have

$$\int_0^\infty P(U)\ dU=\frac{1}{(2\pi mk_BT)^{3/2}}(4\pi 2^{1/2}m^{3/2})(\tfrac{1}{2}\pi^{1/2})(k_BT)^{3/2}$$

$$=1.$$

The algebra has been done correctly and the probability is still normalized.

3.11. EQUIPARTITION OF ENERGY AND BROWNIAN MOVEMENT

A very important application of the Maxwell–Boltzmann distribution is the proof that the average translational kinetic energy per degree of freedom of a particle in thermal contact with a reservoir at temperature T is $k_BT/2$. In three dimensions the average kinetic energy is $3k_BT/2$. This result holds for any term in the total energy that depends on the square of one of the variables (such as a component of the position or the momentum).

The proof is done for the kinetic energy resulting from the x component of momentum. The same procedure can be used for the other components. Rather than keep track of the normalization constant C and the volume in which the particle moves, we will normalize the p_x probability distribution directly:

$$\overline{\left(\frac{p_x^2}{2m}\right)}=\frac{\int_{-\infty}^{\infty}(p_x^2/2m)e^{-p_x^2/2mk_BT}dp_x}{\int_{-\infty}^{\infty}e^{-p_x^2/2mk_BT}dp_x}. \tag{3.51}$$

The integral in the denominator is evaluated in Appendix K and is $(2\pi mk_BT)^{1/2}$. Since the integrand is symmetric, the numerator is

$$\frac{1}{2m}2\int_0^\infty p_x^2e^{-p_x^2/2mk_BT}dp_x.$$

Standard integral tables show that

$$\int_0^\infty x^{2n} e^{-ax^2} dx = \frac{1\times3\times5\times\cdots\times(2n-1)}{2^{n+1}a^n}\sqrt{\frac{\pi}{a}}. \tag{3.52}$$

For the special case $n=1$, this becomes

$$\int_0^\infty x^2 e^{-ax^2} dx = \left(\frac{1}{4a}\right)\left(\frac{\pi}{a}\right)^{1/2}. \tag{3.53}$$

The numerator of Eq. (3.51) is therefore

$$\left(\frac{1}{m}\right)\left(\frac{1}{4}\right)(2mk_BT)(2\pi mk_BT)^{1/2}.$$

Combining this with the denominator gives

$$\overline{\left(\frac{p_x^2}{2m}\right)} = \frac{k_BT}{2}. \tag{3.54}$$

The average value of the kinetic energy corresponding to motion in the x direction is $k_BT/2$, independent of the mass of the particle. The only condition that went into this derivation was that the energy depended on the square of the variable. Any term in the total energy that is a *quadratic function of some variable* will carry through the same way, so that the average energy will be $k_BT/2$ for that variable. This result is called the *equipartition of energy.*

The total translational kinetic energy is the sum of three terms $(p_x^2+p_y^2+p_z^2)/2m$, so the total translational kinetic energy has average value $(\frac{3}{2})k_BT$.

This result is true for particles of *any* mass: atoms, molecules, pollen grains, and so forth. Heavier particles will have a smaller velocity but the same average kinetic energy. Even heavy particles are continually moving with this average kinetic energy. The random motion of pollen particles in water was first seen by a botanist, Robert Brown, in 1827. This *Brownian movement* is an important topic in the next chapter.

3.12. EQUILIBRIUM WHEN PARTICLES CAN BE EXCHANGED: THE CHEMICAL POTENTIAL

Section 3.6 considered two systems that could exchange heat. The most probable or equilibrium state was that in which energy had been exchanged so that the total number of microstates or total entropy was a maximum. This occurred when

$$\frac{1}{\Omega}\left(\frac{\partial \Omega}{\partial U}\right)_{N,V} = \frac{1}{\Omega'}\left(\frac{\partial \Omega'}{\partial U'}\right)_{N',V'}, \tag{3.13}$$

which is equivalent to

$$T=T'.$$

Since $S=k_B \ln \Omega$ this is also equivalent to

$$\left(\frac{\partial S}{\partial U}\right)_{N,V} = \left(\frac{\partial S'}{\partial U'}\right)_{N',V'}.$$

This section considers the case in which the systems can exchange both heat energy and particles; they are in thermal and diffusive contact. The number of particles in each system is not fixed, but their sum is constant:

$$N+N'=N^*. \tag{3.55}$$

Equilibrium will exist for the most probable state, which means that heat has flowed until the two temperatures are the same and Eq. (3.13) is satisfied. The most probable state also requires a maximum in Ω^* or S^* vs N. The arguments used in the earlier section for heat exchange can be applied to obtain the equilibrium condition

$$\frac{1}{\Omega}\left(\frac{\partial \Omega}{\partial N}\right)_{U,V} = \frac{1}{\Omega'}\left(\frac{\partial \Omega'}{\partial N'}\right)_{U',V'}. \tag{3.56}$$

The condition in terms of entropy is

$$\left(\frac{\partial S}{\partial N}\right)_{U,V} = \left(\frac{\partial S'}{\partial N'}\right)_{U',V'}. \tag{3.57}$$

For thermal contact, the temperature was defined in terms of the derivative of S with respect to U, so that equilibrium occurred when $T=T'$. An analogous quantity, the *chemical potential*, is defined by

$$\mu \equiv -T\left(\frac{\partial S}{\partial N}\right)_{U,V}. \tag{3.58}$$

(The reason T is included in the definition will become clear later.) Both thermal and diffusive equilibrium exist when

$$T=T', \quad \mu=\mu'. \tag{3.59}$$

Two systems are in thermal and diffusive equilibrium when they have the same temperature and the same chemical potential.

The units of the chemical potential are energy (J). Since the units of S are $\mathrm{J\,K^{-1}}$ and the units of N are dimensionless,[6] Eq. (3.58) shows that the units of μ are J.

Consider next what happens to the entropy of the total system when particles are exchanged when the system is not in equilibrium. The change of total entropy is

$$\Delta S^* = \left(\frac{\partial S^*}{\partial N}\right)\Delta N = \left(\frac{\partial S}{\partial N}\right)\Delta N + \left(\frac{\partial S'}{\partial N'}\right)\Delta N'.$$

Using the definition of the chemical potential and the fact that $\Delta N' = -\Delta N$, we can rewrite this as

[6]In this book, N represents the number of particles, and the chemical potential has units of energy per particle. In other books it may have units of energy per mole.

$$\Delta S^* = \left(-\frac{\mu}{T} \right) \Delta N - \left(-\frac{\mu'}{T'} \right) \Delta N.$$

If the two temperatures are the same, this is

$$\Delta S^* = \left(\frac{\mu' - \mu}{T} \right) \Delta N. \quad (3.60)$$

We see again that the entropy change will be zero for a small transfer of particles from one system to the other if $\mu = \mu'$. Suppose now that particles flow from A' to A, so that ΔN is positive. If $\mu' > \mu$, that is, the chemical potential of A' is greater than that of A, this will cause an increase in entropy of the system. *If particles move from a system of higher chemical potential to one of lower chemical potential, the entropy of the total system increases.*

3.13. CONCENTRATION DEPENDENCE OF THE CHEMICAL POTENTIAL

The chemical potential of an ideal gas (or a solute in an ideal solution)[7] has the form

$$\Delta \mu = k_B T \ln \left(\frac{C}{C_0} \right) + \Delta(\text{potential energy per particle}). \quad (3.61)$$

We will derive this in Sec. 3.18; for now we show that it is plausible and consistent with the Boltzmann factor.

We know from experience that particles tend to move from a region of higher to lower potential energy, thus increasing their kinetic energy, which can then be transferred as heat to other particles by collision. We also know that particles will spread out to reduce their concentration if they are allowed to. It is the combination of these two factors that causes the Boltzmann distribution of particles in the atmosphere. Both processes, decreasing the potential energy and decreasing the concentration, cause a decrease in the chemical potential and therefore an increase in the entropy. When the atmosphere is in equilibrium, the potential energy term increases with height and the concentration term decreases with height so that the chemical potential is constant.

To see the equivalence between Eq. (3.61) and the Boltzmann factor, suppose that particles can move freely from region 1 to region 2 and that the potential energy difference between the two regions is ΔE_p. The particles will be in equilibrium when $\mu_1 = \mu_2$. From Eq. (3.61) this means that

$$k_B T \ln C_1 + E_{p1} = k_B T \ln C_2 + E_{p_2}.$$

This equation can be rearranged to give

$$\ln C_2 - \ln C_1 = -\frac{E_{p2} - E_{p1}}{k_B T}.$$

If exponentials are taken of each side, the result is

$$\frac{C_2}{C_1} = e^{-(\Delta E_p)/k_B T}.$$

If the temperature of each region is the same, the average kinetic energy will be the same in each system, and $\Delta E_p = \Delta U$. This is then the same as the Boltzmann factor, Eq. (3.29).

There is still another way to look at the concentration dependence. In an ideal gas, the pressure, volume, temperature, and number of particles are related by the equation of state

$$pV = N k_B T.$$

In terms of the particle concentration $C = N/V$, this is

$$p = C k_B T.$$

The work necessary to concentrate the gas from volume V_1 and concentration C_1 to V_2 and C_2 is [see Eq. (1.55)]

$$W_{\text{on gas}} = -\int_{V_1}^{V_2} p(V)\, dV. \quad (3.62)$$

The concentration work at a constant temperature is

$$W = -N k_B T \int_{V_1}^{V_2} \frac{dV}{V} = -N k_B T \ln \frac{V_2}{V_1}.$$

If the final volume is smaller than the initial volume, the logarithm is negative and the concentration work is positive. In terms of the particle concentration $C = N/V$ or the molar concentration $c = n/V$, the concentration work is

$$W_{\text{conc}} = N k_B T \ln \frac{C_2}{C_1} = nRT \ln \frac{c_2}{c_1}. \quad (3.63)$$

The last form was written by observing that $N k_B = nR$ where R is the gas constant per mole.

Comparing Eq. (3.63) with Eq. (3.61), we see that the concentration work at constant temperature is proportional to the change in chemical potential with concentration. It is, in fact, just the number of molecules N times the change in μ: $W_{\text{conc}} = N\, \Delta\mu$.

The concentration work or change of chemical potential can be related to the Boltzmann factor in still another way. Particles are free to move between two regions of different potential energy at the same temperature. The work required to change the concentration is, by Eq. (3.63),

$$W_{\text{conc}} = N\, \Delta\mu = N k_B T \ln \frac{C_2}{C_1}.$$

The concentration ratio is given by a Boltzmann factor:

[7]An ideal solution is defined in Sec. 3.18. This definition of the chemical potential is per molecule, which is consistent with our usage above. Other authors define the chemical potential per mole.

$$\frac{C_2}{C_1} = e^{-(E_{p2}-E_{p1})/k_BT},$$

so that

$$\ln\frac{C_2}{C_1} = \frac{-(E_{p2}-E_{p1})}{k_BT}.$$

Therefore, the concentration work is

$$W_{\text{conc}} = -N(E_{p2}-E_{p1}).$$

If $C_2 < C_1$, W is negative and is equal in magnitude to the increase in potential energy of the molecules. The concentration energy lost by the molecules is precisely that required for them to move to the region of higher potential energy. If $C_2 > C_1$, the loss of potential energy going from region 1 to region 2 provides the energy necessary to concentrate the gas. Alternatively, one may say that the sum of the concentration energy and the potential energy is the same in the two regions. This was, in fact, the statement about the chemical potential at equilibrium: $\mu_2 = \mu_1$.

It was mentioned that the same form for the chemical potential is obtained for a dilute solute. (We will present one way of understanding why in Sec. 3.18.) Therefore, the concentration work calculated for an ideal gas is the same as for an ideal solute. The work required to concentrate 1 mol of substance by a factor of 10 at 310 K is

$$(1\ \text{mol})(8.31\ \text{J mol}^{-1}\ \text{K}^{-1})(310\ \text{K})[\ln(10)]$$

or 5.93×10^3 J. One of the most concentrated substances in the human body is H^+ ion in gastric juice, which has a pH of 1. Since it was concentrated from plasma with a pH of about 7, the concentration ratio is 10^6. The work necessary to concentrate 1 mol is therefore $RT\ln(10^6) = (8.31)(310)(13.82) = 3.56\times10^4$ J.

3.14. SYSTEMS THAT CAN EXCHANGE VOLUME

We have considered two systems that can exchange energy or particles. Now consider the systems shown in Fig. 3.17. They are isolated from the rest of the universe. The vertical line that separates them is a piston that can move and conduct heat, so that energy and volume can be exchanged between the two systems. The piston prevents particles from being exchanged. The constraints are $V^* = V + V'$ and $U^* = U + U'$ from which $dV = -dV'$, $dU = -dU'$. As before, equilibrium exists when the total number of microstates or the total entropy is a maximum. The conditions for maximum entropy are

$$\left(\frac{\partial S^*}{\partial U}\right)_{N,V} = 0, \quad \left(\frac{\partial S^*}{\partial V}\right)_{N,U} = 0.$$

The derivation proceeds as before. For example,

$$\left(\frac{\partial S^*}{\partial V}\right)_{N,U} = \left(\frac{\partial S}{\partial V}\right)_{N,U} + \left(\frac{\partial S'}{\partial V}\right)_{N,U} = \left(\frac{\partial S}{\partial V}\right)_{N,U} - \left(\frac{\partial S'}{\partial V'}\right)_{N',U'}.$$

Equilibrium requires that $T = T'$ so that there is no heat flow. The piston will stop moving and there will be no change of volume when

$$\left(\frac{\partial S}{\partial V}\right)_{N,U} = \left(\frac{\partial S'}{\partial V'}\right)_{N',U'}. \tag{3.64}$$

These derivatives can be evaluated in several ways. The method used here involves some manipulation of derivatives; a more detailed description, consistent with the microscopic picture of energy levels, is found in Reif (1964, pp. 267–273).

For a small exchange of heat and work, the first law can be written as $dU = dQ - dW$. In the present case the only form of work is that related to the change of volume, so $dU = dQ - p\,dV$. It was shown in Eq. (3.23) that $dQ = T\,dS$. Therefore $dU = T\,dS - p\,dV$. This equation can be solved for dS:

$$dS = \left(\frac{1}{T}\right)dU + \left(\frac{p}{T}\right)dV. \tag{3.65}$$

The entropy depends on U, V and N, $S = S(U,V,N)$. If N is not allowed to change, then

$$dS = \left(\frac{\partial S}{\partial U}\right)_{N,V} dU + \left(\frac{\partial S}{\partial V}\right)_{N,U} dV. \tag{3.66}$$

Comparison of this with Eq. (3.65) shows that

$$\left(\frac{\partial S}{\partial U}\right)_{N,V} = \frac{1}{T}, \tag{3.21}$$

$$\left(\frac{\partial S}{\partial V}\right)_{N,U} = \frac{p}{T}. \tag{3.67}$$

The first of these equations was already seen as Eq. (3.21). The second is the one we need to consider equilibrium under volume change. Referring to Eq. (3.64) we see that at equilibrium

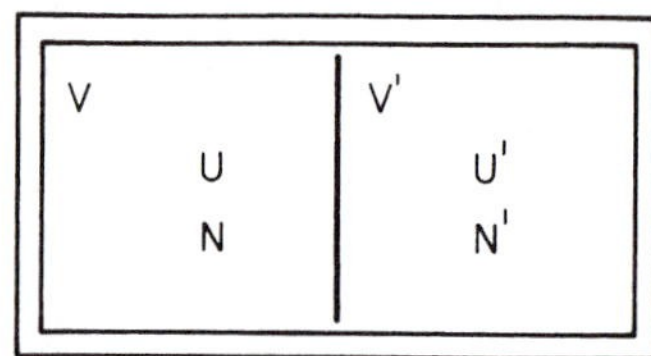

FIGURE 3.17. Two systems can exchange volume because they are separated by a movable piston. Heat can also flow through the piston.

$$\frac{p}{T} = \frac{p'}{T'}.$$

Therefore, equilibrium requires both $T = T'$ and

$$p = p'. \tag{3.68}$$

This agrees with common experience. The piston does not move when the pressure on each side is the same.

3.15. EXTENSIVE VARIABLES AND GENERALIZED FORCES

The number of microstates and the entropy of a system depend on the number of particles, the total energy, and the positions of the energy levels of the system. The positions of the energy levels depend on the volume and may also depend on other macroscopic parameters. For example, they may depend on the length of a stretched muscle fiber or a protein molecule. For charged particles in an electric field, they depend on the charge. For a thin film such as the fluid lining the alveoli of the lungs, the entropy depends on the surface area of the film. The number of particles, energy, volume, electric charge, surface area, and length are all *extensive variables*: if a homogeneous system is divided into two parts, the value of the variable for the total system (volume, charge, etc.) is the sum of the values for each part. A general extensive variable will be called x.

An adiabatic energy change is one in which no heat flows to or from the system. The energy change is due to work done on or by the system as a macroscopic parameter changes, shifting at least some of the energy levels. For each extensive variable x we can define a *generalized force* X such that the energy change in an adiabatic process is

$$dU = -dW = X\ dx. \tag{3.69}$$

(Remember that dU is the increase in energy of the system and dW is the work done *by* the system *on* the surroundings.) Examples of extensive variables and their associated forces are given in Table 3.4.

TABLE 3.4. *Examples of extensive variables and the generalized forces associated with them.*

x	X	$dU = -dW$
Volume V	−Pressure $-p$	$-p\ dV$
Length L	Tension $\mathscr{T}$	$\mathscr{T}\ dL$
Area A	Surface tension $\mathscr{S}$	$\mathscr{S}\ dA$
Charge q	Potential v	$v\ dq$

3.16. THE GENERAL THERMODYNAMIC RELATIONSHIP

Suppose that a system has N particles, total energy U, volume V, and another macroscopic parameter x on which the positions of the energy levels may depend. The number of microstates, and therefore the entropy, will depend on these four variables:

$$S = S(U,N,V,x). \tag{3.70}$$

If each variable is changed by a small amount, there is a change of entropy

$$dS = \left(\frac{\partial S}{\partial U}\right)_{N,V,x} dU + \left(\frac{\partial S}{\partial N}\right)_{U,V,x} dN + \left(\frac{\partial S}{\partial V}\right)_{U,N,x} dV + \left(\frac{\partial S}{\partial x}\right)_{U,N,V} dx. \tag{3.71}$$

Now consider the change of energy of the system. If only heat flow takes place, there is an increase of energy

$$dQ = T\ dS.$$

If an adiabatic process with a constant number of particles takes place, the energy change is

$$-dW = X\ dx - p\ dV.$$

If particles flow into the system without an accompanying flow of heat or work, the energy change is dU_N. It seems reasonable that this energy change, due solely to the movement of the particles, is proportional to dN:

$$dU_N = a\ dN.$$

(It will turn out that the proportionality constant is the chemical potential.) For the total change of energy resulting from all these processes, we can write a statement of the conservation of energy:

$$dU = T\ dS + X\ dx - p\ dV + a\ dN.$$

This is an extension of Eq. (3.5) to the additional variables on which the energy can depend. It can be rearranged as

$$dS = \left(\frac{1}{T}\right) dU - \left(\frac{a}{T}\right) dN + \left(\frac{p}{T}\right) dV - \left(\frac{X}{T}\right) dx. \tag{3.72}$$

Comparison of Eqs. (3.71) and (3.72) shows that

$$\left(\frac{\partial S}{\partial U}\right)_{N,V,x} = \frac{1}{T}, \tag{3.73a}$$

$$\left(\frac{\partial S}{\partial N}\right)_{U,V,x} = -\frac{a}{T}, \tag{3.73b}$$

$$\left(\frac{\partial S}{\partial V}\right)_{U,N,x} = \frac{p}{T}, \tag{3.73c}$$

$$\left(\frac{\partial S}{\partial x}\right)_{U,N,V} = -\frac{X}{T}. \qquad (3.73d)$$

Comparison of Eq. (3.73b) with Eq. (3.58) shows that $a=\mu$. This is why the factor of T was introduced in Eq. (3.58).

Equation (3.72) with the correct value inserted for a is

$$T\,dS = dU - \mu\,dN + p\,dV - X\,dx. \qquad (3.74)$$

This is known as the *thermodynamic identity* or the *fundamental equation of thermodynamics*. It is a combination of the conservation of energy with the relationship between entropy change and heat flow in a reversible process. (A reversible process is one that takes place so slowly that all parts of the system have the same temperature, pressure, etc.) This equation and derivative relations such as Eqs. (3.73) form the basis for the usual approach to thermodynamics.

Finally, let us consider the addition of a particle to a system when the volume is fixed. If we do this without changing the energy, it increases the number of ways the existing energy can be shared and hence the number of microstates. Therefore the entropy increases. If we want to restore the entropy to its original value, we must remove some energy. Exactly the same argument can be made mathematically. We have seen in Eqs. (3.58) and (3.73b) that

$$\mu \equiv -T\left(\frac{\partial S}{\partial N}\right)_{U,V}.$$

Since adding the particle at constant energy increases the entropy, $(\partial S/\partial N)_{U,V}$ is positive and the chemical potential is negative. Next, we rearrange Eq. (3.74) as

$$dU = T dS + \mu dN - p dV$$

and by inspection see that

$$\mu = \left(\frac{\partial U}{\partial N}\right)_{S,V}.$$

Therefore adding a particle at constant volume while keeping the entropy constant requires that energy be removed from the system.

3.17. THE GIBBS FREE ENERGY

A conventional course in thermodynamics develops several functions of the entropy, energy, and macroscopic parameters that are useful in certain special cases. One of these is the *Gibbs free energy*, which is particularly useful in describing changes that occur in a system while the temperature and pressure remain constant. Most changes in a biological system occur under such circumstances.

Imagine a system A in contact with a much larger reservoir as in Fig. 3.18. The reservoir has temperature T' and pressure p'. A movable piston separates A and A'. (At equilibrium, $T=T'$ and $p=p'$.) The reservoir is large enough so that a change of energy or volume of system A does not change T' or p'.

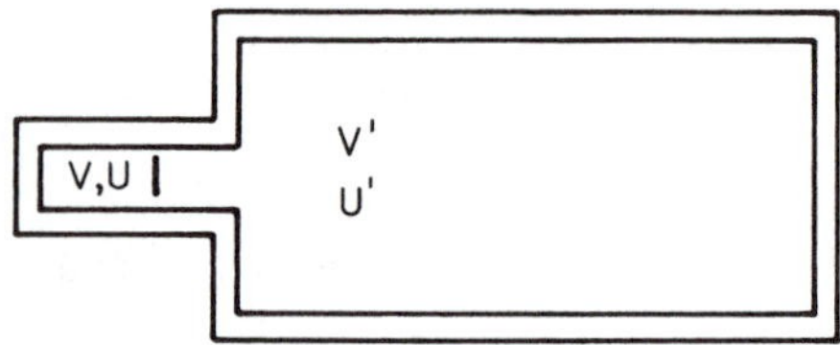

FIGURE 3.18. System A is in contact with reservoir A'. Heat can flow through the piston, which is also free to move. The reservoir is large enough to ensure that anything that happens to system A takes place at constant temperature and pressure.

Consider the change of entropy of the total system that accompanies an exchange of energy or volume between A and A'. Above, this entropy change was set equal to zero to obtain the condition for equilibrium. In this case, however, we will express the total entropy change of system plus reservoir in terms of the changes in system A alone. The total entropy is $S^* = S + S'$, so the total entropy change is

$$dS^* = dS + dS'.$$

If reservoir A' exchanges energy with the system, the energy change is

$$dU' = T'\,dS' - dW' = T'\,dS' - p'\,dV'$$
$$= T'\,dS' + p'\,dV.$$

This can be solved for dS', and the result can be put in the expression for the total entropy change:

$$dS^* = dS + \frac{dU'}{T'} - \frac{p'\,dV}{T'}.$$

We are trying to get dS^* in terms of changes in system A alone. Since A and A' together constitute an isolated system, $dU = -dU'$. Therefore,

$$dS^* = -\frac{-T'\,dS + dU + p'\,dV}{T'}. \qquad (3.75)$$

(Note that a minus sign was introduced in front of this equation.) This expresses the *total* entropy change in terms of changes of S, U, and V in system A and the pressure and temperature of the reservoir with which system A is in contact.

The Gibbs free energy is *defined* to be

$$G \equiv U - T'S + p'V. \qquad (3.76)$$

If the reservoir is large enough so that interaction of the system and reservoir does not change T' and p', then the change of G as system A changes is

$$dG = dU - T'\,dS + p'dV, \tag{3.77}$$

Comparison of Eqs. (3.75) and (3.77) shows that

$$dS^* = -\frac{dG}{T'}. \tag{3.78}$$

The change in entropy of system plus reservoir is related to the change of G, which is a property of the system alone, as long as the pressure and temperature are maintained constant by the reservoir.

To see why G is called a free energy, consider the conservation of energy in the following form:

(work done by the system)

=(energy lost by the system)

+(heat added to the system),

$$dW = -dU + T\,dS.$$

Subtracting $p\,dV$ from both sides of this equation gives

$$dW - p\,dV = -dU + T\,dS - p\,dV = -dG.$$

The right-hand side is the decrease of Gibbs free energy of the system. The work done in any isothermal, isobaric (constant pressure) reversible process, *exclusive of p dV work*, is equal to the decrease of Gibbs free energy of the system. This non–$p\,dV$ work is sometimes called "useful" work. It may represent contraction of a muscle fiber, the transfer of particles from one region to another, the movement of charged particles in an electric field, or a change of concentration of particles. It differs from the change in energy of the system, dU, for two reasons. The volume of the system can change, resulting in $p\,dV$ work, and energy can come from (or be sent to) the reservoir in the form of heat during the process. For example, let the system be a battery at constant temperature and pressure which decreases its internal (chemical) energy and supplies electrical energy. From a chemical energy change dU we subtract $T\,dS$, the heat flow to the surroundings, and $-p\,dV$, the work done on the atmosphere as the liquid in the battery changes volume. What is left is the energy available for electrical work.

As an example of how the Gibbs free energy is used, consider a chemical reaction that takes place in the body at constant temperature and pressure. System A, the region in the body where the reaction takes place, is in contact with a reservoir A' that is large enough to maintain constant temperature and pressure. Suppose that there are four species of particles that interact. Capital letters represent the species and small letters represent the number of atoms or molecules of each that enter in the reaction:

$$aA + bB \leftrightarrow cC + dD.$$

An example is 1 glucose$+6O_2 \leftrightarrow 6CO_2 + 6H_2O$, where $a=1$, $b=6$, $c=6$, $d=6$. The state of the system depends on U, V, N_A, N_B, N_C, and N_D.

We begin with the definition of G, Eq. (3.76), and we call the pressure and temperature of the system and reservoir p and T:

$$G = U - TS + pV.$$

Differentiating, we obtain

$$dG = dU - T\,dS - S\,dT + p\,dV + V\,dp.$$

Generalize Eq. (3.74) for the case of four chemical species:

$$T\,dS = dU - \mu_A dN_A - \mu_B dN_B - \mu_C dN_C - \mu_D dN_D + p\,dV.$$

Insert this in the equation for dG and remember that since the process takes place at constant temperature and pressure, dT and dp are both zero. The result is

$$dG = \mu_A dN_A + \mu_B dN_B + \mu_C dN_C + \mu_D dN_D.$$

In Sec. 3.13 we saw that the concentration dependence of the chemical potential is given by a logarithmic term. Equation (3.61) can be used to write

$$\mu_A = \mu_{A0} + k_B T \ln(C_A/C_0),$$

where μ_{A0} is the chemical potential at a standard concentration (usually 1 mol) and depends on temperature, pH, etc. Note that C_0 is the same reference concentration for all species. As the reaction takes place to the right, we can write the number of molecules gained or lost as

$$dN_A = -a\,dN,$$

$$dN_B = -b\,dN,$$

$$dN_C = c\,dN,$$

$$dN_D = d\,dN,$$

so that we have

$$\begin{aligned} dG = {} & [\mu_{A0} + k_B T \ln(C_A/C_0)](-a\,dN) \\ & + [\mu_{B0} + k_B T \ln(C_B/C_0)](-b\,dN) \\ & + [\mu_{C0} + k_B T \ln(C_C/C_0)](c\,dN) \\ & + [\mu_{D0} + k_B T \ln(C_D/C_0)](d\,dN). \end{aligned}$$

This can be rearranged as (letting $[A]=C_A$, etc.)

$$dG = \left[(c\mu_{C0} + d\mu_{D0} - a\mu_{A0} - b\mu_{B0}) + k_B T \ln\left(\frac{[C]^c [D]^d}{[A]^a [B]^b} \right) - k_B T (a+b-c-d) \ln C_0 \right] dN.$$

Concentrations $[A]$, $[B]$, $[C]$, $[D]$, and C_0 must all be measured in the same units. If these units are such that C_0 is unity (for example 1 mol per liter), then the third term vanishes and we have

$$dG=\left[(c\mu_{C0}+d\mu_{D0}-a\mu_{A0}-b\mu_{B0}) + k_BT \ln\left(\frac{[C]^c[D]^d}{[A]^a[B]^b}\right)\right]dN.$$

Multiplying the expression in square brackets by Avogadro's number converts the chemical potential per molecule to the standard Gibbs free energy per mole, and k_BT to RT. To compensate, the change in number of molecules ΔN is changed to moles Δn:

$$\Delta G=\left[(cG_{C0}+dG_{D0}-aG_{A0}-bG_{B0}) + RT \ln\left(\frac{[C]^c[D]^d}{[A]^a[B]^b}\right)\right]\Delta n. \quad (3.79)$$

The term in small parentheses is the standard free energy change for this reaction, ΔG^0, which can be found in tables. At equilibrium $\Delta G=0$, so

$$0=\Delta G^0+RT \ln\left(\frac{[C]^c[D]^d}{[A]^a[B]^b}\right)=\Delta G^0+RT \ln K_{eq}.$$

The equilibrium constant K_{eq} is related to the standard (1 molar) free-energy change by

$$\Delta G^0=-RT \ln K_{eq},$$

$$K_{eq}=\frac{[C]^c[D]^d}{[A]^a[B]^b}.$$

Many biochemical processes in the body receive free energy from the change of adenosine triphosphate (ATP) to adenosine diphosphate (ADP) plus inorganic phosphorus (P_i). This reaction involves a decrease of free energy. The energy is provided initially by forcing the reaction to go in the other direction to make an excess of ATP. One way this is done is through a very complicated series of chemical reactions known as the *respiration of glucose*. The net effect of these reactions is

$$\text{glucose}+6O_2\rightarrow 6CO_2+6H_2O, \quad \Delta G^0=-680 \text{ kcal},$$

$$36\text{ADP}+36P_i\rightarrow 36\text{ATP}+36H_2O, \quad \Delta G^0=+263 \text{ kcal}.$$

The decrease in free energy of the glucose more than compensates for the increase in free energy of the ATP. The creation of glucose or other sugars is the reverse of the respiration process and is called *photosynthesis*. The free energy required to run the reaction the other direction is supplied as light energy.

3.18. THE CHEMICAL POTENTIAL OF A SOLUTION

We now consider a binary solution of solute and solvent and how the chemical potential changes as these two substances are intermixed.[8] This is a very fundamental process that will lead us to the logarithmic dependence of the chemical potential on solute concentration that we saw in Sec. 3.13, as well as to an expression for the chemical potential of the solvent that we will need in Chap. 5.

To avoid having the subscript s stand for both solute and solvent, we call the solvent water. The distinction between solute and water is artificial; the distinction is usually that the concentration of solute is quite small. We need the entropy change in a solution when N_s solute molecules, which initially were segregated, are mixed with N_w water molecules. We make the calculation for an ideal solution—one in which the total volume of water molecules does not change on mixing and in which there is no heat evolved or absorbed on mixing. This is equivalent to saying that the solute and water molecules are the same size and shape, and that the force between a water molecule and its neighbors is the same as the force between a solute molecule and its neighbors.[9] The resulting entropy change is called the *entropy of mixing*.

To calculate the entropy of mixing, imagine a system with N sites, all occupied by particles. The number of microstates is the number of different ways that particles can be placed in the sites. The first particle can go in any site. The second can go in any of $N-1$ sites, and so forth. The total number of different ways to arrange the particles is $N!$ But if the particles are identical, these states cannot be distinguished, and there is actually only one microstate. The number of microstates is $N!/N!$, where the $N!$ in the numerator gives the number of arrangements and the $N!$ in the denominator divides by the number of indistinguishable states.[10]

Suppose now that we have two different kinds of particles. The total number is $N=N_w+N_s$, and the total number of ways to arrange them is $(N_w+N_s)!$ The N_w water molecules are indistinguishable, so this number must be divided by $N_w!$ Similarly it must be divided by $N_s!$ Therefore, purely because of the ways of arranging the particles, the number of microstates Ω in the mixture is proportional to

$$\frac{(N_w+N_s)!}{N_w!N_s!}.$$

An example of counting microstates is shown in Fig. 3.19.

[8]See also Hildebrand and Scott (1964), p. 17, and Chap. 6.

[9]Extensive work has been done on solutions for which these assumptions are not true. See Hildebrand and Scott (1964); Hildebrand, Prausnitz, and Scott (1970).

[10]The fact that there is only one microstate because of the indistinguishability of the particles is called the Gibbs paradox. For an illuminating discussion of the Gibbs paradox, see Casper and Freier (1973).

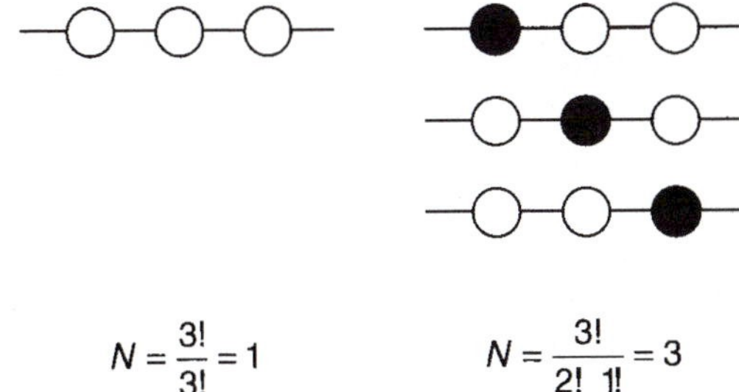

$N = \frac{3!}{3!} = 1$ $N = \frac{3!}{2!\ 1!} = 3$

FIGURE 3.19. The system on the left contains three water molecules. Because they are indistinguishable there is only one way they can be arranged. The system on the right contains two water molecules and one solute molecule. Three different arrangements are possible. In each case the number of arrangements in given by $(N_w+N_s)!/(N_w!N_s!)$.

There may also be dependence on volume and energy; in fact, the dependence on volume and energy may also contain factors of N_w and N_s. However, our assumption that the molecules of water and solute have the same size, shape, and forces of interaction ensures that these dependencies will not change as solute molecules are mixed with water molecules. The only entropy change will be the entropy of mixing.

The entropy change of the mixture relative to the entropy of N_w molecules of pure water and N_s molecules of pure solute is

$$S_{\text{solution}} - S_{\substack{\text{pure water}\\ \text{pure solute}}} = k_B \ln\left(\frac{\Omega_{\text{solution}}}{\Omega_{\substack{\text{pure water}\\ \text{pure solute}}}}\right). \tag{3.80}$$

Since with our assumptions Ω is unity for the pure solute and the pure water, the entropy difference is

$$S_{\text{solution}} - S_{\substack{\text{pure water}\\ \text{pure solute}}} = k_B \ln\left(\frac{(N_w+N_s)!}{N_w!N_s!}\right) = k_B\{\ln[(N_w+N_s)!] - \ln(N_w!) - \ln(N_s!)\}. \tag{3.81}$$

This is symmetric in water and solute, and it is valid for any number of molecules. Since we usually deal with large numbers of molecules and factorials are difficult to work with, let us use Stirling's approximation to write

$$S_{\text{solution}} - S_{\substack{\text{pure water}\\ \text{pure solute}}} = k_B[(N_w+N_s)\ln(N_w+N_s) - N_w \ln N_w - N_s \ln N_s]. \tag{3.82}$$

The next step is to relate the entropy of mixing to the chemical potential. This is done by recalling the definition of the Gibbs free energy, Eq. (3.76):

$$G \equiv U + pV - TS.$$

The sum of the first two terms, $H=U+pV$, is called the *enthalpy*. Any change of the enthalpy is the heat of mixing; in our case it is zero. (The present case is actually more restrictive: p, V, and U are all constant.) Therefore, since T is also constant, the change in Gibbs free energy is due only to the entropy change:

$$\Delta G = -T\ \Delta S = k_B T\left[N_w \ln\left(\frac{N_w}{N_w+N_s}\right) + N_s \ln\left(\frac{N_s}{N_w+N_s}\right)\right]. \tag{3.83}$$

This is still symmetric with water and solute, but it diverges if either N_w or N_s is zero, because of our use of Stirling's approximation.

We now need an expression that relates the change in G to the chemical potential. This can be derived for the general case using the following thermodynamic arguments. We use Eq. (3.76) to write the most general change in G:

$$dG = dU + p\ dV + V\ dp - T\ dS - S\ dT.$$

The fundamental equation of thermodynamics, Eq. (3.74), generalized to two molecular species, is

$$T\ dS = dU - \mu_w dN_w - \mu_s dN_s + p\ dV,$$

so

$$dG = \mu_w dN_w + \mu_s dN_s + V\ dp - S\ dT. \tag{3.84}$$

This can be used to write down some partial derivatives by inspection that are valid in general:

$$\mu_w = \left(\frac{\partial G}{\partial N_w}\right)_{N_s,p,T}, \tag{3.85a}$$

$$\mu_s = \left(\frac{\partial G}{\partial N_s}\right)_{N_w,p,T}, \tag{3.85b}$$

$$V = \left(\frac{\partial G}{\partial p}\right)_{N_s,N_w,T}, \tag{3.85c}$$

$$S = -\left(\frac{\partial G}{\partial T}\right)_{N_s,N_w,p}. \tag{3.85d}$$

To find the chemical potential, we differentiate our expression for G, Eq. (3.83), with respect to N_w and N_s to obtain

$$\mu_w = k_B T \ln x_w, \quad \mu_s = k_B T \ln x_s. \tag{3.86}$$

These have been written in terms of the *mole fractions* or *molecular fractions*

$$x_w = \frac{N_w}{N_w+N_s}, \quad x_s = \frac{N_s}{N_w+N_s}. \tag{3.87}$$

Each chemical potential is zero when the mole fraction for that species is one (i.e., the pure substance). The expressions for μ diverge for x_w or x_s close to zero because of the failure of Stirling's approximation for small values of x.

The last step is to write the chemical potential in terms of the more familiar concentrations instead of mole fractions. We can write the change in chemical potential of the *solute* as the concentration changes from a value C_1 to C_2 as

$$\Delta\mu_s = \mu_s(2) - \mu_s(1) = k_B T \ln(x_2/x_1).$$

As long as the solute is dilute, $N_w + N_s \approx N_w$, so $x_2/x_1 = C_2/C_1$ and

$$\Delta\mu_s = k_B T \ln(C_2/C_1), \tag{3.88}$$

which agrees with Eq. (3.61).

The change in chemical potential of the *water* can be written in terms of the *solute* concentration. Since $x_w + x_s = 1$, $\mu_w = k_B T \ln(1 - x_s)$. For small values of x_s the logarithm can be expanded in a Taylor's series:

$$\ln(1 - x_s) = -x_s - \tfrac{1}{2}x_s^2 - \cdots .$$

The final result is

$$\mu_w = -k_B T x_s = -k_B T N_s/(N_s + N_w)$$

$$\approx -k_B T (N_s/V)/(N_w/V),$$

or

$$\mu_w \approx -\frac{k_B T C_s}{C_w}. \tag{3.89}$$

To reiterate, this is the chemical potential of the water for small solute concentrations. The zero of chemical potential is pure water. The term is negative because the addition of solute decreases the chemical potential of the water, due to the entropy of mixing term.

We now know the concentration dependence of the chemical potential. In Chap. 5 we will be concerned with the movement of solute and water, and we will need to know the depedence of the chemical potentials on pressure. To find this, we use the fact that when the partial derivative of a function is taken with respect to two variables, the results independent of the order of differentiation (Appendix N):

$$\left[\frac{\partial}{\partial p}\left(\frac{\partial G}{\partial N_w}\right)_{T,p,N_s}\right]_{T,N_w} = \left[\frac{\partial}{\partial N_w}\left(\frac{\partial G}{\partial p}\right)_{T,N_w,N_s}\right]_{T,p}$$

From Eqs. (3.85a) and (3.85c), we get

$$\left(\frac{\partial \mu_w}{\partial p}\right)_{T,N_w} = \left(\frac{\partial V}{\partial N_w}\right)_{T,p}, \tag{3.90}$$

For a process at constant temperature, the rate of change of μ_w with p for constant solute concentration is the same as the rate of change of V with N_w when p is fixed.

The quantity $(\partial V/\partial N_w)_{T,p}$ is the rate at which the volume changes when more molecules are added at constant temperature and pressure. For an ideal incompressible liquid it is the molecular volume, $\bar{V}_w$. We can repeat this argument for the solute to obtain

$$\left(\frac{\partial \mu_w}{\partial p}\right)_{T,N_w} = \bar{V}_w, \quad \left(\frac{\partial \mu_s}{\partial p}\right)_{T,N_s} = \bar{V}_s. \tag{3.91}$$

In a solution, the total volume is $V = N_w \bar{V}_w + N_s \bar{V}_s$ where $\bar{V}_w$ and $\bar{V}_s$ are the average volumes occupied by one molecule of water and solute. Dividing by V gives $1 = C_w \bar{V}_w + C_s \bar{V}_s$. If the solution is dilute,

$$\bar{V}_w \approx \frac{1}{C_w}. \tag{3.92}$$

In an ideal solution $\bar{V}_w = \bar{V}_s$. For an ideal dilute solution, we then have

$$\Delta\mu_w = \bar{V}_w(\Delta p - k_B T \Delta C_s) \approx \frac{\Delta p - k_B T \Delta C_s}{C_w}. \tag{3.93}$$

$$\Delta\mu_s = k_B T \ln(C_{s2}/C_{s1}) + \bar{V}_s \Delta p$$

$$\approx k_B T \ln(C_{s2}/C_{s1}) + \bar{V}_w \Delta p. \tag{3.94}$$

We saw this concentration dependence earlier, in Sec. 3.13. If the concentration difference is small, we can write $C_{s2} = C_{s1} + \Delta C_s$ and use the expansion $\ln(1+x) \approx x$ to obtain

$$\Delta\mu_s \approx \frac{k_B T \Delta C_s}{C_s} + \frac{\Delta p}{C_w}. \tag{3.95}$$

3.19. TRANSFORMATION OF RANDOMNESS TO ORDER

When two systems are in equilibrium, the total entropy is a maximum. Yet a living creature is a low-entropy, highly ordered system. Are these two observations in conflict? The answer is no; the living system is not in equilibrium, and it is this lack of equilibrium that allows the entropy to be low. The conditions under which order can be brought to a system—its entropy can be reduced—are discussed briefly in this section.

A car travels with velocity **v** and has kinetic energy $\frac{1}{2}mv^2$. In addition to the random thermal motions of the atoms making up the car, all the atoms have velocity **v** in the same direction (except for those in rotating parts, which have an ordered velocity that is more complicated to describe). If the brake shoes are brought into contact with the brake drums, the car loses kinetic energy, and the shoes and drums become hot. Ordered energy has been converted into disordered, thermal energy; the entropy has increased. Is it possible to heat the drums and shoes with a torch, apply the brakes, and have the car move as the drums and shoes cool off? Energetically, this is possible, but there are only a few microstates in which all the molecules are moving in a manner that constitutes movement of the car. Their number is vanishingly small compared to the number of microstates in which the brake drums are hot. The probability that the car will begin to move is vanishingly small.

An animal is placed in an insulated, isolated container. The animal soon dies and decomposes. Energetically, the animal could form again, but the number of microstates corresponding to a live animal is extremely small compared to all microstates corresponding to the same total energy for all the atoms in the animal.

In some cases, thermal energy can be converted into work. When gas in a cylinder is heated, it expands against a piston that does work. Energy can be supplied to an organism and it lives. To what extent can these processes, which apparently contradict the normal increase of entropy, be made to take place? These questions can be stated in a more basic form.

1. To what extent is it possible to convert internal energy distributed randomly over many molecules into energy that involves a change of a macroscopic parameter of the system? (How much work can be captured from the gas as it expands the piston?)

2. To what extent is it possible to convert a random mixture of simple molecules into complex and highly organized macromolecules?

Both these questions can be reformulated: under what conditions can the entropy of a system be made to decrease?

The answer is that the entropy of a system can be made to decrease if, and only if, it is in contact with one or more auxiliary systems that experience at least a compensating increase in entropy. Then the total entropy remains the same or increases.

One device that can accomplish this process is a heat engine. It operates between two thermal reservoirs at different temperatures, removing heat from the hotter one and injecting heat into the cooler one. Even though less heat goes into the cooler reservoir than was removed from the hotter one (the difference being the mechanical work done by the engine), the overall entropy of the two reservoirs increases. The entropy change of the hot reservoir is a decrease, $-\Delta Q/T$, while the entropy change of the cooler reservoir is an increase, $+\Delta Q'/T'$. Since $T'<T$, the entropy increase more than balances the decrease, even though $\Delta Q'<\Delta Q$. The increase in the log of the number of accessible microstates of the cooler reservoir is greater than the decrease in the log of the number of accessible microstates of the hotter reservoir. The coupled chemical reactions that we saw in Sec. 3.17 are analogous.

SYMBOLS USED IN CHAPTER 3

Symbol	Use	Units	First used on page
a	General variable		63
$\mathbf{a}$	Acceleration	m s^{-2}	42
a	Number of atoms in a molecule		48
a,b,c,d	Number of atoms of species A, B, C, and D to balance a chemical reaction		65
c_j	Concentration (moles/volume)	m^{-3}, l^{-1}	61
e	Electron charge	C	53
e	Base of natural logarithms		52
f	Number of degrees of freedom		48
g	Gravitational acceleration	m s^{-2}	54
h	Planck's constant	J s	55
$\hbar$	Planck's constant divided by 2π	J s	55
k_B	Boltzmann's constant	J K^{-1}	51
m	Mass	kg	42
n	Number of particles in a volume		45
n	Number of "successful" outcomes		44
n,n',n^*	Number of moles in systems A, A' and A^*		61
p	Probability of "success"		45
p	Pressure	Pa	54
$p,\mathbf{p},p_x,p_y,p_z$	Momentum	kg m s^{1}	55
q	Probability of "failure"		45
$\mathbf{r}$	Position vector	m	56
t	Time	s	42
u_i	Energy of the ith energy level	J	47
v,v'	Volume	m^{3}	45
v	Electrical potential	V	53
v,v_x,v_y,v_z	Velocity	m s^{-1}	42
x,y,z	Position coordinate	m	42
x	General variable		49

Symbol	Use	Units	First used on page
x_s, x_w	Mole fractions of solute and solvent (water)		67
y	General variable		51
y	Height	m	54
z	Valence		53
A, A', A^*	Thermodynamic systems		49
A	Surface area	m^2	63
A, B, C, D	Chemically reacting species		65
C_i, C	Concentration (particles per volume)	m^{-3}, l^{-1}	53
C	Normalization constant		55
E_k	Kinetic energy	J	53
E_p	Potential energy	J	53
$\mathbf{F}, F_x, F_y, F_z$	Force	N	42
F	Faraday constant	C mol^{-1}	53
G	Ratio of accessible states in a small system		52
G	Gibbs free energy	J	64
H	Enthalpy	J	67
K_{eq}	Equilibrium constant in chemical reaction		66
L, L_x, L_y, L_z	Length	m	56
M	Number of molecules in a system		47
M	Number of repeated measurements		49
N, N', N^*	Number of particles in a system		44
N_w, N_s	Number of solvent (water) or solute molecules in a system		66
N_A	Avogadro's number	mol^{-1}	53
N_A, N_B, N_C, N_D	Number of moles of species A, B, C, and D consumed or produced in a chemical reaction		65
P	Probability		44
Q	Flow of heat to a system	J	47
R	Ratio of accessible states in a reservoir (Boltzmann's factor)		52
R	Gas constant	J mol^{-1} K^{-1}	53
S	Area	m^2	54
S, S', S^*	Entropy	J K^{-1}	51
T	Kelvin temperature	K	51
U, U', U^*	Total energy of a system	J	47
V	Volume	m^3	45
V_w, V_s	Molecule volume of water, solute	m^3	68
W	Work done by a system on the surroundings	J	47
W_{conc}	Work done on a system to increase the concentration	J	61
X	Generalized force		63
$\mathscr{S}$	Surface tension	N m^{-1}	63
$\mathscr{T}$	Tension	N	63
θ, ϕ	Angles in spherical coordinates		58
τ	$k_B T$	J	51
μ	Chemical potential	J per molecule	60
μ_W, μ_S	Chemical potential of water or solute	J per molecule	67
$\Omega, \Omega', \Omega^*$	Number of accessible microstates for a system when its total energy is in a narrow range		49
$\Gamma(n)$	Gamma (factorial) function of n		59
$-$	A bar over any quantity means that the quantity is averaged over an ensemble of many identically prepared systems		46
$\langle \rangle$	Angular brackets mean an average over time		46

PROBLEMS

Section 3.1

3.1. Use the last column of Table 3.2 to calculate the average value of n, which is defined by

$$\bar{n}=\sum nP(n).$$

Verify that $\bar{n}=Np$ in this case.

3.2. A loose statement is made that "if we throw a coin 1 million times, the number of heads will be very close to half a million." What is the mean number of occurrences of heads in 1 million tries? What is the standard deviation? What does "very close" mean? (You may need to consult Appendices G and H.)

3.3. Write a computer program to simulate measurements of which half of a box a gas molecule is in. Make several measurements with different sets of random numbers, and plot a histogram of the number of times n molecules are found in the left half. Try this for $N=1$, 10, and 100. In BASIC, use the function RND to obtain the random number. Since the numbers are not really random but form a well-defined sequence, a new experiment will require changing the "seed" of the sequence. This is done with the statement RANDOMIZE. Other languages have similar functions.

3.4. Color blindness is a sex-linked defect. The defective gene is located in the X chromosome. Females carry an XX chromosome pair, while males have an XY pair. The trait is recessive, which means that the patient exhibits color blindness only if there is no normal X gene present. Let X_d be a defective gene. Then for a female, the possible gene combinations are

$$XX, \quad XX_d, \quad X_dX_d.$$

For a male, they are

$$XY, \quad X_dY.$$

In a large population about 8% of the males are color-blind. What percentage of the females would you expect to be color-blind?

3.5. A patient with heart disease will sometimes go into ventricular fibrillation, in which different parts of the heart do not beat together, and the heart cannot pump. This is *cardiac arrest*. The following data show the fraction of patients failing to regain normal heart rhythm after attempts at ventricular defibrillation by electric shock [W. D. Weaver *et al.* (1982). *New Engl. J. Med.* **307**: 1101–1106].

Number of Attemps	Fraction persisting in fibrillation
0	1.00
1	0.37
2	0.15
3	0.07
4	0.02

Assume that the probability p of defibrillation on one attempt is independent of other attempts. Obtain an equation for the probability that the patient remains in fibrillation after N attempts. Compare it to the data and estimate p.

3.6. There are N people in a class ($N=80$). What is the probability that no one in the class has a birthday on a particular day? Ignore seasonal variations in birth rate and ignore leap years.

3.7. The death rate for 75-year-old people is 0.089 per year (Commissioners 1941 Standard Ordinary Mortality Table).

(a) What is the probability that an individual aged 75 will die during a 12-hour period? Neglect the fact that some are sick, some are terminally ill, and so on, and assume that the probability is the same for everyone.

(b) Suppose that 10 000 people, all aged 75, are given the flu vaccine at $t=0$. What is the probability that none will die during the next 12 hours? (This underestimates the probability, since very sick people would not be given the vaccine, but they are included in the death rate.)

3.8. This problem is intended to help you understand some of the nuances of the binomial probability distribution.

(a) In a macabre "game" of "roulette" the victim places one bullet in the cylinder of a revolver. There is room for six bullets in the cylinder. The victim spins the cylinder, so there is a probability of $\frac{1}{6}$ that the bullet is in firing position. The victim then places the gun to the head and fires. If the victim survives, the cylinder is spun again and the process is repeated. We can look either at the cumulative probability of "success" (being killed), or the cumulative probability of "failure" (surviving). Make a table for 1000 victims who keep playing the game over and over. Plot the number surviving, the number killed on each try, and the cumulative number killed.

(b) Show that the number surviving can be expressed as $1000e^{-bN}$, where N is the number of tries, and find b.

(c) The data in the following table are from Fédération CECOS, D. Schwartz and M. J. Mayaux (1982). Female fecundity as a function of age, *N. Engl. J. Med.* **306**(7): 404–406. They show the cumulative success rates in different age groups for patients being treated for infertility by artificial insemination from a donor. That is, each month at the time of ovulation each patient who has not yet become pregnant is inseminated artificially. Plot these data. What do they suggest? Make whatever plots can confirm or rule out what you suspect.

Cycle	Fraction pregnant, age≤25	Fraction pregnant, age≥35
0	0	0
1	0.11	0.03
2	0.23	0.14
3	0.30	0.20
4	0.39	0.27
5	0.44	0.35
6	0.51	0.35
7	0.55	0.36
8	0.63	0.39
9	0.65	0.43
10	0.67	0.43
11	0.70	0.46
12	0.74	0.54

Section 3.3

3.9. A thermally insulated ideal gas of particles is confined within a container of volume V. The gas is initially at Kelvin temperature T. The volume of the container is very slowly reduced by moving a piston that constitutes one wall of the container. Give qualitative answers to the following questions.

(a) What happens to the energy levels of each particle?

(b) Is the work done on the gas as its volume decreases positive or negative?

(c) What happens to the energy of the gas?

Section 3.5

3.10. System A has 10^{20} microstates, and system A' has 10^{19} microstates. How many microstates does the combined system have?

3.11. Calculate the Celsius and Kelvin temperatures corresponding to a room temperature of 68 °F, a room temperature of 70 °F, a normal body temperature of 98.6 °F, and a febrile body temperature of 104 °F.

3.12. Calculate and plot Ω, Ω', and Ω^* for Fig. 3.10, thus reproducing the figure. Write down an analytic expression for Ω^* and differentiate to find the value of U for which Ω^* is a maximum.

3.13. We have seen that in general with volume, number of particles, and other parameters that determine the positions of the energy levels held fixed,

$$\frac{1}{\Omega}\frac{d\Omega}{dU}=\frac{1}{k_B T}.$$

Strictly speaking it is not possible to hold the temperature fixed while the energy is changed under these conditions. Suppose that $U=CT$, where C is the heat capacity of the system. Find $\Omega(U)$.

3.14. Systems A and A' are in thermal contact. Show that if $T<T'$, energy flows from A' to A to increase Ω^*, while if $T>T'$, energy flows from A to A'.

3.15. A simple system has *only two* energy levels for each single entity in the system. (The system could, for example, be a collection of ''gates'' in a cell membrane, each with two states, open and closed.) One level has energy u_1, the other has energy u_2. There are N entities in the system. You can answer the following questions without doing any calculations.

(a) What is the minimum energy of the system? How many microstates are there for the minimum energy?

(b) What is the maximum energy of the system? How many microstates are there for which the system has maximum energy?

(c) *Sketch* what $\Omega(U)$ must look like.

(d) Recall the definition of T, Eqs. (3.14) and (3.15). Are there any values of U for which the temperature is negative? Where?

Section 3.6

3.16. Consider the following arrangements of the 26 capital letters of the English alphabet: (a) TWO, (b) any three letters, in any order, that are all different, and (c) any three letters, in any order, which may repeat themselves. For (b) and (c), consider the same letters in a different order to be a different arrangement, and the three ways of selecting AAA, etc., to be different. If each arrangement is a ''microstate,'' find Ω and S in each case.

3.17. Ice and water coexist at 273 K. To melt 1 mol of ice at this temperature, 6000 J are needed. Calculate the entropy difference and the ratio of the number of microstates for 1 mol of ice and 1 mol of water at this temperature. (Do not worry about any volume changes of the ice and water.)

3.18. If a system is maintained at constant volume, no work is done on it as the energy changes. In that case $dU=C(T)\,dT$, where U is the internal energy, C is the heat capacity, and T is the temperature. The specific heat in general depends on the temperature. Suppose that in some temperature region the specific heat varies linearly with temperature: $C(T)=C_0+DT$.

(a) What is the entropy change of the system when it is heated from temperature T_1 to temperature T_2, both of which are in the region where $C(T)=C_0+DT$?

(b) What is the ratio of the number of microstates at T_2 to the number at T_1?

3.19. A substance melts at constant temperature. There are 7 times as many microstates accessible to each molecule of the liquid as there were to each molecule of the solid. Ignore volume changes.

(a) What is the change in entropy of each molecule?

(b) How much heat is required to melt a mole of the substance if the melting temperature is 50 °C?

3.20. The entropy of a monatomic ideal gas at constant energy depends on the volume as $S = Nk_B \ln V + \text{const}$. A gas of N molecules undergoes a process known as a free expansion. Initially it is confined to a volume V by a partition. The partition is ruptured and the gas expands to occupy a volume $2V$. No work is done and no heat flows, so the total energy is unchanged. Calculate the change of entropy and the ratio of the number of microstates after the volume change to the number before.

Section 3.7

3.21. A virus has a mass of 1.7×10^{-14} g. If the virus particles are in thermal equilibrium in the atmosphere, their concentration will vary with height as

$$C(y) = C(0)e^{-y/\lambda}.$$

Evaluate λ. Do you think this answer is reasonable?

3.22. Chemists use Q_{10} to characterize a chemical reaction. It is defined by

$$Q_{10} = \frac{(\text{reaction rate at } T+10)}{(\text{reaction rate at } T)},$$

where T is the Kelvin temperature. If the reaction rate is proportional to the fraction of reacting atoms that have an energy exceeding some threshold ΔU, then, neglecting the $U^{1/2}$ in Eq. (3.49),

$$R \propto \int_{\Delta U}^{\infty} e^{-U/k_BT} dU.$$

(a) Show that

$$R \propto k_B T e^{-\Delta U/k_BT}.$$

(b) Show that

$$\frac{Q_{10}T}{T+10} = \exp\left[\frac{\Delta U}{k_B}\left(\frac{10}{T(T+10)}\right)\right].$$

(c) Estimate ΔU if $Q_{10} = 2$ at $T = 300$ K.

3.23. A pore has three configurations with the energy levels shown. The pore is in thermal equilibrium with the surroundings at temperature T. Find the probabilities p_1, p_2, and p_3 of being in each level. Each level has only one microstate associated with it.

3 ———————— $2U_0$

2 ———————— U_0

1 ———————— 0

3.24. The DNA molecule consists of two intertwined linear monomers. Sticking out from each monomer is one of four bases: adenine (A), guanine (G), thymine (T), or cytosine (C). In the double helix, each base from one strand bonds to a base in the other strand. The correct matches, A–T and G–C, are more tightly bound than are the improper matches. The chain looks something like this, where the last bond shown is an "error." The probability of an error at 300 K is about 10^{-9} per base pair. Assume that this probability is determined by a Boltzmann factor e^{-U/k_BT}, where U is the additional energy required for a mismatch.

(a) Estimate this excess energy.

(b) If such mismatches are the sole cause of mutations in an organism, what would the mutation rate be if the temperature were raised 20 °C?

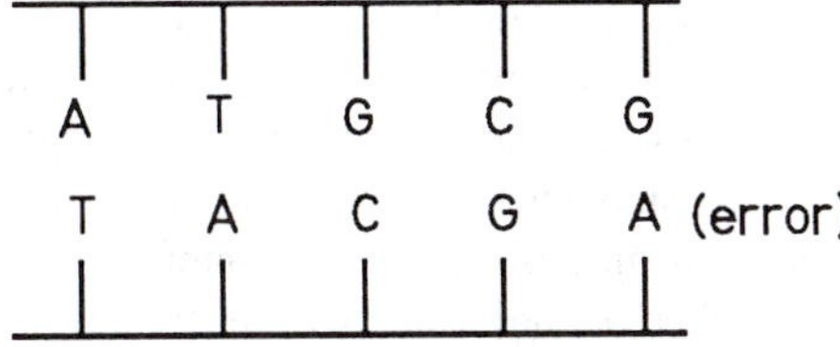

3.25. The data of Problem 2.10 were used to obtain an empirical relationship between the charge integration time τ and the temperature. It might be that τ is determined by a chemical reaction whose rate is given by a Boltzmann factor. Make a new plot based on that assumption and determine the appropriate constants.

3.26. Suppose that particles in water are subjected to an external force $F(y)$ that acts in the y direction. The force is related to the potential energy $E_p(y)$ by $F = -dE_p/dy$. Neglect gravity and buoyancy effects.

(a) Apply Newton's first law to a slice of the fluid in equilibrium to obtain an expression for $p(y)$.

(b) If the particles have a Boltzmann distribution, show that

$$p(y) - p(0) = kT[C(y) - C(0)].$$

Section 3.8

3.27. The concentrations of various ions are measured on the inside and outside of a nerve cell. The following values are obtained when the potential inside the cell is -70 mV with respect to the outside.

Ion	Inside (mmol l^{-1})	Outside (mmol l^{-1})
Na^+	15	145
K^+	150	5
Cl^-	9	125

Comment on which species have concentrations that are consistent with being able to pass freely through the cell wall. Assume $T = 300$ K.

Section 3.10

3.28. Make plots of the classical trajectories in phase space (that is, p_x vs x) for a particle of mass m (a) in free fall, (b) bouncing with momentum $\pm p$ parallel to the x axis between walls at $x=0$ and $x=L$, (c) and undergoing harmonic oscillation along the x axis: $x=x_0 \sin(\omega t)$.

3.29. Suppose that a water molecule in solution has a velocity in the x direction between 200 and 201 m s^{-1}, while it is located within a region of 1 nm along the x axis. In how many cells of phase space of size $\hbar$ might it be located?

3.30. Use the technique described in Sec. 3.10 to show that the probability that a molecule is in one half of the box and has a positive value of v_x is 0.25.

3.31. The vapor pressure of a substance can be calculated using the following model. All molecules in the vapor that strike the surface of the liquid stick. (This number is proportional to the pressure.) Those molecules in the liquid that reach the surface and have enough energy escape. Equilibrium is established when the number sticking per unit area per unit time is equal to the number escaping.

(a) The number of molecules with energy U is proportional to e^{-U/k_BT}. What will be the number with energy greater than the escape energy, U_0? Neglect the factor $U^{1/2}$ in Eq. (3.49).

(b) Use the result of part (a) and values for the vapor pressure of water as a function of temperature to make a plot on semilog paper. From this plot, estimate the escape energy U_0.

(c) The "latent heat of vaporization" of water is 540 calories per gram. Convert the energy per molecule you found in part (b) to calories per gram and compare it with this figure.

3.32. Use the fact that

$$\int_0^\infty e^{-bx^2}dx=\frac{1}{2}\left(\frac{\pi}{b}\right)^{1/2}$$

to show that

$$\int_{-\infty}^{\infty} e^{-p^2/2mk_BT}dp=(2\pi mk_BT)^{1/2}.$$

3.33. The Maxwell–Boltzmann probability distribution $P(U)dU$ can be expressed in the form $Ce^{-U/k_BT}4\pi p^2dp$. Find and expression for the Maxwell–Boltzmann probability distribution for molecular speed, $P(v)dv$, where the speed is $v=(v_x^2+v_y^2+v_z^2)^{1/2}$.

3.34. A macromolecule of density ρ is immersed in an incompressible fluid of density ρ_w at temperature T. The volume v occupied by one macromolecule is known. A dilute solution of the macromolecules is placed in an ultracentrifuge rotating with high angular velocity ω. In the frame of reference rotating with the centrifuge, a particle at rest is acted on by an outward force $m\omega^2 r$, where r is the distance of the particle from the axis.

(a) What is the net force acting on the particle in this frame? Include the effect of buoyancy of the surrounding fluid, of density ρ_w.

(b) Suppose that equilibrium has been reached. Use the Boltzmann factor to find the number of particles per unit volume at distance r.

Section 3.11

3.35. A sensitive balance consists of a weak spring hanging vertically in the earth's gravitational field. The equilibrium position of the end of the spring is $x=0$. When a mass m is added to the spring, it elongates to an average position x_0, around which it vibrates because of thermal energy. In terms of $\Delta x=x-x_0$, the momentum of the mass p_x and the spring constant K, the force which the spring exerts on the mass is Kx_0, and the total energy is

$$U=\frac{p_x^2}{2m}+\tfrac{1}{2}K(\Delta x)^2.$$

(a) What is x_0 in terms of m, g, and K?

(b) Find $\overline{\Delta x^2}=\overline{(x-x_0)^2}$.

(c) What is the smallest mass that can be measured taking a single "snapshot" of the system to find the position of the mass?

3.36. Suppose the total energy of a particle depends on some variable x in the form

$$U=f(\text{other variables})+ax^n,\quad x>0.$$

Show that

$$\bar{U}=\frac{k_BT}{n}.$$

Discuss the case $n=1$, which corresponds to potential energy in a uniform gravitational field. The discussion should consider concentration variation with position.

Section 3.12

3.37. A small system A is in contact with a reservoir A' and can exchange both heat and particles with the reservoir. The number of microstates available to system A does not change. Show that the difference in total entropy when A is in two distinct states is

$$\Delta S^*=-(N_1-N_2)\left(\frac{\partial S}{\partial N}\right)_U-(U_1-U_2)\left(\frac{\partial S}{\partial U}\right)_N$$

so that

$$\frac{P(N_1,U_1)}{P(N_2,U_2)}=\frac{e^{(N_1\mu-U_1)/k_BT}}{e^{(N_2\mu-U_2)/k_BT}}.$$

where T and μ are the temperature and chemical potential of the reservoir. This is called the Gibbs factor, and it reduces to the Boltzmann factor if $N_1=N_2$. Chemists use the notation $\lambda=e^{\mu/k_BT}$, where λ is the absolute activity. Then

$$\frac{P(N_1,U_1)}{P(N_2,U_2)}=\frac{\lambda^{N_1}e^{-U_1/k_BT}}{\lambda^{N_2}e^{-U_2/k_BT}}.$$

3.38. Specialize the results of the previous problem to a series of binding sites on a surface, such as a myoglobin molecule. The two states are

No particle bound at the site	$N_1=0,$	$U_1=0$
One particle bound at the site	$N_2=1,$	$U_2=U_0$

(a) Show that the fraction of sites occupied is

$$f=\frac{\lambda e^{-U_0/k_BT}}{1+\lambda e^{-U_0/k_BT}}.$$

(b) If the sites are in equilibrium with a gas, then $\mu_{\text{gas}}=\mu_{\text{sites}}$ or $\lambda_{\text{gas}}=\lambda_{\text{sites}}$. From the definition $\mu=-T(\partial S/\partial N)_{U,V}$ and the expression for the entropy of a monatomic ideal gas,

$$S(U,V,N)=Nk_B(\ln V+\tfrac{3}{2}\ln U-\tfrac{5}{2}\ln N+\tfrac{5}{2}+\text{const}),$$

where $\text{const}=\frac{3}{2}\ln(m/3\pi\hbar^2)$, show that $f=p/(p_0+p)$, where

$$p_0=\frac{(k_BT)^{5/2}m^{3/2}e^{U_0/k_BT}}{(2\pi\hbar^2)^{3/2}}.$$

This expression fits the data very well. See A. Rossi-Fanelli and E. Antonini (1958). *Archives Biochem. Biophys.* **77**: 478.

Section 3.13

3.39. The entropy of a monatomic ideal gas is

$$S(U,V,N)=Nk_B(\ln V+(\tfrac{3}{2})\ln U-\tfrac{5}{2}\ln N+\tfrac{5}{2}+\text{const}),$$

where $\text{const}=\frac{3}{2}\ln(m/3\pi\hbar^2)$ depends only on the mass of the molecule. Consider two containers of gas at the same temperature and pressure that can exchange particles. Expand the total entropy in a Taylor's series, keep terms to second order, and use the result to find the variance in the fluctuating number of particles in one system. Assume $N\ll N'$. You should obtain the same result obtained from the binomial distribution ($\sigma^2=N$) if you take into account that it is the *temperature* of the gas in the container and not its energy, that should be held fixed. (For a monatomic ideal gas $U=3Nk_BT/2$. Use this result to rewrite the entropy in terms of T, V, and N.)

3.40. Show that the chemical potential of an ideal gas is proportional to the concentration, a result that we have now seen several times for dilute ideal systems. To do so, use the expression for the entropy of a monatomic ideal gas given in the previous problems. Rewrite the thermodynamic identity as

$$dU=T\,dS+\mu\,dN-p\,dV$$

from which we can identify the partial derivative

$$\mu=\left(\frac{\partial U}{\partial N}\right)_{S,V}.$$

The chemical potential is the increase in energy of the system if one particle is added while keeping the entropy and volume fixed. Use the expression for the entropy of the monatomic ideal gas, for the case of N particles with total energy U and $N+1$ particles with total energy $U+\mu$, to show that the chemical potential of the ideal gas is

$$\mu=k_BT\left[\ln\left(\frac{N}{V}\right)-\tfrac{3}{2}\ln(3k_BT/2)-\text{const}\right]$$

or

$$\mu=-k_BT\ln\left[\frac{V}{N}\left(\frac{mk_BT}{2\pi\hbar^2}\right)^{3/2}\right].$$

[A more extensive discussion for other simple systems is given by Cook and Dickerson (1995).]

Section 3.15

3.41. The ear can just hear sound at about 1000 Hz at a level that corresponds to a pressure change of 2×10^{-5} Pa. Atmospheric pressure is 10^5 Pa. Since atmospheric pressure is due to collisions of molecules with the eardrum, there are pressure fluctuations because of fluctuations in the number of collisions in time Δt. We can expect that $\Delta p/p$ is about $1/(\text{number of collisions})^{1/2}$. Suppose that the eardrum has area S and that when detecting a signal at 1000 Hz it averages over a time interval of 0.5 ms. The number of collisions per unit area per unit time is given by $nv/4$, where n is the number of air molecules per unit volume and v is an "average" velocity of 482 m s^{-1}. The radius of the eardrum is 4.5 mm. Find $\Delta p/p$.

Section 3.17

3.42. System A consists of N particles that move from a region where the concentration is C_1 to another where the concentration is C_2, each experiencing a change in chemical potential $\Delta\mu=k_BT\ln(C_2/C_1)$. The process occurs at constant temperature and pressure. What is the ratio of the total number of microstates of system and surroundings after the move to the number before the move? Assume the concentrations do not change.

3.43. If one increases the volume of a liquid at constant p and T, a portion of the liquid evaporates. The amount of liquid decreases as V increases until all the liquid is vaporized. The pressure at which the two phases coexist is called the vapor pressure. The vapor pressure depends on the temperature, as shown. When two phases are in equilibrium, they are in mechanical, thermal, and diffusive equilibrium: $T_l=T_g$, $p_l=p_g$, $\mu_l=\mu_g$. Thus, at any arbitrary point on the vapor-pressure curve,

$$\mu_g(T_0,p_0)=\mu_l(T_0,p_0).$$

Consider some nearby point in the vapor-pressure curve, and expand both chemical potentials in a Taylor's series to show that

$$\frac{dp}{dT}=\frac{(\partial\mu_g/\partial T)_p-(\partial\mu_l/\partial T)_p}{(\partial\mu_l/\partial p)_T-(\partial\mu_g/\partial p)_T},$$

where dp/dT is the slope of the vapor-pressure curve. Use the fact that $G=N\mu(p,T)$, that $(\partial G/\partial T)_{N,p}=-S$, and that $(\partial G/\partial p)_{N,T}=V$, to show that

$$\frac{dp}{dT}=\frac{L}{T\Delta V},$$

where L is the latent heat of vaporization and ΔV is the volume change on vaporization. (Since L and V are both extensive parameters, they can be expressed per mole or per molecule.) This is called the *Clausius–Clapeyron equation.*

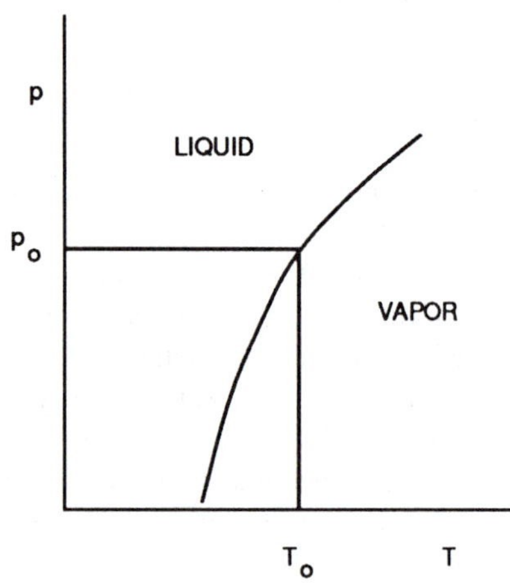

3.44. Use the definition of Gibbs free energy

$$G=U-TS+pV$$

and the thermodynamic identity

$$T\,dS=dU-\mu\,dN+p\,dV$$

to find the partial derivatives of G when N, T, and p are the independent variables. Note that U, S, and V are all extensive variables so that G is proportional to N: $G=N\Phi$. Thereby relate G to the chemical potential.

3.45. (a) Find the change in Gibbs free energy $G=U-TS+pV$ for an ideal gas which changes pressure reversibly from p_1 to p_2 at a constant temperature.

(b) Since $\Delta G=N\Delta\mu$, find $\Delta\mu(T,p)$.

3.46. Use the Clausius–Clapeyron equation for the vapor pressure as a function of temperature (see Problem 3.43),

$$\frac{dp}{dT}=\frac{L}{T\Delta V},$$

and assume an ideal gas so that $\Delta V\approx V_g=Nk_BT/p$ to find the vapor pressure p as a function of temperature.

3.47. The argument leading to the change in G in a chemical reaction can be applied to a single particle moving from a region where the chemical potential is μ_A to a region where the chemical potential is μ_B by letting $dN=-dN_A=dN_B$, in which case $dG=(\mu_B-\mu_A)dN$. We saw in Sec. 3.13 that the chemical potential of a solute in an ideal solution had the form $\Delta\mu=k_BT\ln(C/C_0)+\Delta$(potential energy per particle). Sodium ions of charge $+e$ ($e=1.6\times10^{-19}$ C) are found on one side of a membrane at concentration 145 mmol/L. The electrical potential is zero. On the other side of the membrane the concentration is 15 mmol/L and the potential is -90×10^{-3} V. The change in electrical potential energy is $e\Delta v$. What is the change in Gibbs free energy if a single sodium ion goes from one side to the other? The temperature is 310 K and the pressure is atmospheric.

Section 3.18

3.48. Suppose that a potential energy term as well as a pressure must be added to the chemical potential, as was argued in Sec. 3.13. Consider a column of pure water. What is the difference in chemical potential between the top of the column and the bottom?

3.49. The open circles in the drawing represent water molecules. The solid circles are solute molecules. The vertical line represents a membrane that is permeable to water but not solute. In case (a) there are two water molecules to the right of the membrane. In (b) there is one, and in (c) none. What is the total number of microstates of the combined system in each case?

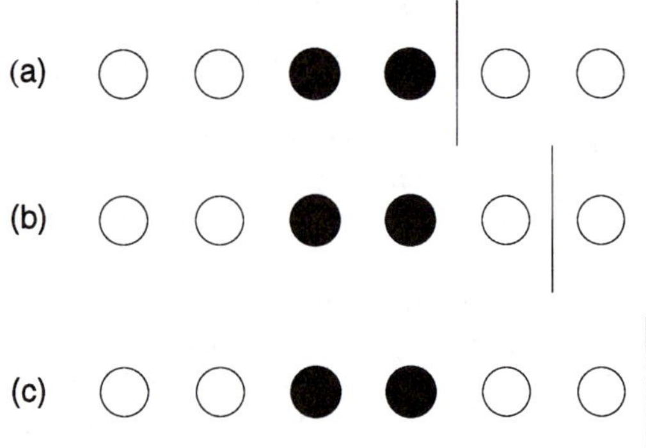

3.50. If we want to apply Eq. (3.95) when there is an appreciable difference in concentration, we can define an average concentration by

$$\Delta\mu_s=k_BT\ln(C_{s2}/C_{s1})\equiv k_BT\frac{\Delta C_s}{\overline{C}_s},$$

$$\overline{C}_s\equiv\frac{\Delta C_s}{\ln(C_{s2}/C_{s1})}=\frac{\Delta C_s}{\ln(1+\Delta C_s/C_{s1})}.$$

Use the Taylor's-series expansion $y=x/\ln(1+x)\approx1+x/2-x^2/12+\cdots$ to find an approximate expression for $\overline{C}_s$.

REFERENCES

Casper, B. M., and S. Freier (1973). ''Gibbs Paradox'' paradox. *Am. J. Phys.* **41**: 509–511.

Cook, G., and R. H. Dickerson (1995). Understanding the chemical potential. *Am. J. Phys* **63** (8): 737–742.

Hildebrand, J. H., J. M. Prausnitz, and R. L. Scott (1970). *The Solubility of Regular and Related Solutions*. New York, Van Nostrand Reinhold.

Hildebrand, J. H., and R. L. Scott (1964). *The Solubility of Nonelectrolytes*, 3rd ed. New York, Dover.

Reif, F. (1964). *Statistical Physics*. Berkeley Physics Course, Vol. 5. New York, McGraw-Hill.

Reif, F. (1965). *Fundamentals of Statistical and Thermal Physics*. New York, McGraw-Hill.

CHAPTER 4

Transport in an Infinite Medium

Chapters 4 and 5 are devoted to one of the most fundamental problems in physiology: the transport of solvent (water) and uncharged solute particles. Chapter 4 develops some general ideas about the movement of solutes in solution. Chapter 5 applies these ideas to movement of water and solute through a membrane.

Section 4.1 defines flux and fluence rate and derives the continuity equation. Section 4.2 shows how to calculate the solute fluence rate when the solute particles are drifting with a constant velocity, as when they are being dragged along by flowing solvent.

The next several sections are devoted to diffusion, the random motion of solute particles. Sections 4.3–4.5 describe random motion in a gas and a liquid. Section 4.6 states Fick's first law, which relates the fluence rate of diffusing particles to the gradient of their concentration. Section 4.7 shows how the proportionality constant in Fick's first law is related to the viscous drag coefficient of the particle in the solution. Section 4.8 shows how Fick's first law and the equation of continuity can be combined to give Fick's second law, a differential equation that tells how the concentration $C(x,y,z,t)$ evolves with time. Section 4.9 discusses various time-independent (steady-state) solutions to the diffusion equation. Section 4.10 discusses steady-state diffusion to or from a cell, including both diffusion through the membrane and in the surrounding medium. Section 4.11 discusses a model of steady-state diffusion of a substance that is being produced at a constant rate inside a spherical cell. Section 4.12 develops a steady-state solution when both drift and diffusion are taking place in one dimension. Section 14.13 provides a very brief introduction to one technique for solving the time-dependent diffusion equation. Section 14.14 describes a simple random-walk model for diffusion.

4.1. FLUX, FLUENCE, AND CONTINUITY

Flow was introduced in Sec. 1.14 of Chap. 1. The *flow rate*, *volume flux*, or *volume current* i is the total volume of material transported per unit time and has units of $m^3\ s^{-1}$. One can also define the mass flux as the total mass transported per unit time or the particle flux as the total number of particles, and so on.

The *particle fluence* is the number of particles transported per unit area across an imaginary surface (m^{-2}). The *volume fluence* is the total volume transported across the surface and has units m^3 times m^{-2}, or m.

The *fluence rate* or *flux density* is the amount of "something" transported across an imaginary surface per unit area per unit time. It can be represented by a vector pointing in the direction the "something" moves and is denoted by $\mathbf{j}$. It has units of "something" $m^{-2}\ s^{-1}$. It is traditional to use a subscript to tell what is being transported: $\mathbf{j}_s$ is solute particle flux density ($m^{-2}\ s^{-1}$), $\mathbf{j}_m$ is mass flux density (kg $m^{-2}\ s^{-1}$), and $\mathbf{j}_v$ is volume flux density ($m^3\ m^{-2}\ s^{-1}$ or $m\ s^{-1}$). In a flowing fluid $\mathbf{j}_v$ is the velocity with which the material moves.

Slightly different nomenclature is used in different fields. The words *flux* and *flux density* are often used interchangeably. Table 4.1 shows some of the names that are encountered. Do not spend much time studying it; it is provided to help you when you must deal with the notation in other books.

As long as we are dealing with a substance that does not appear or disappear (as in a chemical reaction, radioactive decay, or the like), the number of particles or the mass, or in the case of an incompressible liquid, the volume, remains constant or is *conserved*. This conservation leads to a very useful equation called the *equation of continuity*. It will be derived here in terms of the number of particles.

We will first derive it in one dimension. Let the flux density of some species be j particles per unit area per unit time, passing a point. All motion takes place in the x direction along a tube of constant cross-sectional area S. The value of j may depend on position in the tube and on the time:

$$j=j(x,t).$$

The number of particles in the volume shown in Fig. 4.1

TABLE 4.1. *Units and names for j and jS in various fields.*

	j		jS	
Substance	Units	Names	Units	Names
Particles	$m^{-2}\ s^{-1}$	Particle fluence rate Particle current density Particle flux density Particle flux	s^{-1}	Particle flux Particle current Particle flux
Electric charge	$C\ m^{-2}\ s^{-1}$ or $A\ m^{-2}$	Current density	$C\ s^{-1}$ or A	Current
Mass	$kg\ m^{-2}\ s^{-1}$	Mass fluence rate Mass flux density Mass flux	$kg\ s^{-1}$	Mass flux Mass flow
Energy	$J\ m^{-2}\ s^{-1}$ or $W\ m^{-2}$	Energy fluence rate Intensity Energy flux	$J\ s^{-1}$ or W	Energy flux Power

between x and $x+\Delta x$ is $N(x,t)$. At x there may be particles moving both to the right and to the left; the net number to the right in Δt is $j(x,t)$ times the area S times the time Δt. A flux density in the $+x$ direction is called positive. The net number of particles in at x is $j(x,t)S\ \Delta t$.
Similarly, the net number *in* at $x+\Delta x$ is

$$-j(x+\Delta x,t)S\ \Delta t.$$

Combining these gives the net increase in the number of particles in the volume $S\ \Delta x$:

$$\Delta N=[j(x,t)-j(x+\Delta x,t)]S\ \Delta t. \tag{4.1}$$

As $\Delta x \to 0$, the quantity involving j is, by definition, related to the partial derivative of j with respect to x (Appendix N):

$$j(x,t)-j(x+\Delta x,t)=-\frac{\partial j(x,t)}{\partial x}\ \Delta x.$$

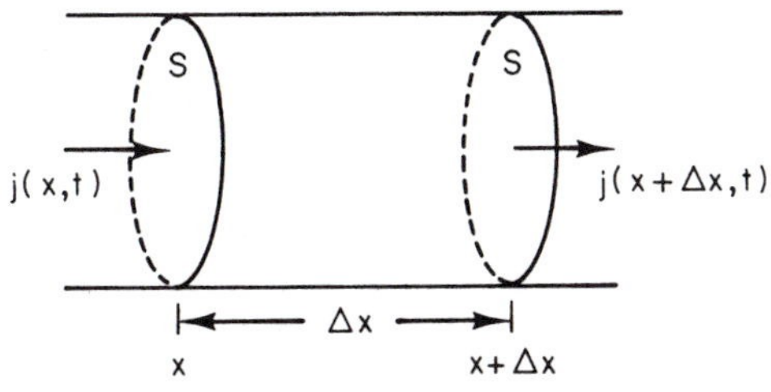

FIGURE 4.1. The fluence rates used to derive the continuity equation in one dimension.

Similarly, the increase in $N(x,t)$ is

$$\Delta N(x,t)=N(x,t+\Delta t)-N(x,t)=\frac{\partial N}{\partial t}\ \Delta t.$$

These two expressions can be substituted in Eq. (4.1) to give

$$\frac{\partial}{\partial t}\ N(x,t)=-(S\ \Delta x)\ \frac{\partial}{\partial x}\ j(x,t).$$

This can be written in terms of the concentration $C(x,t)$ by dividing both sides by the volume $S\ \Delta x$:

$$\frac{\partial C}{\partial t}=-\frac{\partial j}{\partial x}. \tag{4.2}$$

This is the *continuity equation in one dimension.*

In three dimensions the particles are moving in some direction, and it is necessary to consider the fact that $\mathbf{j}$ is a vector with components j_x, j_y, and j_z. The flux across a surface dS oriented at some arbitrary direction with the x,y,z axes is equal to the component of $\mathbf{j}$ perpendicular to the surface times dS. To see this, imagine that $\mathbf{j}$ lies in the xy plane with components j_x and j_y. If $\mathbf{j}$ makes an angle ϕ with the vertical, then

$$j_x=j\ \sin\ \phi, \quad j_y=j\ \cos\ \phi.$$

Consider the small volume shown in Fig. 4.2. If there is no buildup of particles within the volume, the flux in across the two faces parallel to the axes is equal to the flux across dS. The area dS of the slant surface is $dr\ dz$, where dz is the thickness of the volume perpendicular to the paper. The number of particles per second across the face $dy\ dz$ is $j_x dy\ dz=(j\ \sin\ \phi)(dy\ dz)$. Since $dy=dr\ \sin\ \theta$, this may be written as $j\ \sin\ \phi\ \sin\ \theta\ dz\ dr$. Similarly, the number of particles per second in across the bottom face is $j_y dx\ dz=j(\cos\ \phi)(\cos\ \theta)\ dz\ dr$. The sum of these must be equal to the number leaving across the slant face:

$$j(dz\ dr)(\sin\ \phi\ \sin\ \theta+\cos\ \phi\ \cos\ \theta)$$
$$=j(dz\ dr)\cos(\phi-\theta)=j(dS)\cos(\phi-\theta).$$

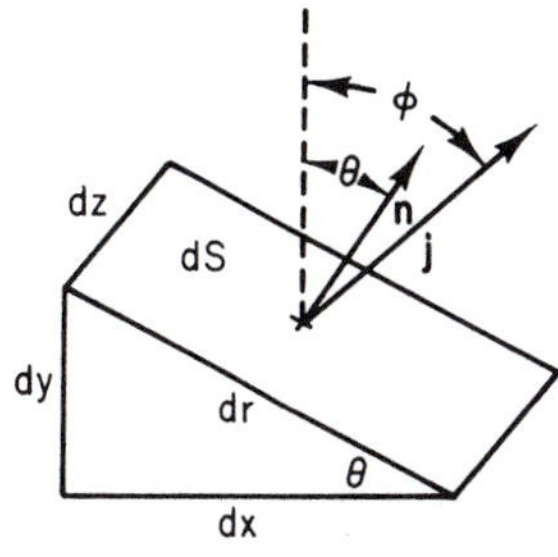

FIGURE 4.2. Volume element used to relate the fluence rate across the slant face to the components of the fluence rate parallel to the x and y axes.

The number of particles per unit area per second across the slant face is, therefore,

$$j \cos(\phi - \theta).$$

Now $\phi - \theta$ is the angle between $\mathbf{j}$ and the vector $\hat{\mathbf{n}}$, which is perpendicular to the surface. If $\hat{\mathbf{n}}$ has unit length, we can write the flux density across dS as

$$j_n$$

(the component of j parallel to $\hat{\mathbf{n}}$), or

$$\mathbf{j} \cdot \hat{\mathbf{n}}$$

(the dot product of j and the normal). The flow per second is sometimes written as

$$(\mathbf{j} \cdot \hat{\mathbf{n}})dS, \quad j_n dS, \quad \text{or} \quad (\mathbf{j} \cdot d\mathbf{S}). \tag{4.3}$$

These are all equivalent: vector $d\mathbf{S}$ is defined to have magnitude dS and to point along the normal to the surface that points outward from the enclosed volume. The same result is obtained (with more algebra) when $\mathbf{j}$ is not in the xy plane.

This result means that if we consider a closed volume as shown in Fig. 4.3, the total number of particles flowing out of the volume can be obtained by adding up the contribution from each element dS. It is

$$(\text{total number of particles out in time } \Delta t) = \left(\iint_{\substack{\text{closed}\\\text{surface}}} j_n dS \right) \Delta t.$$

Since the total number of particles in the volume enclosed by the surface is

$$\iiint_{\substack{\text{enclosed}\\\text{volume}}} C(x,y,z,t) dx\, dy\, dz,$$

we can write[1]

$$\frac{\partial}{\partial t} \iiint_{\text{volume}} C\, dV = - \iint_{\substack{\text{surface}\\\text{enclosing the}\\\text{volume}}} j_n dS. \tag{4.4}$$

The outward flux density or fluence rate of the substance integrated over a closed surface (the net flux through the surface) is equal to the rate of decrease of the amount of substance within the volume enclosed by the surface.

How to evaluate the surface integral is best shown by two examples. First consider a volume defined by two concentric spheres. The smaller sphere has a fixed radius a. Inside the smaller sphere is a lamp radiating light uniformly in all directions. The light enters the volume defined by the two spheres and leaves through the surface of the outer sphere,

[1]We can write dV as $d^3\mathbf{r}$ or $dx\, dy\, dz$.

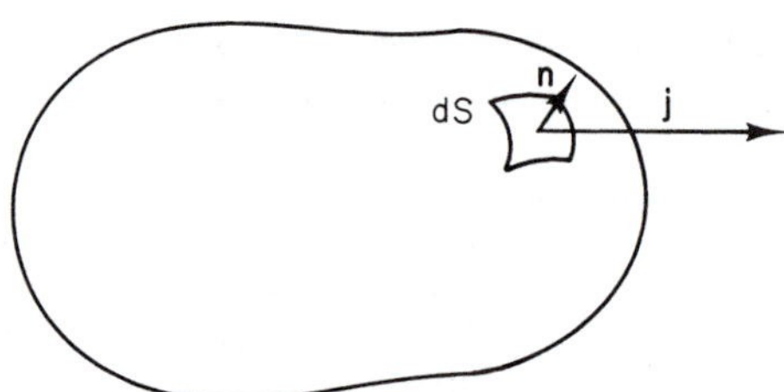

FIGURE 4.3. The total number of particles per second passing through the closed surface (flux) is the sum of the contributions $j_n dS$ from all elements of the surface.

which has radius r. The amount of light energy in the volume defined by the two spheres is not changing, so the rate of energy production by the lamp P is equal to the energy flux through the surface of the outer sphere:

$$P = \iint j_n dS. \tag{4.5}$$

Because of the spherical symmetry, $\mathbf{j}$ is perpendicular to the surface and is the same at all points on the sphere. Therefore,

$$P = j_n \iint dS.$$

Since the integral of dS over the surface of the sphere is $4\pi r^2$,

$$j = j_n = \frac{P}{4\pi r^2}. \tag{4.6}$$

The amount of energy per unit area per unit time crossing the surface of the sphere is the energy fluence rate or the intensity.

The second example is slightly more complicated. Suppose that $\mathbf{j}$ is parallel to the z axis and has the same value everywhere. The net flux through any closed surface will be zero in that case, and we will verify it to show how to evaluate a surface integral. Consider the situation shown in Fig. 4.4. At every point in the shaded strip, $j_n = j \cos\theta$. The area of the shaded strip is (recall the argument accompanying Fig. 3.15) $2\pi r^2 \sin\theta\, d\theta$, so

$$\int j_n dS = \int_0^\pi j \cos\theta\, 2\pi r^2 \sin\theta\, d\theta$$

$$= 2\pi r^2 j \int_0^\pi \cos\theta \sin\theta\, d\theta = 0.$$

The continuity equation can be expressed in terms of derivatives instead of integrals. To derive this form, consider a small rectangular volume located at (x,y,z) and having sides (dx,dy,dz) as shown in Fig. 4.5. Apply Eq. (4.4) to each face of the volume. The rate the substance flows in through the face at x is

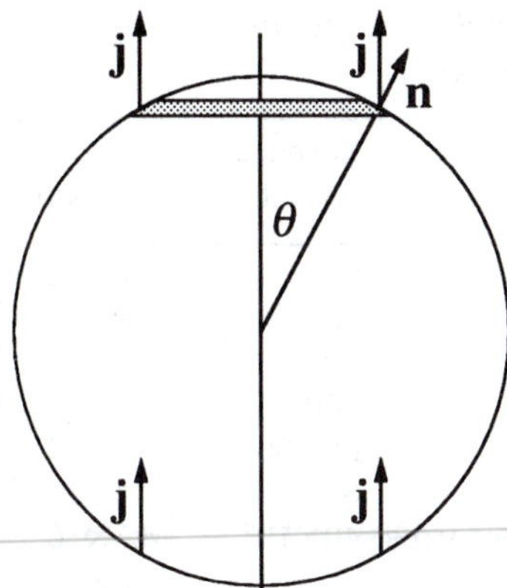

FIGURE 4.4. The fluence rate is the same everywhere. The flux is $\int j_n dS$ over the entire sphere. When the normal component of the fluence rate is outward, the contribution is positive. When it is inward, the contribution is negative.

$$j_x(x)(dy\ dz).$$

At face $x+dx$ it flows out at a rate

$$j_x(x+dx)dy\ dz.$$

There is no contribution to the flow through this face from j_y or j_z, since they are parallel to the face. The net contribution of the two terms is

$$-[j_x(x+dx)-j_x(x)]dy\ dz=-\frac{\partial j_x}{\partial x}dx\ dy\ dz.$$

Similar terms can be written for the faces perpendicular to the y and z axes. The total amount of the substance entering the volume per unit time is the rate of change of the amount within the volume, which is the rate of change of concentration times the volume $dx\ dy\ dz$. Therefore,

$$\frac{\partial C}{\partial t}(dx\ dy\ dz)=-\left(\frac{\partial j_x}{\partial x}+\frac{\partial j_y}{\partial y}+\frac{\partial j_z}{\partial z}\right)(dx\ dy\ dz)$$

or

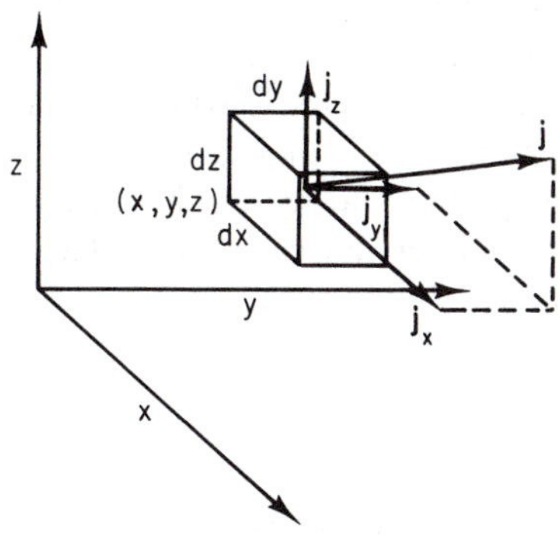

FIGURE 4.5. The small volume used to derive the differential form of the continuity equation.

$$-\frac{\partial C}{\partial t}=\frac{\partial j_x}{\partial x}+\frac{\partial j_y}{\partial y}+\frac{\partial j_z}{\partial z}. \tag{4.7}$$

This is the differential form of the continuity equation. Equation (4.2) was a special case of this when $\mathbf{j}$ was parallel to the x axis.

The combination of derivatives on the right-hand side of Eq. (4.7) occurs frequently enough to warrant a special name. It is called the *divergence* of the vector $\mathbf{j}$:

$$\text{div }\mathbf{j}=\nabla\cdot\mathbf{j}=\frac{\partial j_x}{\partial x}+\frac{\partial j_y}{\partial y}+\frac{\partial j_z}{\partial z}.$$

The continuity equation is therefore

$$\frac{\partial C}{\partial t}=-\text{div }\mathbf{j}. \tag{4.8}$$

This differential form of the continuity equation is completely equivalent to the integral form, Eq. (4.4). It is sometimes more convenient to use Eq. (4.4) and at other times more convenient to use Eq. (4.8).

This derivation assumed that the substance was conserved—neither created nor destroyed. If a chemical reaction is creating the substance at a rate Q particles $\text{m}^{-3}\ \text{s}^{-1}$ (which may depend on position) then the continuity equation becomes

$$\frac{\partial C}{\partial t}=Q-\text{div }\mathbf{j}, \tag{4.9a}$$

$$\frac{\partial}{\partial t}\underset{\text{volume}}{\iiint}C(x,y,z)dV=\underset{\text{volume}}{\iiint}Q(x,y,z)dV-\underset{\substack{\text{surface}\\\text{enclosing the}\\\text{volume}}}{\iint}j_n dS. \tag{4.9b}$$

If particles are being consumed in the chemical reaction, then Q is negative.

The continuity equation says that the rate of decrease of the amount of a conserved substance in a certain region expressed as $-\partial C/\partial t$ is equal to the rate at which it leaves the region expressed as the flow through the surface surrounding the region. The substance may be a certain kind of molecule, electric charge, heat, or mass. If it is electric charge, $\mathbf{j}$ is the electric current per unit area and C is the charge per unit volume. If it is mass, C is the mass per unit volume or density ρ. The continuity equation is found in many contexts; in each it expresses the conservation of some quantity.

In the flow of a liquid, the density of the liquid ρ, the mass M, and volume V are related by

$$M=\rho V.$$

If the liquid is incompressible, a given mass always occupies the same volume, and the density does not change. Therefore,

$$\frac{\partial \rho}{\partial t}=0$$

and the equation of continuity gives

$$\operatorname{div} \mathbf{j}_m=0. \tag{4.10}$$

4.2. DRIFT OR SOLVENT DRAG

One simple way that solute particles can move is to drift with constant velocity. They can do this in a uniform electric or gravitational field if they are also subject to viscous drag, or they can be carried along by the solvent, a process called *drift* or *solvent drag*. (The solute particles are dragged by the solvent.) The solute fluence rate is j_s, with units of particles $m^{-2}\,s^{-1}$ or just $m^{-2}\,s^{-1}$. The number of solute particles passing through a surface is the volume of solution that moves through the surface times the concentration of solute particles. Therefore,

$$\mathbf{j}_s=C\mathbf{j}_v. \tag{4.11}$$

This effect will be explored in greater detail in Sec. 4.12.

4.3. BROWNIAN MOVEMENT

There is also movement of solute molecules when the water is at rest. If the solution is dilute, the solute particles are far apart and hit each other only occasionally. They are struck by water molecules much more often. The result is that they are in continual helter-skelter motion. Each solute molecule is influenced by water molecules, but not by other solute molecules.

In Chap. 3, it was shown that the relative probability for a particle to have energy u when it is in thermal equilibrium with a reservoir at temperature T is given by a Boltzmann factor:[2]

$$P\propto e^{-u/k_BT}.$$

In Chap. 3, the Boltzmann factor was used to show that if any energy term depends on the square of some variable, then the average value of that term is $\frac{1}{2}k_BT$. A particle with kinetic energy of translation $m(v_x^2+v_y^2+v_z^2)/2$ has an average energy $k_BT/2$ for each of the three terms, or a total translational kinetic energy of $3k_BT/2$. This is true regardless of the mass of the particle. Any particle in thermal equilibrium with a reservoir (which can be the surrounding fluid) will move with a mean square velocity given by[3]

$$\overline{v^2}=\frac{3k_BT}{m}. \tag{4.12}$$

[2]The Boltzmann factor provided Perrin with the first means to determine Avogadro's number. The density of particles in the atmosphere is proportional to $\exp(-mgy/k_BT)$, where mgy is the gravitational potential energy of the particles. Using particles for which m was known, Perrin was able to determine k_B for the first time. Since R was already known, Avogadro's number was determined from the relationship $R=N_Ak_B$.

TABLE 4.2. *Values of the rms velocity for various particles at body temperature.*

Particle	Molecular weight	Mass (kg)	v_{rms} (m s^{-1})
H_2	2	3.4×10^{-27}	1940
H_2O	18	3×10^{-26}	652
O_2	32	5.4×10^{-26}	487
Glucose	180	3×10^{-25}	200
Hemoglobin	65 000	1×10^{-22}	11
Bacteriophage	6.2×10^6	1×10^{-20}	1.1
Tobacco mosaic virus	40×10^6	6.7×10^{-20}	0.4
E. coli		2×10^{-15}	0.0025

The square root of $\overline{v^2}$ is called the *root-mean-square*, or *rms*, velocity. It decreases with increasing mass of the particle. The values in Table 4.2 show values of $v_{rms}=(\overline{v^2})^{1/2}$ for different particles at body temperature.

This movement of microscopic-sized particles, resulting from bombardment by much smaller invisible atoms, was first observed by the English botanist Robert Brown in 1827 and is called *Brownian movement*. Solute particles are also subject to this random motion. If the concentration of particles is not uniform, there will be more particles wandering from a region of high concentration to one of low concentration than vice versa. This motion is called *diffusion*.

In the next several sections, we study random motion and diffusion, first for a gas and then for a liquid.

4.4. MOTION IN A GAS: MEAN FREE PATH AND COLLISION TIME

It is possible to define a *mean free path*, which is the average distance a particle travels between successive collisions, and a *collision time*, the average length of time between collisions. Suppose that at $t=0$ there is a collection of N_0 molecules. The number that have moved distance x without suffering a collision is $N(x)$. For short distances dx, the probability that a molecule collides with another molecule is proportional to dx: call it $(1/\lambda)dx$. Then, on the average, the number of molecules having their first collision between x and $x+dx$ is

$$dN=-N(x)\,\frac{1}{\lambda}\,dx.$$

[3]The Maxwell–Boltzmann distribution can be used to show that $\overline{v}_x=0$ (since a particle moves with velocity $+v_x$ or $-v_x$ with the same probability) and that the square of the mean *speed* is $(\overline{v})^2=2.55k_BT/m$. This is slightly less than the mean of the squared velocity, Eq. (4.12).

This is the familiar equation for exponential decay. The number of molecules surviving without any collision is

$$N(x)=N_0e^{-x/\lambda}.$$

To compute the average distance traveled by a molecule between collisions, we multiply each possible value of x by the number of molecules that suffer their first collision between x and $x+dx$. Since $N(x)$ is the number surviving at distance x, and dx/λ is the probability that one of those will have a collision between x and $x+dx$, the mean value of x is

$$\overline{x}=\frac{1}{N_0}\int_0^\infty xN(x)\,\frac{1}{\lambda}\,dx.$$

With the substitutions $s=x/\lambda$ and $N(x)=N_0e^{-s}$, this can be written as

$$\overline{x}=\lambda\int_0^\infty e^{-s}s\,ds=-\lambda[e^{-s}(s+1)]_0^\infty=\lambda. \quad (4.13)$$

Thus λ is the mean free path.

A similar argument can be made for the length of time that each molecule survives before being hit. The probability that a molecule is hit during a short time dt is proportional to dt: call it $(1/t_c)dt$. The number of molecules surviving a time t is given by

$$N=N_0e^{-t/t_c}$$

and the mean time between collisions can be calculated as above. It is t_c, which is called the collision time. The number of collisions per second is the *collision frequency*, $1/t_c$.

It is possible to estimate the mean free path and the collision frequency. Consider a particle of radius a_1 moving through a dilute gas of other particles of radius a_2. For convenience, imagine that particle 1 is moving and that all the other particles are fixed in position. The path of the first particle is shown in Fig. 4.6. If the center of one of these other molecules lies within a distance a_1+a_2 of the moving molecule, there will be a collision. The effect is the same as if the moving particle had radius a_1+a_2 and all the other particles were points. In moving a distance x, the "particle" sweeps out a volume

$$V_x=\pi(a_1+a_2)^2x.$$

On the average, when the particle has traveled a mean free path λ there is one collision. The average number of gas particles in the volume $V_\lambda=\pi(a_1+a_2)^2\lambda$ is therefore 1. The average number of particles per unit volume is Avogadro's number divided by the volume occupied by 1 mole, and the average number in volume V_λ is N_AV_λ/V_M. Since this number is unity,

$$\lambda=\frac{1}{\pi(a_1+a_2)^2(N_A/V_M)}. \quad (4.14)$$

The quantity $\pi(a_1+a_2)^2$ is the area of a circle. It is called the *cross section* for the collision of these particles. The concept of cross section is used extensively in Chap. 14.

This estimation is somewhat crude in its assumption that only one molecule is moving. If all the molecules are of the same kind and have a Maxwell–Boltzmann velocity distribution, then the factor 1 in the numerator is replaced by $2^{-1/2}=0.707$ [Reif (1965, p. 471)].

For a gas at standard temperature and pressure, the volume of 1 mol is 22.4 l$=22.4\times10^3$ cm$^3=22.4\times10^{-3}$ m^3. If $a_1=a_2=1.5\times10^{-8}$ cm$=1.5\times10^{-10}$ m, then Eq. (4.14) can be used to calculate the mean free path:

$$\lambda=\frac{22.4\times10^{-3}\ \text{m}^3}{(3.14)[(3\times10^{-10})^2\ \text{m}^2](6\times10^{23})}=1.3\times10^{-7}\ \text{m}.$$

For a gas at standard temperature and pressure, the mean free path is about 1000 times the molecular diameter, and the assumption of infrequent collisions is justified.

The collision time can be estimated by saying that

$$t_c=\frac{\lambda}{\overline{v}},$$

where $\overline{v}$ is the average speed of the molecules. Ignoring the difference between average speed and rms velocity, we can use Eq. (4.12) to write

$$t_c\approx\lambda\left(\frac{m}{3k_BT}\right)^{1/2}. \quad (4.15)$$

The important feature of this is the dependence on $m^{1/2}$ and on λ. For air at room temperature, $t_c=2\times10^{-10}$ s.

4.5. MOTION IN A LIQUID

The assumptions of the previous section do not hold in a liquid, in which the particle is being continually bombarded by neighbors. Blindly applying Eq. (4.14) to water, we can use the fact that 1 mol is 18 g and occupies 18 cm^3, to obtain $\lambda=1\times10^{-10}$ m, so that $a/\lambda\approx1$, and the assumptions

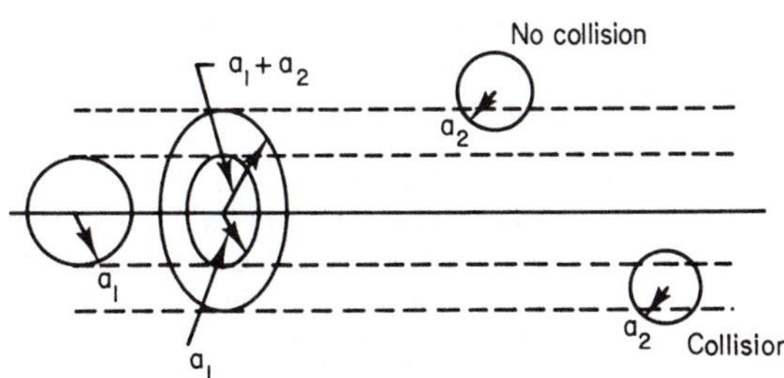

FIGURE 4.6. A particle of radius a_1 moves through a gas of particles of radius a_2. A collision will occur if the center of another particle lies within a distance a_1+a_2 of the trajectory of the particle under consideration.

behind the derivation break down. Estimating the collision time with Eq. (4.15) gives a value that is a factor of 1000 less than for the gas, or 10^{-13} s.

Yet, although these estimates of the mean free path and the collision times are undoubtedly wrong, the concepts appear to be valid. Computer simulations of molecular collisions show that the distribution of free paths is exponential even though the mean free path is only a fraction of a molecular diameter. If diffusion is regarded as a random walk of the diffusing particles (see Sec. 4.12), then the diffusion constant can be related to the mean free path and collision time. Equations (4.14) and (4.15) can then be used to show that the diffusion constant should be inversely proportional to the square of the particle radius. This has been verified experimentally for the diffusion of certain liquids. Evidence for the validity of this random-walk model for diffusion in liquids has been summarized by Hildebrand, Prausnitz, and Scott (1970, pp. 36–39).

A particle in a liquid is subject to a fluctuating force $\mathbf{F}(t)$, which is random in magnitude and direction. The particle begins to move in response to this force. However, once it has begun to move, it suffers more collisions in the front than behind, so the force slows it down. Because the particle can neither stay at rest nor continue to move in the same direction, it undergoes a random, zig-zag motion with average translational kinetic energy $3k_BT/2$. The mean square velocity is not zero, but the mean vector velocity is zero.

For each particle, Newton's second law is

$$m\,\frac{d\mathbf{v}}{dt}=\mathbf{F}(t).$$

This is not very useful as it stands. To make it more tractable, consider a particle with average velocity $\overline{\mathbf{v}}$. (The average means that an ensemble of identically prepared particles is examined.) The particle has more collisions on the front that slow it down. We therefore break up $\mathbf{F}(t)$ into two parts: an average drag force, which will be the same for all the particles in the ensemble, and a rapidly fluctuating part $\mathbf{g}(t)$, which will vary with time and from particle to particle. Newton's second law is then

$$m\,\frac{d\mathbf{v}}{dt}=(\text{drag force})+\mathbf{g}(t),$$

where $\mathbf{g}(t)$ is random in magnitude and direction. The drag force will be zero when $\overline{\mathbf{v}}$ is zero. For average velocities that are not too large it can be approximated by a linear term:

$$(\text{drag force})=-\beta\overline{\mathbf{v}}.$$

With this approximation, Newton's second law is known as the Langevin equation:

$$m\,\frac{d\mathbf{v}}{dt}=-\beta\overline{\mathbf{v}}+\mathbf{g}(t). \tag{4.16}$$

(If the liquid is moving, the drag force will be zero when the particle has the same average velocity as the liquid. So $\overline{\mathbf{v}}$ can be interpreted as the relative velocity of the particle with respect to the liquid.) This equation often has another term in it, which does not average to zero and which represents some external force such as gravity that acts on all the particles. This approximate equation can be solved in some cases, though with difficulty, and has formed the basis for some treatments of the motion of large particles in fluids. With suitable interpretation, it can also form the basis for motion of the fluid molecules themselves.[4] In particular, in dealing with molecular motion, it is necessary to consider the fact that the molecules do not move independently of one another.

In Chap. 1 we discussed fluids for which there was a viscous force proportional to the gradient of the velocity. If a plate perpendicular to the y axis and with area S was moved parallel to the x axis with velocity v_x, the force of the fluid on the plate was given by Eq. (1.32):

$$F_x=-\eta S\,\frac{dv_x}{dy}.$$

This holds true for gases and for many liquids. Such fluids are called Newtonian fluids. (If F_x does not increase proportionally to dv_x/dy, it usually increases less rapidly. The fluid is then called non-Newtonian. There are several different fluids, with different names associated with them, having particular viscous properties.) It was for a Newtonian fluid that we derived the equation for Poiseuille flow, Eq. (1.39).

For a Newtonian fluid, one can show (although it requires some detailed calculation[5]) that the drag force on a spherical particle of radius a is given by

$$\mathbf{F}_{\text{drag}}=-\beta\overline{\mathbf{v}}=-6\pi a\,\eta\overline{\mathbf{v}}. \tag{4.17}$$

This equation is valid when the sphere is so large that there are many collisions of fluid molecules with it and when the velocity is low enough so that there is no turbulence. This result is called Stokes's law.

If the sphere is not moving in an infinite medium but is confined within a cylinder, then a correction must be applied. In that case the viscous drag depends on the velocity of the spherical particle through the fluid, the average velocity of the fluid through the cylinder, and the distance of the particle from the axis of the cylinder.[6]

> Stokes's law is valid for a gas if the mean free path is much less than the molecular radius, so that many collisions with neighboring molecules occur. At the other extreme, a mean free path much greater than the particle radius, the drag force turns out to be

[4]See, for example, Pryde (1966, p. 161).

[5]This is an approximate equation. See Barr (1931, p. 171).

[6]An early correction for particles on the axis of a cylinder is found in Barr (1931), p. 183. More recent work is by Levitt (1975), by Bean (1972), and by Paine and Scherr (1975).

$$F_{\text{drag}} = \alpha \eta a \frac{a}{\lambda} \bar{v}.$$

Although this will not be directly useful to us in considering biological systems, it is mentioned here to show how important it is to understand the conditions under which an equation is valid. Although the dimensions of this new equation are unchanged (we have introduced a factor a/λ, which is dimensionless), the drag force depends on a^2 instead of on a. The reason for the difference is that collisions are now infrequent and that the probability of a collision that imparts some average momentum change is proportional to the projected cross-sectional area of the sphere, πa^2. In the regime of interest to us, in which there are many collisions, we would not expect the force to depend on λ.[7] It is hoped that this will convince you of the danger in taking someone else's equation and using it without understanding it.

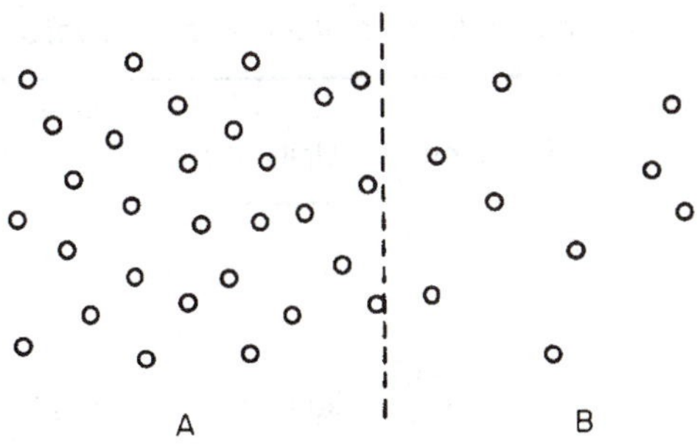

FIGURE 4.7. An example of diffusion. Each molecule at A or B can wander with equal probability to the left or right. There are more molecules at A to wander to the right than there are at B to wander to the left. There is a net flow of molecules from A to B.

4.6. DIFFUSION: FICK'S FIRST LAW

Diffusion is the random movement of particles from a region of higher concentration to a region of lower concentration. The diffusing particles move independently of one another; they may collide frequently with the molecules of the fluid in which they are immersed, but they rarely collide with one another. The surrounding fluid[8] may be at rest, in which case diffusion is the only mechanism for transport of the solute, or it may be flowing, in which case it carries the solute along with it (solvent drag). Both effects can occur together.

We first consider diffusion from a macroscopic point of view and write down an approximate differential equation to describe it. We then obtain a second form of the diffusion equation by combining this with the continuity equation. After discussing briefly some solutions to these equations, we look at the problem from a microscopic point of view, considering the random motion of the particles, and show that we get the same results.

Suppose that the surrounding solvent does not move. If the solute concentration is completely uniform, there is no net flow. As many particles wander to the left as to the right, and the concentration remains the same. There will be local fluctuations in concentration, analogous to those we saw in the preceding chapter for fluctuations in the concentration of a gas, but that is all.

However, if the concentration is higher in region A than in region B to the right of it, there are more particles to wander to the right from A to B than there are to wander to the left from B to A (Fig. 4.7). If the problem is one-dimensional, there is no net flow if $\partial C/\partial x=0$, but there is flow if $\partial C/\partial x \neq 0$. If the concentration difference is small, then the flux density j is linearly proportional to the concentration gradient $\partial C/\partial x$. If the actual process is not linear, this can be thought of as the first term of a Taylor's series expansion (Appendix D). The equation is

$$j_x = -D \frac{\partial C}{\partial x}. \tag{4.18a}$$

Constant D is called the *diffusion coefficient*. The units of D are $\text{m}^2\,\text{s}^{-1}$, as may be seen by noting that the units of j are (something) $\text{m}^{-2}\,\text{s}^{-1}$ and the units of $\partial C/\partial x$ are (something) m^{-4}. This relationship is called *Fick's first law of diffusion*, after Adolf Fick, a German physiologist in the last half of the nineteenth century. The minus sign shows that the flow is in the direction from higher concentration to lower concentration: if $\partial C/\partial x$ is positive, the flow is in the $-x$ direction.

The diffusion equation is one of many forms of the transport equation. Other forms are shown in Table 4.3. The units of the constant are different for the last three entries in the table because the quantity that appears on the right has different units than the quantity on the left. In each case, however, a fluence rate or flux density (of particles, mass, energy, electric charge, or momentum) is related to a rate of change of some other quantity with position. This rate of change is called the *gradient* of the quantity. The gradient is often called the *driving force*. The concentration gradient or driving force "causes" the diffusion of particles; the temperature gradient "causes" the heat flow; the electric voltage gradient "causes" the current flow; the velocity gradient "causes" the momentum flow. All these flows are actually caused by random motion.

The diffusive fluence rate can be related to the gradient of the chemical potential of the solute. Equation (3.61) can be rewritten as

$$\Delta\mu_s = k_B T \ln(C_2/C_1) = k_B T \ln(1+\Delta C_s/C_s)$$
$$\approx k_B T \Delta C_s / C_s,$$

from which $\Delta C_s \approx C_s \Delta\mu_s / k_B T$, so

[7] Kennard (1938, p. 309).

[8] There need not be a surrounding fluid. If the gas within some apparatus is so dilute that more collisions occur between molecules and the walls of the apparatus than between one molecule and another, then their movement is diffusion.

TABLE 4.3. *Various forms of the transport equation.*

Substance flowing	Equation	Units of j	Units of the constant
Particles	$j_s=-D\frac{\partial C}{\partial x}$	$m^{-2}\,s^{-1}$	$m^2\,s^{-1}$
Mass	$j_m=-D\frac{\partial C}{\partial x}$	$kg\,m^{-2}\,s^{-1}$	$m^2\,s^{-1}$
Heat	$j_H=-K\frac{\partial T}{\partial x}$	$J\,m^{-2}\,s^{-1}$ or $kg\,s^{-3}$	$J\,K^{-1}\,m^{-1}\,s^{-1}$
Electric charge	$j_e=-\sigma\frac{\partial V}{\partial x}$	$C\,m^{-2}\,s^{-1}$	$C\,m^{-1}\,s^{-1}\,V^{-1}$ or $\Omega^{-1}\,m^{-1}$
Viscosity (y component of momentum transported in x direction)	$\frac{F}{S}=-\eta\frac{\partial v_y}{\partial x}$	$N\,m^{-2}$ or $kg\,m^{-1}\,s^{-2}$	$kg\,m^{-1}\,s^{-1}$ or Pa s

$$\frac{\partial C_s}{\partial x}=\frac{C_s}{k_BT}\frac{\partial \mu_s}{\partial x}$$

and

$$j_{sx}=-\frac{DC_s}{k_BT}\frac{\partial \mu_s}{\partial x}. \tag{4.18b}$$

The solute flux density is proportional to the diffusion constant, the solute concentration, and the gradient in the chemical potential per solute particle.

In three dimensions, the flow of particles can point in any direction and have components j_x, j_y, and j_z. An equation can be written for each component that is analogous to Eq. (4.18). We can write one vector equation instead of three equations for the three components by defining $\hat{\mathbf{x}}$, $\hat{\mathbf{y}}$, and $\hat{\mathbf{z}}$ to be unit vectors along the axes. Then

$$j_x\hat{\mathbf{x}}+j_y\hat{\mathbf{y}}+j_z\hat{\mathbf{z}}=-D\left(\frac{\partial C}{\partial x}\hat{\mathbf{x}}+\frac{\partial C}{\partial y}\hat{\mathbf{y}}+\frac{\partial C}{\partial z}\hat{\mathbf{z}}\right).$$

We have created a vector that depends on $C(x,y,z,t)$ by performing the indicated differentiations on C and multiplying the results by the appropriate unit vectors. This vector function is the *gradient* of C in three dimensions:

$$\text{grad } C=\boldsymbol{\nabla}C=\frac{\partial C}{\partial x}\hat{\mathbf{x}}+\frac{\partial C}{\partial y}\hat{\mathbf{y}}+\frac{\partial C}{\partial z}\hat{\mathbf{z}}. \tag{4.19}$$

The diffusion equation with this notation is

$$\mathbf{j}=-D\text{ grad } C=-D\boldsymbol{\nabla}C. \tag{4.20}$$

Remember that this is simply shorthand for three equations like Eq. (14.8). If you feel a need to review vector calculus, an excellent text is the one by Schey (1973).

4.7. THE EINSTEIN RELATIONSHIP BETWEEN DIFFUSION AND VISCOSITY

Before we can apply the diffusion equation to real problems, we must determine the value of the diffusion constant D. The experimental determination of D is often based on Fick's second law of diffusion, which combines the first law with the equation of continuity and is discussed in the next section. It is closely related to the viscosity, as was first pointed out by Einstein. This is not surprising, since diffusion is caused by the random motion of the particles under the bombardment of neighboring atoms, and viscous drag is also caused by the bombardment by neighboring atoms. What is remarkable is that a general relationship between them can be deduced quite easily by imagining just the right sort of experiment.

Consider a collection of particles uniformly suspended in a fluid at rest. Imagine that each particle is suddenly subjected to an external force $\mathbf{F}_{\text{ext}}$ (such as gravity) that acts in the $-y$ direction, as shown in Fig. 4.8. The particles will all begin to drift downward, speeding up until the upward viscous force on them balances the external force: $\mathbf{F}_{\text{ext}}-\beta\overline{\mathbf{v}}=\mathbf{0}$. In terms of magnitudes, $F_{\text{ext}}=\beta\overline{v}$.

Because these particles are all moving downward, there is a downward flux density. With reference to Fig. 4.9, the number of particles crossing area S in time Δt will be those within the cylinder of height $\overline{v}\Delta t$. That number is the concentration times the volume $(S\overline{v}\Delta t)$. Dividing by S and Δt gives

$$\mathbf{j}_{\text{drift}}=-\overline{v}C(y)\hat{\mathbf{y}}.$$

As the particles move down, they deplete the upper region of the fluid and cause a concentration gradient. This concentration gradient causes an upward diffusion of particles, with a flux density given by

$$\mathbf{j}_{\text{dif}}=-D\frac{\partial C}{\partial y}\hat{\mathbf{y}}.$$

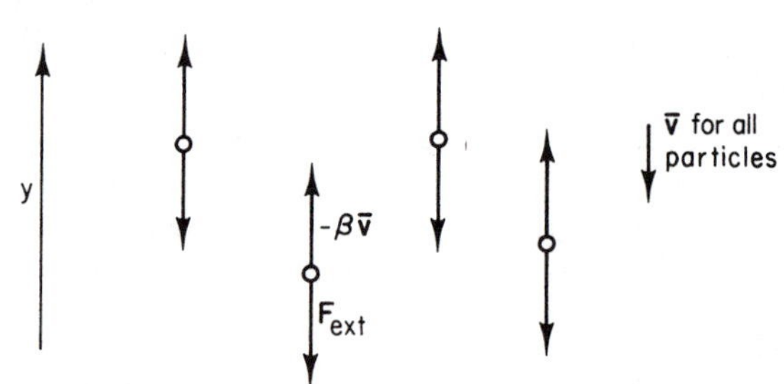

FIGURE 4.8. Particles drifting under the influence of a downward force $\mathbf{F}_{\text{ext}}$.

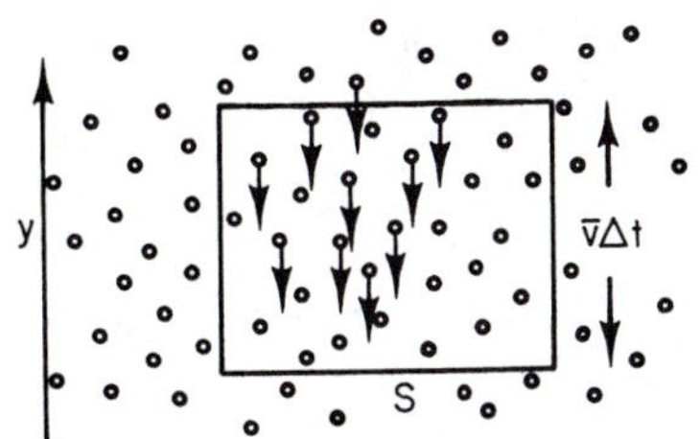

FIGURE 4.9. Calculating the fluence rate of particles drifting downward.

Equilibrium will be established when these two flux densities are equal in magnitude:

$$|\mathbf{j}_{\text{drift}}| = |\mathbf{j}_{\text{dif}}|,$$
$$|\bar{v}C(y)| = \left| D \frac{\partial C}{\partial y} \right|. \tag{4.21}$$

But equilibrium means that the particles have a Boltzmann distribution in y, because their potential energy increases with y (work is required to lift them in opposition to $\mathbf{F}_{\text{ext}}$). For a constant $\mathbf{F}_{\text{ext}}$ independent of y, the energy is

$$u(y) = F_{\text{ext}} y,$$

where F_{ext} is the magnitude of the force. The concentration is

$$C(y) = C(0) e^{-F_{\text{ext}} y / k_B T}.$$

Therefore

$$\frac{\partial C}{\partial y} = -\frac{F_{\text{ext}}}{k_B T} C(y).$$

Inserting this in Eq. (4.21) gives $\bar{v} = D F_{\text{ext}} / k_B T$ or $D = \bar{v} k_B T / F_{\text{ext}}$. In equilibrium the magnitude of $\mathbf{F}_{\text{ext}}$ is equal to the magnitude of the viscous force $\mathbf{f}$. Therefore $D = k_B T \bar{v} / f$. Since the viscous force is proportional to the velocity, $|f| = |\beta \bar{v}|$,

$$D = \frac{k_B T}{\beta}. \tag{4.22}$$

The derivation of this equation required only that the velocities be small enough so that the linear approximations for the diffusion equation and the viscous force are valid. It is independent of the nature of the particle or its size. If in addition the diffusing particles are large enough so that Stokes's law is valid, then

$$\beta = 6 \pi \eta a$$

and[9]

$$D = \frac{k_B T}{6 \pi \eta a}. \tag{4.23}$$

The diffusion coefficient is inversely proportional to the fluid viscosity and the radius of the particle.

Combining Eqs. (4.18b) and (4.22) shows that in terms of the chemical potential,

$$j_{sx} = -\frac{C_s}{\beta} \frac{\partial \mu_s}{\partial x}.$$

The viscosity of water varies rapidly with temperature, as shown in Fig. 4.10. These values of viscosity and Eq. (4.23) have been used to calculate the solid lines for D vs a shown in Fig. 4.11. Various experimental values are also shown. The diffusion coefficient increases rapidly with temperature, so that care must be taken to specify the temperature at which the data are obtained. Since not all the molecules are spherical, there is some uncertainty in the value of the particle radius a. Figure 4.12 is a plot of D at 25 °C (298 K) vs molecular weight M. Considerably more data are available in this form. Although the solid line provides a rough estimate of D if M is known, scatter is considerable because of varying particle shape. Water has an unusually large diffusion coefficient. The diffusion of water is measured by using a few water molecules in which one hydrogen atom is radioactive and measuring how they diffuse. DNA lies a factor of 10 below the curve, presumably because it is partially uncoiled and presents a larger size than other molecules of comparable molecular weight.

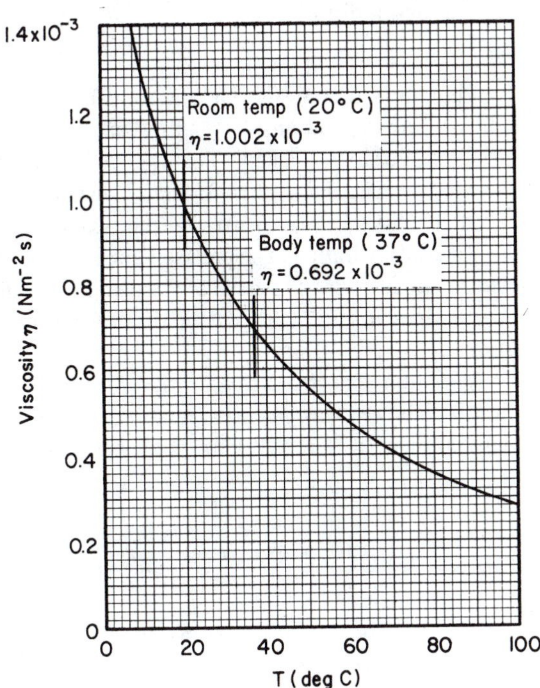

FIGURE 4.10. Viscosity of water at various temperatures. Data are from the *Handbook of Chemistry and Physics*. (1972), 53rd ed., Cleveland, Chemical Rubber, p. F-36.

[9] For self-diffusion (such as radioactively tagged water in water), a hydrodynamic calculation shows that $\beta = 4\pi\eta a$. [R. B. Bird, W. E. Stewart, and E. N. Lightfoot (1960). *Transport Phenomena*, New York, Wiley, p. 514ff.]

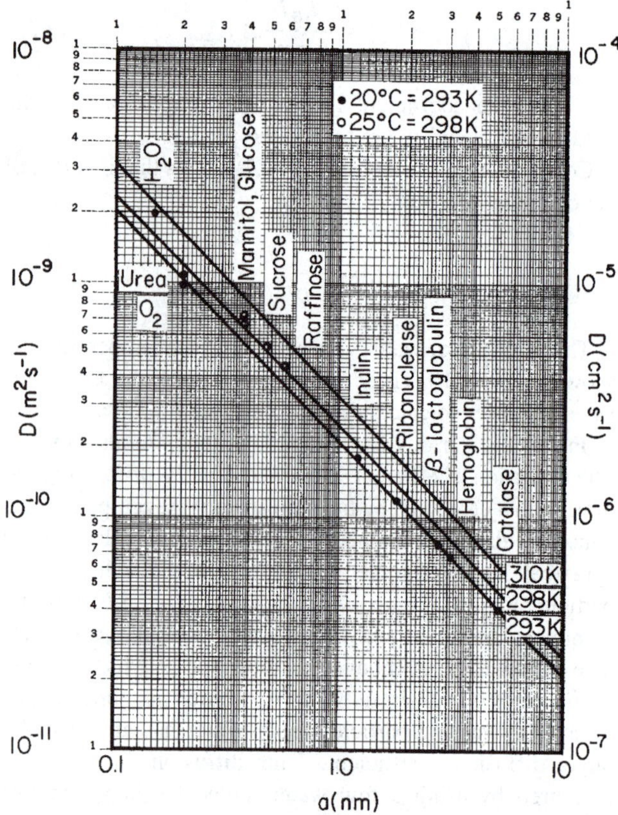

FIGURE 4.11. Diffusion coefficient versus sphere radius *a* for diffusion in water at three different temperatures. Experimental data at 20 °C (293 K) are from Villars and Benedek (1974, p. 2-40). Data at 25 °C (298 K) are from *Handbook of Chemistry and Physics*, (1972), 53rd ed., Cleveland, Chemical Rubber, p. F-47.

If all of the molecules shown had the same density, then their radius would depend on $M^{1/3}$ and the line would have a slope of $-\frac{1}{3}$. The slope is steeper than this, suggesting that the molecules are larger for large M than constant density would predict. This increase in size may be partially attributable to water of hydration. The precise values of diffusion coefficients depend on many details of the particle structure, however the lines in Fig. 4.12 provide an order-of-magnitude estimate.

The assumption that the flux depends linearly on the concentration gradient was an approximation. The diffusion coefficient is found, as a result, to be somewhat concentration dependent.

4.8. FICK'S SECOND LAW OF DIFFUSION

Fick's first law of diffusion, Eq. (4.18), is the observation that for small concentration gradients, the diffusive flux density is proportional to the concentration gradient:

$$j_x = -D\,\frac{\partial C}{\partial x}.$$

If this is differentiated, one obtains

$$\frac{\partial j_x}{\partial x} = -D\,\frac{\partial^2 C}{\partial x^2}.$$

Similar equations hold for the y and z directions. The equation of continuity, Eq. (4.2), is

$$-\frac{\partial C}{\partial t} = \frac{\partial j_x}{\partial x} + \frac{\partial j_y}{\partial y} + \frac{\partial j_z}{\partial z}.$$

If we combine these two equations, we get *Fick's second law of diffusion*:

$$\frac{\partial C}{\partial t} = D\left(\frac{\partial^2 C}{\partial x^2} + \frac{\partial^2 C}{\partial y^2} + \frac{\partial^2 C}{\partial z^2}\right). \tag{4.24}$$

The first law relates the flux of particles to the concentration gradient. The second law tells how the concentration at a point changes with time. It combines the first law and the equation of continuity. The function on the right-hand side of Eq. (4.24),

$$\frac{\partial^2 C}{\partial x^2} + \frac{\partial^2 C}{\partial y^2} + \frac{\partial^2 C}{\partial z^2},$$

is called the *Laplacian* of C. It is often abbreviated as $\nabla^2 C$ (read "del squared C") in American textbooks or ΔC in European books. It is given in other coordinate systems in Appendix L.

In principle, if $C(x,y,z)$ is known at $t=0$, Eq. (4.24) can be solved for $C(x,y,z,t)$ at all later times. (We develop a general, and sometimes useful, equation for doing this below.) We may also look at this equation as a local equation, telling how C changes with time at some point if we know how the concentration changes with position in the neighborhood of that point. The change of concentration with position determines the flux **j**. The changes in flux with position determine how the concentration changes with time.

There is extensive literature on how to solve the diffusion equation (or heat-flow equation, which is the same thing). Instead of discussing a large number of techniques, we show by substitution that a Gaussian or normal distribution function, spreading in a certain way with time, is one solution to Eq. (4.24). In Sec. 4.14 we independently derive the same solution from a random-walk model of diffusion. An important feature of the Gaussian solution is that the center of the distribution of concentration does not move.

For simplicity, we take the distribution to be centered at the origin and find those conditions under which[10]

$$C(x,t) = \frac{C_0}{\sqrt{2\pi}\,\sigma(t)}\,e^{-x^2/2\sigma^2(t)} \tag{4.25}$$

is a solution to the one-dimensional version of Eq. (4.24):

[10]The properties of the Gaussian function, Eq. (4.25), are discussed in Sec. 3.10 and Appendix I.

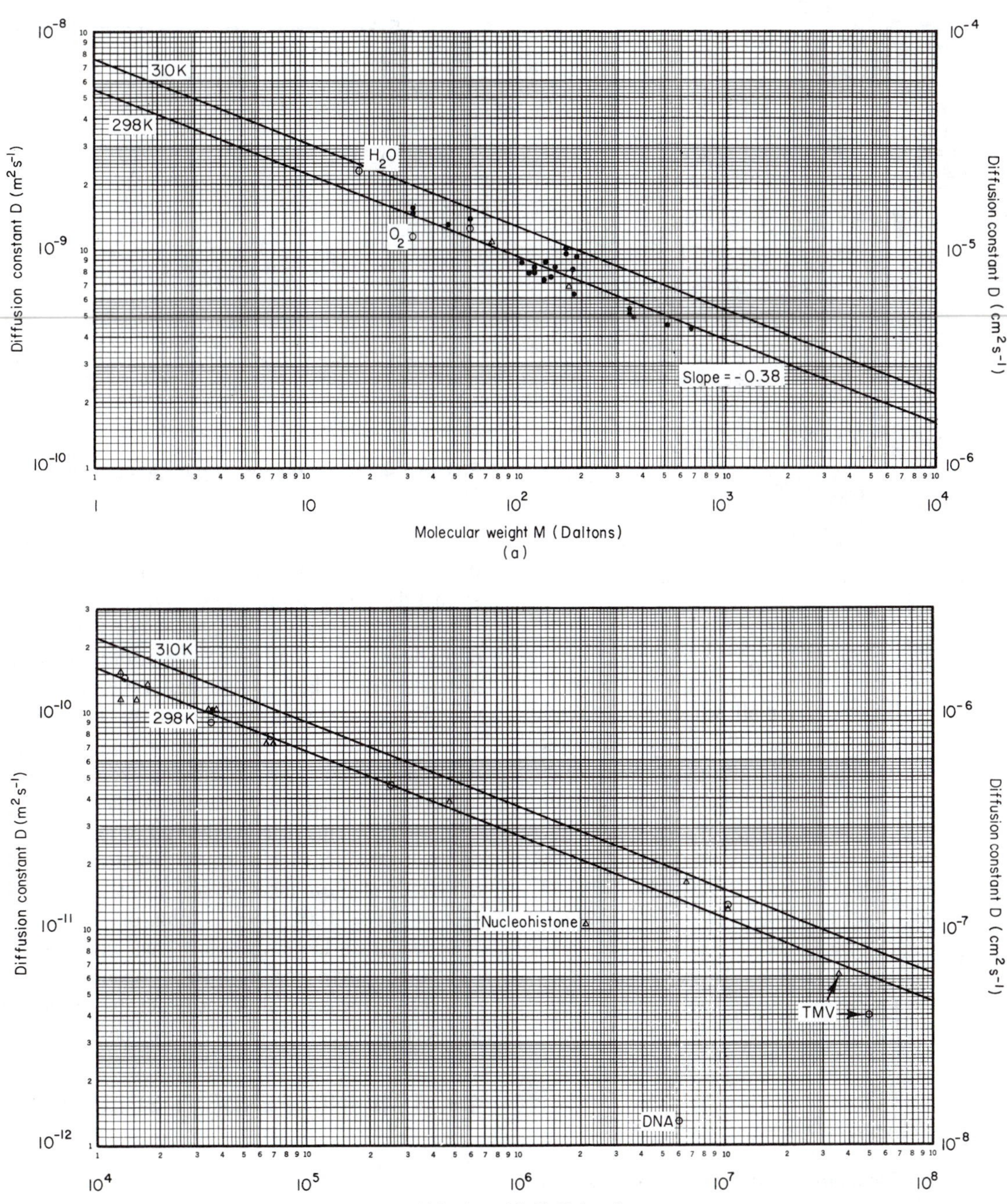

FIGURE 4.12. Diffusion coefficient versus molecular weight in daltons. (One dalton is the mass of one hydrogen atom.) Data are either taken at or corrected to 298 K. The 298-K solid line was drawn by eye through the data; the line at 310 K was drawn parallel to it using the temperature change in Eq. (4.23). Data scatter around the line by about 30%, with occasional larger departures. Solid dots are from *American Institute of Physics Handbook* (1963), 2nd ed., New York, McGraw-Hill, pp. 2-208–2-209. Open circles are from Villars and Benedek (1974), p. 2-40, corrected to 298 K from 293 K by multiplying by 1.15. Triangles are from R. Setlow and E. Pollard (1962), *Molecular Biophysics*. Reading, Mass., Addison-Wesley, pp. 83 and 97, with the same correction.

$$\frac{\partial C}{\partial t} = D \frac{\partial^2 C}{\partial x^2}. \tag{4.26}$$

To do this, we will need various derivatives of Eq. (4.25). They can be evaluated using the chain rule:

$$\frac{\partial C}{\partial t} = C_0 \left(- \frac{1}{\sqrt{2\pi\sigma^2}} e^{-x^2/2\sigma^2} + \frac{1}{\sqrt{2\pi}\sigma} \frac{x^2}{\sigma^3} e^{-x^2/2\sigma^2} \right) \frac{d\sigma}{dt},$$

$$\frac{\partial C}{\partial x} = -\frac{C_0}{\sqrt{2\pi}\sigma} e^{-x^2/2\sigma^2} \frac{x}{\sigma^2},$$

$$\frac{\partial^2 C}{\partial x^2} = -\frac{C_0}{\sqrt{2\pi}\sigma^3} e^{-x^2/2\sigma^2} + \frac{C_0}{\sqrt{2\pi}\sigma} \frac{x}{\sigma^2} e^{-x^2/2\sigma^2} \frac{x}{\sigma^2}.$$

When these are substituted in Eq. (4.26), the result is

$$\frac{C_0}{\sqrt{2\pi}\sigma^2} e^{-x^2/2\sigma^2}\left(-1+\frac{x^2}{\sigma^2}\right)\frac{d\sigma}{dt} = D\frac{C_0}{\sqrt{2\pi}\sigma^3} e^{-x^2/2\sigma^2} \times\left(-1+\frac{x^2}{\sigma^2}\right).$$

We can divide both sides of this equation by

$$\frac{C_0 e^{-x^2/2\sigma^2}}{\sqrt{2\pi}\sigma^2}$$

because this factor is never zero. The result is

$$\left(\frac{x^2}{\sigma^2}-1\right)\frac{d\sigma}{dt} = \frac{D}{\sigma}\left(\frac{x^2}{\sigma^2}-1\right).$$

We can divide by (x^2/σ^2-1) for all values of x except $x=\pm\sigma$. These values of x are where the second derivative of C vanishes; at these points, $\partial C/\partial t=0$ for any value of σ. At all other points, the solution will satisfy the equation only if

$$\sigma\frac{d\sigma}{dt} = D.$$

This can be integrated to give

$$\int \sigma\, d\sigma = \int D\, dt$$

or

$$\tfrac{1}{2}\sigma^2(t) = Dt + \text{const.}$$

Multiply through by 2 and observe that $\sigma^2(0)=\text{const}$, so that

$$\sigma^2(t) = 2Dt + \sigma^2(0). \tag{4.27}$$

If the concentration is initially Gaussian with variance $\sigma^2(0)$, after time t it will still be Gaussian, centered on the same point, with a larger variance given by Eq. (4.27). Figure 4.13 shows this spreading in a typical case. At still earlier times the concentration would have been even more narrowly peaked. In the limit when $\sigma(t)$ is zero, all the particles are at the origin, giving an infinite concentration. This is, of course, impossible. However, all the particles could be very close to the origin, giving a very tall, narrow curve for $C(x)$.

The width of the curve increases as the *square root* of the time. A square-root increase is less rapid than a linear increase, reflecting the fact that as the particles spread out the concentration does not change as rapidly with distance, so that the flux and the rate of spread decrease.

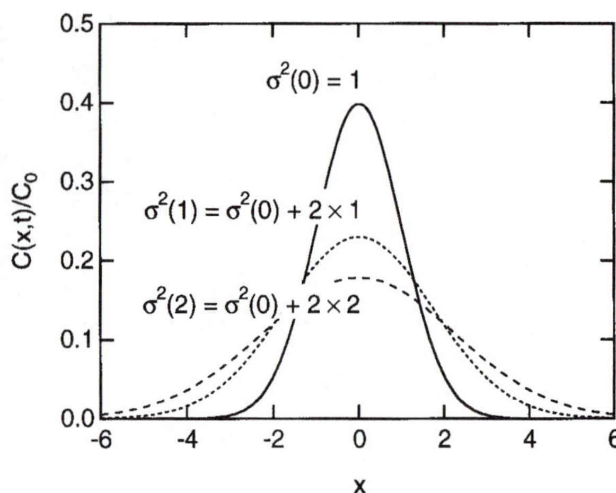

FIGURE 4.13. Spreading of particles by diffusion assuming $D=1$.

Note that the rate of change of concentration with time depends on the second derivative of the concentration with distance. This is because the rate of buildup is the flux into a region at some surface minus the flux out through a nearby surface; each flux is proportional to the gradient of the concentration, so the buildup is proportional to the difference in gradients or the second derivative.

In the Problems listed at the end of this chapter you will discover that diffusion of small particles through water for a distance of 1 μm takes about 1 ms, and diffusion through 100 μm takes 100^2 times as long, or 10 s. The times are even longer for larger particles. Thus, diffusion is an effective mode of transport for distances comparable to the size of a cell, but it is too slow for larger distances. This is why multicelled organisms develop circulatory systems.

4.9. TIME-INDEPENDENT SOLUTIONS

In this section we develop general solutions for diffusion and solvent drag when particles are conserved and the concentration fluence rate are not changing with time. The system is in the steady state. The continuity equation, Eq. (4.8), then becomes div $\mathbf{j}=0$. We consider the solutions for C and $\mathbf{j}$ in one, two, and three dimensions when the symmetry is such that $\mathbf{j}$ depends on only one position coordinate, x or r. These solutions are sometimes appropriate models for limited regions of space. There is always some other region of space serving as a source or sink for the particles that are diffusing, where the model does not apply.

The behavior of $\mathbf{j}$ can be deduced from the continuity equation. In one dimension, such as flow in a pipe or between two infinite planes, the continuity equation is

$$\frac{dj_x}{dx} = 0, \tag{4.28}$$

which has a solution $j_x = b_1$ where b_1 is a constant. (The subscript denotes the constant for the one-dimensional case.) The total flux or current i is constant, so

$$j_x = \frac{i}{S}, \tag{4.29}$$

where S is the area perpendicular to the flow.

In two dimensions, we consider a problem with cylindrical symmetry and consider only flow radially away from or towards the z axis. In that case, the equation in Table L.1 for the divergence becomes

$$\frac{1}{r}\frac{d}{dr}(rj_r) = 0, \tag{4.30}$$

from which

$$\frac{d}{dr}(rj_r) = 0. \tag{4.31}$$

This means that (rj_r) is constant, or

$$j_r = \frac{b_2}{r}. \tag{4.32}$$

This is valid everywhere except along the z axis, where there is a source of particles and the divergence is not zero. The total current leaving a region of length L parallel to the z axis is also constant,

$$j_r = \frac{i}{2\pi Lr}. \tag{4.33}$$

In three dimensions with spherical symmetry, the radial component of the divergence is

$$\frac{1}{r^2}\frac{d}{dr}(r^2 j_r) = 0,$$

from which

$$\frac{d}{dr}(r^2 j_r) = 0, \tag{4.34}$$

so that

$$j_r = \frac{b_3}{r^2} \tag{4.35}$$

or

$$j_r = \frac{i}{4\pi r^2}. \tag{4.36}$$

This is valid everywhere except at the origin, where there is a source of particles.

These results depend only on continuity, time independence, and the assumed symmetry. They are true for diffusion, solvent drag, or any other process. Note the progression in going to higher dimensions: in n dimensions $r^{n-1}j_r$ is constant.

Now consider how the concentration varies in the two limiting cases of pure solvent drag and pure diffusion. (Section 4.12 discusses what happens when both transport modes are important.)

For solvent drag, the velocity of the solvent is the volume flux density $\mathbf{j}_v$ which also satisfies the continuity equation. In one dimension $j_v = i_v/S$. In two dimensions $j_v = i_v/2\pi Lr$, and in three dimensions $j_v = i_v/4\pi r^2$. In each case

$$C_s = \frac{j_s}{j_v} = \frac{i_s}{i_v}. \tag{4.37}$$

Since C_s is constant, there is no diffusion.

For the case of diffusion, $\mathbf{j} = -D\nabla C$. In one dimension this becomes

$$\frac{dC}{dx} = -\frac{i}{SD},$$

which is integrated to give

$$C = -\frac{i}{SD}x + b_1,$$

where b_1 is the constant of integration. The concentration varies linearly in the one-dimensional case. If i is positive (flow in the $+x$ direction), C decreases as x increases. Often the concentration is known at x_1 and x_2, and one wants to know the current. We can write

$$C_1 = -\frac{i}{SD}x_1 + b_1,$$

$$C_2 = -\frac{i}{SD}x_2 + b_1,$$

and solve for i:

$$i = \frac{(C_1 - C_2)SD}{x_2 - x_1}. \tag{4.38a}$$

In two dimensions

$$\frac{dC}{dr} = -\frac{i}{2\pi LD}\frac{1}{r},$$

and the solution is

$$C(r) = -\frac{i}{2\pi LD}\ln r + b_2.$$

We can again solve for the current when the concentrations are known at two different radii:

$$i=\frac{2\pi LD(C_1-C_2)}{\ln(r_2/r_1)}=\frac{2\pi LD(C_2-C_1)}{\ln(r_1/r_2)}. \quad (4.38b)$$

Diffusion in two dimensions with cylindrical symmetry has been used to model the concentration of substances in the region between two capillaries.

In three dimensions, the diffusion equation is

$$\frac{dC}{dr}=-\frac{i}{4\pi Dr^2},$$

which has a solution

$$C(r)=\frac{i}{4\pi Dr}+b_3.$$

The current in terms of the concentration is

$$i=\frac{4\pi D[C(r_1)-C(r_2)]}{1/r_1-1/r_2}. \quad (4.38c)$$

The three-dimensional case is worth further discussion, because it can help us to understand the diffusion of nutrients to a single spherical cell or the diffusion of metabolic waste products away from the cell. Consider the case in which the cell has radius $r_1=R$, the concentration at the cell surface is C_0, and the concentration at infinity is zero. Then

$$i=4\pi DC_0R, \quad (4.39a)$$

$$C(r)=\frac{C_0R}{r}, \quad (4.39b)$$

$$j_r=\frac{C_0DR}{r^2}. \quad (4.39c)$$

The particle current depends on the radius of the cell, not on R^2. This is a very important result and is not what we might naively expect. Diffusion-limited flow of solute in or out of the cell is proportional not to the cell surface area, but to the cell radius. The reason is that the particle movement is limited by diffusion in the region around the cell, and as the cell radius increases, the concentration gradient decreases. (It is possible for the rate of particle migration into the cell to be proportional to the surface area of the cell if some other process, such as transport through the cell membrane, is the rate-limiting step.)

If diffusion is toward the cell, the concentration is C_0 infinitely far away. At the cell surface, every diffusing molecule that arrives is assumed to be captured, and the concentration is zero. The solutions are then

$$i=-4\pi DC_0R, \quad (4.40a)$$

$$C(r)=C_0(1-R/r), \quad (4.40b)$$

$$j_r(r)=\frac{-C_0DR}{r^2}. \quad (4.40c)$$

4.10. EXAMPLE: STEADY-STATE DIFFUSION TO A SPHERICAL CELL AND END EFFECTS

In the preceding section we considered diffusion from infinitely far away to the surface of a spherical cell where the concentration was zero. We now add the effect of steady-state diffusion through a series of pores or channels in the cell membrane. This will lead to a very important result: it does not require very many pores per unit area in the cell membrane to "keep up with" the rate of diffusion of chemicals toward or away from the cell. The result is important for understanding how cells acquire nutrients, how bacteria move in response to chemical stimulation (chemotaxis), and how the leaves of plants function.

To develop the model we need one more result: the current due to diffusion from a disk of radius a where the concentration is C_1 to a plane far away where the concentration is C_2. The disk is embedded in the surface of an impervious plane as shown in Fig 4.14, so particles cannot cross to the region behind the disk. The current is

$$i=4Da(C_1-C_2). \quad (4.41)$$

It is proportional to the radius of the disk, not its surface area. (Obtaining this result requires solving the diffusion equation in three dimensions.)

Consider diffusion through a pore of radius R_p which pierces a membrane of thickness ΔZ, including diffusion in the medium on either side of the membrane (Fig. 4.15). If the material on either side were well stirred, there would be a uniform concentration C_1 on the left and C_4 on the right. Because it is not stirred, there is diffusion in the exterior fluid. Let C_1 and C_4 be measured far away, and call the concentrations at the ends of the pore C_2 on the left and C_3 on the right.

Equation (4.38a) gives the diffusion flux within the pore

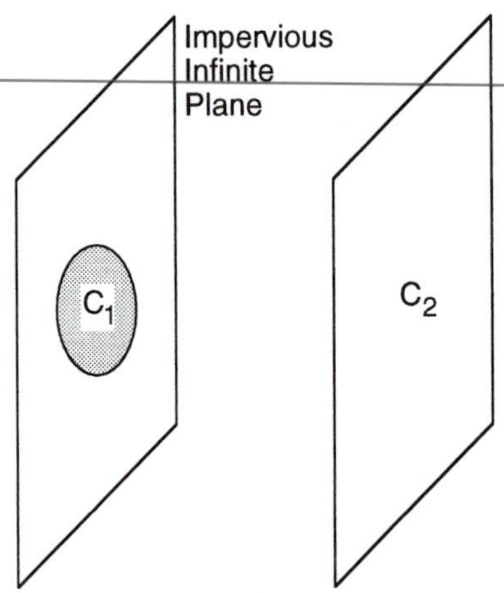

FIGURE 4.14. The diffusion flux from the disk of radius a and concentration C_1 to the infinite sheet where the concentration is C_2 is given by $i=4Da(C_1-C_2)$.

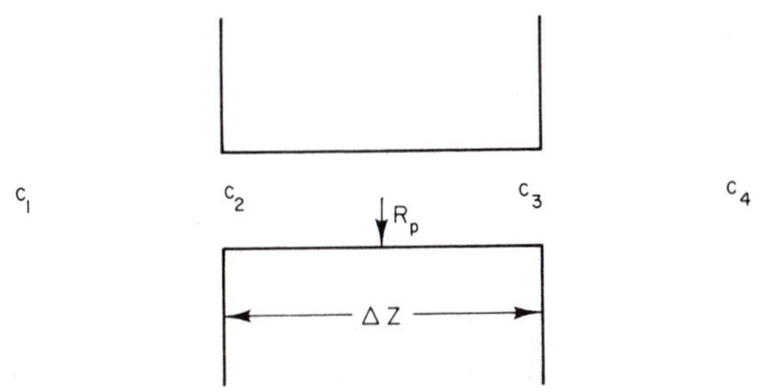

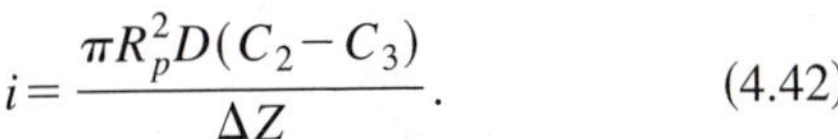

FIGURE 4.15. End effects in diffusion through a pore.

$$i = \frac{\pi R_p^2 D(C_2 - C_3)}{\Delta Z}. \tag{4.42}$$

Diffusion from C_1 to C_2 is given by Eq. (4.41). It is

$$i = 4DR_p(C_1 - C_2), \tag{4.43}$$

while from C_3 to C_4, it is

$$i = 4DR_p(C_3 - C_4). \tag{4.44}$$

In the steady state, there is no buildup of particles and i is the same in each region. We can solve Eqs. (4.42)–(4.44) simultaneously to relate i to concentrations C_1 and C_4:

$$i = \frac{\pi R_p^2 D}{\Delta Z + 2\pi R_p/4}(C_1 - C_4). \tag{4.45}$$

This has the same form as Eq. (4.42), except that the membrane thickness has been replaced by an effective thickness

$$\Delta Z' = \Delta Z + 2\frac{\pi R_p}{4}. \tag{4.46}$$

An extra length $\pi R_p/4$ has been added at each end to correct for diffusion in the unstirred layer on each side of the pore. This correction is important when the pore length is less than two or three times the pore radius.

Now consider diffusion in or out of the spherical cell shown in Fig. 4.16. The radius of the cell is B. The membrane has thickness ΔZ and is pierced by a total of N pores, each of radius R_p. Within the cell we do not know the details of the concentration distribution, since they depend on what sort of chemical reactions are taking place and where. But we will assume that at the radius where diffusion to the pores becomes important, the concentration is C_1. At the inner face of each pore it is C_2, at the outer face it is C_3, and over an approximately spherical surface of radius B' it is C_4. Far away, the concentration is C_5. As a result, there are four separate regions in which we must consider diffusion. The first is from C_1 to the opening of each pore; the second is through the pore; third, there is diffusion from the outer face of each pore to C_4; and, finally, there is diffusion from the spherical object of radius B' to the surrounding medium.

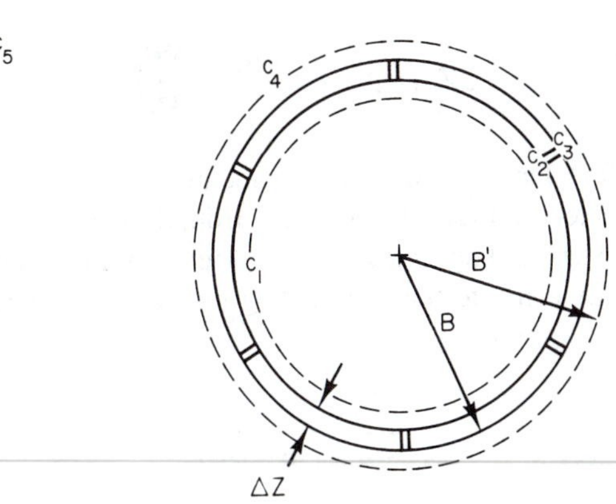

FIGURE 4.16. Diffusive end effects for a spherical cell pierced by pores.

The first three processes are taken into account by applying the end correction to each end of the pores. The flow through one pore is [using Eq. (4.45)]

$$i_{\text{pore}} = \frac{\pi R_p^2 D}{\Delta Z'}(C_1 - C_4), \tag{4.47}$$

where $\Delta Z'$ is given by Eq. (4.46). Since there are N pores in all, the total flow through the cell wall is

$$i_{\text{cell}} = N i_{\text{pore}} = \frac{N\pi R_p^2 D}{\Delta Z'}(C_1 - C_4). \tag{4.48}$$

The diffusion from C_4 to infinity is given by Eq. (4.38c).

$$i_{\text{cell}} = 4\pi DB'(C_4 - C_5), \tag{4.49}$$

where B' is the effective radius for diffusion to the surrounding medium. It is somewhat larger than B. If we equate Eqs. (4.48) and (4.49), we get

$$N\pi R_p^2 DC_1 - N\pi R_p^2 DC_4$$
$$= 4\pi DB'\Delta Z' C_4 - 4\pi DB'\Delta Z' C_5$$

from which

$$C_4 = \frac{NR_p^2 C_1 + 4B'\Delta Z' C_5}{NR_p^2 + 4B'\Delta Z'}.$$

Substituting this back in Eq. (4.49), we get

$$i_{\text{cell}} = \frac{4\pi DB' NR_p^2}{NR_p^2 + 4B'\Delta Z'}(C_1 - C_5), \tag{4.50}$$

This can be rewritten as

$$i_{\text{cell}} = \frac{N\pi R_p^2 D}{\Delta Z_{\text{eff}}}(C_1 - C_5), \tag{4.51}$$

where

$$\Delta Z_{\text{eff}} = \Delta Z + 2\frac{\pi R_p}{4} + \frac{NR_p^2}{4B'}. \qquad (4.52)$$

The first term in ΔZ_{eff} is the membrane thickness. The second term corrects for diffusion from the end of each pore to the surrounding fluid; the last corrects for diffusion away from the cell into the surrounding medium. The third term can be expressed as

$$\frac{NR_p^2}{4B'} = \frac{B}{B'} Bf,$$

where

$$f = \frac{N\pi R_p^2}{4\pi B^2} \qquad (4.53)$$

is the fraction of the cell surface occupied by pores. To estimate B/B', note that B' must be increased by $\pi R_p/4$ because of end effects. The concentration varies near the pores and smooths out further away, so B' must also be increased by an amount roughly equal to l, the spacing of the pores. There are $N/4\pi B^2$ pores per m^2, so $l \approx R_p(\pi/f)^{1/2}$. Assuming both corrections are negligible, the effective pore length is

$$\Delta Z_{\text{eff}} = \Delta Z + 2\left(\frac{\pi R_p}{4}\right) + Bf. \qquad (4.54)$$

Equations (4.51)–(4.54) treat the problem as diffusion through a collection of N pores, corrected for diffusion outside the pore by increasing the length of the pore. It is also useful to write these results as the equation for diffusion around a sphere [Eq. (4.39)], corrected for the diffusion through the cell wall. Writing it in this form gives us insight into how much of the cell wall must be occupied by pores for efficient particle transfer. Solve Eq. (4.53) for NR_p^2 and substitute the result in Eq. (4.50). The result is

$$i_{\text{cell}} = \frac{4\pi DB'B^2 f(C_1 - C_5)}{B^2 f + B'\Delta Z'}$$

$$= 4\pi BD(C_1 - C_5)\frac{B'}{B}\frac{f}{f + \frac{B'}{B}\frac{\Delta Z'}{B}}. \qquad (4.55)$$

This has the form of diffusion to the sphere multiplied by a correction factor. With B'/B again approximated by unity, the correction factor is

$$\frac{f}{f + \Delta Z'/B}.$$

The correction factor is zero when f is zero and becomes nearly unity when the entire cell surface is covered by pores. (Convince yourself that when $f=1$, the correction corresponds to a cell radius $B - \Delta Z'$ instead of B.)

The cell will receive half the maximum possible diffusive flow when the fraction is

$$f = \frac{\Delta Z'}{B}.$$

For a typical cell with $B=5$ μm and $\Delta Z=5$ nm, $f=0.001$. This is a surprisingly small number, but it means that there is plenty of room on the cell surface for different kinds of pores. There are two ways to understand why this number is so small. First, we can regard the ratio of concentration difference to flow as a resistance, analogous to electrical resistance. The total resistance from the inside of the cell to infinity is made up of the resistance from the outside of the cell to infinity plus the resistance of the parallel combination of N pores. Once the resistance of this parallel combination is equal to the resistance from the cell to infinity, adding more pores in parallel does not change the overall resistance very much. The second way to look at it is in terms of the random walks of the diffusing solute molecules. Once a solute molecule has diffused into the neighborhood of the cell, it undergoes many random walks. When it strikes the cell wall, it wanders away again, to return shortly and strike the cell wall someplace else. If the first contact is not at a pore, there is a high probability that the particle will enter a pore on a subsequent contact with the surface.

The same sort of analysis that we have made here can be applied to a plane surface area, such as the underside of a leaf [Meidner and Mansfield (1968)] and to a cylindrical geometry, such as a capillary wall.

The analysis can also be applied to the problem of bacterial chemotaxis—the movement of bacteria along concentration gradients. This problem has been discussed in detail by Berg and Purcell (1977).[11] The cell detects a chemical through some sort of chemical reaction between the chemical and the cell. Suppose that the reaction takes place between the chemical and a binding site of radius R_p on the surface of the cell. We want to know what fraction of the surface area of the cell must be covered by binding sites. This is similar to the diffusion problem of Eq. (4.55), except that if the binding site is on the surface of the cell, there is no diffusion through a pore of length ΔZ. The effective pore length $\Delta Z'$ is just the end correction for one end of the pore, $\pi R_p/4$. Half of the maximum possible flow to the binding site occurs when

$$f = \frac{\pi R_p}{4B}.$$

A typical bacterium might have a radius $B=1$ μm; the binding site might have a radius of a few atoms or 1 nm. With these values $f=7.9\times10^{-4}$. The number of sites would

[11] See also Berg (1975, 1983) and Purcell (1977).

be $f4\pi B^2/\pi R_p^2 = \pi B/R_p = 3000$. There is plenty of room on the cell surface for many different binding sites, each specific for a particular chemical.

An *Escherichia coli* cell typically travels 10–20 body lengths per second. It detects concentration gradients as changes with time. Because of this, Berg and Purcell concluded that a uniform distribution of chemoreceptors over the surface of the cell would be optimum. It would give the highest probability of capture of a chemical molecule that wandered near the cell. However, recent studies of *E. coli* have shown that the receptors are located near the poles of the cell [Maddock and Shapiro (1993); see also the comment by Parkinson and Blair (1993), who point out that the reduced efficiency of sensors could make sense if "eating" or transport into the cell is more important than "smelling"].

The Berg–Purcell model has been extended to provide a time-dependent solution and allow the receptors not to be perfectly absorbing [Zwanzig and Szabo (1991)] and also to have a process in which the molecules attach to the membrane and then diffuse in the two-dimensional membrane surface [Wang, Gou, and Axelrod (1992); Axelrod and Wang (1994)].

4.11. EXAMPLE: A SPHERICAL CELL PRODUCING A SUBSTANCE

Here is a simple model that extends the arguments of Sec. 4.9 to develop a steady-state solution for a spherical cell excreting metabolic products. The cell has radius R. The concentration of some substance inside the cell is $C(r)$, independent of time t and the spherical coordinate angles θ and ϕ. (Spherical coordinates are described in Appendix L and were also used in Sec. 3.10.) The substance is produced at a constant rate Q particles per unit volume per second throughout the cell and leaves through the walls of the cell at a constant fluence rate $j(R)$, independent of t, θ, and ϕ. Assume that all transport is by pure diffusion and the diffusion constant for this substance is D everywhere inside and outside the cell. The material inside the cell is not well stirred. For this model we assume that the cell membrane does not affect the transport process. We could make the model more complicated by introducing the features described in Sec. 4.10. With these assumptions, the cell can be modeled as an infinite homogeneous medium with diffusion constant D that contains a spherical region producing material at rate Q per unit volume per second.

We first find the concentration $C(r)$ inside and outside the cell by using a technique that only works because of the spherical symmetry. We use the continuity equation in the form Eq. (4.9b). Because the concentration is not changing with time, the total amount of material flowing through a spherical surface of radius r is equal to the amount produced within that sphere. For $r<R$

$$4\pi r^2 j(r) = 4\pi r^3 Q/3,$$

$$j(r) = Qr/3.$$

For $r>R$

$$4\pi r^2 j(r) = 4\pi R^3 Q/3,$$

$$j(r) = QR^3/3r^2.$$

Using the fact that $j(r) = -D\,dC/dr$, we obtain for $r<R$

$$\frac{dC}{dr} = -\frac{Q}{3D}r,$$

$$C(r) = -\frac{Qr^2}{6D} + b_1,$$

where b_1 is the constant of integration. For $r>R$,

$$\frac{dC}{dr} = -\frac{QR^3}{3Dr^2},$$

$$C(r) = \frac{QR^3}{3Dr} + b_2.$$

The fact that the concentration must be zero far from the cell means that $b_2=0$. Matching the two expressions at $r=R$ gives

$$-QR^2/6D + b_1 = QR^2/3D,$$

$$b_1 = QR^2/2D,$$

so that

$$C(r) = \begin{cases} \dfrac{Q}{6D}(3R^2 - r^2), & r \leqslant R \\ \dfrac{QR^3}{3Dr}, & r \geqslant R. \end{cases}$$

The other method is more general and can be extended to problems that do not have spherical symmetry. We find solutions to Fick's second law, modified to include the production term Q and with the concentration not changing with time:

$$0 = \frac{\partial C}{\partial t} = D\nabla^2 C + Q,$$

$$\nabla^2 C = -\frac{Q}{D}.$$

In spherical coordinates [Appendix L; Schey (1973)] this is

$$\frac{1}{r^2}\frac{d}{dr}\left(r^2\frac{dC}{dr}\right) + \frac{1}{r^2 \sin\theta}\frac{\partial}{\partial\theta}\left[\sin\theta\left(\frac{\partial C}{\partial\theta}\right)\right] + \frac{1}{r^2\sin^2\theta}\left(\frac{\partial^2 C}{\partial\phi^2}\right) = -\frac{Q}{D}.$$

Since there is no angular dependence, we have separate equations for each domain:

$$\frac{1}{r^2}\frac{d}{dr}\left(r^2\frac{dC}{dr}\right)=\begin{cases}-\frac{Q}{D}, & r<R\\ 0, & r>R.\end{cases}$$

It is necessary to solve each equation in its domain, and then at the boundary require that C be continuous and also that j and therefore dC/dr be continuous. For $r<R$ we get the following (b_1 and b_2 are constants of integration):

$$r^2\frac{dC}{dr}=-\frac{Qr^3}{3D}+b_1,$$

$$\frac{dC}{dr}=-\frac{Qr}{3D}+\frac{b_1}{r^2},$$

$$C(r)=-\frac{Qr^2}{6D}-\frac{b_1}{r}+b_2.$$

Since the concentration is finite at the origin, $b_1=0$:

$$C(r)=b_2-\frac{Qr^2}{6D}, \quad r<R.$$

For $r>R$ we can use the general solution with $Q=0$ and different constants:

$$C(r)=-\frac{b_1'}{r}+b_2'.$$

Far away the concentration is zero, so $b_2'=0$. Matching dC/dr at the boundary gives

$$-\frac{QR}{3D}=\frac{b_1'}{R^2}, \quad b_1'=-\frac{QR^3}{3D}.$$

Matching $C(r)$ at the boundary gives

$$-\frac{QR^2}{6D}+b_2'=-\frac{b_1'}{R}.$$

Putting all of this together gives the same expression for the concentration we had earlier. This technique is a bit more cumbersome, but there are many mathematical tools to extend this technique to cases where there is not spherical symmetry and where Q is a function of position. These advanced techniques can also be used when C is changing with time.

4.12. DRIFT AND DIFFUSION IN ONE DIMENSION

The particle fluence rate due to diffusion on one dimension is $j_{\text{diff}}=-D(\partial C/\partial x)$. That of particles drifting with velocity v is $j_{\text{drift}}=vC$. The total flux density or fluence rate is the sum of both terms:

$$j_s=-D\frac{\partial C}{\partial x}+vC. \tag{4.56}$$

The homogeneous ($j=0$) solution was discussed in Sec. 4.7, where cancellation of these two terms in equilibrium was used to derive the relationship between diffusion constant and viscosity. Using the techniques of Appendix F, we can write the homogeneous solution as

$$C(x)=Ae^{(v/D)x}. \tag{4.57}$$

This can be used to solve the problem of j_s=const when the concentration is C_0 at $x=0$ and C_0' at $x=x_1$, and j_s is constant. $C(x)$ must vary in such a way that the total flux density, the sum of the diffusive and drift terms, is constant. If the concentration is high, then the drift flux density is large and the concentration gradient must be small. If the concentration is small, the diffusive flux, and hence the gradient, must be large. To develop a formal solution, write Eq. (4.56) as

$$\frac{dC}{dx}-\frac{1}{\lambda}C=-\frac{j_s}{D}, \tag{4.58}$$

where $\lambda=D/v$ has the dimensions of length and can be interpreted as the distance over which diffusion is important. If the velocity is zero, diffusion is important everywhere and $\lambda=\infty$. If the velocity is very large, $\lambda\rightarrow 0$. Since v can be either positive or negative, so can λ. A particular solution to Eq. (4.58) is

$$C(x)=\frac{\lambda j_s}{D}=\frac{j_s}{v}.$$

The general solution is the sum of the particular solution and the homogeneous solution, Eq. (4.57):

$$C(x)=Ae^{x/\lambda}+\frac{j_s}{v}. \tag{4.59}$$

The situation is slightly different than what we encountered in Chap. 2. We must determine two constants, A and j_s, given the two concentrations C_0 and C_0'. Writing Eq. (4.59) for $x=0$ and for $x=x_1$, we obtain

$$C_0=A+\frac{j_s}{v},$$
$$C_0'=Ae^{x_1/\lambda}+\frac{j_s}{v}. \tag{4.60}$$

Subtracting these gives

$$C_0'-C_0=A(e^{x_1/\lambda}-1),$$
$$A=(C_0'-C_0)/(e^{x_1/\lambda}-1). \tag{4.61}$$

This can be combined with either of Eqs. (4.60) to give

$$j_s=\frac{C_0e^{x_1/\lambda}-C_0'}{e^{x_1/\lambda}-1}v. \tag{4.62}$$

We can also substitute Eqs. (4.61) and (4.62) in (4.59) to obtain an expression for $C(x)$. The result is

$$C(x)=\frac{C_0(e^{x_1/\lambda}-e^{x/\lambda})+C_0'(e^{x/\lambda}-1)}{e^{x_1/\lambda}-1}. \tag{4.63}$$

We will discuss the implications of this equation below. Let us first determine the average concentration between $x=0$ and $x=x_1$. The average concentration is defined by

$$\bar{C}=\frac{1}{x_1}\int_0^{x_1}C(x)dx. \tag{4.64}$$

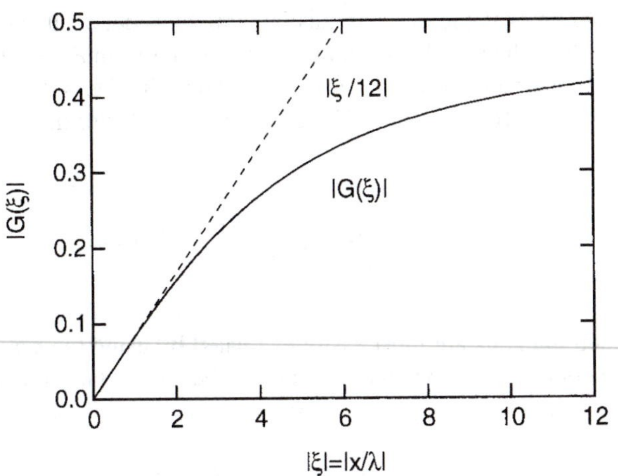

FIGURE 4.17. The correction factor $G(\xi)$ used in Eq. (4.68). The dashed line is the approximation $G(\xi)=\xi/12$, which is valid for small ξ and is used in Eq. (4.67).

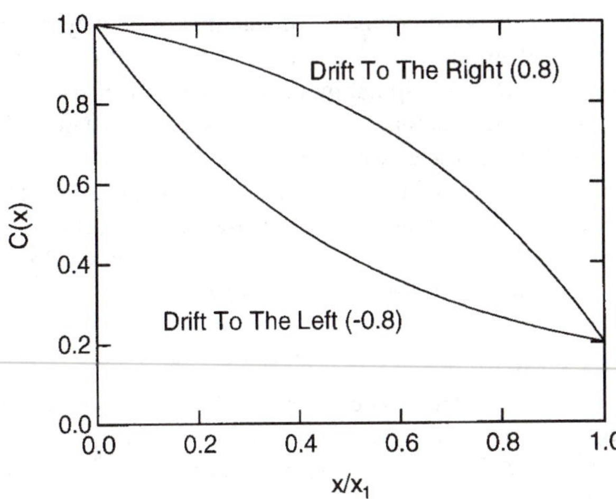

FIGURE 4.18. Concentration profile for combined drift and diffusion. The concentration is 1.0 on the left and 0.2 on the right. For $x_1/\lambda=x_1 v/D=0.8$, drift and diffusion are both to the right. As the concentration falls, the magnitude of the gradient increases. For $x_1/\lambda=x_1 v/D=-0.8$ drift opposes diffusion. As the concentration falls, so does the magnitude of the gradient.

While one could integrate this directly, it is much easier to integrate Eq. (4.56) from 0 to x_1:

$$-D\int_0^{x_1}\left(\frac{dC}{dx}\right)dx+v\int_0^{x_1}C(x)dx=+j_s\int_0^{x_1}dx.$$

The first term is $-D(C_0'-C_0)$. The second is $vx_1\bar{C}$. The third is $j_s x_1$. The equation can therefore be rewritten as

$$v\bar{C}=\frac{D(C_0'-C_0)}{x_1}+j_s. \tag{4.65}$$

Substituting Eq. (4.62) for j_s gives the average concentration

$$\bar{C}=\frac{C_0 e^{x_1/\lambda}-C_0'}{e^{x_1/\lambda}-1}-\frac{\lambda}{x_1}(C_0-C_0'). \tag{4.66}$$

The exponentials can be expanded to give an approximate expression[12]

$$\bar{C}=\tfrac{1}{2}(C_0+C_0')+\frac{x_1}{\lambda}\frac{1}{12}(C_0-C_0'). \tag{4.67}$$

For larger values of x_1/λ, the mean can be written

$$\bar{C}=\frac{1}{2}(C_0+C_0')+(C_0-C_0')G\left(\frac{x_1}{\lambda}\right). \tag{4.68}$$

The correction factor $G(x_1/\lambda)=G(\xi)$, given by

$$G(\xi)=\frac{1}{2}\frac{1+e^{\xi}}{e^{\xi}-1}-\frac{1}{\xi} \tag{4.69}$$

is plotted in Fig. 4.17. The function is odd, and only values for $\xi\geqslant 0$ are shown. For $\xi=0$ ($\lambda=\infty$, pure diffusion), the average concentration is $(C_0+C_0')/2$.

[12]See Levitt (1975, p. 537). For $x_1/\lambda=1.5$, this approximation is within 1%. For $x_1/\lambda=2.5$, the error is about 6%.

Figure 4.18 shows the concentration profile calculated from Eq. (4.63). The concentration is 5 times larger on the left, so diffusion is from left to right. When $x_1/\lambda=x_1 v/D=0.8$, drift is also from left to right. As the concentration falls, the magnitude of the gradient rises, so that the sum of the diffusive and drift fluxes remains the same. When $x_1/\lambda=-0.8$, drift is opposite to diffusion. Therefore, both the concentration and the magnitude of the gradient must rise and fall together to keep total flux density constant.

Equation (4.65) can be rewritten as

$$j_s=\frac{-D(C_0'-C_0)}{x_1}+v\bar{C}. \tag{4.70}$$

This can be interpreted as meaning that the fluence rate is given by the sum of a diffusion term with the average concentration gradient and a drift term with the average concentration. However, the discussion in the preceding paragraph showed that there is actually a continuous change of the relative size of the diffusion and drift terms for different values of x.

4.13. A GENERAL SOLUTION FOR THE PARTICLE CONCENTRATION AS A FUNCTION OF TIME

If $C(x,0)$ is known for $t=0$, it is possible to use the result of Sec. 4.8 to determine $C(x,t)$ at any later time. The key to doing this is that if $C(x,t)dx$ is the number of particles in the region between x and $x+dx$ at time t, it may be be interpreted as the probability of finding a particle in the interval (x,dx) multiplied by the total number of particles. The spreading Gaussian then represents the spread of probability that a particle is between x and $x+dx$.

If a particle is definitely at $x=\xi$ at $t=0$, then $\sigma(0)=0$. The particle cannot remain there because of equipartition of energy: collisions cause it to acquire a mean square velocity $3k_BT/m$ and move. At some later time

$$\sigma(t)=(2Dt)^{1/2}. \tag{4.71}$$

Define $P(\xi,0;x,t)dx$ to be the probability that a particle has diffused to a location between x and $x+dx$ at time t, if it was at $x=\xi$ when $t=0$. This probability is given by Eq. (4.25), except that the distance it has diffused is now $x-\xi$ instead of x. The variance $\sigma^2(t)$ is given by Eq. (4.71). The result is

$$P(\xi,0;x,t)dx=\frac{1}{\sqrt{4Dt\pi}}\,e^{-(x-\xi)^2/4Dt}dx. \tag{4.72}$$

The number of particles initially between $x=\xi$ and $x=\xi+d\xi$ is the concentration per unit length times the length of the interval $C(\xi,0)\;d\xi$, as shown in Fig. 4.19.

The particles can diffuse in either direction. At a later time t, the average number between x and $x+dx$ that came originally from between $x=\xi$ and $x=\xi+d\xi$ is the original number in $(\xi,d\xi)$ times the probability that each one got from there to x. This number is a differential of a differential, $d[C(x,t)\;dx]$, because it is only that portion of the particles in dx that came from the interval $d\xi$:

$$d[C(x,t)\;dx]=C(\xi,0)d\xi\,\frac{1}{\sqrt{4\pi Dt}}\,e^{-(x-\xi)^2/4Dt}dx.$$

To get $C(x,t)\;dx$, it is necessary to integrate over all possible values of ξ:

$$C(x,t)\;dx=\frac{1}{\sqrt{4\pi Dt}}\left(\int_{-\infty}^{\infty}C(\xi,0)e^{-(x-\xi)^2/4Dt}d\xi\right)dx. \tag{4.73}$$

This equation can be used to find $C(x,t)$ at any time, provided that $C(x,t)$ was known at some earlier time. The factor that multiplies $C(\xi,0)$ in the integrand is called the influence function or Green's function for the diffusion problem; it gives the relative weighting of $C(\xi,0)$ in contributing to the later value $C(x,t)$.

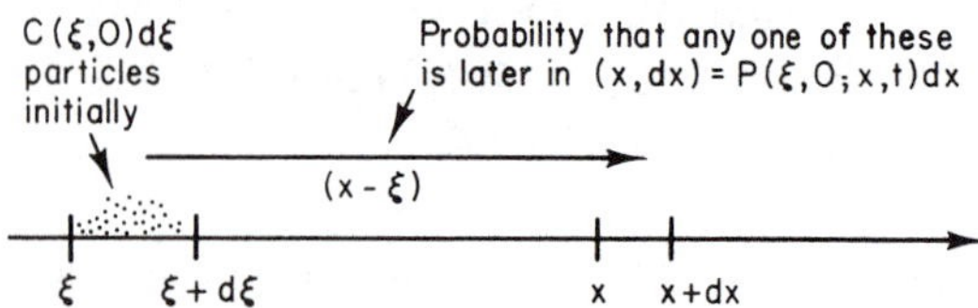

FIGURE 4.19. Diffusion from ξ to x.

As an example of using this integral, consider a situation in which the initial concentration is constant from $\xi=-\infty$ to $\xi=0$ and zero for all positive ξ, as shown in Fig. 4.20. At $t=0$ the diffusion starts. The concentration at later times is given by

$$C(x,t)=\frac{C_0}{\sqrt{4\pi Dt}}\int_{-\infty}^{0}e^{-(x-\xi)^2/4Dt}d\xi.$$

Such integrals are most easily evaluated by using the *error function* that is tabulated in many mathematical tables. It is defined by

$$\text{erf}(z)=\frac{2}{\pi^{1/2}}\int_0^z e^{-t^2}dt. \tag{4.74}$$

The error function is plotted in Fig. 4.21. One must be careful in using tables, for other functions tabulated are related to the error function but differ in normalization constants and in the limits of integration.

To use the error function in evaluating the integral in Eq. (4.73), make the substitution $s=(x-\xi)/(4Dt)^{1/2}$. The integral becomes

$$C(x,t)=\frac{-C_0}{\sqrt{4\pi Dt}}\int_{\infty}^{x/\sqrt{4Dt}}e^{-s^2}\sqrt{4Dt}\;ds.$$

Since

$$\int_A^B f(x)\;dx=\int_0^B f(x)\;dx+\int_A^0 f(x)\;dx$$
$$=\int_0^B f(x)\;dx-\int_0^A f(x)\;dx,$$

this can be written as

$$C(x,t)=\frac{-C_0}{\sqrt{\pi}}\left(\int_0^{x/\sqrt{4Dt}}e^{-s^2}ds-\int_0^{x}e^{-s^2}ds\right)$$

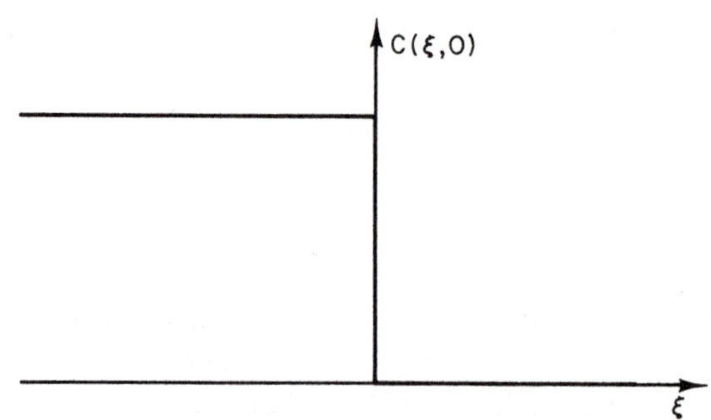

FIGURE 4.20. The initial concentration is constant to the left of the origin and zero to the right of the origin.

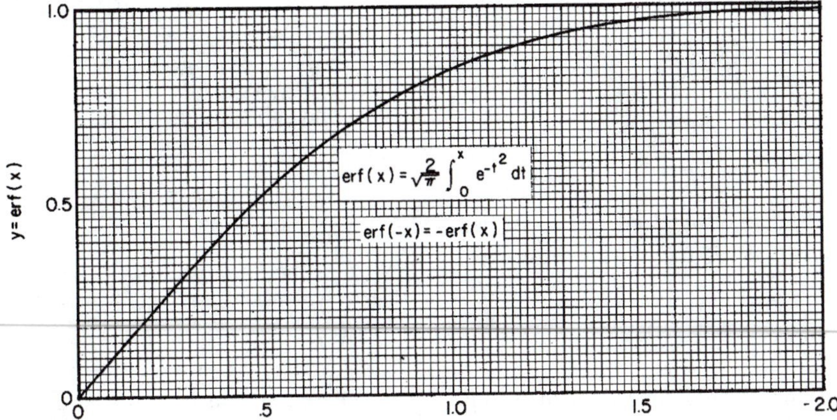

FIGURE 4.21. Plot of the error function erf(x).

$$= \frac{-C_0}{2}\left[\text{erf}\left(\frac{x}{\sqrt{4Dt}}\right)-1\right]=\frac{C_0}{2}\left[1-\text{erf}\left(\frac{x}{\sqrt{4Dt}}\right)\right]. \tag{4.75}$$

The plot in Fig. 4.22 shows how the initially sharp concentration step becomes more diffuse with passing time. Quantitative measurements of the concentration can be used to determine D. Villars and Benedek (1974, pp. 2–57) discuss some experiments to verify the solution we have obtained above and to determine D.

4.14. DIFFUSION AS A RANDOM WALK

The spreading solution to the one-dimensional diffusion equation that we verified can also be obtained by treating the motion of a molecule as a series of independent steps either to the right or to the left along the x axis. (The same treatment can be extended to three dimensions, but we will not do so.) The derivation gives us a somewhat simplified molecular picture of diffusion. The derivation also provides an opportunity to see how the Gaussian distribution approximates the binomial distribution. This section is not necessary to understand the rest of Chaps. 4 and 5, and you should tackle it only if you are familiar with the binomial and Gaussian probability distributions (Appendixes H and I). The model is more restrictive than the diffusion equation derived above, since the latter is the linear approximation to the transport problem.

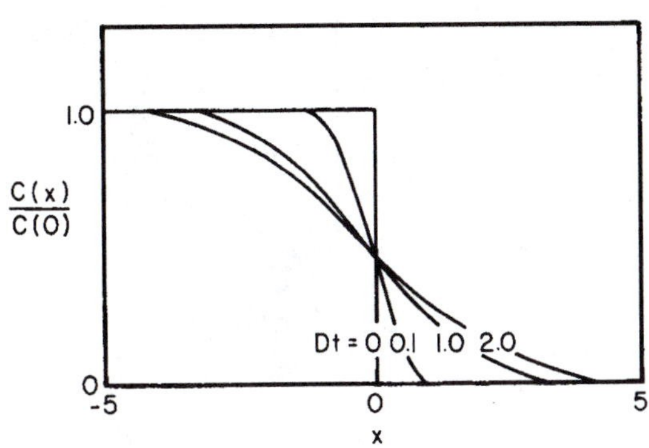

FIGURE 4.22. The spread of an initially sharp boundary due to diffusion.

We use a simplified model in which the diffusing particle always moves in steps of length λ (the mean free path), either in the $+x$ or $-x$ direction. Let the total number of steps taken by the particle be N, of which n are to the right and n' are to the left: $N=n+n'$. Also let $m=n-n'$. The particle's net displacement in the $+x$ direction is then

$$n\lambda - n'\lambda = (n-n')\lambda = m\lambda.$$

Since the steps are independent and a step to the left or right is equally likely ($p=\frac{1}{2}$), the probability of having a displacement $m\lambda$ is given by the binomial probability $P(n;N)$:

$$P(n;N)=\frac{N!}{(n!)(N-n)!}\left(\frac{1}{2}\right)^n\left(\frac{1}{2}\right)^{n'}. \tag{4.76}$$

Since this problem is analogous to throwing a coin, and we know that on the average we get the same number of heads (steps to the left) as tails (steps to the right), we know that the distribution is centered at $n=n'$ or $m=0$. We also know [Eq. (H4)] that the variance in n is given by

$$\overline{n^2}-(\bar{n})^2=Npq=\frac{N}{4}.$$

Since $n=N/2$, this says that

$$\overline{n^2}=\frac{N}{4}+\frac{N^2}{4}.$$

However, we need the variance in m, $\overline{m^2}-(\bar{m})^2$. To obtain it, we write

$$m=2n-N$$

and

$$m^2 = 4n^2 + N^2 - 4nN.$$

Therefore,

$$\overline{m^2} = \overline{4n^2} + N^2 - 4N\bar{n} = N.$$

The variance of the distribution of displacement x is equal to the step length λ times the variance in the number of steps:

$$\sigma^2 = \overline{x^2} = \lambda^2 \overline{m^2} = \lambda^2 N.$$

The number of steps is the elapsed time divided by the collision time $N = t/t_c$. Therefore,

$$\sigma^2 = \frac{\lambda^2 t}{t_c}.$$

Comparing this with Eq. (4.70), we identify $D = \lambda^2/2t_c$, so that

$$\sigma^2 = 2Dt. \qquad (4.77)$$

We have shown that this simple model gives a distribution with fixed mean which spreads with a variance proportional to t. We now must show that the shape is Gaussian. Appendix I shows that the Gaussian is an approximation to the binomial distribution in the limit of large N. Since $\sigma_n^2 = N/4$ and $\bar{n} = N/2$, Eq. (H4) can be used to write

$$P(n) = \left(\frac{2\pi N}{4}\right)^{-1/2} e^{-(n-N/2)^2/(2N/4)}.$$

This can be rewritten in terms of the net number of steps to the right, since $m = n - n' = 2n - N$:

$$P(m) = \left(\frac{2}{\pi N}\right)^{1/2} e^{-m^2/2N}.$$

Note that only every other value of m is allowed. Since $m = 2n - N$, m goes in steps of 2 from $-N$ to N as n goes from 0 to N.

To write the probability distribution in terms of x and t, refer to Fig. 4.23. The spacing between each allowed value of x is 2λ, so that the number of allowed values of m in interval $(x, x+dx)$ is $dx/2\lambda$. Therefore,

$$P(x)dx = P(m)(dx/2\lambda),$$

$$P(x) = \sqrt{\frac{2}{\pi N 4\lambda^2}}\, e^{-m^2/2N}.$$

With the substitutions $m = x/\lambda$ and $N = t/t_c$, this becomes

$$P(x,t) = \sqrt{\frac{t_c}{2\pi\lambda^2 t}}\, e^{-x^2(t_c/2\lambda^2 t)}.$$

With the substitutions $D = \lambda^2/2t_c$ and $C(x,t) = C(0)P(x,t)$, we obtain Eq. (4.25).

The result of Eq. (4.71) is easily extended to two dimensions. Imagine that a total of N steps are taken, half in the x direction and half in the y direction. The $\sigma_x^2 = \sigma_y^2 = \lambda^2(N/2)$. If $r^2 = x^2 + y^2$, $\sigma_r^2 = \sigma_x^2 + \sigma_y^2 = \lambda^2 N$. We still define D in any direction as $\lambda^2/2t_c$, where t_c is the time between steps in that direction. After a total time t, N steps have been taken, but only half of them were in, say, the x direction. Therefore $t_c = 2t/N$. Therefore

$$\sigma_r^2 = \sigma_x^2 + \sigma_y^2 = 4Dt \quad \text{(two dimensions)}. \qquad (4.78)$$

A similar argument in three dimensions gives

$$\sigma_r^2 = \sigma_x^2 + \sigma_y^2 + \sigma_z^2 = 6Dt \quad \text{(three dimensions)}. \qquad (4.79)$$

Figure 4.24 shows the result of a computer simulation of a two-dimensional random walk. A random number is selected to determine whether to step one pixel to the left, up, right, or down—each with the same probability. The trail for 4000 steps is shown in Fig. 4.24(a). The results of continuing for 40 000 steps are shown in Fig. 4.24(b). Note how the particle wanders around one region of space and then takes a number of steps in the same direction to move someplace else. The particle trajectory is "thready." It does not cover space uniformly. A uniform coverage would be very nonrandom. It is only when many particles are considered that a Gaussian distribution of particle concentration results.

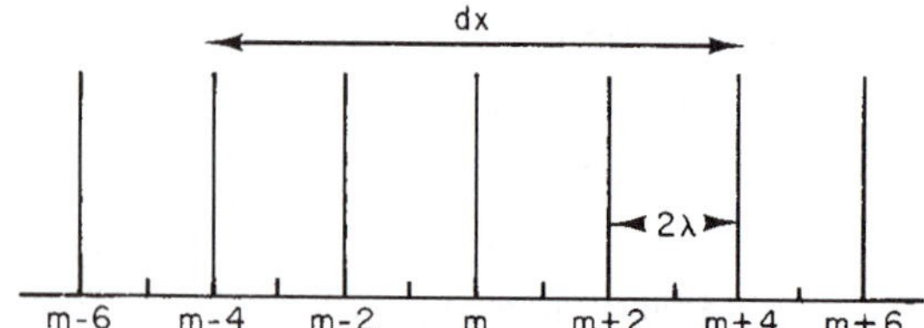

FIGURE 4.23. Relationship between the values of x and the allowed values of m. Every other value of m is missing.

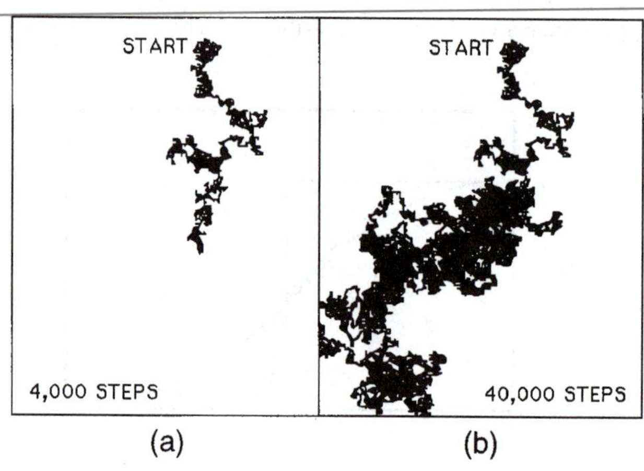

FIGURE 4.24. (a) Trail of a particle after 4000 steps. (b) Trail after additional steps to total 40 000.

Both results in Fig. 4.24 were for the same sequence of random numbers. A computer simulation with 328 runs of 10 000 steps each gave $\overline{x}=-3.3$, $\sigma_x^2=5142$, $\overline{y}=-8.2$, $\sigma_y^2=4773$, and $\overline{x^2+y^2}=10\,027$. The expected values are, respectively, 0, 5 000, 0, 5 000, and 10 000.

SYMBOLS USED IN CHAPTER 4

Symbol	Use	Units	First used on page
a, a_1, a_2	Particle radius	m	83
b_1, b_2, b_3	Constants		91
f	Fraction of cell surface area		94
g	Gravitational acceleration	m s^{-2}	82
$\mathbf{g}$	Force	N	84
i	Particle current	s^{-1}	91
$j, \mathbf{j}, j_s$	Solute fluence rate	m^{-2} s^{-1}	78
j_{drift}	Solute fluence rate due to drift velocity	m^{-2} s^{-1}	86
$\mathbf{j}_m$	Mass fluence rate	kg m^{-2} s^{-1}	78
j_n	Component of **j** normal to surface	m^{-2} s^{-1}	80
$\mathbf{j}_v$	Volume fluence rate	m s^{-1}	78
j_x, j_y, j_z	Components of **j**		79
k_B	Boltzmann's constant	J K^{-1}	82
m	Mass	kg	82
m	$n-n'$		99
$\hat{\mathbf{n}}$	Unit vector normal to a surface		80
n, n'	Number of steps to right, left		99
p, q	Probabilities		99
r	Distance, radius	m	90
s	Dummy variable		83
t	Time	s	78
t_c	Collision time	s	83
u	Energy of a particle	J	82
$v, \mathbf{v}$	Velocity	m s^{-1}	82
x, y, z	Cartesian coordinates	m	78
A	Constant		96
B, B'	Cell radius	m	93
C, C_s	Concentration	m^{-3}	79
D	Diffusion constant	m^2 s^{-1}	85
$F, \mathbf{F}, \mathbf{F}_{\text{ext}}$	Force	N	84
G	Correction factor for average concentration		97
K	Thermal conductivity	J K^{-1} m^{-1} s^{-1}	86
L	Length	m	91
M	Mass	kg	81
M	Molecular weight		88
N, N_0	Number of molecules		79
N	Number of pores on cell surface		93
N	Number of steps in a random walk		99
N_A	Avogadro's number		83
P	Rate of energy production (power)	W	80
P	Probability		82
Q	Rate of creating a substance	m^{-3} s^{-1}	81
R	Gas constant	J K^{-1}	82
R	Radius of a sphere	m	92
R_p	Radius of a pore	m	92
S	Surface area	m^2	78
$d\mathbf{S}$	Vector surface element pointing in the direction of the normal	m^2	80
T	Absolute temperature	K	82
V	Volume	m^3	81
V_M	Volume of 1 mol	m^3	83
ΔZ	Cell membrane thickness	m	92
α	Proportionality constant		85
β	Proportionality constant between force and velocity	N s m^{-1}	84
λ	Mean free path	m	82
λ	Ratio of D/v		96
θ, ϕ	Angles		79
η	Coefficient of viscosity	Pa s	84
σ	Standard deviation		88
σ	Electrical conductivity	$\Omega^{-1} m^{-1}$	86
ξ	Position	m	98
ξ	Dimensionless variable		97
ρ	Mass density	kg m^{-3}	81
μ_s	Chemical potential of solute	J/molecule	85

PROBLEMS

Section 4.1

4.1. A cylindrical pipe with a cross-sectional area $S=1$ cm^2 and length 0.1 cm has $j_s(0)S=200$ s^{-1} and $j_s(0.1)S=150$ s^{-1}.

(a) What is the total rate of buildup of particles in the pipe?

(b) What is the average rate of change of concentration in the section of pipe?

4.2. Write the continuity equation in cylindrical coordinates if $j_\phi=0$ but j_r and j_z can be nonzero.

4.3. Consider two concentric spheres of radii r and $r+dr$. If the particle fluence rate points radially and depends only on r, and the number of particles between r and $r+dr$ is not changing, show that $d(r^2 j)/dr=0$.

Section 4.2

4.4. Suppose that the total blood flow through a region is F ($\mathrm{m^3\,s^{-1}}$). A chemically inert substance is carried into the region in the blood. The total number of molecules of the substance in the region is N. The amount of blood in the region is not changing. Relate dN/dt to the blood flow and the concentrations of substance in the arterial and venous blood (C_A and C_V). This is known as the Fick principle or the Fick tracer method. It is often used with radioactive tracers.

Section 4.3

4.5. Allen *et al.* [(1982). *Science* **218**: 1127–1129] report seeing regular movement of particles in the axoplasm of a squid axon. At a temperature of 21 °C, the following mean drift speeds were observed:

Particle size (μm)	Typical speed ($\mu\mathrm{m\,s^{-1}}$)
0.8–5.0	0.8
0.2–0.6	2

How do these values compare to thermal speeds? (Make a reasonable assumption about the density of particles and assume that they are spherical.)

Section 4.4

4.6. Using the information on the mean free path in the atmosphere and assuming that all molecules have a molecular weight of 30, find the height at which the mean free path is 1 cm.

Section 4.6

4.7. Suppose that

$$C(x,t)=\frac{C_0}{\sqrt{4\pi Dt}}\,e^{-x^2/4Dt}.$$

Find an expression for $j_s(x,t)$.

Section 4.7

4.8. If all macromolecules have the same density, derive the expression for D versus the molecular weight that was used to draw the line in Fig. 4.12.

4.9. For diagnostic studies of the lung, it would be convenient to have radioactive particles that tag the air and that are small enough to penetrate all the way to the alveoli. It is possible to make the isotope ^{99m}Tc into a "pseudogas" by burning a flammable aerosol containing it. The resulting particles have a radius of about 60 nm [W. M. Burch, I. J. Tetley, and J. L. Gras (1984). *Clin. Phys. Physiol. Meas* **5**: 79–85]. Estimate the mean free path for these particles. If it is small compared to the molecular diameter, then Stokes's law applies, and you can use Eq. (4.23) to obtain the diffusion constant. (The viscosity of air at body temperature is about 1.8×10^{-5} Pa s.)

4.10. Figure 4.11 shows that D for O_2 in water at 298 K is $1.2\times10^{-9}\ \mathrm{m^2\,s^{-1}}$ and that the molecular radius of O_2 is 0.2 nm. The diffusion constant of a dilute gas (where the mean free path is larger than the molecular diameter) is $D=\lambda^2/2t_c$, where the collision time is given by Eq. (4.15).

(a) Find a numeric value for the diffusion constant for O_2 in O_2 at 1 atm and 298 K and its ratio to D for O_2 in water. The molecular weight of oxygen is 32.

(b) Assuming that this equation for a dilute gas is valid in water, estimate the mean free path of an oxygen molecule in water.

Section 4.8

4.11. (a) The three-dimensional normalized analog of Eq. (4.25) is

$$C(x,y,z,t)=\frac{C_0}{[2\pi\sigma^2(t)]^{3/2}}\exp\left(-\frac{x^2+y^2+z^2}{2\sigma^2(t)}\right).$$

Find the three-dimensional analog of Eq. (4.27).

(b) Show that $\sigma^2=\overline{x^2}+\overline{y^2}+\overline{z^2}=6Dt$.

4.12. A crude approximation to the Gaussian probability distribution is a rectangle of height P_0 and width $2L$. It gives a constant probability for a distance $\pm L$ either side of the mean.

(a) Determine the value of P_0 and L so that the distribution has the same value of σ as a Gaussian.

(b) Plot $P(x,t)$ if σ is given by Eq. (4.27) and the mean remains centered at the origin for times of 1, 5, 50, 100, and 500 ms. Use D for oxygen diffusing in water at body temperature.

(c) How long does it take for the oxygen to have a reasonable probability of diffusing a distance of 8 μm, the diameter of a capillary?

(d) For $t=100$ ms, plot both the accurate Gaussian and the rectangular approximation.

4.13. Write an equation for Fick's second law in three-dimensional Cartesian coordinates when the diffusion constant depends on position: $D=D(x,y,z)$.

4.14. The heat flow equation in one dimension is

$$j_H=-K\left(\frac{\partial T}{\partial x}\right).$$

One often finds an equation for the "diffusion" of heat:

$$\frac{\partial T}{\partial t}=D\left(\frac{\partial^2 T}{\partial x^2}\right).$$

The units of j_H are J m^{-2} s^{-1}. The internal energy per unit volume is given by $u=\rho CT$, where C is the heat capacity per unit mass and ρ is the density of the material. Derive the second equation from the first and show how D depends on K, C, and ρ.

4.15. A sheet of labeled water molecules starts at the origin in a one-dimensional problem and diffuses in the x direction.

(a) Plot σ vs t for diffusion of water in water.

(b) Deduce a "velocity" versus time. How long does it take for the water to have a reasonable chance of traveling 1 μm? 10 μm? 100 μm? 1 mm? 1 cm? 10 cm?

Section 4.9

4.16. Consider steady-state diffusion through two plane substances as shown in the figure. Show that the diffusion is the same as through a single membrane of thickness $\Delta x_1+\Delta x_2$, with diffusion constant

$$D=\frac{D_1 D_2}{\dfrac{\Delta x_1}{\Delta x_1+\Delta x_2}D_2+\dfrac{\Delta x_2}{\Delta x_1+\Delta x_2}D_1}.$$

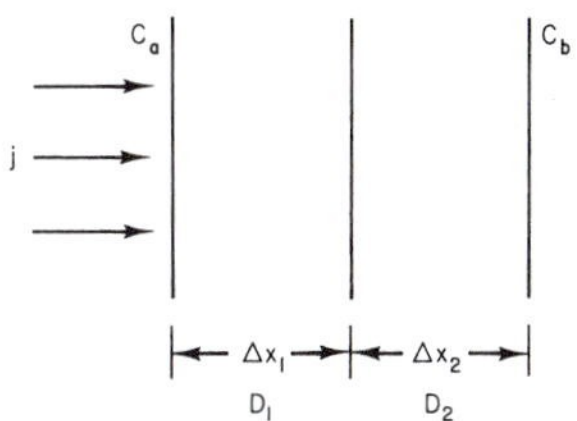

4.17. A fluid on the right of a membrane has different properties than the fluid on the left. Let the diffusion constants on left and right be D_1 and D_2, respectively, and let the pores in the membrane be filled by the fluid on the right a distance xL, where L is the thickness of the membrane.

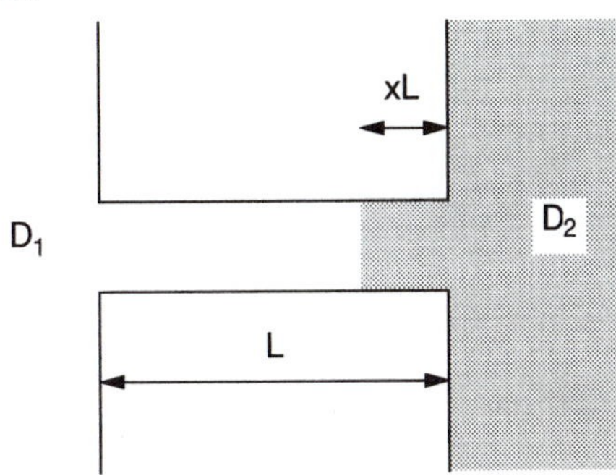

(a) Use the results of Problem 4.16 to determine the effective diffusion constant D for a membrane of thickness L when $D_2=yD_1$, $\Delta x_1=(1-x)L$, and $\Delta x_2=xL$. Neglect end effects.

(b) In the case that oxygen is diffusing in air and water at 310 K, the diffusion constants are $D_1=2.2\times 10^{-5}$ m^2 s^{-1}, $D_2=1.6\times 10^{-9}$ m^2 s^{-1}. Plot D/D_1 vs x.

4.18. Diffusion of water through the skin occurs at a rate of about 350 ml per day. The body surface area is about 1.73 m^2. If the skin is a homogeneous layer about 20 μm thick, estimate the diffusion constant, assuming pure water on the inside of the skin and zero concentration outside.

4.19. A diffusing substance is being consumed by a chemical reaction at a rate Q per unit volume per second. The reaction rate is limited by the concentration of some enzyme, so Q is independent of the concentration of the diffusing substance. For a slab of tissue of thickness b with concentration C_0 at both $x=0$ and $x=b$, solve the equation to find $C(x)$ in the steady state. This is known as the Warburg equation [*Biochem Z.* **142**: 317–350 (1923)]. It is a one-dimensional model for the consumption of oxygen in tissue: points $x=0$ and $x=b$ correspond to the walls of two capillaries side by side.

4.20. Suppose that a diffusing substance disappears in a chemical reaction and that the rate at which it disappears is proportional to the concentration $-kC$. Write down the Fick's second law in this case. Show what the equation becomes if one makes the substitution $C(x,y,z,t)=C'(x,y,z,t)e^{-kl}$.

4.21. A spherical cell has radius R. The total flux through the surface is given by $j_s=-D$ grad C. Suppose that the substance in question has concentration $C(t)$ inside the cell and zero outside. The material outside is removed fast enough so that the concentration remains zero. Using spherical coordinates, find a differential equation for $C(t)$ inside the cell. The thickness of the cell wall is $\Delta r \ll R$.

Section 4.10

4.22. Derive Eq. (4.45).

Section 4.12

4.23. Extend Fick's second law in one dimension $\partial C/\partial t=D(\partial^2 C/\partial x^2)$ to include solvent drag.

Section 4.13

4.24. Write a computer program to model a two-dimensional random walk. Make several repetitions of a random walk of 3600 steps and plot histograms of the displacements in the x and y directions and mean square displacement.

4.25. Write a program to display the motion of 100 particles in two dimensions.

4.26. Particles are released from a point between two perfectly absorbing plates located at $x=0$ and $x=1$. The particles random walk in one dimension until they strike a plate. Find the probability of being captured by the right-hand plate as a function of the position of release, x. (Hint: The probability is related to the diffusive fluence rate to the right-hand plate if the concentration is C_0 at x and is 0 at $x=0$ and $x=1$.)

4.27. The text considered a one-dimensional random-walk problem. Suppose that in two dimensions the walk can occur with equal probability along $+x$, $+y$, $-x$, or $-y$. The total number of steps is $N=N_x+N_y$, where the number of steps along each axis is not always equal to $N/2$.

(a) What is the probability that N_x of the N steps are parallel to the x axis?

(b) What is the probability that the net displacement along the x axis is $m_x\lambda$?

(c) Show that the probability of a particle being at $(m_x\lambda, m_y\lambda)$ after N steps is

$$P(m_x, m_y)=\sum_{N_x} \frac{N!}{N_x!(N-N_x)!}\times\left(\frac{1}{2}\right)^N P(m_x, N_x)P(m_y, N-N_x),$$

where $P(m,N)$ on the right-hand side of this equation is given by Eq. (4.76).

(d) The factor $N!/N_x!(N-N_x)!$ is proportional to a binomial probability. What probability? Where does this factor peak when N is large?

(e) Using the above result, show that

$$P(m_x, m_y)=P\left(m_x, \frac{N}{2}\right)P\left(m_y, \frac{N}{2}\right).$$

(f) Write a Gaussian approximation for two-dimensional diffusion.

REFERENCES

Axelrod, D., and M. D. Wang (1994). Reduction-of-dimensionality kinetics at reaction-limited cell surface receptors. *Biophys. J.* **66**(3), Pt. 1): 588–600.

Barr, G. (1931). *Viscometry*. London, Oxford University Press.

Bean, C. P. (1972). The physics of neutral membranes—neutral pores, in G. Eisenman, ed., *Membranes—A Series of Advances*. New York, Dekker, Vol. 1, pp. 1–55.

Berg, H. C. (1975). Chemotaxis in bacteria. *Ann. Rev. Biophys. Bioeng.* **4**: 119–136.

Berg, H. C. (1983). *Random Walks in Biology*. Princeton, NJ, Princeton University Press.

Berg, H. C., and E. M. Purcell (1977). Physics of chemoreception. *Biophys. J.* **20**: 193–219.

Hildebrand, J. H., J. M. Prausnitz, and R. L. Scott (1970). *Regular and Related Solutions*. New York, Van Nostrand Reinhold.

Kennard, E. H. (1938). *The Kinetic Theory of Gases*. New York, McGraw-Hill.

Levitt, D. (1975). General continuum analysis of transport through pores. I. Proof of Onsager's reciprocity postulate for uniform pore. *Biophys. J.* **15**: 533–551.

Meidner, H., and T. A. Mansfield (1968). *Physiology of Stomata*. New York, McGraw-Hill.

Maddock, J. R., and L. Shapiro (1993). Polar location of the chemoreceptor complex in the *Escherichia coli* cell. *Science* **259**: 1717–1723.

Paine, P. L., and P. Scherr (1975). Drag coefficients for the movement of rigid spheres through liquid-filled cylinders. *Biophys. J.* **15**: 1087–1091.

Parkinson, J. S., and D. F. Blair (1993). Does *E. coli* have a nose? *Science* **259**: 1701–1702.

Pryde, J. A. (1966). *The Liquid State*. London, Hutchinson University Library.

Purcell, E. M. (1977). Life at low Reynolds number. *Am. J. Phys.* **45**: 3–11.

Reif, F. (1965). *Fundamentals of Statistical and Thermal Physics*. New York, McGraw-Hill.

Schey, H. M. (1973). *Div. Grad, Curl, and All That*. New York, Norton.

Villars, F. M. H., and G. Benedek (1974). *Physics with Illustrative Examples from Medicine and Biology*. Reading, MA, Addison-Wesley, Vol. 2.

Wang, D., S.-Y. Gou, and D. Axelrod (1992). Reaction rate enhancement by surface diffusion of adsorbates. *Biophys. Chem.* **43**(2): 117–137.

Zwanzig, R., and A. Szabo (1991). Time dependent rate of diffusion-influenced ligand binding to receptors on cell surfaces. *Biophys. J.* **60**(3): 671–678.

CHAPTER 5

Transport Through Neutral Membranes

The last chapter discussed some of the general features of solute movement in an infinite medium. Solute particles can be carried along with the flowing solution or they can diffuse. This chapter considers the movement of solute and solvent through membranes, ignoring any electrical forces on the particles.

The movement of electrically neutral particles through aqueous pores in membranes has many applications in physiology. They range from flow of nutrients through capillary walls to tissue, to regulation of the amount of fluid in the interstitial space between cells, and to the initial stages of the operation of the kidney.

Sections 5.1 through 5.4 are a qualitative introduction to the flow of water through membranes as a result of hydrostatic pressure differences or osmotic pressure differences. The reader who is not interested in the more advanced material can read just this part of the chapter, culminating in the clinical examples of Sec. 5.4.

Sections 5.5 and 5.6 present phenomenological transport equations that are simple linear relationships between the flow of water and solute particles and the pressure and concentration differences that cause the flows. These relationships are valid for any type of membrane as long as a linear relationship adequately describes the flow and the proportionality constants are regarded as experimentally determined quantities. These equations are applied to the artificial kidney in Sec. 5.7.

Many biological membranes are pierced by pores that admit water and water-soluble molecules up to a certain size. (Lipid-soluble chemicals and some other substances may dissolve in the membrane and then diffuse or may be attached to a carrier molecule or may be transported through the membrane in a vesicle.) Sections 5.8 through 5.11 provide a more advanced treatment of one particular membrane model: a membrane pierced by pores in which electrical forces can be neglected and in which Poiseuille flow takes place. The model leads to expressions for the phenomenological coefficients that are compared to experimental data in Secs. 5.12 and 5.13. Section 5.15 describes measurements on an artificial pore of known geometry, while Sec. 5.13 describes measurements of the properties of pores in the kidney.

Section 5.14 presents a simple model for countercurrent transport, which is important in artificial organs, the kidney, and in conserving heat loss from the extremities.

5.1. MEMBRANES

All cells are surrounded by a membrane 7–10 nm thick. Furthermore, virtually all the physical substructures within the cell are also bounded by membranes. Membranes separate two regions of space; they allow some substances to pass through but not others. The membrane is said to be permeable to a substance that can pass through it; when only certain substances can get through it is called semipermeable. A substance that can pass through is said to be permeant.

The simplest model that one can conceive for a semipermeable membrane is shown in Fig. 5.1(a). A number of pores pierce the membrane. The pores could follow a longer path, as in Fig. 5.1(b). Another simple model is shown in Fig. 5.1(c): there are no pores, but small molecules actually "dissolve" in the membrane and diffuse through. Each example in Fig. 5.1 shows water molecules (open circles), solute molecules (small solid circles), and a large protein molecule that cannot pass through the membrane.

In Figs. 5.1(a) and 5.1(b) the motion of the water molecules is quite different from that of the small solute molecules. Each water molecule is in contact with neighboring water molecules so that when the water molecules move, they flow together. The result is the familiar bulk flow that occurs in a pipe. The solute molecules, on the other hand, are so dilute that they seldom collide with one another. Each one undergoes a random walk *independent of other solute molecules* due to collisions with water molecules.

The motion of each solute molecule is *not* independent of the motion of the surrounding water molecules. If the

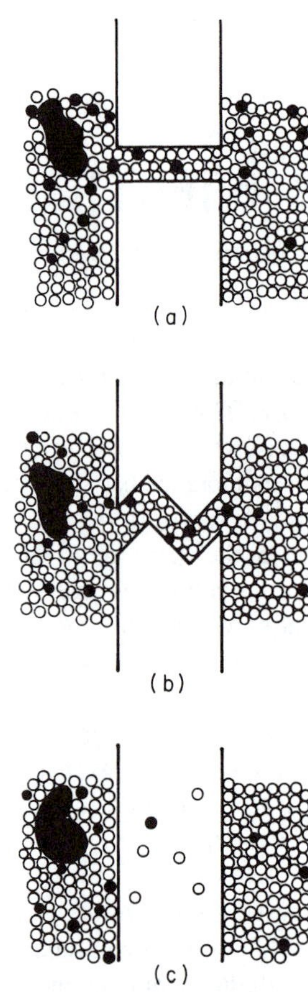

FIGURE 5.1. Simple models for a semipermeable membrane. (a) A straight pore. (b) A pore following a tortuous path. (c) Small molecules dissolve in the membrane and diffuse through.

water is at rest, the movement of the solute molecules is diffusion; if the water is moving, this diffusion is superimposed on a flow of the solute molecules with the moving fluid.

In Fig. 5.1(c), the water and solute molecules are very dilute within the membrane, so that both kinds of molecules diffuse. The water molecules are not in contact with each other, but are in some sort of interstices within the membrane structure, walking randomly in response to thermal agitation of the membrane.

5.2. OSMOTIC PRESSURE IN AN IDEAL GAS

The selective permeability of a membrane gives rise to some striking effects. The flow of water that occurs because solutes are present that cannot get through the membrane is called *osmosis*. Although the phenomena seem strange when they are first encountered, they can be explained quite simply. They are important in a variety of clinical problems that are described in Sec. 5.4. We begin by finding the conditions under which no flow takes place and the direction of flow when it does occur. Later, in Sec. 5.5, we consider the rate of flow in response to a given pressure difference.

It is easiest to understand osmotic pressure by considering the special case of two ideal gases and a membrane that is permeable to one but not the other. This case is simple because the gas molecules do not interact with one another. Then, in Sec. 5.3, we will examine the phenomenon when the substances are liquids.

A box with total volume V^* is shown in Fig. 5.2. It contains N_1^* molecules of gas species 1. If the box is at temperature T, the ideal-gas law relates the pressure, temperature, and the number of molecules:

$$p_1 V^* = N_1^* k_B T. \tag{5.1}$$

This has been written the way physicists like to write it, in terms of the number of molecules N_1^*. It could also have been written in terms of the number of moles n_1^*:

$$p_1 V^* = n_1^* RT.$$

The only difference is that R is per mole instead of per molecule. Since 1 mol contains N_A molecules, where N_A is Avogadro's number, these are related by

$$N_1^* = N_A n_1^*, \quad R = N_A k_B.$$

Numerical values are

$$N_A = 6.022 \times 10^{23} \ \text{mol}^{-1},$$

$$k_B = 1.3807 \times 10^{-23} \ \text{J K}^{-1},$$

$$R = 8.315 \ \text{J mol}^{-1} \text{K}^{-1},$$

$$R = 0.082\,06 \ \text{l atm mol}^{-1} \text{K}^{-1}.$$

It will also be useful to speak of the concentration: the number of molecules or moles per unit volume. We denote mean *molecular* concentration by capital letter C and mean *molar* concentration by lowercase c:

$$C_1 = \frac{N_1^*}{V^*} \ \text{m}^{-3} \text{ or molecules m}^{-3},$$

$$c_1 = \frac{n_1^*}{V^*} \ \text{m}^{-3} \text{ or mol m}^{-3}.$$

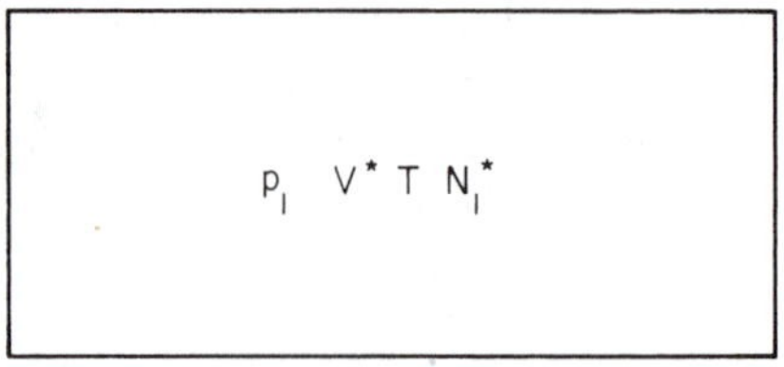

p_1 V^* T N_1^*

FIGURE 5.2. An ideal gas fills a box of volume V^*.

Suppose that a membrane divides volume V^* into two regions of volume V and V' and that the molecules can pass through the membrane (Fig. 5.3). The pressure and temperature are the same in both regions, and we can write

$$p_1 V = N_1 k_B T, \quad p_1 V' = N_1' k_B T.$$

Dividing both sides of each equation by the appropriate volume gives

$$p_1 = C_1 k_B T, \quad p_1 = p_1' = C_1' k_B T. \tag{5.2}$$

This is a statement of equilibrium. When the pressures are the same on both sides of the membrane, no molecules pass through on average. If the pressure is greater on one side than the other, molecules pass through to bring the pressures into equilibrium, as we saw in Chap. 3. Equations (5.2) say nothing about how frequently a molecule that strikes the membrane passes through. It could take hours or days for equilibrium to be attained if the molecules do not pass through very often.

Now, keeping V fixed, introduce species 2 into region V as in Fig. 5.4. Suppose that species 2 cannot pass through the membrane. Bombardment of the membrane by the new molecules on the left causes an additional force on the left side of the membrane. The total pressure in volume V is now the sum of the partial pressures p_1 due to species 1 and p_2 due to the second species:

$$\begin{aligned} p &= p_1 + p_2, \\ p_1 V &= N_1 k_B T, \\ p_2 V &= N_2 k_B T. \end{aligned} \tag{5.3}$$

The ideal-gas law is still obeyed in terms of the total number of molecules in V, $N = N_1 + N_2$:

$$\begin{aligned} pV &= p_1 V + p_2 V = N_1 k_B T + N_2 k_B T \\ &= (N_1 + N_2) k_B T = N k_B T. \end{aligned}$$

In an ideal gas the presence of the second species does not change the partial pressure p_1. The total pressure on the walls and the membrane is increased by p_2 so the membrane is bowed towards the right, but the total pressure is simply the sum of the two partial pressures. The ratio p_1/p is the fraction of the pressure due to collisions of molecules of the first kind with the wall.

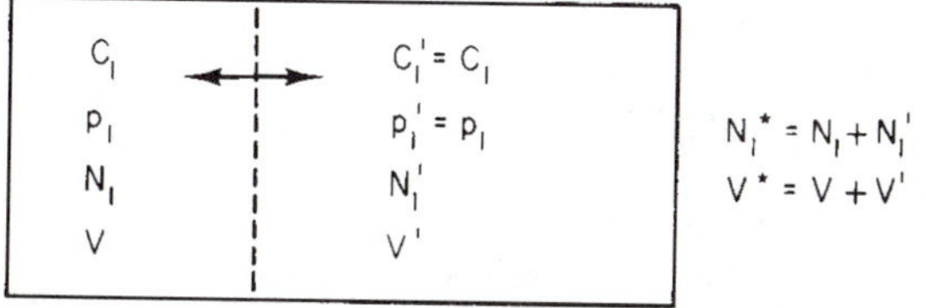

FIGURE 5.3. The introduction of a permeable membrane does not change the pressure or concentration in the gas.

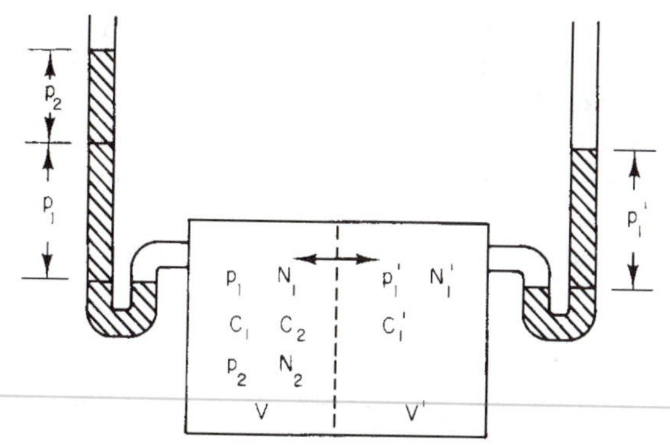

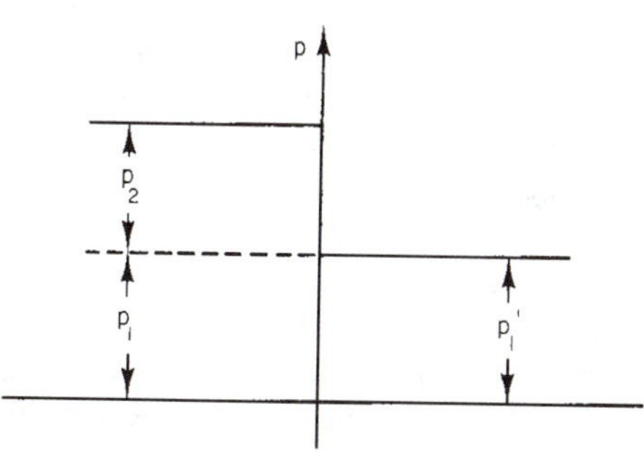

FIGURE 5.4. Species 2, which cannot pass through the membrane, has been introduced in V. The pressure in V is higher than in V' by the partial pressure p_2.

Suppose now that the pressure in V' is raised, either by compressing the gas or by introducing more molecules of type 1, so that instead of $p_1' = p_1$, we have $p_1' = p$. The partial pressure of species 1 is higher in V' than in V. Since these molecules can pass through the membrane, they will flow from V' to V. An identical flow could have been caused without having species 2, simply by raising the pressure in V'. Not every molecule striking the membrane will pass through, but some fraction of all collisions with the wall will result in a molecule passing through. The fraction will depend on the details of the membrane structure. The number going through will be proportional to the number of collisions on one side minus the number of collisions on the other and hence to the difference of partial pressures. If $p_1 > p_1'$, species 1 will flow from V to V'. If $p_1 < p_1'$, the flow will be in the other direction. The details of the membrane will determine how rapid this flow is. *The flow of any species of gas molecule that can pass through the membrane will be from the region of higher partial pressure to lower partial pressure.*

Suppose we start out with only species 1 on each side of the membrane and equal pressures on both sides so that $p = p_1 = p' = p_1'$. There are three ways to make p_1 less than p_1', thereby causing flow from right to left. One is simply to let the gas on the left expand into a larger volume, which lowers $p = p_1$. (Or we could have compressed the gas on the right, raising $p' = p_1'$.) The other two ways involve introducing on the left a species 2 that cannot pass through the membrane. The second way would be to keep the total pressure and volume on the left the same, but remove one molecule of species 1 for every molecule of species 2 that is introduced. The third way would be to increase the volume on the left as each molecule of species 2 is introduced, so that $p=p_1+p_2$ remains the same.

The total partial pressure of all species that cannot pass through the membrane is called the *osmotic pressure* in region V and is usually denoted by π. If the subscript 2 denotes all impermeant species,

$$\pi_2=C_2k_BT. \tag{5.4}$$

The flow through the membrane because of an increase in the osmotic pressure or a decrease in the total pressure is identical. In each case the flow is determined by the difference across the membrane of p_1, the total partial pressure of all the species that can pass through.

The description in the previous paragraphs of partial pressure is easy to visualize, and for the case given it is correct. It is more general, however, to express the condition for equilibrium in terms of the chemical potential. Recall that in Chap. 3 we derived the pressure in terms of volume changes of a system and the chemical potential in terms of the number of particles in the system. Suppose that the membrane separating the two sides is actually a semipermeable piston that is free to move. Equality of the total pressure on both sides of the piston means that the piston will not move and the two systems will not exchange volume. Equality of the chemical potential of a species that can get through the membrane means that the two systems will not exchange particles. It is better, therefore, to say *the flow of any species that can pass through the membrane will be from the region of higher chemical potential to the region of lower chemical potential for that species. If the chemical potentials are the same, there will be no flow.*

The mixture of two ideal gases is a special case of the ideal solution that was described in Sec. 3.18. The chemical potential of species 1 that can pass through the membrane is given by Eq. (3.93):

$$\Delta\mu_1=\bar{V}_1(\Delta p-k_BT\Delta C_2),$$

$$\mu_1-\mu_1'=\bar{V}_1[p-p'-k_BT(C_2-0)],$$

$$\mu_1-\mu_1'=\bar{V}_1(p_1+p_2-p_1'-k_BTC_2).$$

Since $p_2=k_BTC_2$, the chemical potential is the same on both sides of the membrane when $p_1=p_1'$.

5.3. OSMOTIC PRESSURE IN A LIQUID

Imagine now that the two volumes are filled with a solvent, which we will call water. If the pressure of the water is the same in both regions there is no flow of water through the membrane, nor is there exchange of volume if the membrane piston is free to move. Increasing the pressure on one side of the fixed membrane causes water to flow through the membrane from the side with higher pressure to the side with lower pressure. The chemical potential contains a term proportional to the pressure. It was shown in Sec. 3.18 that for an ideal solution[1]

$$\Delta\mu_w=\frac{\Delta p-k_BT\Delta C}{C_w}. \tag{3.93}$$

If there is a solute in the water that can pass freely through the membrane along with the water, the situation is unchanged.

Now let us add some solute on the left that cannot pass through the membrane. We will keep the volume on the left fixed. To add the solute in such a way that the pressure does not change, we must remove some water molecules as we add it. We saw in Chap. 3 that replacing some water molecules with solute increases the entropy of the solution.[2] This means that the Gibbs free energy and the chemical potential are decreased. Water flows from the region on the right, where the chemical potential is higher, to the region on the left, where it is lower. The chemical potential of the water on the left can be increased by increasing the total pressure on the left. The *osmotic pressure* π is the excess pressure that we must apply on the left to prevent the flow of water through the membrane. There is no flow of water when

$$p=p'+\pi.$$

It is more convenient to write all the unprimed quantities on the left:

$$p-\pi=p'.$$

The quantity $p-\pi$ will occur so often in what follows that it is worth a special name. We will define the *driving pressure*

$$p_d\equiv p-\pi. \tag{5.5}$$

As far as I know, it has not been used by other authors. It is a monotonic function of the chemical potential. In an ideal solution it is $C_w\mu_w$. Except in an ideal gas, it is *not* the same as the partial pressure (a concept that is not normally used in a liquid). On the right there is no solute and $p_d'=p'$. *There is no flow when the driving pressure is the same on both sides,*

[1] An ideal solution can be defined in several equivalent ways. One is that it is a solution that obeys Eq. (3.93). Another is that when the separated components are mixed, there is no change of total volume and no heat is evolved or absorbed. See Hildebrand and Scott (1964), Chap. 2.

[2] This, recall, is because the water molecules are indistinguishable. A simple model shows why this happens. Suppose that four water molecules occupy four identical energy levels, and that these are the only four levels available. Because the molecules are indistinguishable, there is only one microstate and the entropy is zero. If one molecule is removed, there are then four separate microstates, corresponding to the empty level being any one of the four. The entropy is $k_BT\ln(4)$.

$$p_d = p'_d, \tag{5.6a}$$

or the chemical potential of the water is the same on both sides,

$$\mu_w = \mu'_w. \tag{5.6b}$$

The water passes through the membrane in the direction from higher p_d to lower p_d (or from higher chemical potential to lower chemical potential). Either the total pressure or the osmotic pressure can be manipulated to change p_d or μ. An increase of total pressure has the same effect as a decrease of osmotic pressure.

Increasing the concentration of the solute increases the osmotic pressure. The fact that $p_d = p - \pi = C_w \mu_w$ means that for ideal solutions obeying Eq. (3.93),

$$\pi = Ck_BT = cRT. \tag{5.7}$$

In many cases this is confirmed by experiment, particularly in dilute solutions. This is known as the van't Hoff law for osmotic pressure.

An *osmole* is the equivalent of a mole of solute particles. The term *osmolality* is used to refer to the number of osmoles per kilogram of *solvent*, while *osmolarity* refers to the number of osmoles per liter of *solution*. The reason for introducing the osmol is that not all impermeant solutes are ideal; their osmotic effects are slightly less than Ck_BT. The osmol takes this correction into account.

5.4. SOME CLINICAL EXAMPLES

As blood flows through capillaries, oxygen and nutrients must leave the blood and get to the cells. Waste products must leave the cells and enter the blood. Diffusion is the main process that accomplishes this transfer. The capillaries are about the diameter of a red cell; the red cells therefore squeeze through the capillary in single file. They move in plasma, which consists of water, electrolytes, small molecules such as glucose and dissolved oxygen or carbon dioxide, and large protein molecules. All but the large protein molecules can pass through the capillary wall.

Outside the capillaries is the interstitial fluid, which bathes the cells. The concentration of protein molecules in the interstitial fluid is much less than it is in the capillaries. Osmosis is an important factor determining the pressure in the interstitial fluid and therefore its volume. The following values (in units[3] of torr) are typical for the osmotic pressure inside and outside the capillary:[4]

$$\text{Inside capillary} \quad \pi_i = 28 \text{ torr}$$

$$\text{Outside capillary, interstitial fluid} \quad \pi_o = 5 \text{ torr}$$

Measurements of the total pressure in the interstitial fluid are difficult, but the value seems to be about -6 torr. It is maintained below atmospheric pressure (taken here to be 0 torr) by the rigidity of the tissues. The driving pressure of water and small molecules outside is therefore

$$p_{w_o} = p_o - \pi_o = -6 - 5 = -11 \text{ torr}.$$

The total hydrostatic pressure within the capillary drops from the arterial end to the venous end, causing blood to flow along the capillary. A typical value at the arterial end is 25 torr; at the venous end, it is 10 torr. If the drop is linear along the capillary, the total pressures versus position is as plotted in Fig. 5.5(a).[5] Subtracting from this the osmotic pressure of the large molecules gives the curve for the driving pressure inside, p_{di}, which is also plotted in Fig. 5.5(a). Figure 5.5(b) shows the total and driving pressures in the interstitial fluid. Figure 5.5(c) compares the driving pressure inside and outside. The pressure is larger inside in the first half of the capillary and larger outside in the second half of the capillary. The result is an outward flow of plasma through the capillary wall in the first half and an inward flow in the second half. There is a very slight excess of outward flow. This fluid returns to the circulation via the lymphatic system.

There are three ways that the balance of Fig. 5.5 can be disturbed, each of which can give rise to edema, a collection of fluid in the tissue. The first is a higher average pressure along the capillary. The second is a reduction in osmotic pressure because of a lower protein concentration in the blood (hypoproteinemia). The third is an increased permeability of the capillary wall to large molecules, which effectively reduces the osmotic pressure. Each is discussed below.

5.4.1. Edema due to Heart Failure

A patient in right heart failure exhibits an abnormal collection of interstitial fluid in the lower part of the body (the legs for a walking patient; the back and buttocks for a patient in bed). This can be understood in terms of the mechanism discussed above. The right heart pumps blood from the veins through the lungs. If it can no longer handle this load, the venous blood is not removed rapidly enough, and the pressure in the veins and the venous end of the capillaries rises. There is a corresponding rise in p_d along the capillary. More fluid flows from the capillary to the interstitial space. The interstitial pressure rises until the net flow is again zero. When the interstitial pressure becomes positive, edema results.

The same process can occur in left heart failure in which the pressure in the pulmonary veins builds up. The patient then has pulmonary edema and may literally drown.

[3] 1 torr = 1 mm Hg = 133.3 Pa = 0.019 34 lb in.$^{-2}$.

[4] A short account of the pressures used here is found in Guyton (1991, Chap. 16). A more detailed discussion is in Guyton, Taylor, and Granger (1975).

[5] This simple discussion uses pressures that compensate for the fact that the surface area of the capillary is larger at the venous end than at the arterial end.

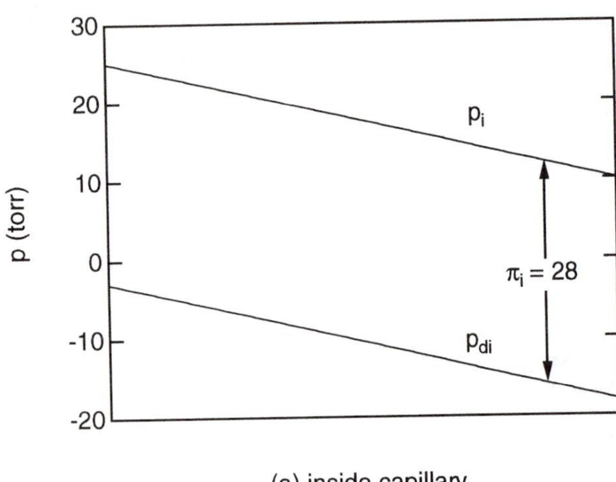

(a) inside capillary

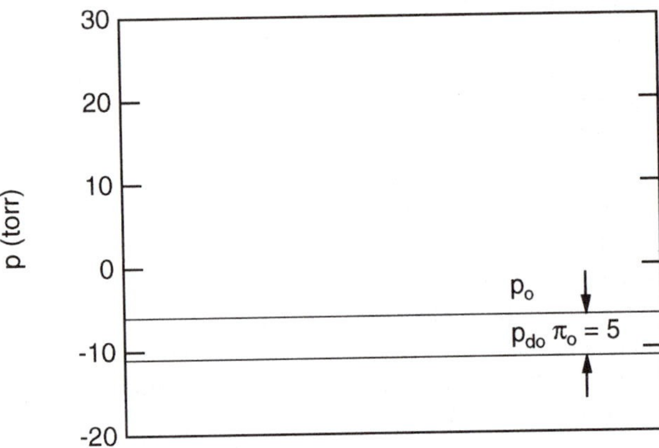

(b) interstitial fluid outside capillary

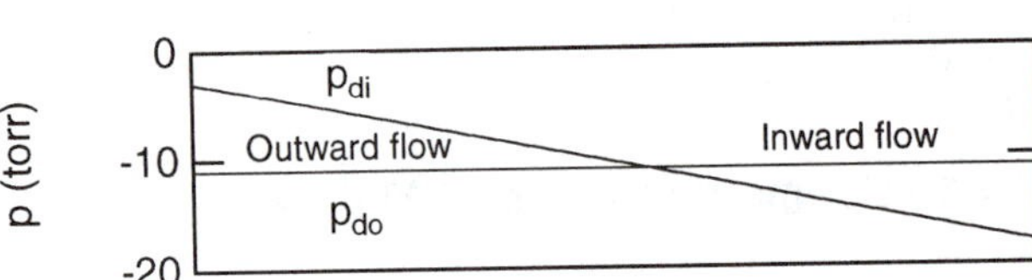

(c) comparison of p_d inside and outside

FIGURE 5.5. Pressures inside and outside the capillary. (a) Inside. (b) Outside. (c) Comparison of the water driving pressure inside and outside.

5.4.2. Nephrotic Syndrome, Liver Disease, and Ascites

Patients can develop an abnormally low amount of protein in the blood serum, hypoproteinemia, which reduces the osmotic pressure of the blood. This can happen, for example, in nephrotic syndrome, a disease of the kidney in which the nephrons (the basic functioning units in the kidney) become permeable to protein, which is then lost in the urine. The lowering of the osmotic pressure in the blood means that the p_d rises. Therefore, there is a net movement of water into the interstitial fluid. Edema can result from hypoproteinemia from other causes, such as liver disease and malnutrition.

The patient with liver disease may suffer a collection of fluid in the abdomen. Fluid collects because the veins of the abdomen flow through the liver before returning to the heart. This allows nutrients absorbed from the gut to be processed immediately and efficiently by the liver. The liver disease may not only decrease the plasma protein concentration, but the vessels going through the liver may become blocked, thereby raising the capillary pressure throughout the abdomen and especially in the liver. A migration of fluid out of the capillaries results. The surface of the liver "weeps" fluid into the abdomen. The excess abdominal fluid is called ascites.

5.4.3. Edema of Inflammatory Reaction

Whenever tissue is injured, whether it is a burn, an infection, an insect bite, or a laceration, a common sequence of events initially occurs that cause edema. They include the following.

1. *Vasodilation.* Capillaries and small blood vessels dilate, and the rate of blood flow is increased. This is responsible for the redness and warmth associated with the inflammatory process.
2. *Fluid exudation.* Plasma, including plasma proteins, leaks from the capillaries because of increased permeability of the capillary wall.
3. *Cellular migration.* The capillary walls become porous enough so that white cells move out of the capillaries and to the site of injury.

5.4.4. Headaches in Renal Dialysis

Dialysis is used to remove urea from the plasma of patients whose kidneys do not function. Urea is in the interstitial brain fluid and the cerebrospinal fluid in the same concentration as in the plasma; however, the permeability of the capillary–brain membrane is low, so equilibration takes several hours.[6] Water, oxygen, and nutrients cross from the capillary to the brain at a much faster rate than urea. As the plasma urea concentration drops, there is a temporary osmotic pressure difference resulting from the urea within the brain. The driving pressure of water is higher in the plasma, and water flows to the brain interstitial fluid. Cerebral edema results, which can cause severe headaches.

The converse of this effect is to inject urea or mannitol, another molecule that does not readily cross the "blood–brain barrier" into the plasma. This lowers the driving pressure of water within the plasma, and water flows from the

[6]Patton *et al.* (1989), Chap. 64.

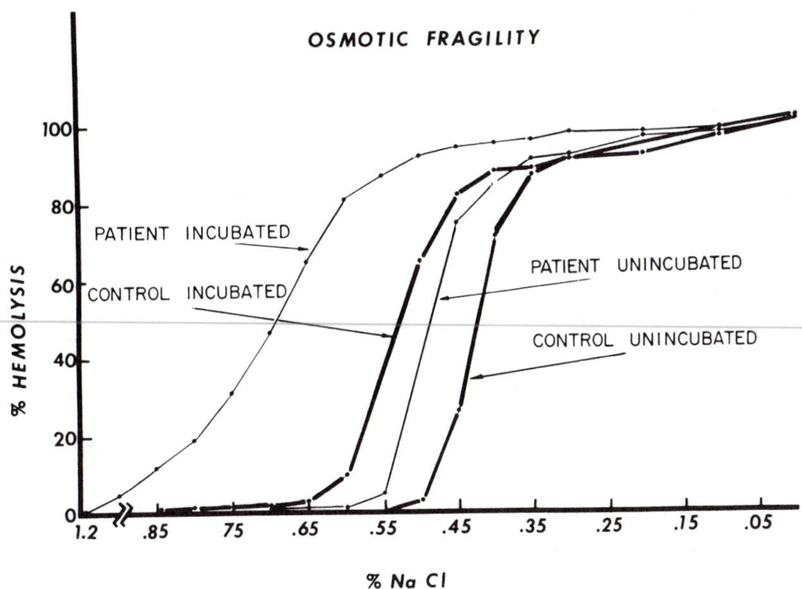

FIGURE 5.6. Osmotic fragility of red cells. The different curves are discussed in the text. From S. I. Rappoport (1971). *Introduction to Hematology*. New York, Harper & Row, p. 99. Reproduced by permission of Harper & Row.

brain into the plasma. Although the effects do not last long, this technique is sometimes used as an emergency treatment for cerebral edema.[7]

5.4.5. Osmotic Diuresis

The functional unit of the kidney is the *nephron*. Water and many solutes pass into the nephron from the blood at the glomerulus. As the urine flows through the rest of the nephron, a series of complicated processes cause a net reabsorption of most of the water and varying amounts of the solutes. Some medium-weight molecules such as mannitol are not reabsorbed at all. If they are present in the nephron, for example, from intravenous administration, the driving pressure of water is lowered and less water is reabsorbed than would be normally. The result is an increase in urine volume and a dehydration of the patient called osmotic diuresis.[8] Similar diuretic action takes place in a diabetic patient who "spills" glucose into the urine.

5.4.6. Osmotic Fragility of Red Cells

Red cells (erythrocytes) are normally disk-shaped, with the center thinner than the rim. In the disease called hereditary spherocytosis the red cells are more rounded. If a red cell is placed in a solution that has a higher driving pressure than that inside the cell, water moves in and the cell swells until it bursts. Since cell membranes (as distinct from the lining of capillaries) are nearly impermeable to sodium, sodium is osmotically active for this purpose.

The osmotic fragility test consists of placing red cells in solutions with different sodium concentrations and determining what fraction of the cells burst. A typical plot of fraction vs sodium concentration is shown in Fig. 5.6. Sodium concentration decreases and p_d increases to the right along the axis.

The patient with hereditary spherocytosis has cells that will be destroyed at a lower external p_d (higher sodium concentration) than normal, because the membrane is more permeable to the sodium.

If the red cells are incubated at body temperature in a sodium solution with the osmolality of plasma for 24 h, the fragility of hereditary spherocytosis cells is markedly increased. During this incubation period the concentration of sodium within the cell *increases*; the sodium cannot escape rapidly when the external concentration is reduced, the driving pressure within the cell is lower than before incubation, and water flows into the cell even more rapidly.

5.5. VOLUME TRANSPORT THROUGH A MEMBRANE

In this section and the next we develop phenomenological equations to describe flow of fluid and flow of solute through a membrane. These are linear approximations to the dependence of the flows on pressure and solute concentration differences. Three parameters are introduced that are

[7] Fishman (1975); White and Likavek (1992).
[8] Gennari and Kassirer (1974); Guyton (1991).

widely used in physiology: the filtration coefficient (or hydraulic permeability), the solute permeability, and the solute reflection coefficient.

The volume fluence rate or volume flow per unit area per second through a membrane is J_v. We will use capital J for the fluence rate through a membrane and lowercase j for fluence rate in bulk solution.

$$J_v = \frac{\text{(total volume per second through membrane area } S)}{S} = \frac{i_v}{S} \text{ m s}^{-1}. \quad (5.8)$$

Consider pure water. The fluence rate depends on the pressure difference across the membrane. When the pressure difference is zero there is no flow. The direction of flow, and therefore the sign of the fluence rate, depends on which side of the membrane has the higher pressure. The simplest relationship that has this property is a linear one:[9]

$$J_v = L_p \Delta p. \quad (5.9)$$

The proportionality constant is called the *filtration coefficient* or *hydraulic permeability*. It depends on the details of the membrane structure, such as the properties of the pores. The SI units for L_p are $\text{m s}^{-1}\text{ Pa}^{-1}$, $\text{m}^3\text{ N}^{-1}\text{ s}^{-1}$, or $\text{m}^2\text{ s kg}^{-1}$. Often in the literature, however, values of L_p are reported in units of (cm/s)/atm. Since 1 atm$=1.01\times10^5$ N m^{-2}, the conversion is

$$1 \text{ cm s}^{-1}\text{ atm}^{-1} = 0.99\times10^{-7} \text{ m s}^{-1}\text{ Pa}^{-1}. \quad (5.10)$$

If a solute is present to which the membrane is completely impermeable, only water will flow, and the flow will depend on Δp_d:

$$\begin{aligned}\Delta p_d &= p_d - p_d' = p - \pi - (p' - \pi') \\ &= p - p' - (\pi - \pi') \\ &= \Delta p - \Delta\pi\end{aligned}$$

so

$$J_v = L_p(\Delta p - \Delta\pi). \quad (5.11)$$

Figure 5.7 shows the pressure relations on each side of the membrane for no flow and for flow in either direction.

When the solute is partially permeant, the volume fluence rate in the linear approximation still depends on both Δp and $\Delta\pi$, but the proportionality constants may be different. Since the solute does not reduce the flow as much as in Eq. (5.11), it is customary to write the two constants as L_p and σL_p:

$$J_v = L_p(\Delta p - \sigma\,\Delta\pi). \quad (5.12)$$

[9]The traditional sign convention has been followed here. There would be a minus sign in the equation if Δp were defined to be $p(x+\Delta x) - p(x)$. However, it is usually defined as $p - p'$. The flow is from the region of higher pressure to the region of lower pressure.

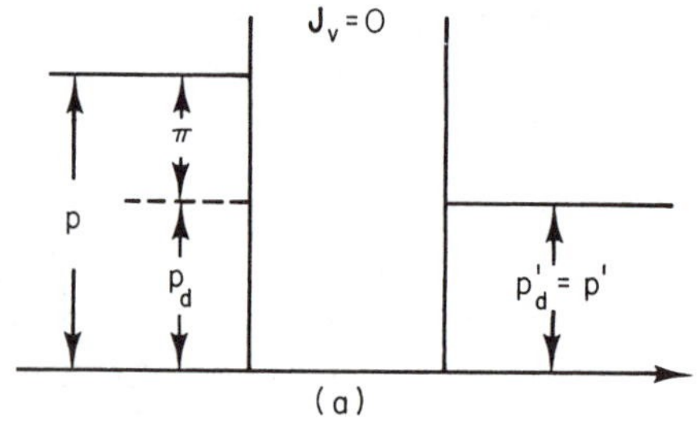

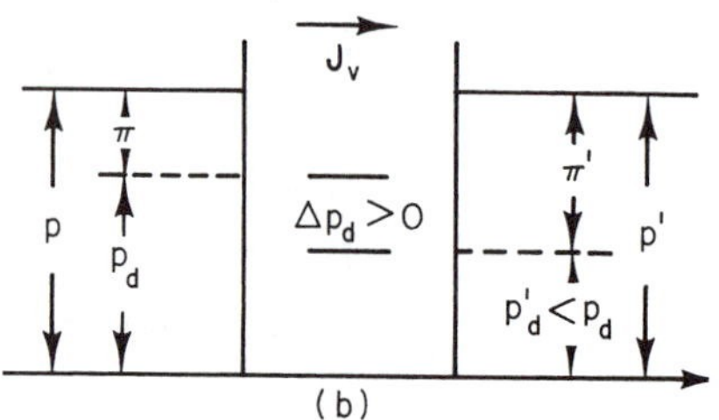

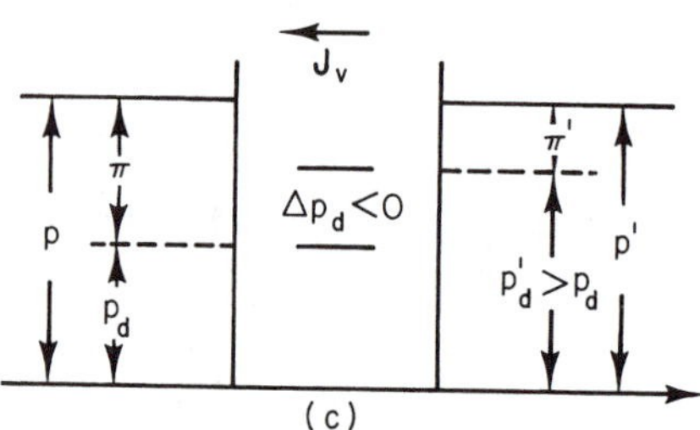

FIGURE 5.7. Different flow possibilities for a completely impermeant solute. (a) $\Delta p_d = 0$, so there is no flow. (b) Flow to the right. (c) Flow to the left.

Parameter L_p is determined by measuring J_v and Δp when $\Delta\pi = 0$, while σ is determined from measurements of Δp and $\Delta\pi$ when $J_v = 0$. Parameter σ is called the *reflection coefficient*. It has different values for different solutes. When $\sigma = 0$ there is no reflection, and the solute particles pass through like water. When $\sigma = 1$ all the solute particles are reflected and Eq. (5.12) is the same as Eq. (5.11).

We can imagine that part of the solute moves freely with the water and part is reflected. We can write

$$p = p_d + \sigma\pi, \quad (5.13)$$

and we can further break this down to a driving pressure for the water p_{dw} and one for the permeant solute:

$$p = \underbrace{p_{dw} + (1-\sigma)\pi}_{\substack{\text{driving pressure}\\ \text{for permeant}\\ \text{molecules} = p_d}} + \underbrace{\sigma\pi}_{\substack{\text{osmotic pressure}\\ \text{of impermeant}\\ \text{molecules}}} . \quad (5.14)$$

where $\overbrace{(1-\sigma)\pi + \sigma\pi}$ = osmotic pressure of all solute molecules.

With this substitution the flow equation becomes

$$J_v = L_p[\Delta p_{dw} + \Delta(1-\sigma)\pi]. \quad (5.15)$$

Figure 5.8 shows the pressure relationships across the membrane.

In the approximation that van't Hoff's law holds, $\pi = k_B TC = RTc$ and Eq. (5.12) can be written as

$$J_v = L_p(\Delta p - \sigma k_B T\ \Delta C), \quad (5.16)$$

$$J_v = L_p(\Delta p - \sigma RT\ \Delta c). \quad (5.17)$$

In Eq. (5.16) the concentration is in molecules m^{-3}; in Eq. (5.17) it is mol m^{-3}. In both cases the units of $k_B T\ \Delta C$ and $RT\ \Delta c$ are pascals.

As an example of volume flow, consider ultrafiltration. *Ultrafiltration* is the process whereby water and small molecules are forced through a membrane by a hydrostatic pressure difference while larger constituents are left behind. An interesting clinical application of ultrafiltration has been proposed. A severely edematous patient (for any of the reasons mentioned in the previous section) must have the extra water removed from the body. This is usually accomplished with diuretics, drugs that increase the renal excretion of water. Some patients may not respond, and in other cases, particularly pulmonary edema, the response may not be fast enough. In the latter case, phlebotomy (bloodletting) is sometimes used to reduce the body water rapidly. This has obvious disadvantages, for example, the removal of blood cells. Silverstein *et al.* (1974) have used ultrafiltration to remove water and sodium from the plasma while leaving the other constituents behind. The apparatus is shown in Fig. 5.9. The ultrafilter consists of a total area $S = 0.2\ m^2$ of membrane, the permeability of which is 1 ml min^{-1} m^{-2} $torr^{-1}$. The pores transport molecules of molecular weight less than 50 000. The filtration rate is set by clamping the ultrafiltrate line (P_F in Fig. 5.9), thereby increasing the pressure on the outside of the ultrafilter and decreasing the pressure drop across the membrane. The pressure was adjusted to give i_v of 32 ml min^{-1} or less, which is equal to that found in a normal kidney.

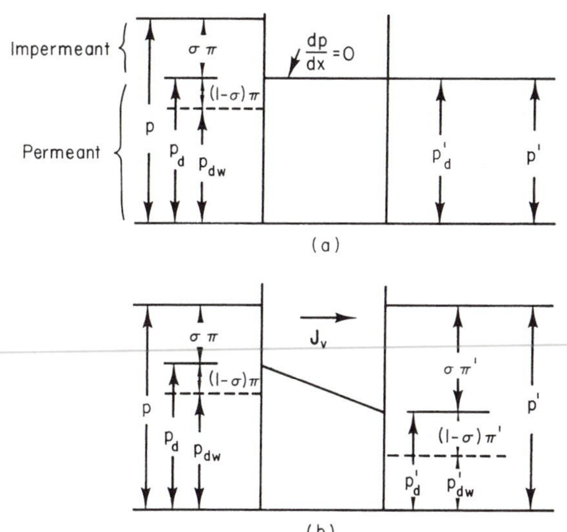

FIGURE 5.8. Pressure relationships on each side of the membrane when $\sigma = \frac{2}{3}$. (d) There is no bulk flow. (b) There is flow to the right.

Figure 5.10 shows the flow vs Δp. The initial slope of this curve determines $L_p S$ and hence L_p. The curve is not

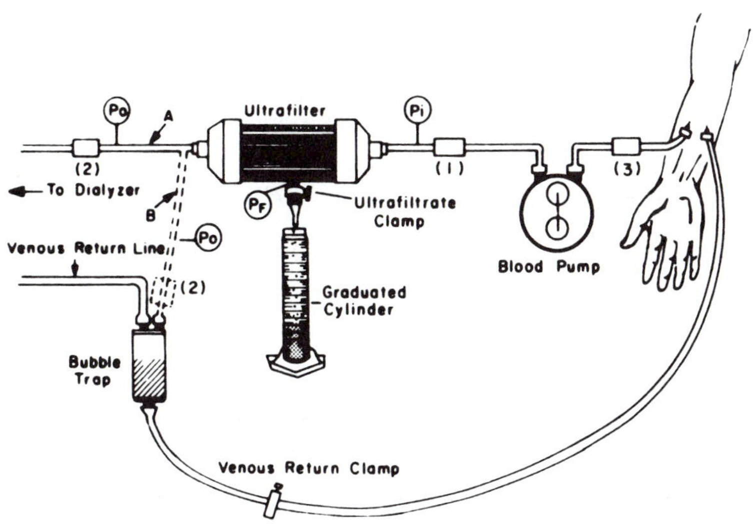

FIGURE 5.9. Apparatus used to treat fluid overload by ultrafiltration. Connection *A* is used for a patient connected to an artificial kidney. Connection *B* is used when the ultrafilter is used by itself. Pressure is monitored at P_i on the input side and P_0 on the output side. Infusion and monitoring sites are (1) anticoagulant infusion site, (2) clotting time in filter, and (3) patient clotting time. From M. E. Silverstein, C. A. Ford, M. J. Lysaght, and L. W. Henderson (1974). Treatment of severe fluid overload by ultrafiltration. Reproduced, by permission, from the *N. Engl. J. Med.* **291**: 747–751. Drawing courtesy of Prof. Henderson.

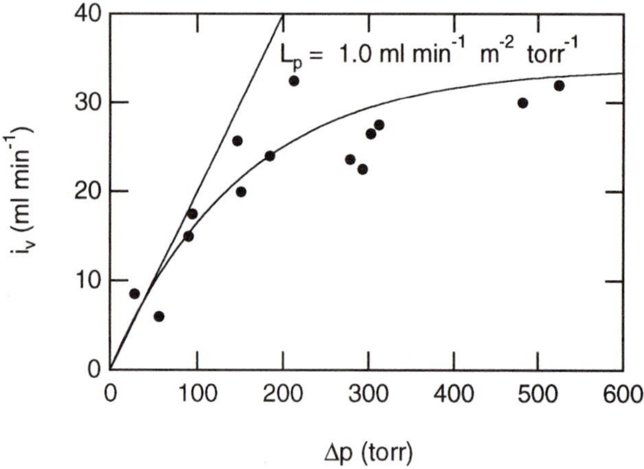

FIGURE 5.10. Filtration rate (flow) i_v, vs transmembrane pressure for a fixed blood flow of 200 ml min^{-1} through the apparatus in Fig. 5.9. The solid straight line shows a value of L_p of 1 ml min^{-1} m^{-2} torr^{-1} as reported by Silverstein *et al.* (see reference in the caption of Fig. 5.9). Modified, by permission, from the *N. Engl. J. Med.* **291**: 747–751 (1974).

linear but saturates at about 32 ml min^{-1} of filtration flow, possibly because of poor mixing within the blood.

Ultrafiltration is sometimes called reverse osmosis. The name is unfortunate, because it suggests some mysterious process unrelated to the principles of this section. Ultrafiltration is often used by campers for purifying water and has been suggested for desalinization of sea water.

5.6. SOLUTE TRANSPORT THROUGH A MEMBRANE

Solute can pass through the membrane in two ways: it can be carried along with flowing water (solvent drag); and it can diffuse.

If there is no reflection ($\sigma=0$) and the solute concentration is the same on both sides of the membrane so there is no diffusion, the flux density or fluence rate is caused by solvent drag and is simply the solute concentration (particles per unit volume) times the volume fluence rate (Sec. 4.2):

$$J_s = C_s J_v .$$

If there is partial reflection with coefficient σ, then

$$J_s = (1-\sigma) C_s J_v .$$

This is consistent with the idea expressed by Eq. (5.14) that a fraction $(1-\sigma)$ of the solute particles can enter the membrane. In that case, C_s is the solute concentration outside the membrane on both sides, and $(1-\sigma)C_s$ is the solute concentration inside the membrane.

If $J_v=0$ there will be no solvent drag but there will be diffusion. The solute flux will be proportional to the concentration gradient and therefore to the concentration difference across the membrane:

$$J_s \propto \Delta C_s .$$

The proportionality constant depends on properties of the membrane. If the membrane is pierced by pores, for example, it depends on pore size, membrane thickness, number of pores per unit area, and the diffusion constant. The dependence will be derived later in this chapter. It is customary to write the proportionality constant as ωRT:

$$J_s = \omega RT \, \Delta C_s .$$

The factor ω is called the *membrane permeability* or *solute permeability*.

In the linear approximation the fluence rate resulting from both processes is the sum of these two terms:

$$J_s = (1-\sigma)\bar{C}_s J_v + \omega RT \, \Delta C_s . \qquad (5.18)$$

Here an average value of $\bar{C}_s$ has been written for the solvent drag term, because the concentration on each side of the membrane is not necessarily the same. The way that this average is taken will become clearer in the discussion of the pore model below.

The solute equation has been written for both fluence rate and concentration in terms of particles. In terms of molar fluence rate and concentration, it is exactly the same:

$$J_s(\text{molar}) = (1-\sigma)\bar{c}_s J_v + \omega RT \, \Delta c_s . \qquad (5.19)$$

Either way, the diffusion proportionality constant is ωRT. It does not change because ΔC_s and J_s (particles) are both written in terms of particles, and Δc_s and J_s (molar) are both written in terms of moles. Referring to Eq. (5.18), the solvent drag term has units of (particles m^{-3}) (m s^{-1})=particles m^{-2} s^{-1}. The term ωRT has units of m s^{-1}. Since the units of RT are joules or N m (per mole), the units of ω are

$$\frac{\text{mol m s}^{-2}}{\text{N m}} = \text{mol N}^{-1}\,\text{s}^{-1}. \qquad (5.20)$$

Further interpretation of ω will be made for specific models.

We have used the same σ in both the solvent drag term and in the preceding section. Although this was made plausible by saying that $1-\sigma$ is the fraction of solute molecules that gets through the membrane, its rigorous proof is more subtle. It has been proved in general using thermodynamic arguments, which can be found in Katchalsky and Curran (1967). It can be proved in detail for specific membrane models.

5.7. EXAMPLE: THE ARTIFICIAL KIDNEY

The artificial kidney provides an example of the use of the transport equations to solve an engineering problem. The problem has been extensively considered by chemical engineers, and we will give only a simple description here. Those interested in pursuing the problem further can begin with reviews by Galetti *et al.* (1995) or Lysaght and Moran (1995). The reader should also be aware that this "high-technology" solution to the problem of chronic renal disease is not an entirely satisfactory one. It is expensive and uncomfortable and leads to degenerative changes in the skeleton and severe atherosclerosis [Lindner *et al.* (1974)]. The alternative treatment, a transplant, has its own problems, related primarily to the immunosuppressive therapy. Anyone who is going to be involved in biomedical engineering or in the treatment of patients with chronic disease should read the account by Calland (1972), a physician with chronic renal failure who had both chronic dialysis and several transplants. The distinction between a high-technology treatment and a real conquest of a disease has been underscored by Thomas (1974, pp. 31–36).

The simplest model of dialysis is shown in Fig. 5.11. Two compartments, the body fluid and the dialysis fluid, are separated by a membrane that is porous to the small molecules to be removed and impermeable to larger molecules. If such a configuration is maintained long enough, then the concentration of any solute that can pass through the membrane will become the same on both sides. The dialysis fluid is prepared with the desired composition of such small molecules as sodium, potassium, and glucose. Volume V' must be larger than V for effective dialysis to take place; otherwise, the concentration of solutes in the dialysis fluid builds up from the initially prepared values. In early work, V' was up to 100 l (since V is about 40 l). Although the fluid was replaced every two hours or so, it was an excellent medium in which to grow bacteria. Although the bacteria could not get through the membrane, they released exotoxins (or, if they died, endotoxins) which diffused back into the patient and caused fever. For the last 30 years a continuous flow system has been used in which the solutes are continually metered into flowing dialysis fluid that is then discarded. Because of this, we will assume that there is no buildup of concentration in the dialysis fluid. (Effectively volume V' is infinite.) We will assume that $\Delta p=0$. (Actually, proteins do cause some osmotic pressure difference, which we will ignore.)

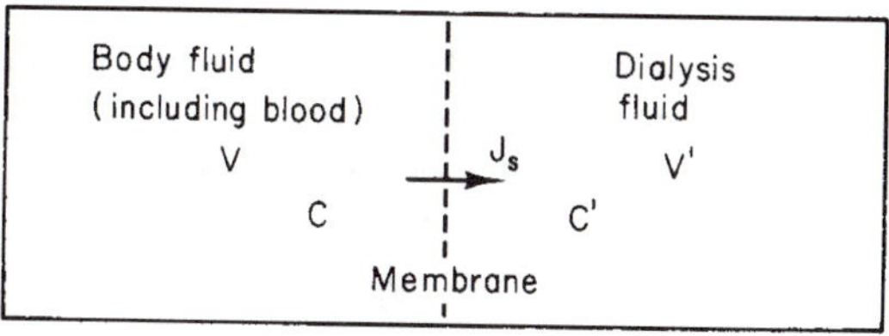

FIGURE 5.11. The simplest model of dialysis. All the body fluid is treated as one compartment; transport across the membrane is assumed to take longer than transport from various body compartments to the blood.

Without solvent drag, the solute transport is by diffusion,

$$J_s=\omega RT(C-C'),$$

where C is the concentration of solute in the blood and C' is the concentration in the dialysis fluid. If the surface area of the membrane is S, then the rate of change of the number of solute molecules is

$$\frac{dN}{dt}=-S\omega RT(C-C').$$

If the solute is well mixed in the body fluid compartment, then $N=CV$, and this equation can be written as

$$\frac{dC}{dt}=-\frac{S\omega RT}{V}(C-C').$$

This is the equation for exponential decay. The steady state solution is $C=C'$. The complete solution is (Appendix F)

$$C(t)=[C(0)-C']e^{-t/\tau}+C', \qquad (5.21)$$

where the time constant τ is

$$\tau=\frac{V}{S\omega RT}. \qquad (5.22)$$

The only things that are adjustable in this equation are the membrane area S and its permeability ω. The size of pores in the membrane is dictated by what solutes are to be retained in the blood. The number of pores per unit area and the thickness of the membrane can be controlled. Typical cellophane membranes have $\omega RT=5\times10^{-6}$ m s^{-1} (with a thickness of 500 μm). The area may be 2 m^2. With a fluid volume $V=40$ l, this gives a time constant

$$\tau=\frac{40\times10^{-3}\ \text{m}^3}{(2\ \text{m}^2)(5\times10^{-6}\ \text{m s}^{-1})}=4\times10^3\ \text{s}=1.1\ \text{h}.$$

Typically, dialysis requires several hours. This longer period is for two reasons. Some of the larger molecules have smaller permeabilities and therefore longer time constants, and rapid dialysis causes cerebral edema and severe headaches.

The actual apparatus is quite complicated. First, it must be sterile, which requires a sterilized, disposable dialysis membrane. Second, the apparatus causes clots, so the blood must be treated with heparin as it enters the machine, and the heparin must be neutralized with protamine as it returns to the patient.

5.8. VOLUME TRANSPORT IN A LARGE PORE

In the next few sections we develop and use a model to predict the values of the phenomenological coefficients of Secs. 5.5 and 5.6. The success of the model depends on its ability to predict behavior, particularly as the size of solute particles is varied. We assume that the membrane has a particularly simple structure.

1. The membrane is pierced by n pores per unit area, all having radius R_p and all being right cylinders. The membrane thickness is ΔZ.
2. The pore and the fluid are electrically neutral. No electrical forces are considered.
3. There is complete mixing on both sides of the pore, so that flow within the liquid on either side can be neglected.
4. The system is in the steady state. There is no variation in flux density (fluence rate) or concentration as a function of time.
5. The pores are large enough so that the bulk flow can be calculated by continuum hydrodynamics.

The quantities considered in these five sections are summarized in Table 5.1.

The results of Chap. 1 can be used when the pore is filled with pure water or water and a solute for which $\sigma=0$. From Eq. (1.39) the flux through a single pore is

$$i_v(\text{single pore}) = \frac{\pi R_p^4}{8\eta}\frac{\Delta p}{\Delta x}. \tag{5.23}$$

The fluence rate through the membrane is obtained by multiplying i_v by nS, the number of pores in area S of membrane, and dividing by S. The result is

$$J_v = \frac{n\pi R_p^4}{8\eta}\frac{\Delta p}{\Delta Z}$$

so that

$$L_p = \frac{n\pi R_p^4}{8\eta\,\Delta Z}. \tag{5.24}$$

TABLE 5.1. *Symbols used for porous membrane.*

Quantity	On left	In pore	On right
Total pressure	p		p'
Solute concentration	C_s	$C(z)$	C_s'
Osmotic pressure	$\pi=k_BTC_s$		$\pi'=k_BTC_s'$
Effectively impermeant part of osmotic pressure	$\sigma\pi$		$\sigma\pi'$
Effectively permeant part of osmotic pressure	$(1-\sigma)\pi$	$p_d(z)$	$(1-\sigma)\pi'$
Water driving pressure	p_{dw}		p_{dw}'

While L_p can be measured fairly easily using Eq. (5.12), it is much more difficult to measure the microscopic quantities needed to test Eq. (5.24). Experimental validation of this equation is considered in a later section. Here we will simply give an example of how calculations are done.

In discussing ultrafiltration we considered a filter (Fig. 5.10) for which $L_p \approx 1$ m min^{-1} m^{-2} torr^{-1}. Since 760 torr$=1\times10^5$ Pa, the hydraulic permeability in SI units is

$$L_p = \frac{1 \text{ ml}}{1 \text{ torr min m}^2}\,\frac{1 \text{ min}}{60 \text{ s}}\times\frac{10^{-6}\text{ m}^3}{1 \text{ ml}}\times\frac{760 \text{ torr}}{1\times10^5 \text{ Pa}}$$
$$=1.27\times10^{-10} \text{ m s}^{-1}\text{ Pa}^{-1}.$$

The manufacturers', literature[10] can be used to estimate

$$R_p \approx 4.5 \text{ nm},$$

$$\Delta Z \approx 10\ \mu\text{m}.^{11}$$

The viscosity of water is 0.9×10^{-3} Pa s at 25 °C. This gives us enough information to estimate n and the fraction of the filter surface that is pores. From Eq. (5.24)

$$n = \frac{8\eta\,\Delta Z\,L_p}{\pi R_p^4} = (8)(0.9\times10^{-3}\text{ Pa s})(10\times10^{-6}\text{ m})$$
$$\times(1.27\times10^{-10}\text{ m s}^{-1}\text{ Pa}^{-1})/\pi(4.5\times10^{-9})^4\text{ m}^4$$
$$=7.10\times10^{15}\ m^{-2}.$$

Since the area of one pore is $\pi R_p^2=6.36\times10^{-17}$ m^2, the total pore area in 1 m^2 is 0.45 m^2, a number that is not unreasonable.

Next consider the volume flow when the reflection coefficient is not zero. The position along the pore is specified by z. The position in a plane perpendicular to the axis of the pore is specified by polar coordinates r and ϕ. Flow of the fluid is described by the vector volume flux density

$$\mathbf{j}_v(r,\phi,z).$$

It is possible to show rigorously that as long as the pore is a right circular cylinder, $\mathbf{j}_v$ points only along z and is independent of ϕ (the fluid does not flow in a spiral and does not flow into or out of the walls):

$$\mathbf{j}_v(r,\phi,z)=j_v(r,z)\hat{\mathbf{z}}. \tag{5.25}$$

The solution is in a steady state and the flow is not changing with time. Therefore the flux density into a volume at z must be the same as the flux density out at $z+dz$:

$$\frac{\partial j_v}{\partial z}=0 \tag{5.26}$$

[10] Amicon XM-50.

[11] This value may not be consistent with the value of L_p quoted. The pore length Δz is not well known, and L_p is variable, depending on experimental conditions.

so that j_v is constant along the z axis (although it can be a function of r). This is just what we saw in Chap. 1 for Poiseuille flow; the variation of j_v with r corresponds to the parabolic velocity profile. A value of $j_v(r)$ that is constant in the z direction requires a constant value of $\partial p/\partial z$ inside the pore.

In the pore, the driving pressure is $p_d(z)$. A typical pressure profile is shown in Fig. 5.12. The symbols are defined in Table 5.1. The pressure in the pore has been drawn with constant slope, since $\partial p_d/\partial z$ is constant. Using Eqs. (5.16) and (5.24), we can write

$$J_v = L_p(\Delta p - \sigma k_B T \,\Delta C_s), \tag{5.27}$$

where

$$L_p = \frac{n\pi R_p^4}{8\eta\,\Delta Z}. \tag{5.28}$$

The value of σ is derived in the next section.

The average value of $j_v(r)$ within the pore will be called $\bar{j}_v$. It is the total flux density through the pore divided by πR_p^2:

$$\bar{j}_v = \frac{i(\text{single pore})}{\pi R_p^2} = \frac{1}{\pi R_p^2}\int_0^{R_p} j_v(r) 2\pi r\, dr$$

$$= \frac{J_v}{n\pi R_p^2} = -\frac{R_p^2}{8\eta}\frac{\partial p_d}{\partial z}. \tag{5.29}$$

5.9. SOLUTE TRANSPORT IN A LARGE PORE

We now consider solute transport in our model pore. The arguments here are very similar to those for combined diffusion and solvent drag that were developed in Sec. 4.12. Those arguments are extended by averaging over the cross section of the pore. We will apply this model when the pore approaches the size of the solute particles. Many objections can be raised to this; in fact, models have been developed that are based on the average frictional force of the wall on the solute particle and the water on the solute particle. Other models consider the fact that in some pores the solute molecules must pass through "single file." We will ignore these more refined models.

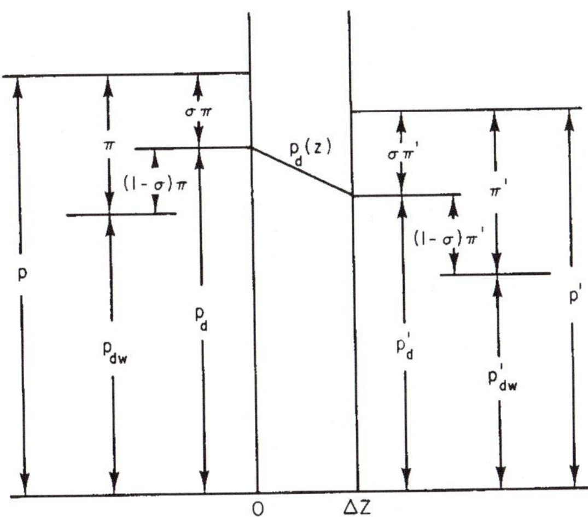

FIGURE 5.12. Pressure within a pore and at the boundaries in the steady state.

Within the pore, the local solute flux is $\mathbf{j}_s(r,\phi,z)$. Arguments similar to those in the preceding section can be offered to show that $\mathbf{j}_s$ points along the z axis and is independent of ϕ:

$$\mathbf{j}_s(r,\phi,z) = j_s(r,z)\hat{\mathbf{z}}. \tag{5.30}$$

The solute concentration does not depend on ϕ, or else there would be diffusion in the ϕ direction and $\mathbf{j}_s$ would have a ϕ component. So

$$C = C(r,z).$$

The r dependence must be kept because the center of a solute molecule of radius a cannot be within a distance a of the wall. Thus $C(r,z)=0$ if $r>R_p-a$. We write[12]

$$C(r,z) = \begin{cases} 0, & R_p - a < r \\ C(z), & 0 \le r \le R_p - a. \end{cases} \tag{5.31}$$

The solute flux due to solvent drag is $C_s j_v$. For diffusion in one dimension the solute flux along the z axis is $-D(\partial C/\partial z)$. For the cylindrical pore we can combine these and write

$$j_s(r,z) = C(r,z) j_v(r,z) - D(r,a,R_p)\frac{\partial C(r,z)}{\partial z}. \tag{5.32}$$

The diffusion constant has been written as a function of r, a, and R_p because in the pore, as distinct from an infinite medium, the constant depends on how close the particle is to the walls. (Remember the relation of D to the viscous drag and the fact that Stokes's law requires modification when the fluid is confined in a tube.)

The preceding section showed that for the steady state j_v is independent of z. A similar argument can be made using the continuity equation for solute particles. Since j_s points in the z direction, the continuity equation is

$$\frac{\partial C}{\partial t} = -\frac{\partial j_s}{\partial z}.$$

When the time derivative vanishes (steady state), j_s is independent of z. Therefore Eq. (5.32) simplifies to

[12]It can be argued that this is the only possible form for $C(r,z)$. See Levitt (1975, p. 535ff.).

$$D(r,a,R_p)\,\frac{\partial C(r,z)}{\partial z}-j_v(r)C(r,z)=-j_s(r). \tag{5.33}$$

The easiest way to write $C(r,z)$ in accordance with Eq. (5.31) is

$$C(r,z)=C(z)\Gamma(r),$$

where

$$\Gamma(r)=\begin{cases}1, & 0\leq r<R_p-a\\ 0, & r>R_p-a.\end{cases} \tag{5.34}$$

With this substitution Eq. (5.33) becomes

$$\Gamma(r)D(r,a,R_p)\,\frac{dC(z)}{dz}-C(z)\Gamma(r)j_v(r)=-j_s(r).$$

This equation can be multiplied by $2\pi r\,dr$ and integrated from $r=0$ to $r=R_p$. The result is

$$\left(\int_0^{R_p}\Gamma(r)D(r,a,R_p)2\pi r\,dr\right)\frac{dC(z)}{dz}-\left(\int_0^{R_p}\Gamma(r)j_v 2\pi r\,dr\right)C(z)=-\int_0^{R_p}j_s(r)2\pi r\,dr. \tag{5.35}$$

The physical meaning of this integration can be understood with the aid of Fig. 5.13, which shows a slab of fluid in the pore between z and $z+dz$. Solute does not cross a surface of constant r but moves parallel to the z axis. Diffusion and solvent drag are considered in each shaded area $2\pi r\,dr$. The integration of Eq. (5.35) establishes an average solute fluence rate, since the right-hand side of the equation is the total flux or current of solute particles per second passing through the pore:

$$i_s=\int_0^{R_p}j_s(r)2\pi r\,dr.$$

As with the volume fluence rate, it is convenient to call the average solute fluence rate $\bar{j}_s$:

$$\bar{j}_s=\frac{i_s}{\pi R_p^2}=\frac{1}{\pi R_p^2}\int_0^{R_p}j_s(r)2\pi r\,dr. \tag{5.36}$$

The first term of Eq. (5.35) is the diffusive flux at z averaged over the entire cross section of the pore. Define an effective diffusion constant

$$D_{\text{eff}}=\frac{1}{\pi R_p^2}\int_0^{R_p}\Gamma(r)D(r,a,R_p)2\pi r\,dr. \tag{5.37}$$

The second term on the left of Eq. (5.35) is the solvent drag flux averaged over the entire cross section of the pore. The integral is

$$\int_0^{R_p}j_v(r)\Gamma(r)2\pi r\,dr=\int_0^{R_p-a}j_v(r)2\pi r\,dr. \tag{5.38}$$

This integral can be evaluated because we know the velocity profile, $j_v(r)$, Eq. (1.38):[13]

$$j_v(r)=\frac{1}{4\eta}\frac{\Delta p}{\Delta x}(R_p^2-r^2).$$

We have already defined the average volume fluence rate to be

$$\bar{j}_v=\frac{1}{\pi R_p^2}\int_0^{R_p}j_v(r)2\pi r\,dr.$$

The desired quantity differs only in the limits of integration. To calculate it, write

$$\int_0^{R_p-a}j_v(r)2\pi r\,dr=\pi R_p^2\bar{j}_v\,\frac{\int_0^{R_p-a}j_v(r)2\pi r\,dr}{\int_0^{R_p}j_v(r)2\pi r\,dr}.$$

The integrals are easily evaluated (see Problems). The result is

$$\int_0^{R_p}j_v(r)\Gamma(r)2\pi r\,dr=\pi R_p^2\bar{j}_v f\left(\frac{a}{R_p}\right), \tag{5.39a}$$

where the function f is

$$f(\xi)=1-4\xi^2+4\xi^3-\xi^4. \tag{5.39b}$$

When Eqs. (5.36), (5.37), and (5.39) are substituted into Eq. (5.35) and each term is divided by πR_p^2, the result is

$$D_{\text{eff}}\left(\frac{dC}{dz}\right)-\bar{j}_v f\left(\frac{a}{R_p}\right)C(z)=-\bar{j}_s \tag{5.40a}$$

or

$$\frac{dC}{dz}-\frac{\bar{j}_v f(a/R_p)}{D_{\text{eff}}}C(z)=-\frac{\bar{j}_s}{D_{\text{eff}}}. \tag{5.40b}$$

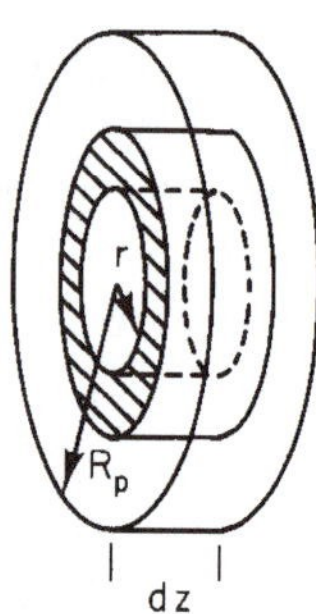

FIGURE 5.13. A slab of fluid in a pore between z and $z+dz$, showing how the integration over r is done.

[13]This ignores the fact that since the walls affect the force on the solute particles, the solute must distort the velocity profile slightly. This point is discussed in the next section.

This is a differential equation for $C(z)$. The right-hand side is the total solute fluence rate, which is constant. On the left-hand side, C varies along the pore so that the diffusive and solvent-drag fluence rates add up to this constant value. If the constant in front of $C(z)$ is written as

$$\frac{1}{\lambda}=\frac{\bar{j}_v f(a/R_p)}{D_{\text{eff}}}, \tag{5.41}$$

this is recognized as Eq. (4.58) for drift plus solvent drag in an infinite medium. The results of Sec. 4.13 can be applied here. It is only necessary to determine values for C_0 and C_0'. Recall that in the pore $C(r,z)=C(z)\Gamma(r)$. The function $\Gamma(r)$ takes into account the reflection that occurs because solute particles cannot be closer to the pore wall than their radius. It was also assumed that the solution on either side of the membrane is well stirred. Therefore, $C_0=C_s$ and $C_0'=C_s'$. Equation (4.70) becomes

$$\bar{j}_s=f\bar{j}_v\bar{C}_s+D_{\text{eff}}\frac{(C_s-C_s')}{\Delta Z}. \tag{5.42}$$

This is an expression for $\bar{j}_s$, the average solute fluence rate in the pore. To get solute fluence rate in the membrane, it must be multiplied by πR_p^2 and the number of pores per unit area. Since $J_v=n\pi R_p^2\bar{j}_v$, we have

$$J_s=f\bar{C}_sJ_v+\frac{n\pi R_p^2D_{\text{eff}}}{\Delta Z}\Delta C_s. \tag{5.43}$$

Comparing this with the general phenomenological equation for solute flow, Eq. (5.18),

$$J_s=(1-\sigma)\bar{C}_sJ_v+\omega RT\,\Delta C_s \tag{5.18}$$

we see that

$$1-\sigma=f,$$

$$\omega RT=\frac{n\pi R_p^2D_{\text{eff}}}{\Delta Z}, \tag{5.44}$$

$$\lambda=\frac{D_{\text{eff}}}{\bar{j}_v(1-\sigma)}=\frac{\omega RT(\Delta Z)}{J_v(1-\sigma)}.$$

The average solute concentration C is obtained from Eq. (4.66) with the substitution of ΔZ for the pore length:

$$\bar{C}_s=\frac{C_se^x-C_s'}{e^x-1}-\frac{1}{x}(C_s-C_s').$$

This can be rearranged as

$$\bar{C}_s=\tfrac{1}{2}(C_s+C_s')+G(x)(C_s-C_s') \tag{5.45a}$$

with

$$G(x)=\frac{1}{2}\left(\frac{1+e^x}{e^x-1}\right)-\frac{1}{x} \tag{5.45b}$$

where $x=\Delta Z/\lambda$. This is the same function we saw in Fig. 4.17. It is odd and is plotted again in Fig. 5.14. An approximate expression for small x is

$$\bar{C}_s=\tfrac{1}{2}(C_s+C_s')+\frac{\Delta Z}{\lambda}\frac{1}{12}(C_s-C_s'). \tag{5.46}$$

The solute concentration away from the sides of the pore is

$$C(z)=\frac{C_s(e^{\Delta Z/\lambda}-e^{z/\lambda})+C_s'(e^{z/\lambda}-1)}{e^{\Delta Z/\lambda}-1}. \tag{5.47}$$

While the concentration profile is not usually measured experimentally, it is useful to plot it to help us visualize the interrelation of diffusion and solvent drag. Call $C_s'/C_s=R$. Equation (5.47) can be rearranged as

$$C(z)=C(0)\left(1-(1-R)\frac{e^{z/\lambda}-1}{e^{\Delta Z/\lambda}-1}\right). \tag{5.48}$$

We can see several things from this equation. First, if the concentration is the same at each end of the pore, $R=1$, the second term in large parentheses vanishes, and the concentration is uniform throughout the pore. If $R\neq1$, then the concentration is that at $z=0$, plus a factor which may be positive or negative, depending on whether R is less than or greater than 1. The ratio of exponentials occurring in that factor is plotted in Fig. 5.15 for different values of $\Delta Z/\lambda$, the ratio of the pore length to the effective diffusion distance.

These curves determine the shape of the concentration profile along the pore. If the flow is zero, $\lambda=D_{\text{eff}}/\bar{j}_v(1-\sigma)$ is infinite and $\Delta Z/\lambda$ is zero. We then have pure diffusion, and the concentration changes uniformly along the pore, corresponding to the straight line in Fig. 5.15. The plots in Fig. 5.16 show what the concentration profiles are like for diffusion to the left and to the right when the flow is to the right. Compare the shape of the concentration

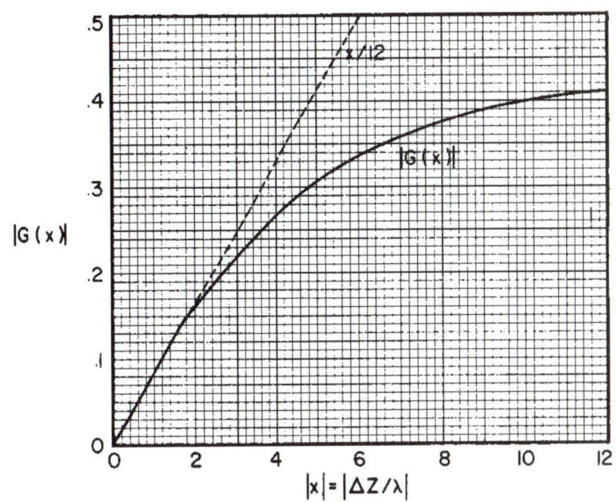

FIGURE 5.14. The correction factor $G(x)$ used in Eq. (5.46). The dashed line is the approximation $G(x)=x/12$, which is valid for small x and is used in Eq. (5.45).

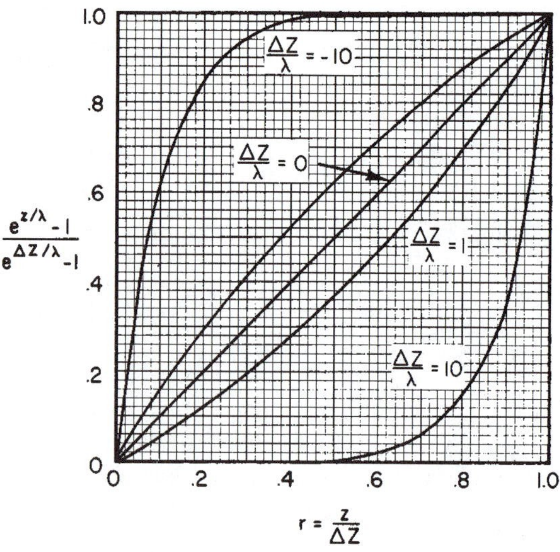

FIGURE 5.15. Plot of the factor $(e^{z/\lambda}-1)/(e^{\Delta Z/\lambda}-1)$, which appears in Eq. (5.48).

profile on the left in Fig. 5.16 with the curve for $\Delta Z/\lambda=1$ in Fig. 5.15. When the concentration is higher on the left, we have to take the mirror image of Fig. 5.15; the curve for $\Delta Z/\lambda=-1$ gives the concentration profile in Fig. 5.16 on the right.

As the pore becomes very long compared to the diffusion length (for example, $|\Delta Z/\lambda|=10$ or more), the concentration along the pore is nearly that carried into the pore by bulk flow from the left, until we get to the far end, where diffusion back up the pore gives a smooth transition to the final concentration on the right.

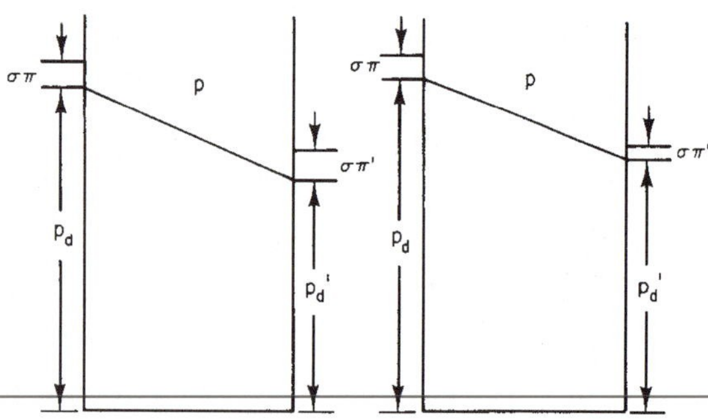

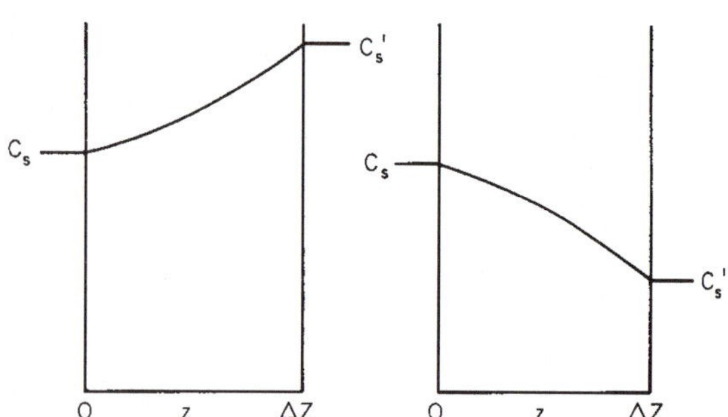

FIGURE 5.16. A possible set of values for p, p_d, and C along a pore for diffusion to the left and diffusion to the right. The fluid on each side of the pore is well stirred and of sufficient volume so that concentrations do not change with time.

Although it is not particularly helpful, we can, if we want, think of the pressure in the pore as being made up of driving pressures due to water and to the solute within the pore:

$$p_d(z)=p_{dw}(z)+p_{ds}(z).$$

Since the effective driving pressure for impermeant solute in the J_v equation is $\sigma k_B T \Delta C$, it would be nice to be able to write

$$p_d(z)=p_{dw}(z)+(1-\sigma)k_B T C(z).$$

This is consistent with the solvent drag flux at position z in the pore, which was given in Eq. (5.40a) by

$$\bar{j}_v f C(z)=\bar{j}_v(1-\sigma)C(z).$$

The "effective" concentration for solvent drag is $(1-\sigma)C(z)$.

Summary

To summarize, the combination of solvent and a solute with reflection coefficient σ has a volume flux

$$J_v=L_p(\Delta p-\sigma k_B T\ \Delta C_s) \tag{5.49}$$

and a solute flux

$$J_s=(1-\sigma)\bar{C}_s J_v+\omega RT\ \Delta C_s. \tag{5.50}$$

The hydraulic permeability is

$$L_p=\frac{n\pi R_p^4}{8\eta\ \Delta Z}. \tag{5.51}$$

The solute permeability is

$$\omega RT=\frac{n\pi R_p^2 D_{\text{eff}}}{\Delta Z}. \tag{5.52}$$

The characteristic length for diffusion is

$$\lambda=\frac{D_{\text{eff}}}{\bar{j}_v(1-\sigma)}=\frac{(\Delta Z)\omega RT}{J_v(1-\sigma)}. \tag{5.53}$$

The average concentration is

$$\bar{C}_s=\tfrac{1}{2}(C_s+C_s')+G(x)\Delta C_s, \tag{5.54}$$

where $G(x)$ is given by Eq. (5.45b). The parameter x is

$$x=\frac{J_v(1-\sigma)}{\omega RT}=\frac{\Delta Z}{\lambda}. \tag{5.55}$$

Notice that the solvent drag term as well as the diffusion term depends on ΔC_s, through the factor $\bar{C}_s$.

There is another form for the solute flow equation that may be useful for calculations, although it mixes up the solvent drag and diffusion terms. Equation (5.55) can be solved for $(1-\sigma)J_v = x\omega RT$.

The result is substituted in Eq. (5.50):

$$J_s = x\omega RT\bar{C}_s + \omega RT(C_s - C_s').$$

Equation (5.54) is then used for $\bar{C}_s$, to get

$$J_s = \omega RT\left(\frac{x}{2}(C_s + C_s') + [xG(x)+1](C_s - C_s')\right).$$

The term $xG(x)+1$ is rewritten with the help of Eq. (5.45b):

$$xG(x)+1 = \frac{x}{2}\frac{e^x+1}{e^x-1}.$$

This is then substituted in the equation for J_s:

$$J_s = \omega RT\left(\frac{x}{2}C_s + \frac{x}{2}C_s' + \frac{x}{2}\frac{e^x+1}{e^x-1}C_s - \frac{x}{2}\frac{e^x+1}{e^x-1}C_s'\right)$$

$$= \frac{x\omega RT}{2}\left[C_s\left(1+\frac{e^x+1}{e^x-1}\right) + C_s'\left(1-\frac{e^x+1}{e^x-1}\right)\right]$$

$$= \omega RT\left(C_s\frac{xe^x}{e^x-1} - C_s'\frac{x}{e^x-1}\right). \qquad (5.56)$$

To use this equation, J_v is calculated and then x is calculated using Eqs. (5.52) and (5.55).

5.10. REFLECTION COEFFICIENT

Previous sections have referred to the fact that not all solute particles can get through the pore. Certainly a particle whose center is off the axis of the pore farther than $R_p - a$ will not get through. The simplest form of correction is the *steric factor*. The ratio of effective area to total area approximates $1-\sigma$. If $\xi = a/R_p$, then

$$1-\sigma \approx \frac{\pi(R_p-a)^2}{\pi R_p^2} = 1 - \frac{2a}{R_p} + \frac{a^2}{R_p^2},$$

$$\sigma = 2\xi - \xi^2.$$

A better calculation was seen in the preceding section. Accept the fact (quoted from thermodynamic results) that the same σ occurs in the equations for J_v and J_s. We saw that the edges of the pore have less bulk flow than the center, so that the steric effect overestimates how many particles are reflected. From Eq. (5.39b),

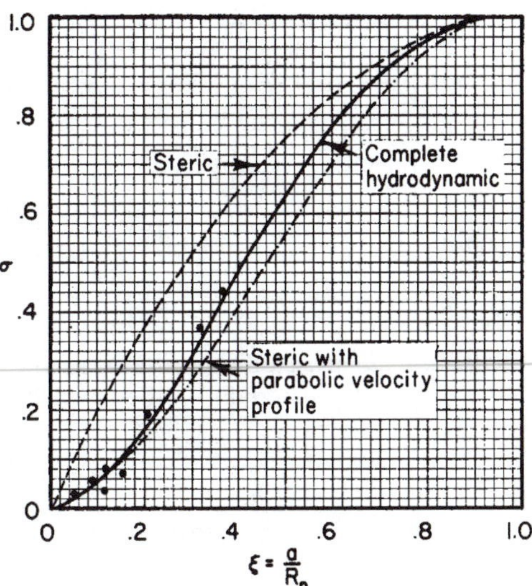

FIGURE 5.17. Calculated values of the reflection coefficient are indicated by the lines. Calculations are shown for the simple steric factor, the steric factor weighted by a parabolic velocity profile, Eq. (5.66), and a more detailed calculation, which takes account of the distortion of the velocity profile by the solute particles. The data points are from Durbin (1960) as reinterpreted by Bean (1972).

$$\sigma = 1 - f = 4\xi^2 - 4\xi^3 + \xi^4. \qquad (5.57)$$

These two approximations to σ are plotted in Fig. 5.17.

It was mentioned in a footnote that the calculation which resulted in Eq. (5.39) neglected the change in velocity profile caused by the solute particles. More rigorous calculations have been done by Levitt (1975) and by Bean (1972, pp. 29–35). Levitt's result is

$$\sigma = \tfrac{16}{3}\xi^2 - \tfrac{20}{3}\xi^3 + \tfrac{7}{3}\xi^4 + 0.35\xi^5. \qquad (5.58)$$

This is valid for $\xi < 0.6$. The three equations for σ are plotted in Fig. 5.17, along with some experimental data from Durbin (1960) using the pore radius assigned by Bean.

The expression for σ is made more useful by having some knowledge of molecular radius for molecules of various molecular weights. Approximate information is given in Fig. 5.18, which is a log–log plot of molecular radius a vs molecular weight M for some common molecules of biological interest. The slope of this curve is 0.38; if the molecules were all of constant density, the dependence of a on M would be $M^{0.33}$; if, instead, these were randomly coiled polymers,[14] the dependence would be on $M^{0.5}$.

[14]See Villars and Benedek (1974, p. 152).

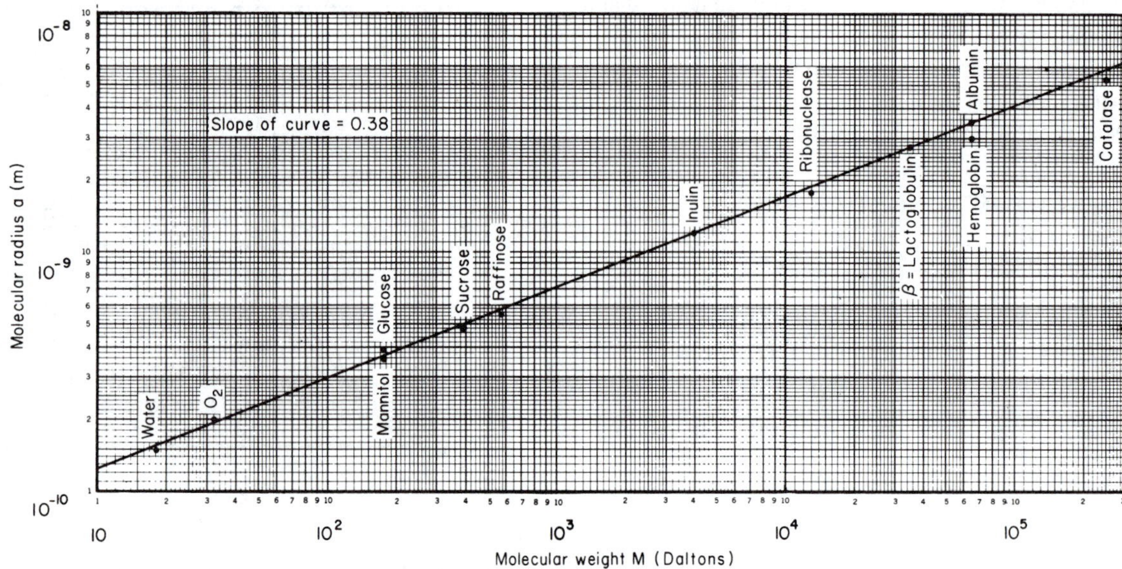

FIGURE 5.18. Plot of the molecular radius vs molecular weight for some common biological molecules. The comments on the slope of the line are discussed in the text.

5.11. THE EFFECT OF PORE WALLS ON DIFFUSION

The solute permeability is given by

$$\omega RT = \frac{n\pi R_p^2 D_{\text{eff}}}{\Delta Z}.$$

An effective diffusion coefficient is used because of the drag on the solute particles by the pore walls. If the pore had an infinitely large diameter, the unrestricted permeability would be

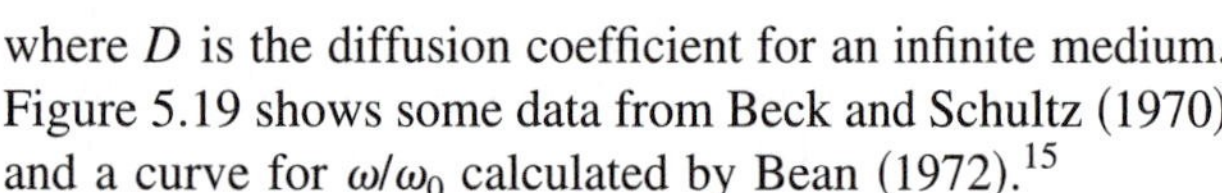

$$\omega_0 RT = \frac{n\pi R_p^2 D}{\Delta Z},$$

where D is the diffusion coefficient for an infinite medium. Figure 5.19 shows some data from Beck and Schultz (1970) and a curve for ω/ω_0 calculated by Bean (1972).[15]

In Europe, filtration rather than dialysis is used to treat kidney patients. There is evidence that some as yet unidentified toxin of medium molecular weight accumulates in the blood. Comparison of $1-\sigma$ from Fig. 5.17 with ω/ω_0 from Fig. 5.19 shows that solvent drag removes medium-sized molecules more effectively. The fluid and electrolytes lost by the patient must be replaced.

[15]Note that the steric factor, which Bean includes separately, is built into D_{eff} through the function $\Gamma(r)$.

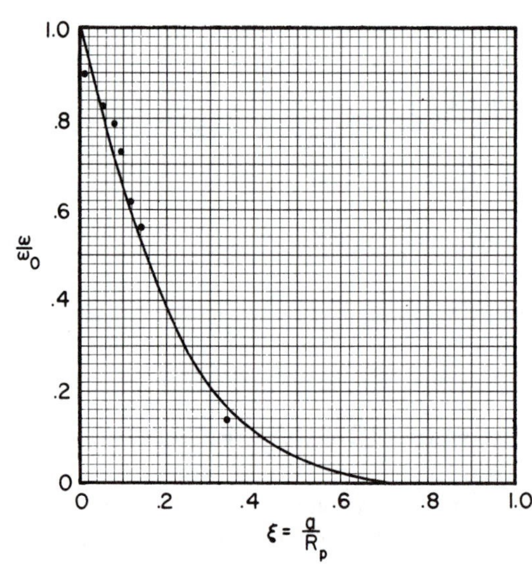

FIGURE 5.19. Plot of ω/ω_0 for experimental data by Beck and Schultz (1970) and a calculation by Bean (1972).

5.12. EXAMPLE: SOLUTE FLOW WHEN THERE IS NO REFLECTION

As an example, we will consider the flow of solute through a membrane when there is no reflection, as a function of the pressure drop across the membrane. Experiments with membranes of this sort have been done using radioactive water for the solute. There is tagged water on the left of the membrane and no tagged water on the right. The solute flux density is most easily calculated using Eq. (5.56). With $C_s'=0$, it gives

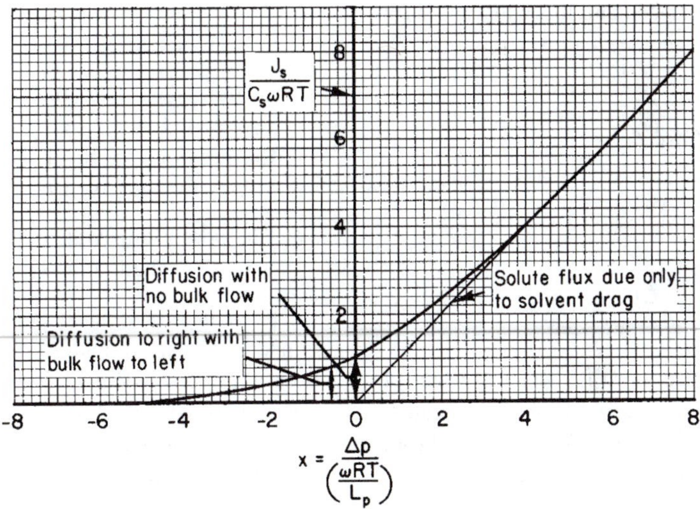

FIGURE 5.20. The flow of solute through a membrane, showing the combination of solvent drag and diffusion. There is no reflection and all solvent is on the left.

$$J_s = \omega RT \frac{xe^x}{e^x - 1} C_s. \tag{5.59a}$$

The value of x is

$$x = \frac{J_v}{\omega RT} = \frac{L_p \Delta p}{\omega RT}. \tag{5.59b}$$

This result is plotted in dimensionless form in Fig. 5.20. When $\Delta p = 0$, there is no solvent drag because there is no bulk flow. For negative Δp, the solvent drag is to the left and subtracts from the diffusive flux, so that $J_s \to 0$. For large positive values of Δp, the flow is dominated by the solvent drag and is proportional to Δp. Figure 5.20 shows these effects and the contribution of diffusion.

We next consider a numerical example. The data are for one of several experiments done by Bean (1969, 1972). The solute was water tagged with tritium (^{3}H). In the experiment we will consider, the pores were diamond-shaped and had a "radius" of about 1.5×10^{-7} m. They were produced by etching fission tracks in mica. The solute flow was measured by observing the radioactivity rise in the right-hand solution, which began with zero concentration of solute. The data are shown in Table 5.2. The temperature was 26.3 °C or 299 K. We will assume that it was 298 K.

The solute flow can be written as follows:

(Total number of molecules through area S per second)

$= J_s S$

$=$ (equivalent flow of bulk fluid)

$\times$(solute concentration)

$= $"$i$"$ C_s$.

From this,

TABLE 5.2. *Experimental data from Bean (1972).*

Pressure (in. of water)	"Flow" "i" ($m^3\ s^{-1}$)
−15	0.19×10^{-12}
−8	0.81×10^{-12}
−4	1.63×10^{-12}
0	3.38×10^{-12}
4	5.94×10^{-12}
8	8.38×10^{-12}

$$``i" = \frac{J_s S}{C_s}.$$

This can be combined with Eq. (5.59a) to obtain

$$``i" = S\omega RT \frac{xe^x}{e^x - 1}. \tag{5.60}$$

To proceed further, we need $x = J_v / \omega RT$ as a function of Δp. Since $\sigma = 0$, the molecules are small and $D_{\text{eff}} = D$. Therefore,

$$\omega RT = \frac{n\pi R_p^2 D}{\Delta Z}.$$

The permeability is

$$L_p = \frac{n\pi R_p^4}{8\eta\, \Delta Z}.$$

Since the pores are diamond-shaped and this equation is for circular pores, we introduce a geometric correction factor g:

$$L_p = \frac{gn\pi R_p^4}{8\eta\,\Delta Z}.$$

With these substitutions, the expression for x is

$$x = \frac{gR_p^2}{8D\eta}\,\Delta p.$$

Bean (1972, p. 15) gives a value for D for water in water of 2.44×10^{-9} $\mathrm{m^2\,s^{-1}}$, at 25 °C. We will use this value, although his experiments were done 1.3 °C higher. From Fig. 4.10, we find that the viscosity at this temperature is $\eta=0.9\times10^{-3}$ Pa s. The pressure data in Table 5.2 are in inches of water. From $p=\rho gh$, we can calculate that 1 in. of water exerts a pressure of

$$(1\times10^3\ \mathrm{kg\,m^{-3}})(9.8\ \mathrm{N\,kg^{-1}})(2.54\times10^{-2}\ \mathrm{m})$$

or 249 Pa. With Δp measured in inches of water and R_p measured in meters, we have

$$x = \frac{249gR_p^2\Delta p}{(8)(2.44\times10^{-9})(0.9\times10^{-3})} = 1.417\times10^{13}\ gR_p^2\Delta p.$$

The coefficient $S\omega RT$ in Eq. (5.60) can be written as

$$S\omega RT = \frac{nSD\pi R_p^2}{\Delta Z}.$$

For the pores in question, $nS=7.54\times10^4$ and $\Delta Z=4.9\times10^{-6}$ m. Therefore,

$$S\omega RT = \frac{(7.54\times10^4)(2.44\times10^{-9})\pi R_p^2}{4.9\times10^{-6}} = 117.9R_p^2.$$

The values for g and for R_p that best fit the data were determined using a computer in the following way.

1. A value for R_p was selected.
2. A value of g was selected.
3. Using these values, x, $F(x)=xe^x/(e^x-1)$, and i were calculated for each of the six pressures.
4. The squares[16] of the residuals (differences between these calculated values and each data point) divided by the experimental value of the flow were summed over all data points. This is called the ''fit.''
5. The value of g was increased by 0.1, and steps 3 and 4 were repeated.
6. If the ''fit'' was improved by step 5, step 5 was repeated. If the ''fit'' did not improve, the results for the previous value of g were printed as the optimum values for the given pore radius.
7. The pore radius was increased and a new, low value of g was selected, and steps 3–6 were repeated.

[16]The procedure for minimizing the sum of the squares of the residuals will be discussed in greater detail in Chap. 11.

The values for g and R_p that best fit the data were determined using the computer program shown in Fig. 5.21. This program is written in C. You should be able to follow the program if you are familiar with any other programming language. Start reading at `main` near the end of the program. The procedures or subroutines that `main` uses come first. `ExpData` places the experimental data points in arrays `p` and `Flow`. The pore radius is varied from 1.0×10^{-7} to 2.2×10^{-7} in steps of 0.1×10^{-7} m. For each pore radius procedure `Loop_On_g` steps the geometric factor g to find the best fit. The fit is calculated in procedure `CalcFit`. If it is an improvement on the previous fit, the values of `g`, `Fit`, and `Flow` are saved permanently by procedure `UseThisFit`. Function `f(x)` is the factor $xe^x/(e^x-1)$ of Eq. (5.60).

The best fit is obtained for $R_p=1.7\times10^{-7}$ m, as shown in Fig. 5.22(a). Flow pressure is plotted in Fig. 5.22(b) for the best value of R_p, along with plots for which R_p is larger and smaller by 0.1×10^{-7} m. The value of g that corresponds to the best fit is 0.7, which means that the hydraulic flow is only 0.7 times what one would expect in a circular pipe. (This correction is discussed from a somewhat different point of view in Bean's papers.)

5.13. EXAMPLE: GLOMERULAR FILTRATION IN THE KIDNEY

The kidney is a vital organ that regulates the composition of the blood and therefore affects the composition of all body fluids. The kidney is a collection of independently functioning units called nephrons that excrete water-soluble wastes and regulate the amount and composition of the body fluids. There are about 10^6 nephrons in each human kidney [Pitts (1974)]. A physicist's view of a nephron is shown in Fig. 5.23. Blood from a renal artery passes first by a membrane in the *glomerulus*, where a large amount of fluid filters through the *basement membrane*. The amount of filtration is regulated by the *afferent* and *efferent arterioles*, which control the pressure of the blood in the glomerulus. The filtrate passes through the *descending* and *ascending tubules*, where most of the filtrate is reabsorbed and other chemicals are secreted into the urine.

While the processes in the tubules involve a great deal of chemistry, including active transport, the initial filtration through the glomerular basement membrane has been explained well using the model we have just developed.

Careful measurements on dog kidneys were interpreted by Verniory *et al.* (1973) in terms of filtration by pores of 5 nm radius in the basement membrane. They used solute molecules of polyvinylpirrolidone labeled with radioactive iodine (^{125}I). These molecules had a distribution of sizes and were separated into fractions with known radii between 1.9 and 3.7 nm. The radii were measured by chromatography.

The measurement of a radioactive solute means that we must determine the solute flux, J_s. It is a combination of solvent drag and diffusion. It is most easily determined using Eq. (5.56). In this case,

```
//Program ModelBean, Figure 5.21
//Produces results in Figure 5.22
//R. K. Hobbie, December 28, 1985; Translated to C January 10, 1994

#include <stdio.h>                  //C input-output routines
#include <math.h>                   //C math routines

const double
   Factor1 = 1.417e13,              //Unit conversion factor for x
   Factor2 = 117.9;                 //Factor for calculating SwRT

double
   p[6],    //Pressure in inches of water
   Flow[6], //Experimental flow in cubic m per s
   FlowCalc[6],                     //Flow calculated this iteration
   FlowPrev[6],                     //Flow in previous iteration
   g, gPrev,                        //Geometric correction factor
   Fit, FitPrev,                    //Sum of weighted square residuals
   x,
   Rp;                              //Pore radius in m

void ExpData(void)                  //Initializes arrays to experimental data
   {
   p[0] = -15;      Flow[0] = 0.19e-12;
   p[1] = -8;       Flow[1] = 0.81e-12;
   p[2] = -4;       Flow[2] = 1.63e-12;
   p[3] = 0;        Flow[3] = 3.37e-12;
   p[4] = 4;        Flow[4] = 5.94e-12;
   p[5] = 8;        Flow[5] = 8.30e-12;
   }

double f (double y)                 //Calculates F(x)
   {
   if (y == 0)
      y = 1;
   else
      y = y * exp(y) / (exp(y) -1);
   return y;
   }                                //end double

void CalcFit (void)                 //Calculates the 'Fit'
   {
   int i;
   Fit = 0;
   for (i = 0;  i <= 5;  ++i)
      {
      x = Factor1 * p[i] * pow (Rp, 2) * g;
      FlowCalc[i] = Factor2 * pow (Rp, 2) * f(x);
      Fit = Fit + pow((Flow[i] - FlowCalc[i]),2) / pow (Flow[i], 2);
      }                             //end for
   }                                //end CalcFit
```

FIGURE 5.21. The computer program used to generate the fit of Fig. 5.22.

```
void UseThisFit(void)                    //sets current data to 'previous' data
   {
   int i;
   gPrev = g;
   FitPrev = Fit;
   for (i = 0;  i <= 5;  ++i)
      FlowPrev [i] = FlowCalc [i];
   }                                     //end UseThisFit

void Loop_On_g(void)                     //calculates g, sets loop for using
Calcfit &
                                         // UseThisFit and displays data
   {
   int i;
   g = 0.2;
   CalcFit();
   UseThisFit();
   g = g + 0.05;
   CalcFit();
   while(FitPrev > Fit)
      {
      UseThisFit();
      g = g + 0.05;
      CalcFit();
      }                                  //end while
   //  Print results
   printf("Rp = %3.2e %4s Fit = %3.2f %4s g = %3.2f \n",
Rp,"",FitPrev,"",gPrev);
   for (i = 0; i <= 5; ++i)
      printf("%6s p = %3.0f %4s Flow = %3.2e \n","",p[i],"", FlowPrev[i]);
   printf("\n");
   scanf("%c");                          //Wait for a character from the
                                         //keyboard to continue
   }                                     //end Loop_On_g

main(void)
   {
   ExpData();
   for(Rp = 1.0e-7; Rp < 2.2e-7; Rp = Rp + 1e-8)
      Loop_On_g();
   }                                     //end main
```

Figure 5.21. (Continued)

$$J_s = \omega RT\left(C_s \frac{xe^x}{e^x - 1} - C_s' \frac{x}{e^x - 1} \right) \quad (5.61)$$

with

$$x = \frac{J_v(1-\sigma)}{\omega RT}.$$

A unique feature of this problem is that the filtrate in the tubule consists entirely of material that has passed through the basement membrane. Since J_s is the solute flux density and J_v the flux density of the fluid in which it is dissolved, the concentration is

$$C_s' = \frac{J_s S\ \Delta t}{J_v S\ \Delta t} = \frac{J_s}{J_v}.$$

From this,

$$J_s = C_s' J_v = \frac{C_s' x \omega RT}{1-\sigma}.$$

This can be put on the left of Eq. (5.56) to give

$$C_s' x = (1-\sigma)\left(C_s \frac{xe^x}{e^x - 1} - C_s' \frac{x}{e^x - 1} \right).$$

The ratio $\phi = C_s'/C_s$ was determined experimentally. To find ϕ, divide each term in the preceding equation by C_s:

$$\phi x = \frac{(1-\sigma)xe^x - (1-\sigma)\phi x}{e^x - 1}.$$

This can be solved for ϕ:

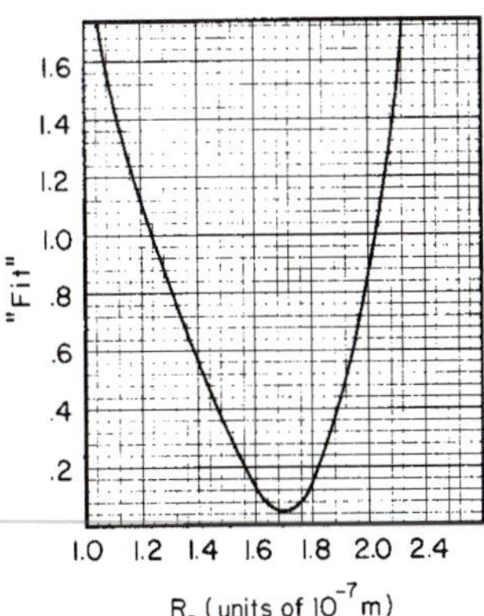

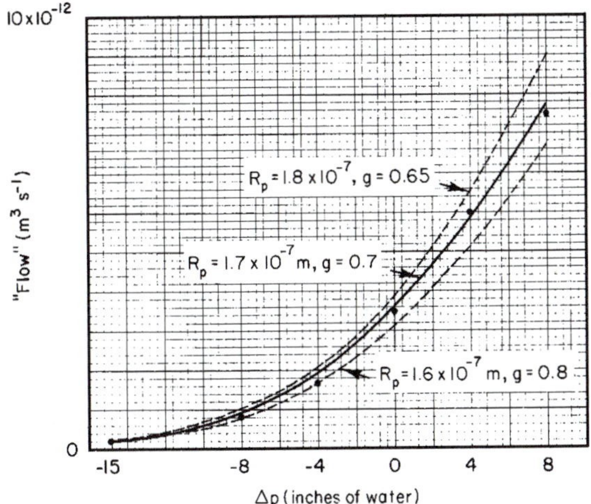

FIGURE 5.22. (a) "Fit" vs R_p. (b) "Flow" vs Δp, showing the experimental data of Table 5.2 and the fits for three values of R_p, with the optimum value of g for each R_p.

$$\phi=\frac{(1-\sigma)e^x}{e^x-\sigma}=\frac{1-\sigma}{1-\sigma e^{-x}}. \tag{5.62}$$

If the solute particles are very small (such as the inulin molecules of Problem 5.42), then $\phi=1$ and $C'_{\text{inulin}}=C_{\text{inulin}}$. Since inulin is not reabsorbed in the tubules, the total amount of inulin excreted in the urine in time Δt can be used to measure $J_{\text{inulin}}S$. If the concentration of inulin in the blood C_{inulin} is measured, then the total flow rate is

$$i_v=J_vS=\frac{J_{\text{inulin}}S}{C_{\text{inulin}}}. \tag{5.63}$$

(Note the units in this equation: J_v is in m s^{-1}; J_{inulin} is in particles m^{-2} s^{-1}.) Verniory *et al.* determined $i_v=3.4\times10^{-7}$ m^3 s^{-1} for dog kidneys. In one day this would amount to 29.4 l. Clearly much of this goes back into the plasma through the tubules. We can use the rectilinear pore model to write the filtration rate as

$$i_v=\frac{n\pi R_p^2 S}{\Delta Z}\frac{R_p^2(\Delta p-\Delta\pi)}{8\eta}. \tag{5.64}$$

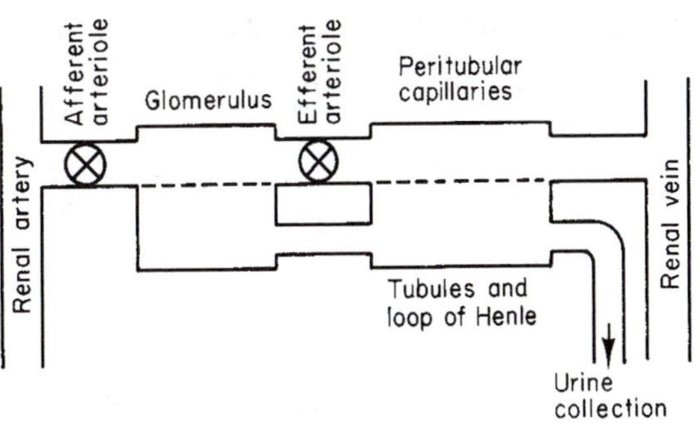

FIGURE 5.23. A schematic view of a nephron. The essential parts are the glomerulus, where initial filtration takes place, and the tubules, where reabsorption and secretion take place. The valves in the afferent and efferent arterioles regulate the pressure difference across the basement membrane in the glomerulus.

The experiment measured ϕ as a function of solute radius a. This requires knowing the ratio $\xi=a/R_p$, which determines the reflection coefficient and ω/ω_0, and the parameter x used in Eq. (5.61). Since $i_v=J_vS$, we can write

$$x=\frac{(i_v/S)[1-\sigma(\xi)]}{[\omega(\xi)/\omega_0]\omega_0RT}=\frac{i_v}{D}\frac{\Delta Z}{n\pi R_p^2S}\frac{1-\sigma(\xi)}{\omega(\xi)/\omega_0}.$$

Using the Stokes–Einstein relation, $D=k_BT/6\pi\eta a$, gives

$$x=\frac{6i_v\eta}{k_BT}\frac{\Delta Z}{n\pi R_p^2S}a\frac{1-\sigma(\xi)}{\omega(\xi)/\omega_0}. \tag{5.65}$$

We need a value of viscosity for plasma. (We do not use the viscosity of whole blood because the cells, with a diameter of about 8000 nm, do not enter the 5-nm pores.) The viscosity is about twice that of water, or 1.4×10^{-3} Pa s at body temperature [Cokelet (1972)].

We have two parameters that are not known: the combination $n\pi R_p^2S/\Delta Z$, which is the ratio of total glomerular pore area per kidney to the membrane thickness, and R_p, which then determines ξ. (Actually, n, R_p, S, and ΔZ are all unknown, but only these two combinations appear.) Verniory *et al.* did a fit varying these parameters, much as we did in the previous section. They obtained the best fit for the values

$$\frac{n\pi R_p^2S}{\Delta Z}=7.5\times10^4 \text{ m}, \quad R_p=5.06 \text{ nm}. \tag{5.66}$$

With these values, and expressing the pressure in torr rather than pascal (760 torr$=10^5$ Pa), we write Eq. (5.64) as

$$i_v=\frac{(7.5\times10^4)(5\times10^{-9})^2(1\times10^5)(\Delta p-\Delta\pi)}{(8)(1.4\times10^{-3})(760)},$$

from which we obtain $\Delta p-\Delta\pi=15.4$ torr. We can also calculate x:

$$x=\frac{(3.4\times 10^{-7}\ \mathrm{m^3\,s^{-1}})(6)(\pi)(1.4\times 10^{-3}\ \mathrm{kg\,m^{-1}\,s^{-1}})}{(7.5\times 10^{4}\ \mathrm{m})(1.38\times 10^{-23})(310)(\mathrm{kg\,m^2\,s^{-2}})}\times a\,\frac{1-\sigma(\xi)}{\omega(\xi)/\omega_0}=(2.8\times 10^{7}\ \mathrm{m^{-1}})(a)\,\frac{1-\sigma(\xi)}{\omega(\xi)/\omega_0}. \quad (5.67)$$

Given values of a and R_p, we can calculate ξ, determine $\sigma(\xi)$ and $\omega(\xi)/\omega_0$ from Eq. (5.58) and Fig. 5.19, and calculate ϕ. This is done in the computer program of Fig. 5.24. The results are shown in Fig. 5.25 along with the data of Verniory *et al.*

```
//   Calculates results for Fig. 5.24
//   R.K. Hobbie, December 28, 1985, Translation to C Jan. 7, 1994

#include <stdio.h>
#include <math.h>

const double
   Rp = 5.066e-9;                        // Pore radius in m

double a,  phi,  x,  xi;

double sigma (double xi)
   {
   double  y;

   if (xi <= 0.6)
      {
      y = 16.0 * pow(xi, 2) - 20.0 * pow(xi, 3) + 7.0 * pow(xi, 4);
      y = y / 3.0 - 0.35 * pow(xi, 5);
      }
   else
      y = 1 - 1.56 * pow( xi - 1, 2);
   return y;
   }                                     // end sigma

double f (double xi)
   {
   double y;
   y = 1 - 2.10443 * xi + 2.08877 * pow(xi, 3);
   y = y - 0.94813 * pow(xi, 5) - 1.372 * pow(xi, 6);
   y = y + 3.87 * pow(xi, 8) - 4.19 * pow(xi, 10);
   y = y * pow(1 - xi, 2);
   return y;
   }                                     //end f

void main ()
   {
   printf("            a          x          phi          xi          sigma       f\n");
   a = 1.0e-9;           //Initial particle radius in m
   do
      {
      xi = a / Rp;
      x = 2.87e7 * a * (1 - sigma(xi)) / f(xi);
      phi = (1 - sigma(xi)) /  (1 - exp(-x) * sigma(xi));
      printf("         %2.1e %3s %4.3f %3s %4.3f %5s",a,"",x,"",phi,"");
      printf("%4.3f %4s %4.3f %1s %4.3f\n",xi,"",sigma(xi),"",f(xi));
      a = a + 2e-10;
      }
   while (a <= 3.6e-9);
   }                                     // end main
```

FIGURE 5.24. The computer program and the results of the calculation for the curve in Fig. 5.25. Equations (5.62) and (5.67) were used. The function for $\sigma(\xi)$ is given in Eq. (5.58), with a parabolic interpolation for $0.6<\xi<1.0$. The function $F=\omega(\xi)/\omega_0$ is calculated using Eqs. (22) and (56) of Bean (1972).

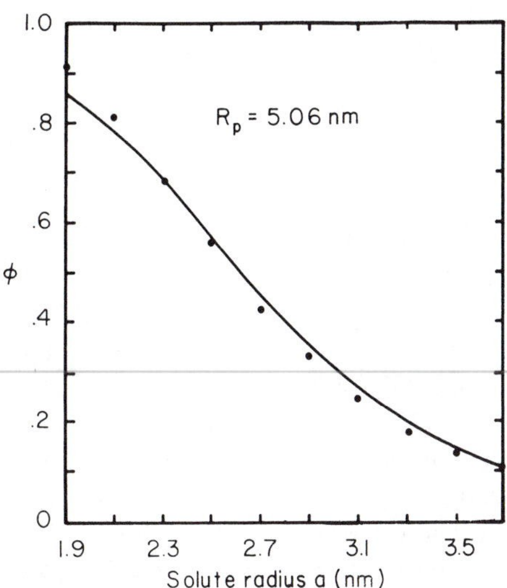

FIGURE 5.25. The data from Verniory *et al.* (1973) with the theoretical fit of Fig. 5.24. The parameters used are given in Eq. (5.58).

5.14. COUNTERCURRENT TRANSPORT

This section considers a problem that demonstrates the principle of *countercurrent* transport. An apparatus (perhaps a dialysis machine or an oxygenator) transports a single solute across a thin membrane of permeability ωRT. On one side of the membrane (the "inside") is a thin layer of solvent that flows along the membrane in the $+x$ direction as shown in Fig. 5.26. On the "outside" is another thin layer of solvent that may be at rest or may flow in either the $+x$ or the $-x$ direction. When it flows in the opposite direction of the fluid inside we have the countercurrent situation.

Suppose that the concentration of solute in the two layers is $C_{\text{in}}(x)$ inside and $C_{\text{out}}(x)$ outside. Solute is transported in the x direction in each fluid layer by pure solvent drag. It passes through the membrane from the side with higher concentration to the other. We develop the model below and show that the steady-state concentration profiles are quite different depending on whether the solvent flows are in the same or opposite directions. The results are shown in Fig. 5.27 for the situation in which the value of C_{in} is 1 and the value of C_{out} is 0 as each solvent starts to flow across the membrane. In Fig. 5.27(a) both flows are to the right; in Fig. 5.27(b) the flows are in opposite directions. The countercurrent case is more effective in reducing C_{in}. The final value of C_{in} is 0.5 in the first case and 0.33 in the second.

To develop the model, we make the following assumptions. The thickness of the fluid layer is h on each side. The concentration of solute in each fluid layer is independent of y, z, and t. The fluid velocity $j_{v\,\text{in}}$ is everywhere constant. The only important mechanism for solute transport within the fluid is solvent drag. Let the length of the slab in the y direction be Y. Inside, the number of particles per second in through the face of the rectangle of area Yh at x is $C_{\text{in}}(x) j_{v\,\text{in}} Yh$. The number out through the face at $x+dx$ is $C_{\text{in}}(x+dx) j_{v\,\text{in}} Yh$. The number through the membrane into

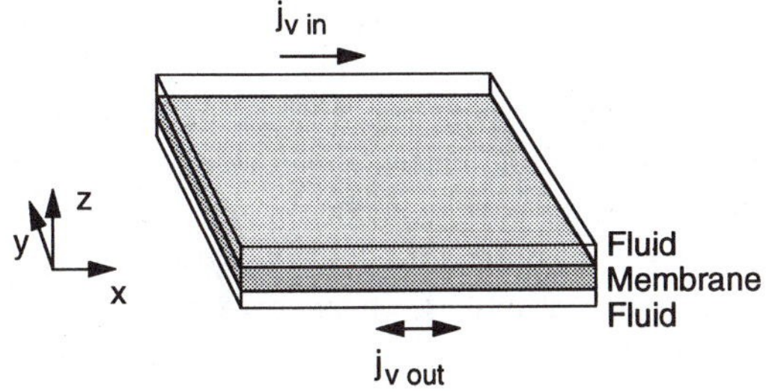

FIGURE 5.26. Layers of fluid containing a solute flow parallel to the x axis on either side of a membrane.

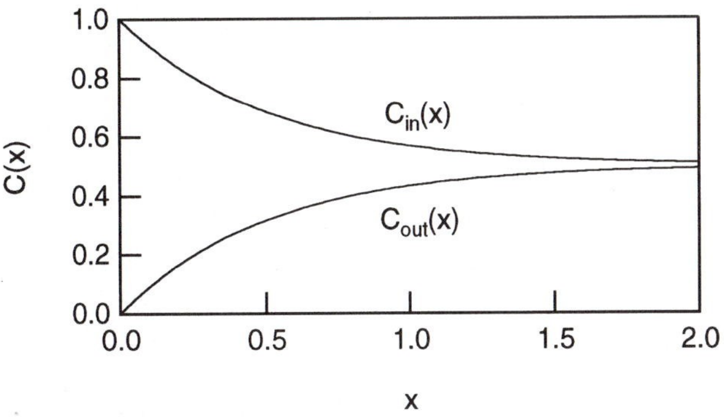

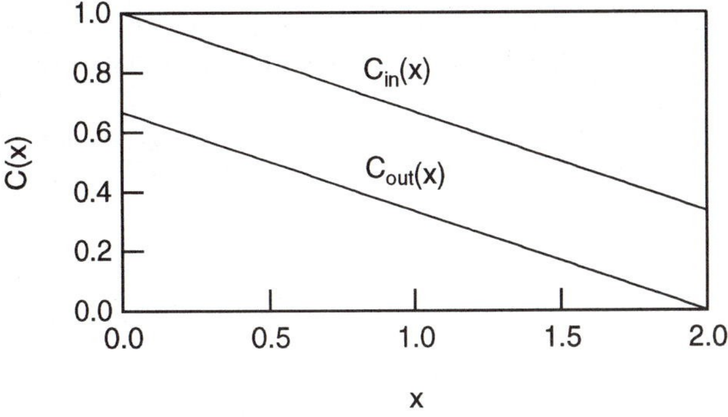

FIGURE 5.27. Solute concentration profiles for two different situations where solvent flows parallel to the membrane surface and solute moves through the membrane from inside to outside. (a) Both fluid layers flow to the right. The concentrations rise and falls exponentially, eventually becoming the same on both sides of the membrane. (b) The countercurrent case, in which the solvent flows are in opposite directions. The solvent outside flows from right to left. The concentrations vary linearly.

the exterior volume is $[C_{in}(x)-C_{out}(x)]\omega RTYdx$. Combining these we get

$$\frac{dC_{in}}{dx}=-\frac{\omega RT}{j_{v\ in}h}[C_{in}(x)-C_{out}(x)]. \tag{5.68}$$

A similar expression can be derived for the exterior:

$$\frac{dC_{out}}{dx}=\frac{\omega RT}{j_{v\ out}h}[C_{in}(x)-C_{out}(x)]. \tag{5.69}$$

In general, h and j_v could be different on each side of the membrane. Our notation allows j_v to have a different direction (sign). Defining $a=\omega RT/j_v h$ we have the coupled differential equations

$$\frac{dC_{in}}{dx}=-a_{in}(C_{in}-C_{out}),$$
$$\frac{dC_{out}}{dx}=+a_{out}(C_{in}-C_{out}). \tag{5.70}$$

We restrict ourselves to the case in which $|a_{in}|=|a_{out}|=a$. Changing the direction of j_v changes the sign of a. Assume h is the same on both sides. The equations show that the slope of $C_{in}(x)$ is minus the slope of $C_{out}(x)$ if both currents are in the same direction, and the two slopes are the same if the currents are in opposite directions. This can be seen in the solutions in Fig. 5.27.

You can verify that Eqs. (5.71) represent a solution of Eqs. (5.70):

$$C_{in}(x)=\frac{c_1}{2}(1+e^{-2ax})+\frac{c_2}{2}(1-e^{-2ax}),$$
$$C_{out}(x)=\frac{c_1}{2}(1-e^{-2ax})+\frac{c_2}{2}(1+e^{-2ax}). \tag{5.71}$$

Setting $c_1=1$ and $c_2=0$ gives a solution for which $C_{in}(0)=1$ and $C_{out}(0)=0$. Figure 5.27(a) shows the concentrations in such a case for $a=1$ and $0<x<2$. If the sign of a is changed in the second differential equation, then the fluid outside is flowing in the opposite direction to the fluid inside. Again you can verify that the most general solution is

$$C_{in}(x)=c_1+(c_2-c_1)ax,$$
$$C_{out}(x)=c_2+(c_2-c_1)ax. \tag{5.72}$$

Figure 5.27(b) is a plot with the constants set so that the concentration inside on the left is 1 and on the outside on the right is zero ($c_1=1$, $c_2=2/3$, $a=1$, $0<x<2$). This configuration is called *countercurrent* flow. We can see from the figure that the transport through the membrane is increased because the concentration difference across the membrane is, on average, greater. To see this quantitatively, note that the total current through the membrane when its length is X is

$$i=\omega RTY\int_0^X[C_{in}(x)-C_{out}(x)]dx.$$

For the first case this is

$$i=\omega RTY(c_1-c_2)\int_0^X e^{-2ax}dx$$
$$=\frac{\omega RTL(c_1-c_2)(1-e^{-2aX})}{2a}.$$

For the second case

$$i=\omega RTY(c_1'-c_2')X.$$

(The primes have been used because the constants are different.) For our example the ratio is

$$\frac{(c_1'-c_2')2aX}{(c_1-c_2)(1-e^{-2ax})}=\frac{(1/3)(2)(1)(2)}{1-e^{-2(1)(2)}}=\frac{1.333}{0.982}=1.36.$$

The countercurrent principle is found in the renal tubules [Guyton (1991), p. 309; Patton *et al.* (1989), p. 1081] and in the villi of the small intestine [Patton *et al.* (1989), p. 915]. The principle is also used to conserve heat in the extremities. If a vein returning from an extremity runs closely parallel to the artery feeding the extremity, the blood in the artery will be cooled and the blood in the vein warmed. As a result, the temperature of the extremity will be lower and the heat loss to the surroundings will be reduced.

SYMBOLS USED IN CHAPTER 5

Symbol	Use	Units	First used on page
a	Solute particle radius	m	117
$a, a_{\text{in}}, a_{\text{out}}$	Parameters	m^{-1}	129
c_1, c_2, c_1', c_2'	Solute concentration	mol m^{-3}	106
f	Temporary function		118
g	Geometric correction factor for diamond-shaped pore		124
g	Gravitational acceleration	m s^{-2}	124
h	Height of column of water	m	124
h	Thickness of fluid layer	m	129
i	Solute current through membrane	s^{-1}	130
i_s	Solute flow	s^{-1}	118
i_v	Volume flow	$\text{m}^3\ \text{s}^{-1}$	112
j_s	Solute fluence rate in pore	$\text{m}^{-2}\ \text{s}^{-1}$	117
$j_v, \mathbf{j}_v$	Volume fluence rate in pore	m s^{-1}	116
k_B	Boltzmann's constant	J K^{-1}	106
n	Number of moles		106
n	Number of pores per unit area	m^{-2}	116
p_1, etc.	Pressure	Pa	106
p_d	"Driving pressure"	Pa	108
p	Total pressure	Pa	107
p_{ds}, p_{dw}	"Driving pressure" of solute or water	Pa	112
r	Radius in cylindrical coordinates	m	116
x	$\Delta Z/\lambda$		119
z	Distance along pore	m	116
$\hat{\mathbf{z}}$	Unit vector in z direction		116
C, C_s, etc.	Molecular concentration of the species indicated by the subscript	m^{-3}	106
D, D_{eff}	Diffusion constant	$\text{m}^2\ \text{s}^{-1}$	117
G	Factor relating solvent drag and diffusion		119
J_s	Solute fluence rate through membrane	$\text{m}^{-2}\ \text{s}^{-1}$	114
J_v	Volume fluence rate through membrane	m s^{-1}	112
L_p	Hydraulic permeability	$\text{m s}^{-1}\ \text{Pa}^{-1}$	112
M	Molecular weight		121
N_1, etc.	Number of molecules		106
N_A	Avogadro's number		106
R	Gas constant	$\text{J mol}^{-1}\ \text{K}^{-1}$	106
R_p	Pore radius	m	116
R	Ratio C_s/C_s'		119
S	Surface area	m^2	112
T	Absolute temperature	K	106
V, V', V^*	Volume	m^3	106
$\overline{V}_s$	Volume of 1 mol of solute	m^3	108
X, Y	Distance	m	129
$\Delta X, \Delta Z$	Pore length	m	116
η	Viscosity	Pa s	116
λ	Effective diffusion distance	m	119
μ	Chemical potential	J molecule^{-1}	108
ξ	a/R_p		118
π	Osmotic pressure	Pa	108
ρ	Density	kg m^{-3}	124
σ	Reflection coefficient		112
τ	Time constant	s	115
ω	Solute permeability	$\text{mol N}^{-1}\ \text{s}^{-1}$	114
ω_0	Solute permeability in an infinite medium	$\text{mol N}^{-1}\ \text{s}^{-1}$	122
ϕ	Angle in cylindrical coordinates		116
ϕ	Ratio C_s/C_s'		126
Γ	Radial dependence of solute concentration		118

PROBLEMS

Section 5.3

5.1. The protein concentration in serum is made up of two main components: albumin (molecular weight 75 000) 4.5 g per 100 ml and globulin (molecular weight 170 000) 2.0 g per 100 ml. Calculate the osmotic pressure due to each constituent. (These results are inaccurate because of electrical effects.)

5.2. If the osmotic pressure in human blood is 7.7 atm at 37 °C, what is the solute concentration assuming that $\sigma=1$? What would be the osmotic pressure at 4 °C?

5.3. Sometimes after trauma the brain becomes very swollen and distended with fluid, a condition known as cerebral edema. To reduce swelling, mannitol may be injected into the bloodstream. This reduces the driving force of water in the blood, and fluid flows from the brain into the blood. If 0.01 mol l^{-1} of mannitol is used, what will be the approximate osmotic pressure?

Section 5.5

5.4. Suppose that L_p is expressed in $m^3\ N^{-1}\ s^{-1}$ or $m\ s^{-1}\ Pa^{-1}$. Find conversion factors to express it in
(a) $ml\ min^{-1}\ cm^{-2}\ torr^{-1}$.
(b) $ml\ s^{-1}\ cm^{-2}$ (in. water)$^{-1}$.
(c) $ml\ s^{-1}\ cm^{-2}$ (lb/in.^{2})$^{-1}$.

5.5. An ideal semipermeable membrane is set up as shown. The membrane surface area is S; the cross-sectional area of the manometer tube is s. At $t=0$, the height of fluid in the manometer is zero. The density of fluid is ρ. Show that the fluid height rises to a final value with an exponential behavior. Find the final value and the time constant. Ignore dilution of the solute.

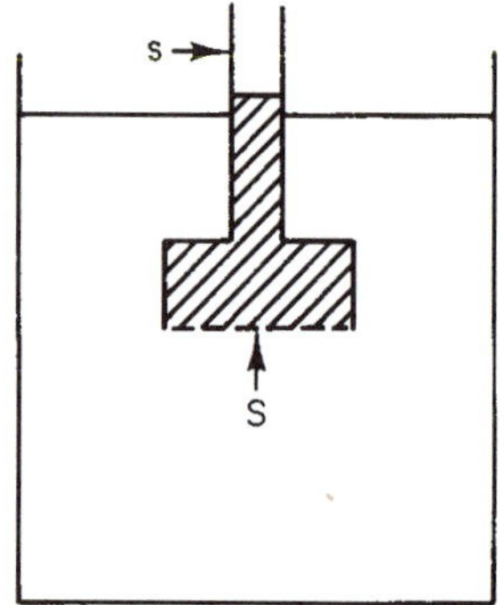

5.6. Consider the design of a lecture demonstration apparatus to show osmotic pressure that uses a commercially available filter as shown in the drawing. Assuming well-stirred fluid on both sides of the membrane and neglecting the change of solute concentration in the manometer tube as water flows in, one finds that height z increases to the equilibrium value exponentially, with a time constant obtained in the previous problem. What would be the time constant if one used the membrane described in Fig. 5.10? For that membrane $L_p=1$ ml $min^{-1}\ m^{-2}\ torr^{-1}$, and the total membrane area is $S=0.2\ m^2$. Suppose that the inner radius of the manometer tube is 1 mm. (One could not use sucrose as a solute, because this particular membrane is permeable to molecules of molecular weight less than 50 000.)

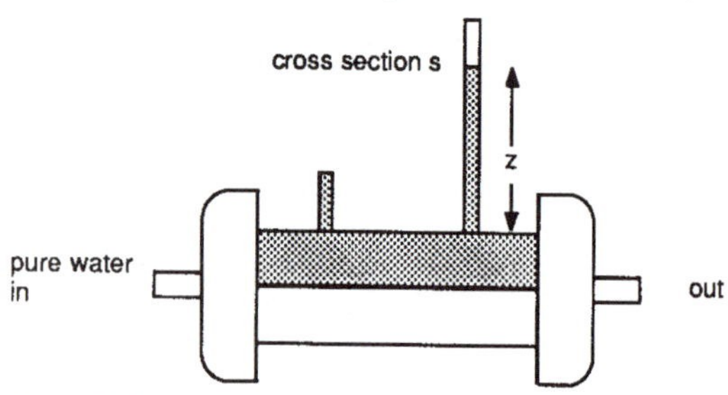

Section 5.6

5.7. Two membranes have permeabilities $\omega_1 RT$ and $\omega_2 RT$. Find the permeability of a two-layered membrane in terms of ω_1 and ω_2.

Section 5.7

5.8. A kidney machine has a membrane permeability $\omega RT=0.5\times10^{-3}$ cm s^{-1}. If the membrane area is 1 m^2, the volume of body fluid is 40 l, and the volume of dialysant is effectively infinite, what is the time constant? How long will it take to reduce the BUN (blood urea nitrogen) concentration from 120 mg per 100 ml to 20 mg per 100 ml?

5.9. Find the pair of coupled differential equations for C and C' for a dialysis machine in which V' is not infinite.

Section 5.9

5.10. Derive Eqs. (5.39a) and (5.39b).

5.11. Write Eq. (5.55) with $J_v=L_p\Delta p$ and $\sigma=0$, and show that it is independent of viscosity.

5.12. Show from Eq. (5.47) that

$$\bar{C}_s=\frac{1}{\Delta Z}\int_0^{\Delta Z} C(z)\,dz$$

$$=\frac{(C_s e^x - C_s') - (1/x)(e^x-1)(C_s-C_s')}{e^x-1},$$

where $x=\Delta Z/\lambda$.

5.13. Expand

$$e^x=1+x+\frac{x^2}{2!}+\frac{x^3}{3!}$$

and show that

$$\bar{C}_s=\tfrac{1}{2}(C_s+C_s')+\frac{x}{12}(C_s-C_s').$$

5.14. Show that Eq. (5.47) gives $C(z)=$const when $\lambda=0$ (pure solvent drag) and gives $dC/dz=$const when $\lambda\to\infty$ (pure diffusion).

5.15. Obtain expressions for J_s when $\lambda=0$ and $\lambda\to\infty$.

5.16. Show that for very large pores when $\sigma=0$ the parameter $x=\Delta Z/\lambda=J_v/\omega RT$ depends only on pore radius, solute particle radius, pressure difference, and temperature and not on viscosity or the number of pores per unit area or the membrane thickness.

5.17. When $C_s'=0$, what are the limiting values of $\bar{C}_s$ as $x\to0$? as $x\to\infty$?

5.18. The calculation of $f(\xi)$ [Eq. (5.39)] assumes that the radius of the solvent molecule is negligible. How would the derivation of Eq. (5.39) change if solute molecules have radius a and solvent molecules have radius b, and what would be the effect?

5.19. A cell has variable volume V and fixed surface area S. The total hydrostatic pressure p is the same inside and outside the cell, and there is complete and instantaneous mixing. Initially the interior and exterior are both pure water. The initial volume of the cell is V_0. At $t=0$ the exterior is bathed in a solution containing an impermeant solute of concentration C_0.

(a) Does the cell shrink to zero volume or expand to its maximum volume, which is a sphere of surface area S?

(b) Derive a differential equation for the volume change and integrate it to find how long it takes for the cell to reach zero or maximum volume.

5.20. A cell has variable volume V and fixed surface area S. The total hydrostatic pressure p is always the same both inside and outside the cell. There is complete and instantaneous mixing both inside and out. An impermeant solute has an initial concentration $C(0)$ both inside and outside. The initial cell volume is V_0. At $t=0$ the exterior solute is removed.

(a) Does the cell shrink to zero volume or expand to its maximum volume, which is a sphere of surface area S?

(b) Derive a differential equation for $V(t)$ and find how long it takes for the cell to reach zero or maximum volume.

5.21. (a) Write J_s in terms of C_s, C_s', J_v, and x.

(b) Specialize the result to the case $C_s'=0$.

5.22. (a) Find the ratio $(1-\sigma)\overline{C}_s J_v/[\omega RT(C_s-C_s')]$ in terms of x, C_s, and C_s'.

(b) Specialize to the case $C_s'=0$ and discuss limiting values for small and large x.

5.23. Consider the following cases for transport of water through a membrane.

1. Water flows by bulk flow through the membrane with $\Delta p=0$. There is an impermeant solute ($\sigma=1$) on the right with concentration C_{big} and zero concentration on the left.
2. There is no volume flow through the membrane ($J_v=0$). Some of the water molecules on the left are tagged with radioactive hydrogen (tritium). The concentration of tagged water molecules is C_s on the left and 0 on the right.
3. There is volume flow, as in case 1, and there are also tagged water molecules on the left.

(a) Find the particle fluence rate of water in the first case in terms of L_p.

(b) Find the particle fluence rate of tagged water in the second case in terms of L_p and ωRT.

(c) Find the particle fluence rate of tagged water in the third case in terms of L_p and ωRT.

(d) Restate the answers in terms of the parameters of a collection of n pores per unit area of radius R_p and length ΔZ.

(e) Estimate the value of x for part (c) if $R_p=10^{-8}$ m and $c_{\text{big}}=c_s=0.1$ mol 1^{-1}.

5.24. Construct diagrams analogous to Fig. 5.16 (a) when the total pressure is the same on both sides and $\pi'=0$ and (b) when $(p-\sigma\pi)<p'$ and $\pi'=0$.

5.25. Discuss the equation

$$J_s=\omega RT\left(C_s\frac{xe^x}{e^x-1}-C_s'\frac{x}{e^x-1}\right),$$

where

$$x=\frac{J_v(1-\sigma)}{\omega RT}$$

for the special case $C_s'=0$ in the two limits $x\to\infty$ and $x\to 0$. (Hint: For small x, $e^x\approx 1+x$.)

5.26. Write Eq. (5.56) in terms of J_v and x instead of ωRT and x.

5.27. A set of large right-cylindrical pores traverses a membrane separating two liquids at the same pressure, $p=p'$. A large solute molecule for which $\sigma=1$ is dissolved on the right at concentration C_0', but not on the left. There are solute molecules of radius a in concentration C_s on the left and 0 on the right. For these smaller molecules, the reflection coefficient is σ. $\sigma C_s(0)<C_0'$.

(a) Does solvent drag augment or hinder the diffusion of the smaller solute molecules?

(b) Write an expression for the initial value of J_v in terms of p, L_p, σ, C_0', and C_s.

(c) Find $x=\Delta Z/\lambda$ in terms of a, R_p, C_s, and C_0', if $a\ll R_p$.

(d) Find J_s when $x\gg 1$.

5.28. Consider the case of water permeability shown in Fig. 5.1(c). Water and solute molecules move through the membrane in the same way. They "dissolve" from solution into the membrane. Assume that the concentration of water molecules just inside the membrane is proportional to the pressure just outside: $C=\alpha p$. The membrane has thickness ΔZ and the diffusion constant for water in the membrane material is D. Under steady-state conditions, derive an expression for L_p.

5.29. Solute is carried through a pipe by solvent drag. The radius of the pipe is b. The average flow along the pipe is $\overline{j}_v$ (independent of r because it has been averaged over r). Assume that *within* the pipe the concentration of solute is independent of radius and can be written as $C(z)$. The solute is carried along purely by solvent drag. Solute concentration outside the pipe is zero. Solute diffuses through the wall of the pipe, which has solute permeability ωRT. In terms of $\overline{j}_v$, b, and ωRT, obtain a differential equation for $C(z)$ and show that C decays exponentially along the pipe. Find the decay constant.

5.30. Consider the case in which solute moves along a tube by a combination of diffusion and solvent drag. Ignore radial diffusion within the tube, but assume that solute is moving out through the walls so that $\overline{j}_s$ is changing with position in the tube. In particular, the *number* of solute particles passing out through the wall in length dz in time dt is $CA2\pi R_p dz\, dt$, where A is related to the permeability of the *wall.* Consider a case in which C does not change with time, but depends only on position along the tube.

(a) Write down the conservation equation for an element of the tube and show that

$$-\frac{\partial j}{\partial z}-\frac{2AC}{R_p}=0.$$

(b) Combine the results of part (a) with Eq. (5.40a) and show that $C(z)$ must satisfy the differential equation

$$\frac{\partial^2 C}{\partial z^2} - \frac{\bar{j}_v f}{D}\frac{\partial C}{\partial z} - \frac{2A}{DR_p}C = 0.$$

Show that this equation will be satisfied if the concentration decreases exponentially along the tube as $C(z) = C_0 e^{-\alpha z}$, where

$$\alpha = \frac{\bar{j}_v f}{2D}\left[-1 + \left(1 + \frac{8AD}{R_p \bar{j}_v^2 f^2}\right)^{1/2}\right].$$

5.31. The volume of a water molecule is V_w and the volume of a solute molecule is V_s. Define a new quantity J_w that is the number of water molecules per unit area per second passing through the membrane. What is J_w in terms of J_v and J_s?

5.32. A membrane of total area S is pierced by n pores per unit area of radius R_p. The pressures in the fluid on each side of the membrane are p and p'. A solute with reflection coefficient σ has concentration C on the left and C' on the right. There are three components to the force exerted by the fluid on the membrane. Force F_1 is the force exerted by the fluid on the nonpore area on each side. Force F_2 is exerted by solute molecules reflected from the pore region. Force F_3 is the viscous drag exerted on the walls of the pores by any fluid flowing through them. Show that for the model used in Secs. 5.8 and 5.9 the sum of these forces is $F_1 + F_2 + F_3 = S(p - p')$.

Section 5.10

5.33. Consider the membrane of Fig. 5.10. $L_p = 1$ ml min^{-1} m^{-2} torr^{-1}. The pores transport no molecules with molecular weight over 50 000. The pore radius corresponds to $\xi = 0.9$ at a molecular weight of 50 000. Find the number of pores per unit area as a function of the membrane thickness. Let $T = 37$ °C. (These data give a different value of R_p than was used in the text.)

5.34. A membrane is pierced by pores with the following properties:

$$\Delta Z = 10^{-6} \text{ m},$$

$$n = 10^8 \text{ pores m}^{-2},$$

$$R_p = 25 \times 10^{-9} \text{ m}.$$

The solute molecules are spheres with radius 10 nm. $T = 25$ °C.

(a) Calculate L_p.
(b) Determine σ.
(c) What fraction of the membrane area is pores?

Section 5.11

5.35. For the membrane of Problem 5.34, determine ωRT.

5.36. In the right-circular-cylinder model, how does ω/ω_0 depend on $\xi = a/R_p$ if the diffusion constant is independent of position in the pore?

5.37. Consider a membrane pierced by uniform right-cylindrical pores of radius R_p. There is transport of water and a single species of solute particle of radius a. If $\sigma = 0.5$, what is ω/ω_0?

5.38. A membrane has the following characteristics:

$$L_p = 1.35 \times 10^{-11} \text{ m s}^{-1} \text{ Pa}^{-1},$$

$$R_p = 2.5 \text{ nm},$$

$$\Delta Z = 10 \ \mu\text{m},$$

$$n = 7.45 \times 10^{15} \text{ pores m}^{-2},$$

$$T = 300 \text{ K } (\eta = 0.84 \times 10^{-3} \text{ Pa s}).$$

The pressure difference across the membrane is $p - p' = 100$ torr. The solution on the left side of the membrane contains mannitol ($a = 0.35$ nm) in concentration $C_s = 3.34 \times 10^{24}$ molecules m^{-3}.

(a) Find σ and ω/ω_0.
(b) Find J_v.
(c) Find the average velocity of fluid in a pore.
(d) Find ωRT.
(e) Find $x = J_v(1 - \sigma)/\omega RT$ and its sign.
(f) Find J_s.

Section 5.12

5.39. From the data shown, estimate L_p and ωRT. The data are for the transport of radioactive water with a concentration of 10^{15} molecules m^{-3} on one side of the membrane and zero on the other side.

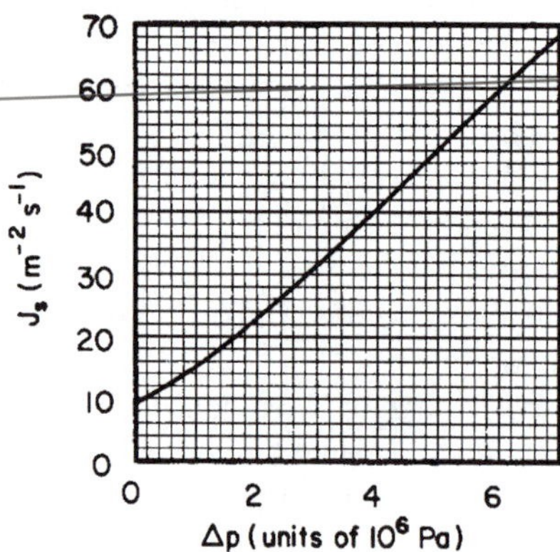

Section 5.13

5.40. The filtration rate of two human kidneys is about 125 ml min^{-1}. The total pore area is 750 cm^2 per 100 g of kidney, and the average mass of each kidney is 150 g. If the pore radius is 3.5 nm and $\Delta Z=50$ nm, find the drop in driving pressure across the glomerular membrane. Use a viscosity $\eta=1.5\times10^{-3}$ Pa s.

5.41. The basement membrane of the glomerulus has the following properties: $R_p=5$ nm, the total glomerular membrane area is 1.5 m^2, the plasma viscosity is 0.002 Pa s, $\Delta p=42$ torr, and $\Delta\pi=32$ torr. The total volume of filtrate is 180 l day^{-1}. Find an expression for the fractional pore area vs ΔZ. If $\Delta Z=60$ nm, what is the fractional pore area? How does this compare with the results of Verniory *et al.* that

$$\frac{n\pi R_p^2 S}{\Delta Z}=7.5\times10^4 \text{ m}$$

$$R_p=5 \text{ nm?}$$

5.42. Show that the glomerular filtration rate is equal to the clearance of inulin. Inulin is a small molecule (1.2 nm radius) which passes through the glomerular basement membrane but is neither reabsorbed nor secreted in the tubules.

REFERENCES

Bean, C. P. (1969). Characterization of cellulose acetate membranes and ultrathin films for reverse osmosis. Research and Development Progress Report No. 465 to the U.S. Department of Interior, Office of Saline Water. Contract No. 14-01-001-1480. Washington, Superintendent of Documents, October 1969.

Bean, C. P. (1972). The physics of porous membranes—Neutral pores, in G. Essenman, ed., *Membranes*. New York, Dekker, Vol. 1, pp. 1–55.

Beck, R. E., and J. S. Schultz (1970). Hindered diffusion in microporous membranes with known pore geometry. *Science* **170**: 1302–1305.

Calland, C. H. (1972). Iatrogenic problems in end-stage renal failure. *N. Engl. J. Med.* **287**: 334–336.

Cokelet, G. R. (1972). The rheology of human blood, in Y. C. Fung *et al.*, eds. *Biomechanics, Its Foundations and Objectives*. Englewood Cliffs, Prentice Hall (especially p. 77, Fig. 4).

Durbin, R. P. (1960). Osmotic flow of water across permeable cellulose membranes. *J. Gen. Physiol.* **44**: 315–326.

Fishman, R. A. (1975). Brain edema. *N. Engl. J. Med.* **293**: 706–711.

Galletti, P. M., C. K. Colton, and M. J. Lysaght (1995). Artificial kidney. In J. D. Bronzino, ed. *The Biomedical Engineering Handbook*. Boca Raton, CRC.

Gennari, F. J., and J. P. Kassirer (1974). Osmotic diuresis. *N. Engl. J. Med.* **291**: 714–720.

Guyton, A. C. (1991). *Textbook of Medical Physiology*, 8th ed. Philadelphia, Saunders.

Guyton, A. C., A. E. Taylor, and H. J. Granger (1975). *Circulatory Physiology II. Dynamics and Control of the Body Fluids*. Philadelphia, Saunders.

Hildebrand, J. H., and R. L. Scott (1964). *The Solubility of Nonelectrolytes*. 3rd ed. New York, Dover.

Katchalsky, A., and P. F. Curran (1967). *Nonequilibrium Thermodynamics in Biophysics*. Cambridge, Harvard University Press.

Levitt, D. G. (1975). General continuum analysis of transport through pores. I. Proof of Onsager's reciprocity postulate for uniform pore. *Biophys. J.* **15**: 533–551.

Lindner, A., B. Charra, D. J. Sherrard, and B. H. Scribner (1974). Accelerated atherosclerosis in prolonged maintenance hemodialysis. *N. Engl. J. Med.* **290**: 697–701.

Lysaght, M. J., and J. Moran (1995). Peritoneal dialysis equipment. In J. D. Bronzino, ed. *The Biomedical Engineering Handbook*. Boca Raton, CRC.

Patton, H. D., A. F. Fuchs, B. Hille, A. M. Scher, and R. F. Steiner, editors (1989). *Textbook of Physiology*, 21st ed. Philadelphia, Saunders.

Pitts, R. F. (1974). *Physiology of the Kidney and Body Fluids*. 3rd ed. Chicago, Yearbook Medical.

Silverstein, M. E., C. A. Ford, M. J. Lysaght, and L. W. Henderson (1974). Treatment of severe fluid overload by ultrafiltration. *N. Engl. J. Med.* **291**: 747–751.

Thomas, Lewis (1974). *Lives of a Cell*. New York, Viking.

Verniory, A., R. DuBois, P. Decoodt, J. P. Gassee, and P. P. Lambert (1973). Measurement of the permeability of biological membranes. Applications to the glomerular wall. *J. Gen. Physiol.* **62**: 489–507.

Villars, F. M. H., and G. B. Benedek (1974). *Physics with Illustrative Examples from Medicine and Biology*, Reading, MA, Addison-Wesley, Vol. 2.

White, R. J., and M. J. Likavek (1992). Current concepts: The diagnosis and initial management of head injury. *N. Engl. J. Med.* **327**: 1502–1511.

CHAPTER 6

Impulses in Nerve and Muscle Cells

A nerve cell conducts an electrochemical impulse because of changes that take place in the cell membrane. These allow movement of ions through the membrane, setting up currents that flow through the membrane and along the cell. Similar impulses travel along muscle cells before they contract. This chapter reviews the basic properties of electric fields and currents that are needed to understand the propagation of the nerve- or muscle-cell impulse.

Section 6.1 introduces the physiology of nerve conduction that is needed. The next eight sections develop the electrostatics and the physics of current flow needed to understand how the action potential propagates along the cell.

The next sections deal with the charge distribution on a resting cell membrane (Sec. 6.10) and the cable model of the axon (Sec. 6.11). If the membrane properties do not change as the voltage across the membrane changes, this leads to electrotonus or passive spread (Sec. 6.12). If the membrane properties do change, a signal can propagate without change of shape. Section 6.13 tells how Hodgkin and Huxley developed equations to describe the membrane changes, and Secs. 6.14 and 6.15 apply their results to the propagation of a nerve impulse. The chapter to this point forms an integrated story of conduction in an unmyelinated axon.

Section 6.16 considers saltatory conduction: the ''jumping'' of an impulse from node to node in a myelinated fiber. Section 6.17 examines the capacitance of a bilayer membrane that has layers with different properties. Section 6.18 shows how minor alterations in the membrane properties can transform the Hodgkin–Huxley model to one that displays repetitive electrical activity.

Section 6.19 shows how tabulated solutions to the electrical capacitance of conductors in different geometries can be used to solve diffusion problems with similar geometric configurations.

6.1. PROPERTIES OF NERVE AND MUSCLE CELLS[1]

A nerve consists of many parallel, independent signal paths, each of which is a nerve cell or fiber. Each cell transmits signals in only one direction; separate cells carry signals to or from the brain. Each cell has an input end (dendrites), a cell body, a long conducting portion or *axon*, and an output end. It is the ends that give the cell its unidirectional character. The input end may be a transducer (stretch receptor, temperature receptor, etc.) or a junction (synapse) with another cell. A threshold mechanism is built into the input end; when an input signal exceeding a certain level is received, the nerve fires and an *impulse* or *action potential* of fixed size and duration travels down the axon. There may be several inputs that can either aid or inhibit each other, depending on the nature of the synapses.

Muscle cells are also long and cylindrical. An electrical impulse travels along a muscle cell to initiate its contraction. This chapter concentrates on the propagation of the action potential in a nerve cell, but the discussion can be regarded as a model for what happens in muscle cells as well.

The axon transmits the impulse without change of shape. The axon can be more than a meter in length, extending from the brain to a synapse low in the spinal cord or from the spinal cord to a finger or toe. Bundles of axons constitute a nerve. The output end branches out in fine nerve endings, which appear to be separated by a gap from the next nerve or muscle cell that they drive.

The long cylindrical axon has properties that are in some ways similar to those of an electric cable. Its diameter may range from less than one micrometer (1 μm) to as much as 500 μm for the giant axon of a squid; in humans the upper

[1]A good discussion of the properties of nerves and the Hodgkin–Huxley experiments is found in Katz (1966). More modern descriptions of nerves and nerve conduction are found in many books, such as Alberts *et al.* (1983 and later editions) and Patton *et al.* (1989).

limit is about 20 μm. Pulses travel along it with speeds ranging from 0.6 to 100 m s^{-1}, depending, among other things, on the diameter of the axon. The axon core may be surrounded by either a membrane (for an *unmyelinated* fiber) or a much thicker sheath of fatty material (*myelin*) that is wound on like tape. A myelinated fiber has its sheath interrupted at intervals and replaced by a short segment of membrane similar to that on an unmyelinated fiber. These interruptions are called *nodes of Ranvier*. A typical human nerve might contain twice as many unmyelinated fibers as myelinated. We will see in Sec. 6.15 that the myelin gives a faster impulse conduction speed for a given axon radius. Myelinated fibers conduct motor information; unmyelinated fibers conduct information such as temperature, for which speed is not important. A typical unmyelinated axon might have a radius of 0.7 μm with a membrane thickness of 5–10 nm. Myelinated fibers have a radius of up to 10 μm, with nodes spaced every 1–2 mm. We will find later that the spacing of the nodes is about 140 times the inner radius of the fiber, a fact that is quite important in the relationship between conduction speed and fiber radius.

A microelectrode inserted inside a resting axon shows an electrical potential that is about 70 mV less than outside the cell. (We will define electrical potential difference in Sec. 6.4.) A typical nerve impulse or action potential or spike in an unmyelinated axon is shown as a function of time in Fig. 6.1. As the impulse passes by the electrode, the potential rises in a millisecond or less to about +40 mV. Then it falls back to about −90 mV and then recovers slowly to its resting value of −70 mV. The membrane is said to *depolarize* and then *repolarize*. The history of recording the action potential has been described by Geddes (1994). The action potential was first measured by Helmholtz around 1850. The measurement technology steadily improved, culminating in the use of a micropipette inserted by Hodgkin and Huxley (1939) into the cut end of the giant axon of the squid, to record the action potential directly.

The information sent along a nerve fiber is coded in the repetition rate of these pulses, all of which are the same shape. Figure 6.2 shows the response of a low-threshold mechanoreceptor in the cornea to a mechanical stimulus. The heavy curve in the bottom panel shows the applied force, and the upper panel shows the impulses.

Comparison of the axoplasm with the *interstitial* or *extracellular* fluid surrounding each axon shows an excess of potassium and a deficit of sodium and chloride ions within the cell, as shown in Fig. 6.3. The regenerative action that produces these sudden changes of membrane potential is caused by changing permeability of the membrane to ions—primarily sodium and potassium ions. These changes are discussed in Secs. 6.13 and 6.14.

The axon can be removed from the rest of the cell and it will still conduct nerve impulses. The speed and shape of the action potential depend on the membrane and the concentration of ions inside and outside the cell. The *intracellular* fluid (*axoplasm*) has been squeezed out of squid giant axons and replaced by an electrolyte solution without altering appreciably the propagation of the impulses—for a while, until the ion concentrations change appreciably. The axoplasm does contain chemicals essential to the long-term metabolic requirements of the cell and to maintaining the ion concentrations.

At the end of a nerve cell the signal passes to another nerve cell or to a muscle cell across a *synapse* or junction. Guyton (1991, Chap. 45) discusses when the transmission

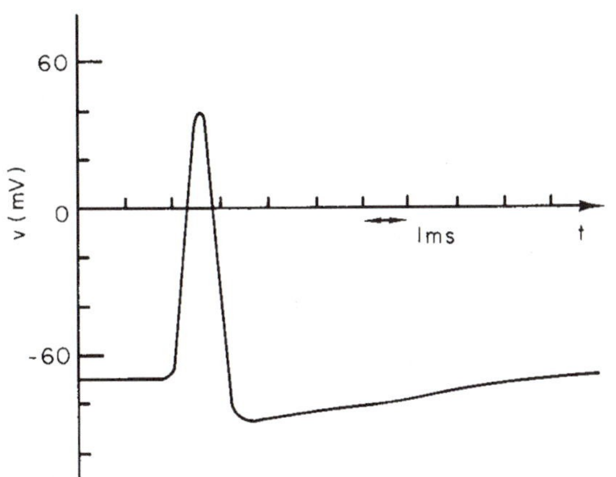

FIGURE 6.1. A typical nerve impulse or action potential, plotted as a function of time.

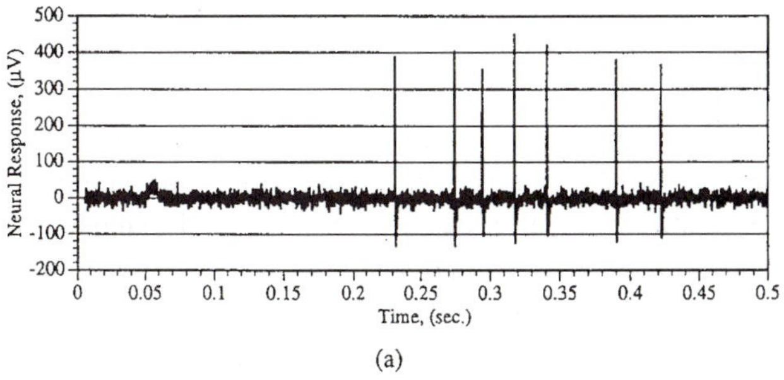

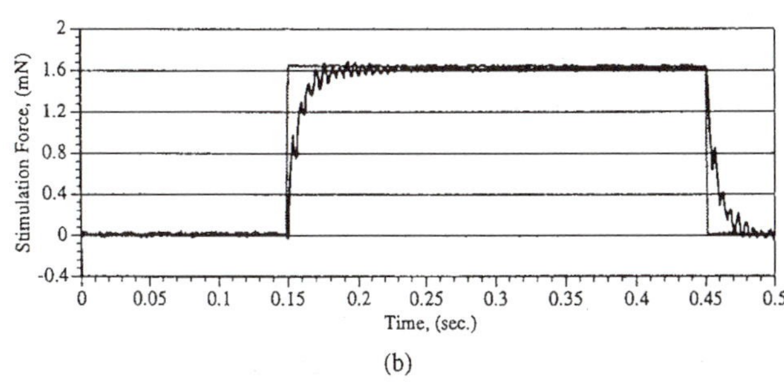

FIGURE 6.2. The response of a mechanical receptor in the cornea to an applied force. (a) The impulses recorded on the surface of the nerve bundle. (b) The applied force. Impulses occur while the force is applied. From B. J. Kane, C. W. Storment, S. W. Crowder, D. L. Tanelian, and G. T. A. Kovacs (1995). Force-sensing microprobe for precise stimulation of mechanoreceptive tissues. *IEEE Trans. Biomed. Eng.* **42**(8):745–750. ©1995 IEEE. Reprinted by permission.

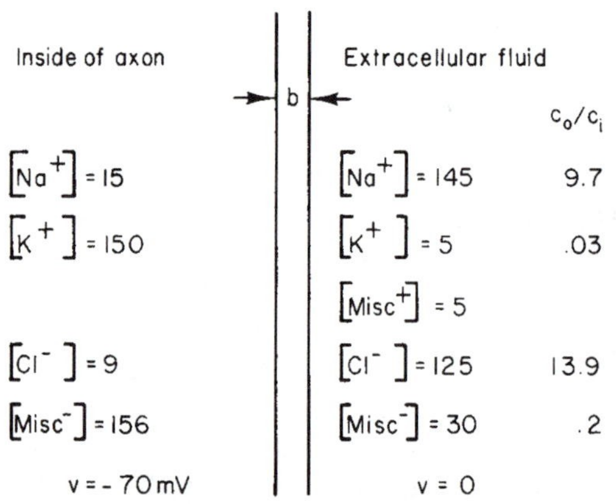

FIGURE 6.3. Ion concentrations in a typical mammalian nerve and in the extracellular fluid surrounding the nerve. Concentrations are in mmol l^{-1}; c_o/c_i is the concentration ratio. The membrane thickness is b

across the synapse is electrical and when it is chemical. There are gaps of 10–20 nm between presynaptic and postsynaptic nerve cells and gaps of 50–100 nm at the neuromuscular junction. Katz (1966) reviews the evidence for the chemical nature of transmission at the usual neuromuscular junction, including a calculation of the small size of any electrical effect [see also Martin (1966)]. Acetylcholine is secreted by these nerve endings when they fire. It increases the permeability of nearby muscle to sodium, which then leaks in and depolarizes the muscle membrane. The process is quantized. Packets of acetylcholine of definite size are liberated [Katz (1966, Chap. 9); Patton *et al.* (1989, Chap. 6)]. A number of other chemical mediators such as norepinephrine, epinephrine, dopamine, serotonin, histamine, glycine, glutamate, aspartate, and gamma-aminobutyric acid, are important elsewhere in the nervous system [Guyton (1991, Chap. 45)].

At the input end of a nerve cell, the response to chemicals from a synapse is often an increase in membrane permeability to sodium ions, which causes depolarization or an increase in the interior potential. In other cases the interior potential becomes more negative (hyperpolarized) and firing is inhibited. If the potential becomes high enough (that is, more positive or less negative), the regenerative action of the membrane takes over, and the cell initiates an impulse. If the input end of the cell acts as a transducer, the interior potential rises when the cell is stimulated. If the input is from another nerve, the signal may cause the potential to increase by a subthreshold amount so that two or more stimuli must be received simultaneously to cause firing, or it may decrease the potential and inhibit stimulation by another nerve at the synapse. This makes possible the logic network that comprises the central nervous system.

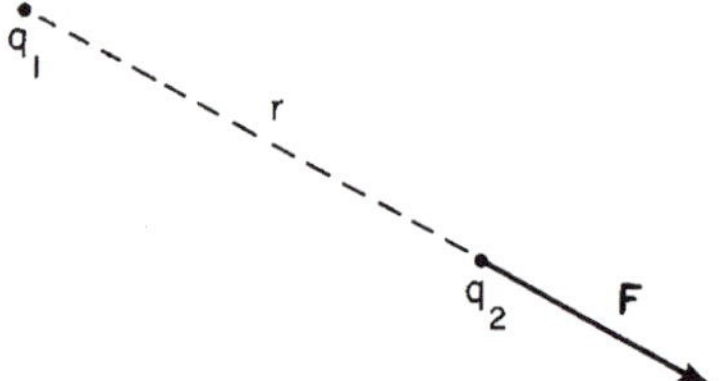

FIGURE 6.4. Force **F** is exerted by charge q_1 on charge q_2. It points along a line between them. An equal and opposite force $-\mathbf{F}$ is exerted by q_2 on q_1.

6.2. COULOMB'S LAW, SUPERPOSITION, AND THE ELECTRIC FIELD

Coulomb's law relates the electrical force between two objects to their electrical charge and separation. For the present purpose, Coulomb's law is a summary of many experiments. If two objects have electrical charge q_1 and q_2, respectively, and are separated by a distance r, then there is a force between them, the magnitude of which is given by

$$|\mathbf{F}| = \left(\frac{1}{4\pi\epsilon_0}\right)\frac{q_1 q_2}{r^2}. \tag{6.1}$$

When the charge is measured in coulombs (C), F in newtons (N), and r in meters (m), the constant has the value

$$\frac{1}{4\pi\epsilon_0} \approx 9\times10^9 \text{ N m}^2\text{ C}^{-2} \tag{6.2}$$

to an accuracy of 0.1%.[2] The direction of the force is along the line between the two charges as shown in Fig. 6.4. If the charges are both positive or both negative, the force is repulsive, which is consistent with assigning a positive sign to **F**. If one is positive and the other negative, then the force is attractive, and **F** has a negative value. Force **F** is exerted by charge q_1 on charge q_2. The force exerted by q_2 on q_1 has the same magnitude but points in the opposite direction. The forces on both charges act to separate them if they have the same sign and to attract them if the signs are opposite.

If two or more charges exert a force on the particular charge being considered, the total force is found by applying Coulomb's law to each charge (paired with the one on which we want to find the force) and adding the vector forces that are so calculated. An example of this is shown in Fig. 6.5. Charges q_1, q_2, and q_3 are $+1.0\times10^{-6}$, -2.0×10^{-6}, and $+3.0\times10^{-6}$ C, respectively. The magnitude of the force that q_1 exerts on q_3 is

$$F_{1\text{ on }3} = \frac{(9\times10^9)(1\times10^{-6})(3\times10^{-6})}{(2\times10^{-2})^2} = 67.5 \text{ N}.$$

[2]The quantity $1/4\pi\epsilon_0$ has been assigned the *exact* value $8.987\,551\,787\,368\,176\,4\times10^9$. This is because in 1983 the velocity of light, c, was defined to be exactly 299 792 458 m s^{-1} and $1/4\pi\epsilon_0 \equiv 10^{-7}c^2$.

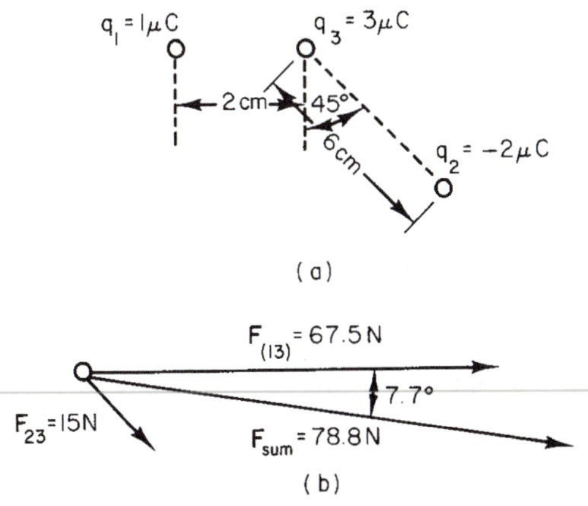

FIGURE 6.5. An example of applying Coulomb's law and adding forces on q_3 due to charges q_1 and q_2. (a) The arrangement of charges. (b) The forces on q_3.

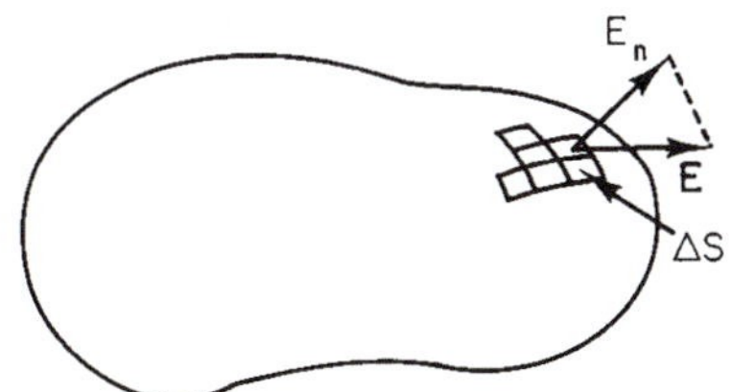

FIGURE 6.6. Calculating the integral of the normal component of **E** through a surface.

Similarly, the force exerted by q_2 on q_3 is

$$F_{2\text{ on }3}=\frac{(9\times10^9)(-2\times10^{-6})(3\times10^{-6})}{(6\times10^{-2})^2}=-15\text{ N}.$$

The minus sign means that the force is attractive, that is, toward q_2. The two forces are shown in Fig. 6.5b, along with their vector sum. The sum can be found by components as in Chap. 1. The result is 78.8 N at an angle of 7.7° clockwise from the direction of $\mathbf{F}_{13}$.

If a collection of charges causes a force to act on some other charge (a "test charge") located somewhere in space, we say that the collection of charges produces an *electric field* at that point in space. One can think, for example, of charge q_1 producing an electric field vector, of magnitude

$$|\mathbf{E}|=\frac{1}{4\pi\epsilon_0}\frac{q_1}{r^2} \tag{6.3}$$

pointing radially away from q_1 (if q_1 is positive) or radially toward q_1 (if q_1 is negative). The force on test charge q_2 placed at the observing point is then

$$\mathbf{F}=q_2\mathbf{E}_1. \tag{6.4}$$

6.3. GAUSS'S LAW

It is possible to derive a theorem about the electric field from a collection of charges, known as *Gauss's law*. Rather than derive it from Coulomb's law, we will state it, and then show that Coulomb's law can be derived from it. Then we will consider some examples of its use.

Divide up *any* closed surface into elements of surface area, such as ΔS in Fig. 6.6. For each element ΔS, calculate the component of **E** normal to the surface, E_n, and multiply it by the magnitude of the surface area ΔS. Add these quantities for the entire closed surface, calling them positive if the normal component of E points out through the surface and negative if E points inward. Gauss's law says that *the resulting sum is equal to the total charge inside the surface, divided by* ϵ_0. In symbols,[3]

$$\int\int E_n dS=\frac{q}{\epsilon_0}=\frac{4\pi q}{4\pi\epsilon_0}. \tag{6.5}$$

This surface integral is exactly the same as the flux of the continuity equation, Eq. (4.4). It is in fact called the electric field flux.

While Gauss's law is always *true*, it is not always *useful*. It is helpful only in cases where **E** is constant over the entire surface of integration, or when the surface can be divided into smaller surfaces, on each of which **E** can be argued to be constant or zero. One of the few cases in which Gauss's law is useful to calculate **E** is the case of a point charge, and another is related to the cell membrane under consideration. In each case, the symmetry of the problem allows the surface of integration to be specified so that **E** perpendicular to the surface is either constant or zero.

The first example is a point charge in empty space. Since such a charge has no preferred orientation (it is a point), and since there is nothing else around to specify a preferred direction in space, the electric field must point radially toward or away from the charge and must depend only on distance from the charge. Therefore, if the surface of integration is a sphere centered on the charge, **E** is equal to E_n and is the same everywhere on the sphere. The term E_n can be taken outside the integral sign in Eq. (6.5) to give

$$\int\int E_n dS=E\int\int dS.$$

The integral of dS over the entire surface of the sphere is just the surface area of the sphere. The surface area of a sphere is $4\pi r^2$ (see Appendix L). Gauss's law gives

$$4\pi r^2E=\frac{q}{\epsilon_0}$$

or

$$E=\frac{q}{4\pi\epsilon_0 r^2}.$$

[3]Some books use one integral sign in this equation and others use two. Strictly speaking the integral over a surface is a two-dimensional integral.

Gauss's law has been used to derive Coulomb's law for the case of a point charge.

If the charge in this problem were not a point charge, nothing would change in the argument as long as the charge distribution was spherically symmetric. As long as the charge is distributed in a spherically symmetric manner, the electric field at a distance r from the center of the distribution is the same as if all the charge within the sphere of radius r were located at the center of the sphere.

Next, consider a problem with cylindrical symmetry rather than spherical symmetry. An example would be an infinitely long line of charge. For such a line of charge, the amount of charge is proportional to the length of the line considered,

$$dq = \lambda\ dx,$$

where λ is the linear charge density in units $\mathrm{C\,m^{-1}}$. Symmetry shows that **E** must point radially outward (or inward) and be perpendicular to the line. Therefore if the Gaussian surface is a cylinder of length L, the axis of which is the line of charge, one can argue that on the end caps $E_n = 0$, while on the wraparound surface of the cylinder $E_n = |\mathbf{E}|$. This is shown in Fig. 6.7. The total integral is therefore the integral for the wraparound surface, which is

$$E \int \int dS.$$

The surface area of the cylinder is its circumference $(2\pi r)$ times its length (L). Therefore Gauss's law becomes

$$2\pi r L E = \frac{\lambda L}{\epsilon_0}$$

or

$$E = \frac{\lambda}{2\pi\epsilon_0 r}. \tag{6.6}$$

Since the constant $1/4\pi\epsilon_0$ is so easily remembered, it is convenient to write this as

$$E = \frac{1}{4\pi\epsilon_0}\frac{2\lambda}{r}. \tag{6.7}$$

Consider next an infinite sheet of charge, with a charge density per unit area of $\sigma\ \mathrm{C\,m^{-2}}$. The symmetry of the situation requires that **E** be perpendicular to the surface.

To see why, suppose that **E** is not perpendicular to the surface. I stand on the surface looking in such a direction that **E** points diagonally off to my left. If I turn around in place, I see **E** pointing diagonally off to my right. Since the charge density is constant and extends an infinite distance in every direction, the charge distribution looks exactly the same as it did before I turned around. The only way to resolve this contradiction (that **E** changed direction while the charge distribution did not change) is to have **E** perpendicular to the surface.

The Gaussian surface can be a cylinder with end caps of area S and sides perpendicular to the surface. Let the end caps be a distance a from the charge sheet on one side and b from the charge sheet on the other, as in Fig. 6.8. **E** is perpendicular to the end caps, so that $\mathbf{E} = E_n$, with no component perpendicular to the sides of the cylinder. Since there is no component of **E** across the cylindrical wall, changing b or a does not change the total flux through the surface. Since the charge inside the volume does not change, **E** must be independent of distance from the sheet of charge. (This is true only because the charge sheet is infinite.) The flux through each end cap is the same, as may be seen from the cross section of the surface in Fig. 6.9. The total flux is therefore $2ES$, while the charge within the volume is σS. Therefore, Gauss's law gives

$$E = \frac{\sigma}{2\epsilon_0} = \frac{1}{4\pi\epsilon_0} 2\pi\sigma. \tag{6.8}$$

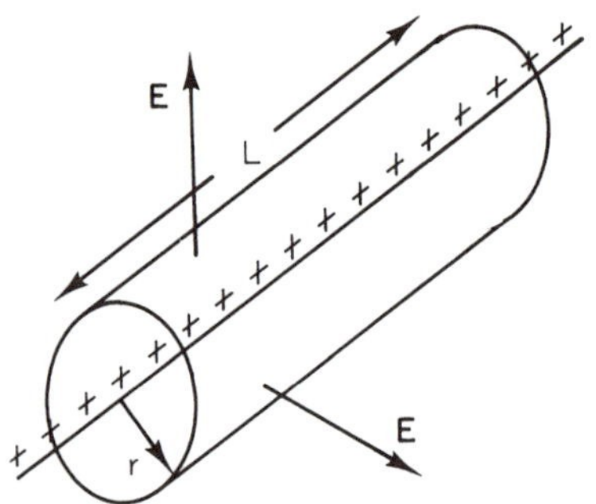

FIGURE 6.7. Gauss's law is used to calculate the electric field from an infinite line of charge. The Gaussian surface is a segment of a cylinder concentric with the line of charge.

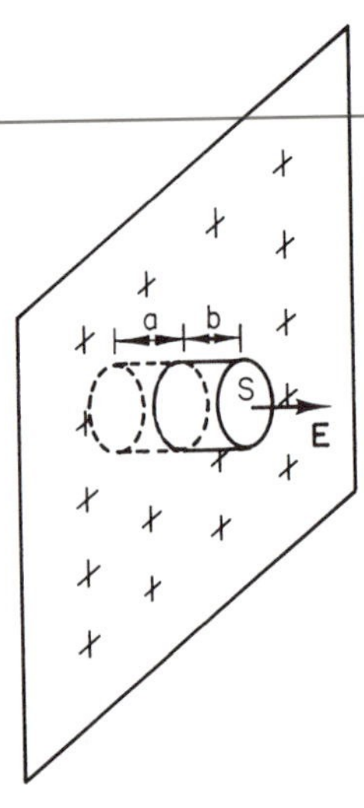

FIGURE 6.8. A portion of an infinite sheet of charge and the appropriate Gaussian surface.

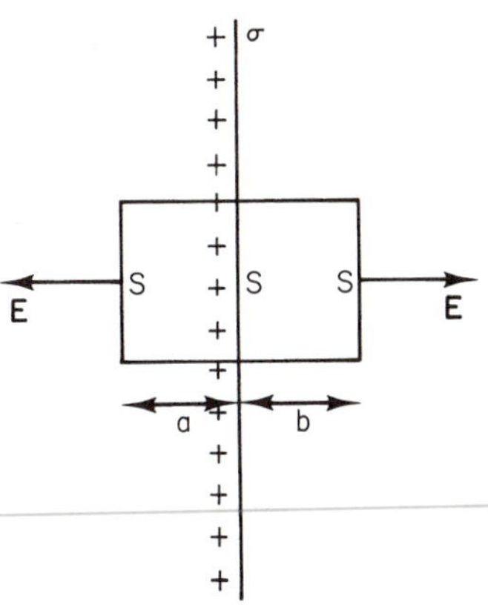

FIGURE 6.9. A side view of the Gaussian surface in Fig. 6.8.

There is, of course, something quite unreal about a sheet of charge extending to infinity. However, it is a good approximation for an observation point close to a finite sheet of charge. If the sheet is limited in extent and the observation point is far away, the distance to all parts of the sheet from the observation point is nearly the same, and the charge sheet may be regarded as a point charge. If one considers a rectangular sheet of charge lying in the xy plane of width $2c$ and length $2b$, as shown in Fig. 6.10, it is possible to calculate exactly the **E** field along the z axis. By symmetry, the field points along the z axis and is 4 times the z component of **E** from the charge from $x=0$ to c and $y=0$ to b. The surface charge density is σ. The distance is $r^2=x^2+y^2+z^2$. The component of E along z is $E\cos\theta=Ez/r$. Therefore, if the charge in element of area $dx\,dy$ is $\sigma\,dx\,dy$, the field is

$$E=\frac{4\sigma z}{4\pi\epsilon_0}\int_0^b dy\int_0^c (x^2+y^2+z^2)^{-3/2}dx.$$

This integral can be evaluated. The result is

$$E=\frac{\sigma}{2\pi\epsilon_0}\left[\frac{\pi}{2}-\sin^{-1}\left(\frac{z^4+(c^2+b^2)z^2-c^2b^2}{z^4+(c^2+b^2)z^2+c^2b^2}\right)\right].$$

This is plotted in Fig. 6.10. for $c=1$, $b=100$. Close to the sheet ($z \ll 1$) the field is constant, as it is for an infinite sheet of charge. Far away compared to 1 m but close compared to 100 m, the field is proportional to $1/r$ as with a line charge. Far away compared to 100 m, the field is proportional to $1/r^2$, as from a point charge.

As a final example, consider two infinite sheets of charge, one with density $-\sigma$ and the other with density $+\sigma$, as shown in Fig. 6.11. This can be solved by using the result for a single sheet of charge, Eq. (6.8), and the principle of superposition. Consider first the region I of Fig. 6.11. There, the negative charge will give an **E** field that has magnitude $\sigma/2\epsilon_0$ and points toward the right, while the positive sheet of charge will give an **E** field of $\sigma/2\epsilon_0$ pointing to the left. The total **E** field in region I is zero. A similar argument can be made in region III with the field of the negative charge pointing left and that of the positive charge pointing to the right. Again the sum is zero. In region II, however, the two **E** fields point in the same direction, and the total field is

$$E=\frac{\sigma}{\epsilon_0}=\frac{1}{4\pi\epsilon_0}4\pi\sigma. \tag{6.9}$$

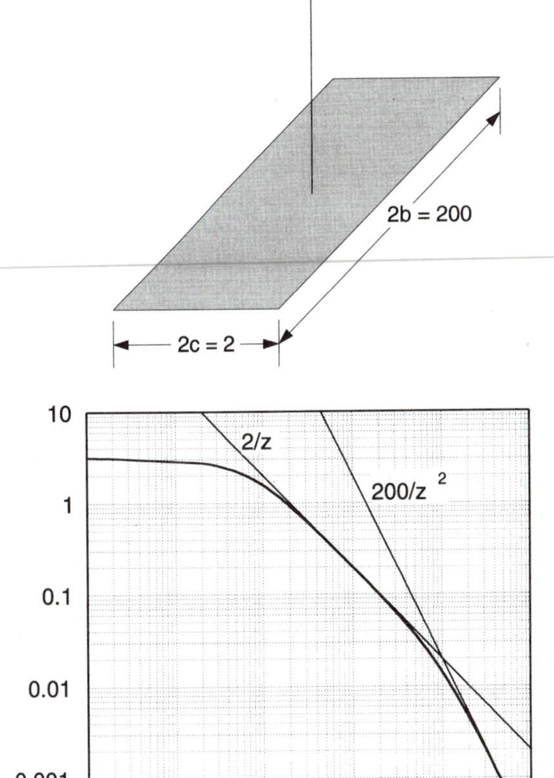

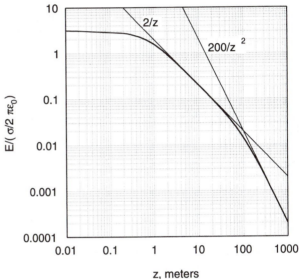

FIGURE 6.10. The electric field from a plate of charge of width 2 m and length 200 m, measured along the perpendicular bisector of the plate. Much closer than 1 m, the field is constant. Around 10 m the field is proportional to $1/r$, the field from a line charge. Farther away than 100 m the field is proportional to $1/r^2$, the field from a point charge.

Notice the factor of 2 difference between Eqs. (6.8) and (6.9). Another way to explain the difference is that there is no **E** in region III of the second case, so that a Gaussian surface can be constructed as shown in Fig. 6.12. Then the flux is zero through every surface except cap A. The charge within the volume is σS, while the flux through cap A is ES. Therefore, $E=\sigma/\epsilon_0$.

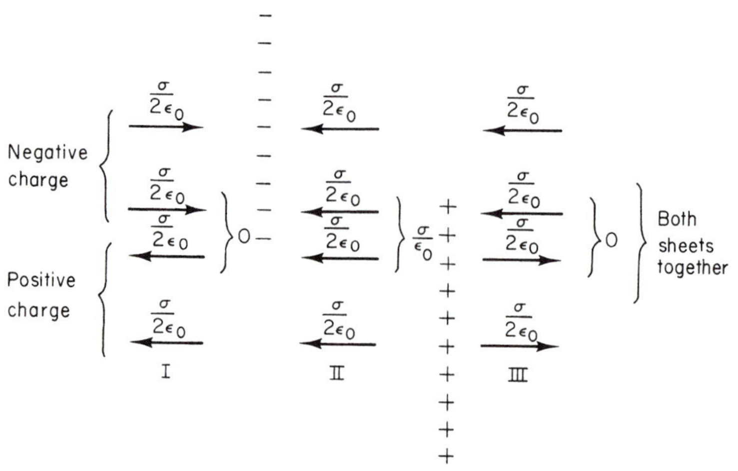

FIGURE 6.11. The electric field due to two infinite sheets of charge. The upper third shows the field due to an infinite sheet of negative charge. The lower third shows the field due to an infinite sheet of positive charge. The middle third shows the field due to both sheets, which is the superposition of the field from each sheet.

Within a nerve cell membrane of 6 nm thickness surrounding a cell of radius 5 μm or 5000 nm, the electric field can be calculated by making the approximation that the sheets of charge are infinite. Suppose that the electric field within the membrane is 1.17×10^7 N C^{-1}. (We will learn how to determine this value later.) From Eq. (6.9) the charge density is

$$\sigma=\frac{E}{4\pi(1/4\pi\epsilon_0)}=\frac{1.17\times10^7}{4\pi(9\times10^9)}=1.03\times10^{-4}\ \mathrm{C\,m^{-2}}.$$

This tells us something about the makeup of the cell. The membrane is in contact with atoms, each of which has a diameter of about 10^{-10} m. Therefore there are approximately 10^{20} atoms (in water molecules, as ions, etc.) in contact with one square meter of the membrane surface. Suppose that the excess charge that causes the electric field in the membrane resides in these atoms and that each atom is either neutral or a monovalent ion. The number of atoms in the square meter which are charged is

$$\frac{1.03\times10^{-4}\ \mathrm{C\,m^{-2}}}{1.6\times10^{-19}\ \mathrm{C\,atom^{-1}}}=6.4\times10^{14}\ \mathrm{atoms\,m^{-2}}.$$

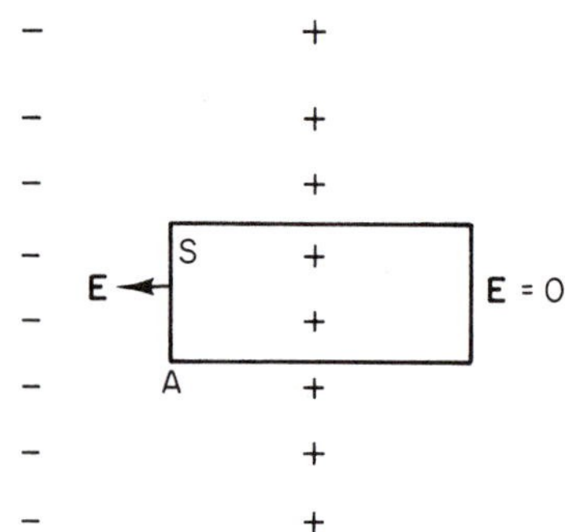

FIGURE 6.12. A Gaussian surface to determine the electric field between two plates.

The fraction of atoms that are charged is $(6.4\times10^{14})/(10^{20})=6.4\times10^{-6}$. Roughly 1 in every 10^5 atoms in contact with the membrane carries an unneutralized charge. (We will have to modify this result later, to account for a partial neutralization of this external charge by charge movement within the membrane.)

6.4. POTENTIAL DIFFERENCE

It is often convenient to talk about the *electrical potential, potential difference*, or *voltage difference* instead of the electric field. The potential is related to the difference in energy of a charge when it is at different points in space. Suppose that an electric field $\mathbf{E}=E_x$ points along the x axis. A positive charge is located at point A. A force F_{ext} must be applied to the charge by something besides the electric field, or else the charge will be accelerated by the force qE_x. If the external force does not exist, the charge will be accelerated to the right by the electric force, and its kinetic energy will increase. The charge can be moved slowly to the right at a constant speed so that its kinetic energy remains fixed, if the external force is always to the left and its magnitude is adjusted so that

$$F_{\mathrm{ext}}=-qE_x.$$

This situation is shown in Fig. 6.13. The external force does work on the charge. One can either say that the total work done on the charge by both forces is equal to zero, or one can ignore the work done by the electric force and invent the idea of potential energy—energy of position—due to the electric field. The increase in potential energy[4] as the charge moves a distance dx is

[4]In earlier chapters the potential energy was called E_{p}, and the total energy was called U. For the next few pages U will be used for potential energy, to avoid confusion with a component of the electric field.

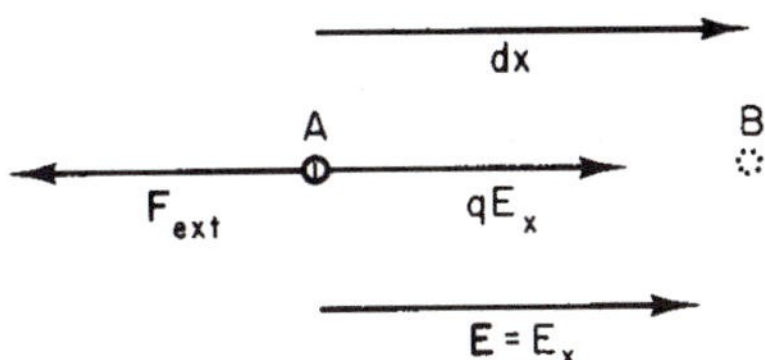

FIGURE 6.13. A charge q is moved from A to B, a distance dx in the x direction. External force $\mathbf{F}_{ext}$ keeps the charge from being accelerated.

$$dU = F_{ext}dx = -qE_x dx.$$

If E_x varies with position, the total change in potential energy when the particle is moved without acceleration from A to B is given by

$$\Delta U = U(B) - U(A) = -q\int_A^B E_x(x)\, dx. \qquad (6.10)$$

For example, in a constant electric field of 1.4×10^7 N C^{-1}, a particle with charge $q = 1.6\times10^{-19}$ C experiences an electric force equal to 2.24×10^{-12} N. If it is moved 5 nm along the x axis, the electric force does 1.12×10^{-20} J of work on it, increasing its kinetic energy. To prevent this increase in kinetic energy, F_{ext} must be applied. The external force does work -1.12×10^{-20} J. We can either say that the total work done by both forces is zero or we can ignore the electrical force and say that the external force increased the potential energy of the particle by -1.12×10^{-20} J as the particle moved from A to B.

If the displacement of the particle is perpendicular to the direction of the electric field, it is also perpendicular to the direction of $\mathbf{F}_{ext}$. Therefore neither force does work on the particle. The potential energy is unchanged when the displacement is perpendicular to the direction of the electric field. This fact can be used to prove [Halliday, Resnick, and Krane (1992, p. 652)] that in three dimensions,

$$\Delta U = U(B) - U(A)$$
$$= -q\left(\int_A^B E_x dx + \int_A^B E_y dy + \int_A^B E_z dz\right). \qquad (6.11)$$

Using the notation of a "dot" or scalar product of two vectors (Sec. 1.8), this can also be written as

$$\Delta U = -q\int \mathbf{E}\cdot d\mathbf{r}. \qquad (6.12)$$

The potential energy difference is measured in joules. It is always proportional to the charge of the particle that is moved in the electric field. It is convenient to define the *potential* difference to be the potential energy difference per unit charge. When the energy difference is in joules and the charge is in coulombs, the ratio is J C^{-1}, which is called a *volt* (V):

$$\Delta v\ \text{(V)} = \frac{\Delta U\ \text{(J)}}{q\ \text{(C)}}. \qquad (6.13)$$

To move a charge of +3 C from point A to point B where the potential is 5 V higher requires that 15 J work be done on the charge. If the charge is then allowed to move back to point A under the influence of only the electric field, its kinetic energy increases by 15 J as the potential energy decreases by the same amount.

This definition of potential, when combined with the definition of potential energy, Eq. (6.10) gives

$$\Delta v = -\int_A^B E_x dx$$

or

$$E_x = -\frac{\partial v}{\partial x}. \qquad (6.14)$$

That is, the component of the electric field in any direction is the negative of the rate of change of potential in that direction. The units of $\mathbf{E}$ were seen earlier to be N C^{-1} (from $\mathbf{F} = q\mathbf{E}$). Equation (6.14) shows that the units of $\mathbf{E}$ may also be called V m^{-1}.

Notice that only differences in potential energy and differences in potential (or colloquially, differences in voltage) are meaningful. We can speak of the potential at a point only if we have previously agreed that the potential at some other point will be called zero. Then we are really speaking of the difference of potential between the reference point and the point in question.

In many cases, it is customary to define the potential to be zero at infinity. Then the potential at point B is

$$v(B) = -\int_\infty^B E_x dx.$$

If you try to apply this equation to the infinite line and sheet of charge, you will discover that it does not work. The reason is that you cannot get infinitely far away from a charge distribution that extends to infinity.

6.5. CONDUCTORS

In some substances, such as metals or liquids containing ions, electric charges are free to move. When all motion of these charges has ceased and static equilibrium exists, there is no net charge within the conductor. To see why there is not, consider a small volume within the conductor. If there were an electric field within that region, the charges there would experience an electric force. Since they are free to move, this force would accelerate them. This force will vanish only when the electric field within the conductor is zero. Therefore, in the static case the electric field within a conductor is zero. Now apply Gauss's law to a small volume within the conductor. Since the electric field in the

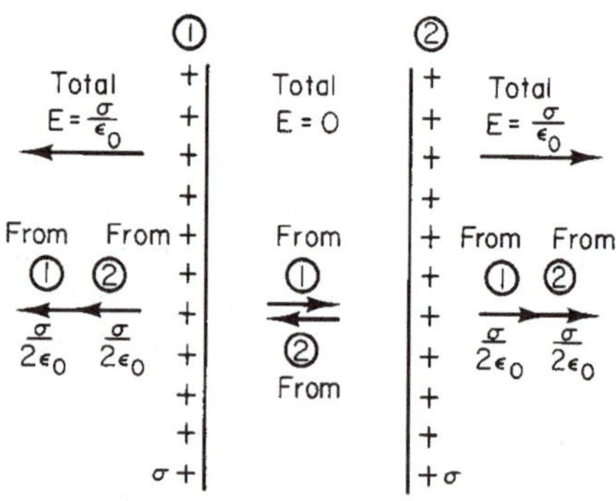

FIGURE 6.14. The electric field in and around a conductor carrying a charge on each surface.

conductor is zero everywhere, the flux through the Gaussian surface is zero, and the net charge within the volume is zero.

At the surface of the conductor, there may well be charge that gives rise to electric fields outside the conductor. Consider, for example, an infinite sheet of metal that has positive charge on it. The positive charge will distribute itself as shown in Fig. 6.14, and either superposition or Gauss's law may be used to show that the electric field outside the conductor is σ/ϵ_0.

Because the electric field is zero throughout a conductor in equilibrium, no work is required to move a charge from one point to another. All parts of the conductor are at the same potential. This statement is true only if the charges are not moving. We will see later that if they are (that is, if a current is flowing), then the electric field in the conductor is not zero and the potential in the conductor is not the same everywhere.

6.6. CAPACITANCE

Suppose that two conductors are fixed in space, with charge $+Q$ on one and $-Q$ on the other. The potential difference between the conductors is proportional to Q. The proportionality constant depends on the geometrical arrangement of the conductors. When the proportionality is written as

$$Q = Cv \tag{6.15}$$

the proportionality constant is called the *capacitance*. The units of capacitance are C V^{-1} or farads (F).

As an example of capacitance, consider two parallel conductors side by side. Let the area of each be S and the separation be b. The charge layers of Fig. 6.12 might be charge on the inner surface of each conductor. The total charge on each conducting plate has magnitude σS. The electric field between the plates is σ/ϵ_0 and the potential difference is $Eb = \sigma b/\epsilon_0$. (Note that the potential difference is proportional to the charge.) The capacitance is

$$C = \frac{Q}{v} = \frac{\sigma S \epsilon_0}{\sigma b} = \frac{\epsilon_0 S}{b}. \tag{6.16}$$

If the plates are separated further with a fixed charge on them, the potential difference increases and the capacitance is decreased. Increasing the area and charge of the plates with fixed σ and fixed b increases Q and C but not v.

6.7. DIELECTRICS

In a conductor, charges rearrange themselves so that there is no electric field within the conductor in the static case. In a dielectric, charges are not free to move far enough to completely cancel the effect of any external electric field, but they can move far enough to cause a partial cancellation.

In some substances, the freedom of charges to move depends on the orientation of the applied field with respect to the crystalline structure of the material. For example, if electrons can shift their positions slightly along a line at 45° to the x and y axes, an electric field along the x axis will cause them to move and induce a field along the y axis. This book deals only with cases in which the induced electric field is parallel to the applied electric field.

The partial neutralization of the external electric field can be understood from the following model. Consider a dielectric in the absence of external fields. The electron distribution of each atom is centered on the nucleus so that there is no electric field (at least when we average over a region containing many atoms). This is shown schematically in Fig. 6.15(a), in which each + sign represents a nucleus and each circle represents a distribution of negative charge in an atom. Figure 6.15(b) shows some external charges producing an electric field. If the dielectric is introduced in the space where this electric field exists, the negative electron clouds are shifted with respect to the nuclei, as shown in Fig. 6.15(b). The result is a polarization electric field $\mathbf{E}_p$, which is in the opposite direction to the external electric field. The total field within the dielectric is the vector sum of these two fields:[5]

$$\mathbf{E}_{\text{tot}} = \mathbf{E}_{\text{ext}} + \mathbf{E}_p . \tag{6.17}$$

When all three vectors are parallel, it is customary to define the electric susceptibility χ by the equation

$$E_p = -\chi E_{\text{tot}} .$$

This can be combined with the previous equation to get

[5]In most textbooks, it is customary to define the polarization by $\mathbf{P} = -\mathbf{E}_p$ or $\mathbf{P} = -\epsilon_0 \mathbf{E}_p$. We have not done that in order to make the phenomenon easier to understand.

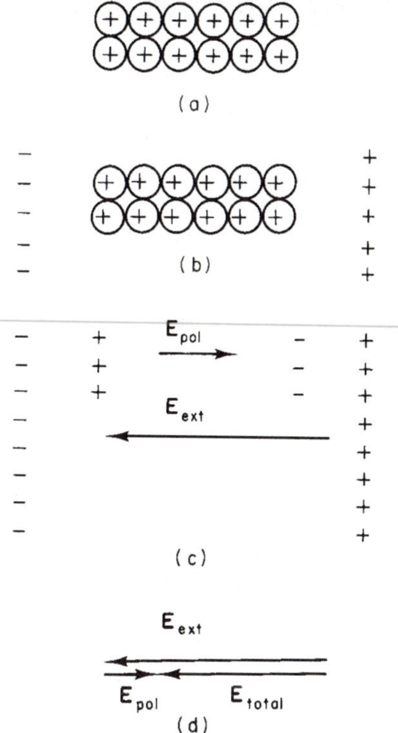

FIGURE 6.15. The polarization of a dielectric by an external electric field. (a) Atoms in the absence of an external field. (b) An external electric field causes a shift of each electron cloud relative to the positively charged nucleus. (c) There is a net buildup of positive charge at the left edge of the dielectric and of negative charge at the right edge. (d) The total electric field within the dielectric is the sum of the external electric field and the polarization electric field induced in the dielectric.

$$E_p=-\frac{\chi}{1+\chi}E_{ext}.$$

The polarization electric field is thus proportional to both the total electric field (proportionality constant $-\chi$) and the external field [proportionality constant $-\chi/(1+\chi)$]. The former relationship is more fundamental, since the field displacing charges in one atom is the total field, due to both external charges and to the charges in neighboring atoms.

The total field within the dielectric is

$$E_{tot}=E_{ext}-\frac{\chi}{1+\chi}E_{ext}=\frac{1}{1+\chi}E_{ext}=\frac{1}{\kappa}E_{ext}. \quad (6.18)$$

The factor $\kappa=1+\chi$ is called the *dielectric constant* of the dielectric. The electric field within the dielectric is reduced by the factor $1/\kappa$ from that which would exist without the dielectric. The dielectric constant for typical nerve membranes is about 7.[6] The dielectric constant of water is quite high (around 80) because the water molecules can easily reorient their charged ends.

[6]This value is high compared to the dielectric constant for a pure lipid, which is between 2 and 3. See the discussion in Sec. 6.17.

The relationship between the applied field, the polarization field, and the total field can be seen in the following example. The electric field between two parallel sheets of charge of density $+\sigma$ and $-\sigma$ per unit area has magnitude $E_{ext}=\sigma/\epsilon_0$. If there is dielectric between them (such as a cell membrane) and if the polarization in the dielectric is uniform, then there is effectively a charge $\pm\sigma'$ induced on the surface of the dielectric that partially neutralizes the external charges. This is shown in Fig. 6.16. The total electric field within the membrane is $\mathbf{E}_{tot}=\mathbf{E}_{ext}+\mathbf{E}_p=\sigma/\epsilon_0-\sigma'/\epsilon_0=\sigma_{net}/\epsilon_0=\mathbf{E}_{ext}/\kappa$.

To recapitulate, in Fig. 6.16 $\mathbf{E}_{ext}$ is σ/ϵ_0 and depends on the external charge distribution; the potential difference between the plates depends on the total field, and its magnitude is E_{tot} times the plate separation.

It is customary to refer to two different kinds of charge. The *free charge* is the charge that we bring into a region. We have some control over it. The *bound charge* is the charge induced in the dielectric by the movement or distortion of atoms and molecules in the dielectric in response to the free charge that has been introduced. Gauss's law can be written either in terms of the total charge (free plus bound)

$$\int\int E_n dS=\frac{q_{tot}}{\epsilon_0}=\frac{q_{free}+q_{bound}}{\epsilon_0} \quad (6.19a)$$

or in terms only of the free charge

$$\int\int \kappa E_n dS=\frac{q_{free}}{\epsilon_0}. \quad (6.19b)$$

The dielectric constant is placed inside the integral sign because the Gaussian surface could pass through materials with different values of the dielectric constant.

As another example of the effect of a dielectric, consider a spherical ion of radius a in which all the charge resides on

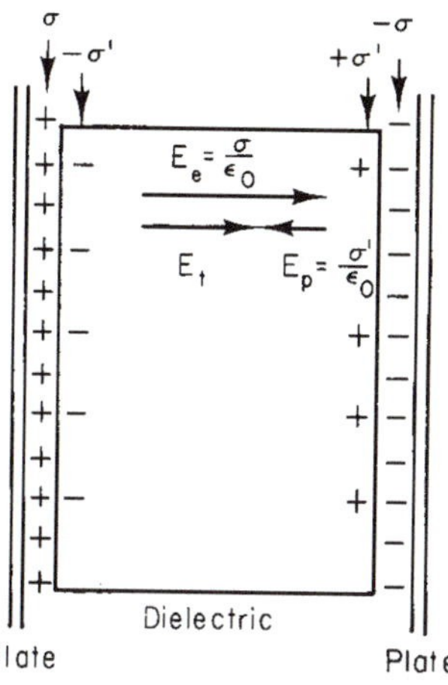

FIGURE 6.16. The polarization electric field reduces the electric field between the plates. The conducting plates could be extracellular and intracellular fluid, and the dielectric could be the cell membrane.

the surface. In a vacuum, the potential at distance r is $v=q/4\pi\epsilon_0 r$, so on the surface of the ion, the potential is $q/4\pi\epsilon_0 a$. The work required to bring to the surface an additional charge dq is $dW=v\ dq=q\ dq/4\pi\epsilon_0 a$. The total work required to place charge Q on the ion is therefore

$$W=\int dW=\frac{1}{4\pi\epsilon_0 a}\int_0^Q q\ dq=\frac{\frac{1}{2}Q^2}{4\pi\epsilon_0 a}.$$

If the sphere is immersed in a uniform dielectric the total electric field and therefore the potential is reduced by a factor κ. The energy required to assemble the ion is then

$$W=\frac{\frac{1}{2}Q^2}{4\pi\epsilon_0 \kappa a}. \tag{6.20}$$

This is called the *Born charging energy*. For an ion of radius 0.2 nm (200 pm) and $Q=1.6\times10^{-19}$ C, the Born charging energy in a vacuum is 5.8×10^{-19} J ion^{-1}. Multiplying by Avogadro's number gives 3.5×10^5 J mol^{-1}. Often in problems involving charges of a few times the electronic charge, it is convenient to use the energy unit electron volt: 1 eV$=1.6\times10^{-19}$ J. For this problem, the Born charging energy is 3.6 eV ion^{-1}.

If the ion is in a dielectric with $\kappa=2$ (a lipid, for example), the Born charging energy is reduced to 1.8 eV ion^{-1}. Water has a very high dielectric constant (about 80) because the water molecules look roughly like that in Fig. 6.17, and the molecules can easily align with an applied electric field. The same ion in water has a Born charging energy of 0.045 eV. At room temperature, the Boltzmann factor for the energy required to create the ion in vacuum is 3.32×10^{-61}. In a lipid, it is 5.76×10^{-31}, and in water, it is 0.175. This explains why it is easy to form ionic solutions in water but not in lipids.

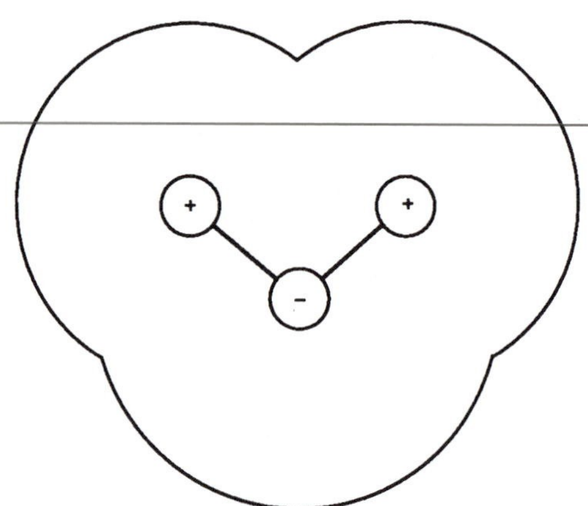

FIGURE 6.17. A schematic diagram of a water molecule. The hydrogen nuclei are 96.5 pm from the oxygen nucleus; the included angle is about 104°. The radius of each hydrogen atom is about 120 pm; the radius of the oxygen atom is about 140 pm. The water molecule has a permanent electric dipole moment.

6.8. CURRENT AND OHM'S LAW

In the electrostatic case, there are no moving charges and no electric field within a conductor. When a current flows in a conductor, charges are moving and there is an electric field.

If the conductor is a wire of cross-sectional area S, the electric current i is the amount of charge per unit time passing a point on the wire. If the amount of charge in time dt is dQ, the current is

$$i=\frac{dQ}{dt}. \tag{6.21}$$

The units of the current are C s^{-1} or amperes (A). The current density j (or j_Q in the notation of Chap. 5), is the current per unit area, i/S. The units are C m^{-2} s^{-1} or A m^{-2}.

In an extended medium, the current density is a vector $\mathbf{j}$ at each point in the medium. The direction of $\mathbf{j}$ is the direction charge is moving at that point.

If there is no electric field in the conductor, there is no average motion of the charges. (There will be random thermal motion, but it will be equally likely in every direction. This random motion of charges is one cause of "noise" in electrical circuits.) To have a current there must be an electrical field in the conductor; this means that there will be a potential difference between two points in the conductor. If there is no potential difference between two points in the conductor, there is no current. For the simple conductor of Fig. 6.18, the current is found to be proportional to the voltage difference between the ends of the conductor. The current is shown flowing from B to A. When $v(B)$ is greater than $v(A)$, v is positive and the current is positive. When v is negative, the current is in the other direction and is also negative.

For the wire of Fig. 6.18, the relationship between current and voltage difference is linear. In that case, we can write

$$i=\frac{1}{R}v=Gv \tag{6.22a}$$

or

$$v=iR. \tag{6.22b}$$

R is called the resistance of the conductor. Since the current is measured in amps and the voltage in volts, its units are V A^{-1} or ohms (Ω). The reciprocal of the resistance is the *conductance* G. Its units are Ω^{-1} or siemens (S).

Ohm's law is not universal. It describes only certain types of conductors. Figure 6.19 shows the current–voltage characteristics of several devices that have nonlinear behavior and that make modern electronic circuits possible. They are shown here not for their own sake, but to emphasize the limited validity of Ohm's law. The nerve cell membrane is not linear.

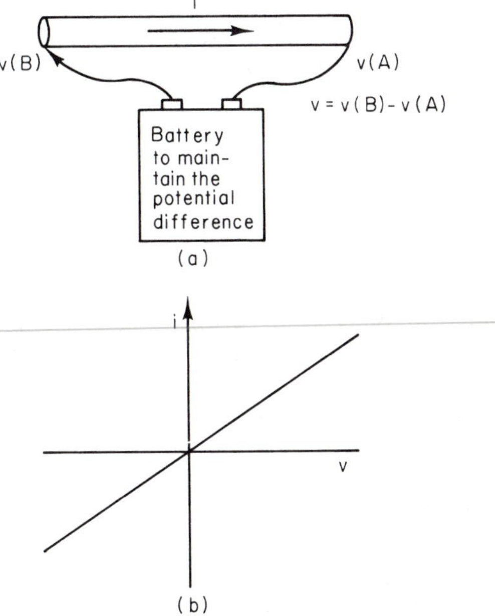

FIGURE 6.18. A current flows in the wire as long as the battery or some other device maintains a potential difference between two points on the wire. The potential difference means that there is an electric field within the wire. If the wire obeys Ohm's law, the current is proportional to the potential difference.

It is possible to write Ohm's law in another form. Placing two identical wires in parallel in the circuit of Fig. 6.18 would cause twice as much current to flow (assuming that the battery maintains the voltage difference at the original level). The current density j remains constant as the cross-sectional area of the wire is changed, when the wire length and voltage difference are held fixed. Similarly, to maintain the same current through a single wire twice as long requires a voltage difference twice as great. Therefore, it is voltage per unit length that determines the current. In this spirit, Ohm's law can be written as

$$j=\frac{i}{S}=\frac{v(B)-v(A)}{RS}.$$

If L is the length of the wire and x the position along it, this can be written as

$$j_x=-\frac{L}{SR}\frac{v(x=L)-v(x=0)}{L}=-\frac{L}{SR}\frac{\partial v}{\partial x}, \tag{6.23a}$$

$$j_x=-\sigma\frac{\partial v}{\partial x}. \tag{6.23b}$$

In three dimensions this alternate statement of Ohm's law becomes

$$\mathbf{j}=\sigma\mathbf{E}. \tag{6.24}$$

The σ in this equation[7] is the electrical conductivity, measured in A m^{-2}/V m^{-1} or S m^{-1}. Its reciprocal is the resistivity of the material, ρ. The units of resistivity are Ω m. For a cylindrical conductor, the resistivity and the resistance are related by

$$\frac{1}{\rho}=\frac{L}{SR}$$

or

$$R=\rho\frac{L}{S}. \tag{6.25}$$

This shows that making the conductor longer increases its resistance, while increasing the cross-sectional area lowers the resistance.

Suppose that a charge is moving through a medium that obeys Ohm's law under the influence of an electric field in the medium. The electric field accelerates the charge, but the energy is continually transferred to the medium by collisions with the other particles in the medium. If a charge Q moves to a lower potential, all the energy it gained is transferred to heat. The rate of energy dissipation is the power

$$P=vi. \tag{6.26}$$

The units of power are J s^{-1} or watts (W). For a material that obeys Ohm's law, Eq. (6.26) can be combined with Ohm's law to give

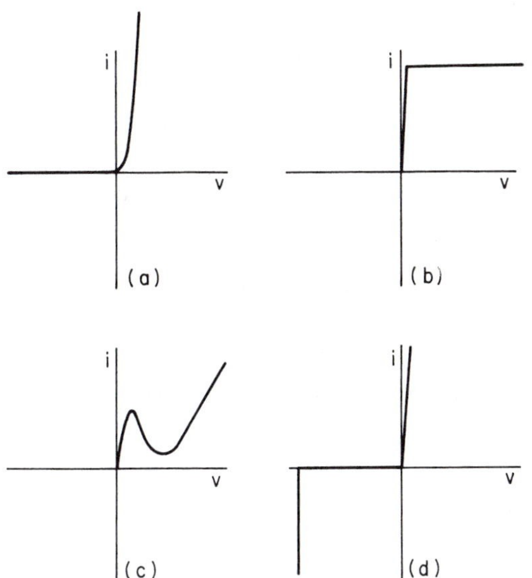

FIGURE 6.19. Current–voltage relationships for some nonlinear devices used in electronic circuits. (a) Diode. (b) Transistor. (c) Tunnel diode. (d) Zener diode.

[7]Note that σ has now been used for two things in this chapter: surface charge per unit area and conductivity. This notation is standard in the literature. You can tell from the context which is meant. Similarly, the symbol ρ is used for charge per unit volume and for resistivity (and for mass density in other chapters). These double usages are found frequently in the literature.

$$P = i^2 R \tag{6.27}$$

or

$$P = \frac{v^2}{R}. \tag{6.28}$$

This type of energy loss has clinical significance. If a patient contacts a source of very high voltage such as an 11 000-V power line, the strong electric fields will cause current to flow throughout the patient's body or limb, because $\mathbf{j} = \sigma \mathbf{E}$. The resistive heating can be enough to boil water within the tissues. If the limb is x rayed, the steam bubbles will look very much like the bubbles that appear in *clostridium* (gas gangrene) infections; if the x ray is deferred a few days, it will be impossible to tell from the x ray whether the bubbles are due to the electrical injury or subsequent infection.

6.9. THE APPLICATION OF OHM'S LAW TO SIMPLE CIRCUITS

The ultimate goal of this chapter is to apply Ohm's law to the axon. Before doing that, however, it is worthwhile to see how it can be applied to some simpler circuits.

The simplest circuit is a resistance R connected across a battery, as shown in Fig. 6.20. The battery voltage of 6 V is the potential difference across the resistor. If the resistance is 3 Ω, the current is

$$i = \frac{v}{R} = \frac{6}{3} = 2 \text{ A}.$$

The rate of heat production in the resistor is

$$P = vi = (6)(2) = 12 \text{ W}.$$

This could also have been calculated from $P = v^2/R = 36/3$, or $P = i^2 R = (4)(3)$. A current of 2 A means that every second 2 C of charge leave the positive terminal of the battery and flow through the resistor. When the charge arrives at the other end of the resistor, it has lost 12 J of energy to heat. The 2 C then travel through the battery back to the positive terminal, gaining 12 J from a chemical reaction within the battery.

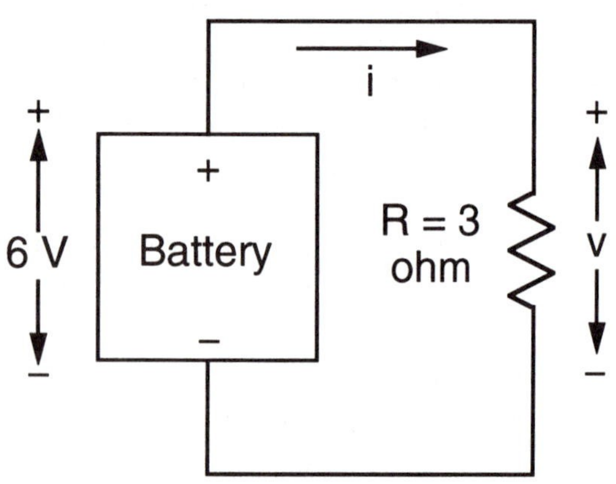

FIGURE 6.20. A resistor connected to a battery.

This example has been stated as though the positive charge moves. In a metallic conductor negative charges (electrons) move from the negative terminal of the battery through the resistor to the positive terminal. From a macroscopic point of view, we cannot tell the difference between the transport of a charge $-q$ from point A to point B, and the transport of a charge $+q$ from point B to point A. Both processes make the total charge at B less positive and the total charge at A more positive by an amount q.

Two fundamental principles used in this discussion have not been explicitly stated. The first is the *conservation of electric charge*: all charge that leaves the battery passes through the resistor. The second is the *conservation of energy*: a charge that starts at some point in the circuit and comes back to its starting point has neither lost nor gained energy. (The energy gained by a charge in the battery is equal to the energy lost by it in passing through the resistor.) These principles become less obvious and more useful in a circuit that is more complicated than the one considered above. They are known as *Kirchhoff's laws*.

In a more complicated circuit, Kirchhoff's first law (conservation of charge) takes the following form. Any junction where the current can flow in different paths is called a *node*. The algebraic sum of all the currents into a node is zero. By algebraic sum we mean that currents into the node are positive, while currents leaving the node are negative, or vice versa. This ensures that no charge will accumulate at the node.

More generally, the node could represent a conductor, such as the plate of a capacitor, on which charge can accumulate. In that case the charge Q changes with time:

$$\frac{dQ}{dt} = \sum \text{ (all currents into or away from the node).}$$

This statement is quite similar to the continuity equation of Sec. 4.1.

As an example of Kirchhoff's first law, consider the node in Fig. 6.21. Conservation of charge requires that $2+3+i=0$ or $i=-5$ A. (In this case positive currents flow into the node; the negative current means that 5 A is flowing out of the node as current i.)

Kirchhoff's second law was used implicitly in the example above to say that the voltage across the resistance

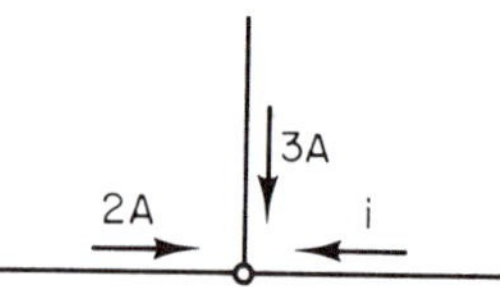

FIGURE 6.21. Conservation of charge means that current i is -5 A.

is 6 V. In general, Kirchhoff's second law says that if one goes around any closed path in a complicated circuit, the total voltage change is zero.

Kirchhoff's laws can be applied to show that the total resistance of a set of resistances in series is

$$R=R_1+R_2+R_3+\cdots .$$

This follows from the definition of resistance, the fact that the same current flows in each resistance, and the total potential difference across the set of resistances is the sum of the potential difference across each one. Kirchhoff's laws can also be used to show that for a collection of resistances in parallel, the total resistance is given by

$$\frac{1}{R}=\frac{1}{R_1}+\frac{1}{R_2}+\frac{1}{R_3}+\cdots ,$$

(see Problem 6.14).

Consider a more complicated example in which two resistors are connected across a battery. The battery voltage is v, and the resistances are R_1 and R_2, as shown in Fig. 6.22. If no current flows out lead A, then conservation of charge requires that the same current i flows in each resistor. The sum of the voltages v_1 and v_2 must also be equal to v. Therefore,

$$i=\frac{v_1}{R_1}=\frac{v_2}{R_2}$$

and

$$v=v_1+v_2=iR_1+iR_2=i(R_1+R_2).$$

The voltage across R_2 is iR_2 or

$$v_2=\frac{R_2}{R_1+R_2}\,v. \tag{6.29}$$

6.10. CHARGE DISTRIBUTION IN THE RESTING NERVE CELL

The axon consists of an ionic intracellular fluid and an ionic extracellular fluid, separated by a membrane. The intracellular and extracellular media are electrical conductors, and when the cell is in equilibrium, there is no current and no electric field in these regions. There will be a field and currents when an impulse is traveling along the axon.

Because the electric field in the resting cell is zero, there is no net charge in the fluid. Positive ions are neutralized by negative ions everywhere except at the membrane. A layer of charge on each surface generates an electric field within the membrane and a potential difference across it.

Measurements with a microelectrode show that the potential within the cell is about 70 mV less than outside. If the potential outside is taken to be zero, then the interior resting potential is −70 mV. Figure 6.23 shows a slice through the cell, showing the membrane on opposite sides of the cell and the charges and electric field. If the potential drops 70 mV as one enters the cell on the left, if the membrane thickness is 6 nm, and if the electric field within the membrane is assumed to be constant, then

$$E=-\frac{dv}{dx}=-\frac{-70\times10^{-3}\ \text{V}}{6\times10^{-9}\ \text{m}}=1.17\times10^{7}\ \text{V m}^{-1}. \tag{6.30}$$

This is how the value of E was determined for use on p. 142.

Except for the layers of charge on the inside and outside of the membrane, which are shown in Fig. 6.23 and which give rise to the electric field and potential difference, the extracellular and intracellular fluids are electrically neutral. However, the ion concentrations are quite different in each. This was shown in Fig. 6.3. There is an excess of sodium

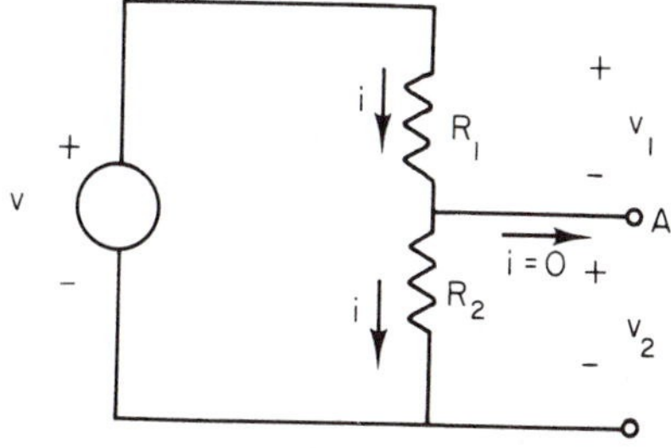

FIGURE 6.22. A more complicated circuit, sometimes called a voltage divider.

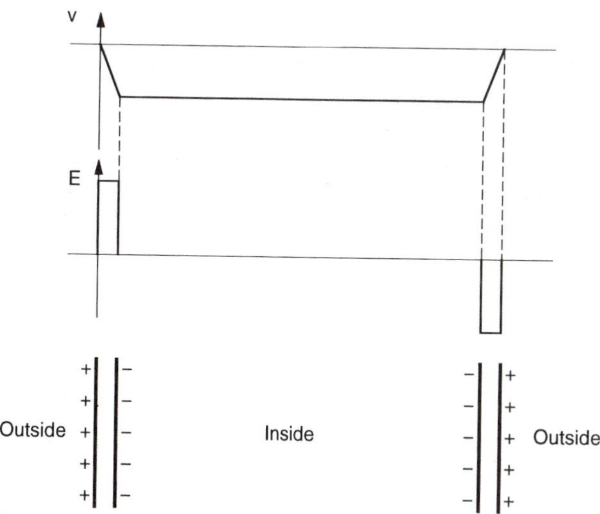

FIGURE 6.23. The potential, electric field, and charge moving across the diameter of a resting nerve cell. Portions of the cell membrane on opposite sides of the cell are shown. Outside the cell the potential and electric field are zero. As one moves to the right into the cell, the electric field in the membrane causes the potential to decrease to −70 mV. Within the cell the field is zero and the potential is constant. Moving out through the right-hand wall the potential rises to zero because of the electric field within the membrane.

ions outside and an excess of potassium ions inside. We will not explain these concentration differences here. (A few models that can cause concentration differences are discussed in Chap. 9.) However, it is possible to see which concentrations (if any) are consistent with the hypothesis that the ions can pass freely through the membrane.

If a particular kind of ion can pass freely through the membrane, equilibrium will be established when the concentration ratio across the membrane is given by a Boltzmann factor or the Nernst equation (see Chap. 3). The potential energy of the ion is zev, where z is the valence of the ion, e the electronic charge (1.6×10^{-19} C), and v the potential in volts. Using subscripts i and o to represent inside and outside the cell, we have

$$\frac{C_i}{C_o}=\frac{e^{-zev_i/k_BT}}{e^{-zev_o/k_BT}}=e^{-ze(v_i-v_o)/k_BT}. \tag{6.31}$$

Here k_B is Boltzmann's constant, 1.38×10^{-23} J K^{-1}. For a situation in which $T=310$ K and $v_i-v_o=-70\times10^{-3}$ V, C_i/C_o is 13.7 for univalent positive ions and $1/13.7=0.073$ for negative ions. The data in Fig. 6.3 are 0.103 for sodium, 30 for potassium, and 0.071 for chloride. From these we see that the chloride concentration ratio is consistent with equilibrium, while the sodium concentration ratio is much too small (too few sodium ions inside) and the potassium concentration ratio is too large (too many potassium ions inside). A potential of -90 mV would bring the potassium concentration ratio into equilibrium, but then chloride would not be in equilibrium and sodium would be even farther from equilibrium. In fact, tracer studies show that potassium leaks out slowly and sodium leaks in slowly. The resting membrane is not completely impermeable to these ions [Hodgkin (1964, Chap. 6); Alberts *et al.* (1983, Chap. 18); Laüger (1991)]. These fluxes are balanced by a process in which energy is expended to pump sodium back out and potassium back in. The usual ratio of sodium to potassium ions in this active transport is 3 sodium to 2 potassium ions [Patton *et al.* (1989, Vol. 1, p. 27)].

The intracellular and extracellular fluid are two conductors separated by a fairly good insulator. The conductors have a capacitance between them. We can estimate this capacitance in two ways. We can either regard the membrane as a plane insulator sandwiched between plane conducting plates (as if the membrane had been laid out flat as in Fig. 6.24), or we can treat it as a dielectric between concentric cylindrical conductors. The text will use the first approximation, while the second will be left to a problem. Suppose that two parallel plates have area S and charge $\pm Q$, respectively, then the charge density on each is $\sigma=\pm(Q/S)$. Equation (6.9) gives the electric field without a dielectric between the conductors:

$$E_e=\frac{\sigma}{\epsilon_0}=\frac{Q}{\epsilon_0 S}.$$

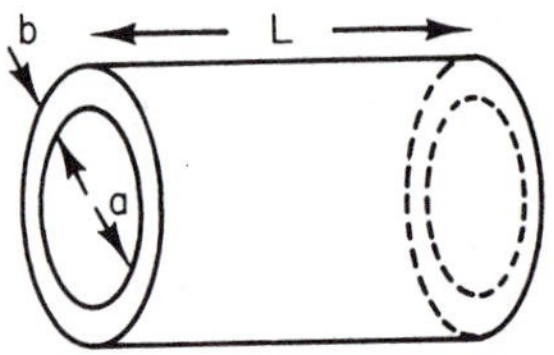

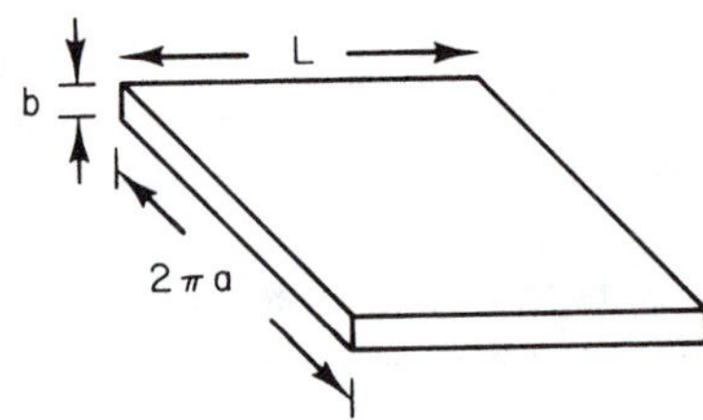

FIGURE 6.24. A portion of a cell membrane of length L, in its original configuration and laid out flat. The membrane thickness is b and the radius of the axon is a. The plane approximation is used to calculate both the capacitance and resistance of the membrane.

With the dielectric of dielectric constant κ, the field is reduced to

$$E=\frac{E_e}{\kappa}=\frac{\sigma}{\kappa\epsilon_0}=\frac{Q}{\kappa\epsilon_0 S}$$

as was seen in Eq. (6.18). The magnitude of the potential difference is E times the plate separation b:

$$v=Eb=\frac{Qb}{\kappa\epsilon_0 S}.$$

The capacitance is $C=Q/v$:

$$C=\frac{Q\kappa\epsilon_0 S}{Qb}=\frac{\kappa\epsilon_0 S}{b}. \tag{6.32}$$

The charge density on the surface of the membrane is obtained from

$$\sigma=\frac{Q}{S}=\frac{Cv}{S}=\frac{\kappa\epsilon_0 v}{b}.$$

Measurements of the dielectric constant κ for axon membrane show it to be about 7. Using values of -70 mV for v and 6×10^{-9} m for b, the capacitance per unit area of membrane can be calculated, as can σ:

$$\frac{C}{S}=\frac{(7)(8.85\times10^{-12})}{6\times10^{-9}}=0.01\ \text{F m}^{-2},$$

$$\sigma=(0.01)(70\times10^{-3})=7\times10^{-4}\ \text{C m}^{-2}. \tag{6.33}$$

This value for the surface charge density is larger by a factor of 7 than that calculated in Sec. 6.3. The reduction of the electric field by polarization of the dielectric has been taken into account in the present calculation. A larger external charge is required to give the same field within the dielectric.

The value of b for myelinated fibers is much greater, typically 2000 nm instead of 6 nm. This reduces the capacitance per unit area by a factor of 300.

6.11. THE APPLICATION OF THE CIRCUIT THEOREMS (KIRCHHOFF'S LAWS) TO A SEGMENT OF AXON: THE CABLE MODEL

We now need to consider the rather complicated flow of charge in the interior of an axon, through the membrane, and in the conducting medium outside the cell. We will model the axon by electric conductors that obey Ohm's law inside and outside the cell and a membrane that has capacitance and also conducts current. We will apply Kirchhoff's laws—conservation of energy and charge—to a small segment of the axon. The result will be a differential equation that is independent of any particular model for the cell membrane. This is called the *cable model* for an axon. We will then apply the cable model in two cases. The first case is when the voltage change does not alter the properties of the membrane. (This can be a small voltage change across the membrane of an unmyelinated axon or a larger voltage change across a myelinated membrane.) The second case is a voltage change that changes the ionic permeability of the membrane, thereby generating a nerve impulse.

Consider the small segment of membrane shown in Fig. 6.25(a). The upper capacitor plate, corresponding to the inside of the membrane, carries a charge Q. The lower capacitor plate (the outside of the membrane) has charge $-Q$. Figure 6.25(a) shows negative ions on the inside and positive ions on the outside of the membrane. This is the usual situation in a resting nerve cell and means that Q is negative in the resting state. The charge on the membrane is related to the potential difference across the membrane by $Q=Cv$. To repeat: when Q is positive the interior potential is greater than outside and $v>0$. In the resting case, Q is negative, $-Q$ is positive, and $v<0$.

If the resistance between the plates of a capacitor is infinite, no current flows, and the charge on the capacitor plates remains constant. However, a membrane is not a perfect insulator; if it were, there would be no nerve conduction. If one side of a membrane has positive charge and the other has negative charge, there are currents, however small, due to each ion species. In the resting state there is a tendency for leakage: for negative ions to move out and positive ions to move in if their concentration ratio is unity. The sum of all the ion currents is the current through the membrane. We call this current i_m and *define* it to be positive when positive charge moves from inside to outside, as in Fig. 6.25(b). Because of this definition, ions moving in the directions shown in Fig. 6.25(a) correspond to a negative value for i_m.

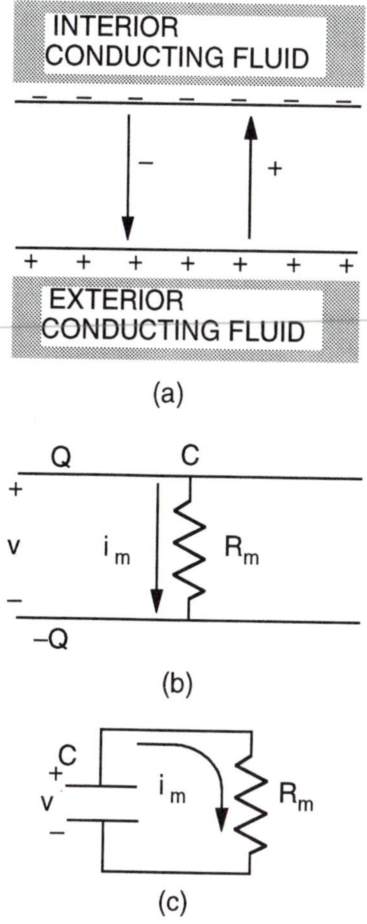

FIGURE 6.25. Leakage currents through the membrane. (a) The flow of positive and negative ions. (b) The membrane capacitance is represented by the parallel plates and the leakage resistance by a single resistor. (c) The capacitance and resistance are usually drawn like this.

Imagine for now that there is no current along the axon, and that there is no mechanism to maintain the charge on the plates of the capacitor. In that case i_m discharges the membrane capacitance, and the charge and potential difference fall to zero as charge flows through the resistor. When i_m is positive, Q and v decrease with time:

$$-i_m=\frac{dQ}{dt}=C\frac{dv}{dt}. \tag{6.34}$$

Let us explore the behavior of this isolated segment of axon a bit further. For now we think of the total leakage current as being through a single effective resistance R_m. This is shown in Fig. 6.25(b). It is customary to draw the resistance separately, as in Fig. 6.25(c). The current is then $i_m=v/R_m$ and

$$C\frac{dv}{dt}=-i_m=-\frac{v}{R_m},$$

$$\frac{dv}{dt} = -\frac{1}{R_m C} v. \qquad (6.35)$$

This is the familiar equation for exponential decay of the voltage (see Chap. 2). If the initial voltage at $t=0$ is v_0, the solution is

$$v(t) = v_0 e^{-t/\tau}, \qquad (6.36)$$

where the time constant τ is given by

$$\tau = R_m C. \qquad (6.37)$$

Referring to Fig. 6.24, we saw that if we have a section of membrane of area S and thickness b the capacitance is given by Eq. (6.32). For a conductor of the same dimensions we saw [Eq. (6.25)] that the resistance is

$$R_m = \frac{\rho_m b}{S},$$

so the time constant is

$$\tau = R_m C = \frac{\rho_m b}{S} \frac{\kappa \epsilon_0 S}{b} = \kappa \epsilon_0 \rho_m. \qquad (6.38)$$

We have the remarkable result that the time constant is independent of both the area and thickness of the membrane. Doubling the area S doubles the amount of charge that must leak off, but it also doubles the membrane current. Doubling b doubles the resistance, but it also makes the membrane capacitance half as large. In each case the factors S and b cancel in the expression for the time constant.

If a very thin lipid membrane is produced artificially, it is found to have a very high resistivity—about 10^{13} Ω m [Scott (1975, p. 493)]. Certain proteins added to the lipid material reduce the resistivity by several orders of magnitude. For natural nerve membrane the resistivity is about

$$\rho_m = 1.6 \times 10^7 \ \Omega \text{ m}. \qquad (6.39)$$

This is the effective resistivity for resting membrane, taking into account all of the ion currents. If ρ_m had this constant value the time constant would be

$$\tau = \kappa \epsilon_0 \rho_m = (7)(8.85 \times 10^{-12})(1.6 \times 10^7) = 1 \times 10^{-3} \text{ s}.$$

(Actually, the resistivity changes drastically as the potential across the membrane changes during the propagation of a nerve impulse.) Since we observe a potential difference across the membrane, there must be a mechanism for renewing the charge on the membrane surface.

The resistance and capacitance of the portion of the axon membrane in Fig. 6.24 can be written in terms of the axon radius a and the length L of the segment by noting that $S = 2\pi a L$. Then one has

$$C = \frac{\kappa \epsilon_0 2\pi a L}{b}, \quad R_m = \frac{\rho_m b}{2\pi a L}.$$

It is convenient to recall that $v = iR$ can be written as $i = Gv$ and introduce the conductance of the membrane segment

$$G_m = \frac{2\pi a L}{\rho_m b}. \qquad (6.40)$$

Both the capacitance and the conductance are proportional to the area of the segment S. It is also convenient to introduce the lowercase symbols c_m and g_m to stand for the membrane capacitance and membrane conductance per unit area:

$$c_m = \frac{C}{S} = \frac{\kappa \epsilon_0}{b}, \qquad (6.41)$$

$$g_m = \frac{G}{S} = \frac{1}{\rho_m b} = \frac{\sigma_m}{b}. \qquad (6.42)$$

(Note that $\sigma_m = 1/\rho_m$ is the electrical conductivity, the reciprocal of the resistivity. It is *not* the charge per unit area. σ is frequently used for both quantities in the literature.)

Both c_m and g_m depend on the membrane thickness as well as the dielectric constant and resistivity of the membrane. The units of c_m and g_m are, respectively, F m^{-2} and S m^{-2}. Be careful: many sources give them per square centimeter instead of per square meter.

We can rewrite Eq. (6.34) in terms of the current density per unit area, j_m, which is proportional to the capacitance per unit area, c_m:

$$-j_m = c_m \frac{dv}{dt}. \qquad (6.43)$$

Table 6.1 shows typical values for the quantities discussed in this section for an unmyelinated axon. These values should not be associated with a particular species. Parameters such as the resistance and capacitance per unit length of the axon are measured directly. Others, such as ρ_m, require an estimate of membrane thickness and are less well known.

Some insight into the magnitude of the charge on the membrane can be obtained from these numbers. The excess charge on the surface of the membrane is 7×10^{-4} C m^{-2} for the unmyelinated fiber. This corresponds to 4.4×10^{15} ions per square meter, if each ion has a charge of 1.6×10^{-19} C. Each atom or ion in contact with the membrane surface occupies an area of about 10^{-20} m^2; thus there are about 10^{20} atoms or ions in contact with a square meter of membrane surface. These may be neutral or positively or negatively charged. If charged, most are neutralized by the presence of a neighbor of opposite charge. The excess charge density that is required can be obtained if $4.4 \times 10^{15}/10^{20}$ or roughly one out of every 20 000 of the atoms in contact with the surface is ionized and not neutralized.

Now let us return to developing the cable model and consider current that flows inside and outside the axon. The

TABLE 6.1. *Properties of a typical unmyelinated nerve.*

a	Axon radius	5×10^{-6} m
b	Membrane thickness	6×10^{-9} m
ρ_i	Resistivity of axoplasm	0.5 Ω m
$r_i = \rho_i/\pi a^2$	Resistance per unit length of axoplasm	$6.4 \times 10^9\ \Omega\ \mathrm{m}^{-1}$
κ	Dielectric constant	7[a]
ρ_m	Resistivity of membrane	$1.6 \times 10^7\ \Omega$ m
$(\kappa\rho)_m$		$112 \times 10^6\ \Omega$ m
$C_m = \kappa\epsilon_0/b$	Capacitance per unit area	10^{-2} F m^{-2}
$2\pi\kappa\epsilon_0 a/b$	Capacitance per unit length	3×10^{-7} F m^{-1}
$g_m = 1/\rho_m b$	Conductance per unit area	10 S m^{-2}
$1/g_m$	Reciprocal of conductance per unit area	0.1 Ω m^2
$2\pi a/\rho_m b$	Conductance per unit length of axon	3.2×10^{-4} S m^{-1}
v	Resting potential inside	-70 mV
$E = v/b$	Electric field in membrane	1.2×10^7 V m^{-1}
$\kappa\epsilon_0 v/b$	Charge per unit area	7×10^{-4} C m^{-2}
	No. of univalent ions per unit area	4.4×10^{15} m^{-2}
	No. of univalent ions per unit length	6.6×10^7 m^{-1}

[a]See Sec. 6.17 for a discussion of this value of the dielectric constant.

cable model assumes that the currents inside are longitudinal, that is, parallel to the axis of the axon. A discussion of departures from this assumption is found in Scott (1975, p. 492). With this assumption, the interior fluid can be regarded as a resistance of length L and radius a as shown in Fig. 6.26. The resistance of such a segment is

$$R_i = \frac{\rho_i L}{S} = \frac{\rho_i L}{\pi a^2}.$$

It is convenient to work with the resistance per unit length:

$$r_i = \frac{R_i}{L} = \frac{\rho_i}{\pi a^2} = \frac{1}{\pi a^2 \sigma_i}. \tag{6.44}$$

The question of resistance of the extracellular fluid for currents outside the axon is more complicated. If the extracellular fluid were infinite in extent, the longitudinal resistance outside the cell would be very small (see Chap. 7). On the other hand, in a nerve fiber or a muscle the axons or muscle cells are packed close together, there is not very much extracellular fluid, and the external resistance per unit length can be significant. There are some very important effects that occur because of this. We will discuss them in Chap. 7.

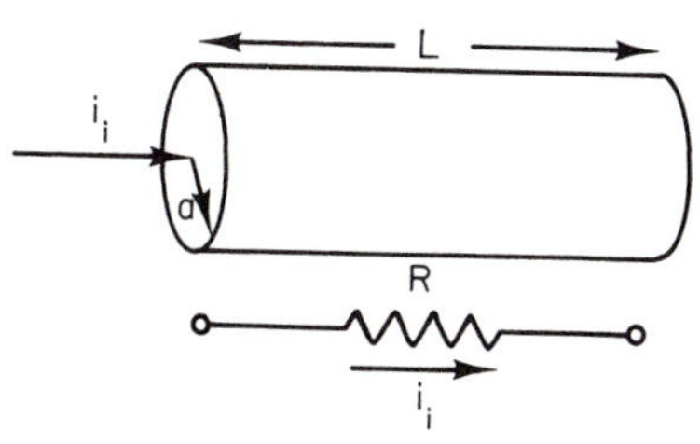

FIGURE 6.26. Axoplasm of length L and radius a can be treated like a simple resistor.

Now consider current flowing along the axon and ignore leakage. One model we can use divides the axon up into many small segments, each with capacitance C, as shown in Fig. 6.27. Each of these small capacitors is separated by a resistance R_i inside and R_o outside. If the nerve is at rest, each capacitor has the resting potential across it, and no current flows in any of the resistors. (We could model this by putting a battery with a potential equal to the resting potential in series with the capacitor.) If we then imagine "shorting out" one capacitor (that is, placing a wire of zero resistance between the two "plates"), all of the neighboring capacitors begin to discharge, and currents flow in the various resistors. This model is explored in the problems. We will use a continuous but equivalent model.

With the aid of Figs. 6.28 and 6.29 we can consider the effect of both membrane and longitudinal currents. Figure 6.28 shows a small region of the axon between x and $x+dx$ and the surrounding membrane. Figure 6.29 shows the same segment of axon more schematically. This diagram represents all of the membrane and the interior and exterior conducting fluid that are between x and $x+dx$. Current i_i flows along the axon on the inside and i_o on the outside.

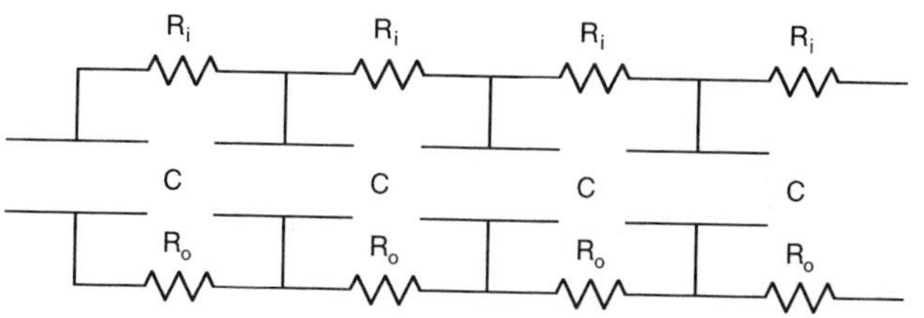

FIGURE 6.27. The membrane can be imagined as being divided into segments. Each segment has capacitance C. The segments are connected by resistances R_i and R_o that carry the current inside and outside the axon. In this drawing the current through the membrane is not shown.

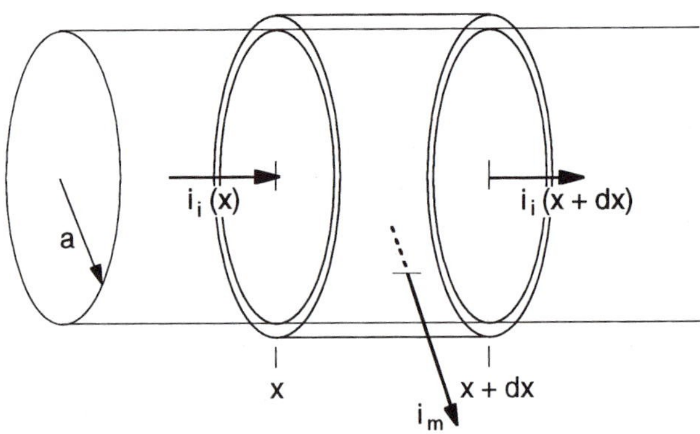

FIGURE 6.28. The membrane surrounding a small portion of axon is shown, along with the longitudinal currents in and out of the segment.

The current through the membrane is i_m. In this section no attempt will be made to relate i_m or j_m to v. Charge Q resides on the inside surface of the membrane and can be either negative or positive. An equal and opposite charge $-Q$ resides on the outer surface of the membrane.

Because the capacitance can charge or discharge, Kirchhoff's law (conservation of charge) does not say that the sum of the currents is zero. Rather, it says that the net current into the volume of axoplasm between x and $x+dx$ changes the charge on the interior surface of the membrane:

$$i_i(x)-i_i(x+dx)-i_m=\frac{dQ}{dt}=C\frac{d(v_i-v_o)}{dt}. \tag{6.45a}$$

When $i_i(x)=i_i(x+dx)=0$ this gives Eq. (6.34). The membrane current i_m represents an average value between x and $x+dx$. It is also a function of x.

We can define

$$di_i=i_i(x+dx)-i_i(x)$$

as the increase in i_i along segment dx, and

$$v_m=v_i-v_o$$

as the transmembrane potential or the potential difference across the membrane. Then we can rewrite Eq. (6.45a) as

$$-di_i=C\frac{dv_m}{dt}+i_m. \tag{6.45b}$$

This is an important equation. It says that when the current flowing inside the axon decreases in a small distance dx, part of the current charges the capacitance of that segment of membrane, and the rest flows through the membrane. In the extracellular medium these currents recombine and become a current $di_o=-di_i$ that flows in the extracellular region. We will discuss the extracellular potential differences arising from this current in Chap. 7. For now, we will assume that v_o is small and can be ignored.

Consider a small segment of axoplasm of length δx, where $\delta x \ll dx$. The voltage at the left end is $v_i(x)$; at the right end it is $v_i(x+\delta x)$. The current along the segment is the voltage difference between the ends divided by the resistance of the segment. The resistance is $R_i=r_i\delta x$. Therefore the current is

$$i_i(x)=\frac{v_i(x)-v_i(x+\delta x)}{r_i\delta x}=-\frac{1}{r_i}\frac{dv_i}{dx}. \tag{6.46}$$

There must be a changing voltage along the axon for a current to flow within it. The minus sign in Eq. (6.46) shows that a current flowing from left to right (in the $+x$ direction) requires a voltage that decreases from left to right, and vice versa. Figure 6.30 shows a hypothetical plot of $v_i(x)$ and the current which would accompany it. Notice that the current is flowing from the region of higher voltage to lower voltage—towards both ends from the region

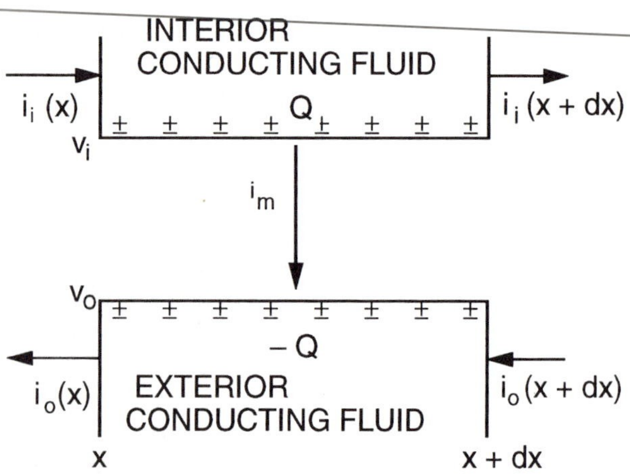

FIGURE 6.29. A more schematic representation of the currents in Fig. 6.28.

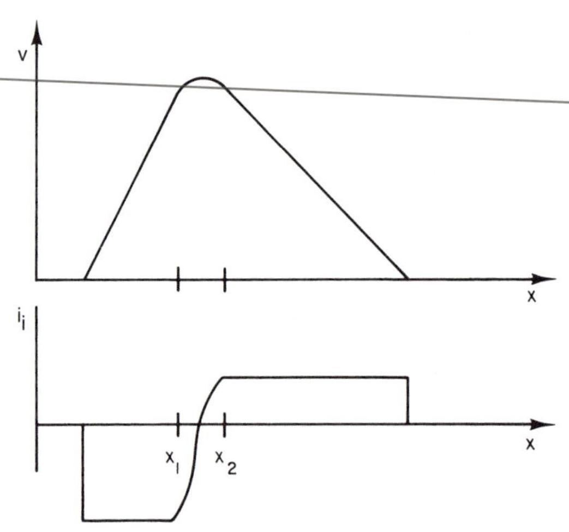

FIGURE 6.30. A hypothetical plot of $v(x)$ and the longitudinal current i_i associated with it.

between x_1 and x_2. In that region either the charge on the membrane is changing or current is flowing through the membrane.

Consider again the cylindrical geometry shown in Fig. 6.28. The surface area of this portion of membrane is $2\pi a\, dx$. Dividing each term of Eq. (6.45a) by the area and remembering the definitions of j_m and c_m we obtain

$$c_m \frac{\partial(v_i - v_o)}{\partial t} = -j_m + \frac{1}{2\pi a}\left[\frac{i_i(x) - i_i(x+dx)}{dx}\right]. \tag{6.47}$$

It is necessary to use partial derivatives because the currents and voltage depend on both x and t as an impulse travels down the nerve. As $dx \rightarrow 0$ the expression in square brackets approaches $-$(the partial derivative of i_i with respect to x):

$$\frac{i_i(x+dx) - i_i(x)}{dx} \rightarrow \frac{\partial i_i}{\partial x}.$$

This can be evaluated using the expression for Ohm's law in the axoplasm, Eq. (6.46):

$$\frac{\partial i_i}{\partial x} = -\frac{1}{r_i}\frac{\partial^2 v_i}{\partial x^2}. \tag{6.48}$$

When this is inserted in Eq. (6.47) the result is

$$c_m \frac{\partial(v_i - v_o)}{\partial t} = -j_m + \frac{1}{2\pi a r_i}\frac{\partial^2 v_i}{\partial x^2}. \tag{6.49}$$

In many cases, when the volume of conducting medium outside the axon is large, $v_o \ll v_i$. If we neglect v_o and drop the subscript on v_i we can rewrite Eq. (6.49) as

$$c_m \frac{\partial v}{\partial t} = -j_m + \frac{1}{2\pi a r_i}\frac{\partial^2 v}{\partial x^2}. \tag{6.50}$$

This is a rather formidable looking equation. It has the form of Fick's second law of diffusion, Eq. (4.24), with the addition of the j_m term. It is worth recalling the origin of each term and verifying that the units are consistent. The term on the left is the rate at which the membrane capacitance is gaining charge per unit area. Therefore all terms in the equation have the units of current per unit area. The first term on the right is the current per unit area through the membrane in the direction that discharges the membrane capacitance. The second term on the right gives the buildup of charge on this area of the membrane because of differences in current along the axon. If $v(x)$ were constant, there would be no current along the inside of the axon. If function $v(x)$ had constant slope, the current along the inside of the axon would be the same everywhere and there would be no charge buildup on the membrane. It is only because $v(x)$ changes slope that i_i is different at two neighboring points in the axon and charge can collect on the membrane. Now, for the units. Since $i = C(dv/dt)$, the units of $c_m \partial v/\partial t$ are current per unit area. The j_m term is by definition current per unit area. Since r_i has the units of $\Omega\, m^{-1}$, the term $2\pi a r_i$, has the units of Ω. When this is combined with $\partial^2 v/\partial x^2$, which has units $V\, m^{-2}$, the result is $A\, m^{-2}$ as required.

This is a very general equation stating Kirchhoff's laws for a segment of the axon. The only assumptions are that the currents depend only on time and position along the axon and that voltage changes on the outside of the axon can be neglected. Particular models for nerve conduction use different relations between j_m and $v(x,t)$.

6.12. ELECTROTONUS OR PASSIVE SPREAD

The simplest membrane model is one that obeys Ohm's law. This approximation is valid if the voltage changes are small enough so that the membrane conductance does not change, or if something has been done to inactivate the normal changes of membrane conductance with voltage. It is also useful for myelinated nerves between the nodes of Ranvier. This is called *electrotonus* or *passive spread.*

In its quiescent state, the voltage all along the inside of the axon has the constant resting value v_r. Both $\partial v/\partial t$ and $\partial^2 v/\partial x^2$ are zero. Equation (6.50) can be satisfied only if $j_m = 0$. Although j_m is zero, it may be made up of several leakage components. In this section we simply assume that j_m is proportional to $v - v_r$:

$$j_m = g_m(v - v_r). \tag{6.51}$$

This simple model does predict that $j_m = 0$ when $v = v_r$. It also predicts that the current will be positive (outward) if $v > v_r$ and negative (inward) if $v < v_r$. It will not explain the propagation of an all-or-nothing nerve impulse. The conductance per unit area, g_m, is assumed to be independent of v and of the past history of the membrane. This is a good assumption only for very small voltage changes. With this assumption, Eq. (6.50) becomes

$$c_m \frac{\partial v}{\partial t} = -g_m(v - v_r) + \frac{1}{2\pi a r_i}\frac{\partial^2 v}{\partial x^2}. \tag{6.52}$$

This equation is usually written in a slightly different form by dividing through by g_m:

$$\frac{1}{2\pi a r_i g_m}\frac{\partial^2 v}{\partial x^2} - v - \frac{c_m}{g_m}\frac{\partial v}{\partial t} = -v_r.$$

It is also customary to make the assignments

$$\lambda^2 = \frac{1}{2\pi a r_i g_m},$$

$$\tau = \frac{c_m}{g_m},$$

so that the equation becomes

$$\lambda^2 \frac{\partial^2 v}{\partial x^2} - v - \tau \frac{\partial v}{\partial t} = -v_r. \tag{6.53}$$

In terms of the primary axon parameters, the parameters in Eq. (6.53) are

$$\lambda^2 = \frac{ab\rho_m}{2\rho_i}, \tag{6.54}$$

$$\tau = \kappa \epsilon_0 \rho_m. \tag{6.55}$$

The time constant was seen before in Eq. (6.38). Equation (6.53) has a steady-state solution $v = v_r$. If a new variable $v' = v - v_r$ is used, it becomes the homogeneous version of the same equation with a steady-state solution $v' = 0$. This homogeneous equation is known as the telegrapher's equation: it was once familiar to physicists and electrical engineers as the equation for a long cable, such as a submarine cable, with capacitance and leakage resistance but negligible inductance [Jeffreys and Jeffreys (1956 p. 602)].

For nerve conduction, the inhomogeneous equation with various exciting terms corresponding to physiological stimuli was discussed by Davis and Lorente de No (1947) and by Hodgkin and Rushton (1946). Their work is summarized by Plonsey (1969, p. 127).

Before considering general solutions to Eq. (6.53), consider some special cases. If $c_m = 0$, so that $\tau = 0$, or if enough time has elapsed so that the voltage is not changing with time and $\partial v / \partial t = 0$, the equation reduces to

$$\lambda^2 \frac{\partial^2 v}{\partial x^2} - v = -v_r.$$

You can verify by substitution that this has a solution

$$v - v_r = \begin{cases} v_0 e^{-x/\lambda}, & x > 0 \\ v_0 e^{+x/\lambda}, & x < 0. \end{cases} \tag{6.56}$$

If the voltage is held at a constant value

$$v = v_r + v_0$$

at some point on the axon, the voltage will decay exponentially to v_r in both directions from that point. This is shown in Fig. 6.31.

Next suppose that $v(x,t)$ does not depend on x, so that there is no longitudinal current in the axon and $\partial^2 v / \partial x^2 = 0$. This is accomplished experimentally by threading a wire axially along the axon. The equation reduces to

$$\tau \frac{\partial v}{\partial t} + v = v_r.$$

This is the familiar equation for exponential decay. If v were held at $v_0 + v_r$ and then the constraint were removed at $t = 0$, the voltage would decay exponentially back to v_r:

$$v - v_r = v_0 e^{-t/\tau}.$$

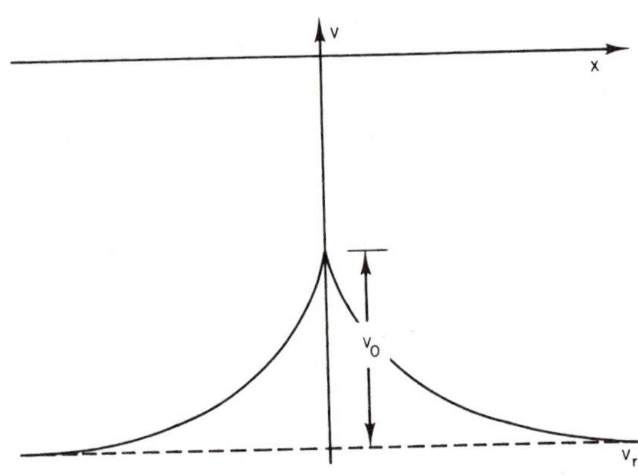

FIGURE 6.31. The voltage distribution along an axon in electrotonus when the membrane capacitance is charged and the voltage is not changing with time.

This represents the discharge of the membrane capacitance through the membrane resistance.

The general case can be simplified by making the substitution

$$v(x,t) = w(x,t) e^{-t/\tau} + v_r$$

in Eq. (6.53). The derivatives are

$$\frac{\partial^2 v}{\partial x^2} = \frac{\partial^2 w}{\partial x^2} \exp\left(\frac{-t}{\tau}\right),$$

$$\frac{\partial v}{\partial t} = -w(x,t)\exp(-t/\tau)/\tau + (\partial w/\partial t)\exp(-t/\tau).$$

Then Eq. (6.53) becomes

$$\left(\lambda^2 \frac{\partial^2 w}{\partial x^2} - \tau \frac{\partial w}{\partial t}\right) e^{-t/\tau} = 0$$

so that

$$\frac{\lambda^2}{\tau} \frac{\partial^2 w}{\partial x^2} = \frac{\partial w}{\partial t}. \tag{6.57}$$

This is the diffusion equation (Fick's second law) which was seen in Chap. 4. The general solution is discussed in the References.

The analytic form will not be reproduced here, but the behavior of $v(x,t) - v_r$ at various times after an excitation is applied is shown in Fig. 6.32. The excitation is a constant current injected at $x = 0$ for all time $t > 0$. After a long time, the curve is identical to that in Fig. 6.31, as the membrane capacitance has fully charged. Only the membrane leakage current attenuates the signal. At earlier times the solution is not precisely exponential; the analytic solution involves error functions (Chap. 4). The change of voltage with time at fixed positions along the cable is also shown. Both the finite propagation time and the attenuation of the signal are evident.

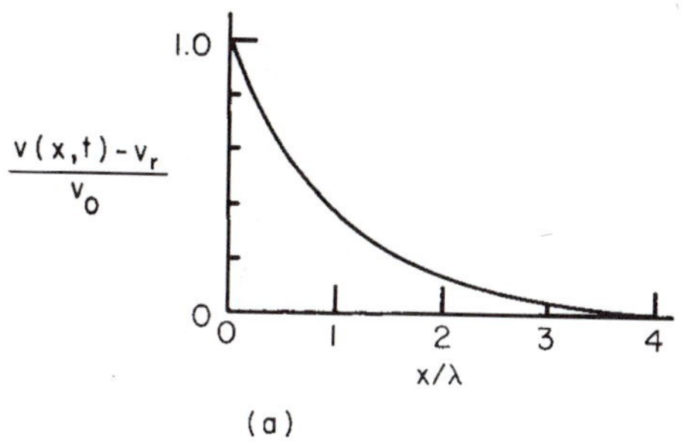

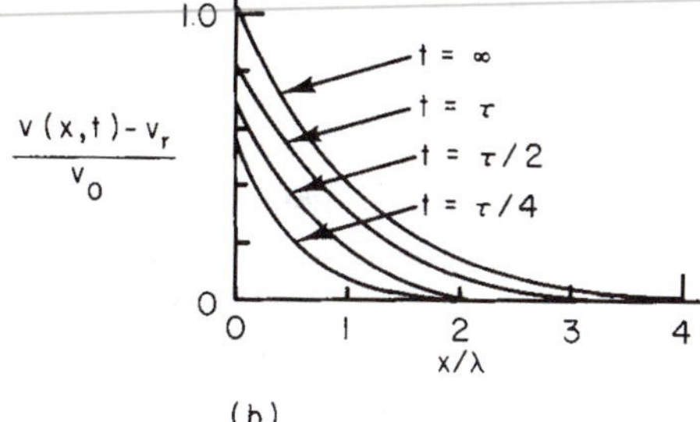

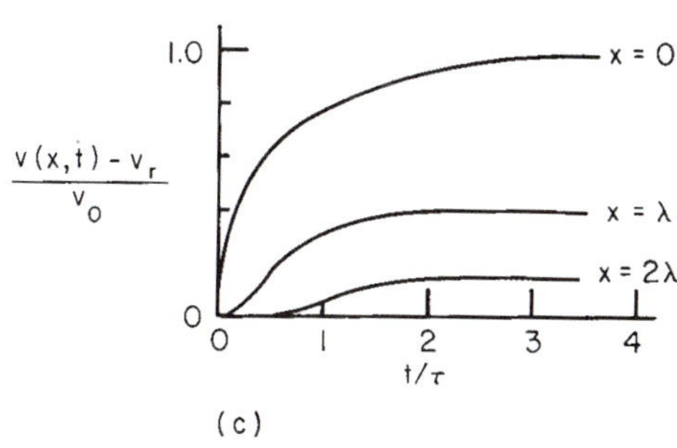

FIGURE 6.32. Some representative solutions to the problem of electrotonus. (a) The voltage along the axon when the membrane capacitance is fully charged and $v-v_r=v_0$ at $x=0$. (b) The voltage along the axon at different times after the application of a constant current at $x=0$. For infinite time this is the same as (a). (c) Voltage at a fixed point on the axon as a function of time, for the same excitation as in (b).

6.13. THE HODGKIN–HUXLEY MODEL FOR MEMBRANE CURRENT

If the voltage at some point on the axon changes by a few millivolts from the resting value, the voltage at other points on the axon is described by electrotonus. However, if the inside voltage rises from the resting value by 20 mV or more, a completely different effect takes place. The potential rises rapidly to a positive value, then falls to about −80 mV, and finally returns to the resting value (Fig. 6.1). This behavior is attributable to a very nonlinear dependence of membrane current on transmembrane voltage.

Considerable work was done on nerve conduction in the late 1940s, culminating in a model that relates the propagation of the action potential to the changes in membrane permeability that accompany a change in voltage. The model [Hodgkin and Huxley (1952)] does not explain why the membrane permeability changes; it relates the shape and conduction speed of the impulse to the observed changes in membrane permeability. Nor does it explain all the changes in current. (For example, the potassium current does fall eventually, and there are some properties of the sodium current that are not adequately described.) Nonetheless, the work was a triumph.

Most of the experiments that led to the Hodgkin–Huxley model were carried out using the giant axon of the squid. This is a single cell several centimeters long and up to 0.5 mm in diameter that can be dissected from the squid. The removal of axoplasm from the preparation and its replacement by electrolytes has shown that the critical phenomena all take place in the membrane. The important results are reviewed in many places [Alberts *et al.* (1983, Chap. 18); Katz (1966, Chaps. 5 and 6); Plonsey (1969, p. 127); Scott (1975, pp. 495–507)].

Voltage-clamp experiments were particularly illuminating. Two long wire electrodes were inserted in the axon and connected to the apparatus shown in Fig. 6.33. The resistance of the wires was so low that the potential at all points along the axon was the same at any instant of time. The potential depended only on time, and not on position. This is called a *space-clamped* experiment. One electrode, paired with an electrode in the surrounding medium, measured the voltage difference across the membrane. The other electrode was used to inject or remove whatever current was necessary to maintain a constant potential difference across the membrane. Measurement of this current allowed calculation of the membrane conductance. This technique is called *voltage clamping*. The experiment was both voltage- and space-clamped.

When the membrane potential was raised abruptly from the resting value to a new value and held there, the resulting current was found to have three components:

1. A current, lasting a few microseconds, that changed the surface charge on the membrane.

2. A current flowing inward which lasted for 1 or 2 ms. Various experiments, such as replacing the sodium ions in the extracellular fluid with some other monovalent ion, showed that this was due to the inward flow of

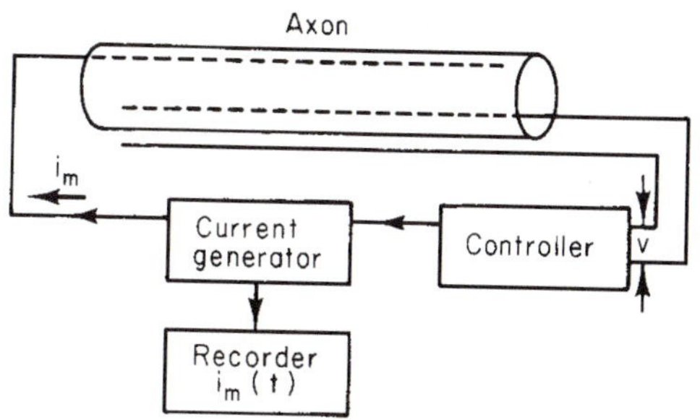

FIGURE 6.33. Apparatus for voltage-clamp measurements.

sodium ions. (Had the potential not been voltage-clamped by the electronic apparatus, this inrush of positive charge would have raised the potential still further.)

3. An outward current that rose in about 4 ms and remained steady for as long as the potential was clamped at this value. Tracer studies showed that this current was due to potassium ions. (Over a time scale of several tens of milliseconds, the potassium current, like the sodium current, does fall back to zero.)

The first current is the $c_m(\partial v/\partial t)$ term of Eq. (6.50); the second and third currents together constitute j_m. Because of the clamping wires, the $\partial^2 v/\partial x^2$ term is zero.

The next step is to develop a model that describes the major ionic constituents of the current. The sodium and potassium contributions to the current will be considered separately; all other contributions will be combined in a "leakage" term:

$$j_m = j_{\mathrm{Na}} + j_{\mathrm{K}} + j_L. \tag{6.58}$$

The leakage includes charge movement due to chloride ions and any other ions that can pass through the membrane.

Consider movement of sodium through the membrane. Similar considerations apply to potassium. The concentrations of sodium inside and out are $[\mathrm{Na}_i]$ and $[\mathrm{Na}_o]$. It will be seen later that the total number of ions moving through the membrane during a nerve pulse in a squid giant axon is too small to change significantly the concentrations inside and outside. Therefore, the concentrations can be taken as fixed. There is no average movement of sodium ions through the membrane, regardless of how permeable it is, when the concentrations and potential are related by the Boltzmann factor or Nernst equation (Sec. 6.10)

$$\frac{[\mathrm{Na}_i]}{[\mathrm{Na}_o]} = e^{-e(v_i - v_o)/k_B T}. \tag{6.31}$$

For given concentrations, the sodium equilibrium or Nernst potential is

$$v_{\mathrm{Na}} = v_i - v_o = \frac{k_B T}{e} \ln\left(\frac{[\mathrm{Na}_o]}{[\mathrm{Na}_i]}\right). \tag{6.59}$$

If $v = v_{\mathrm{Na}}$ there is no current of sodium ions, regardless of the membrane permeability to sodium. If v is greater than v_{Na} (more positive), j_{Na} flows outward. If $v < v_{\mathrm{Na}}$, the sodium current is inward. These currents can be described by

$$j_{\mathrm{Na}} = g_{\mathrm{Na}}(v - v_{\mathrm{Na}}). \tag{6.60}$$

The coefficient g_{Na} is the sodium conductance per unit area. It is not constant but depends on the value of v and, in fact, on the past history of v. Defining the conductance this way makes the functional form of g_{Na} less complex; in particular, it does not have to change sign as v moves through the sodium value v_{Na} and the sodium current reverses direction.

This equation can be multiplied by the membrane area to give a current–voltage relationship. Many authors draw a circuit diagram to represent the current flow through the membrane and along the axon. The sodium voltage–current relationship can be represented by a variable resistance corresponding to g_{Na} in series with a battery at the sodium Nernst potential, as shown in Fig. 6.34(a).

An expression similar to Eq. (6.60) can be written for the potassium current density:

$$j_{\mathrm{K}} = g_{\mathrm{K}}(v - v_{\mathrm{K}}). \tag{6.61}$$

The leakage term will be considered later. To summarize: v is the instantaneous voltage across the membrane. Both v_{K} and v_{Na} are constants depending on the relative ion concentrations inside and outside the cell and the temperature. The conductances per unit area depend on both the present value of v and its past history.

We can now describe the results of the voltage clamp experiments. The voltage in each experiment was changed

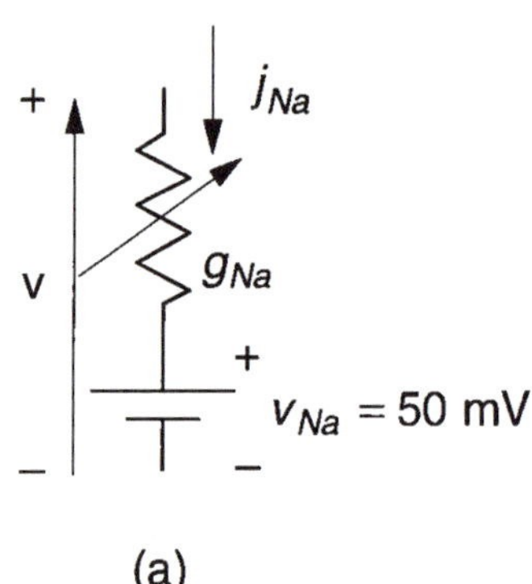

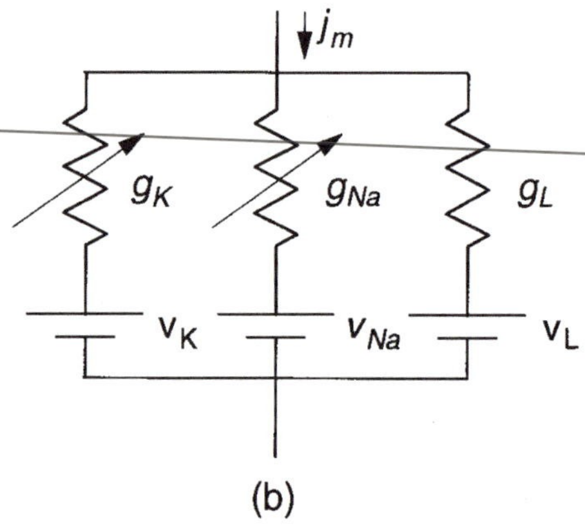

FIGURE 6.34. Equivalent circuits for the membrane current. (a) The sodium current–voltage relationship of Eq. (6.60) is the same as that for a variable resistance in series with a battery at the sodium Nernst potential. (b) The total membrane current can be modeled with three such equivalent circuits.

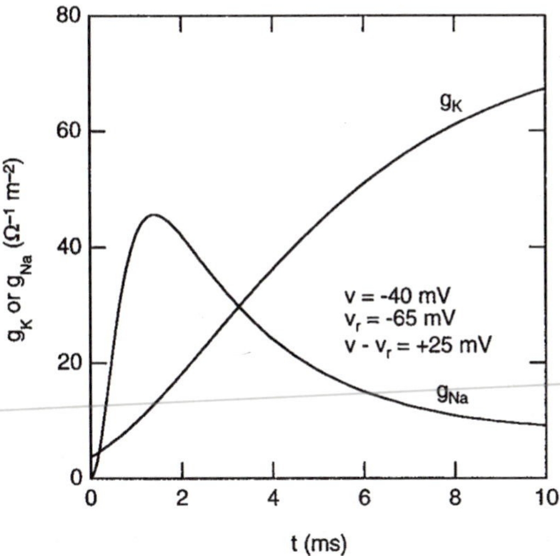

FIGURE 6.35. The behavior of the sodium and potassium conductivities with time in a voltage-clamp experiment. At $t=0$ the voltage was raised by 26 mV from the resting potential. The values are calculated from Eqs. (6.62)–(6.69) and are representative of the experimental data.

from the resting amount by an amount Δv. Therefore, $v-v_{\rm Na}$ and $v-v_{\rm K}$ had constant values after the change, and the changes in current density mirrored the changes in conductivity. Typical results for $\Delta v=26$ mV and $T=6$ °C are shown in Fig. 6.35. [The method of distinguishing sodium from potassium current is described in the original papers, or in Hille (1992, p. 37).] The sodium current rises and falls, while the potassium current rises more slowly. Measurements for longer times show that the potassium current rises to a constant value. Measurements for much longer times show that the potassium current falls after several milliseconds. For other values of Δv the conductance changes are different. Figure 6.36 shows potassium conductances for several values of the clamping voltage.

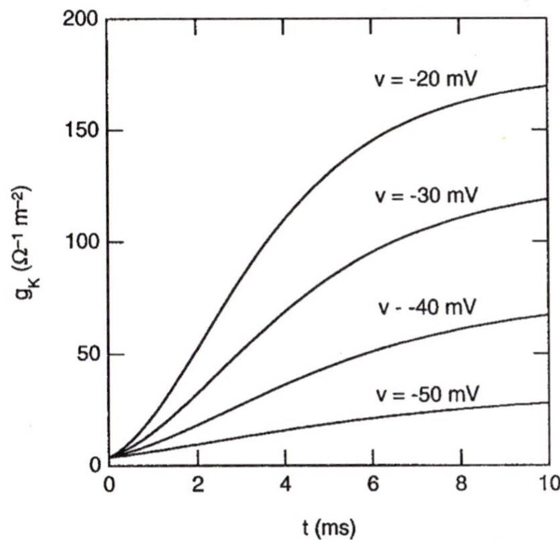

FIGURE 6.36. The behavior of the potassium conductivity for different values of the clamping voltage. These are representative curves calculated from Eqs. (6.62)–(6.64).

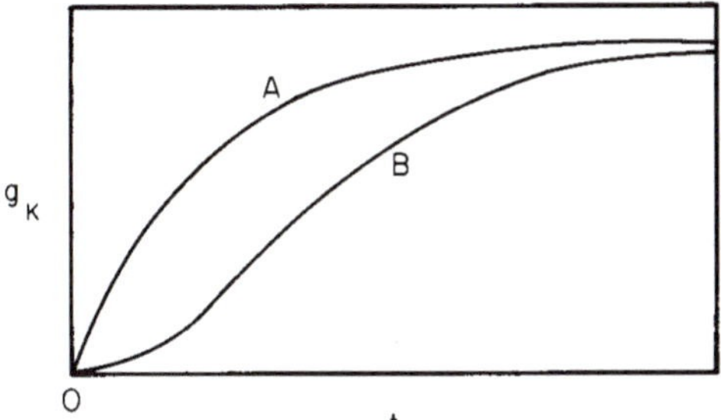

FIGURE 6.37. A comparison of the potassium conductivity data (curve *B*) with an exponential (curve *A*).

Hodgkin and Huxley wanted a way to describe their extensive voltage-clamp data, similar to that in Figs. 6.35 and Fig. 6.36, with a small number of parameters. Assume for now that the initial conductance is zero. The potassium conductance curves of Fig. 6.35 are reminiscent of exponential behavior, such as

$$g_{\rm K}(v,t)=g_{\rm K}(v)(1-e^{-t/\tau(v)}),$$

with both $g_{\rm K}$ and τ depending on the value of the voltage. A simple exponential is not a good fit. Figure 6.37 shows why. Curve *A* has the form of the exponential, while curve *B* has the shape of the conductance data. The conductance data have zero slope at $t=0$ while the exponential has a finite slope. There are many ways to generate a mathematical expression that resembles curve *B* more closely. One is to recall that for small x, $1-e^{-x}$ is approximately x. If we raise x to a power, it will approach the origin with a slope of zero (Fig. 6.38). On the other hand, as $x\to\infty$, $1-e^{-x}\to 1$, which is still one no matter what power it is raised to. This suggests that we try to describe the conductance by

$$g_{\rm K}(v,t)=g_{\rm K\infty}[n_\infty(v)(1-e^{-t/\tau})]^N.$$

In this expression, $g_{\rm K\infty}$ is the largest possible conductance per unit area after a clamp has been applied for a long time. The value of $n_\infty(v)$ varies between 0 and 1 and determines the asymptotic value of the conductivity for a particular value of the voltage step. Hodgkin and Huxley found a

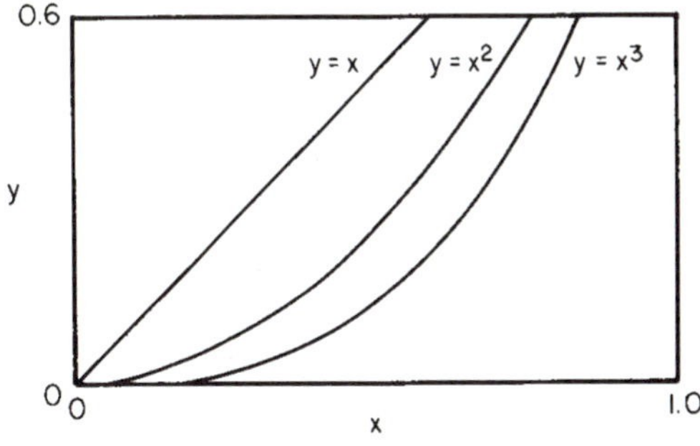

FIGURE 6.38. Graphs of $y=x$, $y=x^2$, and $y=x^3$ to show the behavior of different functions near the origin.

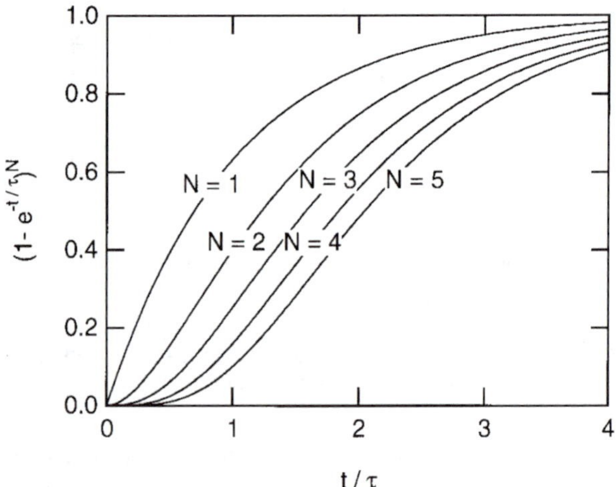

FIGURE 6.39. Plots of $(1-e^{-t/\tau})^N$ for different values of N.

good fit to their data with $N=4$. Plots of $(1-e^{-t/\tau})^N$ are shown in Fig. 6.39 for different values of N.

We now have an empirical fit to the potassium conductance data of the form

$$g_K(v,t)=g_{K\infty}n^4(v,t),$$
$$n(v,t)=n_\infty(v)(1-e^{-t/\tau(v)}). \tag{6.62}$$

The potassium conductance is not initially zero. If the initial value if n_0, the correct expression for $n(v,t)$ is

$$n(v,t)=n_\infty(v)\left(1-\left(\frac{n_\infty-n_0}{n_\infty}\right)e^{-t/\tau(v)}\right).$$

The function n is a solution to the differential equation

$$\frac{dn}{dt}=-\frac{n}{\tau}+\frac{n_\infty}{\tau}. \tag{6.63a}$$

Hodgkin and Huxley wrote this instead in the form

$$\frac{dn}{dt}=\alpha_n(1-n)-\beta_n n. \tag{6.63b}$$

The subscript n on α_n and β_n distinguishes them from similar parameters for the sodium conductance.

The dependence of α_n and β_n on voltage is quite pronounced. With v in mV and α_n and β_n in ms^{-1}, the equations used by Hodgkin and Huxley to describe their experimental values of α_n and β_n are

$$\alpha_n(v)=\frac{0.01[10-(v-v_r)]}{\exp\left(\frac{10-(v-v_r)}{10}\right)-1},$$
$$\beta_n(v)=0.125\exp\left(\frac{-(v-v_r)}{80}\right). \tag{6.64}$$

The quantities α_n and β_n are rate constants in Eq. (6.63b). Like all chemical rate constants, they depend on temperature. The values above are correct when $T=279$ K (6.3 °C). Hodgkin and Huxley assumed that the temperature dependence was described by a Q_{10} of 3. This means that the reaction rate increases by a factor of 3 for every 10 °C temperature rise. The rate at temperature T is obtained by multiplying rates obtained from Eq. (6.64) by

$$3^{(T-6.3)/10}. \tag{6.65}$$

For example, if the temperature is 18.5 °C, the rate must be multiplied by

$$3^{1.22}=3.82.$$

In an actual nerve-conduction process, v is not clamped. The behavior of α_n and β_n was determined from voltage-clamp experiments. Hodgkin and Huxley *assumed* that when v varies with time, the correct value of n can be obtained by integrating Eq. (6.63b). At each instant of time the values of α_n and β_n are those obtained from Eq. (6.64) for the voltage at that instant. This was a big assumption—but it worked. The value of $g_{K\infty}$ that they chose was 360 S m^{-2}.

The sodium conductance was described by two parameters: one reproducing the rise and the other the decay of the conductance. The equation was

$$g_{Na}=m^3hg_{Na\infty}. \tag{6.66}$$

The parameters m and h obeyed equations similar to that for n:

$$\frac{dm}{dt}=\alpha_m(1-m)-\beta_m m, \tag{6.67}$$

$$\frac{dh}{dt}=\alpha_h(1-h)-\beta_h h. \tag{6.68}$$

The v dependences were

$$\alpha_m=\frac{0.1[25-(v-v_r)]}{\exp\left(\frac{25-(v-v_r)}{10}\right)-1},$$
$$\beta_m=4\exp\left(\frac{-(v-v_r)}{18}\right),$$
$$\alpha_h=0.07\exp\left(\frac{-(v-v_r)}{20}\right), \tag{6.69}$$
$$\beta_h=\frac{1}{\exp\left(\frac{30-(v-v_r)}{10}\right)+1}.$$

These values for α and β are also for a temperature of 6.3 °C. The temperature scaling of Eq. (6.65) must be used for other temperatures. The value of $g_{Na\infty}$ is 1200 S m^{-2}. Figure 6.40 plots the time constants and asymptotic values

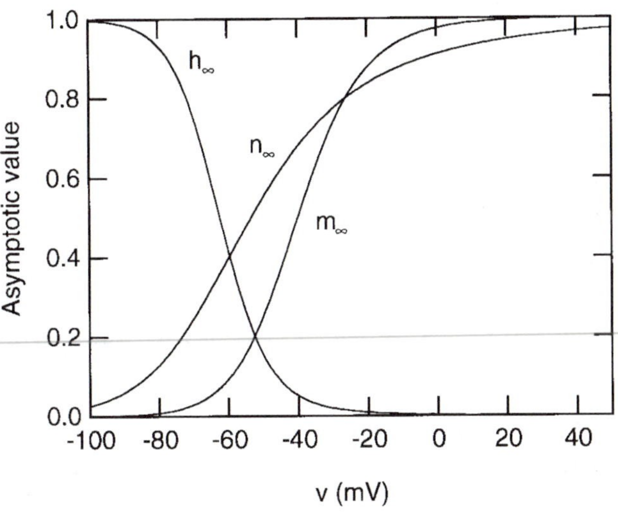

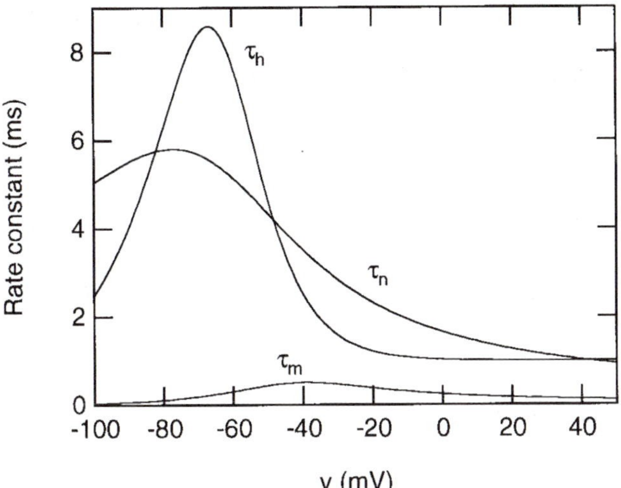

FIGURE 6.40. Plots of the asymptotic values and time constants versus the transmembrane potential.

as a function of membrane potential. These are the parameters for the equations in the form of Eq. (6.63a) rather than Eq. (6.63b).

All other contributions to the current (such as movement of chloride ions) were lumped in the leakage term $j_L = g_L(v - v_L)$. The empirical value for g_L is 3 S m^{-2}. The parameter v_L was adjusted to make the total membrane current equal zero when $v = v_r$. For example, with the data given, zero current is obtained with $v_r = -65$ mV and $v_L = v_r + 10.6$ mV. The three contributions to the membrane current can be thought of as the circuit shown in Fig. 6.34(b).

The Hodgkin–Huxley parameters have been used for a wide variety of nerve and muscle systems, even though they were obtained from measurements of the squid axon. A number of other models have since been developed that incorporate the sodium–potassium pump, calcium, etc. They have also been developed for various muscle and cardiac cells [Dennis *et al.* (1994); Luo and Rudy (1994); Wilders *et al.* (1991)].

6.14. VOLTAGE CHANGES IN A SPACE-CLAMPED AXON

A space-clamped axon has an interior potential $v(t)$ which does not depend on x. If such an axon is stimulated, a voltage pulse is observed. The first test we can make of the Hodgkin–Huxley model is to see if the parameters from the voltage-clamp experiments can also explain this pulse. To do so, it is necessary to insert Eq. (6.58), with all the other equations that are necessary to use it, in Eq. (6.50). Life is made somewhat simpler by the fact that the spatial derivative in Eq. (6.50) vanishes when the wire is in the axon. The result is

$$c_m \frac{\partial v}{\partial t} = -g_{\mathrm{Na}}(v - v_{\mathrm{Na}}) - g_{\mathrm{K}}(v - v_{\mathrm{K}}) - g_L(v - v_L). \tag{6.70}$$

When $v = v_r$ the right-hand side of this equation is zero and v does not change. It is necessary to introduce a stimulus to cause the pulse. This has been done in the computer program of Fig. 6.41, which solves Eq. (6.70). This program is not the most efficient that can be used; it has been written for ease of understanding. A stimulus of 10^{-4} A cm^{-2} is applied between 0.5 and 0.6 ms. This is an additional term in Eq. (6.70), so that in the program, Eq. (6.70) becomes

```
dvdt=(-jMemb+jStim)/Cmemb;
```

In this statement `dvdt` stands for $\partial v/\partial t$, `jMemb` stands for j_m, `Cmemb` for c_m, and `jStim` for the stimulus current. The equation is solved by repeated application of the approximation

```
v=v+dvdt*deltat;
```

which stands for

$$v(t + \Delta t) = v(t) + \frac{\partial v}{\partial t}\Delta t.$$

The program uses $\Delta t = 10^{-6}$ s. The present value of v is used to calculate the rate constants in procedure `Calcab`. These are then used to calculate the present value of each conductivity. The membrane current is then calculated, and the entire process is repeated for the next time step. The results are tabulated in Fig. 6.42 and plotted in Fig. 6.43.

One can see from the plot that j_m is proportional to $\partial v/\partial t$. Note that although g_{Na} is a smooth curve, j_{Na} has an extra wiggle near $t = 2$ ms, caused by the rapid decrease in the magnitude of $v - v_{\mathrm{Na}}$ as the voltage approaches the sodium Nernst potential. The sodium and potassium

```
//program HodgkinHuxley,  Fig. 6.41
//R. K. Hobbie 31 Dec. 1985, Converted to C  Jan. 1994
//Calculates Hodgkin-Huxley space clamped axon at  6.3°C

#include <stdio.h>
#include <math.h>

const float
   deltat =   1e-6,          //deltat for integration
   tPStep =   1e-4,          //print a line this often
   vRest  = -65e-3,          //Resting Potential
   Cmemb  =   1e-6,          //Membrane capacitance
   tMax   =   5e-3,          //Time to quit
   vNa    =  50e-3,          //Sodium Nernst potential
   vK     = -77e-3;          //Potassium Nernst potential

double
   n, m, hh,
   an, am, ah,
   bn, bm, bh,
   dndt, dmdt, dhdt, dvdt,
   gK, gNa,
   jK, jNa, jL, jMemb,
   voltage, t,
   jStim,                    //Stimulus current
   tPrint;                   //Time interval to print

void Calcab (void)
/*  Calculates the alpha and betas for n, m, h, using the
    Hodgkin-Huxley equations. The original equations were
    in mV and ms;  these are in volts and seconds  */
   {
   an = (10 * (-1000 * (voltage - vRest) + 10))
      / (exp((-1000 * (voltage - vRest) + 10) / 10) - 1);
   am = (100 * (-1000 * (voltage - vRest) + 25))
      / (exp((-1000 * (voltage - vRest) + 25) / 10) - 1);
   ah = 70 * exp(-1000 * (voltage - vRest) / 20);
   bn = 125 * exp(-1000 * (voltage - vRest) / 80);
   bm = 4000 * exp(-1000 * (voltage - vRest) / 18);
   bh = 1000/(exp((-1000 * (voltage - vRest) + 30) / 10) + 1);
   }

void Calc_Init_Values(void)   //Calculates initial
                              //values of n, m, hh
   {
   Calcab();
   n = an / (an + bn);
   m = am / (am + bm);
   hh = ah / (ah + bh);
```

FIGURE 6.41. The computer program used to calculate the response of a space-clamped axon to a stimulus. The results are shown in Figs. 6.42 and 6.43.

```
   }

void Calc_Curr(void)
   //   Calculate conductances in siemens per sq cm and
   //   current densities
   {
   gK  = 36e-3 * pow(n, 4);
   gNa = 120e-3 * pow(m, 3) * hh;
   jK  = gK  * (voltage - vK);
   jNa = gNa * (voltage - vNa);
   jL  = 3e-4 * (voltage - vRest - 10.6e-3);
   jMemb = jK + jNa + jL;
   }

main(void)
   {
   //Print Table Headings
   printf("time    v       jMemb     gNa       jNa        gK
jK     jL\n");
   t = 0;
   voltage = vRest;
   tPrint = 0;
   Calc_Init_Values();
   while (t < tMax)              //Step through all the times
      {
      Calc_Curr();               //Calculate membrane current
                                 //from conductances
      if (t >= tPrint)           //Print at certain times
         {
         printf("%4.1f %1s %5.1f %1s ", 1000*t, "",
            1000*voltage, "");
         printf("%8.2e %1s %8.2e %1s %8.2e", jMemb, "",
            gNa, "", jNa );
         printf("%1s %8.2e %1s %8.2e %1s %8.2e\n","", gK, "",
            jK,"",jL);
         tPrint = tPrint + tPStep;
         }                       // end if

      if ((t >= 5e-4) && (t < 6e-4))//Stimulus current at
                                    //beginning
         jStim = 1e-4;
      else
         jStim = 0;              //End of stimulus current
      dvdt = (-jMemb + jStim)/Cmemb;
      voltage = voltage + dvdt * deltat;
      Calcab();                  //Calc alphas, betas for new v
      dndt = an * (1 - n) - bn * n;
      dmdt = am * (1 - m) - bm * m;
      dhdt = ah * (1 - hh) - bh * hh;
  n  = n + dndt * deltat;
  m  = m + dmdt * deltat;
  hh = hh + dhdt * deltat;
  t  = t + deltat;
  }                            //end while
}                              //end main
```

FIGURE 6.41. (Continued.)

t	v	jMemb	gNa	jNa	gK	jK	jL
ms	mV	A/sq cm	S/sq cm	A/sq cm	S/sq cm	A/sq cm	S/sq cm
0.0	-65.0	-3.24e-10	1.06e-05	-1.22e-06	3.67e-04	4.40e-06	-3.18e-06
0.1	-65.0	-3.05e-10	1.06e-05	-1.22e-06	3.67e-04	4.40e-06	-3.18e-06
0.2	-65.0	-2.90e-10	1.06e-05	-1.22e-06	3.67e-04	4.40e-06	-3.18e-06
0.3	-65.0	-2.78e-10	1.06e-05	-1.22e-06	3.67e-04	4.40e-06	-3.18e-06
0.4	-65.0	-2.68e-10	1.06e-05	-1.22e-06	3.67e-04	4.40e-06	-3.18e-06
0.5	-65.0	-2.58e-10	1.06e-05	-1.22e-06	3.67e-04	4.40e-06	-3.18e-06
0.6	-55.3	5.75e-06	1.99e-05	-2.09e-06	3.74e-04	8.11e-06	-2.71e-07
0.7	-55.7	3.13e-06	4.46e-05	-4.72e-06	3.88e-04	8.25e-06	-4.05e-07
0.8	-55.9	5.17e-07	7.08e-05	-7.50e-06	4.02e-04	8.48e-06	-4.59e-07
0.9	-55.9	-1.75e-06	9.55e-05	-1.01e-05	4.16e-04	8.80e-06	-4.40e-07
1.0	-55.6	-3.69e-06	1.19e-04	-1.25e-05	4.31e-04	9.22e-06	-3.58e-07
1.1	-55.1	-5.48e-06	1.43e-04	-1.50e-05	4.46e-04	9.75e-06	-2.19e-07
1.2	-54.5	-7.33e-06	1.69e-04	-1.77e-05	4.62e-04	1.04e-05	-2.79e-08
1.3	-53.7	-9.52e-06	2.02e-04	-2.10e-05	4.80e-04	1.12e-05	2.23e-07
1.4	-52.6	-1.24e-05	2.46e-04	-2.52e-05	5.01e-04	1.22e-05	5.49e-07
1.5	-51.1	-1.66e-05	3.07e-04	-3.11e-05	5.24e-04	1.36e-05	9.79e-07
1.6	-49.2	-2.30e-05	4.02e-04	-3.99e-05	5.52e-04	1.54e-05	1.56e-06
1.7	-46.4	-3.38e-05	5.62e-04	-5.41e-05	5.87e-04	1.80e-05	2.40e-06
1.8	-42.2	-5.34e-05	8.59e-04	-7.92e-05	6.34e-04	2.21e-05	3.67e-06
1.9	-35.1	-9.25e-05	1.50e-03	-1.28e-04	7.02e-04	2.94e-05	5.78e-06
2.0	-22.3	-1.73e-04	3.14e-03	-2.27e-04	8.15e-04	4.45e-05	9.62e-06
2.1	1.0	-2.90e-04	7.91e-03	-3.88e-04	1.03e-03	8.07e-05	1.66e-05
2.2	28.7	-2.08e-04	1.83e-02	-3.89e-04	1.48e-03	1.56e-04	2.49e-05
2.3	39.1	-2.40e-05	2.80e-02	-3.05e-04	2.18e-03	2.53e-04	2.80e-05
2.4	38.7	2.13e-05	3.17e-02	-3.60e-04	3.06e-03	3.53e-04	2.79e-05
2.5	35.8	3.42e-05	3.16e-02	-4.48e-04	4.03e-03	4.55e-04	2.71e-05
2.6	32.0	4.19e-05	2.97e-02	-5.36e-04	5.06e-03	5.52e-04	2.59e-05
2.7	27.5	4.69e-05	2.74e-02	-6.15e-04	6.10e-03	6.37e-04	2.46e-05
2.8	22.7	5.00e-05	2.49e-02	-6.82e-04	7.11e-03	7.09e-04	2.31e-05
2.9	17.6	5.16e-05	2.26e-02	-7.33e-04	8.06e-03	7.63e-04	2.16e-05
3.0	12.4	5.19e-05	2.04e-02	-7.67e-04	8.94e-03	7.99e-04	2.00e-05
3.1	7.2	5.15e-05	1.84e-02	-7.86e-04	9.73e-03	8.19e-04	1.85e-05
3.2	2.1	5.06e-05	1.65e-02	-7.91e-04	1.04e-02	8.25e-04	1.70e-05
3.3	-2.9	4.94e-05	1.48e-02	-7.82e-04	1.10e-02	8.16e-04	1.55e-05
3.4	-7.8	4.80e-05	1.32e-02	-7.63e-04	1.15e-02	7.97e-04	1.40e-05
3.5	-12.5	4.67e-05	1.17e-02	-7.33e-04	1.19e-02	7.68e-04	1.26e-05
3.6	-17.1	4.56e-05	1.04e-02	-6.96e-04	1.22e-02	7.31e-04	1.12e-05
3.7	-21.6	4.47e-05	9.11e-03	-6.52e-04	1.24e-02	6.87e-04	9.84e-06
3.8	-26.1	4.43e-05	7.92e-03	-6.02e-04	1.25e-02	6.38e-04	8.50e-06
3.9	-30.5	4.47e-05	6.79e-03	-5.47e-04	1.26e-02	5.84e-04	7.17e-06
4.0	-35.0	4.64e-05	5.70e-03	-4.85e-04	1.25e-02	5.25e-04	5.81e-06
4.1	-39.8	4.98e-05	4.63e-03	-4.15e-04	1.24e-02	4.61e-04	4.37e-06
4.2	-45.1	5.55e-05	3.54e-03	-3.37e-04	1.22e-02	3.90e-04	2.80e-06
4.3	-51.0	6.30e-05	2.46e-03	-2.48e-04	1.19e-02	3.10e-04	1.02e-06
4.4	-57.6	6.87e-05	1.44e-03	-1.55e-04	1.16e-02	2.24e-04	-9.63e-07
4.5	-64.4	6.46e-05	6.37e-04	-7.29e-05	1.11e-02	1.40e-04	-3.00e-06
4.6	-70.0	4.61e-05	1.95e-04	-2.34e-05	1.07e-02	7.42e-05	-4.69e-06
4.7	-73.5	2.44e-05	4.32e-05	-5.34e-06	1.02e-02	3.55e-05	-5.73e-06
4.8	-75.2	1.04e-05	8.09e-06	-1.01e-06	9.64e-03	1.76e-05	-6.23e-06
4.9	-75.8	3.94e-06	1.62e-06	-2.04e-07	9.15e-03	1.06e-05	-6.43e-06

FIGURE 6.42. Results of the calculation for a space-clamped axon at 6.3 °C.

currents are nearly balanced throughout most of the pulse. It is very misleading to look at curves of g_{Na} and g_K and infer that the potassium current occurs later. The pulse lasts about 2 ms.

If the temperature is raised, the pulse is much shorter. Figure 6.44 shows the impulse when the temperature is 18.5 °C, calculated by multiplying each of the α and β values by 3.82.

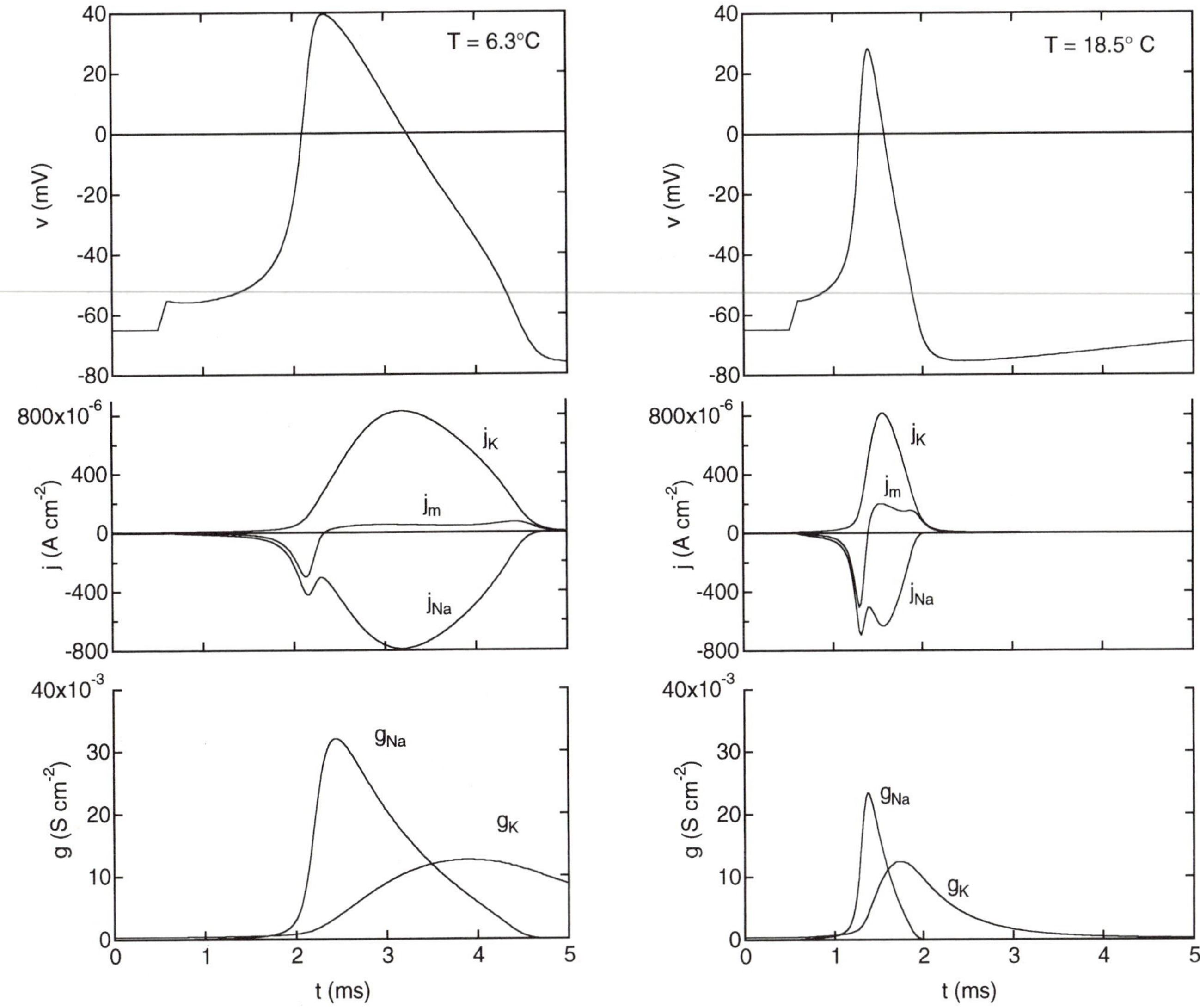

FIGURE 6.43. A plot of the computation presented in Fig. 6.41 for a pulse in a space-clamped squid axon at T=6.3 °C. The axon was stimulated at t=0.5 ms for 0.1 ms.

FIGURE 6.44. A pulse in a space-clamped axon at 18.5 °C. The pulse lasts about 1 ms.

The potassium current is not actually needed to create a nerve impulse because of the leakage current (primarily chloride) and the fact that the sodium conductance decreases with voltage. The potassium current speeds up the repolarization process. It is easy to modify the program of Fig. 6.41 to show this.

6.15. PROPAGATING NERVE IMPULSE

If the wire is not inserted along the axon, the voltage changes in the x direction. A strong enough stimulus at one point results in a pulse that travels along the axon without change of shape. The basic equation that describes it is again Eq. (6.50) with the spatial term and with the Hodgkin–Huxley model for the membrane current:

$$\frac{\partial v}{\partial t} = -\frac{j_m}{c_m} + \frac{1}{2\pi a r_i c_m}\frac{\partial^2 v}{\partial x^2},$$
$$j_m = g_{\mathrm{Na}}(v - v_{\mathrm{Na}}) + g_{\mathrm{K}}(v - v_{\mathrm{K}}) + g_L(v - v_L). \tag{6.71}$$

These can be solved numerically by setting up arrays for values of v, dv/dt, and all the rate constants and conductances at closely spaced discrete values of x along the axon. If index i distinguishes different values of x, then the discrete equation is

```
dvdt[i]=-jMemb[i]/Cmemb
        +(1/(6.28*a*ri*Cmemb*dx*dx))*(v[i+1]
        -2*v[i]+v[i-1]).
```

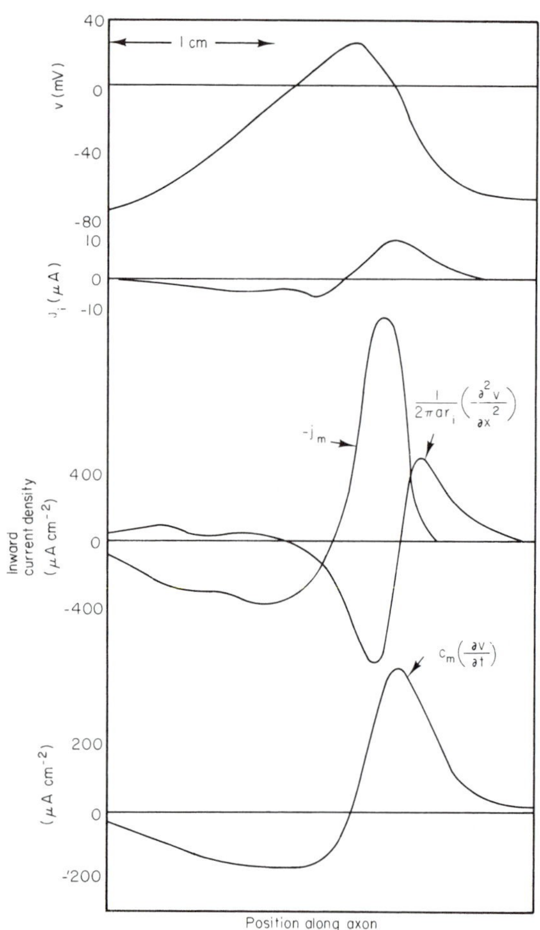

FIGURE 6.45. A propagating pulse plotted against position along the axon at an instant of time. The middle graph shows the two terms comprising the right-hand side of Eq. (6.71). The bottom curve shows the current charging or discharging the membrane.

To recapitulate, Fig. 6.45 shows each term in Eq. (6.71) multiplied through by c_m to have the dimensions of current per unit area. The term

$$c_m \frac{\partial v}{\partial t}$$

is the rate of charge buildup required to change the membrane potential at the rate $\partial v/\partial t$,

$$-j_m = -g_{\mathrm{Na}}(v - v_{\mathrm{Na}}) - g_{\mathrm{K}}(v - v_{\mathrm{K}}) - g_L(v - v_L)$$

is the rate of charge buildup because of current through the membrane, and

$$\frac{1}{2\pi a r_i} \frac{\partial^2 v}{\partial x^2}$$

is the rate of charge buildup on the inner surface of the membrane because the longitudinal current is not uniform.

6.16. MYELINATED FIBERS AND SALTATORY CONDUCTION

We have so far been discussing fibers without the thick myelin sheath. Unmyelinated fibers constitute about two-thirds of the fibers in the human body. They usually have radii of 0.05–0.6 μm. The conduction speed in m s^{-1} is given approximately by $u \approx 1800\sqrt{a}$, where a is the axon radius in meters. (Strictly speaking, in this formula a should be replaced by the outer radius $a+b$ including the membrane thickness, but for an unmyelinated fiber $b \ll a$.)

Myelinated fibers are relatively large, with outer radii of 0.5–10 μm. They are wrapped with many layers of myelin between the nodes of Ranvier, as shown in Fig. 6.46. Typically, the outer radius is $a+b \approx 1.4a$ and the spacing between nodes is proportional to the outer diameter $D = 200(a+b) \approx 280a$ [Deutsch and Deutsch (1993)]. These empirical proportionalities between node spacing and radius and between myelin thickness and radius will be very important to our understanding of the conduction speed. The conduction speed in a myelinated fiber is given approximately by $u \approx 12 \times 10^6(a+b) \approx 17 \times 10^6 a$. The conduction speeds of myelinated and unmyelinated fibers are compared in Fig. 6.47.

In the myelinated region the conduction of the nerve impulse can be modeled by electrotonus because the conductance of the myelin sheath is independent of voltage. At each node a regenerative Hodgkin–Huxley-type (HH-type) conductance change restores the shape of the pulse. Such conduction is called *saltatory conduction* because *saltare* is the Latin verb "to jump."

In the basic model of saltatory conduction, the resistivity of the myelin does not change. The signal is attenuated from one node to the next by the longitudinal and leakage resistance of the myelinated axon and is delayed by the charging of the myelin capacitance—electrotonus. Each node restores the signal to its original size by the regenerative action due to changes in membrane permeability.

We saw that electrotonus is described by

$$\lambda^2 \frac{\partial^2 v}{\partial x^2} - v - \tau \frac{\partial v}{\partial t} = -v_r, \qquad (6.72)$$

where the time constant is

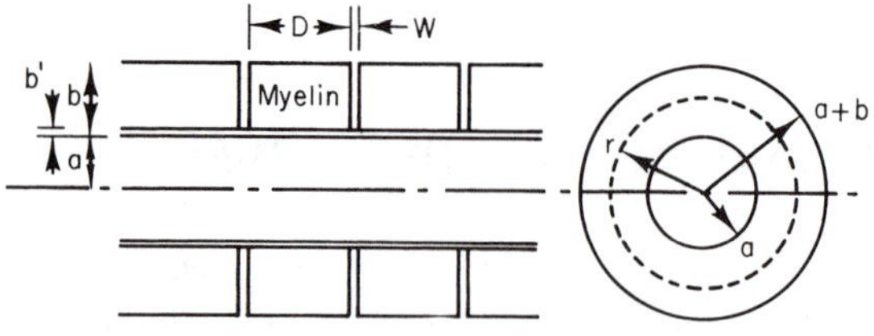

FIGURE 6.46. The idealized structure of a myelinated fiber in longitudinal section and in cross section. The internodal spacing D is actually about 100 times the outer diameter of the axon.

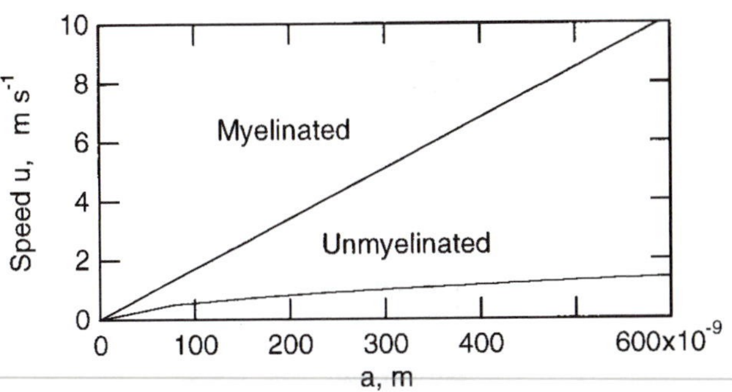

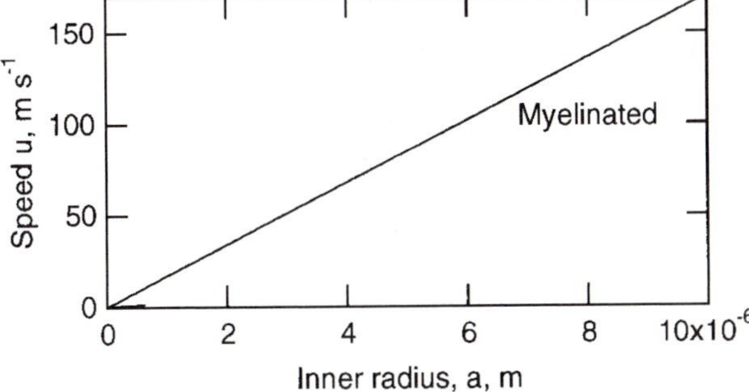

FIGURE 6.47. The conduction speed versus the inner axon radius a for myelinated and unmyelinated fibers. Unmyelinated fibers with $a>0.6$ μm are not found in the body.

$$\tau=\kappa\varepsilon_0\rho_m \tag{6.73}$$

and the space constant is

$$\lambda=\sqrt{\frac{ab\rho_m}{2\rho_i}}. \tag{6.74a}$$

Later in this section we will find that when the myelin thickness is appreciable compared to the inner axon radius, the space constant should be modified:

$$\lambda_{\text{thick}}=\sqrt{\frac{\ln(1+b/a)\rho_m}{2\rho_i}}\,a. \tag{6.74b}$$

For a case in which $a=5$ μm and $b=2$ μm, the change is not very large. The thin membrane version contains the quantity $ab=10\times10^{-12}$ m^2 and the thick myelin version contains $a^2\ln(1+b/a)=8.4\times10^{-12}$ m^2.

We now want to understand the different radial dependence of the conduction speed in the two kinds of fibers. We could do computer modeling for the unmyelinated fiber using Eq. (6.71) with axons of different radii, but this would not provide an equation for $u(a)$. Rather than review the work that has been done (developing equations for the behavior of the foot of the action potential, for example), we will use a simple dimensional argument. This will not give an exact expression for $u(a)$, but it will indicate the functional form it must have.

In either the myelinated or the unmyelinated fiber the signal travels to neighboring regions by electrotonus, where it initiates HH-type membrane conductance changes. In the myelinated case the signal jumps from node to node; in the unmyelinated case the influence is on adjacent parts of the axon. When the neighboring region begins to depolarize, the HH change is much more rapid than that due to electrotonus. (Another way to say this is that during depolarization ρ_m and therefore τ become much smaller.) Therefore the conduction speed is limited by electrotonus. Regardless of the details of the calculation, the speed is proportional to the characteristic length in the problem divided by the characteristic time. For the unmyelinated case it is plausible to assume that the only characteristic length and time are λ and τ, so the speed is

$$u_{\text{unmyelinated}}\propto\frac{\lambda}{\tau}=\sqrt{\frac{b}{2\rho_i\rho_m}\frac{1}{\kappa\varepsilon_0}}\sqrt{a}. \tag{6.75}$$

Since the membrane thickness for an unmyelinated fiber is always about 6 nm, this gives

$$u_{\text{unmyelinated}}\propto270\sqrt{a} \tag{6.76}$$

as shown in Table 6.2.

For myelinated nerves the myelin thickness is $b\approx0.4a$. This means that the space constant is proportional to a:

$$\lambda=\sqrt{\frac{ab\rho_m}{2\rho_i}}=\sqrt{\frac{0.4a^2\rho_m}{2\rho_i}}=a\sqrt{\frac{0.4\rho_m}{2\rho_i}}=1350a. \tag{6.77}$$

The spacing between the nodes, D, is about $280a$. There are two characteristic lengths for the myelinated case, both proportional to a because of the way the myelin is arranged. If we assume that the speed is proportional to D/τ, we obtain

$$u_{\text{myelinated}}\propto4.5\times10^5a. \tag{6.78}$$

If we assume that the speed is proportional to λ/τ, we obtain

$$u_{\text{myelinated}}\propto2.2\times10^6a. \tag{6.79}$$

Table 6.2 compares the space constants, time constants and conduction speeds for myelinated and unmyelinated fibers. The empirical expressions for the conduction speed are 7 or 8 times greater than what we estimate based on λ/τ. We might expect firing at the next node to occur when the signal has risen to about 10% of its final value. This would reduce the time by about a factor of 10 (see Problems).

The internodal spacing is about 20% of the space constant. Suppose that a constant current is injected at one node, as in Fig. 6.31. When the voltage has reached its full value at the next node it is given by

$$\frac{v}{v_0}=e^{-D/\lambda}=e^{-1.4/6.2}=0.8.$$

If for some reason this node does not fire, the signal at the next node will be 0.64 of the original value, and so on. A

TABLE 6.2. *Properties of unmyelinated and myelinated axons of the same radius.*

Quantity	Unmyelinated	Myelinated
Axon inner radius, a	5×10^{-6} m	5×10^{-6} m
Membrane thickness b'	6×10^{-9} m	
Myelin thickness b		2×10^{-6} m
$\kappa\epsilon_0$	6.20×10^{-11} s^{-1} Ω^{-1} m^{-1}	6.20×10^{-11} s^{-1} Ω^{-1} m^{-1}
Axoplasm resistivity ρ_i	$1.1\ \Omega$ m	$1.1\ \Omega$ m
Membrane (resting) or myelin resistivity ρ_m	$10^7\ \Omega$ m	$10^7\ \Omega$ m
Time constant $\tau=\kappa\epsilon_0\rho_m$	6.2×10^{-4} s	6.2×10^{-4} s
Space constant λ	$\lambda=\sqrt{\frac{ab\rho_m}{2\rho_i}}$ $=0.165\sqrt{a}$ $=370\times10^{-6}$ m	$\lambda=\sqrt{\frac{ab\rho_m}{2\rho_i}}=\sqrt{\frac{0.4a^2\rho_m}{2\rho_i}}$ $=a\sqrt{\frac{0.4\rho_m}{2\rho_i}}$ $=1350a$ $=6.8\times10^{-3}$ m
Node spacing D		$D=280a$ $=1.4\times10^{-3}$ m
Conduction speed from model	$u_{\text{unmyelinated}}\propto\frac{\lambda}{\tau}\approx270\sqrt{a}$	$u_{\text{myelinated}}\propto\frac{\lambda}{\tau}\approx2.2\times10^6a$ or $u_{\text{myelinated}}\propto\frac{D}{\tau}=4.5\times10^5a$
Conduction speed, empirical	$u_{\text{unmyelinated}}\approx1800\sqrt{a}$	$u_{\text{myelinated}}\approx17\times10^6a$
Ratio of empirical to model conduction speed	6.7	7.2 or 38
Space constant using thick membrane model		$\lambda=a\sqrt{\frac{\ln(1+b/a)\rho_m}{2\rho_i}}$ $=a\sqrt{\frac{\ln(1.4)\rho_m}{2\rho_i}}$ $=1240a$ $=6.2\times10^{-3}$ m

local anesthetic such as procaine works by preventing permeability changes at the node. It is clear from this discussion that a nerve must be blocked over a distance of several nodes (a centimeter or more) in order for an anesthetic to be effective [Covino (1972)].

We finally develop equations for the resistance and capacitance of a cylindrical membrane whose thickness is appreciable compared to its inner radius. We use Eq. (6.7), which is Gauss's law for cylindrical symmetry, to determine the electric field. Imagine that total charge Q is distributed uniformly over the inner surface of the membrane. If the section of membrane being considered has length D, the charge per unit length is $\lambda=Q/D$. Any charge on the outer surface of the membrane has no effect on the calculation of the electric field between $r=a$ and $r=a+b$ as long as the charge is distributed uniformly on the outer cylindrical surface at $r=a+b$. From Gauss's law the electric field within the membrane of dielectric constant κ is

$$E=\frac{1}{4\pi\varepsilon_0\kappa}\frac{2Q/D}{r}.$$

The potential difference $v(a)-v(a+b)$ is obtained using Eq. (6.14):

$$v(a)-v(a+b)=-\int_{a+b}^{a}E(r)\,dr.$$

Carrying out the integration gives

$$v=v(a)-v(a+b)=\frac{Q}{2\pi\varepsilon_0\kappa D}\ln\left(1+\frac{b}{a}\right). \qquad (6.80)$$

Now place a conducting medium with resistivity $\rho=1/\sigma$ in the region of the membrane. Charge will move. It will be necessary to provide a battery to replenish it. The form of Ohm's law which is most convenient to use is Eq. (6.24): $\mathbf{j}=\sigma\mathbf{E}$. The current per unit area is

$$j=E/\rho=\frac{2Q/D}{\rho4\pi\varepsilon_0\kappa r}.$$

The total current is j times the area of a cylinder of radius r and length D:

$$i = 2\pi r D j = Q/\rho\kappa\varepsilon_0.$$

The resistance of the membrane segment of length D is given by $R = v/i$:

$$R = \frac{\rho}{2\pi D} \ln\left(1 + \frac{b}{a}\right).$$

This can be solved for ρ:

$$\rho = \frac{2\pi R D}{\ln(1+b/a)} \quad \text{(cylinder)}. \tag{6.81}$$

It will be left to the problems to show that the resistivity of a plane resistor of cross sectional area $2\pi a D$ and thickness b is

$$\rho = \frac{2\pi R D}{b/a} \quad \text{(plane)}. \tag{6.82}$$

For a myelinated axon in which $b/a = 0.4$, $\ln(1+b/a) = 0.34$, a difference of about 19%. The capacitance is Q/v:

$$C = \frac{2\pi\kappa\varepsilon_0 D}{\ln(1+b/a)} \quad \text{(cylinder)},$$
$$C = \frac{2\pi\kappa\varepsilon_0 D}{b/a} \quad \text{(plane)}. \tag{6.83}$$

Again, the error in the plane approximation is to use b/a instead of $\ln(1+b/a)$.

6.17. MEMBRANE CAPACITANCE

The value of 7 for the dielectric constant, which has been used throughout this chapter, is considerably higher than the value 2.2, which is known for lipids. The inconsistency arises because part of the membrane is very easily polarized and effectively belongs to the conductor rather than to the dielectric; if the thickness of the lipid alone is considered in calculating the capacitance, then a value of 2.2 for κ is reasonable; if the entire membrane thickness is used, then the much higher dielectric constant for water and the polar groups within the membrane contributes, and $\kappa = 7$ is a reasonable value.

The easiest experiments to understand are those done with artificial bimolecular layers of lipid. The architecture of such a film is shown in Fig. 6.48. Each lipid molecule has a polar head and a hydrophobic tail. The molecules are arranged in a double layer with the heads in the aqueous solution. The dimensions in Fig. 6.48 are consistent with both optical measurements of the film thickness and with the known structure of the lipid molecules. Linear aliphatic hydrocarbons have a bulk dielectric constant of about 2. The polar heads have a much higher dielectric constant, probably about 50. Water has a dielectric constant of about 80.

The capacitance per unit area of bimolecular lipid films is about 0.3×10^{-2} F m^{-2} (0.3 μF cm^{-2}). The simplest way to explain this value is to assume that the polar heads are part of the surrounding conductor. The capacitance per unit area is then

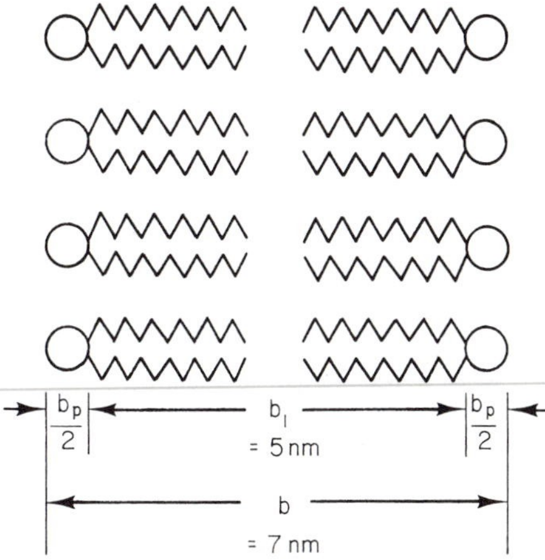

FIGURE 6.48. Structure of a bimolecular lipid membrane.

$$\frac{C}{S} = \frac{\kappa\epsilon_0}{b_1} = \frac{(2.2)(8.85\times 10^{-12})}{5\times 10^{-9}} = 0.4\times 10^{-2}\ \text{F m}^{-2}.$$

A more sophisticated approach is to regard the membrane as made up of three layers: polar, lipid, polar. The same effect can be obtained by considering two layers with all the polar component lumped together, as in Fig. 6.49. Suppose that we put charge $+Q$ on one surface and $-Q$ on the other surface of the membrane. We put no charge on the interface between layers 1 and 2. The charge of zero on the interface can be thought of as a superposition of positive and negative charges as shown in Fig. 6.49. We are referring only to external charge which we place on the membrane; the charges induced by polarization of the dielectric are not shown. They are taken into account by the value of κ. The situation is that of two parallel-plate capacitors in series. Each layer has a capacitance C_i:

$$Q = C_i v_i = \frac{\kappa_i \epsilon_0 S}{b_i} v_i.$$

The total potential across the membrane is

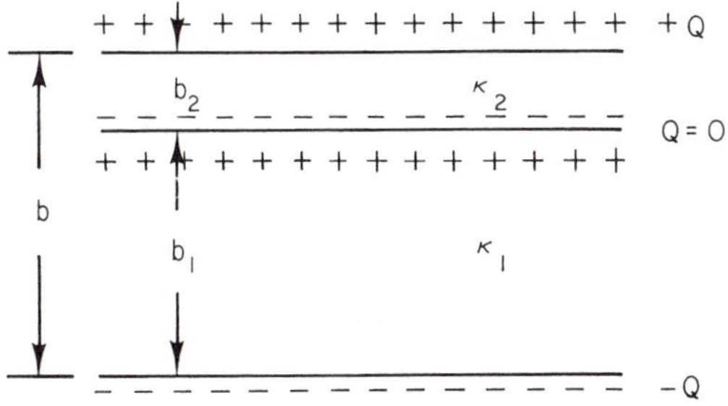

FIGURE 6.49. A membrane composed of two phases. The ith phase has thickness b_i and dielectric constant κ_i. The total thickness is b and the effective dielectric constant is κ. The charges shown are external charge; polarization of the dielectric is not shown but determines the value of κ.

$$v=v_1+v_2=\frac{Q}{C}.$$

The total capacitance is

$$C=\frac{Q}{v_1+v_2}=\frac{Q}{Qb_1/\kappa_1\epsilon_0S+Qb_2/\kappa_2\epsilon_0S}$$

$$=\frac{1}{b_1/\kappa_1\epsilon_0S+b_2/\kappa_2\epsilon_0S}. \tag{6.84}$$

The effective dielectric constant is obtained by equating the total capacitance to $\kappa\epsilon_0S/b$:

$$\kappa=\frac{b}{b_1/\kappa_1+b_2/\kappa_2}. \tag{6.85}$$

Application of these equations to the bimolecular lipid membrane (with $\kappa_1=2.2$, $\kappa_2=50$, $b_1=5$ nm, $b_2=2$ nm) gives

$$\kappa=3.0, \qquad \frac{C}{S}=0.38\times10^{-2}\ \text{F m}^{-2}. \tag{6.86}$$

The capacitance per unit area is nearly that obtained by assuming the polar groups are perfect conductors.

Biological membranes, when hydrated, are about 60% protein and 40% lipid. Current knowledge of the membrane structure is described by Alberts *et al.* (1983). The myelin surrounding a nerve fiber consists of several layers wrapped tightly together. Each repeating layer is made up of two single layers back to back, as shown in Fig. 6.50. The best data on the structure of these layers are from x-ray diffraction experiments. The layers repeat every 17 nm. One model for the structure within a repeat distance is shown in Fig. 6.50 [Worthington (1971, p. 35)]. A single layer of the myelin has a thickness of 8.55 nm. A surprising feature of this model is that the lipid layer is less than half the thickness of that in a bilayer lipid membrane. However, the measured capacitance of a nerve-cell membrane or myelin is greater than for the bilayer lipid membrane; if one is to keep the lipid value for κ, the membrane must be thinner. It is gratifying that the membrane thickness as measured by x-ray diffraction is consistent with the observed membrane capacitance.

To check the consistency, note that Eqs. (6.84) and (6.85) are easily extended to more than two phases. Use the following data:

	κ_i	b_i nm
Water	80	2.2
Lipid	2.2	4.2
Polar	50	10.8

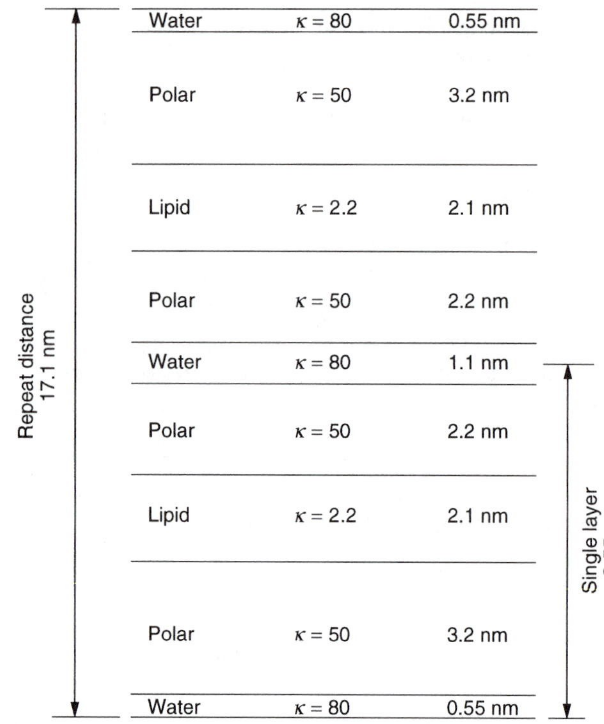

Layer	κ	Thickness
Water	$\kappa=80$	0.55 nm
Polar	$\kappa=50$	3.2 nm
Lipid	$\kappa=2.2$	2.1 nm
Polar	$\kappa=50$	2.2 nm
Water	$\kappa=80$	1.1 nm
Polar	$\kappa=50$	2.2 nm
Lipid	$\kappa=2.2$	2.1 nm
Polar	$\kappa=50$	3.2 nm
Water	$\kappa=80$	0.55 nm

Repeat distance 17.1 nm; Single layer 8.55 nm

FIGURE 6.50. The results of x-ray diffraction measurements of the structure of myelin surrounding frog sciatic nerve. Data are adapted from Worthington (1971, p. 35).

With these values, the effective dielectric constant is

$$\kappa=\frac{17.1}{4.2/2.2+2.1/80+10.8/50}=7.95.$$

If we assume that the membrane on an unmyelinated axon has the same structure as a half-unit of the myelin, then the thickness is 8.55 nm. With a dielectric constant of 7.95, the capacitance per unit area is calculated to be 0.82×10^{-2} F m^{-2}. The measured value is 1.0×10^{-2} F m^{-2}.

When one begins to look at the detailed structure of the membrane as we have done in this section, there is no justification for using the same membrane thickness b for the capacitance and the conductance of the membrane. The capacitance is determined primarily by the thickness of the lipid portion of the membrane; the conductance includes the effect of ions passing through the polar layers. The product, $\kappa\rho$, of the previous section is meaningful only for a membrane that is homogeneous and has the same thickness for both capacitive and conductive effects.

As long as the membrane structure is not being considered, it is safer to express such things as attenuation along the axon in terms of the directly measured parameters: capacitance, conductance, and resistance per unit length of the fiber. Nonetheless, a preliminary formulation in terms of a homogeneous membrane model can be useful to start thinking about the problem.

6.18. RHYTHMIC ELECTRICAL ACTIVITY

Many cells exhibit rhythmic electrical activity. Various nerve transducers produce impulses with a rate of firing that depends on the input to which the transducer is sensitive. The beating of the heart is controlled by the *sino-atrial node* (SA node) that produces periodic pulses that travel throughout the heart muscle.

The mechanism for such repetitive activity is similar to what we have seen in the Hodgkin–Huxley model, though the details of the ionic conductance variations differ. The computer program of Fig. 6.41 can easily be modified to model rhythmic activity. Figure 6.51 shows a plot of the output of a modified program. The only modification was to make jStim be a constant leakage current of 10^{-4} A cm^{-2}. This provides the essential feature: a small inward current between beats that causes the potential inside the cell to increase slowly. When the voltage exceeds a certain threshold, the membrane channels open and the cell produces another impulse.

While this simple change produces repetitive firing, and in fact the shape of the curve in Fig. 6.51 is very similar to that measured in the SA node, the details of ionic conduction are actually very different. The SA node contains no sodium channels. The rapid depolarization is due to an inward calcium current. There are a number of contributions to the current in the SA node, and detailed ionic models of them have been described [Dennis *et al.* (1994); Noble (1995); Noble *et al.* (1989); Wilders *et al.* (1991)]. The slow leakage is a complicated combination of currents, the details of which are still not completely understood [Anumonwo and Jalife (1995); DiFrancesco, Mangoni, and Maccaferri (1995)].

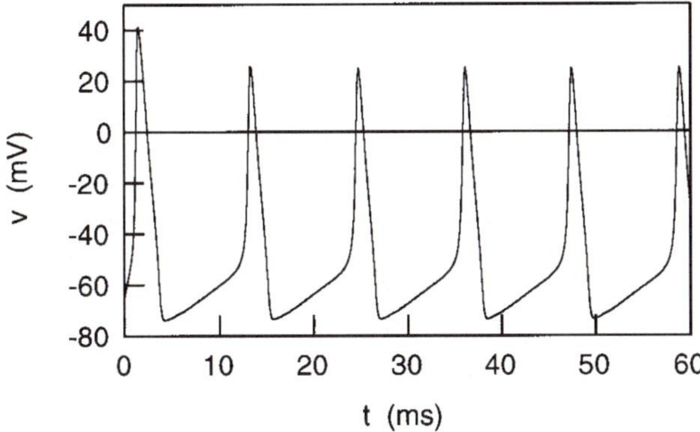

FIGURE 6.51. By changing the leakage current, it is possible to make the Hodgkin–Huxley model display periodic electrical activity.

6.19. THE RELATIONSHIP AMONG THE CAPACITANCE, RESISTANCE, AND DIFFUSION BETWEEN TWO CONDUCTORS

There is a fundamental relationship between the capacitance and resistance between two conductors in a homogeneous conducting dielectric. It is also possible to develop an analogy between capacitance and steady-state diffusion, so that known expressions for the capacitance of conductors in different geometries can be used to solve diffusion problems.

Consider two conductors carrying equal and opposite charge and embedded in an insulating medium with dielectric constant κ. The potential difference between the conductors is Δv, and the magnitude of the charge on each is $Q=C\ \Delta v$. The electric field is $\mathbf{E}(x,y,z)$. In a vacuum Gauss's law applied to a surface surrounding the positively charged conductor gives $\int\int \mathbf{E}\cdot d\mathbf{S}=Q/\epsilon_0$. Polarization in a dielectric surrounding the conductor reduces the electric field by a factor of κ. If E refers to the electric field in the dielectric and Q to the charge on the conductor, Gauss's law becomes

$$\int\int \mathbf{E}\cdot d\mathbf{S}=\frac{Q}{\kappa\epsilon_0}. \tag{6.87}$$

For a given charge on the conductor, the presence of the dielectric reduces E and Δv by $1/\kappa$ and, therefore, increases the capacitance by κ.

Suppose that the dielectric is not a perfect insulator but obeys Ohm's law and has conductivity σ ($\mathbf{j}=\sigma\mathbf{E}$). If some process maintains the magnitude of the charge on each conductor at Q, the current leaving the positive conductor is

$$i=\int\int \mathbf{j}\cdot d\mathbf{S}=\sigma\int\int \mathbf{E}\cdot d\mathbf{S}=\frac{\sigma Q}{\kappa\epsilon_0}. \tag{6.87a}$$

The resistance between the conductors is

$$R=\frac{\Delta v}{i}=\frac{Q/C}{\sigma Q/\kappa\epsilon_0}=\frac{\kappa\epsilon_0}{\sigma C}.$$

This inverse relationship between the resistance and capacitance is independent of the geometry of the conductors, as long as the dielectric constant and conductivity are uniform throughout the medium.

If the charge on the conductors is not replenished, it leaks off with a time constant $\tau=RC=\kappa\epsilon_0/\sigma$. We have seen this result earlier in several special cases; we now understand that it is quite general.

In Chap. 4, we saw that the transport equations for particles, heat, and electric charge all have the same form. We now develop an analogy between these transport equations and the equations for the electric field. The analogy is useful because it relates the diffusion of particles between different regions to the electrical capacitance between conductors with the same geometry; the electrical capacitance in many cases is worked out and available in tables.

Fick's first law of diffusion was developed in Chap. 4:[8]

$$\mathbf{j}_s = -D\nabla c. \tag{4.20}$$

The relationship between fluence rate (particle current density) and particle flux (current) is

$$\iint \mathbf{j}_s \cdot d\mathbf{S} = i_s, \tag{6.88}$$

where i_s is the current of particles out of the volume enclosed by the surface. This equation is very similar to Gauss's law,

$$\iint_{\text{surface}} \mathbf{E} \cdot d\mathbf{S} = \frac{q}{\kappa\epsilon_0}, \tag{6.5}$$

where q is the electric charge. The electric potential and the electric field are related by the three-dimensional version of Eq. (6.14):

$$\mathbf{E} = -\nabla v. \tag{6.89}$$

The similarity between Eqs. (6.89) and (4.20) and between (6.5) and (6.88) suggests that we make the substitutions

$$i_s = \frac{q}{\kappa\epsilon_0},$$

$$c = \frac{v}{D},$$

$$\mathbf{j}_s = \mathbf{E}. \tag{6.90}$$

For any electrostatic configuration in which there are two equipotential surfaces containing charge $+q$ and $-q$, there is an analogous diffusion problem in which there is a flow of particles from one surface to another, each surface having a constant concentration on it. In the electrical case, the charge and potential are related by the capacitance, which is a geometric property of the two equipotential surfaces:

$$q = C\Delta v.$$

An analogous statement can be made for diffusion:

$$i_s = -\frac{C\,\Delta v}{\kappa\epsilon_0} = -\frac{C}{\kappa\epsilon_0} D\,\Delta c. \tag{6.91}$$

We can find the rate of flow of particles if we know the diffusion constant, the concentration difference, and the capacitance for the electrical problem with the same geometry. To see the utility of this method, we will consider some cases of increasing geometrical complexity.

As a first example, suppose that two concentric spheres have radii a and b. You can show (from the work in Problem 6.7, for example) that the capacitance of this configuration is

$$\frac{C}{\kappa\epsilon_0} = \frac{4\pi}{1/a - 1/b}. \tag{6.92}$$

As $b \to \infty$, this becomes

$$\frac{C}{\kappa\epsilon_0} = 4\pi a. \tag{6.93}$$

This can be applied to diffusion into or out of a spherical cell of radius a. If the diffusion is outward, as of waste products, imagine that the outward flow rate is i_s and that the concentration difference between the cell surface and infinity is c_0. Then

$$i_s = -4\pi a D c_0. \tag{6.94}$$

If, on the other hand, the concentration infinitely far away is greater than that at the cell surface by an amount c_0, the number of particles in the cell will increase at a rate

$$i_s = +4\pi a D c_0. \tag{6.95}$$

These results were obtained directly in Chap. 4.

As another example, consider a circular disk of radius a with the other electrode infinitely far away. It is more difficult to calculate the capacitance in this case, but we can look it up [Smythe *et al.* (1957)]. It is $C/\kappa\epsilon_0 = 8a$. But this is the capacitance for the charge on *both sides* of the disk; the lines of **E** and **j** go off in both directions. We want only half of this, since we will use the result to calculate the end correction for a pore. (If we were concerned with diffusion to a disk-shaped cell, we would use the whole thing.) For the half-space

$$i_{s\ \text{half}} = -4Da\,\Delta c \tag{6.96}$$

is proportional to the radius of the disk, not its area.

Still another geometrical situation that may be of interest is the diffusion of particles from one sphere of radius a to another sphere of radius a, when the centers of the spheres are separated by a distance b.

The capacitance between two such spherical electrodes is [Smythe *et al.* (1957, pp. 5–14)]

$$\frac{C}{\kappa\epsilon_0} = 2\pi a \sinh\beta\left(\frac{1}{\sinh\beta} + \frac{1}{\sinh 2\beta} + \frac{1}{\sinh 3\beta} + \cdots\right), \tag{6.97}$$

where

[8] In this section, we will use c for concentration of solute particles and C for the electrical capacitance.

$$\cosh\beta=\frac{b}{2a}.$$

This formula is written in terms of the ''hyperbolic functions''

$$\sinh\beta=\tfrac{1}{2}(e^{\beta}-e^{-\beta}),\qquad(6.98)$$

$$\cosh\beta=\tfrac{1}{2}(e^{\beta}+e^{-\beta}).$$

When the spheres are far apart $b/2a\to\infty$, and

$$\cosh\beta\approx\tfrac{1}{2}e^{\beta},$$

$$\sinh\beta\approx\tfrac{1}{2}e^{\beta}.$$

In that limit,

$$\frac{C}{\kappa\epsilon_0}=2\pi a(\tfrac{1}{2}e^{\beta})\left(\frac{1}{\frac{1}{2}e^{\beta}}+\frac{1}{\frac{1}{2}e^{2\beta}}+\frac{1}{\frac{1}{2}e^{3\beta}}+\cdots\right)=2\pi ae^{\beta}(e^{-\beta}$$

$$+e^{-2\beta}+e^{-3\beta}+\cdots)$$

$$=2\pi a(1+e^{-\beta}+e^{-2\beta}+\cdots)$$

$$=2\pi a\left[1+\left(\frac{a}{b}\right)+\left(\frac{a}{b}\right)^2+\cdots\right].\qquad(6.99)$$

The diffusive flow between two spheres is therefore

$$i_s=-2\pi aD\,\Delta c\qquad(6.100)$$

if they are sufficiently far apart. Note that this is just one-half of the flow from a sphere of radius a to a concentric sphere infinitely far away.

The earlier results in this section show that the electrical resistance between two spherical electrodes sufficiently far apart is $1/2\pi\sigma a$.

SYMBOLS USED IN CHAPTER 6

Symbol	Use	Units	First used on page
a	Distance	m	140
a	Axon inner radius	m	153
a	Radius of spherical cell	m	172
a	Radius of disk	m	172
b,c	Distance	m	140
b	Membrane thickness	m	153
b	Myelin thickness	m	166
b'	Membrane thickness at node of Ranvier	m	166
b_l,b_p	Thickness of membrane lipid and polar layers respectively	m	169
b	Sphere radius	m	172
c	Concentration	mol l^{-1}	172
c_m	Membrane capacitance per unit area	F m^{-2}	152
e	Electronic charge	C	150
g_m,g_K,g_{Na},g_L	Membrane conductance per unit area and conductance for given ion species	S m^{-2}	152
h	Parameter used to describe sodium conductance		160
i	Electric current	A	146
i_i,i_o	Currents along axon	A	153
i_m	Current through a section of membrane	A	151
i_s	Solute current or flux	s^{-1}	172
j	Current per unit area	A m^{-2}	147
j_m	Membrane current per unit area	A m^{-2}	152
j_{Na},j_K,j_L	Membrane current per unit area due to that species	A m^{-2}	158
k_B	Boltzmann constant	J K^{-1}	158
m	Parameter used to describe sodium conductance		160
n	Parameter used to describe potassium conductance		159
q	Electric charge	C	138
r	Distance	m	138
r_i	Resistance per unit length along inside of axon	Ω m^{-1}	153
t	Time	s	146
u	Propagation velocity of a wave or signal	m s^{-1}	167
v	Potential difference	V	143
v_K	Equilibrium (Nernst) potential for potassium	V	158
v_{Na}	Equilibrium (Nernst) potential for sodium	V	158
v_r	Resting membrane potential	V	155
x,z	Distance	m	140

Symbol	Use	Units	First used on page
z	Valence of ion		150
C	Capacitance	F	144
C_i, C_o	Ion concentrations	m^{-3}; mol l^{-1}	150
D	Length of myelinated segment	m	167
$\mathbf{E}, E_x, E_y, E_z$	Electric field and components	$N\,C^{-1}$ or $V\,m^{-1}$	139
E_p	Electric field due to polarization changes	$N\,C^{-1}$ or $V\,m^{-1}$	144
E_e	External electric field	$N\,C^{-1}$ or $V\,m^{-1}$	144
$\mathbf{E}_{tot}, E_t$	Total electric field	$N\,C^{-1}$ or $V\,m^{-1}$	144
$\mathbf{F}$	Force	N	138
$\mathbf{F}_{ext}$	External force	N	142
G, G_m	Conductance (of a section of axon membrane)	Ω^{-1} or S	146
L	Length of cylinder	m	140
$[Na_i], [Na_o]$	Sodium concentration inside and outside axon	m^{-3}	158
P	Power	W	147
Q	Electric charge	C	144
Q_{10}	Factor by which the rate of a chemical reaction increases with a temperature rise of 10 K		160
R	Resistance	Ω	146
R_i	Internal resistance along a segment of axon	Ω	153
R_m	Resistance across a segment of membrane	Ω	151
$S, \Delta S, dS$	Surface area	m^2	139
T	Temperature	K	158
U	Potential energy	J	142
W	Work	J	146
W	Width of node of Ranvier	m	166
$\alpha_m, \beta_m, \alpha_n, \beta_n, \alpha_h, \beta_h$	Rate parameters for Hodgkin–Huxley model	s^{-1}	160
β	Dimensionless variable		172

Symbol	Use	Units	First used on page
ϵ_0	Electrical permittivity of empty space	$N^{-1}\,m^{-2}\,C^2$	138
κ	Dielectric constant		145
λ	Charge per unit length	$C\,m^{-1}$	140
λ	Electrotonus spatial decay constant	m	156
ρ	Resistivity	Ω m	147
ρ_m	Resistivity of membrane	Ω m	152
σ	Charge per unit area	$C\,m^{-2}$	140
σ	Conductivity	$S\,m^{-1}$	147
χ	Electrical susceptibility		144
τ	Time constant	s	152
τ	Electrotonus time constant	s	155

PROBLEMS

Section 6.2

6.1. Two equal and opposite charges $\pm q$ separated a distance a form a dipole. The *dipole moment p* is a vector pointing in the direction from the negative charge to the positive charge of magnitude $p = qa$. In electrochemistry the dipole moment is often expressed in debye: 1 debye (D)$=10^{-18}$ electrostatic units (statcoulomb cm) (1 statcoulomb$=3.3356\times10^{-10}$ C).

(a) Find the relationship between the debye and the SI unit for the dipole moment.

(b) Express the dipole moment of charges $\pm1.6\times10^{-19}$ C separated by 2×10^{-10} m in debye and in the appropriate SI unit.

6.2. Use the principle of superposition to calculate the electric field in regions A, B, C, D, and E in the figure.

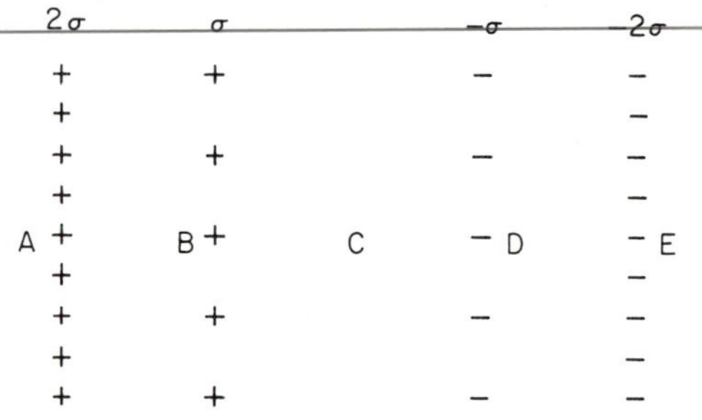

2σ σ $-\sigma$ -2σ

A B C D E

Section 6.3

6.3. An infinite sheet of charge has a thickness $2a$ as shown. The charge density is ρ C m^{-3}. Find the electric field for all values of x.

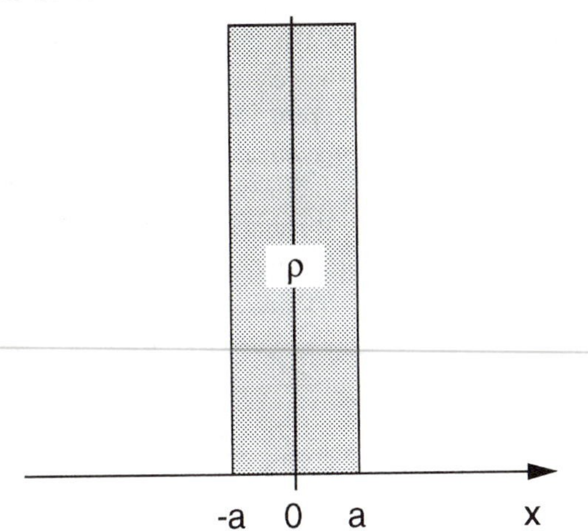

Section 6.6

6.4. Two plane parallel conducting plates each have area S and are separated by a distance b. One carries a charge $+Q$; the other carries a charge $-Q$. Neglect edge effects (i.e., assume the plates are infinite). The plates are electrical conductors.

(a) What is the charge per unit area on each plate? Where does it reside?

(b) What is the electric field between the plates?

(c) What is the capacitance?

(d) As the plate separation is increased what happens to E, v, and C?

(e) If a dielectric is inserted between the plates, what happens to E, v, and C? (See Sec. 6.7.)

6.5. It was shown in the text that the electric field from an infinitely long line of charge, of charge density λ C m^{-1} is $E=\lambda/2\pi\epsilon_0 r$ at a distance r from the line.

(a) Show that if positive charge is distributed with density σ C m^{-2} on the surface of a cylinder of radius a, the electric field is

$$0 \quad \text{for } r<a$$

$$\sigma a/\epsilon_0 r \quad \text{for } r>a.$$

(b) Find the potential difference between a point a distance a from the center of the cylinder and a point a distance d from the center of the cylinder ($d>a$).

(c) Is a or d at the higher potential?

(d) Suppose that another cylinder of radius d is placed concentric with the first. It has a charge σ' per unit area. How will its presence affect the potential difference calculated in part (b)?

(e) Calculate the capacitance between the two cylinders and show that it is $2\pi\epsilon_0 L/\ln(d/a)$, where L is the length of the cylinder.

6.6. The previous problem showed that the capacitance of a pair of concentric cylinders, of radius a and d ($d>a$) is

$$C=\frac{2\pi\epsilon_0 L}{\ln(d/a)}.$$

Suppose now that

$$d=a+b,$$

where b is the thickness of the region separating the two cylinders. (It might, for example, be the thickness of the axon membrane.) Use the fact that

$$\ln(1+x)=x-x^2/2+x^3/3+\cdots$$

to show that, for small b (that is, $b\ll a$), the formula for the capacitance becomes the same as that for a parallel-plate capacitor.

6.7. Find the capacitance of two concentric spherical conductors. The inner sphere has radius a and the outer sphere has radius b.

Section 6.7

6.8. A parallel-plate capacitor has area S and plate separation b. The region between the plates is filled with dielectric of dielectric constant κ. The potential difference between the plates is v.

(a) What is the total electric field in the dielectric?

(b) What is the magnitude of the charge per unit area on the inner surface of the capacitor plates?

(c) What is the magnitude of the polarization charge on the surface of the dielectric?

6.9. For the unmyelinated axon of Table 6.1 and Fig. 6.3,

(a) How many sodium, potassium, and miscellaneous anions are there in a 1-mm segment?

(b) How many water molecules are there in a 1-mm segment?

(c) What is the charge per unit area on the inside of the membrane?

(d) What fraction of all the atoms and ions inside the segment are charged and not neutralized by neighboring ions of the opposite charge?

6.10. (a) Two layers of charge are separated by a distance $b=5\times10^{-9}$ m. The potential difference between them is 70 mV. Find the electric field between the layers, the charge density in each layer, and the pressure exerted by one layer on the other. Compare the pressure to atmospheric. (Hint: The force is calculated by multiplying the charge in a given layer by the electric field due to the charge in the other layer. Think carefully about factors of 2.)

(b) A nerve-cell membrane has positive charge on the outside and negative charge on the inside. These charges attract each other. Assuming a dielectric constant $\kappa=5.7$ for the membrane, an axon radius of 5×10^{-6} m, and a membrane thickness $b=5\times10^{-9}$ m, what is the force per unit area that the charges on one side of the membrane exert on the other? Express the answer in terms of b, the potential across the membrane v, and κ.

6.11. The drawing represents two infinite plane sheets of charge with an infinite slab of dielectric filling part of the space between them. The dashed lines represent cross sections of two Gaussian surfaces. The sides are parallel to the electric field, and the ends are perpendicular to the electric field. Apply the second form of Gauss's law, Eq. (6.19b), to find the electric field within the dielectric using the upper Gaussian surface. Repeat using the lower Gaussian surface.

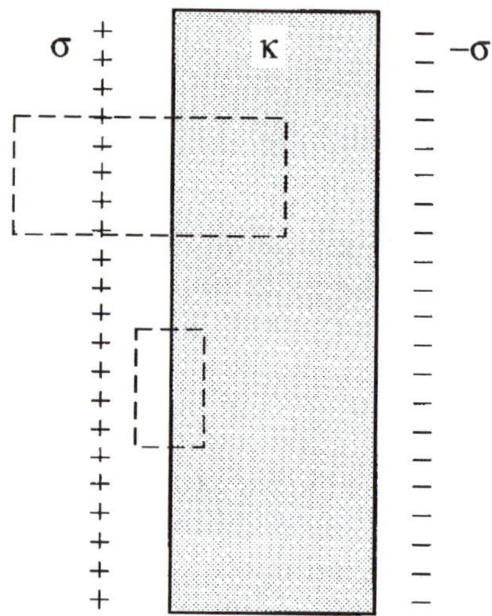

Section 6.8

6.12. This problem will give you some insight into the resistance of electrodes used in neurophysiology. Consider two concentric spherical electrodes; the region between them is filled with material of conductivity σ. The inner radius is a, the outer radius is b.

(a) Imagine that there is a total charge Q on the inner sphere. Find the electric field between the spheres in terms of the potential difference between them and their radii.

(b) The current density in the conducting material is given by $\mathbf{j}=\sigma\mathbf{E}$. Find the total current.

(c) Find the effective resistance, $R=v/i$. What is the resistance as $b\rightarrow\infty$? This is the resistance of a small spherical electrode in an infinite medium; infinite means the other electrode is "far away."

6.13. Find the voltage across each resistor in the figure and the current through it. Also calculate the power dissipation in the resistor.

Section 6.9

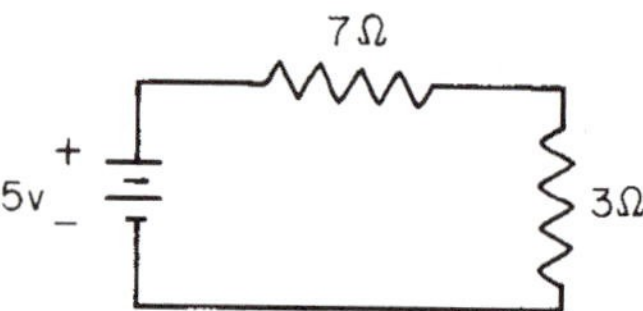

6.14 Derive the equation for the resistance of a set of resistors connected in parallel.

Section 6.11

6.15. The resistivity of the fluid within an axon is 0.5 Ω m. Calculate the resistance along an axon 5 mm long with a radius of 5 μm. Repeat for a radius of 500 μm.

6.16. The voltage along an axon is as shown at some instant of time. The axon radius is 10 μm; the resistivity of the axoplasm is 0.5 Ω m. What is the longitudinal current in the axon as a function of position?

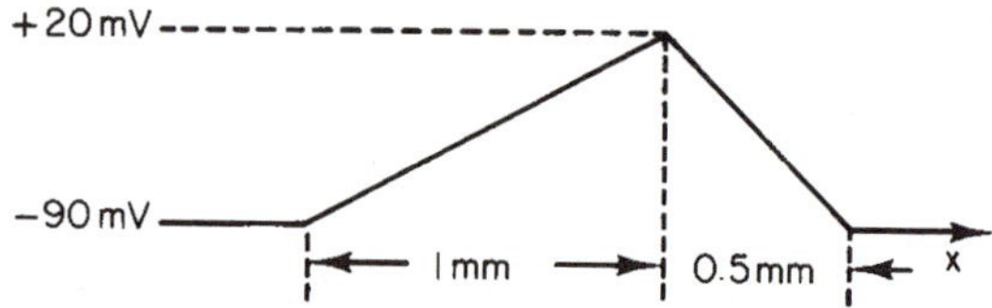

6.17. This problem is designed to show you how a capacitance, such as the cell membrane, charges and discharges. To begin, the switch has been in position B for a long time, so that there is no charge on the capacitor. At $t=0$ the switch is put in position A. It is kept there for 20 s, then thrown back to position B.

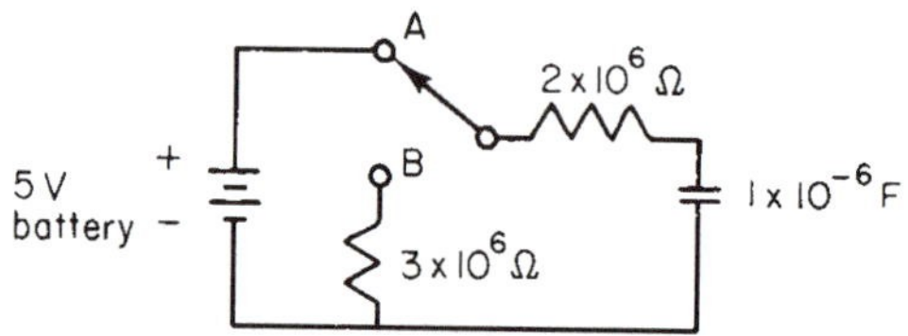

(a) Write a differential equation for the voltage on the capacitor as a function of time when the switch is in position A and solve it.

(b) Repeat when the switch is in position B.

(c) Plot your results.

6.18. Sometimes an organ is lined with a single layer of flat cells. (One example is the lining of the jejunum, the upper portion of the small intestine.) Experimenters may then apply a time-varying voltage across the sheet of cells and measure the resulting current. The cells are packed so tightly together that one model for them is two layers of insulating membrane of dielectric constant κ and thickness b that behave like a capacitor, separated by intracellular fluid of thickness a and resistivity ρ. Find a differential equation or integral equation that relates the total voltage difference across the layer of cells $v(t)$ to the current per unit area through the layer, $j(t)$, in terms of κ, ρ, the membrane thickness b, and the intracellular thickness a.

6.19. The current that appears to go "into" a section of membrane is made up of two parts: that which charges the membrane capacitance and that which is a leakage current through the membrane:

$$i=\frac{v}{R}+C\frac{dv}{dt}.$$

Suppose that the total current is sinusoidal:

$$i=I_0 \cos \omega t.$$

(a) Show that the voltage must be of the form

$$v = I_0 R' \cos \omega t + I_0 X \sin \omega t$$

and that the differential equation is satisfied only if

$$R' = \frac{R}{1+\omega^2(RC)^2},$$

$$X = R\frac{\omega(RC)}{1+\omega^2(RC)^2}.$$

(b) What happens to R' and X as $\omega \to 0$? $\omega \to \infty$? For what value of ω is X a maximum? What is the corresponding value of R'? Plot these points.

(c) Your plot in part *b* should suggest that the locus of X vs R' is a semicircle, centered at $X=0$, $R'=R/2$. Prove that this is so. [Remember that the equation of a circle is $(x-a)^2+(y-b)^2=r^2$.]

6.20. This problem is designed to show you how a "ladder" of resistances can attenuate a signal.

(a) Show that the resistance between points B and G in the circuit is 10 Ω.

(b) Show that the resistance between points A and G in this circuit is also 10 Ω. What will be the result if an infinite number of ladder elements are added to the left of AG?

(c) Assume that v_c (measured with respect to point G) is 6 V. Calculate v_B and v_A. Note that the ratios are the same:

$$\frac{v_B}{v_A} = \frac{v_C}{v_B}.$$

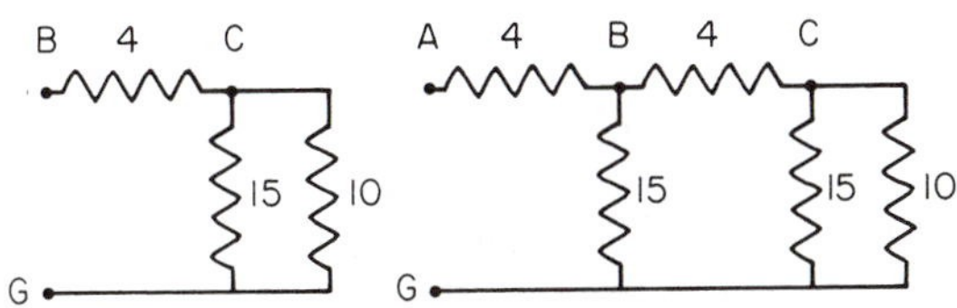

6.21. This is a more general version of the previous problem, which can be applied directly to electrotonus when capacitance is neglected. Consider the ladder shown, which represents an axon. R_0 is the effective resistance between the inside and outside of the axon to the right of the section under consideration. The axon has been divided into small slices; R_i is the resistance along the inside of the axon in the small slice, and R_m is the resistance across the membrane in the slice. The resistance outside the axon is neglected. Note that the resistance looking into the axon to the right of points XX is also R_0.

(a) Show that R_0 is given by a quadratic equation:

$$R_0^2 - R_i R_0 - R_i R_m = 0$$

and that the solution is

$$R_0 = \tfrac{1}{2}[R_i + (R_i^2 + 4R_i R_m)^{1/2}].$$

(b) Show that the ratio of the voltage across one ladder rung to the voltage across the immediately preceding rung is

$$\frac{R_m R_0}{R_m R_0 + R_m R_i + R_i R_0}.$$

(c) Now assume that $R_i = r_i dx$ and $R_m = 1/(2\pi a g_m dx)$. Calculate R_0 and the voltage ratio. Show that the voltage ratio (as $dx \to 0$) is

$$\frac{1}{1+(2\pi a r_i g_m)^{1/2} dx}.$$

(d) The preceding expression is of the form $1/(1+x)$. For sufficiently small x, this is approximately $1-x$. Therefore, show that the voltage change from one rung to the next is

$$dv = -[(2\pi a r_i g_m)^{1/2} dx]v$$

so that v obeys the differential equation

$$\frac{dv}{dx} = -(2\pi a r_i g_m)^{1/2} v.$$

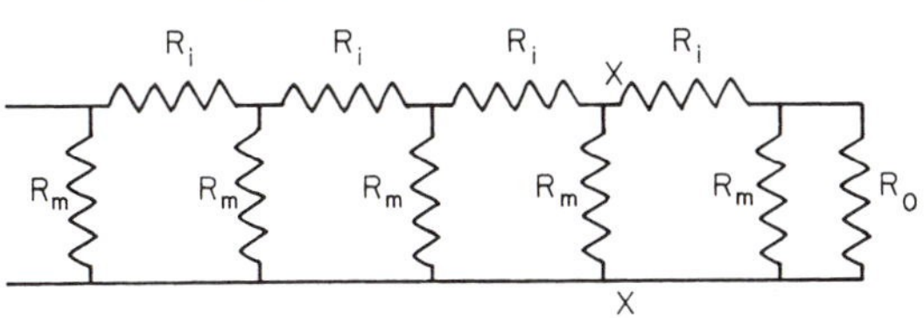

Section 6.12

6.22. Consider the myelinated and unmyelinated axons of Tables 6.1 and 6.2. Compare the decay distance λ for electrotonus in both cases. Neglect attenuation due to the leakage at the node of Ranvier.

6.23. Show by direct substitution that

$$v(x) = v_0 e^{-x/\lambda} + v_{\text{rest}}$$

satisfies the equation

$$\frac{d^2 v}{dx^2} = 2\pi a g_m r_i (v - v_{\text{rest}})$$

if v_{rest} is constant.

6.24. In an electrotonus experiment a microelectrode is inserted in an axon at $x=0$, and a constant current i_0 is injected. After the membrane capacitance has charged, the voltage outside is zero everywhere and the voltage inside is given by Eq. (6.56):

$$v - v_r = \begin{cases} v_0 e^{-x/\lambda}, & x>0 \\ v_0 e^{x/\lambda}, & x<0. \end{cases}$$

(a) Find $i_i(x)$ in terms of v_0, λ, and r_i.

(b) Find $j_m(x)$ in terms of g_m, v_0, and λ.

(c) Find the current i_0 injected at $x=0$ in terms of v_0, λ, and r_i.

(d) Find the input resistance v_0/i_0.

6.25. The equation for electrotonus, as modified in Eq. (6.57), is a diffusion equation. Find the diffusion constant in terms of the axon parameters and evaluate it for a typical case.

Section 6.13

6.26. Use the Hodgkin-Huxley parameters to answer the following questions.

(a) When $v=v_r$, what are α_n and β_n?

(b) Show that $dn/dt=0$ when $n=0.318$. What is the resting value of g_K?

(c) At $t=0$ the voltage is changed to -25 mV and held constant. Find the new values of $\alpha_n, \beta_n, n_\infty$ and the asymptotic value of g_K.

(d) Find an analytic solution for $n(t)$. Plot n and n^4 for $0<t<10$ ms.

6.27. In a voltage-clamp experiment, a wire of radius b is threaded along the interior of an axon of radius a. The axoplasm displaced by the wire is pushed out the end so that the cross-sectional area of the axon containing the wire remains πa^2. The resistivities of wire and axoplasm are ρ_w and ρ_a. Find the wire radius needed so that voltage changes along the axon are reduced by a factor of 100 from what they would be without the wire.

6.28. A wire of resistivity $\rho_w=1.6\times10^{-8}$ Ω m and radius $w=0.1$ mm is threaded along the exact center of an axon segment of radius $a=1$ mm, length $L=1$ cm, and resistivity $\rho_i=0.5$ Ω m. The axon membrane has conductance $g_m=10$ S m^{-2}. Find numerical values for

(a) the resistance along the wire,

(b) the resistance of the axoplasm from the wire to the membrane, and

(c) the resistance of the membrane.

6.29. If the voltage across an axon membrane is changed by 26 mV as in Fig. 6.35, how long will it take for all the potassium to leak out if it continues to move at the constant rate at which it first leaks out? Use the asymptotic value for the potassium conductance from Fig. 6.35, Table 6.1, and Fig. 6.3 for any other values you need.

Section 6.14

6.30. Use the data of Fig. 6.43 to answer the following questions about a nerve impulse in a squid axon of radius $a=0.1$ mm.

(a) Estimate the peak sodium ion flux (ions m^{-2} s^{-1}) and the total number of sodium ions per unit area that pass through the membrane in one pulse.

(b) By what fraction does the sodium concentration in the cell increase during one nerve pulse?

(c) Estimate the peak potassium flux and total potassium transport.

6.31. Show by direct substitution that

$$n(v,t)=n_\infty(v)(1-e^{-t/\tau})$$

satisfies the equation

$$\frac{dn}{dt}=\alpha_n(1-n)-\beta_n n$$

if α_n and β_n are functions of v, but not of time.

6.32. The Hodgkin–Huxley equation for the potassium parameter n is

$$\frac{dn}{dt}=\alpha_n(1-n)-\beta_n n.$$

What is the asymptotic value of n when $t\to\infty$?

6.33. For $t<0$ a squid axon has a resting membrane potential of -65 mV. The sodium Nernst potential is $+50$ mV. The Hodgkin–Huxley parameters are $m=0.05$, $h=0.60$, and $g_{\mathrm{Na}\infty}=1200$ S m^{-2}.

(a) What is j_{Na}?

(b) For $t>0$ a voltage clamp is applied so that $v=-30$ mV. Suppose that $m=0.72(1-e^{-2.2t})$ and $h=0.6e^{-0.63t}$ (here t is in milliseconds). What is the total charge transported across unit area of the membrane by sodium ions?

6.34. A stimulating current of 10^{-4} A cm^{-2} is applied for 100 μs. How much does it change the potential across the membrane?

6.35. Using the resting value of j_K from Fig. 6.41, calculate how long it would take for the concentration of potassium inside an axon of radius 100 μm to decrease by 1%.

6.36. Consider a space-clamped axon for which the resting potential is v_r. Assume that the membrane current density follows a very strange behavior:

$$j_m=\begin{cases} B(v-v_r)^2, & v>v_r \\ 0, & v<v_r. \end{cases}$$

(a) Write a differential equation for $v(t)$.

(b) What are the units of B?

(c) What sign would B have for depolarization to take place after a small positive change of v?

(d) Integrate the equation obtained in (a).

Section 6.15

6.37. A pulse which propagates along the axon with speed u is of the form $v(x,t)=f(x-ut)$.

(a) Use the chain rule to show that this means that

$$\frac{\partial v}{\partial t}=-u\frac{\partial v}{\partial x}, \quad \frac{\partial^2 v}{\partial t^2}=u^2\frac{\partial^2 v}{\partial x^2}.$$

(b) Find an expression for the membrane current per unit area in terms of c_m, ρ_i, ρ_m, a, and the various partial derivatives of f with respect to x.

6.38. An unmyelinated axon has the following properties: radius of 0.25 mm, membrane capacitance of 0.01 F m^{-2}, resistance per unit length along the axon of 2×10^6 Ω m^{-1}, and propagation velocity of 20 m s^{-1}. The propagating pulse passes an observer at $x=0$. The peak of the pulse can be approximated by a parabola, $v(t)=20(1-10t^2)$, where v is in millivolts and t is in milliseconds.

(a) Find the current along the axon at $x=0$, $t=0$.

(b) Find the membrane current per unit area j_m at $x=0$, $t=0$.

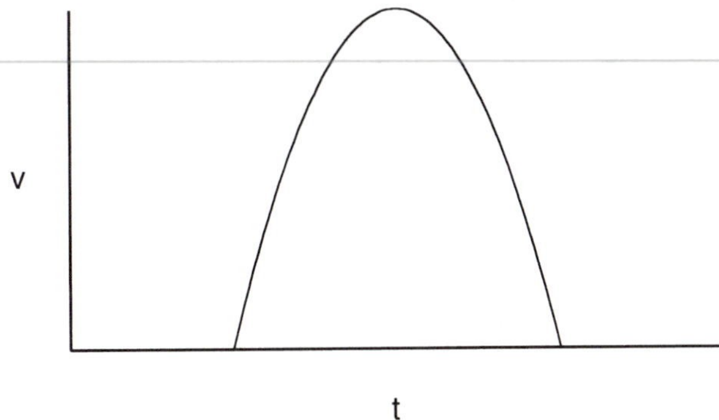

6.39. A space-clamped axon (v independent of distance along axon) has a pulse of the form

$$v(t)-v_r=\begin{cases} 0, & t<-t_1 \\ v_0[1-(t/t_1)^2], & -t_1<t<t_1 \\ 0, & t>t_1 \end{cases}$$

as shown in the previous problem. The axon has radius a, length L, resistivity ρ_i, and membrane capacitance c_m per unit area.

(a) What is the total change in charge on the membrane from $t=-t_1$ to $t=0$?

(b) What is the total change in charge on the membrane from $t=-t_1$ to $t=+t_1$?

(c) What is $j_m(t)$?

(d) If j_m is given by $g_m(v-v_r)$, what is $g_m(t)$? Comment on its behavior.

6.40. A comment was made in the text that the potassium current is not required to generate an action potential. Modify the program of Fig. 6.41 to eliminate the potassium current. (First make sure that you have an unmodified program that reproduces Fig. 6.42 correctly.) Comment on the shape of the resulting pulse. After the pulse there is a new value of the resting potential. Why? Is it significant?

Section 6.16

6.41. Consider a myelinated fiber in which the nodes of Ranvier are spaced every 2 mm. The resistance per unit length along the axon and the capacitance and conductance are as in Table 6.3.

(a) If the voltage difference between nodes is 10 mV, what is the current along the axon? (Assume that the voltages are constant with time, so that the membrane charge does not change. Also neglect leakage current through the membrane.)

(b) If the nerve impulse rises from -70 to $+30$ mV in 0.5 ms, what is the average current required to charge the nodal capacitance?

6.42. Derive Eq. (6.81) and (6.83) for a plane resistor and capacitor.

6.43. A myelinated cylindrical axon has inner radius a and outer radius b. The potential inside is v. Outside it is 0. The myelin is too thick to be treated as a plane sheet of dielectric. Express all answers in terms of a, b, and v.

(a) Give an expression for E for $r<a$.

(b) Give an expression for E for $a<r<b$.

(c) Give an expression for E for $b<r$.

(d) Assuming $\kappa=1$, what is the charge density on the inner surface? The outer surface?

Section 6.18

6.44. Modify the computer program of Fig. 6.41 to have a constant value of `jStim` and run it.

REFERENCES

Alberts, B., D. Bray, J. Lewis, M. Raff, K. Roberts, and J. D. Watson (1983). *Molecular Biology of the Cell*. New York, Garland.

Anumonwo, J. B., and J. Jalife (1995). Cellular and subcellular mechanisms of pacemaker activity initiation and synchronization in the heart. In D. P. Zipes and J. Jalife, eds. *Cardiac Electrophysiology: From Cell to Bedside*, 2nd. ed. Philadelphia, Saunders.

Covino, B. G. (1972). Local anesthesia. *N. Engl. J. Med.* **286**: 975–983.

Davis, Jr., L., and Lorente de No (1947). Contributions to the mathematical theory of electrotonus. *Stud. Rockefeller Inst. Med. Res.* **131**(Part 1): 442–496.

Dennis, S. S, J. W. Clark, C. R. Murphy, and W. R. Giles (1994). A mathematical model of a rabbit sinoatrial node cell. *Am. J. Physiol.* **266**: C832–852.

Deutsch, S., and A. Deutsch (1993). *Understanding the Nervous System*. New York, IEEE.

DiFrancesco, D., M. Mangoni, and G. Maccaferri (1995). The pacemaker current in cardiac cells. In D. P. Zipes and J. Jalife, eds. *Cardiac Electrophysiology: From Cell to Bedside*, 2nd. ed. Philadelphia, Saunders.

Geddes, L. A. (1994). Historical Perspectives 3: Recording of Action Potentials. In J. D. Bronzino, ed. *The Biomedical Engineering Handbook*. Boca Raton, CRC, pp. 1367–1377.

Guyton, A. C. (1991). *Textbook of Medical Physiology*, 8th ed. Philadelphia, Saunders.

Halliday, D., R. Resnick, and K. S. Krane (1992). *Physics*, 4th ed., extended version. New York, Wiley.

Hodgkin, A. L. (1964). *The Conduction of the Nervous Impulse*. Springfield, IL., Thomas.

Hodgkin, A. L., and A. F. Huxley (1939). Action potentials recorded from inside a nerve fiber. *Nature* **144**: 710.

Hodgkin, A. L., and A. F. Huxley (1952). A quantitative description of membrane current and its application to conduction and excitation in nerve. *J. Physiol.* **117**: 500–544.

Hodgkin, A. L., and W. A. Rushton (1946). The electrical constants of a crustacean nerve fiber. *Proc. R. Soc.* B **133**: 444–479.

Jeffreys, H., and B. S. Jeffreys (1956). *Mathematical Physics*. London, Cambridge University Press, p. 602.

Kane, B. J., C. W. Storment, S. W. Crowder, D. L. Tanelian, and G. T. A. Kovacs (1995). Force-sensing microprobe for precise stimulation of mechanoreceptive tissues. *IEEE Trans. Biomed. Eng.* **42**(8): 745–750.

Katz, B. (1966). *Nerve, Muscle and Synapse*. New York, McGraw-Hill.

Laüger, P. (1991). *Electrogenic Ion Pumps*. Sunderland, MA, Sinauer.

Luo, C. H., and Y. Rudy (1994). A dynamic model of the cardiac ventricular action potential. I. Simulations of ionic currents and concentration changes. *Circ. Res.* **74**(6): 1071–1096.

Martin, A. R. (1966). Quantal nature of synaptic transmission. *Physiol. Rev.* **46**: 51–66.

Noble, D., D. DiFrancesco, and J. C. Denyer (1989). Ionic mechanisms in normal and abnormal cardiac pacemaker activity. In Jacklet, J. W., ed. *Cellular and Neuronal Oscillators*. New York, Dekker, pp. 59–85.

Noble, D. (1995). Ionic mechanisms in cardiac electrical activity. In D. P. Zipes and J. Jalife, eds. *Cardiac Electrophysiology: From Cell to Bedside*, 2nd ed. Philadelphia, Saunders.

Patton, H. D., A. F. Fuchs, B. Hille, A. M. Scher, and R. F. Steiner (1989). *Textbook of Physiology*. Philadelphia, Saunders.

Plonsey, R. (1969). *Bioelectric Phenomena*. New York, McGraw-Hill.

Scott, A. C. (1975). The electrophysics of a nerve fiber. *Rev. Mod. Phys.* **47**: 487–533.

Smythe, W. R., S. Silver, J. R. Whinnery, and J. D. Angelakos (1957). Formulas. In D. E. Gray, ed. *American Institute of Physics Handbook*, New York, McGraw-Hill, Chap. 5b.

Wilders, R., H. J. Jongsma, and A. C. G. van Ginnekin (1991). Pacemaker activity of the rabbit sinoatrial node: A comparison of models. *Biophys. J.* **60**: 1202–1216.

Worthington, C. R. (1971). X-ray analysis of nerve myelin, in J. W. Adelman, Jr., ed., *Biophysics and Physiology of Excitable Membranes*, New York, Van Nostrand Reinhold, pp. 1–46.

CHAPTER 7

The Exterior Potential and the Electrocardiogram

In Chap. 6 we assumed that the potential outside a nerve cell is zero. This is only approximately true. There is a small potential that can be measured and has clinical relevance. Before a muscle cell contracts, a wave of depolarization sweeps along the cell. This wave is quite similar to the wave along the axon. Measurement of these external signals gives us the electrocardiogram, the electromyogram, and the electroencephalogram.

In Sec. 7.1 we calculate the potential outside a long cylindrical axon bathed in a uniform conducting medium. Section 7.2 shows that the exterior potential is small compared to the potential inside the cell if there is enough extracellular fluid so that the outside resistance is low. Section 7.3 uses a model in which the action potential is approximated by a triangular pulse to calculate the potential far from the cell. Section 7.4 generalizes this calculation to the case of a pulse of arbitrary shape.

An unusual feature of heart muscle (myocardial) cells is that the cells remain depolarized for 100 ms or so, as described in Secs. 7.5 and 7.6. This means that the potential difference outside the cell is much larger than for other cells, giving rise to the electrocardiogram signals described in Sec. 7.7.

The next two sections discuss electrocardiography and some factors that distort the signal. They make no attempt to consider advanced techniques such as orthogonal leads that are used to reconstruct the electrical activity of the heart from potential difference measurements on the surface of the body. Rather, they are much closer to the way clinicians think about the electrocardiogram, and they can provide a basis from which to learn more complicated techniques.

Section 7.10 talks about improved models that take into account the interaction between the inside and outside of cells and the anisotropies that exist in tissue resistance. Section 7.11 discusses the problem of stimulation: for measurement of evoked responses, for pacemaking, and for defibrillation.

7.1 THE POTENTIAL OUTSIDE A LONG CYLINDRICAL CELL

In Chap. 6 we assumed that the potential is zero everywhere outside an axon. Now we will calculate the exterior potential distribution if the cell is in an infinite uniform conducting medium. We will discover that for the case studied here the exterior potential changes are less than 0.1% of those inside. If the exterior medium is not infinite, the exterior potential changes are larger, as is discussed in Sec. 7.10.

The calculation of electrotonus and the action potential in Chap. 6 was based on the equation

$$c_m \frac{\partial v}{\partial t} + j_m = \frac{1}{2\pi a r_i} \frac{\partial^2 v}{\partial x^2}, \tag{6.50}$$

with the appropriate model for j_m. We made the simplifying assumption that v_o was zero everywhere outside the cell. Equation (6.50) was derived from a statement of the conservation of charge,

$$di_o = -di_i = C \frac{\partial(v_i - v_o)}{\partial t} + i_m, \tag{6.45b}$$

and Ohm's law for the axoplasm,

$$i_i(x) = -\frac{1}{r_i} \frac{\partial v_i}{\partial x}. \tag{6.46}$$

Equation (6.45b) says that if there is an increase (decrease) of current along the inside of the axon, the current must flow in from (out to) the extracellular medium, passing through the membrane or charging or discharging the membrane capacitance on the way.

Consider a single cell stretched along the x axis. We will consider space to be divided into three regions as shown on the left in Fig. 7.1: the interior of the axon (the axoplasm),

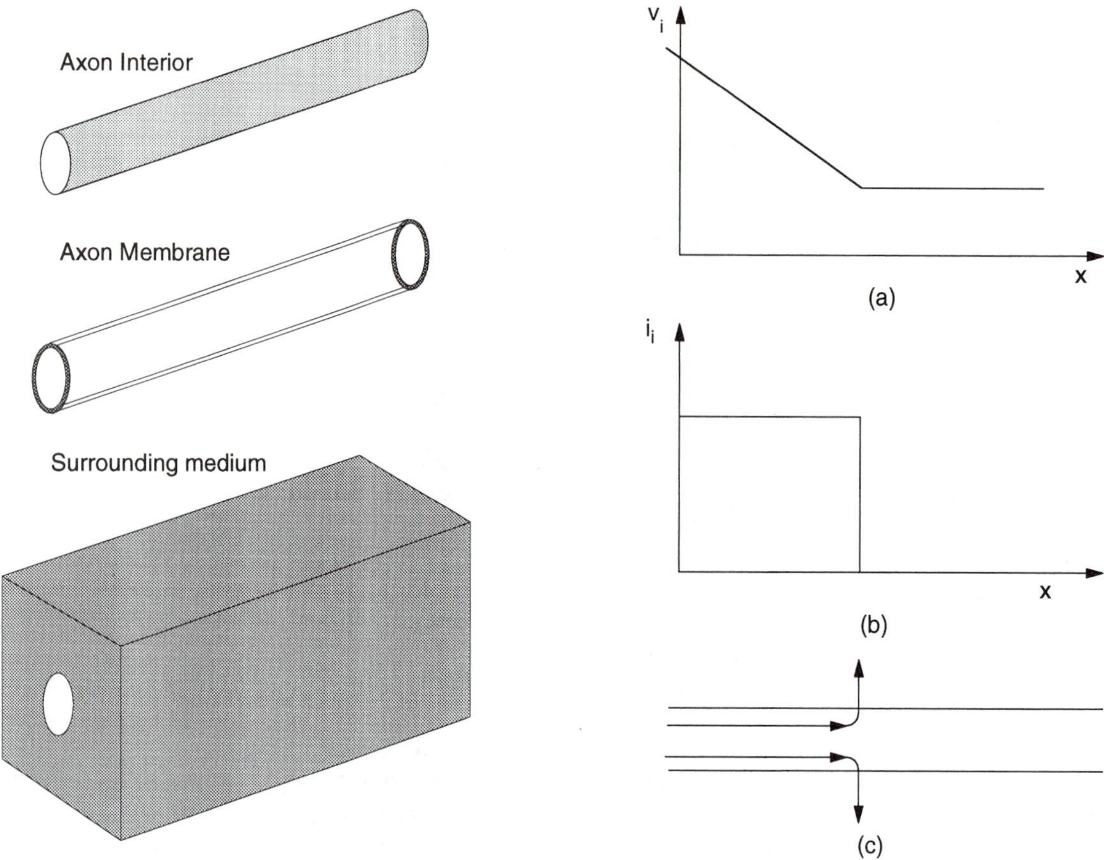

FIGURE 7.1. A conducting cell is stretched along the x axis. (a) A plot of a portion of the interior action potential at one instant of time. (b) A plot of the interior current, proportional to the slope of the action potential because of Ohm's law. (c) Schematic representation of the axon, showing current flowing along the axon and into the exterior conducting medium at the point where the interior current falls to zero.

the axon membrane, and the surrounding medium. Imagine that the current inside the cell is constant to the left of a certain point and zero to the right of that point, as shown on the right in Fig. 7.1(b). Since the material inside the cell obeys Ohm's Law, the interior potential decreases linearly with x as shown in Fig. 7.1(a). Where the current is zero, the interior potential does not change. At the point where the interior current falls to zero, conservation of charge requires that the current pass through the membrane and flow in the external conducting medium, as stated in Eq. (6.45b). Figure 7.1(c) shows the axon with current flowing in the left part of the cell and then flowing into the surrounding medium.

Now consider how the current flows in the surrounding three-dimensional medium. Suppose that the "outside" medium is infinite, homogeneous, and isotropic and has conductivity σ_o. Suppose also that the cell stretched along the x axis is very thin and does not appreciably change the homogeneous and isotropic nature of the extracellular medium, except very close to the x axis. If the current i_o enters the surrounding medium at the origin, the current density $\mathbf{j}$ is directed radially outward and has spherical symmetry. The current density at distance r has magnitude $j=i_o/4\pi r^2$. The magnitude of the electric field is $E=j/\sigma_o=i_o/4\pi\sigma_o r^2$. This has the same form as the electric field from a point charge, for which the electric field is $E=q/4\pi\epsilon_0 r^2$. We speak of i_o as a *point current source.*

We can use the expression for the electric field to calculate the exterior potential. The point current source is shown as the dot in the center of the sphere in Fig. 7.2. To calculate the potential difference between points A and B, it is easiest to integrate Eq. (6.14) along a path from A to B' parallel to the direction of $\mathbf{E}$, and then along $B'B$ where the displacement is always perpendicular to the electric field. (AB' is along a radial line to the charge or current source, and $B'B$ is on the surface of a sphere centered on the charge.) The potential change along $B'B$ is zero. Therefore

$$v(B)-v(A)=-\int_{r_A}^{r_B} E_r dr=-\int_{r_A}^{r_B} \frac{i_o}{4\pi\sigma_o r^2}\,dr$$

$$=\frac{i_o}{4\pi\sigma_o}\left(\frac{1}{r_B}-\frac{1}{r_A}\right).$$

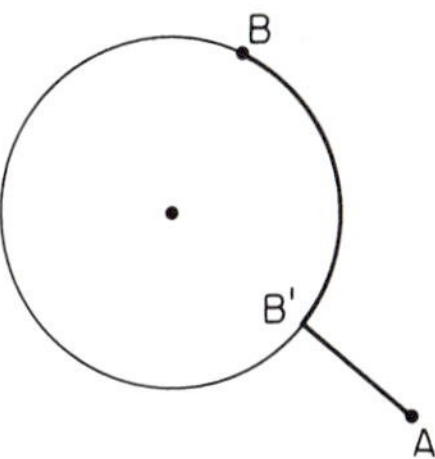

FIGURE 7.2. A point current source is at the center of the circle. The path of integration to calculate the potential difference between points A and B goes first from A to B' and then from B' to B.

Only a difference of potential between two points has meaning. However, it is customary to define the potential to be 0 at $r_A=\infty$ and speak of the potential as a function of position. Then the potential at distance r from a point current source i_o is

$$v(r)=\frac{i_o}{4\pi\sigma_o r}. \tag{7.1}$$

The analogous expression for the potential due to a point charge q is $v(r)=q/4\pi\epsilon_0 r$.

We do not yet have a useful model, because the potential cannot rise forever as we go along the axon to the left. Let us assume that the potential levels off at some point on the left, as shown in Fig. 7.3. This turns out to be a very good model for the electrocardiogram, because the repolarization of myocardial cells does not take place for about 100 ms, so the cell is completely depolarized before repolarization begins. Define the location of the origin so that the transition takes place between $x=0$ and $x=x_2$. The potential change is shown at the top, and the current along the inside of the axon in the middle. The current exists only where there is a voltage gradient, that is between $x=0$ and $x=x_2$. Its magnitude is $i_i=\Delta v_i/R=\Delta v_i\sigma_i\pi a^2/x_2$. This current flows out into the external medium at $x=x_2$ and back into the axon at $x=0$. Such a combination of source and sink of equal magnitude is called a *current dipole*. (A pair of equal and opposite electric charges is called an electric dipole.) The lowest part of the figure shows lines of **j** or **E**. The current is to the right inside the axon (along the axis) and returns outside the axon. The potential at any exterior point is due to two terms: one from the source i_i at $x=x_2$ and the other from the sink $-i_i$ at $x=0$. If r_2 is the distance from the observation point to x_2 and r_0 is the distance to the origin, then

$$v=\frac{\Delta v_i\sigma_i\pi a^2}{4\pi\sigma_o x_2}\left(\frac{1}{r_2}-\frac{1}{r_0}\right)=\frac{\Delta v_i\sigma_i a^2}{4\sigma_o x_2}\left(\frac{1}{r_2}-\frac{1}{r_0}\right). \tag{7.2}$$

Note the similarity of this to the potential from two point charges.

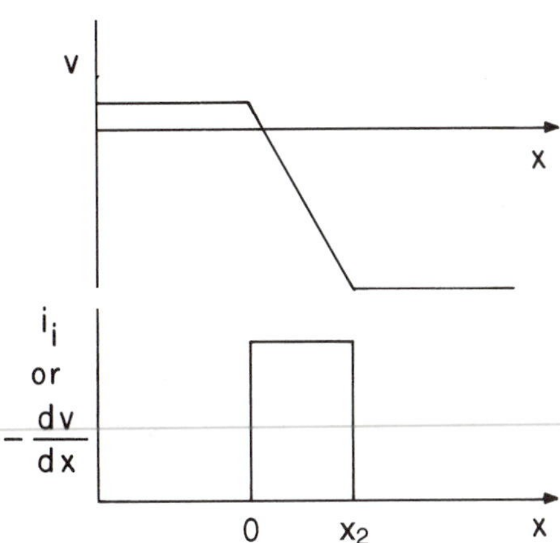

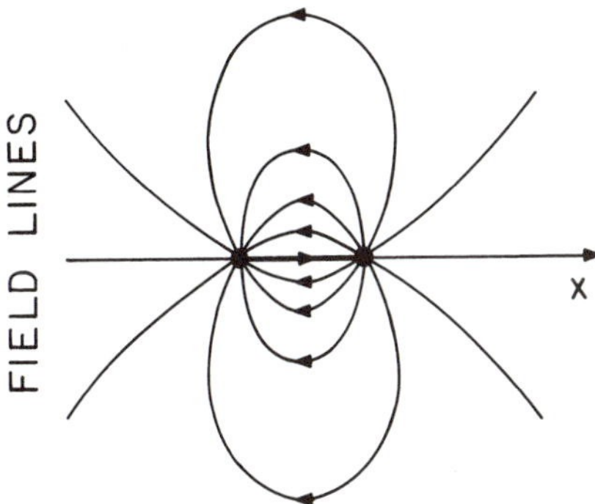

FIGURE 7.3. The potential of Fig. 7.1 is extended to the left in a region of constant (depolarized) potential. The interior current is plotted below the potential. The electric field or current-density lines are plotted at the bottom. The current to the right on the axis is current within the axon; the other lines represent current in the external conducting medium.

Equation (7.2) can be used to estimate the potential at an exterior point for the electrocardiogram because of the long delay before the repolarization of myocardial cells takes place.

To estimate the potential from a nerve impulse, we can approximate the action potential by a triangular potential as shown in Fig. 7.4(a). The potential is zero far to the left. It rises by an amount Δv_i between $x=-x_1$ and $x=0$. It falls linearly to zero at $x=x_2$. The current is plotted in Fig. 7.4(b). In the region just to the left of the origin it is

$$-i_1=-\frac{\Delta v_i\sigma_i\pi a^2}{x_1}. \tag{7.3a}$$

(It is negative because it flows to the left.) To the right of the origin it is

$$i_2=\frac{\Delta v_i\sigma_i\pi a^2}{x_2}. \tag{7.3b}$$

Figure 7.5 shows the outside medium. There is a source of current i_1 at $x=-x_1$, a source i_2 at $x=x_2$, and a sink

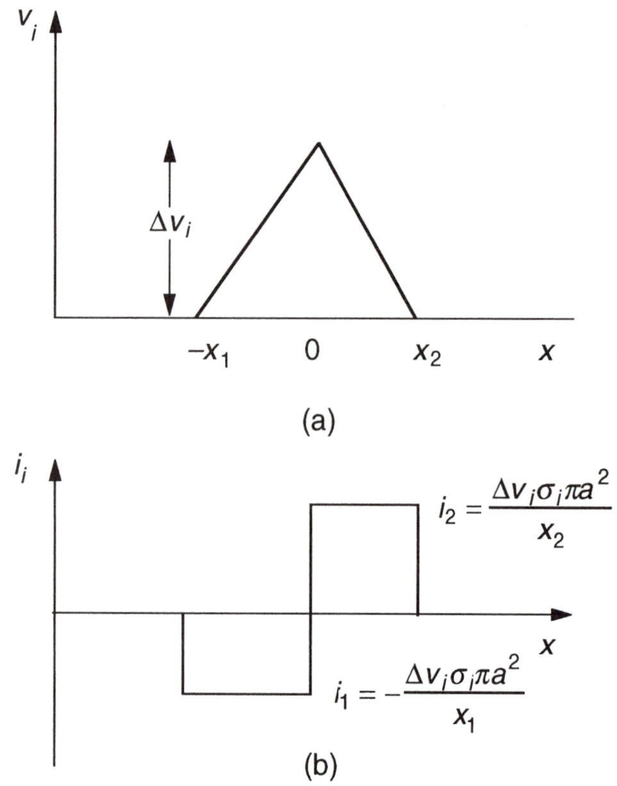

FIGURE 7.4. The action potential is approximated by a triangular wave form. In this piecewise-linear approximation the depolarization and repolarization are both linear. (a) The interior potential. (b) The interior current along the cell.

$-(i_1+i_2)$ at the origin. The potential at observation point P is calculated by repeated application of Eq. (7.1):

$$v=\frac{1}{4\pi\sigma_o}\left(\frac{i_1}{r_1}-\frac{i_1+i_2}{r_0}+\frac{i_2}{r_2}\right). \tag{7.3c}$$

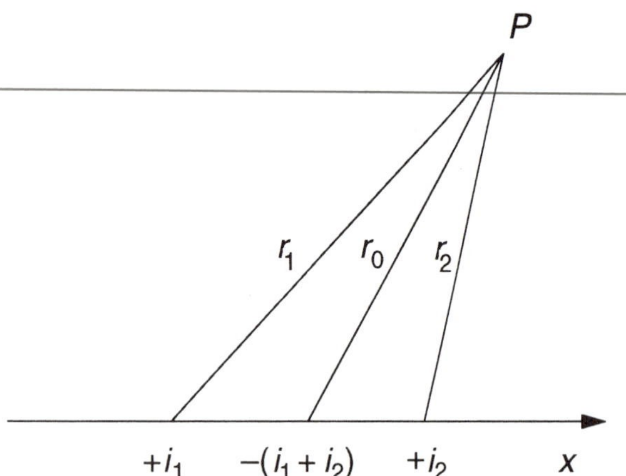

FIGURE 7.5. The cell of Fig. 7.4 is stretched along the x axis. There are current sources at $x=-x_1$ and $x=x_2$, and a current sink at the origin. The distances from each source or sink to the observation point are shown.

Equations (7.3a)–(7.3c) can be combined to give

$$v=\frac{\Delta v_i\sigma_i a^2}{4\sigma_o}\left(\frac{1/x_1}{r_1}-\frac{1/x_1+1/x_2}{r_0}+\frac{1/x_2}{r_2}\right). \tag{7.4}$$

Equations (7.3) and (7.4) are valid at any distance from the cell, as long as we can make the piecewise approximation of the action potential shown in Fig. 7.4.

7.2 THE EXTERIOR POTENTIAL IS SMALL

Let us use Eq. (7.2) for the rising edge of the action potential to estimate the potential outside the cell when it is in an infinite conducting medium. We evaluate Eq. (7.2) close to the surface of the axon where the potential will be largest, say at $x=0$. In that case r_2 is approximately x_2. However, r_0 is not zero. It can never become smaller than $r=a$, the radius of the axon. (The potential would diverge if the model were extended to $r=0$.) We will use an approximate value, $r_0=a$, and call the height of the action potential Δv_i. Then

$$v(0)=\frac{\Delta v_i\sigma_i a^2}{4\sigma_o x_2}\left(\frac{1}{x_2}-\frac{1}{a}\right). \tag{7.5}$$

Since $1/x_2 \ll 1/a$, this becomes

$$v(0)\approx -\frac{\Delta v_i\sigma_i a}{4\sigma_o x_2}. \tag{7.6}$$

Close to $x=x_2$ the potential is

$$v(x_2)=\frac{\Delta v_i\sigma_i a^2}{4\sigma_o x_2}\left(\frac{1}{a}-\frac{1}{x_2}\right)\approx\frac{\Delta v_i\sigma_i a}{4\sigma_o x_2}. \tag{7.7}$$

The potential difference between these two exterior points is

$$\Delta v_o=v(x_2)-v(0)=\frac{\sigma_i}{\sigma_o}\frac{a}{2x_2}\Delta v_i. \tag{7.8}$$

If the conductivities were the same inside and outside, the ratio would be

$$\frac{\Delta v_o}{\Delta v_i}=\frac{a}{2x_2}.$$

The ratio of exterior to interior potential change is proportional to the ratio of the cell radius to the distance along the cell over which the potential changes. From Fig. 6.45 we see that the rising part of the squid action potential has a length $x_2\approx 1$ cm. If $a=0.5$ mm (a quite large axon), then the ratio is $\frac{1}{40}$. For a smaller axon, the ratio is even less.

The same result can be obtained by another argument. The resistance between two points is the ratio of the potential difference between the points to the current flowing between them. Inside the axon **j** and **E** are large because the current is confined to a small region of area πa^2. The resistance inside is $R_i=x_2/\pi a^2\sigma_i$. The same current flows outside, but it is spread out so that **j** and **E** are much less. The resistance between two electrodes in a conducting medium is related to their capacitance (Sec. 6.19). Equations (6.87a) and (6.99) can be used to show that two spherical electrodes of radius a spaced distance x_2 apart ($x_2 \gg a$) have a resistance $R_o = 1/2\pi\sigma_o a$. The voltage ratio is

$$\frac{\Delta v_o}{\Delta v_i} = \frac{R_o}{R_i} = \frac{1}{2\pi\sigma_o a}\,\frac{\pi a^2\sigma_i}{x_2} = \frac{\sigma_i}{\sigma_o}\,\frac{a}{2x_2},$$

the same result as Eq. (7.8).

7.3 THE POTENTIAL FAR FROM THE CELL

In most cases measurements of the potential are made far from the cell—far compared to the distance the action potential spreads out along the cell. If point P is moved far away, Fig. 7.5 looks like Fig. 7.6. The lines r_1, r_0, and r_2 are nearly parallel. If point P is located a distance r_0 from the origin at angle θ with the x axis, then

$$r_2 \approx r_0 - x_2 \cos\theta, \quad r_1 \approx r_0 + x_1 \cos\theta. \tag{7.9}$$

Consider the potential in Eq. (7.2). This is due to the leading edge of the action potential and is a useful model for the electrocardiogram. Substituting Eqs. (7.9) in Eq. (7.2) gives

$$v = \frac{\Delta v_i \sigma_i a^2}{4\sigma_o x_2}\left(\frac{1}{r_0[1-(x_2/r_0)\cos\theta]} - \frac{1}{r_0}\right).$$

You can verify by a Taylor's-series expansion or long division that

$$\frac{1}{1-x} = 1 + x + \cdots, \tag{7.10}$$

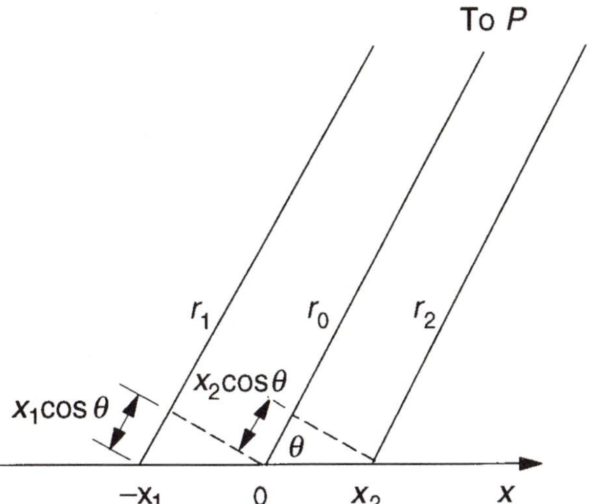

FIGURE 7.6. The observation point P is far away compared to distances x_1 or x_2. The lines to P are nearly parallel.

so that

$$v = \frac{\Delta v_i \sigma_i a^2}{4\sigma_o r_0^2}\cos\theta. \tag{7.11}$$

This is a very important result that will form the basis for our model of the electrocardiogram:

1. The exterior potential depends on Δv_i but not on x_2, the distance the depolarization is spread out along the cell. This is because increasing x_2 decreases the strength of the current at the same time that it increases the potential because the source and sink are further apart.
2. The potential falls off as $1/r^2$ instead of $1/r$ as it would from a point source.
3. The potential varies with angle, being positive to the right of the depolarization region and negative to the left.

It is convenient to define a vector **p** that points along the cell in the direction of the advancing depolarization wave front (the location along the cell where the potential rises). It is called the *activity vector* or *current-dipole moment* for reasons discussed shortly. Its magnitude is

$$p = \pi a^2 \sigma_i \Delta v_i. \tag{7.12}$$

The exterior potential is then (dropping the subscript on **r**)

$$v = \frac{\mathbf{p}\cdot\mathbf{r}}{4\pi\sigma_o r^3}. \tag{7.13}$$

Vector **p** has units of A m. Its magnitude (apart from the conductivity) is the product of the cross-sectional area of the cell and the difference in potential along the cell between the resting and completely depolarized regions. It is called the current-dipole moment because it is the product of the injected and absorbed current and the separation of the source and sink. (The electric-dipole moment is the product of the magnitude of the charges and their separation, with units C m.)

Equation (7.11) can also be written in the form

$$v(r) = \frac{\pi a^2 \cos\theta}{r^2}\,\frac{1}{4\pi}\,\frac{\sigma_i}{\sigma_o}\,\Delta v_i. \tag{7.14}$$

The quantity $\pi a^2 \cos\theta / r^2$ is $\Delta\Omega$, the solid angle[1] subtended at the observation point by a cross section of the cell where the potential changes. In terms of the solid angle

$$v = \frac{\Delta\Omega}{4\pi}\,\frac{\sigma_i}{\sigma_o}\,\Delta v_i. \tag{7.15}$$

Now consider an entire pulse, one where the potential rises and then returns to the resting value. If the approximation of Eq. (7.10) is applied to Eq. (7.4), the result vanishes. It is necessary to make a more accurate approximation, one that takes into account the fact that the vectors

[1]The solid angle is defined in Appendix A.

$\mathbf{r}_1$, $\mathbf{r}_0$, and $\mathbf{r}_2$ are not exactly parallel. Figure 7.7 shows the geometry. We use the law of cosines to write [remember that $\cos(\pi-\theta)=-\cos\theta$]

$$r_1=r_0[1+(2x_1/r_0)\cos\theta+x_1^2/r_0^2]^{1/2},$$

$$r_2=r_0[1-(2x_2/r_0)\cos\theta+x_2^2/r_0^2]^{1/2}.$$

When these are inserted in Eq. (7.4) and a Taylor's-series expansion is done to second order in both x_1/r_0 and x_2/r_0, the result is

$$v=\frac{2\pi a^2}{4\pi r^3}\frac{\sigma_i}{\sigma_o}\frac{\Delta v_i(x_1+x_2)}{2}\frac{3\cos^2\theta-1}{2}. \tag{7.16}$$

The constants have been arranged to show that the term $\Delta v_i(x_1+x_2)/2$ is the area under the impulse when v is plotted as a function of distance along the cell. The angular factor as written with its factor of 2 in the denominator is tabulated in many places as the "Legendre polynomial $P_2(\cos\theta)$." The exterior potential now falls off more rapidly with distance, as $1/r^3$. The angular dependence, shown in Fig. 7.8, is symmetric about $\pi/2$. This shows the angular dependence as one moves around the impulse at a constant distance from it. This is a very different situation and a very different curve from the potential measured at a fixed point near the cell as an impulse travels past. In the latter case r as well as θ is changing. This behavior is discussed in Problems 7.6 and 7.7. The results are shown in Fig. 7.9. The potential from the depolarization is biphasic; that from the complete pulse is triphasic, being positive, then negative, then positive again.

For a single axon in an ionic solution the exterior conductivity is usually higher than in the cell, so $\sigma_i/\sigma_o=0.2$. The conductivity of tissue is considerably less than the conductivity of an ionic solution, and the ratio becomes greater

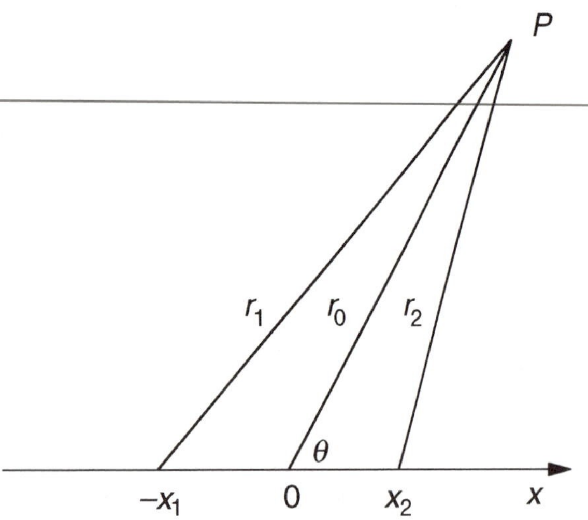

FIGURE 7.7. When the observation point is not so far away, or when a complete nerve impulse is being considered, the law of cosines must be used to relate r_1 and r_2 to r_0.

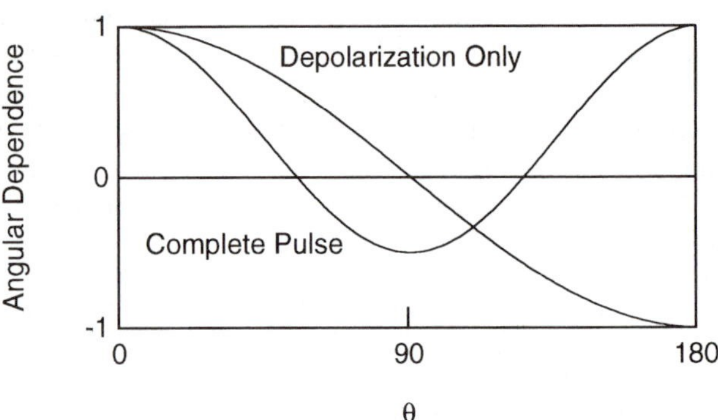

FIGURE 7.8. Plot of the angular dependence of the potential from the entire impulse, Eq. (7.16).

than one. For the electrocardiogram it will be more appropriate to use $\sigma_o=0.33\ \mathrm{S\,m^{-1}}$ (muscle) or $0.08\ \mathrm{S\,m^{-1}}$ (lung), in which case σ_i/σ_o is 6 or 25. We will use an approximate value of 10.

7.4 THE EXTERIOR POTENTIAL FOR AN ARBITRARY PULSE

We have derived the results of the previous sections for an action potential that varies linearly during depolarization and repolarization, a piecewise-linear approximation. In general the action potential does not have sharp changes in slope. We will now consider the general case and find that the results are very similar. For depolarization alone, we will again have a potential depending on the dipole moment. For a complete pulse the potential will depend on the area under the pulse curve.

Again, the axon is stretched along the x axis in an infinite, homogeneous conducting medium. Consider a small segment of axon between x and $x+dx$. If the current entering this segment at x is greater than the current leaving at $x+dx$, the difference must flow into the external medium. From Eq. (6.45b),

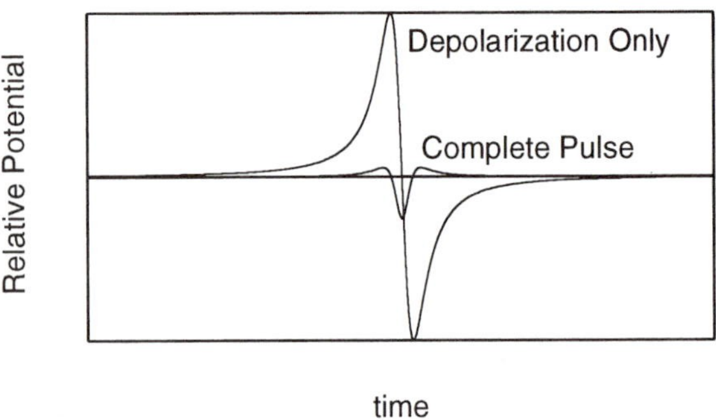

FIGURE 7.9. The potential far from the axon as a function of time as an impulse travels from left to right along the axis. The potential from the complete pulse has been multiplied by a factor of 10 in order to show it.

$$di_o = -di_i = -\frac{\partial i_i(x,t)}{\partial x}\,dx.$$

We can write Ohm's law for the axoplasm as

$$i_i(x,t) = -\pi a^2 \sigma_i \frac{\partial v_i}{\partial x}. \tag{7.17}$$

The current into the exterior medium from length dx of the axon is

$$di_o = \pi a^2 \sigma_i \frac{\partial^2 v_i}{\partial x^2}\,dx. \tag{7.18}$$

It is proportional to the derivative of the current along the axon with respect to x and therefore to the second derivative of the interior potential with respect to x. A small current source di_o generates a potential dv at some point in the external medium given by

$$dv = \frac{di_o}{4\pi\sigma_o r}. \tag{7.19}$$

The potential due to several current sources is obtained by superposition: $v = \int dv$.

If the radius of the cell stretched along the x axis is very small, the cell's influence can be replaced by a current distribution $di_o(x)$ along the x axis. The potential at any point **R** is obtained by integrating Eq. (7.19):

$$v(\mathbf{R}) = \int \frac{di_o}{4\pi\sigma_o r}. \tag{7.20}$$

Vector **R** specifies the point at which the potential is measured, and r is the distance from the measuring point to the point on the x axis where di_o is injected, as shown in Fig. 7.10. Combining this with Eq. (7.18) gives

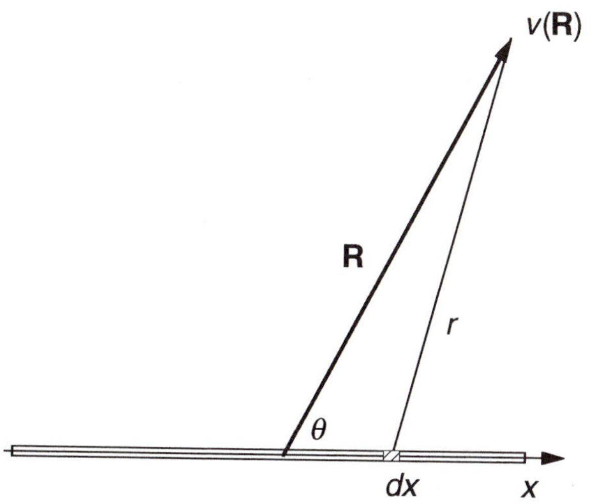

FIGURE 7.10. The potential $v(\mathbf{R})$ is obtained by integrating the potential due to current di_0 from each element dx of the cell.

$$v(\mathbf{R}) = \frac{\pi a^2 \sigma_i}{4\pi\sigma_o} \int \frac{\partial^2 v_i}{\partial x^2} \frac{1}{r}\,dx. \tag{7.21}$$

Although it is difficult to integrate Eq. (7.21) analytically, the integration can be done numerically. Figure 7.11 shows a computer program to carry out this integration for a crayfish lateral giant axon immersed in sea water. The axon radius is 60 μm. The conductivity ratio is $\sigma_i/\sigma_o = 0.2$. The action potential was measured by Watanabe and Grundfest (1961). Clark and Plonsey (1968) showed that it could be well represented by a sum of three Gaussians, with $v_i = 0$ taken to be the resting value. [Since only $\partial^2 v_i/\partial x^2$ enters into Eq. (7.21), the reference level does not matter.] The representation (with v in mV and x in mm) is

$$v_i(x) = 51e^{-[(x-5.4)/1.25]^2} + 72e^{-[(x-6.6)/1.876]^2} + 18e^{-[(x-8.6)/3.003]^2}. \tag{7.22}$$

This function corresponds to an impulse traveling to the *left*. It can be differentiated to obtain an analytic expression for $\partial^2 v_i/\partial x^2$. If the potential is being measured at exterior point (x,y_0), the value of r which is used in Eq. (7.21) is $r = [(x-x_0)^2 + y_0^2]^{1/2}$. The program allows four values of y_0 to be used. The smallest is taken to be a, the radius of the axon.

The results of calculating the exterior potential at $y_0 = a$ are shown in Fig. 7.12. The interior potential, shown in (a), has a peak value of 114 mV. The potential on the surface of the axon (b) ranges from +0.04 to −0.07 mV. In general the exterior potential is less than 0.1% of the interior potential. (This would be different if the extracellular fluid were not infinite.) The original calculation by Clark and Plonsey used much different mathematical techniques; however, the results are very similar. The results of their more accurate calculation are plotted in Fig. 7.13, which was supplied by Roth and Wikswo. Note that the smallest value of y, the distance from the axon, is 0.5 mm.

The approximation that the observer is far from the cell can also be applied to the general case. The physics is exactly the same as in the previous section for the triangular pulse, except that now the pulse has an arbitrary shape so current passes through the membrane at all points along the cell where the second derivative is nonzero. The calculation requires making the same type of approximations in order to evaluate the integral [Eq. (7.21)]. Referring to Fig. 7.10, we again use the law of cosines to write

$$r(x) = R[1 - 2(x/R)\cos\theta + x^2/R^2]^{1/2}.$$

We need to use this in Eq. (7.21). As in the previous section, we make a Taylor's-series expansion of $1/r$. To second order the result is

$$\frac{1}{r} \approx \frac{1}{R}\left(1 + \frac{x}{R}\cos\theta + \frac{1}{2}\frac{x^2}{R^2}(3\cos^2\theta - 1)\right). \tag{7.23}$$

```
//program ClarkAndPlonsey;                    Fig. 7.11
/*R K Hobbie    17 January, 1986*/
/*Converted to C, July 1996*/
/*This program integrates equation 7.21 for the problem which was
  solved by J. Clark and R. Plonsey, The extracellular potential
  field of the single active nerve fiber in a volume conductor.
  Biophys. J. 8:842-864 (1968)*/

/*The nerve pulse is a series of Gaussians used by Clark and
  Plonsey to fit data of Watanabe and Grundfest, J. Gen. Physiol.
  45:267 (1961)*/

/*This program makes use of the Romberg integration function
  qromb, from the Numerical Recipes in C, 2nd ed.*/

#include<stdio.h>
#include<stdlib.h>
#include<math.h>
#include "nr.h"        //N.R. in C header file

/*Global Variables*/
double x0,y0;          //coordinates of observation point

float f(float x)
{
/*Calculates the integrand d2v/dx2 divided by r
   Uses common variables x0 , y0 - - Observation point in m*/
double   xx, r, d2v, temp;

d2v = 0;
xx = (x - 0.0054) / 1.25e-3;
temp = (2 * 51 / (1.25e-3 * 1.25e-3)) * exp(-(xx * xx));
d2v = d2v + temp * (2 * xx * xx - 1);
xx = (x - 0.0066) / 1.876e-3;
temp = (2 * 72 / (1.876e-3 * 1.876e-3)) * exp(-(xx * xx));
d2v = d2v + temp * (2 * xx * xx - 1);
xx = (x - 0.0086) / 3.003e-3;
temp = (2 * 18 / (3.003e-3 * 3.003e-3)) * exp(-(xx * xx));
d2v = d2v + temp * (2 * xx * xx - 1);
r = sqrt((x0 - x) * (x0 - x) + y0 * y0);
return d2v / r;
}

void main()
{
const
   double SigRatio = 0.2, //Interior/exterior conductivity
   a = 6.0e-5;            //axon radius in m
float
```

FIGURE 7.11. The computer program used to calculate the exterior potential by integrating Eq. (7.21) for the problem first solved by Clark and Plonsey (1968). The program uses Romberg integration procedure `qromb` from Press *et al.* (1992).

```
   xstart, xfinish,        //limits of integration
   Integral;
double
   y[4],                   //Calculate at four different distances
   potential, xx           //Exterior potential
  Transmemb;               //Transmembrane potential
int i;
FILE *ofp;        //Outputfile Pointer
if (!(ofp = fopen("Plonsey.out","w")))      //Open output file
{
   printf("cannnot open output file\n");
   exit(1);
}
xstart = 0.0;
xfinish = 0.02;
fprintf(ofp,"    x   \t    v   ");
printf("     x           v    ");
for (i=0; i<4; i++)
{
   y[i] = pow(2,i) * a;
   fprintf(ofp,"\t%9.3e",y[i]);
   printf(" %9.3e ",y[i]);
}
fprintf(ofp,"\n\n");
printf("\n\n");
for(x0 = 0.001; x0 < 0.012; x0 += 0.00025)
{
   xx = (x0 - 0.0054) / 1.25e-3;
   Transmemb = 51 * exp(-xx * xx);
   xx = (x0 - 0.0066) / 1.876e-3;
   Transmemb = Transmemb + 72 * exp(-xx * xx);
   xx = (x0 - 0.0086) / 3.003e-3;
   Transmemb = Transmemb + 18 * exp (-xx*xx);
   fprintf(ofp,"%9.3e\t%10.3e",x0,Transmemb);
   printf("%9.3e %10.3e ",x0,Transmemb);
   for (i=0; i<4 ; i++)
   {
     y0 = y[i];
     Integral = qromb(f,xstart,xfinish); //N.R. in C function Call
     potential = Integral * a * a * SigRatio / 4.0;
     fprintf(ofp,"\t%10.3e",potential);
     printf("%10.3e ",potential);
   }
   fprintf(ofp,"\n");
   printf("\n");
   }
fclose(ofp);
}
```

FIGURE 7.11. (Continued).

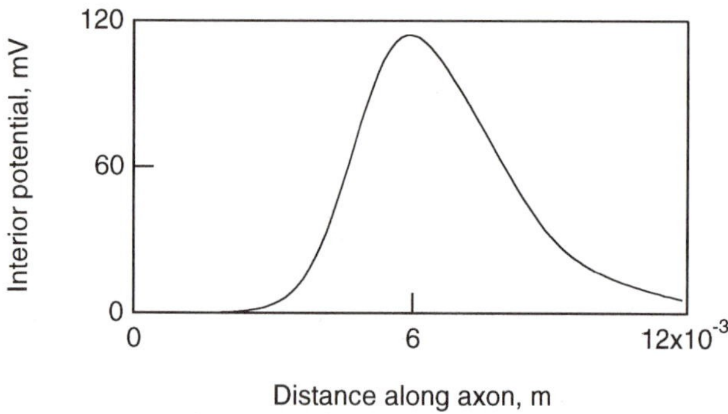

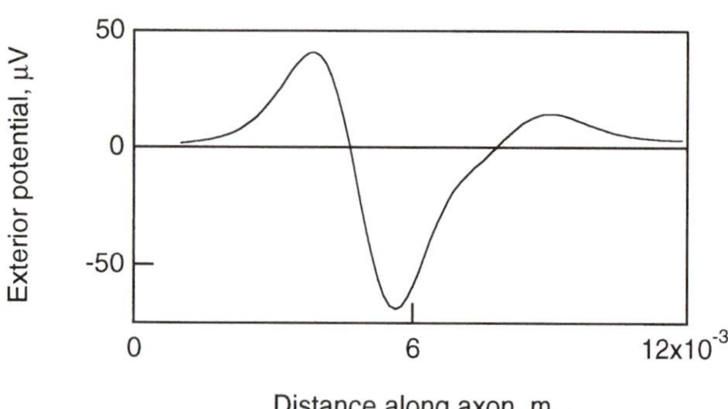

FIGURE 7.12. (a) The transmembrane potential used for the calculation in the program of Fig. 7.11. The impulse is traveling to the left. (b) The exterior potential along the axon calculated by the program.

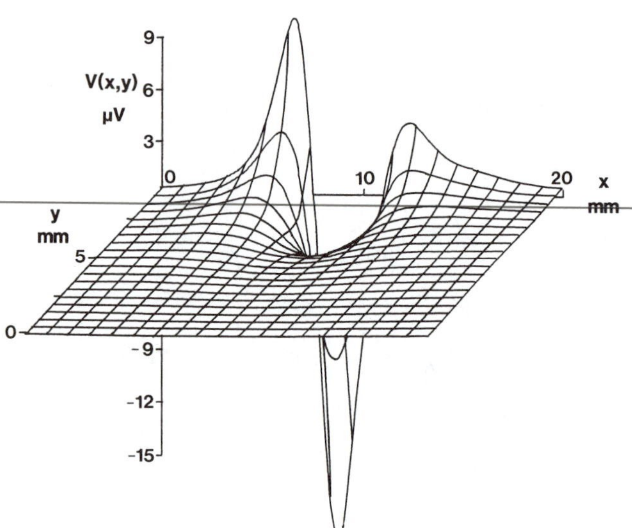

FIGURE 7.13. The exterior potential for the same problem calculated using the more accurate method of Clark and Plonsey (1968). The smallest distance from the axon is $y = 0.5$ mm. The drawing was supplied by B. Roth and J. Wikswo of Vanderbilt University.

The expression for $v(\mathbf{R})$ becomes

$$v(\mathbf{R}) = \frac{\pi a^2 \sigma_i}{4\pi\sigma_o}\left[\frac{1}{R}\int_{x_1}^{x_2}\frac{\partial^2 v_i}{\partial x^2}\,dx + \frac{\cos\theta}{R^2}\int_{x_1}^{x_2}\frac{\partial^2 v_i}{\partial x^2}\,x\,dx + \frac{3\cos^2\theta - 1}{2R^3}\int_{x_1}^{x_2}\frac{\partial^2 v_i}{\partial x^2}\,x^2 dx\right]. \tag{7.24}$$

There are three integrals inside the square brackets that we must evaluate. Take limits of integration x_1 and x_2 to be points where $\partial v_i/\partial x = 0$. The first integral is $\partial v_i/\partial x$ and vanishes. The second integral can be integrated by parts. Since $\partial v_i/\partial x = 0$ at the end points the second integral is $v_i(x_1) - v_i(x_2)$. The third integral is integrated by parts twice and is

$$\int_{x_1}^{x_2}\frac{\partial^2 v_i}{\partial x^2}\,x^2 dx = \left[x^2\frac{\partial v_i}{\partial x}\right]_{x_1}^{x_2} - 2[xv(x)]_{x_1}^{x_2} + 2\int_{x_1}^{x_2} v_i(x)\,dx. \tag{7.25}$$

The first term of this vanishes because of the way the end points were chosen.

We now apply these results to Eq. (7.24) in two cases. The first is the case of depolarization only, which is useful in considering the electrocardiogram. Set up the coordinate system so the origin is someplace in the impulse where $\partial v_i/\partial x = 0$. The total change in v_i is Δv_i. Then $x_1 = 0$, $v_i(x_1) = \Delta v_i$, $v_i(x_2) = 0$. The first nonvanishing term of Eq. (7.24) requires only the second integral:

$$v(\mathbf{R}) = \frac{\pi a^2 \sigma_i}{4\pi\sigma_o}\frac{\cos\theta}{R^2}\Delta v_i. \tag{7.26}$$

We obtained this result in a special case as Eq. (7.11).

In the second case we consider the complete pulse, and we take x_1 to the left of the pulse and x_2 to the right. The first integral in Eq. (7.24) still vanishes. Now the second integral also vanishes because $v_i(x_1) = v_i(x_2) = v_{\text{rest}}$ and $\Delta v_i = 0$. It is necessary to use the third integral, Eq. (7.25). The first term in Eq. (7.25) vanishes. The second and third terms must be considered together. Rewrite the potential in terms of departures from the resting potential:

$$v_i(x) = v_{\text{rest}} + v_{\text{depol}}(x).$$

The second term in Eq. (7.25) is $-2v_{\text{rest}}(x_2 - x_1)$. The third term of Eq. (7.25) is

$$2\int_{x_1}^{x_2} v_{\text{depol}}(x)dx + 2(x_2 - x_1)v_{\text{rest}}.$$

Adding these gives

$$v(\mathbf{R}) = \frac{2\pi a^2 \sigma_i}{4\pi\sigma_o}\frac{1}{R^3}\frac{3\cos^2\theta - 1}{2}\int_{x_1}^{x_2}[v_i(x) - v_{\text{rest}}]dx. \tag{7.27}$$

Again, we saw a special case of this as Eq. (7.16).

Note the progression in these results. When we are looking at one corner of a depolarization pulse, we have a current source or sink, and the potential is proportional to $1/R$ [Eq. (7.1)]. We do not find this situation in physiology, because the potential would have to keep rising forever. When we consider the entire depolarization portion of the wave form, the potential is proportional to $1/R^2$, as in Eqs. (7.11) or (7.26). This is a good model for the electrocardiogram because the repolarization does not commence until the entire heart is depolarized. When the entire pulse is considered, the potential is proportional to $1/R^3$ as in Eqs. (7.16) or (7.27). This is a good model for nerve conduction. The potential is considerably less in this case because of the $1/R^3$ dependence. This is an example of a technique called a multipole expansion. Generally, defining $\xi = x/R$, one can make the expansion

$$\frac{1}{(1-2\xi\cos\theta+\xi^2)^{1/2}} = P_0 + \xi P_1 + \xi^2 P_2 + \xi^3 P_3 + \cdots, \tag{7.28}$$

where the P_n are functions of $\cos\theta$ and are called *Legendre polynomials*. The first few Legendre polynomials are

$$\begin{aligned} P_0 &= 1, \\ P_1 &= \cos\theta, \\ P_2 &= \tfrac{1}{2}(3\cos^2\theta - 1), \\ P_3 &= \tfrac{1}{2}(5\cos^3\theta - 3\cos\theta). \end{aligned} \tag{7.29}$$

All of these calculations are based on a model in which the current flows parallel to the axis of the cell, passes through the membrane, and then returns in the extracellular conducting medium. This model is called the line approximation. It is, of course, impossible for current to pass through the membrane if it always flows parallel to the axis of the cell. It is possible to do an exact calculation in which $\mathbf{j}$ has a radial component everywhere as well as one parallel to the axis of the cell. (See Sec. 7.11 for a description of how this is done.) Trayonava, Henriquez, and Plonsey (1990) have compared the exact solution with two approximations, one of which is the line approximation. The line approximation is quite good if the radius of the cell is much smaller than the distance along the cell over which the depolarization takes place.

7.5 ELECTRICAL PROPERTIES OF THE HEART

There are many similarities between myocardial cells and nerve cells: a membrane separates extracellular and intracellular fluids; the concentrations of the principal ions are about the same; except for a small amount of charge on the membrane, the extracellular and intracellular fluids are electrically neutral; and selective ion channels are responsible for the initiation and propagation of the action potentials. There are also major differences: myocardial cells in mammals are about 100 μm long and 10 μm in diameter. The interiors of neighboring cells are connected through *gap junctions*, so current and ions flow directly from one cell to another [Beyer *et al.* (1995)]. This continuum of cells is called a *syncytium*. There are also important differences in the details of the ion currents. We continue for now to use the simple model of long, one-dimensional cells. Refinements to this model are discussed in Sec. 7.10.

In the resting state, the cytoplasm (fluid inside the cell) of an atrial cell is at about -70 mV, while that in a ventricular cell is at about -90 mV. When a cell depolarizes, the action potential lasts for 100–300 ms, depending on the species. A ''typical'' pulse is shown in Fig. 7.14. There are variations in pulse shape between species and also in different parts of the heart. The initial rapid depolarization in heart muscle cells is caused by an inward sodium current (phase 0 on the curve). The fall at phase 1 is caused by a transient outward potassium current. This current is small in *endocardium* (near the *inside* of the heart) but is prominent enough in the outer layers of the heart (*epicardium*) so that there can be a ''spike and dome'' shape to the potential [Campbell *et al.* (1995); Antzelevitch *et al.* (1995)]. This is followed by a Ca^{2+} influx that maintains the plateau (phase 2) of the action potential. The ''slow'' potassium channels finally open [Kass (1995)], and potassium efflux causes repolarization (phase 3). During phase 4 the original ion concentrations are restored.

The heart can beat in isolation. If it is removed from an animal and bathed in nutrient solution, it continues to beat spontaneously. With each beat, a wave of depolarization sweeps over the heart, and it contracts. The wave is initiated by some specialized fibers located in the right atrium called the *sinoatrial node* (SA node). As was mentioned in Sec. 6.18, the SA node does not have the usual sodium channels, and the depolarization is due to calcium. The shape of the SA node potential is much more like Fig. 6.51 than Fig.

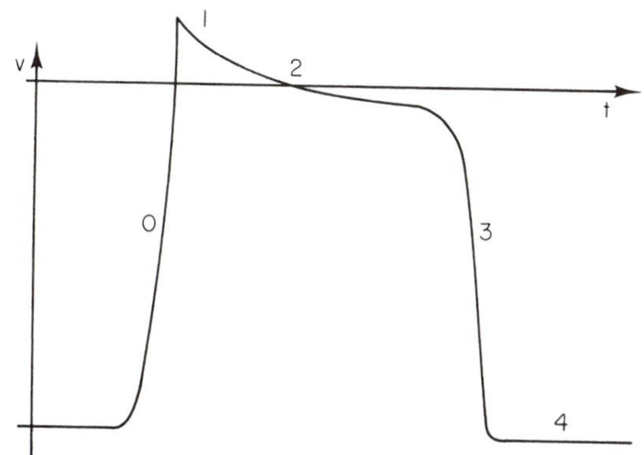

FIGURE 7.14. Depolarization and repolarization of a myocardial cell. The depolarization lasts 200–300 ms. The resting potential is about -90 mV.

7.14. Depolarization of the SA node causes rapid depolarization in the rest of the heart. The rate at which the SA node repetitively "fires" is increased by the sympathetic nerves to the heart (which release norepinephrine) and decreased by the parasympathetic nerves (which release acetylcholine). Devices that produce such periodic firing are common in physics and engineering. They are called "free-running relaxation oscillators."

Figure 7.15 shows how the depolarization progresses through the heart. Once the SA node has fired, the depolarization sweeps across both atria (a,b). When the atria are completely depolarized (c) there is no depolarization wavefront. The atria are separated from the ventricles by fibrous connective tissue that does not transmit the impulse. The only connection between the atria and the ventricles is some conduction tissue called the *atrioventricular node* (AV node). After passing through the AV node, the depolarization spreads rapidly over the ventricles on the conduction system—a set of specialized muscle cells on the inner walls of the ventricles (d,e) and finally through the myocardium of each ventricle to the outer wall (e,f,g). The network consists of the common bundle (or bundle of His), the left and right bundles, and the fine network of Purkinje fibers. The AV node will spontaneously depolarize at a rate of about 50 beats min^{-1}; it usually does not because it is triggered by the more rapid beating of the atria. In well-trained athletes, the resting pulse rate can be so low that the AV node fires spontaneously, giving rise to what are called nodal escape beats. These are physiologic and no cause for concern.

Sodium conductivity increases as the transmembrane potential rises. This means that electrotonus assists in the propagation of the signal. The potassium permeability also increases with voltage. Moreover, as the voltage falls, it approaches the potassium Nernst potential, and the current falls for this reason as well. Therefore electrotonus does not assist in the propagation of the repolarization, which is a local phenomenon.

Normally, the depolarization progresses through the myocardium in an orderly fashion. It is followed by repolarization, and after a brief *refractory period* the heart is ready to beat again. During the refractory period the cells do not respond to a stimulus. It is possible in abnormal situations for a wave of depolarization to travel in a closed path through the myocardium. This closed path, called a *reentrant circuit*, surrounds an obstacle such as scar tissue, the aorta, or the pulmonry artery. It can also surround an area that simply has different conduction properties. If travel around the reentrant circuit takes long enough for the refractory period to be over, the cells depolarize again, and the wave can continue to travel on its closed path. Reentrant excitation is thought to be responsible for several kinds of heart disease, including most life-threatening ventricular tachycardias Another type of reentrant excitation is spiral waves that occur because of the non-linear nature of the myocardium [Davidenko (1995)]. (Such nonlinear behavior will be discussed a bit more in Chap. 10.) It is also possible for a reentrant wave to leave behind a refractory state that blocks normal conduction.

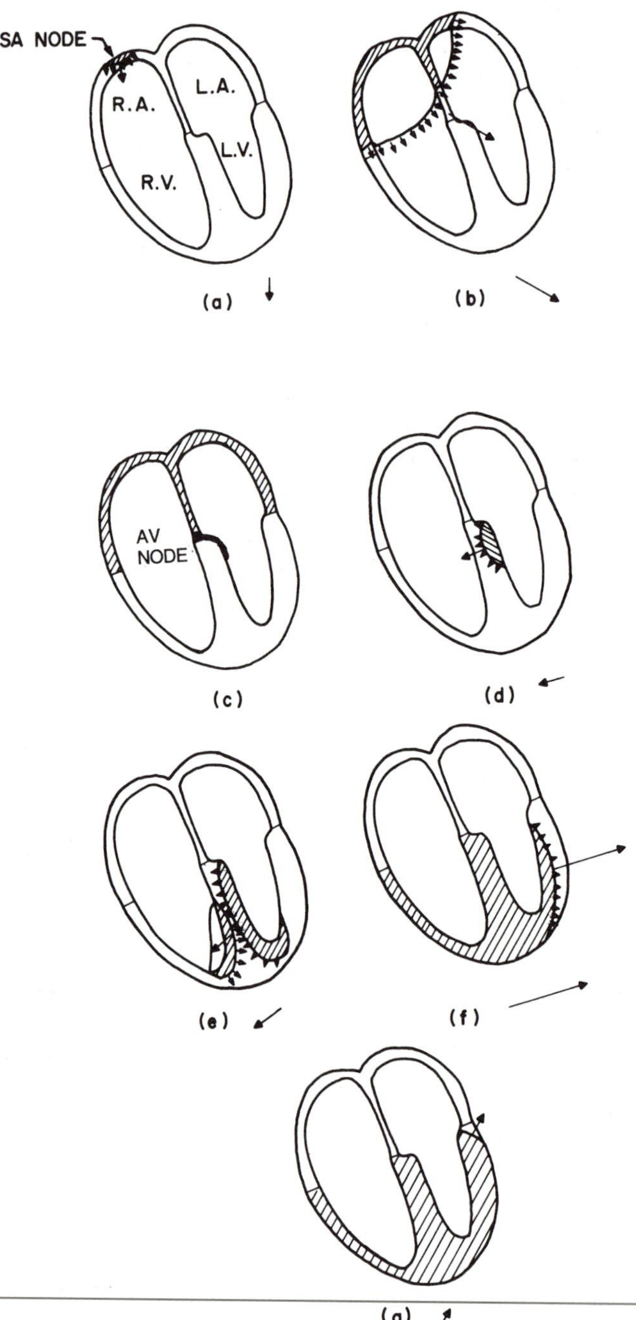

FIGURE 7.15. The wave of depolarization sweeping over the heart. Atrial and ventricular muscle are not connected except through the AV node. (a) Depolarization beginning at the SA node. (b) Atria nearly depolarized. (c) The AV node is conducting. (d) Beginning of depolarization of the left ventricle. (e, f) Continuing ventricular depolarization. (g) Ventricular depolarization nearly complete. From R. K. Hobbie (1973). The electrocardiogram as an example of electrostatics. *Am. J. Phys.* **41**: 824–831.

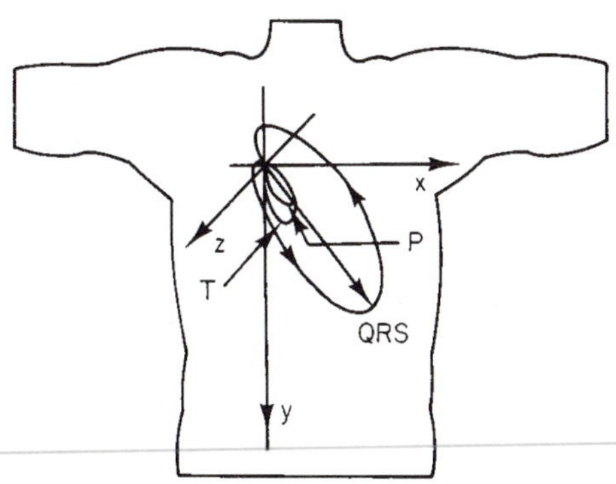

FIGURE 7.16. Locus of the tip of the total current-dipole vector during the cardiac cycle.

7.6 THE CURRENT-DIPOLE VECTOR OF THE HEART AS A FUNCTION OF TIME

Each myocardial cell depolarizes and repolarizes during the cardiac cycle. The total current-dipole vector at any instant is the sum of the vectors for all the cells in the heart. This section considers how the total current-dipole vector changes with time as the myocardium depolarizes and then repolarizes. Initially, all the cells are completely polarized (resting) and there is no net dipole moment. The cells begin to depolarize near the SA node, and a wave of depolarization sweeps across the atria. For each myocardial cell, the dipole vector points in the direction that the wave of depolarization is traveling[2] and moves along the cell with the depolarization wave. These vectors for all the cells that are depolarizing constitute an advancing wave that moves across the heart.

The potential at the point of observation can be calculated by applying Eq. (7.13) for each cell. Vector **r** is the vector from the cell to the point of observation and is different for each cell. However, we will assume for now that the observation point is so far from the heart that all points in the myocardium are nearly equidistant from it. This is a terrible assumption; later we will be more realistic. It allows us to speak of the *instantaneous total current-dipole moment*, which is the sum of the dipole moments of all depolarizing cells at that instant.

The locus of the tip of the total dipole moment during the cardiac cycle is shown in Fig. 7.16 for a typical case. The x axis points to the patient's left, the y axis to the patient's feet, and the z axis from back to front. The small loop labeled P occurs during atrial depolarization. The loop labeled QRS is the result of ventricular depolarization. Ventricular repolarization gives rise to the "T wave." Atrial repolarization is masked by ventricular depolarization. A plot of the x, y, and z components of **p** is shown in Fig. 7.17. These components are typical; there can be considerable variation in the directions of the loops in Fig. 7.16.

[2]If one takes into account the anisotropies in the conductivities of myocardial tissue discussed in Sec. 7.10, the depolarization does not travel in the direction that **p** points. We ignore this for now.

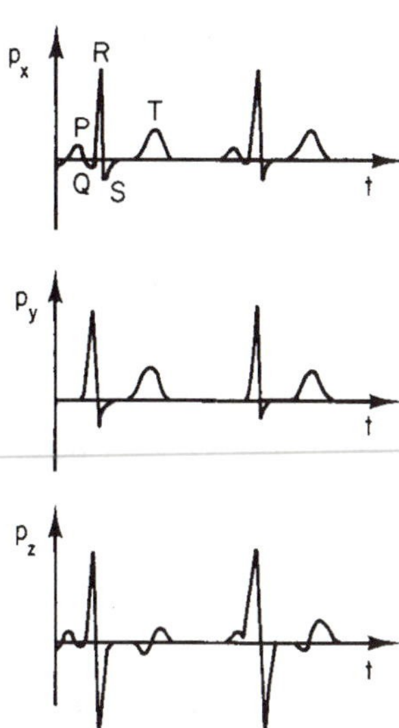

FIGURE 7.17. The three components of the total current-dipole vector as a function of time.

7.7 THE ELECTROCARDIOGRAPHIC LEADS

We turn next to how the electrocardiographic measurements are made. We will use a model in which the torso is modeled as an infinite homogeneous conductor and will continue to assume that every myocardial cell is the same distance from each electrode. Both assumptions are wrong, of course, and later we will improve upon them.

The potential at **r** from vector **p** is given by Eq. (7.13). The potential difference at two points at positions $\mathbf{r}_1$ and $\mathbf{r}_2$, each a distance r from the cell, is therefore (see Fig. 7.18)

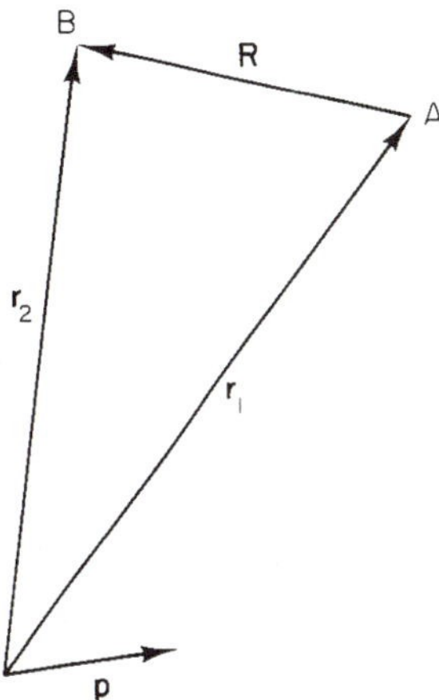

FIGURE 7.18. Geometry for calculating the potential difference of **p** between points *A* and *B*.

$$v(\mathbf{r}_2,\mathbf{r}_1)=\frac{\mathbf{p}\cdot(\mathbf{r}_2-\mathbf{r}_1)}{4\pi\sigma_0 r^3}.$$

Denoting $\mathbf{r}_2-\mathbf{r}_1$ by $\mathbf{R}$, we have

$$v=\frac{\mathbf{p}\cdot\mathbf{R}}{4\pi\sigma_0 r^3}. \tag{7.30}$$

The potential difference between two electrodes separated by a displacement **R** *and equidistant from the current-dipole vector* **p** *measures the instantaneous projection of the vector* **p** *on* **R**.

If the depolarization can be described by a single current-dipole vector, only three measurements are needed in principle, corresponding to the projections on three perpendicular axes. The standard electrocardiogram (ECG) records 12 potential differences using nine electrodes. There are many reasons for this. The body is not an infinite, homogeneous conductor, and the relationship between cellular dipole moments and the potential is more complicated than our model; to convert the three perpendicular components to the instantaneous values of **p** would require a mathematical reconstruction; and the electrodes are not far away compared to the size of the heart. With 12 recorded potential differences, it is fairly easy to interpret the electrocardiogram by inspection.

The first three electrodes are placed on each wrist and the left leg. The limbs serve as extensions of the wires, so that the potential is measured where the limbs join the body. This is a major correction to our crude model that the heart is in an infinite conducting medium. If the subject were immersed in a conducting medium such as sea water, movement of the arms would change the size of the ECG signal because it would change **R**. In air, however, movement of the arms does not change the size of the signal. The simplest correction to explain this is to say that **R** for the two arm electrodes goes from shoulder to shoulder. These three electrodes measure potential differences between three points located approximately as shown in Fig. 7.19. The dimensions are for a typical adult. The three potential differences are called *limb leads I, II,* and *III:*

$$\begin{aligned}\mathrm{I}&=v_B-v_A,\\ \mathrm{II}&=v_C-v_A,\\ \mathrm{III}&=v_C-v_B.\end{aligned} \tag{7.31}$$

In the approximation used here, the voltage difference I is proportional to the projection of **p** on $\mathbf{R}_\mathrm{I}$, and so forth. These leads measure the projections of **p** on the three vectors $\mathbf{R}_\mathrm{I}$, $\mathbf{R}_\mathrm{II}$, and $\mathbf{R}_\mathrm{III}$ of Fig. 7.19.

It is customary also to combine these three potentials in a slightly different way to obtain projections of **p** on three other directions. These combinations are called the *augmented limb leads*. They contain no information that was not already present in the limb leads, but the six signals are easier to interpret by inspection. The combinations are

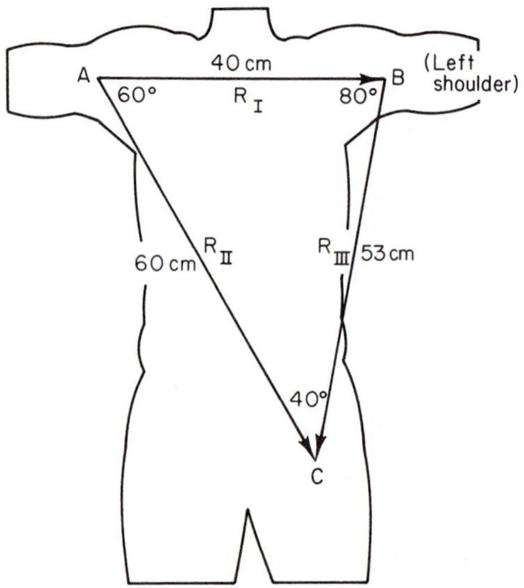

FIGURE 7.19. Vectors connecting the three electrodes for a typical patient. The limbs are extensions of the leads of the electrocardiograph machine.

$$\begin{aligned}aVR&=v_A-\tfrac{1}{2}(v_B+v_C)=-\tfrac{1}{2}(\mathrm{I}+\mathrm{II}),\\ aVL&=v_B-\tfrac{1}{2}(v_A+v_C)=\tfrac{1}{2}(\mathrm{I}-\mathrm{III}),\\ aVF&=v_C-\tfrac{1}{2}(v_A+v_B)=\tfrac{1}{2}(\mathrm{II}+\mathrm{III}).\end{aligned} \tag{7.32}$$

These are proportional to the projections of **p** on vectors $\mathbf{R}_L$, $\mathbf{R}_R$, and $\mathbf{R}_F$ of Fig. 7.20. The subscripts refer to the fact that the vectors point toward the left shoulder, right shoulder, and foot, respectively.

The six lines in Fig. 7.20 are spaced approximately every 30° in the frontal plane. Many texts argue that the leads are

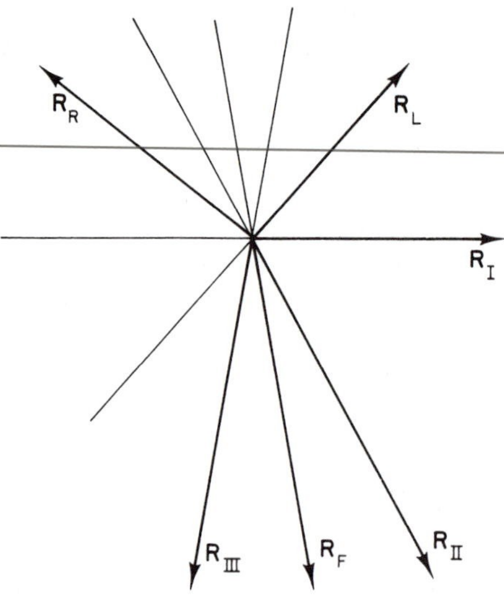

FIGURE 7.20. The six directions in the frontal plane defined by the limb leads and the augmented limb leads. The angles are for the same subject as in Fig. 7.19.

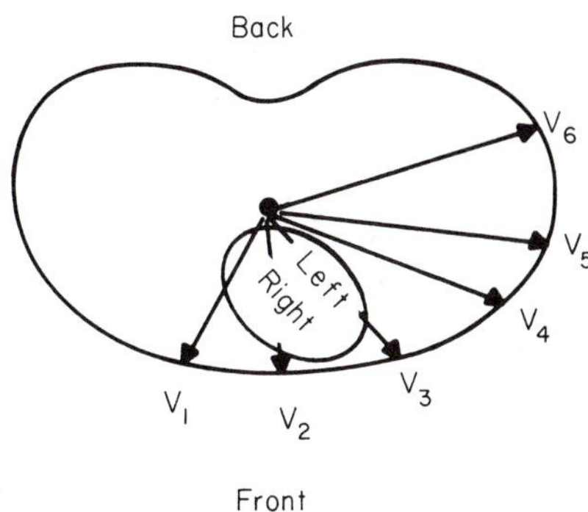

FIGURE 7.21. The location of the precordial leads and the directions of the components of **p** which they measure. From R. K. Hobbie (1973). The electrocardiogram as an example of electrostatics. *Am. J. Phys.* **41**: 824–831.

spaced exactly every 30° and that the triangle of Fig. 7.19 is an equilateral triangle (Einthoven's triangle). While the directions are not far from 30°, this assumption is not really necessary. Physicians often want to know the direction of **p** at some point during the cardiac cycle, or the average direction of **p** during the QRS wave (ventricular depolarization). With six directions measured, this can be determined by inspection.

These six leads measure projections in the frontal plane. It is also necessary to have at least one projection in a plane perpendicular to the frontal plane. It is customary to place six leads across the chest wall in front of the heart; they are called the *precordial leads*. Their locations are shown in Fig. 7.21. The potential difference is measured between each precordial electrode and the average of v_A, v_B, and v_C. A lead therefore measures the projection of **p** on a vector from the center of triangle ABC to the electrode for that lead. This fact is not obvious, and in fact is true only if differences in $1/r^2$ are neglected. To see that it is true with the appropriate approximation, pick an arbitrary point O and from it construct vectors $\mathbf{R}_A$, $\mathbf{R}_B$, $\mathbf{R}_C$, and $\mathbf{R}_D$ to the points A, B, and C of Fig. 7.22 and to the precordial electrode at D. The desired potential is

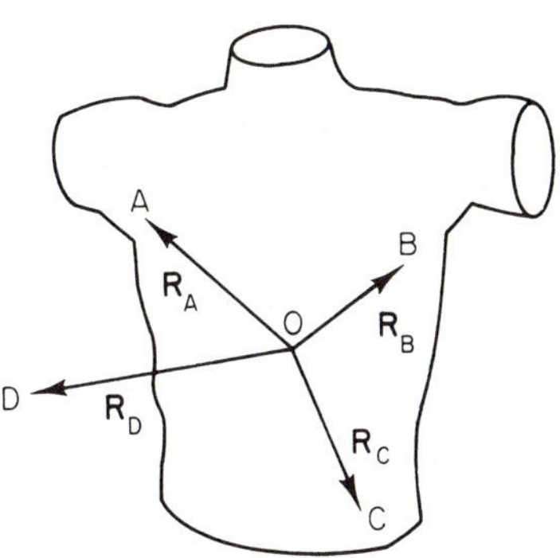

FIGURE 7.22. A perspective drawing of the vectors used to calculate the potential in a precordial lead.

$$v = v_D - \frac{v_A + v_B + v_C}{3}.$$

It can be calculated using Eq. (7.30) for each term:

$$v = \frac{1}{4\pi\sigma_0}\left[\frac{\mathbf{p}\cdot\mathbf{R}_D}{R_D^3} - \frac{1}{3}\left(\frac{\mathbf{p}\cdot\mathbf{R}_A}{R_A^3} + \frac{\mathbf{p}\cdot\mathbf{R}_B}{R_B^3} + \frac{\mathbf{p}\cdot\mathbf{R}_C}{R_C^3}\right)\right].$$

So far, the location of O is arbitrary. If it is picked to be at the center of the triangle, then $\mathbf{R}_A + \mathbf{R}_B + \mathbf{R}_C = \mathbf{0}$. (This is the definition of "center.") It will also be true that $R_A \approx R_B \approx R_C$ so that the term in large parentheses will vanish. The desired potential difference is then

$$v = \frac{1}{4\pi\sigma_0}\frac{\mathbf{p}\cdot\mathbf{R}_D}{R_D^3}.$$

In this approximation, each precordial lead measures the projection of **p** on a vector from the center of the triangle ABC to the electrode. The amplitude of the signal will be larger than for the limb leads, because $R_D < R_A$. Some of the precordial leads are quite close to the heart. The assumption that r is the same for all parts of the myocardium is not valid. Because of the factor $1/r^2$, the greatest contribution to the potential comes from the closest regions of myocardium. A lead is said to "look at" the myocardium closest to it.

7.8 SOME ELECTROCARDIOGRAMS

A normal electrocardiogram is shown in Fig. 7.23. When **p** has its greatest magnitude during the QRS wave, it is nearly parallel to $\mathbf{R}_{\mathrm{II}}$. There is almost no signal in aVL, which is perpendicular to $\mathbf{R}_{\mathrm{II}}$.

Compare this to Fig. 7.24, which shows the electrocardiogram for a patient with *right ventricular hypertrophy*, an

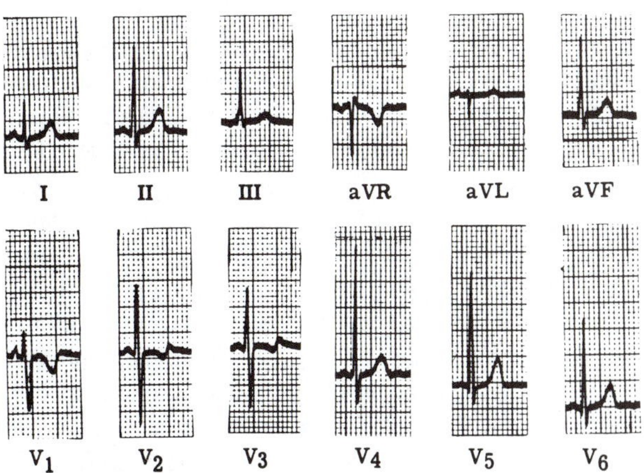

FIGURE 7.23. A normal electrocardiogram. The large divisions are 0.5 mV vertically and 0.2 s horizontally. From R. K. Hobbie (1973). The electrocardiogram as an example of electrostatics. *Am. J. Phys.* **41**: 824–831. The electrocardiogram was supplied by Prof. James H. Moller, M.D.

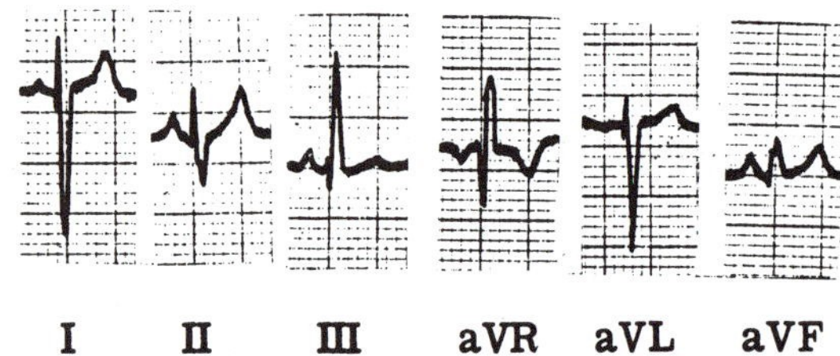

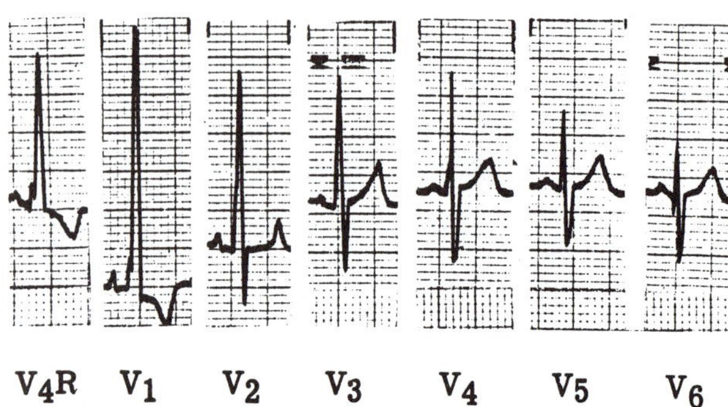

FIGURE 7.24. The electrocardiogram of a patient with right ventricular hypertrophy. Reproduced by permission from R. K. Hobbie (1973). The electrocardiogram as an example of electrostatics. *Am. J. Phys.* **41**: 824–831. The electrocardiogram was supplied by Prof. James H. Moller, M.D.

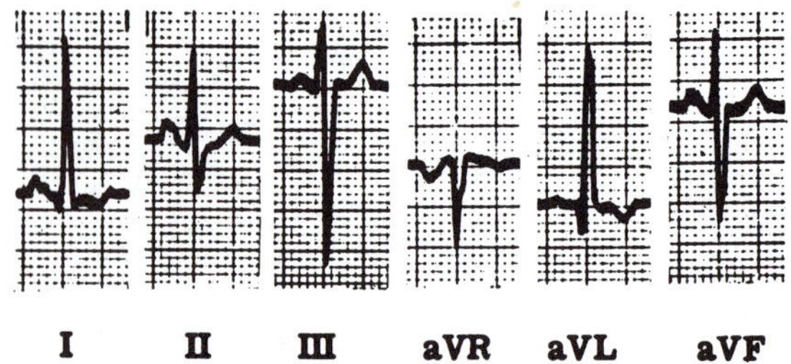

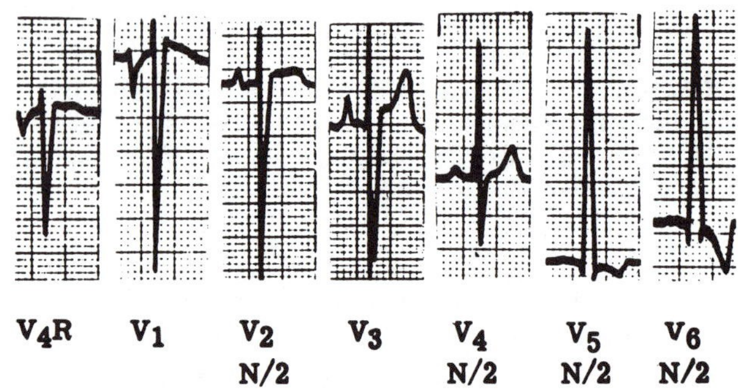

FIGURE 7.25. The electrocardiogram for a patient with left ventricular hypertrophy. The electrocardiogram was supplied by Prof. James H. Moller, M.D.

enlargement and thickening of the right ventricle. Because of the greater right ventricular volume **p** points to the right during the QRS wave, so that the QRS signal is negative in lead I. Lead aVF shows that there is very little vertical component of **p** during the QRS wave. The precordial leads V_1 and V_2 show the strongest signals, because the right ventricle faces the front of the body. In this case an extra lead V_4R has been used, which is symmetrical with V_4 but on the right side of the body.

The electrocardiogram in Fig. 7.25 from a patient with left ventricular hypertrophy. The thicker left ventricular wall causes the QRS dipole to point to the left. As a result, lead I has an abnormally high peak, aVL is large and positive, V_2 is negative, and V_4, V_5, and V_6 have very large positive peaks. These last four leads are shown at half scale.

A fault in the conduction system known as a *bundle branch block* causes the depolarization wave to travel through the myocardium rather than over the conduction system. Since the speed of propagation in myocardium is slower than that in the conduction system, the depolarization takes longer than usual. An electrocardiogram for a patient with *right bundle branch block* (a block in the bundle for the right ventricle) is shown in Fig. 7.26. The effect is most striking in leads that are most sensitive to the right ventricle: precordial leads 1 and 2. In V_1 the early part of the QRS wave has the usual biphasic, up–down pattern as the left ventricle depolarizes. This is followed by a large and prolonged vector pointing to the right, as the right ventricle slowly depolarizes. Lead V_2 shows a strong and prolonged bipolar signal as the right ventricle depolarizes.

7.9 WHY LEAD I IS NOT EXACTLY HORIZONTAL

Specialists in the interpretation of electrocardiograms are aware that $\mathbf{R}_\mathrm{I}$ is not horizontal and in the frontal plane, but points posteriorly, up, and to the left. This can be explained with our model by the fact that $R_A \neq R_B$. The potential for lead I is

$$v_B - v_A = \frac{1}{4\pi\sigma_0}\left[\frac{\mathbf{p}\cdot\mathbf{R}_B}{R_B^3} - \frac{\mathbf{p}\cdot\mathbf{R}_A}{R_A^3}\right].$$

Adding and subtracting the quantity $\mathbf{p}\cdot\mathbf{R}_B/R_A^3$ gives

$$v_B - v_A = \frac{1}{4\pi\sigma_0}\left[\frac{\mathbf{p}\cdot(\mathbf{R}_B-\mathbf{R}_A)}{R_A^3} - \mathbf{p}\cdot\mathbf{R}_B\left(\frac{1}{R_B^3} - \frac{1}{R_A^3}\right)\right].$$

The second term can be multiplied by R_A^3/R_A^3, and we get

$$v_B - v_A = \frac{1}{4\pi\sigma_0 R_A^3}\left[\mathbf{p}\cdot\mathbf{R}_\mathrm{I} + \left(\frac{R_A^3}{R_B^3} - 1\right)\mathbf{p}\cdot\mathbf{R}_B\right]. \quad (7.33)$$

The expression in square brackets is of the form

$$\mathbf{p}\cdot\mathbf{R}_\mathrm{I}',$$

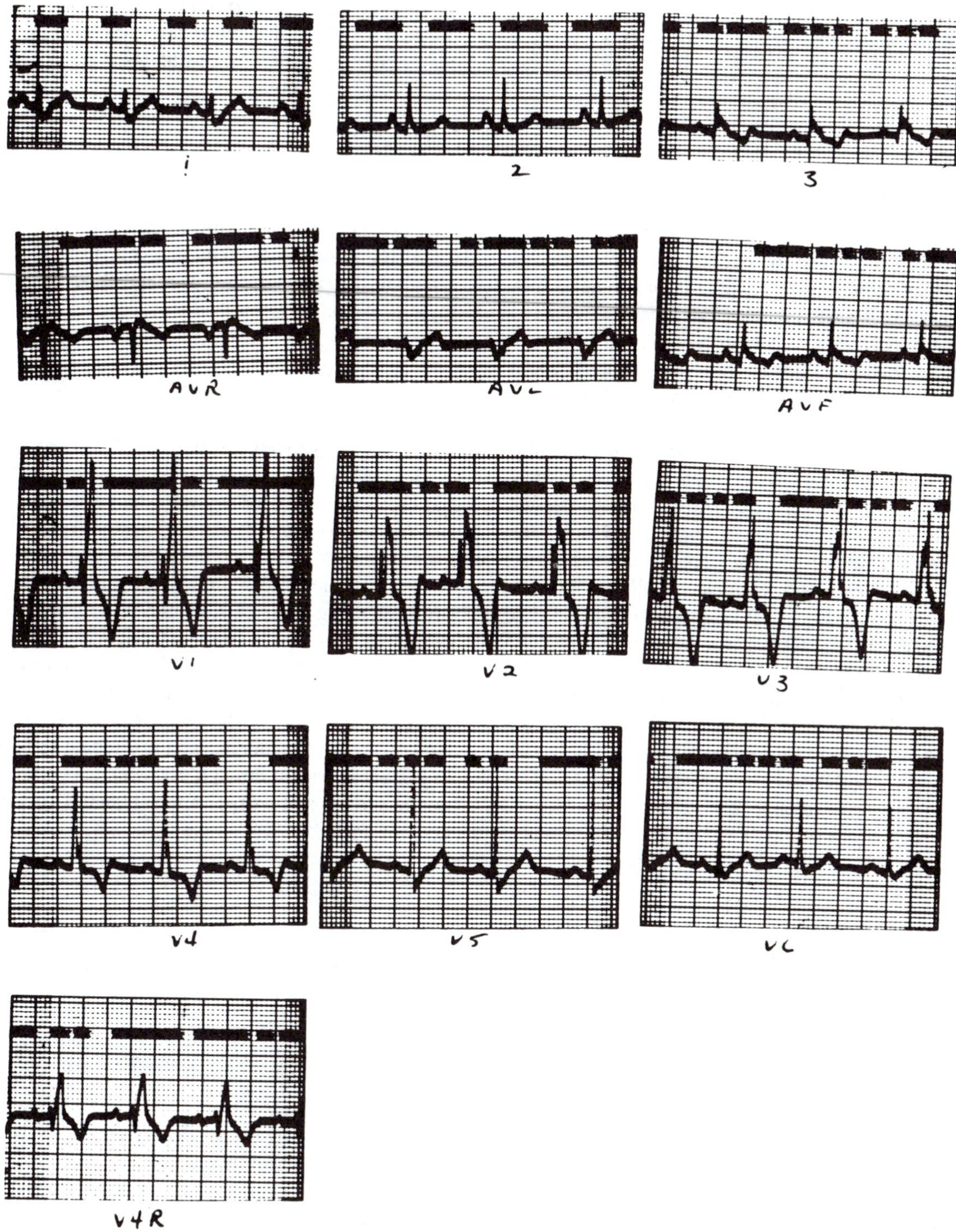

FIGURE 7.26. The electrocardiogram for a patient with right bundle branch block. The electrocardiogram was supplied by Prof. James H. Moller, M.D.

where

$$\mathbf{R}_{\mathrm{I}}' = \mathbf{R}_{\mathrm{I}} + \left(\frac{R_A^3}{R_B^3} - 1 \right) \mathbf{R}_B . \tag{7.34}$$

For the same patient used earlier, with point O taken at the center of the cardiac outline [Goss, (1954, p. 75)], we get $R_A = 0.24$ m, $R_B = 0.21$ m, and $R_C = 0.41$ m. Then $R_A^3/R_B^3 - 1 = 0.5$, and the axis is deviated 8°.

Similar calculations can be done for the other axes. The results are

$$\mathbf{R}'_{\mathrm{III}}=\left(\frac{R_A}{R_C}\right)^3\mathbf{R}_{\mathrm{III}}+\left[\left(\frac{R_A}{R_C}\right)^3-\left(\frac{R_A}{R_B}\right)^3\right]\mathbf{R}_B, \quad (7.35)$$

$$\mathbf{R}'_{\mathrm{II}}=\mathbf{R}_{\mathrm{II}}+\left[\left(\frac{R_A}{R_C}\right)^3-1\right]\mathbf{R}_C. \quad (7.36)$$

7.10 REFINEMENTS TO THE MODEL

Our model for the potential outside a nerve or muscle cell has been a long single conducting fiber in an infinite, homogeneous medium. We will consider four ways to extend and improve the model. The first is to recognize that current must also flow radially inside the cell. If it did not, it could never leave the cell. At the same time we will abandon the assumption that the presence of the cell along the x axis does not perturb the current outside the cell. The third improvement is to recognize that the conductivity may depend on position. This is particularly important outside the cell, where we have muscle, fat, lungs, etc. Finally, the conductivity at a given point may depend on which direction the current flows—for example, parallel or perpendicular to the cells.

In order to make these refinements to the model, we must develop a different formulation of the problem. Consider some region of space containing a conducting material described by Ohm's law. The electric field is related to the potential by the three-dimensional extension of Eq. (6.14):

$$E_x=-\frac{\partial v}{\partial x},$$

$$E_y=-\frac{\partial v}{\partial y}, \quad (7.37)$$

$$E_z=-\frac{\partial v}{\partial z},$$

or

$$\mathbf{E}=-\operatorname{grad} v=-\boldsymbol{\nabla} v. \quad (7.38)$$

If the material is isotropic and obeys Ohm's law, then from Eq. (6.24)

$$\mathbf{j}=\sigma\mathbf{E}=-\sigma\boldsymbol{\nabla} v. \quad (7.39)$$

We now apply the equation of continuity or conservation of charge, casting Eq. (4.8) in terms of the electric current density $\mathbf{j}$ and the electric charge per unit volume, ρ:

$$\frac{\partial\rho}{\partial t}=-\boldsymbol{\nabla}\cdot\mathbf{j}. \quad (7.40)$$

Combining these two equations gives

$$\frac{\partial\rho}{\partial t}=\operatorname{div}(\sigma\operatorname{grad} v)=\boldsymbol{\nabla}\cdot(\sigma\boldsymbol{\nabla} v). \quad (7.41)$$

Leaving the conductivity inside the divergence term allows the conductivity to depend on position. If the conductivity is the same everywhere it can be taken outside the divergence operator to give

$$\frac{\partial\rho}{\partial t}=\sigma\nabla^2 v=\sigma\left(\frac{\partial^2 v}{\partial x^2}+\frac{\partial^2 v}{\partial y^2}+\frac{\partial^2 v}{\partial z^2}\right). \quad (7.42a)$$

We can write this in cylindrical coordinates which are more useful for modeling a cylindrical cell. From Appendix L, assuming that the potential does not depend on the polar angle θ, we have

$$\frac{\partial\rho}{\partial t}=\sigma\nabla^2 v=\sigma\left[\frac{1}{r}\frac{\partial}{\partial r}\left(r\frac{\partial v}{\partial r}\right)+\frac{\partial^2 v}{\partial z^2}\right]. \quad (7.42b)$$

These are very general equations, applicable to any volume of space where the material is homogeneous and isotropic and obeys Ohm's law. They were derived using Ohm's law and the conservation of charge. Equation (7.42a) is actually the same result we had in Eq. (6.49). To see this, apply it in a one-dimensional case to the interior of a single cell stretched along the x axis. Consider the charge in a small cylindrical region of axoplasm of length h and radius a, the cylindrical surface of which is surrounded by cell membrane. The total charge Q within the axoplasm is related to the interior charge density ρ_i by $Q=\pi a^2 h\rho_i$. Any charge that accumulates in this region due to changes in $\mathbf{j}$ either flows out through the membrane or accumulates on the surface and charges a membrane of capacitance c_m per unit area.

$$\frac{\partial Q}{\partial t}=\pi a^2 h\frac{\partial\rho_i}{\partial t}=C\frac{\partial v_m}{\partial t}+i_m=2\pi ah\left(c_m\frac{\partial v_m}{\partial t}+j_m\right).$$

Combining this with the one-dimensional version of Eq. (7.42a) applied to the inside of the cell, we obtain

$$c_m\frac{\partial v_m}{\partial t}+j_m=\frac{\pi a^2 h}{2\pi ah}\sigma_i\frac{\partial^2 v_i}{\partial x^2}=\frac{\sigma_i a}{2}\frac{\partial^2 v_i}{\partial x^2}. \quad (7.43)$$

This is the same as Eq. (6.49), except that it is written in terms of σ_i, a, and h instead of a and r_i.

Now we can make the first two improvements. Except at the cell membrane, where charge on the membrane capacitance is changing as the membrane potential changes, $\partial\rho/\partial t=0$. If we assume that the transmembrane potential v_m is known, then Eq. (7.42b) can be applied separately to the extracellular and the intracellular fluid for a long straight axon to determine the potential everywhere outside (or inside). This was first done by Clark and Plonsey (1968). In the extracellular and intracellular fluids, Eq. (7.42b) becomes

$$\frac{1}{r}\frac{\partial}{\partial r}r\frac{\partial v_o(r,z)}{\partial r}+\frac{\partial^2 v_o(r,z)}{\partial z^2}=0, \quad r>a$$

$$\frac{1}{r}\frac{\partial}{\partial r}\, r\,\frac{\partial v_i(r,z)}{\partial r}+\frac{\partial^2 v_i(r,z)}{\partial z^2}=0, \quad r<a \tag{7.44}$$

$$v_m(z)=v_i(a,z)-v_o(a,z).$$

With v_m known, these equations were solved for the potential distribution inside and outside the cell. This is the calculation that was done to obtain Fig. 7.13. The result of this type of calculation has been compared to the line-source model by Trayanova, Henriquez, and Plonsey (1990).

To make the next improvement, consider again Eq. (7.43). It can also be written as

$$\beta\left(c_m\frac{\partial v_m}{\partial t}+j_m\right)=\sigma_i\frac{\partial^2 v_i}{\partial x^2}, \tag{7.45}$$

where $\beta=2\pi ah/\pi a^2 h=2/a$ is the ratio of surface area to volume of the cell. Our cell was cylindrical. With other geometrical configurations, such as a cubic or a spherical cell, β would have a different value, but it always has the dimensions of $(\text{length})^{-1}$. In the general three-dimensional case, we have

$$\underbrace{\beta\left(c_m\frac{\partial v_m}{\partial t}+j_m\right)}_{\text{zero, except at the cell membrane}}=\text{div}(\sigma_i\ v_i)=\nabla\cdot(\sigma_i\nabla v_i). \tag{7.46}$$

Both σ_i and v_i are functions of position. The left-hand side is zero except at the cell membrane. The main theme of this chapter has been that current stops flowing inside the cell must flow outside the cell. We can write an analogous equation for the region outside the cell:

$$\underbrace{-\beta\left(c_m\frac{\partial v_m}{\partial t}+j_m\right)}_{\text{zero, except at the cell membrane}}=\text{div}(\sigma_o\ v_o)=\nabla\cdot(\sigma_o\nabla v_o). \tag{7.47}$$

Calculations have been done for over 20 years in which the potential on the surface of the thorax was obtained by solving Eq. (7.47) numerically, with the conductivity varying from organ to organ (or the mathematical equivalent). Examples are found in Stanley *et al.* (1986) and Purcell *et al.* (1988). See also some of the references for Sec. 8.5.

The final improvement recognizes that the material in which the current flows is generally not isotropic. If it is still described by Ohm's law, then we can write $\mathbf{j}=\tilde{\boldsymbol{\sigma}}\cdot\mathbf{E}$ where $\tilde{\boldsymbol{\sigma}}$ is a matrix or tensor. In Cartesian coordinates

$$\begin{pmatrix} j_x \\ j_y \\ j_z \end{pmatrix}=\begin{pmatrix} \sigma_{xx} & \sigma_{xy} & \sigma_{xz} \\ \sigma_{yx} & \sigma_{yy} & \sigma_{yz} \\ \sigma_{zx} & \sigma_{zy} & \sigma_{zz} \end{pmatrix}\begin{pmatrix} E_x \\ E_y \\ E_z \end{pmatrix}. \tag{7.48}$$

This is a compact notation for

$$j_x=\sigma_{xx}E_x+\sigma_{xy}E_y+\sigma_{xz}E_z,$$

with similar equations for j_y and j_z. It can be shown that the conductivity matrix must be symmetric, so there are actually six conductivity coefficients, not nine. It is often possible to make some of the matrix elements zero by suitable choice of a coordinate system and suitable orientation of the axes.

In terms of the conductivity matrix, Eq. (7.46) becomes

$$\underbrace{\beta\left(c_m\frac{\partial v_m}{\partial t}+j_m\right)}_{\text{zero, except at membrane surface}}=\text{div}(\tilde{\boldsymbol{\sigma}}_i\cdot\ v_i)$$
$$=\nabla\cdot(\tilde{\boldsymbol{\sigma}}_i\cdot\nabla v_i). \tag{7.49}$$

Myocardial cells are typically about 10 μm in diameter and 100 μm long. They have the added complication that they are connected to one another by gap junctions, as shown schematically in Fig. 7.27. This allows currents to flow directly from one cell to another without flowing in the extracellular medium. The *bidomain* (two-domain) model is often used to model this situation. It considers a region, small compared to the size of the heart, that contains many cells and their surrounding extracellular fluid. It simplifies the problem by assuming that each small volume element contains two domains, intracellular and extracellular. Think of the volume element as the entire region shown in Fig. 7.27. There are two potentials in each small volume element: $v_i(\mathbf{r},t)$ and $v_o(\mathbf{r},t)$. These potentials are averages over the intracellular and extracellular domains contained in the volume element. The transmembrane potential is the difference between these two potentials: $v_m(\mathbf{r},t)=v_i(\mathbf{r},t)-v_o(\mathbf{r},t)$. Charge can pass freely between the two domains, but the total charge within a volume

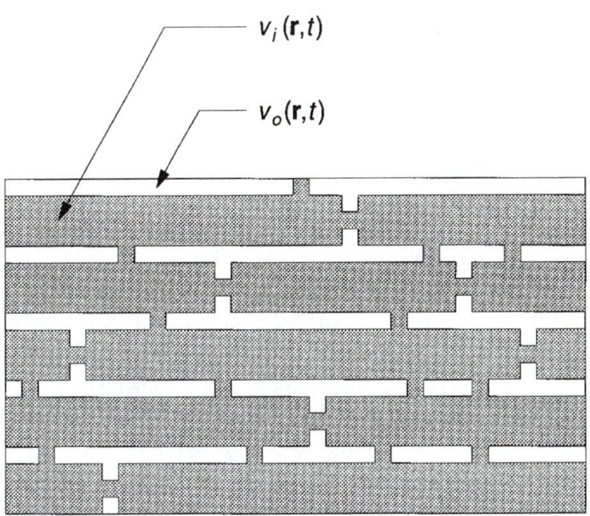

FIGURE 7.27. The interior of myocardial cells (shaded) is connected to adjoining cells by gap junctions. The bidomain model assumes that in a small region of space (large compared to a cell) there are two potentials: the interior potential and outside potential that are functions of position and time.

element is conserved. If the current densities in each domain are $\mathbf{j}_i$ and $\mathbf{j}_o$, then the divergence of the sum is zero: $\nabla\cdot(\mathbf{j}_i+\mathbf{j}_o)=0$. The divergence of each current individually passes through or charges the membrane. The anisotropic analogs of Eqs. (7.46) and (7.47) are now

$$\beta\left(c_m\frac{\partial v_m}{\partial t}+j_m\right)=\mathrm{div}(\tilde{\boldsymbol{\sigma}}_i\cdot v_i)=\boldsymbol{\nabla}\cdot(\tilde{\boldsymbol{\sigma}}_i\cdot\boldsymbol{\nabla}v_i),$$

$$-\beta\left(c_m\frac{\partial v_m}{\partial t}+j_m\right)=\mathrm{div}(\tilde{\boldsymbol{\sigma}}_o\cdot v_o)=\boldsymbol{\nabla}\cdot(\tilde{\boldsymbol{\sigma}}_o\cdot\boldsymbol{\nabla}v_o),\tag{7.50}$$

Since in this model every small volume element contains both intracellular and extracellular space, the restriction on the left-hand side of these equations has been removed: the left-hand side is now also a function of position. The quantity β is still the surface-to-volume ratio, but now it is the membrane surface area per unit volume of the entire bidomain—both intracellular and extracellular volumes. For example, if we consider that the cells are all cylindrical of length h and radius a, then the surface area of a cell is $2\pi ah$. If the fraction of the total volume occupied by cells is f, then the total volume associated with this cell is $\pi a^2h/f$, so

$$\beta=\frac{2f}{a}.\tag{7.51}$$

The membrane current j_m can be modeled by either a passive (Ohm's law—electrotonus) membrane or with one of the models for an active membrane.

The bidomain can also be placed in a surrounding bath in which there is a current density $\mathbf{j}_e$ and a potential v_e. (It would be on the surface of this region that one would measure the normal electrocardiogram.) In this region we would have the equations

$$\mathbf{j}_e=-\sigma_e\boldsymbol{\nabla}v_e,\quad \boldsymbol{\nabla}\cdot\mathbf{j}_e=0.\tag{7.52}$$

Various authors have applied the bidomain concept to models in one, two, and three dimensions. This has been done with v_m assumed known [an anisotropic extension of Eqs. (7.44)] or with various models for j_m.

The reader interested in learning how to make calculations using the bidomain model might start with a paper by Roth (1991) that treats a one-dimensional bidomain model that can be solved in closed form. The paper discusses the boundary conditions that must be applied at the surface of the bidomain, an issue that has been ignored here. Another clear example of the use of the bidomain model with cylindrical symmetry is also given by Roth (1988). He allows the conductivities in the r and z directions to be different, in both the inside and outside regions. This leads to the differential equation

$$\frac{1}{r}\frac{\partial}{\partial r}r\frac{\partial}{\partial r}[\sigma_{ir}v_i(r,z)+\sigma_{or}v_o(r,z)]$$
$$+\frac{\partial^2}{\partial z^2}[\sigma_{iz}v_i(r,z)+\sigma_{oz}v_o(r,z)]=0.\tag{7.53}$$

Roth then discusses the changes of variables and assumptions that must be made in order to obtain an analytic solution, as well as the very complicated paths that the return current follows. Most of the important features of the bidomain results occur because the anisotropy ratio is different in the inside and outside domains: $\sigma_{ir}/\sigma_{iz}\neq\sigma_{or}/\sigma_{oz}$. Roth also describes how the geometry of the cells and the gap junctions give rise to these anisotropy ratios. The paper also shows that for a cylindrical bidomain, measurement of the exterior potential far away is given accurately by the dipole model. Whether the bidomain model gives this same result for a large depolarizing wave front has not yet been determined. A good recent review has been provided by Henriquez (1993).

Henriquez and Plonsey (1990) showed that for a simple fiber v_o was negligible (about 0.04 mV). Using the bidomain model increased v_o to about 2 mV. Restricting the volume of the extracellular return space made v_o as large as 34 mV. The larger extracellular resistance also slowed the conduction velocity. The extracellular and intracellular voltages have also been measured experimentally, for example by Knisley *et al.* (1991). Various one-dimensional bidomain models have also been developed that treat the cells and gap junctions as having a periodic variation in conductivity [Cartree and Plonsey (1992); Trayanova (1994)]. Other examples of calculations using the bidomain model are discussed in the next section and in Chap. 8.

7.11 ELECTRICAL STIMULATION

The information that has been developed in this chapter can also be used to understand some of the features of stimulating electrodes. These may be used for electromyographic studies, for stimulating muscles to contract, for a cochlear implant to partially restore hearing, for cardiac pacing, and even for defibrillation. The electrodes may be inserted in cells, placed in or on a muscle, or placed on the skin.

A pulse of current is sent to the stimulating electrode. The current required to produce a response depends on the shape and size of the electrode, its placement, the kind of cell being stimulated, and the duration of the pulse. For a given electrode geometry the shorter the pulse, the larger the current required for a tissue response. For very long pulses there is a minimum current required to stimulate that is called *rheobase*. The strength-duration curve was first discovered by G. Weiss in 1901. He expressed it in terms of total charge in the stimulating pulse. A description of the strength-duration curve and its history has been given by Geddes and Bourland (1985). They also describe some techniques for making accurate measurements. The strength-duration curve for current was first described by Lapicque in 1909 as

$$i = i_R \left(1 + \frac{t_C}{t} \right), \tag{7.54}$$

where i is the current required for stimulation, i_R is the minimum current or rheobase, t is the duration of the pulse, and t_C is *chronaxie*, the duration of the pulse that requires a doubling of the rheobase current.

Equation (7.54) provides an empirical fit to the experimental data. We can develop a model to explain it using information from Chap. 6. A nerve fires after a certain departure from the resting potential. Subthreshold behavior can be modeled by electrotonus. Suppose that we inject a stimulating current into a cell at the origin. Equation (6.56) gave the voltage along the axon for a current injected in the cell at the origin after an infinitely long time:

$$v - v_r = v_0 e^{-|x|/\lambda}.$$

The solution to Problem 6.24 shows that the current injected is

$$i_0 = 2 v_0 / \lambda r_i. \tag{7.55}$$

The quantities λ and r_i are defined in Chap. 6. The factor of 2 arises because current flows both ways along the cell. The rheobase current is

$$i_R = \frac{2 v_{\text{threshold}}}{\lambda r_i}. \tag{7.56}$$

If we assume that the threshold voltage is independent of pulse duration, we can use the curve for $x=0$ in Fig. 6.32(c) to relate the minimum current to the pulse duration. As long as the pulse is applied, the voltage will rise along this curve. When the current is turned off, the voltage will start to fall. If the voltage had reached threshold, the cell will fire. This curve is the solution of Eq. (6.53). The solution is [Plonsey (1969, p. 132)]

$$v(0,t) - v_r = v_0 \operatorname{erf}\left(\sqrt{\frac{t}{\tau}} \right), \tag{7.57}$$

where τ is the membrane time constant, $\kappa \epsilon_0 \rho_m$. The error function is defined in Eq. (4.74) and is plotted in Fig. 4.21. The current required for stimulation with an intracellular electrode at the origin is therefore

$$i = \frac{2 v_{\text{threshold}}}{\lambda r_i \operatorname{erf}(\sqrt{t/\tau})} = \frac{i_R}{\operatorname{erf}(\sqrt{t/\tau})}. \tag{7.58}$$

Chronaxie can be related to the time constant τ by setting $i = 2 i_R$:

$$2 i_R = \frac{i_R}{\operatorname{erf}\left(\sqrt{\frac{t_C}{\tau}} \right)}. \tag{7.59}$$

From a table of values of the error function, we find

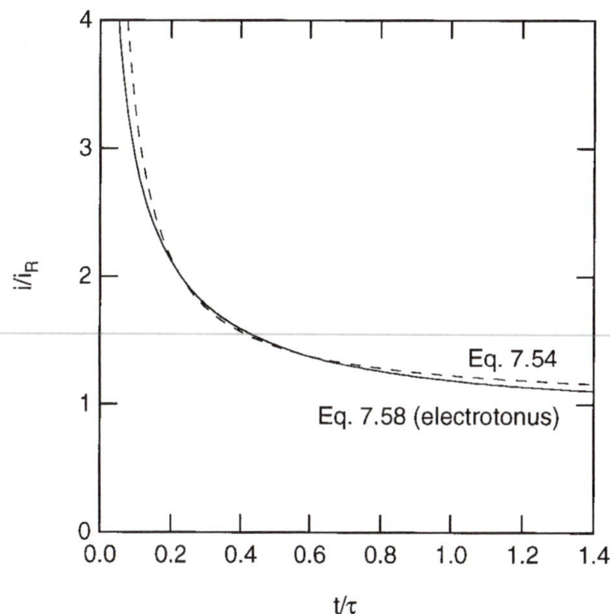

FIGURE 7.28. The stimulus strength-duration curve plotted for the chronaxie–rheobase model, Eq. (7.54) and for electrotonus, Eq. (7.58).

$$t_C = 0.228 \tau. \tag{7.60}$$

Figure 7.28 compares the standard empirical curve, Eq. (7.54), with this model. The curves are experimentally indistinguishable.

Equation (7.54) is also used for surface electrodes. Table 7.1 shows some experimental values for rheobase and chronaxie. The further the electrode from the tissue being stimulated, the greater the rheobase current that is required.

An electrode that is transferring positive charge to the medium is called an *anode*. One that is collecting positive charge is called a *cathode*. If the stimulating electrode is inside the cell, a positive current leaving the electrode will increase the positive charge within the cell and depolarize

TABLE 7.1. *Comparison of values for rheobase and chronaxie for different stimulations.*

Stimulation	Rheobase (mA)	Chronaxie (ms)
Intracellular, from Table 6.1, $v_{\text{threshold}} = 15$ mV	6.7×10^{-6}	0.23
Myocardium, from best pacing electrodes (Roth, private communication)	0.1	
Motor nerves for inspiration, from stimulation of chest wall [Voorhees *et al.* (1992)]	49	0.17
Myocardium, from stimulation of chest wall [Voorhees *et al.* (1992)]	204	1.82

it. Another way to say it is that current from the electrode flows out through the membrane, so the inside of the membrane will be made more positive than the outside. On the other hand, an anodic electrode just outside the cell will send positive current in through the membrane near the electrode, as shown in Fig. 7.29. This lowers the potential inside and hyperpolarizes the membrane near the electrode. Further away from the stimulation point will be a region where current flows out through the membrane, thus depolarizing the cell. However, the outward current is in general spread out over more membrane, so the current density and hence the depolarization is less than the hyperpolarization near the anode. The situation is, of course, reversed for a cathodic electrode. Figure 7.29 is conceptual; to draw the field lines accurately would require taking into account the conductivities of the extracellular and intracellular fluid as well as the membrane.

The electrotonus model also helps us understand another effect that is observed: the *virtual cathode*. The point of origin for a stimulus can be measured by placing sensing electrodes in or on the heart at different distances from the stimulating electrode and plotting the time required for the depolarization wave front to reach the electrode vs its position. Extrapolation to the time of stimulus gives the size of the region of initial depolarization. Imagine a stimulating electrode inside a one-dimensional cell. When the stimulus current is just above rheobase, the region of depolarization is very small and surrounds the electrode. As the stimulating current is increased, the size of the initial depolarized region grows. From Eqs. (6.56) and (7.59) we obtain

$$v_{\text{threshold}} = \frac{i_0 \lambda r_i}{2} e^{-x_{\text{vc}}/\lambda}$$

or

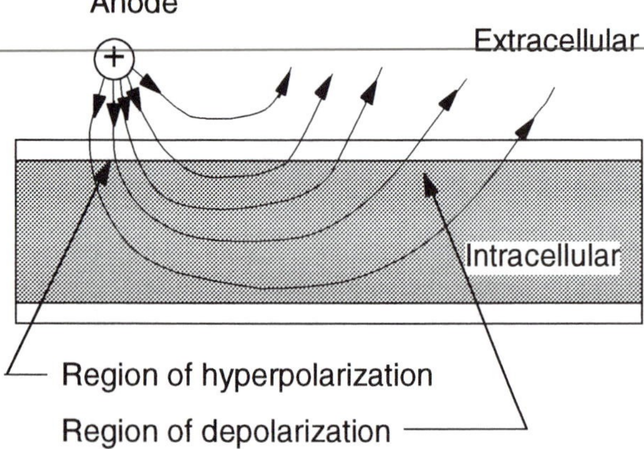

FIGURE 7.29. A schematic drawing showing why there is a region of hyperpolarization near a stimulating anode (positive electrode) with a region of weaker depolarization further away.

$$x_{\text{vc}} = \lambda \ln\left(\frac{i_0 \lambda r_i}{2 v_{\text{threshold}}}\right) = \lambda \ln\left(\frac{i_0}{i_R}\right), \tag{7.61}$$

where x_{vc} is the size of the virtual cathode.

Cardiac pacemakers are a useful treatment for certain heart diseases [Moses *et al.* (1991); Barold (1985)]. The most frequent are an abnormally slow pulse rate (*bradycardia*) associated with symptoms such as dizziness, fainting (*syncope*), or heart failure. These may arise from a problem with the SA node (*sick sinus syndrome*) or with the conduction system (*heart block*). A pacemaker can be used temporarily or permanently. The pacing electrode can be threaded through a vein from the shoulder to the right ventricle (transvenous pacing) or placed directly in the myocardium during heart surgery. Sometimes two pacing electrodes are used, one in the atrium and one in the ventricle. The pacing electrode can be unipolar or bipolar. With a unipolar electrode, the stimulation current flows into the myocardium and returns to the case of the pacemaker, which is often placed in a pocket in the muscle of the chest wall near the shoulder. The return current in a bipolar electrode goes to a ring electrode a few centimeters back along the pacing lead from the electrode at the tip. The surface area of a typical tip is about 10 mm^2 (10^{-5} m^2). The current density required to initiate depolarization depends on the spatial distribution of the current and may be in the range 100–1000 A m^{-2}. The resistance of the tissue is 150–500 Ω. Thus, in this model the rheobase current is about[3] 1 mA. After the pacing electrode is implanted, the size of the voltage pulse required to initiate ventricular activity rises because inflammatory tissue grows around the electrode. It is conducting, but the myocardium is further away, and the inflammatory tissue effectively increases the size of the electrode, thereby reducing the current density. After six months or so, the inflammation has been replaced by a small fibrous capsule, resulting in an effective electrode size larger than the bare electrode but smaller than the region of inflammation. Recently, electrodes that elute steroids have been used to reduce the inflammation.

The bidomain model has been used to understand the response of cardiac tissue to stimulations [Roth and Wikswo (1994); Roth (1994); Wikswo (1995)]. The former simulation explains a remarkable experimental observation. Although the speed of the wave front is greater along the fibers than perpendicular to them, if the stimulation is well above threshold, the wavefront originates farther from the cathode in the direction perpendicular to the fibers—the direction in which the speed of propagation is slower. The simulations show that this is due to the anisotropy in conductivity. This is called the "dog-bone" shape of the virtual cathode. It can rotate with depth in the myocardium because the myocardial fibers change orientation. The difference in anisotropy accentuates the effect of a region of

[3]Acute implants of smaller electrodes where the electrode resistance is low, as well as computer simulations have shown simulation with currents as small as 18 μA (Roth, private communication).

hyperpolarization surrounding the depolarization region produced by a cathodic electrode. Strong point stimulation can also produce reentry waves that can interfere with the desired pacing effect.

One of the fundamental problems with research in this area can be seen in equations like (7.54). The variable on the left is the transmembrane potential v_m. The variable on the right is the potential inside or outside the cell. Measurement of v_m requires measurement or calculation of the difference $v_i - v_o$. Experimental measurements of the transmembrane potential often rely on the use of a voltage-sensitive dye whose fluorescence changes with the transmembrane potential [Knisley, Hill, and Ideker (1994); Neunlist and Tung (1995)].

SYMBOLS USED IN CHAPTER 7

Symbol	Use	Units	First used on page
a	Axon radius	m	181
c_m	Membrane capacitance per unit area	$\mathrm{F\,m^{-2}}$	181
i	Current	A	183
i_i	Current along inside of axon	A	181
i_m, i_o	Current through membrane	A	181
$j, \mathbf{j}$	Current density	$\mathrm{A\,m^{-2}}$	183
j_m	Current density through membrane	$\mathrm{A\,m^{-2}}$	181
$p, \mathbf{p}$	Activity vector or current dipole moment	$\mathrm{V\,m^2}$	185
q	Charge	C	183
r_i	Resistance per unit length	$\Omega\,\mathrm{m^{-1}}$	181
$r, \mathbf{r}$	Distance	m	182
t	Time	s	181
v	Potential difference	V	181
v_i	Potential inside axon	V	181
x, y, x_0, x_1, x_2, y_0	Distance	m	181
C	Capacitance	F	198
E	Electric field	$\mathrm{V\,m^{-1}}$	182
R	Resistance	Ω	183
$R, \mathbf{R}$	Distance	m	187
β	Ratio of surface area to volume	$\mathrm{m^{-1}}$	199
ϵ_0	Permittivity of free space	$\mathrm{N^{-1}\,m^{-2}\,C^2}$	183
$\sigma, \sigma_i, \sigma_o$	Electrical conductivity	$\mathrm{S\,m^{-1}}$	182
ρ	Charge density	$\mathrm{C\,m^{-3}}$	198
θ	Angle		185
Ω	Solid angle		185

PROBLEMS

Section 7.1

7.1. A single nerve or muscle cell is stretched along the x axis and embedded in an infinite homogeneous medium of conductivity σ_0. Current i_0 leaves the cell at $x=b$ and enters the cell again at $x=-b$. Find the current density **j** at distance r from the axis in the $x=0$ plane.

7.2. An axon is stretched along the x axis. At one instant of time an impulse traveling along the axon has the form shown in the graph. The electrical conductivity inside the axon of radius a is σ_i. In the infinite external medium it is σ_0. Find an expression for the potential at point (x_0, y_0).

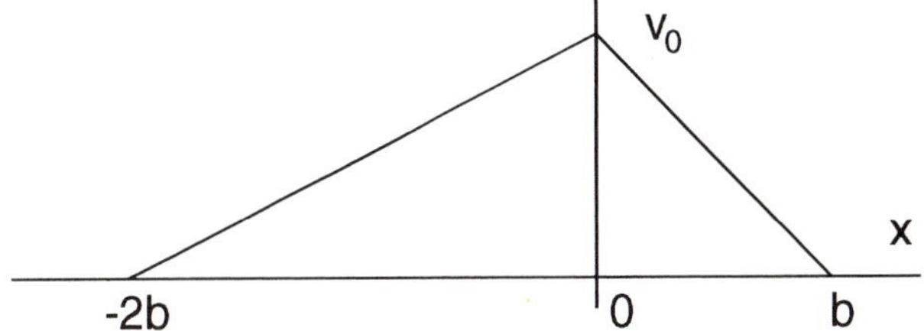

7.3. The interior potential of a cylindrical cell is plotted at one instant of time. Distances along the cell are given in terms of length b. The cell has radius a and electrical conductivity σ_i. The resting potential is 0 and the depolarized potential is v_0. The conductivity of the external medium is σ_0.

(a) Find expressions for, and plot, the current along the cell in the four regions ($x<0$, $0<x<2b$, $2b<x<3b$, $3b<x$).

(b) Find the potential at a point (x,y) outside the cell in terms of the parameters given in the problem. The point is not necessarily far from the cell.

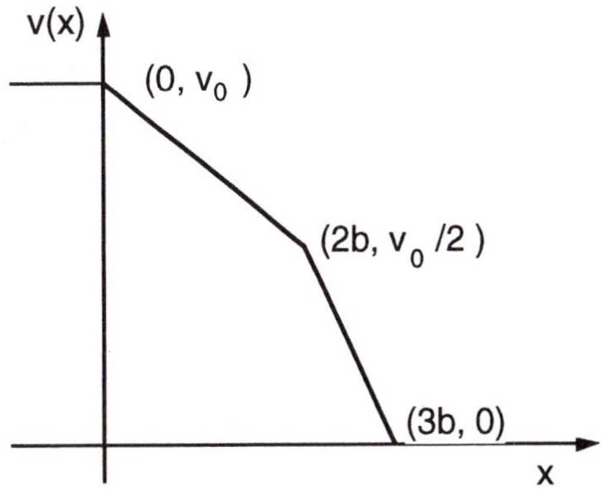

Section 7.2

7.4. Modify the closing argument of Sec. 7.2 by considering electrodes that are disks rather than spheres. (Hint: The capacitance you will need is given in Sec. 6.19.)

Section 7.3

7.5. What would be the current-dipole moment of a nerve cell of radius 2 μm when it depolarizes? Would myelination make any difference? Does the result depend on the rise time of the depolarization? If the impulse lasts 1 ms and the conduction speed is 5 m s^{-1}, how far apart are the rising and falling edges of the pulse?

7.6. An axon or muscle cell is stretched along the x axis on either side of the origin. As it depolarizes, a constant current dipole **p** pointing to the right sweeps along the axis with velocity u. An electrode at $(x=0, y=a)$ measures the potential with respect to $v=0$ at infinity. Ignore repolarization. Find an expression for v at the electrode as a function of time and sketch it. Assume that at $t=0$, **p** is directly under the electrode at $x=0$.

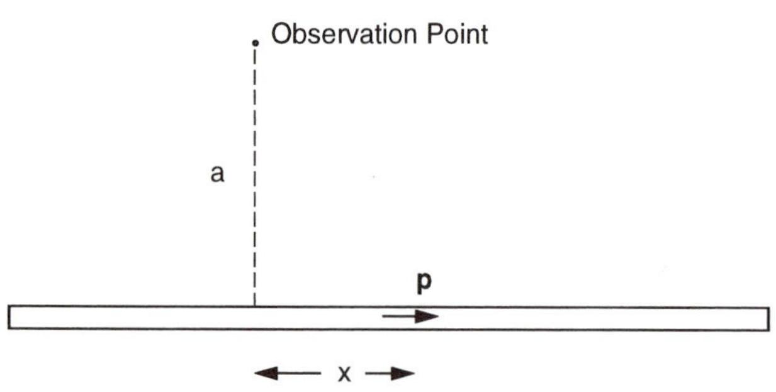

7.7. An electrode at $(x=0, y=a)$ measures the potential outside an axon with respect to $v=0$ at infinity. A nerve impulse is at point x along the axon, measured from the perpendicular from the electrode to the axon. At $x+b$ is a current dipole pointing to the right, which represents the depolarization wave front. At $x-b$ is a vector of the same magnitude pointing to the left, which represents repolarization. Obtain an expression for v as a function of x, b, p, and a. Plot it in the case $a=1$, $b=0.05$.

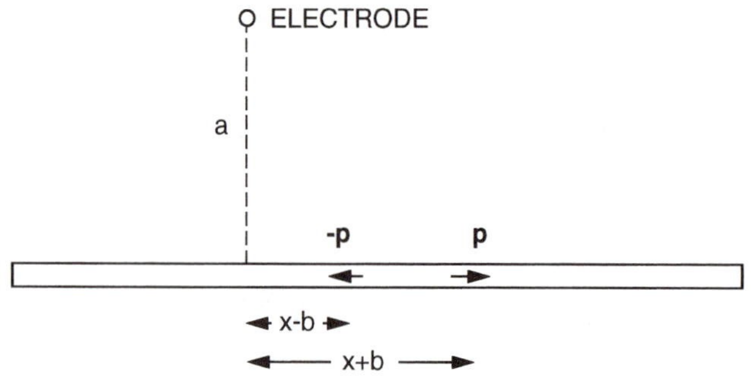

Section 7.4

7.8. Run the program of Fig. 7.12 and plot the potential for different distances from the axon.

7.9. Modify the program of Fig. 7.12 to calculate the potential from a single Gaussian action potential and plot the potential.

Section 7.7

7.10. Two electrodes are placed in a uniform conducting medium 10 cm from a cell of radius 5 μm and 10 cm from each other, so that the two electrodes and the cell form an equilateral triangle. When the cell depolarizes the potential rises 90 mV. What will be the potential difference between the two electrodes when the cell orientation is optimum? How many cells would be needed to give a potential difference of 1 mV between the electrodes? Assume $\sigma_i/\sigma_o=10$.

7.11. Guess whatever parameters you need to predict the voltage at the peak of the QRS wave in lead II. Compare your results to the electrocardiogram of Fig. 7.23.

7.12. At a particular instant of the cardiac cycle, **p** is located at the midpoint of a line connecting two electrodes that are 50 cm apart, and **p** is parallel to that line. At that instant the magnitude of the potential difference between the electrodes is 1.5 mV. Upon depolarization, the potential change within the cells has magnitude 90 mV.

(a) What is the magnitude of **p**?

(b) If $\sigma_i/\sigma_o=10$, what is the cross-sectional area of the advancing region of depolarization?

7.13. A semi-infinite slab of myocardium occupies the region $z>0$. A hemispherical wave of depolarization moves radially away from the origin through the slab. At some instant of time the radius of the hemispherical depolarizing wavefront is R. Assume that $\mathbf{p}=\int d\mathbf{p}$, that $d\mathbf{p}$ is everywhere perpendicular to the advancing wavefront, and that the magnitude of $d\mathbf{p}$ is proportional to the local area of the wavefront. Find **p**. Assume that the observation point is very far away compared to R.

7.14. Make measurements on yourself and construct Fig. 7.19.

7.15. Experiments have been done in which a dog heart was stimulated by an electrode deep within the myocardium. No external potential difference was detected until the spherical wave of depolarization grew large enough so that part of it intercepted one wall of the heart. Why?

7.16. Prove directly from Eq. (7.31) that I−II+III=0. (It is sometimes said that the equilateral nature of Einthoven's triangle is necessary to prove this.)

7.17. Derive Eqs. (7.32).

Section 7.9

7.18. Make a scale drawing to determine R_I' [Eq. (7.34)].

7.19. Derive Eqs. (7.35) and (7.36).

Section 7.10

7.20. Ohm's law says that $\mathbf{j}=\sigma\mathbf{E}$. Draw what $\mathbf{j}$ and $\mathbf{E}$ look like (a) in a circuit consisting of a battery and a resistor; (b) for the current flowing when a nerve cell depolarizes.

7.21. Obtain the values for β for a cube of length a on a side, for a cylinder of radius a and length h, and for a sphere of radius a.

Section 7.11

7.22. Verify Eq. (7.56).

7.23. Verify the values given for rheobase and chronaxie in Table 7.1 that are based on Table 6.1.

7.24. Find the equivalent of Eq. (7.54) in terms of the charge required for the stimulation.

7.25. If the medium has a constant resistance, find the energy required for stimulation as a function of pulse duration.

7.26. Another equation that is used to describe strength-duration curves is $i_0=i_R/(1-e^{-\alpha t})$. Express this in terms of chronaxie. Plot this curve on Fig. 7.28.

7.27. A typical pacemaker electrode has a surface area of 10 mm^2. What is its resistance into an infinite medium if it is modeled as a sphere? If it is modeled as a disk? (You will have to use results from Chap. 6 and assign a value for σ_o.)

REFERENCES

Antzelevitch, C., S. Sicouri, A. Lukas, V. V. Nesterenko, D-W. Liu, and J. M. Di Diego (1995). Regional differences of electrophysiology in ventricular cells: physiological and clinical implications. In D. P. Zipes and J. Jalife, eds. *Cardiac Electrophysiology: From Cell to Bedside*, 2nd ed. Philadelphia, Saunders, pp. 228–245.

Barold, S. S. (1985). *Modern Cardiac Pacing*. Mount Kisco, NY, Futura.

Beyer, E. C., R. D. Veenstra, H. L. Kanter, and J. E. Saffitz (1995). Molecular structure and patterns of expression of cardiac gap junction proteins. In D. P. Zipes and J. Jalife, eds. *Cardiac Electrophysiology: From Cell to Bedside*, 2nd ed. Philadelphia, Saunders, pp. 31–38.

Campbell, D. L., R. L. Rasmusson, M. B. Comer, and H. C. Strauss (1995). The cardiac calcium-independent transient outward potassium current: kinetics, molecular properties, and role in ventricular repolarization. In D. P. Zipes and J. Jalife, eds. *Cardiac Electrophysiology: From Cell to Bedside*, 2nd ed. Philadelphia, Saunders, pp. 83–96.

Cartee, L. A., and R. Plonsey (1992). The effect of cellular discontinuities on the transient subthreshold response of a one-dimensional cardiac model. *IEEE Trans. Biomed. Eng.* **39**(3): 260–270.

Clark, J. and R. Plonsey (1968). The extracellular potential field of a single active nerve fiber in a volume conductor. *Biophys J.* **8**: 842–864.

Davidenko, J. M. (1995). Spiral waves in the heart: Experimental demonstration of a theory. In D. P. Zipes and J. Jalife, eds. *Cardiac Electrophysiology: From Cell to Bedside*, 2nd ed. Philadelphia, Saunders, pp. 478–488.

Goss, C. M. (1954). *Gray's Anatomy of the Human Body*, 26th ed. Philadelphia, Lea & Febiger, p. 75.

Geddes, L. A., and J. D. Bourland (1985). The strength-duration curve. *IEEE Trans. Biomed. Eng.* **32**: 458–459.

Guyton, A. C. (1991). *Textbook of Medical Physiology*, 8th ed. Philadelphia, Saunders.

Henriquez, C. S. (1993). Simulating the electrical behavior of cardiac tissue using the bidomain model. *Crit. Rev. Biomed. Eng.* **21**(1): 1–77.

Henriquez, C. S., and R. Plonsey (1990). Simulation of propagation along a cylindrical bundle of cardiac tissue—II: Results of simulation. *IEEE Trans. Biomed. Eng.* **37**: 861–875.

Kass, R. S. (1995). Delayed potassium channels in the heart: Cellular, molecular and regulatory properties. In D. P. Zipes and J. Jalife, eds. *Cardiac Electrophysiology: From Cell to Bedside*, 2nd ed. Philadelphia, Saunders, pp. 74–82.

Knisley, S. B., B. C. Hill, and R. E. Ideker (1994). Virtual electrode effects in myocardial fibers. *Biophys. J.* **66**: 719–728.

Knisley, S. B., T. Maruyama, and J. W. Buchanan, Jr. (1991). Interstitial potential during propagation in bathed ventricular muscle. *Biophys. J.* **59**: 509–515.

Lapicque, L. (1909). Definition experimentale de l'excitabilite. Comptes Rendus Acad. Sci. **67**(2): 280–283.

Moses, H. W. *et al.* (1991). *A Practical Guide to Cardiac Pacing*. 3rd ed. Boston, Little, Brown.

Neunlist, M. and L. Tung (1995). Spatial distribution of cardiac transmembrane potentials around an extracellular electrode: Dependence on fiber orientation. *Biophys. J.* **68**: 2310–2322.

Patton, H. D. *et al.* (1989). *Textbook of Physiology*. 21st ed. Philadelphia, Saunders, Vol. 2.

Plonsey, R. (1969). *Bioelectric Phenomena*. New York, McGraw-Hill.

Press, W. H., S. A. Teukolsky, W. T. Vetterling, and B. P. Flannery (1992). *Numerical Recipes in C: The Art of Scientific Computing*, 2nd ed., reprinted with corrections, 1995. New York, Cambridge University Press.

Purcell, C. J., G. Stroink, and B. M. Horacek (1988). Effect of torso boundaries on electrical potential and magnetic field of a dipole. *IEEE Trans. Biomed. Eng.* **35**(9): 671–678.

Roth, B. J. (1988). The electrical potential produced by a strand of cardiac muscle: A bidomain analysis. *Ann. Biomed. Eng.* **16**: 609–637.

Roth, B. J. (1991). A comparison of two boundary conditions used with the bidomain model of cardiac tissue. *Ann. Biomed. Eng.* **19**: 669–678.

Roth, B. J. (1994). Mechanisms for electrical stimulation of excitable tissue. *Crit. Rev. Biomed. Eng.* **22**(3/4): 253–305.

Roth, B. J. and J. P. Wikswo, Jr. (1994). Electrical stimulation of cardiac tissue: A bidomain model with active membrane properties. *IEEE Trans. Biomed. Eng.* **41**(3): 232–240.

Stanley, P. C., T. C. Pilkington, and M. N. Morrow (1986). The effects of thoracic inhomogeneities on the relationship between epicardial and torso potentials. *IEEE Trans. Biomed. Eng.* **BME-33**(3): 273–284.

Trayanova N. (1994). An approximate solution to the periodic bidomain equations in one dimension. *Math. Biosc.* **120**(2): 189–210.

Trayanova, N., C. S. Henriquez, and R. Plonsey (1990). Limitations of approximate solutions for computing extracellular potential of single fibers and bundle equivalents. *IEEE Trans. Biomed Eng.* **37**(1): 22–35.

Voorhees, C. R., W. D. Voorhees III, L. A. Geddes, J. D. Bourland, and M. Hinds (1992). The chronaxie for myocardium and motor nerve in the dog with surface chest electrodes. *IEEE Trans. Biomed. Eng.* **39**(6): 624–628.

Watanabe, A., and H. Grundfest (1961). Impulse propagation at the septal and commisural junctions of crayfish lateral giant axons. *J. Gen. Physiol.* **45**: 267–308.

Wikswo, J. P., Jr. (1995). Tissue anisotropy, the cardiac bidomain, and the virtual cathode effect. In D. P. Zipes and J. Jalife, eds. *Cardiac Electrophysiology: From Cell to Bedside*, 2nd ed. Philadelphia, Saunders, pp. 348–361.

CHAPTER 8

Biomagnetism

The field of biomagnetism has exploded in the last two decades. Magnetic signals have been detected from the heart, brain, skeletal muscles, and isolated nerve and muscle preparations. Measurements of the magnetic susceptibility of the lung show the effect of dust inhalation. Susceptibility measurements of the heart can determine blood volume, while the susceptibility of the liver can measure iron stores in the body. Bacteria and some animals contain aggregates of magnetic particles, often attached to neural tissue. Many bacteria use these magnetic particles to determine which way is down. There is evidence that magnetism is used for orientation by other animals.

Sections 8.1 and 8.2 review the basics of magnetism, assuming that the reader has already studied this material. Section 8.3 calculates the magnetic field of an axon in an infinite conducting medium. This result, which shows that the field is due primarily to the current dipole in the interior of the axon, is also approximately true for the magnetocardiogram and evoked responses from the brain, described in Secs. 8.4 and 8.5.

Section 8.6 reviews electromagnetic induction. Section 8.7 introduces diamagnetic, paramagnetic, and ferromagnetic materials and describes biomagnetic effects that depend on magnetic materials. Section 8.8 reviews instrumentation for measuring these weak magnetic signals. Section 8.9 describes the use of varying magnetic fields to stimulate nerves or muscles.

8.1. THE MAGNETIC FORCE ON A MOVING CHARGE

Lodestone, compass needles, and other forms of magnetism have been known for centuries, but it was not until 1820 that Hans Christen Oersted showed that an electric current could deflect a compass needle. We now know that magnetism results from electric forces that moving charges exert on other moving charges and that the appearance of the magnetic force is a consequence of special relativity. The properties of magnetic forces can be derived from what we know of electric forces, combined with two assumptions: (1) the laws of nature are identical in two reference systems moving with a constant relative velocity, and (2) there is no preferred direction in space. An excellent development of magnetism from these assumptions is found in Purcell (1985). The development here is more traditional and is incomplete.

Suppose that a beam of electrons is accelerated in a cathode ray tube (as in an oscilloscope, computer display, or television receiver) and causes a spot of light to be emitted where it strikes a fluorescent screen. The electron source is cathode C in Fig. 8.1. The accelerating electrode is E. The fact that the beam is accelerated toward a positively charged electrode confirms that the electrons are negatively charged. The beam normally strikes the screen at point X. Placing a battery between plates A and B creates an electric field that deflects the beam as it passes between the plates. If plate A is positively charged, the beam is deflected upward to point Y. If the battery is removed and the north pole of a bar magnet is brought to the position shown, the beam is deflected to point Z.

We say that a *magnetic field* exists in the space surrounding the bar magnet and that the direction of the magnetic field at any point is the direction a small compass needle located there would point. This experiment shows that the force is at right angles to both the direction of the magnetic field and the velocity of the charged particle. Detailed experiments show that the magnitude of the force $\mathbf{F}$ is proportional to the charge, the magnitude of the velocity $\mathbf{v}$, and the strength of magnetic field $\mathbf{B}$. (In fact, modern definitions of the magnetic field are based on this proportionality.) The magnitude of the force is greatest when $\mathbf{v}$ and $\mathbf{B}$ are perpendicular. We have seen a relationship like this between three vectors before: the vector product or cross product, which was associated with torque and defined in Sec. 1.3. We write

$$\mathbf{F}=q(\mathbf{v}\times\mathbf{B}). \tag{8.1}$$

The SI unit of $\mathbf{B}$ is the tesla or T. An earlier name was the

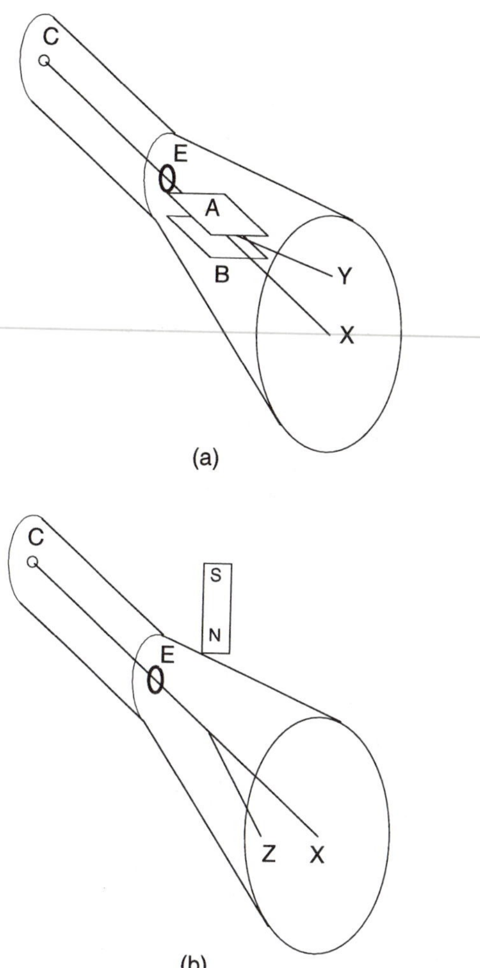

FIGURE 8.1. An electron beam generated at cathode *C* and accelerated through electrode *E* strikes the fluorescent screen on the right. (a) A positive charge on plate *A* and negative charge on plate *B* deflects the beam from *X* to *Y*. (b) A bar magnet brought close as shown deflects the beam to point *Z*.

weber per square meter. The cgs unit is the gauss or G: $1\ \mathrm{T}=10^4\ \mathrm{G}$. If a coordinate system is set up so that $\mathbf{v}$ is along the x axis and $\mathbf{B}$ is along the y axis, then $\mathbf{v}\times\mathbf{B}$ and the force on a positive charge are along the $+z$ axis. For negatively charged electrons $\mathbf{F}$ is in the opposite direction. Combining Eq. (8.1) with the electric force gives the full expression for the electromagnetic force, often called the *Lorentz force*:

$$\mathbf{F}=q(\mathbf{E}+\mathbf{v}\times\mathbf{B}). \tag{8.2}$$

Since current in a wire is the result of moving charges, there is a force on a segment of wire carrying a current. Suppose that there are C particles per unit volume, each with charge q, drifting with speed v along a segment of wire of length ds and cross sectional area S. In time dt the total charge passing a given plane is $CvqS\ dt$ [see Eq. (4.11)] so that the current is $i=CvqS$. If there is a magnetic field perpendicular to the wire, the magnitude of the force on each particle is qvB and the total force is $CS\ ds\ qvB=iB\ ds$. If vector $d\mathbf{s}$ is defined along the wire in the direction of the positive current, then the contribution to the magnetic force from this segment of the wire is

$$d\mathbf{F}=i(d\mathbf{s}\times\mathbf{B}). \tag{8.3}$$

If a small square loop of wire is placed in a uniform magnetic field and a current is made to flow in the wire, there is a magnetic force on each arm of the square. (The current can be led to and from the rectangle by two parallel closely spaced wires, in which the forces cancel because the currents are in opposite directions. Forces not considered here maintain the position of the loop.) Figure 8.2 shows the orientation of the loop in the magnetic field. Each side has length a. The magnetic moment $\mathbf{m}$ is perpendicular to the loop and makes an angle θ with the direction of $\mathbf{B}$. Sides HE and FG are perpendicular to the field. The force on side EF has magnitude $iBa\ \sin\phi$ and is directed as shown. Side GH has a force of equal magnitude in the opposite direction. On side FG the force is down and on side HE it is up, both with magnitude iBa. The vector sum of all the forces is zero. There is a torque, however. If the torque is taken about the center of the loop, the FG force and HE force each exert a torque of magnitude $(iBa)(a/2)\cos\phi$. The total torque is therefore $iBa^2\cos\phi$. The loop is said to have a magnetic moment $\mathbf{m}$ of magnitude iS, where S is the area of the loop. Vector $\mathbf{m}$ is defined to point perpendicular to the loop in the direction of the thumb of the right hand when the fingers curl in the direction of the current around the loop. The units of m are $\mathrm{A\ m^2}$ or $\mathrm{J\ T^{-1}}$. In terms of angle θ between $\mathbf{m}$ and $\mathbf{B}$, the torque exerted by the magnetic field on the magnetic moment has magnitude $ia^2B\ \sin\theta$, so

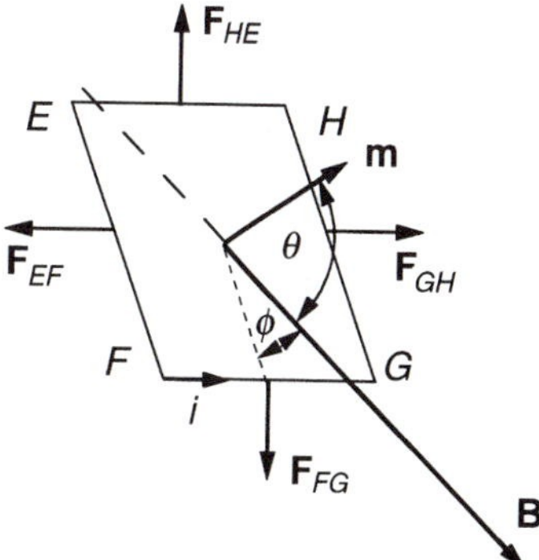

FIGURE 8.2. A current-carrying loop in a uniform magnetic field is shown in perspective. The dashed line from the center of the loop to the center of edge *FG*, vector **B** and vector **m** all lie in the same plane. The sum of angles θ and ϕ is $\pi/2$. The forces on opposite sides add to zero. There is a torque on the loop unless its plane is perpendicular to the field ($\phi=\pi/2$). The magnetic moment **m** is perpendicular to the plane of the loop.

$$\tau = \mathbf{m} \times \mathbf{B}. \tag{8.4}$$

The torque is zero when **m** and **B** are parallel or antiparallel. When they are parallel the equilibrium is stable: if there is a small displacement of **m** the torque acts to return it to equilibrium. When they are antiparallel, the equilibrium is unstable.

A small current loop can be used to test for the presence of a magnetic field. At equilibrium **m** points in the same direction that a small compass needle would point and gives the direction of **B**. Measuring the torque for a known displacement of m from this direction gives the magnitude of **B**.

8.2. THE MAGNETIC FIELD OF A MOVING CHARGE OR A CURRENT

With a compass needle or small sensing coil we can in principle map the magnetic field surrounding a bar magnet or a wire carrying a current. If we examine the field near a long straight wire carrying current i, we find that **B** is always at right angles to the wire and at distance r has magnitude

$$B = \frac{\mu_0 i}{2\pi r}. \tag{8.5}$$

The constant μ_0 is analogous to ϵ_0 in the electrostatic case and is exactly $4\pi \times 10^{-7}$ T m A^{-1} (or Ω s m^{-1}). Figure 8.3 shows the direction of **B** at various locations around a wire. The direction of the force is consistent with Eq. (8.2) if the direction of **B** is defined to be the direction in which the fingers of the right hand curl when the thumb points along the wire in the direction of the (positive) current.

Close to the wire **B** is always at right angles to the wire, in contrast to the electric field, which close to a charge always points toward or away from it. In the electric case, the flux of **E** through a closed surface is proportional to the charge within the volume enclosed by the surface (Gauss's law). In contrast, the flux of **B** through a closed surface is always zero. In the notation of Sec. 4.1,

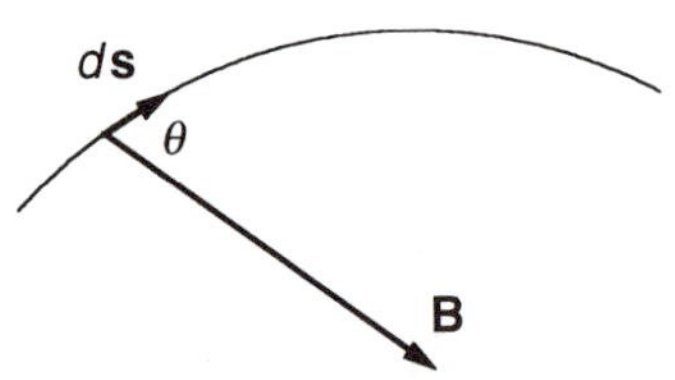

FIGURE 8.4. The line integral of $\mathbf{B}\cdot d\mathbf{s}$ is calculated by multiplying $d\mathbf{s}$ by the component of **B** parallel to $d\mathbf{s}$, that is $B\cos\theta$.

$$\iint_{\text{closed surface}} B_n dS = \iint_{\text{closed surface}} \mathbf{B}\cdot d\mathbf{S} = 0. \tag{8.6}$$

If single magnetic charges (magnetic monopoles) existed, the flux would be proportional to the magnetic charge within the volume. Magnetic monopoles have never been observed, in spite of considerable effort to find them.

As in the electric case, we can construct lines of **B**. The tangent to the line always points in the direction of **B**. For the long wire, the lines of **B** are circles. One can show from Eq. (8.6) that lines of **B** always close on themselves.

Equation (8.6) has the form of the continuity equation, Eq. (4.4), with **B** substituted for **j** and with $C=0$. The differential version of Eq. (8.6) can therefore be obtained from Eq. (4.8). It is

$$\text{div } \mathbf{B} = \nabla \cdot \mathbf{B} = 0. \tag{8.7}$$

It is also interesting to consider the line integral of **B** around a closed path. That is, for any element of the path $d\mathbf{s}$ shown in Fig. 8.4 take the projection of **B** in the direction of $d\mathbf{s}$, $B\cos\theta$. Sum up all the contributions $B\cos\theta\, ds$ along the entire closed path. For path $ABCD$ in Fig. 8.5(a), the result is zero. The reason is that $B\, ds\cos\theta$ is zero on segments AB and CD. On segment DA it is $(\mu_0 i/2\pi a) \times (a\phi) = \mu_0 i\phi/2\pi$, while on segment BC it is $-(\mu_0 i/2\pi b)(b\phi) = -\mu_0 i\phi/2\pi$. In Fig. 8.5(b) the path is circular with the wire at the center, and the line integral is $B(2\pi a) = \mu_0 i$. This result is general:

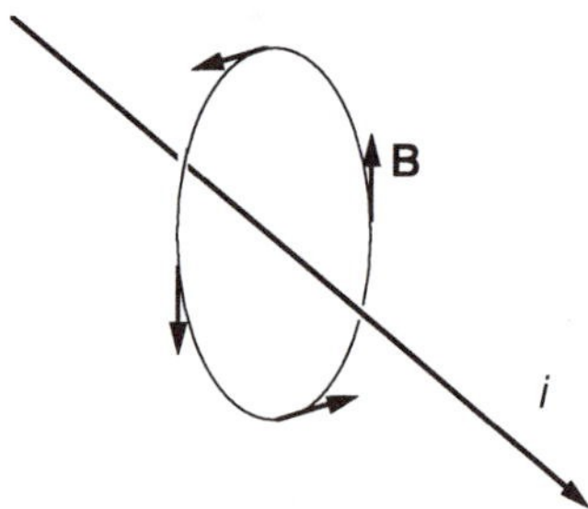

FIGURE 8.3. The magnetic field around a current-carrying wire is at right angles to the wire and the perpendicular from the observation point to the wire. The magnitude is inversely proportional to the distance from the wire.

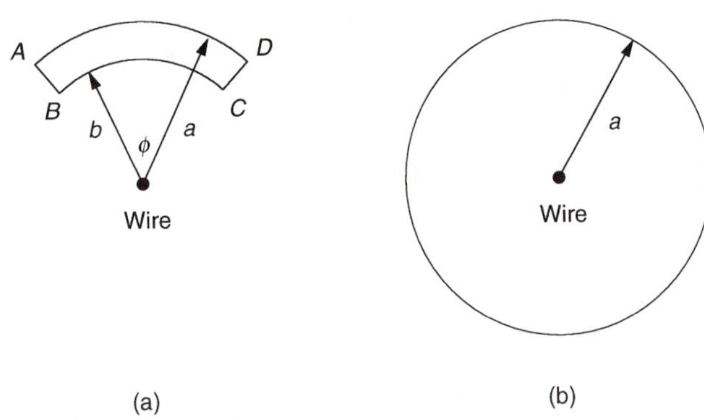

FIGURE 8.5. Two paths of integration. In (a) the path does not encircle the wire carrying the current, and $\oint \mathbf{B}\cdot d\mathbf{s} = 0$. In (b) the path encircles the wire and $\oint \mathbf{B}\cdot d\mathbf{s} = \mu_0 i$.

$$\oint B \cos \theta \, ds = \oint \mathbf{B} \cdot d\mathbf{s} = \mu_0 i. \qquad (8.8)$$

The circle on the integral sign means that the integral is taken around a closed path. The line integral of the magnetic field around a closed path is equal to μ_0 times the current through a circuit enclosed by that path. If two wires carrying equal and opposite currents are enclosed by the path of the line integral, the integral is zero. It does *not* mean that $\mathbf{B}$ is zero everywhere on the path.

A more general statement is that for steady currents the line integral of $\mathbf{B}$ around a closed path is equal to the integral of the current density $\mathbf{j}$ through any surface enclosed by the path:

$$\oint \mathbf{B} \cdot d\mathbf{s} = \mu_0 \int \int \mathbf{j} \cdot d\mathbf{S}. \qquad (8.9)$$

This is known as *Ampere's circuital law*. Like Gauss's law, it is always true but not always useful. It is true for currents that do not vary with time, but it can be used to calculate the magnetic field only if symmetry can be used to argue that $\mathbf{B}$ is always parallel to the path and has the same magnitude at all points on the path.

The surface used to calculate the right-hand side can be any surface bounded by the path used on the left. Since we are dealing with steady currents for which there is no charge accumulation, the continuity equation, Eq. (4.4), shows that the flux of $\mathbf{j}$ (the total current) through any closed surface is zero. Two surfaces S and S', both bounded by the path, form a closed surface as shown in Fig. 8.6. The total current through surface S is the same as the total current through S'.

In situations where the symmetry of the problem does not allow the field to be calculated from Ampere's law, it is possible to find the field due to a steady current in a closed circuit using the *Biot–Savart law*. The contribution $d\mathbf{B}$ to the magnetic field from current i flowing along a line element $d\mathbf{s}$ is

$$d\mathbf{B} = \frac{\mu_0 i}{4\pi} \frac{d\mathbf{s} \times \mathbf{r}}{r^3}. \qquad (8.10)$$

Vector $\mathbf{r}$ is from the current element to the point where the field is to be calculated. The field is found by integrating over the *entire circuit*.

Figure 8.7 shows how this integration is done for an infinitely long straight wire along the x axis. The contribution at point P is obtained by dropping a perpendicular from P to the wire to define $x=0$. The distance from P to the wire is a. The contribution from an element dx at point x is

$$dB = \frac{\mu_0 i}{4\pi} \frac{dx \sin \theta}{r^2} = \frac{\mu_0 i}{4\pi} \frac{a dx}{r^3}.$$

Since $r^2 = a^2 + x^2$ the total field is

$$B = \frac{\mu_0 i}{4\pi} \int_{-\infty}^{\infty} \frac{a \, dx}{(a^2+x^2)^{3/2}} = \frac{\mu_0 i a}{4\pi} \left[\frac{x}{a^2(x^2+a^2)^{1/2}} \right]_{-\infty}^{\infty}$$

$$= \frac{\mu_0 i}{2\pi a}.$$

This agrees with Eq. (8.5) and the result obtained using Ampere's circuital law.

A steady current from a point source which spreads uniformly in all directions generates no magnetic field. To see why consider Fig. 8.8. The source of current is at O. The magnetic field at P can be calculated using the Biot–Savart law. For any element $d\mathbf{s}$ a symmetric element $d\mathbf{s}'$ can be selected, such that $d\mathbf{s} \times \mathbf{r} = -d\mathbf{s}' \times \mathbf{r}'$. Associated with each element is a small area dA, and the current along $d\mathbf{s}$ is $i = j \, dA$. We can set $dA = dA'$ so i is the same in each case. Therefore, $\mathbf{B} = \mathbf{0}$. (This can also be shown using Ampere's law; see Problem 8.7.)

Derivation of Ampere's law requires that there be no charge buildup, so that the total current through a closed surface is zero. However, we will consider an action potential in which the membrane capacitance charges and discharges. To see how this affects Ampere's law, consider current i charging the two shaded capacitor plates in Fig.

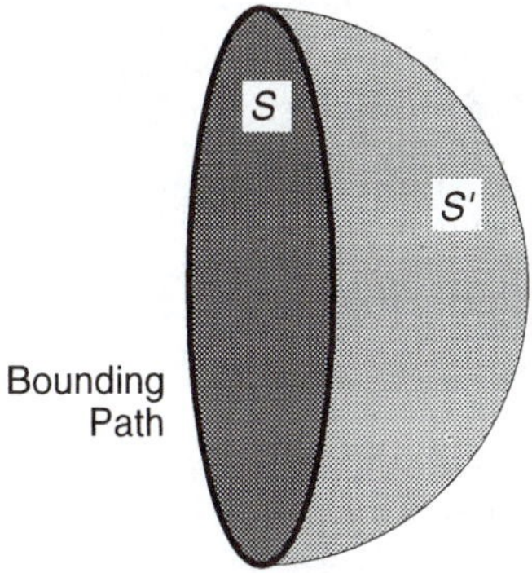

FIGURE 8.6. Since the total current or flux of **j** through any closed surface is zero, the current through surface S is equal to the current through surface S'.

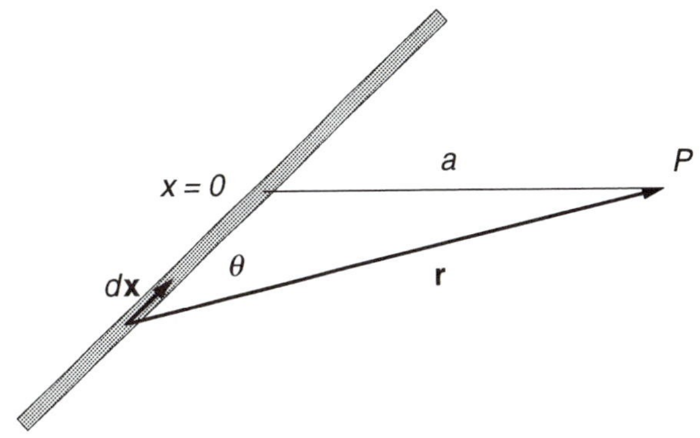

FIGURE 8.7. The Biot–Savart law is used to calculate the magnetic field at point P due to an infinite wire.

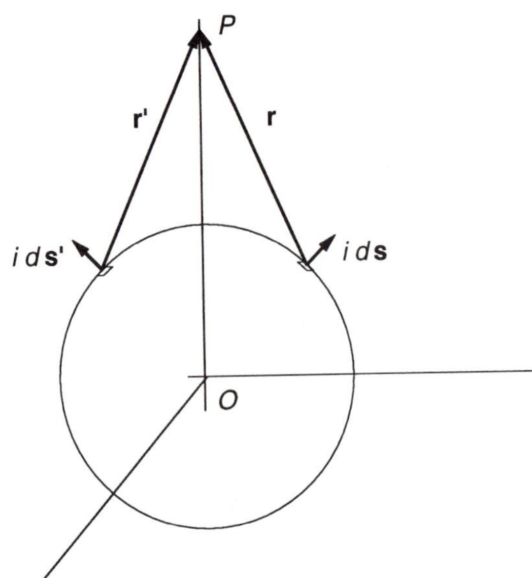

FIGURE 8.8. The magnetic field from a spherically symmetric radial distribution of current is zero. The source at O sends current uniformly in all directions. P is the observation point. For any element ds there is a corresponding ds' such that $d\mathbf{s}\times\mathbf{r}=-d\mathbf{s}'\times\mathbf{r}'$. The current through a small area $d\mathbf{A}$ around ds is i. The same current flows through a corresponding area around $d\mathbf{s}'$. Can you obtain the same result by a symmetry argument?

8.9. The area of each capacitor plate is A. The region between the plates, of thickness b, is filled with dielectric of dielectric constant κ. The integral $\int\int \mathbf{j}\cdot d\mathbf{S}$ is i for surface S and zero for surface S'. Because of the current, the charge density σ on the left-hand plate is increasing at a rate given by $i=A\ d\sigma/dt$, while on the right-hand plate the charge is decreasing because $i=-A\ d\sigma/dt$. Since the electric field between the plates is $E=\sigma/\kappa\epsilon_0$ we can say that $i=Ad(\kappa\epsilon_0 E)/dt$. The quantity $\mathbf{D}=\kappa\epsilon_0\mathbf{E}$ is called the *electric displacement*, and

$$\mathbf{j}_d=\frac{\partial \mathbf{D}}{\partial t}$$

is called the *displacement current density*. More careful consideration shows that Ampere's law is valid when there is charge buildup if we replace $\mathbf{j}$ by $\mathbf{j}+\mathbf{j}_d$:

$$\oint \mathbf{B}\cdot d\mathbf{s}=\mu_0\int\int(\mathbf{j}+\mathbf{j}_d)\cdot d\mathbf{S}. \tag{8.11}$$

With this change, if S and S' are circles of radius a, Ampere's law gives $B=\mu_0 i/2\pi a$ for either one. (The radius of the circle must be very large; see the discussion in the next paragraph.)

What current should be used in the Biot–Savart law? A very surprising answer is that as long as the fields are relatively slowly varying (so that the emission of radio waves is not important), the displacement current contributes nothing. We are free to include it or ignore it. Purcell (1985, p. 328) and Shadowitz (1975, p. 416) discuss why this is so. It is not always easy to calculate the entire displacement current. For example, Fig. 8.10 shows how the conduction current and displacement current vary when current charges a capacitor. Notice that the displacement current flows to and from the back sides of the capacitor plates. This is why we said in the previous paragraph that the radius of the curve defining surfaces S and S' must be very large in order that one surface have no net flux of displacement current and the other have all of it. Whatever their size, however, Eq. (8.11) is valid.

It was mentioned above that a steady current from a point source that spreads uniformly in all directions generates no magnetic field according to the Biot–Savart law. Yet any circular loop has current flowing through it, so Ampere's law suggests that there is a field. The discrepancy is

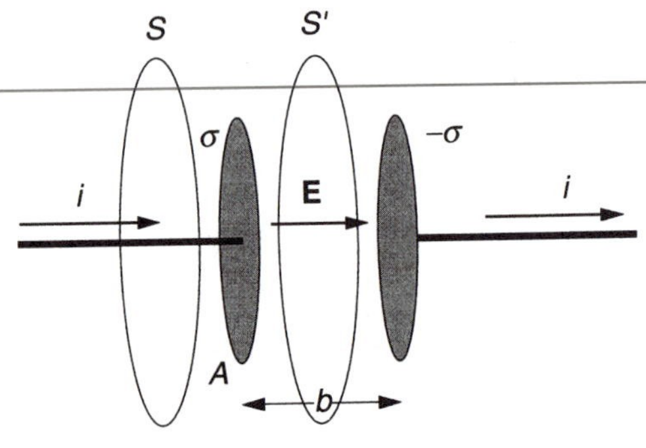

FIGURE 8.9. A wire and capacitor plates are shown. The integral of the current density through surface S, which is pierced by the wire, is i. Through surface S', which is between the capacitor plates, the integral is zero. If the displacement current density is included, both surface integrals are the same. (If surfaces S and S' are not large enough, there is also a net displacement current through S, as can be seen from Fig. 8.10.)

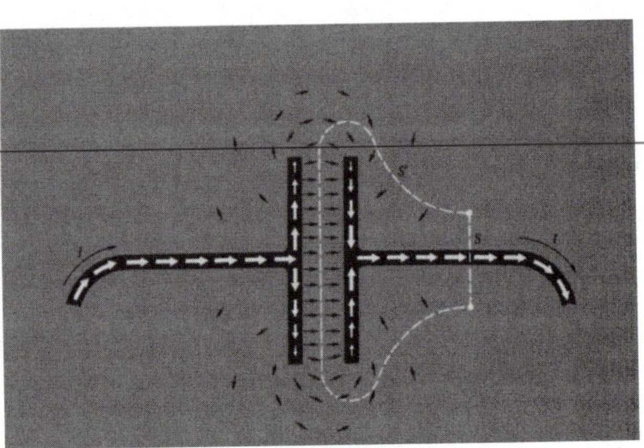

FIGURE 8.10. The conduction current (white arrows) and displacement current (black arrows) in a discharging capacitor are shown. The conduction current decreases with distance out the capacitor plates. The displacement current includes the fringing field. [From E. Purcell (1984). *Electricity and Magnetism*, 2nd ed. Berkeley Physics Series, Vol. 2. New York, McGraw-Hill. Used by permission.)

resolved by noting that the current comes from a charge q at the origin that is being drained off by $i=-dq/dt$. This gives rise to a displacement current $\mathbf{j}_d$ that cancels $\mathbf{j}$ (see Problem 8.7).

8.3. THE MAGNETIC FIELD OF AN AXON IN AN INFINITE, HOMOGENEOUS CONDUCTING MEDIUM

We can use the Biot–Savart law to calculate the magnetic field due to an action potential propagating down an infinitely long axon stretched along the x axis and embedded in an infinite homogeneous conducting medium. Section 7.1 showed that there are three components to the current: i_i along the interior of the axon, di_o out through the membrane (including both displacement current and conduction current), and current in the surrounding medium.

The principle of superposition allows us to calculate the field due to the exterior current by finding the magnetic field $d\mathbf{B}$ from current di_o into the surrounding medium from axon element dx, and then integrating along the axon. We saw in Chap. 7 that the current in the external medium from a small element dx flows uniformly in all directions from a point source. We learned in the preceding section that the magnetic field generated by a spherically symmetric radial current is zero. Therefore, in the approximation that the axon is very thin, we can ignore the *external* current from each element dx. We can do this only because the medium is infinite, homogeneous, and isotropic. When the exterior conductor has boundaries or structure, the symmetry is broken and the external currents contribute to the magnetic field. Our calculation breaks down very close to the axon. Distortions from the field due to the external current because the axon is not infinitely thin are about 1% close to the axon. The current through the cell wall gives a very small contribution to the magnetic field—roughly 1 part in 10^6.

The major contribution is therefore from i_i. We use the law of Biot–Savart, Eq. (8.10). The observation point is in the xy plane at $(x_0, y_0, 0)$ and the axon lies along the x axis so that $d\mathbf{s}=\hat{\mathbf{x}}\,dx$, as shown in Fig. 8.11. The product $d\mathbf{s}\times\mathbf{r}$ can be evaluated using Eq. (1.8) or (1.9):

$$d\mathbf{s}\times\mathbf{r}=\begin{vmatrix} \hat{\mathbf{x}} & \hat{\mathbf{y}} & \hat{\mathbf{z}} \\ dx & 0 & 0 \\ x_0-x & y_0 & 0 \end{vmatrix}=dx\,y_0\hat{\mathbf{z}}.$$

The term in the denominator is $r^3=[(x_0-x)^2+y_0^2]^{3/2}$. The magnetic field in the xy plane is in the z direction and has magnitude

$$B_z=\frac{\mu_0 y_0}{4\pi}\int\frac{i_i(x)\,dx}{[(x_0-x)^2+y_0^2]^{3/2}}.$$

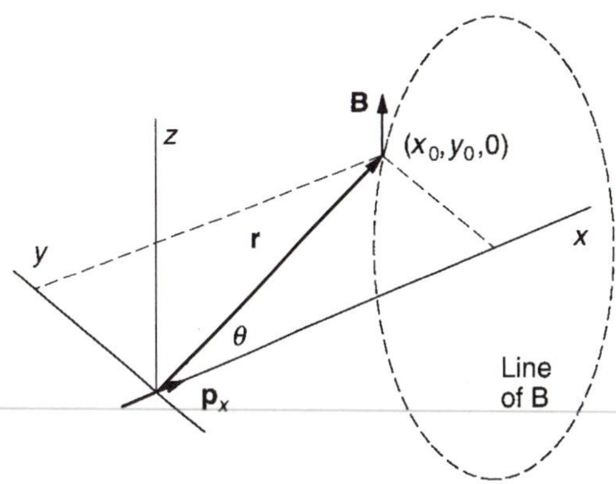

FIGURE 8.11. The geometry for calculation of the magnetic field due to a current element idx or current dipole p_x stretched along the x axis.

It was shown in Eqs. (7.3) that $i_i=-\pi a^2\sigma_i(dv_i/dx)$ and that the interior potential v_i is very nearly equal to the transmembrane potential v. The final expression for B_z is

$$B_z=-\frac{\mu_0 a^2\sigma_i y_0}{4}\int\frac{[dv_i(x)/dx]\,dx}{[(x_0-x)^2+y_0^2]^{3/2}}. \tag{8.12a}$$

The computer program in Fig. 8.12 evaluates the field for the same crayfish axon whose external potential was studied in Sec. 7.1. The field at a distance $2a$ from the axon is plotted in Fig. 8.13. Again, the results agree well with more sophisticated calculations. [Swinney and Wikswo (1980); Woosley, Roth, and Wikswo (1985)]. The latter reference is particularly clear and should be accessible to those who have studied the convolution integral in Chap. 12. A three-dimensional plot of their results is shown in Fig. 8.14.

It is worth repeating that a calculation this simple succeeds only because the axon and the exterior medium are infinite. If there are boundaries, or if there are regions in the external medium where the conductivity changes, then current in the external medium does contribute to the magnetic field. For example, an isolated nerve preparation in air would have the external current flowing in a thin layer of ionic solution along the outside of the axon, where it would generate a field that almost completely cancels that from i_i.

An approximation valid at large distances can be obtained from Eq. (8.12a) by expanding the denominator in much the same way we did to obtain Eqs. (7.26) and (7.27). The observation point is (R,θ) in the xy plane. In this case we need the expansion of

$$\frac{1}{r^3}=\frac{1}{R^3}\left(1-2\frac{x}{R}\cos\theta+\frac{x^2}{R^2}\right)^{-3/2}\approx\frac{1}{R^3}\left(1+\frac{3x\cos\theta}{R}+\cdots\right).$$

The final result is

```
/*R K Hobbe 7 May, 1986. Converted to C July, 1996
This program integrates equation 8.12 for a problem which was
first sovled by K. R. Swinney and J. P. Wilkswo, the extracellular
magnetic field of the single active nerve fiber in a volume
conductor. Biophys. J. 32:719-732 (1980). The nerve pulse is a
series of gaussians used by Clark and Plonsey to fit the data of
Watanabe and Grundfest, J. Gen. Physiol. 45:267 (1961).
Uses Romberg integration routine qromb from W. H. Press et al.
Numerical Recipes in C.*/

#include <stdio.h>
#include <stdlib.h>
#include <math.h>
#include "nr.h"              //Press, et al. Numerical Recipes in C

/*Global Variables*/
const double pi = 3.14159265359;
double x0, y0,                    //coordinates of observation point
A[3], B[3], C[3];                        //parameters for gaussian

float f(float x)
{
   /*Calculates the integrand dv/dx divided by r^3*/
   double xx, r, dv;
   int i;
   dv = 0;
   for (i=0; i<3; i++)
   {
      xx = B[i] * (x- C[i]);
      dv = dv-2 * A[i] * B[i] * xx * exp(-xx * xx);
   }
   r = sqrt((x0 - x) * (x0 - x) + y0 * y0);
   return dv / (r * r * r);
}
void main()
{
   FILE *ofp;                               //output file pointer
   const double Mu0 = 4.0e-7 * pi,    //permeability of free space
   axonradius = 6.0e-5,                        //axon radius in m
   xstart = 0.0, xfinish = 0.020;          //limits of integration
   double integral,
      MagField,                               //Exterior potential
      y[4];       //Calculate at four different distances from axon
   int i;

   if (!(ofp = fopen("OutputFile","w")))
   {
      printf("cannnot open output file\n");
      exit(1);
```

FIGURE 8.12. The program used to calculate the magnetic field outside an axon in an infinite homogeneous conductor using Eq. (8.12). It uses the Romberg integration routing `qromb` from Press *et al.* (1992).

```
    }
    A[0] = 0.051;
    A[1] = 0.072;
    A[2] = 0.018;
    B[0] = 800;
    B[1] = 533;
    B[2] = 333;
    C[0] = 0.0054;
    C[1] = 0.0066;
    C[2] = 0.0086;
    y[0] = 2 * axonradius;
    y[1] = 0.001;
    y[2] = 0.003;
    y[3] = 0.01;

    printf("%12.5g %12.5g %12.5g %12.5g\n",y[0],y[1],y[2],y[3]);
    fprintf(ofp, "%12.5g\t%12.5g\t%12.5g\t%12.5g\n",
            y[0],y[1],y[2],y[3]);
    for(x0 = xstart; x0 < xfinish; x0 += 0.00025)
    {
      printf("%12.5f",x0);
      fprintf(ofp,"%12.5f",x0);
      for(i=0; i<4; i++)
      {
        y0 = y[i];
        integral = qromb(f, xstart, xfinish);
        MagField = -Mu0 * axonradius * axonradius * y0 * integral /
               4.0;
        printf(" %12.5e",MagField);
        fprintf(ofp,"\t%12g",MagField);
      }
      printf("\n");
      fprintf(ofp,"\n");
    }
    fclose(ofp);
}
```

FIGURE 8.12. (Continued.)

$$B_z = \frac{\mu_0 \pi a^2 \sigma_i \sin\theta}{4\pi R^2}(v_i(x_1) - v_i(x_2)) + \frac{\mu_0 \pi a^2 \sigma_i \, 3\sin\theta\cos\theta}{4\pi R^3}\int_{x_1}^{x_2}(v_i(x) - v_{\text{resting}})dx. \tag{8.12b}$$

The first term is the same as what we have in terms of **p**. For a complete pulse the first term vanishes and the second term is used.

8.4. THE MAGNETOCARDIOGRAM

It is now feasible to measure magnetic fields arising from the electrical activity of the heart and brain. The models developed in Sec. 8.3 and in Chap. 7 can be used to compare the electric and magnetic signals from a current dipole **p**. The instrumentation for these measurements is described in Sec. 8.8.

For a single cell at the origin in a homogeneous conducting medium, the exterior potential at observation point **r** is

$$v = \frac{\mathbf{p}\cdot\mathbf{r}}{4\pi\sigma_o r^3}. \tag{7.13}$$

The current dipole **p** points along the cell in the direction of the advancing depolarization wave and has magnitude

$$p = \pi a^2 \sigma_i \Delta v_i. \tag{7.12}$$

An expression analogous to Eq. (7.13) describes the magnetic field of a depolarizing cell. We consider the field due to current along the x axis and then generalize the

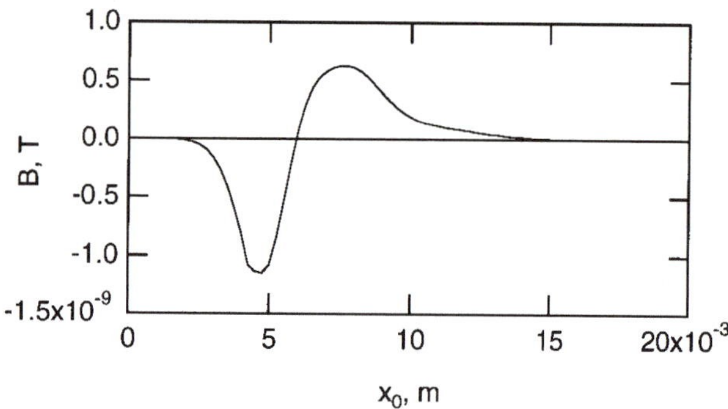

FIGURE 8.13. The magnetic field B_z 0.12 mm from a crayfish axon in an infinite homogeneous conducting medium is shown. The field was calculated using the program of Fig. 8.12. The exterior potential for this configuration was calculated in Sec. 7.4.

result. The derivation begins with Eq. (8.12b) and uses the geometry of Fig. 8.11. The region of depolarization occupies only a millimeter or so along the cell. Since the measurements are made much farther away, the denominator can be removed from the integral, which is then just $\int (dv/dx)\, dx$. If the depolarization is at the origin, then the expression for B_z for $z=0$ is

$$B_z=-\frac{\mu_0 a^2 \sigma_i y_0[v(x_2)-v(x_1)]}{4(x_0^2+y_0^2)^{3/2}}=\frac{\mu_0}{4\pi}\frac{p y_0}{(x_0^2+y_0^2)^{3/2}}. \tag{8.13}$$

Figure 8.11 shows that $y_0=r \sin\theta$, so that $py_0=pr \sin\theta=|\mathbf{p}\times\mathbf{r}|$. The direction of **B** is also consistent with the cross product. Generalizing, we have for a single cell,

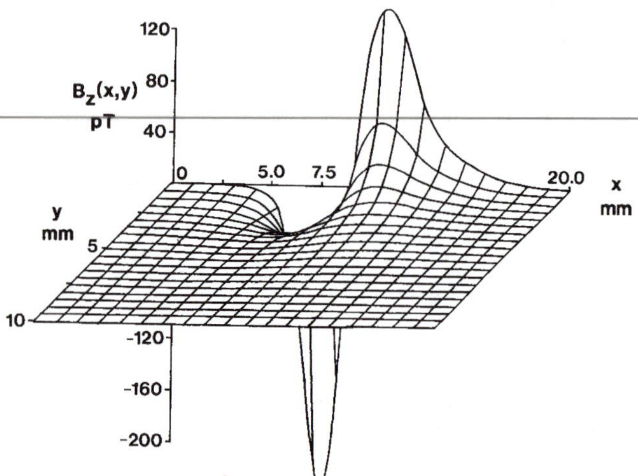

FIGURE 8.14. A three-dimensional plot of the magnetic field around the crayfish axon. The minimum distance from the wire is 0.5 mm. The drawing was supplied by B. Roth and J. P. Wikswo, Vanderbilt University.

$$\mathbf{B}=\mu_0\frac{\mathbf{p}\times\mathbf{r}}{4\pi r^3}. \tag{8.14}$$

Note the remarkable similarity between Eqs. (7.13) and (8.14). One involves the dot product, and the other the cross product. For both, the field falls as $1/r^2$. If we are considering the cardiogram, either field from the entire heart is the superposition of the field from many cells. As with the electrocardiogram, the first approximation for the magnetocardiogram is to ignore changes in $1/r^2$ and speak of the total current-dipole vector.

Measurements of either the potential or the magnetic field can be used to determine the location of **p**. Assume that **p** is at the origin. We will adopt the coordinate system usually used for the magnetocardiogram. The x axis points to the patient's left, the y axis points up, and the z axis points toward the front of the patient, roughly perpendicular to the chest wall. The anterior chest surface is considered to be the xy plane at some fixed value of z. We ignore distortions to the field which arise because no current can flow in the region beyond the body, and we assume that the conductivity of the body is homogeneous and isotropic. From Eq. (8.14), we obtain the three components of **B** along the line $(x,0,z)$:

$$B_x=\frac{\mu_0 p_y z}{4\pi r^3},$$

$$B_y=\frac{\mu_0(p_z x - z p_x)}{4\pi r^3},$$

$$B_z=-\frac{\mu_0 p_y x}{4\pi r^3}. \tag{8.15}$$

Compare these results to the lines of **B** in Fig. 8.15, which were drawn for the three components of **p** using the right-hand rule. Along the line being considered ($y=0$, $z=\text{const}$), p_x contributes only to B_y, and B_y is always negative. Component p_y contributes to both B_x and B_z; the latter changes sign while the former does not, as we change the value of x. Component p_z gives only a y component of **B** that changes sign as x changes sign. The component normal to the body surface, B_z is given by

$$B_z(x,0,z)=-\frac{\mu_0}{4\pi}\frac{p_y x}{(x^2+z^2)^{3/2}}.$$

Figure 8.16 plots contours for the potential and the magnetic field component B_z perpendicular to the body surface when **p** points along the y axis. Again, distortions because of changes in conductivity are ignored. The similarity of the two sets of contours is clear. The contours of constant potential are proportional to $p_y y/r^3$, while the contours for B_z are proportional to $-p_y x/r^3$. Either contour map can be used to determine the location and depth of **p**. To be specific, consider the contours for B_z. The field is proportional to the function $x/(x^2+z^2)^{3/2}$, which changes

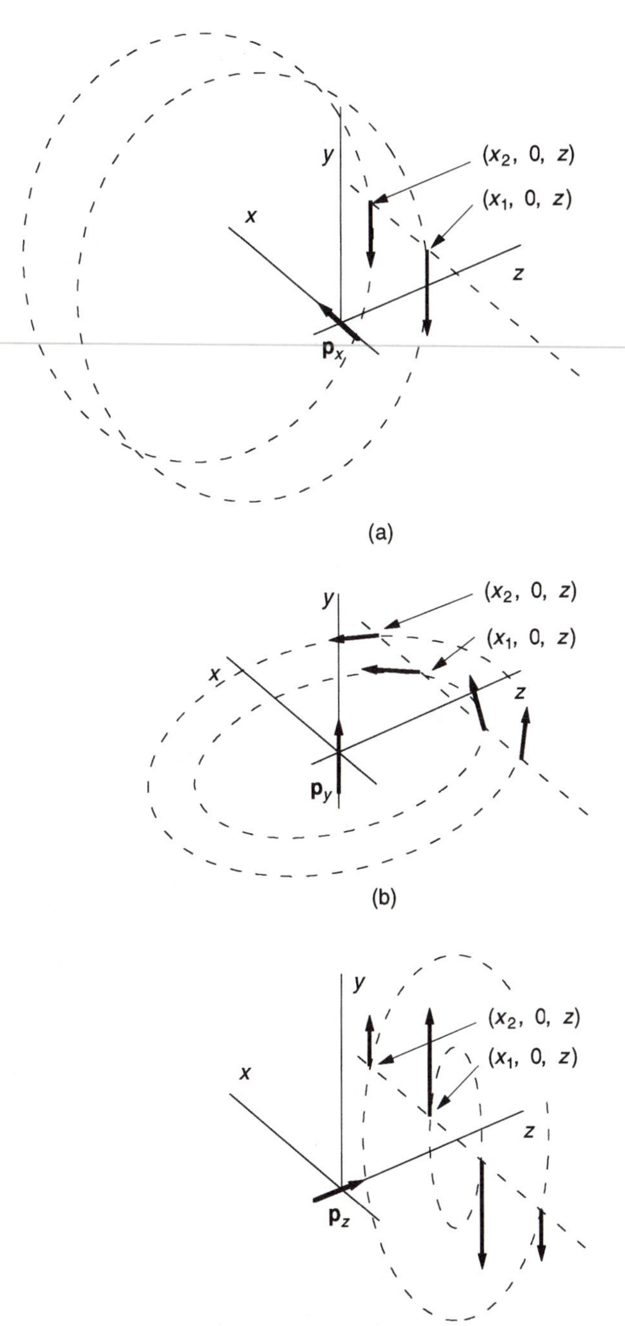

FIGURE 8.15. The magnetic field produced by the three components of a current dipole at the origin. The coordinate system is that customarily used for magnetocardiography. The *x* axis points toward the subject's left, the *y* axis is vertical, and the *z* axis points forward through the subject's chest. The coordinate system is viewed over the subject's right shoulder.

sign right over the source and has a maximum and a minimum at $x=\pm z/\sqrt{2}$. The depth of the source z is related to the spacing Δx along the x axis between the maximum and minimum by

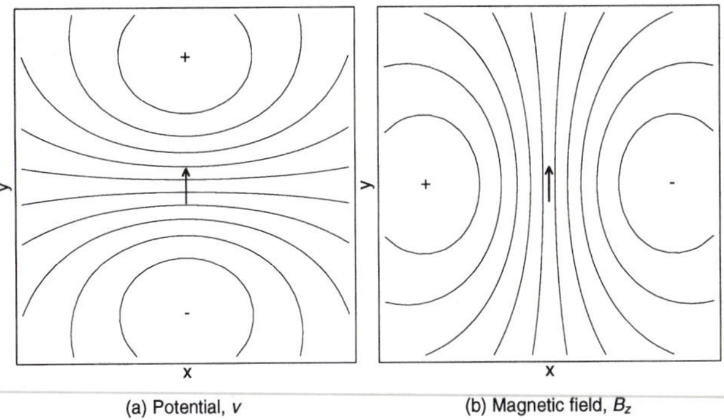

FIGURE 8.16. Contour plots in the *xy* plane for (a) the potential and (b) the *z* component of the magnetic field from a current dipole **p** pointing along the *y* axis, calculated for an infinite, isotropic conducting medium.

$$z=\frac{\Delta x}{\sqrt{2}}. \tag{8.16}$$

The source is located directly beneath the point on the axis where $B_z=0$, and its strength is related to the maximum value of B_z by

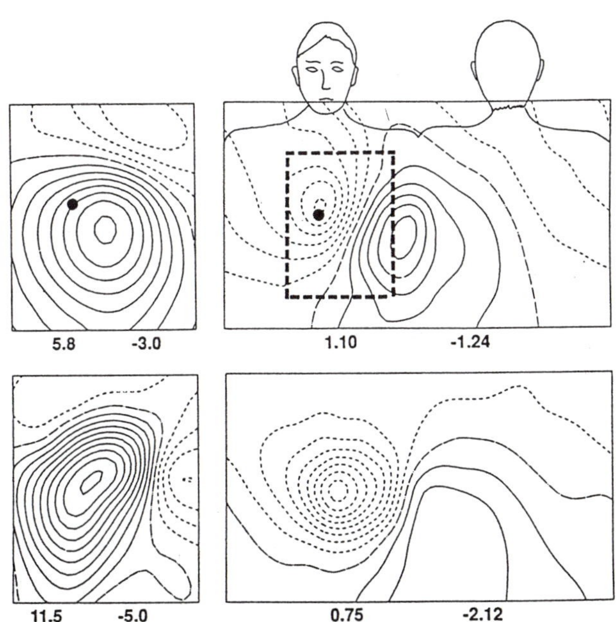

FIGURE 8.17. Maps of the magnetic field perpendicular to the body (left) and the body surface potential (right) in two patients. The upper row is for a normal patient, and the lower row for a patient with an anterior myocardial infarction. The dashed area in the potential map corresponds to the area for which the magnetic field was measured. The dot in the upper row shows where the midline intersects the level of the fourth intercostal space. Note how the constant contours for the magnetic field are oriented at right angles to the isopotential lines, as in the previous figure. From G. Stroink (1992). Used by permission.

$$B_z(\max) = \frac{\mu_0 p_y}{6\pi\sqrt{3}z^2}. \qquad (8.17)$$

Figure 8.17 shows real maps of the potential and the magnetic field on the surface of the chest. While the basic features are described by the simple current dipole model, the exact shape of the contours in Fig. 8.17 differs from the shape in Fig. 8.16. This is due to variations in conductivity of the body. The surface potential is distorted by conductivity differences throughout the thorax; the magnetic field is particularly susceptible to return currents flowing just below the surface of the body. Hosaka *et al.* (1976) did an early calculation of the effect of currents at the surface of the torso on the magnetocardiogram. They found that the component of **B** perpendicular to the body surface is modified by about 30% by the return current. Tangential components of **B** are influenced more; this is why the normal component B_z is usually measured. Other calculations have been done by Purcell *et al.* (1988). Tan *et al.* (1992) show that using a model of the conductivities that matches the geometry of the patient's thorax allows accurate localization of the current dipole source from the surface measurements. Stroink (1992) reviews magnetic field maps; surface potential maps are reviewed by Flowers and Horan (1995) and by Stroink (1996).

The magnetic field close to the heart is affected by the anisotropy of the tissue conductivity. Figure 8.18 shows measurements made 1.5 mm from a 1-mm-thick slice of canine myocardium by Staton *et al.* (1993). Panel A shows the time course of simultaneous recordings from three pickup coils 3 mm in diameter and separated by 4 mm. There are striking differences over 4 mm. Panel B shows a magnetic field contour map during stimulus from another experiment. Instead of having one peak and one valley as in Figs. 8.16 and 8.17, it shows a cloverleaf or *quatrefoil* pattern. Panels C and D show the field contours and the current flow in a third experiment, 6 ms after stimulation. This field and current pattern is predicted by bidomain calculations [Wikswo (1995)].

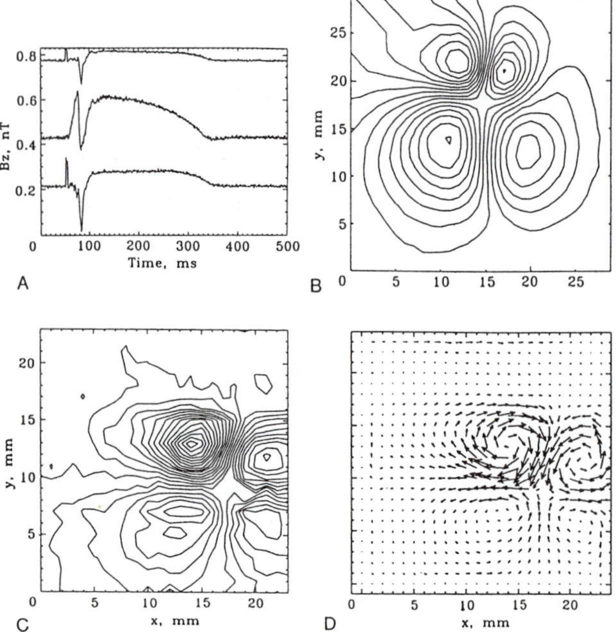

FIGURE 8.18. The results of magnetic field measurements very close to a slice of canine myocardium. The panels are described in the text. Portions from J. P. Wikswo, Jr. (1995). Tissue anisotropy, the cardiac bidomain, and the virtual cathode effect. In D. P. Zipes and J. Jalife, eds. *Cardiac Electrophysiology: From Cell to Bedside*, 2nd ed. Philadelphia, Saunders, pp. 348–361. Portions from D. J. Staton, R. N. Friedman, and J. P. Wikswo, Jr. (1993). High resolution SQUID imaging of octupolar currents in anisotropic cardiac tissue. *IEEE Trans. Appl. Superconduct.* **3**(1):1934–1936. © 1993 IEEE.

8.5. MAGNETIC FIELDS FROM NERVES

The magnetic signals from a nerve action potential are weaker than those from the heart for two reasons. First, the current-dipole vector associated with the repolarization follows close behind the depolarization and reduces the field. (The largest unmyelinated axons in the body have a conduction speed of about 1 m s^{-1} and the pulse is about 1 mm long. Myelinated fibers have a pulse length up to 8–10 times longer.) Second, the cross-sectional area of the advancing wavefront is much smaller. However, action potentials have been measured in nerve [Barach *et al.* (1985); Roth and Wikswo (1985); Woosley *et al.* (1985)] and in muscle [Gielen *et al.* (1991)]. They have also been measured in green algae [Trontelj *et al.* (1994)].

We saw in Sec. 6.1 that nerve cells have an input end (dendrites), a cell body, and an axon. The signal that propagates from a synapse through the dendrites to the cell body and axon is much smaller (about 10 mV) and longer (10 ms) than an action potential that travels along the axon. The cells at the surface of the cerebral cortex have dendrites that are like the trunk of a tree perpendicular to the surface of the cortex, with branches from several directions coming to the trunk. The signal from the trunk is the primary contributor to the magnetoencephalogram (MEG) and electroencephalogram (EEG). The problems show that the magnetic field associated with the rise of the post-synaptic potential is more easily observed outside the brain than is the action potential. One can see from Fig. 8.19 that in a spherically symmetric conducting medium the radial component of **p** does not contribute to the radial magnetic field. Therefore the MEG is most sensitive to detecting activity in the fissures of the cortex, where the ''trunk'' of the post-synaptic dendrite is perpendicular to the surface of the fissure. Since the skull is not a perfect sphere, there is some effect of radial components of **p** on the MEG. The EEG is sensitive to both radial and tangential components of **p**.

Measurements of the magnetoencephalogram are often

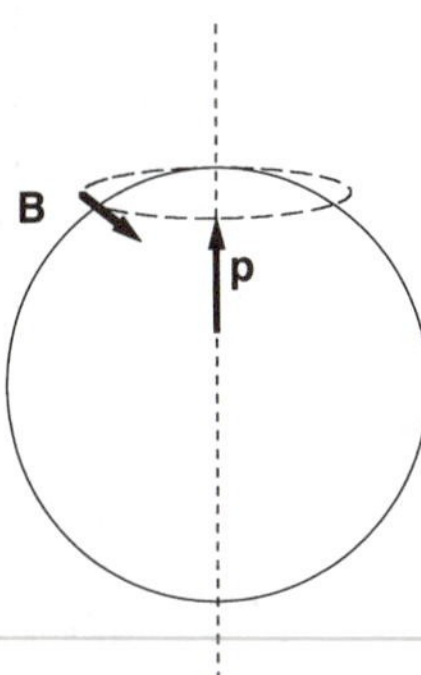

FIGURE 8.19. The lines of **B** around a current dipole **p** are such that the dipole is oriented along the radius of a sphere, there are no components of the field perpendicular to the surface of the sphere.

based on evoked responses. A repetitive stimulus is presented to the subject—audible, visual, or tactile—or the subject is asked to perform a repetitive task such as flexing a finger. Signal-averaging techniques are used to identify the associated changes in magnetic field (see Chap. 11). Figure 8.20 shows averaged magnetic field contours measured over the scalp of a subject who was listening to a string of words presented in random order every 2.3 s. Sometimes the subject was asked to read and ignore the words. At other times the subject was asked to count how many of the words were on a list. The first peak, 100 ms after presentation of the word, was the same in both cases.

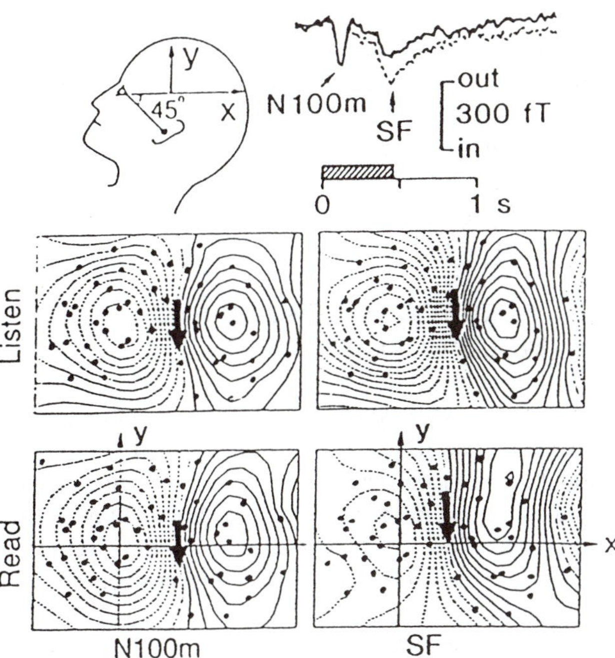

FIGURE 8.20. Magnetic field maps recorded over the scalp of a subject who was either ignoring a series of words by reading or counting how many of the words were in a predetermined list. The features are discussed in the text. From Hämäläinen *et al.* (1993). Used by permission.

The sustained field peak, SF, was considerably stronger when the subject was paying attention to the list. Magnetic contours and the equivalent current dipole source are also shown.

The information available from the EEG and MEG has been reviewed by Wikswo, Gevins, and Williamson (1993).

Magnetic measurements have also been made of slower signals. Grimes *et al.* (1985) found ion currents in the human leg that change over an hour or so. Thomas *et al.* (1993) found a quasistatic ionic current in the chick embryo, probably associated with active-transport pumps. Richards *et al.* (1995a) have measured a magnetic signal associated with slow currents in the small intestine of the rabbit and its supporting mesentery. The signal changes appreciably if the blood supply to the intestine is cut off. These measurements could be clinically useful, because mesenteric ischemia is difficult to diagnose early. Measurements have recently been made in humans [Richards *et al.* (1995b), Petrie *et al.* (1996)].

8.6. ELECTROMAGNETIC INDUCTION

In 1831 Faraday discovered that a *changing* magnetic field causes an electric current to flow in a circuit. It does not matter whether the magnetic field is from a permanent magnet moving with respect to the circuit or the changing current in another circuit. The results of many experiments can be summarized in the *Faraday induction law*:

$$\oint \mathbf{E}\cdot d\mathbf{s} = -\frac{d}{dt}\int\int \mathbf{B}\cdot d\mathbf{S} = -\frac{d\Phi}{dt}. \tag{8.18}$$

It states that the line integral of **E** around a closed path is equal to the negative of the rate of change of the magnetic flux through any surface bounded by the path. The relationship between the direction of **S** and $d\mathbf{s}$ is given by a right-hand rule: if the fingers of the right hand curl around the circuit in the direction of $d\mathbf{s}$, the thumb of the right hand points in the direction of a positive normal to **S**. The units of magnetic flux $\Phi = \int\int \mathbf{B}\cdot d\mathbf{S}$ are T m^2 or weber (Wb).

The differential form of the Faraday induction law is (see Problem 8.14)

$$\text{curl } \mathbf{E} = \nabla \times \mathbf{E} = -\frac{\partial \mathbf{B}}{\partial t}. \tag{8.19}$$

The result of the vector operation curl is another vector. In Cartesian coordinates the components of $\nabla \times \mathbf{E}$ are

$$(\nabla \times \mathbf{E})_x = \frac{\partial E_z}{\partial y} - \frac{\partial E_y}{\partial z},$$

$$(\nabla \times \mathbf{E})_y = \frac{\partial E_x}{\partial z} - \frac{\partial E_z}{\partial x},$$

$$(\nabla \times \mathbf{E})_z = \frac{\partial E_y}{\partial x} - \frac{\partial E_x}{\partial y}.$$

These can be abbreviated by using determinant notation as

$$\nabla\times\mathbf{E}=\begin{vmatrix}\hat{\mathbf{x}} & \hat{\mathbf{y}} & \hat{\mathbf{z}}\\ \frac{\partial}{\partial x} & \frac{\partial}{\partial y} & \frac{\partial}{\partial z}\\ E_x & E_y & E_z\end{vmatrix}. \tag{8.20}$$

The integral form of the Faraday induction law can be used to determine **E** only if the symmetry is such that **E** is always parallel to $d\mathbf{s}$ and has the same magnitude all along the path. One situation where it can be used is a circular loop of wire in the xy plane centered at the origin. The radius of the loop is a and its normal is along the $+z$ axis. Suppose that everywhere in the xy plane within the boundary of the circle the field points along z and depends only on time: $\mathbf{B}(x,y,z,t)=B(t)\hat{\mathbf{z}}$. Symmetry shows that **E** has the same magnitude everywhere and is always tangent to the loop. Equation (8.18) gives

$$E=-\frac{a}{2}\frac{dB}{dt}. \tag{8.21}$$

If the loop is made of material that obeys Ohm's law, there is a current of density $j=\sigma E=-(\sigma a/2)(dB/dt)$. If the radius of the wire is b, then $i=-(\pi ab^2\sigma/2)(dB/dt)$. Figure 8.21 shows the direction of the induced current if dB/dt is positive. The induced current sets up its own magnetic field which points in the $-z$ direction within the loop, opposing the primary field increase within the loop. The induced current always opposes the *change* of magnetic field that produces it. This is called *Lenz's law*. If it were not true, the induced current, once started, would increase indefinitely.

This result does not require that the ring be hollow; it can be part of a much larger conductor. The larger the conductor, the greater the radius of the path along which the induced current can flow. The currents that changing magnetic fields induce in conductors are called *eddy currents* and cause heating losses in the conductor. Iron, which is a conductor, is often used as a core in transformer windings to increase the intensity of the magnetic field. To reduce the eddy-current losses, the cores are made of thin layers of iron insulated from one another by varnish. This limits the radius of the path in which the eddy currents can flow. Some coils and transformers are wound on cores of powdered iron dispersed in an insulating binder. Rooms with thick conducting walls (aluminum, about 2 cm thick) have been used to shield against 60 Hz magnetic fields from power wiring. The eddy currents induced in the aluminum attenuate the field by about a factor of 200 [Stroink *et al.* (1981)].

Rapidly changing magnetic fields can induce currents large enough to trigger nerve impulses. This is discussed in Sec. 8.9.

The quantity $\int_a^b \mathbf{E}\cdot d\mathbf{s}$ is the work done per unit charge in moving from a to b and is called the *electromotive force along the path from a to b*. Terminology is not always consistent; see the discussion by Page (1977). The details of how a changing magnetic field causes a current to flow were shown above for a circular conductor. The force on a moving charge due to the induced electric field is balanced by the drag force as the charge drifts through the conductor. Energy supplied by the changing magnetic field is dissipated as heat. If a voltmeter is attached to two points on the circle, the voltmeter reading may seem paradoxical, until one realizes that there may be changing flux in the voltmeter leads as well. An additional complication is that when there is any region of space in which $\nabla\times\mathbf{E}\neq\mathbf{0}$, then it is possible for $\int_a^b \mathbf{E}\cdot d\mathbf{s}$ to depend on the path (rather than just the end points), even if the magnetic field is zero at all points on the path. This is described clearly and in detail by Romer (1982).

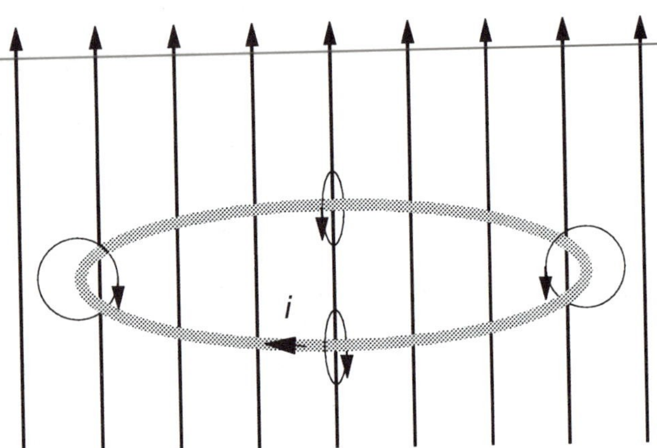

FIGURE 8.21. A magnetic field increasing in the direction shown induces a current in the loop. This current generates a magnetic field in the opposite direction, opposing the change in the magnetic field.

8.7. MAGNETIC MATERIALS AND BIOLOGICAL SYSTEMS

Just as the electric field can be altered by the polarization of a dielectric, the magnetic field can be altered by matter. Biological measurements can be based on alterations of the field by an organ in the body. Some cells exhibit permanent magnetism, which is important for measuring direction in some bacteria and probably in some higher organisms.

The effects of magnetic fields on material are more complicated than those of electric fields. Since there are no known magnetic charges (monopoles), we must consider the effect of magnetic fields on current loops or magnetic dipoles. Figure 8.22 shows a current loop in a magnetic field that decreases as z increases. As a result the lines of **B** spread apart. The loop has radius a, carries current i, and

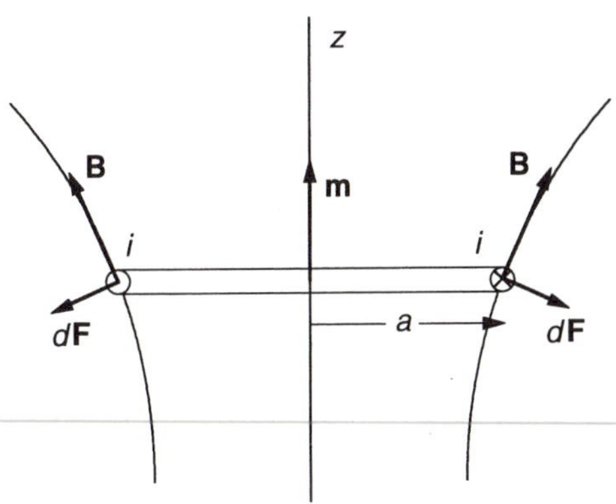

FIGURE 8.22. A current loop in an inhomogeneous magnetic field experiences a force toward the region of stronger magnetic field.

has magnetic moment[1] **m**. For the orientation shown, there is a force on the loop in the $-z$ direction that is toward the region where the field is stronger. If the magnetic moment of the loop were not parallel to **B**, there would also be a torque on the loop. For ease in calculation, imagine that the loop has been placed in the field in such a way that along the axis of the loop, **B** points in the z direction. Then the spreading of the lines of **B** means that **B** has a component radially outward all around the loop. Because of the symmetry B_r has a constant magnitude everywhere around the loop, and the force on the loop is $-2\pi a i B_r(a)$. Field $B_r(a)$ is found by considering the fact that the total magnetic flux through all surfaces of the pillbox in Fig. 8.23 is zero. The net outward flux is

$$[B_z(z+dz)-B_z(z)]\pi a^2+B_r(a)2\pi a dz=0.$$

This can be rearranged to give

$$\left(\frac{\partial B_z}{\partial z}+\frac{2}{a}B_r(a)\right)\pi a^2 dz=0,$$

from which

$$B_r(a)=-\frac{a}{2}\frac{\partial B_z}{\partial z}.$$

The force on the loop is therefore

$$F_z=\pi a^2 i\frac{\partial B_z}{\partial z}=m_z\frac{\partial B_z}{\partial z}. \qquad (8.22)$$

If **m** is parallel to **B** the force is toward the region of stronger field; if **m** is antiparallel to **B** the force is toward the region of weaker field.

[1]Be careful. We are talking about two different kinds of dipoles in this chapter. The current dipole **p** is a source and sink of current and has units A m. The magnetic dipole **m**, equivalent to a small magnet with north and south poles, has units A m^2. The magnetic field from a magnetic dipole falls off as $1/r^3$.

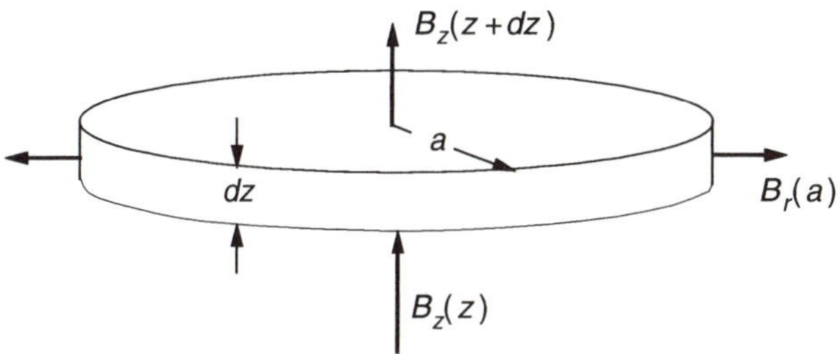

FIGURE 8.23. Gauss's law for **B** is applied to a pillbox of radius a and thickness dz.

An atom can have a magnetic moment because of two effects.[2] The motion of the electrons in orbit about the nucleus constitutes a current; as a result of which there may be an orbital magnetic moment. The intrinsic spin of each electron gives rise to a spin magnetic moment, independent of any orbital motion. In most atoms, the orbital magnetic moments average to zero, and most of the electrons are arranged in pairs whose spins cancel. The atom therefore usually has no net magnetic moment.

Most substances placed in an inhomogeneous field experience a weak force *away* from the region of strong field, and the force is roughly proportional to the square of the field strength, an effect called *diamagnetism.* It can be understood with a simple classical model. As the atom is moved into the magnetic field, the Faraday induction effect distorts the orbits of the electrons to induce a magnetic dipole moment proportional to **B** and in the opposite direction, consistent with Lenz's law. The force is therefore proportional to $B_z(\partial B_z/\partial z)$. Purcell (1985) has a quantitative treatment of this model.

A few substances are *attracted* to the region of stronger field, again with a force that is often proportional to the square of the field. Each atom of these *paramagnetic* substances has a permanent magnetic moment associated with the spin of an unpaired electron. Thermal motion normally keeps the magnetic moments of different atoms oriented randomly. As the substance is brought into the magnetic field the spin magnetic moments of different atoms begin to align with the magnetic field. A magnetic dipole moment is induced in the substance, but this time it is in the direction of **B**, and the substance is attracted to the magnet.

Some substances placed in an inhomogeneous magnetic field experience much stronger attraction than do paramagnetic substances. In these substances some of the atomic moments are aligned even in the absence of an external field. They are permanent magnets. Further alignment of the atomic moments may take place in an external field, but complete alignment often takes place in relatively weak external fields. These substances are called *ferromagnets.* The individual atoms have magnetic moments, and there are

[2]Much weaker magnetic moments of the atomic nucleus are considered in Chap. 17.

forces between atoms which cause the spins to align. Section 14-4 of Eisberg and Resnick (1985) provides a relatively simple explanation of the quantum-mechanical effects underlying this spin alignment. *Ferrimagnets* are similar to ferromagnets, but the crystals contain two different kinds of ions with different magnetic moments.

The *magnetization* **M** is the average magnetic moment per unit volume. It is defined by considering volume ΔV that has total magnetic moment $\Delta\mathbf{m}=\Sigma\mathbf{m}_i$, where the summation is taken over all atoms in the volume, and taking the ratio

$$\mathbf{M}=\frac{\Delta\mathbf{m}}{\Delta V}. \tag{8.23}$$

We have seen that a current loop possesses a magnetic moment of magnitude $m=iS$. One can imagine a current giving rise to any magnetic moment, even one associated with electron spin. Such currents are called *bound* currents and must be included in Ampere's law. The currents that flow due to conduction—that we can control by changing the conductivity of the material or throwing a switch—are called *free currents*. One can show that if we define the new vector

$$\mathbf{H}=\mathbf{B}/\mu_0-\mathbf{M}, \tag{8.24}$$

it depends only on the free currents:

$$\oint \mathbf{H}\cdot d\mathbf{s}=\int\int \mathbf{j}_{\text{free}}\cdot d\mathbf{S}. \tag{8.25}$$

Vector **H** is called the *magnetic field intensity*. It has units A m^{-1}. It does not have the physical significance of **B** (it does not appear in the Lorentz force, and its divergence is not always zero, for example). However, it often simplifies computations, because we control free current in the laboratory.

In a vacuum, $\mathbf{B}=\mu_0\mathbf{H}$. It has been traditional to define the *magnetic permeability* of a medium in which **B**, **M**, and **H** are all proportional to one another by the equation

$$B=\mu H, \tag{8.26}$$

in which case

$$\frac{\mu}{\mu_0}=1+\frac{M}{H}=1+\chi_m. \tag{8.27}$$

In diamagnetic materials the *magnetic susceptibility* χ_m is less than 0 and $\mu<\mu_0$. A typical diamagnetic susceptibility is $\approx -1\times10^{-5}$. In paramagnetic materials $\chi_m>0$ and $\mu>\mu_0$. A typical paramagnetic susceptibility is $\approx 1\times10^{-4}$.

The relationship between B and H in ferromagnetic substances is nonlinear and is characterized by a BH curve. A "typical" curve is shown in Fig. 8.24. The fact that the curves for increasing and decreasing H do not coincide is called *hysteresis*. The arrows show the direction in which H changes on each branch of the curve. Saturation takes place beyond points W and Y. When $H=0$ there is a *remanent* magnetic field (points X and Z). If the temperature of the sample is raised above a critical temperature called the *Curie temperature*, the magnetism is destroyed.

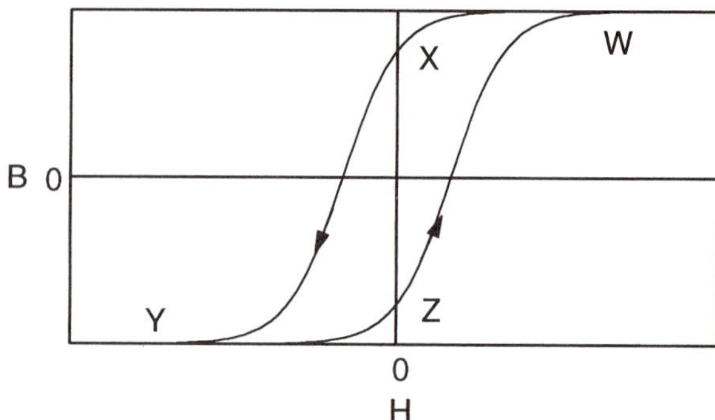

FIGURE 8.24. A typical curve of B vs H for a ferromagnetic material. The curve shows hysteresis, and the arrows show the direction of travel around the curve $WXYZ$. Points W and Y show where B saturates. Points X and Z show the remanent magnetic field when $H=0$.

Several kinds of measurements can be based on magnetic effects in materials. A common component of dust inhaled by miners and industrial workers is magnetite, Fe_3O_4, which is ferrimagnetic. By placing the thorax in a fixed magnetic field for a few seconds, the particles can be aligned. The field is turned off and the remanent field measured. The use of *magnetopneumography* in occupational health is described by Stroink (1985). Cohen *et al.* (1984) have modeled the process by which the particles are magnetized, as well as the relaxation process by which the magnetization disappears after the external field is removed. Relaxation curves are used to estimate intracellular viscosity and the motility of macrophages (scavenger white cells) in the alveoli [Stahlhofen and Moller (1993)].

The magnetic susceptibility of blood and myocardium is different from the susceptibility of surrounding lung tissue. An externally applied magnetic field induces a field that changes as the volume of the heart changes. It can be measured externally. The theory and experiments have been described by Wikswo (1980).

Susceptibility measurements can also be used to measure the total iron stores in the body. Normally the body contains 3–4 g of iron. About a quarter of it is stored in the liver. The amount of iron can be elevated from a large number of blood transfusions or in certain rare diseases such as hemochromatosis and hemosiderosis. The liver is an organ whose susceptibility can easily be measured. The susceptibility varies linearly with the amount of iron deposited. Magnetic susceptometry has been used to estimate body iron stores [Brittenham *et al.* (1983); Nielsen *et al.* (1995)].

Magnetism is used for orientation by some organisms. Several species of bacteria contain linear strings of up to 20 particles of magnetite, each about 50 nm on a side encased in a membrane [Frankel *et al.* (1979); Moskowitz (1995)]. Over a dozen different bacteria have been identified that

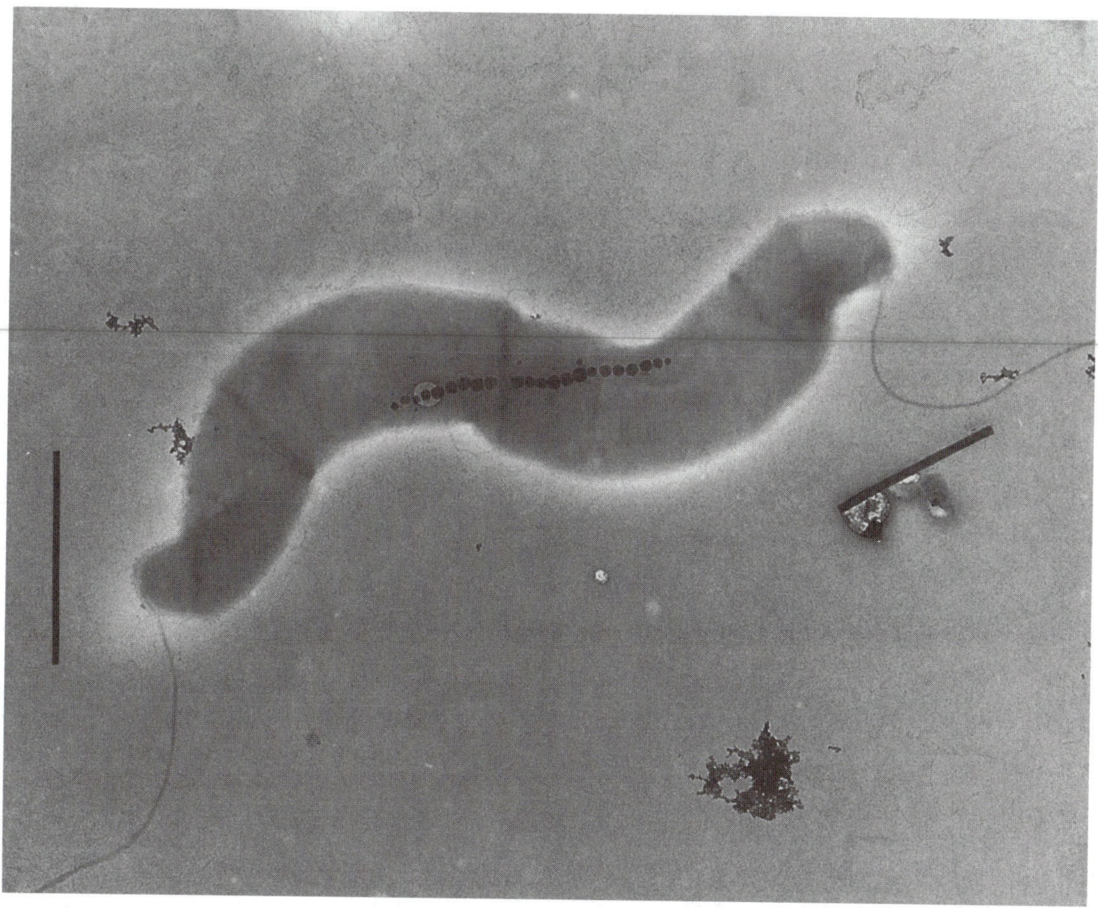

FIGURE 8.25. The magnetosomes, small particles of magnetite in the magnetotactic bacterium *Aquaspirillum magnetotacticum* are shown. The bar is 1 μm long. The photograph was taken by Y. Gorby and was supplied by N. Blakemore and R. Blakemore, University of New Hampshire.

synthesize these intracellular, membrane-bound particles or magnetosomes (Fig. 8.25). Bacteria in the northern hemisphere have been shown to seek the north pole. Because of the tilt of the earth's field, they burrow deeper into the environment in which they live. Similar bacteria in the southern hemisphere burrow down by seeking the south pole. In the laboratory the bacteria align themselves with the local field. In the problems you will learn that there is sufficient magnetic material in each bacterium to align it with the earth's field just like a compass needle. Other bacteria that live in oxygen-poor, sulfide-rich environments contain magnetosomes composed of greigite, Fe_3S_4 rather than magnetite (Fe_3O_4). In aquatic habitats, high concentrations of magnetotactic bacteria are usually found near the *oxic–anoxic transition zone* (OATZ), which contains a very high concentration of both kinds of bacteria. In freshwater environments the OATZ is usually at the sediment–water interface. In marine environments it is displaced up into the water column. Since some bacteria prefer more oxygen and others prefer less, and they both have the same kind of propulsion and orientation mechanism, one wonders why one kind of bacteria is not swimming out of the favorable environment. Frankel and Bazylinski (1994) have proposed that the magnetic field and the magnetosomes keep the organism aligned with the field, and that they change the direction in which their flagellum rotates to move in the direction that leads them to a more favorable concentration of some desired chemical.

Magnetosomes are found in other species and are likely also to be used for orientation. One species of algae contains about 3000 magnetic particles, each of which is about $40 \times 40 \times 140$ nm [de Araujo *et al.* (1986)]. Bees, pigeons, and fish contain magnetic particles. It is more difficult to demonstrate their function, because of the variety of other sensory information available to these animals. For example, homing pigeons with magnets attached to their heads could orient well on sunny days but not on cloudy ones [Walcott *et al.* (1979)]. There is evidence that bees orient in a magnetic field. These experiments are summarized in Frankel (1984). The net magnetic moment in the bees is oriented transversely in the body [Gould *et al.* (1978)]. In pigeons the magnetic material is located in the dura (the outer covering of the brain) or skull. In all of these cases, the material has been identified as magnetite. In the yellowfin tuna, data are compatible with about 8.5×10^7 magnetic particles, each of which is a single domain of magnetite in the shape of an approximately 50-nm cube [Walker *et al.* (1984)].

There is now evidence that birds may actually have three compasses. Since the magnetic and geographic poles are fairly far apart, migratory birds must correct their magnetic compasses as they fly. The Savannah sparrow is known to have a magnetic compass and a star compass and to take visual cues from the sky at sunset. Able and Able (1995) have shown that adult Savannah sparrows that are subjected to a field pointing in a different direction than the earth's field will at first trust their magnetic compasses, but over a few days they recalibrate their magnetic compasses with their star compasses. An accompanying editorial [Gould (1995)] places their work in context.

The fact that the magnetite particles seem to be about 50 nm on a side is physically significant. Frankel (1984) summarizes arguments that if the particles are smaller than about 35 nm on a side, thermal effects can destroy the alignment of the individual particles. If they are larger than about 76 nm, multiple domains can form within a particle, decreasing the magnetic moment. Chains of bacterial magnetosomes have recently been used to calibrate a magnetic force microscope [Proksch *et al.* (1995)].

8.8. DETECTION OF WEAK MAGNETIC FIELDS

The detection of weak fields from the body is a technological triumph. The field strength from lung particles is about 10^{-9} T; from the magnetocardiogram it is about 10^{-10} T; from the brain it is 10^{-12} T for spontaneous (α wave) activity and 10^{-13} T for evoked responses. These signals must be compared to 10^{-4} T for the earth's magnetic field. Noise due to spontaneous changes in the earth's field can be as high as 10^{-7} T. Noise due to power lines, machinery, and the like can be 10^{-5}–10^{-4} T.

If the signal is strong enough, it can be detected with conventional coils and signal-averaging techniques that are described in Chap. 11. Barach *et al.* (1985) used a small detector through which a single axon was threaded. The detector consisted of a toroidal magnetic core wound with many turns of fine wire (Fig. 8.26). Current passing through the hole in the toroid generated a magnetic field that was concentrated in the ferromagnetic material of the toroid. When the field changed, a measurable voltage was induced in the surrounding coil.

The signals from the body are weaker, and their measurement requires higher sensitivity and often special techniques to reduce noise. Hämäläinen *et al.* (1993) present a detailed discussion of the instrumentation problems. Sensitive detectors are constructed from superconducting materials. Some compounds, when cooled below a certain critical temperature, undergo a sudden transition and their

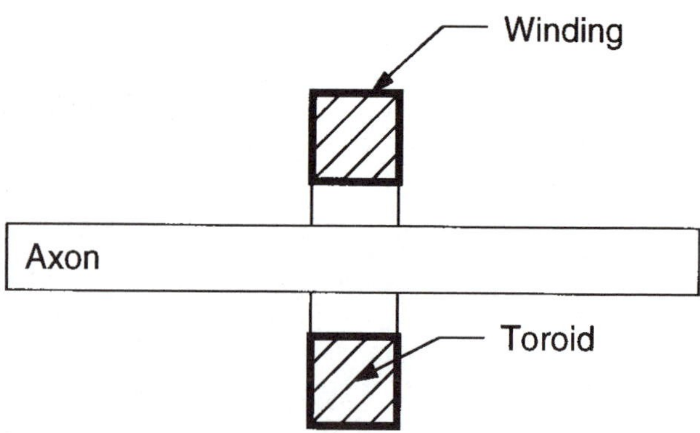

FIGURE 8.26. A nerve cell preparation is threaded through the magnetic toroid to measure the magnetic field. The changing magnetic flux in the toroid induces an electromotive force in the winding. Any external current that flows through the hole in the toroid diminishes the magnetic field.

electrical resistance falls to zero. A current in a loop of superconducting wire persists for as long as the wire is maintained in the superconducting state. The reason there is a superconducting state is a well-understood quantum-mechanical effect that we cannot go into here. It is due to the cooperative motion of many electrons in the superconductor [Eisberg and Resnick (1985), Sec. 14.1; Clarke (1994)]. The integral $\oint \mathbf{E}\cdot d\mathbf{s}$ around a superconducting ring is zero, which means that $d\Phi/dt$ is zero, and the magnetic flux through a superconducting loop cannot change. If one tries to change the magnetic field with some external source, the current in the superconducting circuit changes so that the flux remains the same.

The detector is called a superconducting quantum interference device (SQUID). The operation of a SQUID and biological applications is described in the *Scientific American* article by Clarke (1994). Wikswo (1995a) surveys the use of SQUIDs for applications in biomagnetism and nondestructive testing. A technical discussion is also available [Hämäläinen *et al.* (1993)]. The dc SQUID requires a superconducting circuit with two branches, each of which contains a very thin nonsuperconducting "weak link" known as a Josephson junction (Fig. 8.27). As the

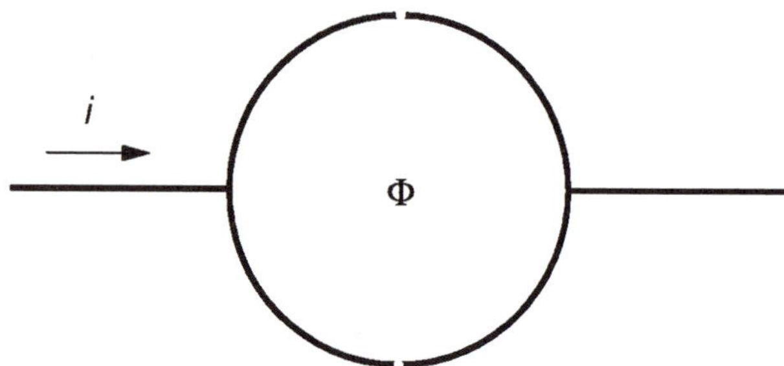

FIGURE 8.27. A dc SQUID is shown. The solid lines represent superconducting wires, broken by Josephson junctions at the top and bottom. The total current through both wires depends on Φ, the magnetic flux through the circle.

magnetic field is changed, these weak links allow the flux in the loop to change. The phase of the quantum mechanical wave function of the collectively moving electrons differs in the two branches by an amount depending on the magnetic flux linked by the circuit. The total current depends on the interference of these two wave functions and is of the form $I = 2I_0 \cos(\pi\Phi/\Phi_0)$, where Φ is the flux through the circuit. The quantity $\Phi_0 = h/2e$, where h is Planck's constant (see Chap. 13) and e is the electron charge, is the quantum-mechanical unit of flux and has a value equal to 2.068×10^{-15} T m^2. Because interference changes corresponding to a small fraction of this can be measured, the SQUID is very sensitive. The SQUID must be operated at temperatures where it is superconducting. It used to be necessary to keep a SQUID in a liquid-helium bath, which is expensive to operate because of the high evaporation rate of liquid helium. With the advent of high-temperature superconductors, SQUIDS have the potential to operate at liquid-nitrogen temperatures, where the cooling problems are much less severe.

A typical magnetometer for biomagnetic research contains a *flux transporter*, a superconducting detector coil d a centimeter or so in radius, coupled to a very small multiturn output coil o that matches the size of the SQUID and is placed right next to it. This is shown schematically in Fig. 8.28. The wires between the two loops are close together and have negligible area between them. The total flux, which is constant because the entire circuit is superconducting, is $\Phi = \Phi_d + \Phi_o$. The large area of the detecting coil increases its sensitivity. Any change in the magnetic field at the detector causes an opposite change in the flux and magnetic field at the output coil.

Because ambient natural and artificial background magnetic fields are so high, measurements are often made in special shielded rooms. These can be built of ferromagnetic materials or of conductors to take advantage of eddy current attenuation, or they may have active circuits to cancel the background fields. It has proven possible in some cases to eliminate the need for these rooms by using specially designed flux transporter that are less sensitive

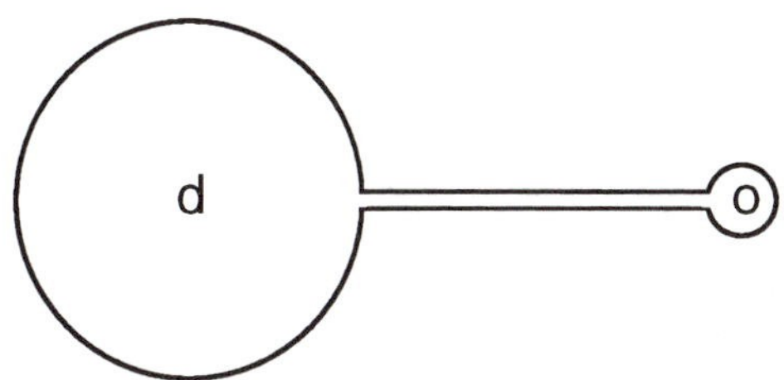

FIGURE 8.28. A superconducting loop shaped as shown becomes a flux transporter. Because the total flux in the loop is constant, a change of flux in the detecting loop *d* is accompanied by an equal and opposite flux change in the output loop *o*. The diameter of the output loop is matched to the size of the SQUID. Sensitivity is increased because the detecting loop has a larger area.

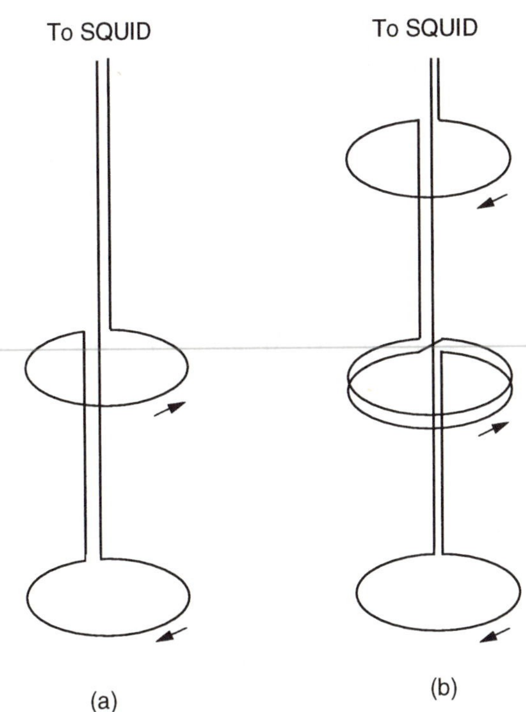

FIGURE 8.29. Gradiometers are sensitive to nearby sources of the magnetic field but are much less sensitive to distant sources. (a) A first-order gradiometer. (b) A second-order gradiometer.

to distant sources but measure the nearby source with almost the same sensitivity as a single loop. If a distant background source can be represented by a magnetic dipole, the field falls as $1/r^3$. The signal in a magnetometer (Fig. 8.28) would be proportional to this. Problem 8.25 shows that the signal from distant a dipole detected by a first-order gradiometer [Fig. 8.29(a)] is proportional to $1/r^4$ and that the signal in a second order gradiometer [Fig. 8.29(b)] is proportional to $1/r^5$. Both gradiometers are insensitive to background that does not vary with position. Yet the loop closest to the nearby signal source detects a much stronger signal than the loops that are further away. With modern multi-channel detector systems, one need not use gradiometer coils. A large number of coils are used at different locations and the signals from them are combined to give the same suppression of background from distant sources.

8.9. MAGNETIC STIMULATION

Since a changing magnetic field generates an induced electric field, it is possible to stimulate nerve or muscle cells without using electrodes. The advantage is that for a given induced current deep within the brain, the currents in the scalp that are induced by the magnetic field are far less than the currents that would be required for electrical stimulation. Therefore magnetic stimulation is relatively painless. Magnetic stimulation can be used to diagnose central nervous system diseases that slow the conduction velocity in motor nerves without changing the conduction velocity in sensory nerves. It could be used to map motor brain function and to monitor motor nerves during spinal cord surgery.

One of the earliest investigations was reported in 1985 by Barker *et al.* who used a solenoid in which the magnetic field changed by 2 T in 110 μs to apply a stimulus to different points on a subject's arm and skull. The stimulus made a subject's finger twitch after the delay required for the nerve impulse to travel to the muscle. For a region of radius $a=10^{-2}$ m in material of conductivity 1 S m^{-1}, the induced current density for the field change in Barker's solenoid was 90 A m^{-2}. (This is for conducting material inside the solenoid; the field falls off outside the solenoid, so the induced current is less.) This current density is large compared to current densities in nerves (Chap. 6).

Brasil-Neto *et al.* (1992) studied the effects of coil orientation, stimulus intensity, and shape of the induced current pulse on the amplitudes of motor evoked potentials stimulated magnetically. Several theoretical analyses and models have been made, including those by Heller and van Hulsteyn (1992) and Eselle and Stuchly (1995). Magnetic stimulation is included in the review by Roth (1994).

SYMBOLS USED IN CHAPTER 8

Symbol	Use	Units	First used on page
a,b	Distance	m	209
i	Current	A	207
j, $\mathbf{j}$	Current density	A m^{-2}	209
j_d	Displacement current density	A m^{-2}	209
m, $\mathbf{m}$	Magnetic moment	A m^2	207
$\mathbf{p}$	Current dipole moment	A m	213
q	Charge	C	206
r	Distance	m	208
s, $\mathbf{s}$	Linear displacement	m	207
t	Time	s	217
v, $\mathbf{v}$	Velocity	m s^{-1}	206
v	Electrical potential	V	211
x,y,z	Coordinates	m	209
$\hat{\mathbf{x}},\hat{\mathbf{y}},\hat{\mathbf{z}}$	Unit vectors		211
A	Area	m^2	209
B, $\mathbf{B}$	Magnetic field	T	206
D, $\mathbf{D}$	Electric "displacement"	C m^{-2}	209
E, $\mathbf{E}$	Electric field	V m^{-1}	207
F	Force	N	206
H	Magnetic field intensity	A m^{-1}	220
M	Magnetization	A m^{-1}	220
S, $\mathbf{S}$	Surface area	m^2	208
ϵ_0	Electric permittivity of space	N^{-1} m^{-2} C^2	210
θ	Angle		207
κ	Dielectric constant		210
μ	Magnetic permeability	Ω s m^{-1}	220
μ_0	Magnetic permeability of space	Ω s m^{-1}	208
σ	Charge per unit area	C m^{-2}	210
σ_i,σ_o	Electrical conductivity	S m^{-1}	211
τ, $\boldsymbol{\tau}$	Torque	N m	208
ϕ	Angle		207
χ_m	Magnetic susceptibility		220
Φ	Magnetic flux	T m^2 or Wb	217

PROBLEMS

Section 8.1

8.1. An electric dipole consists of charges $\pm q$ separated a distance b. Show that the torque on an electric dipole in a steady electric field is given by $\boldsymbol{\tau}=\mathbf{p}\times\mathbf{E}$, where $\mathbf{p}$ is the dipole moment of magnitude qb, pointing in the direction from $-q$ to $+q$.

8.2. Show that the units of $\mathbf{m}$, A m^2 or J T^{-1}, are equivalent.

8.3. Show that the units of μ_0, T m A^{-1}, are equivalent to Ω s m^{-1}.

Section 8.2

8.4. A very long solenoid of radius a has current i in the windings. The windings are closely spaced and there are N turns per meter. What is the magnetic field in the solenoid? (Hint: if the solenoid is very long, the field inside is uniform and the field outside is zero.)

8.5. Show that $d\mathbf{D}/dt$ has the dimensions of current density.

8.6. A circular loop of radius a and area S carries current i. The loop is at the origin and lies in the xy plane. Find the magnetic field at any point on the z axis and show that it is proportional to the magnetic moment of the loop, $|\mathbf{m}|=iS$.

8.7. Show that a point source of current discharges at such a rate that the displacement current density everywhere cancels the current density, so that Ampere's law also predicts that the magnetic field is zero.

Section 8.3

8.8. The current along an axon is $i_i(x)=i_0$, $0<x<x_1$ and is zero everywhere else. The axon is in an infinite homogeneous conducting medium.

(a) What is $v_i(x)$?

(b) Find **B** at a point (x_0,y_0).

8.9. One can obtain a very different physical picture of the source of a magnetic field using the Biot–Savart law than one gets using Ampere's law, even though the field is the same. A ring of radius a is perpendicular to the x axis and centered at x_0. Current flows along the x axis from $x=0$ to $x=x_2$. There is a spherically symmetric current in at $x=0$ and a spherically symmetric current out from x_2. Calculate the magnetic field at a point on the ring using Ampere's law and using the Biot–Savart law. Discuss the difference in interpretation; your expression for the field should be the same both ways. [A more extensive discussion of three different ways the source of the magnetic field can be viewed is given by Barach (1987).]

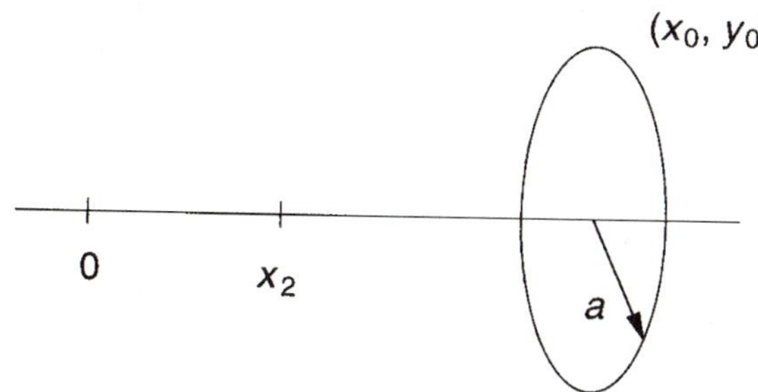

8.10. Suppose that $i_i(t)$ is determined by measurement of the magnetic field around an axon. Numerical differentiation of the data gives derivatives of i_i also. Use the arguments of Sec. 6.11 and Problem 6.37 to show that for an action potential traveling without change of shape, one can determine the membrane current density from

$$j_m=\frac{1}{2\pi a u}\frac{\partial i_i}{\partial t}-c_m u r_i i_i.$$

For an application of this technique, see Barach *et al.* (1985).

Section 8.4

8.11. Use the same technique as in Chap. 7 to estimate the magnitude of the magnetocardiogram signal.

8.12. (a) Derive Eqs. (8.16) and (8.17).

(b) What effect will y and z components of **p** have for measurements taken along an axis with $y=0$?

Section 8.5

8.13. Consider two cylindrical cells of radius 1 μm. One is an axon with an action potential lasting 1 ms and traveling at 1 m/s with a depolarization amplitude of 100 mV. The other is a dendrite with a post-synaptic potential depolarization of 10 mV. The conductivity within both cells is 1 S m^{-1}.

(a) Compare the magnetic field 5 cm away from the dendrite with depolarization only and the axon with a complete pulse.

(b) If the minimum magnetic field that can be detected is 100×10^{-15} T, how many dendrites must be simultaneously excited to detect the signal?

(c) Pyramidal cells in the cortex are aligned properly to generate this kind of signal. Assume the dendrite is 2 mm long. There are about 50 000 neurons/mm^3 in the cortex, of which 70% are pyramidal cells. Find the volume of the smallest excited region that could be detected if all the pyramidal cells in the volume simultaneously had a post-synaptic depolarization of 10 mV.

Section 8.6

8.14. Consider a current loop with one corner at (0,0,0) and the diagonally opposite corner at $(dx,dy,0)$, in a changing magnetic field that has components $(0,0,dB/dt)$. Show that for this configuration Eq. (8.19) follows from Eq. (8.18).

8.15. Derive Eq. (8.21).

8.16. It is possible that the Faraday induction law allows marine sharks, skates, and rays to orient in a magnetic field [Frankel (1984)]. If a shark can detect an electric field strength of 0.5 μV m^{-1}, how fast would it have to swim through the earth's magnetic field to experience an equivalent force on a charged particle? The earth's field is about 5×10^{-5} T.

Section 8.7

8.17. Magnetite, Fe_3O_4, has a density of 5.24 g cm^{-3} and a magnetic moment of 3.75×10^{-23} A m^2 per molecule. If a cubic sample 50 nm on a side is completely magnetized, what is the total magnetic moment? What is the magnitude of M?

8.18. The magnetic moment of a magnetosome, one of the small particles of magnetite in a bacterium, is about 6.40×10^{-17} A m^2. Assume that the magnetic activity in all the species listed is due to a collection of magnetosomes of this size. The table shows values given in the references cited in the text. The earth's magnetic field is about 5×10^{-5} T. Fill in the remaining entries in the table.

Organism	Number of magnetosomes	Total magnetic moment	mB_{earth}/k_BT
Bacterium	20		
Bees		1.2×10^{-9}	
Pigeons		5.0×10^{-9}	
Tuna	8.5×10^7		

8.19. In this problem you will work out the orientation of a bacterium if the entire organism simply aligns like a compass needle in the earth's field of 5×10^{-5} T.

(a) Show that $\boldsymbol{\tau}=\mathbf{m}\times\mathbf{B}$ implies an orientation energy $U=-mB\cos\theta$.

(b) The bacterium has a single flagellum that causes it to swim in the direction of its long axis with speed v_0. The component of its velocity in the direction of the earth's field is $v_x=v_0\cos\theta$. In the absence of the magnetic torque, the probability that a bacterium is at an angle between θ and $\theta+d\theta$ with the earth's field is $d\Omega=2\pi\sin\theta\,d\theta$. With the torque, the probability that a bacterium is at angle θ with the earth's field is modified by a Boltzmann factor $\exp(-U/k_BT)$. Find the average velocity in the directionof the earth's field. Use $m=1.28\times10^{-15}$ A m^2.

8.20. Suppose that the bacterium of the preceding problem is swimming in a tank aligned with the earth's field. An external coil suddenly reverses the direction of the field but leaves the magnitude unchanged. Assume that the bacterium is a sphere of radius a. A torque τ on a small sphere of radius a in a medium of viscosity η causes the sphere to rotate at a rate $d\theta/dt$, such that $\tau=8\pi a^3\eta(d\theta/dt)$. For simplicity, assume that all motion takes place in a plane.

(a) Show that $d\theta/dt=\sin\theta/t_0$, where $t_0=8\pi a^3\eta/mB$.

(b) Evaluate t_0 for a bacterium of radius 20 μm in the earth's magnetic field. Use $m=1.28\times10^{-15}$ A m^2

(c) The velocity component perpendicular to the field is $v_y=v_0\sin\theta$. Show that when the bacterium rotates from angle θ_1 to θ_2 it has moved a distance $y=v_0t_0(\theta_2-\theta_1)$.

(d) Show that the time required to change from angle ϵ to $\pi-\epsilon$ is $t_0\ln[(1+\cos\epsilon)/(1-\cos\epsilon)]$.

Section 8.8

8.21. The earth's magnetic field is about 5×10^{-5} T. A typical magnetoencephalogram is about 10^{-13} T. How much angular variation in the detector (due to vibration, for example) is tolerable?

8.22. The spatial gradient in the earth's field is about 10^{-11} T m^{-1}. How much lateral movement can be tolerated in measuring a magnetoencephalogram of about 10^{-13} T?

8.23. Show that the units of $h/2e$ are V s, and that this is also a unit of magnetic flux.

8.24. Suppose that a SQUID of area 0.1 cm^2 can resolve a magnetic flux change $\Delta\Phi=10^{-3}\Phi_0$. What is the corresponding change in B?

8.25. The first difference of B is $B(x+a)-B(x)$. What is the second difference? Compare the first and second differences to what is detected by a first-order and second-order gradiometer. Assume that B is constant over the area of each gradiometer loop. Use these results to determine the signal resulting from a distant but unwanted dipole source with a magnetic field that falls as $1/r^3$.

8.26. A first-order gradiometer is used to measure the magnetic field at a point $(x_0,\ 0,\ z_0)$from a current dipole described by Eq. (8.15). The gradient is measured at position $z=x_0/\sqrt{2}$. The coils are at x_0 and x_0+a and are perpendicular to the z axis. Find the net flux in the gradiometer in terms of x_0 and a and the radius b of the coils. Assume B is uniform across each coil.

8.27. Figure 8.29(a) shows a gradiometer for measuring $\partial B_z/\partial z$. Sketch a gradient coil for measuring $\partial B_z/\partial x$.

Section 8.9

8.28. Suppose that a magnetic stimulator consists of a single-turn coil of radius $a=2$ cm. It is desired to have a magnetic field of 2 T on the axis of the coil and a distance $b=2$ cm away.

(a) Calculate the current required, using symmetry and the Biot–Savart law. (Hint: Use the results of Problem 8.6.)

(b) Assume that the magnetic field rises from 0 to 2 T in 100 μs. Assume also that the flux through the coil is equal to the field at the center of the coil multiplied by the area of the coil. Calculate the emf induced in the coil.

REFERENCES

Able, K. P., and M. A. Able (1995). Interactions in the flexible orientation system of a migratory bird. *Nature* **375**: 230–223.

Barach, J. P. (1987). The effect of ohmic return current on biomagnetic fields. *J. Theor. Biol.* **125**: 187–191.

Barach, J. P., B. J. Roth, and J. P. Wikswo (1985). Magnetic measurements of action currents in a single nerve axon: A core conductor model. *IEEE Trans. Biomed. Eng.* **BME-32**: 136–140.

Barker, A. T., R. Jalinos, and I. L. Freeston (1985). Non-invasive magnetic stimulation of the human cortex. *Lancet* **1**(8437): 1106–1107.

Brasil-Neto J. P., L. G. Cohen, M. Panizza, J. Nilsson, B. J. Roth, and M. Hallett (1992). Optimal focal transcranial magnetic activation of the human motor cortex: effects of coil orientation, shape of the induced current pulse, and stimulus intensity. *J. Clin. Neurophysiol.* **9**(1): 132–136.

Brittenham, G. M., D. E. Farrell, J. W. Harris, E. S. Feldman, E. H. Danish, W. A. Muir, J. H. Tripp, J. N. Brennan, and E. M. Bellon (1983). Diagnostic assessment of human iron stores by measurement of hepatic magnetic susceptibility. *Nuovo Cimento* **2D**: 567–581.

Clarke, J. (1994). SQUIDS. *Sci. Am.* Aug. 1994: 46–53.

Cohen, D., and I. Nemoto (1984). Ferrimagnetic particles in the lung part I: The magnetizing process. *IEEE Trans. Biomed. Eng.* **BME-31**: 261–273.

Cohen, D., I. Nemoto, L. Kaufman, and S. Arai (1984). Ferrimagnetic particles in the lung part II: The relaxation process. *IEEE Trans. Biomed. Eng.* **BME-31**: 274–285.

de Araujo, F. F., M. A. Pires, R. B. Frankel, and C. E. M. Bicudo (1986). Magnetite and magnetotaxis in algae. *Biophys. J.* **50**: 375–378.

Eisberg, R., and R. Resnick (1985). *Quantum Physics of Atoms, Molecules, Solids, Nuclei and Particles*, 2nd ed. New York, Wiley.

Eselle, K. P., and M. A. Stuchly (1995). Cylindrical tissue model for magnetic field stimulation of neurons: Effects of coil geometry. *IEEE Trans. Biomed. Eng.* **42**(9): 934–941.

Flowers, N. C., and L. G. G. Horan (1995). Body surface potential mapping. In D. P. Zipes and J. Jalife, eds. *Cardiac Electrophysiology: From Cell to Bedside*, 2nd ed., Philadelphia, Saunders, pp. 1049–1067.

Frankel, R. B. (1984). Magnetic guidance of organisms. *Ann. Rev. Biophys. Bioeng.* **13**: 85–103.

Frankel, R. B., R. P. Blakemore, and R. S. Wolfe (1979). Magnetite in freshwater magnetotactic bacteria. *Science* **203**: 1355–1356.

Frankel, R. B., and D. A. Bazylinski (1994). Magnetotaxis and magnetic particles in bacteria. *Hyperfine Interactions* **90**: 135–142.

Gielen, F. L., R. N. Friedman, and J. P. Wikswo, Jr. (1991). In vivo magnetic and electric recordings from nerve bundles and single motor units in mammalian skeletal muscle. Correlations with muscle force. *J. Gen. Physiol.* **98**(5): 1043–1061.

Gould, J. L. (1995). Constant compass calibration. *Nature* **375**:184.

Gould, J. L., J. L. Kirschvink, and K. S. Deffeyes (1978). Bees have magnetic remanence. *Science* **201**: 1026–1028.

Grimes, D. I. F., R. F. Lennard, and S. J. Swithenby (1985). Macroscopic ion currents within the human leg. *Phys. Med. Biol.* **30**: 1101–1112.

Hämäläinen, M., R. Harri, R. J. Ilmoniemi, J. Knuutila, and O. V. Lounasmaa (1993). Magnetoencephalography—theory, instrumentation, and applications to noninvasive studies of the working human brain. *Rev. Mod. Phys.* **65**(2): 413–497.

Heller, L. and D. B. van Hulsteyn (1992). Brain stimulation using magnetic sources: Theoretical aspects. *Biophys. J.* **63**: 129–138.

Hosaka, H., D. Cohen, B. N. Cuffin, and B. M. Horacek (1976). The effect of torso boundaries on the magnetocardiogram. *J. Electrocardiogr.* **9**: 418–425.

Moskowitz, B. M. (1995). Biomineralization of magnetic minerals. *Rev. Geophys. Suppl.* **33**(Part 1 Suppl. S): 123–128.

Nielsen, P., R. Fischer, R. Englehardt, P. Tondury, E. E. Gabbe, and G. E. Janka (1995). Liver iron stores in patients with secondary haemosiderosis under iron chelation therapy with deferoxamine or deferiprone. *Brit. J. Hematology* **91**: 827–833 (1995).

Page, C. H. (1977). Electromotive force, potential difference, and voltage. *Am. J. Phys.* **45**: 978–980.

Petrie, R. J., P. van Leeuwen, B. Brandts, G. Turnbull, S. J. O. Veldhuyzen van Zanten, and G. Stroink (1996). Single and multichannel measurements of gastrointestinal activity in the fasted and fed states. In C. Aine, E. Flynn, Y. Okada, G. Stroink, S. Swithenby, and C. Woods, eds. *Biomag.96: Advances in Biomagnetism Research.* Berlin, Springer-Verlag.

Press, W. H., S. A. Teukolsky, W. T. Vetterling, and B. P. Flannery (1992). *Numerical Recipes in C: The Art of Scientific Computing*, 2nd ed. reprinted with corrections, 1995. New York, Cambridge University Press.

Proksch, R. B., T. E. Schäffer, B. M. Moskowitz, E. D. Dahlberg, D. A. Bazylinski, and R. B. Frankel (1995). Magnetic force microscopy of the submicron magnetic assembly in a magnetotactic bacterium. *Appl. Phys. Lett.* **66**(19): 2582–2584.

Purcell, E. M. (1985). *Electricity and Magnetism*, 2nd ed. Berkeley Physics Course. New York, McGraw-Hill, Vol. 2.

Purcell, C. J., G. Stroink, and B. M. Horacek (1988). Effect of torso boundaries on electrical potential and magnetic field of a dipole. *IEEE Trans. Biomed. Eng.* **35**(9): 671–678.

Richards, W. O., C. L. Garrard, S. H. Allos, L. A. Bradshaw, D. J. Staton, and J. P. Wikswo, Jr. (1995a). Noninvasive diagnosis of mesenteric ischemia using a squid magnetometer. *Ann. Surg.* **221**(6): 696–704.

Richards, W. O., D. Staton, J. Golzarian, R. N. Friedman, and J. P. Wikswo, Jr. (1995b). Non-invasive SQUID magnetometer measurement of human gastric and small bowel electrical activity. In C. Baumgertner *et al.*, eds. *Biomagnetism: Fundamental Research and Clinical Applications.* Elsevier, IOS.

Romer, R. H. (1982). What do "voltmeters" measure?: Faraday's law in a multiply connected region. *Am. J. Phys.* **50**: 1089–1093.

Roth, B. J. (1994). Mechanisms for electrical stimulation of excitable tissue. *Crit. Rev. Biomed. Eng.* **22**(3/4): 253–305.

Roth, B. J., and J. P. Wikswo, Jr. (1985). The magnetic field of a single axon: A comparison of theory and experiment. *Biophys. J.* **48**: 93–109.

Shadowitz, A. (1975). *The Electromagnetic Field.* New York, McGraw-Hill.

Stahlhofen, W., and W. Moller (1993). Behaviour of magnetic microparticles in the human lung. *Rad. Env. Biophys.* **32**(3): 221–238.

Staton, D. J., R. N. Friedman, and J. P. Wikswo, Jr. (1993). High resolution SQUID imaging of octupolar currents in anisotropic cardiac tissue. *IEEE Trans. Appl. Superconduct.* **3**(1): 1934–1936.

Stroink, G. (1985). Magnetic measurements to determine dust loads and clearance rates in industrial workers and miners. *Med. & Biol. Eng. & Computing* **23**: 44–49.

Stroink, G. (1992). Cardiomagnetic Imaging. In R. A. Dunn and A. S. Berson, eds. *Frontiers in Cardiovascular Imaging.* New York, Raven, pp. 161–177.

Stroink, G., M. J. R. Lamothe, and M. J. Gardner (1996). Magnetocardiographic and electrocardiographic mapping studies. In H. Weinstock, ed. SQUID sensors: *Fundamentals, Fabrication and Applications.* NATO ASI Series, Dordrecht, The Netherlands, Kluwer.

Stroink, G., B. Blackford, B. Brown, and M. Horacek (1981). Aluminum shielded room for biomagnetic measurements. *Rev. Sci. Instrum.* **52**(3): 463–468.

Swinney, K. R., and J. P. Wikswo, Jr. (1980). A calculation of the magnetic field of a nerve action potential. *Biophys. J.* **32**: 719–732.

Tan, G. A., F. Brauer, G. Stroink, and C. J. Purcell (1992). The effect of measurement conditions on MCG inverse solutions. *IEEE Trans. Biomed. Eng.* **39**(9): 921–927.

Thomas, I. M., M. Freake, S. J. Swithenby, and J. P. Wikswo, Jr. (1993). A distributed quasi-static ionic current in the 3–4 day old chick embryo. *Phys. Med. Biol.* **38**: 1311–1328.

Trontelj, Z., R. Zorec, V. Jabinsek, and S. N. Erné (1994). Magnetic detection of a single action potential in *Chara corallina* internodal cells. Biophys. J. **66**: 1694–1696.

Walcott, C., J. L. Gould, and J. L. Kirschvink (1979). Pigeons have magnets. *Science* **205**: 1027–1029.

Walker, M. M., J. L. Kirschvink, S.-B. R. Chang, and A. E. Dizon (1984). A candidate magnetic sense organ in the yellowfin tuna. *Science* **224**: 751–753.

Wikswo, J. P., Jr. (1980). Noninvasive magnetic detection of cardiac mechanical activity: Theory. *Med. Phys.* **7**: 297–306.

Wikswo, J. P., Jr. (1995a). SQUID magnetometers for biomagnetism and nondestructive testing: Important questions and initial answers. *IEEE Trans. Appl. Superconduct.* **5**(2): 74–120.

Wikswo, J. P., Jr. (1995b). Tissue anisotropy, the cardiac bidomain, and the virtual cathode effect. In D. P. Zipes and J. Jalife, eds. *Cardiac Electrophysiology: From Cell to Bedside*, 2nd ed., Philadelphia, Saunders, pp. 348–361.

Wikswo, J. P., Jr., A. Gevins, and S. J. Williamson (1993). The future of the EEG and MEG. *Electroencephalogr. Clinical Neurophysiol.* **87**: 1–9.

Woosley, J. K., B. J. Roth, and J. P. Wikswo, Jr. (1985). The magnetic field of a single axon; A volume conductor model. *Math. Biosci.* **76**: 1–36.

CHAPTER 9

Electricity and Magnetism at the Cellular Level

This chapter describes a number of topics related to charged membranes and the movement of ions through them. These range from the basics of how the presence of impermeant ions alters the concentration ratios of permeant ions, to the movement of ions under the combined influence of an electric field and diffusion, and to simple models for gating in ion channels in cell membranes. It also discusses mechanisms for the detection of weak electric and magnetic fields and the possible effects of weak low-frequency electric and magnetic fields on cells.

Section 9.1 discusses Donnan equilibrium in which the presence of an impermeant ion on one side of a membrane, along with other ions that can pass through, causes a potential difference to build up across the membrane. The potential difference across the membrane exists even though the bulk solution on each side of the membrane is electrically neutral. Section 9.2 examines the Gouy–Chapman model for the charge buildup at each surface of the membrane that gives rise to this potential difference. This same model is extended in three dimensions to the cloud of counterions surrounding each ion in solution—the Debye–Hückel model of Sec. 9.3.

Since water molecules have a net dipole moment, they align themselves so as to nearly cancel the electric field of each ion. Very close to the ion the electric field is so strong that even complete alignment is insufficient to cancel the ion's field. This saturation of the dielectric is described in Sec. 9.4.

Ions move in solution by *diffusion* if there is a concentration gradient and by *drift* if there is an applied electric field. The Nernst–Planck equation (Sec. 9.5) describes this motion. When several ion species are moving through a membrane, there can be zero total electric current, even though there is a flow of each species through the membrane. A constant-field model for this situation leads to the Goldman equation of Sec. 9.6.

The next two sections describe channels in active cell membranes. Section 9.7 describes a simple model for gating—the opening and closing of channels, as well as limitations to the conductance of each channel imposed by diffusion to the mouth of the channel. Section 9.8 introduces noise—the fluctuations in channel current that limit measurement accuracy but also can be used to determine properties of the channels.

Section 9.9 describes how channels can detect very small mechanical motions, as in the ear, and how certain fish can detect very small electric fields in sea water. Both of these processes are working near the limit of sensitivity set by random thermal motion.

Section 9.10 introduces an area of great interest and controversy: whether rather weak, low-frequency electric and magnetic fields can have any effect on cells. We discuss some of the physical aspects of the problem and conclude that such effects are highly unlikely.

9.1. DONNAN EQUILIBRIUM

There is usually an electrical potential difference across the wall of a capillary. There is also a potential difference across the cell membrane, and the concentration of certain ion species is different in the intracellular and extracellular fluid. In Chap. 3 we saw that if the potential difference across the membrane is $v' - v$, an ion of valence z is in equilibrium when

$$C'/C = e^{-ze(v'-v)/k_BT}. \tag{3.30}$$

With this concentration ratio there is no current, even if the membrane is permeable to the species. This result is a special case of the Boltzmann factor, more familiar in physiology as the Nernst equation:

$$v' - v = -\frac{RT}{zF}\ln\left(\frac{C'}{C}\right). \tag{3.34}$$

It is often said—incorrectly—that the Nernst equation shows how the concentration of an ion species *causes* the potential difference across the membrane. We saw in Chap. 6 that the potential difference across the membrane is caused by layers of charge on each side of the membrane that create an electric field in the membrane. The solutions on each side of the membrane are electrically neutral except at the boundary with the membrane. (If there were an electric field in the solution, ions would move until the field was zero; then Gauss's law could be used to show that any volume contains zero charge.) We will learn in Sec. 9.2 the typical distance from the membrane occupied by the charged layer, and in Sec. 9.3 we will find the distance scale over which there are microscopic departures from neutrality in a bulk ionic solution.

The concentration differences do not *directly* cause the potential difference. However, if the concentration of an ion species on one side of the membrane is varied, the potential often changes in a manner that is approximated by Eq. (3.34) over a wide range of concentrations. We will now explore one mechanism by which this can happen. This is particularly important for the walls of capillaries, where charged proteins in the blood are too large to pass through the gaps between cells in the capillary walls, but is it also applicable to the cell membrane.

In Donnan equilibrium, the potential difference arises because one ion species cannot pass through the membrane at all. Consider the hypothetical case of Fig. 9.1. Let potassium ions exist on either side of the membrane in concentrations [K] and [K′]. The membrane is also permeable to chloride ions, which exist in concentrations [Cl] and [Cl′]. In addition, there are large charged molecules $[M^+]$ and $[M^-]$ that cannot pass through the membrane. Their concentrations are $[M^+]$, $[M^{+\prime}]$, $[M^-]$, and $[M^{-\prime}]$. For simplicity, we assume they are monovalent. The potential on the left is 0; on the right it is v'. Assume that the concentrations of the large molecules are fixed. The potassium concentration on one side of the membrane will be

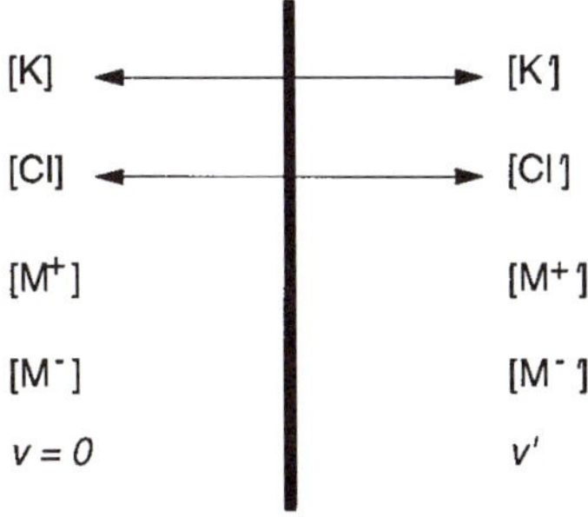

FIGURE 9.1. Ion concentrations on either side of a membrane. Species that can pass through the membrane are indicated by double-headed arrows.

assumed known, and we must solve for four variables: [K′], [Cl], [Cl′], and v'. Therefore, four equations are needed.

The first two equations state that the solutions on either side are electrically neutral:

$$[M^+]+[\mathrm{K}]=[\mathrm{Cl}]+[M^-], \tag{9.1}$$

$$[M^{+\prime}]+[\mathrm{K}']=[\mathrm{Cl}']+[M^{-\prime}]. \tag{9.2}$$

Equation (9.1) can be solved for *Cl*. It will be convenient to define $[M]=[M^+]-[M^-]$ and $[M']=[M^{+\prime}]-[M^{-\prime}]$:

$$Cl=K+(M^+-M^-)=K+M. \tag{9.3}$$

Note that adding any amount of KCl to the solution on the left automatically satisfies this equation, since any increase in [K] is accompanied by a like increase in [Cl]. The other two equations state that the concentrations of potassium and chloride on the two sides of the membrane are related by a Boltzmann factor. Since the valence $z=+1$ for [K] and -1 for [Cl] we have

$$\frac{[\mathrm{K}']}{[\mathrm{K}]}=\frac{[\mathrm{Cl}]}{[\mathrm{Cl}']}=e^{-ev'/k_BT}. \tag{9.4}$$

The chloride concentration on the right is $[\mathrm{Cl}']=[\mathrm{Cl}]([\mathrm{Cl}']/[\mathrm{Cl}])=[\mathrm{Cl}]([\mathrm{K}]/[\mathrm{K}'])$, so that from Eq. (9.2) $[\mathrm{K}']+[M']=[\mathrm{Cl}]([\mathrm{K}]/[\mathrm{K}'])$. This can be rewritten as a quadratic equation in [K′], since all the other concentrations are known:

$$[\mathrm{K}']^2+[M'][\mathrm{K}']-[\mathrm{K}][\mathrm{Cl}]=0.$$

The solution is

$$[\mathrm{K}']=\frac{-[M']+\sqrt{[M']^2+4[\mathrm{K}][\mathrm{Cl}]}}{2}. \tag{9.5}$$

(The negative square root is discarded because it would give a negative potassium concentration.) Recall that [Cl] is known from Eq. (9.3). Once [K′] and [Cl′] are known, v' is determined from Eq. (9.4). Solutions for different values of [K] are shown in Table 9.1 and Fig. 9.2 and 9.3 for the conditions

$$[M^+]=145\ \mathrm{mmol\,l^{-1}},$$

$$[M^-]=30\ \mathrm{mmol\,l^{-1}},$$

$$[M]=115\ \mathrm{mmol\,l^{-1}},$$

$$[M^{+\prime}]=15\ \mathrm{mmol\,l^{-1}},$$

$$[M^{-\prime}]=156\ \mathrm{mmol\,l^{-1}},$$

$$[M']=-141\ \mathrm{mmol\,l^{-1}},$$

$$T=310\ \mathrm{K}.$$

(At this temperature $k_BT/e=26.75$ mV.)

TABLE 9.1. *Variation of concentrations and voltage as K is varied.*

[K]	[Cl]	[K′]	[Cl′]	[C]/[Cl′]=[K′]/[K]	v' (mV)
0.01	115.01	141.01	0.01	14101	−255.57
0.10	115.10	141.08	0.08	1410.8	−193.99
0.20	115.20	141.16	0.16	705.8	−175.46
0.50	115.50	141.41	0.41	282.8	−151.00
1.00	116.00	141.82	0.82	141.8	−132.53
2.00	117.00	142.64	1.64	71.32	−114.15
5.00	120.00	145.13	4.13	29.03	−90.10
10.00	125.00	149.37	8.37	14.94	−72.33
20.00	135.00	158.08	17.08	7.904	−55.30
50.00	165.00	185.48	44.48	3.710	−35.07
100.00	215.00	233.20	92.20	2.332	−22.65
200.00	315.00	331.21	190.21	1.656	−13.49
500.00	615.00	629.49	488.49	1.259	−6.16

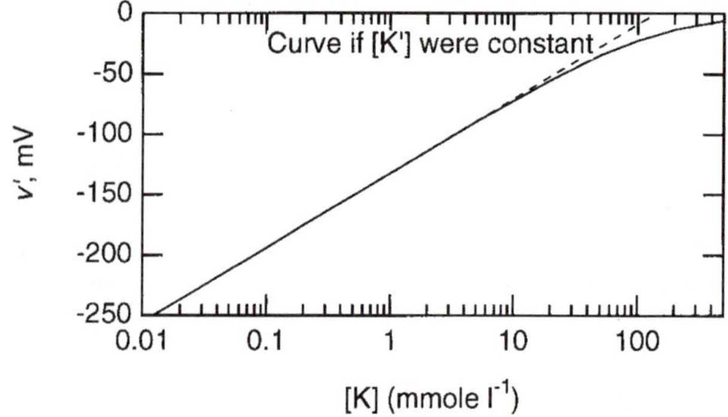

FIGURE 9.3. Membrane potential v' vs [K] for the example of Donnan equilibrium. For [K]<10 m*M* the curve is like the Nernst equation because [K′] has a nearly constant value of 141 m*M*. The dashed line shows the relationship if [K′] were constant.

Several features of this solution are worth noting. First, changing [K] does change the potential, but the mechanism is indirect. The Boltzmann factor still applies; minuscule changes in concentration are sufficient to provide layers of charge on the membrane surface that generate a potential difference such that these concentrations are at equilibrium. Figure 9.3 shows that over a wide range of *K*, *K′* does not change significantly, and therefore the curve of v' vs ln[K] is nearly a straight line. We could equally well have regarded *Cl* as the independent variable. The impermeable ions enter the equation only as their net charge, $[M]=[M^+]-[M^-]$ and $[M']=[M^{+\prime}]-[M^{-\prime}]$. As the concentrations [K] and [Cl] get larger, the impermeant ions become less important, the potential approaches zero, and the ratios [K′]/[K] and [Cl′]/[Cl] approach unity.

Donnan equilibrium may well explain the potential that exists across the capillary wall, which is impermeable to negatively charged proteins but is permeable to other ions. There is evidence that it does not adequately explain the potential across a cell membrane. For example, the membrane is known to be slightly permeable to sodium, although the sodium concentration is nowhere near what it would be if the sodium were in equilibrium. In the next section we examine one model for how the ions are distributed at the interface in Donnan equilibrium.

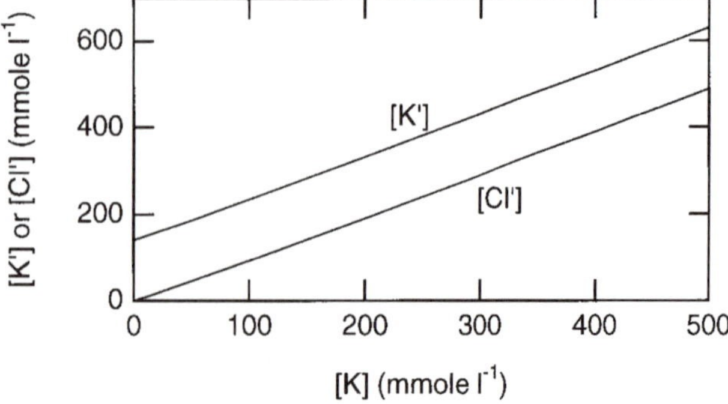

FIGURE 9.2. Variation of [K′] and [Cl′] with [K] in the example of Donnan equilibrium.

9.2. POTENTIAL CHANGE AT AN INTERFACE: THE GOUY–CHAPMAN MODEL

In this section we study a model that was used independently by Gouy and by Chapman to study the interface between a metal electrode and an ionic solution. They investigated the potential changes along the x axis perpendicular to a large plane electrode. The same model was used by Schottky to study the charge distribution in a semiconductor. Biological applications were described by Mauro (1962). We show the features of the model by examining the transition region for the Donnan equilibrium example described in the preceding section.

An infinitely thin membrane at $x=0$ is assumed to be permeable to potassium and chloride ions. Their concentrations are $K(x)$ and $Cl(x)$. An impermeant positive cation has concentration $M(x)$ for $x>0$. For negative x, $M(x)=0$. Impermeant anions are ignored. Far to the left the potential is zero and the concentrations are [K] and [Cl]. Far to the right they are v', [K′], [Cl′], and $[M']$.

The first step is to relate the charge distribution to the potential. If v and E change only in the x direction, then Gauss's law can be applied to a slab of cross-sectional area S between x and $x+dx$ as shown in Fig. 9.4. The net flux out through the surface at $x+dx$ is $E_x(x+dx)S$. The net outward flux at x is $-E_x(x)S$. There is no contribution to the flux through the other surfaces. The total ionic charge in the volume is $\rho_{\text{ext}}(x)S\,dx$. We include the effect of water polarization by using the dielectric constant for water,

variable at the opening, the change in the output variable is the open-loop gain times the change in the input variable:

$$\text{OLG}=G_1G_2=\left(\frac{\partial x}{\partial y}\right)_{\text{box } g}\left(\frac{\partial y}{\partial x}\right)_{\text{box } f}=\frac{\partial g}{\partial y}\frac{\partial f}{\partial x}. \tag{10.7}$$

The open-loop gain can be calculated by taking the derivatives in either order, which corresponds to breaking the loop after either box (Fig. 10.6).

If the relationships between the derivatives have been plotted as in Fig. 10.5, it may be easiest to evaluate the derivatives graphically. In that case, it is easiest to work with $\partial y/\partial x$ for box g. But $\partial y/\partial x = 1/(\partial x/\partial y)$. Therefore,

$$\text{OLG}=G_1G_2=\frac{(\partial y/\partial x)_{\text{box } f}}{(\partial y/\partial x)_{\text{box } g}}. \tag{10.8}$$

It is important to calculate the gain in the direction that causality operates. Going around the loop the wrong way gives the reciprocal of the open-loop gain.

We can now calculate how much feedback reduces the change in x, compared to the case in which there is no feedback and the value of y going into box g is held fixed. For box g, where $x=g(y,p)$, we can write for small changes in p and y

$$\Delta x=\left(\frac{\partial x}{\partial p}\right)_{y,\text{box } g}\Delta p+\left(\frac{\partial x}{\partial y}\right)_{p,\text{box } g}\Delta y=\left(\frac{\partial x}{\partial p}\right)_{y,\text{box } g}\Delta p + G_1\Delta y. \tag{10.9}$$

When there is no feedback, Δy is zero and

$$\Delta x=\left(\frac{\partial x}{\partial p}\right)_{y,\text{box } g}\Delta p.$$

When there is feedback, there is a value of Δy to be included. If the change in x with feedback is $\Delta x'$, the change in y can be calculated from the second box:

$$\Delta y = G_2\Delta x'. \tag{10.10}$$

This can be combined with Eq. (10.9):

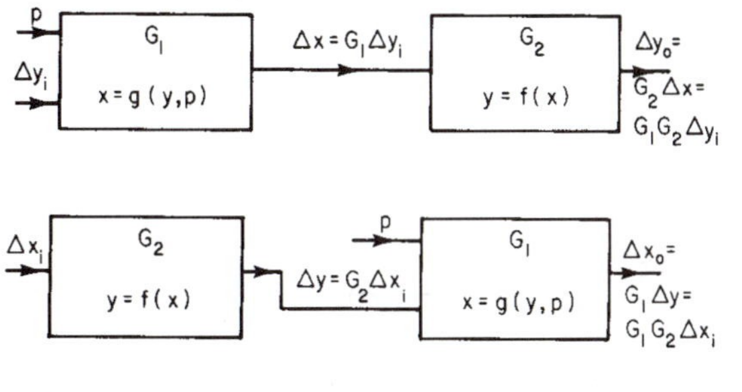

$$\Delta x'=\left(\frac{\partial x}{\partial p}\right)_y\Delta p+G_1(G_2\Delta x')=\Delta x+G_1G_2\Delta x'.$$

This is solved for $\Delta x'$:

$$\Delta x'=\frac{\Delta x}{1-G_1G_2}=\frac{\Delta x}{1-\text{OLG}}. \tag{10.11}$$

The effect of feedback is to cause a change in y which reduces the change in x by the factor $1-\text{OLG}$. When the feedback is negative, the open-loop gain is negative, $1-\text{OLG}$ is greater than one, and there is a reduction in Δx. If the feedback is positive and the open-loop gain is less than one, $\Delta x'$ is larger than Δx.

For the respiration example, the equations for each box are

$$x=g(y,p)=\frac{15.47p}{y-2.07},$$
$$y=f(x)=\begin{cases}10, & x\leqslant 40\\ 10+2.5(x-40), & x>40.\end{cases} \tag{10.12}$$

The derivatives are

$$\left(\frac{\partial g}{\partial p}\right)_y=\frac{15.47}{y-2.07},$$
$$G_1=\left(\frac{\partial g}{\partial y}\right)_p=-\frac{15.47p}{(y-2.07)^2},$$
$$G_2=\left(\frac{\partial f}{\partial x}\right)=2.5.$$

At operating point A in Fig. 10.5, the values are

$$x=45.07,\quad p=60,\quad y=22.67,$$
$$\left(\frac{\partial g}{\partial p}\right)_y=0.757, \tag{10.13}$$
$$G_1=-2.19,\quad G_2=2.5,\quad \text{OLG}=-5.48.$$

If p changes from 60 to 62, then without feedback $\Delta x=(0.757)(2)=1.5$. With feedback, $\Delta x'=1.5/(1+5.48)=0.23$.

10.4. APPROACH TO EQUILIBRIUM WITHOUT FEEDBACK

The technique described in the preceeding section allows us to determine the equilibrium state or operating point of a system if we can measure the functions f and g. It does not tell us how the system behaves when it is not at the equilibrium point, nor does it tell us how the system moves from one point to another when parameter p is changed. To learn that, we need an equation of motion for each process or box

FIGURE 10.6. The open loop gain is calculated by opening the loop at any point. (a) Loop opened in y. (b) Loop opened in x.

in the feedback loop. The equation of motion is usually a differential equation. In real systems the differential equation is often nonlinear and difficult to solve. We first consider models described by linear differential equations, and then we consider some of the behaviors of nonlinear systems.

The response of a system cannot be infinitely fast. At equilibrium the rate of exhaling carbon dioxide is the same as the rate of production throughout the body. If the rate of production rises in a certain muscle group, the extra carbon dioxide enters the blood and is distributed throughout the body, and the carbon dioxide concentrations in the blood and alveoli rise gradually. To develop a quantitative model, assume that all the carbon dioxide in the body is stored in a single well-stirred compartment of volume V_c. This assumption of uniform concentration is certainly an oversimplification. The total number of moles is n and the concentration is n/V_c. The concentration in the blood is related to the partial pressure in the alveoli by a solubility constant α: $n/V_c=\alpha x$. Therefore $dn/dt=\alpha V_c dx/dt$. Moreover, dn/dt is equal to the rate of production [Eq. (10.2)] minus the rate of removal [Eq. (10.3)]:

$$\frac{dx}{dt}=\frac{F\dot{o}}{\alpha V_c}-\frac{x(y-b)}{\alpha V_c RT}.$$

We change the definition of F to take account of the fact that $\dot{o}$ and p are both the rate of oxygen consumption in slightly different units ($\dot{o}$ is in mol s^{-1} and p is in mmol min^{-1}):

$$\frac{dx}{dt}=\frac{Fp}{\alpha V_c}-\frac{x(y-b)}{\alpha V_c RT}. \qquad (10.14)$$

This differential equation depends on both x and y and in fact is nonlinear since the variables are multiplied together in the last term. At equilibrium $dx/dt=0$ and Eq. (10.14) gives Eq. (10.4).

If y is constant (a constant breathing rate, which could be accomplished by placing the subject on a respirator), then there is no feedback and Eq. (10.14) is a linear differential equation with constant coefficients:

$$\frac{dx}{dt}+\frac{y_0-b}{RT\alpha V_c}x=\frac{Fp}{\alpha V_c}.$$

It can be solved using the techniques of Appendix F. Suppose that for $t\leq 0$, $p=p_0$, $x=x_0$, and $y=y_0$. For $t>0$ the subject exercises, so that $p=p_0+\Delta p$, $x=x_0+\xi$, and y is unchanged. The equation then becomes

$$\frac{d\xi}{dt}+\frac{y_0-b}{RT\alpha V_c}\xi=\frac{F\,\Delta p}{\alpha V_c}. \qquad (10.15)$$

The homogeneous equation is

$$\frac{d\xi}{dt}+\frac{1}{\tau_1}\xi=0, \qquad (10.16)$$

where the time constant is

$$\tau_1=\frac{RT\alpha V_c}{y_0-b}. \qquad (10.17)$$

The homogeneous solution is $\xi=Ae^{-t/\tau_1}$. The particular solution is

$$\xi=\frac{FRT}{y_0-b}\Delta p=a\,\Delta p,$$

so the complete solution is $\xi=a\,\Delta p+Ae^{-t/\tau_1}$. We now use the initial condition to determine A. At $t=0$ $\xi=0$, so $A=-a\,\Delta p$. The complete solution without feedback that matches the initial condition is

$$x-x_0=a\,\Delta p(1-e^{-t/\tau_1}). \qquad (10.18)$$

Figure 10.7 shows how x changes with time on a plot of x vs. t and a plot of x vs y. The dots are spaced at equal times.

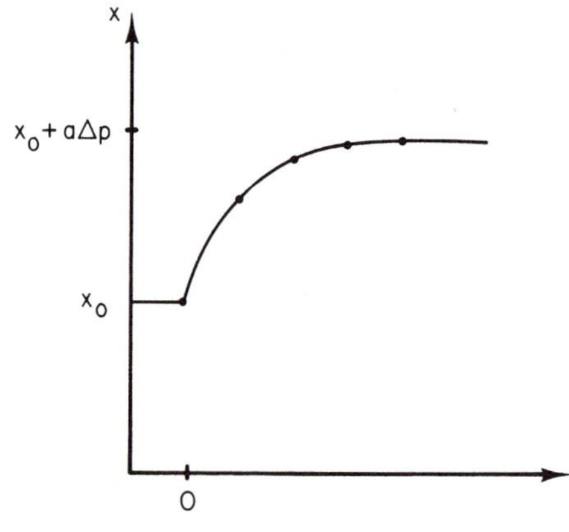

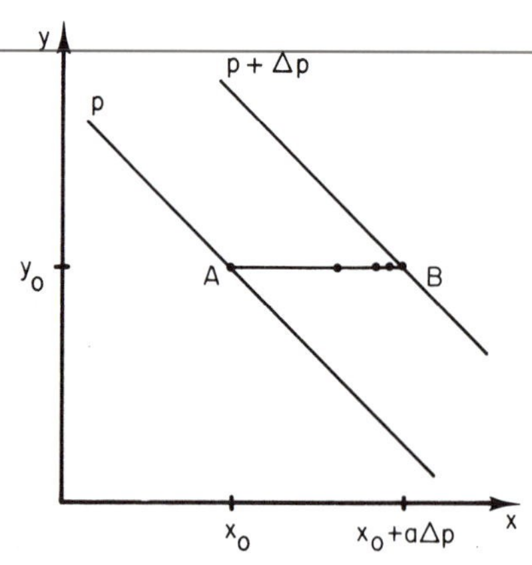

FIGURE 10.7. The change in x without feedback in response to a step change in parameter p. (a) Plot of x vs t. (b) Plot of x vs y.

10.5. APPROACH TO EQUILIBRIUM IN A FEEDBACK LOOP WITH ONE TIME CONSTANT

Suppose now that y is allowed to change and that $\eta = y - y_0$. We can write the equation for the change in x, Eq. (10.14) as

$$\frac{d\xi}{dt} = \frac{dx}{dt} = \frac{Fp_0}{\alpha V_c} + \frac{F\Delta p}{\alpha V_c} - \frac{(x_0+\xi)(y_0-b+\eta)}{\alpha V_c RT}$$

$$= \underbrace{\frac{Fp_0}{\alpha V_c} - \frac{x_0(y_0-b)}{\alpha V_c RT}}_{=0} + \frac{F\Delta p}{\alpha V_c} - \frac{\xi(y_0-b)}{\alpha V_c RT} - \frac{x_0\eta}{\alpha V_c RT} - \frac{\xi\eta}{\alpha V_c RT}.$$

Multiplying all terms by τ_1 as defined in Eq. (10.17), and identifying

$$G_1 = \left(\frac{\partial g}{\partial y}\right)_p = -\frac{x_0}{y_0-b},$$

we obtain

$$\tau_1 \frac{d\xi}{dt} = a\ \Delta p - \xi + G_1 \eta - \frac{\xi\eta}{y_0-b}. \qquad (10.19)$$

The product $\xi\eta$ in the last term makes the equation nonlinear. If we assume that the last term can be neglected, we have a linear differential equation

$$\tau_1 \frac{d\xi}{dt} + \xi = a\ \Delta p + G_1 \eta. \qquad (10.20)$$

Now assume that the response of the second box is linear and instantaneous, so that

$$\eta = G_2 \xi. \qquad (10.21)$$

If this is substituted in the linearized equation, Eq. (10.20), the result is

$$\tau_1 \frac{d\xi}{dt} + (1 - G_1 G_2)\xi = a\ \Delta p. \qquad (10.22)$$

The steady-state solution before $t=0$ is

$$x_0 = \frac{a p_0}{1 - G_1 G_2}.$$

At $t=0$ the oxygen demand is changed to $p_0 + \Delta p$. The new steady-state (inhomogeneous) solution is $\xi = a\Delta p/(1 - G_1 G_2)$ and the homogeneous solution is $\xi = A e^{-t/\tau}$ where the time constant is

$$\tau = \frac{\tau_1}{1 - G_1 G_2}. \qquad (10.23)$$

[You can show this by dividing each term in Eq. (10.22) by τ_1 and comparing it to the equation for exponential decay.]

After combining the homogeneous and inhomogeneous solutions and using the initial condition $\xi(0)=0$ to determine A, we obtain the final result:

$$\xi = x - x_0 = \frac{a\Delta p}{1 - G_1 G_2}(1 - e^{-t/\tau}). \qquad (10.24)$$

This solution has the same form as Eq. (10.18). Both the total change in x and the time constant have been reduced by the factor $1/(1 - G_1 G_2)$. The change in y can be determined from $\eta = G_2 \xi$. The new solution is plotted along with the old solution in Fig. 10.8. This plot is for a system in which the open-loop gain is $G_1 G_2 = -1.3$. The time constant and the change in x are both reduced by 1/2.3.

It is important to realize that although the feedback reduced the time constant, it has not made x change faster. The curve of $x(t)$ with feedback has always changed less than the curve without feedback, and it has always changed more slowly. The reduction in time constant occurs because x does not change as much with feedback present, so it reaches its asymptotic value more quickly.

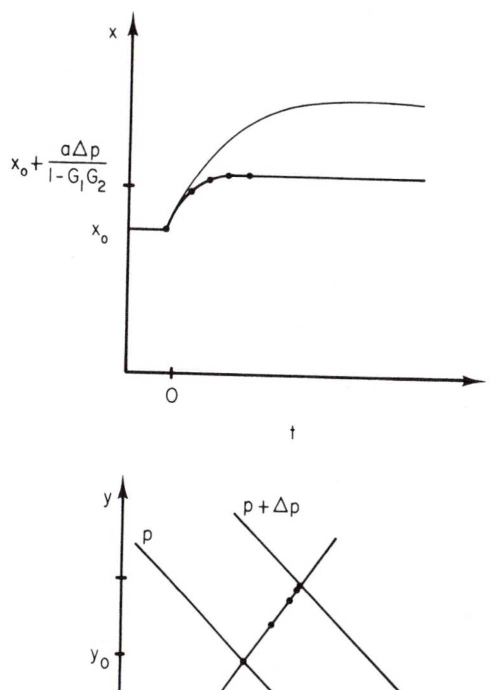

FIGURE 10.8. The change in x with feedback in response to a step change on parameter p. (a) Plot of x vs t. The change in x without feedback is shown for comparison. (b) Plot of x vs y.

This result assumes that box f has a negligible time constant. Applied to the respiratory example, it means that the carbon dioxide–sensing system responds rapidly compared to the time it takes for carbon dioxide levels within the blood to change after a change in p. Figure 10.9(a) repeats Fig. 10.8 and shows the changes in x and y resulting from a step change in p. When the second time constant is negligible, y is always proportional to x and the system moves back and forth along line AB.

The CO_2 sensors actually take a while to respond. To see what effect this might have, imagine the extreme case where the sensors are very slow compared to the change of carbon dioxide concentration in the blood. In that case, when p changes, y does not change right away. The system behaves at first as if there were no feedback, moving from point A to point C in Fig. 10.9(b). As the feedback slowly takes effect, the system moves from C to B. When the exercise ends, the system moves to point D because the subject is breathing too hard. Then it finally moves from D back to A. The actual system behaves in a manner somewhere between these two extremes, as we will see in the next section.

Consider a third possibility, that a regulatory mechanism anticipates the increased metabolic demand. This might happen if we took deep breaths before we began to exercise, or if additional muscle movement signaled the respiratory control center before the carbon dioxide concentration had a chance to change. Suppose that such anticipation is the only feedback mechanism. With the initiation of exercise, y changes to its final value. The level of carbon dioxide has not yet built up, so the increased ventilation reduces x below its normal value. We can approximate this by point D in Fig. 10.9(c). As the increased activity drives x up, the system moves at constant y to point B. When the exercise stops, y drops immediately to the resting value, though carbon dioxide is still coming out of the muscles. The result is that x rises to point C before finally falling back to point A.

Figure 10.10 shows what actually happens in the control of respiration. There is a fast neurological control and a slower chemical control. The result is a combination of the processes in Figs. 10.9(a)–10.9(c). Note that $x = P_{CO_2}$ is in the upper graph and the ventilation rate y is in the lower graph, the opposite of Fig. 10.9.

If we had not made the linear approximation we would not have been able to solve the equation, but the behavior would have been very similar. The nonlinear equation is obtained by substituting the equation for the second box, Eq. (10.21), in Eq. (10.19) instead of Eq. (10.20). The resulting equation is

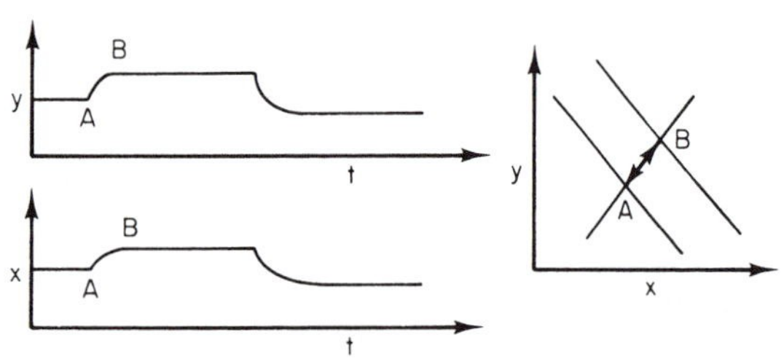

(a) Second time constant negligible

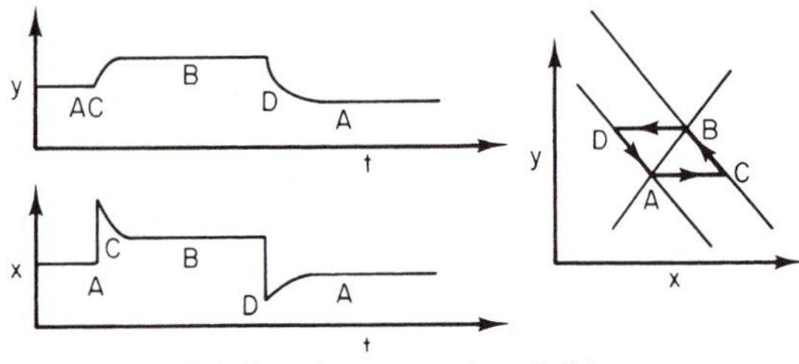

(b) First time constant negligible

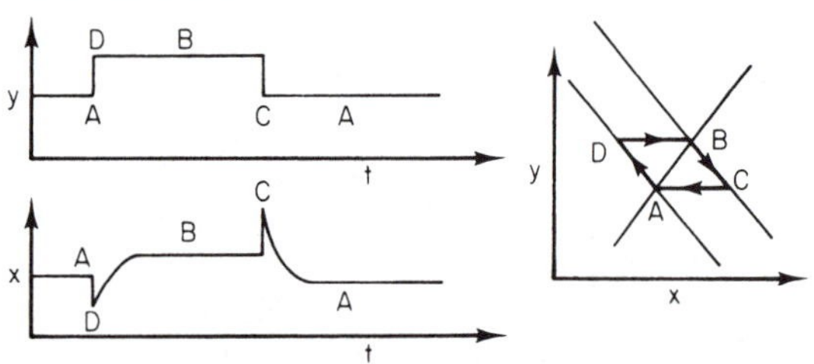

(c) Anticipation in second stage

FIGURE 10.9. Changes in x and y after a step change in parameter p. (a) The second time constant is negligible compared to the first; x and y move exponentially to their new equilibrium values. (b) The first time constant is negligible; the slow change in y means that there is no feedback at first. (c) The second stage anticipates the change in y that will be required; there is too much feedback at first.

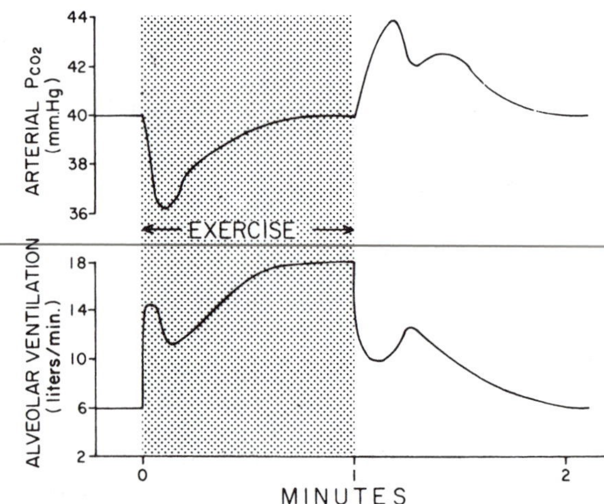

FIGURE 10.10. Change of arterial P_{CO_2} and alveolar ventilation in response to exercise. Note that $x = P_{CO_2}$ is in the upper graph and the ventilation rate y is in the lower graph, the opposite of Fig. 10.9. Reproduced from A. C. Guyton (1966). *Textbook of Medical Physiology*, 9th ed. Philadelphia, Saunders, with permission of the publisher. Data are extrapolated to humans from dogs. The dog experiments are described in C. R. Bainton (1972). *J. Appl. Physiol.* **33**:778.

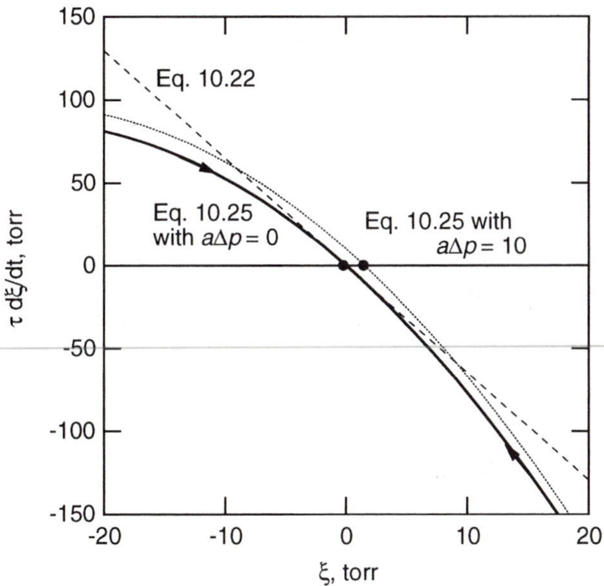

FIGURE 10.11. Plots of $\tau_1(d\xi/dt)$ vs ξ. The straight dashed line is the linear approximation. The parabolas are plots of the nonlinear equation, Eq. (10.25), for two different values of parameter p. The closed circles show stable fixed points.

$$\tau_1 \frac{d\xi}{dt} = a\ \Delta p - (1 - G_1 G_2)\xi - \frac{G_2 \xi^2}{y_0 - b}. \qquad (10.25)$$

Both this and the linear version are plotted in Fig. 10.11 for $a\ \Delta p = 0$. In each case $d\xi/dt$ is positive when $\xi < 0$ and negative when $\xi > 0$, so ξ approaches zero as time goes on. The direction of evolution of ξ is shown by the arrows on the nonlinear curve. This is often called a *one-dimensional flow*. The variable ξ "flows" to the origin, which is called a *stable fixed point* of the flow. If we change $a\ \Delta p$ to 10, the curve shifts as indicated by the dotted line, and the fixed point moves to a slightly different value of ξ.

This is a particular case of a differential equation in one dependent variable of the form

$$\frac{dx}{dt} = f(x).$$

A great deal about the solution to the general equation can be learned by graphing it as we have done above. When the derivative is positive the function increases with time, and when it is negative it decreases. Figure 10.12(a) shows a more complicated function, with arrows showing the direction of the flow. The stable fixed point is indicated by a solid circle. There are two *unstable fixed points*, indicated by open circles. If x has precisely the value of an unstable fixed point, it remains there because $dx/dt = 0$. However, if it is displaced even a small amount, it flows away from the unstable point. Figure 10.12(b) shows just the x axis with the fixed points and the arrows. Stable fixed points are often called *attractors* or *sinks*. The unstable fixed points are called *repellers* or *sources*. Chapter 2 of Strogatz (1994) has an excellent and detailed discussion of one-dimensional flows.

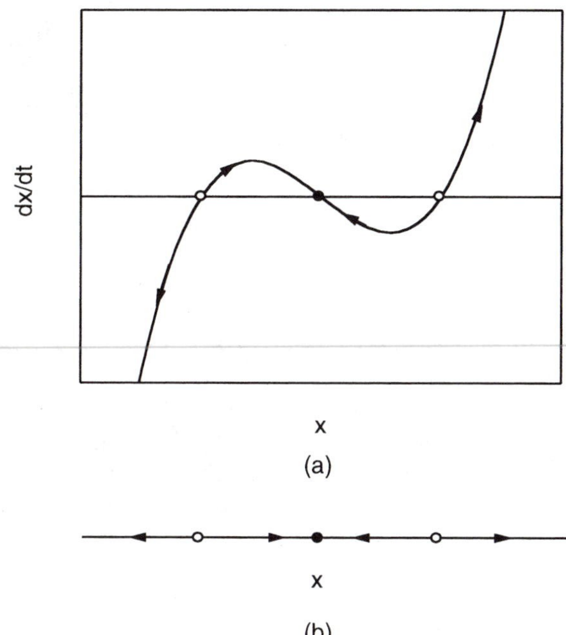

FIGURE 10.12. A more complicated flow on the line is shown. (a) Plot of dx/dt vs x. The arrows show the direction that x changes. The open circles show unstable fixed points, and the filled circle is a stable fixed point. (b) The fixed points and the direction of flow are shown on the x axis.

10.6. A FEEDBACK LOOP WITH TWO TIME CONSTANTS

In the preceeding section we considered a feedback loop in which only one process had a significant time constant. The other process responded "instantaneously;" its time constant was much shorter. Here we consider the case in which both processes have comparable time constants. We will see that it is possible for such a system to exhibit damped sinusoidal behavior in response to an abrupt change in one of the parameters. Whether it does or not depends on the relative values of the two time constants and the open-loop gain. We consider both graphical and analytical techniques for solving this problem.

In earlier sections we discussed control of breathing. Equation (10.20) was the linear model for the departure of one variable from equilibrium:

$$\tau_1 \frac{d\xi}{dt} = -\xi + G_1 \eta + a\ \Delta p. \qquad (10.20)$$

For the second process, instead of $\eta = G_2 \xi$ we assume that the behavior is given by an analogous equation

$$\tau_2 \frac{d\eta}{dt} = G_2\xi - \eta. \qquad (10.26)$$

For negative feedback either G_1 or G_2 must be negative.

We have a special case of a pair of first-order differential equations

$$\frac{dx_1}{dt} = f_1(x_1, x_2),$$
$$\frac{dx_2}{dt} = f_2(x_1, x_2). \qquad (10.27)$$

(Here x_1 and x_2 are general variables and have no relationship to the breathing problem considered earlier.)

We first combine the two first-order equations to make a second-order equation which, because we are using linear equations, can be solved exactly. To do this, differentiate Eq. (10.20):

$$\tau_1 \frac{d^2\xi}{dt^2} + \frac{d\xi}{dt} = a \frac{dp}{dt} + G_1 \frac{d\eta}{dt}.$$

Substitute Eq. (10.26) in this and obtain

$$\tau_1 \frac{d^2\xi}{dt^2} + \frac{d\xi}{dt} = -\frac{G_1}{\tau_2}\eta + \frac{G_1G_2}{\tau_2}\xi + a\frac{dp}{dt}.$$

To eliminate the η in this equation, solve Eq. (10.20) for $G_1\eta$ and substitute it in this equation. The result is

$$\tau_1 \frac{d^2\xi}{dt^2} + \frac{d\xi}{dt} = -\frac{\tau_1}{\tau_2}\frac{d\xi}{dt} - \frac{1}{\tau_2}\xi + \frac{a}{\tau_2}p + \frac{G_1G_2}{\tau_2}\xi + a\frac{dp}{dt}.$$

After like terms are combined, the result is

$$\frac{d^2\xi}{dt^2} + \left(\frac{1}{\tau_1} + \frac{1}{\tau_2}\right)\frac{d\xi}{dt} + \frac{1-G_1G_2}{\tau_1\tau_2}\xi = \frac{a}{\tau_1\tau_2}p(t) + \frac{a}{\tau_1}\frac{dp}{dt}. \qquad (10.28)$$

This is another linear differential equation with constant coefficients. The right-hand side is a known function of time, since $p(t)$ is known. The homogeneous equation is very common in physics and is called the *harmonic oscillator* equation. It is usually written in the form

$$\frac{d^2\xi}{dt^2} + 2\alpha\frac{d\xi}{dt} + \omega_0^2\xi = 0,$$

with the identifications

$$2\alpha = \frac{1}{\tau_1} + \frac{1}{\tau_2} = \frac{\tau_1+\tau_2}{\tau_1\tau_2},$$
$$\omega_0^2 = \frac{1-G_1G_2}{\tau_1\tau_2}. \qquad (10.29)$$

It is shown in Appendix F that as long as $\alpha \geqslant \omega_0$, the system is critically damped or overdamped and there will be no oscillation or "ringing." This will be the case if

$$\frac{(\tau_1+\tau_2)^2}{4\tau_1^2\tau_2^2} \geqslant \frac{1-G_1G_2}{\tau_1\tau_2}$$

or

$$\frac{(\tau_1+\tau_2)^2}{4\tau_1\tau_2} \geqslant 1 - G_1G_2. \qquad (10.30)$$

This equation is symmetric in τ_1 and τ_2. The important parameter is $x = \tau_1/\tau_2$. There is no ringing when

$$\frac{(1+x)^2}{4x} \geqslant 1 - G_1G_2.$$

Since the feedback is negative, $G_1G_2 = -|G_1G_2|$. Then there is no ringing if

$$|G_1G_2| < \frac{x}{4} + \frac{1}{4x} - \frac{1}{2}, \quad G_1G_2 < 0. \qquad (10.31)$$

If the two time constants are equal ($x=1$), the right-hand side of Eq. (10.31) is zero. There will be ringing if the open-loop gain has a magnitude greater than zero. For large values of x (say $x>10$), the equation is approximately $|G_1G_2| < x/4$. If the magnitude of the open-loop gain is larger than this, there will be ringing.

We can see the general behavior of Eqs. (10.20) and (10.26) by examining the behavior of the derivatives. Both derivatives are zero and there is a fixed point when

$$\xi = \frac{a\,\Delta p}{1-G_1G_2}, \quad \eta = \frac{G_2 a\,\Delta p}{1-G_1G_2}.$$

For $\Delta p = 0$ the fixed point is at the origin. Figures 10.13 and 10.14 show plots of ξ and η for different values of the gain and damping. The plots of η vs ξ are called *state-space* plots or *phase-plane* plots. Those shown spiral to the fixed point. Depending on the values of the gains and time constants (try positive feedback) there can also be exponentially growing solutions. An extensive literature exists analyzing stability for both Eqs. (10.27) and their linearized versions. See Chaps. 5 and 6 of Strogatz (1994) or Chap. 3 of Hilborn (1995).

10.7. MODELS USING NONLINEAR DIFFERENTIAL EQUATIONS

We have used many models in this book. In Chap. 2 we introduced a linear differential equation that leads to exponential growth or decay, and we used it to model tumor and bacterial growth and the movement of drugs through the body. We briefly examined some nonlinear extensions of this model. In Chap. 4 we modeled diffusion processes with linear equations, Fick's first and second laws, and we used a linear model to describe solvent drag. In Chap. 5 we used the model of a right-cylindrical pore. In Chap. 6 we used both a linear model, electrotonus, and a nonlinear model,

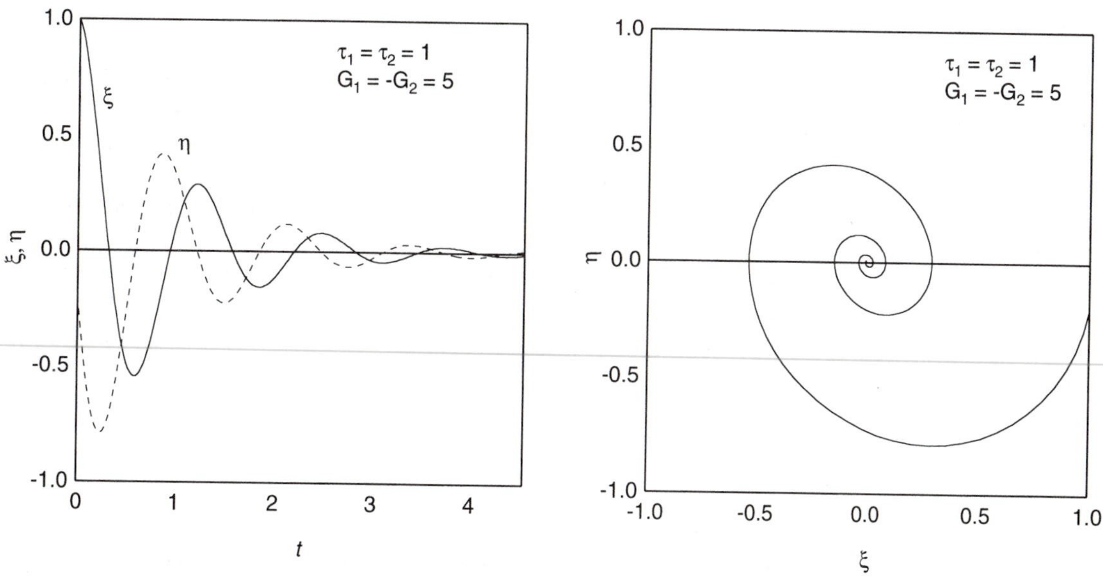

FIGURE 10.13. A solution to Eq. (10.28) is plotted that has a value 1 and time derivative zero when $t=0$. The variable η is obtained from ξ by using Eq. (10.26). Plots of ξ and η vs t are shown on the left. A state-space plot of η vs ξ is shown on the right.

the Hodgkin–Huxley equations. In this chapter we introduced a linear model for feedback, and we saw how two linear processes in a feedback loop could lead to oscillations, the linear harmonic oscillator.

Linear models have one advantage: they can be solved exactly. But most processes in nature are not linear. Jules-Henri Poincaré realized at the turn of the century that systems described exactly by the completely deterministic equations of Newton's laws could exhibit wild behavior. Poincaré was studying the three-body problem in astronomy (such as sun–earth–moon). While we are all familiar with the fact that the motion of the sun–earth–moon system is evolving smoothly with time and that eclipses can be predicted centuries in advance, this smooth behavior does not happen for all systems. For certain ranges of parameters (such as the masses of the objects) and initial positions and velocities, the solutions can exhibit behavior that is now termed *chaotic*. If we consider the motion that results from two sets of initial conditions that differ from each other only by an infinitesimal amount in one of the variables, we find that in chaotic behavior there can be solutions that diverge exponentially from each other as time

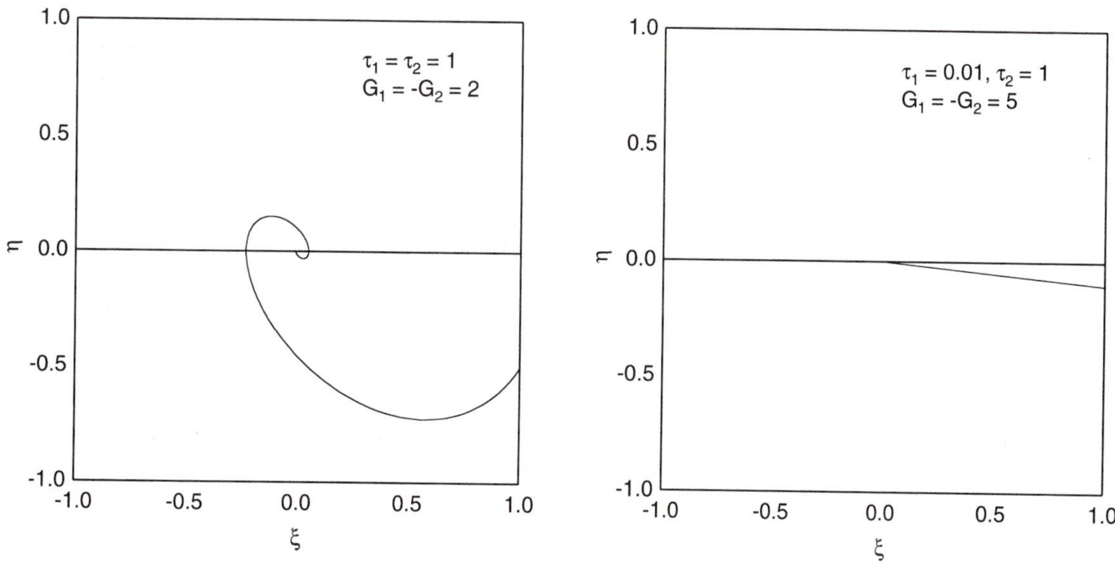

FIGURE 10.14. Additional state-space plots for the same initial conditions as in Fig. 10.13 with different values of the parameters.

goes on, even though the solutions remain bounded. Poincaré developed some geometrical techniques for studying the behavior of such systems. Thorough study of such systems requires the use of a digital computer. As a result, it has only been since the 1970s that we have realized how often chaotic behavior can occur in a system governed by deterministic equations. With computers we have gained more insight into the properties of chaotic behavior.

Just as the harmonic oscillator provides a model for behavior seen in many contexts from electric circuits to shock absorbers in automobiles to the endocrine system, certain features of nonlinear models have wide applicability. These include period doubling, the ability to reset the phase (timing) of a nonlinear oscillator, and deterministic chaos.

Some have said that Newtonian physics has been overthrown by chaos. This is not true. The same equations hold; predictable motions with which we have long been familiar still take place. Much of our current technology is based on them. We build television sets and send a spacecraft to explore several planets in succession. With chaos, we have come to understand a rich set of solutions to these same equations that we were not equipped to study before.

Many books about nonlinear systems have been written. A particularly interesting one for this audience is by Kaplan and Glass (1995). It is written for biologists and has many clear and relevant examples. Others are by Glass and Mackey (1988), by Hilborn (1995), and by Strogatz (1994).

Space limitations prevent more than a brief hint at some of the features of nonlinear dynamics, here and in Chap. 11. In this section we will discuss some one- and two-dimensional nonlinear differential equations. These will not lead to chaos, but will allow us to describe a very simple model for *phase resetting*. In Sec. 10.8 we will discuss equations that exhibit chaotic behavior.

Suppose that a nonlinear system with N variables can be described by a set of first-order differential equations:

$$\begin{aligned}
\frac{dx_1}{dt} &= f_1(x_1, x_2, \ldots, x_N), \\
\frac{dx_2}{dt} &= f_2(x_1, x_2, \ldots, x_N), \\
&\ldots, \\
\frac{dx_N}{dt} &= f_N(x_1, x_2, \ldots, x_N).
\end{aligned} \tag{10.32}$$

[These are an extension of the pair of differential equations we saw as Eqs. (10.27). Our model of breathing had two variables. It would be more realistic to use a breathing model with more variables, since alveolar ventilation also depends on arterial pH, weakly on oxygen partial pressure, and on the nervous factors that were described earlier.]

If the equations are cast in this form with N variables, then N initial conditions are required, corresponding to the constant of integration required for each equation. It is customary to say that there are *N degrees of freedom.* This is the language of *system dynamics.* This definition of degrees of freedom is different from what we used in Chap. 3, where each degree of freedom was represented by a second-order differential equation ($d^2x/dt^2 = F_x/m$, for example) and two initial conditions were required for each degree of freedom.

We can put Newton's second law in this form by writing two first-order differential equations instead of one second-order equation. For motion in one dimension, instead of

$$m\frac{d^2x}{dt^2} = F(x,v),$$

we write a pair of first-order equations:

$$\frac{dv}{dt} = \frac{F(x,v)}{m}, \quad \frac{dx}{dt} = v.$$

This system has two degrees of freedom in our new terminology. In either description, two initial conditions are required.

In many situations the force [or more generally the function on the right-hand side of Eqs. (10.32)] is time dependent. In standard form, the functions on the right do not depend on time. This is remedied by introducing one more variable, $x_{N+1} = t$. The additional differential equation is $dx_{N+1}/dt = 1$.

The evolution or "motion" of the system can be thought of as a trajectory in N-dimensional space, starting from the point that represents the initial conditions. Time is a parameter. We have seen an example of this for two dimensions in Figs. 10.13 and 10.14. It is possible to prove that two distinct trajectories cannot intersect in a finite period of time and that a single trajectory cannot cross itself at a later time [see Hilborn (1995, p. 76) or Strogatz (1994, p. 149)].[1] This is true in the full N-dimensional space; if we were to measure only two variables, we could see apparent intersections in the state plane that we were observing. This means that chaotic behavior, in which variables appear to change wildly and two-dimensional trajectories appear to cross, does not occur for a pair of differential equations of the form in Eqs. (10.32). At least three variables are required. A system with two degrees of freedom that is externally driven[2] can exhibit chaotic behavior because of the additional variable x_{N+1} that is introduced.

In Chap. 2 we studied the logistic differential equation

[1]In Figs. 10.13 and 10.14 we saw three different trajectories decaying exponentially towards the origin.

[2]That is, one of the functions on the right-hand side of the set of equations depends on time.

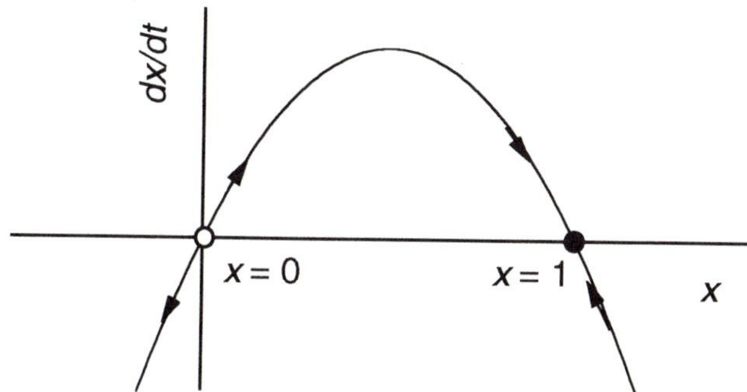

FIGURE 10.15. Plot of dx/dt vs x for the logistic differential equation.

$$\frac{dy}{dt} = by\left(1 - \frac{y}{y_\infty}\right). \tag{2.27}$$

It is convenient to rewrite the logistic equation in terms of the dimensionless variable $x = y/y_\infty$:

$$\frac{dx}{dt} = bx(1-x). \tag{10.33}$$

This separates the scale factor y_∞ from the dynamic factor b that tells how rapidly y and x are changing.[3] A plot of dx/dt vs x is shown in Fig. 10.15. There is an unstable fixed point at $x=0$ and a stable fixed point at $x=1$. The logistic equation is one of a whole class of nonlinear first-order differential equations for which dx/dt as a function of x has a maximum. It has been studied extensively because of its relative simplicity, and it has been used for population modeling. (Better models are available.[4] The logistic model assumes that the population is independent of the populations of other species, that the growth of the species does not affect the carrying capacity y_∞, and that the population increases smoothly with time.)

Many of the important features of nonlinear systems do not occur with one degree of freedom. We can make a very simple model system that displays the properties of systems with two degrees of freedom by combining the logistic equation for variable r with an angle variable θ that increases at a constant rate:

$$\frac{dr}{dt} = ar(1-r), \quad \frac{d\theta}{dt} = 2\pi. \tag{10.34}$$

This has the form of Eqs. (10.32). We can interpret (r, θ) as the polar coordinates of a point in the xy plane. When t has increased from 0 to 1 the angle has increased from 0 to 2π, which is equivalent to starting again with $\theta=0$. This system has been used by many authors. Glass and Mackey (1988) have proposed that it be called the *radial isochron clock*. Typical behavior is shown in Fig. 10.16(a). If $r=1$ there is a circular orbit corresponding to the stable fixed point of Eq. (10.33). Such a stable orbit is called a *stable limit cycle*.[5] There is an unstable limit cycle, $r=0$, corresponding to the unstable fixed point of Eq. (10.33). Any initial conditions except $r=0$ give trajectories that move toward the stable circular limit cycle as time progresses. The set of points in the xy plane lying on orbits that move to the limit cycle as $t \to \infty$ is called the *basin of attraction* for the limit cycle. In this case the basin of attraction includes all points except the origin. If we look at the time behavior, Fig. 10.16(b) shows the behavior of $x = r\cos\theta$ on the limit cycle. The oscillator might provide timing information as the phase moves through some value. Figure 10.16(c) shows a series of pulses every time θ is a multiple of 2π.

In many cases the differential equations contain one or

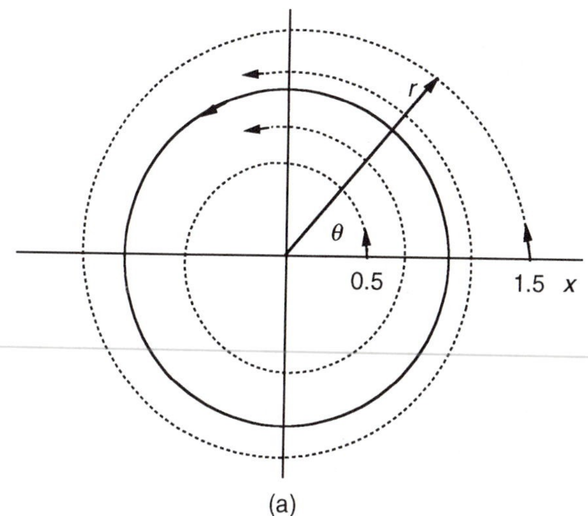

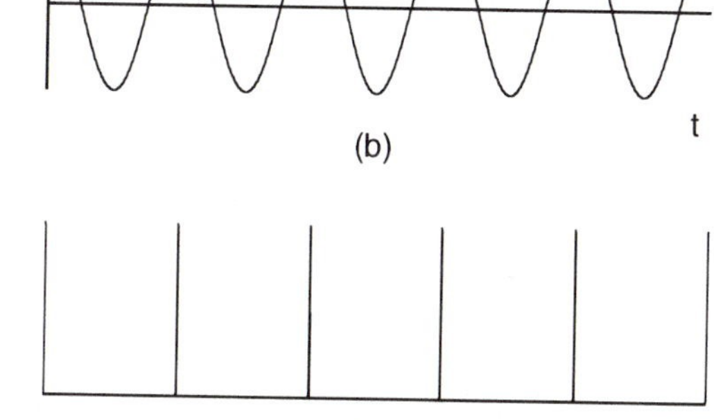

FIGURE 10.16. A system with two degrees of freedom. (a) The limit cycle is represented by the solid circle. Systems starting elsewhere in the plane have trajectories that approach the limit cycle as $t \to \infty$, as shown by the dashed lines. (b) The value of $x = r\cos\theta$ is plotted as a function of time. (c) A timing pulse is generated every time θ is a multiple of 2π.

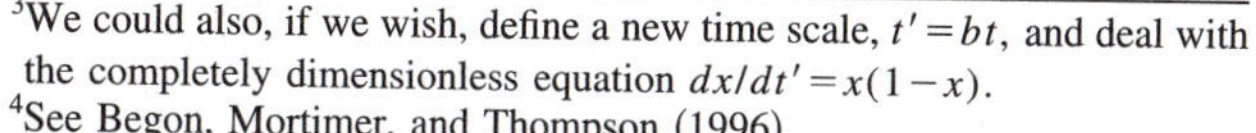

[3] We could also, if we wish, define a new time scale, $t' = bt$, and deal with the completely dimensionless equation $dx/dt' = x(1-x)$.

[4] See Begon, Mortimer, and Thompson (1996).

[5] A *stable limit cycle* is an oscillation in the solutions to a set of differential equations that is always reestablished following any small perturbation.

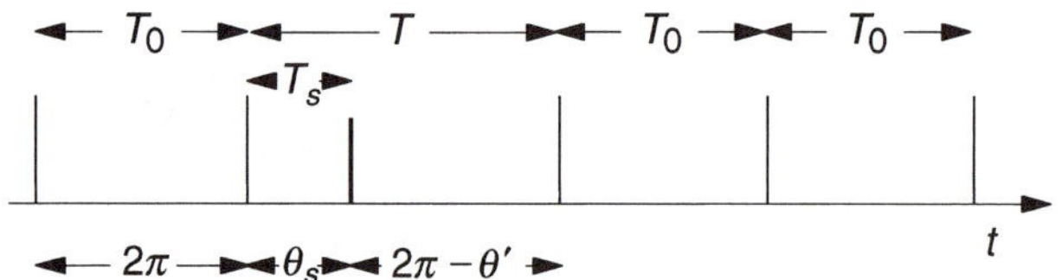

FIGURE 10.17. Resetting the phase of an oscillation. The oscillator fires regularly with period T_0. A stimulus a time T_s after it has fired causes a period of length T, after which the periods are again T_0.

more parameters that can be varied, and the number and shape of the limit cycles change as the parameters are changed. A point in parameter space at which the number of limit cycles changes or their stability changes is called a *bifurcation*. We will see examples of bifurcations in the next section. See the references for a much more extensive discussion.

One important characteristic of nonlinear oscillators is that a single pulse can reset their phase. If they are subject to a series of periodic pulses they can be entrained to oscillate at the driving frequency. (The nonlinear oscillators that sweep the electron beam across the screen of a television tube are entrained by synchronization pulses in the television signal.) Our simple two-dimensional model exhibits phase resetting that is very similar to that exhibited by cardiac tissue.[6]

Suppose that a cardiac pacemaker depolarizes every T_0 seconds and that it can be modeled by our radial isochron clock. Assume that depolarization occurs when $\theta=0$ or a multiple of 2π. A stimulus is applied at time T_s after the beginning of the cycle, as shown in Fig. 10.17. As a result the time from the previous depolarization to the next one is changed to T, after which the period reverts to T_0. (In a real experiment, it may be necessary to wait several cycles before measuring so that any transient behavior has time to decay, and then extrapolate back to find the value of T.) Often a stimulus early in the cycle is found to delay the next depolarization, while a stimulus late in the cycle advances it. Our model provides a simple geometric interpretation of this behavior, independent of any knowledge of the detailed dynamics.

A delayed depolarization is shown in Fig. 10.17. Pulses are occurring every T_0 seconds when the phase is a multiple of 2π (that is, 0). A stimulus is applied at a time T_s after the previous pulse, at which time the phase is θ_s. Since the phase advances linearly, we have the proportion

$$\frac{T_s}{T_0}=\frac{\theta_s}{2\pi}.$$

Suppose the stimulus causes the system to move to a new state with a phase θ' which we do not yet know. Since in our model $d\theta/dt$ is constant, the phase advances after the stimulus at the same rate as it would have without the stimulus. The next pulse occurs when the phase again reaches 2π. This occurs at a time T after the previous pulse, or a time $T-T_s$ after the stimulus, when the phase has increased from θ' to 2π. Therefore

$$\frac{T-T_s}{T_0}=\frac{2\pi-\theta'}{2\pi}$$

and

$$\frac{T}{T_0}=\frac{2\pi+\theta_s-\theta'}{2\pi}. \tag{10.35}$$

We use our limit cycle model to relate θ_s and θ' as shown in Fig. 10.18. The system has been moving on a circle of unit radius representing the stable limit cycle. Assume that the only effect of the stimulus is to shift the value of x by a distance b along the $+x$ axis. For the angles shown in Fig. 10.18(a) this results in a point with $\theta' < \theta_s$, a delay in the phase or $T>T_0$. For an initial angle in

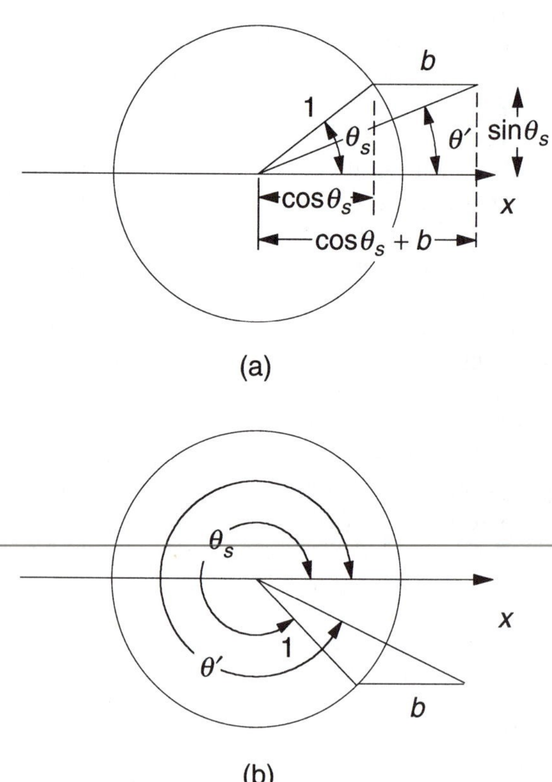

FIGURE 10.18. The limit-cycle oscillator model for resetting the phase of an oscillation. At phase θ_s a stimulus changes the value of x by an amount $+b$. (a) For angles $0<\theta_s<\pi$, this places the system on a trajectory with a larger value of r and a smaller phase θ', delaying the next pulse. (b) For angles $\pi<\theta_s<2\pi$, the stimulus results in a larger phase and the next pulse occurs earlier. The system returns to the limit cycle while θ continues to increase at a constant rate.

[6]This discussion is based on Glass and Mackey (1988), p. 104ff. See also the works by Winfree (1980, 1987, 1994, 1995).

the lower half plane [Fig. 10.18(b)] it results in $\theta' > \theta_s$ and $T < T_0$. The relation between the two angles can be obtained from the triangles:

$$\cos\theta' = \frac{\cos\theta_s + b}{[(\cos\theta_s + b)^2 + \sin^2\theta_s]^{1/2}}$$

$$= \frac{\cos\theta_s + b}{(1 + b^2 + 2b\cos\theta_s)^{1/2}}.$$

The stimulus changed both θ and r. After the stimulus each evolves independently according to its own differential equation. The trajectory returns to the limit cycle as r returns to its attractor, but the phase is forever altered. Figure 10.19(a) is a plot of θ' vs θ_s for two values of b. When $b<1$, θ' takes on all values, while for $b>1$, θ' is restricted to values near 0 and 2π. The first case is called a *type-1 phase resetting*, and the second is called a *type-0 phase resetting*. Figure 10.19(b) combines these results with Eq. (10.35) to determine T/T_0 as a function of T_s/T_0.

Figure 10.20 shows experimental data for electrotonically stimulated Purkinje fibers from the conduction system of a dog. The fibers were undergoing spontaneous oscillation with $T_0 = 1.575$ s. Stimuli of two different amplitudes were applied at different parts of the cycle, T_s/T_0. Two different curves were obtained. The one with larger current looks like the curve with $b = 1.05$ in Fig. 10.19(b), while the one with smaller current looks like the curve with $b = 0.95$.

It is theoretically possible to apply a stimulus that would put the system at the point $r=0$ in the state space. In that case it would not oscillate, though for this model $r=0$ is an unstable equilibrium point and any slight perturbation would lead the system back to the stable limit cycle. In more complicated models it is possible to have a region of state space corresponding to no oscillation and a basin of attraction that leads to it. Figure 10.21 shows the results of a calculation by Winfree (1995) of the effect of stimuli on resetting the phase of the Hodgkin–Huxley equations adjusted to oscillate spontaneously. The abscissa is the

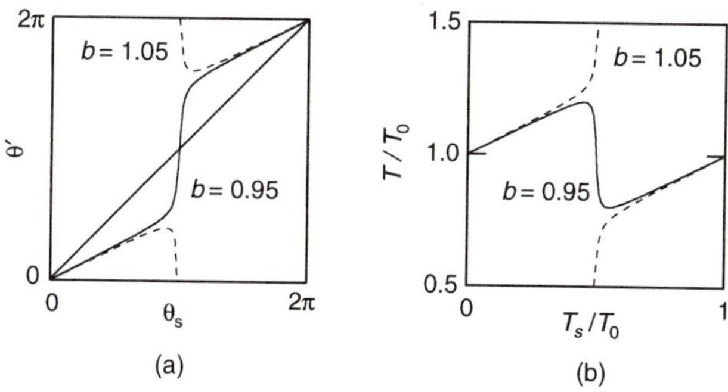

FIGURE 10.19. Plots of (a) the new phase vs the old phase and (b) the length of the period vs the time when the stimulus is applied.

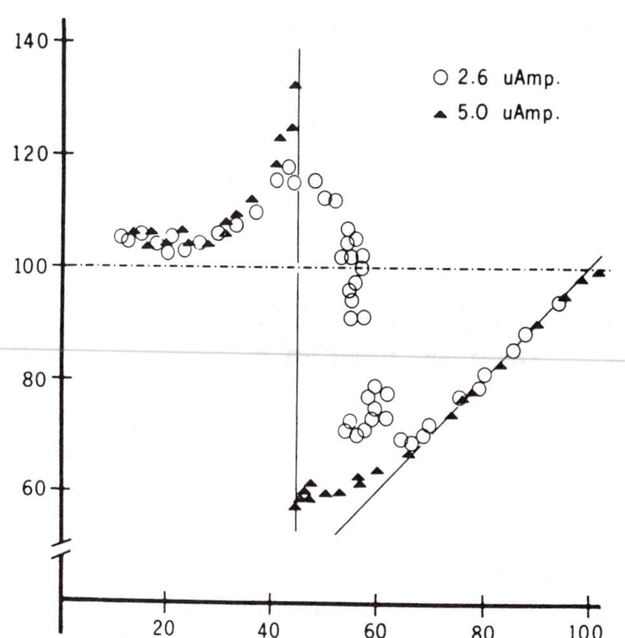

FIGURE 10.20. Phase resetting of a spontaneously oscillating Purkinje fiber by stimulation with an electrical impulse. The abscissa is T_s/T_0 expressed as a percentage. The ordinate is T/T_0 expressed as a percentage. Two different stimulus strengths were used. Compare the smaller stimulus (the open circles) to $b=0.95$ and the larger stimulus to $b=1.05$ in Fig. 10.19(b). Reproduced with permission from J. Jalife and G. K. Moe (1976). Effect of electrotonic potential on pacemaker activity of canine Purkinje fibers in relation to parasystole. *Circ. Res.* **39**: 801–808. Copyright 1976 American Heart Association.

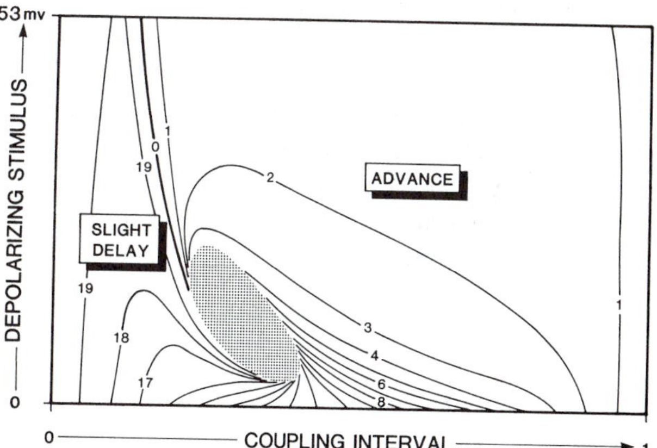

FIGURE 10.21. Phase resetting in a Hodgkin–Huxley model. The coupling interval is the delay from the previous pulse to the stimulation pulse in fractions of a period. The ordinate shows the size of the stimulus pulse in mV. The contours show the latency or time from the stimulus to the next pulse, measured in twentieths of a period. From A. T. Winfree (1987), *When Time Breaks Down*, Princeton, NJ, Princeton University Press. Copyright ©1987. Reproduced by permission of Princeton University Press.

coupling interval or the time after the previous pulse at which the stimulus is delivered. The ordinate is the height of the depolarizing pulse in mV. The contour lines show different values of the latency—the time in twentieths of a cycle period from the stimulus to the next pulse. The region of state space where annihilation of the oscillation occurs is shaded and has been called a ''black hole'' by Winfree (1995).[7]

10.8. DIFFERENCE EQUATIONS AND CHAOTIC BEHAVIOR

We have alluded to the possibility of chaotic behavior, but we have not yet seen it. Chaotic behavior of nonlinear differential equations requires three degrees of freedom. It is possible to see chaotic behavior in difference equations with a single degree of freedom because the restriction that the trajectory cannot cross itself or another trajectory no longer applies. It arose from the continuous nature of the trajectories for a system of differential equations.

We considered the logistic differential equation as a model for population growth. The differential equation assumes that the population changes continuously. For some species each generation is distinct, and a difference equation is a better model of the population than a differential equation. An example might be an insect population where one generation lays eggs and dies, and the next year a new generation emerges. A model that has been used for this case is the logistic difference equation or *logistic map*

$$y_{j+1}=ay_j\left(1-\frac{y_j}{y_\infty}\right)$$

with $a>0$ and j the generation number. It can again be cast in dimensionless form by defining $x_j=y_j/y_\infty$:

$$x_{j+1}=ax_j(1-x_j). \tag{10.36}$$

While superficially this looks like the logistic differential equation, it leads to very different behavior. The stable points are not even the same. A plot of x_{j+1} vs x_j is a parabola, from which we can immediately see the following properties of the logistic map:

$$x_j<0,\quad x_{j+1}<0,$$

$$x_j=0,\quad x_{j+1}=0,$$

$$0<x_j<1,\quad x_{j+1}>0,$$

$$x_j=1,\quad x_{j+1}=0,$$

$$x_j>1,\quad x_{j+1}<0.$$

If we are to use this as a population model, we must restrict x to values between 0 and 1 so the values do not go to $-\infty$. In order to keep successive values of the map within the interval (0,1) we also make the restriction $a<4$.

For the logistic differential equation, $x=1$ was a point of stable equilibrium. However, for the logistic map, if $x_j=1$ the next value is $x_{j+1}=0$. The equilibrium value x^* can be obtained by solving Eq. (10.36):

$$x^*=ax^*(1-x^*)$$

$$=1-1/a. \tag{10.37}$$

This can be interpreted graphically as the intersection of Eq. (10.36) with the equation $x_{j+1}=x_j$ as shown in Fig. 10.22. You can see from either the graph or from Eq. (10.37) that there is no solution for positive x if $a<1$. For $a=1$ the solution occurs at $x^*=0$. For $a=3$ the equilibrium solution is $x^*=2/3$. Figure 10.23 shows how, for $a=2.9$ and an initial value $x_0=0.2$, the values of x_j approach the equilibrium value $x^*=0.655$. This equilibrium point is called an *attractor*.

Figure 10.23 also shows the remarkable behavior that results when a is increased to 3.1. The values of x_j do not come to equilibrium. Rather, they oscillate about the former equilibrium value, taking on first a larger value and then a smaller value. This is called a *period-2 cycle*. The behavior of the map has undergone *period doubling*. What is different about this value of a? Nothing looks strange about Fig. 10.22. But it turns out that if we consider the slope of the graph of x_{j+1} vs x_j at x^*, we find that for $a>3$ the slope of the curve at the intersection has a magnitude greater than 1. Many books explore the implications of this.

The period doubling continues with increasing a. For $a>3.449$ there is a cycle of period 4. A plot of the period-4

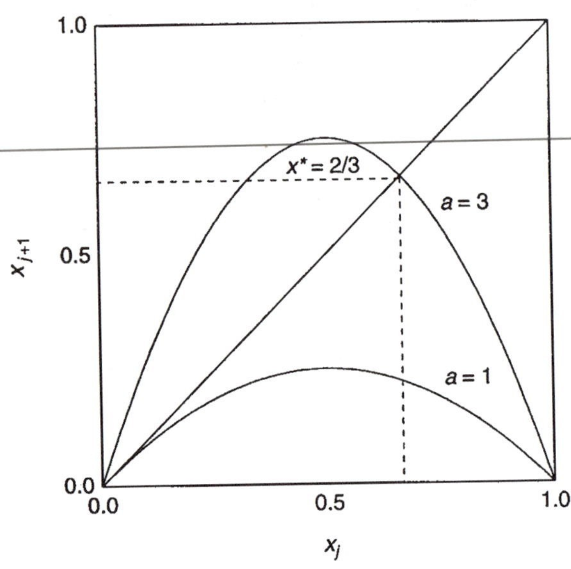

FIGURE 10.22. Plot of x_{j+1} vs x_j for the logistic difference equation or logistic map, for two values of parameter a.

[7]See Winfree (1987), especially Chaps. 3 and 4, or Glass and Mackey (1988), pp. 93–97.

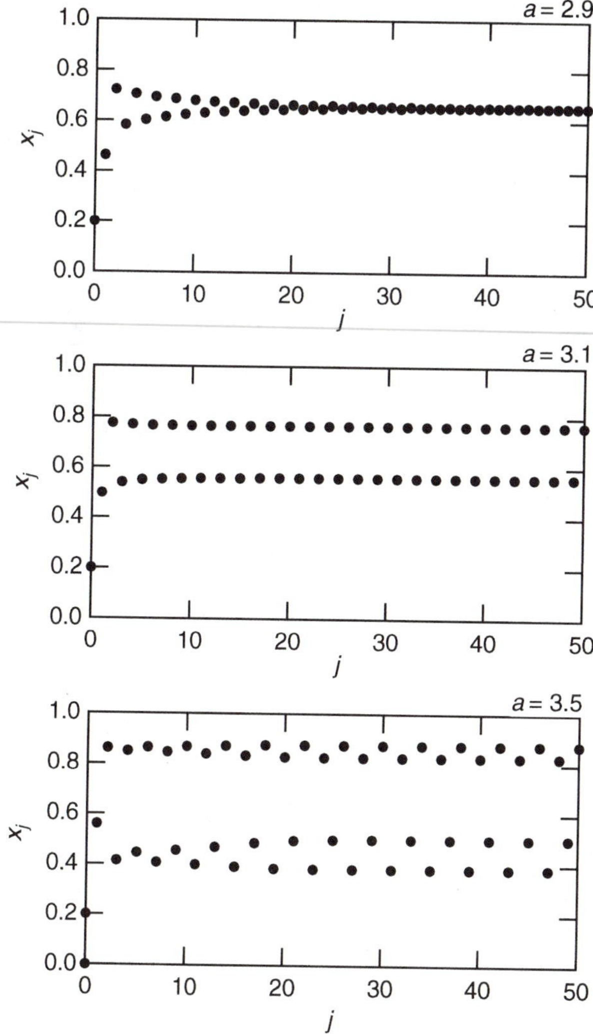

FIGURE 10.23. Plots of x_j vs j for different values of a, showing how the sequence of values converges to one, two, or four values of x called the attractors.

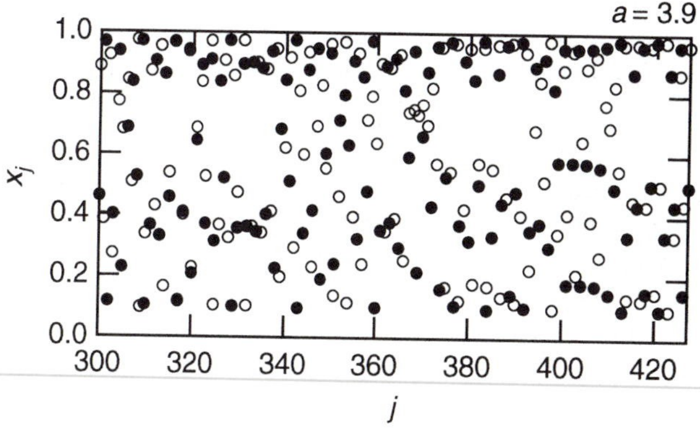

FIGURE 10.24. For this value of a the solution is aperiodic. There is no attractor.

cycle for $a=3.5$ is also shown in Fig. 10.23. For $a>3.54\,409$ there is a cycle of period 8. The period doubling continues, with periods 2^N occurring at more and more closely spaced values of a. When $a>3.569\,946$, for many values of a the behavior is aperiodic, and the values of x_j never form a repeating sequence. Remarkably, there are ranges of a in this region for which a repeating sequence again occurs, but they are very narrow. The details of this behavior are found in many texts. In the context of ecology they are reviewed in a classic paper by May (1976).

For $a<3.569\,946$, starting from different initial values x_0 leads after a number of iterations to the same set of values for the x_j. For values of a larger than this, starting from slightly different values of x_0 usually leads to very different values of x_j, and the differences become greater and greater for larger values of j. This is shown in Fig. 10.24 for $a=3.9$. The sequence is plotted from $j=301$ to $j=350$. The solid circles represent the sequence starting with $x_0=0.20$; the open circles represent the sequence for $x_0=0.21$.

This is an example of chaotic behavior or *deterministic chaos*. Deterministic chaos has four important characteristics:

1. The system is deterministic, governed by a set of equations that define the evolution of the system.
2. The behavior is bounded. It does not go off to infinity.
3. The behavior of the variables is aperiodic in the chaotic regime. The values never repeat.
4. The behavior depends very sensitively on the initial conditions.

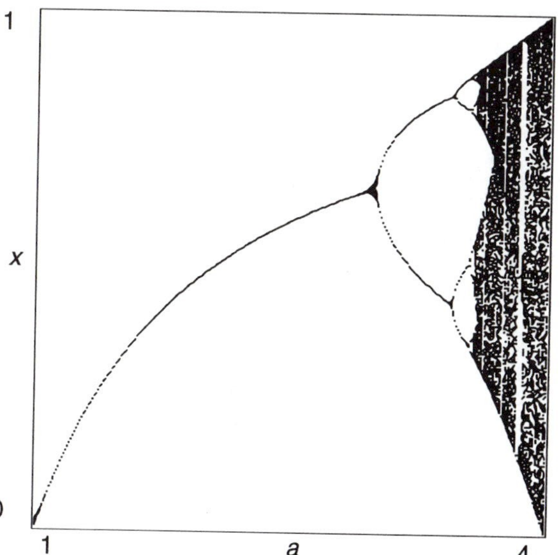

FIGURE 10.25. A bifurcation diagram for the logistic map, showing 300 values of x_j for values of a between 1 and 4. The plot was made using the Macintosh software *A Dimension of Chaos* by Matthew A. Hall.

Figure 10.25 shows the values of x_j that occur after any transients have died away for different values of parameter a. The diagram was made by picking a value of a. Then many different values of x_0 were selected at random and the iterations were made. After 50 iterations, the next 300 values of x_j were plotted. Then a was incremented slightly and the process was repeated. This is called a *bifurcation diagram*. The figure shows the range $1<a<4$. The asymptotic value of x_j rises according to $x^*=1-1/a$ until period doubling occurs at $a=3$. A four-cycle appears for $a>3.449$, and for $a>3.569\,946$ chaos sets in. Within the chaotic region are very narrow bands of finite periodicity.

Figure 10.26 shows a feature of many chaotic systems called *self-similarity*. The bifurcation diagram is plotted for two ranges of a: 3.4–4.0 and 3.743–3.745. The x scale is expanded in the second diagram. Note the similarity to the two bifurcation diagrams.

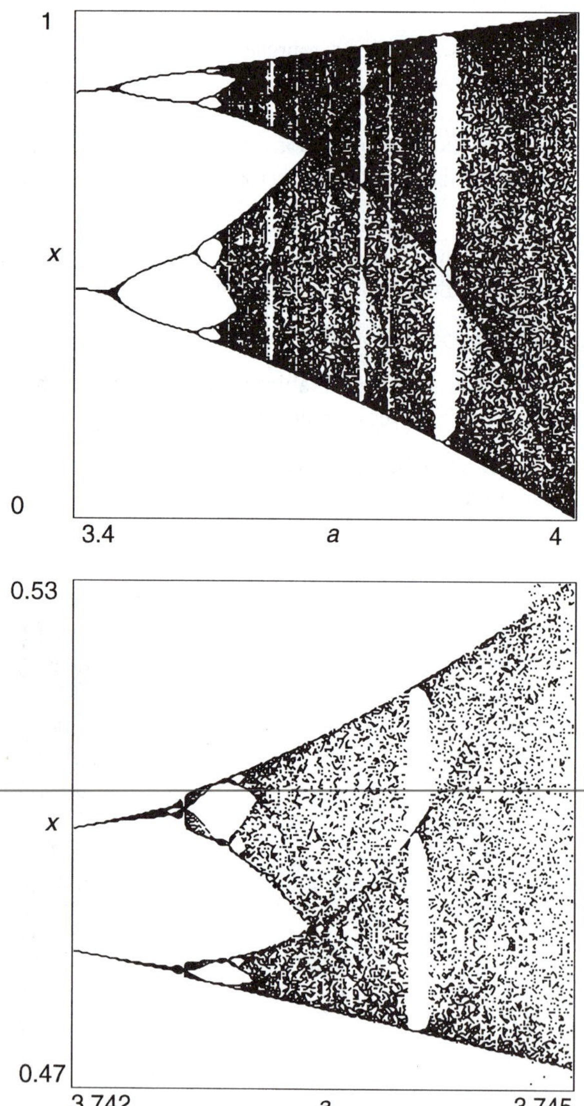

FIGURE 10.26. An example of self-similarity. The top curve shows $3.4<a<4$. The bottom curve shows $3.742<a<3.745$. Note its similarity to the top curve. The plot was made using the Macintosh software *A Dimension of Chaos* by Matthew A. Hall.

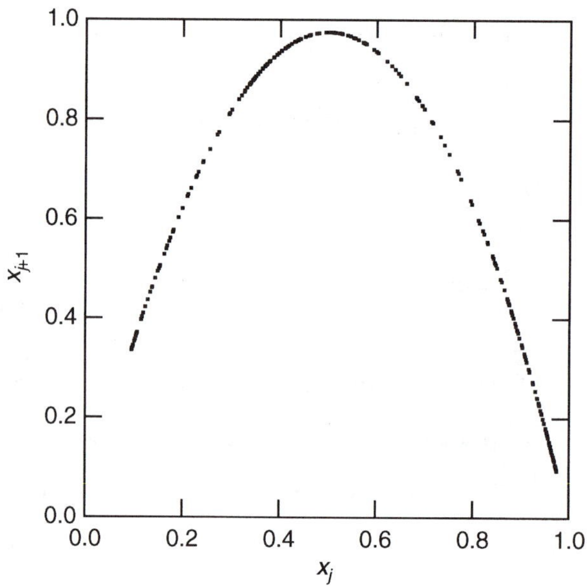

FIGURE 10.27. A plot of x_{j+1} vs x_j for the data of Fig. 10.24 recovers the logistic map.

Even though the plot of x_j vs j in Fig. 10.24 has no obvious pattern, the values of x_j were obtained from the logistic map. When we plot x_j vs x_j+1 the points fall on the map (Fig. 10.27).

The simplest systems in which chaotic behavior can be seen are first-order difference equations in which x_{j+1} is a function of x_j. The function is peaked and "tunable" by some parameter. Chaotic behavior occurs for some values of the parameter. In fact, it appears that the ratios of the parameter values involved in the period doubling and approach to chaos may be independent of the particular shape of the curve.[8]

Some systems exhibit *quasiperiodicity*. Consider the map $x_{j+1}=x_j+b$ where b is a fixed parameter. Wrap the function back on itself so that x remains in the interval (0,1). This is done by using the modulo or remainder function.[9] The map is

$$x_{j+1}=x_j+b \pmod 1). \tag{10.38}$$

The function is plotted in Fig. 10.28 for $b=0.3$. The map is plotted in (a). The apparent discontinuities are due to the wrapping. A sequence of 50 points is plotted in Fig. 10.28(b). Because $b=3/10$ is a rational fraction, the points repeat themselves exactly every 10 steps. This can be seen in Fig. 10.28(c), which plots 128 consecutive points on a

[8]See Hilborn (1995), Chap. 2, Kaplan and Glass (1995), p. 30, or Strogatz (1994), p. 370.

[9]The function x mod n gives the number that remains after subtracting n from x enough times so that the result is less than 1. For example, 1.5742 mod 1=0.5742; 7.5 mod 1=0.5.

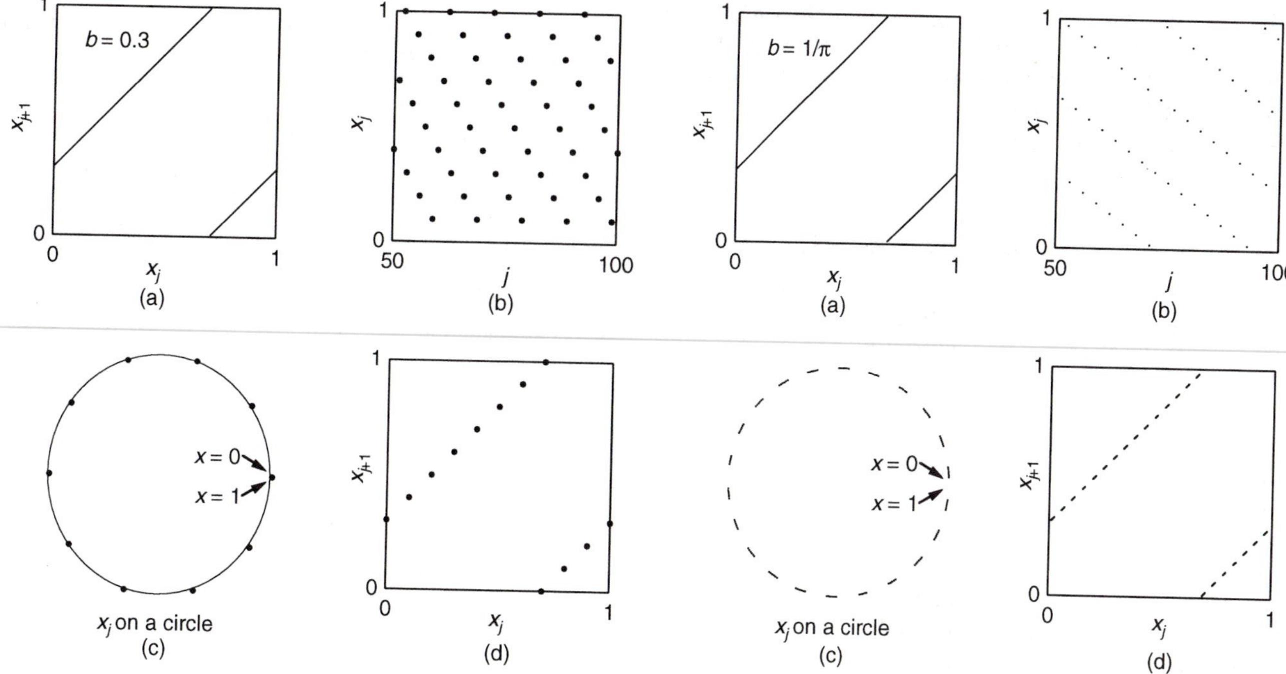

FIGURE 10.28. The linear map $x_{j+1}=x_j+0.3$ (mod 1). (a) Plot of the map. (b) Plot of x_j vs n for 50 points. (c) The map plotted on a circle for 128 points. (d) Plot of x_{j+1} vs x_j gives 128 points on the map.

FIGURE 10.29. The linear map $x_{j+1}=x_j+1/\pi$ (mod 1). (a) Plot of the map. (b) Plot of x_j vs n for 50 points. (c) The map plotted on a circle for 128 points. (d) Plot of x_{j+1} vs x_j gives 128 points on the map.

circle. The angle counterclockwise from the horizontal axis is $\theta_j=2\pi x_j$. The 128 values all fall at 10 points on the circle. The plot of x_{j+1} vs x_j in Fig. 10.28(d) has 10 points that fall on the map.

Compare this with Fig. 10.29, which is a plot of the same map for an irrational value of the parameter, $b=1/\pi$. The curve in *a* looks very similar. However, the values of x_j never repeat. This is difficult to see from Fig. 10.29(b), but can be seen in (c), where the 128 points are all at different values of θ. If more points were plotted, the circle would be completely filled. All of the points plotted in (d) are also different, but of course they lie on the map function. If we were to make a bifurcation diagram the values of x_j would fill all points on the graph, unless b were a rational fraction, when there would be a finite number of points. This appears at first sight to be chaotic behavior, but it is not. The function is deterministic, it is bounded, and the values of x never repeat. But it does not satisfy the last criterion: sensitive dependence on initial conditions. In chaotic behavior, two trajectories that start from initial points that are very close diverge in time. If a slightly different value of x_0 is used for this map, all of the values in the new sequence are shifted from the original sequence by the same amount. There is no divergence of the two solutions. In quasiperiodicity, the trajectories for two points that are initially close remain close.

10.9. A FEEDBACK LOOP WITH A TIME CONSTANT AND A FIXED DELAY

In Sec. 10.6 we saw that if both processes in a two-stage feedback system had comparable time constants, there was the possibility for damped oscillations or "ringing." Another possibility is that a portion of the system may respond to values of a state variable at some earlier time. The fixed time delay could be the time it takes a signal to travel along a nerve or the time it takes for a chemical to pass through a blood vessel.

We will consider a linear model for such a system, as shown in Fig. 10.30:

$$\tau_1 \frac{dy}{dt}+y=G_1 x+p_1,$$
$$x=G_2 y(t-t_d)+p_2. \tag{10.39}$$

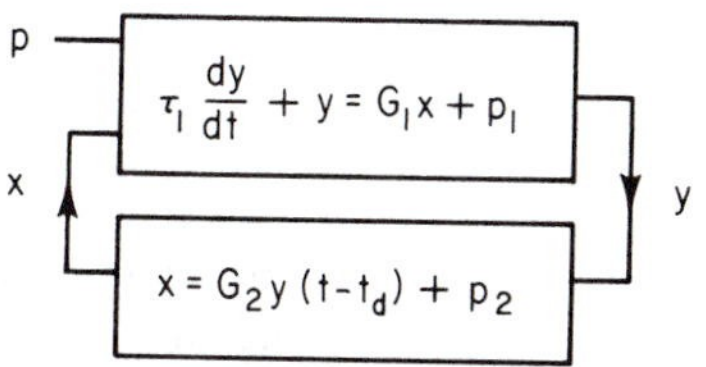

FIGURE 10.30. A two-stage feedback loop. The upper process is described by a single time constant; the lower one introduces a fixed time delay.

The first equation is like those in the preceding section, except that the factor a multiplying p_1 is set equal to unity. The second equation says that $x(t)$ is proportional to the value of y at the earlier time $t-t_d$, plus some other parameter p_2. These can be combined to give a *delay-differential equation*:

$$\tau_1 \frac{dy}{dt} = -y + G_1 G_2 y(t-t_d) + p_1 + G_1 p_2$$

or, defining $p = p_1 + G_1 p_2$ to eliminate clutter,

$$\tau_1 \frac{dy}{dt} = -y + G_1 G_2 y(t-t_d) + p. \qquad (10.40)$$

This equation can give rise to sustained as well as damped oscillations. It is not hard to see why. Suppose that y is above some equilibrium value and that $G_1 G_2 < 0$. The first term on the right causes y to decrease toward equilibrium. But when it is nearly at equilibrium the second term, responding to an earlier positive value of y, continues to make y decrease so y goes negative. Now y is below the equilibrium value and the same arguments can be applied as y increases. This paragraph could go on for a long time.

Why do we now have oscillations for a system with apparently only one degree of freedom? The reason is the delay term. In order to specify the initial state of the system at $t=0$, we must specify the value of y for all times $-t_d < t < 0$. This is effectively an infinite number of values of y. Delay differential equations have an infinite number of degrees of freedom.

The mathematics for such a system become quite involved (even for the linear system we discuss here). The techniques for solving the equation were first described by Hayes (1950). The equation has been considered for biological examples by Glass and Mackey (1988).

The derivative is zero and the equation has a fixed point y_f when $y_f = p/(1 - G_1 G_2)$. It is convenient to work with the new variable $w = y - y_f$ and rewrite Eq. (10.40) as

$$\tau_1 \frac{dw}{dt} = -w + G_1 G_2 w(t-t_d).$$

We make another simplifying assumption; that the magnitude of the open-loop gain $G_1 G_2$ is so much greater than 1 that the $-w$ term can be neglected.[10] Then the equation becomes

$$\frac{dw}{dt} = \frac{G_1 G_2}{\tau_1} w(t-t_d).$$

Now recall that since $G_1 G_2 \ll -1$, this coefficient is approximately the negative of the reciprocal of the time constant with no delay and *with* feedback [see Eq. (10.23)]. Therefore the equation we will solve is

$$\frac{dw}{dt} = -\frac{1}{\tau} w(t-t_d). \qquad (10.41)$$

If the delay time is zero, this is the familiar equation for exponential decay. As we argued above, a delay can allow oscillation. One can show by substitution that for certain values of the parameters one possible solution has the form $w(t) = w_0 e^{-\gamma t} \cos \omega t$. We will find the conditions for a steady oscillation of the form $w(t) = w_0 \cos \omega t$. The left-hand side of Eq. (10.41) is $dw/dt = -\omega w_0 \sin \omega t$. The right-hand side is

$$-(1/\tau) w_0 \cos(\omega t - \omega t_d) = -(1/\tau) w_0 \cos \omega t \cos \omega t_d$$
$$-(1/\tau) w_0 \sin \omega t \sin \omega t_d.$$

Therefore the proposed solution will satisfy Eq. (10.41) only if

$$-\omega w_0 \sin \omega t = -(1/\tau) w_0 \sin \omega t_d \sin \omega t$$

and

$$0 = -(1/\tau) w_0 \cos \omega t_d \cos \omega t$$

from which we get $\omega = 1/\tau$ and $\cos \omega t_d = 0$ or $\omega t_d = \pi/2$. Combining these gives $t_d/\tau = \pi/2$. From these we see exactly how the sustained oscillation occurs. The delay time and frequency are such that the shift is exactly one-quarter cycle. This is the same shift that would be obtained by taking the second time derivative of the undelayed function, which would lead to the undamped harmonic oscillator equation.

10.10. FEEDBACK LOOPS: A SUMMARY

The last several sections have been mathematically complex. However, you do not need to memorize a large number of equations to carry away the heart of what is in them. The essential features are as follows:

1. If the equations relating the input and output variables of each process of a feedback loop are known, then their simultaneous solution gives the equilibrium or steady-state values of the variables. (In a biological system it may be very difficult to get these equations.) The solution is called the operating point or a fixed point of the system of equations.
2. If a single process in the feedback loop determines the time behavior, and the rate of return of a variable to equilibrium is proportional to the distance of that variable from equilibrium, then the return to equilibrium is an exponential decay and the system can be characterized by a time constant.
3. In a feedback system one variable changes to stabilize another variable. The amount of stabilization and the accompanying decrease in time constant depend on the open-loop gain.

[10] If you are considering a problem where this is not a reasonable assumption, see the Appendix of Glass and Mackey (1988).

4. It is possible to have oscillatory behavior with damped or constant amplitude if the two processes have comparable time constants and sufficient open-loop gain, or if one of the processes depends on the value of its input variable at an earlier time or if the process has three or more degrees of freedom.
5. A nonlinear system oscillating on a limit cycle can have its phase reset by an external stimulus.
6. Nonlinear systems of difference equations with one or more degrees of freedom or nonlinear systems of differential equations with three or more degrees of freedom may exhibit bifurcations and chaotic behavior.

10.11. ADDITIONAL EXAMPLES

This section provides some additional examples of the principles we have seen above. The details of the experiments and modeling are given in the references.

10.11.1. Cheyne–Stokes Respiration

We have seen how the body responds to CO_2 levels in the blood by controlling the rate and amplitude of breathing to maintain the CO_2 concentration within a narrow range. The frequency and amplitude of breathing can also undergo oscillation. Some patients almost stop breathing for a minute or so and then breathe with much greater amplitude than normal. This is called Cheyne–Stokes breathing. Guyton, Crowell, and Moore (1956) showed that diverting carotid artery blood in dogs through a long length of tubing increased the transit time between heart and brain and caused Cheyne–Stokes respirations. Cheyne–Stokes respirations have been modeled with a nonlinear delay-differential equation by Mackey and Glass. Their results are shown in Fig. 10.31.

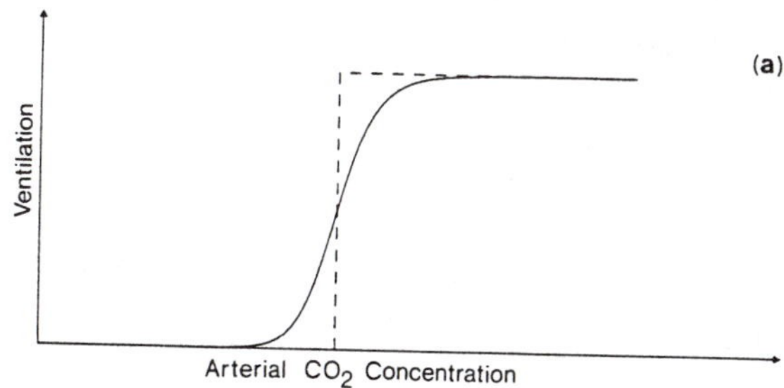

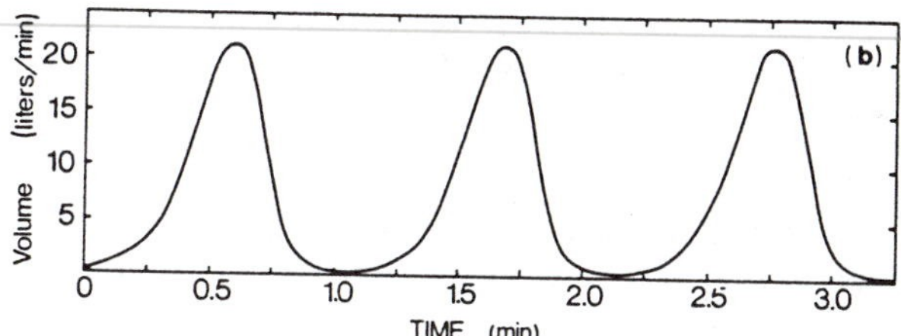

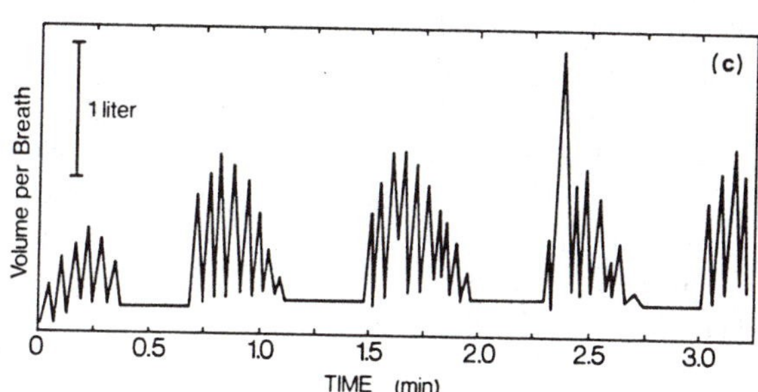

FIGURE 10.31. Cheyne–Stokes respirations. (a) The curve used to model $y=g(x(t-t_d))$. (b) The results of the model calculation. (c) Ventilation during Cheyne–Stokes respiration. From L. Glass and M. C. Mackey (1988). *From Clocks to Chaos*. Princeton, NJ, Princeton University Press. Copyright © 1988 by Princeton University Press. Reproduced by permission. Panels (b) and (c) reprinted with permission from M. C. Mackey and L. Glass (1977), Oscillation and chaos in physiological control systems. *Science* **197**: 287–289. Copyright 1977 American Association for Advancement of Science.

10.11.2. Hot Tubs and Heat Stroke

Problems 10.10 and 10.11 discuss how the body perspires in order to prevent increases in body temperature. At the same time blood flows through vessels near the surface of the skin, giving the flushed appearance of an overheated person. The cooling comes from the evaporation of the perspiration from the skin. If the perspiration cannot evaporate or is wiped off, the feedback loop is broken and the cooling does not occur. If a subject in a hot tub overheats, the same blood flow pattern and perspiration occur, but now heat flows into the body from the hot water in the tub. The feedback has become positive instead of negative, and heat stroke and possibly death occurs. This has been described in the physics literature by Bartlett and Braun (1983).

10.11.3. Pupil Size

The pupil changes diameter in response to the amount of light entering the eye. This is one of the most easily studied feedback systems in the body, because it is possible to break the loop and to change the gain of the system. Let the variables be as follows: x is the amount of light striking the retina, p is the light intensity, and y the pupil area. In the normal case, x is proportional to y and p: $x=Apy$. The body responds to increasing x by decreasing y so $y=f(x)$. These processes are shown in Fig. 10.32.

The reason this system can be studied so easily is that shining a very narrow beam of light into the pupil means that the change of pupil radius no longer affects x; the loop is broken in the upper box of Fig. 10.32. Shining a light into the eye so that it is on the edge of the pupil increases the

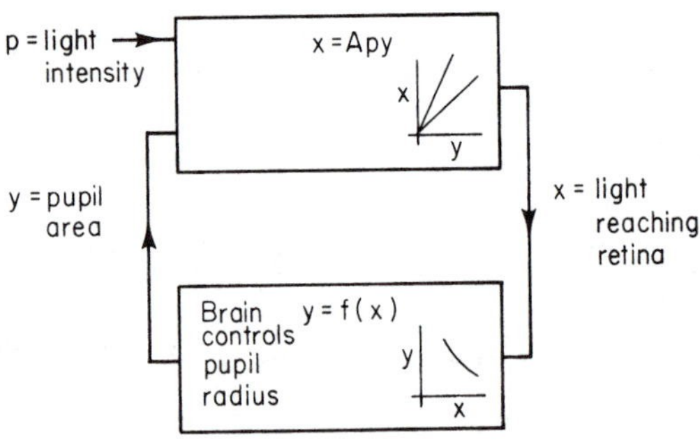

FIGURE 10.32. The feedback system for controlling the size of the pupil.

gain in the upper box. These schemes are shown in Fig. 10.33. Furthermore, it has been discovered experimentally that the process in the lower box controls the size of both pupils, even though light is directed at only one eye.

The properties of $y=f(x)$ have been studied extensively by Stark [see Stark (1957, 1968, 1984)]. The results are consistent with a feedback loop having several time constants and also a fixed delay. Increasing the open loop gain as in Fig. 10.33(c) causes the pupil to oscillate at a frequency of about 1.3 Hz (cycles per second). Stark (1984) reviews this work, including the use of noise to analyze the system and nonlinearities.

10.11.4. Oscillating White-Blood-Cell Counts

A delay-differential equation has been used to model the production of red and white blood cells. Figure 10.34 shows the actual white count for a patient with chronic granulocytic leukemia as well as the results of a model calculation. The striking feature of the model is the emergence of an aperiodic pattern when the delay time is increased from 6 to 20 days.

10.11.5. Waves in Excitable Media

The propagation of an action potential is one example of the propagation of a wave in excitable media. We saw in Chap. 7 that waves of depolarization sweep through cardiac tissue. The circulation of a wave of contraction in a ring of cardiac tissue was first demonstrated by Mines in 1914. It was first thought that such a wave had to circulate around an anatomic obstacle, but it is now recognized that no obstacle is needed.

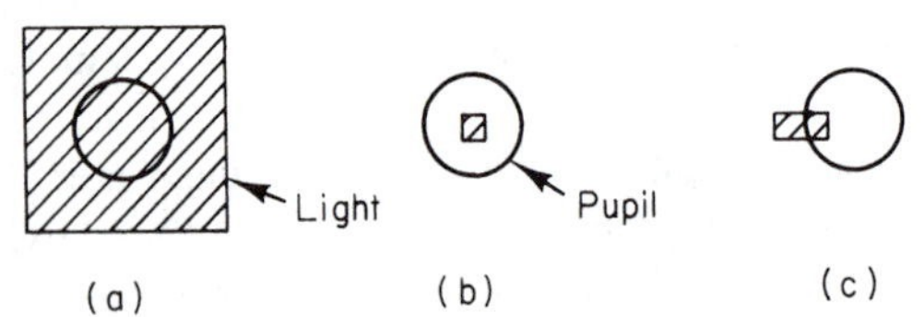

FIGURE 10.33. The feedback loop for pupil size can be changed by changing the way in which light strikes the eye. (a) In the normal situation $x=Apy$. (b) When the spot of light is smaller than the pupil, the feedback loop is broken. (c) When the spot of light strikes the edge of the pupil, the gain is increased.

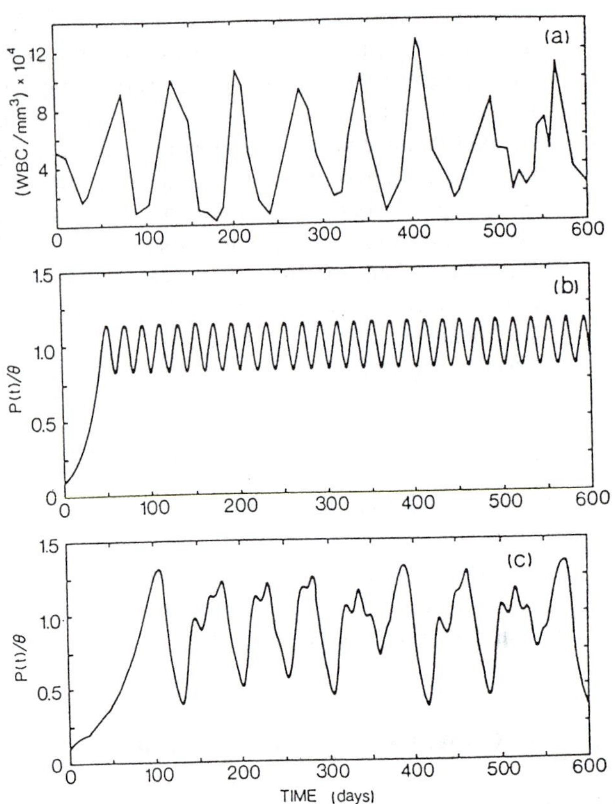

FIGURE 10.34. A nonlinear model for white-blood-cell production. (a) White-blood-cell count from a patient with chronic granulocytic leukemia. (b) The results of a nonlinear delay-differential equation model with a delay time of 6 days are an oscillation with a period of 20 days. (c) The results of using the same model with a delay time of 20 days are aperiodic. Reprinted with permission from M. C. Mackey and L. Glass (1977). Oscillation and chaos in physiological control systems. *Science* **197**: 287–289. Copyright 1977 American Association for Advancement of Science.

Waves in thin slices of cardiac tissue often have the shape of spirals, very similar to simulations produced by a model similar to a two-dimensional Hodgkin–Huxley model [for example, see Pertsov and Jalife (1995)]. These waves turn out to occur in many contexts beside the heart. They have also been seen in the Belousov–Zhabotinsky chemical reaction,[11] in social amoebae, in the retina of the eye, and as calcium waves in oocytes. Beautiful photographs of all of these are found in Winfree (1987).

These spiral waves seem to be another ubiquitous phenomenon (like period doubling) that depends only on

[11]There are many references. See Mielczarek *et al.* (1983); Epstein *et al.* (1983); and Winfree (1987).

the coarse features of the model. They can be generated with simple computer models called *cellular automata*. The rules for such an automaton and photographs of the resulting spiral waves are shown in Chap. 2 of Kaplan and Glass (1995).

The study of three-dimensional spiral waves in the heart is currently a very active field [Keener and Panfilov (1995); Mercader *et al.* (1995); Pertsov and Jalife (1995); Winfree (1994, 1995)]. They can lead to ventricular tachycardia, they can meander, much as a tornado does, and their conversion to turbulence is a possible mechanism for the development of ventricular fibrillation (see the next example).

10.11.6. Period Doubling and Chaos in Heart Cells

Guevara *et al.* (1981) have subjected small aggregates of chick heart-cell aggregates to periodic stimulation. The stimulation frequency was slightly less than the natural frequency of oscillation. The behavior of the preparation is shown in Fig. 10.35 and can best be seen by examining the *bottom* of the leading edge of the sharp positive pulse. The top strip on the left is phase locking. This is followed on the right in the top strip by an alternation characteristic of frequency doubling. The middle strip shows a variation of period 4. The bottom strip shows irregularity that is consistent with deterministic chaos.

Garfinkel *et al.* (1992) have also observed period doubling in a stimulated preparation of rabbit heart. Arrhythmias were induced by adding drugs to the solution

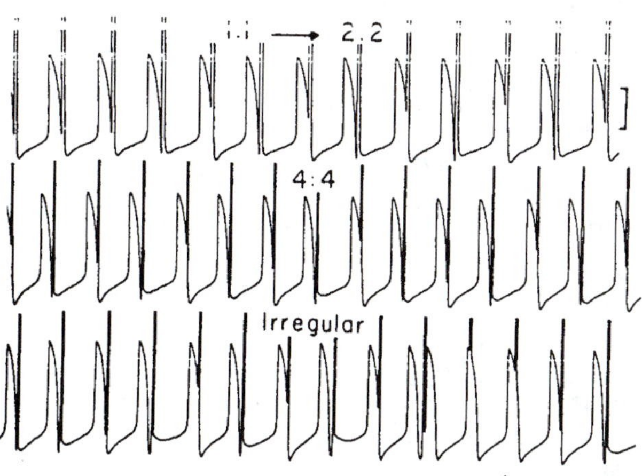

FIGURE 10.35. An aggregate of chick heart cells was periodically stimulated. Follow the bottom of the beginning of each sharp spike. The left part of the top strip shows phase locking. The right-hand portion of the top strip shows period doubling. The middle strip shows a period-4 behavior. The bottom strip shows irregular behavior consistent with deterministic chaos. Reprinted with permission from M. R. Guevara, L. Glass, and A. Shrier (1981). Phase locking, period-doubling bifurcations and irregular dynamics in periodically stimulated cardiac cells. *Science* **214**: 1350–1353. Copyright 1981 American Association for Advancement of Science.

perfusing the preparation. Figure 10.36 shows plots of the recorded action potentials and a plot of the map of I_n vs I_{n-1}, where I is the interval between beats. In panels A and B there is a constant interbeat interval and one point on the map. Panels C and D show period doubling. Panels E and F show a period-4 pattern. Panels G and H are completely aperiodic. Garfinkel *et al.* describe how to apply perturbations to restore periodicity. However, Christini and Collins (1995) have reported that similar techniques used by Schiff *et al.* (1994) to control a hippocampal preparation can be used to control a simple nonchaotic random system, which raises questions about whether the system is really being controlled.

Ventricular fibrillation is "the rapid, disorganized, and asynchronous contraction of ventricular muscle. ... it represents the final common pathway for death in most patients who experience out-of-hospital cardiac arrest, and its rate of recurrence is on the order of 30% in the first year in successfully resuscitated patients." [Epstein and Ideker (1995)]. It appears to be due to meandering waves, and it does not occur unless the myocardium is sufficiently thick [Winfree (1994, 1995)].

Witkowski *et al.* (1993, 1994) have made electrode arrays with a spacing of about 200 μm that can be placed directly on the myocardium. The membrane current i_m can be estimated from the spatial derivatives of the extracellular (interstitial) potential. Recently [Witkowski *et al.* (1995)] this technique has provided evidence that ventricular fibrillation has a component with simpler dynamics than had previously been thought.

SYMBOLS USED IN CHAPTER 10

Symbol	Use	Units	First used on page
a, b	Arbitrary parameter		263
a	Parameter in logistic map		270
a,b	Constant in logistic equation	s^{-1}	267
b	Dead space in lungs	m^3 or l	257
b	Amplitude of stimulus		268
f, g, h, i	Functions		258
j	Index for successive values in difference equation		270
m	Mass	kg	266
n	Number of moles of dissolved carbon dioxide	mol	260
$\dot{o}$	Rate of oxygen consumption	m s^{-1}	257
p	Rate of oxygen consumption	mmol min^{-1}	257
p	Light intensity		275
r	Variable		267
t	Time	s	266

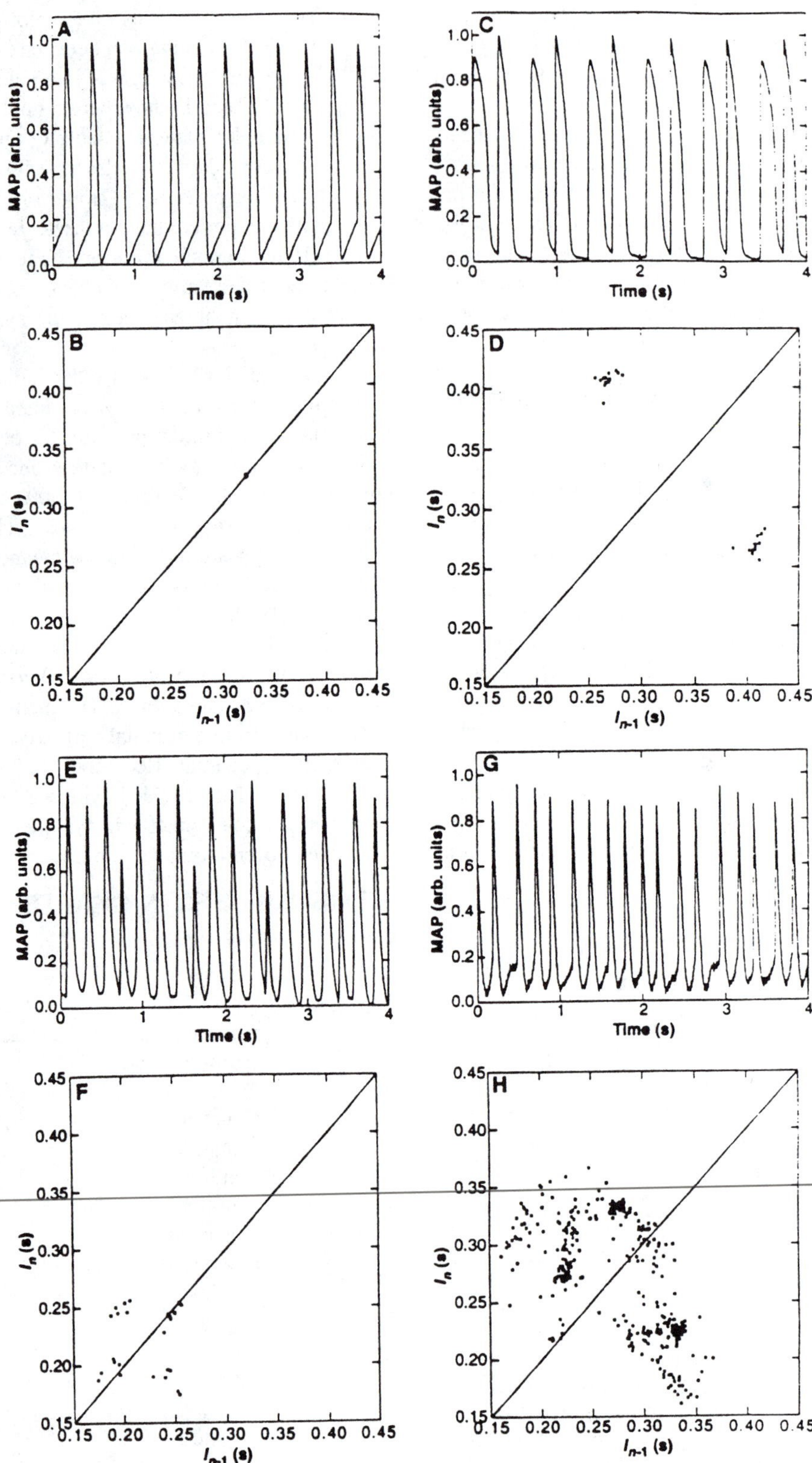

FIGURE 10.36. The results of experiments on a preparation consisting of intraventricular septum from a rabbit heart. Plots show the recorded action potentials and the map of I_n vs I_{n-1} where I is the interval between beats. In A and B there is a constant interbeat interval and one point on the map. Panels C and D show period doubling. Panels E and F show a period-4 pattern. Panels G and H are completely aperiodic. Reprinted with permission from A. Garfinkel, M. L. Spano, W. J. Ditto, and J. N. Weiss (1992). Controlling cardiac chaos. *Science* **257**: 1230–1235. Copyright 1992 American Association for Advancement of Science.

Symbol	Use	Units	First used on page
t_d	Delay time		273
w,x,y,z	General variables		273
x	Partial pressure of carbon dioxide	torr	257
x	Amount of light striking retina		275
x^*	Equilibrium value of x		270
y	Ventilation rate	$\mathrm{l\ min^{-1}}$	257
y	Pupil area		275
y_∞	Constant (carrying capacity) in logistic equation		267
A	Proportionality constant		260
F	Respiratory quotient		257
F_x	x component of force		266
G_1,G_2	Gain		258
R	Gas constant	$\mathrm{J\ K^{-1}\ mol^{-1}}$	257
T	Temperature	K	257
T,T_0,T_s	Time	s	268
$\dot{V}$	Rate that air flows through the alveoli in the lungs	$\mathrm{m^3\ s^{-1}}$	257
V_c	Compartment in which carbon dioxide is distributed throughout the body	$\mathrm{m^3}$ or l	260
α	Solubility constant	$\mathrm{mol\ l^{-1}\ torr^{-1}}$	260
θ,θ',θ_s	Angle		267
τ,τ_1,τ_2	Time constant	s	260
ω	Angular frequency	$\mathrm{rad\ s^{-1}}$	264
ξ,η	Variables		260

PROBLEMS

Section 10.1

10.1. Make the unit conversions to show that Eq. (10.4) is equivalent to Eq. (10.1).

Section 10.2

10.2. The level of the thyroid hormone thyroxine (T4) in the blood is regulated by a feedback system. Thyroid stimulating hormone (TSH) is released by the pituitary. The thyroid responds to increased levels of TSH by producing more T4. The T4 then acts through the hypothalamus and pituitary to reduce the amount of TSH.

(a) On a graph of T4 vs TSH, plot hypothetical curves showing these two processes and indicate the equilibrium or operating point.

(b) T4 contains four iodine atoms. If the body has an insufficient supply of dietary iodine, the thyroid cannot make enough T4. What changes in the graphs will result? (This causes iodine deficiency goiter or thyroid hyperplasia. With the advent of iodized table salt and the use of iodine by bakers in bread dough to make their equipment easier to clean, the disease has almost disappeared.)

10.3. For the feedback system

$$x=\left(\frac{y-p}{3}\right)^{1/2}, \quad y=4-x^2$$

assume that the independent variable is on the right in each equation.

(a) Plot y vs x for each process.

(b) Find the operating point when $p=0$.

(c) Find the operating point when $p=0.1$.

Section 10.3

10.4. Find the open-loop gain for the system described in Problem 10.3.

10.5. Find the open-loop gain for the system shown.

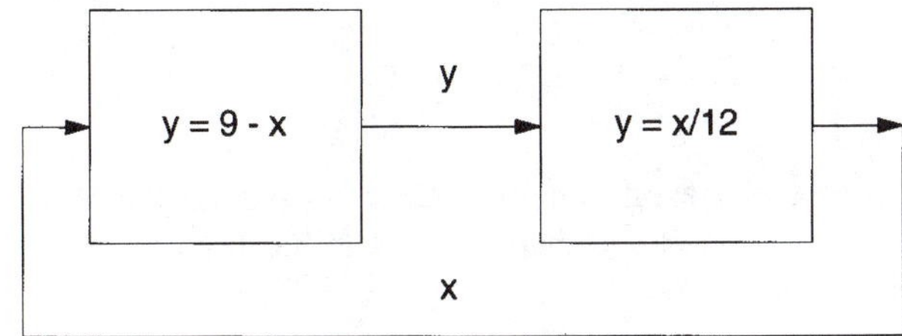

10.6. A feedback loop has the three stages shown. Find the operating point and the open-loop gain if these variables are all positive.

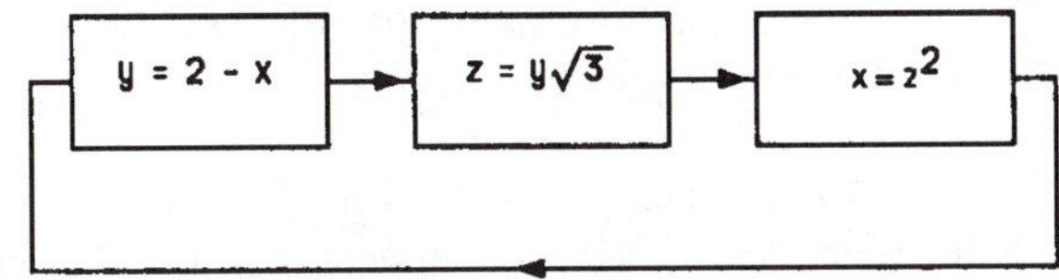

10.7. Consider how thyroid hormone is removed from the body by the kidneys. The variables are V, the total plasma volume (l); C, the plasma concentration of thyroid hormone ($\mathrm{mol\ l^{-1}}$); y, the total amount of hormone (mol); and R, the rate of hormone production ($\mathrm{mol\ s^{-1}}$). In the steady state, removal is at rate $dy/dt=R-KC=0$. Then $R=KC$ and Q is not changing with time (see Chap. 2). The clearance is a measure of the kidney's ability to remove hormone, since the removal process depends on the concentration.

(a) Plot K vs C for two different values of R. Show on your graph what happens if K remains fixed as R changes.

(b) It has been found experimentally [D. S. Riggs (1952). *Pharmacol. Rev.* **4**: 284–370] that K increases as C increases: $K=aC$. Plot this on your graph, too.

(c) Draw a block diagram showing the proper cause and effect relationship between C and K.

(d) Calculate the open-loop gain. Show how changes in C are altered by the feedback mechanism.

10.8. A substance is produced in the body and removed at rate R. The concentration is C. The clearance is defined to be K. In the steady state $0=dy/dt=R-KC$, or $K=R/C$. It is found experimentally that the clearance depends on the concentration as $K=aC^n$, where C is the independent variable. Find the open-loop gain, eliminating K and a from your answer.

10.9. The kidney excretes phosphate in the following way. The total plasma volume V_p contains phosphate at concentration C_p: $Q_p=C_pV_p$. A volume of plasma $\dot{V}_f$ is filtered through the renal glomeruli into the nephrons each second. Within the nephron, phosphate is either reabsorbed into the plasma or excreted into the urine. Experiments show that virtually all phosphate is reabsorbed up to some rate $\dot{Q}_{\max}$:

$$\left(\frac{dQ}{dt}\right)_{\text{reabs}}=\begin{cases} C_p\dot{V}_f, & C_p\dot{V}_f<\dot{Q}_{\max} \\ \dot{Q}_{\max}, & C_p\dot{V}_f\geqslant\dot{Q}_{\max}. \end{cases}$$

As in Problem 10.7(a), at equilibrium the clearance of phosphate from the plasma is defined as

$$K=\frac{(dQ/dt)_{\text{excreted into urine}}}{C_p}.$$

Suppose that exogenous phosphate is entering the plasma at a fixed rate R and that steady state has been reached so that

$$R=\left(\frac{dQ}{dt}\right)_{\text{excreted into urine}}.$$

(a) What value for reabsorption does this imply?

(b) Determine two equations relating K and C_p and plot them.

(c) Calculate the open-loop gain of the feedback loop.

10.10. With considerable simplification, consider the body to have a constant temperature T throughout and a total heat capacity C. The total amount of thermal energy in the body is U. The heat capacity is defined so that $dU=C\,dT$. The source of the thermal energy is the body's metabolism:

$$\left(\frac{dU}{dt}\right)_{\text{in}}=M.$$

If sweating is ignored, the rate of loss of energy by convection and radiation is approximately proportional to the amount by which the body temperature exceeds the ambient or surrounding temperature:

$$\left(\frac{dU}{dt}\right)_{\text{loss}}=K(T-T_a).$$

(a) What is the steady-state temperature as a function of M and T_a?

(b) Write a differential equation for T as a function of time. Suppose that M suddenly jumps by a fixed amount. What is the time constant?

10.11. When the body temperature is above 37 °C, sweating becomes important. The rate of energy loss is proportional to the amount of water evaporated. If all the perspiration evaporates, sweating loss can be approximated by

$$\left(\frac{dU}{dt}\right)_{\text{sweat}}=L(T-37).$$

(a) Modify the differential equation of the previous problem to include $(dU/dt)_{\text{sweat}}$ as the input variable with T as the output variable. Combine it with this new equation to make a feedback loop. Determine the new equilibrium temperature and the time constant.

(b) Make numerical comparisons for the previous problem and this one when $M=71$ kcal h^{-1}, $C=70$ kcal °C^{-1}, $K=25$ kcal h^{-1} °C^{-1}, $L=750$ kcal h^{-1} °C^{-1}, $T_a=38$ °C (high enough to ensure sweating).

10.12. A simplified model of the circulation is shown. Normally we know that the arterial pressure is the same as that in the carotid sinus: $p_{\text{art}}=p_{\text{sinus}}$. In experiments on dogs whose vagus nerves were cut, the carotid arteries were isolated and perfused by a separate pump. This broke the feedback loop and allowed the curve on the accompanying graph to be obtained. The empirical equation shown [based on the work of M. Scher and A. C. Young, (1963). Serroanalysis of carotid sinus reflex effects on peripheral resistance. *Circulation Res.* **12**: 152–165, summarized in Riggs (1970)] is

$$p_{\text{art}}=90+\frac{120}{1+\exp[(p_{\text{sinus}}-165)/5]}.$$

(a) Draw a block diagram of the complete feedback system. Label the blocks, show the functional relationship for each one, and indicate the proper cause-and-effect relationship.

(b) Find the operating point.

(c) Find the open-loop gain.

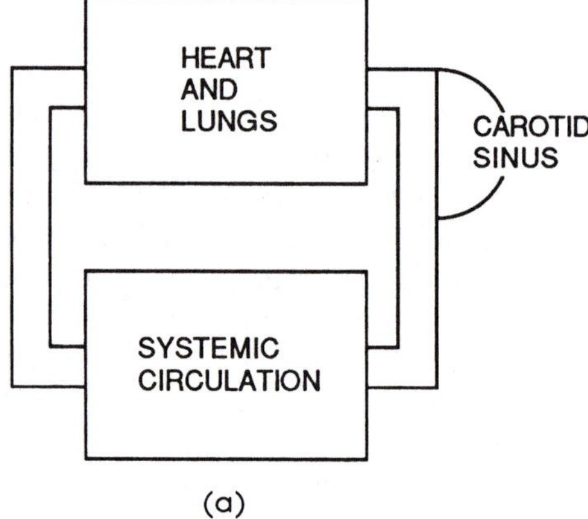

(a)

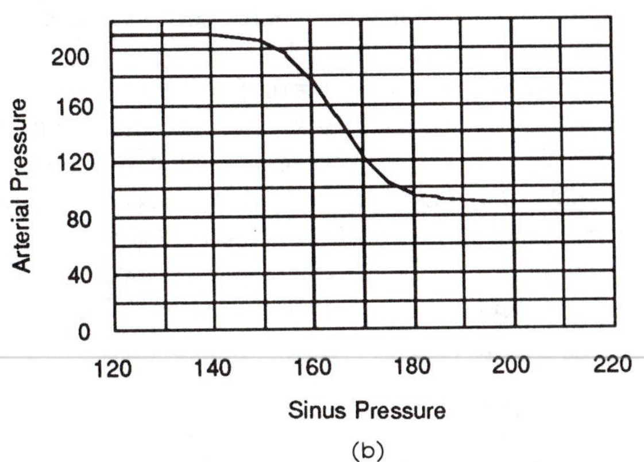

(b)

10.13. Consider the following special case of linear feedback:

$$\Delta x = G_1(\Delta p + \Delta y), \quad \Delta y = G_2 \Delta x.$$

Find the ratio $\Delta x/\Delta p$ when $G_1 \ll -1$, $G_2 < 1$.

10.14. Differentiate Eq. (10.4) and show that the expression for G_1 is the same as in Eq. (10.19).

10.15. For the thyroid problem, Problem 10.7, write a differential equation that can be solved to give C as a function of time. Suppose that at $t=0$, R suddenly becomes 0. What is the differential equation then? Solve the equation: note that it is not linear.

Section 10.5

10.16. The following is a vastly oversimplified model of calcium regulation in dogs. Calcium is stored in body fluids and bones. Experiments show that the calcium concentration in the blood of a dog obeys approximately the equation [Riggs (1970, p. 491)]

$$3.9\frac{dC}{dt} + 1.4C = 81.2 + \dot{Q}_{\text{iv}} + \dot{Q}_r(t),$$

where C is the plasma concentration in mg l^{-1}, t is the time in h, $\dot{Q}_{\text{iv}}$ is the rate of intravenous infusion of calcium in mg h^{-1}, and $\dot{Q}_r$ is the rate of reabsorption of calcium from bone into the blood in mg h^{-1}. (The numerical constants are consistent with these units.) The rate of reabsorption depends on the level of parathyroid hormone (PTH) concentration in the blood, which in turn depends on the calcium concentration. Instead of measuring the PTH concentration, experimenters found that $\dot{Q}_r$ and C are related empirically by $\dot{Q}_r = 188 - 1.34C$, where C is the independent variable.

(a) Draw a block diagram with variables $\dot{Q}_r$ and C.

(b) Write equations to describe the steady state and find steady-state values of $\dot{Q}_r$ and C when $\dot{Q}_{\text{iv}} = 0$.

(c) Find the open-loop gain.

(d) Find the time constant for the change of C when the parathyroid glands have been removed, in response to a step change in $\dot{Q}_{\text{iv}}$.

(e) Find the time constant for the change in C in response to a step change in calcium infusion when the parathyroid glands are intact, so that the feedback loop is closed.

10.17. This problem is a simplification by the author suggested by the data of Chick *et al.* (1977). Artificial pancreas using living beta cells: Effects on glucose homeostasis in diabetic rats. *Science* **197**: 780–781. Experimental data on diabetic rats show that the insulin level is 0 and the glucose level is 500. When an artificial pancreas is installed, a new operating point is reached for which $i=40$ and $g=100$.

(a) Make the simplest assumption possible: glucose level responds to insulin level according to $g = A + G_1 i$, while insulin responds to glucose as $i = G_2 g$. Find the open-loop gain.

(b) The same series of experiments showed that when the feedback loop is closed, the time constant for glucose to fall is 1.67 h. When the artificial pancreas is removed, the glucose level rises with a time constant of 10.67 h. Estimate the open-loop gain, assuming that the insulin level changes instantaneously.

Section 10.6

10.18. Multiply Eq. (10.28) by τ_2 and show that it reduces to Eq. (10.25) when $\tau_2 \ll \tau_1$.

10.19. For the two-stage feedback loop with equal time constants τ, show that oscillation results with a frequency

$$\omega = \frac{(|\text{OLG}|)^{1/2}}{\tau}.$$

10.20. Consider two substances in the plasma with concentrations X and Y. (They might be glucose and insulin.) Assume that experiment has established the following facts.

(a) The steady-state values of each concentration are X_0 and Y_0. Departures from them are $x = X - X_0$ and $y = Y - Y_0$.

(b) When $y=0$, X is removed from the body at a rate proportional to x. This is true for both positive and negative values of x:

$$\frac{dx}{dt} = -\frac{1}{\tau_1}x.$$

(c) When $x=0$, Y influences the rate at which X changes in an approximately linear fashion. An increase of Y above Y_0 $(y>0)$ increases the rate of disappearance of x.

(d) When $x=0$, y is cleared at a rate proportional to y:

$$\frac{dy}{dt} = -\frac{1}{\tau_2}y.$$

(e) When $y=0$ and x is nonzero, a positive value of x stimulates the production of Y, while a negative value of x inhibits the production of Y.

Assume that the rate of production is a linear function of x. Write down two linear differential equations to model these observations. That is, add a term to each of the equations given that describes observations (c) and (e).

10.21. Combine the two equations obtained in the previous problem into a single differential equation in x. Show that it has the form

$$\frac{d^2x}{dt^2}+\left(\frac{1}{\tau_1}+\frac{1}{\tau_2}\right)\frac{dx}{dt}+\frac{1-\text{OLG}}{\tau_1\tau_2}x=0.$$

Use the result of Problem 10.17 to obtain $1-\text{OLG}$ and suppose that $\tau_1=50$ min. For what value of τ_2 will critical damping occur? (If you find two values of τ_2, which seems more reasonable?) If τ_2 is greater than the value you select, will the system be overdamped or underdamped? (Do not take these results too seriously.)

10.22. This problem explores the response of a simple linear system from the point of view of the system's response to sinusoidal signals of various frequencies.

(a) The differential equation describing a system with time constant τ and gain G is

$$\frac{dx}{dt}=\frac{1}{\tau}x+\frac{G}{\tau}y.$$

Show by substitution that if $y=Y\sin\omega t$, then $x=X\sin(\omega t+\phi)$, where $\tan\phi=\omega\tau$ and $X(\omega\tau\sin\phi+\cos\phi)=GY$.

(b) Use the relation $\tan\phi=\omega\tau$ to establish the triangle shown, and use it to show that $X=GY/(1+\omega^2\tau^2)^{1/2}$. These two relations give the response of the system in the frequency domain.

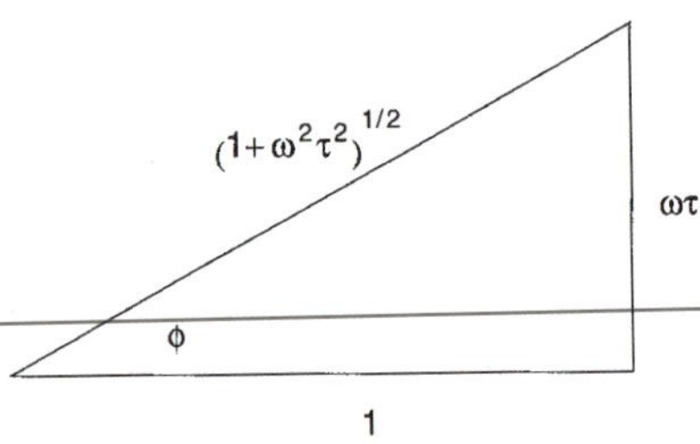

10.23. The following model for the attrition of troops in battle was developed by F. W. Lanchester and has been found to work reasonably well in several battles. The number of friendly troops is $F(t)$ and the number of enemy troops is $E(t)$. The rates of change are given by

$$\frac{dF}{dt}=-aE,\quad \frac{dE}{dt}=-bF,$$

where a and b are the "effectiveness" of each side. The initial number of troops on each side is F_0 and E_0.

(a) What are the initial values of dF/dt and dE/dt?

(b) Obtain a differential equation for F.

(c) Find the most general solution to this differential equation and determine the coefficients from the initial conditions.

(d) Plot F and E for $a=b=0.05$ and $E_0=2F_0$.

10.24. The equation $dF/dt=-aE$ could also be thought of as describing a predator–prey situation if a represents the number of animals that the "enemy" eats per unit time. Ignoring latent periods such as gestation and infancy, what is the simplest way the equation could be modified to take account of reproduction and other ways of dying?

Section 10.7

10.25. Make a phase-space plot and discuss stability for $dy/dt=by$, $dy/dt=-by$, and $dy/dt=a-by$.

10.26. Make a drawing similar to Fig. 10.15 for the differential equation

$$\frac{dx}{dt}=x(c-x^2)$$

for different values of c (positive and negative) and describe the stability of the fixed points as a function of c.

10.27. (a) Make drawings of the tip of the vector that defines θ' in Fig. 10.18 to show that when $b<1$, θ' takes on all values, while for $b>1$, θ' is restricted to values near 0 and 2π.

(b) Redraw Fig. 10.19(a) in the case that the angles are not reset to zero when they reach 2π.

10.28. Consider the undamped harmonic oscillator in the form

$$\frac{dx}{dt}=v,\quad \frac{dv}{dt}=-\omega_0^2x.$$

(a) Make a phase-plane plot.

(b) Is the closed trajectory a limit cycle? Why or why not?

(c) Add a damping force proportional to $-v$ and redraw the phase-plane plot.

Section 10.8

10.29. Show that for the logistic difference equation, the slope dx_{n+1}/dx_n at x^* is given by $2-a$, so that for $a>3$ the slope has magnitude >1.

10.30. Use a spreadsheet to plot x_n for different values of a and explore the period doubling.

Section 10.9

10.31. By substitution show that $w(t)=w_0e^{-\gamma t}\cos\omega t$ can be a solution of the delay-differential equation, Eq. (10.41) if

$$y = \frac{1}{\tau} e^{\gamma t_d} \cos \omega t_d,$$

$$\omega = \frac{1}{\tau} e^{\gamma t_d} \sin \omega t_d.$$

Introduce the dimensionless variables $\alpha = t_d/\tau$, $\xi = \omega t_d$, and $\eta = \gamma t_d$ and show that the result is the simultaneous equations

$$\xi = \alpha e^{\eta} \sin \xi, \quad \eta = \alpha e^{\eta} \cos \xi.$$

From these obtain the equivalent equations $\eta = \xi \cot \xi$. and $\xi^2 = \alpha^2 e^{2\eta} - \eta^2$. Show how these can be solved graphically if α is known.

Section 10.11

10.32. Find an equation relating L, the total amount of light energy per second reaching the retina, I, the intensity of the light (W m^{-2}), and R, the radius of the pupil. Calculate $G = \partial L/\partial R$. Consider the two cases shown in the figure. In the first there is uniform illumination of the pupil. In the second a rectangle of illumination partially overlaps the pupil so that the area within the pupil is ab.

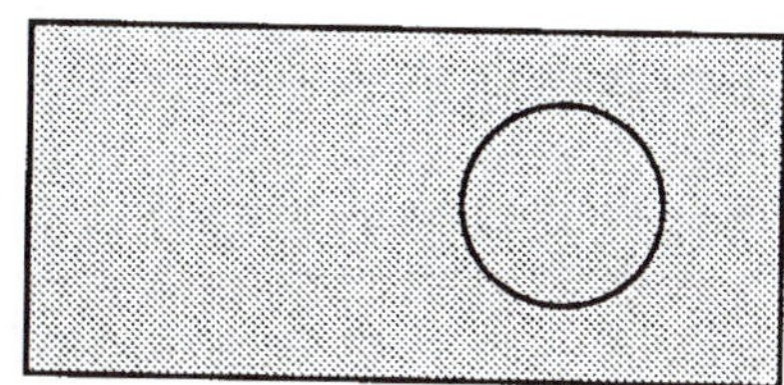

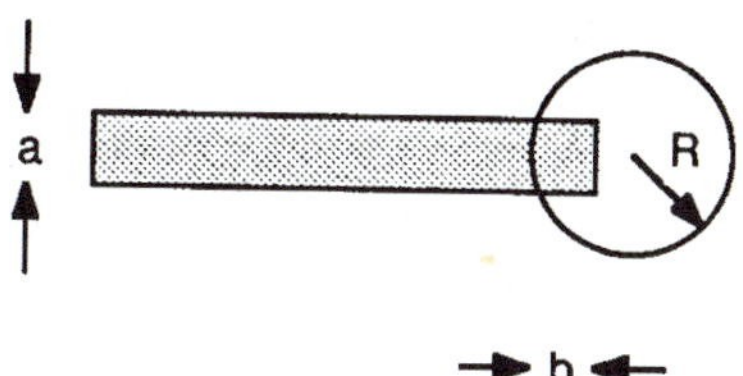

REFERENCES

Bartlett, A. A., and T. J. Braun (1983). Death in a hot tub: the physics of heat stroke. *Am. J. Phys.* **51**(2): 127–132.

Begon, M., M. Mortimer, and D. J. Thompson (1996). *Population Ecology: A Unified Study of Animals and Plants*, 3rd ed. Cambridge, MA, Blackwell Science.

Christini, D. J. and J. J. Collins (1995). Controlling nonchaotic neuronal noise using chaos control techniques. Phys. Rev. Lett. **75**(14): 2782–2785.

Epstein, A. E., and R. E. Ideker (1995). Ventricular fibrillation. In D. P. Zipes and J. Jalife, eds. *Cardiac Electrophysiology: From Cell to Bedside*, 2nd ed. Philadelphia, Saunders, pp. 927–933.

Epstein, I. R., K. Kustin, P. De Kepper, and M. Orbán (1983). Oscillating chemical reactions. *Sci. Am.* March: 112–123.

Garfinkel, A., M. L. Spano, W. L. Ditto, and J. N. Weiss (1992). Controlling cardiac chaos. *Science* **257**: 1230–1235.

Glass, L., and M. C. Mackey (1988). *From Clocks to Chaos*. Princeton, NJ, Princeton University Press.

Guevara, M. R., L. Glass, and A. Shrier (1981). Phase-locking, period-doubling bifurcations and irregular dynamics in periodically stimulated cardiac cells. *Science* **214**: 1350–1353.

Guyton, A. C., J. W. Crowell, and J. W. Moore (1956). Basic oscillating mechanism of Cheyne–Stokes breathing. *Am. J. Physiol.* **187**: 395–398.

Hayes, N. D. (1950). Roots of the transcendental equation associated with a certain difference-differential equation. *J. London Math. Soc.* **25**: 226–232.

Hilborn, R. C. (1995). *Chaos and Nonlinear Dynamics*. New York, Oxford University Press.

Kaplan, D., and L. Glass (1995). *Understanding Nonlinear Dynamics*. New York, Springer.

Keener, J. P., and A. V. Panfilov (1995). Three-dimensional propagation in the heart: The effects of geometry and fiber orientation on propagation in the myocardium. In D. P. Zipes and J. Jalife, eds. *Cardiac Electrophysiology: From Cell to Bedside*, 2nd ed. Philadelphia, Saunders, pp. 335–348.

Mackey, M. C., and L. Glass (1977). Oscillation and chaos in physiological control systems. *Science* **197**: 287–289.

May, R. M. (1976). Simple mathematical models with very complicated dynamics. *Nature (London)* **261**: 459–467.

Mercader, M. A., D. C. Michaels, and J. Jalife (1995). Reentrant activity in the form of spiral waves in mathematical models of the sinoatrial node. In D. P. Zipes and J. Jalife, eds. *Cardiac Electrophysiology: From Cell to Bedside*, 2nd ed. Philadelphia, Saunders, pp. 389–403.

Mielczarek, E. V., J. S. Turner, D. Leiter, and L. Davis (1983). Chemical clocks: Experimental and theoretical models of nonlinear behavior. *Am. J. Phys.* **51**(1): 32–42.

Mines, G. R. (1914). On circulating excitation of heart muscles and their possible relation to tachycardia and fibrillation. *Trans. R. Soc. Canada* **4**: 43–53.

Patton, H. D., A. F. Fuchs, B. Hille, A. M. Scher, and R. F. Steiner, eds. (1989). *Textbook of Physiology*, 21st ed. Philadelphia, Saunders.

Pertsov, A. M., and J. Jalife (1995). Three-dimensional vortex-like reentry. In D. P. Zipes and J. Jalife, eds. *Cardiac Electrophysiology: From Cell to Bedside*, 2nd ed. Philadelphia, Saunders, pp. 403–410.

Riggs, D. S. (1970). *Control Theory and Physiological Feedback Mechanisms*. Baltimore, Williams and Wilkins.

Schiff, S. J., K. Kerger, D. H. Duong, T. Chang, M. L. Spano, and W. L. Ditto (1994). Controlling chaos in the brain. *Nature* **370**(6491): 615–620.

Stark, L. (1957). A servoanalytic study of consensual pupil reflex to light. *J. Neurophys.* **20**: 17–26.

Stark, L. (1968). *Neurological Control Systems: Studies in Bioengineering*. New York, Plenum, pp. 73–84.

Stark, L. W. (1984). The pupil as a paradigm for neurological control systems. *IEEE Trans. Biomed. Eng.* **31**: 919–924.

Strogatz, S. H. (1994). *Nonlinear Dynamics and Chaos*. Reading, MA, Addison-Wesley.

Winfree, A. T. (1980). *The Geometry of Biological Time*. New York, Springer.

Winfree, A. T. (1987). *When Time Breaks Down*. Princeton, NJ, Princeton University Press.

Winfree, A. T. (1994). Electrical turbulence in three-dimensional heart muscle. *Science* **266**: 1003–1006; Persistent tangled vortex rings in generic excitable media. *Nature* **371**: 233–236.

Winfree, A. T. (1995). Theory of spirals. In D. P. Zipes and J. Jalife, eds. *Cardiac Electrophysiology: From Cell to Bedside*, 2nd ed., Philadelphia, Saunders, pp. 379–389.

Witkowski, F. X., K. M. Kavanagh, P. A. Penkoske, and R. Plonsey (1993). In vivo estimation of cardiac transmembrane current. *Circulation Res.* **72**(2): 424–439.

Witkowski, F. X., R. Plonsey, P. A. Penkoske, and K. M. Kavanagh (1994). Significance of inwardly directed transmembrane current in determination of local myocardial electrical activation during ventricular fibrillation. *Circulation Res.* **74**(3): 507–524.

Witkowski, F. X., K. M. Kavanagh, P. A. Penkoske, R. Plonsey, M. L. Spano, W. L. Ditto, and D. T. Kaplan (1995). Evidence for determinism in ventricular fibrillation. *Phys. Rev. Lett.* **75**(6): 1230–1233.

CHAPTER 11

The Method of Least Squares and Signal Analysis

This chapter deals with three common problems in experimental science. The first is fitting a discrete set of experimental data with a mathematical function. The function usually has some parameters that must be adjusted to give a ''best'' fit. The second is to detect a periodic change in some variable—a signal—which may be masked by random changes—noise—superimposed on the signal. The third is to determine whether sets of apparently unsystematic data are from a random process or a process governed by deterministic chaotic behavior.

Though not the exclusive province of physics, these techniques are used in many fields, including physiology and biophysics. The fitting techniques lead naturally to Fourier series, which are used extensively in image reconstruction and image analysis. Using least squares or Fourier series normally requires extensive computation. Commercial packages for making these calculations are readily available. The Problems at the end of the chapter are often artificially designed for simple computation, rather than being ''real.'' I hope that the chapter will help you develop some intuition for the techniques before you use the commercial packages.

This chapter is a self-contained discussion of signal analysis. It is a prerequisite to Chap. 12 on image reconstruction.

We will find that a periodic signal can be built up of sine waves of different frequencies, and that it is possible to speak of the *frequency spectrum* of a signal. The first five sections of the chapter show how to adjust the parameters in a polynomial or in a sum of sines and cosines to fit experimental data. Sections 11.6 and 11.7 discuss sine and cosine expansions for continuous periodic functions. Sections 11.8 and 11.9 introduce the cross-correlation and autocorrelation functions and their relation to the power spectrum. Sections 11.12 and 11.13 extend these techniques to pulses. Sections 11.14 and 11.15 introduce noise and the use of correlation functions to detect signals that are masked by noise.

Many linear feedback systems are most easily studied by how they respond to sinusoidal stimuli at various frequencies, and there are techniques using impulse or noise stimuli that provide the same information. Section 11.16 explains the frequency response of a linear system, and the next section describes the effect of a simple linear system on the power spectrum of Johnson noise. The last section introduces some of the concepts involved in testing data for chaotic behavior.

11.1. THE METHOD OF LEAST SQUARES AND POLYNOMIAL REGRESSION

In this section we show how to approximate or ''fit'' a set of discrete data y_j with a polynomial function

$$y(x_j)=\sum_k a_k x_j^k .$$

Several criteria can be used to determine the ''best'' fit [Press *et al.* (1992)]; the one described in this chapter is called the *method of least squares*. Instead of immediately deriving the general polynomial result, we first consider the simple (and rather useless) fit $y=x+b$ (the coefficient of x is unity), then the more useful linear fit $y=ax+b$.

Suppose that we wish to describe the data in Table 11.1 by a fitting function $y(x)$. A plot of the data suggests that a straight line will be a reasonable approximation to the data. For mathematical simplicity, we first try a line with unit slope but adjustable intercept:

$$y(x_j)=x_j+b. \tag{11.1}$$

Figure 11.1(a) plots y vs x for different values of b. It is clear by inspection that the curves for $b=1$ and $b=2$ are

TABLE 11.1. *Sample data.*

x	y
1	2
4	6
5	7

closer to the points than those for $b=0$ or $b=3$. For a quantitative measure of how good the fit is we will use the quantity

$$Q=\frac{1}{N}\sum_{j=1}^{N}[y_j-y(x_j)]^2, \qquad (11.2)$$

which is called the *mean square error*. It is the square of the residuals [the differences between the measured value of y and the values of y calculated from the approximation to the data, $y_j-y(x_j)$] summed over all N data points and divided by N. It is reminiscent of the variance, with the mean replaced by the fitting function $y(x_j)$. The least-squares technique adjusts the parameters in the function $y(x_j)$ to make Q a minimum. Table 11.2 shows the steps in the calculation of Q for various values of b. Figure 11.1(b) shows how Q changes as b is changed. It is tedious to calculate Q for many different values of b; instead we can treat this as a maximum–minimum problem in calculus. We write

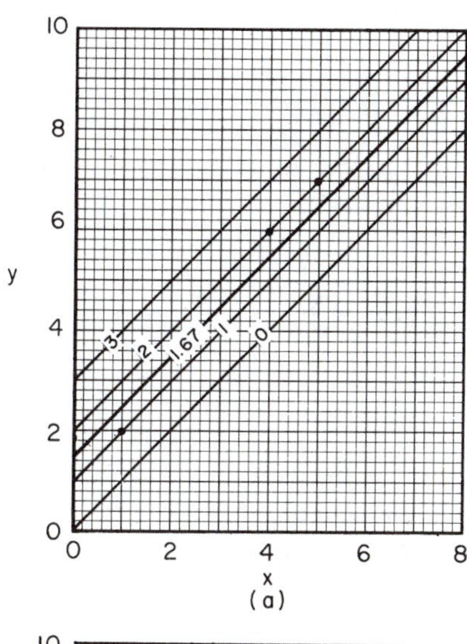

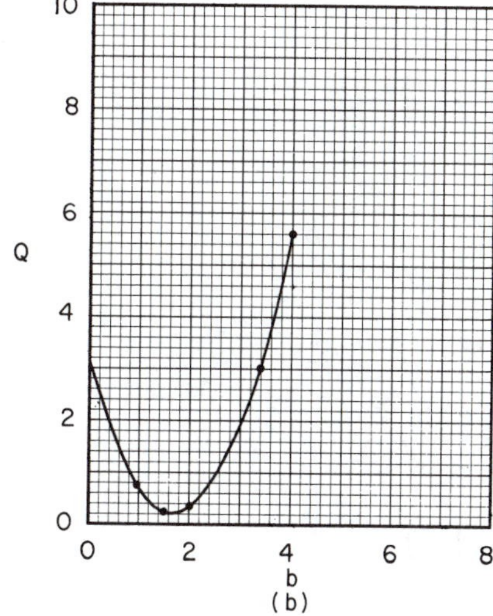

FIGURE 11.1. Fits to the data of Table 11.1 by a curve of the form $y=x+b$. (a) Plots of y vs x. (b) Plot of Q vs b. Q is defined in Eq. (11.2).

$$Q=\frac{1}{N}\sum_{j=1}^{N}(y_j-x_j-b)^2=\frac{1}{N}[(y_1-x_1-b)^2+(y_2-x_2-b)^2+\cdots].$$

The derivative is

$$\frac{dQ}{db}=-\frac{1}{N}\sum_{j=1}^{N}2(y_j-x_j-b)=\frac{1}{N}[-2(y_1-x_1-b)-2(y_2-x_2-b)+\cdots].$$

Setting this equal to zero to find the extremum gives

$$\sum_{j=1}^{N}y_j=\sum_{j=1}^{N}x_j+\sum_{j=1}^{N}b$$

or, not bothering to show explicitly that the index ranges over all the data points,

$$\sum_j y_j=\sum_j x_j+Nb.$$

Using this result for the example above gives $15=10+3b$, or $b=1.67$ for the smallest value of Q.

This problem is rather artificial, because for simplicity we did not allow the slope of the line to vary. The maximum–minimum procedure is easily extended to two or more parameters. If the fitting function is given by $y=ax+b$, then Q becomes

$$Q=\frac{1}{N}\sum_{j=1}^{N}(y_j-ax_j-b)^2.$$

At the minimum, both $\partial Q/\partial a=0$ and $\partial Q/\partial b=0$. The former gives

$$\frac{\partial Q}{\partial a}=\frac{2}{N}\sum_{j=1}^{N}(y_j-ax_j-b)(-x_j)=0$$

or

$$\sum_j x_jy_j-a\sum_j x_j^2-b\sum_j x_j=0. \qquad (11.3)$$

The latter gives

$$\frac{\partial Q}{\partial b}=\frac{2}{N}\sum_{j=1}^{N}(y_j-ax_j-b)(-1)=0$$

or

$$\sum_j y_j-a\sum_j x_j-Nb=0. \qquad (11.4)$$

TABLE 11.2. *Calculation of Q for the example of Eq. 11.1.*

Index			b=0		b=1		b=1.25	
j	x_j	y_j	$y(x_j)$	$[y_j-y(x_j)]^2$	$y(x_j)$	$[y_j-y(x_j)]^2$	$y(x_j)$	$[y_j-y(x_j)]^2$
1	1	2	1	1	2	0	2.25	0.0625
2	4	6	4	4	5	1	5.25	0.5625
3	5	7	5	4	6	1	6.25	0.5625
Sum				9		2		1.1875
Q				3		0.67		0.395

For this example $\Sigma_j x_j=10$, $\Sigma_j y_j=15$, $\Sigma_j x_j^2=42$, and $\Sigma_j x_j y_j=61$. Therefore, Eqs. (11.3) and (11.4) become

$$42a+10b=61,$$

$$10a+3b=15.$$

These can be easily solved.

A general expression for the solution to Eqs. (11.3) and (11.4) is

$$a=\frac{N\left(\sum_j x_j y_j\right)-\left(\sum_j x_j\right)\left(\sum_j y_j\right)}{N\left(\sum_j x_j^2\right)-\left(\sum_j x_j\right)^2}, \qquad (11.5a)$$

$$b=\frac{\sum_j y_j}{N}-\frac{a\left(\sum_j x_j\right)}{N}\equiv\bar{y}-a\bar{x}, \qquad (11.5b)$$

where $\bar{x}$ and $\bar{y}$ are the means. In doing computations where the range of data is small compared to the mean, better numerical accuracy can be obtained by forming the sums

$$S_{xx}=\sum_j (x_j-\bar{x})^2,$$

$$S_{xy}=\sum_j (x_j-\bar{x})(y_j-\bar{y}),$$

in terms of which

$$a=S_{xy}/S_{xx}. \qquad (11.5c)$$

For our example,

$$a=\frac{(3)(61)-(10)(15)}{(3)(42)-100}=1.27$$

$$b=\frac{15}{3}-\frac{(1.27)(10)}{3}=0.77.$$

The best straight-line fit to the data of Table 11.1 is $y=0.77+1.27x$. The value of Q, calculated from Eq. (11.2), is 0.013. The best fit is plotted in Fig. 11.2. It is considerably better than the fit with the slope constrained to be one.

This method can be extended to a polynomial of arbitrary degree. The only requirement is that the number of adjustable parameters (which is one more than the degree of the polynomial) be less than the number of data points. If this requirement is not met, the equations cannot be solved uniquely; see Problem 11.9. If the polynomial is written as

$$y(x_j)=a_0+a_1x_j+a_2x_j^2+\cdots+a_nx_j^n=\sum_{k=0}^{n} a_kx_j^k, \qquad (11.6)$$

then the expression for the mean square error is

$$Q=\frac{1}{N}\sum_{j=1}^{N}\left(y_j-\sum_{k=0}^{n} a_kx_j^k\right)^2. \qquad (11.7)$$

Index j ranges over the data points; index k ranges over the terms in the polynomial. This expression for Q can be differentiated with respect to one of the parameters, say, a_m:

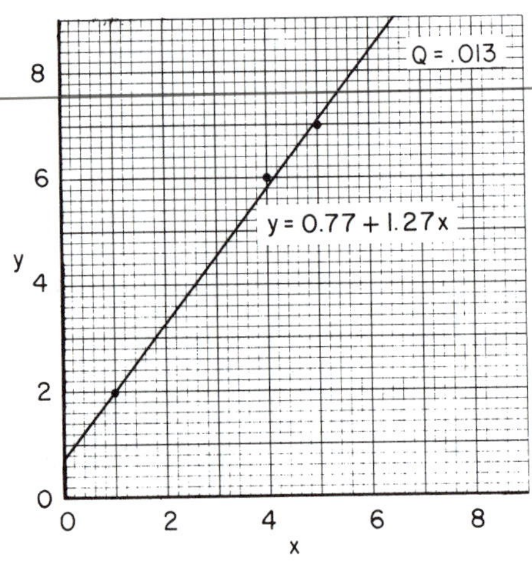

FIGURE 11.2. The best fit to the data of Table 11.1 with the function $y=ax+b$. Both the slope and the intercept have been chosen to minimize Q.

$$\frac{\partial Q}{\partial a_m}=\frac{2}{N}\sum_{j=1}^{N}\left[\left(y_j-\sum_{k=0}^{n}a_kx_j^k\right)(-x_j^m)\right].$$

Setting this equal to zero gives

$$\sum_j y_jx_j^m=\sum_k\sum_j a_kx_j^kx_j^m=\sum_k a_k\sum_j x_j^{k+m}.$$

This is one of the equations we need. Doing the same thing for all values of m, $m=0,1,2,\ldots,n$, gives $n+1$ equations that must be solved simultaneously for the $n+1$ parameters $a_0,a_1,\ldots,a_n$.

For $m=0$:

$$\sum_j y_j=a_0\sum_j x_j^0+a_1\sum_j x_j+a_2\sum_j x_j^2+\cdots+a_n\sum_j x_j^n.$$

For $m=1$:

$$\sum_j x_jy_j=a_0\sum_j x_j+a_1\sum_j x_j^2+a_2\sum_j x_j^3+\cdots$$
$$+a_n\sum_j x_j^{n+1}. \qquad (11.8)$$

For $m=n$:

$$\sum_j x_j^ny_j=a_0\sum_j x_j^n+a_1\sum_j x_j^{n+1}+a_2\sum_j x_j^{n+2}+\cdots$$
$$+a_n\sum_j x_j^{2n}.$$

Solving these equations is not as formidable a task as it seems. Given the data points (x_j,y_j), the sums are all evaluated. When these numbers are inserted in Eqs. (11.8), the result is a set of $n+1$ simultaneous equations in the $n+1$ unknown coefficients a_k. This technique is called *linear least-squares fitting of a polynomial* or *polynomial regression.* Routines for solving the simultaneous equations or for carrying out the whole procedure are readily available.

The least-squares technique described here gives each data point the same weight. If some points are measured more accurately than others, they should be given more weight. This can be done in the following way. If there is an associated error δy_j for each data point, then one can weight each data point inversely as the square of the error and minimize

$$Q=\frac{1}{N}\sum_{j=1}^{N}\frac{[y_j-y(x_j)]^2}{(\delta y_j)^2}. \qquad (11.9)$$

It is easy to show that the effect is to add a factor of $(1/\delta y_j)^2$ to each term in the sums in Eqs. (11.8) [Gatland (1993)]. This assumes that errors exist only in the y values. If there are errors in the x values as well, it is possible to make an approximate correction based on an effective error in the y values [Orear (1982)] or to use an iterative but exact least-squares method [Lybanon (1984)]. The treatment of unequal errors has been discussed by Gatland (1993) and by Gatland and Thompson (1993).

11.2. NONLINEAR LEAST SQUARES

This same technique can be used to fit the data with a single exponential $y=ae^{-bx}$, where a and b are to be determined. Take logarithms of each side of the equation:

$$\log y=\log a-bx\log e,$$

$$v=a'-b'x.$$

This can be fitted by the linear equation, determining constants a' and b' by the techniques described above.

Things are not so simple if there is reason to believe that the function might be a sum of exponentials:

$$y=a_1e^{-b_1x}+a_2e^{-b_2x}+\cdots.$$

When the derivatives of this function are set equal to zero, the equations in a_1,a_2, etc., will be linear because if the b_k are known, e^{-b_kx} can be calculated. The equation for determining the b's will be transcendental equations that are quite difficult to solve. With a sum of two or more exponentials, taking logs does not avoid the problem. The problem can be solved using the technique of *nonlinear least squares.* One makes an initial guess for each parameter[1] $b_{10},b_{20},\ldots,b_{k0}$ and says that the correct value of each b is given by $b_k=b_{k0}+h_k$. The calculated value of y is written as a Taylor's-series expansion with all the derivatives evaluated for $b_{10},b_{20},\ldots$:

$$y(x_j;b_1,b_2,\ldots)=y(x_j;b_{10},b_{20},\ldots)+\frac{\partial y}{\partial b_1}h_1+\frac{\partial y}{\partial b_2}h_2$$
$$+\cdots.$$

Since y and its derivatives can be evaluated using the current guess for each b, the expression is linear in the h_k, and the linear least-squares technique can be used to determine the values of the h_k that minimize Q. After each h_k has been determined, the revised values $b_k=b_{k0}+h_k$ are used as the initial guesses and the process is repeated. The process is repeated until a minimum value of Q is found. The technique is not always stable; it can overshoot and give too large a value for h_k. There are many ways to improve the process to ensure more rapid convergence. The most common is called the Levenberg–Marquardt method [see Bevington and Robinson (1992) or Press *et al.* (1992)].

[1]The parameters a_k can either be included in the parameter list, or the values of a_k for each trial set b_k can be determined by linear least squares.

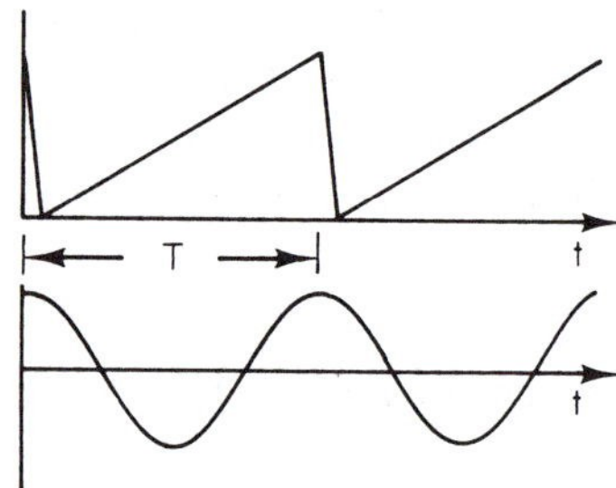

FIGURE 11.3. Two different periodic functions.

11.3. THE PRESENCE OF MANY FREQUENCIES IN A PERIODIC FUNCTION

A function y that repeats itself after a time[2] T is said to be *periodic* with period T. The mathematical description of this periodicity is

$$y(t+T)=y(t). \tag{11.10}$$

Two examples of functions with period T are shown in Fig. 11.3. One of these functions is a sine wave, $y(t)=A\sin(\omega_0 t-\phi)$, where A is the amplitude, ω_0 is the angular frequency, and ϕ is the phase of the function. Changing the amplitude changes the height of the function. Changing the phase shifts the function along the time axis. The sine function repeats itself when the argument shifts by 2π radians. It repeats itself after time T, where $\omega_0 T = 2\pi$. Therefore the angular frequency is related to the period by

$$\omega_0=\frac{2\pi}{T}\ \mathrm{s}^{-1}. \tag{11.11}$$

(The units are rad s^{-1}, but radians are dimensionless.) It is completely equivalent to write the function in terms of the frequency as $y(t)=A\sin(2\pi f_0 t-\phi)$. The frequency f_0 is the number of cycles per second. Its units are s^{-1} or hertz (Hz) (hertz is not used for angular frequency):

$$f_0=\frac{1}{T}=\frac{\omega_0}{2\pi}\ \mathrm{Hz}. \tag{11.12}$$

It is possible to write function y as a sum of a sine term and a cosine term instead of using phase ϕ:

$$\begin{aligned} y(t) &= A\sin(\omega_0 t-\phi)=A(\sin\omega_0 t\cos\phi-\cos\omega_0 t\sin\phi) \\ &= (A\cos\phi)\sin(\omega_0 t)-(A\sin\phi)\cos(\omega_0 t) \\ &= S\sin(\omega_0 t)-C\cos(\omega_0 t). \end{aligned} \tag{11.13}$$

[2]Although we speak of t and time, the technique can be applied to any independent variable if the dependent variable repeats as in Eq. (11.10). Zebra stripes are (almost) periodic functions of position.

The upper function graphed in Fig. 11.3 also has period T. But it has a single frequency only if it is a sine wave.

Harmonics are integer multiples of the fundamental frequency. They have the time dependence $\cos(k\omega_0 t)$ or $\sin(k\omega_0 t)$, where $k=2,3,4,\ldots$ and also have period T. (They also have shorter periods, but they still satisfy the definition [Eq. (11.10)] for a function of period T.)

We can generate periodic functions of different shapes by combining various harmonics. Different combinations of the fundamental, third harmonic, and fifth harmonic are shown in Fig. 11.4. In this figure, (a) is a pure sine wave, (b) and (c) have some third harmonic added with a different phase in each case, and (d) and (e) show the addition of a fifth harmonic term to (b) with different phases.

An *even function* is one for which $y(-t)=y(t)$. For an *odd function*, $y(-t)=-y(t)$. The cosine is even, and the sine is odd. A sum of sine terms gives an odd function. A sum of cosine terms gives an even function.

11.4. FOURIER SERIES FOR DISCRETE DATA

The ability to adjust the amplitude of sines and cosines to approximate a specific shape suggests that discrete periodic data can be fitted by a function of the form

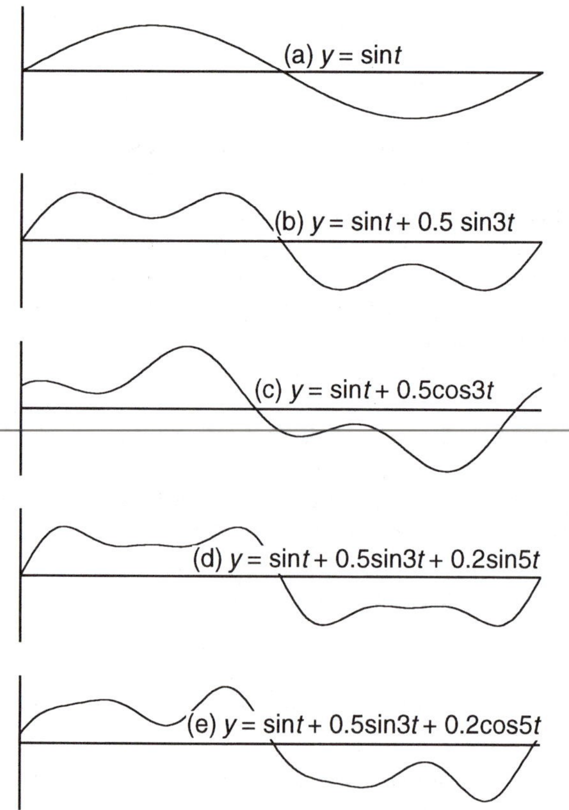

FIGURE 11.4. Various periodic signals made by adding sine waves that are harmonically related. Each signal has an angular frequency $\omega_0=1$ and a period $T=2\pi$.

$$y(t_j)=a_0+\sum_{k=1}^{n} a_k \cos(k\omega_0 t_j)+\sum_{k=1}^{n} b_k \sin(k\omega_0 t_j)$$

$$=a_0+\sum_{k=1}^{n} a_k \cos(k2\pi f_0 t_j)+\sum_{k=1}^{n} b_k \sin(k2\pi f_0 t_j). \tag{11.14}$$

The period $T=2\pi/\omega_0$ is a characteristic of the calculated function, not the function being fitted. There are $2n+1$ parameters $(a_0;a_1,\ldots,a_n;b_1,\ldots,b_n)$. We will talk about restrictions on n later. If the least-squares criterion is used to determine the parameters, Eq. (11.14) is a *Fourier-series* representation of the data.

Why adopt a least-squares criterion? Why minimize the average square error (mean of the square of the residuals) instead of minimizing something else like the average of the absolute value of the error? In some cases a good justification for the least-squares criterion can be made. For example, in an electric circuit $v^2(t)/R$ or $i^2(t)R$ is the instantaneous power dissipation in a resistor, and the least-squares criterion minimizes the average power in the residuals. However, other criteria are possible, and in some situations they may be preferable.

Using the least-squares criterion to determine the coefficients to fit N data points requires minimizing the mean square residual

$$Q=\frac{1}{N}\sum_{j=1}^{N}\left(y_j-a_0-\sum_{k=1}^{n} a_k \cos(k\omega_0 t_j)-\sum_{k=1}^{n} b_k \sin(k\omega_0 t_j)\right)^2. \tag{11.15}$$

The derivatives that must be set to zero are

$$\frac{\partial Q}{\partial a_0}=-\frac{2}{N}\sum_{j=1}^{N}\left[\left(y_j-a_0-\sum_{k=1}^{n} a_k \cos(k\omega_0 t_j)-\sum_{k=1}^{n} b_k \sin(k\omega_0 t_j)\right)(1)\right],$$

$$\frac{\partial Q}{\partial a_m}=-\frac{2}{N}\sum_{j=1}^{N}\left[\left(y_j-a_0-\sum_{k=1}^{n} a_k \cos(k\omega_0 t_j)-\sum_{k=1}^{n} b_k \sin(k\omega_0 t_j)\right)\cos(m\omega_0 t_j)\right],$$

and

$$\frac{\partial Q}{\partial b_m}=-\frac{2}{N}\sum_{j=1}^{N}\left[\left(y_j-a_0-\sum_{k=1}^{n} a_k \cos(k\omega_0 t_j)-\sum_{k=1}^{n} b_k \sin(k\omega_0 t_j)\right)\sin(m\omega_0 t_j)\right].$$

Setting each derivative equal to zero and interchanging the order of the summations give $2n+1$ equations analogous to Eq. (11.8). The first is

$$\sum_{j=1}^{N} y_j=Na_0+\sum_{k=1}^{n} a_k\sum_{j=1}^{N} \cos(k\omega_0 t_j)+\sum_{k=1}^{n} b_k\sum_{j=1}^{N} \sin(k\omega_0 t_j). \tag{11.16}$$

There are n equations of the form

$$\sum_{j=1}^{N} y_j \cos(m\omega_0 t_j)=a_0\sum_{j=1}^{N} \cos(m\omega_0 t_j)+\sum_{k=1}^{n} a_k\sum_{j=1}^{N} \cos(k\omega_0 t_j)\cos(m\omega_0 t_j)+\sum_{k=1}^{n} b_k\sum_{j=1}^{N} \sin(k\omega_0 t_j)\cos(m\omega_0 t_j) \tag{11.17}$$

for $m=1,\ldots,n$, and n more of the form

$$\sum_{j=1}^{N} y_j \sin(m\omega_0 t_j)=a_0\sum_{j=1}^{N} \sin(m\omega_0 t_j)+\sum_{k=1}^{n} a_k\sum_{j=1}^{N} \cos(k\omega_0 t_j)\sin(m\omega_0 t_j)+\sum_{k=1}^{n} b_k\sum_{j=1}^{N} \sin(k\omega_0 t_j)\sin(m\omega_0 t_j). \tag{11.18}$$

Since the t_j are known, each of the sums over the data points (index j) can be evaluated independent of the y_j.

If the data points are equally spaced, the equations become much simpler. There are N data points spread out over an interval T: $t_j=jT/N=2\pi j/N\omega_0$, $j=1,\ldots,N$. The arguments of the sines and cosines are of the form $(2\pi jk/N)$. One can show that

$$\sum_{j=1}^{N} \cos\left(\frac{2\pi jk}{N}\right)=\begin{cases} N, & k=0 \text{ or } k=N \\ 0 & \text{otherwise,}\end{cases} \tag{11.19}$$

$$\sum_{j=1}^{N} \sin\left(\frac{2\pi jk}{N}\right)=0, \quad \text{for all } k, \tag{11.20}$$

$$\sum_{j=1}^{N} \cos\left(\frac{2\pi jk}{N}\right)\cos\left(\frac{2\pi jm}{N}\right)=\begin{cases} N/2, & k=m \text{ or } k=N-m \\ 0 & \text{otherwise,}\end{cases} \tag{11.21}$$

$$\sum_{j=1}^{N} \sin\left(\frac{2\pi jk}{N}\right)\sin\left(\frac{2\pi jm}{N}\right)=\begin{cases} N/2, & k=m \\ -N/2, & k=N-m \\ 0 & \text{otherwise,}\end{cases} \tag{11.22}$$

$$\sum_{j=1}^{N} \sin\left(\frac{2\pi jk}{N}\right)\cos\left(\frac{2\pi jm}{N}\right)=0 \quad \text{for all } k. \tag{11.23}$$

Because of these properties, Eqs. (11.16)–(11.18) become a set of independent equations when the data are equally spaced:

$$a_0=\frac{1}{N}\sum_{j=1}^{N} y_j \tag{11.24a}$$

$$a_m=\frac{2}{N}\sum_{j=1}^{N} y_j \cos\left(\frac{2\pi jm}{N}\right) \tag{11.24b}$$

$$b_m=\frac{2}{N}\sum_{j=1}^{N} y_j \sin\left(\frac{2\pi jm}{N}\right). \tag{11.24c}$$

It is customary to change the notation to make the equations more symmetric. Figure 11.5 shows the four different times corresponding to $N=4$ with $j=1,2,3,4$. Because of the periodicity of the sines and cosines, $j=N$ gives exactly the same value of a sine or cosine as does $j=0$. Therefore, if we reassign the data point corresponding to $j=N$ to have the value $j=0$ and sum from 0 to $N-1$, the sums will be unchanged:

$$a_0=\frac{1}{N}\sum_{j=0}^{N-1} y_j, \tag{11.25a}$$

$$a_m=\frac{2}{N}\sum_{j=0}^{N-1} y_j \cos\left(\frac{2\pi jm}{N}\right), \tag{11.25b}$$

$$b_m=\frac{2}{N}\sum_{j=0}^{N-1} y_j \sin\left(\frac{2\pi jm}{N}\right). \tag{11.25c}$$

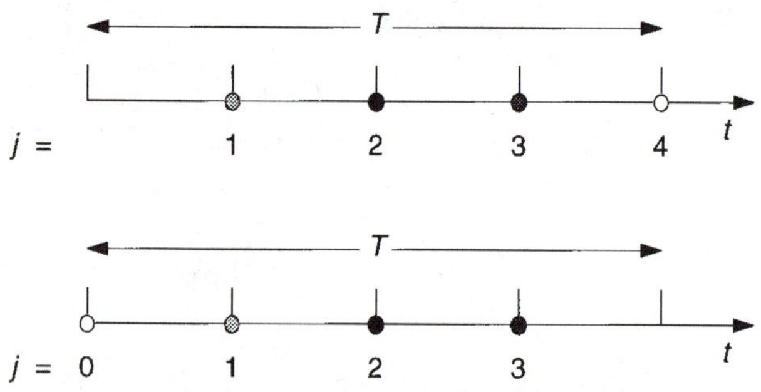

FIGURE 11.5. A case where $n=4$. The values of time are spaced by T/n and distributed uniformly. In the top case the values of j range from 1 to n. In the lower case they range from 0 to $n-1$. The values of all the trigonometric functions are the same for $j=0$ and for $j=n$.

For equally spaced data the function can be written as

$$y_j=y(t_j)=a_0+\sum_{k=1}^{n} a_k \cos\left(\frac{2\pi jk}{N}\right)+\sum_{k=1}^{n} b_k \sin\left(\frac{2\pi jk}{N}\right). \tag{11.25d}$$

Since there are N independent data points, there can be at most N independent coefficients. Therefore $2n+1\leq N$, or

$$n\leq\frac{N-1}{2}. \tag{11.26}$$

This means that there must be at least two samples per period at the highest frequency present. This is known as the *Nyquist sampling criterion.*

You can show (see the problems at the end of this chapter) that the symmetry and antisymmetry in Eqs. (11.21) and (11.22) for $k=N-m$ means that Eqs. (11.24) and (11.25) for $k>N/2$ repeat those for $k<N/2$. We can use this fact to make the equations more symmetric by changing the factor in front of the summations in Eqs. (11.25b) and (11.25c) to be $1/N$ instead of $2/N$ and extending the summation in Eq. (11.25d) all the way to $n=N-1$. Since $\cos(0)=1$ and $\sin(0)=0$, we can include the term a_0 by including $k=0$ in the sum. We then have the set of equations

$$y_j=y(t_j)=\sum_{k=0}^{N-1} a_k \cos\left(\frac{2\pi jk}{N}\right)+\sum_{k=0}^{N-1} b_k \sin\left(\frac{2\pi jk}{N}\right), \tag{11.27a}$$

$$a_k=\frac{1}{N}\sum_{j=0}^{N-1} y_j \cos\left(\frac{2\pi jk}{N}\right), \tag{11.27b}$$

$$b_k=\frac{1}{N}\sum_{j=0}^{N-1} y_j \sin\left(\frac{2\pi jk}{N}\right). \tag{11.27c}$$

This set of equations is the usual form for the *discrete Fourier transform.* We will continue to use our earlier form, Eqs. (11.25), in the rest of this chapter.

The Fourier transform is usually written in terms of complex exponentials. We have avoided using the notation of complex exponentials. It is not necessary for anything done in this book. The sole advantage of complex exponentials is that they simplify the notation. The actual calculations must be done with real numbers. Since you will undoubtedly see complex notation in other books, the notation is included here for completeness. The numbers that we have been using are called real numbers. The number $i=\sqrt{-1}$ is called an *imaginary number.* A combination of a real and imaginary number is called a *complex number.* The remarkable property of imaginary numbers that make them useful in this context is that

$$e^{i\theta}=\cos\theta+i\sin\theta. \tag{11.28}$$

If we define the complex number $Y_k=a_k-ib_k$, we can write Eqs. (11.27) as

$$Y_k=\frac{1}{N}\sum_{j=0}^{N-1} y_j e^{-i2\pi jk/N} \tag{11.29a}$$

and

$$y_j=\sum_{k=0}^{N-1} Y_k e^{i2\pi jk/N}. \tag{11.29b}$$

Since our function y is assumed to be real, in the second equation we keep only the real part of the sum. To repeat:

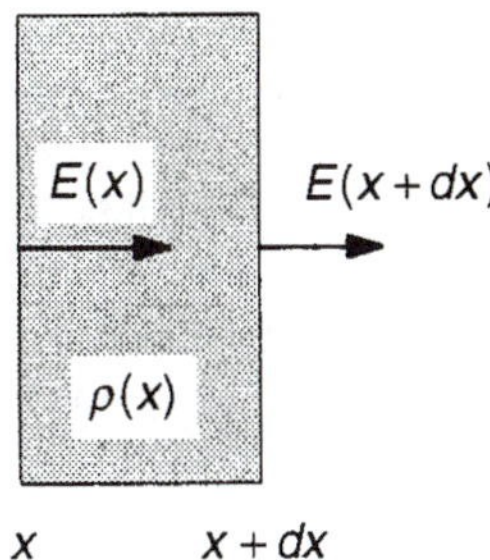

FIGURE 9.4. Gauss's law is applied to the shaded volume to derive Poisson's equation in one dimension.

which is about 80. Applying Gauss's law in the form Eq. (6.19b), we obtain[1]

$$E_x(x+dx)-E_x(x)=\frac{4\pi\rho_{\text{ext}}(x)\ dx}{4\pi\kappa\epsilon_0},$$

$$\frac{dE_x}{dx}=\frac{4\pi\rho_{\text{ext}}(x)}{4\pi\kappa\epsilon_0}.$$

Finally, since $E_x=-(\partial v/\partial x)$, we have the one-dimensional Poisson equation,

$$\frac{d^2v}{dx^2}=-\frac{4\pi\rho_{\text{ext}}(x)}{4\pi\kappa\epsilon_0}. \tag{9.6}$$

This equation was derived in much the same way that the equation of continuity was combined with Fick's first law to derive Fick's second law. The same procedure can be used in three dimensions to derive the general form of Poisson's equation:

$$\nabla^2 v=-\frac{4\pi\rho_{\text{ext}}(\mathbf{r})}{4\pi\kappa\epsilon_0}. \tag{9.7}$$

Had we chosen to include the water polarization charge explicitly in a total charge density, we would have obtained

$$\nabla^2 v=-\frac{4\pi\rho_{\text{tot}}(\mathbf{r})}{4\pi\epsilon_0}. \tag{9.8}$$

For the model being considered the ions are all univalent, so the ionic charge density at x is related to the concentrations by

$$\rho_{\text{ext}}(x)=e[K(x)+M(x)-Cl(x)]. \tag{9.9a}$$

More generally, for a series of ion species each with concentration C_i and valence z_i,

$$\rho_{\text{ext}}(\mathbf{r})=e\sum_i z_iC_i(\mathbf{r}). \tag{9.9b}$$

The next step is to assume that the concentrations of all ions are given by Boltzmann factors and are therefore related to the potential by

$$K(x)=[\text{K}]e^{-ev(x)/k_BT}\quad\text{for all } x$$

$$Cl(x)=[\text{Cl}]\ e^{ev(x)/k_BT}\quad\text{for all } x \tag{9.10a}$$

$$M(x)=[M']e^{-e(v(x)-v')/k_BT},\quad x>0.$$

(Remember that $M(x)=0$ to the left of the origin.) An equivalent general expression is

$$\rho_{\text{ext}}(\mathbf{r})=e\sum_i z_iC_i\ \exp\left(\frac{-z_iev(\mathbf{r})}{k_BT}\right), \tag{9.10b}$$

where C_i is the concentration in the region where $v=0$.

Combining Eqs. (9.7) and (9.10b) gives the *Poisson–Boltzmann equation* for a dielectric:

$$\nabla^2 v=-\frac{4\pi e}{4\pi\epsilon_0\kappa}\sum_i z_iC_i\ \exp\left(\frac{-z_iev(\mathbf{r})}{k_BT}\right). \tag{9.11}$$

For the specific problem at hand the Poisson–Boltzmann equation takes the form

$$\frac{d^2v}{dx^2}=\frac{-4\pi e}{4\pi\epsilon_0\kappa}([\text{K}]e^{-ev(x)/k_BT}-[\text{Cl}]\ e^{ev(x)/k_BT}).$$

This applies for $x<0$ only. While it is possible to solve this exactly using numerical techniques [see Mauro (1962)], we will confine ourselves to the case in which $\xi=ev/k_BT\ll 1$ and we can make the approximation $e^{\xi}\approx 1+\xi$. (This is accurate to 0.5% for $\xi=0.1$, to 10% for $\xi=0.5$, and to 25% for $\xi=0.8$.) With this approximation

$$\rho_{\text{ext}}=e\sum C_iz_i\left(1-\frac{z_iev}{k_BT}\right)=e\sum C_iz_i-e^2\sum\frac{C_iz_i^2v}{k_BT}. \tag{9.12}$$

Far from the membrane the solution is electrically neutral, so the first term vanishes. We are left with the *linear Poisson–Boltzmann equation:*

$$\nabla^2 v(\mathbf{r})=\frac{4\pi e^2\sum C_iz_i^2}{4\pi\epsilon_0\kappa k_BT}\ v(\mathbf{r}). \tag{9.13}$$

The coefficient of $v(\mathbf{r})$ on the right has the dimensions of $1/(\text{length})^2$. It will also appear in other contexts. It is known as the *Debye length*:

$$\frac{1}{\lambda_D^2}=\frac{4\pi e^2\sum C_iz_i^2}{4\pi\epsilon_0\kappa k_BT}. \tag{9.14}$$

The linearized Poisson–Boltzmann equation is

[1]Throughout this section I keep 4π in both numerator and denominator that could be canceled. I did this for two reasons. First, it is easiest for me to remember that the quantity $1/4\pi\epsilon_0$ is about 9×10^9; second, for those who do not use SI units, the factor $1/4\pi\epsilon_0$ does not appear, but the other factor of 4π remains.

$$\nabla^2 v = \frac{v}{\lambda_D^2}. \tag{9.15}$$

For the one-dimensional problem and $x<0$, it is

$$\frac{d^2 v}{dx^2} = \frac{v}{\lambda_D^2}, \tag{9.16}$$

where

$$\frac{1}{\lambda_D^2} = \frac{4\pi e^2([\text{K}]+[\text{Cl}])}{4\pi\epsilon_0 \kappa k_B T}. \tag{9.17}$$

The methods of Appendix F can be applied to solve this equation.[2] The characteristic equation (F.4) is $s^2 = 1/\lambda_D^2$, so the solution for $x<0$ is $v(x) = Ae^{-x/\lambda_D} + Be^{x/\lambda_D}$. The potential is zero far to the left, so $A=0$. Therefore, the solution for $x<0$ is

$$v(x) = Be^{x/\lambda_D}, \quad x<0. \tag{9.18}$$

It is most convenient to write the concentrations for $x>0$ in terms of the concentrations far to the right. It is now necessary to include the impermeant ions.

$$K(x) = [\text{K}']e^{-e[v(x)-v']/k_B T},$$

$$Cl(x) = [\text{Cl}']e^{e[v(x)-v']/k_B T},$$

$$M(x) = [M']e^{-e[v(x)-v']/k_B T}. \tag{9.19}$$

The linearized Poisson-Boltzmann equation for $x>0$ is then

$$\frac{d^2 v}{dx^2} = -\frac{4\pi e}{4\pi\epsilon_0 \kappa}\left([K'] - \frac{[\text{K}']ev(x)}{k_B T} + \frac{[\text{K}']ev'}{k_B T} - [\text{Cl}'] - \frac{[\text{Cl}']ev(x)}{k_B T} + \frac{[\text{Cl}']ev'}{k_B T} + [M'] - \frac{[M']ev(x)}{k_B T} + \frac{[M']ev'}{k_B T}\right). \tag{9.20}$$

Since neutrality requires that $[\text{K}'] + [M'] - [\text{Cl}'] = 0$, the constant terms cancel. With the definition

$$\frac{1}{\lambda_D'^2} = \frac{4\pi e^2([\text{K}']+[\text{Cl}']+[M'])}{4\pi\epsilon_0 \kappa k_B T}, \tag{9.21}$$

this can be written as

$$\frac{d^2 v}{dx^2} - \frac{v(x)}{\lambda_D'^2} = -\frac{v'}{\lambda_D'^2}. \tag{9.22}$$

This is an inhomogeneous linear differential equation with constant coefficients. As pointed out in Appendix F, the most general solution is the sum of the solution to the homogeneous equation (i.e., with the right hand side equal to 0) and any solution of the inhomogeneous equation, with the constants adjusted to satisfy whatever boundary conditions exist. In this case $v(x) = v'$ satisfies the inhomogeneous equation, so the most general solution is

$$v(x) = A'e^{-x/\lambda_D'} + B'e^{x/\lambda_D'} + v'.$$

Far to the right $v = v'$ so $B' = 0$. Therefore, the solution we need is

$$v(x) = A'e^{-x/\lambda_D'} + v'. \tag{9.23}$$

This solution for $x>0$ must be combined with the solution for $x<0$, Eq. (9.18). At $x=0$ the potential must be continuous. Therefore $B = A' + v'$. Also at $x=0$ the electric field, and therefore dv/dx, is continuous. (If dv/dx were not continuous, the second derivative and ρ_{ext} would be infinite.) This requirement gives the equation $B/\lambda_D = -A'/\lambda_D'$. Solving these two equations, we obtain

$$A' = \frac{-v'\lambda_D'}{\lambda_D' + \lambda_D}, \quad B = \frac{v'\lambda_D}{\lambda_D' + \lambda_D}. \tag{9.24}$$

Figures 9.5 and 9.6 show the potential, concentration, and charge density for the case $[\text{K}] = 100$ and $[M'] = 50$ mmol l^{-1}. The other parameters are given in Table 9.2. Since the radii of ions are about 0.2 nm, the Debye length is several ionic diameters, and the continuous model we have used is reasonable. The Debye length becomes smaller for higher concentrations.

9.3. IONS IN SOLUTION: THE DEBYE–HÜCKEL MODEL FOR THE SURROUNDING ION CLOUD

In an ionic solution, ions of opposite charge attract one another. A model of this neutralization was developed by Debye and Hückel a few years after Gouy and Chapman developed the model in the previous section. The Debye–

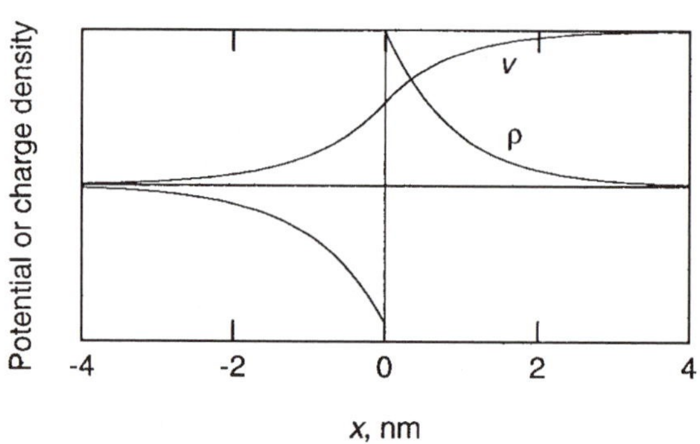

FIGURE 9.5. The potential and charge density in the vicinity of the Donnan membrane. There is a layer of negative charge on the left of the membrane and of positive charge on the right. Each decays with the Debye length given by the ion concentrations far from the membrane.

[2] We have seen this equation before in electrotonus when the membrane capacitance is fully charged.

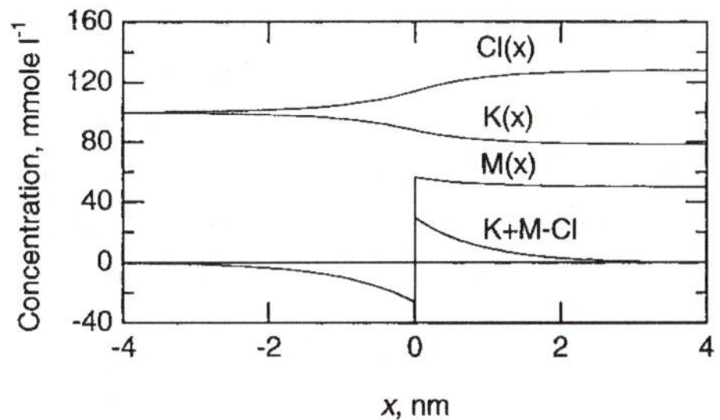

FIGURE 9.6. Concentration profiles across the Donnan membrane. The concentration $K(x)+M(x)-Cl(x)$ is proportional to the charge density.

Hückel model singles out a particular ion and assumes that the average concentration of the counterions surrounding it is given by a Boltzmann factor. Screening by the counterions causes the potential to fall much more rapidly than $1/r$. One major difficulty with this assumption is that each counterion is also a central ion; therefore, the notion of a continuous cloud of counterions represents some sort of average.

We consider a situation in which the electric field, potential, and charge distribution are spherically symmetric. We could begin with Eq. (9.8) and look up the Laplacian operator in spherical coordinates. However, it is instructive to derive Poisson's equation for the spherically symmetric case. Consider two concentric spheres of radius r and radius $r+dr$. Apply Gauss's law to the volume contained between the two surfaces. If $\mathbf{E}$ is spherically symmetric, the flux through the inner sphere is $4\pi r^2 E(r)$. It points into the sphere and is therefore negative. The outward flux at $r+dr$ is

$$4\pi(r+dr)^2 E(r+dr)$$
$$=4\pi[r^2+2rdr+(dr)^2]\left[E(r)+\left(\frac{dE}{dr}\right)dr\right].$$

If we keep only terms of order dr or less, the outward flux through the outer sphere is

$$4\pi r^2 E(r)+8\pi rE(r)dr+4\pi r^2\frac{dE}{dr}dr.$$

TABLE 9.2. *Parameters for the Donnan interface when [K]=100, [M]=0, and [M′]=50 mmol l^{-1} at T=310 K.*

[Cl]	100 mmol l^{-1}
[K]	100 mmol l^{-1}
[M]	0 mmol l^{-1}
[K′]	78.1 mmol l^{-1}
[Cl′]	128.1 mmol l^{-1}
[M′]	50 mmol l^{-1}
v'	6.617 mV
λ_D	0.991 nm
λ_D'	0.875 nm

The net flux out of the volume is $8\pi rE(r)dr+4\pi r^2(dE/dr)dr$. The total charge in the shell is $\rho_{ext}(r)$ times the volume of the shell, $4\pi r^2 dr$. Therefore, Gauss's law is

$$8\pi rE(r)dr+4\pi r^2\frac{dE}{dr}dr=\frac{\rho_{ext}(r)4\pi r^2}{\kappa\epsilon_0}dr$$

or

$$\frac{1}{r^2}\frac{d}{dr}r^2E(r)=\frac{4\pi\rho_{ext}(r)}{4\pi\epsilon_0\kappa}. \tag{9.25}$$

Since $E(r)=-dv/dr$, the final equation for the potential is

$$\frac{1}{r^2}\frac{d}{dr}r^2\frac{dv}{dr}=-\frac{4\pi\rho_{ext}(r)}{4\pi\epsilon_0\kappa}. \tag{9.26}$$

The Poisson–Boltzmann equation in spherical coordinates, the analog of Eq. (9.11), is

$$\frac{1}{r^2}\frac{d}{dr}r^2\frac{dv}{dr}=-\frac{4\pi e}{4\pi\epsilon_0\kappa}\sum z_iC_i\exp\left(\frac{-z_iev(r)}{k_BT}\right). \tag{9.27}$$

We again make a linear approximation to the Boltzmann factor to obtain the linear Poisson–Boltzmann equation for spherical symmetry:

$$\frac{1}{r^2}\frac{d}{dr}r^2\frac{dv}{dr}=\frac{1}{\lambda_D^2}v(r). \tag{9.28}$$

The Debye length λ_D is defined in Eq. (9.14). With the substitution $v(r)=u(r)/r$, the equation becomes

$$\frac{d^2u}{dr^2}=\frac{1}{\lambda_D^2}u(r), \tag{9.29}$$

which is the same as Eq. (9.16). Therefore the solution is

$$v(r)=\frac{u(r)}{r}=\frac{Ae^{-r/\lambda_D}+Be^{r/\lambda_D}}{r}.$$

Requiring that $v(r)$ approach 0 as $r\to\infty$ means that $B=0$. For small r, the electric field $(-dv/dr)$ is that of an unshielded ion of charge ze. Therefore $A=ze/4\pi\epsilon_0\kappa$, and the final solution is

$$v(r)=\frac{ze\exp(-r/\lambda_D)}{4\pi\epsilon_0\kappa r}. \tag{9.30}$$

This is the potential of a point charge ze in a dielectric, modified by an exponential decay over the Debye length. From Eq. (9.14) one sees that the greater the concentration of counterions, the shorter the Debye length.

Table 9.3 shows values of $v(r)$, $\xi = ev/k_BT$, and the potential from an unscreened point charge in water of dielectric constant 80, when the ion concentrations are those given in Fig. 6.3. A typical ion radius is about 0.2 nm. We will discover in the next section that the dielectric constant saturates for $r<0.25$ nm. Therefore, values are given in Table 9.3 only for $r>0.3$ nm. The table shows that the assumption $e^{\xi} \approx 1+\xi$ is reasonable only for $r>0.5$ nm. The Debye length is $\lambda_D = 0.77$ nm.

The charge density of the ion cloud can be obtained from Eqs. (9.26) and (9.30). The result is

$$\rho_{\text{ext}}(r) = \frac{-ze}{4\pi\lambda_D^2 r} e^{-r/\lambda_D}. \tag{9.31}$$

Ignoring for the moment the region over which the linearization is valid, the total charge in the counterion cloud inside a sphere of radius a is

$$\int_0^a 4\pi r^2 \rho_{\text{ext}}(r)\, dr.$$

Adding to this a point charge ze at the origin gives the total charge due to both the ion and the counterion cloud inside radius a:

$$q(a) = ze\left(1+\frac{a}{\lambda_D}\right) e^{-a/\lambda_D}. \tag{9.32}$$

This function approaches ze, the charge of the point ion, as $a\to 0$, and it approaches 0 as $a\to\infty$. Table 9.3 also shows the values of $q(a)/e$. Ninety percent of the counterion charge resides within 3 nm of the central ion. The charge on the central ion is half neutralized by charge in a sphere of radius 1.3 nm, about six ionic radii. Figure 9.7 shows schematically an ion of radius 0.2 nm. Since a monovalent ion will be neutralized by a single counterion, it is clear that the assumption of a continuous charge distribution equal to the average is a bit strained. The shaded circle of radius 0.25 nm represents the region in which the water molecules are completely polarized and the dielectric constant is less than 80; this is discussed in the next section.

If a neutral ionic fluid containing mutually screened ions and counterions is placed in a constant electric field pointing to the right, the positive ions will begin to drift to the right and the negative ions to the left. This is discussed further in Sec. 9.5.

TABLE 9.3. *The Debye–Hückel potential for a monovalent ion in a solution of ions of the concentration given in Fig. 6.2 for the interior of an axon. Also shown are the parameter zev/k_BT, the unscreened potential, and the charge inside a sphere of radius r.*

r (nm)	$v(r)$ (V)	ev/k_BT	$e/(4\pi\epsilon_0\kappa r)$ (V)	$q(r)/e$
0.3	0.040 62	1.52	0.059 92	0.94
0.4	0.026 76	1.00	0.044 94	0.90
0.5	0.018 81	0.70	0.035 95	0.86
0.6	0.013 77	0.51	0.029 96	0.82
0.7	0.010 37	0.39	0.025 68	0.77
0.8	0.007 97	0.30	0.022 47	0.72
0.9	0.006 22	0.23	0.019 97	0.67
1.0	0.004 92	0.18	0.017 98	0.63
1.2	0.003 16	0.12	0.014 98	0.54
1.4	0.002 09	0.08	0.012 84	0.46
1.6	0.001 41	0.05	0.011 23	0.39
1.8	0.000 97	0.04	0.009 99	0.32
2.0	0.000 67	0.03	0.008 99	0.27
2.2	0.000 47	0.02	0.008 17	0.22
2.4	0.000 33	0.01	0.007 49	0.18
2.6	0.000 24	0.01	0.006 91	0.15
2.8	0.000 17	0.01	0.006 42	0.12
3.0	0.000 12	0.00	0.005 99	0.10

9.4. SATURATION OF THE DIELECTRIC

The electric field in vacuum at distance r from a point charge q is $E = q/(4\pi\epsilon_0 r^2)$. If the charge is in a dielectric, the field is reduced by a factor $1/\kappa$, except at very small distances, where the electric field is so strong that the polarization of the dielectric is saturated.

A molecule of water appears schematically as shown in Fig. 6.17. The radius of each hydrogen atom is about 120 pm; the radius of the oxygen is about 140 pm. Each hydrogen nucleus is 96.5 pm from the oxygen; the angle between them is 104°. The hydrogen atoms share their electrons with the oxygen in such a way that each hydrogen

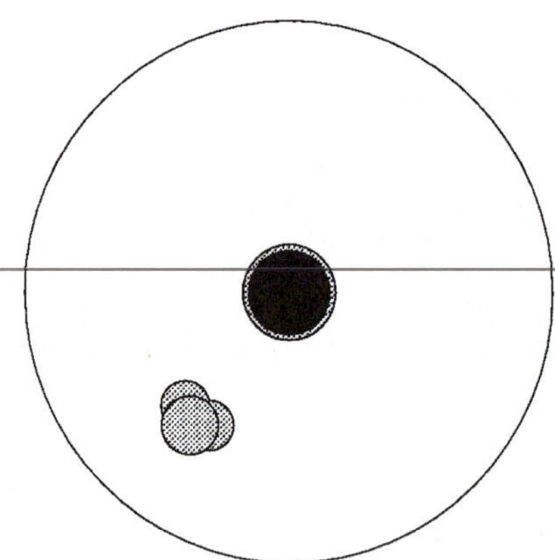

FIGURE 9.7. Schematic picture of the regions surrounding an ion. The solid circle in the center represents the ion of radius 0.2 nm. The shaded circle shows the region in which the polarization of the water is saturated. The outer circle of radius 1.3 nm represents the region within which the cloud of counterions has neutralized half of the charge on the ion, which means that on the average a counterion will be in this region half of the time. This radius depends on the ion concentrations that are those for the interior of a squid axon. A scale drawing of a water molecule is also shown.

atom has a net positive charge and the oxygen has a net negative charge. A pair of charges $\pm q$ separated by distance b has an *electric dipole moment* $\mathbf{p}_e$ of magnitude $p_e = qb$. The vector points from the negative to the positive charge. The magnitude of the dipole moment of a water molecule is 6.237×10^{-30} C m.

Each molecule of a dielectric in an applied electric field has an induced dipole moment that reduces the field. This dipole moment can be caused by a displacement of the electron cloud with respect to the nucleus, or it can represent (as for a polar molecule like water) an average molecular alignment against the tendency of thermal motion to orient the water molecules randomly.

The average induced dipole moment gives rise to the polarization field E_{pol} [Eqs. (6.17)–(6.18)]. To see the relationship, consider a small volume in the dielectric with N molecules per unit volume. Each molecule has a dipole moment $p_e = qb$. Far from this volume, the potential is due primarily to the dipole moment of each molecule. This can be shown by arguments like those in Secs. 7.3 and 7.4. The potential depends on the total dipole moment of the volume. The total number of dipoles in the volume is $NS\,dx$, so $p_{\mathrm{tot}} = p_e NS\,dx$. This is equivalent to a charge $q' = p_e NS$ on the ends of the volume element, or a surface of charge density

$$\sigma_q' = \frac{q'}{S} = p_e N. \tag{9.33}$$

Now consider a parallel-plate capacitor as shown in Fig. 9.8. Imagine a series of small volume elements in the dielectric. The induced charges $\pm \sigma_q$ on adjacent surfaces of each row of volume elements cancel except at the end of each row. The polarization field is therefore due entirely to the induced charge of surface density $\pm \sigma_q'$ at each end of the dielectric. The magnitude of the field is

$$E_{\mathrm{pol}} = \frac{\sigma_q'}{\epsilon_0} = \frac{N p_e}{\epsilon_0}. \tag{9.34}$$

The quantity $N p_e$ is the dipole moment per unit volume and is called the *polarization* P.

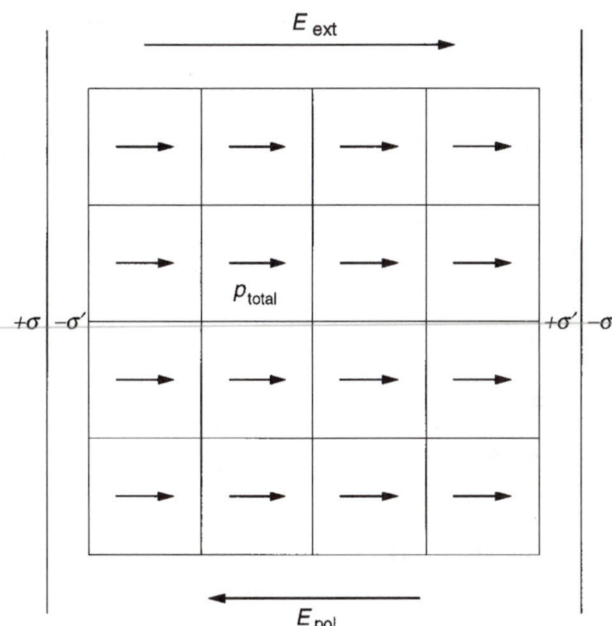

FIGURE 9.8. A dipole moment p_{tot} is induced in each volume element of the dielectric. The total effect is the same as a charge density $\pm\sigma'$ induced on the surfaces of the dielectric.

As the external electric field is increased, E_{pol}, which points in the opposite direction, also increases. This corresponds to the water molecules becoming more and more aligned.[3] From the definition of susceptibility and the dielectric constant in Sec. 6.7, the magnitudes are related by

$$E_{\mathrm{pol}} = -\frac{\chi}{1+\chi} E_{\mathrm{ext}} = -\left(1 - \frac{1}{\kappa}\right) E_{\mathrm{ext}}.$$

For a monovalent ion in water, $E_{\mathrm{pol}} = (79/80) E_{\mathrm{ext}} = (79/80) e/(4\pi\epsilon_0 r^2)$. When the dipoles are completely aligned, E_{pol} saturates at its maximum value, given by Eq. (9.34) with the molecular dipole moment substituted for p_e. The number of water molecules per unit volume is obtained from the fact that 1 mol has a mass of 18 g, occupies 1 cm^3/g, and contains N_A molecules:

$$E_{\mathrm{pol}}(\max) = \frac{(N_A \text{ molecule/mol})(1 \text{ g/cm}^3)(10^6 \text{ cm}^3/\text{m}^3)(6.237\times 10^{-30} \text{ C m/molecule})}{(18 \text{ g/molecule})\epsilon_0 \text{ C V}^{-1}\text{ m}^{-1}} = 2.36\times 10^{10} \text{ V m}^{-1}.$$

Figure 9.9 shows the fields E_{ext} and E_{pol} around a monovalent ion. As E_{pol} saturates, E_{tot} rises toward the value it would have without a dielectric. The dielectric constant falls from 80 to 1 at about 0.23 nm. The fall is actually more gradual than this, because thermal motion smears the alignment of the dipoles. Close to an ion the potential is larger than $q/(4\pi\epsilon_0 r)$. This changes the Born charging energy [Eq. (6.20)] and the free energy change as an ion dissolves in a solvent [Bockris and Reddy (1970), Chap. 2].

[3]A more sophisticated model for the alignment of the electric dipoles in the electric field is analogous to that for magnetic moments in Sec. 18.3.

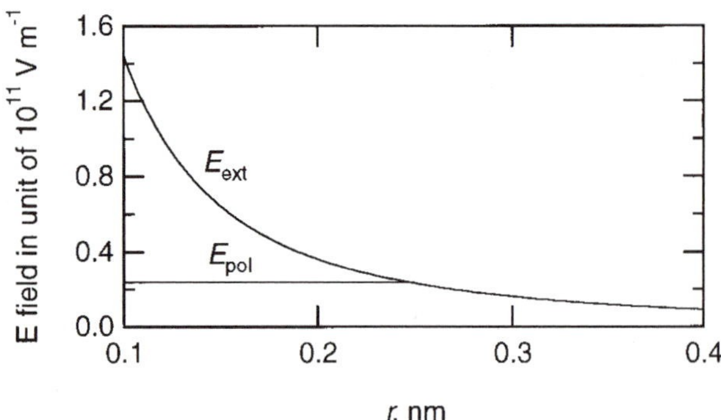

FIGURE 9.9. The electric field around a monovalent point charge and the polarization electric field due to the water. The polarization field saturates for $r<0.25$ nm.

9.5. ION MOVEMENT IN SOLUTION: THE NERNST–PLANCK EQUATION

Solute particles can move by diffusion. They can also move if they have an average velocity v_{solute}.[4] One way they can have an average velocity is if they are at rest on average with respect to a moving solution. This is *solvent drag*. The solute particles can also be dragged through the solution by an external force that acts on them, such as gravity or an electric force. The number per unit area per unit time crossing a place is Cv_{solute}. The solute particle fluence rate (particle current density) due to both diffusion and the solute velocity in the x direction is[5]

$$j_s = -D\frac{dC}{dx} + C(x)v_{\text{solute}}. \tag{9.35}$$

Suppose that an external force $\mathbf{F}=ze\mathbf{E}$ acts on the solute particles in the x direction. They will be accelerated until the viscous drag on them is equal to the magnitude of F. But we saw in Chap. 4 that the viscous drag is $f=-\beta(v_{\text{solute}}-v_{\text{solvent}})$ where $v_{\text{solute}}-v_{\text{solvent}}$ is the relative velocity of the solute through the solvent. Coefficient β is related to the diffusion constant by $\beta=k_BT/D$. Therefore, the particles are no longer accelerated when

$$zeE=\beta(v_{\text{solute}}-v_{\text{solvent}}), \quad v_{\text{solute}}-v_{\text{solvent}}=zeE/\beta. \tag{9.36}$$

Equation (9.35) can be rewritten as

$$j_s=-D\frac{dC}{dx}+C(x)[v_{\text{solvent}}+(v_{\text{solute}}-v_{\text{solvent}})].$$

[4]We have an unfortunate notation problem. We have been using v for both velocity and potential. They will now occur in the same equation. All velocities will have a subscript, such as solute or solvent.

[5]We use x for the distance in the direction parallel to E because z is used for valence.

Now v_{solvent} is the volume of solvent that flows per unit area per unit time and is just j_v. With this substitution and using Eq. (9.36), the particle current density is

$$j_s=-D\frac{dC}{dx}+C(x)j_v+C(x)zeE\frac{D}{k_BT}. \tag{9.37}$$

The first term represents solute motion due to diffusion, the second represents solute dragged along with the bulk flow of the solution (solvent drag), and the third represents drift due to the applied electric field.

We will consider only the case in which there is no bulk flow of solution, so $j_v=0$. The equation then reduces to

$$j_s=-D\frac{dC}{dx}+\frac{zeE}{k_BT}DC. \tag{9.38}$$

It is called the *Nernst–Planck equation.* Diffusion is always toward the region of lower concentration, while for positive charge the movement with constant velocity is in the direction of $\mathbf{E}$. For negative charges it is in the opposite direction.

Consider the current density in bulk solution between planes at $x=0$ where $v(x)=0$ and $x=L$ where $v(x)=v$. If there is no concentration gradient and the potential changes uniformly, then $E=-dv/dx=-v/L$ points in the $-x$ direction, and the particle current density is

$$j_s=-\frac{zeDCv}{k_BTL}.$$

The electrical current density j is obtained by multiplying j_s by the charge on each particle, ze:

$$j=-\frac{z^2e^2DCS}{k_BTL}\frac{v}{S}=-\frac{G(C)}{S}v. \tag{9.39}$$

If $v(L)>v(0)$, the current is to the left and is negative. Recalling that $G=\sigma S/L=1/R=S/\rho L$, we obtain the conductivity in bulk solution

$$\sigma=\frac{1}{\rho}=\frac{z^2e^2DC}{k_BT}. \tag{9.40}$$

If several ion species carry current and can be assumed to move independently, then the total conductivity is the sum of the conductivities for each ion. Table 9.4 shows contributions to the conductivity for various species at typical concentrations.

This model is satisfactory for material such as the inside of an axon where the concentrations are constant and the material is electrically neutral, so that the ions themselves do not on average contribute to the electric field. We have assumed that the ions move independently, which will happen only if the electric field of other ions can be ignored.

TABLE 9.4. *Conductivities of ions at various concentrations at 25 °C. Diffusion constants for each ion are from Hille (1992, p. 268). (For each species $\sigma = z^2 e^2 DC/k_BT$, where C is in ion m^{-3}, the conductivities add, and $\rho = 1/\sigma$.*

	D ($m^2\,s^{-1}$)	C (mmol l^{-1})	σ ($S\,m^{-1}$)	ρ (Ω m)
		Extracellular squid axon		
Na	1.33×10^{-9}	145	0.723	
K	1.96×10^{-9}	5	0.037	
Cl	2.03×10^{-9}	125	0.951	
			1.711	0.584
		Intracellular squid axon		
Na	1.33×10^{-9}	15	0.075	
K	1.96×10^{-9}	150	1.102	
Cl	2.03×10^{-9}	9	0.069	
			1.246	0.803

We can model ions flowing from a region of one concentration to another (such as crossing the axon membrane) with the Nernst–Planck equation. Writing it for the electric current density and using the fact that $E(x) = -dv/dx$, we have

$$j = -zeD\frac{dC}{dx} - \frac{z^2e^2D}{k_BT}\frac{dv}{dx}C. \tag{9.41}$$

It is simpler to use the dimensionless variable $u(x) = zev(x)/k_BT$, which is the ratio of an ion's energy to thermal energy:

$$j = -zeD\left(\frac{dC}{dx} + C\frac{du}{dx}\right). \tag{9.42}$$

If we assume that dv/dx is constant throughout the region, $v(0)=0$ and $v(L)=v$, then the gradient is $dv/dx = v/L$, and Eq. (9.41) becomes

$$\frac{dC}{dx} - \frac{1}{\lambda}C = -\frac{j}{zeD}, \tag{9.43}$$

where the characteristic length for this model (*not* the Debye length) is

$$\lambda = -\frac{L}{u} = -\frac{k_BTL}{zev}. \tag{9.44}$$

This is the same as Eq. (4.58), except for the denominator of the term involving j. Here the denominator is zeD because j is the electric current density instead of the particle current density. The solution analogous to Eq. (4.62) is

$$j = \frac{zeD}{\lambda}\frac{C_0e^{L/\lambda} - C_0'}{e^{L/\lambda} - 1}$$

or

$$j = \frac{zeD}{\lambda}\frac{C_0e^{-u} - C_0'}{e^{-u} - 1}, \tag{9.45}$$

where C_0 is the ion concentration at $x=0$ and C_0' is the concentration at $x=L$.

The current vanishes if $C_0e^{L/\lambda} - C_0' = 0$, or

$$\frac{C_0'}{C_0} = e^{L/\lambda} = e^{-zev/k_BT} = e^{-u}.$$

This is the Boltzmann factor.

Eq. (9.45) can be written in terms of the original variables:

$$j = -\frac{z^2e^2Dv}{k_BTL}\frac{C_0e^{-zev/k_BT} - C_0'}{e^{-zev/k_BT} - 1} = -\frac{zeDu}{L}\frac{C_0e^{-u} - C_0'}{e^{-u} - 1}. \tag{9.46}$$

It is interesting to compare this to Eq. (9.39). Since G depends on concentration, it is useful to factor out a factor C_0 and write

$$j = -\frac{z^2e^2DC_0}{k_BTL}\frac{e^{-zev/k_BT} - C_0'/C_0}{e^{-zev/k_BT} - 1}v$$
$$= -\frac{G(C_0)}{S}\frac{e^{-zev/k_BT} - C_0'/C_0}{e^{-zev/k_BT} - 1}v. \tag{9.47}$$

If $C_0 = C_0'$, we recover Eq. (9.39). Figure 9.10 shows the current density in A m^{-2} for a situation where $C_0 = 145$ mmol l^{-1}, $C_0' = 15$, and $L = 1$ m. The diffusion constant for sodium from Table 9.4 has been used. Because $C_0 > C_0'$, equilibrium occurs when $v = +57.3$ mV at 20 °C.

Note the nonlinearity of the current–voltage relationship that arises because $C_0 \neq C_0'$. For very negative potentials the flow is almost entirely from left to right and the current

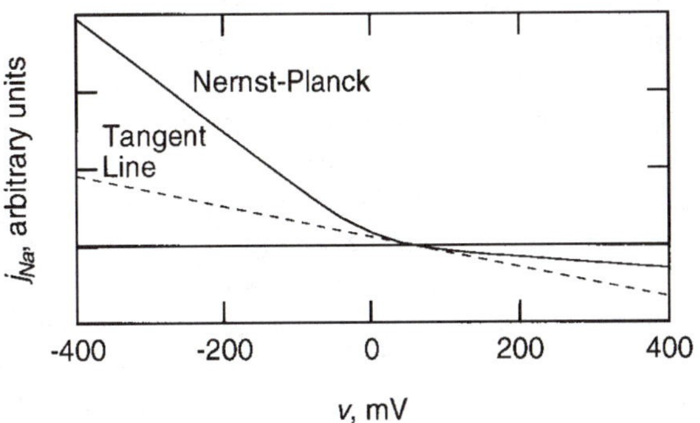

FIGURE 9.10. Sodium current versus applied potential for the constant field Nernst–Planck model when the sodium concentration is 145 m*M* on the left and 15 m*M* on the right. The calculation was done using Eq. (9.46) for $T = 293$ K. The tangent line was calculated using Eq. (9.48). The nonlinearity or rectification occurs because of the different ion concentrations on each side.

density approaches $G(C_0)v/S$ while for very positive potentials the flow is from right to left and the current density approaches $G(C_0')v/S$. This asymmetry is fundamental. It occurs because there are different numbers of charge carriers on the left and right. When this behavior is seen in channels in cell membranes, they are often called *rectifier channels*. This same asymmetry in differences in the concentration of charge carriers is responsible for rectification in semiconductors.

Near the Nernst potential the current density has the form $j=-(G/S)(v-v_{\text{Nernst}})$ if

$$\frac{G}{S}=\frac{G(C_0)(zev_{\text{Nernst}}/k_BT)e^{-zev_{\text{Nernst}}/k_BT}}{S(e^{zev_{\text{Nernst}}/k_BT}-1)}. \qquad (9.48)$$

This equation was used to derive the tangent line shown in Fig. 9.10.

The constant-field model is an oversimplification. The field can be distorted by fixed charges located in the region. Moreover, the model is internally inconsistent. There are electric fields generated by the flowing ions, which become important at high concentrations. The fact that $j=0$ when the potential is equal to the Nernst potential is fundamental and holds for any ion or model for conduction. It can be derived in the general case from Eq. (9.43) (see the Problems). A self-consistent analytic solution for the case of a single ion species has been known for 50 years. The solution has been extended by many workers and has been generalized by Leuchtag and Swihart (1977) to the case in which all the ions have the same charge.

9.6. ZERO TOTAL CURRENT IN A CONSTANT-FIELD MEMBRANE: THE GOLDMAN EQUATION

The Nernst-Planck equation can be used to calculate the current due to movement of ions through a membrane in which there is a constant electric field. We assume a constant field because it leads to an analytic solution and because we have no knowledge of internal structure or the behavior of counterions which could change the field. The resulting equations, showing the conditions under which there is no net current, are called the *Goldman* or the *Goldman–Hodgkin–Katz* (GHK) equations.

The GHK equations can be derived by assuming either a homogeneous membrane, in which case the Nernst–Planck equation is simply applied to each species, or cylindrical pores of constant cross section. Since we know that the pores do not have a constant electric field (Sec. 9.7) and it is quite unlikely that they have constant cross section, the GHK equation is an approximation. Nevertheless, it has been used widely in the study of excitable membranes.

We will show the derivation for a cylindrical pore that has a constant circular cross section. We use cylindrical coordinates (r,ϕ,x), where x is the axis of the cylinder. (Again, z denotes the valence of the ions.) Let the outside of the membrane be at $x=0$ and the inside at $x=L$, where the potential is v and $u=zev/k_BT$. The arguments of Sec. 5.9 about the r and x dependence can be applied to Eq. (9.42). The analog of Eq. (5.33) is

$$j(r)=-zeD(r,a,R_p)\left(\frac{\partial C(r,x)}{\partial x}+\frac{u}{L}\,C(r,x)\right). \qquad (9.49)$$

Again the concentration can be written as $C(r,x)=C(x)\Gamma(r)$. Equation (9.49) becomes

$$j(r)=-ze\Gamma(r)D(r,a,R_p)\left(\frac{\partial C(x)}{\partial x}+\frac{u}{L}\,C(x)\right). \qquad (9.50)$$

This can be multiplied by $2\pi r\,dr$ and integrated over the pore area. There are two integrals to consider. The first defines the average current density for a particular species:

$$\int_0^{R_p} j(r)2\pi r\,dr=\pi R_p^2\bar{j}. \qquad (9.51)$$

The second defines an effective diffusion constant:

$$\int_0^{R_p}\Gamma(r)D(r,a,R_p)2\pi r\,dr=\pi R_p^2 D_{\text{eff}}. \qquad (9.52)$$

The integrated current density equation is, therefore,

$$\bar{j}=-zeD_{\text{eff}}\left(\frac{dC(x)}{dx}+\frac{u}{L}\,C(x)\right). \qquad (9.53)$$

Consideration of the r dependence in the pore has given an equation exactly like Eq. (9.42), but with D_{eff} instead of D. Equations (9.43) and (9.44) are still valid. The form of λ is unchanged: $\lambda=-k_BTL/zev$. As in Sec. 5.9, conversion from a single pore to unit area of the membrane requires multiplying $\bar{j}$ by $n\pi R_p^2$. We define $\omega_sRT=n\pi R_p^2D_{\text{eff}}/L$ and call the concentration outside C_1 and the concentration inside C_2. The electric current density per unit area of membrane is

$$\begin{aligned}J'&=\frac{z^2e^2\omega_sRTv}{k_BT}\,\frac{C_1e^{-zev/k_BT}-C_2}{1-e^{-zev/k_BT}}\\&=z^2e^2v\,\omega_sN_A\,\frac{C_1e^{-zev/k_BT}-C_2}{1-e^{-zev/k_BT}}.\end{aligned} \qquad (9.54)$$

Suppose that three species can pass through the membrane: sodium, potassium, and chloride. Equation (9.54) can be applied separately to each species to obtain the following three equations:

$$J'_{\text{Na}}=e^2v\,\omega_{\text{Na}}N_A\,\frac{[\text{Na}_1]e^{-ev/k_BT}-[\text{Na}_2]}{1-e^{-ev/k_BT}}, \qquad (9.55\text{a})$$

$$J'_{\text{K}}=e^2v\,\omega_{\text{K}}N_A\,\frac{[\text{K}_1]e^{-ev/k_BT}-[\text{K}_2]}{1-e^{-ev/k_BT}}, \qquad (9.55\text{b})$$

$$J'_{\text{Cl}}=e^2 v\,\omega_{\text{Cl}}N_A\,\frac{[\text{Cl}_1]e^{+ev/k_BT}-[\text{Cl}_2]}{1-e^{+ev/k_BT}}. \qquad (9.55\text{c})$$

Instead of saying that each fluence rate is zero, let us investigate the consequences if the *sum* of the three fluence rates is zero and no other charges pass through the membrane. This will ensure that the amount of charge within the cell does not change with time, but it will not keep the concentration of each species within the cell fixed with time. This less stringent requirement becomes

$$J'_{\text{Na}}+J'_{\text{K}}+J'_{\text{Cl}}=0.$$

Adding Eqs. (9.55) together and factoring out $N_A e^2 v/(1-e^{-ev/k_BT})$ gives

$$(\omega_{\text{Na}}[\text{Na}_1]+\omega_{\text{K}}[\text{K}_1]+\omega_{\text{Cl}}[\text{Cl}_2])e^{-ev/k_BT}$$
$$=\omega_{\text{Na}}[\text{Na}_2]+\omega_{\text{K}}[\text{K}_2]+\omega_{\text{Cl}}[\text{Cl}_1],$$

or

$$v=\frac{k_BT}{e}\ln\left(\frac{\omega_{\text{Na}}[\text{Na}_1]+\omega_{\text{K}}[\text{K}_1]+\omega_{\text{Cl}}[\text{Cl}_2]}{\omega_{\text{Na}}[\text{Na}_2]+\omega_{\text{K}}[\text{K}_2]+\omega_{\text{Cl}}[\text{Cl}_1]}\right). \qquad (9.56)$$

As an example of the use of the GHK equation, consider the behavior when the concentration of some external ion is changed. We will use the concentrations of Fig. 6.2, except for the ion whose concentration is being changed. The particle concentrations are in mmol l^{-1} (any units can be used since ratios are taken):

$$[\text{Na}_1]=145,$$
$$[\text{K}_1]=5,$$
$$[\text{Cl}_1]=[\text{Na}_1]+[\text{K}_1]-25,$$
$$[\text{Na}_2]=15,$$
$$[\text{K}_2]=150,$$
$$[\text{Cl}_2]=[\text{Na}_2]+[\text{K}_2]-156.$$

The permeabilities are not known. However, only the ratio to ω_{K} matters. If we take the ratio $\omega_{\text{K}}:\omega_{\text{Na}}:\omega_{\text{Cl}}$ to be 1.0:0.04:0.45 and use $T=300$ K, then Eq. (9.56) is (in mV)

$$v=25.88\ \ln\left(\frac{[\text{K}_1]+0.04[\text{Na}_1]+0.45([\text{Na}_2]+[\text{K}_2]-156)}{[\text{K}_2]+0.04[\text{Na}_2]+0.45([\text{Na}_1]+[\text{K}_1]-25)}\right).$$

This has been plotted in Fig. 9.11. for the case that $[\text{K}_1]$ is varied and for the case that $[\text{Na}_1]$ is varied. In each case Cl ions are also added to the external solution in an equal amount. There is a region of potassium concentration over which the behavior is nearly exponential, and one could be misled into thinking that the potential–concentration relation was given either by the Nernst equation alone or by Donnan equilibrium. The potential change with sodium concentration is much less because of the low permeability of the membrane to sodium.

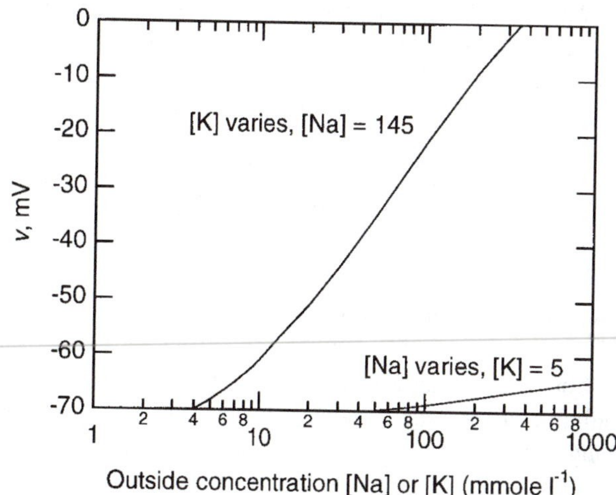

FIGURE 9.11. The potential difference across a cell membrane as a function of changes in the exterior concentration of KCl or NaCl, calculated using the Goldman equation.

The assumption that the total current through the membrane is zero guarantees that there will be no charge buildup inside the cell; however, the individual currents are not zero, so there may be concentration changes with time. We will next investigate the magnitude of this effect. Equation (9.54) can be converted to particle flux instead of charge flux by dividing by ze. The result for ion s is

$$J_s=zev\,\omega_s\,\frac{C_1e^{-zev/k_BT}-C_2}{1-e^{-zev/k_BT}}.$$

The concentrations are converted from mmol l^{-1} to particles/m^3 by multiplying by Avogadro's number. (The factors of 10 in the conversion happen to cancel out.) Consider the previous example at $T=300$ K, $[\text{K}_1]=5$, $[\text{Na}_1]=145$, and $v=-68.17$ mV. The exponential factor for the positive ions is $e^{-ev/k_BT}=13.929$, while for the chloride ion it is the reciprocal, 0.0718. If we write $\omega_{\text{Na}}=0.04\omega_{\text{K}}$ and $\omega_{\text{Cl}}=0.45\omega_{\text{K}}$, then the fluxes for the three ions are

$$J_{\text{K}}=+(6.55\times10^3)\omega_{\text{K}}(6.215),$$
$$J_{\text{Na}}=-(6.55\times10^3)\omega_{\text{K}}(6.202),$$
$$J_{\text{Cl}}=-(6.55\times10^3)\omega_{\text{K}}(0.013),$$

and the total current is zero.

Although the GHK equations are widely used because of their simplicity, some cautions are in order. Their derivation assumed independence of the moving ions. We know that this is an oversimplification for several reasons. Experiments show that the currents saturate for high concentrations. The distortion of the electric field by other ions was ignored. The permeability (diffusion constant) was assumed to be constant. The pore was assumed to have a constant cross section and constant electric field. A somewhat less restrictive model for the reversal potential can be derived for a pair of ions with the same valence if we assume that any variations in $D(x)$ for the two ions are similar (Problem 9.12). With that assumption the reversal potential is

$$v_{\text{rev}} = \frac{k_B T}{ze} \ln\left(\frac{\omega_a C_{a1} + \omega_b C_{b1}}{\omega_a C_{a2} + \omega_b C_{b2}}\right). \tag{9.57}$$

9.7. MEMBRANE CHANNELS

In Chap. 6 we described some of the properties of the sodium and potassium channels in a squid axon. There are many other kinds of channels. Variations exist not only from one organism to another, but in different kinds of cells in the same organism. The classic monograph on ionic channels is the book by Hille (1992).

There are several different kinds of potassium channels. Most open after depolarization; a few open after hyperpolarization. Potassium channels in axons (like the ones we encountered in Chap. 6) are called *delayed rectifiers* because of their delay in opening after a voltage step. The properties of sodium channels are more uniform from one cell type to another. Calcium channels pass much smaller currents than sodium or potassium channels because calcium concentrations are much smaller; the calcium current density is usually about $\frac{1}{10}$ the current density for sodium or potassium. Calcium channels usually activate with depolarization. Since the concentration of calcium inside cells is usually very small, the interior calcium concentration can increase 20-fold in response to depolarization. This increase in concentration often initiates a chemical reaction, for example, to cause contraction of a muscle cell. Chloride channels often have a large conductivity. The chloride concentration ratio in some muscle is such that the resting potential is close to the chloride Nernst potential. As a result, small changes in the potential cause relatively large chloride currents, which tend to stabilize the resting potential.

The earliest voltage-clamp measurements were difficult to sort out. Hodgkin and Huxley changed the concentration of extracellular sodium, substituting impermeant choline ions, to determine what part of the current was due to sodium and what was due to potassium. Figure 9.12(a) shows typical currents. In the mid-1960s various drugs were found that at very small concentrations selectively block conduction of a particular ion species. We now know that these drugs bind to the channels that conduct the ions. The next big advance was patch-clamp recording in 1976. Micropipettes were sealed against a cell membrane that had been cleaned of connective tissue by treatment with enzymes. A very-high-resistance seal resulted [$(2-3) \times 10^7\ \Omega$] that allowed one to see the opening and closing of individual channels. For this work Erwin Neher and Bert Salmon received the Nobel Prize for physiology in 1992. Around 1980 Neher's group found a way to make even

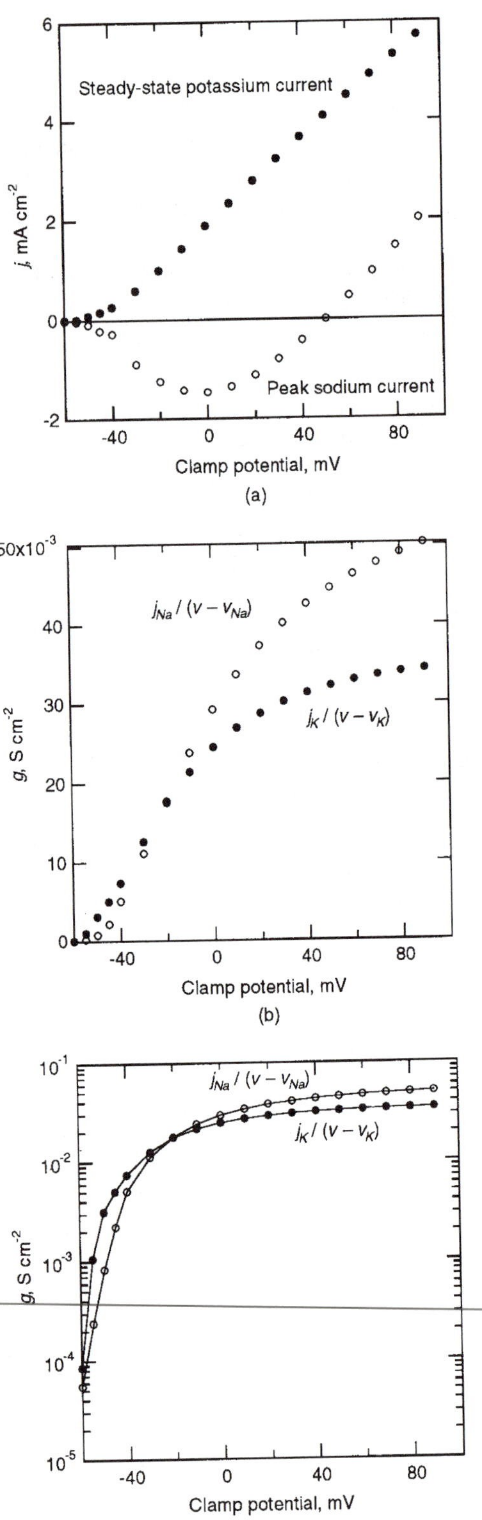

FIGURE 9.12. Steady-state potassium current and peak sodium current for a squid axon subject to a voltage clamp vs the transmembrane potential during the clamp. These are not real data, but were generated using the Hodgkin–Huxley model. (a) Current density. (b) Current density divided by the difference between the potential and the Nernst potential, to give the conductance per unit area [see Eq. (6.60)]. (c) The same data as in (b) plotted on semi-log paper.

higher-resistance (10^{10}–10^{11} Ω) seals that reduced the noise even further and allowed patches of membrane to be torn from the cell while adhering to the pipette [Hamill, *et al.* (1981)]. The relationship of noise to resistance will be discussed below.

It was found that the pores open and close randomly, as shown in Fig. 9.13. Thus, the Hodgkin–Huxley model describes the average behavior of many pores, not the kinetics of single pores. Note how the current through an open pore changes as a function of the applied potential. These direct patch-clamp measurements show that a single open pore can pass at least 1 pA of current or 6×10^6 monovalent ions per second. Most can pass much more. While no perfectly selective channel is known, most channels are quite selective: for example, some potassium channels show a 100:1 preference for potassium over sodium.

During the last decade, gene splicing combined with patch-clamp recording has provided a wealth of new information. The portion of the DNA responsible for synthesizing the membrane channel has been identified. One example that has been extensively studied is a potassium channel from the fruit fly, *Drosophila melanogaster*. The *Shaker* fruit fly mutant shakes its legs under anesthesia. It was possible to identify exactly the portion of the fly's DNA responsible for the mutation. It was then possible to place *Shaker* DNA in other cells that do not normally have potassium channels. They immediately made functioning channels.

The current view is that the *Shaker* potassium channel consists of four identical subunits, each containing six α helices that span the membrane. The pore presumably runs along the four fold-symmetry axis, as shown in end view in Fig. 9.14. Sodium and calcium channels are very similar, but the four subunits are not identical. Recent reviews of voltage-gated channels have been prepared by Sigworth (1993) and by Keynes (1994).

Let us now explore some of the physics of channels. Combining the macroscopic current density with the current in a single channel shows that there are not many channels per unit area of the membrane (see Problems 9.13 and 9.14). It is illuminating to consider what effect currents of this magnitude and duration have on the transmembrane potential. The capacitance per unit area of biological membranes is about 0.01 F m^{-2} (1 μF cm^{-2}). A channel conducting 1 pA for 1 ms allows 10^{-15} C to pass. This is enough charge to change the potential 100 mV on an area of 10^{-12} m^2 or 1 μm^2. This charge transfer corresponds to about 6000 monovalent ions per μm^2.

Figure 9.12(a) shows the steady-state potassium and peak sodium current densities for a squid axon. The ion concentrations are known, and we saw in Chap. 6 that the Nernst potentials at 6.3 °C were +50 mV for sodium and −77 mV for potassium. Figure 9.12(b) shows the conductance per unit area, obtained by dividing the current by $v - v_{\mathrm{Nernst}}$. Figure 9.12(c) shows a semilogarithmic plot of the conductance per unit area.

The sodium current density changes sign at the sodium Nernst potential. While a measured zero crossing is an accurate way to determine the Nernst potential, extrapolation to find the zero-crossing can be quite misleading. The potassium current density appears to be linear over a large region, and it is tempting to extrapolate to find v_K. The extrapolation shows zero current at about −40 mV, which is far from v_K. The reason can be seen in Fig. 9.12(b), which shows that g_K is varying considerably over the region where j_K appears to be linear; this distorts the slope and changes the extrapolated intersection.

A simple two-state model can explain the shape of the curves in Fig. 9.12(c). The conductance per unit area of a membrane is the product of the conductance of an open pore and the average number of pores per unit area that are open. The model assumes that each channel has a gate that

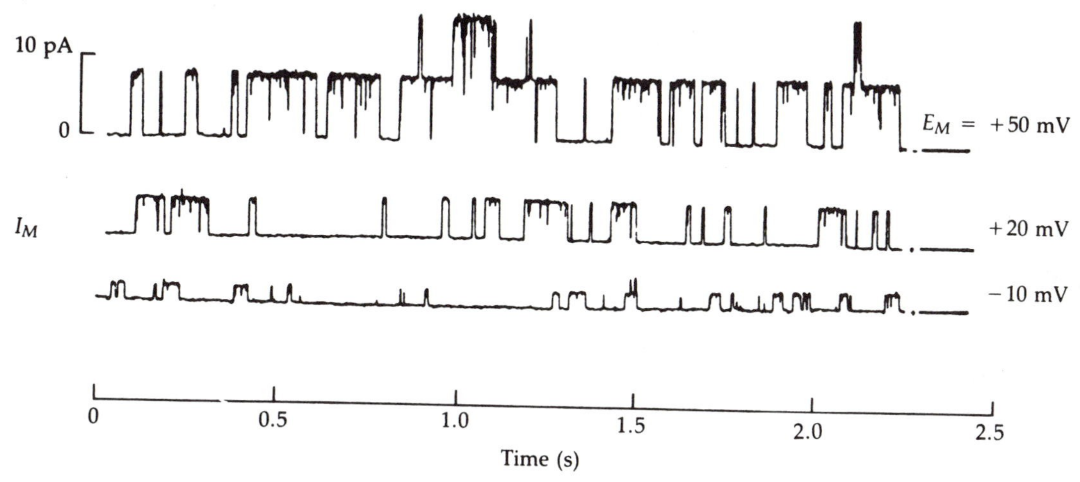

FIGURE 9.13. Opening of single K(Ca) channels. From B. S. Pallotta, K. L. Magleby, and J. N. Barrett (1981). Single channel recordings of Ca^{2+}-activated currents in rat muscle cell culture. *Nature* **293**: 471–474. Reprinted with permission from *Nature* (London).

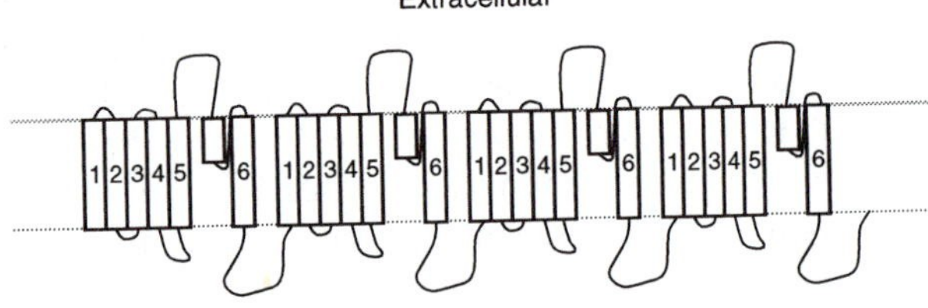

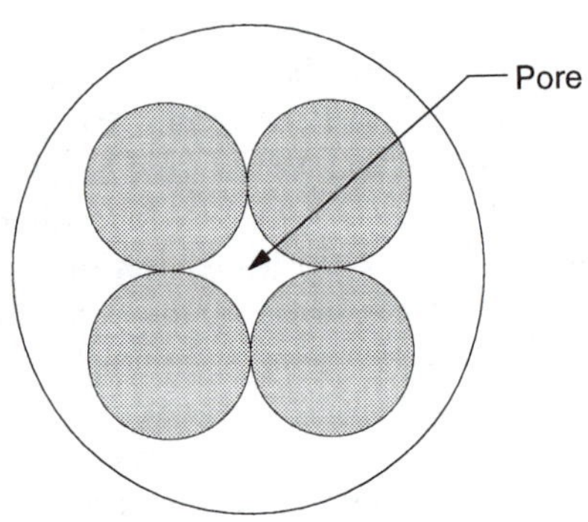

FIGURE 9.14. Current understanding of the structure of a *Shaker* potassium channel. There are four identical subunits, each consisting of six α-helices that traverse the membrane. At the bottom is a schematic end view, showing how the four subunits create a pore at their center.

is either open or closed. When the gate is open, the channel has a conductance determined by the passive properties of the rest of the channel. The rapid increase of conductance between -60 and -30 mV corresponds to a rapidly increasing probability that the gate is open.

Suppose that each channel has a gate with two states: open and closed. When there is no average electric field in the membrane ($v=0$) the energy of the open state is $w=u_o k_B T$ greater than the closed state. Suppose also that as the gate opens and closes, a charge q associated with the gate moves a small distance parallel to the axis of the pore. When there is a potential v across the membrane, the charge moves through a potential difference αv, where $\alpha<1$. The total energy change when the gate opens with potential v across the membrane is then $w+\alpha q v$. The quantity αq is often written as ze and called the *equivalent gating charge.* In terms of $k_B T$ the energy change when the pore opens is $u=u_o+zev/k_B T$.

Let p_o be the probability that a pore is open and p_c be the probability that it is closed. The probabilities are related by a Boltzmann factor: $p_o=p_c e^{-u}$. Since $p_o+p_c=1$, $p_o=e^{-u}/(1+e^{-u})=1/(1+e^u)$,

$$p_o=\frac{1}{1+e^{u_o+zev/k_B T}}. \tag{9.58}$$

For very large values of u (small values of p_o),

$$p_o \approx e^{-(u_o+zev/k_B T)}. \tag{9.59}$$

The conductance per unit area of the membrane is the conductance of an open pore times by the number of pores per unit area (that is g_∞), multiplied by p_o. Figure 9.15 shows plots of the "data" and lines generated from Eq. (9.58). The fit parameters have been adjusted to provide good fits at the lowest conductances. For sodium $u_0=-10.5$ and $z=-7$; for potassium $u_0=-19$ and $z=-10$. The fact that u_0 is very negative means that when $v=0$ the energy of an open gate is much less than the energy of a closed gate. Nearly all of the pores are open, as can be seen from the $v=0$ point in Fig. 9.15. The fact that $z=-7$ or -10 means that when the pore opens or closes the equivalent of 7 (or 10) electron charges must move through the full transmembrane potential difference. Many more charges could be displaced a much smaller distance and experience a much smaller potential change. More sophisticated multilevel models are discussed by Sigworth (1993).

This charge movement constitutes a very small current called the *gating current.* It is different from the current to charge the membrane capacitance. We saw above that during a 1-pA pulse lasting 1 ms, about 6000 monovalent ions flow through the membrane. The gating charge is about 10 monovalent charges, a ratio of about 600. The gating current is so small that it has not yet been measured in a single channel, but it can be measured by manipulating the ions bathing the membrane in a patch-clamp experiment.

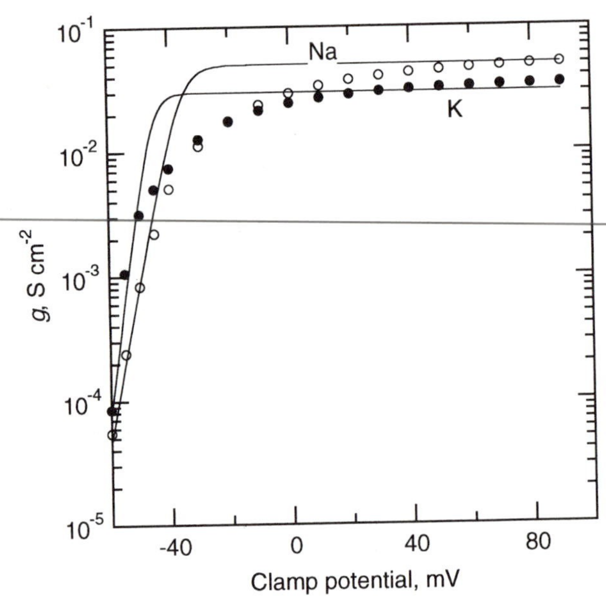

FIGURE 9.15. Plot of sodium and potassium conductivities from Fig. 9.12(c) with fits by Eq. (9.58). For sodium $u_0=-10.5$ and $z=-7$; for potassium $u_0=-19$ and $z=-10$.

Figure 9.16 shows the results of a set of experiments with *Shaker* potassium channels. Panel A shows the macroscopic depolarizations to +20 and +80 mV for a patch with about 400 channels. The peak current at +80 mV is 1.25 pA per channel. Panel B shows the gating current recorded from another patch containing about 8000 channels. Potassium was removed from the solution bathing the interior surface of the membrane. The gating current lasts slightly less than 1 ms and peaks at about 4.5×10^{-15} A, about 300 times less than the channel current. The agreement with our first estimate of 600 times less is satisfactory, given the accuracy of the data. Panels C and D show recordings similar to panel A, but with many fewer channels in the patch. The results from three successive depolarizing pulses are shown in each case. The channel openings are similar to those in Fig. 9.13, but are recorded at a much shorter time scale. The increased current through an open channel and the higher probability of being open for a clamp of +80 mV are both apparent.

A very simple approximate calculation shows that there is not much ion–ion interaction in a channel. A current of 1 pA is 6.25×10^{6} monovalent ions per second, so that the average time between the passage of successive ions through the channel is 1.6×10^{-7} s. Now we need the transit time. While no perfectly selective channel is known, most channels are quite selective: a 100:1 preference for potassium over sodium in some potassium channels. In a uniform electric field giving 80 mV across the membrane, a monovalent ion would have a drift velocity of 0.6 m s^{-1} based on the bulk diffusion constant. [See the discussion surrounding Eq. (4.22).] Because ions in the pore are confined, let us use $\frac{1}{10}$ of this, or 0.06 m s^{-1}. (The diffusion constant is proportional to the solute permeability; see Sec. 5.11. Ignoring electric forces, we see from Fig. 5.19 that $\omega/\omega_0=0.1$ corresponds to $a/R_p=0.4$. So this is probably still a high drift velocity.) Then the time it takes the ion to pass through the channel is its length (assume 6 nm) divided by the average speed, or 10^{-7} s. The fraction of the time there is an ion in the channel is 0.625.

We can make some other estimates of channel parameters. Over some part of its length, the channel must be narrow enough so the wall can interact directly with the ion that is passing through, not shielded by water molecules. The pore must therefore narrow to a radius of 0.3 to 0.7 nm in some region. Let us assume a cylindrical pore of radius 0.7 nm and length 6 nm. The average number of water molecules in the channel is 308; the concentration of ions is 113 mmol l^{-1}, which is certainly in the right region. The resistance of a channel while it is open is $R=v/i$ $=80$ mV/1 pA$=8\times10^{7}$ Ω. (We should actually use $v-v_{\text{Nernst}}$, but this is a rough estimate. If we were going to be more accurate, we should also use the Nernst–Planck equation, recognizing that the ions move by diffusion as well as drift.)

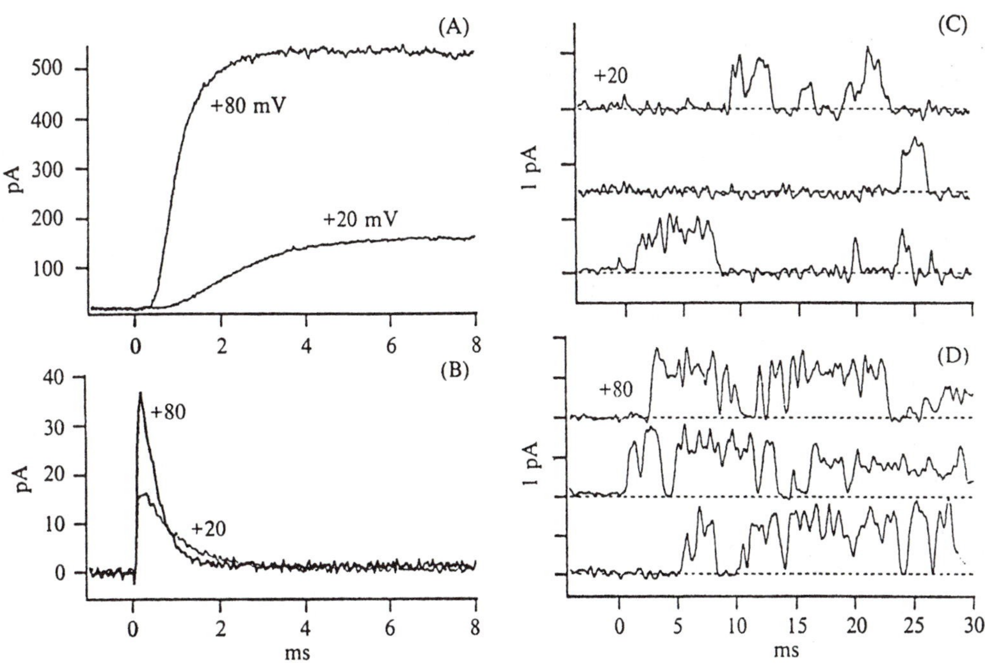

FIGURE 9.16. The results of a set of experiments with *Shaker* potassium channels. Panel A shows the macroscopic depolarizations to +20 and +80 mV for a patch with about 400 channels. Panel B shows the gating current recorded from another patch containing about 8000 channels. Potassium was removed from the solution bathing the interior surface of the membrane. Panels C and D show recordings similar to panel A, but with many fewer channels in the patch. The results from three successive depolarizing pulses are shown in each case. From F. J. Sigworth (1993). Voltage gating of ion channels. *Quart. Rev. Biophys.* **27**: 1–40. Reprinted with permission of Cambridge University Press.

Finally, let us examine diffusion effects at the mouth of the pore. We have assumed that the solution on either side of the membrane is well stirred, so that the concentration is C_0 or C_0' everywhere. In fact there is no stirring, and as the current through a pore increases, the concentration at the membrane surface falls on the side where ions enter the membrane and rises on the side where ions leave. The rate of diffusion to and from the pore limits the current, and the current saturates as the voltage is increased. Also, if the concentrations are high enough, the ions no longer move independently and the current for a fixed voltage is not proportional to the concentration. Simple examples of these two effects are discussed below.

We explore voltage saturation for a uniform pore of radius a and length L using the constant-field model. The concentration in the solution far from the membrane is C_0 on the left and C_0' on the right. If the ions flow from left to right, a gradient forms near the mouth of the pore as ions diffuse to the pore mouth from the bulk solution. Therefore the concentration at the pore mouth is $C_0 - \delta C$. (We assume that there are other impermeant ions that keep the bulk solution electrically neutral.) Similarly, ions traveling through the pore build up the concentration at the exit until the diffusion current away from the exit equals the current through the pore. The concentration at the exit is then $C_0' + \delta C'$.

In Sec. 6.19 we saw that the particle diffusion current between a disk of radius a and infinity is $4D_0 a\,\delta C$. (The subscript denotes the diffusion constant for bulk medium.) The electric current to the pore entrance is therefore $4D_0 aze\,\delta C$, while the current diffusing away from the exit is $4D_0 aze\,\delta C'$. In the steady state these are equal, so $\delta C' = \delta C$.

The current in the pore is πa^2 times the current density. From Eq. (9.46)

$$i = -\frac{\pi a^2 zeDu}{L}\,\frac{(C_0 - \delta C)e^{-u} - (C_0' + \delta C)}{e^{-u} - 1}. \quad (9.60)$$

This can be rearranged as

$$i = -\frac{\pi a^2 zeDu}{L}\,\frac{C_0 e^{-u} - C_0' - \delta C(e^{-u} + 1)}{e^{-u} - 1}$$

and further rearranged to give

$$i = -\frac{\pi a^2 zeDu}{L}\left[\frac{C_0 e^{-u} - C_0'}{e^{-u} - 1} + \delta C\,\frac{1 + e^{-u}}{1 - e^{-u}}\right].$$

The algebra is simpler with the temporary substitutions $\alpha = C_0'/C_0$, $f = (1 + e^{-u})/(1 - e^{-u})$, and $g = (e^{-u} - \alpha)/(e^{-u} - 1)$. The current through the pore equals the diffusion current into the pore in the steady state:

$$i = -(\pi a^2 zeDu/L)(C_0 g + \delta C f) = 4D_0 aze\,\delta C. \quad (9.61)$$

This can be solved for δC:

$$\delta C = -\frac{C_0 g}{f + 4D_0 L/\pi Dau}.$$

Inserting this in Eq. (9.61) we obtain our final expression for the current:

$$i = -\frac{\pi a^2 zeDu}{L}\,\frac{C_0 e^{-u} - C_0'}{e^{-u} - 1}\,\frac{1}{1 + \dfrac{\pi Dau(1 + e^{-u})}{4D_0 L(1 - e^{-u})}}. \quad (9.62)$$

This is the original constant-field expression for the current multiplied by a correction factor. The correction factor depends on the dimensionless ratios D/D_0 and a/L, as well as the potential through $u = zev/k_B T$. It does not depend on the concentration.

This derivation assumed that the concentration just inside the pore is equal to the concentration just outside the pore mouth. One could also assume a partition fraction β, so that $C(0) = \beta C_0$ and $C(L) = \beta C_0'$. The effect would be the same as if the diffusion constant in the pore were changed to βD.

The only concentration dependence in Eq. (9.62) is the factor $C_0 e^{-u} - C_0'$, which depends linearly on concentration. The current for a fixed voltage saturates as the concentration is increased. The conductance can often be described by an equation of the form

$$G = \frac{\gamma C}{1 + C/K}, \quad (9.63)$$

which rises linearly as γC for small concentrations and saturates at γK for large concentrations. Values of K are typically in the range 40–200 mM. Saturation means that there is a maximum rate at which ions can pass through the pore. It has been explained by a model in which ions must temporarily bind to a certain site to move through the pore [reviewed in Hille (1992)] and also in terms of the Nernst–Planck model if only one ion can occupy the selective region (Levitt, 1986).

9.8. NOISE

The current fluctuates while a channel is open, as can be seen in Figs. 9.13 and 9.16. Some of the fluctuation is due to noise in the measurement apparatus. However, there are some fundamental physical lower limits to the fluctuations resulting from noise in the membrane patch itself. We discuss these briefly here, with a more extensive discussion in Chap. 11. DeFelice (1981) wrote an excellent book on noise in membranes.

The first (and smallest) limitation is called *shot* noise. It is due to the fact that the charge is transported by ions that move randomly and independently through the channels. Imagine a single open conducting channel with an average current $\bar{i}$ of monovalent ions. During time Δt (which can be any interval shorter than the time the channel is open) the average charge flow is $\bar{i}\,\Delta t$ and the average number of ions is $\bar{n}=\bar{i}\,\Delta t/e$. Since there are a very large number of ions that might flow through the channel (occurrences) and the probability that any one ion moves through the channel during Δt is very small, we have the Poisson limit of the binomial distribution. The variance in the number of ions is $\sigma_n^2=\bar{n}=\bar{i}\,\Delta t/e$. Since the average charge transported is $\bar{q}=\bar{n}e$, the variance in the charge is $\sigma_q^2=e^2\sigma_n^2=e\bar{i}\,\Delta t$. When many samples of length Δt are measured, the variance in the current is $\sigma_i^2=\sigma_q^2/(\Delta t)^2$. The standard deviations are

$$\sigma_n=\left(\frac{\bar{i}\,\Delta t}{e}\right)^{1/2},$$

$$\sigma_q=(e\bar{i}\,\Delta t)^{1/2},$$

$$\sigma_i=\left(\frac{e\bar{i}}{\Delta t}\right)^{1/2}, \tag{9.64}$$

and the fractional standard deviations are

$$\frac{\sigma_n}{\bar{n}}=\left(\frac{e}{\bar{i}\,\Delta t}\right)^{1/2},$$

$$\frac{\sigma_q}{\bar{q}}=\left(\frac{e}{\bar{i}\,\Delta t}\right)^{1/2},$$

$$\frac{\sigma_i}{\bar{i}}=\left(\frac{e}{\bar{i}\,\Delta t}\right)^{1/2}. \tag{9.65}$$

For a current of 1 pA, the fractional standard deviation is 0.013 when the sampling time is 1 ms and 0.04 when the sampling time is 0.1 ms. These are much smaller than what is observed in the figures.

The next source of noise is called *Johnson noise*. It arises from thermal fluctuations or Brownian movement of the ions. It can be derived from a microscopic model of conduction (either in an ionic solution or a metal), but we will do it using the equipartition of energy.

First, we need an expression for the energy contained in a charged capacitor. To obtain it, imagine that an amount of charge $+dq$ is transferred from the negative to the positive conductor. This increases the amount of positive charge on the positive conductor and also increases the amount of negative charge on the negative conductor. The work required to transfer the charge when the potential difference between the conductors is v is $v\,dq$. The energy stored in the capacitor is the total work required to charge the conductor from 0 to q. Remembering that $q=Cv$, we have

$$U=\int_0^q v\,dq=\frac{1}{C}\int_0^q q\,dq=\frac{q^2}{2C}=\frac{Cv^2}{2}. \tag{9.66}$$

If the capacitor is completely isolated, there can be a constant charge on each conductor with no fluctuations. If the capacitor is in thermal contact with its surroundings and is in equilibrium, then the equipartition theorem applies. The capacitor can be brought into thermal equilibrium with its surroundings by connecting a resistance R between the conductors. This will discharge the capacitor so $\bar{q}=\bar{v}=0$. There will be fluctuations around these zero values. Because the expression for the energy depends on the square of the variables, the mean square value is given by the equipartition of energy theorem, Eq. (3.54). We will assume that when the capacitor is charged, thermal fluctuations give the same variances:

$$\sigma_v^2=(\overline{v^2}-\bar{v}^2)=\overline{v^2}=k_BT/C, \tag{9.67a}$$

$$\sigma_q^2=(\overline{q^2}-\bar{q}^2)=\overline{q^2}=Ck_BT. \tag{9.67b}$$

In a simple RC circuit, $i=v/R$, so

$$\sigma_i^2=\sigma_v^2/R^2=k_BT/R^2C. \tag{9.67c}$$

Since changes in current or voltage in an RC circuit occur with time constant $\tau=RC$, we can also write these as

$$\sigma_v^2=Rk_BT/\tau, \quad \sigma_i^2=k_BT/R\tau. \tag{9.68}$$

These are special cases of a more general relationship that will be discussed in Chap. 11.

We can use these to determine some of the requirements for patch-clamp recording. In order to see the current from a single channel with some accuracy, let us require that the standard deviation of the current fluctuation be less than $\frac{1}{8}$ of the signal we want to measure. (This signal-to-noise ratio, SNR, of 8 is arbitrary.). First consider the limitation due to Johnson noise. We want $\sigma_i<\bar{i}/8$ or $\sigma_i^2<\bar{i}^2/64$. From this we obtain

$$R>\left(\frac{k_BT}{C}\right)^{1/2}\frac{8}{\bar{i}}. \tag{9.69}$$

The capacitance of a patch of membrane of 1 μm radius is 3.1×10^{-14} F. At a temperature of 300 K and for an average current of 1 pA, this gives $R>3\times10^9$ Ω. Larger values of R will give an even higher SNR. There are several sources of thermal noise in a recording electrode, all discussed in the paper by Hamill *et al.* (1981). These are order-of-magnitude results; one must determine carefully which capacitances and resistances provide the dominant effects.

We can also see when shot noise is important. The ratio of Johnson noise to shot noise is

$$\frac{\sigma_i^2(\text{Johnson})}{\sigma_i^2(\text{shot})}=\frac{k_BT/R\tau}{e\bar{i}/\Delta t}=\frac{k_BT}{Re\bar{i}}. \tag{9.70}$$

This ratio is less than 1 and shot noise is important when $R>k_BT/e\bar{i}=2.6\times10^{10}$ Ω. Shot noise has been detected in channel gating currents and subjected to very sophisticated analysis. See the paper by Crouzy and Sigworth (1993) and the references therein.

9.9. SENSORY TRANSDUCERS

Animals have very sensitive senses. We saw (Problem 3.41) that the ear can hear sounds at 1000 Hz that are just greater than the pressure fluctuations due to molecular collisions on the ear drum. An eye that is adopted to the dark can detect flashes of light corresponding to a few photons (Chap. 13). Many animals can smell chemicals when only a few molecules strike their sensory organs. The electric skate can detect extremely small electric fields. In each case there is a transducer that converts the sensory stimulus into a series of nerve impulses. The transducer must have sufficient sensitivity, and it must also absorb an amount of energy greater than that from random thermal bombardment (Brownian movement).[6] We describe here two transducers: the mechanoreceptors (hair cells) of the inner ear and the electric organ of the skate.

Various transduction mechanisms are reviewed in Chap. 8 of Hille (1992). The mechanoreceptors of the bullfrog inner ear have been studied for many years. The hair-cell current rises from 0 to 100 pA with an 0.5-μm displacement. Each hair cell is cylindrical. On its end face are found about 60 very small *stereocilia*, each 1–50 μm long and with a 100–500-nm radius. The tips of these stereocilia are linked by thin filaments. The hair cell and stereocilia that detect sound in the ear are attached to the basilar membrane in the cochlea of the ear and move in a very viscous fluid as the basilar membrane vibrates. Hair cells detecting accelerations of the entire animal are attached to a suspended dense body. It is believed that as the stereocilia move, these filaments pull open flaps at the end of ion channels, allowing ions to enter the cell and initiate the conduction process. This is shown schematically in Fig. 9.17. Denk and Webb (1989) have used a laser interferometer to measure the motion of the hair cells. They found that the spontaneous motion consists primarily of thermal excitation (Brownian motion). Fluctuations in the intracellular voltage were also measured. They often correlated with the motion of the hair cells.

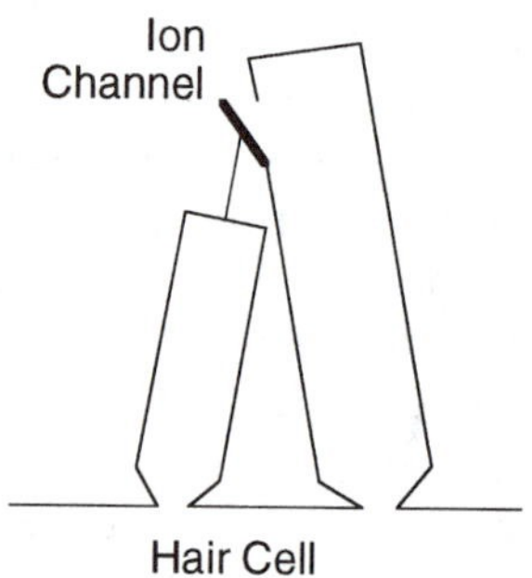

FIGURE 9.17. A schematic diagram of two stereocilia linked by a filament that opens a channel as the cillia move back and forth.

Freshwater catfish respond to electric fields as low as 10^{-4} V m^{-1} and saltwater sharks and rays can detect fields of 5×10^{-7} V m^{-1}. A brief review has been given by Bastian (1994); Kalmijn (1988) provides a very complete review. The saltwater fish have a more complicated sensory apparatus than the freshwater fish, known as the *ampullae of Lorenzini.* Kalmijn and his colleagues discovered that the ocean flounder generates a current dipole of 3×10^{-7} A m. In sea water of resistivity 0.23 Ω m this gives an electric field of 2×10^{-5} V m^{-1} at a distance 10 cm in front of the flounder. They were able to show in a beautiful series of behavioral experiments that dogfish (a small shark) could detect the electric field 0.4 m from a current dipole of 4×10^{-7} A m, corresponding to an electric field of 5×10^{-7} V m^{-1}. The fish would bite at the electrodes, ignoring a nearby odor source. A field of 10^{-4} V m^{-1} would elicit the startle response. A field $\frac{1}{10}$ as large caused a physiologic response. The animals responded to a constant field or a sinusoidally alternating fields up to 4 Hz. At 8 Hz the threshold increased by a factor of 2.

In a recent series of experiments Lu and Fishman (1994) have dissected out the ampulla of Lorenzini and measured its response in the laboratory. They found that the resting rate of firing of the organ is about 35 Hz (impulses per second) and that an applied electric potential increases or decreases the firing rate by about 1 Hz μV^{-1}, depending on its sign. The firing rate saturated for potential differences of 100 μV.

The behavioral experiments showed sensitivity to an electric field of 0.5 μV m−1. The anatomy of the ampulla is such that the organ senses the potential difference between the surface of the fish and deep in its interior. Pickard (1988) shows that for a spherical fish of radius a, this gives a potential of $3aE/2$, where E is the external electric field. The amplitude of the potential difference oscillation of a fish of length $\frac{1}{3}$ m is therefore 0.25 μV. This is enough to cause a 1% change in firing rate, which could be detected by the following neuronal circuits.

The Johnson noise is somewhat smaller than the signal detected. To estimate it, use Eq. (9.67a) with the ampullary capacitance of 0.15 μF measured by Lu and Fishman. The standard deviation of the noise is 0.17 μV.

9.10. POSSIBLE EFFECTS OF WEAK EXTERNAL ELECTRIC AND MAGNETIC FIELDS

There is a controversy over whether radio-frequency and power-line-frequency electric or magnetic fields can cause cancer. While the effect, if any, is quite small, the literature is vast, involving both epidemiological and laboratory studies. Results are conflicting, and the mechanisms by which such an effect might occur are not yet understood,

[6]For the detection of light, the amount of energy per photon is so much greater than k_BT that shot noise dominates.

though models have been proposed, some of which are inconsistent with basic physical principles such as the Boltzmann factor, the mean free path of ions, and thermal fluctuations. It is beyond our scope to do more than discuss some very basic underlying physics and provide pointers to the field.

We have seen that electric charges give rise to electric fields and moving electric charges (currents) generate magnetic fields. The electric field lines start and end on charges, and the magnetic field lines surround the current generating the field. We will see in Chap. 12 that accelerated charges generate electromagnetic *radiation*, in which the electric and magnetic fields are interrelated and the field lines close on themselves far from the source charges. Energy is radiated: it leaves the source and never returns. This radiated energy is in the form of discrete packets or *photons*, whose energy is related to the frequency of oscillation of the fields (which is determined by the frequency of oscillation of the charges that radiated the field). The energy of each photon is $E=h\nu$, where h is Planck's constant and ν the frequency of oscillation. At room temperature, we have seen that the important energy size for random thermal motion is $k_BT=4\times10^{-21}$ J. At 60 Hz, the energy in each photon is much smaller: 4×10^{-32} J. At 100 MHz it is 7×10^{-26} J. For electromagnetic radiation in the ultraviolet and beyond, which certainly can harm cells, the photon energy is 5×10^{-19} J or greater, quite large compared to k_BT. At the very low frequencies we are considering, it is the strength of the electric or magnetic field that is important, not the energy of individual photons. A more detailed discussion of the distinction between these low-frequency "near fields" and radiation fields is found in Polk (1996).

While the possible induction or promotion of cancer has been the reason for recent concern, there are many possible biological effects of electric or magnetic fields besides causing cancer. We have already seen electrical burns, cardiac pacing, and nerve and muscle stimulation by electric or rapidly changing magnetic fields. Even stronger electric fields increase membrane permeability. This is believed to be due to the transient formation of pores (*electroporation*). Pores can be formed, for example, by microsecond-length pulses with a field strength of about 10^8 V m^{-1} [Freeman *et al.* (1994); Weaver (1994)]. Microwaves are used to heat tissue. Nerve stimulation requires a few millivolts across the cell membrane, or about 10^5-10^6 V m^{-1}. Both electric and magnetic fields are used to promote bone healing, with field strengths in tissue in the fracture region of 10^{-1} V m^{-1} [Tenforde (1995)].

Much weaker fields are produced by power lines. Barnes (1995) reviews power-line fields and finds in a typical home average electric fields in air next to the body of about 7 V m^{-1} with peak values of 200 V m^{-1}. We will find below that the fields within the body are much less. Average magnetic fields are about 0.1 μT, with peaks up to 4 times as large. Within the body they are about the same. Tenforde reviews both power-line- and radio-frequency field intensities.

Epidemiological studies have been very valuable in tracing the cause of infectious outbreaks. They have also indicated that smoking increases the probability of developing lung cancer by 3000%—a factor of 30. There are difficulties with epidemiological studies when the effect is small. These range from problems of statistical fluctuations when the number of patients is small to the fact that associations do not prove causality and there may be unsuspected variables that are confusing the picture. The problem is exacerbated when positive findings receive widespread publicity and negative findings are ignored by the press. Though there is still some debate, a recent review by Moulder and Foster[7] (1995) finds that the association between power-frequency fields and cancer is weak. Two other recent articles by Carstensen (1995) and by Bren (1995) reach similar conclusions. One recent review by two epidemiologists [Poole and Ozonoff (1996)] argues that the risk of childhood cancer shows an increase with dose. However, a review by another epidemiologist [Heath (1996)] concludes, "The weakness and inconsistent nature of epidemiologic data, combined with the continued dearth of coherent and reproducible findings from experimental laboratory research, leave one uncertain and rather doubtful that any real biological link exists between EMF [electromagnetic field] exposure and carcinogenicity." A recent report by a committee of the National Academy of Sciences concludes that "the current body of evidence does not show that exposure to these fields presents a human-health hazard."..."The committee reviewed residential exposure levels to electric and magnetic fields, evaluated the available epidemiological studies, and examined laboratory investigations that used cells, isolated tissues, and animals."

There are now a large number of laboratory studies. They, too, are ambiguous. In some cases when positive effects have not been replicated it appears to be due to better technique: more careful controls or a "blind" study in which the experimenter recording the effect truly does not know whether the field is on or off.[8] In other cases, those whose results were not replicated feel that the attempt

[7]Moulder and Foster were also guest editors for the July/August 1996 issue of *IEEE Engineering in Medicine and Biology*. It has eleven articles reviewing EMF Health effects, plus a Foreword and an Afterword by the guest editors. One of the articles describes how to access the literature. See also the electronic document *Powerlines—Cancer FAQ*, available via ftp at //ftp.mcw.edu/pub/emf-and-cancer and at many other sites.

[8]Foster (1996) has reviewed many of the laboratory studies and describes cases where subtle cues meant the observers were not making truly "blind" observations. Though not directly relevant to the issue under discussion here, a classic study by Tucker and Schmitt (1978) at the University of Minnesota is worth noting. They were seeking to detect possible human perception of 60-Hz magnetic fields. There appeared to be an effect. For five years they kept providing better and better isolation of the subject from subtle auditory clues that might bias the subjects. With their final isolation chamber, none of the 200 subjects could reliably perceive whether the field was on or off. Had they been less thorough and persistent, they would have reported a positive effect that does not exist.

to replicate did not adhere closely enough to the original experimental design. Misakian *et al.* (1993) provide an extensive review of the biological, physical, and electrical parameters that must be considered in such experiments. Goodman *et al.* (1995) review the results to date and state some of the difficulties that must be overcome for definitive experiments. (They feel that fields can produce significant biological responses, even at low levels.) A recent book by Blank (1995) reviews many aspects of the problem.

We will review a few of the basic principles that must be kept in mind. One of the important principles is the relationship between the electric field in air and the field within the body, which is a conductor. A simple model that shows how this coupling takes place is the one-dimensional problem shown in Fig. 9.18. An infinite slab of tissue has dielectric constant κ and conductivity σ. In the air perpendicular to the surface of the slab is an external oscillating electric field $E(t)=E_0 \cos \omega t$. We assume that the dielectric constant is independent of frequency and accounts for the polarization of the tissue. An ionic current flows and causes free charge per unit area $\pm\sigma_q$ to accumulate on the surfaces of the slab. Within the slab the field is $E_1(t)$ and the current density is $j=\sigma E_1$. Applying Gauss's law [Eq. (6.19b)] to either surface gives

$$-\epsilon_0 E_0 \cos \omega t+\kappa\epsilon_0 E_1(t)=\sigma_q(t). \qquad (9.71)$$

Conservation of free charge at the surface requires that[9]

$$\sigma E_1=j=-\frac{d\sigma_q}{dt}. \qquad (9.72)$$

If we differentiate Eq. (9.71) and combine it with Eq. (9.72), we obtain

$$\frac{dE_1}{dt}+\frac{\sigma}{\kappa\epsilon_0}E_1=\frac{\omega}{\kappa}E_0 \sin \omega t. \qquad (9.73)$$

The factor $\sigma/\kappa\epsilon_0$ is characteristic of the tissue and has the dimensions of frequency. We will call it ω_t. Typical tissue conductivity is about 0.1 Ω^{-1} m^{-1}. We must be careful about the value of the dielectric constant. We have used a value of 80 for water. However, tissue is much more complex than pure water and there are several effects that alter the dielectric constant [Foster and Schwan (1996)]. It takes time for both the polarization charges and conducting ions to move. As a result, both the conductivity and the dielectric constant of tissue depend on the frequency of the applied electric field and in fact are not independent of one another (see Foster and Schwan, especially pp. 31–41). Several effects change the conductivity and dielectric constant as a function of frequency. At power-line frequencies the dominant effect is the slight movement of the counterions and charge in the double layer at a cell membrane in response to the applied electric field. As a result, $\kappa\approx 10^6$ and $\omega_t=1.1\times 10^4$ s^{-1}.

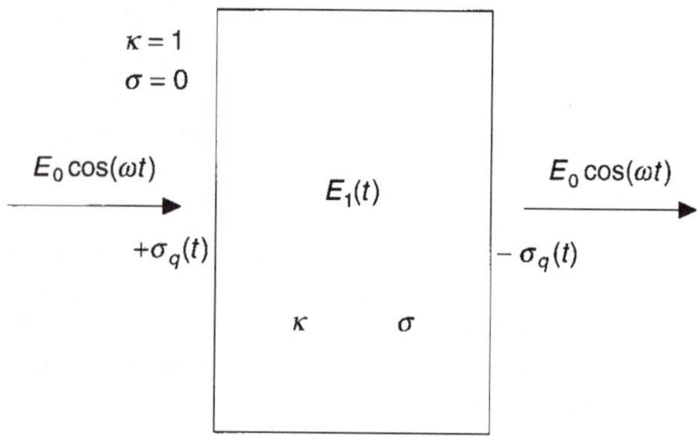

FIGURE 9.18. An infinite slab of tissue is immersed in an oscillating electric field of amplitude E_0 in air.

We try a solution to Eq. (9.73) of the form $E_1(t)=A \sin \omega t+B \cos \omega t$. It satisfies the equation if

$$A=\frac{\omega_t/\omega}{\kappa(1+\omega_t^2/\omega^2)}E_0\approx\frac{\omega\epsilon_0}{\sigma}E_0,$$

$$B=-\frac{\omega}{\omega_t}A=-\frac{1}{\kappa(1+\omega_t^2/\omega^2)}E_0\approx 0. \qquad (9.74)$$

The exact values for 60 Hz and a dielectric constant of 10^6 are $A=3.3\times 10^{-8}E_0$, $B=1.1\times 10^{-9}E_0$. The amplitude of the field in tissue is $E_1\approx A$. The field in air is reduced by a factor of about 3×10^{-8} in tissue because the tissue is a good conductor. The total reduction is nearly the same for a dielectric constant of 80, as can be seen from the fact that the approximate form for A does not depend on κ.

Another important factor is the electric fields that exist in and near a cell. One model that has been used extensively is a spherical cell with inner radius a and membrane thickness b immersed in an infinite conducting medium in which is an electric field E_1 far from the cell. This problem can be solved exactly by solving Poisson's equation in the three regions and matching boundary conditions much as we did to obtain Eq. (9.72). The results, for a case in which a 60-Hz field is applied, the conductivities of the extracellular and intracellular fluids are the same, $a=10$ μm and $b=6$ nm, and $\sigma_{\text{membrane}}=2.4\times 10^{-8}\sigma$ are indicated[10] in Fig. 9.19. Only the amplitude of the electric field is shown. The important features of this solution are that the field just outside the cell is roughly the same as the field far away, the field inside the membrane is magnified by a large factor (a/b), and the field inside the cell is multiplied by a very

[9]Readers who are familiar with the concepts of reactance and complex impedance must be frustrated because I have not used them. The reason is pedagogic. Because many in my intended audience may have had only one year of calculus, I want to avoid the use of complex numbers. In Chap. 11 I introduce them as a parallel notation. They are widely used in the image reconstruction described in Chap. 12.

[10]Calculated using equations in Polk (1995), p. 62.

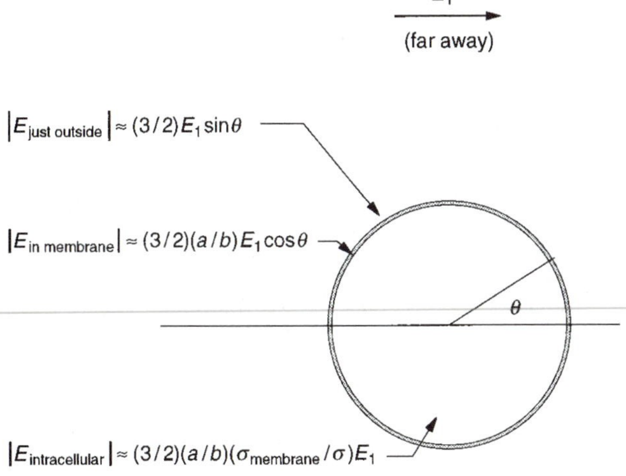

FIGURE 9.19. The electric fields in and around a spherical cell. The cell has radius a and membrane thickness b. The field far from the cell has amplitude E_1.

small number $(a\sigma_{\text{membrane}}/b\sigma)$. It is the low conductivity and thinness of the membrane that are responsible for these effects.

The external field that causes a biological effect is a signal that must be detected by the process that gives rise to the effect. We can also examine the signal-to-noise ratio. The noise can be either thermal (Johnson) noise, shot noise, or noise from the electric currents that normally flow in the body due to nerve conduction and muscle contraction. To have a signal that is not masked by Johnson noise, we must have an electric field E such that

$$\frac{zev}{k_BT} = \frac{zeEd}{k_BT} \geqslant 1, \tag{9.75}$$

where z is the valence of an effective charge that moves a distance d in the electric field E. From this,

$$zd \geqslant \frac{k_BT}{eE}. \tag{9.76}$$

Table 9.5 shows the result of a calculation using a field in air of 300 V m^{-1}. We see that a process occurring within

TABLE 9.5. *Comparison of signal to thermal noise for an applied electric field in air of 300 V m^{-1}. $T = 300$ K. Spacing d is the minimum displacement of a charge ze that will satisfy Eq. (9.76).*

Model	Just outside the cell	In the cell membrane	Inside the cell
E (V m^{-1})	1.01×10^{-5}	1.62×10^{-02}	5.40×10^{-10}
k_BT/eE (m)	$2.57\times10^{+3}$	1.59	$4.79\times10^{+7}$
Assumed z	10	10	10
d (m)	$2.57\times10^{+2}$	1.59×10^{-1}	$4.79\times10^{+6}$

the cell membrane is most sensitive, requiring the smallest value of d. Even in that case, the distance over which a charge $10e$ would have to move is about 0.1 m. It seems unlikely that interactions of the electric field with a single cell cause any effects.

One proposal to overcome this signal-to-noise problem is that the biological effect is due to the averaging of the field over many cells or over time. This was first proposed by Weaver and Astumian (1990), and a specific model has been formulated by Astumian, Weaver, and Adair (1995). The model applies Eq. (9.58) and shows that, if the concentration of some substance outside the cell is much larger than inside, the response to an oscillating v is ''rectification'' or a net inward current. This would allow an accumulation of the substance within the cell. The averaging times in their model are 1–3 h. Weaver and Astumian (1995) review the entire causality problem, including the effects of shot noise.

Magnetic interactions have also been proposed. The magnetic field is not attenuated at the body surface like the electric field is. In 1990 Kirschvink reported that the human brain contains several million magnetosomes per gram. Recently Kobayashi, Kirschvink, and Nesson (1995) have reported that contamination with magnetic particles could affect laboratory experiments with cell cultures, even if the cells being studied do not normally contain magnetosomes. Commercial disposable, presterilized plastic laboratory ware used in tissue culture experiments were found to contain ferromagnetic particles smaller than 100 nm that are readily taken up by white blood cells.

What about the signal-to-noise ratio for magnetic effects? The situation is somewhat more favorable than for the electric case. We saw in Chap. 8 that a single magnetosome has appreciable alignment with the earth's magnetic field, even in the presence of thermal bombardment. The earth's field is about 5×10^{-5} T. For a single magnetosome

$$\frac{mB_{\text{earth}}}{k_BT} = \frac{(6.4\times10^{-17})(5\times10^{-5})}{(1.38\times10^{-23})(300)} = 0.77. \tag{9.77}$$

For a larger magnetosome of radius 100 μm $m = 2\times10^{-15}$ A m^2 and the energy ratio in the earth's field is 24. The field due to a typical power line is about 100 times smaller: about 2×10^{-7} T.

Kirschvink proposed a model whereby a magnetosome in a field of 10^{-4}–10^{-3} T could rotate to open a membrane channel. As an example of the debate that continues in this area, Adair (1991, 1992, 1993, 1994) argued that a magnetic interaction cannot overcome thermal noise in a 60-Hz field of 5×10^{-6} T. However, Polk (1994) argues that more biologically realistic parameters, including a large number of magnetosomes in a cell, could allow an interaction at 2×10^{-6} T.

The essential features of all the models are like this. Imagine a particle with magnetic moment **m** in the earth's field. It will tend to align with the field as shown in Fig. 9.20(a). The direction of the magnetic moment with the earth's field is θ. Apply an alternating field $B_0 \cos \omega t$ at right angles to the earth's field, as shown in Fig. 9.20(b). [The parameter ω is the angular frequency, $\omega = 2\pi\nu$, where ν is the number of cycles per second the field oscillates. The units of ω are radians per second; the units of ν are cycles per second or hertz (Hz). For a 60-Hz power frequency, $\omega = 377\ \mathrm{s}^{-1}$.] There are three torques on the particle. The first is viscous drag, which is proportional to the angular velocity of the particle $d\theta/dt$ but in the opposite direction. The second is the torque tending to align **m** with the earth's field, $-mB_{\text{earth}} \sin \theta$. The third tends to align **m** with the alternating field, $mB_0 \cos \omega t \cos \theta$. Assume that the acceleration is so small that the particle is in rotational equilibrium. (This is not necessary, but it simplifies the math.) Then, from Eq. (1.14),

$$-\beta \frac{d\theta}{dt} + mB_{\text{earth}} \sin \theta - mB_0 \cos \omega t \cos \theta = 0. \tag{9.78}$$

In order to linearize the equation, assume that θ is small enough so that $\sin \theta \approx \theta$ and $\cos \theta \approx 1$. The linearized equation is

$$\beta \frac{d\theta}{dt} - mB_{\text{earth}}\theta = -mB_0 \cos \omega t. \tag{9.79}$$

This is a linear differential equation with constant coefficients that can be solved by the techniques of Appendix F. Consider only the particular solution and try a solution of the form

$$\theta = \theta_1 \cos \omega t + \theta_2 \sin \omega t. \tag{9.80}$$

Substitution of this in the equation shows that

$$\theta_1 = \frac{m^2 B_0 B_{\text{earth}}}{(\omega\beta)^2 + (mB_{\text{earth}})^2},$$

$$\theta_2 = \frac{\omega\beta m B_0}{(\omega\beta)^2 + (mB_{\text{earth}})^2},$$

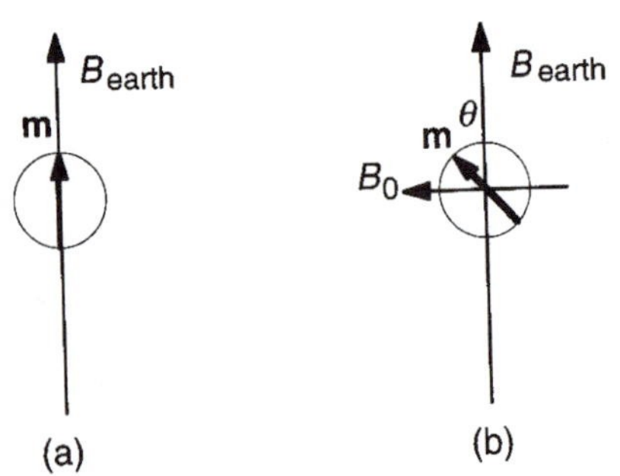

FIGURE 9.20. A particle with magnetic moment **m** (a) aligned with the earth's magnetic field and (b) at an angle θ with the earth's field because of an applied field B_0.

$$\theta_m = \frac{mB_0}{[(\omega\beta)^2 + (mB_{\text{earth}})^2]^{1/2}}, \tag{9.81}$$

where the maximum amplitude is $\theta_m^2 = \theta_1^2 + \theta_2^2$. We saw in Chap. 4 (Stokes's law) that the translational viscous drag on a spherical particle is $6\pi\eta a v$. The viscous torque on a rotating sphere is $8\pi\eta a^3 (d\theta/dt)$. The measured values for viscosity inside a cell range from 0.003 to 0.015 N s m^{-2} [Polk (1994)]. Using the average of these, $\beta = 0.009a^3$. The magnetic moment of a single-domain magnetosome is also proportional to volume: $m = 2\times 10^6 a^3$. This leads to a maximum amplitude that is independent of a:

$$\theta_m = \frac{2\times 10^6 a^3 B_0}{[(377)^2(0.23)^2 a^6 + (2\times 10^6)^2 (5\times 10^{-5})^2 a^6]^{1/2}}$$
$$= 1.5\times 10^4 B_0. \tag{9.82}$$

Kirschvink originally argued from data about hair-cell deformation that a deflection of 16° or 0.3 rad is needed. This would require $B_0 = 5\times 10^{-5}$ T. (He had a slightly different value because he used a different viscosity. He also included the torque due to the force on the channel gate.)

In the absence of the applied field, the thermal fluctuations in angle can be estimated as follows. In the linear approximation, the work required to displace the particle an amount θ from the direction of the earth's field is

$$W = \int \tau\, d\theta = \int mB_{\text{earth}}\theta\, d\theta = mB_{\text{earth}} \frac{\theta^2}{2}. \tag{9.83}$$

Equipartition of energy again gives us

$$\overline{\theta^2_{\text{thermal}}} = \frac{k_B T}{mB_{\text{earth}}} = \frac{k_B T}{(2\times 10^6)(5\times 10^{-5})a^3} = \frac{k_B T}{100a^3}. \tag{9.84}$$

For a 50-nm magnetosome, this gives $\theta_{\text{rms}} = 0.58$ rad. For a 100-nm magnetosome it is 0.2 rad, comparable to the maximum angles of deduced from the model in the preceding paragraph.

SYMBOLS USED IN CHAPTER 9

Symbol	Use	Units	First used on page
a	Radius	m	234
a	Particle radius	m	238
b	Spacing	m	235
d	Displacement of charge	m	249
e	Electron charge	C	228
f	Force	N	236
f,g	Temporary functions		243

Symbol	Use	Units	First used on page
g_K	Potassium conductance per unit area	$\Omega^{-1}\ m^{-2}$ or S	241
i	Electric current	A	243
$j, \bar{j}$	Electric current density	$A\ m^{-2}$	237
j_s	Particle current density	particle $m^{-2}\ s^{-1}$	236
k_B	Boltzmann constant	$J\ K^{-1}$	228
$\mathbf{m}$	Magnetic moment	$A\ m^2$	250
n	Number of ions		245
p, p_c, p_o	Probability		242
p_e, p_{tot}	Electric dipole moment	C m	235
q	Charge	C	245
r	Position	m	231
r	Radius in spherical coordinates		233
t	Time	s	250
u	$rv(r)$	V m	233
u, u_0, w	Energy (normalized to k_BT)		237
v, v'	Potential	V	228
v_{Nernst}	Nernst potential	V	238
v_{solute}	Velocity of solute	$m\ s^{-1}$	236
$v_{solvent}$	Velocity of solvent	$m\ s^{-1}$	236
x	Position	m	231
z	Valence		228
A, B, A', B'	Constants	V	232
B_{earth}	Earth's magnetic field	T	249
B_0	Amplitude of applied oscillating magnetic field	T	249
C, C'	Concentration	particle m^{-3}	228
C_i	Concentration of species i	particle m^{-3}	231
[Cl], [Cl′]	Chloride concentration	particle m^{-3}	229
C	Capacitance	F	245
D, D_{eff}, D_0	Diffusion constant	$m^2\ s^{-1}$	236
E, E_x, E_0, E_1	Electric field	$V\ m^{-1}$	233
E_{ext}	External electric field	$V\ m^{-1}$	235
E_{pol}	Polarization electric field	$V\ m^{-1}$	235
F	Faraday constant	C/mol	229
F	Force	N	236
G	Conductance	S	237
J	Current per limit area of membrane	$A\ m^{-2}$	239

Symbol	Use	Units	First used on page
K	Saturation concentration	particle m^{-3}	244
$[K], [K']$	Potassium concentration	particle m^{-3}	231
L	Separation	m	236
$[M^+], [M^{+\prime}]$	Concentration of impermeant cations	particle m^{-3}	231
$[M^-], [M^{-\prime}]$	Concentration of impermeant anions	particle m^{-3}	231
$[M], [M']$	Net concentration of impermeant ions	particle m^{-3}	231
N	Number per unit volume	m^{-3}	235
N_A	Avogadro's number	mol^{-1}	235
R	Gas constant	$J\ K^{-1}\ mol^{-1}$	229
R	Resistance	Ω	245
R_p	Pore radius	m	238
S	Area	m^2	237
T	Temperature	K	228
U, W	Energy	J	245
β	Linear viscous drag coefficient	$N\ s\ m^{-1}$	236
β	Rotational viscous drag coefficient	N m s	250
β	Partition fraction		244
γ	Proportionality constant	$\Omega^{-1}\ m^3$	244
ϵ_0	Electrical permittivity of vacuum		231
κ	Dielectric constant		231
λ_D	Debye length	m	232
λ	Length	m	237
ρ, ρ_{ext}	Charge density	$C\ m^{-3}$	231
ρ	Resistivity	Ω m	236
σ_q, σ_q'	Charge per unit area	$C\ m^{-2}$	235
σ	Conductivity	$\Omega^{-1}\ m^{-1}$	236
σ_i	Standard deviation of current	A	245
σ_n	Standard deviation of number of ions		245
σ_q	Standard deviation of charge	C	245
σ_v	Standard deviation of voltage	V	245
τ	Time constant	s	245
τ	Torque	Nm	250
θ	Angle		250
χ	Susceptibility		235

Symbol	Use	Units	First used on page
$\omega_s, \omega, \omega_0$	Solute permeability	$N^{-1}\,s^{-1}$	238
ω	Angular frequency	s^{-1}	248
ω_t	Characteristic angular frequency of tissue	s^{-1}	248
Γ	Radial concentration factor		238

PROBLEMS

Section 9.1

9.1. The chloride ratio between plasma and interstitial fluid is 0.95. Plasma protein has a valence of about -18 and a concentration of about 10^{-3} mol l^{-1}. In the interstitial fluid, $[Na']=[Cl']=155$ mmol l^{-1}. Find the sodium and chloride concentrations in the plasma and the potential difference across the capillary wall, assuming Donnan equilibrium.

9.2. Suppose that there are two compartments with equal volume $V=1$ l, separated by a membrane that is permeable to K and Cl ions. Impermeant positive ions have a concentration 0 on the left and $[M']=[M^+]=10$ mmol l^{-1} on the right. The initial concentration of potassium is $[K_0]=30$ mmol l^{-1} on the left. $T=310$ K.

(a) Find the initial concentrations of potassium and chloride on both sides and the potential difference.

(b) A fixed amount of potassium chloride (10 mmole) is added on the left. After things have come to equilibrium, find the new concentrations and potential difference.

Section 9.2

9.3. Derive the Poisson equation from Gauss's law in Cartesian coordinates in three dimensions.

9.4. Consider ions uniformly dispersed in a solution. Find the average linear separation of the ions for concentrations of 1, 10, 100, and 1000 mmol l^{-1}.

9.5. Verify Eq. 9.20.

9.6. Verify the parameters presented in Table 9.2. How accurate is the approximation $e^x=1+x$ in this case?

Section 9.3

9.7. The value of A used to obtain Eq. (9.30) was determined by saying that as $r\rightarrow 0$, the electric field must approach $ze/4\pi\epsilon_0\kappa r^2$. An elaboration of the model would be to say that the central ion has radius a and that the electric field at $r=a$ must be the same as the field at the surface of the ion, $ze/4\pi\epsilon_0\kappa a^2$. How does this change the expression for $v(r)$?

Section 9.5

9.8. Find an expression for the slope of the Nernst–Planck constant-field curve in Fig. 9.10 when v is equal to the Nernst potential, v_0. Hint: expand the exponentials in Eq. (9.46) around v_0.

9.9. Show that when $j=0$, Eq. (9.43) gives $C(x)=C_0 e^{-zev(x)/k_BT}$, as we already know must be true in equilibrium. Hint: solve for dv/dx.

9.10. The discussion surrounding Eqs. (9.35)–(9.42) was for a model of ions in a pore with constant electric field. It is also possible to write an integral version of the Nernst–Planck equation. Consider a single channel in which the current is the same for all values of x, the distance along the channel. If the diffusion constant and cross-sectional area of the channel are allowed to vary, and with the usual substitution $u(x)=zev(x)/k_BT$, Eq. (9.42) becomes

$$i=j(x)S(x)=-zeD(x)S(x)\left(\frac{dC}{dx}+C(x)\,\frac{du}{dx}\right).$$

(a) Show that if each term is multiplied by e^u, this can be written as

$$\frac{ie^{u(x)}}{D(x)S(x)}=-ze\left(e^{u(x)}\,\frac{dC}{dx}+C(x)e^{u(x)}\,\frac{du}{dx}\right).$$

(b) Show that if the integration is carried from x_1 to x_2, then the current in the channel is

$$i=-\frac{ze[C(x_2)e^{u(x_2)}-C(x_1)e^{u(x_1)}]}{I},$$

where the integral

$$I=\int_{x_1}^{x_2}\frac{e^{u(x)}\,dx}{S(x)D(x)},$$

contains all the information about the channel.

Section 9.6

9.11. Apply Eq. (9.43) to two species, a and b, for which D is a function of x.

(a) Show that since

$$d(Ce^{zev/k_BT})/dx=e^{zev/k_BT}(dC/dx)+Ce^{zev/k_BT}(ze/k_BT)\times(dv/dx)$$

each equation can be written in the form

$$j_a=-z_aeD_a(x)e^{-zev/k_BT}d[C_a(x)e^{zev/k_BT}]/dx.$$

(b) Consider the functions I_a and I_b, where

$$I_a = \frac{-1}{z_a e} \int D_a^{-1}(x) e^{zev/k_BT} dx.$$

Show that since j_a and j_b are constants,

$$j_a I_a = C_{a2} e^{zev/k_BT} - C_{a1}, \quad j_b I_b = C_{b2} e^{zev/k_BT} - C_{b1}.$$

(c) Show that if $D_b(x)/D_a(x)$ is a constant equal to ω_a/ω_b, then Eq. (9.57) results.

9.12. Direct measurements of the reversal potential to obtain permeability ratios are difficult because the concentrations inside the axon are not known. One can overcome this by measuring how the reversal potential changes as outside concentrations are varied. Obtain an equation for the shift of reversal potential if two measurements are made: one in which the concentration $C_{a1}=0$, the other with $C_{b1}=0$.

Section 9.7

9.13. A patch–clamp experiment shows that the conductance of a single Ca^{2+} channel is $G=25\times 10^{-12}$ S. The membrane thickness is $b=6\times 10^{-9}$ m. Use $v=50$ mV.

(a) Assuming that the resistivity of the fluid in the channel is $\rho=0.5\ \Omega$ m, find an expression and numerical value for the channel radius a.

(b) If the conductance per unit area is 1200 S m^{-2}, find the number of pores per unit area.

(c) The current is $i=Gv$, where v is the applied voltage. Find an expression for n, the number of calcium ions per second passing through the channel, in terms of whichever of parameters G, v, b, and a are necessary.

(d) How many calcium ions are in the channel at one time, if the calcium concentration is C mmol l^{-1}?

9.14. A potassium channel might have a radius of 200 pm and a length of 6 nm. If it contained potassium at a concentration of 150 mmol l^{-1}, how many potassium ions on average would be in the channel?

9.15. How long does it take for a sodium ion to drift in the electric field (assumed constant) through a membrane of thickness L and applied potential Δv? How long does it take to move by pure diffusion? Find numerical values when the membrane is 6 nm thick and potential difference is 70 mV.

9.16. Suppose that a sodium pore when open passes 10 pA and $j_{Na}=2\times 10^{-5}$ A cm^{-2}. Calculate the number of open pores per unit area and the average linear spacing between them.

9.17. Calculate the current density of sodium ions in a region of length 6 nm due to (a) pure diffusion when there is no potential difference and the concentrations are 145 and 15 mmol l^{-1}, (b) pure drift when the concentration is 145 mmol l^{-1} and the potential difference is 70 mV, and (c) both diffusion and drift if the electric field is constant.

9.18. Patch–clamp recording is done with a micropipette of radius 1 μm.

(a) If the pipette encircles a single channel with conductance 20 pS, what is the channel current when the channel is open and the voltage across the membrane is 20 mV away from the Nernst potential for the ion in question? Make a simple estimate using Ohm's law.

(b) Assuming a capacitance of 0.01 F m^{-2}, what current charges the capacitance of the membrane patch under the micropipette if a 20-mV change occurs linearly in 5 μs?

9.19. The following circuit illustrates the effects that must be considered when an electrode is used to measure the properties of a patch of membrane. R_1 is the resistance of the electrode. R_2 and C are properties of the membrane. The applied voltage $v_0(t)$ is a step at $t=0$. The electrode current is $i(t)$. The voltage across the membrane patch is $v(t)$.

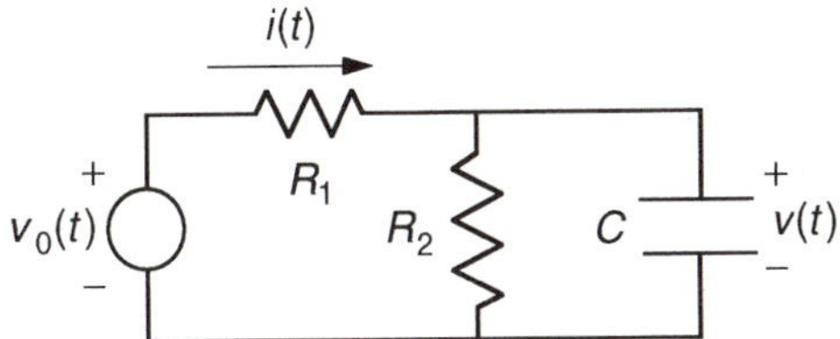

(a) Show that

$$v_0(t) = R_1 C \frac{dv}{dt} + \frac{R_1+R_2}{R_2} v(t).$$

(b) Show that the time constant is

$$\tau = \frac{R_1 R_2 C}{R_1 + R_2}$$

and that $\tau \to R_1 C$ if $R_1 \ll R_2$, $\tau \to R_2 C$ if $R_1 \gg R_2$.

(c) If $v_0(t)$ is a step of height v_0 at $t=0$, show that

$$v(t) = v_0 \frac{R_2}{R_1+R_2} (1 - e^{-t/\tau})$$

and

$$i(t) = v_0 \frac{1}{R_1+R_2} \left(1 + \frac{R_2}{R_1} e^{-t/\tau}\right).$$

(d) Plot $v(t)$ and $i(t)$.

(e) The case $R_1 \ll R_2$ is called *voltage-clamped.* Find expressions for $v(t)$ and $i(t)$ in that case and plot them. Where does the transient current flow? For fixed R_2, what is the time constant?

(f) In the *current-clamped* case, $R_1 \gg R_2$ and $i_0 = v_0/R_1$. Find expressions for $v(t)$ and $i(t)$ and plot them. For fixed R_2, what is the time constant?

(g) Make numerical plots of $v(t)$ and $i(t)$ when $v_0 = 150$ mV, $R_1=10^6\ \Omega$, $C=5\times 10^{-12}$ F, and $R_2=10^{11}\ \Omega$.

9.20. A patch–clamp experiment is done with a micropipette having a resistance of $10^6\ \Omega$. When 150 mV is applied across the membrane, the current is 0 when the pores are closed and 1 pA when one channel is open. The membrane capacitance is 4×10^{-3} F m^{-2}. The microelectrode tip has an inner radius of 20 μm. What is the time constant for voltage changes? Does it depend on whether the channel has opened or closed?

9.21. Obtain expressions for the current in the constant-field model with and without the correction for end effects when $C_0 = C_0'$. Find the limiting value for the current as $v \rightarrow \pm\infty$.

Section 9.9

9.22. In some nerve membranes a region of "negative resistance" is found, in which the current decreases as the voltage is increased.

(a) Where have we seen this behavior before?

(b) To see why it happens, consider two cases. The current through the membrane is given by $j = g(v) \times (v - v_0)$, where $g(v)$ is a property of the membrane, and the Nernst potential v_0 depends on the ion concentration on either side of the membrane. For this problem let $v_0 = +50$ mV. Calculate j as a function of v for two cases: (a) $g(v) = 1$; (b) the conductance increases rapidly with voltage: $g(v) = (5.6 \times 10^{-7}) e^{0.288v}$ (v in mV).

(c) Negative resistance increases the sensitivity of the ampullae of Lorenzini, as measured by Lu and Fishman. To see why, calculate the output voltage in a two-resistance voltage divider network (as in Fig. 6.22) and discuss what happens if R_2 is negative.

Section 9.10

9.23. (a) Here is one way that signal-to-noise ratio can be improved. Suppose that there are N receptors, connected in the nervous system in such a way that an output response requires a logical AND between all N of receptors. Whether or not there is a response is sampled every T seconds. If the signal exists, all N receptors respond. If the signal does not exist, each receptor responds to thermal noise with a probability p (which might be $p = e^{-U/k_B T}$, where U is an activation energy). Assume that p is the same for each receptor, and that whether a receptor has responded to thermal noise is independent of the response of all other receptors and also independent of its response at any other time. What is the signal-to-noise (*S/N*) ratio as a function of N? Suppose that $N = 8$. Plot S/N as a function of p.

(b) Find $U/k_B T$ vs N for $S/N = 4$.

9.24. Here is another way to look at the signal-to-noise ratio.

(a) Show that the energy of a charged parallel-plate capacitance can be written as $\kappa\epsilon_0 E^2 V/2$, where $V = Sd$ is the volume between the plates. This is a special case of a general relationship that the energy per unit volume associated with an electric field is $\kappa\epsilon_0 E^2/2$.

(b) Use the information about the magnitude of the electric field in the cell membrane from Fig. 9.19 to calculate the total electrostatic energy in the membrane.

(c) Compare the ratio of the total electrostatic energy to $k_B T$ when the air field is 300 V m^{-1}. This overestimates the ratio, because the energy is spread over the entire membrane and is not available to interact in one place.

9.25. Obtain Eq. (9.84) from the expression $U = -mB \cos\theta$ that was derived in Problem 8.19, by making a suitable expansion for small angles.

9.26. Compare the electric field in tissue at 60 Hz from an air field of 300 V m^{-1} with the equivalent field due to eddy currents (Faraday induction law) by a magnetic field of 10^{-7} T. Use a 10-cm radius.

REFERENCES

Adair, R. (1991). Constraints on biological effects of weak extremely-low-frequency electromagnetic fields. *Phys Rev. A* **43**: 1039–1048.

Adair, R. (1992). Reply to "Comment on 'Constraints on biological effects of weak extremely-low-frequency electromagnetic fields.'" *Phys. Rev. A* **46**: 2185–2187.

Adair, R. (1993). Effects of ELF magnetic fields on biological magnetite. *Bioelectromagnetics* **14**: 1–4.

Adair, R. (1994). Constraints of thermal noise on the effects of weak 60-Hz magnetic fields acting on biological magnetite. *Proc. Nat. Acad. Sci. USA* **91**: 2925–2929.

Astumian, R. D., J. C. Weaver, and R. K. Adair (1995). Rectification and signal averaging of weak electric fields by biological cells. *Proc. Nat. Acad. Sci. USA* **92**: 3740–3743.

Barnes, F. S. (1995). Typical electric and magnetic field exposures at power-line frequencies and their coupling to biological systems. In M. Blank, ed. *Electromagnetic Fields: Biological Interactions and Mechanisms.* Washington, American Chemical Society, pp. 37–55.

Bastian, J. (1994). Electrosensory organisms. *Physics Today* **47**(2): 30–37.

Blank, M., ed. (1995). *Electromagnetic Fields: Biological Interactions and Mechanisms.* Washington, American Chemical Society.

Bockris, J. O'M., and A. K. N. Reddy (1970). *Modern Electrochemistry.* New York, Plenum, Vol. 1.

Bren, S. P. A. (1995). 60 Hz EMF health effects—a scientific uncertainty. *IEEE Eng. Med. Biol.* **14**: 370–374.

Carstensen, E. L. (1995). Magnetic fields and cancer. *IEEE Eng. Med. Biol.* **14**: 362–369.

Crouzy, S. C., and F. J. Sigworth (1993). Fluctuations in ion channel gating currents: Analysis of nonstationary shot noise. *Biophys. J.* **64**: 68–76.

DeFelice, L. J. (1981). *Introduction to Membrane Noise.* New York, Plenum.

Denk, W., and W. W. Webb (1989). Thermal-noise-limited transduction observed in mechanosensory receptors of the inner ear. *Phys. Rev. Lett.* **63**(2): 207–210.

Foster, K. R. (1996). Electromagnetic field effects and mechanisms: In search of an anchor. *IEEE Eng. Med. Biol.* **15**(4): 50–56.

Foster, K. R., and H. P. Schwan (1996). Dielectric properties of tissues. In C. Polk and E. Postow, eds. *Handbook of Biological Effects of Electromagnetic Fields.* Boca Raton, CRC, pp. 25–102.

Freeman, S. A., M. A. Wang, and J. C. Weaver (1994). Theory of electroporation of planar bilayer membranes: Predictions of the aqueous area, change in capacitance, and pore-pore separation. *Biophys. J.* **67**: 42–56.

Goodman, E. M., B. Greenebaum, and M. T. Marron (1995). Effects of electromagnetic fields on molecules and cells. *Intl. Rev. Cytology* **158**: 279–338.

Hamill, O. P., A. Marty, E. Neher, B. Salmon, and F. J. Sigworth (1981). Improved patch-clamp techniques for high-resolution current recording from cells and cell-free membrane patches. *Pflügers Arch.* **391**: 85–100.

Heath, C. W., Jr. (1996). Electromagnetic field exposure and cancer: A review of epidemiologic evidence. *Cancer J. Clin.* **65**: 29–44.

Hille, B. (1992). *Ionic Channels of Excitable Membranes,* 2nd ed. Sunderland, MA, Sinauer Associates.

Kalmijn, Ad. J. (1988). Detection of weak electric fields. In J. Aetma, *et al.*, eds. *Sensory Biology of Aquatic Animals.* New York, Springer, pp. 151–186.

Keynes, R. D. (1994). The kinetics of voltage-gated ion channels. *Quart. Rev. Biophys.* **27**(4): 339–434.

Kirschvink, J. L. (1992). Comment on "Constraints on biological effects of weak extremely-low-frequency electromagnetic fields." *Phys. Rev. A* **46**: 2178–2184.

Kobayashi, A. K., J. L. Kirschvink, and M. H. Nesson (1995). Ferromagnetism and EMFs. *Nature* **374**: 123.

Leuchtag, H. R., and C. J. Swihart (1977). Steady-state electrodiffusion. Scaling, exact solutions for ions of one charge, and the phase plane. *Biophys. J.* **17**: 27–46.

Levitt, D. G. (1986). Interpretation of biological ion channel flux data: Reaction-rate versus continuum theory. *Ann. Rev. Biophys. Biophys. Chem.* **15**: 29–57.

Lu, J., and H. M. Fishman (1994). Interaction of apical and basal membrane ion channels underlies electroreception in ampullary epithelia of skates. *Biophys. J.* **67**: 1525–1533.

Mauro, A. (1962). Space charge regions in fixed charge membranes and the associated property of capacitance. *Biophys. J.* **2**: 179–198.

Misakian, M., A. R. Sheppard, D. Krause, M. E. Frazier, and D. L. Miller (1993). Biological, physical, and electrical parameters for in vitro studies with ELF magnetic and electric fields: A primer. *Bioelectromagnetics (Suppl.)* **2**: 1–73.

Moulder, J. E., and K. R. Foster (1995). Biological effects of power-frequency fields as they relate to carcinogenesis. *Proc. Soc. Expt. Biol. Med.* **209**: 309–323.

National Research Council (1996). Committee on the Possible Effects of Electromagnetic fields on Biologic Systems. *Possible Health Effects of Exposure to Residential Electric and Magnetic Fields.* Washington, DC. National Academy Press, Prepublication Copy.

Pickard, W. F. (1988). A model for the acute electrosensitivity of cartilaginous fishes. *IEEE Trans. Biomed. Eng.* **35**(4): 243–249.

Polk, C. (1994). Effects of extremely-low-frequency magnetic fields on biological magnetite. *Bioelectromagnetics* **15**: 261–270.

Polk, C. (1995). Bioelectromagnetic dosimetry. In M. Blank, ed. *Electromagnetic Fields: Biological Interactions and Mechanisms.* Washington, American Chemical Society, pp. 57–78.

Polk, C. (1996). Introduction. In C. Polk and E. Postow, eds. *Handbook of Biological Effects of Electromagnetic Fields.* Boca Raton, CRC, pp. 1–23.

Poole, C., and D. Ozonoff (1996). Magnetic fields and childhood cancers: An investigation of dose response analyses. *IEEE Eng. Med. Biol.* **15**(4): 41–49.

Sigworth, F. J. (1993). Voltage gating of ion channels. *Quart. Rev. Biophys.* **27**(1): 1–40.

Tenforde, T. S. (1995). Spectrum and intensity of environmental electromagnetic fields from natural and man-made sources. In M. Blank, ed. *Electromagnetic Fields: Biological Interactions and Mechanisms.* Washington, American Chemical Society, pp. 13–35.

Tucker, R. D., and O. H. Schmitt (1978). Tests for human perception of 60 Hz moderate strength magnetic fields. *IEEE Trans. Biomed. Eng.* **BME25**: 509–518.

Weaver, J. C. (1994). Molecular basis for cell membrane electroporation. *Ann. N.Y. Acad. Sci.* **720**: 141–152.

Weaver, J. C., and R. D. Astumian (1990). The response of living cells to very weak electric fields: The thermal noise limit. *Science* **247**: 459–462.

Weaver, J. C., and R. D. Astumian (1995). Issues relating to causality of bioelectromagnetic fields. In M. Blank, ed. *Electromagnetic Fields: Biological Interactions and Mechanisms.* Washington, American Chemical Society, pp. 79–96.

CHAPTER 10

Feedback and Control

We now turn to the way in which the body regulates such things as temperature, oxygen concentration in the blood, cardiac output, number of red or white cells, and calcium concentrations in the blood. All of these are regulated by a *feedback loop*. A feedback loop exists if variable x determines the value of variable y, and variable y in turn determines the value of variable x.

Suppose that x is the deviation of a bullet from its desired path. A bullet has no feedback; after it has left the gun, its deviation from the desired path is determined by the initial aim of the gun, fall due to gravity, drift caused by the wind, and air turbulence. An accuracy of one part in 10^4 (a tenth of an inch in 50 ft) is quite good. A car, on the other hand, is steered by the driver. If deviation x becomes appreciable, the driver changes y, the position of the steering wheel. The value of y determines x through the steering mechanism and the tires. It is possible to have a car deviate less than 1 ft from the desired position within a lane after driving 3000 miles, an accuracy of one part in 10^7 or 10^8. This is an example of *negative feedback*. If x gets too large, the factors in the feedback loop tend to reduce it.

Negative-feedback systems can generate oscillations of their variables. We see oscillations in physiological systems on many different time scales, from the rhythmic activity of the heart, to breathing in and out, to changes in the rate of breathing, to daily variations in body temperature, blood pressure, and hormone levels, to monthly variations such as the menstrual cycle, to annual variations such as hibernation, coloring, fur growth, and reproduction.

It is also possible to have positive feedback. Two bickering children can goad each other to new heights of anger. Blood pressure is regulated in part by sensors in the kidney. A patient with high blood pressure may suffer damage to the blood vessels, including those feeding the kidneys, which reduces the blood pressure at the sensors. The sensors then ask for still higher blood pressure, which accelerates the damage, which leads to still higher blood pressure, and so on.

The simplest feedback loop consists of two processes: one in which y depends on x and another in which x depends on y. The loop can have many more variables. Steering the car, in addition to the variables of lane position and steering-wheel position, involves vision, neuromuscular processes, all of the variables in the automobile's steering mechanism, and the Newtonian mechanics of the car's motion—with external variables such as the behavior of other drivers continually bombarding the system.

Sections 10.1–10.3 deal with the relationships between the feedback variables when the system is in *equilibrium* or in the *steady state* and none of the variables are changing with time. The techniques for determining the *operating point*, the steady-state values of the variables, are graphical and can be applied to any system if the relationship among the variables is known.

When the system is not at equilibrium, it returns to the equilibrium point if the system is stable. Although the equations describing this return to equilibrium are usually not linear, Secs. 10.4–10.6 discuss how linear systems behave when they are not at the operating point. A linear system may "decay" exponentially to the steady-state values or it may exhibit damped oscillations.

Most systems are not linear. Section 10.7 discusses systems described by nonlinear equations in one or two dimensions, introducing some of the vocabulary and graphical techniques of nonlinear systems analysis. It closes with an example of resetting the phase of a biological oscillator. Section 10.8 introduces the ideas of period doubling and chaotic behavior through difference equations and the logistic map. It then describes a linear map that appears to be chaotic but is not. Section 10.9 shows how a linear differential equation that depends on a fixed delay in the variable can exhibit either damped or continuous oscillations. Section 10.10 summarizes the earlier sections, and Sec. 10.11 gives several biological examples.

10.1. STEADY-STATE RELATIONSHIPS AMONG VARIABLES

Any feedback loop can be broken down, conceptually at least, into separate processes that relate a dependent variable to an independent variable and possibly to some other parameters. Figure 10.1 shows an example. In the first process the thyroid gland, in response to thyroid stimulating hormone (TSH) from the pituitary, produces the thyroid hormones thyroxine (T4) and tri-iodothyronine (T3). An increase of TSH increases production of T3 and T4. These processes depend on other parameters, such as the amount of iodine available in the body to incorporate into the T3 and T4. In the second process, the pituitary increases the production of TSH if the concentration of T3 in the blood falls. It may also respond to T4 and other variables as well. (This is an oversimplification. The pituitary actually responds to hormones secreted by the hypothalamus. The hypothalamus is responding to the levels of T3 and T4.)

For a quantitative example, consider a simple model relating the amount of carbon dioxide in the alveoli (air sacs of the lung) and the rate of breathing (ventilation rate). If the body is producing CO_2 at a constant rate, a given ventilation rate corresponds to a definite value of P_{CO_2}, the partial pressure of carbon dioxide in the alveoli. In the steady state the amount of CO_2 exhaled (the volume of gas leaving the lungs per minute times P_{CO_2}) is just equal to the amount produced in the body. Figure 10.2 shows this relationship when the pH and P_{O_2} of the blood are fixed. As ventilation rate rises, P_{CO_2} falls. We are ignoring several other feedback loops [Riggs (1970, pp. 401–418)]. If the metabolic rate rises, P_{CO_2} also rises. Experiments show that ventilation rate y and alveolar P_{CO_2} (which we will call x) are related by (Riggs, op. cit.)

$$y = 2.07 + 15.47p/x, \quad x = \frac{15.47p}{y - 2.07}. \tag{10.1}$$

In these equations y is measured in l min^{-1}, x is in torr, and parameter p is the body's oxygen consumption in mmol min^{-1}. In these relationships y is the independent variable and x is the dependent variable. A typical resting person requires $p = 15$ mmol min^{-1}.

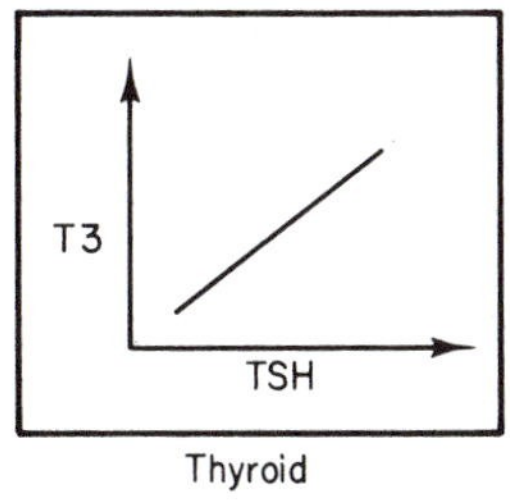

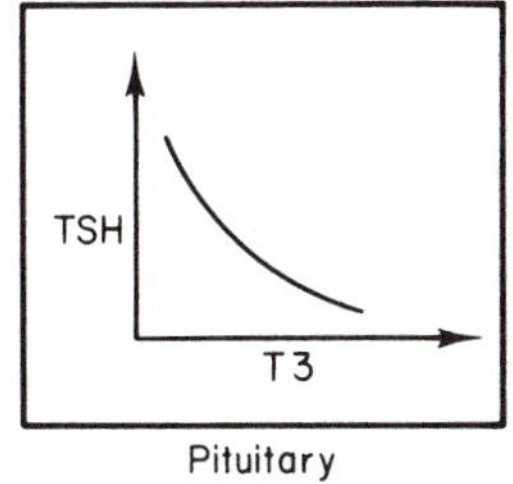

FIGURE 10.1. Schematic curves of the relationship between thyroid hormone (T3) and thyroid stimulating hormone (TSH) in the thyroid gland and in the pituitary.

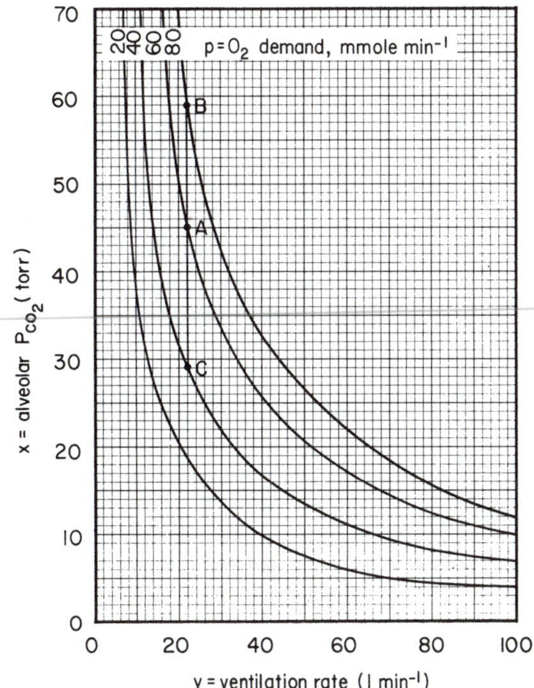

FIGURE 10.2. The pressure of CO_2 in the alveoli of the lungs decreases as the ventilation rate is increased. The different curves correspond to different total metabolic rates.

Equation 10.1 can be derived using a simple model for respiration. Let the metabolic rate of the body be described by $\dot{o}$, the rate of oxygen consumption in mol s^{-1}. (The overdot denotes a time derivative or rate.) The *respiratory quotient* F relates $\dot{o}$ to the rate of CO_2 production, so

$$(\text{rate of production}) = F\dot{o}. \tag{10.2}$$

A typical value of F is 0.8.

Carbon dioxide is removed from the body by breathing. If the rate at which air flows through the alveoli is $\dot{V}$ m^3 s^{-1}, then the rate of removal is obtained from the ideal-gas law:

$$(\text{rate of removal}) = \frac{x\dot{V}}{RT}.$$

The rate $\dot{V}$ is less than the ventilation rate y because air in the trachea and bronchi does not exchange oxygen carbon dioxide with the blood: $\dot{V} = y - b$. Therefore

$$(\text{rate of removal}) = \frac{x(y-b)}{RT}. \tag{10.3}$$

In equilibrium the rate of production is equal to the rate of removal, so

$$F\dot{o} = \frac{x(y-b)}{RT}$$

or

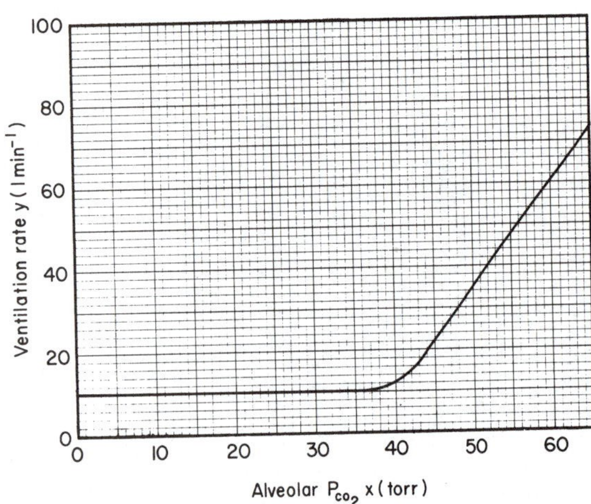

FIGURE 10.3. When P_{CO_2} in the blood rises above 40 torr, the breathing rate increases. P_{CO_2} in the blood is almost the same as in the alveoli.

$$x = \frac{RTF\dot{o}}{y-b}. \tag{10.4}$$

With the proper conversion of units from p to $\dot{o}$, this is Eq. (10.1).

If the metabolic rate were to change without a change in breathing rate, $x = P_{CO_2}$ would change drastically. Suppose that someone exercises moderately so that $p = 60$ mmol min^{-1}, $y = 21$ l min^{-1}, and $x = 45$ torr, point A in Fig. 10.2. If y remained constant while p rose to 80 mmol min^{-1}, x would soar to about 59; if p fell to 40, x would drop to 29. Feedback ensures that this does not happen. One of the feedback mechanisms consists of sensors that measure x and cause y to change. Figure 10.3 shows a typical curve for a 70-kg male [Patton *et al.* (1989, p. 1034)]. (The concentration of CO_2 in blood is nearly the same as in the alveoli.)

10.2. DETERMINING THE OPERATING POINT

We now have two processes relating the steady-state values of x and y. For alveolar gas exchange, we know x as a function of y: $x = g(y,p)$. For the regulatory mechanism, we know $y = f(x)$. Together, these constitute a feedback loop, Fig. 10.4. To find the operating point, these two equations must be solved simultaneously. The easiest way to do this is to plot them on the same graph as in Fig. 10.5. When $p = 60$ mmol min^{-1}, the operating point is at A. In a plot like this the horizontal axis represents the independent variable for one process and the dependent variable for the other.

If the feedback loop includes several variables, for example

$$x = f(w), \quad y = g(x), \quad z = h(y), \quad w = i(z),$$

we can combine three of these equations to get $x = F(y)$ and plot it with $y = g(x)$.

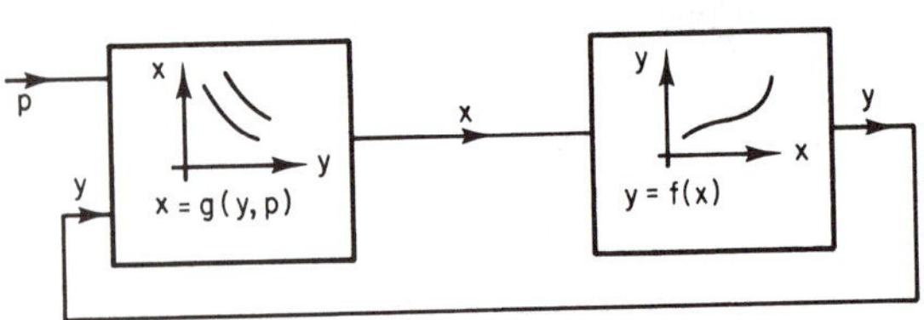

FIGURE 10.4. A general feedback loop. Either box may involve some parameters.

10.3. REGULATION OF A VARIABLE AND OPEN-LOOP GAIN

We can also see from Fig. 10.5 how feedback causes y to change in response to a change in parameter p to reduce the change in x. If y does not change, a change of p from 60 to 80 causes the operating point to go from A to B. In fact, y increases so that the new operating point is at D. The feedback loop is said to regulate the value of x.

The *gain* of a box is the ratio of the change in the output variable to the change in the input variable. For the first box in Fig. 10.4,

$$G_1 = \left(\frac{\Delta x}{\Delta y}\right)_{\substack{\text{box } g \\ p \text{ fixed}}} = \left(\frac{\partial g}{\partial y}\right)_p. \tag{10.5}$$

For the second box,

$$G_2 = \left(\frac{\Delta y}{\Delta x}\right)_{\text{box } f} = \frac{\partial f}{\partial x}. \tag{10.6}$$

The product $G_1 G_2$ is called the *open-loop gain* (OLG). Its name comes from the fact that if the feedback loop is opened at any point and a small change is made in the input

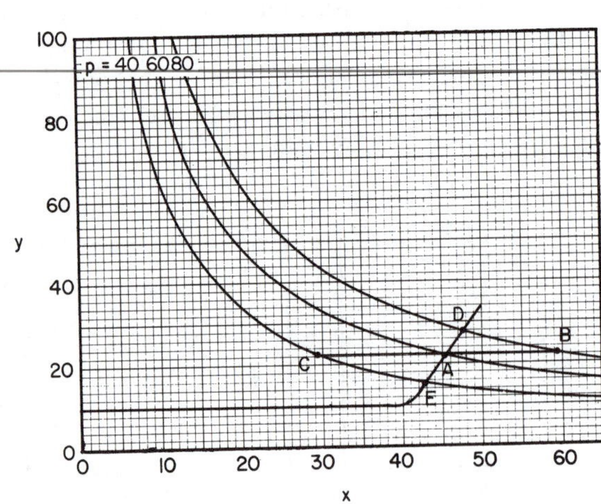

FIGURE 10.5. Regulation of the breathing rate. A change of metabolic rate (parameter p) causes a change in ventilation rate y, so that $x = P_{CO_2}$ does not change as much.

TABLE 11.3. *Fourier coefficients obtained for the square wave fit.*

Term k	a_k	b_k
0	0.000	
1	−0.031	1.273
2	0.000	0.000
3	−0.031	0.424
4	0.000	0.000
5	−0.031	0.253
6	0.000	0.000
7	−0.031	0.181

this gives only a more compact notation. It does not save in the actual calculations.

Figures 11.6–11.9 show fits to a square wave with 128 data points. The function is $y_j=1$, $j=0,\ldots,63$ and $y_j=-1$, $j=64,\ldots,127$. This is an odd function of t. Therefore, the series should contain only sine terms; all a_k should be zero. The calculated coefficients are shown in Table 11.3. Some of the a_k values are small but not exactly zero. This is due to the finite number of data points; the a_k become smaller as N is increased. The even values of the b_k vanish. We will see why below.

Figure 11.6 shows the square wave as dots and $y(x)$ as a smooth curve when $b_1=1.273$ and all the other coefficients are zero. This provides the minimum Q obtainable with a single term. Figure 11.7 shows why Q is larger for any other value of b_1. Figure 11.8 shows the terms for $k=1$ and

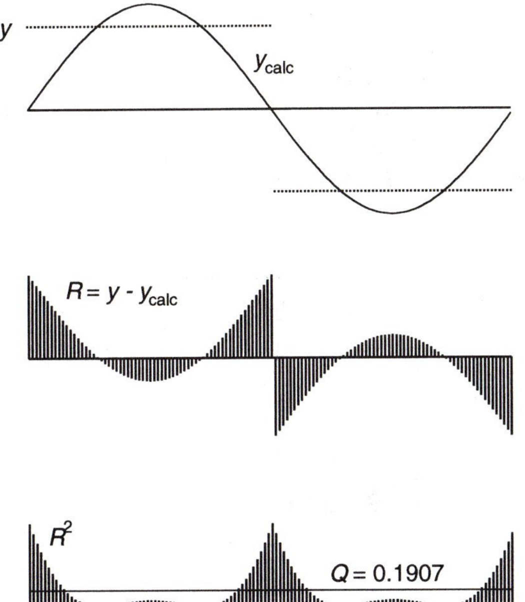

FIGURE 11.6. A square wave $y(t_j)$ and the calculated function $y(t)=b_1\sin(\omega_0 t)$ are shown, along with the residuals and the squares of the residuals for each point. The value of b_1 is $4/\pi$, which minimizes Q for that term.

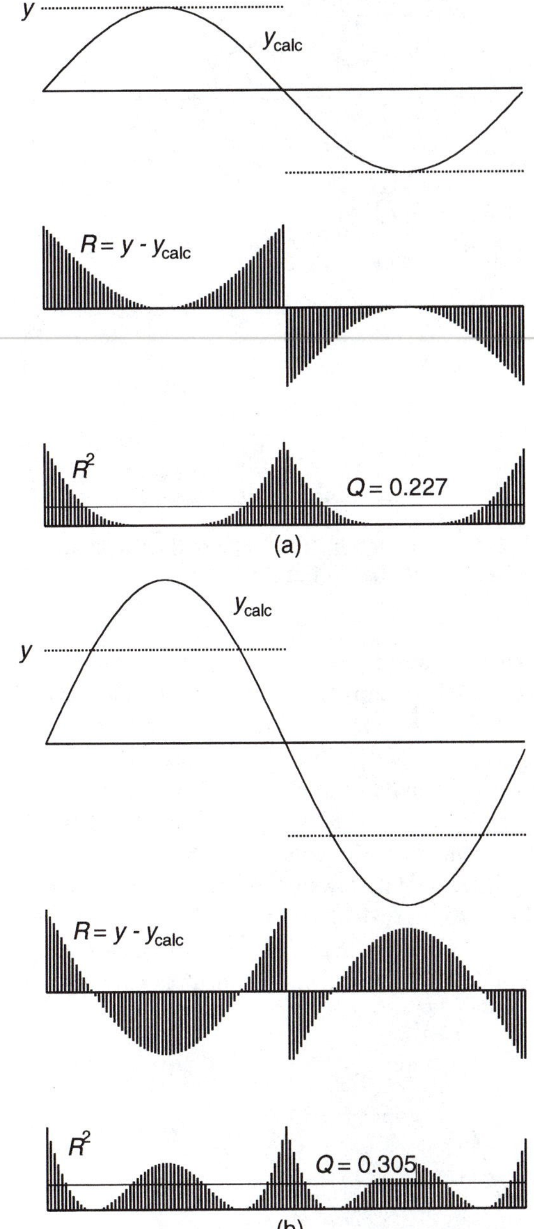

FIGURE 11.7. A single term is used to approximate the square wave. (a) $b_1=1.00$, which is too small a value. (b) $b_1=1.75$, which is too large. In both cases Q is larger than the minimum value for a single term, shown in Fig. 11.6.

$k=3$. The value of Q is further reduced. Figure 11.9 shows why even terms do not reduce Q. In this case $b_2=0.5$ has been added to b_1. There is improvement for $0<t<T/4$ and $3T/4<t<T$, but between those regions the fit is made worse.

If the data points y_j are a sine or cosine with exact frequency ω_0 or a harmonic, and if no random errors superimposed on the data, then only the coefficients corresponding to those terms will be nonzero. The reason is that if the function is exactly periodic, then by sampling for one period we have effectively sampled for an infinite time. If

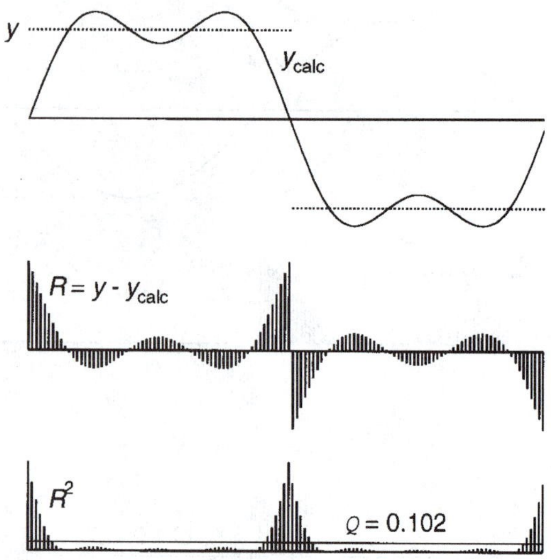

FIGURE 11.8. Terms b_1 and b_3 have their optimum values. Q is smaller than that in Fig. 11.6.

the frequency of y is not an exact harmonic of ω_0, then the Fourier series contains terms at several frequencies. This is shown in Figs. 11.10 and 11.11 for the data $y_j = \sin[6.6\pi(j+1)/N]$. These figures also show the effect of increasing the number of samples. Figure 11.10 shows the y_j for $N=20$ and $N=80$ samples during the period of the measurement. For 20 samples, $n=9$; for 80 samples, $n=39$. Figure 11.11 shows $(a_k^2+b_k^2)^{1/2}$ for both sample sets for $k=0$ to 9, calculated using Eqs. (11.25). For 80 samples, the value of $(a_k^2+b_k^2)^{1/2}$ is very small for $k>9$ and is not plotted. The frequency spectrum is virtually independent of the number of samples. The largest amplitude occurs for $k=3$. If one imagines a smooth curve drawn through the histogram, its peak would be slightly above $k=3$. (Note the asymmetry between $k=2$ and $k=4$.) We will see later that the width of this curve depends on the duration of the measurement, T. [There are slight differences if the function is shifted one point, $y(t_j) = \sin(6.6\pi j/N)$.]

Figures 11.12 and 11.13 show the number of spontaneous births per hour vs local time of day for several hundred thousand live births in various parts of the world. Figure 11.12 shows the best fit for $k=1$ ($T=24$ h); in Fig. 11.13 the $k=3$ term ($T=8$ h) has been added.

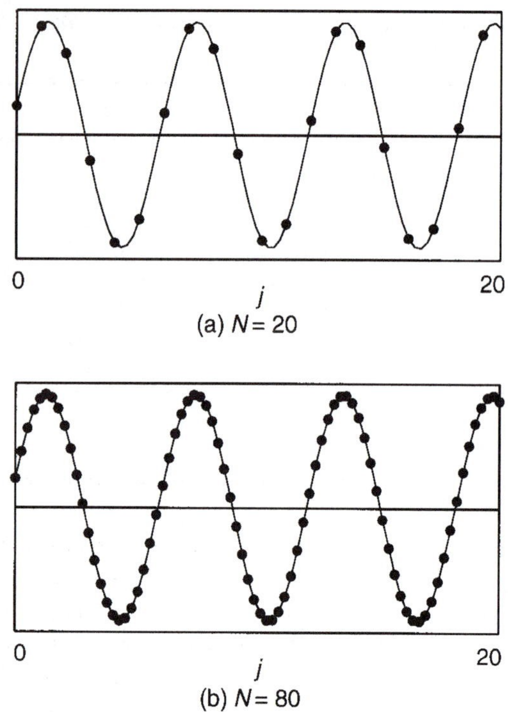

FIGURE 11.10. Sine wave $y_j = \sin[6.6\pi(j+1)/N]$ (a) with 20 data points and (b) with 80 data points.

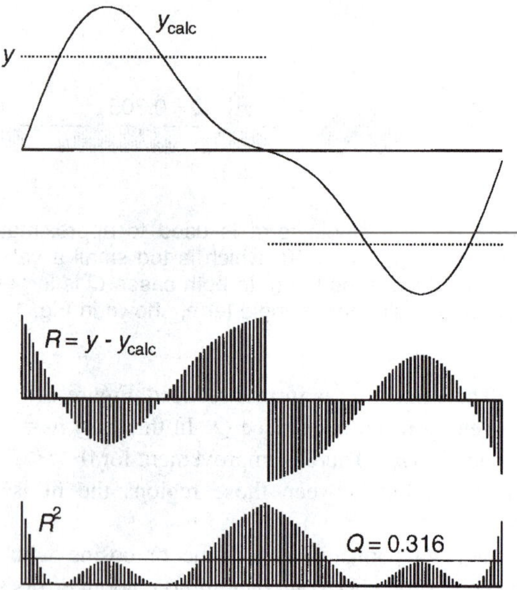

FIGURE 11.9. This shows why even terms do not contribute. A term $b_2 = 0.5$ has been added to a term with the correct value of b_1. It improves the fit for $T < T/4$ and $T > 3T/4$ but then makes it worse between $T/4$ and $3T/4$.

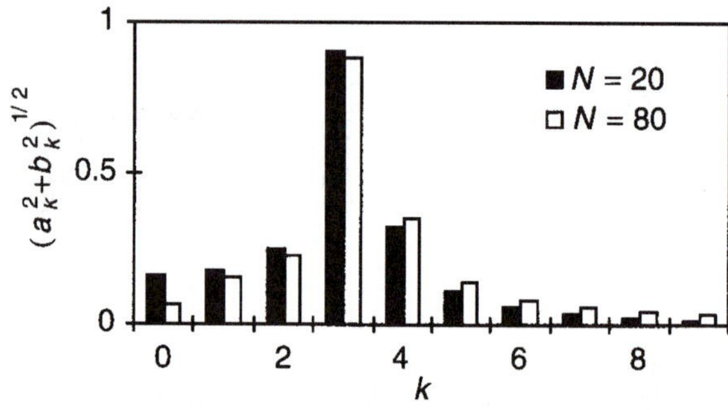

FIGURE 11.11. The amplitude of the mixed sine and cosine coefficients $(a_k^2+b_k^2)^{1/2}$ vs k for the function $y_j = \sin[6.6\pi(j+1)/N]$. The signal is sampled for 20 or 80 data points. The amplitude spectrum is independent of the number of samples.

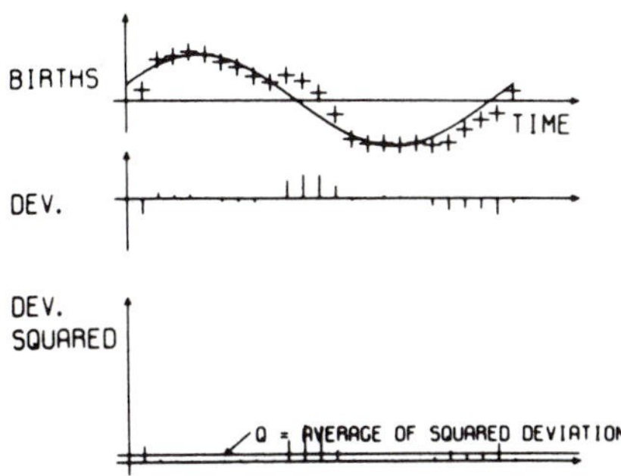

FIGURE 11.12. The rate of spontaneous births is plotted vs time of day. The amplitude and phase have been adjusted to give the smallest value of Q. From R. K. Hobbie and F. Halberg (1974). Rhythmometry made easy. In M. Ferin *et al.*, eds. *Biorhythms in Human Reproduction.* Wiley, New York, pp. 37–41. Reproduced by permission of John Wiley & Sons.

The calculation of the Fourier coefficients using our equations involves N evaluations of the sine or cosine, N multiplications, and N additions for each coefficient. There are N coefficients, so that there must be N^2 evaluations of the sines and cosines. This uses a lot of computer time. It is possible to reduce sharply the number of sine or cosine evaluations by using the formulas

$$\cos(kx)=\cos[(k-1)x]\cos(x)-\sin[(k-1)x]\sin(x),$$

$$\sin(kx)=\sin[(k-1)x]\cos(x)+\cos[(k-1)x]\sin(x).$$

However, about N^2 multiplications are still required. Cooley and Tukey (1965) showed that it is possible to group the data in such a way that the number of multiplications is about $(N/2)\log_2 N$ instead of N^2, a technique known as the *fast Fourier transform* (FFT). For example, for $1024=2^{10}$ data points, $N^2=1\,048\,576$, while $(N/2)\log_2 N=(512)(10)=5120$. This speeds up the calculation by a factor of 204. The techniques for the FFT are discussed by many authors [see Press *et al.* (1992) or Visscher (1996)]. Bracewell (1990) has written an interesting review of all the popular numerical transforms. He points out that the grouping used in the FFT dates back to Gauss in the early nineteenth century.

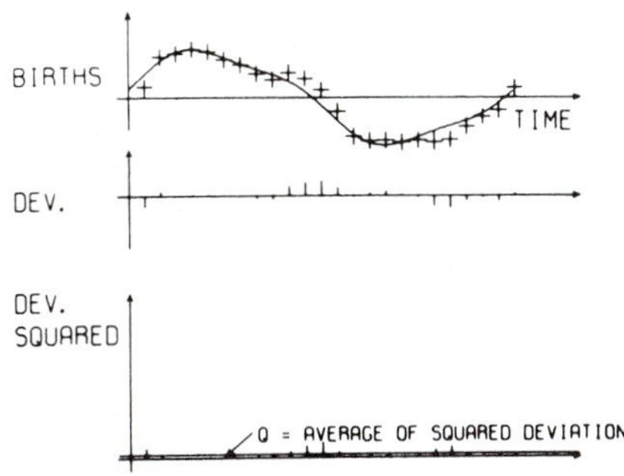

FIGURE 11.13. An additional term with a period of 8 h (the third harmonic) has been added to the expression used in Fig. 11.12. Q is decreased further. From R. K. Hobbie and F. Halberg (1974). Rhythmometry made easy. In M. Ferin *et al.*, eds. *Biorhythms in Human Reproduction.* Wiley, New York, pp. 37–41. Reproduced by permission of John Wiley & Sons.

11.5. PITFALLS OF DISCRETE SAMPLING: ALIASING

We saw in the preceding section that N samples in time T allow the determination of unique Fourier coefficients only for the terms from $k=0$ to $n=(N-1)/2$. This means that for a sampling interval T/N, the maximum angular frequency is $(N-1)\omega_0/2$. The period of the highest frequency that can be determined is $T_{\min}=2T/(N-1)$. This is approximately twice the spacing of the data points. One must sample at least twice per period to determine the coefficient at a particular frequency.

If a frequency component is present whose period is less than twice the sampling interval, it will appear in the analysis at a much *lower* frequency. This is the familiar stroboscopic effect in which the wheels of the stagecoach appear to rotate backward because the samples (movie frames) are not made rapidly enough. In signal analysis, this is called *aliasing*. It can be seen in Fig. 11.14, which shows a sine wave sampled at regularly spaced intervals that are longer than half a period.

This phenomenon is inescapable if frequencies greater than $(N-1)\omega_0/2$ are present. They must be removed by analog or digital techniques *before* the sampling is done. For a more detailed discussion, see Blackman and Tukey (1958) or Press *et al.* (1992). An example of aliasing is found in a later section, in Fig. 11.42. Maughan *et al.* (1973) have pointed out how researchers have been "stung" by this problem in hematology.

11.6. FOURIER SERIES FOR A PERIODIC FUNCTION

It is also possible to define the Fourier series for a periodic function when the function is given rather than discrete values of y. The function need only be *piecewise continuous*, that is, with a finite number of discontinuities. The calculated function is given by the analog of Eq. (11.14):

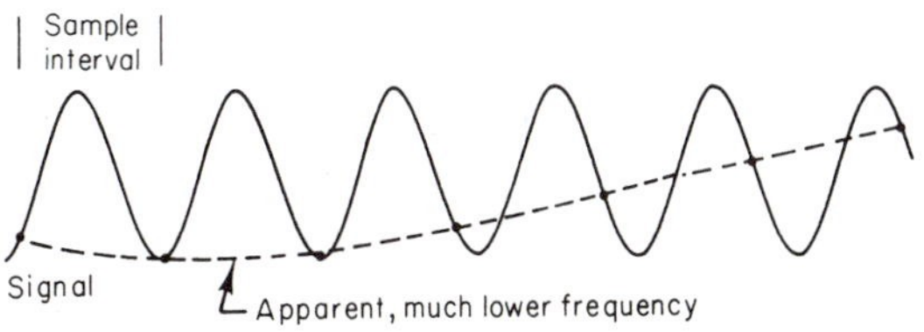

FIGURE 11.14. An example of aliasing. The data are sampled less often than twice per period and appear to be at a much lower frequency.

$$y_{\text{calc}}(t)=a_0+\sum_{k=1}^{n} a_k \cos(k\omega_0 t)+\sum_{k=1}^{n} b_k \sin(k\omega_0 t). \tag{11.30}$$

The quantity to be minimized is still the mean square error, in this case

$$Q=\frac{1}{T}\int_{t'}^{t'+T}[y(t)-y_{\text{calc}}(t)]^2 dt. \tag{11.31}$$

When Q is a minimum, $\partial Q/\partial a_m$ and $\partial Q/\partial b_m$ must be zero for each coefficient. For example,

$$\frac{\partial Q}{\partial a_m}=\frac{1}{T}\frac{\partial}{\partial a_m}\int_0^T\left(y(t)-a_0-\sum_{k=1}^{n}[a_k\cos(k\omega_0 t)+b_k\sin(k\omega_0 t)]\right)^2 dt$$

$$=-\frac{2}{T}\int_0^T\left[\left(y(t)-a_0-\sum_{k=1}^{n}[a_k\cos(k\omega_0 t)+b_k\sin(k\omega_0 t)]\right)\cos(m\omega_0 t)\right]dt.$$

This equation must be satisfied for each value of m from 1 to n. If the order of integration and summation is interchanged, the result is

$$-\frac{2}{T}\left(\int_0^T y(t)\cos(m\omega_0 t)\,dt-a_0\int_0^T\cos(m\omega_0 t)\,dt-\sum_{k=1}^{n}a_k\int_0^T\cos(k\omega_0 t)\cos(m\omega_0 t)\,dt-\sum_{k=1}^{n}b_k\int_0^T\sin(k\omega_0 t)\cos(m\omega_0 t)\,dt\right)=0.$$

The integral of $\cos(m\omega_0 t)$ over a period vanishes. The last two integrals are of the form given in Appendix E, Eqs. (E.4) and (E.5):

$$\int_0^T\cos(k\omega_0 t)\cos(m\omega_0 t)\,dt=\begin{cases}0, & k\neq m\\ T/2, & k=m\end{cases}$$

$$\int_0^T\sin(k\omega_0 t)\cos(m\omega_0 t)\,dt=0. \tag{11.32}$$

These results are the *orthogonality relations* for the trigonometric functions. Inserting these values, we find that only one term in the summation over k remains, and we have

$$-\frac{2}{T}\left(\int_0^T y(t)\cos(m\omega_0 t)\,dt-a_m\int_0^T\cos^2(m\omega_0 t)\,dt\right)=0.$$

This equation is satisfied only if the term in large parentheses vanishes, or

$$a_m=\frac{2}{T}\int_0^T y(t)\cos(m\omega_0 t)\,dt. \tag{11.33a}$$

Minimizing with respect to b_m gives

$$b_m=\frac{2}{T}\int_0^T y(t)\sin(m\omega_0 t)\,dt, \tag{11.33b}$$

and minimizing with respect to a_0 gives

$$a_0=\frac{1}{T}\int_0^T y(t)\,dt. \tag{11.33c}$$

These equations are completely general. Because of the orthogonality of the integrals, the coefficients are independent, just as they were in the discrete case for equally spaced data. This is not surprising, since the continuous case corresponds to an infinite set of uniformly spaced data.

Note the similarity of these equations to the discrete results, Eqs. (11.24). In each case a_0 is the average of the function over the period. The other coefficients are twice the average of the signal multiplied by the sine or cosine whose coefficient is being calculated. The integrals can be taken over any period. Sometimes it is convenient to make the interval $-T/2$ to $T/2$. As we would expect, the integrals involving sines vanish when y is an even function, and those involving cosines vanish when y is an odd function.

For the square wave $y(t)=1$, $0<t<T/2$; $y(t)=-1$, $T/2<t<T$, we find

$$a_k=0$$
$$b_k=\begin{cases}0, & k \text{ even}\\ 4/\pi k, & k \text{ odd.}\end{cases} \tag{11.34}$$

Table 11.4 shows the first few coefficients for the Fourier series representing the square wave, obtained from Eq. (11.34). They are the same as those for the discrete data in Table 11.3. Figure 11.15 shows the fits for $n=3$ and $n=39$. As the number of terms in the fit is increased, the value of Q decreases. However, spikes of constant height (about 18% of the amplitude of the square wave or 9% of the discontinuity in y) remain. These are seen in Fig. 11.15. These spikes appear whenever there is a discontinuity in y and are called the *Gibbs phenomenon*. A discussion of

TABLE 11.4. *Value of the kth coefficient and the value of Q when terms through the kth are included from Eq. (11.34).*

k	b_k	Q
1	1.2732	0.19
3	0.4244	0.10
5	0.2546	0.07
7	0.1819	0.05
9	0.1415	0.04

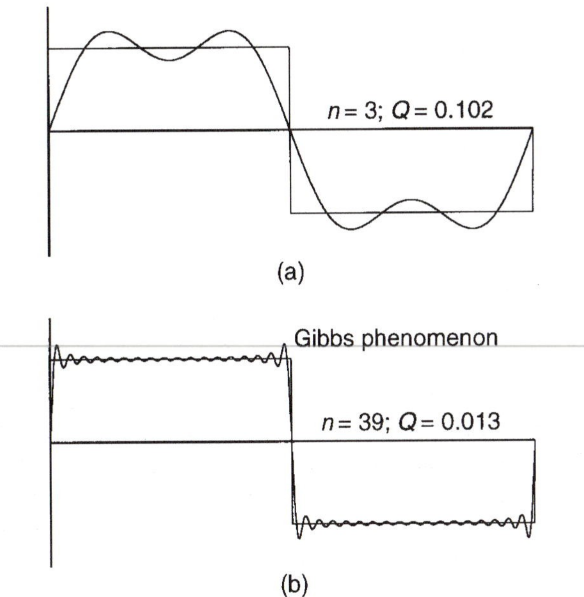

FIGURE 11.15. Fit to the square wave. (a) Fit with the terms for $k=1$ and $k=3$. The value of Q is 0.1. (b) Fit with terms through $k=39$. Q is very small, but the Gibbs phenomenon—spikes near the discontinuity—is apparent.

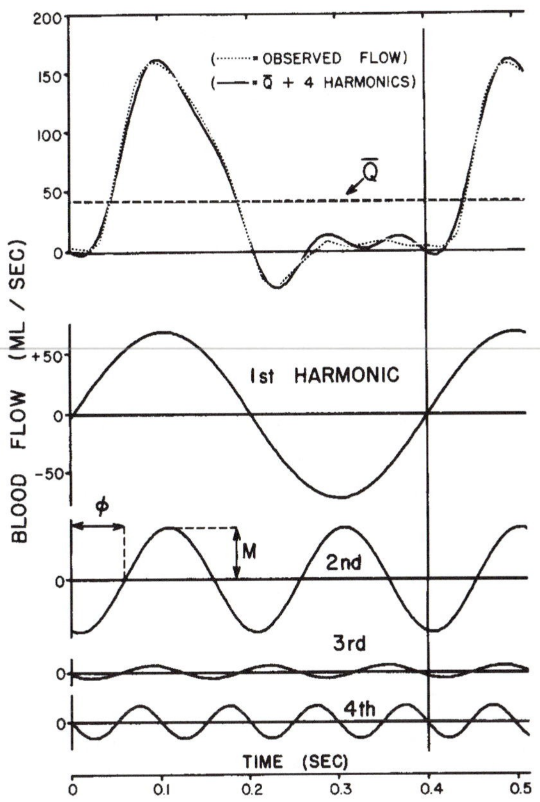

FIGURE 11.16. Analysis of the pulmonary arterial blood flow in a dog, in terms of a Fourier series. From W. R. Milnor (1972). Reproduced by permission from the *New England Journal of Medicine*. Drawing courtesy of Professor Milnor.

their origin can be found in Guillemin (1949, p. 436) and in Jeffreys and Jeffreys (1956, p. 455). They are another indication that least-squares may not always be the most desirable criterion for an approximation.

Figure 11.16 shows the blood flow in the pulmonary artery of a dog as a function of time. It has been fitted by a mean and four terms of the form $M_k \sin(k\omega_0 t - \phi_k)$. The technique is useful because the elastic properties of the arterial wall can be described in terms of sinusoidal pressure variations at various frequencies [Milnor (1972)].

11.7. THE POWER SPECTRUM

Since the power dissipated in a resistor is v^2/R, the square of any function (or signal) is often called the power. For a periodic signal $y(t)$, we have seen that the signal can be written as

$$y(t) = a_0 + \sum_{k=1}^{n} [a_k \cos(k\omega_0 t) + b_k \sin(k\omega_0 t)] + \epsilon_n(t), \tag{11.35}$$

where the error $\epsilon_n(t)$ is the difference between the signal and the sum over n terms. [This equation defines $\epsilon_n(t)$.] The coefficients are given by Eqs. (11.33).

The average "power" in the signal is defined to be[3]

$$\langle y^2 \rangle = \lim_{T' \to \infty} \frac{1}{2T'} \int_{-T'}^{T'} y^2(t)\, dt. \tag{11.36}$$

For a periodic signal, the same result can be obtained by integrating over one period:

$$\langle y^2 \rangle = \frac{1}{T} \int_0^T y^2(t)\, dt. \tag{11.37}$$

To calculate this using Eq. (11.35) for $y(t)$, we have to write the sum twice and multiply both sums together:

$$\frac{1}{T}\int_0^T y^2(t)\, dt = \frac{1}{T}\int_0^T \left(a_0 + \sum_{k=1}^{n} [a_k \cos(k\omega_0 t) + b_k \sin(k\omega_0 t)] + \epsilon_n(t) \right) \left(a_0 + \sum_{j=1}^{n} [a_j \cos(j\omega_0 t) + b_j \sin(j\omega_0 t)] + \epsilon_n(t) \right) dt.$$

When these terms are multiplied together and written out, we have

[3]The time average of a variable will be denoted by $\langle\ \rangle$ brackets.

$$\langle y^2\rangle=\frac{1}{T}\int_0^T dt\Bigg(\overset{\text{(i)}}{a_0^2}+\overset{\text{(ii)}}{2a_0\sum_{k=1}^{n}[a_k\cos(k\omega_0 t)}+\overset{\text{(iii)}}{b_k\sin(k\omega_0 t)]}+\overset{\text{(iv)}}{2a_0\epsilon_n(t)}+\sum_{k=1}^{n}[\overset{\text{(v)}}{a_k^2\cos^2(k\omega_0 t)}+\overset{\text{(vi)}}{b_k^2\sin^2(k\omega_0 t)}]$$

$$+\sum_{k=1}^{n}\sum_{j\neq k}[\overset{\text{(vii)}}{a_k a_j\cos(k\omega_0 t)\cos(j\omega_0 t)}+\overset{\text{(viii)}}{b_k b_j\sin(k\omega_0 t)\sin(j\omega_0 t)}]+\overset{\text{(ix)}}{2\sum_{k=1}^{n}\sum_{j=1}^{n}a_k b_j\cos(k\omega_0 t)\sin(j\omega_0 t)}$$

$$+\overset{\text{(x)}}{2\epsilon_n(t)\sum_{k=1}^{n}[a_k\cos(k\omega_0 t)+b_k\sin(k\omega_0 t)]}+\overset{\text{(xi)}}{\epsilon_n^2(t)}\Bigg).$$

Each term has been labeled (i) through (xi). Assume that the function y is sufficiently well behaved so that the order of integration and summation can be interchanged. Term (i) gives a_0^2. Terms (ii) and (iii) are integrals of the cosine or sine over an integral number of cycles and vanish. Term (iv) gives[4]

$$2a_0\frac{1}{T}\int_0^T \epsilon_n(t)\,dt=0.$$

Terms (v) and (vi) give $a_k^2/2$ and $b_k^2/2$. Terms (vii), (viii), and (ix) all vanish because of Eq. (11.32). Terms like (x) vanish because $\epsilon_n(t)$ can be approximated by a sum of sine and cosine terms extending from $k=n+1$ to infinity that are orthogonal to terms for $1<k<n$. Term (xi) is $\langle\epsilon_n^2\rangle$. We finally have for the average power

$$\langle y^2(t)\rangle=a_0^2+\tfrac{1}{2}\sum_{k=1}^{n}(a_k^2+b_k^2)+\langle\epsilon_n^2\rangle=\sum_{k=0}^{n}\Phi_k+\langle\epsilon_n^2\rangle. \tag{11.38}$$

The coefficients are defined by Eqs. (11.33). We could have made a similar argument for the discrete Fourier series of Eqs. (11.24) or (11.25) and obtained the same result. In both cases the average power is a sum of terms Φ_k that represent the average power at each frequency. The term $\Phi_0=a_0^2$ is the average of the square of the zero-frequency or dc (direct-current) term; $\Phi_k=\frac{1}{2}(a_k^2+b_k^2)$ is the average of the squares of the terms $a_k\cos(k\omega_0 t)$ and $b_k\sin(k\omega_0 t)$; and $\langle\epsilon_n^2\rangle$ is the average of the square of the error term. Figure 11.17 shows the power spectrum of the square wave that was used in the example.

11.8. CORRELATION FUNCTIONS

The correlation function is useful to test whether two functions of time are correlated, that is, whether a change in one is accompanied by a change in the other. Let the two variables to be tested be $y_1(t)$ and $y_2(t)$. The change in y_2 may occur earlier or later than the change in y_1; therefore the correlation must be examined as one of the variables is shifted in time. Examples of pairs of variables that may be correlated are wheat price and rainfall, urinary output and fluid intake, and the voltage changes at two different points along the nerve cell axon. The variables may or may not be periodic. Exhibiting a correlation does not establish a cause-and-effect relationship. (The height of a growing tree may correlate for several years with an increase in the stock market.)

To calculate the cross-correlation function of y_1 and y_2, advance $y_2(t)$ by an amount τ, multiply y_1 by the shifted y_2, and integrate the product. Figure 11.18 shows the process for two pulses. The second pulse occurs 2 s later than the first. As the second pulse is advanced the pulses begin to overlap. When the second pulse has been advanced by 2 s the overlap is greatest; as it is advanced more, the overlap falls to zero. The cross-correlation function depends on τ and is plotted in Fig. 11.18(c). The mathematical statement of this procedure for a pulse is

$$\phi_{12}(\tau)=\int_{-\infty}^{\infty}y_1(t)y_2(t+\tau)dt. \tag{11.39}$$

The integrand makes a positive contribution to the integral if $y_1(t)$ and $y_2(t+\tau)$ are both positive at the same time or both negative at the same time. It makes a negative contribution if one function is positive while the other is negative.

If the signals are not pulses, then the cross-correlation integral is defined as

$$\phi_{12}(\tau)=\langle y_1(t)y_2(t+\tau)\rangle. \tag{11.40}$$

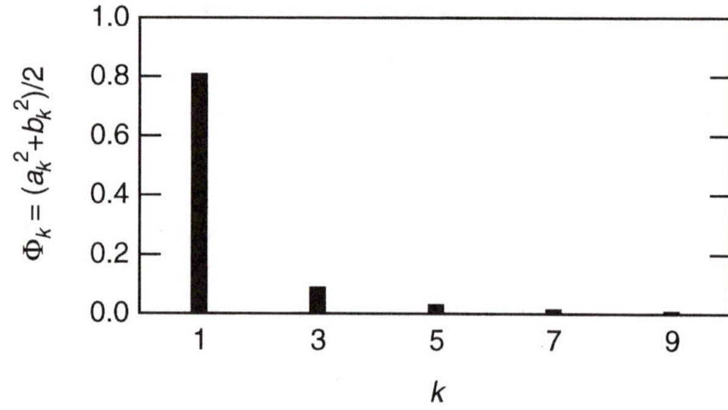

FIGURE 11.17. The power spectrum Φ_k for the square wave of Fig. 11.6 or Fig. 11.16, calculated using the values of b_k from Table 11.4.

[4]The quantity $y(t)-a_0$ has an average of zero. Since all the sine and cosine terms have an average of zero, ϵ_n also has an average of zero.

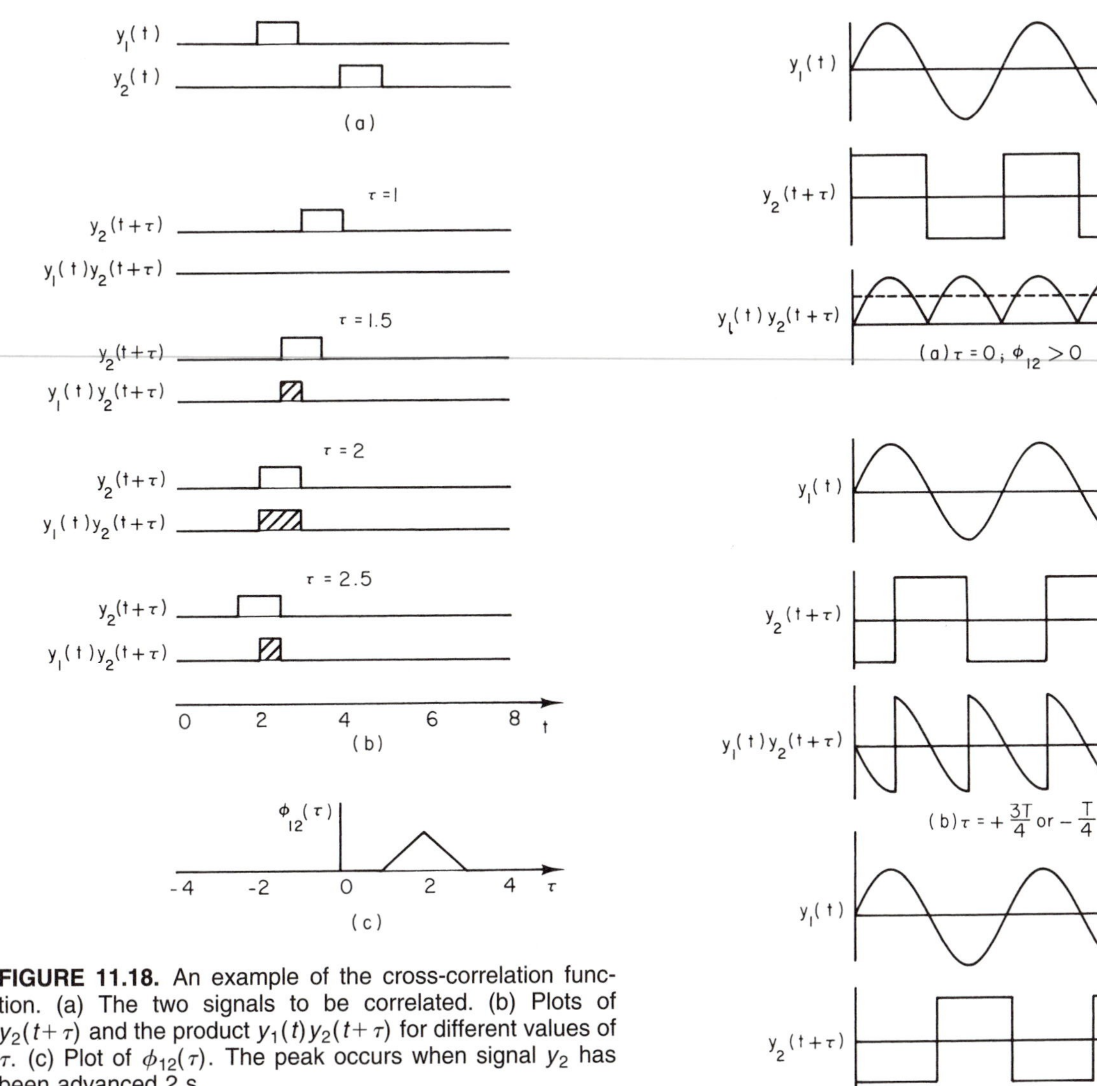

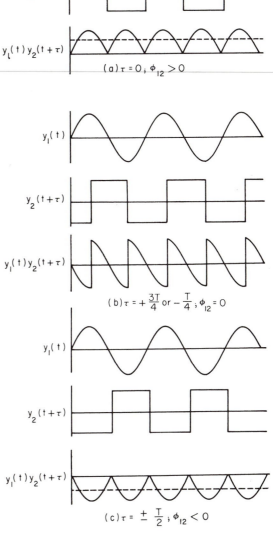

FIGURE 11.18. An example of the cross-correlation function. (a) The two signals to be correlated. (b) Plots of $y_2(t+\tau)$ and the product $y_1(t)y_2(t+\tau)$ for different values of τ. (c) Plot of $\phi_{12}(\tau)$. The peak occurs when signal y_2 has been advanced 2 s.

FIGURE 11.19. Cross correlation of a square wave and a sine wave of the same period.

As before, the average is the integral over a long time divided by the time interval:

$$\phi_{12}(\tau) = \lim_{T\to\infty} \frac{1}{2T} \int_{-T}^{T} y_1(t) y_2(t+\tau)\, dt. \quad (11.41)$$

If the signals have period T, the average can be taken by integrating over a single period:

$$\phi_{12}(\tau) = \frac{1}{T} \int_{t'}^{t'+T} y_1(t) y_2(t+\tau)\, dt. \quad (11.42)$$

Note the difference in units between ϕ_{12} as defined for pulses in Eq. (11.39) where the units of ϕ are the units of y^2 times time, and ϕ_{12} defined in Eq. (11.40) or (11.41) where the units are those of y^2.

As an example of the cross correlation, consider a square wave that has value ± 1 and a sine wave with the same period (Fig. 11.19). When the square wave and sine wave are in phase, the product is always positive and the cross correlation has its maximum value. As the square wave is shifted the product is sometimes positive and sometimes negative. When they are 1/4 period out of phase, the average of the integrand is zero, as shown in Fig. 11.19(b). Still more shift results in the correlation function becoming positive again, with a shift of one full period giving the same result as no shift.

The cross correlation depends only on the relative shift of the two signals. It does not matter whether y_2 is advanced by an amount τ or y_1 is delayed by the same amount:

$$\phi_{21}(-\tau) = \phi_{12}(\tau). \quad (11.43)$$

The *autocorrelation* function is the correlation of the signal with itself:

$$\phi_{11}(\tau)=\begin{cases} \int y_1(t)y_1(t+\tau)dt & \text{(pulse)} \quad (11.44) \\ \langle y_1(t)y_1(t+\tau)\rangle & \text{(nonpulse).} \quad (11.45) \end{cases}$$

Since the signal is correlated with itself, advancing one copy of the signal is the same as delaying the other. The autocorrelation is an even function of τ:

$$\phi_{11}(\tau)=\phi_{11}(-\tau). \tag{11.46}$$

The autocorrelation function for a sine wave can be calculated analytically. If the amplitude of the sine wave is A,

$$\begin{aligned}\phi_{11}(\tau) &= \frac{A^2}{T}\int_0^T \sin(\omega t)\sin(\omega t+\omega\tau)\ dt \\ &= \frac{A^2}{T}\int_0^T \sin(\omega t)[\sin(\omega t)\cos(\omega\tau) \\ &\quad +\cos(\omega t)\sin(\omega\tau)]\ dt \\ &= A^2\cos(\omega\tau)\left(\frac{1}{T}\int_0^T \sin^2(\omega t)\ dt\right)+A^2\sin(\omega\tau) \\ &\quad \times\left(\frac{1}{T}\int_0^T \sin(\omega t)\cos(\omega t)\ dt\right).\end{aligned}$$

It is shown in Appendix E that the first term in large parentheses is $\frac{1}{2}$ and the second is 0. Therefore the autocorrelation of the sine wave is

$$\phi_{11}(\tau)=\frac{A^2}{2}\cos(\omega\tau). \tag{11.47}$$

The same result can be obtained using the averaging notation.

As a final example, consider the autocorrelation of a square wave of unit amplitude. One period is drawn in Fig. 11.20 showing the wave, the advanced wave, and the product. The average is the net area divided by T. The area above the axis is $(2)(T/2-\tau)(1)$ since there are two rectangles of height 1 and width $T/2-\tau$. From this must be subtracted the area of the two rectangles of height 1 and width τ that are below the axis. The net area is $T-4\tau$. The autocorrelation function is

$$\phi_{11}(\tau)=1-4\tau/T, \quad 0<\tau<T/2. \tag{11.48}$$

The plot of the integrand in Fig. 11.20 is only valid for $0<\tau<T/2$. We can use the fact that the autocorrelation is an even function to draw ϕ for $-T/2<\tau<0$. We then have ϕ for the whole period. It is plotted in Fig. 11.21.

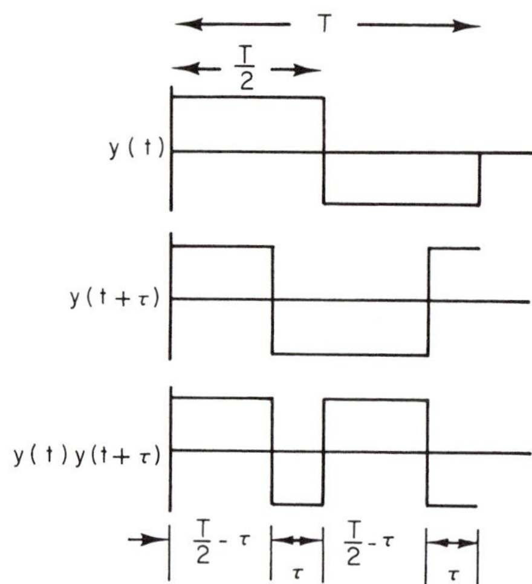

FIGURE 11.20. Plots of $y(t)$, $y(t+\tau)$, and their product for a square wave.

11.9. THE AUTOCORRELATION FUNCTION AND THE POWER SPECTRUM

We saw that the power spectrum of a periodic signal is related to the coefficients in its Fourier transform:

$$\langle y^2(t)\rangle=a_0^2+\tfrac{1}{2}\sum_{k=1}^{n}(a_k^2+b_k^2), \tag{11.38}$$

with the term for each value of k representing the amount of power carried in the signal component at that frequency. The Fourier series for the autocorrelation function carries the same information. To see this, calculate the autocorrelation function of

$$y_1(t)=a_0+\sum_{k=1}^{n}[a_k\cos(k\omega_0 t)+b_k\sin(k\omega_0 t)].$$

We can write

$$\begin{aligned}\langle y_1(t)y_1(t+\tau)\rangle &= \left\langle\left(a_0+\sum_{k=1}^{n}[a_k\cos(k\omega_0 t)\right.\right. \\ &\quad \left. +b_k\sin(k\omega_0 t)]\right)\left(a_0\right. \\ &\quad +\sum_{j=1}^{n}\{a_j\cos[j\omega_0(t+\tau)] \\ &\quad \left.\left. +b_j\sin[j\omega_0(t+\tau)]\}\right)\right\rangle.\end{aligned}$$

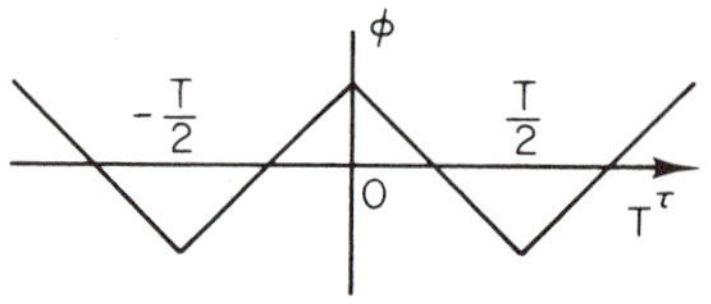

FIGURE 11.21. Plot of $\phi_{11}(\tau)$ for the square wave.

The next step is to multiply out all the terms as we did when deriving Eq. (11.38). We then use the trigonometric identities[5]

$$\cos(x+y)=\cos x \cos y-\sin x \sin y,$$

$$\sin(x+y)=\cos x \sin y+\sin x \cos y.$$

For many of the terms, either the averages are zero or pairs of terms cancel. We finally obtain

$$\phi_{11}(\tau)=a_0^2+\sum_{k=1}^{n} \tfrac{1}{2}(a_k^2+b_k^2)\cos(k\omega_0\tau). \quad (11.49)$$

This has only cosine terms, since the autocorrelation function is even.

For zero shift,

$$\phi_{11}(0)=a_0^2+\sum_{k=1}^{n} \tfrac{1}{2}(a_k^2+b_k^2).$$

Comparison with Eq. (11.38) shows that this is the power in the signal. We can get this result directly from Eq. (11.36). The integral is the same as the definition of the autocorrelation function when $\tau=0$.

The Fourier series for the autocorrelation function is particularly easy to obtain. We need only pick out the coefficients in Eq. (11.49). Write the Fourier expansion of the autocorrelation function as

$$\phi_{11}(\tau)=\alpha_0+\sum_{k=1}^{n} \alpha_k \cos(k\omega_0\tau). \quad (11.50)$$

Comparing terms in Eqs. (11.49) and (11.50) shows that $\alpha_0=a_0^2$ and $\alpha_k=\frac{1}{2}(a_k^2+b_k^2)$. We can also compare these with the definition of Φ_k in Eq. (11.38) and say that

$$\Phi_0=[\text{average dc (zero-frequency) power}]=\alpha_0=a_0^2, \quad (11.51)$$

$$\Phi_k=(\text{average power at frequency } k\omega_0)=\alpha_k =\tfrac{1}{2}(a_k^2+b_k^2).$$

The autocorrelation function contains no phase information about the signal. This is reflected in the fact that $\alpha_k =\frac{1}{2}(a_k^2+b_k^2)$; the sine and cosine terms at a given frequency are completely mixed together.

There are two ways to find the power Φ_k at frequency $k\omega_0$. Both are shown in Fig. 11.22. The function $y(t)$ and its Fourier coefficients are completely equivalent, and one can go from one to the other. Squaring the coefficients and adding them gives the power spectrum. This is a one-way process; once they have been squared and added, there is no way to separate them again. The autocorrelation function also involves squaring and adding and is a one-way process.

[5] The virtue of the complex notation is that these addition formulae become the standard rule for multiplying exponentials: $e^{i(x+y)}=e^{ix}e^{iy}$.

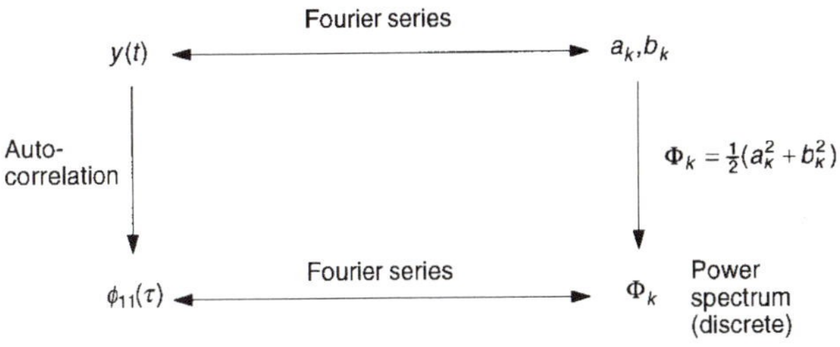

FIGURE 11.22. The power spectrum of a periodic signal can be obtained from either the squares of the Fourier coefficients of the signal or from the Fourier coefficients of the autocorrelation function.

The autocorrelation function and the power spectrum are related by a Fourier series and can be calculated from each other.

11.10. NONPERIODIC SIGNALS AND FOURIER INTEGRALS

Sometimes we have to deal with a signal that is a pulse that occurs just once. Several pulses are shown in Fig. 11.23; they come in an infinite variety of shapes. Noise is another signal that never repeats itself and is therefore not periodic. The *Fourier integral* or *Fourier transform* is an extension of the Fourier series that allows us to deal with nonperiodic signals.

The Fourier series expansion of a periodic function $y(t)$ was seen earlier to be

$$y(t)=a_0+\sum_{k=1}^{n} a_k \cos(k\omega_0 t)+\sum_{k=1}^{n} b_k \sin(k\omega_0 t), \quad (11.30)$$

with the coefficients given by

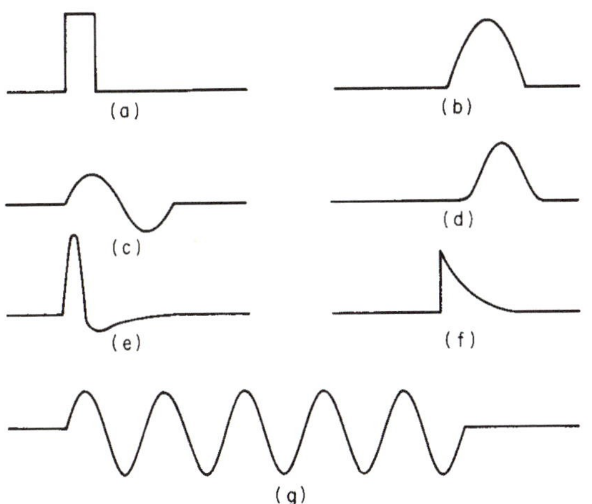

FIGURE 11.23. Various pulses. The common feature is that they occur once. (a) Square pulse. (b) Half cycle of a sine wave. (c) One cycle of a sine wave. (d) Gaussian. (e) Nerve pulse. (f) Exponentially decaying pulse. (g) Gated sine wave.

$$a_0=\frac{1}{T}\int_0^T y(t)\,dt,$$

$$a_m=\frac{2}{T}\int_0^T y(t)\cos(m\omega_0 t)\,dt,$$

$$b_m=\frac{2}{T}\int_0^T y(t)\sin(m\omega_0 t)\,dt. \qquad (11.33)$$

Since $y(t)$ has period T, the integrals in Eqs. (11.33) can be over any interval that is one period long. Let us therefore make the limits of integration $-T/2$ to $T/2$ and also remember that $1/T=\omega_0/2\pi$. With these substitutions, Eqs. (11.33) become

$$a_0=\frac{\omega_0}{2\pi}\int_{-T/2}^{T/2} y(t)\,dt,$$

$$a_k=\frac{\omega_0}{\pi}\int_{-T/2}^{T/2} y(t)\cos(k\omega_0 t)\,dt,$$

$$b_k=\frac{\omega_0}{\pi}\int_{-T/2}^{T/2} y(t)\sin(k\omega_0 t)\,dt. \qquad (11.52)$$

Now allow k to have negative as well as positive values. If the coefficients for negative k are also defined by Eqs. (11.52), they have the properties [since $\cos(k\omega_0 t)=\cos(-k\omega_0 t)$ and $\sin(k\omega_0 t)=-\sin(-k\omega_0 t)$],

$$a_{-k}=a_k, \quad b_{-k}=-b_k.$$

Therefore the terms $a_k\cos(k\omega_0 t)$ and $b_k\sin(k\omega_0 t)$ in Eq. (11.30) are the same function of t whether k is positive or negative. By introducing negative values of k we can make the coefficients in front of the integrals for a_k and b_k in Eqs. (11.52) become $\omega_0/2\pi$. This is the same trick used to obtain the discrete equations, Eqs. (11.27). With negative values of k allowed, we have

$$y(t)=a_0'+\sum_{\substack{k=-\infty\\k\neq 0}}^{\infty}[a_k'\cos(k\omega_0 t)+b_k'\sin(k\omega_0 t)],$$

$$a_0'=\frac{\omega_0}{2\pi}\int_{-T/2}^{T/2} y(t)\,dt,$$

$$a_k'=\frac{\omega_0}{2\pi}\int_{-T/2}^{T/2} y(t)\cos(k\omega_0 t)\,dt,$$

$$b_k'=\frac{\omega_0}{2\pi}\int_{-T/2}^{T/2} y(t)\sin(k\omega_0 t)\,dt.$$

Since $\cos(0\omega_0 t)=1$ and $\sin(0\omega_0 t)=0$, we can incorporate the definition of a_0' into the definition of a_k' and introduce b_0 which is always zero. The sum can then include $k=0$:

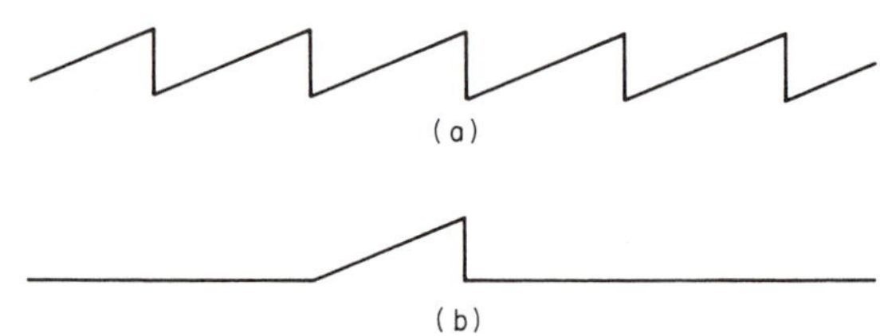

FIGURE 11.24. (a) A periodic signal. (b) A nonperiodic signal.

$$y(t)=\sum_{k=-\infty}^{\infty}[a_k'\cos(k\omega_0 t)+b_k'\sin(k\omega_0 t)],$$

$$a_k'=\frac{\omega_0}{2\pi}\int_{-T/2}^{T/2} y(t)\cos(k\omega_0 t)\,dt,$$

$$b_k'=\frac{\omega_0}{2\pi}\int_{-T/2}^{T/2} y(t)\sin(k\omega_0 t)\,dt. \qquad (11.53)$$

A final change of variables defines $C_k=2\pi a_k'/\omega_0$ and $S_k=2\pi b_k'/\omega_0$. With these changes the Fourier series and its coefficients are

$$y(t)=\frac{\omega_0}{2\pi}\sum_{k=-\infty}^{\infty}[C_k\cos(k\omega_0 t)+S_k\sin(k\omega_0 t)]$$

$$C_k=\int_{-T/2}^{T/2} y(t)\cos(k\omega_0 t)\,dt,$$

$$S_k=\int_{-T/2}^{T/2} y(t)\sin(k\omega_0 t)\,dt. \qquad (11.54)$$

To recapitulate, there is nothing fundamentally new in Eq. (11.54). Negative values of k were introduced so that the sum goes over each value of k twice (except for $k=0$). This allowed the coefficients to be made half as large.

These equations can be used to calculate the Fourier series for a periodic signal such as that shown in Fig. 11.24(a). Suppose that instead we want to find the coefficients for the nonperiodic signal shown in Fig. 11.24(b). This signal can be approximated by another periodic signal shown in Fig. 11.25. The approximation to Fig. 11.24(b) becomes better and better as T is made longer. As T becomes infinite, the fundamental angular frequency ω_0 approaches 0. Define $\omega=k\omega_0$. The frequencies ω are discrete with spacing ω_0. Consider a small frequency interval encompassing many values of k, $\Delta\omega=\omega_0\Delta k$, as shown in Fig. 11.26. Since ω_0 is approaching zero, there

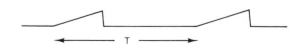

FIGURE 11.25. An approximation to the nonperiodic signal shown in Fig. 11.24(b).

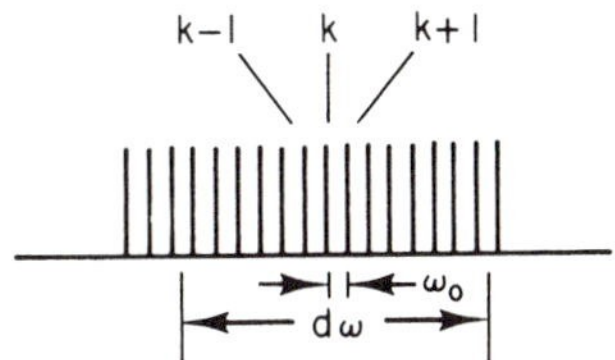

FIGURE 11.26. A histogram of C_k vs k.

can be many values of ω between k and $k+\Delta k$. The frequencies will be nearly the same, so the values of C_k will be nearly the same. All of the terms in the sum in Eq. (11.54) can be replaced by an average value of C_k or S_k multiplied by the number of values of k in the interval, which is Δk. But $\Delta k = \Delta\omega/\omega_0$. Finally, we set $C_k = C(\omega)$ and $\Delta\omega = d\omega$. The sum becomes an integral:

$$y(t)=\frac{\omega_0}{2\pi}\int_{-\infty}^{\infty}[C(\omega)\cos(\omega t)+S(\omega)\sin(\omega t)]\,\frac{d\omega}{\omega_0},$$

or finally, since $d\omega = 2\pi\, df$,

$$y(t)=\frac{1}{2\pi}\int_{-\infty}^{\infty}[C(\omega)\cos(\omega t)+S(\omega)\sin(\omega t)]\,d\omega$$

$$=\int_{-\infty}^{\infty}[C(f)\cos(2\pi f t)+S(f)\sin(2\pi f t)]\,df$$

$$C(\omega)=\int_{-\infty}^{\infty}y(t)\cos(\omega t)\,dt, \tag{11.55}$$

$$S(\omega)=\int_{-\infty}^{\infty}y(t)\sin(\omega t)\,dt.$$

These equations constitute a *Fourier integral pair* or *Fourier transform pair*. They are completely symmetric in the variables f and t and symmetric apart from the factor 2π in the variables ω and t. One obtains $C(\omega)$ or $S(\omega)$ by multiplying the function $y(t)$ by the appropriate trigonometric function and integrating over time. One obtains $y(t)$ by multiplying C and S by the appropriate trigonometric function and integrating over frequency.

Using complex notation, we define

$$Y(\omega)=C(\omega)-iS(\omega) \tag{11.56}$$

and write

$$y(t)=\frac{1}{2\pi}\int_{-\infty}^{\infty}Y(\omega)e^{i\omega t}\,d\omega=\int_{-\infty}^{\infty}Y(\omega)e^{i\omega t}\,df,$$
$$Y(\omega)=\int_{-\infty}^{\infty}y(t)e^{-i\omega t}dt. \tag{11.57}$$

As an example, consider the function

$$y(t)=\begin{cases}0, & t\leq 0\\ Ae^{-at}, & t>0.\end{cases} \tag{11.58}$$

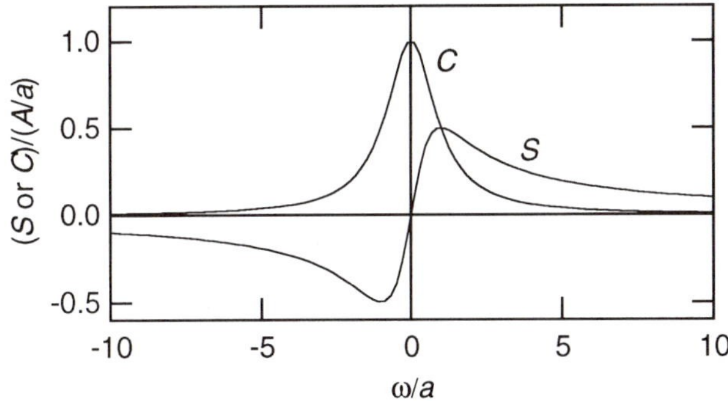

FIGURE 11.27. The sine and cosine coefficients in the Fourier transform of an exponentially decaying pulse.

The functions C and S are evaluated using Eqs. (11.55). Since $y(t)$ is zero for negative times, the integrals extend from 0 to infinity. They are found in all standard integral tables:

$$C(\omega)=A\int_0^{\infty}e^{-at}\cos(\omega t)\,dt=\frac{A/a}{1+(\omega/a)^2},$$
$$S(\omega)=A\int_0^{\infty}e^{-at}\sin(\omega t)\,dt=\frac{(A/a)(\omega/a)}{1+(\omega/a)^2}. \tag{11.59}$$

These are plotted in Fig. 11.27. Function C is even, while S is odd. The functions are plotted on log–log graph paper in Fig. 11.28. Remember that only positive values of ω/a can be shown on a logarithmic scale, so the origin and negative frequencies cannot be shown. It is apparent from the slopes of the curves that C falls off as $(\omega/a)^{-2}$ while S falls more slowly, as $(\omega/a)^{-1}$. One way of explaining this difference is to note that the function $y(t)$ can be written as a sum of even and odd parts as shown in Fig. 11.29. The odd function, which is given by the sine terms in the integral, has a discontinuity, while the even function does not. A more detailed study of Fourier expansions shows that a function with a discontinuity has coefficients that decrease as $1/\omega$ or $1/k$, while the coefficients of a function without a discontinuity decrease more rapidly. (Recall that the coefficients of the square wave were $4/\pi k$.)

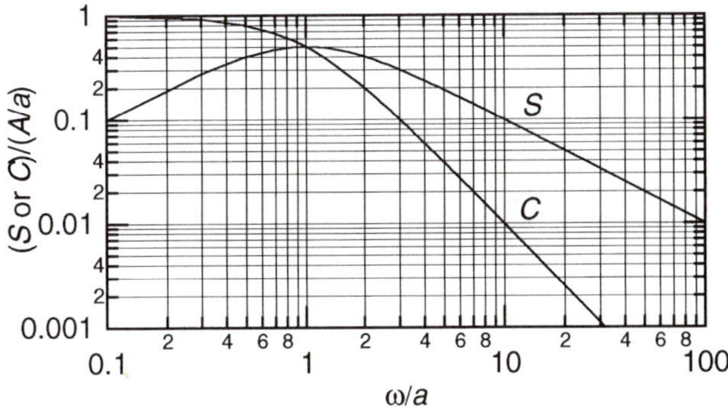

FIGURE 11.28. Log–log plot of the coefficients in Fig. 11.27.

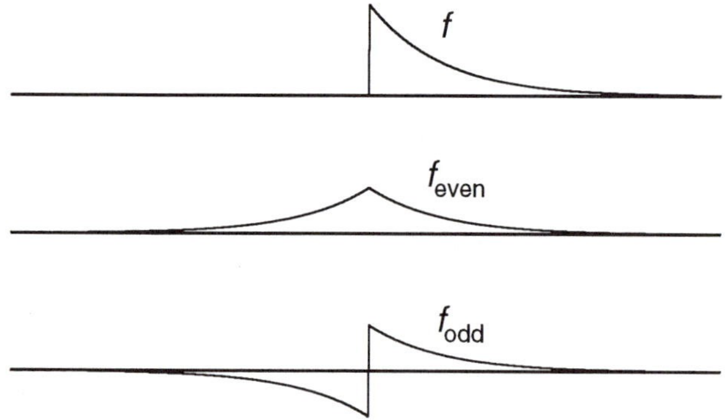

FIGURE 11.29. Function $f(t)$ and its even and odd parts.

11.11. THE DELTA FUNCTION

It will be useful in the next sections to introduce a pulse that is very narrow, very tall, and has unit area under its curve. Physicists call this function the *delta function* $\delta(t)$. Engineers call it the *impulse function* $u_0(t)$.

The δ function is defined by the equations

$$\delta(t)=0, \quad t\neq 0$$

$$\int_{-\epsilon}^{\epsilon} \delta(t)\, dt=\int_{-\infty}^{\infty} \delta(t)\, dt=1. \tag{11.60}$$

The δ function can be thought of as a rectangle of width a and height $1/a$ in the limit $a\to 0$, or as a Gaussian function (Appendix I) as $\sigma\to 0$. Many other functions have the same limiting properties. The δ function is not like the usual function in mathematics because of its infinite discontinuity at the origin. It is one of a class of "generalized functions" whose properties have been rigorously developed by mathematicians[6] since they were first used by the physicist P. A. M. Dirac.

Since integrating across the origin picks up this spike of unit area, the integral of the δ function is a step of unit height at the origin. The δ function and its integral are shown in Fig. 11.30. The δ function can be positioned at $t=a$ by writing $\delta(t-a)$ because the argument vanishes at $t=a$.

Multiplying any function by the δ function and integrating picks out the value of the function when the argument of the δ function is zero:

$$\int_{-\infty}^{\infty} y(t)\delta(t)\, dt=y(0)\int_{-\infty}^{\infty} \delta(t)\, dt=y(0),$$

$$\int_{-\infty}^{\infty} y(t)\delta(t-a)\, dt=y(a)\int_{-\infty}^{\infty} \delta(t-a)\, dt=y(a). \tag{11.61}$$

[6]A rigorous but relatively elementary mathematical treatment is given by Lighthill (1958).

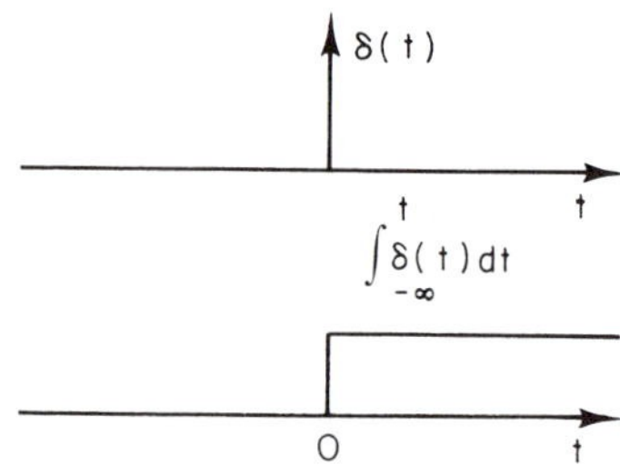

FIGURE 11.30. The δ function and its integral.

The second integral on each line is based on the fact that $y(t)$ has a constant value when the δ function is nonzero so it can be taken outside the integral.

The δ function has the following properties that are proved in Problem 11.19:

$$\delta(t)=\delta(-t),$$

$$t\ \delta(t)=0,$$

$$\delta(at)=\frac{1}{a}\ \delta(t). \tag{11.62}$$

11.12. THE ENERGY SPECTRUM OF A PULSE AND PARSEVAL'S THEOREM

For a periodic signal, the average power is

$$\lim_{T\to\infty} \frac{1}{2T}\int_{-T}^{T} y^2(t)\, dt=a_0^2+\sum_k \tfrac{1}{2}(a_k^2+b_k^2).$$

It is necessary to use the average power because the integral is infinite. For a pulse the integral is finite and the average power vanishes. In that case we use the integral without dividing by T. It is called the energy in the pulse.

Since $y(t)$ is given by a Fourier integral, the energy in the pulse can be written as

$$\begin{aligned}\int_{-\infty}^{\infty} y^2(t)dt &= \left(\frac{1}{2\pi}\right)^2 \int_{-\infty}^{\infty} dt\int_{-\infty}^{\infty} d\omega\int_{-\infty}^{\infty} d\omega'[C(\omega)\cos(\omega t)\\ &\quad +S(\omega)\sin(\omega t)][C(\omega')\cos(\omega t)+S(\omega')\sin(\omega t)].\end{aligned} \tag{11.63}$$

If the terms are multiplied out, this becomes

$$\begin{aligned}\int_{-\infty}^{\infty} y^2(t)dt=\left(\frac{1}{2\pi}\right)^2 \int dt\int d\omega\int d\omega'[C(\omega)C(\omega')\\ \times\cos(\omega t)\cos(\omega' t)\\ +C(\omega)S(\omega')\cos(\omega t)\sin(\omega' t)\end{aligned}$$

$$+S(\omega)C(\omega')\sin(\omega t)\cos(\omega' t)$$

$$+S(\omega)S(\omega')\sin(\omega t)\sin(\omega' t)]. \qquad (11.64)$$

To simplify this expression, we interchange the order of integration, carrying out the time integration first. [It is necessary to assume that the function $y(t)$ is sufficiently well behaved so that this can be done.]

Changing the order gives three integrals to consider:

$$\int_{-\infty}^{\infty} dt \, \cos(\omega t)\cos(\omega' t)$$

$$\int_{-\infty}^{\infty} dt \, \cos(\omega t)\sin(\omega' t)$$

$$\int_{-\infty}^{\infty} dt \, \sin(\omega t)\sin(\omega' t).$$

These are analogous to the trigonometric integrals of Appendix E, except that they extend over all time instead of just one period. Therefore we might expect that an integral such as $\int_{-\infty}^{\infty} dt \, \cos(\omega t)\sin(\omega' t)$ would vanish for all possible values of ω and ω'. We might expect that the integrals $\int_{-\infty}^{\infty} dt \, \cos(\omega t)\cos(\omega' t)$ and $\int_{-\infty}^{\infty} dt \, \sin(\omega t)\sin(\omega' t)$ would vanish when $\omega \neq \omega'$ and be infinite when $\omega = \omega'$. Such is indeed the case. This is reminiscent of the δ function, but it does not tell us the exact relationship of these integrals to it. To find the exact values of the integrals we use the following trick. *Assume* that $y(t)$ is such that $C(\omega) = \delta(\omega - \omega')$. Then, using Eqs. (11.55) and (11.61) we get

$$y(t) = \frac{1}{2\pi} \int_{-\infty}^{\infty} \delta(\omega - \omega')\cos(\omega t) \, d\omega = \frac{1}{2\pi} \cos(\omega' t).$$

The inverse equation for $C(\omega)$ is

$$C(\omega) = \int_{-\infty}^{\infty} y(t)\cos(\omega t) \, dt$$

$$= \frac{1}{2\pi} \int \cos(\omega' t)\cos(\omega t) \, dt.$$

But $C(\omega) = \delta(\omega - \omega')$. Therefore

$$\int_{-\infty}^{\infty} dt \, \cos(\omega t)\cos(\omega' t) = 2\pi \, \delta(\omega - \omega'). \qquad (11.65a)$$

A similar argument shows that

$$\int_{-\infty}^{\infty} dt \, \sin(\omega t)\sin(\omega' t) = 2\pi \, \delta(\omega - \omega'). \qquad (11.65b)$$

The fact that both the sine and cosine integrals are the same should not be surprising, since a sine curve is jut a cosine curve shifted along the axis and we are integrating from $-\infty$ to ∞.

These integrals can be used to evaluate Eq. (11.64). The result is

$$\int_{-\infty}^{\infty} y^2(t) \, dt = \frac{1}{2\pi} \int_{-\infty}^{\infty} d\omega \int_{-\infty}^{\infty} d\omega' [C(\omega)C(\omega')\delta(\omega - \omega') + S(\omega)S(\omega')\delta(\omega - \omega')]$$

$$= \frac{1}{2\pi} \int_{-\infty}^{\infty} d\omega [C^2(\omega) + S^2(\omega)]. \qquad (11.66)$$

This result is known as *Parseval's theorem*. If we define the function

$$\Phi'(\omega) = C^2(\omega) + S^2(\omega), \qquad (11.67a)$$

then Eq. (11.66) takes the form

$$\int_{-\infty}^{\infty} y^2(t) \, dt = \int_{-\infty}^{\infty} \Phi'(\omega) \frac{d\omega}{2\pi} = \int_{-\infty}^{\infty} \Phi'(f) \, df. \qquad (11.67b)$$

The prime is to remind us that this is energy and not power. The left-hand side is the total energy in the signal, and $y^2(t) \, dt$ is the amount of energy between t and $t + dt$. This suggests that we call $\Phi'(\omega) \, d\omega/2\pi = \Phi'(\omega) \, df$ the amount of energy in the angular frequency interval between ω and $\omega + d\omega$ or the frequency interval between f and $f + df$.

This is very similar to what we had for the discrete case in Eq. (11.38). In the discrete case there was a factor of $\frac{1}{2}$ in the relationship of Φ_κ to the Fourier coefficients that does not appear here. The reason is that our equations for the discrete case did not include negative frequencies, while our continuous case does. The discrete coefficients were twice as large. If we had taken the sum to include negative frequencies in the discrete case, the factor of 1/2 would not appear.

The energy spectrum of the exponential pulse that was used earlier as an example is

$$\Phi'(\omega) = C^2(\omega) + S^2(\omega) = \left(\frac{A}{a}\right)^2 \frac{1}{1 + (\omega/a)^2}. \qquad (11.68)$$

It is plotted in Fig. 11.31.

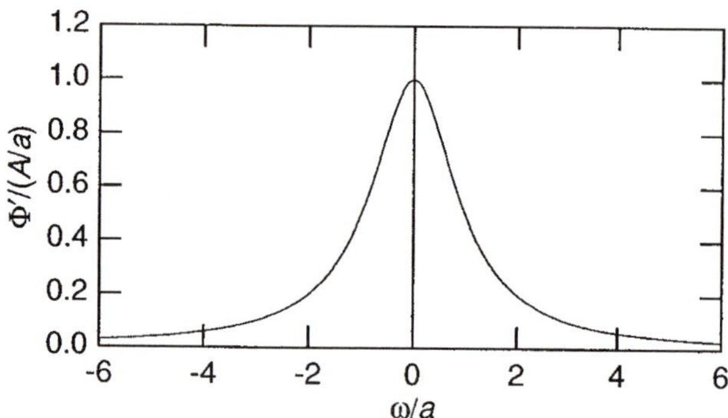

FIGURE 11.31. The energy spectrum $\Phi'(\omega)$ for an exponential pulse.

11.13. THE AUTOCORRELATION OF A PULSE AND ITS RELATION TO THE ENERGY SPECTRUM

The correlation functions for pulses are defined as integrals instead of averages:

$$\phi_{12}(\tau)=\int_{-\infty}^{\infty} y_1(t)y_2(t+\tau)\,dt,$$
$$\phi_{11}(\tau)=\int_{-\infty}^{\infty} y_1(t)y_1(t+\tau)\,dt. \tag{11.69}$$

Consider the autocorrelation function of the exponential pulse, Eq. (11.58). Figure 11.32 shows the functions involved in calculating the autocorrelation for a typical positive value of τ. Since the autocorrelation function is even, negative values of τ need not be considered. The product of the function and the shifted function is $(Ae^{-at})(Ae^{-a(t+\tau)})=A^2e^{-a\tau}e^{-2at}$. It can be seen from Fig. 11.32 that the limits of integration are from zero to infinity. Thus,

$$\phi_{11}(\tau)=A^2e^{-a\tau}\int_0^{\infty} e^{-2at}dt=\frac{A^2e^{-a\tau}}{2a}, \quad \tau>0.$$

Because ϕ_{11} is even, the full autocorrelation function is

$$\phi_{11}(\tau)=\frac{A^2}{2a}e^{-a|\tau|}. \tag{11.70}$$

This is plotted in Fig. 11.33.

The autocorrelation function has a Fourier transform. Only the cosine term appears, since ϕ is even:

$$\Phi'(\omega)=\frac{A^2}{2a}\int_{-\infty}^{\infty} e^{-a|t|}\cos(\omega t)\,dt$$
$$=\frac{A^2}{a}\int_0^{\infty} e^{-at}\cos(\omega t)\,dt.$$

We have seen this integral before, in conjunction with Eq. (11.59). The result is

$$\Phi'(\omega)=\left(\frac{A}{a}\right)^2\frac{1}{1+(\omega/a)^2}.$$

Comparing this with Eq. (11.67), we again see that

$$\Phi'(\omega)=C^2(\omega)+S^2(\omega). \tag{11.71}$$

This relationship between the autocorrelation and Φ' can be proved in general by representing each function in the definition of the autocorrelation function by its Fourier transform, using the trigonometric addition formulas, carrying out the time integration first, and using the δ-function definitions. The result is

$$\phi_{11}(\tau)=\frac{1}{2\pi}\int_{-\infty}^{\infty}[C^2(\omega)+S^2(\omega)]\cos(\omega\tau)\,d\omega$$

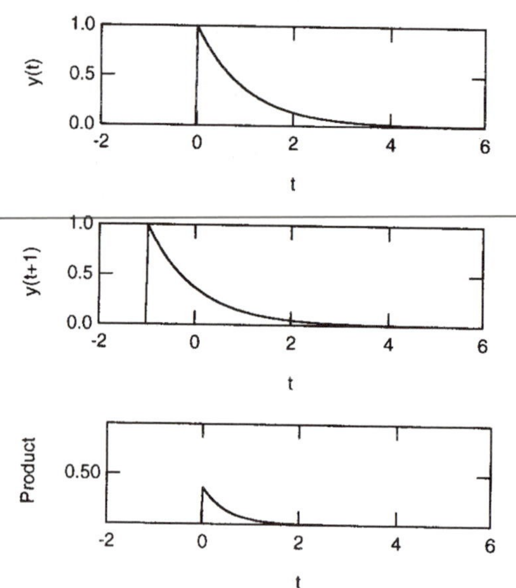

FIGURE 11.32. Calculation of the autocorrelation of the exponential pulse. The figure shows $y(t)$, $y(t+\tau)$, and their product, for $\tau=1$.

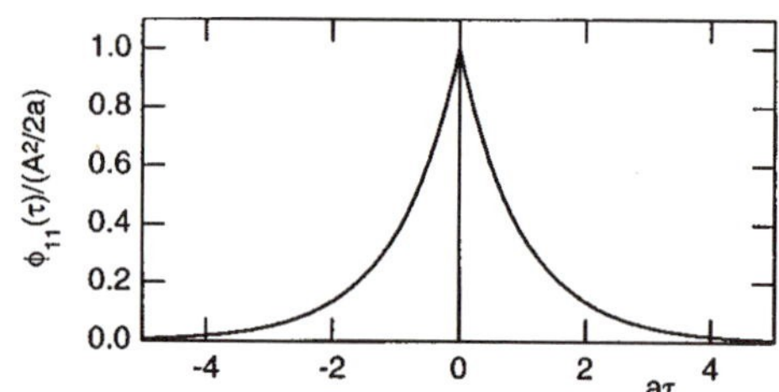

FIGURE 11.33. The autocorrelation function for an exponentially decaying pulse.

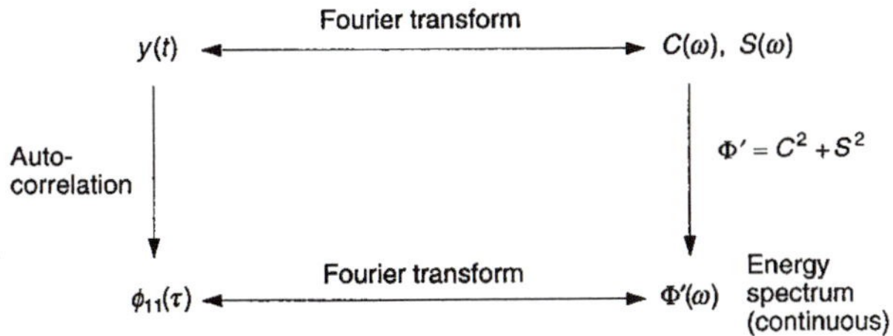

FIGURE 11.34. Two ways to obtain the energy spectrum of a pulse signal.

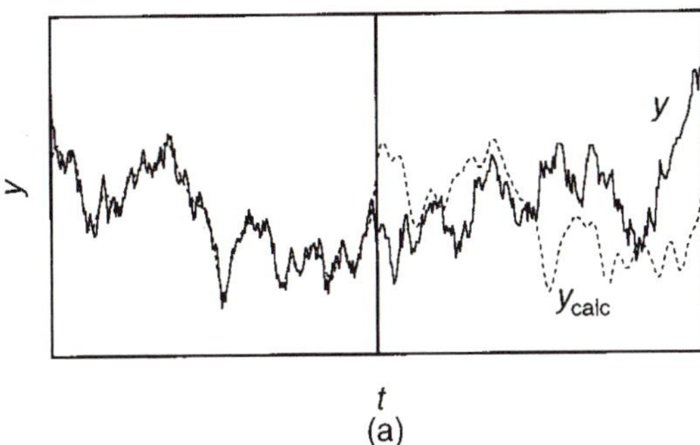

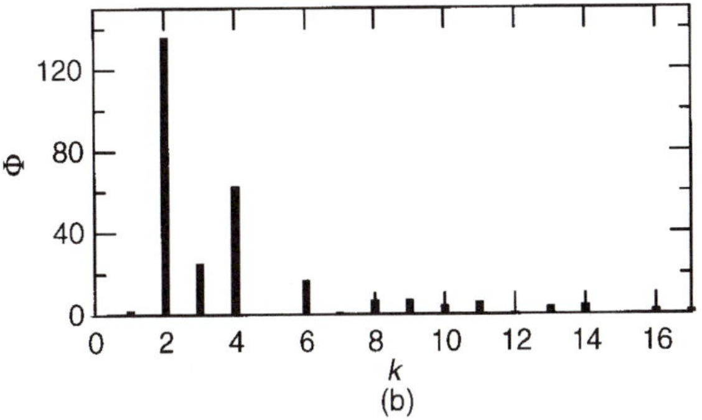

FIGURE 11.35. (a) The function y_j that was calculated from a one-dimensional random walk with a Gaussian-distributed step length. The dashed line shows the function calculated from the Fourier coefficients based on the first half. It does not fit the second half of the function. This is characteristic of random functions. (b) The power spectrum calculated from the first half of y_j. The dc component has been suppressed because it depends on the starting value of y.

$$= \frac{1}{2\pi} \int_{-\infty}^{\infty} \Phi'(\omega) \cos(\omega\tau)\, d\omega. \tag{11.72}$$

As with the periodic signal, there are two ways to go from the signal to the energy spectrum. The Fourier transform is taken of either the original function or the autocorrelation function. Squaring and adding is done either in the time domain to $y(t)$ to obtain the autocorrelation function, or in the frequency domain by squaring and adding the coefficients. The Fourier transforms can be taken in either direction. Squaring and adding is one-directional and makes it impossible to go from the energy spectrum back to the original function. These processes are illustrated in Fig. 11.34.

11.14. NOISE

The function $y(t)$ we wish to study is often the result of a measurement of some system: the electrocardiogram, the electroencephalogram, blood flow, etc., and is called a *signal.* Most signals are accompanied by *noise.* Random noise fluctuates in such a way that we cannot predict what its value will be at some future time. Instead we must talk about the probability that the noise has a certain value. A key problem is to learn as much as one can about a signal which is contaminated by noise. The techniques discussed in this chapter are often useful.

A very important property of noise can be seen from the data shown in Fig. 11.35(a). The data consist of 460 discrete values that appear to have several similar peaks. A discrete Fourier transform of the first 230 values gives fairly large values for the first few coefficients a_k and b_k. Yet these values of a_k and b_k fail to describe subsequent values of y_j. The reason is that the y_j are actually random. In this case they are the net displacement after j steps in a random walk in which each step length is Gaussian distributed with standard deviation $\sigma = 5$. A segment of the procedure used to generate the y_j is shown in Fig. 11.36. The Fourier transform of a random function does not exist. We can apply the recipe to the data and calculate the coefficients. But if we apply the same recipe to some other set of data points from the random function we get different values of the coefficients, although the sum of their squares, $(a_k^2 + b_k^2)^{1/2}$ would be nearly the same. The sum of the squares of the coefficients is plotted in Fig. 11.35(b). It is the phases that change randomly, while the amount of energy at a particular frequency remains constant.

Noise is not periodic, but neither is it a pulse. It has finite power, but it will have infinite energy if the noise goes on "forever." To describe noise we must use averages, calculated over a time interval that is "long enough" so that the average does not change. Suppose that we are measuring the electrical potential between a pair of electrodes on the scalp. Suppose that there is no obvious periodicity, and we think it is noise. If we measure the potential for only a few milliseconds, we will get one average value. If we measure for the same length of time a few minutes later, we may get a different average. But if we average for two or three minutes, then a repetition gives the same average.

In general, random signals may vary with time in such a way that this average changes. (If we repeat the measurements on the scalp in a few hours, the averages may be different.) We will *assume* that properties such as the mean and standard deviation and power spectrum do not change with time, so that if we average over a "long enough" interval and repeat the average at a later time, we get the same result. Processes that generate data with these prop-

```
//Figure 11.36 Code Segment
#include <time.h>
#include <stdio.h>
#include <stdlib.h>
#include <math.h>


void main()
{
   #define IMAX 100
   const double pi = 3.14159265359;
   double Stdev, Normrand1, Normrand2, step,
      max = RAND_MAX;       //maximum value returned by rand() call
   int i;
   double y[IMAX];
   //...
   srand ((unsigned int) (time(NULL) / 2)); //Randomly initialize
             //rand() function by using srand and the system time
   Stdev = 5;
   y[1] = 200;
   for (i=2; IMAX; i++)
   {
      // Get 2 random numbers uniformly distributed on (0,1)
      Normrand1 = rand() / max;
      Normrand2 = rand() / max;

      //Box and Mueller algorithm
      step = Stdev * sqrt(-2 * log(Normrand1)) * cos(2 * pi *
         Normrand2);

      //Generate total displacement of random walk at step i
      y[i] = y[i-1] + (int) (step + 0.5);
   }
   scanf("%i",i);
}
```

FIGURE 11.36. This shows the segment of the computer program that generated the random walk shown in Fig. 11.24. Function `RandomX` returns a random number distributed uniformly on the interval $(1,2^{31}-2)$. The division by the `Scalb` function converts this to a random number distributed uniformly on the interval (0,1). Conversion to a Gaussian distribution is done by a technique described by G. E. P. Box and M. E. Mueller (1958). A note on the generation of normal deviates. *Ann. Math. Stat.* **28**: 610–611.

erties are called *stationary*. We limit our discussion to stationary random processes.

The correlation functions are not particularly useful for well-defined periodic signals, but they are very useful to describe noise or a signal that is contaminated by noise. (In fact, they allow us to detect a periodic signal that is completely hidden by noise. The technique is described in the next section.) Space limitations require us to state some properties of the autocorrelation function of noise without proof, though the results are plausible. Many discussions of noise are available. An excellent one with a biological focus is by DeFelice (1981).

The autocorrelation function is given by Eq. (11.45):

$$\phi_{11}(\tau)=\langle y_1(t)y_1(t+\tau)\rangle=\lim_{T\to\infty}\frac{1}{2T}\int_{-T}^{T}y_1(t)y_1(t+\tau)\,dt.$$

The properties of the autocorrelation function depend on the details of the noise. Some possible shapes for the autocorrelation function are shown in Fig. 11.37.

The following properties of the autocorrelation function can be proved:

1. The autocorrelation function is an even function of τ. This follows from the definition.

2. The autocorrelation function for $\tau=0$, $\phi_{11}(0)$, measures the average power in the signal. This also follows from the definition.

3. For a random signal with no constant or periodic components, the autocorrelation function goes to zero as $\tau\to\infty$. This is plausible, since for large shifts, if the signal is completely random, there is no correlation.

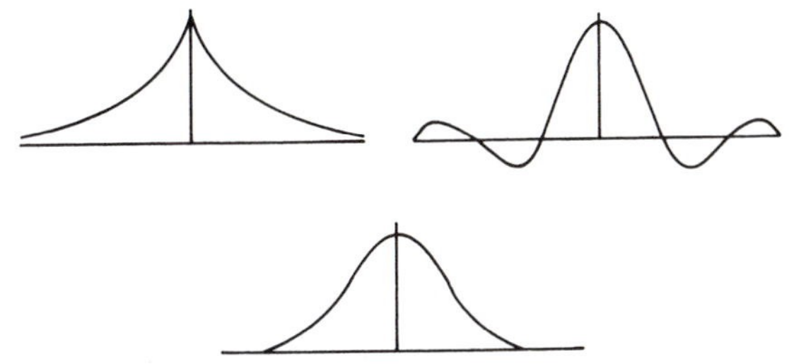

FIGURE 11.37. Some possible autocorrelation functions of noise.

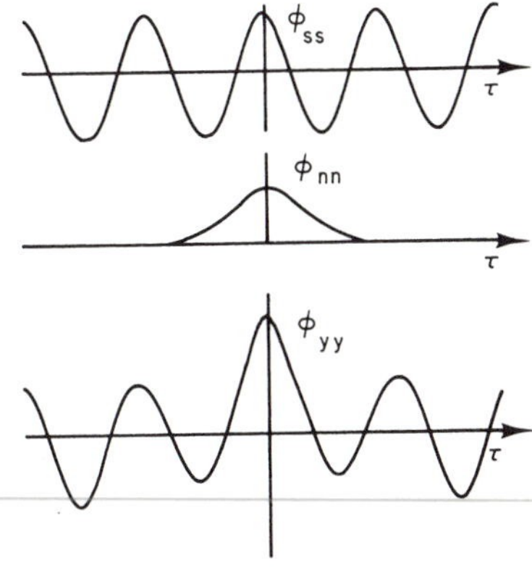

FIGURE 11.38. The autocorrelation functions of a sine wave signal ϕ_{ss}, noise ϕ_{nn}, and signal plus noise ϕ_{yy}.

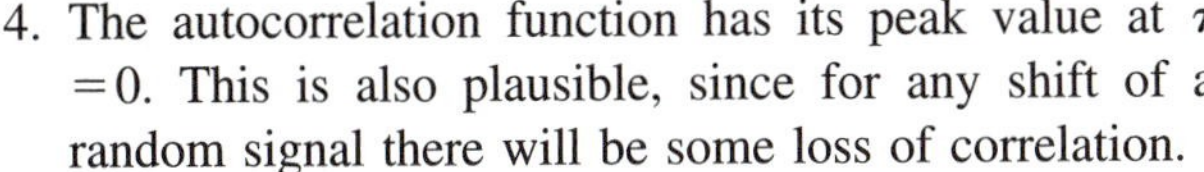

4. The autocorrelation function has its peak value at $\tau = 0$. This is also plausible, since for any shift of a random signal there will be some loss of correlation.

11.15. CORRELATION FUNCTIONS AND DETECTING SIGNALS

The autocorrelation function is useful for detecting a periodic signal in the presence of noise. Suppose that the periodic signal is $s(t)$, the random noise is $n(t)$, and the average of both is zero. The combination of signal and noise is

$$y(t)=s(t)+n(t). \tag{11.73}$$

The autocorrelation of the combination is

$$\begin{aligned}\phi_{yy}(\tau)&=\langle[s(t)+n(t)][s(t+\tau)+n(t+\tau)]\rangle\\&=\langle s(t)s(t+\tau)\rangle+\langle s(t)n(t+\tau)\rangle+\langle n(t)s(t+\tau)\rangle\\&\quad+\langle n(t)n(t+\tau)\rangle.\end{aligned}$$

Each term in the average can be identified as a correlation function:

$$\phi_{yy}(\tau)=\phi_{ss}(\tau)+\phi_{sn}(\tau)+\phi_{ns}(\tau)+\phi_{nn}(\tau).$$

Since the noise is random, the cross correlations ϕ_{ns} and ϕ_{sn} should be zero if the averages were taken over a sufficiently long time. Therefore,

$$\phi_{yy}(\tau)=\phi_{ss}(\tau)+\phi_{nn}(\tau). \tag{11.74}$$

The autocorrelation of a periodic signal is periodic in τ, while the autocorrelation of the noise approaches zero if τ is long enough. The three terms in Eq. (11.74) will look something like the plots in Fig. 11.38.

If we suspect that a periodic signal is masked by noise, we can calculate the autocorrelation function. If the autocorrelation function shows periodicity that persists for long shift times τ, a periodic signal is present. The period of the correlation function is the same as that of the signal. Acquisition of the data and calculation of the correlation function are done with digital techniques. Press *et al.* (1992) have an excellent discussion of the techniques and pitfalls.

If the period of a signal is known to be T, perhaps from the autocorrelation function or more likely because one is looking for the response evoked by a periodic stimulus, it is possible to take consecutive segments of the combined signal plus noise of length T, place them one on top of another, and average them. The signal will be the same in each segment, while the noise will be uncorrelated. After N sampling periods, the noise is reduced by $1/\sqrt{N}$.

Examples of this are the *visual or auditory evoked response*. The signal in the electroencephalogram (EEG) or magnetoencephalogram is measured in response to a flash of light or an audible click. (In other experiments the subject may perform a repetitive task.) The stimulus is repeated over and over while the signal plus noise is recorded and averaged.

The averaging procedure can be described in terms of a cross correlation. Suppose a local signal $l(t)$ is produced in synchrony with the stimulus. The cross correlation of $l(t)$ with $y(t)$ is

$$\phi_{yl}(\tau)=\langle[s(t)+n(t)]l(t+\tau)\rangle=\phi_{sl}+\phi_{nl}.$$

Whatever the local signal is, its cross correlation with the noise approaches zero for long averaging times, so

$$\phi_{yl}(\tau)=\phi_{sl}(\tau). \tag{11.75}$$

If the local signal is a series of narrow spikes approximated by δ functions, then

$$l(t)=\delta(t)+\delta(t-T)+\delta(t-2T)+\cdots.$$

Since both $s(t)$ and $l(t)$ are periodic with the same period, the average can be taken over a single period. The integral then contains one δ function:

$$\phi_{yl}(-\tau)=\phi_{sl}(-\tau)=\frac{1}{T}\int_0^T s(t)\,\delta(t-\tau)\,dt=\frac{s(\tau)}{T}.$$

The correlation function reproduces the signal. Figure 11.39 shows an example of signal averaging for an evoked response in the EEG for increasing values of N.

We have already seen that the Fourier transform of a random signal does not exist. Because the phases of a random signal are continually changing, we were unable to predict the future behavior of a time series in Fig. 11.35. If the signal is stationary, averages, including the average power, do not change with time and have meaning. The autocorrelation function of a random signal does exist, and so does its Fourier transform. If $\Phi(\omega)$ is the Fourier transform of the autocorrelation function of a random signal, then

$$\lim_{T\to\infty} \frac{1}{2T} \int_{-T}^{T} y^2(t)dt = \int_{-\infty}^{\infty} \Phi(\omega) \frac{d\omega}{2\pi} \qquad (11.76)$$

and we can think of Φ as giving the power between frequencies f and $f+df$. This is called the *Wiener theorem for random signals*. The quantity Φ is often called the *power spectral density* or PSD. Figure 11.40 summarizes how the power or energy spectrum can be obtained for a periodic signal, a pulse, and a random signal.

In the digital realm there are several ways to calculate the

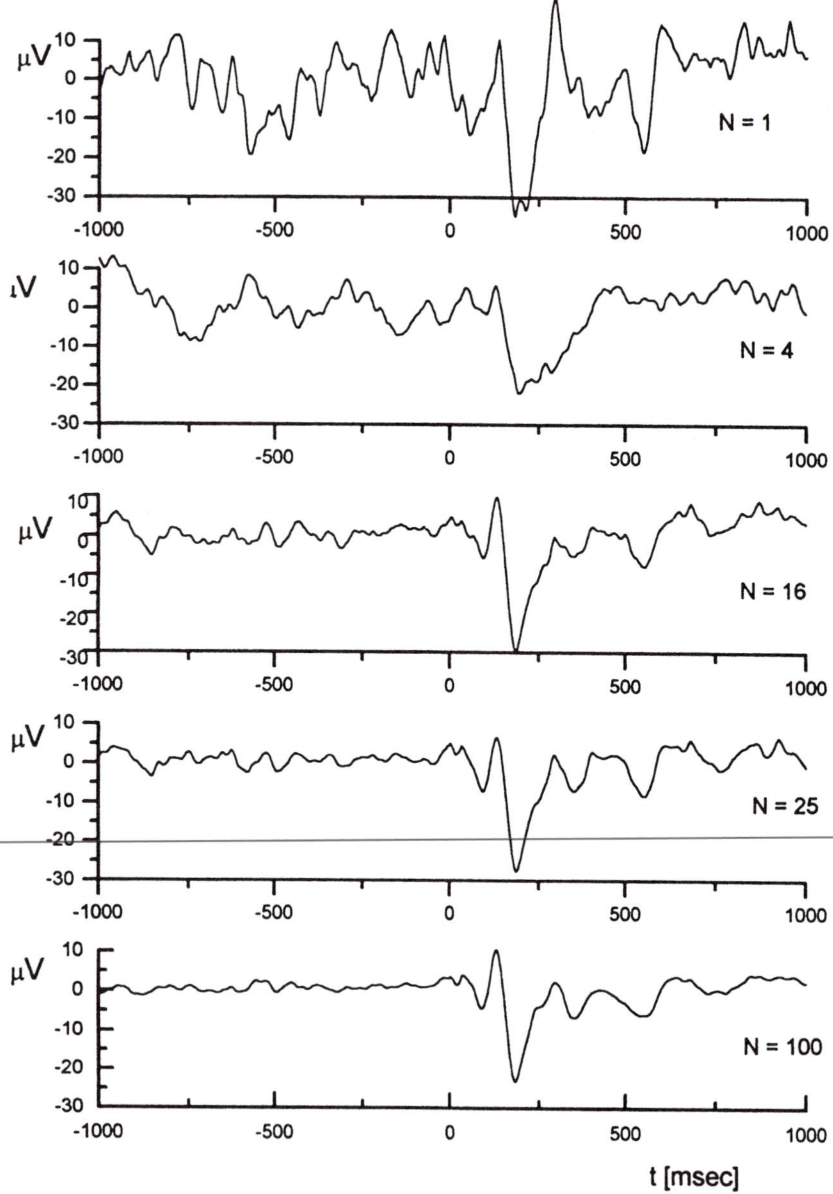

FIGURE 11.39. An example of signal averaging. An evoked response is recorded along with the EEG from a scalp electrode. As the number of repetitions N is increased, the EEG background decreases and the evoked response stands out. Reprinted with permission from L. C. Mainardi, A. M. Bianchi, and S. Cerutti (1995). Digital biomedical signal acquisition and processing, in S. Bronzino, ed. *Biomedical Engineering Handbook*. Boca Raton, FL, CRC. Copyright © 1995 CRC Press.

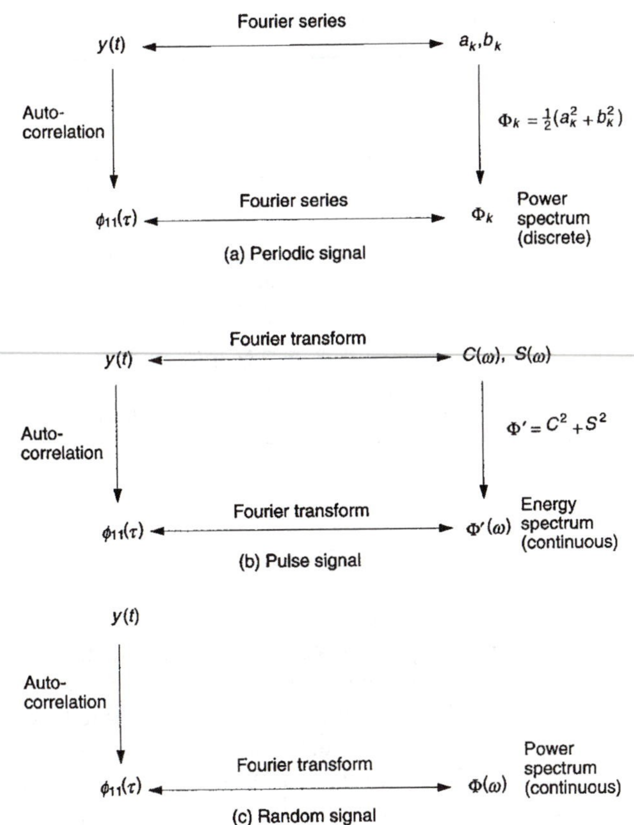

FIGURE 11.40. The relationships between the power spectrum or energy spectrum and (a) a periodic signal, (b) a pulse, (c) a random signal. The Fourier transform and series are bidirectional; the other processes are not.

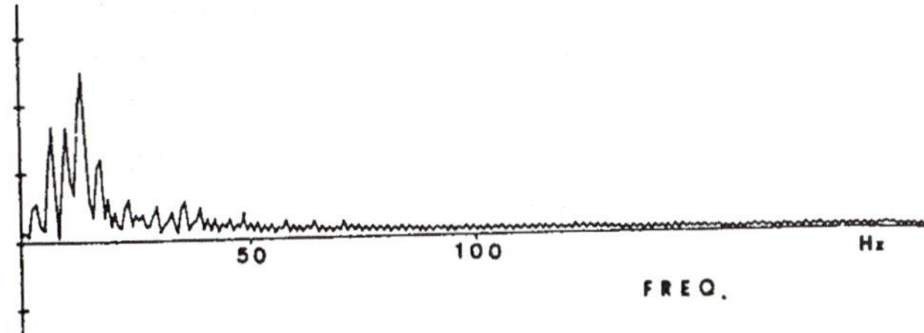

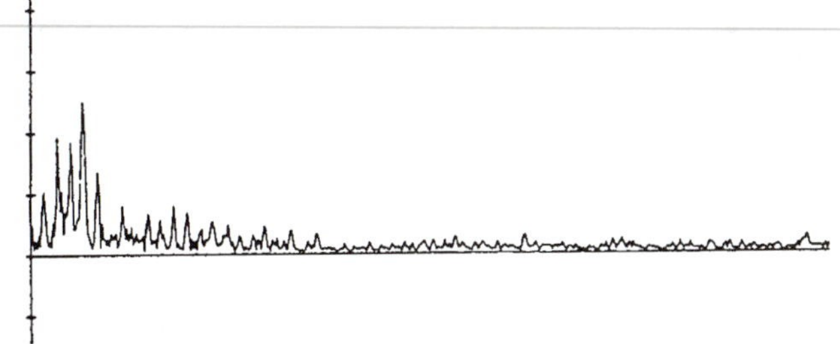

FIGURE 11.41. The power spectrum from a surface electromyogram calculated two different ways. The top panel shows the Blackman–Tukey method, which is a fast Fourier transform of a digital estimate of the autocorrelation function. The lower panel is the sum of the squares of the coefficients in a direct fast Fourier transform of the discrete data. Reprinted, with permission, from A. Cohen (1995). Biomedical Signals: Origin and Dynamic Characteristics; Frequency-Domain Analysis, in S. Bronzino, ed. *Biomedical Engineering Handbook*. Boca Raton, FL, CRC. Copyright © 1995 CRC Press.

power spectral density.[7] The Blackman–Tukey method makes a digital estimate of the correlation function and takes its discrete Fourier transform, as described in Fig. 11.40(c). The periodogram uses the discrete Fourier transform directly. Though the Fourier transform of a random signal does not exist because of the randomly changing phases, the sum of the squares of the coefficients is stable. In fact, we plotted Φ_k calculated from the discrete Fourier transform in Fig. 11.35(b). Figure 11.41 shows both ways of calculating $\Phi(f)$ for a surface electromyogram—the signal from a muscle measured on the surface of the skin. Slight differences can be seen, but they are not significant.

Figure 11.42 shows the power spectrum of an EEG signal and also the effect of aliasing. The original signal has no frequency components above 40 Hz. Sampling was done at 80 Hz. A 50-Hz power frequency signal was added, and the analysis caused a spurious response at 30 Hz. The right-hand panel also shows the mirror-image power spectrum from 40 to 80 Hz that should be thought of as occurring at negative frequencies (the factor of 2 again).

It is worth pausing to review the units of the various functions we have introduced. They become confusing because we have three different cases: a periodic signal that is infinite in extent, a pulse signal that is of finite duration, and a random-noise signal that is also infinite in extent but not periodic. For both signals that are infinite in extent we must use the "power," and for the pulse we must use "energy." Often in signal analysis the units of "power" and "energy" may not be watts or joules. If the signal is a voltage, then the power dissipated in resistance R is v^2/R in watts. Our "power" defined from the equations above would be just v^2.

Suppose that the signal y is measured in "units." Then the "power" is in (units)^2 and the "energy" for a pulse is in (units)^2 s. The correlation functions for the infinite signals are in (units)^2 while those for pulses are in (units)^2 s. Table 11.5 summarizes the situation.

11.16. FREQUENCY RESPONSE OF A LINEAR SYSTEM

Chapter 10 discussed feedback in a linear system in terms of the solution of a differential equation that described the response of the system as a function of time. The simplest system treated there was described by Eq. (10.20):

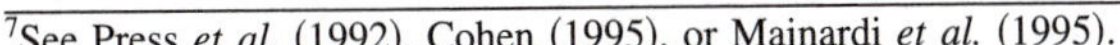

[7] See Press *et al.* (1992), Cohen (1995), or Mainardi *et al.* (1995).

TABLE 11.5. *Units used in the various functions in this chapter, assuming that y is measured in (unit).*

Type of function Signal	Expansion coefficients	Correlation functions	Power or energy
Discrete periodic y (unit)	a_k, b_k, (unit)	ϕ [(unit)^2]	Power [(unit)^2] Φ_k [(unit)^2]
Pulse y (unit)	C, S [(unit s)]	ϕ [(unit)^2 s]	Energy [(unit)^2 s] $\Phi'(\omega)$ [(unit)^2 s^2] $\Phi'(\omega)\,d\omega$ [(unit)^2 s]
Random y (unit)		ϕ [(unit)^2]	Power [(unit)^2] $\Phi(\omega)$ [(unit)^2 s] $\Phi(\omega)\,d\omega$[(unit)^2]

$$\tau_1 \frac{dx}{dt} + x = ap(t) + G_1 y(t). \tag{11.77}$$

Function $p(t)$ is the *input* parameter. This equation was combined with Eq. (10.21) to obtain

$$\tau_1 \frac{dx}{dt} + (1 - G_1 G_2)x = ap(t). \tag{11.78}$$

It is often useful to characterize the behavior of a system by its response to sine waves of different frequencies instead of by its time response. The most familiar example is the audio amplifier: the output signal $x(t)$ is some function of an input signal $p(t)$ that is seldom a pure sinusoid. An equation analogous to Eq. (11.78) relates x and p. The amplifier is usually described as having "a frequency response of -0.5 dB at 10 Hz and 30 kHz." It is easy to feed a sinusoidal signal of different frequencies into the amplifier and measure the amplitude ratio of the output sine wave to the input sine wave.[8] To describe the amplifier completely, it is also necessary to measure the phase delay or the time delay at each frequency. The combination of amplitude and phase response is called the *transfer function* of the amplifier.

In principle, once the properties of a linear system are known, either in terms of a differential equation or the transfer function, its response to any input can be calculated. In the time domain, one solves the differential equation with input $p(t)$ on the right-hand side. In the frequency domain, one computes the Fourier transform of the input, makes the appropriate changes in amplitude and phase at every frequency according to the transfer function, and takes the inverse Fourier transform of the result. The inverse transform gives the output response as a function of time. Sometimes the differential equation may be impossible to solve analytically or the inverse Fourier transform cannot be obtained, and numerical solutions are all that can be obtained.

The frequency-response technique may be particularly useful if the system has several stages (a microphone, an amplifier, one or more loudspeakers); one can multiply the amplitudes and add the phases of each stage.

If the differential equation is known, the frequency response can be calculated. Conversely, if the frequency

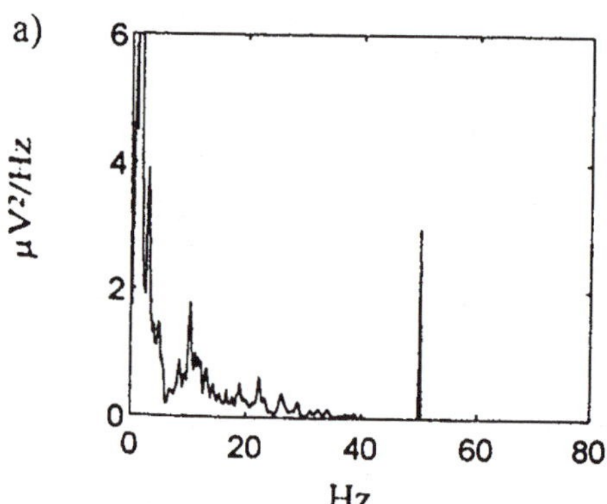

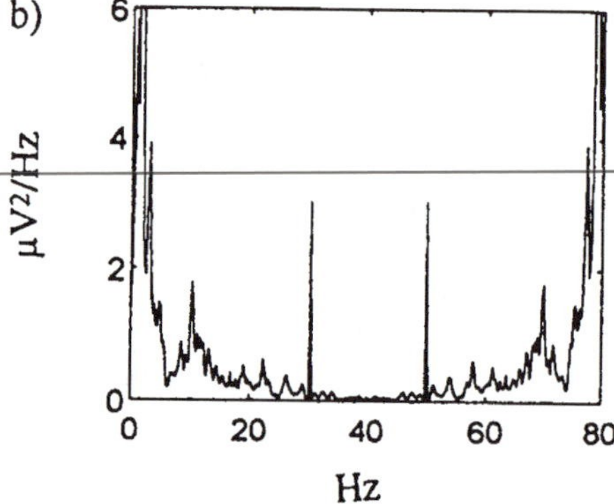

FIGURE 11.42. The power spectrum of an electroencephalogram signal showing the problem with aliasing, and also the presence of negative frequencies appearing as a spectrum above the Nyquist frequency. Reprinted with permission from L. C. Mainardi, A. M. Bianchi, and S. Cerutti. Digital biomedical signal acquisition and processing, in S. Bronzino, ed. *Biomedical Engineering Handbook*. Boca Raton, FL, CRC. Copyright © 1995 CRC Press.

[8]The technique works only for a linear system. If the system is not linear, the output will not be sinusoidal.

and phase responses are known, the differential equation can be deduced. We give an example of the former approach in this section. The latter technique requires more mathematics than we have developed.

As an example of the frequency response method of describing the system, consider Eq. (11.78). With $G_2=0$, the results apply to the case without feedback, Eq. (11.77). Let $p(t)=\cos(\omega t)$ and $a=1$. We want a solution of the form

$$x(t)=G(\omega)\cos(\omega t-\theta), \tag{11.79}$$

where $G(\omega)$ is the overall "gain" or amplitude ratio at frequency ω. We can show by substitution that Eq. (11.79) is a solution of Eq. (11.78) if

$$G(\omega)=\frac{1}{1-G_1G_2}\left(\frac{1}{1+\omega^2\tau_1^2/(1-G_1G_2)^2}\right)^{1/2}, \tag{11.80}$$

$$\tan\,\theta=\frac{\omega\tau_1}{1-G_1G_2}.$$

The behavior of the gain is plotted in Fig. 11.43, both without feedback $(1-G_1G_2=1)$ and with feedback $(1-G_1G_2=3)$. At low frequencies the gain is constant. It falls at high frequencies $(\omega\tau\gg 1)$ as ω^{-1}. When $\omega=1/\tau_1$ (without feedback) or $\omega=3/\tau_1$ (with feedback), the gain is $1/\sqrt{2}$ times its value at zero frequency. This frequency is called the *half-power frequency* because the power is proportional to the square of the signal and its value at the half-power frequency is $\frac{1}{2}$ times its value at zero frequency.

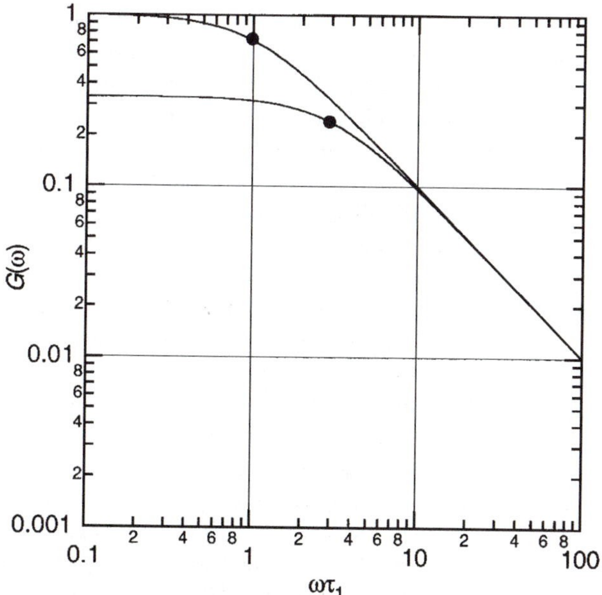

FIGURE 11.43. Plot of $G(\omega)$ for a system described by Eq. (11.80). Two cases are shown: without feedback $(1-G_1G_2=1)$ and with feedback $(1-G_1G_2=3)$. The dots mark the half-power frequencies (see text).

Negative feedback reduces the gain and also raises the half-power frequency from $1/\tau_1$ to $(1-G_1G_2)/\tau_1$. The time constant is reduced by the feedback from τ_1 to $\tau_1/(1-G_1G_2)$. [Recall Eq. (10.23).]

The gain is often expressed in decibels[9] (dB):

$$[\text{gain (dB)}]=20\log_{10}G(\omega). \tag{11.81}$$

A gain ratio of unity is equivalent to 0 dB; a gain of 1000 is $20\log_{10}(1000)=60$ dB. One advantage to expressing gain in decibels is that the gains in dB for several stages add. If the first process has a gain of 2 (6 dB) and the second has a gain of 100 (40 dB), the overall gain is 200 (46 dB). For the amplifier whose gain has fallen by 0.5 dB at 10 Hz and 30 kHz, the ratio $G(\omega)/G_{max}$ is given by solving

$$-0.5=20\log_{10}(G/G_{max}),$$

$$G/G_{max}=10^{-0.025}=0.944.$$

The gain has fallen to 94.4% of its maximum value at 10 Hz and 30 kHz. If the maximum gain were 1000 (60 dB), then the gain would have fallen to 944 (59.5 dB) at 10 Hz and 30 kHz.

The fall in gain is called the *roll-off*, in this case the high-frequency roll-off. At high frequencies the gain is proportional to $1/\omega$, so it drops by a factor of 2 (6 dB) when the frequency doubles (1 *octave*). Therefore the gain has a high-frequency roll-off of 6 dB per octave. A roll-off of 6 dB per octave is characteristic of systems with a single time constant, as in Eq. (11.78).

As an example we show that the response of the system to a δ function calculated in the time domain is consistent with the frequency response. Let the input be

$$p(t)=\delta(t).$$

The Fourier transform of the input is

$$C_{in}(\omega)=\int_{-\infty}^{\infty}\delta(t)\,\cos(\omega t)\,dt=1,$$

$$S_{in}(\omega)=\int_{-\infty}^{\infty}\delta(t)\sin(\omega t)\,dt=0.$$

The δ function contains constant power at all frequencies. The sine coefficients are zero because a δ function at $t=0$ is an even function. The gain and phase delay are applied to $C(\omega)$ to get the Fourier transform of the output signal. Although we started with a purely even function (only cosine terms) the phase shift means that the output contains both sine and cosine terms. To calculate the output, we write Eq. (11.79) as

$$x(t)=\int[G(\omega)\cos\,\theta\,\cos(\omega t)+G(\omega)\sin\,\theta\,\sin(\omega t)]\,d\omega,$$

[9]The bel is the logarithm to the base 10 of the *power* ratio. The decibel is one-tenth as large as the bel. Since the power ratio is the square of the voltage ratio, the factor in Eq. (11.81) is 20.

from which

$$C_{\text{out}}(\omega)=G(\omega)\cos\theta,$$

$$S_{\text{out}}(\omega)=G(\omega)\sin\theta.$$

From Eq. (11.80) we get (letting $G_2=0$ and doing a fair amount of algebra)

$$C_{\text{out}}(\omega)=\frac{1}{1+\omega^2\tau_1^2},$$
$$S_{\text{out}}(\omega)=\frac{\omega\tau_1}{1+\omega^2\tau_1^2}. \tag{11.82}$$

It is easier to solve the differential equation, take the Fourier transform of the solution, and compare it to Eq. (11.82) than it is to find the inverse transform with the mathematical tools at our disposal. For $G_2=0$ the equation to be solved is

$$\tau_1\frac{dx}{dt}+x=\delta(t).$$

For all positive t a steady-state solution is $x(t)=0$. The solution of the homogeneous equation is $x(t)=Ae^{-t/\tau_1}$. The value A is obtained by integrating the equation from $-\epsilon$ to $+\epsilon$ as $\epsilon\rightarrow 0$:

$$\tau_1\int_{-\epsilon}^{\epsilon}\frac{dx}{dt}\,dt+\int_{-\epsilon}^{\epsilon}x\,dt=\int_{-\epsilon}^{\epsilon}\delta(t)\,dt.$$

The first term is $x(\epsilon)-x(-\epsilon)\rightarrow x(0)-0$. The second term vanishes in the limit, since x is finite and the width goes to zero. From the definition of the δ function the right-hand side of the equation is 1. Therefore

$$x=\begin{cases}0, & t<0\\ (1/\tau_1)e^{-t/\tau_1}, & t>0.\end{cases} \tag{11.83}$$

The Fourier coefficients of this function were calculated in Eqs. (11.59). They are

$$C(\omega)=\frac{(1/\tau_1)/(1/\tau_1)}{1+\omega^2\tau_1^2},$$

$$S(\omega)=\frac{\omega\tau_1(1/\tau_1)/(1/\tau_1)}{1+\omega^2\tau_1^2}.$$

These agree with Eqs. (11.82). We have demonstrated that the response of this particular linear system to a δ function is the Fourier transform of the transfer function of the system.

Although the system is not linear, one can see the frequency response of a physiological system in Figure 11.44. Glucose was administered intravenously to two subjects in a sinusoidal fashion with a period of 144 min, as shown in the top panel. The middle panel shows the resulting insulin secretion rate in a normal subject. Insulin secretion adjusts rapidly to the changing glucose level, and the spectral power density has a peak at a period of 144 min. (Note that the spectrum is plotted vs period, not frequency.) The bottom panel shows the results for a subject with impaired glucose tolerance (diabetes). There are oscillations in the insulin secretion rate, but they are irregular and at a shorter period, as can be seen in the PSD.

11.17. THE FREQUENCY SPECTRUM OF NOISE

In Sec. 9.8 we introduced Johnson noise and shot noise. Both are inescapable. Johnson noise arises from the Brownian motion of charge carriers in a conductor; shot noise arises from fluctuations due to the discrete nature of the charge carriers. When we introduced Johnson noise we said nothing about its frequency spectrum. We used the equipartition theorem to argue that since the energy on a capacitor depends on the square of the voltage, there would be fluctuations in a capacitor whose average voltage is zero given by (in the notation of this chapter)

$$\tfrac{1}{2}C\langle v^2\rangle=\tfrac{1}{2}k_BT. \tag{11.84}$$

(In this section we will use T both for time and, when immediately following the Boltzmann constant, for temperature. We also have, briefly, C for capacitance as well as for the Fourier cosine coefficient. We will eliminate the use of C for capacitance as much as possible.)

If the capacitor is completely isolated the charge on its plates, and hence the voltage between them, cannot fluctuate. The equipartition theorem applies to the capacitor only when it is in thermal equilibrium with its surroundings. This thermal contact can be provided by a resistor R between the plates of the capacitor. It is actually the Brownian movement of the charge carriers in this resistor that cause the Johnson noise. In analyzing the noise in electric circuits, it is customary to imagine that the noise arises in an *ideal voltage source*: a ''battery'' that maintains the voltage across its terminals—fluctuating randomly with time—regardless of how much current flows through it. It is placed in series with the resistor. This is not a real source. It is a fictitious source that gives the correct results in circuit analysis. We call the voltage across this noise source $e(t)$ and we want to learn about its properties.

Imagine that we place the noise source and its associated resistor across the plates of a capacitor, as shown in Fig. 11.45. We want to relate the voltage across the capacitor,

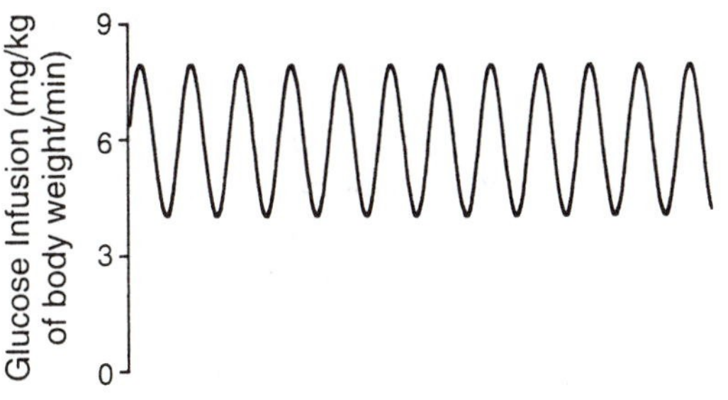

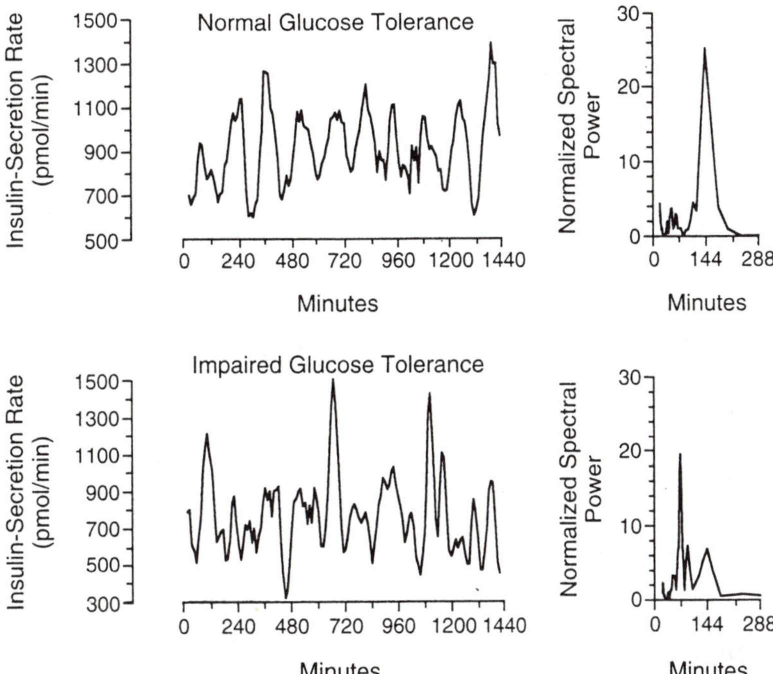

FIGURE 11.44. An example of the frequency response of a system. Glucose was administered intravenously to two subjects in a sinusoidal fashion with a period of 144 min, as shown in the top panel. The middle panel shows the insulin secretion rate in a normal subject. Insulin secretion adjusts rapidly to the changing glucose level, and the spectral power density, which is plotted versus period, has a peak at $T=144$ min, as shown in the spectral plot on the right. The bottom panel shows the results for a subject with impaired glucose tolerance (diabetes). There are oscillations in the insulin secretion rate, but they are irregular and at a lower period, as can be seen in the PSD. From K. S. Polonsky, J. Sturis, and G. I. Bell (1996). Non-insulin-dependent diabetes mellitus—A genetically programmed failure of the beta cell to compensate for insulin resistance. *New Engl. J. Med.* **334**(12): 777–783. Modified from M. M. O'Meara *et al.* (1993). Lack of control by glucose of ultradian insulin secretory oscillations in impaired glucose tolerance and in non-insulin-dependent diabetes mellitus. *J. Clin. Invest.* **92**: 262–271. Used by permission of the *New England Journal of Medicine* and the *Journal of Clinical Investigation.*

v, to the voltage across the noise source, e. We know that $e(t)=v(t)+Ri(t)$, and that $i=C\,dv/dt$. [See the discussion surrounding Eqs. (6.34) and (6.35).] Therefore

$$e(t)=v(t)+RC\,\frac{dv}{dt}=\tau_1\,\frac{dv}{dt}+v. \qquad (11.85)$$

(By introducing $\tau_1=RC$ we eliminate the need to use C for capacitance until the very end of the argument. We use the subscript on τ_1 to distinguish it from the argument of the correlation function.)

Even though the voltage is random, let us assume we can write it as a Fourier integral. Our final results depend only on the power spectrum and not on the phases. We write

$$v(t)=\frac{1}{2\pi}\int_{-\infty}^{\infty}[C(\omega)\cos(\omega t)+S(\omega)\sin(\omega t)]\,d\omega. \qquad (11.86)$$

Differentiating this gives an expression for dv/dt:

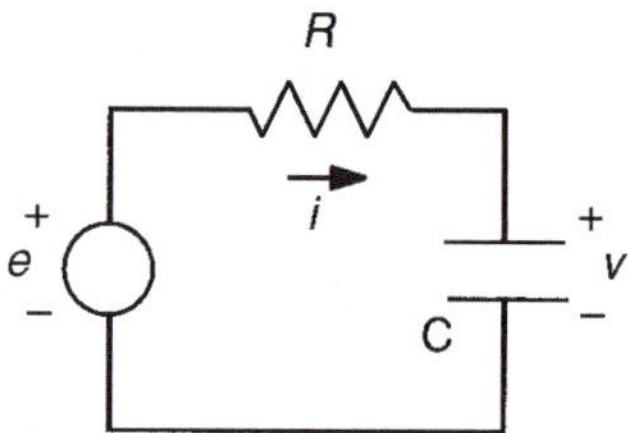

FIGURE 11.45. The circuit for analyzing the noise produced by a resistance R connected to capacitance C. The circuit assumes that the noise is generated in a voltage source $e(t)$ in series with the resistance. The voltage across the capacitance is v.

$$\frac{dv}{dt}=\frac{1}{2\pi}\int_{-\infty}^{\infty}[-\omega C(\omega)\sin(\omega t)+\omega S(\omega)\cos(\omega t)]\,d\omega. \tag{11.87}$$

Combining these with Eq. (11.85) gives us the Fourier transform of $e(t)$:

$$\begin{aligned}e(t)&=\frac{1}{2\pi}\int_{-\infty}^{\infty}\{[C(\omega)+\omega\tau_1 S(\omega)]\cos(\omega t)\\&\quad+[S(\omega)-\omega\tau_1 C(\omega)]\sin(\omega t)\}\,d\omega\\&=\frac{1}{2\pi}\int_{-\infty}^{\infty}\{[\alpha(\omega)]\cos(\omega t)+[\beta(\omega)]\sin(\omega t)\}\,d\omega.\end{aligned}$$

We now need to calculate $\langle v^2(t)\rangle$ and $\langle e^2(t)\rangle$. The calculation is exactly the same as what we did to derive Parseval's theorem, in Eqs. (11.63)–(11.66), except that we are dealing with random signals instead of pulses and we have to introduce

$$\lim_{T\to\infty}\frac{1}{2T}$$

on each side of the equation. When we do this, we find

$$\begin{aligned}\langle v^2(t)\rangle&=\frac{1}{2\pi}\int_{-\infty}^{\infty}[C^2(\omega)+S^2(\omega)]\,d\omega\\&=\frac{1}{2\pi}\int_{-\infty}^{\infty}\Phi_v(\omega)\,d\omega,\\\langle e^2(t)\rangle&=\frac{1}{2\pi}\int_{-\infty}^{\infty}[\alpha^2(\omega)+\beta^2(\omega)]\,d\omega\\&=\frac{1}{2\pi}\int_{-\infty}^{\infty}\Phi_e(\omega)\,d\omega.\end{aligned} \tag{11.88}$$

If we expand Φ_e, we find that

$$\begin{aligned}\Phi_e(\omega)&=\alpha^2(\omega)+\beta^2(\omega)\\&=[C^2(\omega)+S^2(\omega)](1+\omega^2\tau_1^2)\end{aligned}$$

$$\Phi_e(\omega)=\Phi_v(\omega)(1+\omega^2\tau_1^2). \tag{11.89}$$

Johnson noise was discovered experimentally by J. B. Johnson in 1926. The next year Nyquist explained its origin using thermodynamic arguments and showed that until one reaches frequencies high enough so that quantum-mechanical effects are important, Φ_e is a constant independent of frequency [Nyquist (1928)]. We will not reproduce his argument; rather we will assume that Φ_e is a constant and find the value of Φ_e for which the mean square voltage across the capacitor satisfies the equipartition theorem, Eq. (11.84).

The expression for Φ_v becomes

$$\Phi_v(\omega)=\frac{\Phi_e}{1+\omega^2\tau_1^2}, \tag{11.90}$$

and from the first of Eqs. (11.88),

$$\begin{aligned}\langle v^2(t)\rangle&=\frac{1}{2\pi}\int_{-\infty}^{\infty}\Phi_v(\omega)\,d\omega=\frac{\Phi_e}{2\pi}\int_{-\infty}^{\infty}\frac{d\omega}{1+\omega^2\tau_1^2}\\&=\frac{\Phi_e}{2\pi\tau_1}\int_{-\infty}^{\infty}\frac{dx}{1+x^2}\\&=\frac{\Phi_e}{2\pi\tau_1}[\tan^{-1}(\infty)-\tan^{-1}(-\infty)]=\frac{\Phi_e}{2\tau_1}.\end{aligned} \tag{11.91}$$

Putting this expression in the equipartition statement, Eq. (11.84), and remembering that $\tau_1=RC$, we obtain

$$\begin{aligned}\tfrac{1}{2}C\langle v^2(t)\rangle&=\tfrac{1}{2}C\,\frac{\Phi_e}{2RC}=\tfrac{1}{2}k_BT,\\\Phi_e&=2Rk_BT.\end{aligned} \tag{11.92}$$

The units of Φ are V^2 s or $V^2\ \mathrm{Hz}^{-1}$. This is for frequencies that extend from $-\infty$ to ∞. If we were dealing with only positive frequencies, we would have

$$\Phi_e=4Rk_BT \quad \text{(using positive frequencies only)}. \tag{11.93}$$

Either way, this says that the power spectrum for the fictitious source $e(t)$ is constant so there is equal power at all frequencies (up to the limits imposed by quantum mechanical effects). For this reason, Johnson noise is called *white noise* in analogy with white light that contains all frequencies. The voltage fluctuations across the capacitor have the power spectrum

$$\Phi_v(\omega)=\begin{cases}\dfrac{2Rk_BT}{1+\omega^2\tau_1^2}, & \text{using positive and negative frequencies}\\[2ex] \dfrac{4Rk_BT}{1+\omega^2\tau_1^2}, & \text{using positive frequencies only.}\end{cases} \quad (11.94)$$

Figure 11.46 shows the Johnson-noise power spectra and rms voltage spectra plotted vs frequency. These are based on $T=300$ K, $R=10^6\ \Omega$, $C=10^{-9}$ F, and $\tau_1=RC=10^{-3}$ s. The labels on the ordinates are worth discussion. On the left we have Φ/R, which from Eq. (11.94) is in joules, which is W s or W Hz^{-1}. The units for the graph on the right that are consistent with this are W$^{1/2}$ s$^{1/2}$=W$^{1/2}$ Hz$^{-1/2}$=V $\Omega^{-1/2}$ Hz$^{-1/2}$. The resistance has been included to make the units V Hz$^{-1/2}$. The $1/f^2$ falloff at high frequencies is due to the frequency response of the RC circuit and is not characteristic of the noise. Figure 11.47 shows an example: the spectral density of the magnetic field from an article on the magnetoencephalogram. The units are femtotesla (hertz)$^{-1/2}$ (1 femtotesla=1 fT=10^{-15} T).

We can determine the autocorrelation functions $\phi_{ee}(\tau)$ and $\phi_{vv}(\tau)$. Equation (11.72) gave the Fourier transform of the autocorrelation function for a pulse. For a random signal the autocorrelation is very similar but involves the power instead of the energy:

$$\phi_{ee}(\tau)=\frac{1}{2\pi}\int_{-\infty}^{\infty}\Phi_e(\omega)\cos(\omega\tau)\ d\omega,$$

$$\phi_{vv}(\tau)=\frac{1}{2\pi}\int_{-\infty}^{\infty}\Phi_v(\omega)\cos(\omega\tau)\ d\omega. \quad (11.95)$$

For the voltage source the autocorrelation function is

$$\phi_{ee}(\tau)=\frac{2Rk_BT}{2\pi}\int_{-\infty}^{\infty}\cos(\omega\tau)\ d\omega. \quad (11.96)$$

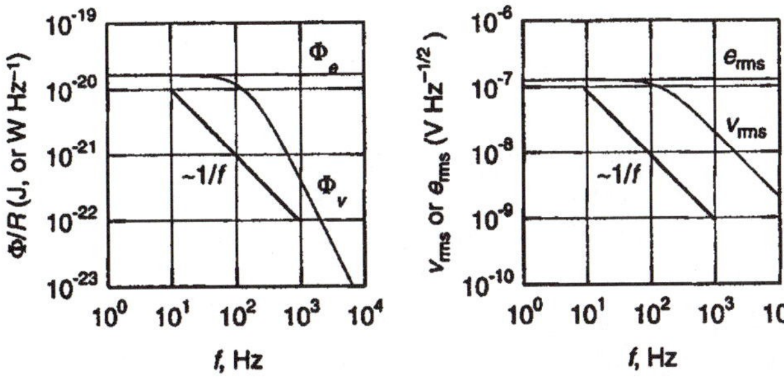

FIGURE 11.46. The power spectrum of the noise source e and the voltage across the capacitor v. The left panel plots Φ/R vs f. The right panel plots v_{rms} in each frequency interval. The parameters are described in the text.

To evaluate this, consider Eq. (11.65a), which shows the Fourier transform of the δ function. The integral there is over time. Interchange the time and angular frequency variables to write

$$\int_{-\infty}^{\infty}\cos(\omega\tau)\cos(\omega\tau')\ d\omega=2\pi\ \delta(\tau-\tau'). \quad (11.97)$$

Let $\tau'=0$:

$$\int_{-\infty}^{\infty}\cos(\omega\tau)\ d\omega=2\pi\ \delta(\tau). \quad (11.98)$$

The final expression for the autocorrelation function of the noise source is

$$\phi_{ee}(\tau)=2Rk_BT\ \delta(\tau). \quad (11.99)$$

To find $\phi_{vv}(\tau)$, consider the discussion surrounding Eqs. (11.69) and (11.70). There we discussed the Fourier transform pair (letting $a=1/\tau_1$)

$$\frac{A^2}{1+\omega^2\tau_1^2}\ \xleftrightarrow{\text{Fourier transform}}\ \frac{A^2}{2\tau_1}e^{-|\tau|/\tau_1}, \quad (11.100)$$

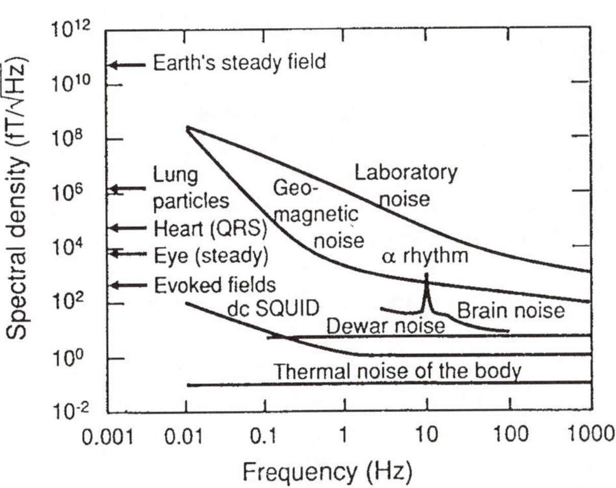

FIGURE 11.47. Spectral density of various sources of the magnetic field, expressed in terms of the magnetic field in femtotesla (1 fT=10^{-15} T). From M. Hämäläinen, R. Harri, R. J. Ilmoniemi, J. Knuutila, and O. V. Lounasmaa (1993). Magnetoencephalography—theory, instrumentation, and applications to noninvasive studies of the working human brain. *Rev. Mod. Phys.* **65**(2): 413–497. Used by permission.

from which we obtain the autocorrelation function for the voltage across the capacitor:

$$\phi_{vv}(\tau)=\frac{Rk_BT}{\tau_1}\,e^{-|\tau|/\tau_1}. \tag{11.101}$$

Let us compare these two results. The autocorrelation of the noise source is a δ function. Any shift at all destroys the correlation. The noise equivalent voltage source and resistor, isolated from anything else, respond instantaneously to random noise changes, the correlation function is infinitely narrow, and all frequencies are present. When the source and resistor are connected to a capacitor, the voltage across the capacitor cannot change instantaneously. There is a high- frequency roll-off, and the voltage at one time is correlated with the voltage at surrounding times. As the time constant of the circuit becomes smaller, $\phi_{vv}(\tau)$ becomes narrower and taller, approaching the δ function.

The power spectrum across the capacitor has the same form as the square of the magnitude of the gain (transfer function) of Eq. (11.80). This is the transfer function for an RC circuit, as can be seen by comparing Eq. (11.78) with Eq. (11.85). This is a special case of a general result, that linear systems can be analyzed by measuring how they respond to white noise.

Chapter 9 also mentioned shot noise, which occurs because the charge carriers have a finite charge, so the number of them passing a given point in a circuit in a given time fluctuates about an average value. One can show that shot noise is also white noise.

Johnson noise and shot noise are fundamental and independent of the details of the construction of the resistance. The former depends on the Brownian motion of the charge carriers, and the latter depends on the number of charge carriers required to transport a given amount of charge. They are irreducible lower limits on the noise (for a given resistance and temperature). If one measures the noise in a real resistor in a circuit, one finds additional or ''excess'' noise that can be reduced by changing the materials or construction of the resistor. This excess noise often has a $1/f$ frequency dependence. For white noise the power in every frequency interval is proportional to the width of the interval, so there is 10 times as much power in the frequency decade from 10 to 100 Hz as in the decade from 1 to 10 Hz. For $1/f$ noise, on the other hand, there is equal power in each frequency decade. This kind of noise is sometimes called ''pink noise'' in allusion to the fact that pink light has more power in the red (lower frequency) part of the spectrum than the rest.

Noise with a $1/f$ spectrum had been discovered in many places: resistors, transistors, and the fluctuations in the flow of sand in an hourglass, in traffic flow, in the heartbeat, and even in human cognition. It is thought that there must be some universal principle underlying $1/f$ noise, possibly related to chaos, but this is still an area of active investigation.

11.18. TESTING DATA FOR CHAOTIC BEHAVIOR

A major problem in data analysis is to find the meaningful signal due to the physical or biological process in the presence of noise. We have introduced some of the analysis techniques in this chapter. A problem that has only become important in recent years is to determine whether a variable that is apparently random is due to truly random behavior in the underlying process or whether the process is displaying chaotic behavior. The techniques for determining this are still under development and are beyond the scope of this book. An excellent introduction is found in Chap. 6 of Kaplan and Glass (1995). We close by mentioning two of the tools used in this analysis: embedding and surrogate data.

One of the problems in analyzing data from complex systems is that we may not be able to measure all of the variables. For example, we may have the electrocardiogram or even an intracellular potential recording but have no information about the details of the ionic currents of several species through the membrane that change the potential. We may measure the level of thyroid hormones T3 and T4 but have no information about the other hormones in the thyroid–hypothalamus–pituitary feedback system. Fortunately, we do not need to measure all the variables. There is a data-reduction technique that can be applied to a few of the variables that shows the dynamics of the full system.

To see how embedding works, consider a system with two degrees of freedom described by a set of nonlinear differential equations with the form of Eqs. (10.32). In order to make the subscript on x available to index measurements of the variable at different times, we write the variables as x and y instead of x_1 and x_2:

$$\frac{dx}{dt}=f_1(x,y), \quad \frac{dy}{dt}=f_2(x,y).$$

A phase-space plot would be in the xy plane. Suppose we only measure variable x, and that we obtain a sequence of measurements $x_j=x(t_j)$. The time derivative is approximately

$$\frac{x_{j+h}-x_j}{t_{h+j}-t_j}\approx\frac{dx}{dt}=f_1(x,y).$$

A series of measurements at different times gives us information about how function f_1 depends on x. A remarkable result that we state without proof is that it also gives information about the entire system. [See Kaplan and Glass (1995) for a more detailed discussion and references to the literature.] Figure 11.48 shows this in a specific case. It is a calculation using the van der Pol oscillator. This nonlinear oscillator has been used to model many systems since it was first proposed in the 1920s. It can be written as the pair of first-order equations

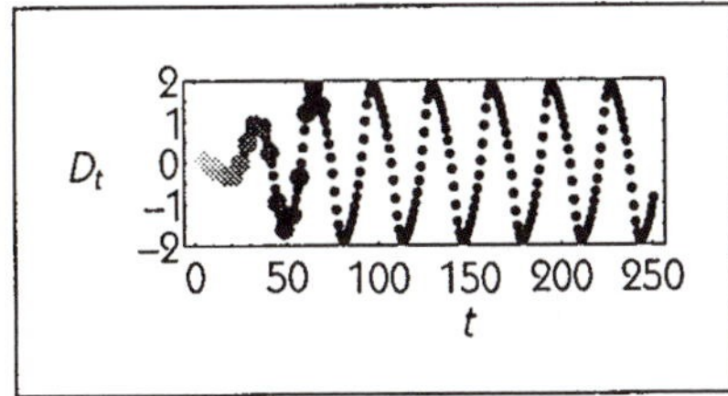

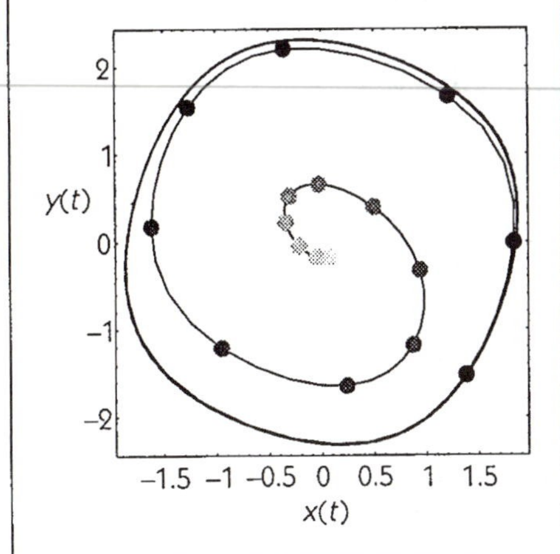

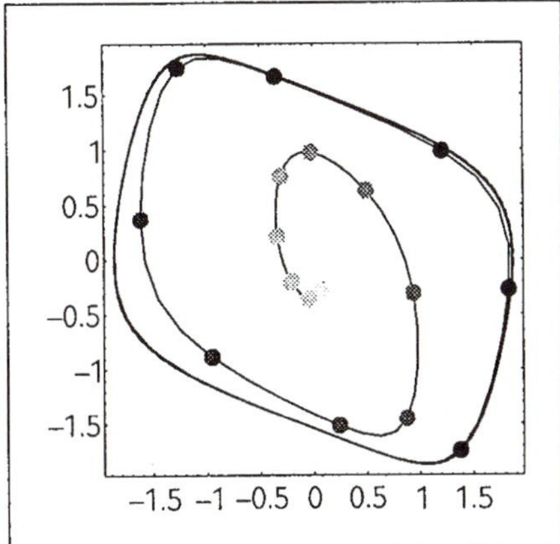

FIGURE 11.48. Plots of the solution to the van der Pol equation with a certain set of initial conditions. The top panel shows values of x_j vs j (labeled as D_t vs t). The middle panel is a phase-plane plot of y vs x. The bottom panel plots x_{j+10} vs x_j. Shading is used to identify some of the early data points in all three panels. The trajectory in the bottom panel has the same characteristics as the phase-plane plot. From Kaplan and Glass (1995). Used by permission of Springer-Verlag.

$$\frac{dx}{dt}=\frac{1}{a}\left(y-\frac{x^3}{3}+x\right),\quad \frac{dy}{dt}=-ax,$$

where a is a very small positive number. The top panel of Fig. 11.48 shows values of x_j vs j (labeled as D_t vs t). The middle panel shows a phase-plane plot of y vs x. The bottom panel plots x_{j+10} vs x_j. Shading is used to identify some of the early data points in all three panels. The trajectory in the bottom panel has all the same characteristics as the phase-plane plot.

This is an example of a general technique called *time-lag embedding*. The set of differential equations with two degrees of freedom has been converted into a nonlinear map in one degree of freedom.

For a system with three degrees of freedom, we could make a three-dimensional plot by creating sets of three numbers from the n measured values, which we can think of and plot as the three components of a vector

$$\mathbf{x}_j=(x_j,x_{j-h},x_{j-2h}),\quad j=2h,2h+1,\ldots,n-1.$$

In general, we can construct a p-dimensional set of vectors

$$\mathbf{x}_j=(x_j,x_{j-h},\ldots x_{j-ph}),\quad j=ph,\ldots,n-1.$$

We call p the *embedding dimension* and the *embedding lag*. There are a number of further calculations that can be done to the embedded vector to help decide on the behavior of the underlying system. These are described in Kaplan and Glass (1995).

In general, a fully conclusive answer to the question of whether the data are due to a random process or a chaotic process cannot be obtained, though strong indications can be. The most rigorous way to test for the presence of chaotic behavior is to make the hypothesis—called the *null hypothesis*—that the data are explained by a linear process plus random noise. One then develops a test statistic (several standard tests are used) and compares the value of the test statistic for the real data to its value for sets of data that are consistent with the null hypothesis. These sets are called *surrogate data*. We examined one linear system with noise: the random walk of Fig. 11.35. The next value in the sequence was the previous value plus random noise. We saw that the power spectrum was defined, but the phases changed randomly. We can think of any linear system driven by random noise as having a defined transfer function $G(\omega)$ with random phases. Therefore one can generate sets of surrogate data by taking the transform of the original data in the form of an amplitude and phase, related to C and S by Eq. (11.13). One then randomizes the phases and calculates the inverse Fourier transform of the randomized coefficients to generate the surrogate data sequence. The surrogate data have the same power spectrum and autocorrelation function as the original data. One then applies the various test statistics. If we were to do this to the data from Fig. 11.35, we would find the tests the same for the original data and the sets of surrogate data, because the original data set is consistent with the null hypothesis.

SYMBOLS USED IN CHAPTER 11

Symbol	Use	Units	First used on page
a	Coefficient in polynomial fit		286
a	Slope		285
a	Coefficient of even (cosine) term		289
a	Parameter in exponential		287
a	Arbitrary constant		302
b	Intercept		284
b	Parameter in exponential		287
b	Coefficient of odd (sine) term		289
e	Noise voltage source	V	312
f,f_0	Frequency	Hz	288
h	Small quantity		287
h	Shift index		317
i	$\sqrt{-1}$		290
j	Index, usually denoting data point		284
k	Index denoting terms in sum		284
k_B	Boltzmann constant	J K^{-1}	312
l,m	A particular value of index k		287
n	Maximum value of index k		286
n	Noise		307
p	Parameter		310
p	Dimension of vector		317
s	Signal		307
t	Time	s	294
v	log y		287
v	Voltage	V	312
x	Independent variable		284
$\mathbf{x}$	Vector of data points		317
y	Dependent variable		284
A	Amplitude		298
C,C_k	Amplitude of cosine term		288
C	Capacitance	F	312
G	Gain		310
N	Number of data points		285
Q	Goodness of fit or mean square residual		285
R	Resistance	Ω	313
S,S_k	Amplitude of sine term		288
S_{xx},S_{xy}	Sums of residuals and their products		286
T	Period	s	288
T	Temperature	K	312
Y,Y_k	Complex Fourier transform or series of y		290
α	Fourier coefficient in autocorrelation function		299
δy	Uncertainty in y		287
δ	Delta function		302
ϵ	Error		295
ϵ	Small number (limit of integration)		302
ϕ,θ	Phase		288
ϕ	Correlation function		296
τ	Shift time	s	296
τ_1	Time constant	s	310
ω,ω_0	Angular frequency	s^{-1}	288
Φ_k	Power at frequency $k\omega_0$		299
$\Phi(\omega)$	Power in frequency interval		308
$\Phi'(\omega)$	Energy in frequency interval		303

PROBLEMS

Section 11.1

11.1. Find the least squares straight line fit to the following data:

x	y
0	2
1	5
2	8
3	11

11.2. Suppose that you wish to pick one number to characterize a set of data $x_1,x_2,\ldots,x_N$. Prove that the mean $\bar{x}$ defined by

$$\bar{x}=\frac{1}{N}\sum_{j=1}^{N} x_j ,$$

minimizes the mean square error

$$Q=\frac{1}{N}\sum_{j=1}^{N} (x_j-\bar{x})^2 .$$

11.3. Derive Eqs. (11.5).

11.4. Suppose that the experimental values $y(x_j)$ are exactly equal to the calculated values plus random noise for each data point: $y(x_j)=y_{\text{calc}}(x_j)+n_j$. What is Q?

11.5. You wish to fit a set of data (x_j, y_j) with an expression of the form $y=Bx^2$. Differentiate the expression for Q to find an equation for B.

11.6. Obtain equations for the linear least-squares fit of $y=Bx^m$ to data by making a change of variables. (Hint: Review Sec. 2.10.)

11.7. Apply the results of Problem 11.6 to the case of Problem 11.5. Why does it give slightly different results?

11.8. Carry out a numerical comparison of Problems 11.5 and 11.7 with the data points

x	y
1	3
2	12
3	27

Repeat with

x	y
1	2.9
2	12.1
3	27.1

11.9. This problem is designed to show you what happens when the number of parameters exceeds the number of data points. Suppose that you have two data points:

x	y
0	1
1	4

Find the best fits for one parameter (the mean) and two parameters ($y=ax+b$). Then try to fit the data with three parameters (a quadratic). What happens when you try to solve the equations?

Section 11.4

11.10. Write a computer program to verify Eqs. (11.19)–(11.23).

11.11. Consider Eqs. (11.16)–(11.18) when $n=N$ and show that all equations for $m>N/2$ reproduce the equations for $m<N/2$.

11.12. Analyze the following data, assuming that a sine and cosine term with a period of 24 h are the only terms present:

Time	y
0000	10.0
0300	14.1
0600	10.0
0900	0.0
1200	−10.0
1500	−14.1
1800	−10.0
2100	0.0
2400	same as 0000

Section 11.6

11.13. Use Eqs. (11.33) to derive Eq. (11.34).

Section 11.8

11.14. Suppose that $y(x,t)=y(x-vt)$. Calculate the cross correlation between signals $y(x_1)$ and $y(x_2)$.

Section 11.9

11.15. Consider a square wave of amplitude A and period T.

(a) What are the coefficients in a Fourier-series expansion?

(b) What is the power spectrum?

(c) What is the autocorrelation of the square wave?

(d) Find the Fourier-series expansion of the autocorrelation function and compare it to the power spectrum.

11.16. The series of pulses shown are an approximation for the concentration of follicle-stimulating hormone (FSH) released during the menstrual cycle.

(a) Determine a_0, a_k, and b_k in terms of d and T.

(b) Sketch the autocorrelation function.

(c) What is the power spectrum?

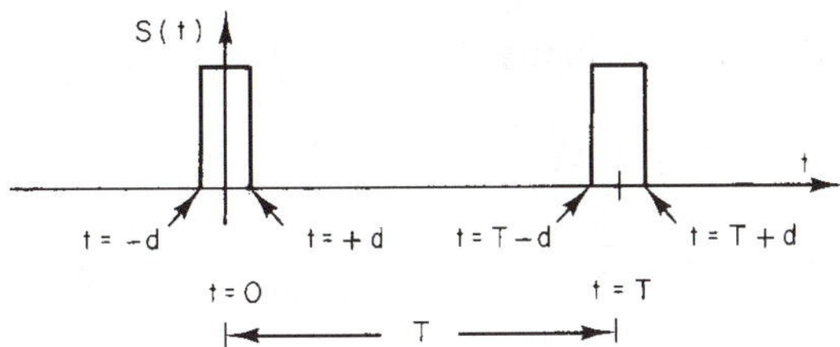

11.17. Consider the following simplified model for the periodic release of follicle-stimulating hormone (FSH). At $t=0$ a substance is released so the plasma concentration rises to value C_0. The substance is cleared so that $C(t) = C_0 e^{-t/\tau}$. Thereafter the substance is released in like amounts at times T, $2T$, and so on. Furthermore, $\tau \ll T$.

(a) Plot $C(t)$ for two or three periods.

(b) Find general expressions for a_0, a_k, and b_k. Use the fact that integrals from 0 to T can be extended to infinity because $\tau \ll T$. Use the following integral table:

$$\int_0^\infty e^{-ax} dx = \frac{1}{a},$$

$$\int_0^\infty e^{-ax} \cos(mx)\, dx = \frac{a}{a^2+m^2}$$

$$\int_0^\infty e^{-ax} \sin(mx)\, dx = \frac{m}{a^2+m^2}.$$

(c) What is the ''power'' at each frequency?

(d) Plot the ''power'' for $k=1,10,100$ for two cases: $\tau/T=0.1$ and 0.01. Compare the results to the results of Problem 11.16.

(e) Discuss qualitatively the effect that making the pulses narrower has on the power spectrum. Does the use of Fourier series seem reasonable in this case? Which description of the process is easier—the time domain or the frequency domain?

(f) It has sometimes been said that if the transform for a given frequency is written as $A_k \cos(k\omega_0 t - \phi_k)$ that ϕ_k gives timing information. What is ϕ_1 in this case? ϕ_2? Do you agree with the statement?

11.18. Calculate the autocorrelation function and the power spectrum for the previous problem.

Section 11.11

11.19. Prove that

$$\delta(t) = \delta(-t),$$

$$t\delta(t) = 0,$$

$$\delta(at) = \frac{1}{a}\,\delta(t).$$

Section 11.12

11.20. Rewrite Eqs. (11.59) in terms of an amplitude and a phase. Plot them.

11.21. Find the Fourier transform of

$$f(t) = \begin{cases} 1, & -a \leqslant t \leqslant a \\ 0 & \text{everywhere else.} \end{cases}$$

11.22. Find the Fourier transform of

$$y = \begin{cases} e^{-at} \sin(\omega_0 t), & t \geqslant 0 \\ 0, & t < 0. \end{cases}$$

Determine $C(\omega)$, $S(\omega)$, and $\Phi'(\omega)$ for $\omega > 0$ if the term that peaks at negative frequencies can be ignored for positive frequencies.

Section 11.15

11.23. Here are some data.

(a) Plot them.

(b) If you are told that there is a signal in these data with a period of 4 s, you can group them together and average them. This is equivalent to taking the cross correlation with a series of δ functions. Estimate the signal shape.

t	y	t	y	t	y
1	5	13	4	25	3
2	4	14	2	26	10
3	5	15	6	27	3
4	3	16	0	28	0
5	10	17	3	29	3
6	11	18	2	30	11
7	10	19	6	31	1
8	8	20	0	32	0
9	11	21	7	33	9
10	4	22	7	34	4
11	7	23	4	35	1
12	0	24	2	36	1

Section 11.16

11.24. Verify that Eqs. (11.79) and (11.80) are solutions of Eq. (11.78).

11.25. Equation (11.80) is plotted on log–log graph paper in Fig. 11.43. Plot it on linear graph paper.

11.26. If the frequency response of a system were proportional to

$$\frac{1}{1+(\omega/\omega_0)^3},$$

what would be the high frequency roll-off in decibels per octave for $\omega \gg \omega_0$?

11.27. Consider a signal $y = A\cos(\omega t)$. What is the time derivative? For a fixed value of A, how does the derivative compare to the original signal as the frequency is increased? Repeat these considerations for the integral of $y(t)$.

Section 11.17

11.28. Show that integration of Eq. (11.101) over all shift times is consistent with the integration of the δ function that is obtained in the limit $\tau_1 \to 0$.

REFERENCES

Bevington, P. R., and D. K. Robinson (1992). *Data Reduction and Error Analysis for the Physical Sciences*, 2nd ed. New York, McGraw-Hill.

Blackman, R. B., and J. W. Tukey (1958). *The Measurement of Power Spectra*. AT&T. New York, Dover, pp. 32–33.

Bracewell, R. N. (1990). Numerical transforms. *Science* **248**: 697–704.

Cohen, A. (1995). Biomedical signals: Origin and dynamic characteristics; frequency-domain analysis. In J. D. Bronzino, ed. *The Biomedical Engineering Handbook*. Boca Raton, FL, CRC, pp. 805–827.

Cooley, J. W., and J. W. Tukey (1965). An algorithm for the machine calculation of complex Fourier series. *Math. Comput.* **119**: 297–301.

DeFelice, L. J. (1981). *Introduction to Membrane Noise*. New York, Plenum.

Gatland, I. R. (1993). A weight-watcher's guide to least-squares fitting. *Comput. Phys.* **7**(3): 280–285.

Gatland, I. R., and W. J. Thompson (1993). Parameter bias estimation for log-transformed data with arbitrary error characteristics. *Am. J. Phys.* **61**(3): 269–272.

Guillemin, E. (1949). *The Mathematics of Circuit Analysis*. Cambridge, MA, MIT Press, p. 341.

Jeffreys, H., and B. S. Jeffreys (1956). *Mathematical Physics*. Cambridge, England, Cambridge University Press, p. 455.

Kaplan, D., and L. Glass (1995). *Understanding Nonlinear Dynamics*. New York, Springer.

Lighthill, M. J. (1958). *An Introduction to Fourier Analysis and Generalized Functions*. Cambridge, England, Cambridge University Press.

Lybanon, M. (1984). A better least-squares method when both variables have uncertainties. *Am. J. Phys.* **52**: 22–26.

Mainardi, L. T., A. M. Bianchi, and S. Cerutti (1995). Digital biomedical signal acquisition and processing. In J. D. Bronzino, ed. *The Biomedical Engineering Handbook*. Boca Raton, FL, CRC, pp. 828–852.

Maughan, W. Z., C. R. Bishop, T. A. Pryor, and J. W. Athens (1973). The question of cycling of blood neutrophil concentrations and pitfalls in the analysis of sampled data. *Blood*. **41**: 85–91.

Milnor, W. R. (1972). Pulsatile blood flow. *New Eng. J. Med.* **287**: 27–34.

Nyquist, H. (1928). Thermal agitation of electric charge in conductors. *Phys. Rev.* **32**: 110–113.

Orear, Jay (1982). Least squares when both variables have uncertainties. *Am. J. Phys.* **50**: 912–916.

Press, W. H., S. A. Teukolsky, W. T. Vetterling, and B. P. Flannery (1992). *Numerical Recipes in C: The Art of Scientific Computing*, 2nd ed., reprinted with corrections, 1995. New York, Cambridge University Press.

Visscher, P. B. (1996). The FFT: Fourier transforming one bit at a time. *Comput. Phys.* **10**(5): 438–443.

CHAPTER 12

Images

Images are very important in the remainder of this book. They may be formed by the eye, a camera, an x-ray machine, a nuclear medicine camera, magnetic resonance imaging, or ultrasound. The concepts developed in Chap. 11 can be used to understand and describe image quality. The same concepts are also used to reconstruct computed tomographic or magnetic resonance slice images of the body.

The convolution integral of Sec. 12.1 shows how the response of a linear system can be related to the input to the system and the impulse (δ function) response of the system. It forms the basis for the rest of the chapter. The Fourier-transform properties of the convolution are also described in this section. Section 12.2 extends the convolution integral and Fourier transform to two dimensions. Section 12.3 introduces some concepts from photometry that are needed to describe the energy per unit time in an image. Section 12.4 discusses three common ways to form images: a pinhole, a lens, and a shadow. Section 12.5 introduces quantitative ways to relate the object to the image, using the techniques developed in Chap. 11 to describe the blurring that occurs. Section 12.6 shows the importance of different spatial frequencies in an image and their effect on the quality of the image.

Sections 12.7 and 12.8 pose the fundamental problem of reconstructing slices from projections and introduce two techniques for solving it: the Fourier transform and filtered back projection. Section 12.9 provides a numerical example of filtered back projection for a circularly symmetric object.

12.1. CONVOLUTION

We are now going to apply the techniques developed in Chap. 11 to describe the formation of images. An image is a function of position, usually in two dimensions. Functions of time are easier to think about, so we start with a one-dimensional example that is a function of time: a high-fidelity sound system. A hi-fi system is (one hopes) linear, which means that the relationship between the output response and a complicated input function can be written as a superposition of responses to more elementary input functions. The output might be the instantaneous air pressure at some point in the room; the input might be the air pressure at a microphone or the magnetization on a strip of tape.

It takes a certain amount of time for the signal to propagate through the system. In the simplest case the response at the ear would exactly reproduce the response at the input a very short time earlier. In actual practice the response at time t may depend on the input at a number of earlier times, because of limitations in the electronic equipment or echoes in the room. If the entire system is linear, the output $g(t)$ can be written as a superposition integral, summing the weighted response to inputs at other times. If $f(t')$ is the input function and h is the weighting, the output $g(t)$ is

$$g(t)=\int_{-\infty}^{\infty} f(t')h(t,t')\, dt'. \tag{12.1}$$

Variable t' is a dummy variable. The integration is over all values of t' and it does not appear in the final result, which depends only on the functional form of f and h. Note also that if f and g are expressed in the same units, then h has the dimensions of s^{-1}.

If input f is a δ function at time t_0', then

$$g(t)=\int_{-\infty}^{\infty} \delta(t'-t_0')h(t,t')dt'=h(t,t_0'). \tag{12.2}$$

We see that $h(t,t')$ is the *impulse response* of the system to an impulse at time t'. If the impulse response of a linear system is known, it is possible to calculate the response to any arbitrary input.

If, in addition to being linear, the system responds to an impulse the same way regardless of when it occurs, the system is said to be *stationary*. In the hi-fi example, this means that no one is adjusting the volume or tone controls. For a stationary system the impulse response depends only on the *time difference* $t-t'$:

$$h(t,t')=h(t-t'), \tag{12.3}$$

and the superposition integral takes the form

$$g(t)=\int_{-\infty}^{\infty} f(t')h(t-t')dt'. \tag{12.4a}$$

This is called the *convolution integral.* It is often abbreviated as

$$g(t)=f(t)\otimes h(t). \tag{12.4b}$$

For the hi-fi system the function $h(t-t')$ is zero for all t' larger (later) than t; the response does not depend on future inputs. For the images we will be considering shortly, where the variables represent positions in the object and image, h can exist for negative arguments.

We saw an example of the impulse response in Sec. 11.16, where we found that the solution of the differential equation for the system was a step exponential, Eq. (11.83). For that simple linear system we can write

$$h(t-t')=\begin{cases} 0, & t<t' \\ (1/\tau_1)e^{-(t-t')/\tau_1}, & t>t' \end{cases}. \tag{12.5}$$

We have seen superposition integrals before: for one-dimensional diffusion [Eq. (4.73)] and for the potential [Eq. (7.21)] and magnetic field [Eq. (8.12)] outside a cell.

There is an important relationship between the Fourier transforms of the functions appearing in the convolution integral, which was hinted in Sec. 11.16. If the sine and cosine transforms of function h are denoted by $C_h(\omega)$ and $S_h(\omega)$, with similar notation for f and g, the relationships can be written

$$\begin{aligned} C_g(\omega)&=C_f(\omega)C_h(\omega)-S_f(\omega)S_h(\omega), \\ S_g(\omega)&=C_f(\omega)S_h(\omega)+S_f(\omega)C_h(\omega). \end{aligned} \tag{12.6a}$$

This is called the *convolution theorem.*[1]

Equations (12.6a) are similar to the addition formulas for sines and cosines, which are of course used in the derivation. To derive them, we take the Fourier transforms of f and h:

$$f(t')=\frac{1}{2\pi}\int_{-\infty}^{\infty}[C_f(\omega)\cos(\omega t')+S_f(\omega)\sin(\omega t')]d\omega,$$

$$h(t-t')=\frac{1}{2\pi}\int_{-\infty}^{\infty}\{C_h(\omega)\cos[\omega(t-t')] + S_h(\omega)\sin[\omega(t-t')]\}\,d\omega.$$

[1]If we were using complex exponential notation, the Fourier transforms would be related by.

$$G(\omega)=F(\omega)H(\omega). \tag{12.6b}$$

Then

$$\begin{aligned} g(t)&=\int_{-\infty}^{\infty} f(t')h(t-t')\,dt' \\ &=\left(\frac{1}{2\pi}\right)^2\int_{-\infty}^{\infty}dt'\left[\int_{-\infty}^{\infty}d\omega[C_f(\omega)\cos(\omega t') \right. \\ &\quad +S_f(\omega)\sin(\omega t')]\int_{-\infty}^{\infty}d\omega'\{C_h(\omega')\cos[\omega'(t-t')] \\ &\quad \left. +S_h(\omega')\sin[\omega'(t-t')]\}\right]. \end{aligned}$$

The trigonometric addition formulas and the fact that $\sin(-\omega' t')=-\sin(\omega' t')$ are used to rewrite this and expand it, much as we did in the last chapter. Carrying out the integration over t' first and using the properties of integrals of the δ function gives

$$\begin{aligned} g(t)&=\frac{1}{2\pi}\int_{-\infty}^{\infty}d\omega[C_f(\omega)C_h(\omega)-S_f(\omega)S_h(\omega)]\cos(\omega t) \\ &\quad+\frac{1}{2\pi}\int_{-\infty}^{\infty}d\omega[C_f(\omega)S_h(\omega) \\ &\quad+S_f(\omega)C_h(\omega)]\sin(\omega t). \end{aligned}$$

Comparison of this with Eqs. (11.55) proves Eq. (12.6a).

Fourier techniques need not be restricted to frequency and time. The quality and resolution of the image on the retina, an x-ray film, or a photograph are best described in terms of *spatial frequency.* The distance across the image in some direction is x, and a sinusoidal variation in the image would have the form $A(k)\sin(kx-\phi)$. The *angular spatial frequency* k has units of radians per meter. It is $k=2\pi/\lambda$, where λ is the wavelength, in analogy to $\omega=2\pi/T$. Alternatively, we can use the *spatial frequency* $1/\lambda$, with units of cycles per meter or cycles per millimeter.

12.2. THE FOURIER TRANSFORM AND CONVOLUTION IN TWO DIMENSIONS

The convolution and Fourier transform in two dimensions are needed to analyze the response of a system that forms a two-dimensional image of a two-dimensional object. The object can be represented by function $f(x',y')$. In general the image is given by

$$g(x,y)=\int_{-\infty}^{\infty}\int_{-\infty}^{\infty} f(x',y')h(x,x';y,y')\,dx'dy'. \tag{12.7}$$

If the contribution of object point (x',y') to the image at (x,y) depends only on the relative distances $x-x'$ and y

$-y'$, then the two-dimensional impulse response is $h(x-x',y-y')$, and the image is obtained by the two-dimensional convolution

$$g(x,y)=\int_{-\infty}^{\infty}\int_{-\infty}^{\infty} f(x',y')h(x-x',y-y')\,dx'\,dy' \tag{12.8a}$$

or

$$g(x,y)=f(x,y)\otimes\otimes h(x,y). \tag{12.8b}$$

The Fourier transform in two dimensions is defined by

$$f(x,y)=\left(\frac{1}{2\pi}\right)^2\int_{-\infty}^{\infty}dk_x\int_{-\infty}^{\infty}dk_y[C(k_x,k_y)\cos(k_xx+k_yy)+S(k_x,k_y)\sin(k_xx+k_yy)]. \tag{12.9a}$$

The coefficients are given by

$$C(k_x,k_y)=\int_{-\infty}^{\infty}dx\int_{-\infty}^{\infty}dy\ f(x,y)\cos(k_xx+k_yy), \tag{12.9b}$$

$$S(k_x,k_y)=\int_{-\infty}^{\infty}dx\int_{-\infty}^{\infty}dy\ f(x,y)\sin(k_xx+k_yy). \tag{12.9c}$$

The Fourier transforms of the functions in the convolution are related by equations similar to those for the one-dimensional convolution[2]

$$C_g(k_x,k_y)=C_f(k_x,k_y)C_h(k_x,k_y)-S_f(k_x,k_y)S_h(k_x,k_y), \tag{12.10}$$

$$S_g(k_x,k_y)=C_f(k_x,k_y)S_h(k_x,k_y)+S_f(k_x,k_y)C_h(k_x,k_y).$$

12.3. RADIOMETRY

This section develops some of the concepts and vocabulary of *radiometry*, the measurement of radiant energy. We will be considering four types of radiant energy in the next five chapters: infrared radiation, visible light, ultraviolet radiation, and x rays. The notation for the quantities used in radiometry differs by discipline. Radiometry will be revisited in Sec. 13.10.

The total amount of energy being considered is the *radiant energy R*, measured in joules. It can be the energy *emitted* by a source, *transferred* from one region to another, or *received* by a detector. We will often use subscripts *s* and *d* to refer to the source and detector. In optics the radiant energy is electromagnetic radiation. In radiological physics we will also consider energy transported by charged and neutral particles such as electrons and neutrons.

The rate at which the energy is radiated, transferred, or received is the *radiant power P* (watts).

The simplest source is a point that radiates uniformly in all directions. The *radiant intensity* or radiant power per unit solid angle (Appendix A) leaving a point source radiating uniformly in all directions is

$$\frac{dP}{d\Omega}=\frac{P}{4\pi}\ (\mathrm{W\ sr^{-1}}).$$

The power per unit area falls as $1/r^2$, while the power per unit solid angle is independent of r.[3] The point source need not be uniform. For example, it might be a searchlight 1 m in diameter viewed from a point several kilometers away so that it appears to be a point. The light might be confined to a cone with a half-angle of 1°. Then a plot of $dP/d\Omega$ might look like Fig. 12.1. The total power radiated by the point source is

$$P=\int\frac{dP}{d\Omega}\,d\Omega. \tag{12.12}$$

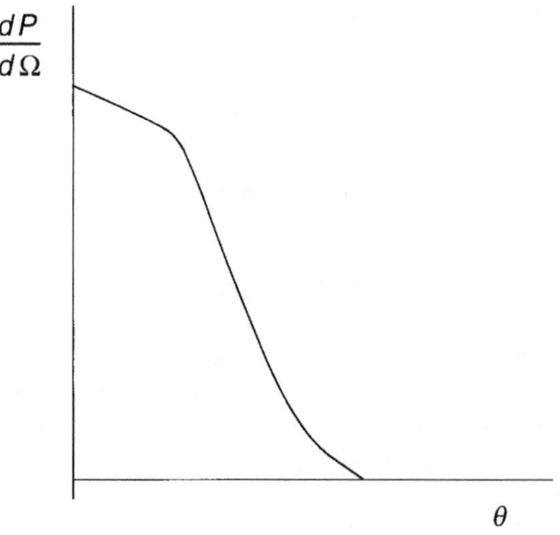

FIGURE 12.1. A plot of power per unit solid angle as a function of angle from the axis of a hypothetical searchlight.

[2] With complex notation we would define the two-dimensional Fourier transform pair by

$$F(k_x,k_y)=\int\int f(x,y)e^{i(k_xx+k_yy)}\,dx\,dy,$$

$$f(x,y)=\left(\frac{1}{2\pi}\right)^2\int\int F(k_x,k_y)e^{-i(k_xx+k_yy)}\,dk_x\,dk_y, \tag{12.11a}$$

and the convolution theorem would be

$$G(k_x,k_y)=F(k_x,k_y)H(k_x,k_y). \tag{12.11b}$$

[3] The lighting industry calls $dP/d\Omega$ the intensity, while in physical optics intensity is used for power per unit area. We will try to avoid using the word intensity.

If the power per unit solid angle is symmetric about the axis of the beam and θ is the angle with respect to the beam axis, then (see Appendix L)

$$P=\int_0^\pi \frac{dP}{d\Omega} 2\pi \sin\theta\, d\theta.$$

Now consider the energy striking a surface. The *irradiance E* is the power per unit area incident on a surface. The strict definition is the ratio of the power incident on an infinitesimal element of detector surface dS_d to the area projected perpendicular to the direction the radiant energy is traveling. If θ_d is the angle between a normal to the surface and the direction of propagation, the irradiance is

$$E=\frac{dP}{\cos\theta_d\, dS_d}. \quad (12.13)$$

For a point source radiating uniformly in all directions, the power at distance r is spread uniformly over a sphere of area $4\pi r^2$, so

$$E=\frac{P}{4\pi r^2}. \quad (12.14)$$

For an extended source the power emitted by the surface is proportional to both the size of the emitting area dS_s and the solid angle of the cone $d\Omega$ into which the energy is radiated, as shown in Fig. 12.2. The solid angle subtended by a small element of area on the detector is $d\Omega$, as shown by the dashed lines. The amount of power radiated into $d\Omega$ from dS_s is

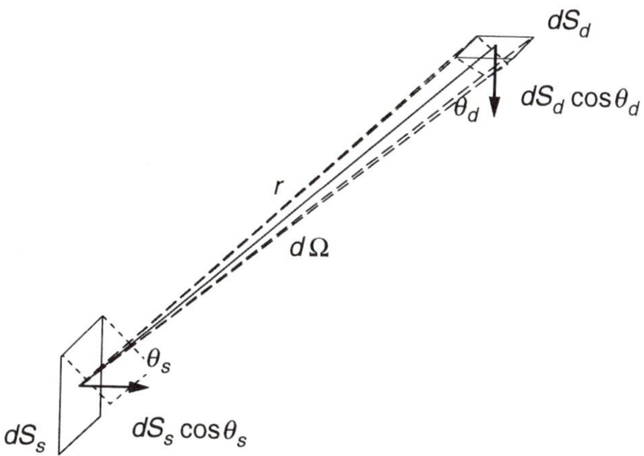

FIGURE 12.2. Radiant energy is emitted from an element of surface area dS_s into a cone of solid angle $d\Omega$. The direction of emission is at an angle θ_s with the normal to the surface. A detecting surface has an element of area dS_d oriented at a direction θ_d to the direction of travel of the radiation from source to detector.

$$L dS_s d\Omega=\frac{d^2P}{\cos\theta_s\, dS_s\, d\Omega} dS_s d\Omega, \quad (12.15)$$

where the *radiance L* depends on the direction of emission as well as the location on the surface. The radiance is the *power emitted from the source per unit solid angle per unit area of surface projected perpendicular to the emerging beam.* This equation is valid whether the energy is emitted directly from the source (as in a glowing object) or is scattered by the surface (as from this page). The total power emitted is

$$P=\int\int L\, dS_s\, d\Omega. \quad (12.16)$$

The distinction between angles and areas for the source and the detector is shown in Fig. 12.2. Note that the solid angle subtended at the source by dS_d is $d\Omega = dS_d \cos\theta_d/r^2$. The power into an area dS_d of the detector from area dS_s of the source is therefore

$$d^2P=\frac{L\cos\theta_s\cos\theta_d\, dS_s dS_d}{r^2}. \quad (12.17)$$

12.4. FORMING IMAGES

There are many ways to form an image. We consider three that are important either in human vision or in radiology.

12.4.1. Pinhole

The simplest is a pinhole, shown in Fig. 12.3. If radiant energy from the object travels in straight lines, that which passes through the pinhole in the barrier falls on a small region in the image plane. There is an inverted (upside-down) image for any value of the object distance u or image distance v. We continue using a notation in which the input

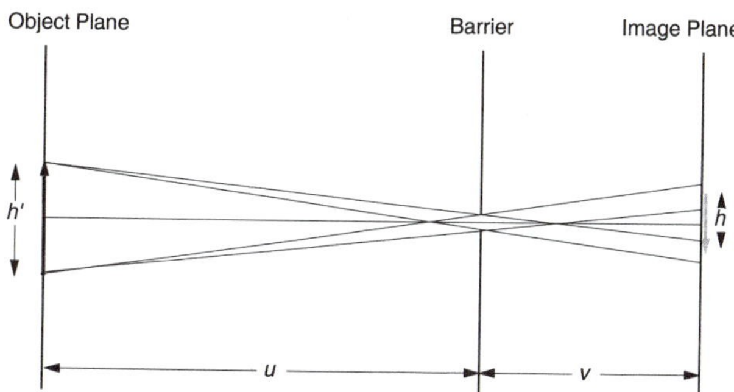

FIGURE 12.3. The image formed by a pinhole in the barrier. The object height is h' and the image height is h. The image is inverted. There is blurring because of the size of the pinhole. Making the pinhole smaller reduces this blurring, lets through less light, and may also cause diffraction effects that blur the image.

or object variables are primed and the output or image variables are unprimed. If the object has a height h' we can draw lines from the extremes of the object through the center of the pinhole to define the center of the corresponding image points and determine the image height h. By similar triangles, the *magnification* is

$$m=\frac{h}{h'}=\frac{v}{u}, \tag{12.18}$$

where u is the object distance (object to screen) and v the image distance, as shown in Fig. 12.3. The image is blurred. The height of the blurred image spot w depends on the diameter $2a$ of the pinhole:

$$w=\frac{2a(v+u)}{u}. \tag{12.19}$$

Decreasing the size of the pinhole makes the image sharper, until the size becomes so small that diffraction effects begin to blur things again.[4] Pinhole images are often used to verify the size of the focal spot (source of x rays) in an x-ray tube.

The drawback of a pinhole is that not much radiant energy reaches the image. We can calculate the amount of energy per unit area in the image. The radiant power striking the image from a small area of the object is given by Eq. (12.17), where the solid angle is the area of the pinhole divided by the square of the object distance, and both angles are close to zero:

$$(\text{power in image})=L\ dS'\ d\Omega=L\ \frac{dS'\ \pi a^2}{u^2}, \tag{12.20}$$

This energy falls on a certain area of the image. We are usually interested in the case where the pinhole is small, but large enough so that diffraction effects can be ignored. The power falls on an image area dS related to the object area dS' by[5] $dS=m^2dS'=(v/u)^2dS'$. The irradiance in the image is

$$E_{\text{image}}=\frac{L\ dS'\ d\Omega}{dS}=\frac{L\ dS'(\pi a^2/u^2)}{dS'(v^2/u^2)}=\frac{L\pi a^2}{v^2}. \tag{12.21}$$

If we define $d\Omega'$ to be the solid angle that the pinhole subtends at the position of the image, then we can write this as

$$E_{\text{image}}=L\ d\Omega'. \tag{12.22}$$

[4]We will not discuss diffraction effects. See, for example, Williams and Becklund (1972).

[5]The m^2 appears in this expression because the image is magnified both in the plane of the paper and perpendicular to it. Imagine dS' to be a square of side h' and dS to be a square of side h.

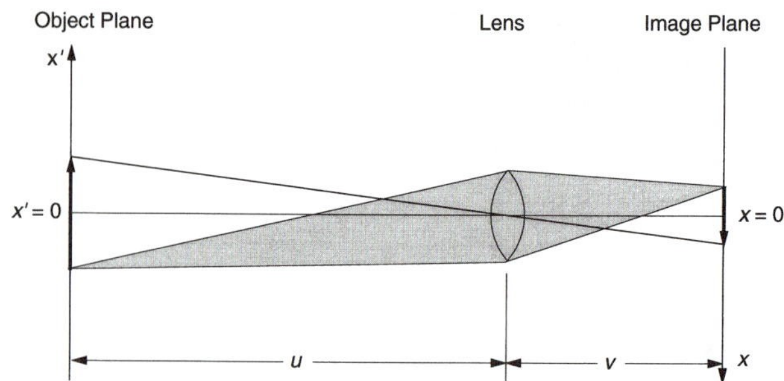

FIGURE 12.4. The image formed by an ideal lens. All light emerging from a very small region on the object in the solid angle that enters the lens is focused into a corresponding small region in the image plane. This solid angle is represented by the shading. Calculations are done using coordinate systems that point in opposite directions because of the inversion of the image.

Equations (12.21) and (12.22) tell us that the power per unit area or irradiance in the image is the radiance of the object times the area of the pinhole divided by the image distance squared, which is the same as the object energy radiance times the solid angle that the pinhole subtends at the image.

12.4.2. Ideal Lens

The main difference between a pinhole and an ideal lens is that the lens captures radiant energy emitted in a much greater solid angle and focuses it back to a point in the image. A lens is shown in Fig. 12.4. Radiant energy again emerges perpendicular to the surface and is captured by the lens. If it is an ideal lens, the radiant energy from any point on the object is focused to a point in the image. A ray can be drawn from a point on the object through the center of the lens to the corresponding point on the image.[6] The magnification is again given by Eq. (12.18). In fact, Eqs. (12.20)–(12.22) apply to the lens as well as the pinhole. The only difference is that the radius of the lens can now be made much larger, to gather more radiant energy, without blurring the image.

Recall from geometrical optics that for a thin lens the object distance, image distance, and focal length are all related by

$$\frac{1}{u}+\frac{1}{v}=\frac{1}{f}. \tag{12.23}$$

The image distance has its smallest value, f, when the object is infinitely far away. As the object is brought closer, the image distance increases, but it does not increase very much until the object distance is only a few times the focal length. As long as the object is farther away, the image

[6]This is true in the approximation that we have a ''thin'' lens. For a thick lens there is some point, usually near the geometric center of the lens, through which the same ray can be drawn.

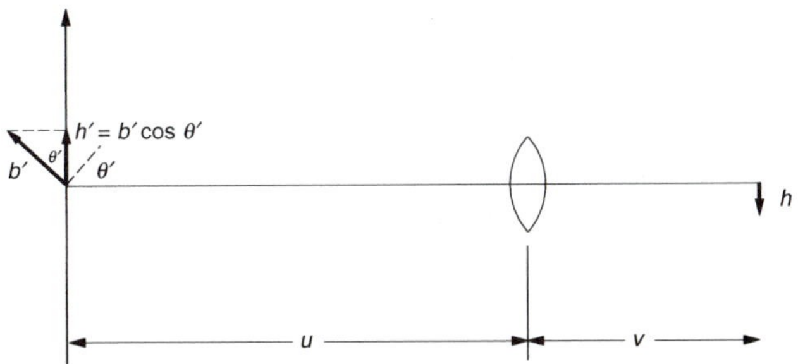

FIGURE 12.5. A surface oblique to the axis of the lens emits radiation from area dS'. The relation between dS' and dS depends on angle θ'. The apparent height of the object is $h'=b' \cos\theta$.

distance is approximately equal to f, and the power per unit area or energy fluence rate in the image is nearly independent of the object distance. The reason that the energy fluence rate in the image is independent of object distance is that the power entering the lens decreases as $1/u^2$ when the object is moved farther away, but at the same time the area of the image in which the power is concentrated becomes smaller, being nearly proportional[7] to $1/u^2$.

If the object surface is not perpendicular to the axis of the lens, the results must be modified as in Fig. 12.5. The fact that the radiant energy may not be emitted uniformly by the surface is taken into account by $L(\theta')$. To find the relationship between the area dS' of a patch on the surface of the object and the corresponding patch dS in the image, let the object have a length l' perpendicular to the page and length b' as in Fig. 12.5. The "height" of element b' perpendicular to the axis of the lens is $h'=b' \cos\theta'$. Since Eq. (12.15) still defines the magnification, we have

$$dS=lh=l'h'(v/u)^2=dS' \cos\theta' \, (v/u)^2.$$

The irradiance in the image is then

$$E_{\text{image}}=\frac{L(\theta')\cos\theta' \, \pi a^2}{v^2 \cos\theta'}=\frac{L(\theta')\pi a^2}{v^2}. \tag{12.24}$$

The angular dependence of L must be measured. In some cases where the reflection from the surface is "perfectly diffuse"[8] $L(\theta')$ has the form

$$L(\theta')=L_0. \tag{12.25}$$

This is called *Lambert's law of illumination* or Lambert's cosine law. A surface described by Lambert's law will have equal power per unit area in the image regardless of the viewing angle. Look at surfaces around you. Do similar surfaces illuminated the same way appear to have the same brightness when they are oblique to your line of vision?

12.4.3. Shadows with a Point Source

The third example is important in making x-ray pictures of the body. A point source of radiant energy projects a shadow of the object on the image plane, as shown in Fig. 12.6. The magnification is

$$m=\frac{u+v}{u}, \tag{12.26}$$

and the blurring occurs because the source has a finite extent. Figure 12.6 is an idealization because the object is actually thick, with components at several different distances from the source. Moreover, the object is not completely opaque and only part of the radiant energy is absorbed.

12.5. THE RELATIONSHIP BETWEEN THE OBJECT AND THE IMAGE

We can use the discussion of photometry to develop a mathematical expression relating an object to its image. We will find that with certain simplifying approximations, it is given by a two-dimensional convolution. This means that we can learn a great deal from the Fourier-transform properties of the convolution.

Suppose that an object is in the $x'y'$ plane and has a radiance $L(x',y')$ that varies from place to place on the object. From a patch of object $dx'dy'$ an amount of power $L(x',y')dx'dy'd\Omega$ passes through the pinhole or lens and strikes the image plane. In an ideal situation all of this power would be focused in area $dx\,dy$ and the power per

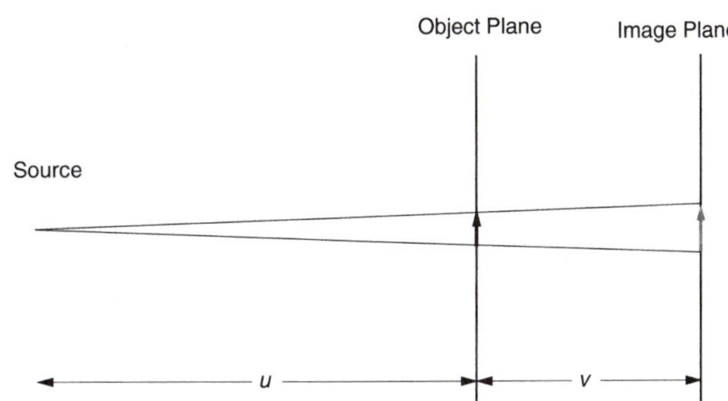

FIGURE 12.6. A point source casts a shadow of an object in the image plane.

[7]The quantity that determines how much power per unit area is in the image is $\pi a^2/v^2$, which is approximately $\pi a^2/f^2$. Camera buffs will recognize that this is inversely proportional to the square of the *f number*, the ratio of the lens focal length to diameter. The notation $f/8$ means that $f/(2a)=8$, or $2a=f/8$. The smaller the f number, the more power per unit area in the image. For the case of an $f/8$ lens, $\pi a^2/f^2=\pi/(4\times 8^2)$. Those familiar with taking pictures of close objects will also recognize that the "effective f number" that is used to calculate exposures for close objects is the f number times the image distance divided by the focal length. This serves to put v^2 back in the denominator of the original expression instead of approximating it by f^2.

[8]We will discuss what this means in the next chapter. Sometimes Eq. (12.16) is defined without the factor $\cos\theta_s$, in which case Lambert's law has the form $L(\theta_s)=L_0 \cos\theta_s$.

unit area in the image would be given by Eq. (12.22). In a more general case aberrations (blurring) in the lens, diffraction, or scattering from dust or other objects between the object and image spread the radiant power out over a broader area. The relationship between the object radiance and the image illuminance is given by Eq. (12.7):

$$E_{\text{image}}(x,y)=\int\int L(x',y')h(x,y;x',y')\,dx'\,dy'. \tag{12.27}$$

Function h is called the *point-spread function*. The point-spread function tells how power from a point source at (x',y') spreads out over the image plane. It receives its name from the following. If we imagine that the object is a point described by $L(x',y')=L\;\delta(x'-x_0')\delta(y'-y_0')$, then integration shows that

$$E_{\text{image}}=h(x,y;x_0',y_0').$$

The point-spread function has the same functional form as the image from a point source, just as did the impulse response in one dimension.

For the ideal lens that ignores diffraction effects, you can verify that the point-spread function is

$$h(x,y;x_0',y_0')=\frac{m^2\pi a^2}{v^2}\,\delta(x-mx')\delta(y-my'). \tag{12.28}$$

This assumes that the origins of the xy and $x'y'$ planes are on the axis of the lens and that the object is small enough so that v^2 and the solid angle are constant for all points on the object. It also assumes that the directions of the image axes are opposite to the directions of the axes in the object plane to account for the inversion of the image (see Fig. 12.4). The δ functions pick out the values $(x'=x/m,\; y'=y/m)$ in the object plane to contribute to the image at (x,y). You can make the verification by substituting Eq. (12.28) in Eq. (12.27) and using the properties of the δ function from Eq. (11.62). Carrying out the integrations gives our earlier expression for the power per unit area in the image, Eq. (12.21).

This discussion assumes that *intensities* add. This is true when the oscillations of the radiant energy (such as the electric field for light waves) have random phases lasting for a time short compared to the measurement time. Such radiant energy is called *incoherent*.[9]

We have already seen that when the impulse response in a one-dimensional system depends on coordinate differences such as $t-t'$ (or $x-x'$ or $x-mx'$), the system is stationary. In this case it is also said to be *space invariant*: changing the position of the object changes the position of the image but not its functional form. Stationarity is easier to obtain in a system such as a hi-fi system than in an imaging system, but we usually assume that it holds in an imaging system as well. For example, we obtained space invariance for our ideal lens by ignoring any change in solid angle with position. For a space-invariant system

$$E_{\text{image}}(x,y)=\int\int L(x',y')h(x-mx',y-my')\,dx'\,dy'. \tag{12.29}$$

This is a two-dimensional convolution, which we first saw in Sec. 12.2. The convolution theorem is

$$\begin{aligned}C_{\text{image}}(k_x,k_y)&=C_{\text{object}}(k_x,k_y)C_h(k_x,k_y)\\&\quad-S_{\text{object}}(k_x,k_y)S_h(k_x,k_y),\end{aligned} \tag{12.30}$$

$$\begin{aligned}S_{\text{image}}(k_x,k_y)&=C_{\text{object}}(k_x,k_y)S_h(k_x,k_y)\\&\quad+S_{\text{object}}(k_x,k_y)C_h(k_x,k_y).\end{aligned}$$

The *optical transfer function* (OTF) is the Fourier transform of the point-spread function, $C_h(k_x,k_y)$ and $S_h(k_x,k_y)$. It is analogous to the transfer function for an amplifier (Sec. 11.16). The *modulation transfer function* (MTF) is the amplitude of the OTF:

$$\text{MTF}(k_x,k_y)=[C_h^2(k_x,k_y)+S_h^2(k_x,k_y)]^{1/2}. \tag{12.31}$$

The *phase transfer function* is

$$\text{PTF}(k_x,k_y)=\tan^{-1}\left(\frac{S_h(k_x,k_y)}{C_h(k_x,k_y)}\right). \tag{12.32}$$

Often the transfer functions are normalized by dividing them by their value at zero spatial frequency.

The modulation transfer function can be measured by using a set of objects whose radiance varies sinusoidally at different spatial frequencies. Since we are dealing with incoherent systems, the radiance cannot be negative and must be offset by a zero-frequency component:

$$L(x,y)=a+b\,\cos(k_x x+k_y y),\quad 0<b<a. \tag{12.33}$$

The image is described by

$$\begin{aligned}E=\;&\text{MTF}(0,0)a+\text{MTF}(k_x,k_y)b\,\cos[k_x x+k_y y\\&+\phi(k_x,k_y)].\end{aligned} \tag{12.34}$$

The *modulation* of the object is defined to be

[9]These arguments also work for coherent radiation, where the phases are important, but the point-spread function is for the amplitude of the wave instead of the square of the amplitude (intensity), and the calculation gives rise to interference and diffraction effects.

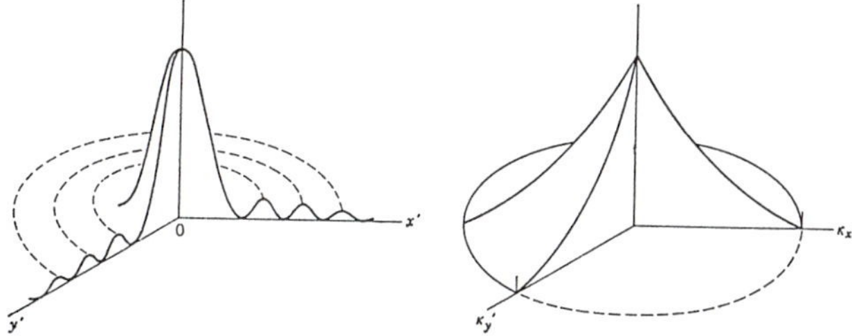

FIGURE 12.7. The point-spread function and modulation transfer function for a diffraction-limited circular aperture. From C. S. Williams and O. A. Becklund (1972). *Optics: A Short Course for Engineers and Scientists*. New York, Wiley. Used by permission of the authors.

$$(\text{modulation}) = \frac{L_{\max} - L_{\min}}{L_{\max} + L_{\min}} = \frac{(a+b)-(a-b)}{(a+b)+(a-b)} = \frac{b}{a}. \tag{12.35}$$

A similar expression defines the modulation of the image. The modulation transfer function is the ratio of the modulation of the image divided by the modulation of the object. The phase of the optical transfer function describes shifts of the phase of the image at each angular frequency along the appropriate axis. It is fully as important as the amplitude, since it describes the evenness or oddness of the image about its stated origin.

The modulation transfer function of an ideal system would be flat for all spatial frequencies. However, there is an upper limit imposed by diffraction, if nothing else. Figure 12.7 shows the point-spread function and MTF for a diffraction-limited case. Figure 12.8 shows three possible modulation transfer functions for an imaging system. The upper one represents the diffraction limit. It has the same general shape as in Fig. 12.7. Curves 1 and 2 might be for real systems. While the second system transmits more of the highest spatial frequencies, it transmits less of the midrange frequencies, and its image would not have as much "punch" as the first system. Figure 12.9 shows the modulation transfer functions of several photographic films, with (a) being the most sensitive and (e) the least sensitive but with the highest resolution. Photographers are well aware of the tradeoff between speed and resolution in film. Fast films are "more grainy" than slow films.

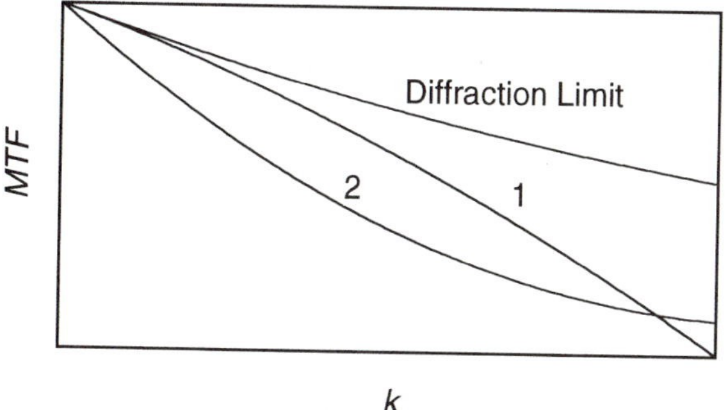

FIGURE 12.8. Three possible modulation transfer functions. The top one is the diffraction limit for monochromatic light. (Compare it with Fig. 12.7.) Curve 2 is higher than curve 1 at the highest value of k shown, but an image produced by system 2 would not have as much "punch." It has less content at the middle spatial frequencies.

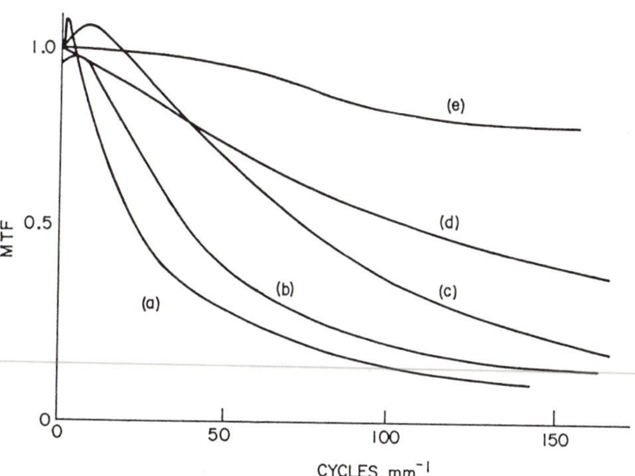

FIGURE 12.9. Some representative modulation transfer functions for various photographic films, showing how the resolution decreases as the film sensitivity increases. Film (a) has the greatest sensitivity and worst resolution. Film (e) is the least sensitive ("slowest") and has the best resolution. From R. Shaw (1979). Photographic Detectors. Chap. 5 of *Applied Optics and Engineering*, New York, Academic, Vol. 7., pp. 121–154.

A complex imaging system may have several components, just as the hi-fi system did. If the system is linear, the modulation transfer function for the combined system is the product of the modulation transfer functions for each component. The optical transfer functions combine according to equations like Eq. (12.10).

The *line-spread function* is the response of a system to a line object in a plane perpendicular to the axis of the lens. In general, the system is not isotropic and the line-spread function depends on the direction of the line. The Fourier transform of the line-spread function along the y axis is $C_h(k_x,0)$ and $S_h(k_x,0)$. Figures 12.10 and 12.11 show a geometrical interpretation of the point-spread function and the line-spread function. The *edge spread function* is the response to an object that has a step in the radiance. All of these functions are interrelated. A discussion of how one can be obtained from another is found in many places, including Chap. 9 of Gaskill (1972).

12.6. SPATIAL FREQUENCIES IN AN IMAGE

There are some universal relationships between the spatial frequencies present in an image and the character of the image. These relationships hold whether the image is a photograph, an x-ray film, a computed tomographic scan, an ultrasound or nuclear medicine image, of a magnetic reso-

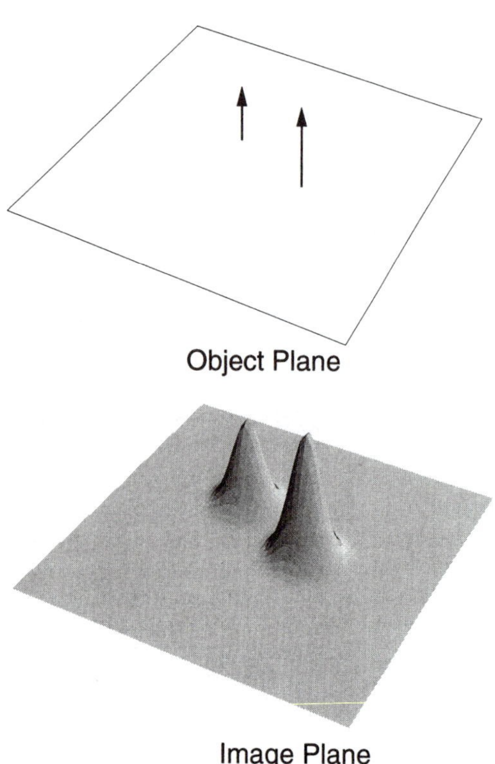

FIGURE 12.10. The point-spread function. Two impulse sources of different height are shown in the object plane. The response to them is shown in the image plane.

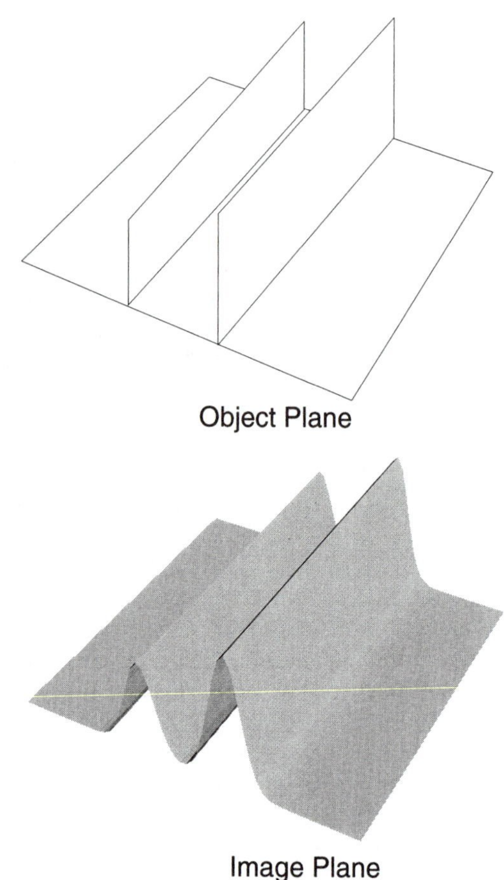

FIGURE 12.11. The line-spread function. Two line sources are shown in the object plane. The response to them is shown in the image plane.

nance image. In this section we describe these general relationships, which we will use throughout the rest of the book.

The first general relationship concerns the size of an image and the spatial frequencies present. The object is nonperiodic. An image must be represented by a Fourier series, not a Fourier integral, because we cannot have an infinite number of samples. The Fourier series representing the image is periodic. We saw in Chap. 11 that if the lowest angular frequency present is ω_0, the period is $T = 2\pi/\omega_0$. For simplicity, consider only the x direction and the corresponding spatial frequencies k. Figure 12.12 shows an image of length D represented by a Fourier series of period L. The lowest spatial frequency present is $k_0=2\pi/L$. The series has harmonics with separation $\Delta k = k_0$. If the images are not to overlap, we must have $D<L$. This leads to the fundamental relationship

$$D<\frac{2\pi}{\Delta k}. \tag{12.36}$$

The lowest spatial frequency present (which equals the separation of the spatial frequencies) determines the size of the image (the "field of view" or FOV).

Sampling the signal N times or reconstructing the signal with N discrete values gives a spatial resolution $\Delta x = D/N$. This allows (or requires) the determination of $N/2$ cosine coefficients and $N/2$ sine coefficients. The highest spatial frequency present is $k_{max}=N\Delta k/2$. We obtain

$$\Delta x=\frac{D}{2}\frac{\Delta k}{k_{max}}=\frac{\pi}{k_{max}}. \tag{12.37}$$

The resolution is inversely proportional to the highest spatial frequency present. As we saw for the Fourier series representing a square wave, the higher harmonics give fine

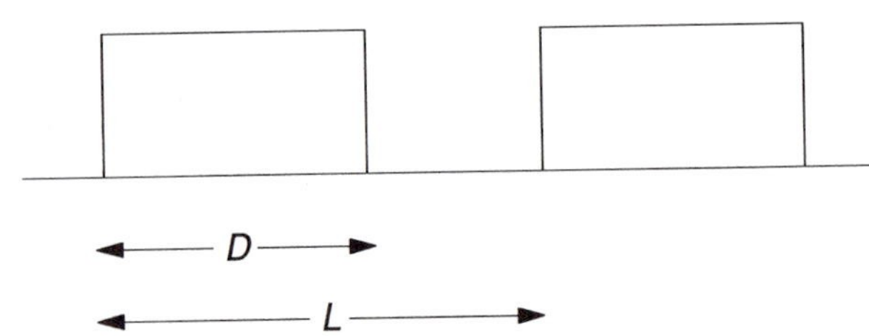

FIGURE 12.12. The relationship between the length of an object and the periodicity of a Fourier series representing its image. If the entire object is to be imaged, the period L must be greater than the length of the object D.

detail and sharpness to the image. The xerographic technique used in copying machines tends to accentuate the higher spatial frequencies. Large dark areas are difficult to reproduce.

To reiterate: *The lowest spatial frequency in the image determines the field of view. The highest spatial frequency in the image determines the resolution. The lower the minimum spatial frequency, the larger the field of view. The higher the maximum spatial frequency, the finer the resolution.*

Here are a number of pictures that show how changing the coefficients in certain regions of k space affect an image. Figure 12.13(b) shows a transverse scan of a head by magnetic resonance imaging. This is a normal image to compare with the following figures. It has a spatial resolution of 256×256 points. The magnitude of its Fourier transform is shown in Fig. 12.13(a). Figure 12.14 shows the cosine and sine coefficients in the expansion.

Figures 12.15 and 12.16 show what happens when the high-frequency Fourier components are removed. In the first case they have been removed above $k_{x\ \max}/2$ and $k_{y\ \max}/2$. In the second they are removed above $k_{x\ \max}/4$ and $k_{y\ \max}/4$. Compare the blurring in these figures with the original image.

When the low-frequency coefficients are set to zero as in Fig. 12.17, only the high-frequency edges remain. In this case the Fourier components below $k_{x\ \max}/4$ and $k_{y\ \max}/4$ have been set to zero. (Keeping the same values of $k_{\max}$ and Δk and removing the information on those coefficients keeps the field of view the same.)

Figure 12.18 shows the aliasing that results from setting every other Fourier coefficient to zero. This has the effect of doubling Δk in Eq. (12.36). Since the width of the image has not been changed, this leads to aliasing, which shows up as the ''ghost'' images. In the first case alternate Fourier coefficients have been removed in k_x space, in the second in both k_x and k_y.

In summary: *Low spatial frequencies provide shape, contrast, and brightness. High spatial frequencies provide resolution, edges, and sharp detail.*

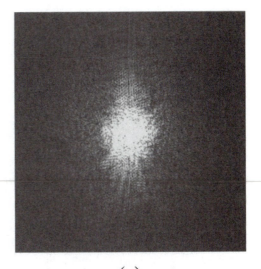

(a)

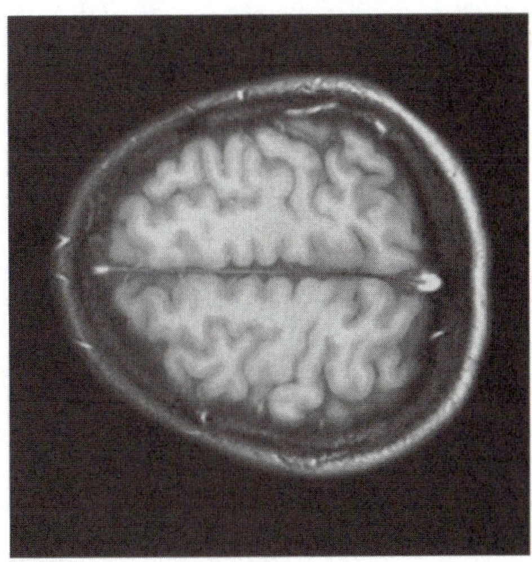

(b)

FIGURE 12.13. A magnetic resonance imaging head scan: (a) The amplitude C^2+S^2 in k space. (b) The image. This is a normal image to compare with the following figures. Prepared by Mr. Tuong Huu Le, Center for Magnetic Resonance Research, University of Minnesota. Thanks also to Professor Xiaoping Hu.

12.7. TWO-DIMENSIONAL IMAGE RECONSTRUCTION FROM PROJECTIONS BY FOURIER TRANSFORM

The reconstruction problem can be stated as follows. A function $f(x,y)$ exists in two dimensions. Measurements are made that give *projections*: the integral of $f(x,y)$ along some line as a function of displacement perpendicular to that line. For example, integration parallel to the y axis gives a function of x,

$$F(x)=\int_{-\infty}^{\infty} f(x,y)\,dy, \qquad (12.38)$$

as shown in Fig. (12.19). The scan is repeated at many different angles θ with the x axis, giving a set of functions $F_\theta(x')$. The problem is to reconstruct $f(x,y)$ from the set of functions $F_\theta(x')$. Several different techniques can be used. A very detailed reference is the book by Cho *et al.* (1993).

We will consider two of these techniques: reconstruction by Fourier transform, where the Fourier coefficients are obtained from projections [in a magnetic resonance imaging (MRI) machine they are measured directly] and filtered back projection (Sec. 12.8).

The Fourier transform technique is easiest to understand. Consider Eqs. (12.9). If $k_y=0$ in Eq. (12.9b), the result is

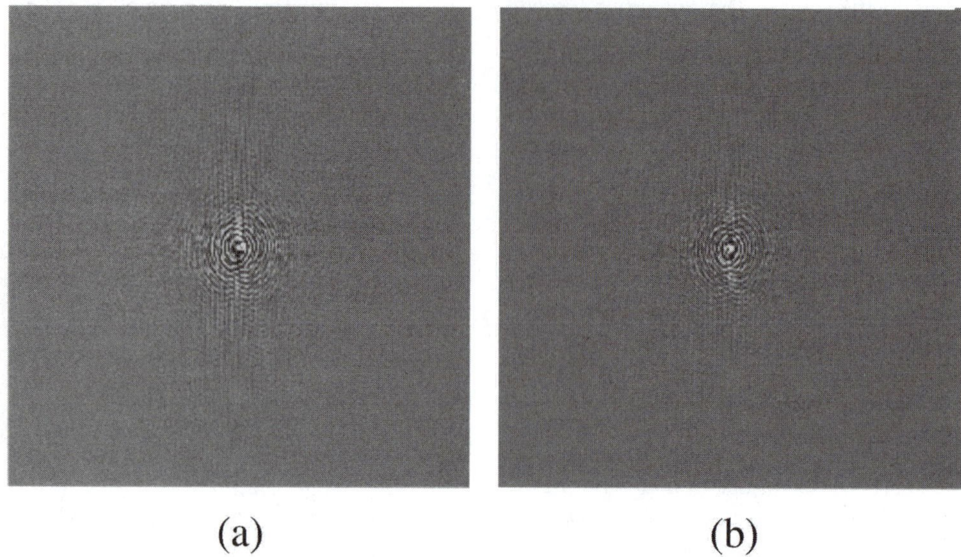

FIGURE 12.14. The sine and cosine coefficients for the image in Fig. 12.13. (a) $C(k_x,k_y)$. (b) $S(k_x,k_y)$. Prepared by Mr. Tuong Huu Le, Center for Magnetic Resonance Research, University of Minnesota. Thanks also to Professor Xiaoping Hu.

$$C(k_x,0)=\int_{-\infty}^{\infty} dx\ \cos(k_x x)\int_{-\infty}^{\infty} dy\ f(x,y)$$
$$=\int_{-\infty}^{\infty} dx\ \cos(k_x x)F(x). \tag{12.39}$$

Similarly

$$S(k_x,0)=\int_{-\infty}^{\infty} dx\ \sin(k_x x)F(x). \tag{12.40}$$

To state this in words: the Fourier transform of $F(x)$ determines the sine and cosine transforms of $f(x,y)$ along the line $k_y=0$ (the k_x axis) in the spatial frequency plane. This is shown in Fig. 12.20. A scan F_θ in another direction can be Fourier-transformed to give C and S at an angle θ with the k_x axis. The transforms of a set of projections at many different angles provide values of C and S throughout the k_x-k_y plane that can be used in Eq. (12.9a) to calculate $f(x,y)$. In Chap. 17 we will find that the data from an MRI scan give the functions $C(k_x,k_y)$ and $S(k_x,k_y)$ directly.

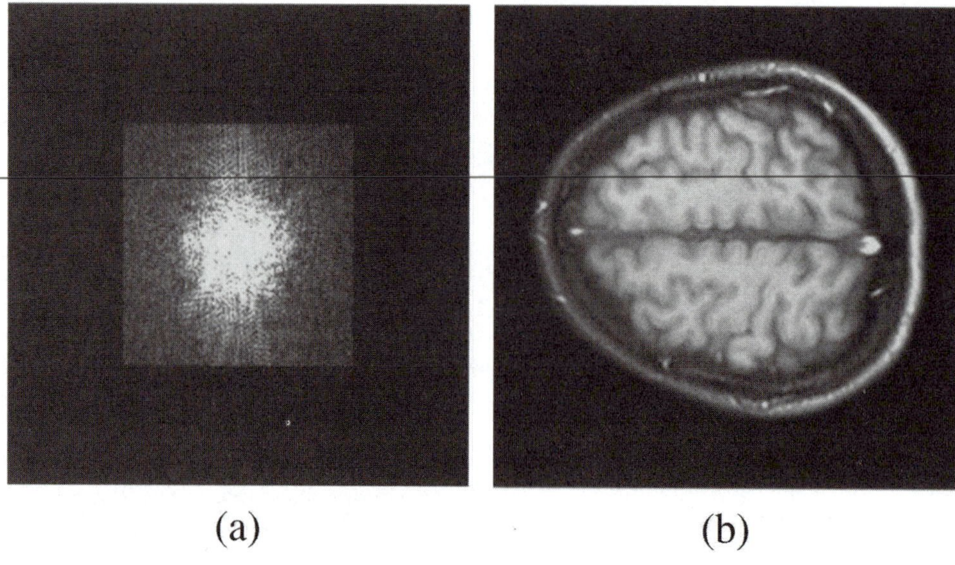

FIGURE 12.15. The image that results when the high-frequency Fourier components above $k_{x\,\max}/2$ and $_{y\,\max}/2$ are removed. Note the blurring compared to Fig. 12.13. Prepared by Mr. Tuong Huu Le, Center for Magnetic Resonance Research, University of Minnesota. Thanks also to Professor Xiaoping Hu.

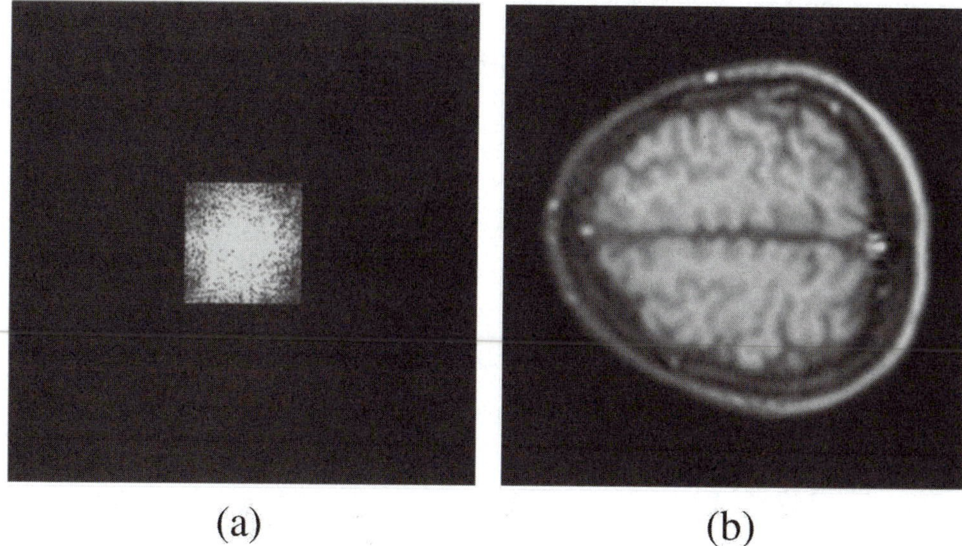

FIGURE 12.16. The image that results when the high-frequency Fourier components above $k_{x\,\max}/4$ and $k_{y\,\max}/4$ are removed. The blurring is even greater. Prepared by Mr. Tuong Huu Le, Center for Magnetic Resonance Research, University of Minnesota. Thanks also to Professor Xiaoping Hu.

In practice, the transforms are discrete. Using the notation that includes the redundant frequencies above $N/2$ and makes the coefficients half as large [Eqs. (11.27)], the two-dimensional discrete Fourier transform (DFT) is[10]

$$f_{jk}=\sum_{l=0}^{N-1}\sum_{m=0}^{N-1} C_{lm}\cos\frac{2\pi(jl+km)}{N}$$

$$+\sum_{l=0}^{N-1}\sum_{m=0}^{N-1} S_{lm}\sin\frac{2\pi(jl+km)}{N}. \quad (12.41a)$$

The coefficients are given by

$$C_{lm}=\frac{1}{N^2}\sum_{j=0}^{N-1}\sum_{k=0}^{N-1} f_{jk}\cos\frac{2\pi(jl+km)}{N}, \quad (12.41b)$$

$$S_{lm}=\frac{1}{N^2}\sum_{j=0}^{N-1}\sum_{k=0}^{N-1} f_{jk}\sin\frac{2\pi(jl+km)}{N}. \quad (12.41c)$$

Making a DFT of the projections gives values for C and S that lie on the circles in Fig. 12.21. Taking the inverse transform to calculate the reconstructed image requires values at the lattice points. They are obtained by interpolation. The details of how the interpolation is made are crucial when using the Fourier transform reconstruction technique.

[10]In this notation the low frequencies occur for low values of the indices l and m. Usually, as in Figs. 12.13–12.18, the indices are shifted so $k=0$ occurs in the middle of the sum.

12.8. RECONSTRUCTION FROM PROJECTIONS BY FILTERED BACK PROJECTION

Filtered back projection is more difficult to understand than the direct Fourier technique. It is easy to see that every point in the object contributes to some point in each projection. The converse is also true. In a back projection every point in each projection contributes to some point in the reconstructed image. This can be seen from Figure 12.22, which shows two points A and B and three projections. For point A, which is at the center of rotation, the relevant value of x' is the same in each projection, while for point B the value of x' is different in each projection.

A very simple procedure would be to construct an image by back-projecting every projection. The back projection $f_b(x,y)$ at point (x,y) is the sum of $F_\theta(x')$ for every projection or scan, using the value of x' that corresponds to the original projection through that point. That is, for Fig. 12.22, the back projection at point A would be the sum of the three values for which the solid projection lines intersect the scans, while for point B it would be the sum of the values where the three dashed lines strike the scans. This gives a rather crude image, but we will see how to refine it.[11]

Figure 12.23 shows how to relate the values of x' and

[11]To see why it is crude, suppose the original object is a disk at the origin. Every projection will be the same because of the symmetry in angle. Every back projection will lay down a contribution to the image along a stripe. Even though the reconstructed image will be largest where the original circle was, the image will have nonzero values throughout the image plane. We will see this example in Sec. 12.9.

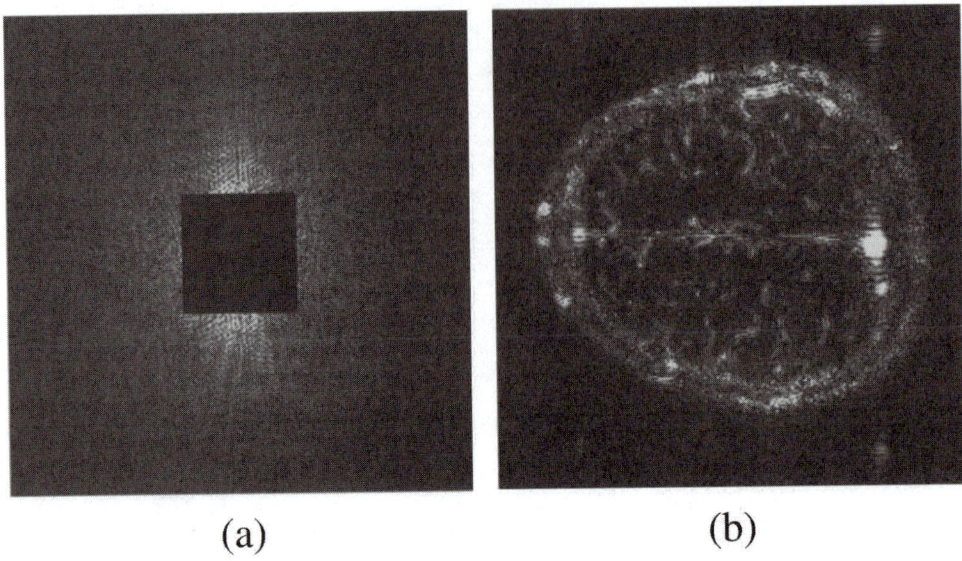

FIGURE 12.17. The image that results when the low-frequency Fourier components below $k_{x\,\max}/4$ and $k_{y\,\max}/4$ are removed. Prepared by Mr. Tuong Huu Le, Center for Magnetic Resonance Research, University of Minnesota. Thanks also to Professor Xiaoping Hu.

y' for a projection at angle θ to the object or image coordinates x and y. The transformations are

$$x' = x \cos\theta + y \sin\theta,$$
$$y' = -x \sin\theta + y \cos\theta, \tag{12.42}$$

and the inverse transformations are

$$x = x' \cos\theta - y' \sin\theta,$$
$$y = x' \sin\theta + y' \cos\theta. \tag{12.43}$$

Mathematically, the projection at angle θ is integrated along the line y':

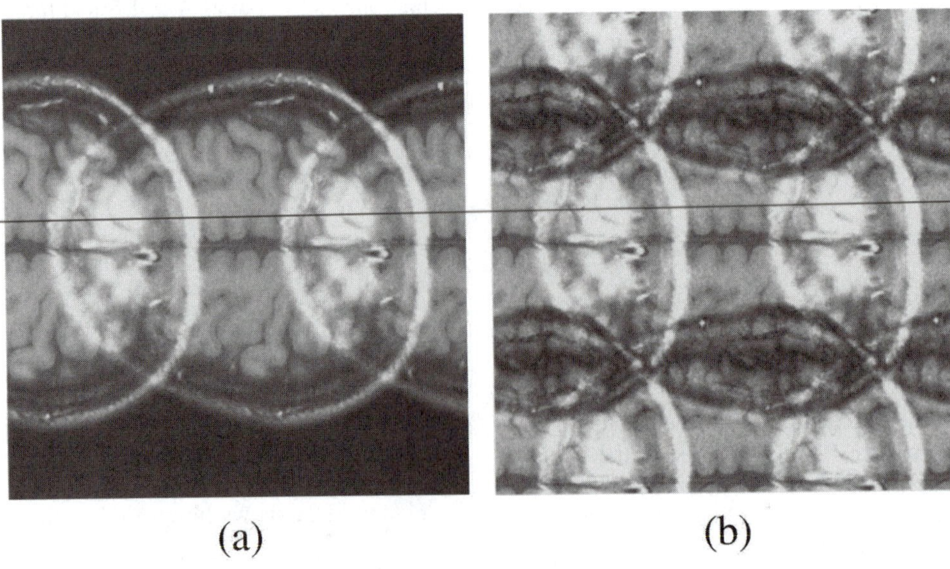

FIGURE 12.18. The Fourier coefficients for every other value of k have been set to zero. This has the effect of doubling Δk in Eq. (12.36). Since the width of the image has not been changed, this leads to aliasing, which shows up as the "ghost" images. (a) Every other value of k_x has been removed. (b) Every other value of both k_x and k_y has been removed. Prepared by Mr. Tuong Huu Le, Center for Magnetic Resonance Research, University of Minnesota. Thanks also to Professor Xiaoping Hu.

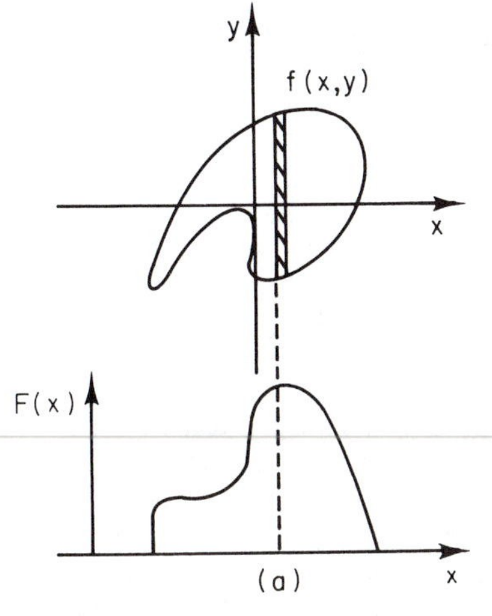

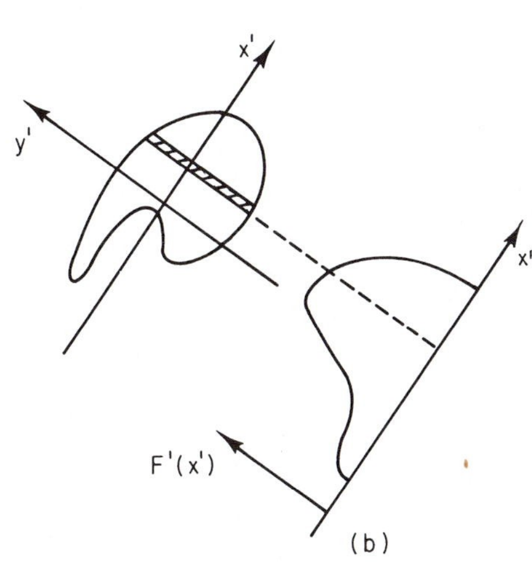

FIGURE 12.19. (a) Function $F(x)$ is the integral of $f(x,y)$ over all y. (b) The scan is repeated at angle θ with the x axis.

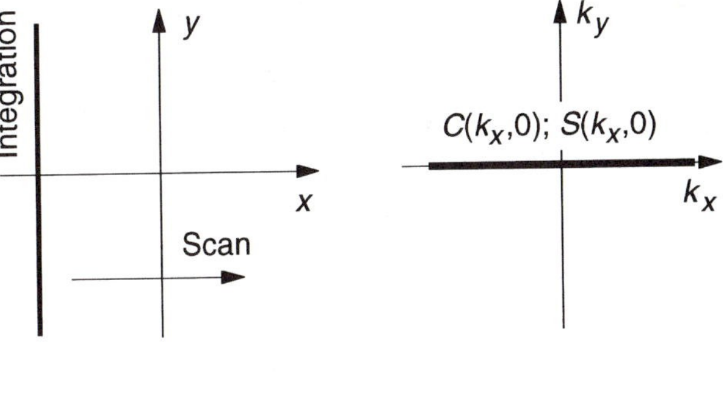

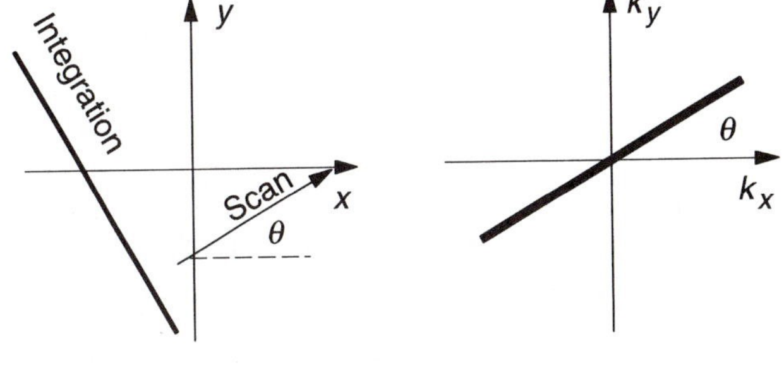

FIGURE 12.20. The Fourier transform of $F(x)=\int f(x,y)\,dy$ gives Fourier coefficients C and S along the k_x axis ($k_y=0$). Scans at other angles give C and S along corresponding lines in the k_xk_y plane.

$$F_\theta(x')=\int f(x,y)\,dy'=\int f(x'\cos\theta-y'\sin\theta,x'\sin\theta+y'\cos\theta)\,dy'. \tag{12.44}$$

The definition of the back-projected function is

$$f_b(x,y)=\int_0^\pi F_\theta(x')\,d\theta, \tag{12.45}$$

where x' is determined for each projection by using Eq. (12.42). The limits of integration are 0 and π since the projection for $\pi+\theta$ repeats the projection for angle θ.

We will now show that the image $f_b(x,y)$ obtained by taking projections of the object $F_\theta(x')$ and then back-projecting them is equivalent to taking the convolution of the object with the function $h(x-x',y-y')=1/r$, where r is the distance in the xy plane from the object point to the image point.

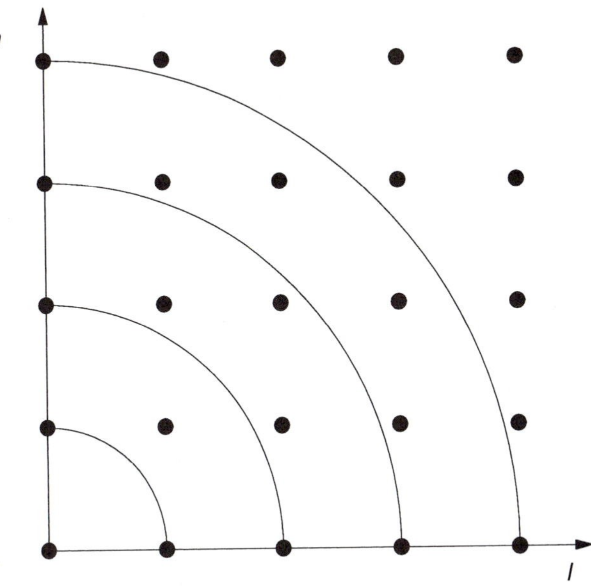

FIGURE 12.21. The two-dimensional Fourier reconstruction requires values of C and S at the lattice points shown. The Fourier transforms of the projections F_θ give the coefficients along the circular arcs. Interpolation is necessary to do the reconstruction.

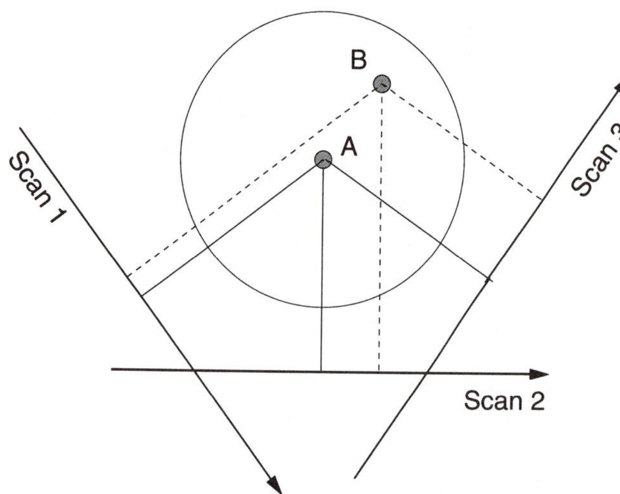

FIGURE 12.22. The principle of back projection. Each point in the image is generated by summing all values of $F_\theta(x')$ that projected through that point. For point A at the center of rotation, the appropriate value of x' is the same at each angle. For other points such as B, the value of x' is different at each angle.

To simplify the algebra, we will find the back projection at the origin. We want the set of projections for $x'=0$ as a function of scan angle θ. They are, from Eq. (12.44),

$$F_\theta(0)=\int_{-\infty}^{\infty} f(-y' \sin \theta, y' \cos \theta)\, dy'. \qquad (12.46)$$

In terms of angle $\theta'=\theta+\pi/2$ which is the angle from the x axis to the y' axis,

$$F_\theta(0)=\int_{-\infty}^{\infty} f(y' \cos \theta', y' \sin \theta')\, dy'.$$

The arguments of f look very much like components of a vector, which suggests expressing the integral in polar coordinates. Since y' is a dummy variable, call it r'. In terms of r' and θ' the projection is

$$F_\theta(0)=\int_{-\infty}^{\infty} f(r',\theta')\, dr'. \qquad (12.47)$$

Inserting this expression in Eq. (12.45) gives for the back projection

$$f_b(0,0)=\int_{-\infty}^{\infty}\int_0^{\pi} f(r',\theta')\, dr'\, d\theta'. \qquad (12.48)$$

Figure 12.24(a) shows how y' (that is, r') is integrated from $-\infty$ to ∞ while θ' goes from 0 to π. For the purposes of Eq. (12.48) the limits of integration can be changed as in Fig. 12.24(b). Variable r' can range from 0 to ∞ while θ' goes from 0 to 2π. Then the expression for f_b looks even more like an integration in polar coordinates:

$$f_b(0,0)=\int_0^{\infty}\int_0^{2\pi} f(r',\theta)\, dr'\, d\theta'.$$

There is still one difference between this and polar coordinates. The element of area, which is $dx'\, dy'$ in Cartesian coordinates, is $r'\, dr'\, d\theta'$ in polar coordinates. Therefore, let us rewrite this as

$$f_b(0,0)=\int_0^{\infty}\int_0^{2\pi}\left(\frac{f(r',\theta')}{r'}\right) r'\, dr'\, d\theta'. \qquad (12.49)$$

We now change to the Cartesian variables x' and y'. The back-projected image at the origin is

$$f_b(0,0)=\int_{-\infty}^{\infty}\int_{-\infty}^{\infty}\frac{f(x',y')}{(x'^2+y'^2)^{1/2}}\, dx'\, dy'. \qquad (12.50)$$

For an arbitrary point (x,y) the result is similar:

$$f_b(x,y)=\int_{-\infty}^{\infty}\int_{-\infty}^{\infty}\frac{f(x',y')}{[(x-x')^2+(y-y')^2]^{1/2}}\, dx'\, dy'. \qquad (12.51)$$

We have shown that the image obtained by taking projections of the object $F_\theta(x')$ and then back projecting them is

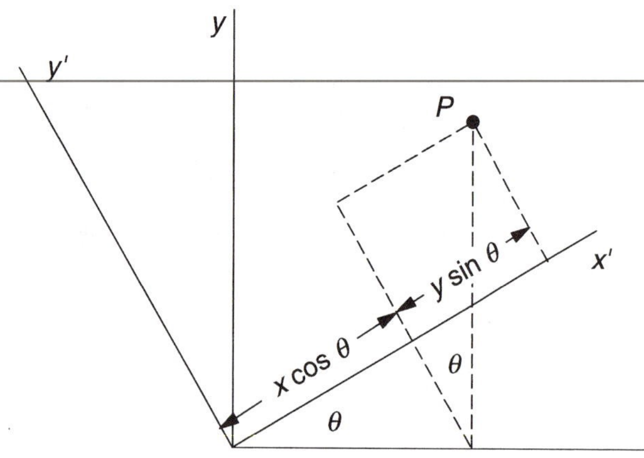

FIGURE 12.23. By considering components of the coordinates of point P in both coordinate systems, one can derive the transformation equations, Eqs. (12.42) and (12.43).

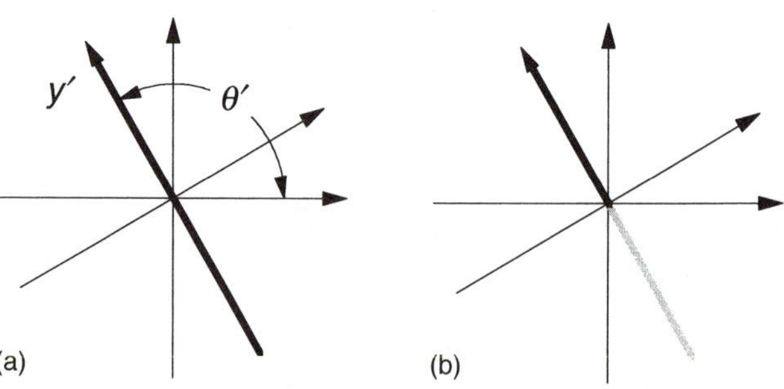

FIGURE 12.24. Integration for the back projection is over y' from $-\infty$ to $+\infty$ and from $\theta=0$ to π, as shown in (a). This can be converted to an integral from 0 to ∞ if the angular integration is taken from 0 to 2π, as shown in (b).

equivalent to taking the convolution of the object with the function $h(x-x',y-y')=1/r$, where r is the distance in the xy plane from the object point to the image point.

The back-projected image is not a faithful reproduction of the object. But it is possible to manipulate the projections $F_\theta(x')$ to produce a function $G_\theta(x')$ whose back projection is the desired $f(x,y)$. This is the process of *filtering* before making the back projection. To find the relationship between F and the desired function G, note that there is some function $g(x,y)$ that we do not know, but which, when projected and then back projected, yields the desired function $f(x,y)$. That is,

$$f(x,y)=g_b(x,y)$$

$$=\int_{-\infty}^{\infty}\int_{-\infty}^{\infty}\frac{g(x',y')}{[(x-x')^2+(y-y')^2]^{1/2}}\,dx'\,dy'. \tag{12.52}$$

Equations (12.10) relate the Fourier coefficients of f, g, and $h(r)=1/r$:

$$C_f(k_x,k_y)=C_g(k_x,k_y)C_h(k_x,k_y)-S_g(k_x,k_y)S_h(k_x,k_y),$$

$$S_f(k_x,k_y)=C_g(k_x,k_y)S_h(k_x,k_y)+S_g(k_x,k_y)C_h(k_x,k_y).$$

These can be solved for

$$S_g=\frac{C_hS_f-S_hC_f}{C_h^2+S_h^2} \tag{12.53a}$$

and

$$C_g=\frac{C_hC_f+S_hS_f}{C_h^2+S_h^2}. \tag{12.53b}$$

One can show by direct integration (see problem 12.17) that the Fourier transform of $h(r)=1/r$ is

$$C_h(k_x,k_y)=2\pi(k_x^2+k_y^2)^{-1/2},$$
$$S_h(k_x,k_y)=0, \tag{12.54}$$

so that

$$C_g(k_x,k_y)=\frac{1}{2\pi}(k_x^2+k_y^2)^{1/2}C_f(k_x,k_y),$$
$$S_g(k_x,k_y)=\frac{1}{2\pi}(k_x^2+k_y^2)^{1/2}S_f(k_x,k_y). \tag{12.55}$$

If function $g(x,y)$ were known and were projected to give $G_\theta(x')$, then back-projecting G_θ would give the desired $f(x,y)$. The final step is to relate $G_\theta(x')$ and $F_\theta(x')$ so that we do not have to know $g(x,y)$. To establish this relationship, consider a projection on the x axis. Equations (12.39) and (12.40) show that

$$F(x)=\frac{1}{2\pi}\int_{-\infty}^{\infty}[C_f(k_x,0)\cos(k_xx)$$
$$+S_g(k_x,0)\sin(k_xx)]\,dk_x,$$

while

$$G(x)=\frac{1}{2\pi}\int_{-\infty}^{\infty}[C_g(k_x,0)\cos(k_xx)$$
$$+S_g(k_x,0)\sin(k_xx)]\,dk_x.$$

Equations (12.55) relate the Fourier coefficients for F and G. For $k_y=0$, $(k_x^2+k_y^2)^{1/2}=|k_x|$. Therefore

$$G(x)=\frac{1}{(2\pi)^2}\int_{-\infty}^{\infty}[C_f(k_x,0)\cos(k_xx)$$
$$+S_f(k_x,0)\sin(k_xx)]|k_x|\,dk_x. \tag{12.56}$$

This result is independent of the choice of axis, so it must be true for any projection. There is a function $h(x)$ which can be convolved with any $F_\theta(x)$ to give the desired function $G_\theta(x)$. Equation (12.56) shows that

$$C_g(k_x,0)=C_f(k_x,0)|k_x|/2\pi,$$

$$S_g(k_x,0)=S_f(k_x,0)|k_x|/2\pi.$$

Comparison with Eqs. (12.9) shows that

$$C_h=|k_x|/2\pi,$$

$$S_h=0.$$

Therefore

$$h(x)=\frac{1}{(2\pi)^2}\int_{-\infty}^{\infty}|k_x|\cos(k_xx)\,dk_x.$$

Because the integrand is an even function, we can multiply by 2 and integrate from zero to infinity. The integral to infinity does not exist. However, there is some maximum spatial frequency, roughly the reciprocal of the resolution we want, which we call $k_{x\,\max}$. We can therefore cut the integral off at this maximum spatial frequency and obtain

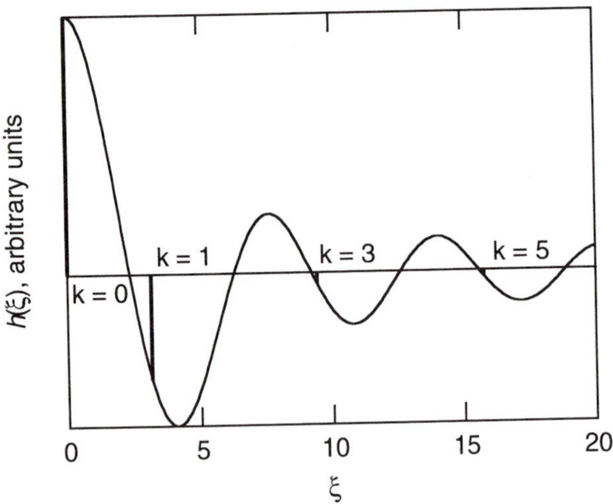

FIGURE 12.25. The weighting function $h(x)$ of Eq. (12.57). The bars show the nonzero values for the example in Section 12.9.

$$h(x)=\frac{1}{(2\pi)^2}\int_0^{k_{x\,\max}} k_x \cos(k_x x)\,dk_x$$

$$=\frac{1}{(2\pi)^2}\left[\frac{\cos(k_x x)}{x^2}+\frac{k_x \sin(k_x x)}{x}\right]_0^{k_{x\,\max}}$$

$$=\frac{k_{x\,\max}^2}{(2\pi)^2}\,[2\,\mathrm{sinc}(\xi)-\mathrm{sinc}^2(\xi/2)], \qquad (12.57)$$

where $\xi=k_{x\,\max}x$ and $\mathrm{sinc}(\xi)=\sin(\xi)/\xi$. The function $h(x)$ is plotted in Fig. 12.25. Using a sharp high-frequency cutoff introduces some problems, which are described below and which are easily overcome.

To summarize: *If each projection F_θ is convolved with the function h of Eq. (12.57) and then back-projected, the back-projected image is equal to the desired image.*

12.9. AN EXAMPLE OF FILTERED BACK PROJECTION

It is not difficult to write a computer program to perform filtered back projection if execution speed is not a concern. For our example we will use an object with circular symmetry, so that every projection is equivalent and only one projection needs to be convolved with the weighting function h. Because of the circular symmetry the back projection need be done only along one diameter. The program shown in Fig. 12.26 was used to reconstruct the image.

The "top-hat" function used is used as the object:

$$f(x,y)=\begin{cases}1, & x^2+y^2<a^2\\ 0 & \text{otherwise}\end{cases}. \qquad (12.58)$$

The projection is $F(x)=2(a^2-x^2)^{1/2}$ for $x^2<a^2$. Procedure `CalcF` evaluates $F(x)$ for 100 points. Variables x and i are related by $x=i/(N/2)-1$, so that x ranges from -1 to 1 as index i goes from 0 to 100. The value of a is 0.5.

The convolution is done by procedure `Convolve`, which uses convolving function h to operate on function F to produce G. The discrete form of h is obtained from Eq. (12.57) by the following argument, originally due to Ramachandran and Lakshminarayanan (see Cho, *et al.* (1993), p. 80). Variable x is considered on the interval $(-1,1)$, so the period is 2 and $\omega_0=\pi$. The maximum spatial frequency is $k_{x\,\max}=N\pi/2$. The value of x in the weighting function $h(x)$ depends on the value of index $k=i-j$: $x_i-x_j=2(i-j)/N=2k/N$. Therefore $\xi=k_{x\,\max}x=(N\pi/2)(2/N)k=\pi k$, where k is an integer. From Eq. (12.57) we obtain

$$h(k=0)=N^2/16,$$

$$h(k\ \text{even})=0,$$

$$h(k\ \text{odd})=(N^2/4)(-1/k^2\pi^2). \qquad (12.59)$$

Procedure `Convolve` replaces the integral of Eq. (12.4a) by a sum. The factor dx in the integral becomes $1/N$ in the sum.

Procedure `BackProject` forms the image from G. One hundred eighty projections are done in 1° increments from 0° to 179°. The value of x is determined from $x=i\cos\theta$, but it is shifted so that the rotation takes place about $i=50$. Unless x is at the end points, the value of G is obtained by linear interpolation. The value of $\Delta\theta$ used to convert the integral to a sum is $\pi/180$.

Procedure `PrintData` writes the data for the plots shown in Fig. 12.27. One can see from inspection of Fig. 12.27 how the convolution converts the semicircular projection F into a function G that is flat-topped over the region of nonvanishing f and has a negative contribution in the wings. Figure 12.28 shows what the image looks like if the back projection is done without first performing the convolution.

One can also see from Fig. 12.27 that ringing is introduced at the sharp edges. This is characteristic of the sharp high-frequency cutoff at $k_{x\,\max}$ (similar to the Fourier series representation of a square wave with only a finite number of terms). Early computer tomography (CT) scans created with the convolution function presented here showed a dark band just inside the skull where there was an abrupt change in $f(x,y)$ upon going from bone to brain (Fig. 12.29). A gradual high-frequency cutoff changes the details of $h(k)$ and eliminates this ringing (Fig. 12.30).

```
/*Fig. 12.16. R Hobbie, 7/6/86. Converted to C by James Roberts, July, 1996.*/

#include <stdio.h>
#include <stdlib.h>
#include <math.h>

#define N 100                                         //Number of data points
const int n = N;
const double pi = 3.141592654;

double F[N], Image[N],        //Projection of object and image along a line [i]
       G[N];                                          //Convolved Projection

void CalcF(double *F)
/*Calculates the projection of a circle of radius a centered at N/2.*/
{
   int i,i1,i2;
   double a = 0.5, x;

   a = 0.5;
   i1 = (int) (50 - a * ((float) n) / 2.0);
   i2 = (int) (50 + a * ((float) n) / 2.0);
   for (i=0;i<n;i++)
   {
      x = 2 * (i+1) / (float) n - 1.0;
      F[i] = 0;
      if (i+1 > i1 && i+1< i2)
         F[i] = F[i] + 2 * sqrt(pow(a,2) - pow(x,2));
   }
}

double H(int k)                         /*This is the function of Eq. 12.27 */
{
   if (k==0)
      return  pow((float) n,2)/16.0;
   else if(k%2 == 1 || (-k)%2 == 1)
      return  -pow(((float) n)/(pi*(float) k),2) / 4.0;
   else
      return  0;
}

void Convolve(double * y, double * G)
{
   int i,j;
   double temp;

   for (i=0;i<n;i++)
   {
      temp = 0;
      for (j=0;j<n;j++)
        if (y[j] != 0)
         temp = temp + H(i-j) * y[j];
      G[i] = temp/((float) n);
   }
}

void BackProject(double * G, double * Image)
{
```

FIGURE 12.26. The program used to make a filtered back projection of a circularly symmetric function.

```
    const int MaxProj = 180;

    int i,j,l;
    double x,xinterp,
           temp,c;

    for(i=0;i<n;i++)
       Image[i] = 0;
    for (j=0;j<MaxProj;j++)
    {
       c = cos(pi*((float) j)/180.0);
       for (i=0;i<n;i++)
       {
          x = ((float) n) / 2.0 + (i+1-((float) n) / 2.0) * c;
          l = (int) x;
          xinterp = x-l;
          if (l<=1)
             temp = G[0];
          else if(l>=n)
             temp = G[n-1];
          else
             temp = G[l-1] +xinterp * (G[l] - G[l-1]);
          Image[i] = Image[i] + temp;
       }
    }
    for(i=0;i<n;i++)
       Image[i] = Image[i] * pi / ((float) MaxProj);
}

void PrintData(int n1, int n2, double * x, FILE *fp, char *title)
{
    int i,j;
    fprintf(fp,"\n\n%s\n",title);

    j = 0;
    for (i = n1 -1; i < n2; i++)
    {
       if (j%10 == 0)
          fprintf(fp,"\n%2i",i+1);
       fprintf(fp,"\t%8.3f",x[i]);
       j++;
    }
    fprintf(fp,"\t%8.3f",x[i]);
}

void main()
{
    FILE *ofp;
    if (!(ofp = fopen("OutputFile","w")))
    {
       printf("cannot open output file\n");
       exit(1);
    }

    fprintf(ofp,"\n");

    CalcF(F);
    PrintData(1,n,F,ofp,"Projected Object F");
```

FIGURE 12.26. (Continued.)

```
    Convolve(F,G);
    PrintData(1,n,G,ofp,"Convolved Projection G");

    BackProject(G,Image);
    PrintData(1,n,Image,ofp,"Back-Projected Image");

    fclose(ofp);
}
```

FIGURE 12.26 (Continued.)

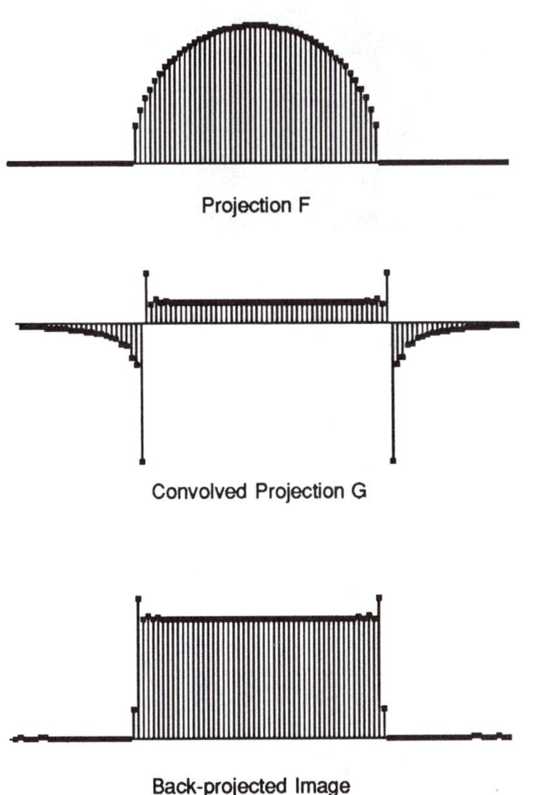

FIGURE 12.27. Reconstruction of a circularly symmetric image by filtered back projection. (a) The projection $F(x)$. (b) The convolved projection $G(x)$. (c) The image from back-projecting the convolved data.

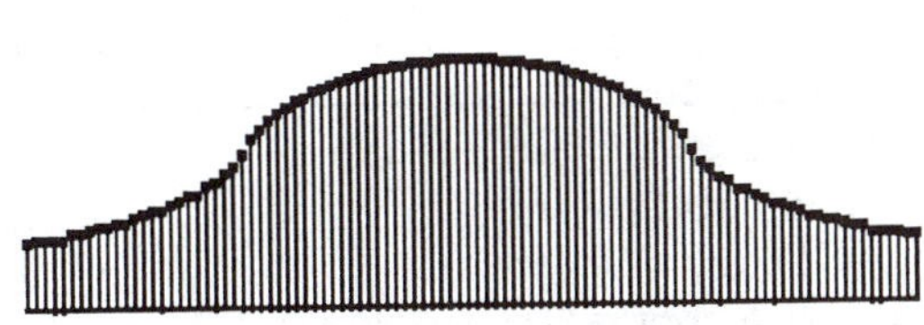

FIGURE 12.28. Reconstruction by simple back projection without convolution. The object is the same as in Fig. 12.27.

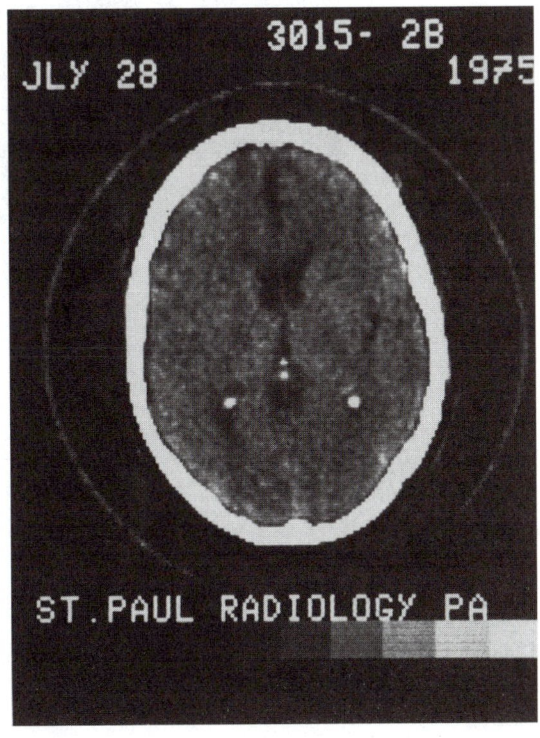

FIGURE 12.29. An early CT brain scan, showing ringing inside the skull. Photograph courtesy of St. Paul Radiology Associates, St. Paul, MN.

SYMBOLS USED IN CHAPTER 12

Symbol	Use	Units	First used on page
a	Radius of pinhole	m	326
a,b	Constants		328
a	Radius of "top-hat" function	m	338
b'	Length of object	m	327
f,g	Arbitrary functions		322
f	Focal length of lens	m	326
f_b,g_b	Back-projected images of f,g		335

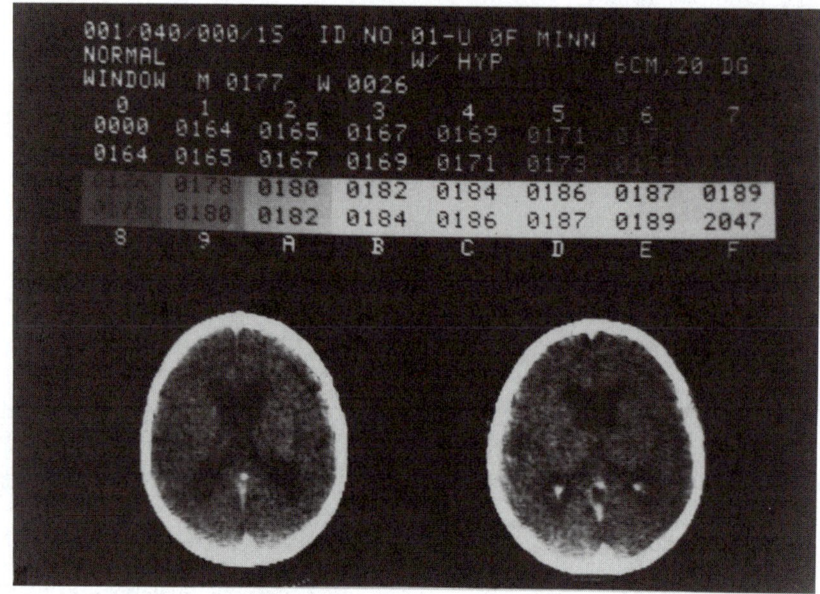

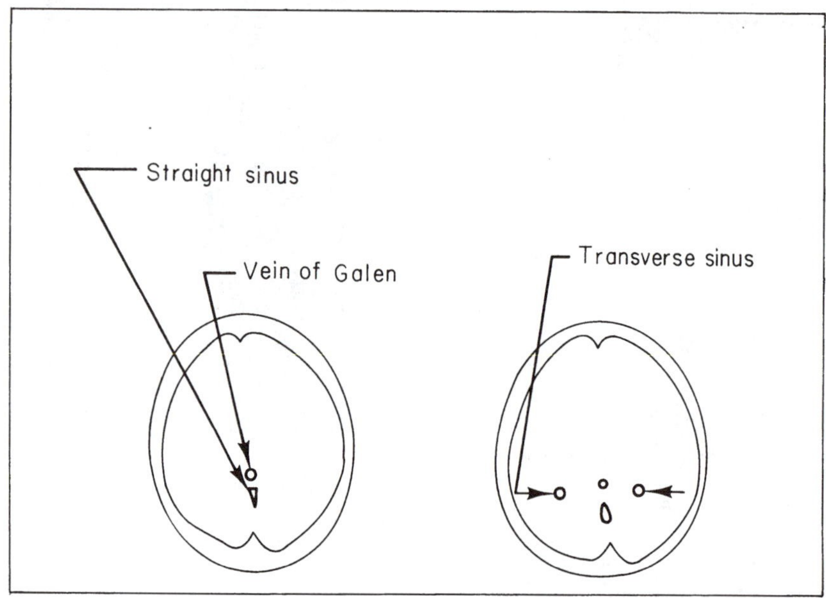

FIGURE 12.30. A set of brain scans using a gradual high-frequency cutoff to eliminate ringing. Photograph courtesy of Professor J. T. Payne, Department of Diagnostic Radiology, University of Minnesota.

Symbol	Use	Units	First used on page
h	Point-spread function; impulse		322
	Response for convolution		327
h,h'	Image, object height	m	324
i	$\sqrt{-1}$		324
j,k	Subscript indices for data		333
k,k_x,k_y	Spatial frequency	m^{-1}	324
l'	Length of object	m	327
l,m	Subscript indices for Fourier coefficients		333

Symbol	Use	Units	First used on page
m	Magnification		326
r, r'	Coordinate variable	m	325
t, t'	Time or arbitrary variable		322
u	Object distance	m	325
v	Image distance	m	325
w	Height of blurred image	m	326
x, y, x', y'	Distance; coordinates in image or object plane; rotated coordinate system for image reconstruction	m	323
A	Amplitude		323
C_f	Fourier cosine transform of function f		323
D	Length of image	m	330
E	Irradiance	W m^{-2}	326
F or F_θ	Projection of function f		331
F, G, H	Complex Fourier transforms of f, g, h		324
L	Radiance	W m^{-2} sr^{-1}	325
L	Spatial period of Fourier series for reconstructing an image	m	330
N	Total number of data points; number of discrete values; on one dimension of an image		330
P	Radiant power	W	324
R	Radiant energy	J	324
S	Surface area	m^2	325
S_f	Fourier sine transform of function f		323
T	Period	s	323
δ	Dirac delta function		322
λ	Wavelength	m	323
ϕ	Phase		328
θ, θ'	Angle		325
θ_d, θ_s	Angle between the direction of travel of radiant energy and a normal to the surface of the source or detector		325
θ'	Projection angle with other axis		336
τ_1	Parameter	s	323
ω, ω_0	Angular frequency	s^{-1}	323
ξ	Dummy variable		338
Ω	Solid angle	sr	324

PROBLEMS

Section 12.1

12.1. Compare Eq. (12.4a) to Eqs. (4.73) and (7.21) and deduce the impulse response for those two systems.

12.2. Except for the minus sign, Eq. (12.4a) is the same integral that defines the cross-correlation function. There are some important differences, however. Show that the convolution function is commutative—interchanging the order of variables gives the same result—but that the cross-correlation function is not.

12.3. Fill in the details in the derivation of Eq. (12.6a).

Section 12.2

12.4. If you are familiar with complex variables, use the definition of the Fourier transform in Eq. (12.9) to prove the convolution theorem, Eq. (12.10).

12.5. What are the two-dimensional images whose Fourier transforms are shown?

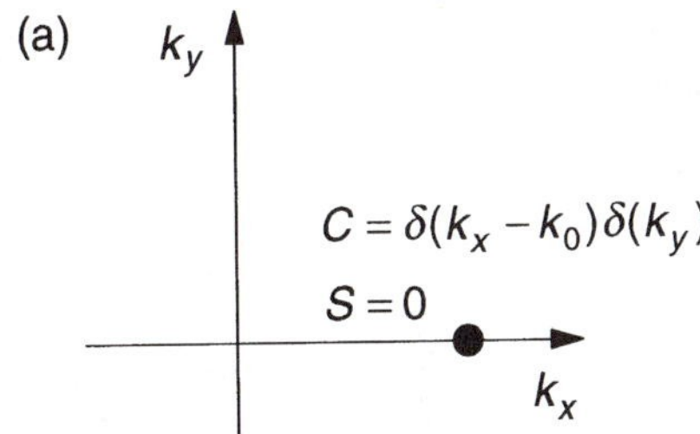

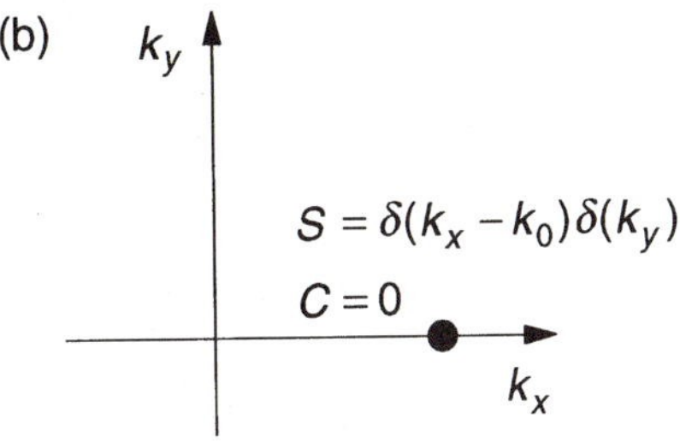

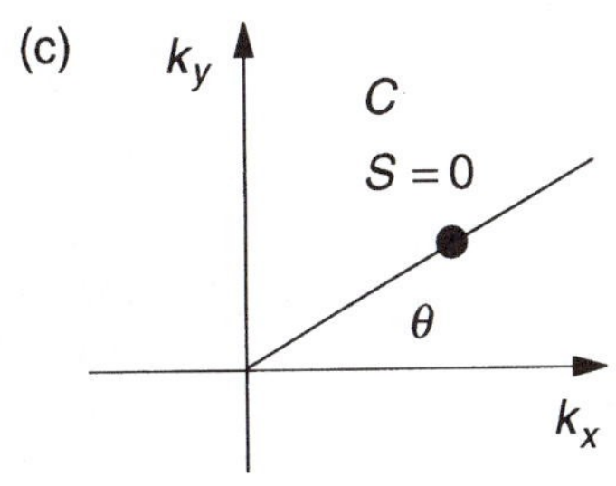

Section 12.4

12.6. Suppose that a sphere radiates uniformly from its surface according to Lambert's cosine law: $L=L_0$. By considering area $dS=2\pi r^2 \sin\theta\, d\theta$ on the surface of a sphere, find the power radiated per steradian in the direction of the z axis and the total power radiated.

12.7. Show that the total power per unit area radiated from a surface obeying Lambert's cosine law is $W_r = \pi L_0$. This quantity is called the exitance.

Section 12.5

12.8. Complete the verification of Eq. (12.28) suggested in the text.

12.9. Find the Fourier transform of the point-spread function for the ideal lens, Eq. (12.28). This does not include diffraction effects.

12.10. Use Eq. (12.30) to show that the sum of the squares of the Fourier coefficients of the image is equal to the sum of the squares of the Fourier coefficients of the object times the square of the modulation transfer function, for a given set of spatial frequencies k_x, k_y.

12.11. Write the modulation of the image in terms of the variables in Eq. (12.35).

12.12. How does magnification m change the spatial frequencies in going from object to image? Since one is concerned about seeing detail in the object, resolution and spatial frequencies are usually converted to object coordinates in medical imaging, while they are left in terms of the detector coordinates in photography.

Section 12.6

12.13. This problem shows how increasing the detail in an image introduces high-frequency components. Find the continuous Fourier transform of the two functions

$$f_1(x)=\begin{cases} 0, & x<0 \\ 1, & 0<x<1, \\ 0, & x>1 \end{cases}$$

$$f_2(x)=\begin{cases} 0, & x<0 \\ \sqrt{3/2}, & 0<x<\frac{1}{3} \\ 0, & \frac{1}{3}<x<\frac{2}{3}. \\ \sqrt{3/2}, & \frac{2}{3}<x<1 \\ 0, & x>1 \end{cases}$$

Plot $a(k_x)=[C^2(k_x)+S^2(k_x)]^{1/2}$ for each function using a spreadsheet or plotting package, for the range $-45<k_x<45$. Compare the features of each plot. Both functions have the same value of $\int_{-\infty}^{\infty} f^2(x)dx$.

Section 12.7

12.14. Suppose that $f(x,y)$ is independent of y. Find expressions for $C(k_x,k_y)$ and $S(k_x,k_y)$ and insert them in the expression for $f(x,y)$ to verify that $f(x,y)$ is recovered. You will need Eqs. (11.65).

12.15. Suppose that the object is a point at the origin, so that $f(x,y)=\delta(x)\delta(y)$. Find the projection $F(x)$ and the transform functions $C(k_x,0)$ and $S(k_x,0)$. Use these results to reconstruct the image using the Fourier technique.

Section 12.8

12.16. Derive Eqs. (12.42) and (12.43).

12.17. Consider the Fourier transform of $1/r$. The coefficients are given by

$$C(k_x,k_y)=\int_{-\infty}^{\infty}\int_{-\infty}^{\infty}\frac{dx\, dy\, \cos(k_x x+k_y y)}{(x^2+y^2)^{1/2}},$$

$$S(k_x,k_y)=\int_{-\infty}^{\infty}\int_{-\infty}^{\infty}\frac{dx\, dy\, \sin(k_x x+k_y y)}{(x^2+y^2)^{1/2}}.$$

Transform to polar coordinates ($x=r\cos\theta$, $y=r\sin\theta$). Show from symmetry considerations of the angular integral that $S=0$. Use the fact that

$$\int_0^{2\pi}\cos(kr\cos\theta)\, d\theta=2\pi J_0(kr)$$

and

$$\int_0^{\infty} J_0(kr)\, dr=\frac{1}{k}$$

to derive Eqs. (12.54). The function $J_0(x)$ is a *Bessel function of order zero*. It is tabulated and has known properties, similar to a trigonometric function. [See Abramowitz and Stegun (1972, p. 360)].

Section 12.9

12.18. Verify Eqs. (12.59).

12.19. Modify the program of Fig. 12.30 and run it without the convolution.

12.20. Modify the program of Fig. 12.30 to reconstruct an annulus instead of a top-hat function.

REFERENCES

Abramowitz, M. and I. A. Stegun. (1972). *Handbook of Mathematical Functions With Formulas, Graphs and Mathematical Tables*. Washington, U.S. Government Printing Office.

Cho, Z.-h., J. P. Jones, and M. Singh (1993). *Foundations of Medical Imaging*. New York, Wiley.

Gaskill, J. D. (1972). *Linear Systems, Fourier transforms, and Optics*. New York, Wiley.

Ramachandran, G. N., and A. V. Lakshminarayanan (1971). Three-dimensional reconstruction from radiographs and electron micrographs: Application of convolutions instead of Fourier transforms. *Proc. Nat. Acad. Sci. U.S.* **68**: 2236–2240.

Shaw, R. (1979). Photographic Detectors. Ch. 5 of *Applied Optics and Optical Engineering*, New York, Academic, Vol. 7, pp. 121–154.

Williams, C. S., and O. A. Becklund (1972). *Optics: A Short Course for Engineers and Scientists*. New York, Wiley.

CHAPTER 13

Atoms and Light

This chapter describes some of the biologically important properties of infrared, visible, and ultraviolet light. X rays are discussed in Chaps. 14 and 15. The chapter makes no attempt to discuss interference or diffraction. A brief discussion of geometrical optics accompanies the description of image formation in the eye and errors of refraction.

This chapter considers light to be photons (Sec. 13.1). Photons can be emitted or absorbed when single atoms change energy levels, and they have certain frequencies characteristic of the atom, as described in Sec. 13.2. Molecules have additional energy levels shown in Sec. 13.3. Biological examples include spectrophotometry, photodissociation, immunofluorescence, infrared spectroscopy, and Raman scattering. There is such an extensive literature about these that the discussion here is quite brief. (The discussion of Raman scattering is in Sec. 13.7.)

Section 13.4 describes the scattering and absorption of radiation, processes that are important in the rest of this chapter and in Chaps. 14–16. The probability of scattering or absorption is measured by the cross section, which is also introduced here.

Photons can be absorbed and emitted by some substances in a continuous range of frequencies or wavelengths. This happens when many atoms interact with each other and blur the energy levels, as in liquids and solids. This leads to the concept of thermal radiation described in Sec. 13.5. Examples of thermal radiation are infrared radiation by the skin and ultraviolet radiation by the sun. The former is discussed in Sec. 13.6.

Infrared and optical photons may scatter many times in tissue without being absorbed. In some cases the process can be modeled accurately with the diffusion approximation developed in Sec. 13.7, along with measurements based on these techniques.

Blue and ultraviolet light are used for therapy, as described in Sec. 13.8. They can also be harmful, particularly to skin and eyes.

Lasers are used to heat tissue, often rapidly enough to do surgery as water in the tissue suddenly boils. Models of this process include the laser bioheat equation that is developed in Sec. 13.9.

Section 13.10 returns to the problem of radiometry—measuring radiation—that was introduced in Sec. 12.4. All of the important quantities are defined, and the corresponding photometric and actinometric quantities are also introduced.

Section 13.11 describes how the eye focuses an image on the retina and the correction of simple errors of refraction. A final example of the photon nature of light is given in Sec. 13.12: the statistical limit to dark-adapted vision—shot noise—which is important when the eye is operating in its most sensitive mode.

13.1. THE NATURE OF LIGHT: WAVES VERSUS PHOTONS

Light travels in a vacuum with a velocity $c=3\times10^8$ m s^{-1} (to an accuracy of 0.07%). When light travels through matter, its velocity is less than this and is given by

$$c' = \frac{c}{n}, \qquad (13.1)$$

where n is the *index of refraction* of the substance. The value of the index of refraction depends on both the composition of the substance and the color of the light.

A controversy over the nature of light existed for centuries. Sir Isaac Newton explained many properties of light with a particle model in the seventeenth century. In the early nineteenth century, Thomas Young described some interference experiments that could be explained only by assuming that light is a wave. By the end of the nineteenth century nearly all known properties of light, including many of its interactions with matter, could be explained by assuming that light consists of an electromagnetic wave. By an electromagnetic wave, we mean that

1. Light can be produced by accelerating an electric charge.
2. Light has an electric and a magnetic field associated with it; the force that the light exerts on a charged particle is given by Eq. (8.2), $\mathbf{F}=q(\mathbf{E}+\mathbf{v}\times\mathbf{B})$. The force due to the magnetic field is usually very small.
3. The velocity of light traveling in a vacuum is given by electromagnetic theory in terms of the parameters ϵ_0 and μ_0 measured in the laboratory for "ordinary" electric and magnetic fields.

In the early twentieth century, light was discovered to have *both* particle properties and electromagnetic wave properties at the same time. This rather disconcerting discovery was followed in a few years by the discovery that matter, which had been thought to consist of particles, also has wave properties.

A traveling wave of light can be described by a function of the form $f(x-c't)$, which represents a disturbance traveling along the x axis in the positive direction. (To keep a particular value for the argument of f, x must increase as time increases.) If the wave is sinusoidal, then the period, frequency,[1] and wavelength can be defined:

$$\nu=\frac{1}{T}, \quad c'=\lambda\nu. \tag{13.2}$$

As light moves from one medium into another where it travels with a different speed, the frequency remains the same. The wavelength changes as the speed changes.

Each particle of light or *photon* has energy E. The energy of each photon (a "particle" concept) is related to the frequency ν (a "wave" concept) by the relation

$$E=h\nu=\frac{hc'}{\lambda}. \tag{13.3}$$

The proportionality constant h is called *Planck's constant.* It has the numerical value[2]

$$h=6.63\times10^{-34}\ \text{J s}=4.14\times10^{-15}\ \text{eV s}. \tag{13.4}$$

It is sometimes useful to use the number "h stroke" or "h bar":

$$\hbar=\frac{h}{2\pi},$$
$$\hbar=1.05\times10^{-34}\ \text{J s}=0.66\times10^{-15}\ \text{eV s}. \tag{13.5}$$

In terms of the angular frequency $\omega=2\pi\nu$,

$$E=\hbar\omega. \tag{13.6}$$

The electromagnetic spectrum includes radio waves, microwaves, infrared, visible, and ultraviolet light, x rays, and γ rays. Table 13.1 shows the wavelengths that separate these arbitrary regions, together with the frequencies and the energies of the photons. Visible-light photons have an energy of a few electron volts. X rays are 10^4-10^7 times more energetic, while γ rays, which come from atomic nuclei, are often even more energetic but may have energies overlapping x-ray energies. The only difference between x rays and γ rays is their source.

The property of light that we associate with color is the frequency or the energy of each photon. Visible light covers a narrow range of frequencies, about an octave (a factor of 2). Table 13.2 shows the wavelengths and frequencies dividing the colors of the visible spectrum.

Most of the effects discussed in this chapter can be explained by assuming that light is made up of photons.

13.2. ATOMIC ENERGY LEVELS AND ATOMIC SPECTRA

The simplest system that can emit or absorb light is an isolated atom. An atom is isolated if it is in a monatomic gas. Isolated atoms are found to exist with specific internal energies, not including translational kinetic energy. An atom can change from one energy level to another by emitting or absorbing a photon with an energy equal to the energy difference between the levels. Let the energy levels be labeled by $i=1,2,3,\ldots$, with the energy of the ith state being E_i. It is found that there is a lowest possible internal energy for the atom; when the atom is in this state, no

TABLE 13.1. *The regions of the electromagnetic spectrum and their boundaries.*

Name	Wavelength (m)	Frequency (Hz)	Energy (eV)
Radio waves			
	1	3×10^{8}	1.24×10^{-6}
Microwaves			
	1×10^{-3}	3×10^{11}	1.24×10^{-3}
Extreme infrared			
	15×10^{-6}	2×10^{13}	0.083
Far infrared			
	6×10^{-6}	5×10^{13}	0.207
Middle infrared			
	3×10^{-6}	1×10^{14}	0.414
Near infrared			
	0.75×10^{-6}	4×10^{14}	1.65
Visible			
	0.4×10^{-6}	7.5×10^{14}	3.1
Ultraviolet			
	12×10^{-9}	2.4×10^{16}	100
X rays, γ rays			

[1]We used f for frequency in earlier chapters because this is customary when discussing noise. Here we adopt the notation most often used in atomic physics.

[2]The *electron volt* (eV) is a unit of energy. 1 eV $=1.6\times10^{-19}$ J. It is the energy acquired by an electron that moves through a potential difference of 1 V.

TABLE 13.2. *The visible electromagnetic spectrum.*

Color	Wavelength (nm)	Frequency (Hz)	Energy (eV)
	750	4.0×10^{14}	1.65
Red			
	610	4.9×10^{14}	2.03
Orange			
	590	5.1×10^{14}	2.10
Yellow			
	570	5.26×10^{14}	2.17
Green			
	500	6.0×10^{14}	2.48
Blue			
	450	6.7×10^{14}	2.76
Violet			
	400	7.5×10^{14}	3.1

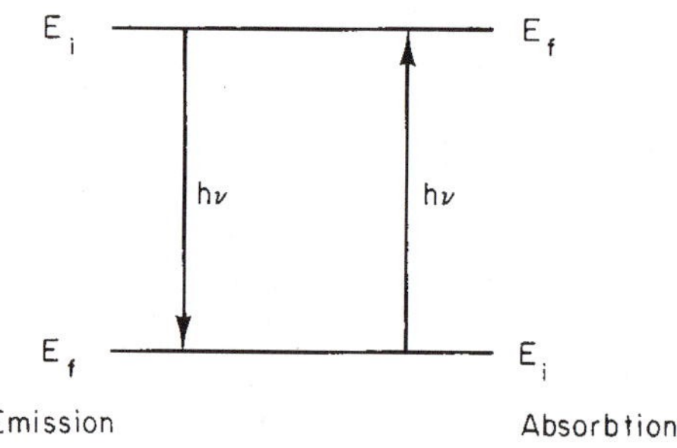

FIGURE 13.1. A system can change from one energy to another by emitting or absorbing a photon. The photon has an energy equal to the difference in energies of the two levels.

further energy loss can take place. If E_i is greater than the lowest energy, then the atom can lose energy by emitting a photon of energy $E_i - E_f$ and exist in a lower-energy state E_f (Fig. 13.1).

It is possible, using techniques of quantum mechanics, to calculate the energies of the levels with reasonable accuracy (and in some cases with spectacular accuracy). For our purposes, we need only recognize that energy levels exist and know their approximate values. You may be familiar with the hydrogen atom, in which the energy of the nth level is given by

$$E_n = -\left(\frac{1}{4\pi\epsilon_0}\right)^2 \frac{m_e e^4}{2\hbar^2 n^2}, \quad n=1,2,3,\ldots . \tag{13.7}$$

The energy is in joules when the electron mass m_e is in kilograms, the electronic charge e is in coulombs, and $\hbar$ is in J s. The Coulomb's law constant $1/4\pi\epsilon_0$ is given in Eq. (6.2). Dividing the energy in joules by e gives the energy in electron volts:

$$E_n = -\frac{13.6}{n^2} \quad \text{(in eV)}. \tag{13.8}$$

The energy-level diagram in Fig. 13.2 shows these energies and some transitions between them. In other cases, the energy depends not only on the integer $n=1,2,3,4,\ldots$, but on other quantum numbers as well.

Figure 13.3 plots the spectrum for hydrogen vs wavelength, along with some of the energy levels of hydrogen. Letters $a,b,c,\ldots$ mark lines in the spectrum and the associated transitions.

In general, the energy of an atom depends on the values of five quantum numbers for each electron in the atom. The quantum numbers are

$n=1,2,3,\ldots$	the principal quantum number
$l=0,1,2,\ldots,n-1$	the orbital angular momentum quantum number
$s=\frac{1}{2}$	the spin quantum number
$m_l=-l,-(l-1),\ldots,l-1,l$	"z component" of the orbital angular momentum
$m_s=-\frac{1}{2}, \frac{1}{2}$	"z component" of the spin

Sometimes the last two quantum numbers, m_l and m_s, are replaced by two other quantum numbers, j and m_j. The allowed values of j and m_j are

$j=l-\frac{1}{2}$ *or* $l+\frac{1}{2}$, except that $j=\frac{1}{2}$ when $l=0$	total angular momentum quantum number
$m_j=-j,-(j-1),\ldots,j-1,j$	"z component" of the total angular momentum

Whether one uses m_l and m_s or j and m_j, in either case each electron is described by five quantum numbers, one of which is always $\frac{1}{2}$. There are four quantum numbers that can change, corresponding to the three space degrees of freedom and the "spin" associated with m_s. The internal energy of the atom is the sum of the kinetic and potential energies of each electron. The energy of each electron depends on the values of its quantum numbers. It is influenced by the elec-

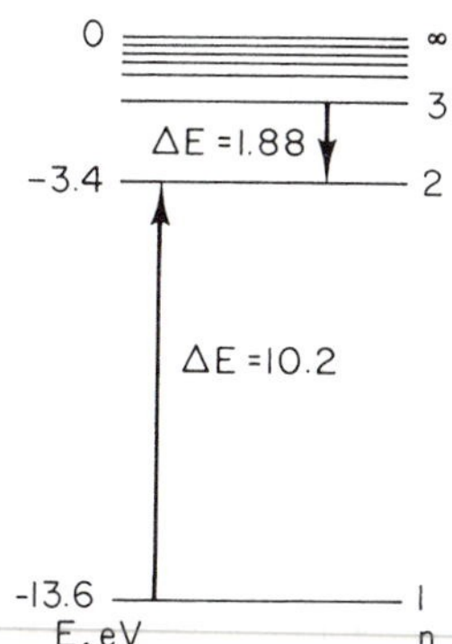

FIGURE 13.2. Energy levels in a hydrogen atom. Transitions are shown corresponding to the emission and absorption of light.

tric field generated by the nucleus and all the other electrons. There are also magnetic interactions between electrons and between each electron and the nucleus, because the moving charges generate magnetic fields. No two electrons in an atom can have the same values for all their quantum numbers, a fact known as the *Pauli exclusion principle*.

The *ionization energy* is the smallest amount of energy required to remove an electron from the atom when the atom is in its ground state. For hydrogen the ionization energy is 13.6 eV. In contrast, it takes only 5.1 eV to remove the least-tightly-bound electron from a sodium atom.

An atom can receive energy from an external source, such as a collision with another atom or some other particle. It can also absorb a photon of the proper energy. Absorbing just the right amount of energy allows one of its electrons to move to a higher energy level, as long as that level is not already occupied. The atom can then get rid of this excess energy by radiating a photon, with the excited electron falling to an unoccupied state with lower energy. This change is usually consistent with the following *selection rules*, which can be derived using quantum mechanics:

$$\Delta l = \pm 1, \quad \Delta j = 0, \pm 1. \tag{13.9}$$

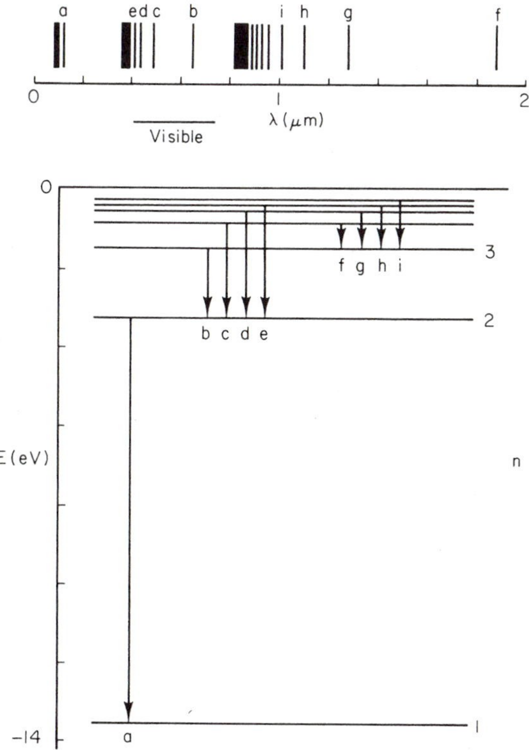

FIGURE 13.3. The spectrum for hydrogen plotted vs wavelength and the energy levels for hydrogen. Some spectral lines and the corresponding transitions have been labeled.

13.3. MOLECULAR ENERGY LEVELS

In addition to internal energy, an atom can have kinetic energy of translation with three degrees of freedom, as described in Chap. 3. The translational kinetic energy is also quantized, but as long as the atom is not confined to a very small volume, the levels are so closely spaced that the translational kinetic energy can be regarded as continuous.

Two atoms together have six degrees of translational freedom, because each can move in three-dimensional space. However, if the atoms are bound together, their motions are not independent One can speak of the three degrees of freedom for translation of the molecule as a whole (center-of-mass motion) and also the vector displacement of one atom from the other. This is shown in Fig. 13.4. Vector **r** locates the center of mass of the two atoms. It is located at a point such that $m_1\mathbf{R}_1 = -m_2\mathbf{R}_2$.

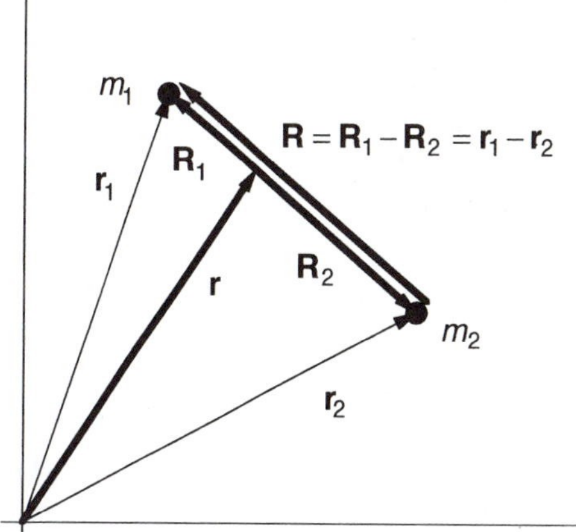

FIGURE 13.4. A diatomic molecule. Vectors $\mathbf{r}_1$ and $\mathbf{r}_2$ are the positions of the atoms measured in the laboratory. Vectors $\mathbf{R}_1$ and $\mathbf{R}_2$ are coordinates in the center-of-mass system.

Consider two particles of mass m_1 and m_2. Their positions with respect to some fixed origin are $\mathbf{r}_1$ and $\mathbf{r}_2$. The velocity of each particle is $\mathbf{v}_i = d\mathbf{r}_i/dt$. The kinetic energy of the ith particle is $T_i = \frac{1}{2}m_i(\mathbf{v}_i \cdot \mathbf{v}_i)$. Define the center of mass by

$$\mathbf{r} = \frac{m_1\mathbf{r}_1 + m_2\mathbf{r}_2}{m_1+m_2}$$

and the vectors from the center of mass to each particle by

$$\mathbf{R}_1 = \mathbf{r}_1 - \mathbf{r} = \frac{m_2(\mathbf{r}_1 - \mathbf{r}_2)}{m_1+m_2} = \frac{m_2\mathbf{R}}{m_1+m_2},$$

$$\mathbf{R}_2 = \frac{-m_1\mathbf{R}}{m_1+m_2}.$$

The total kinetic energy is $T = \frac{1}{2}m_1(\mathbf{v}_1 \cdot \mathbf{v}_1) + \frac{1}{2}m_2(\mathbf{v}_2 \cdot \mathbf{v}_2)$. Since $\mathbf{v}_i = \mathbf{v} + \mathbf{V}_i$, we have

$$2T = (m_1+m_2)(\mathbf{v}\cdot\mathbf{v}) + m_1(\mathbf{V}_1 \cdot \mathbf{V}_1) + m_2(\mathbf{V}_2 \cdot \mathbf{V}_2) + 2\mathbf{v}\cdot(m_1\mathbf{V}_1 + m_2\mathbf{V}_2).$$

The last term vanishes because $m_1\mathbf{R}_1 + m_2\mathbf{R}_2 = \mathbf{0}$. Consider the second term. Differentiating $\mathbf{R}_1 = m_2\mathbf{R}/(m_1+m_2)$ shows that

$$\mathbf{V}_1 \cdot \mathbf{V}_1 = \left(\frac{m_2}{m_1+m_2}\right)^2 V^2,$$

$$\mathbf{V}_2 \cdot \mathbf{V}_2 = \left(\frac{m_1}{m_1+m_2}\right)^2 V^2.$$

Therefore,

$$T = \tfrac{1}{2}(m_1+m_2)v^2 + \frac{1}{2}\frac{m_1m_2}{m_1+m_2}V^2.$$

The first term is the kinetic energy of a point mass $m_1 + m_2$ traveling at the speed of the center of mass. The second is the kinetic energy of a particle having the "reduced mass" $m_1m_2/(m_1+m_2)$ and the velocity of relative motion of the two particles, $V = |\mathbf{V}| = |d\mathbf{R}/dt|$. If $\mathbf{R}$ changes magnitude, the particles are *vibrating*. If $\mathbf{R}$ has a fixed magnitude the molecule can *rotate*. If the molecule is rotating in some plane with angular velocity ω, then

$$\frac{1}{2}\frac{m_1m_2}{m_1+m_2}V^2 = \frac{1}{2}\frac{m_1m_2}{m_1+m_2}R^2\omega^2 = \tfrac{1}{2}I\omega^2.$$

The quantity $I = [m_1m_2/(m_1+m_2)]R^2 = m_1R_1^2 + m_2R_2^2$ is the *moment of inertia* of the two objects [Halliday, Resnick, and Krane, (1992, p. 245ff)]. The angular momentum of a mass about some point is sometimes called the "moment of the momentum" about that point, in the same sense that the torque is the moment of a force about some point. In this case the angular momentum is

$$L = R_1(m_1v_1) + R_2(m_2v_2) = m_1R_1^2\omega^2 + m_2R_2^2\omega^2 = I\omega.$$

These two equations can be combined to give the rotational kinetic energy in terms of the angular momentum about the center of mass:

$$T = \frac{L^2}{2I}.$$

Quantum mechanically, the angular momentum cannot take on any arbitrary value. The square of the angular momentum is restricted to the values

$$L^2 = r(r+1)\hbar^2, \quad r = 0,1,2,\ldots\,.$$

Since there is no potential energy, the total energy of rotation of the molecule is

$$E_r = \frac{r(r+1)\hbar^2}{2I}, \quad r = 0,1,2,\ldots\,. \tag{13.10}$$

The spacing of the rotational levels is shown in Fig. 13.5. A detailed calculation using quantum mechanics shows that when a photon is emitted or absorbed, r must change by ± 1. Therefore the photon energy is

$$\Delta E_r = E_r - E_{r-1} = \frac{\hbar^2}{I}r, \quad r = 1,2,\ldots\,. \tag{13.11}$$

The problems at the end of the chapter show that these photons have low energies, so that rotational spectra lie in the far-infrared region (far meaning far from the visible region, i.e., very long wavelengths).

The other possibility is that the atoms in the molecule vibrate back and forth along their line of centers. If two masses have an equilibrium position a certain distance apart, work must done either to push them closer together or to pull them farther apart. In either case the potential energy is increased. At the equilibrium separation the potential energy is a minimum. Figure 13.6 shows the potential energy E_p of a sodium ion and a chloride ion as a function

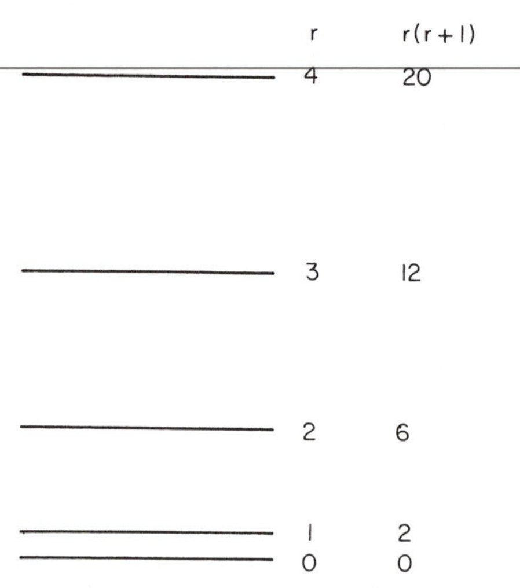

FIGURE 13.5. Energy levels of a rotating molecule.

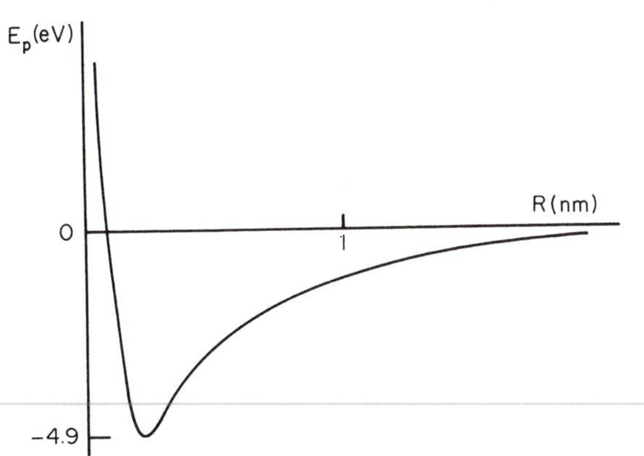

FIGURE 13.6. The potential energy of a sodium ion and a chloride ion as a function of their nuclear separation.

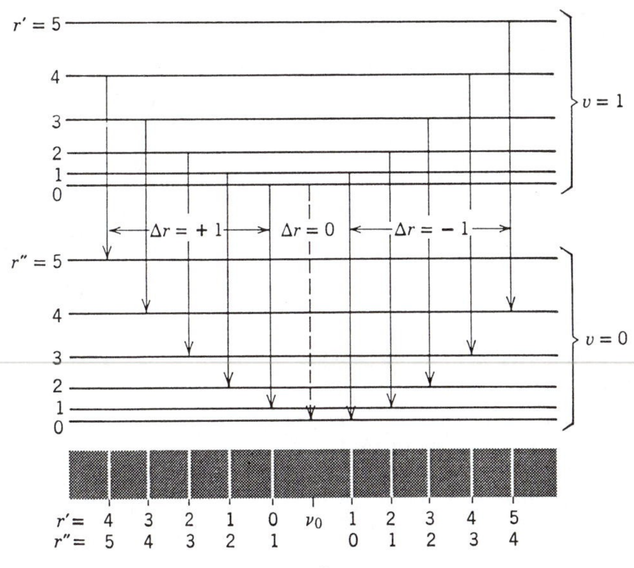

FIGURE 13.7. Transitions for vibrational–rotational spectra. From R. M. Eisberg and R. Resnick (1985). *Quantum Physics of Atoms, Molecules, Solids, Nuclei and Particles*, 2nd ed. New York, Wiley, p. 428. Reproduced by permission of John Wiley & Sons.

of their separation. The potential has a minimum at a separation of about 0.2 nm. The simplest function that has a minimum is a parabola. A parabola can be used to approximate the minimum in Fig. 13.6: $E_p(R)=\frac{1}{2}k(R-R_0)^2$. R_0 is the equilibrium separation. Since (see Sec. 6.4) $dE_p=-F dr$, the force between the ions is $F=-dE_p/dR=-k(R-R_0)$, which is the linear approximation to the force between the two ions. The force is attractive if $R>R_0$ and repulsive if $R<R_0$.

A mass subject to a linear restoring force is called a *harmonic oscillator* (Appendix F). A mass m subject to a linear restoring force $-kx$ oscillates with an angular frequency $\omega^2=k/m$. Classically, the energy of the oscillating mass depends on the amplitude of the motion and can have any value. Quantum mechanically, it is restricted to values

$$E_v=\hbar\omega(v+\tfrac{1}{2}), \quad v=0,12,\ldots . \tag{13.12}$$

This is the total energy, including both kinetic and potential energy. The levels are spaced equally by an amount $\hbar\omega$. The spacing is usually greater than that for rotational levels, often in the infrared. The transitions that give rise to the emission or absorption of photons require a change in the rotational quantum numbers as well as the vibrational ones. The selection rules are

$$\Delta r=\pm 1, \quad \Delta v=\pm 1. \tag{13.13}$$

Some of these vibrational–rotational transitions are shown in Fig. 13.7.

Finally, there can be transitions involving v, r, and the electronic quantum numbers as well. When the electronic quantum numbers change, the shape of the interatomic potential changes, as shown in Fig. 13.8. The details of molecular spectra are fairly involved and are summarized in many texts. Transitions of biological importance are discussed in Grossweiner (1994, pp. 33–38). If the electron selection rules are satisfied, the transition is fairly rapid (typically 10^{-8} s), a process called *fluorescence*. Sometimes the electron becomes trapped in a state where it cannot decay according to the electronic selection rules of Eq. (13.9). It may then have a lifetime up to several seconds before decaying, a phenomenon called *phosphorescence*.

13.4. SCATTERING AND ABSORPTION OF RADIATION; CROSS SECTION

Photons in a vacuum travel in a straight line. When they travel through matter they are apparently[3] slowed down, leading to an index of refraction greater than unity; they may also be scattered or absorbed. Visible light does not pass through a building wall, but it does pass through a glass window. The absorption may depend on the frequency or wavelength of the light. The window can be made of colored glass. The light can also be scattered. This leads to the blue of the sky or to the white of clouds. If there is absorption as well as scattering, the clouds may appear gray instead of white. How light is scattered or absorbed in tissue has become very important in biophysics. Infrared light absorption can be used to measure chemical composition of the body. Light is also used for therapy and for laser surgery. This section shows how scattering and absorption are measured and how light behaves in tissue.

Imagine that we have a distant source of photons that travel in straight lines, and that we collimate the beam (send it through an aperture) so that a nearly parallel beam of

[3]Individual photons travel at speed c, yet the light wave travels at speed c/n. The slowing down of light in a medium is due to interference between the primary beam and scattered photons. This is discussed in Sherwood (1996), in Milonni (1996), and the references cited in these papers.

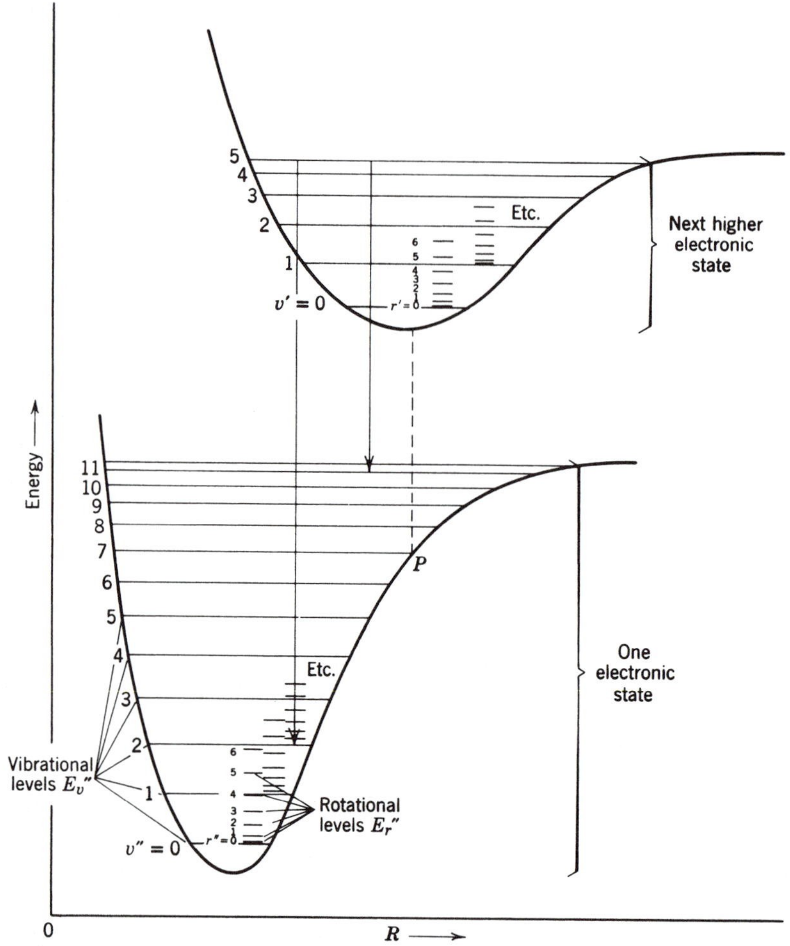

FIGURE 13.8. A combination of changes in electronic quantum numbers within an atom and of vibrational and rotational quantum numbers within the molecule. From R. M. Eisberg and R. Resnick (1985). *Quantum Physics of Atoms, Molecules, Solids, Nuclei and Particles*, 2nd ed. New York, Wiley, p. 430. Reproduced by permission of John Wiley & Sons.

photons is available to us. Imagine also that we can see the tracks of the N photons in the beam, as in Fig. 13.9. When a thin sample of material of thickness dz is placed in the beam, a certain number of photons are scattered and a certain number are absorbed. If we repeat the experiment many times, we find that the number of photons scattered fluctuates about an average value that we call dN_s and the number absorbed fluctuates about an average value dN_a. When we vary the thickness of the absorber, we find that if it is sufficiently thin, the average number of photons scattered and absorbed is proportional to the thickness as well as the number of incident photons:

$$dN_s = \mu_s N\,dz, \quad dN_a = \mu_a N\,dz. \tag{13.14}$$

The total number of unscattered photons N changes according to

$$dN = -(dN_s + dN_a) = -N(\mu_s + \mu_a)dz$$

with solution

$$N(z) = N_0 e^{-\mu z} = N_0 e^{-(\mu_s + \mu_a)z}. \tag{13.15}$$

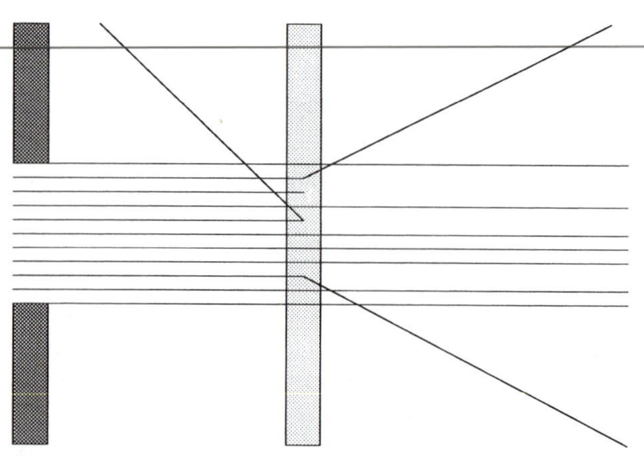

FIGURE 13.9. A collimated beam of photons passes from left to right through a thin slice of material. Some photons pass through, some are scattered, and some are absorbed.

The quantity μ is the *total linear attenuation coefficient.* Quantities μ_s and μ_a are the linear scattering and absorption coefficients. Both depend on the material and the energy of the photons. This kind of exponential absorption is known as *Beer's law.*

The interaction of photons with matter is statistical. The *cross section* σ is an effective area proportional to the probability that an interaction takes place. The interaction takes place with a "target entity." It is sometimes convenient to define the target to be a single molecule, at other times an atom, and still other times one of the electrons within an atom. We can visualize the meaning of the cross section by considering either a single target entity interacting with a beam of photons or a single photon interacting with a thin foil of targets. Both are shown in Fig. 13.10. For the single target in Fig. 13.10(a), consider a beam of N photons passing through the area S with a uniform number per unit area N/S. Let the average number of interactions be $\bar{n}$. The cross section per target entity is defined by saying that the fraction of photons that interact is equal to the fraction of the area occupied by the cross section:

$$\frac{\bar{n}}{N}=\frac{\sigma}{S}. \tag{13.16}$$

We denote the number of photons per unit area by Φ and write Eq. (13.16) as $\bar{n}=\sigma\Phi$. This is the average number of scatterings per target entity or the *probability of interaction per target entity when the beam has* Φ *photons per unit area*:

$$p=\sigma\Phi. \tag{13.17}$$

Strictly speaking, $\bar{n}$ is dimensionless, σ has the dimensions m^2 and Φ has dimensions m^{-2}. However, it is often helpful to think of $\bar{n}$ as being interactions per target entity and σ as being m^2 per target entity.

Alternatively, imagine sending a beam of photons at the target of area S' shown in Fig. 13.10(b). There are N_T target entities per unit area in the path of the beam, each having an associated area σ. The fraction of the photons that interact is again the fraction of the area that is covered:

$$\frac{\bar{n}}{N}=\frac{\sigma S' N_T}{S'}=\sigma N_T \tag{13.18}$$

This is the *probability that a single photon interacts when there are* N_T *target entities per unit area.* Note the symmetry with Eq. (13.17). In the first case there is one target entity and a certain number of photons per unit area. In the second case there is one photon and a certain number of target entities per unit area.

If a number of mutually exclusive interactions can take place (such as absorption and scattering), we can define a cross section for each kind of interaction. The probabilities and the cross sections add:

$$\sigma=\sum_i \sigma_i. \tag{13.19}$$

The second interpretation we had above can be used to relate the cross section to the attenuation coefficient. The number of target entities per unit area is equal to the number per unit volume times the thickness of the target along the beam. To obtain the number of target atoms per unit volume, recall that 1 mol of atoms contains Avogadro's number N_A atoms. If A is the mass of a target containing 1 mol of atoms and the target has mass density ρ, then volume V has mass ρV and contains $\rho V/A$ mol and $N_A\rho V/A$ atoms. Therefore the number of atoms per unit volume is $N_A\rho/A$, and the number of atoms per unit area is

$$N_T=\frac{N_A\rho}{A}\,dz, \tag{13.20}$$

The linear coefficients are related to their corresponding cross sections by

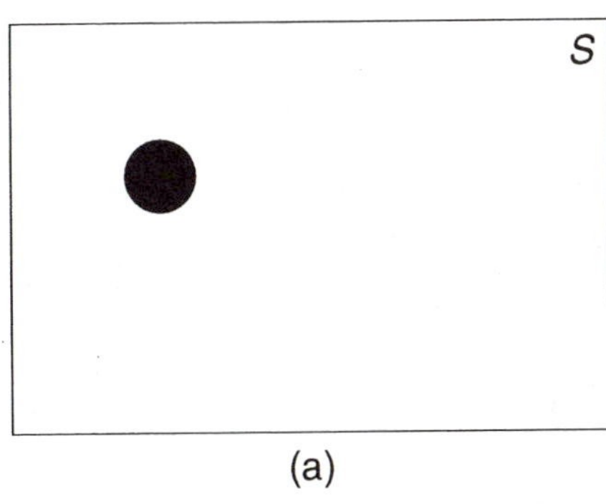

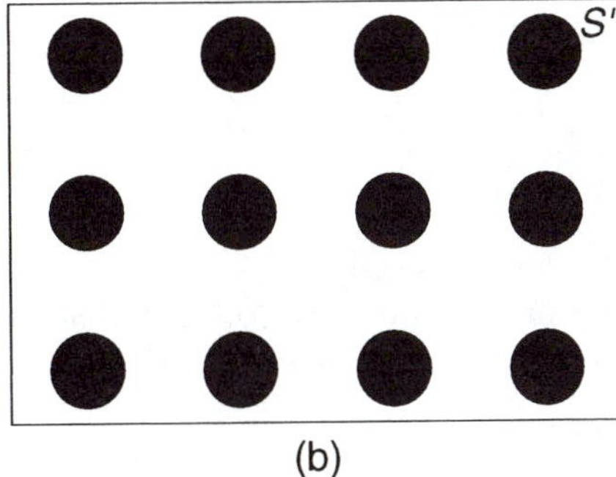

FIGURE 13.10. Each circle represents the cross section σ associated with a target entity such as an atom. (a) There is one atom in area S. (b) There are T target atoms per unit area in area S'.

$$\mu_s = \frac{N_A \rho}{A} \sigma_s,$$

$$\mu_a = \frac{N_A \rho}{A} \sigma_a, \qquad (13.21)$$

$$\mu = \frac{N_A \rho}{A} (\sigma_s + \sigma_a) = \frac{N_A \rho}{A} \sigma_{\text{tot}},$$

where σ_{tot} is the sum of all the interaction cross sections.

Be careful with units! Avogadro's number is defined to be 6.022137×10^{23} entities per mole, which is the number in a *gram* atomic weight. For carbon, $A = 12.01 \times 10^{-3}$ kg mol^{-1} and $\rho = 2.0 \times 10^3$ kg m^{-3}.

We may wish to know the probability that particles (in this case photons) are scattered in a certain direction. We have to consider the probability that they are scattered into a small solid angle $d\Omega$. In this case σ is called the *differential scattering cross section* and is often written as

$$\frac{d\sigma}{d\Omega} d\Omega \quad \text{or} \quad \sigma(\theta) d\Omega. \qquad (13.22)$$

The units of the differential scattering cross section are m^2 sr^{-1}. The differential cross section depends on θ, the direction of travel of the incident and scattered particles. In a spherical coordinate system in which the incident particle moves along the z axis, the solid angle is $d\Omega = \sin\theta \, d\theta \, d\phi$ (Appendix L). If the cross section has no ϕ dependence, then the integration over ϕ can be carried out and $d\Omega = 2\pi \sin\theta \, d\theta$. These solid angles are shown in Fig. 13.11.

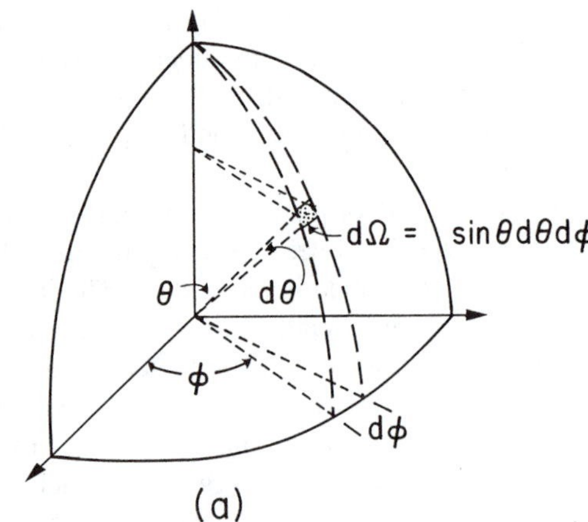

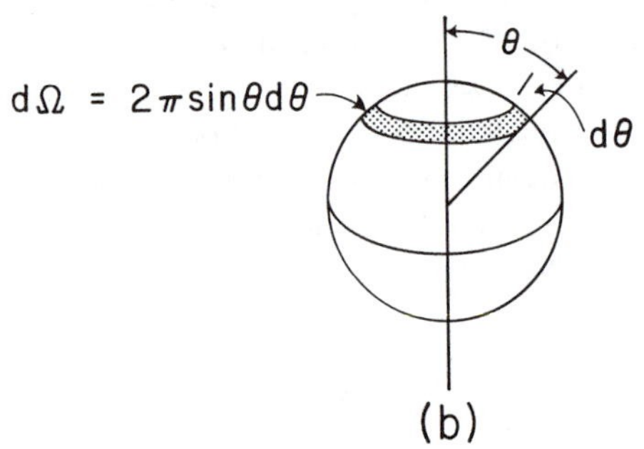

FIGURE 13.11. (a) A small solid angle $d\Omega = \sin\theta \, d\theta \, d\phi$ surrounds the direction defined by angles θ and ϕ. (b) The solid angle $d\Omega = 2\pi \sin\theta \, d\theta$ results from integrating over ϕ.

There are three ways to interpret the exponential decay of the beam that has not undergone any interactions. First, the number of particles remaining in the beam that have undergone no interaction decreases as the target becomes thicker, so that the number of particles available to interact in the deeper layers is less. Second, the exponential can be regarded as taking into account the fact that in a thicker sample some of the target atoms are hidden behind others and are therefore less effective in causing new interactions. The third interpretation is in terms of the Poisson probability distribution (Appendix J). Each layer of thickness dz provides a separate chance for the beam particles to interact. The probability of interacting in any one layer dz is small, $p = \sigma_{\text{tot}} N_A \rho dz / A$, while the total number of "tries" is z/dz. The average number of interactions is $m = p \times$ number of tries. The probability of no interaction is $e^{-m} = \exp(-\sigma_{\text{tot}} N_A \rho z / A) = e^{-\mu z}$.

When the cross section for scattering is large, things can become quite complicated. For example, photons may scatter many times and be traveling through the material in all directions. Various approximations have been used to model photon transport such a case. We will examine some of them shortly. One simple correction that is often made is to consider the average direction a scattered photon travels, for example, the average value of the cosine of the scattering angle, $g = \overline{\cos\theta}$, where θ is the angle of a single scattering. If the average angle of scattering is very small, g is nearly 1. If the photon is scattered backward, $g = -1$, and if the scattering is isotropic, $g = 0$. Formally,

$$g = \frac{\int_0^\pi \sigma(\theta) \cos\theta \, 2\pi \sin\theta \, d\theta}{\int_0^\pi \sigma(\theta) 2\pi \sin\theta \, d\theta}. \qquad (13.23)$$

The *reduced scattering coefficient*

$$\mu_s' = (1 - g)\mu_s \qquad (13.24)$$

is what is usually measured.

The values of the absorption and scattering coefficients vary widely. For infrared light at 780 nm, values are roughly[4]

$$\mu_s' = 1500 \ \text{m}^{-1}, \quad \mu_a = 5 \ \text{m}^{-1}.$$

[4] These are "eyeballed" from data for various tissues reported in the article by Yodh and Chance (1995). Values are up to ten times larger at other wavelengths. See Table 5.2 in Grossweiner (1994).

13.5. THERMAL RADIATION

Any atomic gas emits light if it is heated to a few thousand kelvins. The light consists of a line spectrum. The famous yellow line of sodium has

$$\lambda = 589.2 \text{ nm},$$

$$\nu = 5.092 \times 10^{14} \text{ Hz},$$

$$E = h\nu = 3.38 \times 10^{-19} \text{ J} = 2.11 \text{ eV}.$$

These photons are emitted when sodium atoms lose 2.11 eV and return to their ground state. If the sodium atoms are excited by thermal collisions, the probability that a sodium atom is in the excited state, relative to the probability that it is in the ground state, is given by the Boltzmann factor

$$\frac{P_{\text{excited}}}{P_{\text{ground}}} = e^{-E/k_B T}.$$

At room temperature $k_B T = 4.14 \times 10^{-21}$ J, so $e^{-E/k_B T} = e^{-81.5} = 3.8 \times 10^{-36}$. The number of atoms in the excited state is utterly negligible. If the temperature is raised to 1500 K, $e^{-E/k_B T}$ is 8×10^{-8}, and enough atoms are excited to give off light as they fall back to the ground state.

If a piece of iron is heated to 1500 K, it glows with a white–yellow color. The approximate temperatures at which the hot metal appears various colors are given in Table 13.3. If the light from the glowing metal is analyzed with a spectroscope, it is found to consist of a continuous range of colors rather than discrete lines.

The difference between the spectra of single atoms and the spectra of solids and liquids can be understood from the following argument. If we have N isolated identical atoms, each atom has an energy level at the energy shown in Fig. 13.12(a). There are a total of N levels, one for each atom. When two of these atoms are brought close together, the levels shift slightly and split into two closely spaced levels because of interaction between the atoms. The two levels for a pair of atoms are shown in Fig. 13.12(b). If three atoms are brought close together the level splits into three levels as shown in Fig. 13.12(c). If a large number of atoms are brought close together, the N level spreads out into a *band*, Fig. 13.12(d). Transitions from one band to another

TABLE 13.3. *Approximate color temperatures.*[a]

Color	T (K)
Red, just visible in daylight	750–800
Cherry red	975–1175
Yellow	1200–1505
White	1425–1800
Dazzling (bluish) white	1900

[a]The range of values reflects the differences between scales established by different observers.

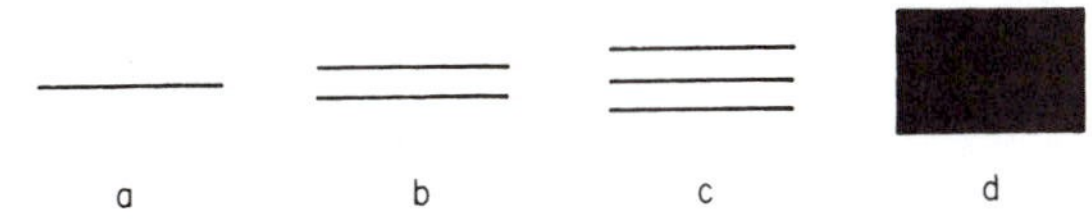

FIGURE 13.12. The splitting of energy levels as many atoms are brought together. (a) A single atom. (b) Two atoms. (c) Three atoms. (d) Many atoms.

can have many different energies, and photons with a continuous range of energies can be emitted or absorbed.

The relative number of photons of different energies that will be emitted or absorbed depends on the nature of the substance. Glass and sodium chloride crystals are nearly transparent in the visible spectrum because the spacing of the levels is such that no photons of these energies are absorbed. When such substances are heated enough to populate the higher energy levels, no photons of these energies are emitted.

A substance that has so many closely spaced levels that it can absorb every photon that strikes it appears black. It is called a *blackbody*. It is difficult if not impossible to make a surface that is completely absorbing; the absorption can be improved by making a cavity, as in Fig. 13.13. Photons entering the hole in the cavity bounce from the walls many times before chancing to pass out through the hole again, and they therefore have a greater chance of being absorbed. Such a hole in a cavity is a better approximation to a blackbody than is the absorbing material lining the cavity.

If the surface is not completely absorbing, we define the *emissivity* $\epsilon(\lambda)$, which is the fraction of light absorbed at wavelength λ. (Why emission and absorption are closely related is discussed below.) If the light all passes through some transparent material or is completely reflected, then $\epsilon = 0$; if it is all absorbed, $\epsilon = 1$. A blackbody has $\epsilon(\lambda) = 1$ for all wavelengths. An object for which $\epsilon(\lambda)$ is constant but less than 1 is called a gray body.

When a blackbody is heated, the light given off has a continuous spectrum because the energy levels are so closely spaced. By imagining two different cavities in contact, one can argue[5] that the amount of energy coming out of a blackbody cavity depends only on the temperature of the walls and not on the nature of the surfaces.

The spectrum of power per unit area emitted by a completely black surface in the wavelength interval between λ and $\lambda + d\lambda$ is

$$W_\lambda(\lambda, T)\, d\lambda,$$

a universal function called the *blackbody radiation function*. It has units of W m^{-3}, although it is often expressed as W cm^{-2} μm^{-1}. The value of W_λ is plotted for several different temperatures in Fig. 13.14. As the black surface or cavity walls become hotter, the spectrum shifts toward shorter wavelengths, which is consistent with the observa-

[5]For a brief discussion see Halliday, Resnick, and Krane (1992, p. 1021ff). A more detailed treatment is found in Bramson (1968, Chap. IV).

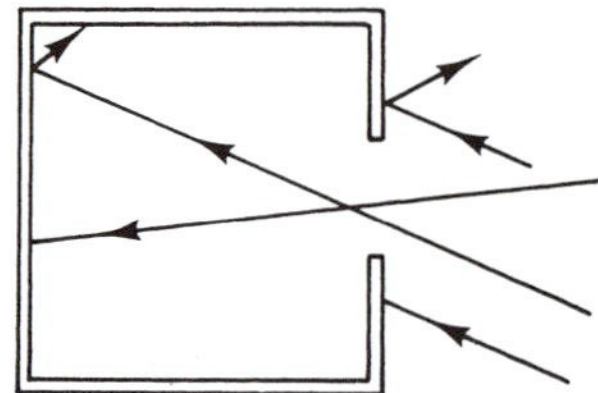

FIGURE 13.13. A small hole in the wall of a cavity is a better blackbody than the walls of the cavity are. Any light that enters the hole must be reflected several times before emerging. It can be absorbed by the wall at any reflection. If the walls appear black, the hole appears even blacker. (The walls are highly absorbing diffuse reflectors.)

tions in Table 13.3. The visible region of the spectrum is marked on the abscissa in Fig. 13.14; even at 1600 K when the radiating surface appears white, most of the energy is radiated in the infrared.

Figure 13.15 plots $W_\lambda(\lambda,T)$ for two temperatures near body temperature (37 °C=310 K). Compare the scales of Figs. 13.14 and 13.15 and note how much more energy is emitted by a blackbody at the higher temperature and how it is shifted to shorter wavelengths.

Much work was done on the properties of blackbody or thermal or cavity radiation in the late 1800s and early 1900s. While some properties could be explained by classical physics, others could not. The description of the function $W_\lambda(\lambda,T)$ by Planck is one of the foundations of quantum mechanics. We will not discuss the history of these developments, but will simply summarize the properties of the blackbody radiation function that are important to us.

The value of $W_\lambda(\lambda,T)$ is given by

$$W_\lambda(\lambda,T)=\frac{2\pi c^2 h}{\lambda^5(e^{hc/\lambda k_B T}-1)}. \tag{13.25}$$

This equation has several interesting properties. First, consider the factor $e^{hc/\lambda k_B T}$ in the denominator. Since light consists of photons of energy $E=h\nu=hc/\lambda$, the factor in parentheses in the denominator is $e^{E/k_B T}-1$. For very large energies (short wavelengths) the 1 can be neglected and the effect of the denominator on Eq. (13.25) is like a Boltzmann factor.

Second, let us write the radiation function in terms of frequency. Let λ_1 and $\lambda_2=\lambda_1+d\lambda$ be two slightly different wavelengths, with power $W_\lambda(\lambda,T)d\lambda$ emitted per unit surface area between λ_1 and λ_2. The same power must be emitted[6] between frequencies $\nu_1=c/\lambda_1$ and $\nu_2=c/\lambda_2$:

$$W_\nu(\nu,T)d\nu=W_\lambda(\lambda,T)d\lambda. \tag{13.26}$$

Since $\nu=c/\lambda$, $d\nu/d\lambda=-c/\lambda^2$, and

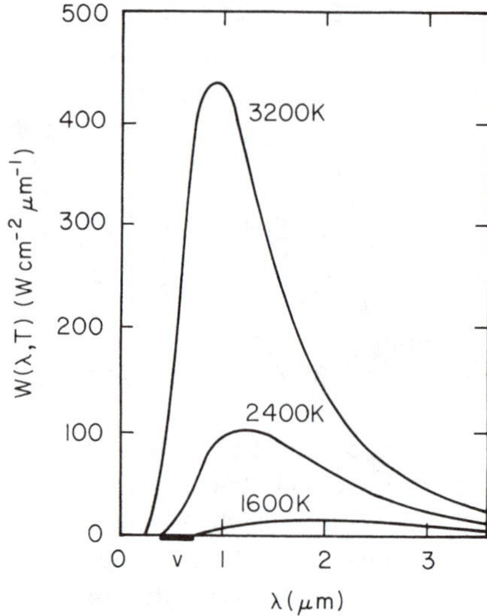

FIGURE 13.14. The blackbody radiation function for several temperatures. The visible spectrum is marked by v.

$$|d\nu|=+\frac{c}{\lambda^2}|d\lambda|. \tag{13.27}$$

Equations (13.25)–(13.27) can be combined to give

$$W_\nu(\nu,T)=\frac{2\pi\nu^2(h\nu)}{c^2(e^{h\nu/k_B T}-1)}. \tag{13.28}$$

Next, we can show that W_λ has a peak that shifts to shorter wavelengths with higher temperature. Equation (13.25) can be differentiated to show that for a fixed temperature the peak in W_λ occurs at wavelength

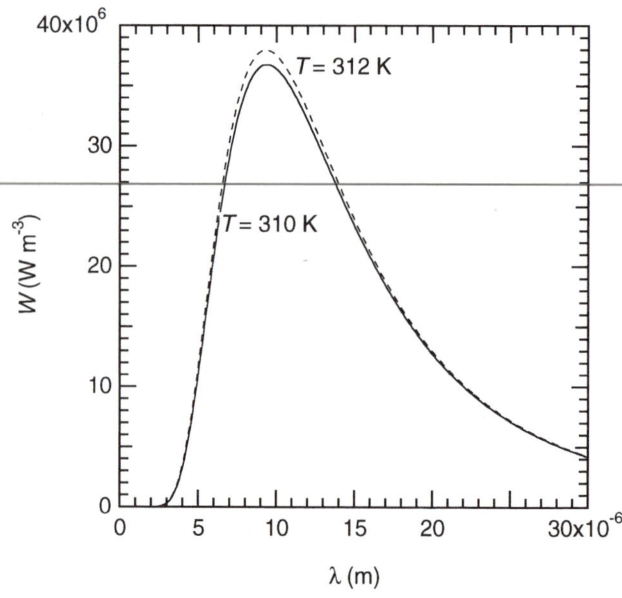

FIGURE 13.15. The blackbody radiation function $W(\lambda,T)$ for T=310 K and T=312 K.

[6] $W_\lambda(\lambda,T)$ and $W_\nu(\nu,T)$ do not have the same functional form. In fact, they have different units. The units of $W_\lambda(\lambda,T)$ are W m^{-3}, while those of $W_\nu(\nu,T)$ are W m^{-2} s.

$$\lambda_{max}T = \frac{hc}{4.9651 k_B} \tag{13.29}$$

or

$$\lambda_{max}T = 2.9 \times 10^{-3} \text{ m K}.$$

Equations (13.25) and (13.29) can be combined to give the power radiated per unit area at the peak of the spectrum:

$$W_\lambda(\lambda_{max}, T) = (1.29 \times 10^{-15})T^5 \text{ W cm}^{-2}\ \mu\text{m}^{-1}$$
$$= (1.29 \times 10^{-5})T^5 \text{ W m}^{-3}. \tag{13.30}$$

We can find the total amount of power emitted per unit surface area by integrating[7] Eq. (13.26):

$$W_{tot}(T) = \int_0^\infty W_\lambda(\lambda, T)\,d\lambda = \int_0^\infty W_\nu(\nu, T)\,d\nu$$
$$= \frac{2\pi^5 k_B^4}{15c^2h^3} T^4 = \sigma T^4. \tag{13.31}$$

This is the *Stefan–Boltzmann* law. The Stefan–Boltzmann constant σ, which has no relationship to cross section, was known empirically before Planck's work. It has the numerical value

$$\sigma = 5.67 \times 10^{-8} \text{ W m}^{-2}\text{ K}^{-4}. \tag{13.32}$$

All this is true for a blackbody. Thermodynamic arguments can be made to show that if a body does not completely absorb light at some wavelength, that is $\epsilon(\lambda) < 1$, then the power emitted at that wavelength is

$$\epsilon(\lambda) W_\lambda(\lambda, T). \tag{13.33}$$

This is the same $\epsilon(\lambda)$ that was introduced earlier in this section. It is called the *emissivity* of the surface. This implies that a surface that appears blackest when it is absorbing radiation will be brightest when it is heated. Figure 13.16 shows a small hole in a piece of tungsten that has been heated. The hole forms the opening to a cavity and is therefore more absorbing than is the tungsten surface. When heated, the hole emits more light than the tungsten surface.

13.6. INFRARED RADIATION FROM THE BODY

The body radiates energy in the infrared, and this is a significant source of energy loss. Infrared radiation has been used for over 40 years to image the body, but the value of the technique is still a matter of debate. In a later section we will see how the scattering of infrared radiation by the body can be used to learn information about tissue beneath the surface.

[7]This is not a simple integration. See Gasiorowicz (1974, p. 6).

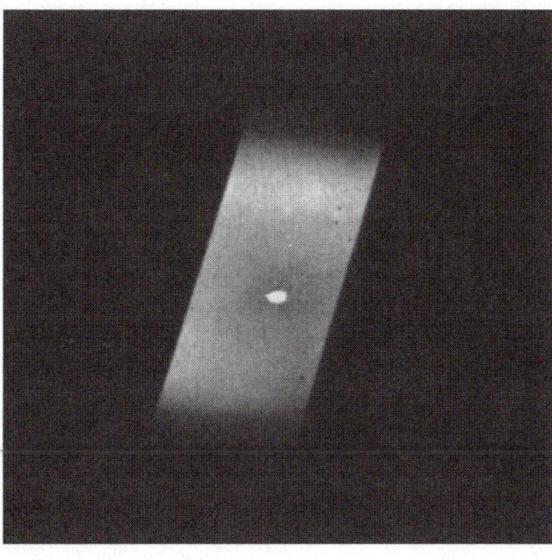

FIGURE 13.16. A photograph of an incandescent tungsten tube with a small hole drilled in it. The radiation emerging from the hole is brighter than that from the tungsten surface. From D. Halliday, R. Resnick, and K. S. Krane (1992). *Fundamentals of Physics*, 4th ed. extended. New York, Wiley, Vol. 2, p. 1022. Reproduced by permission of John Wiley & Sons.

Measurements of the emissivity of human skin have shown that for $1\ \mu\text{m} < \lambda \leq 14\ \mu\text{m}$, $\epsilon(\lambda) = 0.98 \pm 0.01$. This value was found for white, black, and burned skin [Steketee (1973)]. In the infrared region in which the human body radiates, it is very nearly a blackbody. Let us apply Eq. (13.31) to see what the blackbody radiation from the human body is. The total surface area of a typical adult male is about 1.73 m^2. The surface temperature is about 33 °C = 306 K (this is less than the core temperature of 310 K). Therefore the total power radiated is $w_{tot} = SW_{tot} = S\sigma T^4$ = 875 W. This is a large number, nearly 9 times the basal metabolic rate of 100 W. The reason it is so large is that it assumes the surroundings are at absolute zero, or that the subject is radiating in empty space with no surroundings. When there are nearby surfaces, radiation from them is received by the subject, and the net radiation is considerably less than 875 W. The easiest arrangement for which to calculate the net heat loss is a blackbody at temperature T surrounded by a surface at temperature T_s (Fig. 13.17). At equilibrium the temperature of both objects is the same, $T = T_s$. The power emitted by the body is equal to the power absorbed. Increasing T increases the power emitted

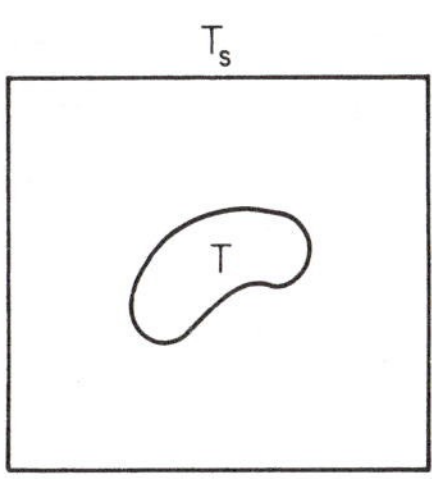

FIGURE 13.17. A blackbody at temperature T within a container with wall temperature T_s.

according to $W = \sigma T^4$. The body then emits more power than it absorbs. Equilibrium is restored when the body has cooled or the surroundings have warmed so that the temperatures are again the same. Thermodynamic arguments can be made to show that the net power radiated by the body is

$$w_{\text{tot}} \approx S\sigma(T^4 - T_s^4). \tag{13.34}$$

If the object is not a blackbody or the wall temperature is not uniform, the net power loss is more complicated. However, this model represents a considerable improvement over our previous calculation. Suppose that the surroundings are at a temperature $T_s = 293$ K (20 °C). The net loss is

$$w_{\text{tot}} = (1.73)(5.67 \times 10^{-8})(306^4 - 293^4) = 137 \text{ W}.$$

This says that a nude subject surrounded by walls at 20 °C would have to exercise to maintain body temperature, even if the air temperature were warm enough so that heat conduction and convection losses were zero.

If you have lived in a cold climate, you have probably felt cold in a room at night when the drapes are open, even though a thermometer records air temperatures that should be comfortable. This is because of radiation from you to the cold window. The glass is transparent only in the visible range; for infrared radiation it is opaque and has a high emissivity. The radiation of the cold window back to you is much less than your radiation to it, and you feel cold.

This same problem can occur with a premature infant in an incubator. If the incubator is placed near a window, one wall of the incubator can be cooled by radiation to the window. The infant can be cooled by radiation to the wall of the incubator, even though a shiny (low-emissivity) thermometer in the incubator records a reasonable air temperature. One solution is to be careful where an incubator is placed and insulate its walls; another has been to redesign incubators with a feedback loop controlling the infant's temperature.

Infrared radiation can be used to image the body. Two types of infrared imaging are used. In infrared photography the subject is illuminated by an external source with wavelengths from 700 to 900 nm. The difference in absorption between oxygenated and nonoxygenated hemoglobin allows one to view veins lying within 2 or 3 mm of the skin. Either infrared film or a solid-state camera can be used.

Thermal imaging detects thermal radiation from the skin surface. Significant emission from human skin occurs in the range 4–30 μm, with a peak at 9 μm (Fig. 13.15). The detectors typically respond to wavelengths below 6–12 μm. Thermography began about 1957 with a report that skin temperature over a breast cancer was slightly elevated. There was great hope that thermography would provide an inexpensive way to screen for breast cancer, but there have been too many technical problems. There is more variability in vascular patterns in normal breasts than was first realized, so that differences of temperature at corresponding points in each breast is not an accurate diagnostic criterion. The thermal environment in which the examination is done is extremely important. The sensitivity (ability to detect breast cancer) is too low to use it as a screening device. Thermography has also been proposed to detect and to diagnose various circulatory problems. Thermography is generally not accepted at this time [Cotton (1992); Blume (1993)], though it still has its proponents. One group has reported that measurements as a function of time can detect oscillations in skin temperature that are part of the normal thermal regulatory process, and that changes in these oscillations may occur in certain diseases [Anbar (1994)]. Although the physics of the design of a thermographic camera is a nice example of radiometry, I have removed the discussion from this edition.

Infrared radiation from the tympanic membrane (eardrum) and ear canal is used to measure body temperature. One instrument is based on pyroelectric sensors, which were developed for the mass market of motion detectors [Fraden (1991)]. They have a permanent electric dipole moment whose magnitude changes with temperature.

13.7. THE DIFFUSION APPROXIMATION TO PHOTON TRANSPORT

13.7.1. General Theory

In a number of cases a portion of the body is illuminated with infrared, visible, or ultraviolet light for diagnosis or treatment. Understanding how the photons interact is important both for measurements and for determining the proper dose for treatment. There is so much scattering that Beer's law (exponential decay) is not a valid approximation. The most accurate studies are done with "Monte Carlo" computer simulations, in which probabilistic calculations are used to follow a large number of photons as they repeatedly interact in the tissue being simulated. There are various approximations that are often useful. The field is reviewed in Chap. 5 of Grossweiner (1994). One of the approximations, the diffusion approximation, is described here.

If the photons have undergone enough scattering in a medium, all memory of their original direction is lost. In that case the movement of the photons can be modeled by the diffusion equation. In Sec. 4.11 we wrote Fick's second law as

$$\frac{\partial C}{\partial t} = D\nabla^2 C + Q.$$

The left-hand side of the equation is the rate at which the concentration, the number of particles per unit volume, is increasing. The term $D\nabla^2 C$ is the net diffusive flow into the small volume, the particle current being given by

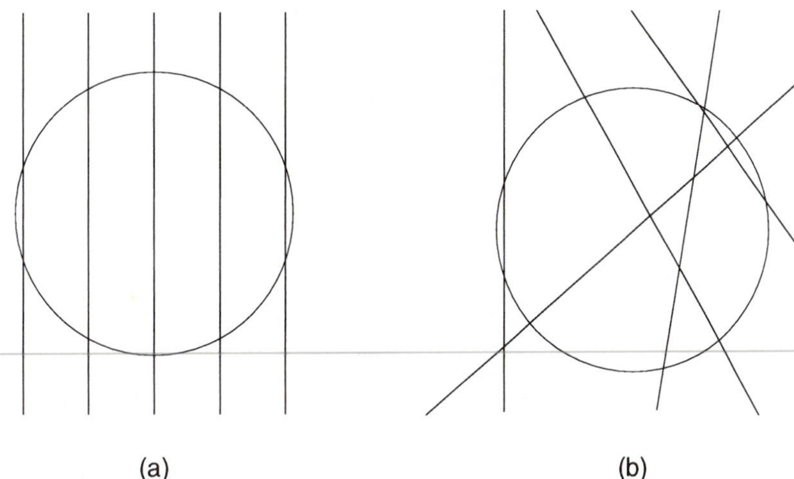

FIGURE 13.18. The particle fluence is the ratio of the expectation or average value of the number of particles passing through the sphere to the area of a great circle of the sphere, πa^2. It depends on the total number of particles passing through the sphere, regardless of the direction they travel.

$\mathbf{j} = -D\boldsymbol{\nabla} C$. The last term is the rate of production or loss of particles within the volume by other processes, depending on whether Q is positive or negative.

Let us suppose that we can apply this to photons. The concentration must be the number of *diffusing* photons per unit volume. Any in the incident beam are still traveling in the original direction and are not diffusing. Therefore there may be a source term, which we will call s, due to the incident photons. But photons are also being absorbed. They are traveling with a speed $c' = c/n$, where n is the index of refraction of the medium. In time dt they travel a distance $dx = c'\,dt$, and the probability that they are absorbed is $\mu_a dx = \mu_a c'\,dt$. Therefore the diffusion equation for photons is

$$\frac{\partial C}{\partial t} = D\nabla^2 C + s - \mu_a c' C. \tag{13.35}$$

Each term has the units of photons $\mathrm{m}^{-3}\,\mathrm{s}^{-1}$.

In photon transfer, it is customary to make two changes in this equation. The first is to divide all terms by the speed of the photons in the medium,[8] c'. The result is

$$\frac{1}{c'}\frac{\partial C}{\partial t} = D'\nabla^2 C - \mu_a C + \frac{s}{c'},$$

where $D' = D/c'$ is referred to in the photon transfer literature as the *photon diffusion constant.* It has dimensions of length.

Two important quantities in radiation transfer are the *photon* or *particle fluence* and the *photon fluence rate*. The International Commission on Radiation Units (ICRU) defines the particle fluence as follows for any kind of particle, including photons: At the point of interest construct a small sphere of radius a. Let the number of particles striking the surface of the sphere during some time interval have an *expectation value* N. (The expectation value is the mean of a set of measurements in the limit as the number of measurements becomes infinite.) The *particle fluence* Φ is the ratio $N/\pi a^2$, where πa^2 is the area of a great circle of the sphere, that is, the area of a circle having the same radius as the sphere. This is shown in Fig. 13.18 and is a generalization of our earlier use of Φ as the number of particles per unit area. It neatly avoids having to introduce obliquity factors, since for any direction the particles travel, one can construct a great circle on the sphere that is perpendicular to their path. The particle fluence rate is

$$\varphi = \frac{d\Phi}{dt}.$$

We saw in Chap. 4 that for a group of particles all traveling with the same speed, the number transported across a plane per unit area per unit time is equal to their concentration times their speed. The photon concentration is related to the photon fluence rate by $C = \varphi/c'$, and the photon diffusion equation becomes

$$\frac{1}{c'}\frac{\partial \varphi}{\partial t} = D'\nabla^2 \varphi - \mu_a \varphi + s. \tag{13.36}$$

This is the form that is usually found in the literature. The units of each term are photons $\mathrm{m}^{-3}\,s^{-1}$. One can show that[9]

[8]Most papers in this field use c as the velocity of light in the medium. I prefer to reserve c for the fundamental constant, the velocity of light in vacuum.

[9]See, for example, Duderstadt and Hamilton (1976, pp. 133–136).

$$D' = \frac{1}{3[\mu_a + (1-g)\mu_s]} = \frac{1}{3(\mu_a + \mu_s')}. \quad (13.37)$$

13.7.2. Continuous Measurements

If the tissue is continuously irradiated with photons at a constant rate, the term containing the time derivative vanishes. If in addition we use a broad beam of photons so that we have a one-dimensional problem and we are far enough into the tissue so that the source term can be ignored, the model is

$$D' \frac{d^2\varphi}{dx^2} = \mu_a \varphi. \quad (13.38)$$

This has an exponential solution $\varphi = \varphi_0 e^{-\mu_{\text{eff}} x}$, where $\mu_{\text{eff}} = \{3\mu_a(\mu_a + (1-g)\mu_s)\}^{1/2}$. It is interesting to see what these numbers mean. Using the ''typical'' values from Sec. 13.4, the mean depth for the unattenuated beam is

$$\lambda_{\text{unatten}} = \frac{1}{\mu} = \frac{1}{\mu_a + \mu_s'} = \frac{1}{1505} = 0.66 \text{ mm}.$$

For the diffuse beam the mean depth is about 10 times this:

$$\lambda_{\text{diffuse}} = \frac{1}{\mu_{\text{eff}}} = \frac{1}{\sqrt{(3)(5)(1505)}} = 6.7 \text{ mm}.$$

These values are for a wavelength where the tissue is fairly transparent. The diffusion equation can be solved for other geometries that model the light source being used.[10] One problem with these measurements is that it gives only μ_{eff}, which is a combination of μ_a and μ_s'. Also, the path length may be ambiguous because the geometry cannot be modeled accurately.

13.7.3. Pulsed Measurements

A technique made possible by ultrashort light pulses from a laser is time-dependent diffusion. It allows determination of both μ_s and μ_a. A very short (150-ps) pulse of light strikes a small region on the surface of the tissue. A detector placed on the surface and typically about 4 cm away records the arriving photons. A typical plot of the reflected photon fluence rate is shown in Fig. 13.19. Patterson *et al.* (1989) have shown that the reflected fluence rate after a pulse is approximately

$$R(r,t) = (4\pi D'c't)^{-3/2} z_0 t^{-1} \exp(-\mu_a c't) \times \exp\left(-\frac{r^2 + z_0^2}{4D'c't}\right). \quad (13.39)$$

Here r is the distance of the detector from the source along the surface of the skin, $c't$ is the total distance the photon

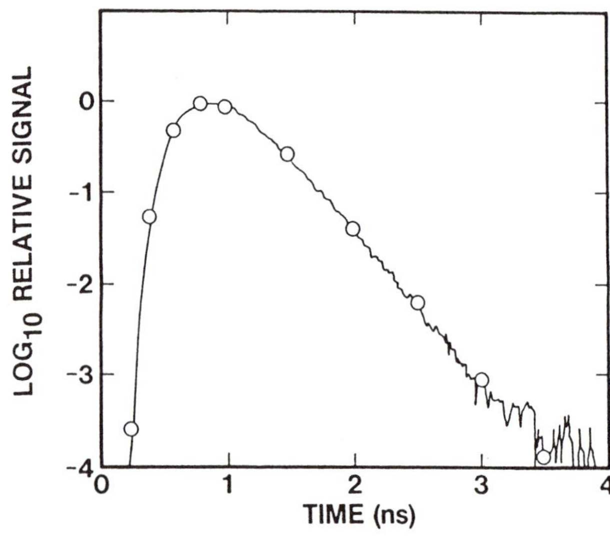

FIGURE 13.19. Time-resolved infrared spectroscopy. The line is a measurement of the reflected photons from the calf of a human volunteer at a distance of 4 cm from the pulsed source. The wavelength is 760 nm. The circles are calculated using Eq. (13.39) and normalized to the peak value. From M. S. Patterson, B. Chance, and B. C. Wilson (1989). Time resolved reflectance and transmittance for the non-invasive measurement of tissue optical properties. *Appl. Opt.* **28**(12): 2331–2336. Copyright by the Optical Society of America.

has traveled before detection, and $z_0 = 1/[(1-g)\mu_s]$ is the depth at which all the incident photons are assumed to scatter and become part of the diffuse photon pool. This curve fits Fig. 13.19 well and can be used to determine μ_a and $(1-g)\mu_s$. We can understand the various terms in Eq. (13.39). The last term is a Gaussian spreading in the r direction away from the z axis where the photons were injected. This is a two-dimensional problem. Compare this with Eq. (4.77), which shows that in two dimensions $\sigma_r^2 = 4Dt$, and recall that $D = D'c'$. The next-to-last term is the fraction of the photons in the pulse that are absorbed, $\exp(-\mu_a x)$, where x is the total distance the photons have traveled. The first term is the normalization that reduces the amplitude of the Gaussian as it spreads.

A related technique is to apply a continuous laser beam whose amplitude is modulated at various frequencies between 50 and 800 MHz. The Fourier transform of Eq. (13.39) gives the change in amplitude and phase of the detected signal. Their variation with frequency can also be used to determine μ_a and μ_s'.[11]

13.7.4. Refinements to the Model

The diffusion equation, Eq. (13.36), is an approximation, and the solution given, Eq. (13.39), requires some unrealistic assumptions about the boundary conditions at the

[10]See, for example, Grossweiner (1944, p. 98).

[11]See, for example, Sevick *et al.* (1991) or Pogue and Patterson (1994).

surface of the medium ($z=0$). Hielscher *et al.* (1995) have recently compared experiment, Monte Carlo calculations, and solutions to the diffusion equation with three different boundary conditions. They found that Eq. (13.39) was the easiest to use but leads to errors in the estimates of the coefficients that become worse when the detector and source are close together. Their Monte Carlo calculations fit the data quite well.

We have ignored the reflections that occur when light goes from one medium into another with a different index of refraction. These are discussed in the literature.

Near-infrared spectroscopy is an active research area. Recent papers include the determination of optical properties and blood oxygenation of tissue using continuous sources [Liu *et al.* (1995)]. The scattering and absorption coefficients are such that measurements of brain oxygenation can be made through the skull; see Okada *et al.* (1995).

13.7.5. The Integrating-Sphere Photometer

We now discuss the theory of the integrating-sphere photometer, which can be used to measure diffusely scattered light. Its operation is based on the fact that when light is diffusely reflected from any point on the inside surface of a hollow sphere, the amount of reflected light per unit area striking all other points on the inside of the sphere is the same. This remarkable result is easily proved. The radius of the sphere is R. Let the light be reflected from a point along the z axis at the bottom of the sphere, as shown in Fig. 13.20. Because of the spherical symmetry, we need only consider light striking some other point on the surface that is in the plane of the paper and at polar angle θ. The isosceles triangle shows that $\theta_s=\theta_d=\theta/2$ and $r=2R\cos\theta/2$. Equation (12.17) gives the power per unit area striking dS_d. If the surface obeys Lambert's law $L(\theta_s)=L_0$, the power striking dS_d is

$$E\,dS_d=\frac{L_0\cos(\theta_s)\cos(\theta_d)\,dS_s\,dS_d}{r^2}$$

$$=\frac{L_0\cos(\theta_s)\cos(\theta_d)\,dS_s\,dS_d}{(2R\cos\theta/2)^2}=\frac{L_0\,dS_s\,dS_d}{4R^2}.$$

The power and the irradiance are independent of θ and are therefore the same everywhere inside the sphere.

Figure 13.21(a) shows an integrating sphere with an entrance aperture of area $S_{\rm in}$ and an exit aperture of area $S_{\rm exit}$. The total surface area of the sphere is S. Suppose that an amount of radiant energy $R_{\rm in}$ enters the sphere through the entrance aperture and strikes the opposite surface. A fraction α is diffusely reflected and spreads out uniformly over the inside surface of the sphere. The energy out through the exit aperture after the first scattering is $R_1=R_{\rm in}\alpha(S_{\rm exit}/S)$. Some energy also escapes through the entrance aperture. The fraction of the energy striking a

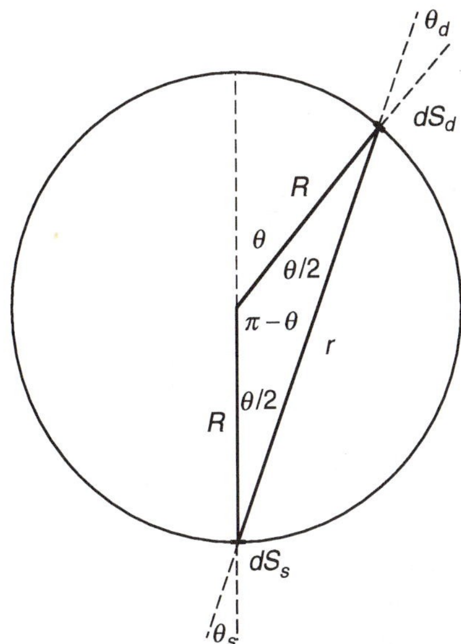

FIGURE 13.20. Light diffusely reflected from an element of area dS_s on the inside of a sphere uniformly illuminates the surface of the sphere. The derivation considers light striking an area dS_d at angle θ with the axis of the sphere passing through the reflecting element.

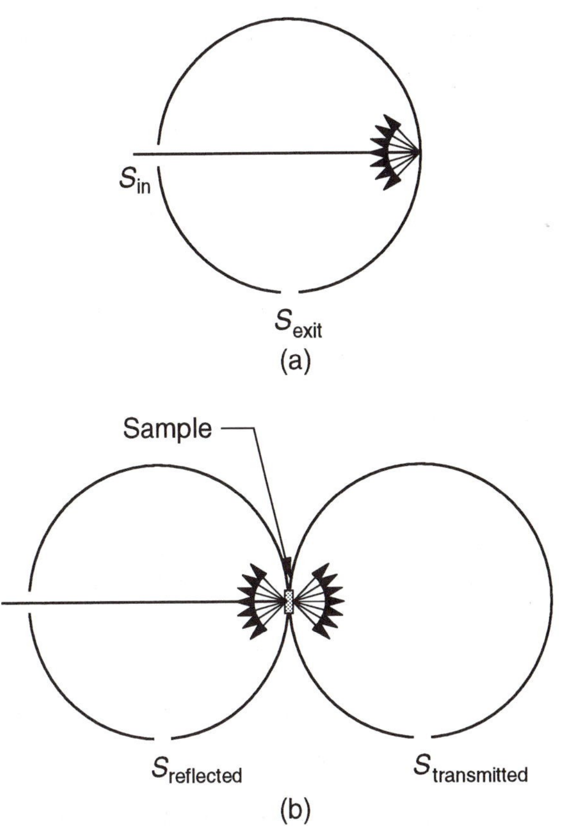

FIGURE 13.21. (a) An integrating sphere. (b) A double integrating sphere used to measure both reflected and transmitted light.

reflecting surface in the sphere is $\beta=(S-S_{exit}-S_{in})/S$. This again scatters uniformly over the inner surface. The energy escaping through the exit aperture on the second scattering is $R_2=R_{in}\alpha^2\beta(S_{exit}/S)$. The total amount of energy leaving through the exit aperture is

$$R_{out}=R_1+R_2+R_3+\cdots$$
$$=R_{in}\alpha(S_{exit}/S)[1+\alpha\beta+(\alpha\beta)^2+\cdots]$$
$$=\frac{R_{in}\alpha(S_{exit}/S)}{1-\alpha\beta}. \tag{13.40}$$

A double integrating sphere, shown in Fig. 13.21(b), can be used to measure both the reflected and transmitted light from a sample placed at the junction of the spheres.

13.7.6. Biological Applications of Infrared Scattering

We now mention some of the applications of infrared probes, applying infrared light to the body and measuring the reflected or transmitted radiation.

A great deal of work has been done to measure the oxygenation of the blood as a function of time by determining the absorption at two different wavelengths. Figure 13.22 shows the absorption coefficients for oxygenated and deoxygenated hemoglobin and water. The greater absorption of blue light in oxygenated hemoglobin makes oxygenated blood red. The wavelength 800 nm at which both forms of hemoglobin have the same absorption is called the *isosbestic point*. Measurements of oxygenation are often made by comparing the absorption at two wavelengths on either side of this point.

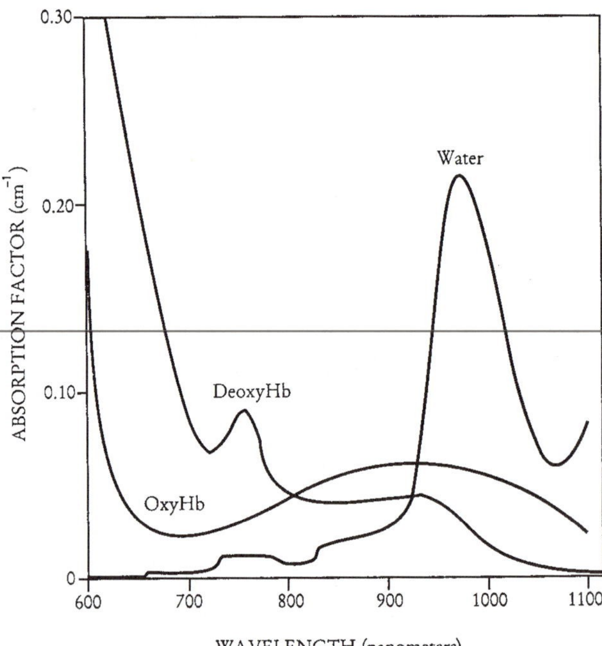

FIGURE 13.22. The absorption coefficient μ_a for water, oxyhemoglobin, and deoxyhemoglobin. From A. Yodh and B. Chance (1995). Spectroscopy and imaging with diffusing light. *Phys. Today* March: 34–40. Used by permission of the American Institute of Physics.

As was mentioned above, one of the difficulties with these measurements is knowing the path length, since photons undergo many scatterings before being absorbed or reaching the detector. Scattering from many tissues besides hemoglobin distorts the signal. Nonetheless, this technique is now widely used with "pulse oximeters" that fit over a finger. An historical review and discussion of the measurement problems has been provided by Mendelson (1992). Other uses of infrared probes are described by Mendelson (1995) and by Flewelling (1995).

Infrared probes are used extensively in the laboratory. Since the vibrational and rotational levels depend on the masses, separations, and forces between the various atoms bound in a molecule, it is not surprising that infrared spectroscopy can be used to identify specific bonds. This is a useful technique in chemistry. Biological applications are made difficult by the fact that the absorption coefficients are large and thin samples must be used, particularly in an aqueous environment.

One way around this is *Raman scattering*. Raman scattering is the scattering of visible light (for which a biological preparation is much more likely to be transparent), in which the scattered photon does not have its original energy, but has lost or gained energy corresponding to a rotational or vibrational transition. The effect was discovered in 1928, and its measurement became practical with the development of laser technology. Lasers provide a source of visible light with sufficient stability so that the Raman lines can be resolved. An idealized example is shown in Fig. 13.23. This technique allows the "catalog" of energy levels known from infrared spectroscopy to be applied to systems that are strongly absorptive in the infrared but transparent to visible light. Many discussions of Raman spectroscopy are available. A fairly theoretical one by Berne and Pecora (1976) relies heavily on autocorrelation functions and spectral analysis that we saw in Chap. 11.

13.8. BLUE AND ULTRAVIOLET RADIATION

The energy of individual photons of blue and ultraviolet light is high enough to trigger chemical reactions in the body. These can be both harmful and beneficial. A beneficial effect is the use of blue light to treat neonatal jaundice. The most common harmful effect is the development of sunburn, skin cancer, and premature aging of the skin.

Neonatal jaundice occurs when bilirubin builds up in the blood. Bilirubin is a toxic waste product of the decomposition of the hemoglobin that is released when red blood cells die. It is insoluble in water and cannot be excreted through either the kidney or the gut. It is excreted only after being

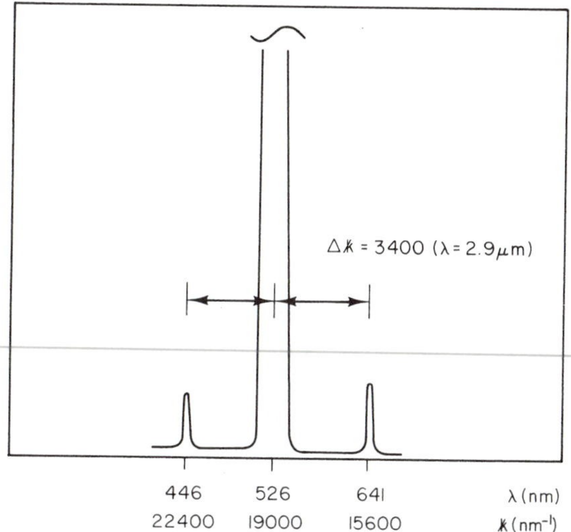

FIGURE 13.23. In Raman scattering, a photon gains or loses energy due to a change in the vibrational or rotational state of the scattering molecule. An idealized example for water is shown. The very intense line has no energy change; the weak lines are Raman scattering.

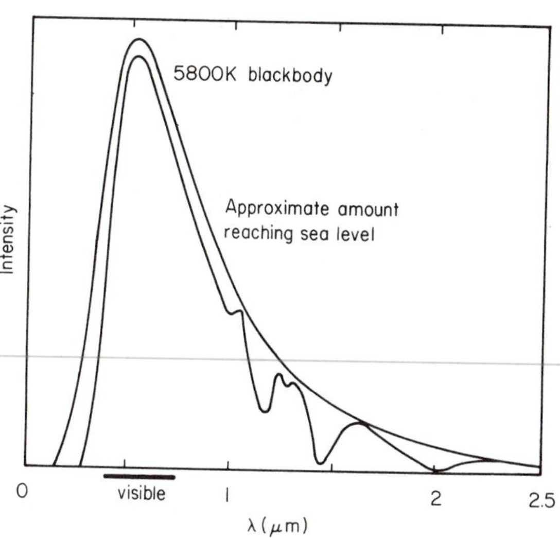

FIGURE 13.24. The solar spectrum and the approximate spectrum reaching the earth after atmospheric attenuation.

conjugated with glucuronic acid in the liver. Bilirubin monoglucuronate and bilirubin diglucuronate are both water soluble. They are excreted in the bile and leave via the gut. Some newborns have immature livers that cannot carry out the conjugation. In other cases there is an increased death rate of red blood cells (hemolysis) and the liver cannot keep up. The serum bilirubin level can become quite high, leading to a series of neurological symptoms known as kernicterus. The abnormal yellow color of the skin called jaundice is due to bilirubin in the capillaries under the skin.

It was discovered accidentally that when the skin of a newborn with jaundice was exposed to bright light, the jaundice color went away. Photons of blue light have sufficient energy to convert the bilirubin molecule into more soluble and apparently less harmful forms [McDonagh (1985)]. Photons of longer wavelength have less energy and are completely ineffective. The standard form of phototherapy used to be to place the baby "under the lights." Since the lights were bright and also emitted some ultraviolet, it was necessary to place patches over the baby's eyes. Also, since the baby's skin had to be exposed to the lights, it had to be placed in an incubator to keep it warm. Recently, a fiberoptic blanket has been developed that can be wrapped around the baby's torso under clothing or other blankets. The optical fibers conduct the light from the source directly to the skin. Eye patches are not needed, and the baby can be fed and handled. Typical energy fluence rates are $(4-6)\times 10^{-2}$ W m^{-2} nm^{-1} in the range 425–475 nm. Acceptance by nursing staff and parents is very high [Murphy and Oellrich (1990)]. The blanket can even be used for home treatment, though this is still controversial. Volume 24, Issue 4 of *Health Devices* (August, 1995) is devoted to an assessment of fiberoptic phototherapy systems.

Ultraviolet light can come from the sun or from lamps. The maximum intensity of solar radiation is in the green, at about 500 nm. The sun emits approximately like a thermal radiator at a temperature of 5800 K. Figure 13.24 shows a 5800-K thermal radiation curve. The power per unit area from the sun at all wavelengths striking the earth's outer atmosphere, the solar constant, can be calculated by regarding the sun as a thermal radiator. The calculated value is 1390 W m^{-2} (2 cal cm^{-2} min^{-1}). Satellite measurements give 1372 W m^{-2} [Madronich (1993)]. Because of reflection, scattering, atmospheric absorption, and so forth, the amount actually striking the earth's surface is about 1000 W m^{-2}. Figure 13.24 also shows the effect of absorption of sunlight in the atmosphere. The sharp cutoff at 320 nm is due to atmospheric ozone (O_3), which absorbs strongly from 200 to 320 nm. It absorbs more weakly at wavelengths as long as 360 nm. Molecular oxygen absorbs strongly below 180 nm.

The ultraviolet spectrum is qualitatively divided into the following regions:

UVA	315–400 nm
UVB	280–315 nm[12]
UVC or middle UV	200–280 nm
Vacuum UV	<240 nm
Far UV	120–200 nm
Extreme UV	10–120 nm

Only the first three are of biological significance, because the others are strongly absorbed in the atmosphere.

Madronich (1993) gives a detailed discussion of the

[12] In Europe the range of UVB radiation is taken to be 290–320 nm.

various factors that reduce the ultraviolet energy reaching the earth's surface. The sensitivity of DNA decreases as the wavelength increases. Figure 13.25 shows the solar radiation reaching the ground when the sun is at different angles from the zenith (directly overhead), weighted for DNA sensitivity. Biological effects of ultraviolet light are reviewed by Diffey (1991). Computer programs are available that calculate the total dose during a day at different latitudes and altitudes.[13]

There are several responses of the skin to ultraviolet light. In order to understand them one must know something about the anatomy and physiology of skin. The outer layer of the skin, the *epidermis*, consists of three sublayers (Fig. 13.26). A single layer of *basal cells* is on the inside. Most of these cells produce keratin, a protein that gives the outer layers of skin its strength. About 10% of the cells are *melanocytes* that produce the pigment *melanin*. Next comes a sublayer of about seven cells, called the *prickle layer*. On top of this is a two- or three-cell layer called the *stratum granulosum* or granular layer. The surface is a layer of dead cells, primarily keratin but also cellular debris, called the *stratum corneum* or horny layer. Basal cells are constantly produced in the basal layer, migrate outward, become the stratum corneum, and are sloughed off.

In order to discuss injury to tissue, both here by ultraviolet light and in later chapters by x rays, we need to introduce some specialized terms. The body's immediate (*acute*) response to an injury, whether it is an infection, a bump, a cut or a burn, is the *inflammatory response* described in Sec. 5.4. Prolonged (*chronic*) irritation may result in abnormal cell growth. The abnormalities of cell growth that result in organs or tissues that are larger than normal are *hypertrophy*, an enlargement of existing cells, and *hyperplasia*, an enlargement due to the formation of new cells. The aberrations in cell growth patterns are shown in Table 13.4. They are *metaplasia*, *dysplasia*, and *anaplasia*. Metaplasia is reversible and goes away if the stimulus or irritant is removed. Dysplasia is sometimes reversible and sometimes progresses to become cancerous. Anaplastic changes are present in nearly all forms of cancer. Anaplasia may result from dysplasia, or it may arise directly from normal cells.

The acute effect of ultraviolet radiation is reddening of the skin or *erythema* due to increased blood flow in the *dermis*, the layer beneath the epidermis. This is part of the inflammatory reaction. The amount of energy that just produces detectable erythema is called the *minimum erythemal dose*. It is difficult to measure in an objective manner. New instrumentation allows quantitative measurements [Diffey and Farr (1997)]. The 1987 *reference action spectrum* adopted by the CIE[14] shows the relative sensitivity of the skin versus wavelength for the production of erythema. It is

$$\epsilon(\lambda)=\begin{cases}1.0, & 250\leq\lambda\leq 298 \text{ nm}\\ 10^{0.094(298-\lambda)}, & 298\leq\lambda\leq 328 \text{ nm.}\\ 10^{0.015(139-\lambda)}, & 328\leq\lambda\leq 400 \text{ nm}\end{cases} \tag{13.41}$$

This is plotted in Fig. 13.27. The minimum erythemal dose at 254 nm is about 6×10^7 J m^{-2} [Diffey and Farr (1991, Table 2)]. There is considerable scatter from one experiment to another. When the degree of erythema is plotted vs energy per unit area, the slope of the curve depends on the wavelength. Early effects on skin include sunburn, tanning (now thought to be an injury response), and thickening. Daily exposure for two to seven weeks causes a three to fivefold thickening of the stratum corneum.

Some patients have an abnormally high sensitivity to ultraviolet exposure. They may exhibit abnormal photosensitivity because of various diseases or from taking drugs such as phenothiazines (one of the classes of major tranquilizers), sulfa drugs, dimethylchlortetracycline, the antidiabetic sulfonureas, and thiazide diuretics and even from drinking quinine water. Photocontact dermatitis is

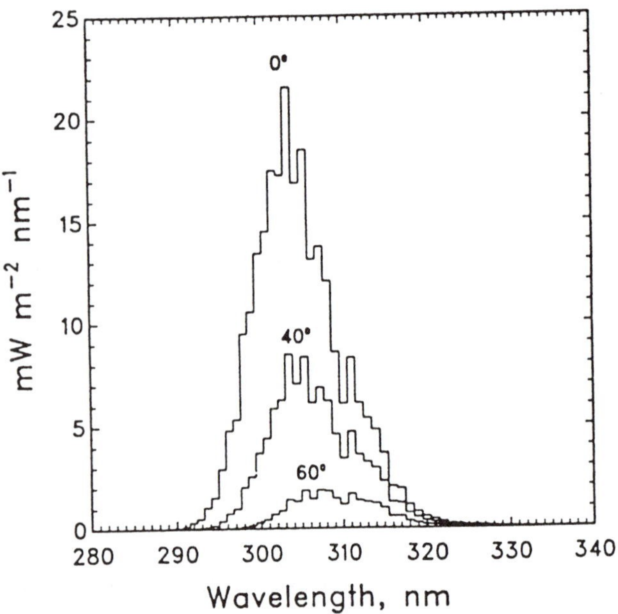

FIGURE 13.25. Spectral dose rates weighted for ability to damage DNA for three different angles of the sun from overhead. The calculation assumes clear skies and an ozone layer of 300 Dobson units (1 DU=2.69 $\times10^{20}$ molecule m^{-2}. From S. Madronich (1993). The atmosphere and UV-B radiation at ground level. In A. R. Young *et al.*, eds. *Environmental UV Photobiology.* New York, Plenum, pp. 1–39.

[13]A computer program for evaluating solar UV exposure has been provided, with a listing, by Schaefer (1993). See also Diffey and Cameron (1984).

[14]*Commission International de l'Eclairage* or International Commission on Illumination.

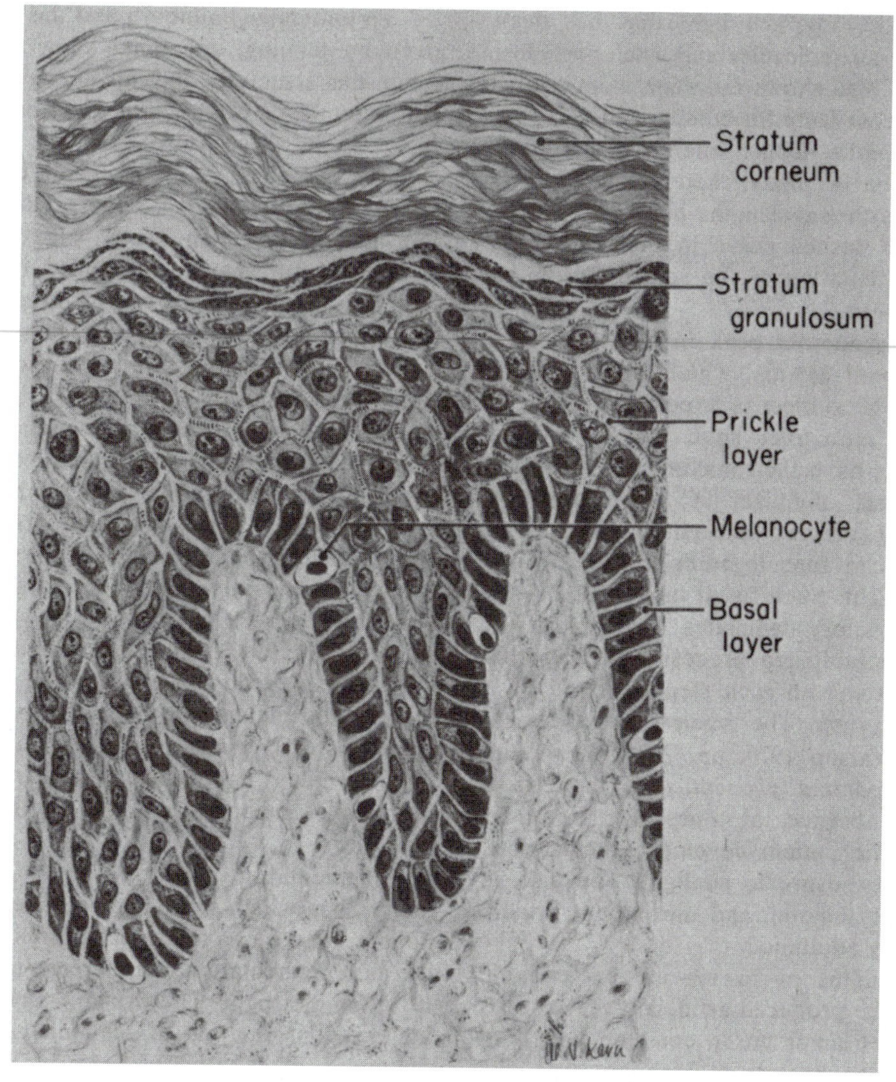

FIGURE 13.26. The epidermis. The basal layer contains the cells from which the other layers are derived. As the cells move toward the surface they become the prickle layer and the stratum granulosum (granular layer). The stratum corneum is dead cellular debris. The melanocytes, which produce melanin granules, are in the basal layer. From D. M. Pillsbury (1971). *A Manual of Dermatology.* Philadelphia, Saunders, p. 5. Reproduced by permission of the W. A. Saunders Co.

caused by interaction of photons with substances placed on the skin, such as perfumes containing furocoumarins, lime peel, fungi, and fluorescein dye used in lipsticks.

Chronic exposure to ultraviolet radiation causes premature aging of the skin. The skin becomes leathery and wrinkled and loses elasticity. The characteristics of photoaged skin are quite different from skin with normal aging [Kligman (1989)]. UVA radiation was once thought to be harmless. We now understand that UVA radiation contributes substantially to premature skin aging because it penetrates into the dermis.[15] Moseley (1994) reviews the safety issues involving ultraviolet light and lasers.

There are three types of skin cancer. Basal-cell carcinoma (BCC) is most common, followed by squamous-cell carcinoma (SCC). These are together called nonmelanoma or nonmelanocytic skin cancer (NMSC). Basal-cell carcinomas can be quite invasive (Fig. 15.45) but rarely metastasize or spread to distant organs. Squamous-cell carcinomas are more prone to metastasis. Melanomas are much more aggressive and frequently metastasize.

Armstrong and Kricker (1995) review the epidemiology of skin cancer. This summary is based on their paper. There are geographic differences in *incidence*, the number of newly diagnosed cases per 100 000 population per year. Estimates of incidence for the three types of skin cancer for whites in the United States are given in Table 13.5. Melanoma incidence rates for whites in the United States are approximately 10 times higher than those for blacks living

[15]There has been at least one report of skin cancer associated with UVA radiation from a cosmetic tanning bed [Lever and Lawrence (1995)].

TABLE 13.4. *Abnormal changes in tissue.*

Metaphasia	A reversible change in which one cell type is replaced by another
Dysplasia	Variation in size, shape, and organization of the cells. Literally, "deranged development"
Anaplasia	A marked, irreversible, and regressive change from adult cells that are differentiated in form to more primitive, less differentiated cells

Differences between benign and malignant tumors

Characteristic	Benign	Malignant
Histological differentiation (microscopic appearance)	Often typical of the tissue of origin	Not well differentiated; atypical cells
Mode of growth	Expands inside a capsule	Expansive; also infiltrative, with no capsule
Rate of growth	Progressive; usually slow; few cells undergoing mitosis (division) at any one time	May be rapid, with many cells undergoing mitosis
Metastasis (distant spread)	Absent	Frequently present

in the same geographic area, and 2–6 times higher than those for Hispanics living in the same area. An increase in melanoma incidence in professional or indoor occupations has been observed, probably related to an increase in recreational sun exposure. Also, the incidence is higher for people born in countries with lots of sunlight than for people who migrate to those countries, suggesting that the number of years of exposure or the age at exposure is important. The incidence of melanoma increases with age until about age 50 and slows somewhat in older people. The incidence of NMSC increases steadily with age. The incidence of melanoma from the early 1960s to the late 1980s has increased at a rate of about 5% per year in populations of European origin, while increasing much less or not at all in other populations. Most of this increase has been on the trunk, particularly in men. Similar increases in NMSC have been seen in the United States, primarily for BCC on the trunk. (It is much more difficult to obtain accurate figures for NMSC than for melanoma, because NMSC data are not typically kept since it has a much lower mortality.) The epidemiological data suggest an association between skin cancer and exposure to sunlight. There are also laboratory studies of the damage to cells caused by ultraviolet light. Armstrong and Kricker conclude that the evidence "leaves little room for doubt that sun exposure causes both melanoma and NMSC." There is some epidemiological evidence that a pattern of infrequent, intense exposure to ultraviolet light is more likely to lead to melanoma than relatively continuous exposure.

Sun protection can reduce skin cancer. Protection is most important in childhood years, both because children receive 3 times the annual sun exposure of adults and because the skin of children is more susceptible to cancer-causing changes [Truhan (1991)]. A number of chemical sunscreens are available, and there is good evidence that sunscreens with a sun protection factor (SPF) of 15 or more are quite helpful. It has been argued that some sunscreens do not adequately protect against UVA radiation. However, formulations are changing, and some experts feel that the UVA protection in any sunscreen with an SPF of 15 or more is adequate [Urbach (1993)]. Because the skin of chil-

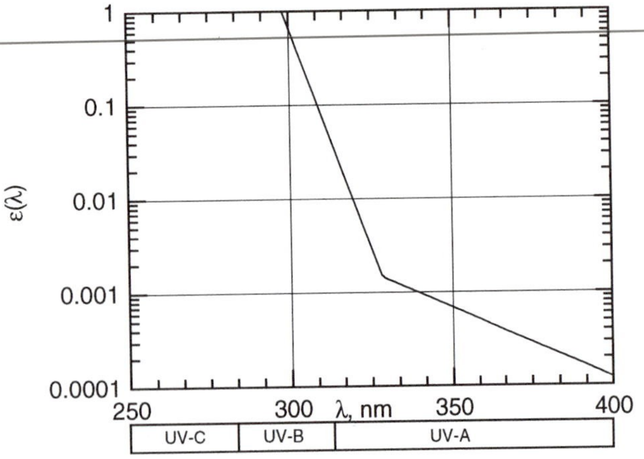

FIGURE 13.27. The erythema action spectrum $\epsilon(\lambda)$ for ultraviolet light, as adopted by the CIE in 1987.

TABLE 13.5. *Estimates of skin cancer incidence rates per 100 000.*[a]

Cancer type	Population	Males	Females
Melanoma	White, New Orleans, 1983–87	6.9	5.3
	White, Hawaii, 1983–87	22.2	14.9
SCC	White, U.S., 1994 (rough est.)	100	45
BCC	White, U.S., 1994 (rough est.)	400	200

[a]Simplifications made by the author from data in Armstrong and Kricker (1995).

dren during the first six months of life may absorb these chemicals, the Food and Drug Administration (FDA) recommends that infants under 6 months should be kept out of the sun or physically shaded from it. Because of the high reflectivity of sand and snow, beach umbrellas provide at most a factor of 2 protection. Hats need to have a brim that is at least 7.5 cm wide [Diffey and Cheeseman (1992)].

The effect of ultraviolet light on the eye has been reviewed by Bergmanson and Söderberg (1995). Acute effects include *keratitis* (inflammation of the cornea, the transparent portion of the eyeball) and *conjunctivitis* (inflammation of the conjunctiva, the mucous membrane covering the eye), also known as snow blindness or welder's flash. Laboratory studies show that ultraviolet-light exposure causes thickening of the cornea and disrupts corneal metabolism. UVC radiation is absorbed by the cornea. The crystalline lens absorbs UVB and, in older persons, UVA and visible light. Only a little UVA light reaches the retina. The retina is also susceptible to trauma from blue light. Low doses cause photochemical changes in tissues, while high doses also cause thermal damage.

Chronic low exposure to ultraviolet light causes permanent damage to the cornea, known as *droplet keratopathy* or *spheroid degeneration*. UVA radiation is a significant factor in the development of a *pterygium*, a hyperplasia of the conjunctiva that may grow over the cornea and impair sight. Rarely, it can cause blindness.

Properly designed spectacles and contact lenses can protect the eye against ultraviolet light. However, both must be designed to absorb ultraviolet. Soft contacts are larger and provide more protection than rigid gas-permeable contacts. Protection from high ultraviolet light-intensity requires sunglasses or welding goggles. Wide-brimmed hats also help protect the eye from ultraviolet light.

Ultraviolet light is used in therapy, primarily for the treatment of a skin disease called *psoriasis*. Psoriasis is an inflammatory disorder in which the basal cells move out to the stratum corneum in much less than the normal 28 days. The skin is red and has thick scaling. UVB radiation, often in conjunction with coal tar applied to the skin, has been used as a treatment for psoriasis since the 1920s. In the 1960s a treatment was developed that uses UVA and a chemical either applied to the skin or administered systemically (photochemotherapy or PUVA—psoralen UVA). The chemical is a psoralen derivative. It affects DNA, and when the affected DNA, is irradiated with ultraviolet light, cross-links form, preventing replication. The treatment works, but it has also been found to cause cancer. There was a 20-year scientific controversy about this, with different results being reported in the United States and Europe. There are now well-defined guidelines for the use of PUVA [Studniberg and Weller (1993)]. Details of PUVA therapy are found in Grossweiner (1994, pp. 162–167).

PUVA therapy is also useful in cutaneous T-cell lymphoma, a disease that first becomes apparent on the skin and then moves to internal organs. Another treatment, *extracorporeal photopheresis*, involves removing the patient's blood, extracting the red blood cells, irradiating the plasma and white blood cells with UVA light outside the body, and returning the red blood cells and the irradiated white blood cells and plasma to the patient [Grossweiner (1944, pp. 167f)].

13.9. HEATING TISSUE WITH LIGHT

Sometimes tissue is irradiated in order to heat it; in other cases tissue heating is an undesired side effect of irradiation. In either case, we need to understand how the temperature changes result from the irradiation. Examples of intentional heating are *hyperthermia* (heating of tissue as part of cancer therapy) or *laser surgery* (tissue ablation[16]). Tissue is ablated when sufficient energy is deposited to vaporize the tissue. Heating may be a side effect of phototherapy.

The temperature changes are often modeled by a heat-flow equation containing a source term for the deposition of photon energy and a term representing flow of energy away from the site in warmed blood. This is one form of the *bioheat equation*, which can include additional terms in more complicated models.

The linear equation for heat conduction was mentioned as one form of the transport equation in Table 4.3:

$$j_H = -K\frac{dT}{dx},$$

with the units of the thermal conductivity K being $\mathrm{J\,K^{-1}\,m^{-1}\,s^{-1}}$. When extended to three dimensions and combined with the equation of continuity (conservation of energy), this gives a heat-conduction equation with the same form as Fick's second law for diffusion:

$$\rho_t C_t \frac{\partial T}{\partial t} = K\nabla^2 T. \qquad (13.42)$$

Here ρ_t is the density of the tissue ($\mathrm{kg\,m^{-3}}$) and C_t the tissue heat capacity ($\mathrm{J\,kg^{-1}}$). The left-hand side of the equation is the rate of energy increase in the tissue per unit volume, and the right-hand side is the net rate of heat flow into that volume by conduction—energy flowing because warmer molecules with more kinetic energy transfer energy to cooler neighbors in a collision process analogous to a random walk. This model is for solids; in liquid one must also consider convection.

We now add a term for energy carried away by flowing blood. In the linear approximation it is proportional to the temperature difference between the tissue and the blood supply and also to the rate of blood flow. Units for this term can be quite confusing and need to be examined in detail. Blood flow is usually defined by physiologists as the *perfusion* P, which is the volume flow of blood per unit mass of tissue. The SI units P are

[16]In surgery, *ablation* means the excision or amputation of tissue.

$$P\,\frac{\text{m}^3\ \text{(blood)}}{[\text{kg (tissue)}]\text{s}}.$$

Its product with the tissue density is the volume flow of blood per unit volume of tissue:

$$\rho_t P=\frac{[\text{kg(tissue)}][\text{m}^3\ \text{(blood)}]}{[\text{m}^3\ \text{(tissue)}][\text{kg (tissue)}]\text{s}}=\frac{\text{m}^3\ \text{(blood)}}{[\text{m}^3\ \text{(tissue)}]\text{s}}=\text{s}^{-1}.$$

The quantity is analogous to clearance (Chap. 2). Its inverse is the time it takes for a volume of blood equal to the tissue volume to flow through the tissue. Each term of our heat-flow equation has units of energy per unit volume of tissue per second. If we assume that the blood enters the tissue at temperature T_0 and leaves at temperature T, the energy lost by the volume is the heat capacity of blood, C_b, times its mass per unit volume times the temperature rise. The new term in the heat-flow equation is

$$C_b\,\frac{\text{J}}{\text{K [kg (blood)]}}\times\rho_b\,\frac{\text{kg (blood)}}{\text{m}^3\ \text{(blood)}}\times\rho_t P\,\frac{\text{m}^3\ \text{(blood)}}{[\text{m}^3\ \text{(tissue)}]\ \text{s}}$$
$$\times[(T-T_0)\ \text{K}]$$

or

$$C_b\rho_b\rho_t P(T-T_0)\,\frac{\text{J}}{[\text{m}^3\ \text{(tissue)}]\ \text{s}},$$

so the heat-flow equation with blood flow added is

$$\rho_t C_t\,\frac{\partial T}{\partial t}=K\nabla^2 T-C_b\rho_b\rho_t P(T-T_0).$$

The last term we consider is the energy deposited by the photon beam. In Sec. 13.7 we defined the particle fluence and particle fluence rate for photons. The definition can be used for both collimated beams and diffuse radiation. In a similar way we define the *energy fluence* Ψ as the ratio of the expectation value of the amount of photon energy traversing a small sphere of radius a divided by the area of a great circle of the sphere, πa^2. The *energy fluence rate* is

$$\psi=\frac{d\Psi}{dt}. \tag{13.43}$$

The energy per unit volume lost by a beam with energy fluence rate ψ can be determined by the following argument. Consider only the fluence rate due to photons traveling in a certain direction. Orient the z axis is in that direction and consider a small volume $dS\,dz$. The rate at which energy flows into the volume is $\psi\,dS$, and the rate at which it is absorbed is $(\psi\,dS)(\mu_a dz)$. Therefore the rate of absorption per unit volume is $\mu_a\psi$, independent of the direction the photons travel. The final heat-flow equation is

$$\rho_t C_t\,\frac{\partial T}{\partial t}=K\nabla^2 T-C_b\rho_b\rho_t P(T-T_0)+\mu_a\psi. \tag{13.44}$$

For monoenergetic photons the photon energy fluence rate is related to the photon fluence rate by

$$\psi=h\nu\varphi. \tag{13.45}$$

In general, one must first solve Eq. (13.36) to determine ψ and then solve Eq. (13.44). We could add other terms, such as one for the thermal energy produced by metabolism within the tissue.

Sometimes Eq. (13.44) is written with all terms divided by $\rho_t C_t$, and sometimes with all terms divided by K. If we divide by $\rho_t C_t$ the equation is similar in form to the diffusion equation in Chap. 4:

$$\frac{\partial T}{\partial t}=D\nabla^2 T-\frac{C_b}{C_t}\,\rho_b P(T-T_0)+\frac{\mu_a}{\rho_t C_t}\,\psi, \tag{13.46}$$

where

$$D=\frac{K}{\rho_t C_t}. \tag{13.47}$$

Values of D are in the range $(0.5–2.5)\times10^{-7}\ \text{m}^2\,\text{s}^{-1}$ depending on the tissue type [Grossweiner (1994), pp. 127–129)]. We saw in Chap. 4 that for a spreading Gaussian solution to the diffusion equation the variance is $\sigma_x^2=\sigma_y^2=\sigma_z^2=2Dt$. The thermal relaxation time, that is, the average time for the temperature rise to spread a distance x, is therefore $x^2/2D$ in one dimension, $x^2/4D$ in two dimensions, and $x^2/6D$ in three dimensions.

There is an interplay between the thermal conductivity term and the blood-flow term. The thermal penetration depth δ_{th} is the distance at which the two terms are comparable. For larger distances blood flow is more important. To estimate the penetration depth, assume that $T-T_0$ changes over this distance. Then the Laplacian is approximated by $\nabla^2 T\approx(T-T_0)/\delta_{\text{th}}^2$. Equating the diffusive and blood flow terms gives

$$D\,\frac{T-T_0}{\delta_{\text{th}}^2}=\frac{C_b}{C_t}\,(\rho_b P)(T-T_0),$$

so

$$\delta_{\text{th}}^2=D\,\frac{C_t}{C_b}\,\frac{1}{\rho_b P}=\frac{K}{\rho_t C_b\rho_b P}. \tag{13.48}$$

Grossweiner (1994) discusses values for the various tissue parameters, their temperature dependence, and simple models for tissue heating and ablation.

13.10. RADIOMETRY AND PHOTOMETRY

Section 12.3 introduced some of the concepts of radiometry, the measurement of radiant energy. Concepts for the measurement of radiant energy were developed simultaneously in different disciplines and even in different wavelength regions, depending on the purpose and the measurement techniques that were originally available. It is recommended that the term *photometry* be reserved for measurement of the ability of electromagnetic radiation to produce a human visual sensation, that *radiometry* be used to describe the measurement of radiant energy independent of its effect on a particular detector, and that *actinometry* be used to denote the measurement of photon flux or photon dose (total number of photons) independent of any subsequent photoactivated process [Zalewski (1995, p. 24.7)]. This section reviews radiometric units and introduces a few of the related units from photometry and actinometry. Nomenclature is slightly different for x rays and charged particles.

Section 12.3 introduced the radiant energy, the radiant power, the radiance and the irradiance. Section 13.7 described two more quantities, the photon fluence and the photon fluence rate. The energy fluence and energy fluence rate were introduced in Sec. 13.9. These are reviewed and compared here so that all the definitions are in one place. The definitions are summarized in Table 13.6. Symbols are shown for quantities used in this text. The third column shows symbols that have recently been recommended by the American Association of Physicists in Medicine [AAPM 57 (1996)]. They often differ from the usage in this book.

The total amount of energy being considered is the *radiant energy* R, measured in joules. It can be the energy emitted by a source, transferred from one region to another, or received by a detector.[17] The rate at which the energy is radiated, transferred, or received is the *radiant power* P (watts).

The radiant energy leaving a source can travel in many different directions. The radiation striking a surface can come from many different directions. If we consider any small area in space there will generally be radiation passing through that area traveling in many different directions. In each case the radiant energy or the radiant power is proportional to the magnitude of the small area projected perpendicular to the direction the energy is traveling, and to the size of the solid angle being considered. The *radiance* L is the amount of radiant power per unit solid angle per unit surface area projected perpendicular to the direction of the radiant energy. The radiance of radiation traveling through a small area in space is sometimes difficult to visualize. Figures 13.28 and 13.29 may help. Figure 13.28 shows radiation leaving three points on a surface at the left. Some of it passes through the surface represented by the thicker vertical line. The energy passing through that surface has components from each point on the radiating surface. Figure 13.29 shows radiation in a very narrow cone of solid angles passing through surface dS whose normal is at an angle θ with the beam direction. The radiance is the power per unit solid angle divided by $dS \cos \theta$.

[17]In optics the radiant energy is electromagnetic radiation. In radiological physics we will have radiant energy transported by x rays and also charged and neutral particles such as electrons, protons, and neutrons.

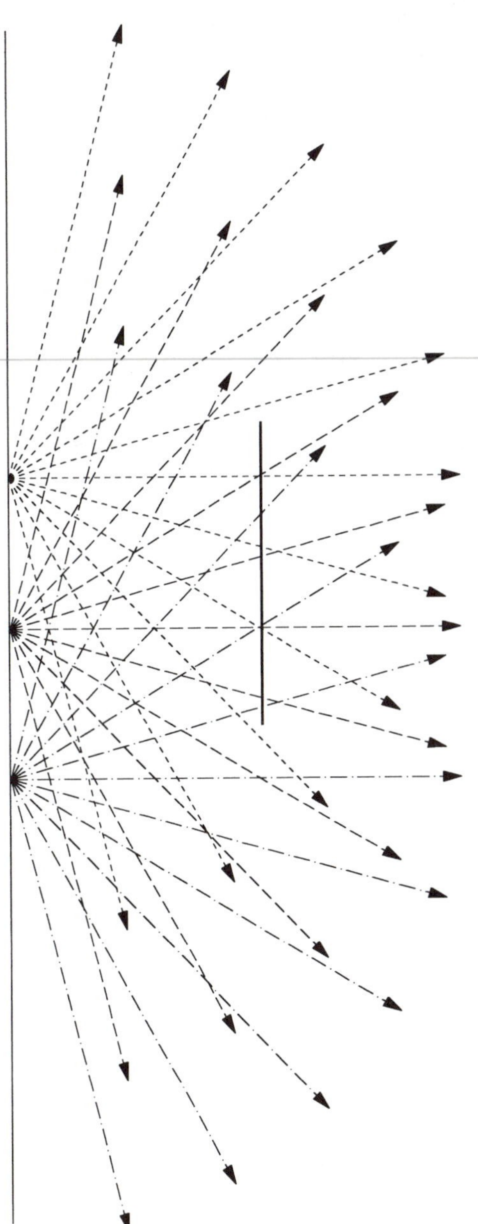

FIGURE 13.28. Radiation emitted from different points of the surface on the left strikes the surface on the right.

We have already seen the *energy fluence* Ψ, which is a measure of the total radiation entering or leaving a small volume of space. It is the total amount of energy striking a small sphere of radius a divided by the area of a great circle πa^2 in the limit as the radius approaches zero. Strictly speaking, if we repeat the experiment many times, the amount of energy striking the sphere fluctuates. The energy fluence is defined in terms of the expectation value of this fluctuating quantity. Figure 13.30 shows two examples. In Fig. 13.30(a), a parallel beam with energy R passes through a circular area πa^2 for a time Δt. In Fig. 13.30(b), a total amount of energy R strikes a sphere of radius a from many different directions. In both cases $\Psi = R/\pi a^2$. Notice that

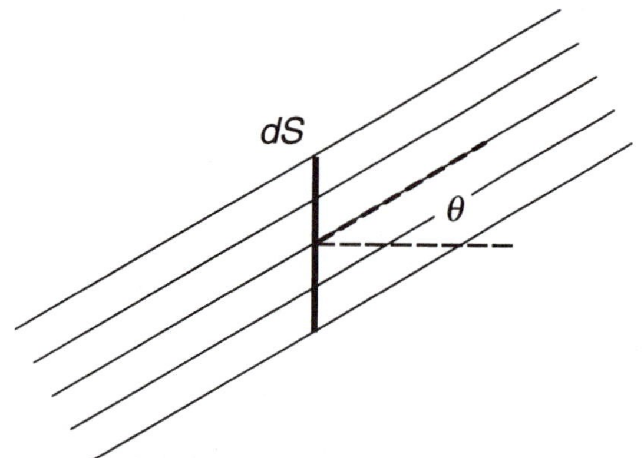

FIGURE 13.29. Surface area dS, projected perpendicular to the direction of the radiation, has projected area $dS \cos\theta$.

some of the energy passing through the sphere passes outside a great circle that is not perpendicular to the direction in which the radiation is traveling, but it does pass through a great circle constructed perpendicular to its direction of travel. The energy fluence rate is the amount of energy fluence per unit time (which for the small sphere is $P/\pi a^2$):

$$\psi = \frac{d\Psi}{dt}. \tag{13.49}$$

The *radiant intensity* is defined only for a point source. It is the radiant power or radiant flux emitted by a point source in a given direction per unit solid angle. Because intensity is often used in physics for power per unit area, we will avoid using the term.

The *exitance* W_r is the radiant power or flux emitted per unit area of a surface. The *irradiance* E is the power per unit area incident on a surface.

We can derive some useful relationships for a beam of collimated radiation all traveling in one direction (a *plane wave*). Imagine that the collimated beam comes from a point source radiating power P. The energy fluence rate at distance r from the source is the power through a sphere of radius a divided by πa^2:

$$\psi = \frac{\pi a^2 P}{4\pi r^2}\frac{1}{\pi a^2} = \frac{P}{4\pi r^2}.$$

This is also the power per unit area incident on a circle of radius a oriented perpendicular to the beam. Therefore, for a collimated beam,

$$\psi = E \quad \text{(collimated beam)}. \tag{13.50}$$

For isotropic (Lambertian) radiation, $L = L_0$. The power incident on a small element of surface area dS_d from angle $d\Omega$ is $L_0 dS_d \cos\theta d\Omega$, where θ is the angle that the incident radiation makes with the normal to the surface. The solid angle is $2\pi \sin\theta d\theta$ [see Fig. 13.11(b)]. The irradiance is

$$E = \frac{dS_d 2\pi L_0 \int_0^{\pi/2} \cos\theta \sin\theta \, d\theta}{dS_d} = \pi L_0. \tag{13.51}$$

The same geometry is used with dS_s to show that for isotropic radiation the exitance is

$$W_r = \pi L_0. \tag{13.52}$$

To determine the energy fluence rate for isotropic radiation consider a small sphere of radius a and the radiation arriving in a small solid angle $d\Omega$ about a line perpendicular to a great circle of the sphere. The power is $L_0 \pi a^2 d\Omega$. This argument applies for any direction of the radiation. Integrating over all directions gives the total power $L_0 \pi a^2 4\pi$. Therefore, for isotropic (Lambertian) radiation,

$$\psi = 4\pi L_0 = 4E \quad \text{(isotropic radiation)}. \tag{13.53}$$

When the energy is not monochromatic, we define the amount of energy per unit wavelength interval as R_λ, with units J m^{-1} or J nm^{-1}. The total energy between wavelengths λ_1 and λ_2 is

$$\int_{\lambda_1}^{\lambda_2} R_\lambda(\lambda) \, d\lambda. \tag{13.54}$$

For the photometric units we also need to know the sensitivity of the eye. The eye contains two types of light receptors: *rods*, which have no color discrimination but are most sensitive, and *cones*, which are less sensitive and can discriminate color. *Photopic* vision is normal vision at high levels of illumination in which the eye can distinguish colors. *Scotopic* vision occurs at low light levels with a dark-adapted eye. The CIE has established the spectral efficiency function V for the eye of a standard observer for both photopic vision [$V(\lambda)$] and scotopic vision [$V'(\lambda)$]. Both are normalized to unity at their peak (Fig. 13.31).

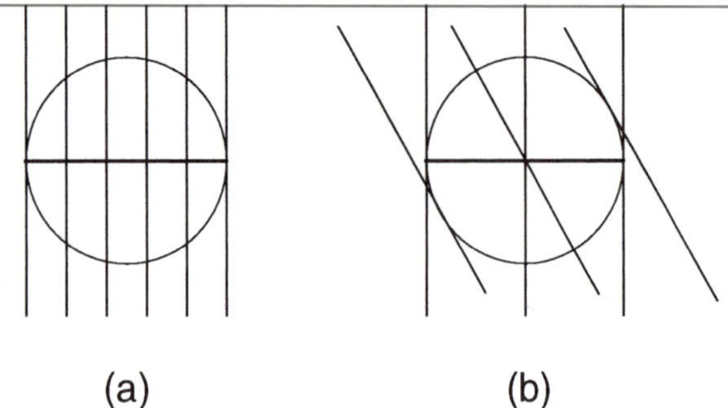

FIGURE 13.30. The energy fluence is the amount of radiant energy striking the sphere of radius a divided by the area of a great circle, πa^2, in the limit as the radius approaches zero. (a) Radiation traveling in one direction. (b) Radiation traveling in two directions.

TABLE 13.6. *A comparison of radiometric, photometric, and actinometric quantities. Symbols are given for those quantities used in this text. The column "Symbol somtimes used" gives an alternate symbol that is often found. See, for example, AAPM 57 (1996).*

Radiometric quantity	Symbol used here	Symbol sometimes used	Units	Photometric quantity	Symbol	Units	Actinometric quantity	Symbol	Units
General quantities									
Radiant energy emitted, transferred, or received.	R	Q	J	*Luminous energy.*	R_v	lm s	*Number of photons* emitted, transferred, or received.	N	
Radiant flux or radiant power emitted, transferred, or received.	P	P or Φ or $\dot{R}$	W	*Luminous flux.*	P_v	lm	*Photon flux.*		s^{-1}
Radiance: the radiant power per unit solid angle per unit area of surface projected perpendicular to the radiant energy. It can be defined on the surface of a source or detector or at any point on the path of a ray of radiation.	L	r	W m^{-2} sr^{-1}	*Luminance.*	L_v	candela m^{-2} (cd m^{-2})	*Photon flux radiance.*		m^{-2} sr^{-1}
Energy fluence: the ratio of the expectation value of the radiant energy striking a small sphere to the area of a great circle of the sphere.	Ψ	H_0	J m^{-2}				*Photon (or particle) fluence:* the ratio of the expectation value of the number of photons striking a small sphere to the area of a great circle of the sphere.	Φ	m^{-2}
Energy fluence rate: the energy fluence per unit time.	ψ	E_0	W m^{-2}				*Photon fluence rate*: the photon fluence per unit time	φ	m^{-2} s^{-1}
Quantities emitted from a surface									
Radiant intensity: radiant power or flux emitted by a *point source* in a given direction per unit solid angle.		I	W sr^{-1}	*Luminous intensity.*		lm sr^{-1} or candela (cd)	*Photon flux intensity.*		sr^{-1}
Exitance: radiant power or flux emitted or reflected per unit area.		W_r	W m^{-2}	*Luminous exitance.*		lm m^{-2}	*Photon exitance.*		m^{-2}
Quantities incident on a surface									
Irradiance: the power per unit area incident on a surface.		E	W m^{-2}	*Illuminance.*		lm m^{-2} or lux	*Photon flux irradiance.*		m^{-2} s^{-1}
Radiant exposure: radiant energy arriving per unit area.		H	J m^{-2}	*Luminous exposure.*		lm s m^{-2}	*Photon flux exposure.*		m^{-2}

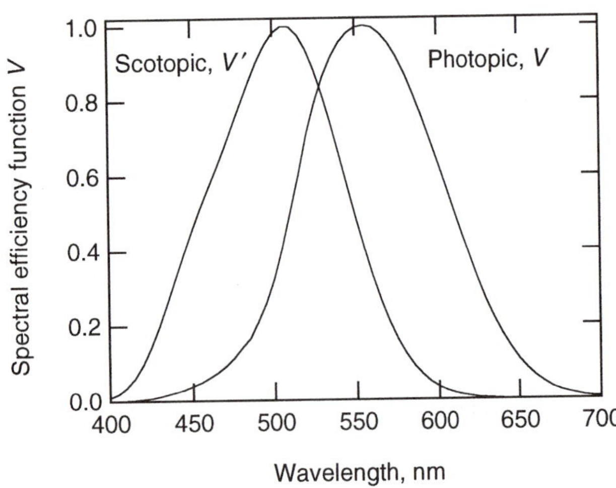

FIGURE 13.31. The spectral efficiency functions for the CIE standard eye. Plotted from data in Table 2 of Zalewski (1995).

The *luminous flux* P_v in lumens (lm) is the analog of the energy flux P. The peak sensitivity for photopic vision is for green light, $\lambda = 555$ nm. At that wavelength the relationship between P and P_v is

$$P = 1 \text{ W} \Leftrightarrow P_v = 683 \text{ lm}, \tag{13.55a}$$

$$P_v = 1 \text{ lm} \Leftrightarrow P = 1.464 \times 10^{-3} \text{ W}.$$

The ratio P_v/P at 555 nm is the *luminous efficacy* for photopic vision, $K_m = 683$ lm W^{-1}. For a distribution of wavelengths,

$$P_v = K_m \int_{400 \text{ nm}}^{700 \text{ nm}} V(\lambda) P_\lambda(\lambda)\, d\lambda. \tag{13.55b}$$

An analogous relationship holds for scotopic vision, with $K_m' \cong 1700$ lm W^{-1}:

$$P_v(\text{scotopic}) = K_m' \int_{400 \text{ nm}}^{700 \text{ nm}} V'(\lambda) P_\lambda(\lambda)\, d\lambda. \tag{13.55c}$$

If P were spread uniformly over the visible spectrum, the overall conversion efficiency would be about 200 lm W^{-1}. A typical incandescent lamp has an efficiency of 10–20 lm W^{-1}, while a fluorescent lamp has an efficiency of 60–80 lm W^{-1}. The number of lumens per steradian is the *luminous intensity*, in lm sr^{-1}. The lumen per steradian is also called the *candle*. Other units are shown in Table 13.6.

The actinometric quantities count the number of photons. For monochromatic photons the energy is the number of photons times $h\nu$. Therefore an actinometric quantity is easily obtained when the radiometric quantity is known. The units are shown in Table 13.6.

13.11. THE EYE

This section presents a simple model for the eye, sufficient for us to understand how refractive errors are corrected and to see how photons strike the retina, so that the sensitivity of the eye can be determined in the next section.

A simplified cross section of the eye is shown in Fig. 13.32. The principal components through which the light passes are the curved, thin, transparent *cornea*, the *aqueous*, the *lens*, the *vitreous*, and the *retina*. The *iris* defines the area of the *pupil*, the opening in front of the lens through which light passes.

When light passes through a surface from one medium into another, part is reflected and part is transmitted. The transmitted light usually changes direction, a process called *refraction*. Figure 13.33 shows the angles involved, all measured with respect to the dashed line, which is normal to the surface at the point where the light ray strikes. The angle the reflected light makes with the normal is the same as the angle of incidence, $\theta_r = \theta_1$. The refracted light is described by Snell's law, $n_1 \sin\theta_1 = n_2 \sin\theta_2$.

When light from an object strikes the eye, it must be bent or refracted to form an image on the retina. Most of the refraction takes place at the surface between the air and the cornea. The cornea is very thin, and a light ray is deflected only a very small distance before it strikes the aqueous. Thus, most of the refraction occurs because of the difference between the index of refraction of the air ($n = 1.00$) and the aqueous ($n = 1.33$). The light then passes through the crystalline lens ($n = 1.42$) and the vitreous ($n = 1.33$). The lens changes shape to provide the adjustable part of the overall refraction.

A number of models at varying levels of sophistication are used to describe the formation of the image on the retina. The most detailed take into account the refraction at each surface where the index of refraction changes,

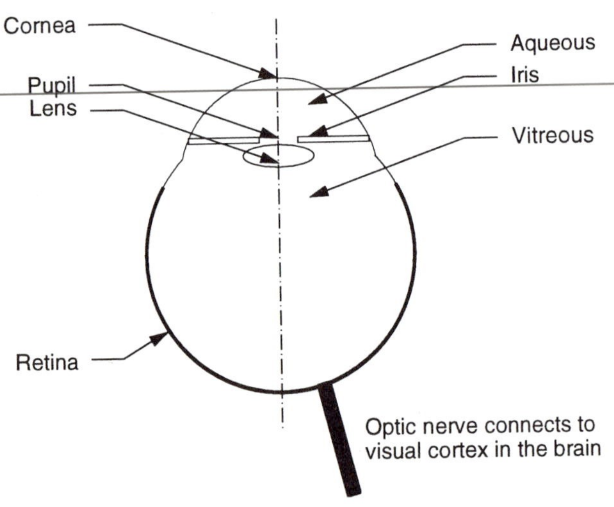

FIGURE 13.32. A simplified cross section of the left eye, viewed from above.

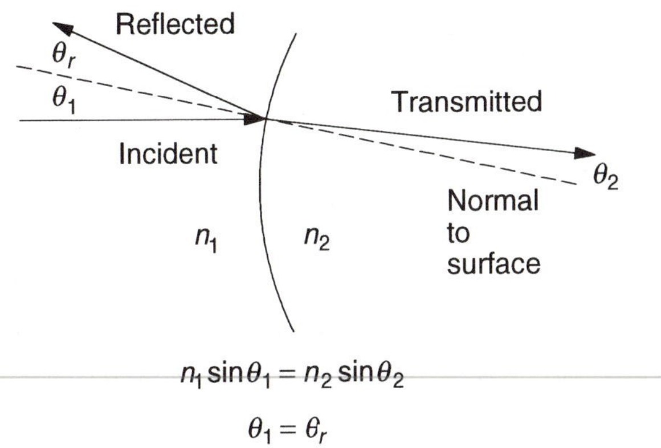

FIGURE 13.33. When light passes from one medium to another with a different index of refraction, it is bent so that $n_1 \sin \theta_1 = n_2 \sin \theta_2$. The angle of incidence is equal to the angle of reflection. All angles are measured with respect to the normal to the surface.

including variations in different layers of the lens itself. Others treat only the refraction at the air–cornea, aqueous–lens, and lens–vitreous interfaces. The simplest model, and the one we will use, treats the eye as a thin lens of adjustable focal length f, with object distance u and fixed image distance v, as shown in Fig. 13.34. The object and image distances and focal length are related by the *thin-lens equation* found in any general physics book:

$$\frac{1}{u}+\frac{1}{v}=\frac{1}{f}. \tag{13.56}$$

When the object is infinitely far away the image distance is equal to the focal length of the lens, $v=f$. As the object is brought closer to the eye the image distance cannot change, but the lens changes to increase the focal length. A typical value for v is 1.7 cm.

In ophthalmology and optometry it is customary to describe the refraction of the eye in terms of the *vergence*. When light rays are emanating from a point they are diverging, and the vergence is negative. When they are coming toward a point the vergence is positive and they are

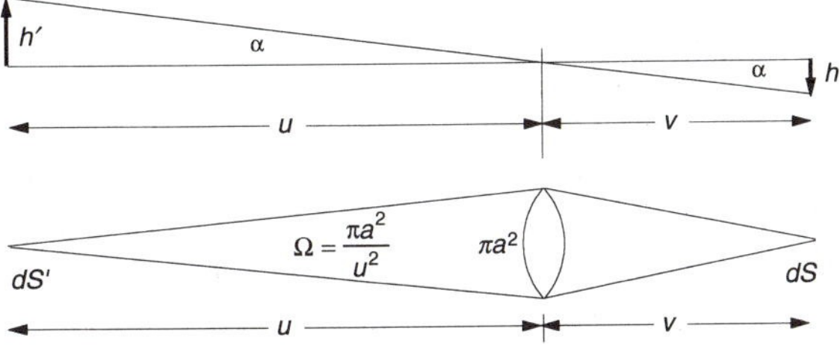

FIGURE 13.34. A source of height h' emits light in all directions. Some of this light is intercepted by a lens and focused in an image. (a) Relation between object and image distances and sizes. (b) Collection of light by the lens.

converging. When they are parallel, the vergence is zero. Quantitatively, the vergences for the geometry shown in Fig. 13.34 are

$$U=-\frac{1}{u} \quad \text{(diverging from the object)},$$

$$V=\frac{1}{v} \quad \text{(converging to the image)}, \tag{13.57}$$

$$F=\frac{1}{f} \quad \text{(a converging lens)}.$$

The relationship between the vergences is

$$V=F+U. \tag{13.58}$$

When the distances are in meters, the vergences are in *diopters*. A given eye requires a particular value of V to form the image. The converging power of all the refracting surfaces in the eye must be $F=V$ in order to focus on an object infinitely far away. Closer objects require more convergence from the eye, which is provided by the lens. Table 13.7 shows typical values for the converging power of the eye. Most of the convergence is provided by the front surface. Using the values in Table 13.7, we see that when the eye is relaxed, $F=V=59$ diopters, $U=0$, and the eye is focused on an object infinitely far away. With $F=69$ diopters, $U=-10$, and the eye is focused on an object 0.1 m away. This ability of the lens to change shape and provide additional converging power is called *accommodation.*

In the normal, or *emmetropic* eye, the length of the eye is such that when the lens is relaxed, rays with no vergence (parallel rays from a source infinitely far away) are focused on the retina ($V=F$). In nearsightedness or *myopia,* parallel rays come to a focus in front of the retina. The eye is slightly too long for the shape of the cornea. The total converging power of the eye is too great, and the relaxed eye focuses at some closer distance, from which the rays are diverging. Myopia can be corrected by placing a diverging spectacle or contact lens in front of the eye, so that incoming parallel rays are diverging when they strike the cornea. The farsighted or *hypermetropic* eye does not have enough converging power. The subject can focus on distant objects by providing some additional converging power

TABLE 13.7. *Convergence power of the eye.*

Refracting structure	Relaxed normal eye (diopters)	Most converging eye (age 25) (diopters)
Air–cornea surface	45	45
Lens	14	24
Entire eye	59	69

from the lens, but then the lens cannot provide enough converging power to focus on nearby objects. The corrective lens provides additional convergence.

When the eye is not symmetric about an axis through the center of the lens, the images from objects oriented at different angles in the plane perpendicular to the axis form at different distances from the lens. This is called *astigmatism*, and it can be corrected with a spectacle lens that is not symmetric about the axis. The lack of symmetry usually occurs at the surface of the cornea, so a contact lens can restore the symmetry.

Surgery to change the radius of curvature of the cornea can also be used to correct errors of refraction.

As we age the accommodation of the eye decreases, as shown in Fig. 13.35. A normal viewing distance of 25 cm or less requires 4 diopters or more of accommodation. The graph shows that this limit is usually reached in the early 40s. Bifocals provide a portion of the spectacle lens that has increased converging power, usually in the bottom part of the lens. This can be done either by grinding the lower portion of the lens with a different radius of curvature or by fusing glass with a different index of refraction into the lens.

The sharpness of the image is reduced by two other effects: *chromatic aberration* and *spherical aberration*. Chromatic aberration occurs because the index of refraction varies with wavelength. There is nearly a 2-diopter change in overall refractive power from the red to the blue. Spherical aberration occurs because the refractive power changes with distance from the axis of the eye. This is different from astigmatism, which is a departure from symmetry at different angles about the axis.

A concept important in both vision and photography is *depth of field*. It can be understood by referring to Fig. 13.36. The retina is behind the plane in which the image is in focus. In dim light, the pupil of the eye is fully open and

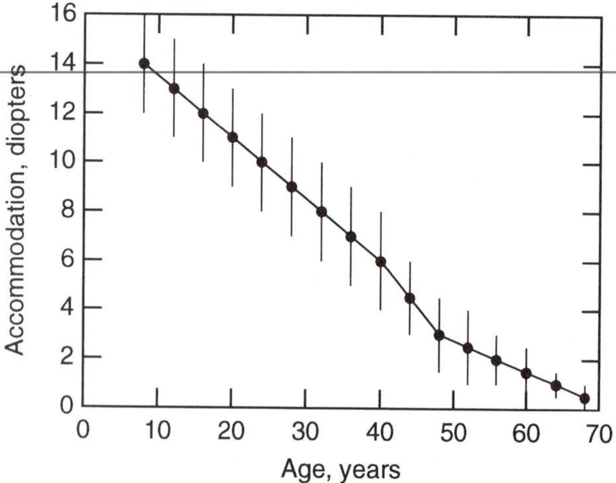

FIGURE 13.35. Accommodation vs age. There is considerable variation between individuals, shown by the error bars.

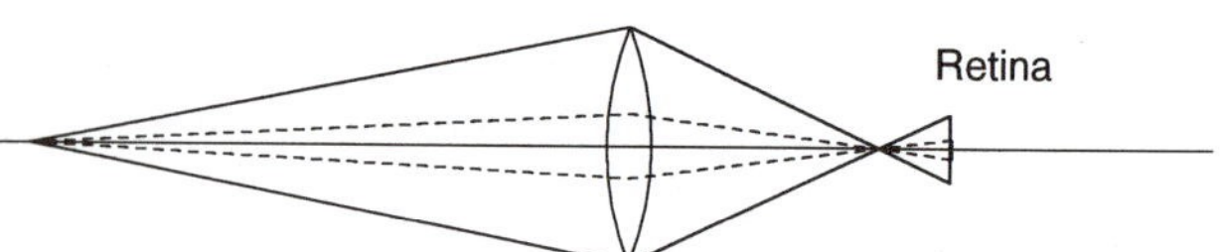

FIGURE 13.36. Depth of field is illustrated by this ray diagram. The retina is slightly behind the plane of focus. In dim light, the pupil of the eye is fully open and light from a point object is spread out over the larger circle on the retina. When the light is brighter and the pupil is smaller, light from the same point object is confined to the smaller circle defined by the dashed lines.

light from a point object is spread out over the larger circle on the retina defined by the solid rays. In brighter light the pupil is smaller, and light from the same point object is confined to the smaller circle defined by the dashed lines. This is why we can see better in brighter light. An older person whose accommodation is less and who is trying to avoid bifocals often finds that bright light makes it easier to see nearby objects.

Point-spread functions and modulation transfer functions can be used to describe the image. [See, for example, Charman (1995) or Grievenkamp *et al.* (1995).] A simpler model describes the image by a Gaussian with a certain standard deviation, equal to the square root of the sum of the variances due to various effects. The maximum photopic (bright-light) resolution of the eye is limited by four effects: diffraction of the light passing through the circular aperture of the pupil (5–8 μm), spacing of the receptors ($\approx$3 μm), chromatic and spherical aberrations (10–20 μm) and noise in eyeball aim (a few micrometers) [Stark and Theodoris (1973)].[18] The total standard deviation is $(6^2+3^2+15^2+5^2)^{1/2}=17$ μm in the image on the retina. Since the diameter of the eyeball is about 2 cm, this corresponds to an angular size (α in Fig. 13.34) of $(17\times10^{-6})/(2\times10^{-2})=8.5\times10^{-4}$ rad$=0.048°=2.9$ minutes of arc. [For further discussion, see Cornsweet (1970, Chap. 3).]

13.12. QUANTUM EFFECTS IN DARK-ADAPTED VISION

The visual process involves two steps. First, the eye creates an image of an external object on the retina as described above. Then the photon stimulus is transduced into neurological signals that are interpreted by the central nervous system. The discussion here is limited to a classic experiment on scotopic vision that show the importance of quantum effects (shot noise) in human vision in dim light.

The experiment was performed by Hecht, Schlaer, and Pirenne in 1942. The experiment has been described in

[18]We could express all of these effects in terms of the modulation transfer functions for each process.

many places. A detailed nonmathematical description is that by Cornsweet (1970). A more mathematical review by Pirenne (1962) is also available.

The retina can be divided into two regions. The *fovea*, the area of greatest visual discrimination, is composed entirely of cones. The percentage of rods is highest a few millimeters away from the fovea, and this part of the retina is most sensitive to faint light. The experiment was done by having the subject look directly at a very dim red fixation point while a green light was flashed in such a place that its image fell on the most sensitive part of the retina. The dark-adapted eye increases sensitivity by a factor of about 5000.

Experiments on the sensitivity of the dark-adapted eye to flashes of weak light have shown that if the flash duration is less than 100 ms and the light on the retina covers a "receptor field" less than 10 minutes of arc in size, the scotopic response of the eye depends on the total amount of energy or the total number of photons in the flash. Photons striking anywhere within the field during this time have the same effect; the eye must combine the effects occurring in all receptors in the receptor field in a tenth of a second. A scotopic receptor field is shown in Fig. 13.37. [This scotopic field size (10 minutes of arc) cannot be compared to the 2.9 minutes for maximum resolution, which is for photopic vision on a different part of the retina.]

In the Hecht–Schlaer–Pirenne experiment the flashes were short enough and small enough so that only the total number of photons was important. The fraction of flashes that the subject recognized was measured as a function of the total flash energy. A typical response curve is shown in Fig. 13.38. Let q be the number of photons striking the cornea in front of the pupil in each flash, which is the total energy in the flash divided by the energy of each photon. For the 510-nm green light used, the photon energy is $hc/\lambda = 3.89\times10^{-19}$ J. The number of photons striking the cornea can be determined as follows. Let Lt be the radiance times the duration of the flash. From Eq. (12.24) the number

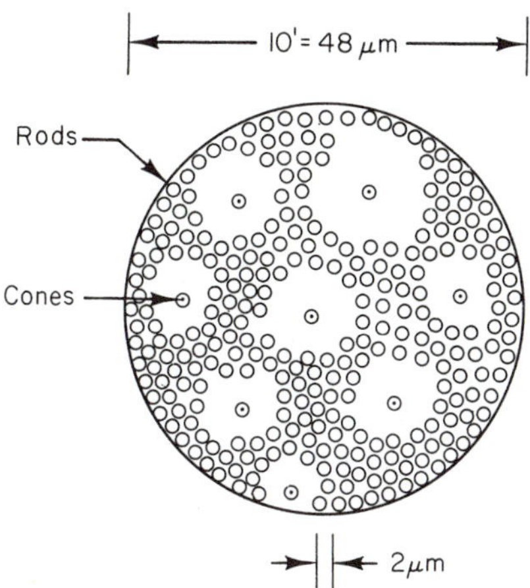

FIGURE 13.37. An example of a 10-minute-of-arc field superimposed on the rods and cones in the retina in the region of greatest sensitivity.

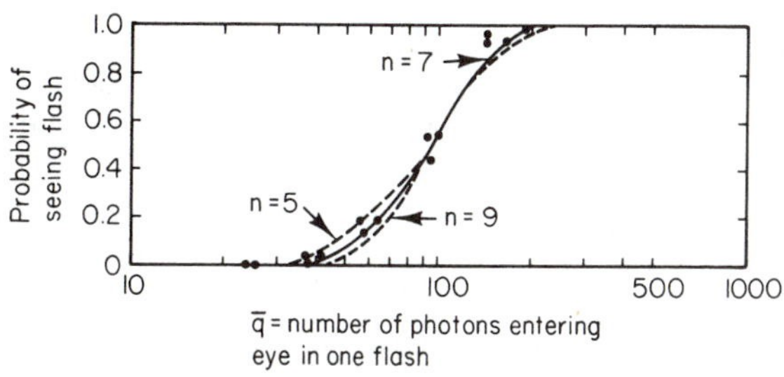

FIGURE 13.38. Typical response in the experiments of Hecht, Schlaer, and Pirenne. Curves are calculated using Eq. (13.61). Data are from S. Hecht, S. Schlaer, and M. H. Pirenne (1942). Used with permission of the Journal of General Physiology.

of photons striking the cornea that would be in the image if there were no losses is [see Eq. (12.24)]

$$q = \frac{(Lt)(\pi a^2)\, dS'}{h\nu u^2} = \frac{(Lt)(\pi a^2)\, dS}{h\nu v^2}. \qquad (13.59)$$

The number of photons fluctuates from flash to flash. Therefore we should speak of $\bar{q}$, the average number of photons striking the cornea per flash. Of these, only some fraction f actually reach the retina and are absorbed by a visual pigment molecule. The average number absorbed is

$$m = f\bar{q}. \qquad (13.60)$$

Let us next postulate that some minimum number of quanta n must be absorbed during the flash in order for the subject to see it. If the average number absorbed per flash is m, there will sometimes be more and sometimes less than n photons absorbed per flash. The probability of absorbing x photons per flash is given by the Poisson distribution $P(x;m)$ (Appendix J). The probability of seeing the flash[19] is the probability that x is greater than or equal to n:

$$\begin{aligned} P(\text{seeing}) &= \sum_{x=n}^{\infty} P(x;m) \\ &= 1 - P(0;m) - P(1;m) - \cdots - P(n-1;m) \\ &= 1 - e^{-m}\left(1 + m + \frac{m^2}{2!} + \cdots + \frac{m^{n-1}}{(n-1)!}\right). \end{aligned} \qquad (13.61)$$

This function is plotted in Fig. 13.39 as a function of m for various values of n, with both a linear and a logarithmic scale for m.

Hecht, Schlaer, and Pirenne used an ingenious method to determine n. They plotted their data vs. the logarithm of $\bar{q}$. Since $m = f\bar{q}$, $\log m = \log f + \log \bar{q}$; different values of f correspond to shifting the curve along the axis. They then compared the experimental data to the various theoretical curves for the probability of seeing a flash, plotted against

[19] See also the discussion of target theory in Chap. 15.

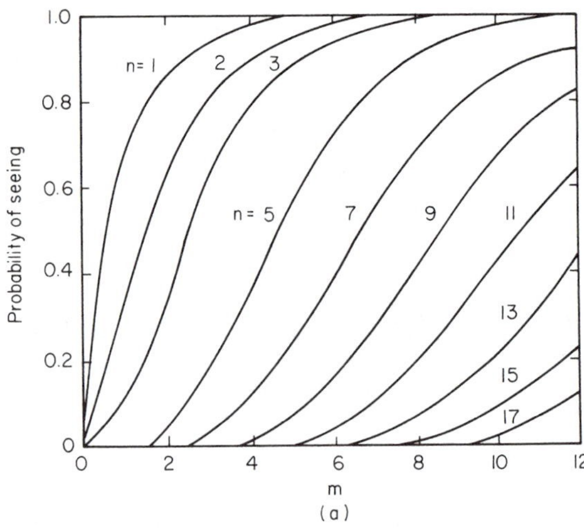

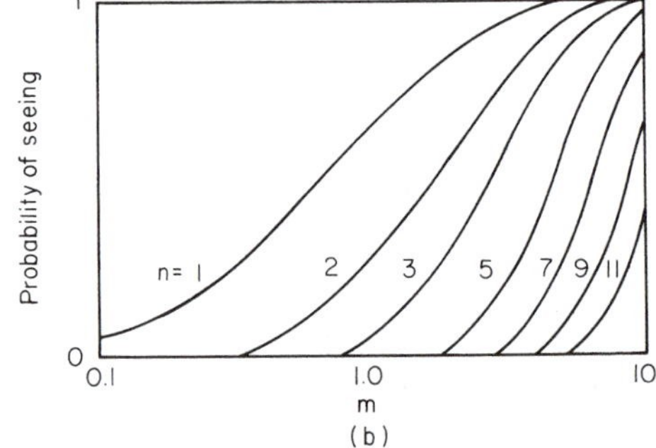

FIGURE 13.39. The probability of seeing a flash, plotted vs (a) *m*; (b) log *m*.

log *m*. Sliding the paper containing the data along the log *m* axis is equivalent to trying different values of f. The data in Fig. 13.38 are shown along with the curves for $n=5$, 7, and 9. For these data, $n=7$ gives the best fit. From Fig. 13.38, a 55% chance of detecting the flash corresponds to 100 photons for $\bar{q}$ while being consistent with $m=7$. Therefore, $f=0.07$.

Hecht, Schlaer, and Pirenne deduced that about seven photons must be absorbed by the rods in the area of integration shown in Fig. 13.37 within 0.1 s in order for the brain to detect the flash of light. Their data were consistent with the hypothesis that the photons arrived at random, with the actual number in each flash obeying a Poisson distribution. This is a very simple model. The visual system is quite complicated, and there are sources of noise at the retina and in the processing centers in the brain that integrate the signals from the various rods within a visual field. The subjects can also adjust their response, responding to fewer photons but having a greater rate of false positives—saying there is a response when there really is not. Models that account for these effects still have about 100 photons at the cornea, but a higher value of f (0.2) and a larger value of n (26) because of the additional noise sources [Teich *et al.* (1982)].

If the light intensity is increased, m increases. There will be shot-noise fluctuations with a standard deviation equal to $m^{1/2}$, and the eye should be unable to detect brightness changes smaller than this. Measurements by H. B. Barlow in 1957 showed that as long as short flashes spanning only one visual field are used, the minimum detectable intensity depends on the square root of the light intensity. This statistical limit to detecting intensity changes is a lower limit; for larger sources and longer exposure times the minimum detectable brightness change is larger and is more nearly proportional to the intensity than to the square root of the intensity [Rose (1973)].

The phototransduction mechanism is quite complicated. See the reviews by Baylor (1987) and by Yau (1994). Photoreceptors have a dark current that is reduced when light falls on them. In other words, the light hyperpolarizes the cell. This lowers the rate of release of a neurotransmitter, cyclic GMP. Single photon response has been detected.

SYMBOLS USED IN CHAPTER 13

Symbol	Use	Units	First used on page
a	Radius	m	369
c	Speed of light in a vacuum	m s^{-1}	346
c'	Speed of light in a medium	m s^{-1}	347
e	Charge on electron	C	348
f	Fraction of photons reaching retina		375
f	Focal length	m	373
g	Scattering anisotropy factor		354
h	Planck's constant	J s	347
$\hbar$	Planck's constant divided by 2π	J s	347
j	Total angular momentum quantum number		348
j	Energy transport in heat flow	W m^{-2}	367
k	Spring constant	N m^{-1}	351
k_B	Boltzmann constant	J K^{-1}	355
l	Orbital angular momentum quantum number		348
m	Mass	kg	348
m	Average number		375
m_e	Mass of electron	kg	348
m_i	Mass of ith particle	kg	349

Symbol	Use	Units	First used on page
m_j, m_l, m_s	z quantum number for angular momentum		348
n	Index of refraction		346
n	Principal quantum number		348
$\bar{n}$	Average number of photons that interact		353
n	Minimum number of photons to trigger a response		375
p	Probability		353
q	Electric charge	C	347
q	Number of photons entering the eye in a flash		375
$\bar{q}$	Average value of q		375
r	Rotational quantum number		350
$r, \mathbf{r}$	Cordinate		349
s	Spin quantum number		348
s	Source term in diffusion equation	$m^{-3}\ s^{-1}$	359
t	Time		347
$\mathbf{v}$	Velocity	ms^{-1}	347
v	Vibrational quantum number		351
u, v	Object and image distances	m	373
w_{tot}	Net power radiated	W	357
x, z	Distance	m	347
z_0	Depth of first scattering	m	360
A	Molar mass	kg	353
$\mathbf{B}$	Magnetic field	T	347
C	Concentration	m^{-3}	358
C_b, C_t	Heat capacity of blood, tissue	$J\ kg^{-1}\ K^{-1}$	367
D	Diffusion constant	$m^2\ s^{-1}$	358
D'	Photon diffusion constant	m	359
D	Thermal diffusion constant	$m^2\ s^{-1}$	358
$\mathbf{E}$	Electric field	$V\ m^{-1}$	347
E	Total energy	J	348
E_r	Rotational energy	J	350
E_v	Vibrational energy	J	351
E	Irradiance	$W\ m^{-2}$	361

Symbol	Use	Units	First used on page
$F, \mathbf{F}$	Force	N	347
F	Converging power of a lens	diopter (m^{-1})	373
I	Moment of inertia	$kg\ m^2$	350
K	Thermal conductivity	$W\ K^{-1}\ m^{-1}$	367
K_m	Luminous efficacy, photopic	$lm\ W^{-1}$	372
K'_m	Luminous efficacy, scotopic	$m\ W^{-1}$	372
L	Angular momentum	$kg\ m^2\ s^{-1}$	350
L	Radiance	$W\ m^{-2}\ sr^{-1}$	361
N	Number of photons		352
N_a	Number absorbed		352
N_s	Number scattered		352
N_A	Avogadro's number		353
N_T	Number of target entities per unit area along beam	m^{-2}	353
P	Probability		375
P	Tissue perfusion	$m^3\ kg^{-1}\ s^{-1}$	368
P	Radiant power	W	369
P_v	Luminous flux	lm	372
Q	Rate of production	$m^{-3}\ s^{-1}$	358
$R, \mathbf{R}$	Coordinate of atom distance	m	349
R	Reflected fluence rate	$m^{-2}\ s^{-1}$	360
R	Radiant energy	J	361
R_λ	Radiant energy per unit wavelength interval	$J\ m^{-1}$ or $J\ nm^{-1}$	371
S, S'	Surface area	m^2	353
T	Kinetic energy	J	350
T	Temperature	K	355
U	Object vergence	diopter (m^{-1})	373
$V, \mathbf{V}$	Velocity	$m\ s^{-1}$	350
V	Photopic spectral efficiency function		371
V'	Scotopic spectral efficiency function		371
V	Image vergence	diopter (m^{-1})	373
W_λ	Blackbody radiation function	$W\ m^{-3}$ or $W\ m^{-2}\ nm^{-1}$	355
W_ν	Blackbody radiation function	$W\ m^{-2}\ s^{-1}$	356
W_r	Exitance	$W\ m^{-2}$	369
α	Angle		373
α	Fraction of radiant energy reflected		362

Symbol	Use	Units	First used on page
β	Fraction of surface area of sphere that reflects		362
δ_{th}	Thermal penetration depth	m	368
ϵ_0	Electrical permittivity of empty space	$N^{-1}\,C^2\,m^{-2}$	348
$\epsilon(\lambda)$	Reference action spectrum		357
θ, ϕ	Angles		354
φ	Particle fluence rate	$m^{-2}\,s^{-1}$	359
λ	Wavelength	m	347
μ	Total linear attenuation coefficient	m^{-1}	352
μ_a	Linear absorption coefficient	m^{-1}	352
μ_s	Linear scattering coefficient	m^{-1}	352
μ_s'	Reduced linear scattering coefficient	m^{-1}	354
μ_{eff}	Effective linear attenuation coefficient	m^{-1}	360
ρ, ρ_b, ρ_t	Density, density of blood, density of tissue	$kg\,m^{-3}$	353
$\sigma, \sigma_i, \sigma_a, \sigma_s, \sigma_{tot}$	Cross section	m^2	353
$\sigma(\theta)$, $d\sigma/d\Omega$	Differential scattering cross section	$m^2\,sr^{-1}$	354
σ	Stefan–Boltzmann constant	$W\,m^{-2}\,K^{-4}$	357
$\sigma_r^2, \sigma_x^2, \sigma_y^2, \sigma_z^2$	Variance for diffusion or heat flow	m^2	368
ν	Frequency	s^{-1}	347
ω	Angular frequency	radian s^{-1}	347
ψ	Energy fluence rate	$W\,m^{-2}$	368
Ψ	Energy fluence	$J\,m^{-2}$	368
Φ	Particle fluence	m^{-2}	359

359

PROBLEMS

Section 13.1

13.1. An einstein is 1 mol of photons. Derive an expression for the energy in an einstein as a function of wavelength λ. Express the answer in kilocalories and the wavelength in nanometers.

Section 13.2

13.2. Use Eq. (13.7) to derive Eq. (13.8).

Section 13.3

13.3. Estimate $\hbar^2/2I$ for an HCl molecule. What would the spacing of rotational levels be?

13.4. An inulin molecule has a molecular weight of 4000 dalton (that is, 1 mol has a mass of 4000 g). Assume that it is spherical with a radius of 1.2 nm. What is the angular frequency ω of a photon absorbed when its rotational quantum number changes from 10 to 11? The moment of inertia of a sphere rotating about an axis through its center is $I=(2/5)mR^2$.

13.5. The rotational spectrum of HCl contains lines at 60.4, 69.0, 80.4, 96.4, and 120.4 μm. What is the moment of inertia of an HCl molecule?

13.6. Consider a combined rotational–vibrational transition for which r goes from 1 to 0 while v goes from v to $(v-1)$. Find the frequencies of the photons emitted in terms of the moment of inertia of the molecule I, the angular frequency of vibration of the atoms in the molecule ω, and the quantum number v.

13.7. A rotating molecule emits photons. Find the ratio of the angular frequency of the photons, ω_{phot}, to the angular frequency of rotation of the molecule ω_{rot}, as a function of the orbital angular momentum quantum number r.

Section 13.4

13.8. A beam with 200 particles per square centimeter passes by an atom. The particles are uniformly and randomly distributed in the area of the beam.

(a) Fifty particles are scattered. What is the total scattering cross section?

(b) Ten particles are scattered in a cone of 0.1 sr solid angle about a particular direction. What is the differential cross section in $m^2\,sr^{-1}$?

13.9. The differential scattering cross section for a beam of x-ray photons of a certain energy from carbon at an angle θ is $50\times10^{-30}\,m^2\,sr^{-1}$. A beam of 10^5 photons strikes a pure carbon target of thickness 0.3 cm. The density of carbon is $2\,g\,cm^{-3}$, and the atomic weight is 12. The detector is a circle of 1-cm radius located 20 cm from the target. How many scattered photons enter the detector?

13.10. In photochemistry one often uses the extinction coefficient e, defined by $\mu_a=eC$, where C is the concentration in moles per liter. This assumes the substance being measured is dissolved in a completely transparent solvent.

(a) What are the units of the extinction coefficient?

(b) What is the conversion between the extinction coefficient and the absorption cross section?

13.11. Suppose that the absorption coefficient in some biological substance is $5\,m^{-1}$. Make the very crude assumption that the substance has the density of water and a molecular weight of 18. What is the absorption cross section?

Section 13.5

13.12. Sodium is introduced into a flame at 2500 K. What fraction of the atoms are in their first excited state? In their ground state? (Remember that the characteristic sodium line is yellow.) If the flame temperature changes by 10 K, what is the fractional change in the population of each state? Which method of measuring sodium concentration is more stable to changes in flame temperature: measuring the intensity of an emitted line or measuring the amount of absorption?

13.13. Show that the maximum of the thermal radiation function $W_\lambda(\lambda,T)$ occurs at a wavelength such that $e^x(5-x)=5$, where $x=hc/(\lambda_{\max}k_BT)$. Verify that $x=4.9651$ is a solution of this transcendental equation, so that

$$T\lambda_{\max}=\frac{hc/k_B}{4.9651}.$$

13.14. Two parallel surfaces of area S have unit emissivity and are at temperatures T_1 and T_2 [$T_1>T_2$, panel (a)]. They are large compared to their separation, so that all radiation emitted by one surface strikes the other. Assume that radiation is emitted and absorbed only by the two surfaces that face each other. Let P_0 be the energy lost per unit time by body 1. A new sheet of perfectly absorbing material is introduced between bodies 1 and 2, as shown in panel (b). It comes to equilibrium temperature T. Let P be the net energy lost by surface 1 in this case. Find P/P_0 in terms of T_1 and T_2.

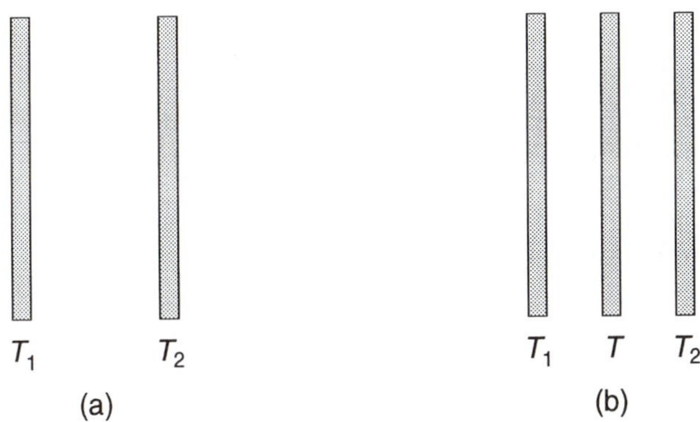

13.15. The sun has a radius of 6.9×10^8 m. The earth is 149.5×10^9 m from the sun. Treat the sun as a thermal radiator at 5800 K and calculate the energy from the sun per unit area per unit time striking the upper atmosphere of the earth. State the result in W m^{-2} and cal cm^{-2} min^{-1}.

13.16. If all the energy received by the earth from the sun is lost as thermal radiation (a poor assumption because a significant amount is reflected from cloud cover), what is the equilibrium temperature of the earth?

Section 13.6

13.17. Show that an approximation to Eq. (13.34) for small temperature differences is $w_{\text{tot}}=SK_{\text{rad}}(T-T_s)$. Deduce the value of K_{rad} at body temperature. [Hint: Factor $T^4-T_s^4=(T-T_s)(\cdots)$.] You should get $K_{\text{rad}}=6.76$ W m^{-2} K^{-1}.

13.18. What fractional change in $W_\lambda(\lambda,T)$ for thermal radiation from the human body results when there is a temperature change of 1 K at 5 μm? 9 μm? 15 μm?

Section 13.7

13.19. (a) Find the slope of the log R vs t in Eq. (13.39). What is its value for large times? (b) What can be determined from the time when R has its maximum value? (Hint: log R has a maximum when R has a maximum.)

13.20. The result of one set of infrared measurements, in human calf (leg) muscle, gave a total scattering coefficient $\mu_s=8.3$ cm^{-1} and an absorption $\mu_a=0.176$ cm^{-1}.

(a) What fraction of the photons have not scattered in passing through a layer that is 8 μm thick? (This corresponds roughly to the size of a cell.)

(b) On average, how many scattering events take place for each absorption event?

(c) What is the cross section for scattering per molecule? For this estimate, assume the muscle consists entirely of water, with molecular weight 18 and density 10^3 kg m^{-3}.

13.21. Consider the ratio of the radiant energy out of an integrating sphere to the radiant energy in, from Eq. (13.40). Discuss the stability of the ratio as $\alpha\beta$ approaches 1. (In practice reflecting surfaces with α about 0.9 are better than more reflecting surfaces.)

13.22. Use Eq. (13.40) to discuss why a hole in a cavity such as that in Fig. 13.13 looks blacker than the surface, even if the surface is a good absorber (that is, the value of α for the surface is very small).

13.23. An infrared transition involves an energy of 0.1 eV. What are the corresponding frequency and wavelength? If the Raman effect is observed with light at 550 nm, what will be the frequencies and wavelengths of each Raman line?

13.24. A Raman spectrum has a line at 500 nm with subsidiary lines at 400 and 667 nm. What is the wavelength of the corresponding infrared line?

Section 13.8

13.25. (a) Suppose that the threshold for erythema caused by sunlight with $\lambda=300$ nm is 30 J m^{-2}. Does this suggest that the result is thermal (an excessive temperature increase) or something else, like the photoelectric effect or photodissociation? Make some reasonable assumptions to estimate the temperature rise.

(b) The energy in sunlight at all wavelengths reaching the earth is 2 cal cm^{-2} min^{-1}. Suppose that the total body area exposed is 0.6 m^2. What would be the temperature rise per minute for a 60-kg person if there were no heat-loss mechanisms? Compare the rate of energy absorption to the basal metabolic rate, about 100 W.

13.26. Suppose that the energy fluence rate of a parallel beam of ultraviolet light that has passed through thickness x of solution is given by $\psi=\psi_0 e^{-\mu_a x}$. (Scattering is ignored.) The absorption coefficient μ_a is related to the concentration C (molecules cm^{-3}) of the absorbing molecules in the solution by $\mu_a=aC$. Biophysicists working with ultraviolet light define the dose rate to be the power absorbed per molecule of absorber. (This is a different definition of dose than is used in Chap. 14.) Calculate the dose rate for a thin layer ($\mu_a x \ll 1$).

13.27. A beam of photons passes through a monatomic gas of molecular weight A and absorption cross section σ. Ignore scattering. The gas obeys the ideal gas law, $pV=Nk_BT$.

(a) Find the attenuation coefficient in terms of σ, p, and any other necessary variables.

(b) Generalize the result to a mixture of several gases, each with cross section σ_i, partial pressure p_i, and N_i molecules.

13.28. The attenuation of a beam of photons in a gas of pressure p is given by $d\Phi=-\Phi(\sigma p/k_BT)dx$, where σ is the cross section, k_B the Boltzmann constant, T the absolute temperature, and x the path length. Suppose that the pressure is given as a function of altitude y by $p=p_0 e^{-mgy/k_BT}$. What is the total attenuation by the entire atmosphere?

13.29. Consider a beam of photons incident on the atmosphere from directly overhead. The atmosphere contains several species of molecules, each with partial pressure p_i. The absorption coefficient is $\mu_a=(1/k_BT)\Sigma_i\sigma_i p_i$. If each constituent of the atmosphere varies with height y as $p_i(y)=p_{0i}\exp(-m_i g y/k_BT)$, show that the fluence rate striking the earth is given by an expression of the form $e^{-\alpha}$ and find α.

Section 13.9

13.30. Consider a tissue with a heat capacity of $3.6\ \mathrm{J\,kg^{-1}\,K^{-1}}$, a density of $1000\ \mathrm{kg\,m^{-3}}$, and a thermal conductivity of $0.5\ \mathrm{W\,m^{-1}\,K^{-1}}$. Assume the heat capacity of blood is the same, and that the tissue perfusion is $4.17\times10^{-6}\ \mathrm{m^3\,kg^{-1}\,s^{-1}}$. Find the thermal diffusivity, the time for the heat to flow 1 cm, and the thermal penetration depth.

Section 13.10

13.31. How many photons per second correspond to 1 lm at 555 nm for photopic vision? At 510 nm for scotopic vision?

Section 13.11

13.32. A person is nearsighted, and the relaxed eye focuses at a distance of 50 cm. What is the strength of the desired corrective lens in diopters?

13.33. What is the distance of closest vision for an average person with normal vision at age 20? Age 40? Age 60?

13.34. A person of age 40 and is fitted with bifocals with a +1 diopter strength. What are the closest and farthest distances of focus without the bifocal lens and with it? By the time the person is age 50, what are they with and without the same lens?

Section 13.12

13.35. How many photons per 0.1 s enter the eye from a 100-W light bulb 1000 ft away? Assume the pupil is 6 mm in diameter. How far away can a 100-W bulb be seen if there is no absorption in the atmosphere? Use a luminous efficiency of $17\ \mathrm{lm\,W^{-1}}$ and then assume an equivalent light source at 555 nm.

13.36. The table below shows the radiance of some extended sources. Without worrying about obliquity factors (assume that all the light is at normal incidence), calculate the number of photons entering a receptive field of 0.17° diameter when the pupil diameter is 6 mm and the integration time is 0.1 s. Assume a conversion efficiency of $100\ \mathrm{lm\,W^{-1}}$ and then assume that all the photons are at 555 nm.

Source	Radiance ($\mathrm{lm\,m^{-2}\,sr^{-1}}$)
White paper in sunlight	25000
Clear sky	3200
Surface of the moon	2900
White paper in moonlight	0.03

13.37. A piece of paper is illuminated by dim light so that its radiance is $0.01\ \mathrm{lm\,m^{-2}\,sr^{-1}}$ in the direction of a camera. A camera lens 1 cm in diameter is 0.6 m from the paper. The sheet of paper is 10×10 cm. The shutter of the camera is open for 1 ms. Assume all the light is at 555 nm. How many photons from the paper enter the lens of the camera while the shutter is open?

13.38. If three or more photons must be absorbed by a visual receptor field for the observer to see a flash, how often will the flash be seen if the average number of photons absorbed in a receptor field per flash is four?

13.39. Assume that an average of d photons are detected and that the photons are Poisson distributed. What must d be to detect a signal that is a 1% change in d, if the signal-to-noise ratio must be at least 5?

13.40. Suppose that the average number of photons striking a target during an exposure is m. The probability that x photons strike during a similar exposure is given by the Poisson distribution. What is the probability that an organism responds to an exposure of radiation in each of the following cases?

(a) The response of the organism requires that a single target within the organism be hit by two or more photons.

(b) The response of the organism requires that two targets within the organism each be struck by one or more photons during the exposure.

REFERENCES

AAPM 57 (1996). *Recommended Nomenclature for Physical Quantities in Medical Applications of Light.* American Association of Physicists in Medicine (AAPM) Report No. 57. College Park, MD.

Anbar, M. (1994). *Quantitative Telethermometry in Medical Diagnosis and Management.* Boca Raton, FL, CRC.

Armstrong, B. K., and A. Kricker (1995). Skin cancer. *Dermatolog. Clinics* **13**(3): 583–594.

Barlow, H. B. (1957). Incremental thresholds at low intensities considered as signal/noise discriminations. *J. Physiol.* **136**: 469–488.

Baylor, D. A. (1987). Photoreceptor signals and vision. *Invest. Ophthalmol. Visual Sci.* **28**(1): 34–49.

Bergmanson, J. P. G., and P. G. Söderberg (1995). The significance of ultraviolet radiation for eye diseases. A review with comments on the efficacy of UV-blocking contact lenses. *Ophthalm. Physiolog. Opt.* **15**(2): 83–91.

Berne, B. J., and R. Pecora (1976). *Dynamic Light Scattering: With Applications to Chemistry, Biology, and Physics.* New York, Wiley.

Blume, S. (1993). Social process and the assessment of a new imaging technique. *Intl. J. Technol. Assess. Health Care* **9**(3): 335–345.

Bramson, M. A. (1968). *Infrared Radiation.* New York, Plenum.

Charman, W. N. (1995). Optics of the eye, Chap. 24 in M. Bass, editor in chief, *Handbook of Optics*, 2nd ed. Sponsored by the Optical Society of America. New York, McGraw-Hill.

Cornsweet, T. N. (1970). *Visual Perception*, New York, Academic.

Cotton, P. (1992). AMA's council on scientific affairs to take a fresh look at thermography. *J. Am. Med. Assoc.* **267**(14): 1885–1887.

Diffey, B. L. (1991). Solar ultraviolet radiation effects on biological systems. *Phys. Med. Biol.* **36**(3): 299–328.

Diffey, B. L., and J. Cameron (1984). A microcomputer program to predict sunburn exposure. *Med. Phys.* **11**(6): 869–870.

Diffey, B. L., and J. Cheeseman (1992). Sun protection with hats. *Brit. J. Dermatol.* **127**(1): 10–12.

Diffey, B. L., and P. M. Farr (1991). Quantitative aspects of ultraviolet erythema. *Clin. Phys. Physiol. Meas.* **12**(4): 311–325.

Duderstadt, J. J., and L. J Hamilton (1976). *Nuclear Reactor Analysis.* New York, Wiley.

Flewelling, R. (1995). Noninvasive Optical Monitoring. In J. D. Bronzino, ed. *The Biomedical Engineering Handbook.* Boca Raton, FL, CRC, pp. 1346–1356.

Fraden, J. (1991). Noncontact temperature measurements in medicine. In D. L. Wise, ed. *Bioinstrumentation and Biosensors.* New York, Dekker, pp. 511–549.

Gasiorowicz, S. (1974). *Quantum Physics.* New York, Wiley.

Grievenkamp, J. E., J. Schwiegerling, J. M. Miller, and M. D. Mellinger (1995). Visual acuity modeling using optical raytracing of schematic eyes. *Am. J. Ophthalmol.* **120**(2): 227–240.

Grossweiner, L. (1994). *The Science of Phototherapy.* Boca Raton, CRC.

Halliday, D., R. Resnick, and K. S. Krane (1992). *Fundamentals of Physics*, 4th ed. New York, Wiley.

Hecht, S., S. Schlaer, and M. H. Pirenne (1942). Energy, quanta, and vision. *J. Gen. Physiol.* **25**: 819–840.

Hielscher, A., S. L. Jacques, L. Wang, and F. K. Tittel (1995). The influence of boundary conditions on the accuracy of diffusion theory in time-resolved reflectance spectroscopy of biological tissues. *Phys. Med. Biol.* **40**: 1957–1975.

Kligman, L. H. (1989). Photoaging: manifestations, prevention, and treatment. *Clin. Geriatr. Med.* **5**: 235–251.

Lever, L. R., and C. M. Lawrence (1995). Nonmelanoma skin cancer associated with the use of a tanning bed. *New Engl. J. Med.* **332**(21): 1450–1451.

Liu, H., D. A. Boas, Y. Zhang, A. G. Yodh, and B. Chance (1995). Determination of optical properties and blood oxygenation in tissue using continuous NIR light. *Phys. Med. Biol.* **40**: 1983–1993.

Madronich, S. (1993). The atmosphere and UV-B radiation at ground level. In A. R. Young *et al.*, eds. *Environmental UV Photobiology.* New York, Plenum, pp. 1–39.

McDonagh, A. F. (1985). Light effects on transport and excretion of bilirubin in newborns. In R. J. Wortman, M. J. Baum, and T. J. Potts, Jr., eds. *The Medical and Biological Effects of Light. Ann. N.Y. Acad. Sci.* **453**: 65–72.

Mendelson, Y. (1992). Pulse oximetry: Theory and applications for noninvasive monitoring. *Clin. Chem.* **38**(9): 1601–1607.

Mendelson, Y. (1995). Optical Sensors. In J. D. Bronzino, ed. *The Biomedical Engineering Handbook.* Boca Raton, FL, CRC, pp. 764–778.

Milonni, P. W. (1996). Answer to question on Snell's law in quantum mechanics. *Am. J. Phys.* **64**(7): 842.

Moseley, H. (1994). Ultraviolet and laser radiation safety. *Phys. Med. Biol.* **39**: 1765–1799.

Murphy, M. R., and R. G. Oellrich (1990). A new method of phototherapy: Nursing perspectives. *J. Perinatol.* **10**(3): 249–251.

Okada, E., M. Firbank, and D. T. Delpy (1995). The effect of underlying tissue on the spatial sensitivity profile of near-infrared spectroscopy. *Phys. Med. Biol.* **40**: 2093–2108.

Patterson, M. S., B. Chance, and B. C. Wilson (1989). Time resolved reflectance and transmittance for the non-invasive measurement of tissue optical properties. *Appl. Opt.* **28**(12): 2331–2336.

Pirenne, M. H. (1962). Chapters 5 and 6 of H. Davidson, ed., *The Eye.* Vol. 2. *The Visual Process.* New York, Academic.

Pogue, B. W., and M. S. Patterson (1994). Frequency-domain optical absorption spectroscopy of finite tissue volumes using diffusion theory. *Phys. Med. Biol.* **39**: 1157–1180.

Rose, A. (1973). *Vision: Human and Electronic.* New York, Plenum.

Schaefer, B. E. (1993). Suntan and the ozone layer. *Sky and Telescope* **86**(1): 83–86, July.

Sevick, E. M., B. Chance, J. Leigh, S. Nioka, and M. Maris (1991). Quantitation of time- and frequency-resolved optical spectra for the determination of tissue oxygenation. *Analyt. Biochem.* **195**: 330–351.

Sherwood, B. A. (1996). Answer to question on Snell's law in quantum mechanics. *Am. J. Phys.* **64**(7): 840–842.

Stark, L. C., and G. C. Theodoris (1973). Information theory in physiology, in J. H. U. Brown and D. S. Gans. eds. *Engineering Principles in Physiology*, New York, Academic, Vol. 1, pp. 13–31.

Steketee, J. (1973). Spectral emissivity of skin and pericardium. *Phys. Med. Biol.* **18**: 686–694.

Studniberg, H. M., and P. Weller (1993). PUVA, UVB, psoriasis, and nonmelanoma skin cancer. *J. Am. Acad. Dermatol.* **29**: 1013–1022.

Teich, M. C., P. R. Prucnal, G. Vannucci, M. E. Breton, and W. J. McGill (1982). Multiplication noise in the human visual system at threshold: 1. Quantum fluctuations and minimal detectable energy. *J. Opt. Soc. Am.* **72**(4): 419–431.

Truhan, A. P. (1991). Sun protection in childhood. *Clin. Pediatrics* **30**(12): 676–681.

Urbach, F. (1993). Ultraviolet A transmission by modern sunscreens: is there a real risk? [Editorial review] *Photoderm. Photoimmunol. and Photomed.* **9**: 237–241.

Yau, K.-W. (1994). Phototransduction mechanism in retinal rods and cones. *Investigative Ophthalmol. Visual Sci.* **35**(1): 9–32.

Yodh, A., and B. Chance (1995). Spectroscopy and imaging with diffusing light. *Phys. Today.* March: 34–40.

Zalewski, E. F. (1995). Radiometry and Photometry, Chap. 24 of M. Bass, editor in chief, *Handbook of Optics*, 2nd ed. Sponsored by the Optical Society of America. New York, McGraw-Hill.

CHAPTER 14

Interaction of Photons and Charged Particles with Matter

An x-ray image records variations in the passage of x rays through the body because of scattering and absorption. A side effect of the making the image is the absorption of some x-ray energy by the body. Radiation therapy depends on the absorption of large amounts of x-ray energy by a tumor. Diagnostic procedures in nuclear medicine (Chap. 16) introduce a small amount of radioactive substance in the body. Radiation from the radioactive nuclei is then detected. Some of the energy from the photons or charged particles emitted by the radioactive nucleus is absorbed in the body. To describe all of these effects requires that we understand the interaction of photons and charged particles with matter.

In Chap. 13 we discussed the transport of photons of ultraviolet and lower energy—a few electron volts or less. Now we will discuss the transport of photons of much higher energy—10 keV and above. We will also discuss the movement through matter of charged particles such as electrons, protons, and heavier ions. These high energy photons and charged particles are called *ionizing radiation*, because they produce ionization in the material through which they pass. The distinction is blurred, since ultraviolet light can also ionize.

A charged particle moving through matter loses energy by local ionization, disruption of chemical bonds, and increasing the energy of atoms it passes near. It is said to be *directly ionizing*. Photons passing through matter transfer energy to charged particles, which in turn affect the material. These photons are *indirectly ionizing*.

Photons and charged particles interact primarily with the electrons in atoms. Section 14.1 describes the energy levels of atomic electrons. Section 14.2 describes the various processes by which photons interact; these are elaborated in the next four sections, leading in Sec. 14.7 to the concept of a photon attenuation coefficient. Attenuation is extended to compounds and mixtures in Sec. 14.8.

An atom is often left in an excited state by a photon interaction. The mechanisms by which it loses energy are covered in Sec. 14.9. The energy that is transferred to electrons can cause radiation damage. The transfer process is described in Secs. 14.10 and 14.15–14.17.

Section 14.11 introduces the charged-particle stopping power, which is the rate of energy loss by a charged particle as it passes through material. Extensions of this concept, which are important in radiation damage, are the linear energy transfer and the restricted collision stopping power, introduced in Sec. 14.12. A charged particle travels a certain distance through material as it loses its kinetic energy. This leads in Sec. 14.13 to the concept of range. Charged particles also lose energy by emitting photons. The radiation yield is also discussed in Sec. 14.13. Insight into the process of radiation damage is gained by examining track structure in Sec. 14.14.

The last three sections return to the movement of energy from a photon beam to matter. The discussion requires both an understanding of both photon interactions and charged-particle stopping power and range.

14.1. ATOMIC ENERGY LEVELS AND X-RAY ABSORPTION

A neutral atom has a nuclear charge $+Ze$ surrounded by a cloud of Z electrons. As was described in Chap. 13, each electron has a definite energy, characterized by a set of five quantum numbers: n, l, s (which is always $\frac{1}{2}$), j, and m_j. (Instead of j and m_j, the numbers m_l and m_s are sometimes used.) There are restrictions on the values of the numbers:

$n=1,2,3,\ldots$	the principal quantum number
$l=0,1,2,\ldots,n-1$	the orbital angular momentum quantum number
$s=\frac{1}{2}$	the spin quantum number
$j=l-\frac{1}{2}$ or $l+\frac{1}{2}$, except that $j=\frac{1}{2}$ when $l=0$	total angular momentum quantum number
$m_j=-j,-j-1,\ldots,j-1,j$	"z component" of the total angular momentum

(14.1)

The dependence of the electron energy on m_j is very slight, unless the atom is in a magnetic field.

In each atom, only one electron can have a particular set of values of the quantum numbers. Since the atoms we are considering are not in a magnetic field, electrons with different values of m_j but the same values for n, l, and j will all be assumed to have the same energy. Electrons with different values of n are said to be in different *shells*. The shell for $n=1$ is called the K shell; those for $n=2,3,4,\ldots$ are labeled $L,M,N,\ldots$. Different values of l and j for a fixed value of n are called *subshells*, denoted by roman numeral subscripts on the shell labels. The maximum number of electrons that can be placed in a subshell is $2(2l+1)$. Each electron bound to the atom and has a certain negative energy, with zero energy defined when the electron is just unbound, that is, infinitely far away from the atom. Table 14.1 lists the energy levels of electrons in tungsten. Some of the levels in Table 14.1 are shown in Fig. 14.1. The scale is logarithmic. Since the energies are negative, the magnitude increases in the downward direction.

The *ionization energy* is the energy required to remove the least-tightly-bound electron from the atom. For tungsten, it is about 6 eV. If one plots the ionization energy or the chemical valence of atoms as a function of Z, one finds abrupt changes when the last electron's value of n or l changes.

In contrast to this behavior of the outer electrons, the energy of an inner electron with fixed values of n and l varies smoothly with Z. To a first approximation, the two innermost K electrons are attracted by a nuclear charge Ze. The energy of the level can be estimated using Eq. (13.7) for hydrogen, with the nuclear charge e replaced by Ze:

TABLE 14.1. *Energy levels for electrons in a tungsten atom ($Z=74$).*

n	l	j	Number of electrons	X-ray label	Energy (eV)
1	0	$\frac{1}{2}$	2	K	−69 525
2	0	$\frac{1}{2}$	2	L_{I}	−12 100
	1	$\frac{1}{2}$	2	L_{II}	−11 544
	1	$\frac{3}{2}$	4	L_{III}	−10 207
3	0	$\frac{1}{2}$	2	M_{I}	−2 819
	1	$\frac{1}{2}$	2	M_{II}	−2 575
	1	$\frac{3}{2}$	4	M_{III}	−2 281
	2	$\frac{3}{2}$	4	M_{IV}	−1 872
	2	$\frac{5}{2}$	6	M_{V}	−1 809
4	0	$\frac{1}{2}$	2	N_{I}	−595
	1	$\frac{1}{2}$	2	N_{II}	−492
	1	$\frac{3}{2}$	4	N_{III}	−425
	2	$\frac{3}{2}$	4	N_{IV}	−259
	2	$\frac{5}{2}$	6	N_{V}	−245
	3	$\frac{5}{2},\frac{7}{2}$	14	$N_{VI,VII}$	−35
5	0	$\frac{1}{2}$	2	O_{I}	−77
	1	$\frac{1}{2}$	2	O_{II}	−47
	1	$\frac{3}{2}$	4	O_{III}	−36
	2	$\frac{3}{2},\frac{5}{2}$	4	$O_{IV,V}$	−6
6			2	P_{I}	

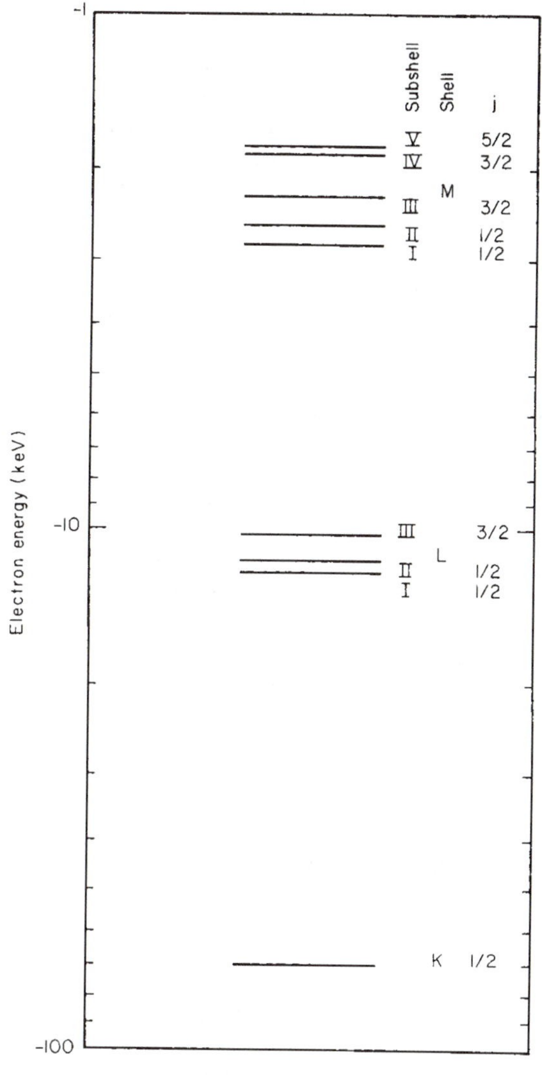

FIGURE 14.1. Energy levels for electrons in tungsten.

$$E_K = -\frac{13.6Z^2}{1^2}. \tag{14.2}$$

The two electrons also repel each other and experience some repulsion by electrons in other shells. This effect is called *charge screening*. Experiment (measuring values of E_K) shows that the effective charge seen by a K electron is *approximately* $Z_{\text{eff}} \approx Z-3$ for heavy elements, so that for K electrons ($n=1$),

$$E_K \approx -13.6(Z-3)^2 \quad (\text{in eV}). \tag{14.3}$$

The screening is greater for electrons with larger values of n, which may be thought of as being in ''orbits'' of larger radius.

14.2. PHOTON INTERACTIONS

There are a number of different ways in which a photon can interact with an atom. The more important ones will be considered here. It is convenient to adopt a notation (γ,bc) where γ represents the incident photon, and b and c are the results of the interaction. For example (γ,γ) represents initial and final photons having the same energy; in a (γ,e) interaction the photon is absorbed and only an electron emerges. This section describes the common interactions and the energy balance for each case.

In the *photoelectric effect* (γ,e), the photon is absorbed by the atom and a single electron is ejected. The initial photon energy $h\nu_0$ is equal to the final energy. The recoil kinetic energy of the atom is very small because its mass is large, so the final energy is the kinetic energy of the electron, T_{el}, plus the excitation energy of the atom. The excitation energy is equal to the binding energy of the ejected electron, B. The energy balance is therefore

$$h\nu_0 = T_{\text{el}} + B. \tag{14.4}$$

The atom subsequently loses its excitation energy. The deexcitation process is described below and involves the emission of additional photons or electrons. The photoelectric cross section is τ.

In Compton scattering, $(\gamma,\gamma' e)$, the original photon disappears and a photon of lower energy and an electron emerge. The statement of energy conservation is

$$h\nu_0 = h\nu + T_{\text{el}} + B.$$

Usually the photon energy is high enough so that B can be neglected, and this is written as

$$h\nu_0 = h\nu + T_{\text{el}}. \tag{14.5}$$

The Compton cross section for scattering from a single electron is σ_C. *Incoherent scattering* is Compton scattering from all the electrons in the atom, with cross section σ_{incoh}.

Coherent scattering is a (γ,γ) process in which the photon is elastically scattered from the entire atom. That is, the internal energy of the atom does not change. The recoil kinetic energy of the atom is very small (see Problem 14.9), and it is a good approximation to say that the energy of the incident photon equals the energy of the scattered photon:

$$h\nu_0 = h\nu. \tag{14.6}$$

The cross section for coherent scattering is σ_{coh}.

It is also possible for the final photon to have a different energy from the initial photon (γ,γ') without the emission of an electron. The internal energy of the target atom or molecule increases or decreases by a corresponding amount. Again, the recoil kinetic energy of the atom is negligible. Examples are fluorescence and Raman scattering. [In fluorescence, if one waits long enough, additional photons are emitted, in which case the reaction could be denoted as $(\gamma,\gamma'\gamma'')$, or $(\gamma,2\gamma)$, or even $(\gamma,3\gamma)$.]

At high energies *pair production* takes place. This is a (γ,e^+e^-) reaction. Since it takes energy to create the electron and positron, their rest energies must be included in the energy balance equation:

$$h\nu_0 = T_+ + m_ec^2 + T_- + m_ec^2 = T_+ + T_- + 2m_ec^2. \tag{14.7}$$

The cross section for pair production is κ.

Figure 14.2 shows the cross section for interactions of photons with carbon for photon energies from 10 to 10^{11} eV. At the lowest energies the photoelectric effect dominates. Between 10 keV and 10 MeV Compton scattering is most important. Above 10 MeV pair production takes over. There is a small bump at about 20 MeV due to nuclear effects, but its contribution to the cross section is only a few percent of that due to pair production. The four important effects are discussed in the next four sections.

14.3. THE PHOTOELECTRIC EFFECT

In the photoelectric effect a photon of energy $h\nu_0$ is absorbed by an atom and an electron of kinetic energy $T_{\text{el}} = h\nu_0 - B$ is ejected. B is the magnitude of the binding energy of the electron and depends on which shell the electron was in. Therefore it is labeled B_K, B_L, and so forth. The cross section for the photoelectric effect, denoted by τ, is a sum of terms for each shell:

$$\tau = \tau_K + \tau_L + \tau_M + \cdots . \tag{14.8}$$

As the energy of a photon beam is decreased, the photoelectric cross section increases rapidly. For photon energies too small to remove an electron from the K shell, the cross section for the K-shell photoelectric effect is zero. Even though photons do not have enough energy to remove an electron from the K shell, they may have enough energy to remove L-shell electrons. The cross section for L electron photoelectric effect is much smaller than that for K elec-

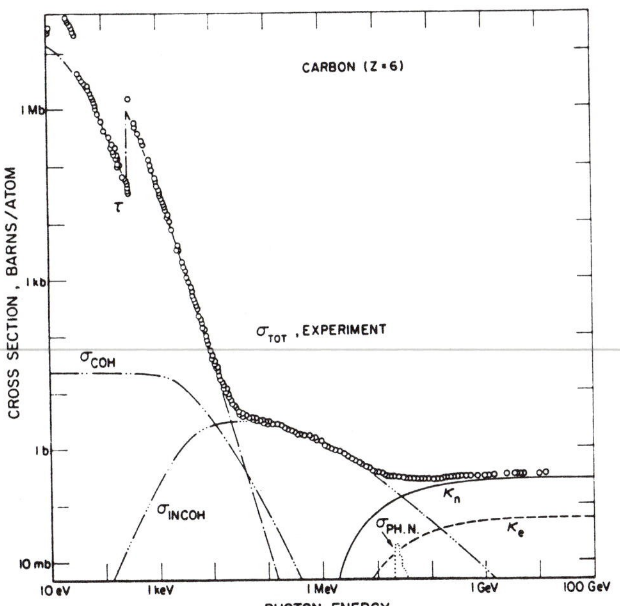

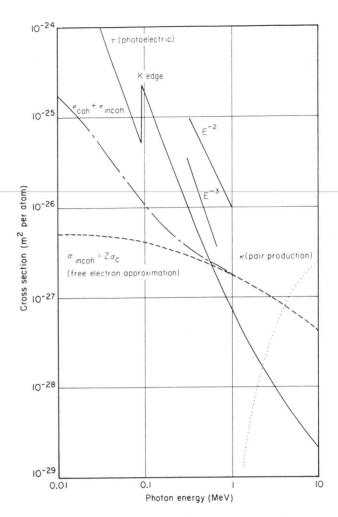

FIGURE 14.2. Total cross section for the interactions of photons with carbon versus photon energy. The photoelectric cross section is τ, the coherent scattering cross section σ_{coh}, the total Compton cross section σ_{incoh}, and the nuclear and electronic pair production are κ_n and κ_e. The photonuclear scattering cross section σ_{PHN} is also shown. The cross section is given in barns: 1 b$=10^{-28}$ m^2. From Hubbell, Gimm, and Øverbø (1980). Reproduced by permission of the American Institute of Physics. Photograph courtesy of J. H. Hubbell.

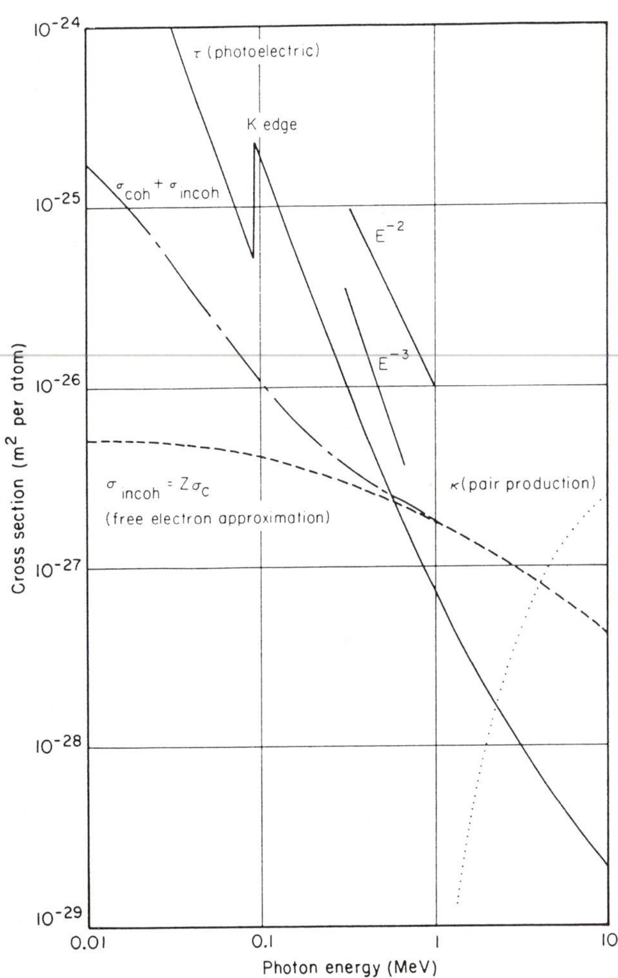

FIGURE 14.3. Cross sections for the photoelectric effect and incoherent and coherent scattering from lead. The binding energies of the K and L shells are 0.088 and 0.0152 MeV. Plotted from Table 3.22 of Hubbell (1969).

trons, but it increases with decreasing energy until its threshold energy is reached. This energy dependence is shown for lead in Fig. 14.3, which plots the cross section for the photoelectric effect, incoherent Compton scattering, and coherent scattering. The K absorption edge for the photoelectric effect is seen. The photoelectric effect below the K absorption edge is due to $L,M,\ldots$ electrons; above this energy the K electrons also participate. Above 0.8 MeV in lead Compton scattering becomes more important than the photoelectric effect.

The energy dependence of the photoelectric effect cross section is between E^{-2} and E^{-4}. An approximation to the Z and E dependence of the photoelectric cross section τ near 100 keV is

$$\tau \propto Z^4 E^{-3}. \tag{14.9}$$

Once an atom has absorbed a photon and ejected a photoelectron, it is in an excited state. The atom will eventually lose this excitation energy by capturing an electron and returning to its ground state. The deexcitation processes are described in Sec. 14.9.

14.4. COMPTON SCATTERING

Compton scattering is a $(\gamma,\gamma' e)$ process. The equations that are used to relate the energy and angle of the emerging photon and electron, as well as the equations that give the cross section for the scattering, are usually derived assuming that the electron is free and at rest. We turn first to the kinematics. A photon has energy E and momentum p, related by

$$E = h\nu = pc. \tag{14.10}$$

This is a special case of a more general relationship from special relativity:

$$E^2 = (pc)^2 + (m_0c^2)^2. \tag{14.11}$$

In these equations E is the total energy of the particle, p its momentum, m_0 the "rest mass" of the particle (measured

FIGURE 14.4. Momentum relationships in Compton scattering. (a) Before. (b) After. The photon emerges at angle θ, the electron at angle ϕ.

when it is not moving), and m_0c^2 is the "rest energy."[1] For a photon, which can never be at rest, $m_0=0$. [Equation (14.10) can also be derived from the classical electromagnetic theory of light.]

The conservation of energy and momentum can be used to derive the relationship between the angle at which the scattered photon emerges and its energy. A detailed knowledge of the forces involved is necessary to calculate the relative number of photons scattered at different angles; in fact, this calculation must be done using quantum mechanics. Figure 14.4 shows the geometry involved in the scattering. The electron emerges with momentum p, kinetic energy T, and total energy $E=T+m_ec^2$. It emerges at an angle ϕ with the direction of the incident photon. The scattered photon emerges with a frequency ν', which is lower than ν_0 because it has lost energy, at angle θ. Conservation of momentum in the direction of the incident photon gives

$$\frac{h\nu_0}{c}=\frac{h\nu'}{c}\cos\theta+p\cos\phi,$$

while conservation of momentum at right angles to that direction gives

$$\frac{h\nu'}{c}\sin\theta=p\sin\phi.$$

Conservation of energy gives

$$h\nu_0=h\nu'+T.$$

The equation $E=T+m_ec^2$ can be combined with Eq. (14.11) to give

$$(pc)^2=T^2+2m_ec^2T.$$

The last four equations can then be combined and solved for various unknowns.

The wavelength of the scattered photon is given by

$$\lambda'-\lambda_0=\frac{c}{\nu'}-\frac{c}{\nu_0}=\frac{h}{m_ec}(1-\cos\theta). \quad (14.12)$$

The wavelength shift (but not the frequency or energy shift) is independent of the incident wavelength. The quantity h/m_ec has the dimensions of length and is called the *Compton wavelength of the electron.* Its numerical value is

$$\lambda_C=\frac{h}{m_ec}=2.427\times10^{-12}\ \text{m}=2.427\times10^{-3}\ \text{nm}. \quad (14.13)$$

If Eq. (14.12) is solved for the energy of the scattered photon, the result is

$$h\nu'=\frac{m_ec^2}{1-\cos\theta+1/x}, \quad (14.14)$$

where x is the energy of the incident photon in units of $m_ec^2=511$ keV:

$$x=\frac{h\nu_0}{m_ec^2}. \quad (14.15)$$

The energy of the recoil electron is $T=h\nu_0-h\nu'$:

$$T=\frac{h\nu_0(2x\cos^2\phi)}{(1+x)^2-x^2\cos^2\phi}=\frac{h\nu_0x(1-\cos\theta)}{1+x(1-\cos\theta)}. \quad (14.16)$$

Figure 14.5 shows the energy of the scattered photon and the recoil electron as a function of the angle of emergence of the photon. The sum of the two energies is 1 MeV, the energy of the incident photon.

The inclusion of dynamics, which allows us to determine the relative number of photons scattered at each angle, is fairly complicated. The quantum-mechanical result is known as the *Klein–Nishina* formula. The result depends on the polarization of the photons. For unpolarized photons, the cross section per unit solid angle for a photon to be scattered at angle θ is

$$\frac{d\sigma_c}{d\Omega}=\frac{r_e^2}{2}\left[\frac{1+\cos^2\theta+\dfrac{x^2(1-\cos\theta)^2}{1+x(1-\cos\theta)}}{[1+x(1-\cos\theta)]^2}\right], \quad (14.17)$$

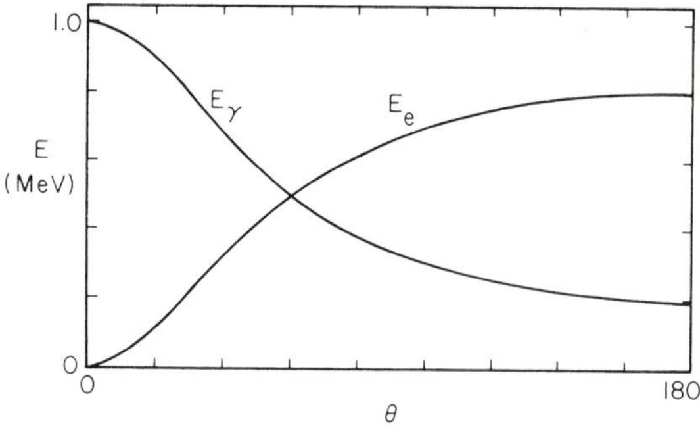

FIGURE 14.5. Energy of emerging photon and recoil electron as a function of θ, the angle of the emerging photon, for an incident photon of energy 1 MeV.

[1] Since this is the only relativistic result we will need, it is not developed here. A discussion can be found in any book on special relativity.

where $r_e = e^2/4\pi\epsilon_0 m_e c^2$ is the "classical radius" of the electron. The cross section is plotted in Fig. 14.6. It is peaked in the forward direction at high energies. As $x \rightarrow 0$ (long wavelengths or low energy) it approaches

$$\frac{d\sigma_c}{d\Omega} = \frac{r_e^2(1+\cos^2\theta)}{2}, \tag{14.18}$$

$$r_e = \frac{e^2}{4\pi\epsilon_0 m_e c^2} = 2.818\times 10^{-15}\ \text{m},$$

which is symmetric about 90°.

Equation (14.17) can be integrated over all angles to obtain the total Compton cross section for a single electron:

$$\sigma_c = 2\pi r_e^2\left[\frac{1+x}{x^2}\left(\frac{2(1+x)}{1+2x} - \frac{\ln(1+2x)}{x}\right) + \frac{\ln(1+2x)}{2x} - \frac{1+3x}{(1+2x)^2}\right]. \tag{14.19}$$

As $x \rightarrow 0$, this approaches

$$\sigma_c \rightarrow \frac{8\pi r_e^2}{3} = 6.652\times 10^{-29}\ \text{m}^2. \tag{14.20}$$

Figure 14.7 shows σ_C as a function of energy.

The classical analog of Compton scattering is Thomson scattering of an electromagnetic wave by a free electron. The electron experiences the electric field **E** of an incident plane electromagnetic wave and therefore has an acceleration $-e\mathbf{E}/m$. Accelerated charges radiate electromagnetic waves, and the energy radiated in different directions can be calculated, giving Eqs. (14.18) and (14.20). [See, for example, Rossi (1952, Chap. 8).]

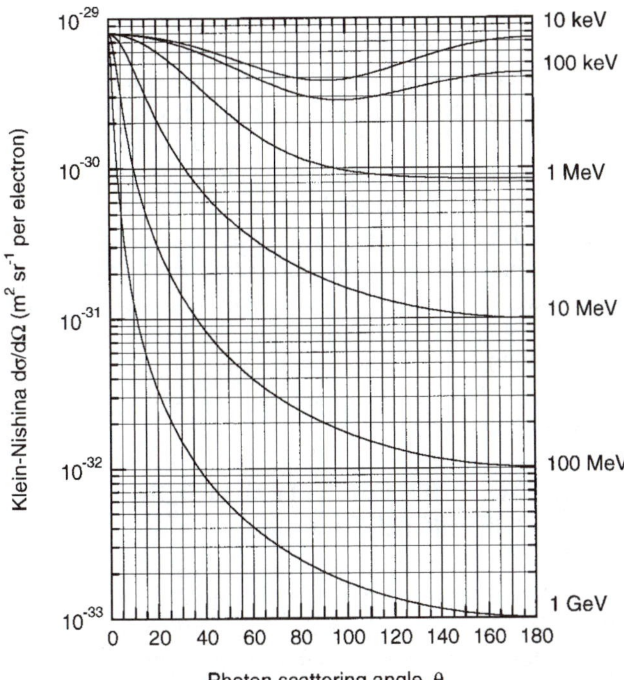

FIGURE 14.6. Differential cross section for Compton scattering of unpolarized photons from a free electron, calculated from Eq. (14.17).

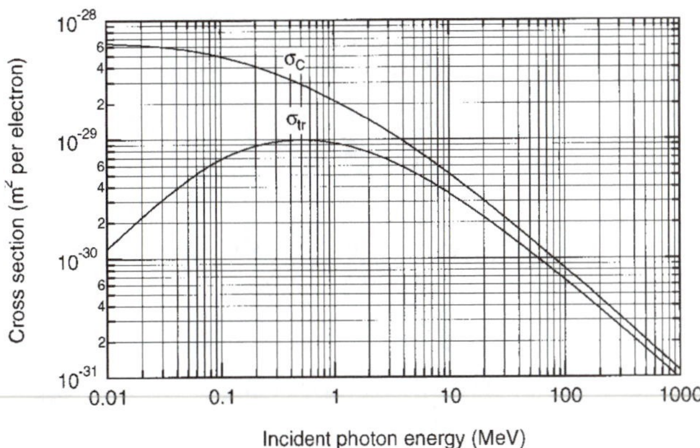

FIGURE 14.7. The total cross section σ_C for Compton scattering by a single electron and the cross section for energy transfer $\sigma_{tr} = f_C\sigma_C$.

The Compton cross section is for a single electron. For an atom containing Z electrons, the maximum value of the incoherent cross section occurs if all Z electrons take part in the Compton scattering:

$$\sigma_{\text{incoh}} \leqslant Z\sigma_C.$$

At low energies σ_{incoh} falls below this maximum value because the electrons are bound and not at rest. The falloff can be seen in Fig. 14.2. For carbon $Z\sigma_C = 4.0 \times 10^{-28}$ m^2. This value is approached by σ_{incoh} near 10 keV. For lower photon energies, σ_{incoh} is much less. This fall off is appreciable for energies as high as 7–8 keV, even though the K-shell binding energy in carbon is only 283 eV. The electron motion and binding in the target atom cause a small spread in the energy of the scattered photons [Carlsson and Carlsson(1982)].

Departures of the angular distribution and incoherent cross section from Z times the Klein–Nishina formula are discussed by Hubble *et al.* (1975) and by Jackson and Hawkes (1981).

We will need to know the average energy transferred to an electron in a Compton scattering. Equation (14.16) gives the ratio of electron kinetic energy to original photon energy as a function of photon scattering angle. The *transfer cross section* is defined to be

$$\sigma_{\text{tr}} = \int_0^\pi \frac{d\sigma_C}{d\Omega}\frac{T}{h\nu_0} 2\pi \sin\theta\, d\theta = f_C\sigma_C. \tag{14.21}$$

This can be integrated. The result is [see Attix (1986, p. 134)]

$$\sigma_{\mathrm{tr}}=(2\pi r_e^2)\left[\frac{2(1+x)^2}{x^2(1+2x)}-\frac{1+3x}{(1+2x)^2}\right.$$
$$-\frac{(1+x)(2x^2-2x-1)}{x^2(1+2x)^2}-\frac{4x^2}{3(1+2x)^3}$$
$$\left.-\left(\frac{1+x}{x^3}-\frac{1}{2x}+\frac{1}{2x^3}\right)\ln(1+2x)\right]. \quad (14.22)$$

This quantity is also plotted in Fig. 14.7. Equation (14.22) is a rather nasty equation to evaluate, particularly at low energies, because many of the terms nearly cancel.

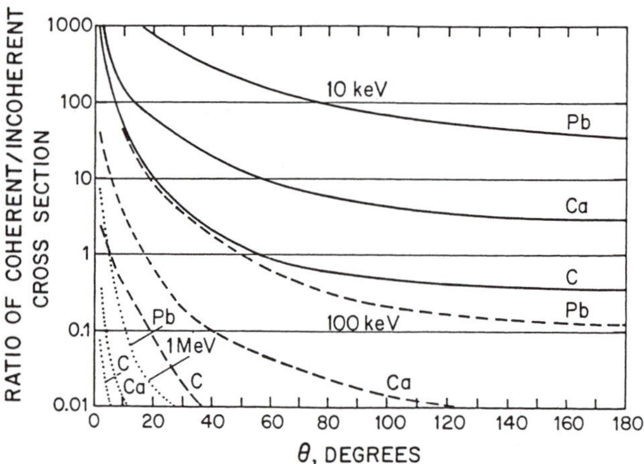

FIGURE 14.8. The ratio of coherent to incoherent scattering cross section as a function of angle, for various target atoms and photon energies. Calculated from tables in Hubbell *et al.* (1975). Reprinted, with permission, from R. K. Hobbie, Interaction of Photons and Charged Particles in Matter, in L. Williams, ed. *Nuclear Medical Physics.* Copyright © 1987 CRC Press, Inc. Boca Raton, FL.

14.5. COHERENT SCATTERING

A photon can also scatter elastically from an atom, with none of the electrons leaving their energy levels. This (γ,γ) process is called *coherent scattering* (sometimes called Rayleigh scattering), and its cross section is σ_{coh}. The entire atom recoils; if one substitutes the atomic mass in Eqs. (14.15) and (14.16), one finds that the atomic recoil kinetic energy is negligible.

The primary mechanism for coherent scattering is the oscillation of the electron cloud in the atom in response to the electric field of the incident photons. There are small contributions to the scattering from nuclear processes. The cross section can be calculated classically as an extension of Thomson scattering, or it can be done using various degrees of quantum-mechanical sophistication [see Kissel *et al.* (1980) or Pratt (1982)].

The coherent cross section is peaked in the forward direction because of interference effects between electromagnetic waves scattered by various parts of the electron cloud. The peak is narrower for elements of lower atomic number and for higher energies. The ratio of coherent to incoherent scattering cross sections is shown in Fig. 14.8.

If the wavelength of the incident photons is large compared to the size of the atom, then all Z electrons behave like a single particle with charge $-Z_e$ and mass Zm_e. From Eqs. (14.18) and (14.20), one can see that the cross section in this limit is Z^2 times the single-electron value: $Z^2\sigma_C$. The limiting value for carbon is 2.39×10^{-27} m^2, which can be compared to the low energy limit for σ_{coh} in Fig. 14.2.

14.6. PAIR PRODUCTION

A photon with energy above 1.02 MeV can produce a particle–antiparticle pair: a negative electron and a positive electron or *positron*. Conservation of energy requires that

$$h\nu_0=\underbrace{T_-+m_ec^2}_{\text{electron}}+\underbrace{T_++m_ec^2}_{\text{positron}}=T_++T_-+2m_ec^2. \quad (14.23)$$

One can show, using $h\nu_0=pc$ for the photon, that momentum is not conserved by the positron and electron if Eq. (14.23) is satisfied. However, pair production always takes place in the Coulomb field of another particle (usually a nucleus) that recoils to conserve momentum. Since the nucleus has a large mass and its kinetic energy is $p^2/2m$, its energy is small compared to the terms in Eq. (14.23). Since the rest energy (m_ec^2) of an electron or positron is 0.51 MeV, pair production is energetically impossible for photons below $2m_ec^2=1.02$ MeV. The cross section for this (γ,e^+e^-) reaction is κ_n.

Pair production with excitation or ionization of the recoil atom can take place at energies that are only slightly higher; however, the cross section does not become appreciable until the incident photon energy exceeds $4m_ec^2=2.04$ MeV, the threshold for pair production in which a free electron (rather than a nucleus) recoils to conserve momentum. Because ionization and free-electron pair production are $(\gamma,e^-e^-e^+)$ processes, this is usually called *triplet production*. Extensive data are given in Hubbell *et al.* (1980). The cross section for both processes is $\kappa=\kappa_n+\kappa_e$. The energy dependence of κ can be seen in Figs. 14.2 and 14.3.

14.7. THE PHOTON ATTENUATION COEFFICIENT

Consider the arrangement shown in Fig. 14.9(a), in which a beam of photons is collimated so that a narrow beam strikes a detector. A scattering material is then introduced in the beam. Some of the photons pass through the material without interaction. Others are scattered. Still others disappear because of photoelectric effect or pair-production interactions. If we measure only photons that remain in the

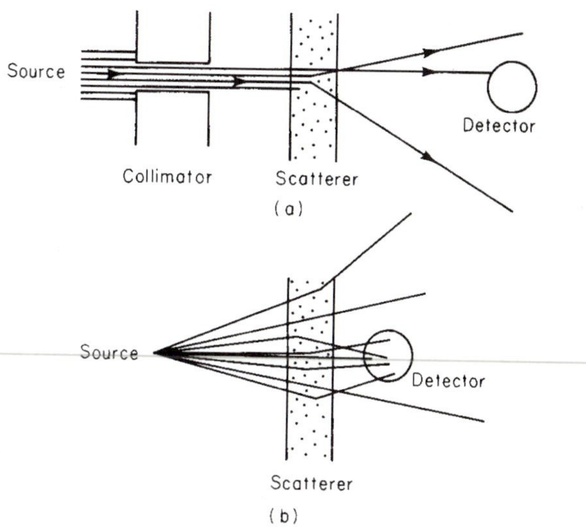

FIGURE 14.9. Measurements with (a) narrow-beam geometry and (b) broad-beam geometry.

unscattered beam, the loss of photons is called *attenuation* of the beam. Attenuation includes both scattering and absorption. We record as still belonging to the beam only photons that did not interact; they still travel in the forward direction with the original energy. This is called a *narrow beam geometry* measurement. It is an idealization, because photons that undergo Compton or coherent scattering through a small angle can still strike the detector. Figure 14.9(b) shows a source, scatterer, and detector geometry that is much more difficult to interpret. In this case photons that are initially traveling in a different direction are scattered into the detector. These are called *broad beam geometry* experiments.

In narrow-beam geometry, the total cross section is related to the total number of particles that have interacted in the scatterer. Let N be the number of particles that have not undergone any interaction in passing through scattering material of thickness z. We saw in Sec. 13.4 that the number of particles that have not interacted decreases in thickness dz by

$$dN = -\frac{\sigma_{tot} N_A \rho}{A} N\, dz,$$

so that

$$\frac{dN}{dz} = -\mu_{atten} N,$$

where

$$\mu_{atten} = \frac{N_A \rho \sigma_{tot}}{A}. \tag{14.24}$$

In these equations ρ is the mass density of the target material and A is its atomic weight. The number of particles that have undergone no interaction decays exponentially with distance:

$$N(z) = N_0 e^{-\mu_{atten} z}. \tag{14.25}$$

The quantity μ_{atten} is called the *total linear attenuation coefficient.*

In a broad-beam geometry configuration the total number of photons reaching the detector includes secondary photons and is larger than the value given by Eq. (14.25).

The units in Eqs. (14.25) are worth discussing. Avogadro's number is $6.022\,045\times 10^{23}$ entities mol^{-1}. If the density ρ is in kg m^{-3} and σ_{tot} is in m^2, then A must be expressed in kg mol^{-1} and μ_{atten} is in m^{-1}. On the other hand, it is possible to express ρ in g cm^{-3}, σ_{tot} in cm^2, and A in g mol^{-1}, so that μ_{atten} is in cm^{-1}. As an example, consider carbon, for which $A = 12.011\times 10^{-3}$ kg mol^{-1} $= 12.011$ g mol^{-1}. If $\sigma_{tot} = 1.269\times 10^{-28}$ m^2 atom^{-1} $= 1.269\times 10^{-24}$ cm^2 atom^{-1}, then either

$$\mu_{atten} = \frac{(6.022\times 10^{23}\ \text{atom mol}^{-1})(2.000\times 10^{3}\ \text{kg m}^{-3})(1.269\times 10^{-28}\ \text{m}^2\ \text{atom}^{-1})}{12.011\times 10^{-3}\ \text{kg mol}^{-1}} = 12.7\ \text{m}^{-1}$$

or

$$\mu_{atten} = \frac{(6.022\times 10^{23}\ \text{atom mol}^{-1})(2.000\ \text{g cm}^{-3})(1.269\times 10^{-24}\ \text{cm}^2\ \text{atom}^{-1})}{12.011\ \text{g mol}^{-1}} = 0.127\ \text{cm}^{-1}$$

The total cross section for photon interactions is

$$\sigma_{tot} = \sigma_{coh} + \sigma_{incoh} + \tau + \kappa. \tag{14.26a}$$

In many situations the coherently scattered photons cannot be distinguished from those unscattered, and σ_{coh} should not be included:

$$\sigma_{tot} = \sigma_{incoh} + \tau + \kappa. \tag{14.26b}$$

Tables usually include total cross sections and attenuation coefficients both with and without coherent scattering.

It is possible to regroup the terms in Eqs. (14.24) and (14.25) in a slightly different way:

$$dN = -N \frac{N_A \sigma_{tot}}{A} \rho \, dz.$$

The quantity $N_A \sigma_{tot}/A$ is the *mass attenuation coefficient*, $\mu_{atten}/\rho(\mathrm{m^2\ kg^{-1}})$:

$$\frac{\mu_{atten}}{\rho} = \frac{N_A \sigma_{tot}}{A}. \quad (14.27)$$

The exponential attenuation is then

$$N(\rho z) = N_0 e^{-(\mu_{atten}/\rho)(\rho z)}. \quad (14.28)$$

The mass attenuation coefficient has the advantage of being independent of the density of the target material, which is particularly useful if the target is a gas. It has an additional advantage if Compton scattering is the dominant interaction. If $\sigma_{tot} = Z\sigma_C$, then

$$\frac{\mu_{atten}}{\rho} = \frac{Z\sigma_C N_A}{A}.$$

Since Z/A is nearly 1/2 for all elements except hydrogen, this quantity changes very little throughout the periodic table. This constancy is *not* true for the photoelectric effect or pair production. Figure 14.10 plots the mass attenuation coefficient vs energy for three elements spanning the periodic table. It is nearly independent of Z around 1 MeV where Compton scattering is dominant. The K and L absorption edges can be seen for lead; for the lighter elements they are below 10 keV. Figure 14.11 shows the contributions to μ_{atten}/ρ for air from the photoelectric effect, incoherent scattering, and pair production.

14.8. COMPOUNDS AND MIXTURES

The usual procedure for dealing with mixtures and compounds is to assume that each atom scatters independently. If the cross section for element i summed over all the interaction processes of interest is denoted by σ_i, then Eq. (13.18) is replaced by

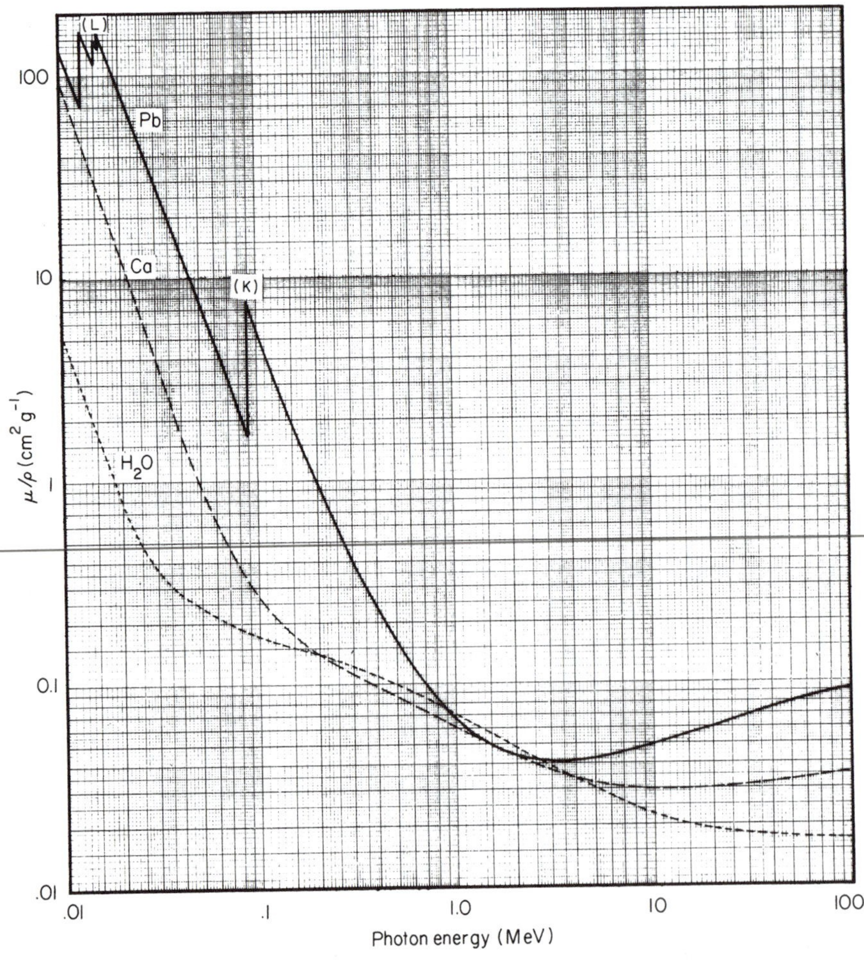

FIGURE 14.10. Mass attenuation coefficient vs energy for lead, calcium, and water. Near 1 MeV the mass attenuation coefficient is nearly independent of Z. Plotted from Tables 3-15, 3-22, and 3-24 of Hubbell (1969).

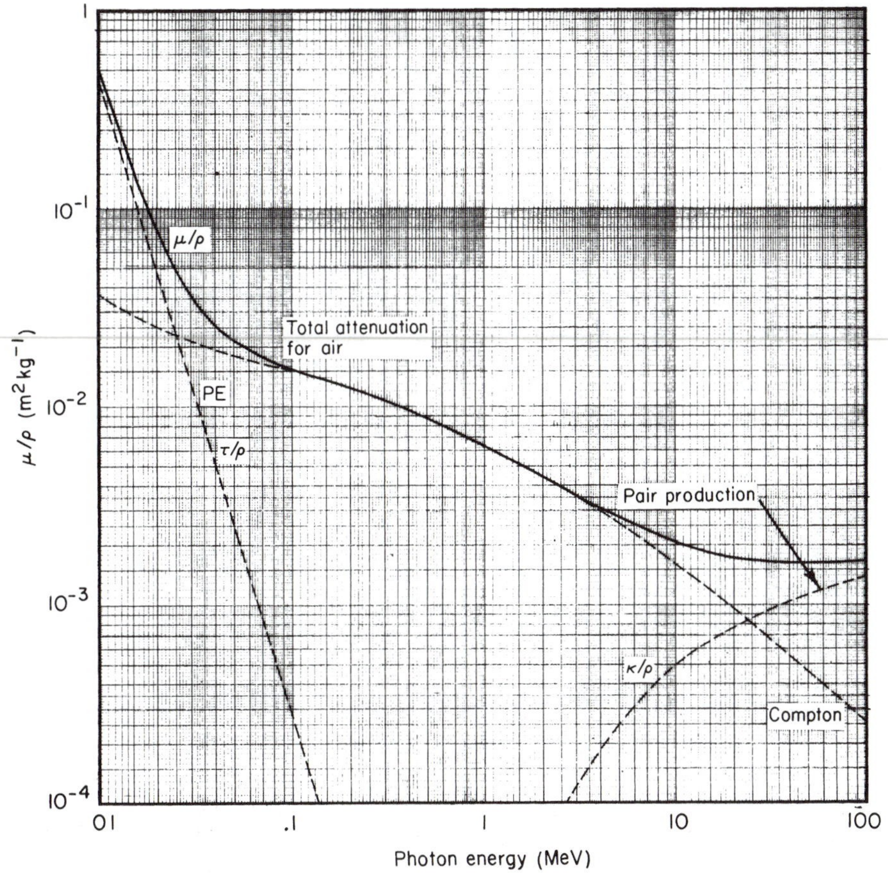

FIGURE 14.11. Mass attenuation coefficient vs energy for air. Plotted from Table 3-27 of Hubbell (1969). Coherent scattering is not included.

$$\frac{\bar{n}}{N} = \sum_i \sigma_i (N_T)_i = \left(\sum_i \sigma_i (N_{TV})_i \right) dz, \tag{14.29}$$

where $(N_T)_i$ is the number of target atoms of species i *per unit projected area* of the target and $(N_{TV})_i$ is the number of target atoms *per unit volume*. The sum is taken over all elements in the compound or mixture.

It is possible to replace the sum by the product of the cross section per molecule multiplied by the number of molecules per unit volume. The cross section per molecule is the sum of the cross sections for all the atoms in the molecule. To see that this is so, note that a volume of scatterer V contains a total mass $M = \rho V$ and that the mass of each element is M_i. The mass fraction of each element is $w_i = M_i / M$. The total number of atoms of species i in volume V is the number of moles times Avogadro's number:

$$(N_{TV})_i = \frac{M_i N_A}{A_i V} = \frac{w_i}{A_i} \rho N_A . \tag{14.30}$$

The mass fraction of element i in a compound containing a_i atoms per molecule with atomic mass A_i is

$$w_i = \frac{a_i A_i}{A_{\text{mol}}}, \tag{14.31}$$

where A_{mol} is the molecular weight. Therefore

$$\sum_i \sigma_i (N_{TV})_i = \left(\sum_i \frac{a_i \sigma_i}{A_{\text{mol}}} \right) \rho N_{\text{A}}$$
$$= \left(\sum_i a_i \sigma_i \right) \frac{\rho N_A}{A_{\text{mol}}} = \sigma_{\text{mol}} (N_{TV})_{\text{mol}} . \tag{14.32}$$

The factor $(N_{TV})_{\text{mol}} = \rho N_A / A_{\text{mol}}$ is the number of molecules per unit volume. When a target entity (molecule) consists of a collection of subentities (atoms), we can say that in this approximation (all subentities interacting independently), the cross section per entity is the sum of the cross sections for each subentity. For example, for the molecule CH_4, the total molecular cross section is $\sigma_{\text{carbon}} + 4\sigma_{\text{hydrogen}}$ and the molecular weight is $[(4 \times 1) + 12 = 16] \times 10^{-3}$ kg mol^{-1}.

14.9. DEEXCITATION OF ATOMS

After the photoelectric effect, Compton scattering, or triplet production, an atom is left with a hole in some electron shell. An atom can be left in a similar state when an electron is knocked out by a passing charged particle or by certain transformations in the atomic nucleus that are discussed in Chap. 16.

The hole in the shell can be filled by two competing processes: a *radiative transition*, in which a photon is emitted as an electron falls into the hole from a higher level, or a nonradiative or radiationless transition, such as the emission of an *Auger electron* from a higher level as a second electron falls from a higher level to fill the hole. Both processes are shown in Fig. 14.12. In the radiative transition, the energy of the photon is equal to the difference in binding energies of the two levels. For the example of Fig. 14.12, the photon energy is $B_K - B_L$. The emission of an L-shell Auger electron is shown in Fig. 14.12(c): its energy is $T=(B_K-B_L)-B_L=B_K-2B_L$. Table 14.2 shows the energy changes that occur after a hole is created in an atom by photoelectric excitation. It is worth understanding each table entry in detail. Two different paths for deexcitation are shown: one for photon emission and one for ejection of an Auger electron. The sum of the total energy, which includes photons, electrons, and excitation energy of the atom, does not change.

The photon that is emitted is called a *characteristic photon* or a *fluorescence photon*. Its energy is given by the difference of two electron energy levels in the atom. There is an historical nomenclature for these photons. Because a hole moving to larger values of n corresponds to a decrease in the total energy of an atom, it is customary to draw the energy levels for holes instead of electrons, as in Fig. 14.13. Transitions in which the hole is initially in the $n=1$ state give rise to the K series of x rays, those in which the initial hole is in the $n=2$ state give rise to the L series, and so on.

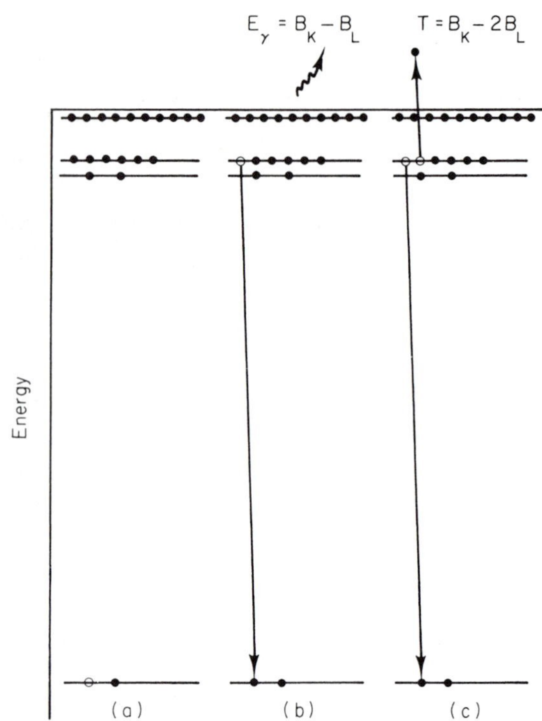

FIGURE 14.12. Two possible mechanisms for the deexcitation of an atom with a hole in the K shell. (a) The atom with the hole in the K shell. (b) An electron has moved from the L shell to the K shell with emission of a photon of energy $B_K - B_L$. (c) An electron has moved from the L shell to the K shell. The energy liberated is taken by another electron from the L shell, which emerges with energy $B_K - 2B_L$. This electron is called an Auger electron.

Greek letters (and their subscripts) are used to denote the shell (and subshell) of the final hole. The transitions shown in Fig. 14.13 are consistent with certain *selection rules*, which can be derived using quantum theory:

TABLE 14.2. *Energy changes in the photoelectric effect and in subsequent deexcitation.*

Process	Total photon energy	Total electron energy	Atom excitation energy	Sum
Before photon strikes atom	$h\nu$	0	0	$h\nu$
After photoelectron is ejected [Fig. 14.12(a)]	0	$h\nu - B_K$	B_K	$h\nu$
Case 1: Deexcitation by the emission of a K and an L photon				
Emission of K fluorescence photon [Fig. 14.12(b)]	$B_K - B_L$	$h\nu - B_K$	B_L	$h\nu$
Emission of L fluorescence photon	$B_K - B_L, B_L$	$h\nu - B_K$	0	$h\nu$
Case 2: Deexcitation by emission of an Auger electron from the L shell.				
Emission of Auger electron [Fig. 14.12(c)]	0	$h\nu - B_K$, $B_K - 2B_L$	$2B_L$	$h\nu$
First L-shell hole filled by fluorescence	B_L	$h\nu - B_K$, $B_K - 2B_L$	B_L	$h\nu$
Second L-shell hole filled by fluorescence	B_L, B_L	$h\nu - B_K$, $B_K - 2B_L$	0	$h\nu$

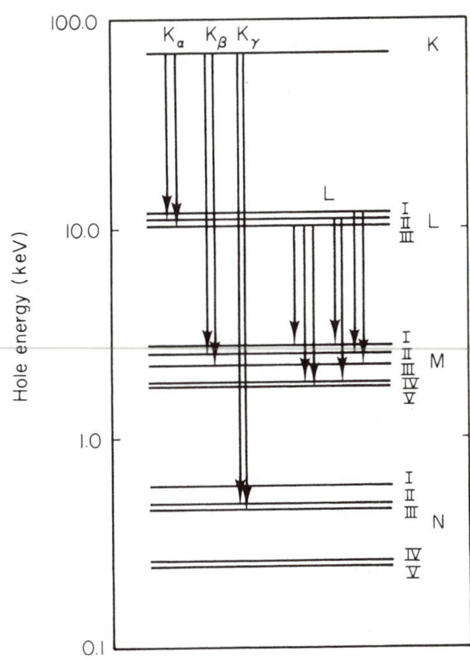

FIGURE 14.13. Energy-level diagram for holes in tungsten, and some of the x-ray transitions.

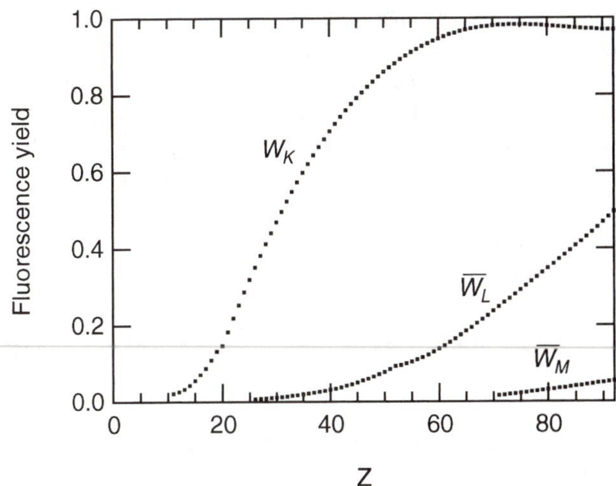

FIGURE 14.14. Fluorescence yields for *K*-, *L*-, and *M*-shell vacancies as a function of atomic number *Z*. Points are calculated using the polynomial coefficients in Table 8 of Hubbell *et al.* (1994).

$$\Delta l = \pm 1, \quad \Delta j = 0, \pm 1. \tag{14.33}$$

We saw in Eqs. (14.2) and (14.3) that the position of a level could be estimated by the Bohr formula corrected for screening. The energy of the K_α line (which depends on screening for two values of n) can be fitted empirically by

$$E_{K_\alpha} = \left(\frac{3}{4}\right)(13.6)(Z-1)^2. \tag{14.34}$$

After creation of a hole in the K shell, whether the atom deexcites by emitting a photon or an Auger electron is random. The probability of photon emission is called the *fluorescence yield*, W_K. The *Auger yield* is $A_K = 1 - W_K$. For a vacancy in the L or higher shells, one must consider the fluorescence yield for each subshell, defined as the number of photons emitted with an initial state corresponding to a hole in a subshell, divided by the number of holes in that subshell. The situation is further complicated by the fact that radiationless transitions can take place within the subshell, thereby altering the number of vacancies in each subshell. These are called *Coster–Kronig* transitions, and they are also accompanied by the emission of an electron. For example, a hole in the $L_{\rm I}$ shell can be filled by an electron from the $L_{\rm III}$ shell with the ejection of an M-shell electron. A *super–Coster–Kronig* transition would involve electrons all within the same shell, for example, a hole in the $M_{\rm I}$ shell filled by an electron from the $M_{\rm II}$ shell with the ejection of an electron from the $M_{\rm IV}$ shell.

One can define an average $\overline{W}_L, \overline{W}_M$, etc., but it is not a fundamental property of the atom, since it depends on the vacancy distribution in the subshells. Bambynek *et al.* (1972) review the physics of atomic deexcitations and present theoretical and experimental data for the fundamental parameters. They show that $\overline{W}_L$ is less sensitive to the initial vacancy distribution than one might expect, because of the rapid changes in hole distribution caused by the Coster–Kronig transitions. Hubbell *et al.* (1994) provide a more recent review and propose empirical polynomials to fit the fluorescent yield data as a function of atomic number. Figure 14.14 shows values for W_K, $\overline{W}_L$, and $\overline{W}_M$ as a function if Z. The points are calculated using the polynomials suggested by Hubbell *et al.* (1994). One can see from this figure that radiationless transitions are much more important (the fluorescent yield is much smaller) for the L shell than for the K shell. They are nearly the sole process for higher shells. The deexcitation is often called the *Auger cascade*. The Auger cascade produces many vacancies in the outer shells of the atom, and some of these may be filled by electrons from other atoms in the same molecule. This process can break molecular bonds. Moreover, the Auger and Coster–Kronig electrons from the higher shells can be very numerous. They are of such low energy that they travel only a fraction of a cell diameter. This must be taken into consideration when estimating cell damage from radiation. The effect of radiationless transitions is quite important for certain radioactive isotopes that are administered to a patient, particularly when they are bound to the cellular DNA. We will discuss them further in Chap. 16.

14.10. ENERGY TRANSFER FROM PHOTONS TO ELECTRONS

The attenuation coefficient gives the rate at which photons interact as they pass through material. If a beam of monoenergetic photons of energy $E=h\nu$ and particle fluence Φ passes through a thin layer dx of material, the number of particles per unit area that interact in the layer, $-d\Phi$, is proportional to the fluence and the attenuation coefficient: $-d\Phi=\Phi\mu_{\text{atten}}dx$. The energy fluence is $\Psi=h\nu\Phi$. The reduction of energy fluence of unscattered photons is $-d\Psi=-h\nu\, d\Phi$. For a thick absorber we can say that the number of unscattered photons and the energy carried by unscattered photons decay as

$$\Phi_{\text{unscatt}}=\Phi_0 e^{-\mu_{\text{atten}}x}, \quad \Psi_{\text{unscatt}}=\Psi_0 e^{-\mu_{\text{atten}}x}. \tag{14.35}$$

The total energy flow is much more complicated. Every photon that interacts contributes to a pool of secondary photons of lower energy and to a pool of electrons and positrons. Figure 14.15 shows the processes by which energy can move between the photon pool and the electron–positron pool. Energy that remains as secondary photons, such as those resulting from fluorescence or Compton scattering, can carry energy long distances from the site of the initial interaction. Ionizing particles (photoelectrons, Auger electrons, Compton recoil electrons, and electron–positron pairs) usually lose their energy relatively close to where they were produced. We will see in Sec. 14.13 that for primary photons below 10 MeV, the mean free path of the secondary electrons is usually short compared to that of the photons. Damage to cells is caused by local ionization or excitation of atoms and molecules. This damage is done much more efficiently by the electrons than by the photons.

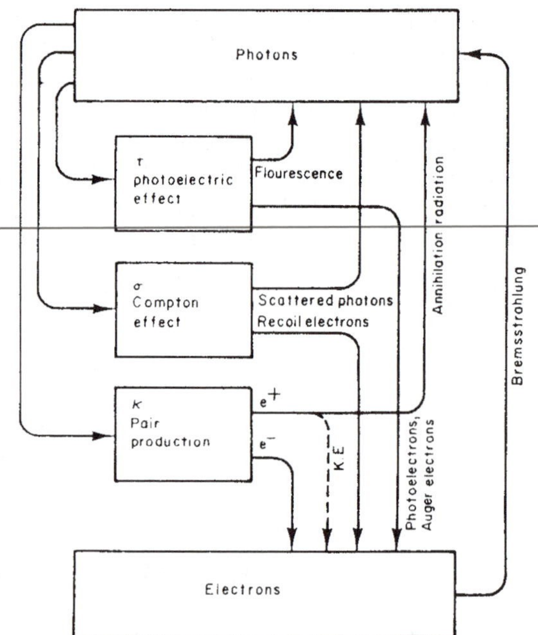

FIGURE 14.15. Routes for the transfer of energy between photons and electrons.

The *mass energy transfer coefficient* μ_{tr}/ρ is a measure of the energy transferred from primary photons to charged particles in the interaction. If N monoenergetic photons of energy E strike a thin absorber of thickness dx, the amount of energy transferred to charged particles is defined to be

$$\overline{dE_{\text{tr}}}=NE\mu_{\text{tr}}dx,$$

so that

$$\frac{\mu_{\text{tr}}}{\rho}=\frac{1}{\rho NE}\frac{\overline{dE_{\text{tr}}}}{dx}. \tag{14.36}$$

We can relate μ_{tr} to μ_{atten}. Suppose the material contains a single atomic species and that f_i is the average fraction of the photon energy that is transferred to charged particles in process i. (Different values of i denote the photoelectric effect, incoherent scattering, coherent scattering, and pair production.) Multiplying the number of photons that interact by their energy E and by f_i gives the energy transferred. Comparison with Eq. (14.24) shows that

$$\frac{\mu_{\text{tr}}}{\rho}=\frac{N_A}{A}\sum_i f_i\sigma_i. \tag{14.37}$$

Coherent scattering produces no charged particles, so

$$\frac{\mu_{\text{tr}}}{\rho}=\frac{N_A}{A}(\tau f_\tau+\sigma_{\text{incoh}}f_C+\kappa f_\kappa). \tag{14.38}$$

Fraction f_τ for the photoelectric effect can be written in terms of δ, the average energy emitted as fluorescence radiation per photon absorbed. The quantity δ is calculated taking into account all atomic energy levels and the fluorescence yield for each shell. The average electron energy is $h\nu-\delta$, so

$$f_\tau=\frac{h\nu-\delta}{h\nu}=1-\frac{\delta}{h\nu}. \tag{14.39}$$

We can estimate δ by assuming that τ_K is the dominant term in Eq. (14.8). The probability that the hole in the K shell is filled by fluorescence is W_K. The energy of the photon is B_K-B_L or B_K-B_M, and so on. A hole is left in a higher shell, which may decay by photon or Auger-electron emission. The latter is much more likely for the higher shells. Therefore nearly all of the photons emitted have energy B_K-B_L, so we have the approximate relationship

$$\delta\approx W_K(B_K-B_L). \tag{14.40}$$

For Compton scattering, the fraction of the energy transferred to electrons is implicit in Eqs. (14.21) and (14.22). The transfer cross section $f_C\sigma_C$, is plotted in Fig. 14.7.

For pair production, energy in excess of $2m_ec^2$ becomes kinetic energy of the electron and positron. The fraction is

$$f_\kappa=1-\frac{2m_ec^2}{h\nu}. \tag{14.41}$$

All of these can be combined to estimate μ_{tr}.

We will see in Sec. 14.11 that charged particles traveling through material can radiate photons through a process known as *bremsstrahlung*. The *mass energy-absorption coefficient* μ_{en} takes this additional effect into account. It is defined as

$$\frac{\mu_{en}}{\rho}=\frac{\mu_{tr}}{\rho}(1-g), \tag{14.42}$$

where g is the fraction of the energy of secondary electrons that is converted back into photons by bremsstrahlung in the material. The fraction of the energy converted to photons depends on the energy of the electrons. Since the average electron energy is different in the three processes, we can write (again assuming noninteracting atoms in the target material)

$$\frac{\mu_{en}}{\rho}=\frac{N_A}{A}\sum_i f_i\sigma_i(1-g_i). \tag{14.43}$$

In addition to bremsstrahlung, there is another process that converts charged-particle energy back into photon energy. Positrons usually come to rest and then combine with an electron to produce *annihilation radiation*. Occasionally, a positron annihilates while it is still in flight, thereby reducing the amount of positron kinetic energy that is available to excite atoms. While not mentioned in the ICRU Report 33 (1980) definition, this effect has been included in the tabulations of μ_{en}/ρ by Hubbell (1982). Seltzer (1993) has recently reviewed the calculation of μ_{tr}/ρ and μ_{en}/ρ.

The energy-transfer and energy-absorption coefficients differ appreciably when the kinetic energies of the secondary charged particles are comparable to their rest energies, particularly in high-Z materials. The ratio μ_{en}/μ_{tr} for carbon falls from 1.00 when $h\nu=0.5$ MeV to 0.96 when $h\nu=10$ MeV. For lead at the same energies it is 0.97 and 0.74. Tables are given by Attix (1986). The difference between the attenuation and the energy-absorption coefficients is greatest at energies where Compton scattering predominates, since the scattered photon carries away a great deal of energy. Figure 14.16 compares μ_{atten}/ρ and μ_{en}/ρ for water.

A short table of attenuation and energy-transfer coefficients is found in Appendix O. More extensive tables are found in Hubbell and Seltzer (1996). Another data source is a computer program provided by Boone and Chavez (1996).

We will return to these concepts in Sec. 14.15 to discuss the dose, or energy per unit mass deposited in tissue or a detector. First, we must discuss energy loss by charged particles.

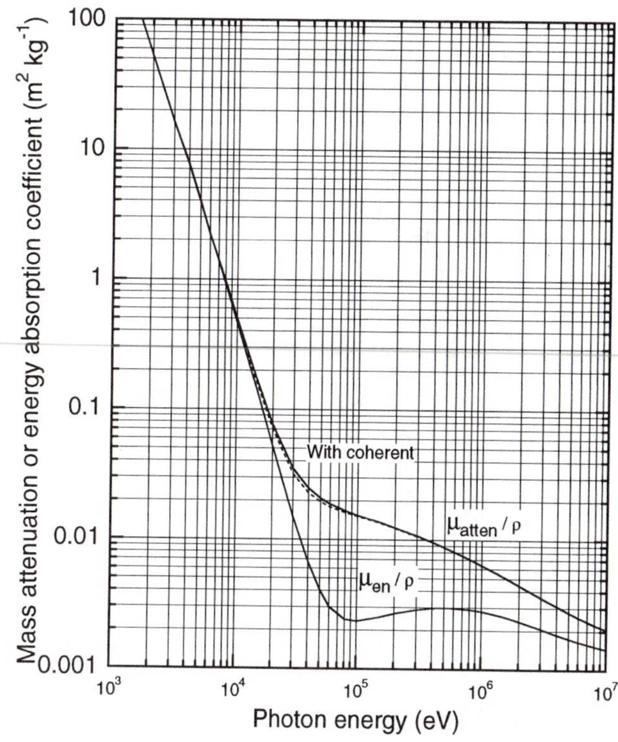

FIGURE 14.16. Coherent and incoherent attenuation coefficients and the mass energy absorption coefficient for water. Plotted from data in Hubbell (1982).

14.11. CHARGED-PARTICLE STOPPING POWER

The behavior of a particle with charge ze and mass M_1 passing through material is very different from the behavior of a photon. When a photon interacts, it usually disappears, either being completely absorbed as in the photoelectric effect or pair production or being replaced by a photon of different energy traveling in a different direction as in Compton scattering. The exception to this statement is coherent scattering, where a photon of the same energy travels in a different direction. A charged particle has a much larger interaction cross section than a photon—typically 10^4-10^5 times as large. Therefore the ''unattenuated'' charged-particle beam falls to zero almost immediately.

Each interaction usually causes only a slight decrease in the particle's energy, and it is convenient to follow the charged particle along its path. Figure 14.27 shows the tracks of some α particles (helium nuclei) in photographic emulsion. The length of the fiducial marks is 10 μm. Each particle entered at the bottom of the figure and stopped near the top. Figures 14.28 and 14.29 show the tracks of electrons. Figure 14.27 is in photographic emulsion, while Fig. 14.29 is in water. We will be discussing these tracks in detail in Sec. 14.14. For now, we need only note that the α particle tracks are fairly straight, with some deviation near the end of the track. The electrons, being lighter, show considerably more scattering.[2]

It is convenient to speak of how much energy the charged particle loses per unit path length, the *stopping power*, and its *range*—roughly, the total distance it travels before losing all its energy. The stopping power is the expectation value of the amount of kinetic energy T lost by the projectile per unit path length. (The term *power* is historical. The units of stopping power are J m^{-1} not J s^{-1}.) The *mass stopping power* is the stopping power divided by the density of the stopping material and is analogous to the mass attenuation coefficient (often we will say stopping power when we actually mean mass stopping power):

$$S=-\frac{dT}{dx}, \quad \frac{S}{\rho}=-\frac{1}{\rho}\frac{dT}{dx}. \tag{14.44}$$

In the energy-loss process, the projectile interacts with the target atom. The projectile loses energy W, which becomes kinetic energy or internal excitation energy of the target atom. Internal excitation may include ionization of the atom. If the atoms in the material act independently, the cross section per atom for an interaction that results in an energy loss between W and $W+dW$ is $(d\sigma/dW)\, dW$. The results of Sec. 13.4 can be used to write the probability that a projectile loses an amount of energy between W and $W+dW$ while traversing a thickness dx of a substance of atomic mass number A and density ρ:

$$(\text{probability})=\frac{\bar{n}}{N}=\frac{N_A\rho}{A}\,dx\,\frac{d\sigma}{dW}\,dW. \tag{14.45}$$

The average total energy loss is

$$dT=\frac{N_A\rho}{A}\,dx\int_0^{W_{\max}} W\,\frac{d\sigma}{dW}\,dW, \tag{14.46}$$

and the mass stopping power is

$$\frac{S}{\rho}=\frac{N_A}{A}\int_0^{W_{\max}} W\,\frac{d\sigma}{dW}\,dW. \tag{14.47}$$

The integral is sometimes called the *stopping cross section* ϵ. Its units are J m^2.

Figure 14.17 shows the mass stopping power for protons, α particles ($z=2$, $M_\alpha=4M_p$), and electrons and positrons ($z=\pm1$) in carbon as a function of energy. We see a number of features of these curves:

1. All of the stopping power curves have roughly the same shape, rising with increasing energy, reaching a peak, and then falling.
2. There is a region where the stopping power falls approximately as $1/T$.

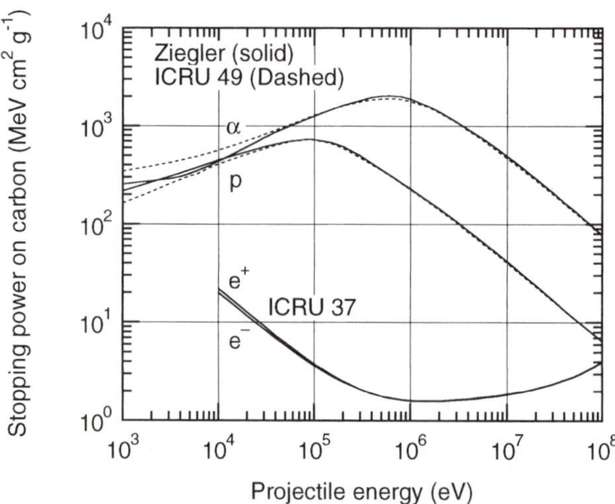

FIGURE 14.17. The mass stopping power for electrons, positrons, protons, and α particles in carbon vs kinetic energy. Plotted from data in ICRU 37, ICRU 49, and the program SRIM (Stopping and Range of Ions in Matter), version 96.04 [see Ziegler, Biersack and Littmark (1985)].

3. At still higher energies the curves rise again. This can be seen for the electron and positron curves above 1 MeV. Similar increases occur in the proton and α-particle curves at higher energies than are plotted here.

The similarities suggest that the stopping power curves for different projectiles may be related. Figure 14.18 shows the similarities more clearly. The stopping powers are plotted vs particle speed in the form $\beta=v/c$. At low energies β is related to kinetic energy by

$$\beta=\left(\frac{2T}{Mc^2}\right)^{1/2}. \tag{14.48}$$

This equation is correct for small β but allows v to become greater than c as the kinetic energy is increased. The relativistically correct expression,

$$\beta=\left[1-\left(\frac{1}{T/Mc^2+1}\right)^2\right]^{1/2}, \tag{14.49}$$

for which $\beta\to1$ as T becomes infinite, was used to convert Fig. 14.17 to Fig. 14.18. The α-particle stopping power in Fig. 14.17 has been divided by the square of the α-particle charge number $z^2=4$. All three curves of $(1/z^2)S/\rho$ vs β are described by very similar functions for $\beta>0.04$, though the electron and α-particle curves are about 10% below the proton curve.[3] At low speeds the scaled α-particle curve falls significantly below the proton curve. The reason, the formation of an electron cloud on the α particle, is discussed below.

It is not difficult to understand the basic shape of the stopping power curve. Most of the energy loss is from the

[2]This distinction between photons and charged particles represents two extremes on a continuum, and we must be careful not to adhere to the distinction too rigidly. A photon may be coherently scattered through a small angle with no loss of energy, while a charged particle may occasionally lose so much energy that it can no longer be followed.

[3]A value $\beta=0.04$ corresponds to a kinetic energy of 400 eV for electrons, 800 keV for protons, and 3.2 MeV for α particles.

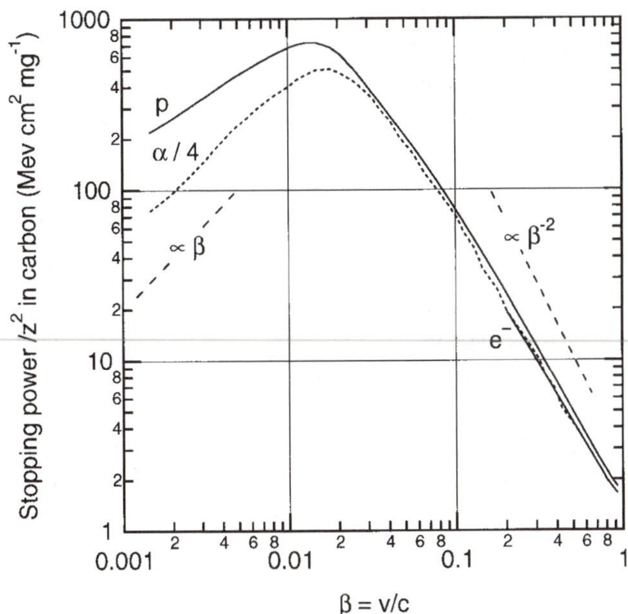

FIGURE 14.18. The scaled stopping power. The stopping power is plotted vs the speed $\beta = v/c$ of the projectile for electrons, protons, and α particles. The α-particle stopping power has been divided by 4, the square of the α particle charge z. Proton and α-particle stopping powers are from the program SRIM (see caption for Fig. 14.17). The electron stopping power is from ICRU Report 37 (1984).

projectile to the electrons of the target atom. Since the electrons are bound to the target nucleus, the speed with which the projectile passes the target is important. Imagine pushing slowly on a swing with a force that gradually increases and then decreases. The net force on the swing is the vector sum of the external force exerted F_{ext}, the vertical pull of gravity, and the tension in the ropes and equals the swing's mass times acceleration. For small horizontal displacements x from equilibrium, the vector sum of the weight and the tension in the string is horizontal and nearly proportional to x. It points toward the equilibrium position, and for small displacements is approximately a linear restoring force. If the proportionality constant is k,

$$ma = F_{ext} - kx.$$

This is the equation of motion for an undamped harmonic oscillator (Chap. 10 and Appendix F). If the force builds up slowly, there is a very small acceleration, and the swing angle changes so that $F_{ext} \approx kx$. As the force decreases the swing returns to its resting position. All of the work that was done to displace the swing is now returned as work by the swing on the source of the external force. No net energy has been imparted to the swing. This is called an *adiabatic* process or approximation, a slightly different use of the term than in Chap. 3.

At the other extreme, the force could be applied for a very short time, building up to a peak and falling quickly. In this case, the swing does not have time to move and $F_{ext} = ma$. This can be integrated to give

$$\int F_{ext}\,dt = m \int a\,dt = m(v_{final} - v_{initial}). \quad (14.50)$$

The swing acquires a velocity and hence some kinetic energy. The integral of force with respect to time is called the *impulse*, and this limit is the *impulse approximation.*

The two limits depend on whether the duration of the force is long or short compared to the natural period of the swing. The atomic electrons are bound, and they have a natural period that is the circumference of their orbit divided by their speed $v_{electron}$. The length of time that a projectile exerts a force on the electrons is roughly the diameter of the atom divided by the projectile speed. Ignoring factors of 2π, we see that the passage of the projectile will be adiabatic if

$$\frac{d_{atom}}{v_{projectile}} \gg \frac{d_{atom}}{v_{electron}}$$

or $v_{projectile} \ll v_{electron}$. The impulse approximation will be valid if $v_{projectile} \gg v_{electron}$.

This is sufficient to explain the shape of the stopping-power curves in Fig. 14.18. When the projectile has very low energy it moves past the atom so slowly that the electrons have time to rearrange themselves[4] and then return to their original state as the projectile leaves, restoring to the projectile the energy that they received while rearranging. As the projectile speed increases, the process is no longer adiabatic, first for the more slowly moving outer electrons and then for more and more of the inner atomic electrons as the speed increases. At the other extreme, when the projectile speed becomes high enough we can think of the process in terms of the impulse approximation. The faster the projectile moves by, the shorter the time the force is applied and the smaller the energy transfer. The energy transfer is most effective, and the peak of the stopping power occurs, when the speed of the projectile is about equal to the speed of the atomic electrons in the target.

The cross section in Eqs. (14.45)–(14.47) is the sum of cross sections for three possible processes. We have already described the stopping power due to interactions of the projectile with the target electrons, S_e. There is another contribution to the stopping power from interactions of the projectile with the target nucleus, S_n. It is also possible for the energy loss to involve the radiation of a photon, so we also have radiative stopping power, S_r. Because these are independent processes, the total stopping power and the cross section are each the sum of three terms:

[4]Classically, if the electrons go around the nucleus many times while the projectile moves by, the shape of their orbits can change in response to the projectile. Quantum mechanically, the shape of the wave function can change, but the quantum numbers do not change.

$$\frac{S}{\rho}=\frac{S_e}{\rho}+\frac{S_n}{\rho}+\frac{S_r}{\rho},$$
$$\frac{d\sigma}{dW}=\left(\frac{d\sigma}{dW}\right)_e+\left(\frac{d\sigma}{dW}\right)_n+\left(\frac{d\sigma}{dW}\right)_r. \tag{14.51}$$

To compare these processes, we need to consider the maximum energy that can be transferred, as well as the relative probability of each process. The maximum possible energy transfer W_{max} can be calculated using conservation of energy and momentum. For a collision of a projectile of mass M_1 and kinetic energy T with a target particle of mass M_2 which is initially at rest, a nonrelativistic calculation gives

$$W_{max}=\frac{4TM_1M_2}{(M_1+M_2)^2}. \tag{14.52}$$

The analogous relativistic equation (needed, for example, when the projectile is an electron) is

$$W_{max}=\frac{2(2+T/M_1c^2)TM_1M_2}{M_1^2+2(1+T/M_1c^2)M_1M_2+M_2^2}. \tag{14.53}$$

The values of W_{max} for representative projectiles and targets are shown in Table 14.3, along with the percentage of the stopping power due to nuclear collisions. For electrons, the table also shows the percentage of the stopping power due to radiative transitions. The percentages are calculated from ICRU Report 49 (1993). Electrons can scatter from nuclei, but the amount of recoil energy transferred to the nucleus is very small. Although electrons undergo a great deal of nuclear scattering, which results in a tortuous path through material, they lose very little energy in a nuclear scattering. The heavier projectiles can lose relatively more energy in each nuclear collision than in each electron collision. For a given kind of projectile, nuclear stopping is more important at lower energies, because less energy can be transferred to an electron. The heavier the projectile for a given energy, the more important the nuclear term becomes, for the same reason.

The collision of electrons with electrons is a special case. Equation (14.52) or (14.53) gives $W_{max}=T$. Consider the collision of two billiard balls of the same mass. If the projectile misses the target, it continues straight ahead with its original energy and $W=0$. If it hits the target head on, it comes to rest and the target travels in the same direction with the same energy that the projectile had—a situation indistinguishable from the complete miss. It is customary (but arbitrary) in the case of identical particles to say that the particle with higher energy is the projectile, so $W_{max}=T/2$. This adjustment has been made in Table 14.3 for electrons on electrons and protons on protons.

TABLE 14.3. *Maximum energy transfer and relative importance of nuclear and radiative interactions for various projectiles and targets.*

Projectile	Target	Nuclear W_{max} (eV)	Electron W_{max} (eV)	S_n/S	S_r/S
Electron, 100 keV	Hydrogen	240	50 000		0.01%
	Carbon	20	50 000		0.09%
	Lead	1	50 000		2.2%
Electron, 1 MeV	Hydrogen	4 300	500 000		0.13%
	Carbon	360	500 000		0.65%
	Lead	20	500 000		11.5%
Proton, 10 keV	Hydrogen	5 000	20	1.7%	
	Carbon	2 800	20	1.6%	
	Lead	200	20	1.5%	
Proton, 100 keV	Hydrogen	50 000	220	0.17%	
	Carbon	28 400	220	0.15%	
	Lead	1 900	220	0.24%	
Proton, 1 MeV	Hydrogen	500 000	2 200	0.11%	
	Carbon	280 000	2 200	0.07%	
	Lead	19 000	2 200	0.09%	
α particle, 10 keV	Hydrogen	6 400	5	27%	
	Carbon	7 500	5	12%	
	Lead	700	5	10%	
α particle, 100 keV	Hydrogen	64 000	50	1.6%	
	Carbon	75 000	50	1.1%	
	Lead	7 400	50	1.8%	
α particle, 1 MeV	Hydrogen	640 000	500	0.13%	
	Carbon	750 000	500	0.12%	
	Lead	74 000	500	0.20%	

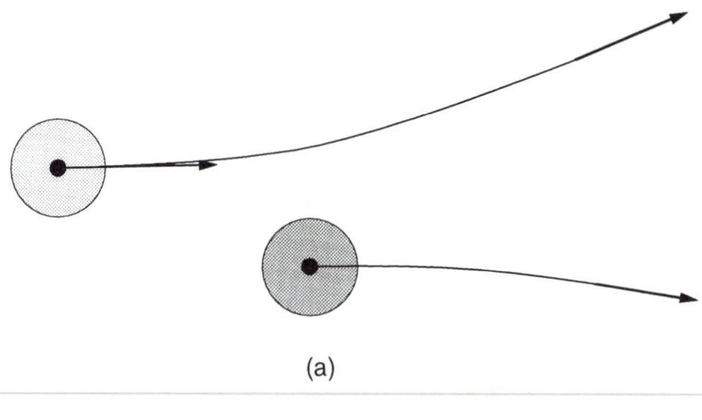

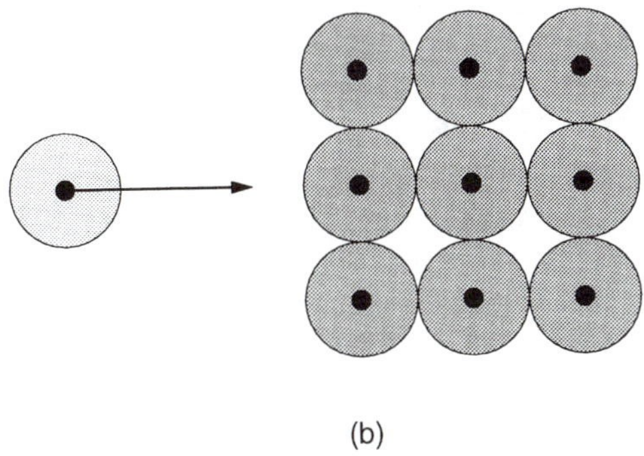

FIGURE 14.19. A projectile, which may or may not carry an electron cloud, moves past a target atom. (a) In a gas the projectile interacts with one atom at a time. (b) In a liquid or a solid, neighboring atoms may influence the interaction.

Radiation is only important for electrons and occurs in a certain fraction of the elastic electron scatterings from the target nucleus. Nuclear scattering gives the electron a fairly large acceleration. Classically, an accelerated charged particle radiates electromagnetic waves. This process is called *bremsstrahlung*—braking or deceleration radiation. The energy radiated is proportional to the square of the acceleration, so bremsstrahlung is only important for light projectiles. There is also a contribution from electron–electron or positron–electron scattering. The electron–electron contribution vanishes at low energies, although the positron–electron bremsstrahlung does not.[5] We will see in Chap. 15 that bremsstrahlung is an important component of the x-ray spectrum produced when a beam of electrons strikes a target. Even so, the fraction of the electron energy that is converted to radiation is small.

An atom has a radius of a few times 10^{-10} m. The nucleus of the atom is much smaller, about 10^{-15} m, and contains most of the atom's mass. The atom's size is determined by the electron cloud around the nucleus. Figure 14.19(a) shows a projectile entering at the left and traveling to the right through a gas. It interacts with one target atom at a time. The solid black dots represent the nuclei of the projectile and the target atom. The shaded circles represent the electron clouds. The projectile may or may not have an electron cloud, which is shown with lighter shading. Figure 14.19(b) shows the interaction with a solid or liquid in which the target atoms are tightly packed, and it may not be accurate to say that the projectile interacts with only one atom at a time.

Classically, the motion of a charged projectile past a charged target depends on the charges and masses of the particles, the initial velocity or kinetic energy of the projectile, and the *impact parameter* b, which is the perpendicular distance from a line through the initial velocity of the projectile to the target, as shown in Fig. 14.20. The classical cross section for having an impact parameter between b and $b+db$ is the area of the ring, $2\pi b\, db$. If we could relate b to the energy loss W, we would have the cross section $d\sigma/dW$ of Eq. (14.47).

The energy-loss process is quite complicated, and the cross section cannot be calculated exactly. A great deal of experimental and theoretical work on stopping powers has been done, extending from 1899 to the present time. Different models are used for the low energy regime and the high energy regime. The history is nicely reviewed by Ziegler, Biersack, and Littmark (1985). Much of the recent work on stopping powers has been motivated by the use of ion implantation to make semiconductors, the analysis of materials using ion beams, and medical applications. Currently stopping powers of low-energy heavy ions can be calculated with an accuracy of better than 10%. For high-speed light ions the accuracy is better than 2%. Current references are found in Ziegler, Biersack, and Littmark (1985).

14.11.1. Interaction with Target Electrons

We first consider the interaction of the projectile with a target electron, which leads to the *electronic stopping power*, S_e. Many authors call it the *collision stopping power*, S_{col}. There can be interactions in which a single electron is ejected from a target atom or interactions with the electron cloud as a whole (a "plasmon" excitation).

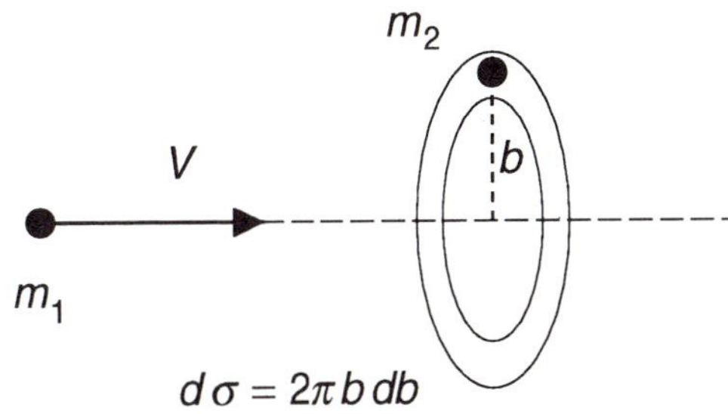

FIGURE 14.20. The impact parameter is the perpendicular distance from the target particle to a line extended from the projectile in the direction of its velocity before the interaction.

[5]This difference can be understood classically. In the first approximation, the radiation by a charge is proportional to the product of the charge times its acceleration, qa. For two interacting electrons, $a_1=-a_2$, $q_1=q_2$, and the sum of these two terms vanishes. For an electron and a positron $a_1=-a_2$, $q_1=-q_2$, and the two terms add.

The stopping power at higher energies, where it is nearly proportional to β^{-2}, has been modeled by Bohr, by Bethe, and by Bloch [see the review by Ahlen (1980)]. The Bethe–Bloch model is also valid for relativistic energies. A nonrelativistic model for high energies was developed by J. Lindhard and his colleagues [see references in Ziegler, Biersack, and Littmark (1985)]. It allows more accurate calculations of which electrons in the target receive energy from the projectile.

We can gain considerable insight into the high-energy loss process by making a classical calculation of the cross section for transferring energy to an electron using the impulse approximation. This is a simplification of the Bethe–Bloch model. In our model, a heavy projectile passes by a free electron that is at rest. Momentum is transferred from the projectile to the electron. Because of its large mass, the projectile's velocity does not change appreciably, but the lighter electron acquires an appreciable velocity and kinetic energy. If the momentum transferred to the electron is p, its kinetic energy is $p^2/2m_e$. That kinetic energy must have been lost by the projectile.

Figure 14.21 shows a particle of mass M, charge ze, and velocity $V=\beta c$ moving past a stationary electron. The impact parameter b is the perpendicular distance from the electron to the path of the projectile. The distance from the projectile to the electron is r, and the distance along the path to the point of closest approach is ξ. The momentum transferred to the electron is $\mathbf{p}=\int \mathbf{F}\,dt=-e\int \mathbf{E}\,dt$. By symmetry, there is no component of $\mathbf{p}$ parallel to the path of the projectile. The reason is shown in Fig. 14.22. For each location of the projectile that gives a parallel component of $\mathbf{p}$ in one direction, there is a position an equal distance on the other side of the point of closest approach that gives a component of $\mathbf{p}$ with the same magnitude but in the opposite direction. The perpendicular component of $\mathbf{p}$ is the same for both locations, so there is a net perpendicular component of momentum transfer. The perpendicular component of $\mathbf{E}$ is

$$E_\perp = E\sin\theta = \frac{ze\sin\theta}{4\pi\epsilon_0 r^2} = \frac{zeb}{4\pi\epsilon_0 r^3} = \frac{ze}{4\pi\epsilon_0}\frac{b}{(\xi^2+b^2)^{3/2}}.$$

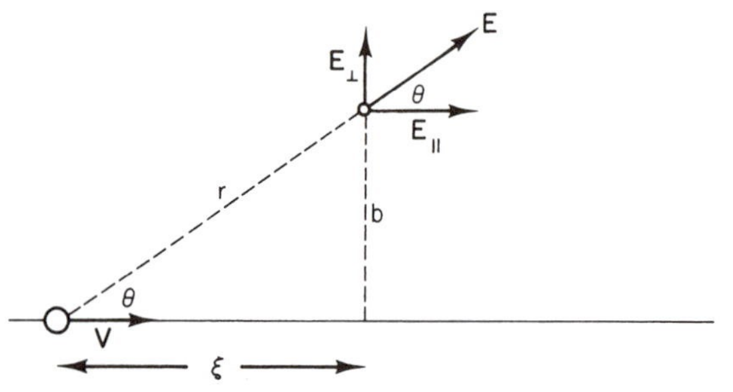

FIGURE 14.21. A heavy particle of charge ze, mass M, and velocity **V** moves past a stationary electron.

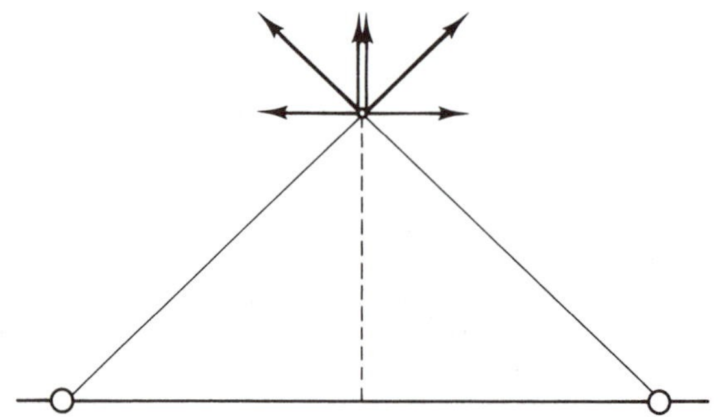

FIGURE 14.22. Why the parallel component of **p** is zero. For every point where the projectile gives a particular $\mathbf{E}_\parallel$, there is a symmetric point where $\mathbf{E}_\parallel$ is equal but opposite. The components $\mathbf{E}_\perp$ are in the same direction in both places, so the perpendicular component of **p** does not vanish.

The perpendicular impulse is $\int F_\perp dt = -e\int E_\perp (dt/d\xi)d\xi$. If the fraction of energy lost by the projectile is small, then $dt/d\xi = 1/\beta c$ does not change during the collision. The impulse is therefore

$$p = -\frac{e}{V}\int E_\perp d\xi = -\frac{ze^2 b}{4\pi\epsilon_0 \beta c}\int_{-\infty}^{\infty}\frac{d\xi}{(\xi^2+b^2)^{3/2}}$$

$$= -\frac{ze^2 b}{4\pi\epsilon_0\beta c}\lim_{x\to\infty}\left[\frac{\xi}{b^2(\xi^2+b^2)^{1/2}}\right]_{-x}^{x}$$

$$= -\frac{2ze^2}{4\pi\epsilon_0\beta c b}.$$

The smaller the impact parameter, the greater the momentum transfer to the electron. The kinetic energy acquired by the electron is

$$W = \frac{p^2}{2m_e} = \frac{2z^2e^4}{(4\pi\epsilon_0)^2 m_e c^2\beta^2 b^2}.$$

The factor $e^4/[(4\pi\epsilon_0)^2 m_e c^2]$ depends only on the charge and mass of the electron. It can also be written as $r_e^2 m_e c^2$, where r_e is the classical radius of the electron [Eq. (14.18)]. The factor has the numerical value

$$r_e^2 m_e c^2 = 6.50\times10^{-43}\ \text{J m}^2 = 4.06\times10^{-24}\ \text{eV m}^2.$$

Using this notation the energy transfer is

$$W = \frac{2z^2 r_e^2 m_e c^2}{\beta^2 b^2}. \tag{14.54}$$

Here z is the charge on the projectile, βc is its speed, and b is the impact parameter. Note that W does not depend on the mass of the heavy projectile, but only on its speed. As the speed becomes less, the energy transfer becomes greater, because the projectile takes longer to move past the electron and the force is exerted for a longer time (as long as the time is still short enough so that the impulse approximation remains valid).

If the electrons are uniformly distributed, the cross section for each electron is $d\sigma=(d\sigma/dW)\,dW=2\pi b\,db$. This can be written, with the help of Eq. (14.54), in terms of W:

$$\frac{d\sigma}{dW}dW=\frac{4\pi z^2 r_e^2 m_e c^2}{2\beta^2}\frac{dW}{W^2}. \tag{14.55}$$

This expression diverges as W approaches zero, corresponding to very large impact parameters. However, the assumption that the target electrons are free fails in this limit, so that there is some effective lower limit $W_{\min}$. Also, the greater the impact parameter, the longer the electron will experience the force exerted by the projectile (though it will be weaker). If the time is too long, the electron can move in response to the force and not absorb as much energy. The impulse approximation is no longer valid. We have already seen that there is a maximum energy transfer $W_{\max}$. Multiplying the cross section by W and integrating from $W_{\min}$ to $W_{\max}$ gives

$$\frac{S_e}{\rho}=\frac{4\pi N_A r_e^2 m_e c^2}{\beta^2}\frac{Z}{A}z^2\ln\left(\frac{W_{\max}}{W_{\min}}\right). \tag{14.56}$$

The factor $4\pi N_A r_e^2 m_e c^2$ has the value 30.707 eV m^2 mol^{-1}=0.307 07 MeV cm^2 mol^{-1}.

A quantum-mechanical calculation gives a result of essentially the same form as Eq. (14.56). The logarithmic term includes both ionization and plasmon excitation[6] and is called the *stopping number per atomic electron* $L(\beta,z,Z)$:

$$\frac{S_e}{\rho}=\frac{4\pi r_e^2 m_e c^2}{\beta^2}N_A\frac{Z}{A}z^2 L(\beta,z,Z). \tag{14.57}$$

For heavy charged particles L has the form

$$L(\beta,z,Z)=L_0+zL_1+z^2L_2,$$
$$L_0=\ln\left(\frac{\beta^2}{1-\beta^2}\right)+\ln\left(\frac{2m_e c^2}{I(z)}\right)-\beta^2-\frac{C}{Z}-\frac{\delta}{2}. \tag{14.58}$$

Equation (14.57) with $L=L_0$ is often called the *Bethe–Bloch formula.* The second term in L_0 depends on $I(Z)$, the ionization potential of the atoms in the absorber, averaged over all the electrons in the atom. Values of $I(Z)$ have been calculated theoretically and also derived from measurements of the stopping power. They range from 14.8 eV for hydrogen to 884 eV for uranium. The value 14.8 eV is greater than the ground-state energy of hydrogen, 13.6 eV, because the ejected electron has some average kinetic energy. Values of I can vary considerably, depending on whether the other correction terms are present. For example, values of I in the literature for hydrogen range from 11 to 20 eV. Discussions of values for I and the various terms in L can be found in ICRU Report 49 (1993), in Ahlen (1980), and in Attix (1986). The term $\delta/2$ corrects for the *density effect.* The calculation above assumed that the electron experienced the full electric field of the projectile. However, other electrons in the absorber move slightly, polarizing the absorber and reducing the field. This effect becomes important at high energies as the electric field is distorted by relativistic effects. It also depends on the density of the absorber. A small density effect persists in conductors even at low energies; however, it is usually incorporated into the value of $I(Z)$. For the projectile energies we are considering, the density effect is most important for electrons.

An alternative nonrelativistic treatment by Lindhard and colleagues allows the use of accurate atomic electron density distributions and also considers the effect of electrons in neighboring atoms.[7] In the Lindhard model the stopping power is

$$\frac{S_e}{\rho}=\frac{N_A}{A}\int z^2 I(V,\rho_e)\rho_e 4\pi r^2 dr, \tag{14.59}$$

where z is the projectile charge, I is the *stopping interaction strength* in J m^2 (more often in eV pm^2),[8] ρ_e is the electron density in the atom (in units of the electron charge), and $4\pi r^2\,dr$ is the volume element. Integration of ρ_e over all volume gives Z, the atomic number of the target. Comparison of Eqs. (14.59) and (14.47) shows that the integral in Eq. (14.59) is the stopping cross section per target atom.

Figure 14.23 shows how the Lindhard model explains why the stopping power falls below the $1/\beta^2$ curve at lower projectile velocities. Each panel shows the electron density in copper, $4\pi r^2\rho_\epsilon$, and the interaction strength I. Their product, the solid line, is the integrand in Eq. (14.59). The integral is taken from 0 to 0.14 nm (1.4 Å). The K, L, and M shells of copper can be seen in the electron density curve. Figure 14.23(a) is for a 10-MeV proton or some other heavy ion with the same speed. The projectile is moving fast enough so that all electrons except those in the K shell interact with it. Contrast this with Fig. 14.23(b), which is for a 100-keV proton. The projectile speed is much less, and the interaction is almost exclusively with the outer electrons.[9]

Both the Bethe–Bloch and Lindhard models fail at low energy, because the electrons are not free and many of the interactions are adiabatic. Some models [reviewed by Ziegler, Biersack, and Littmark (1985)] predict a stopping power proportional to projectile velocity. This has been found to be true in general, though not for all elements. The

[6]A plasmon excitation is due to the interaction of the projectile with the entire electron cloud of the atom.

[7]The electron density functions are calculated using quantum mechanics. The problem is to find the electron distribution by solving Schrödinger's equation with the potential distribution due to the nucleus and the potential due to the electron charge distribution for which one is solving. This self-consistent computation is called the Hartree–Fock approximation.

[8]I is not the same as the average ionization energy of Eq. (14.58).

[9]The solid line representing the integrand does not fall to zero at 0.12 nm=1.2 Å because of the effect of electrons from neighboring atoms. In a solid there are no regions where the electron density is zero.

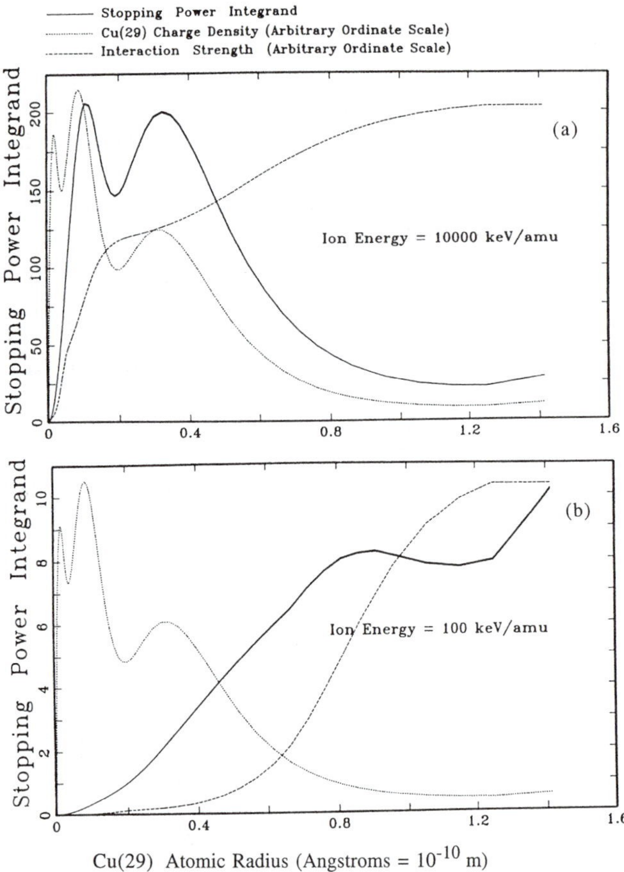

FIGURE 14.23. Calculation of the stopping power at low energies involves integrating the product of the electron charge distribution in the target atom and the interaction strength function, which depends on the projectile speed. The dotted line shows the electron charge density for copper. The solid line shows the integrand. (a) For 10-MeV protons, all electrons but those in the *K* shell contribute. (b) For 100-keV protons the interaction function has changed, and only the outermost electrons contribute. Note the much different ordinate scales in (a) and (b). Provided by J. F. Ziegler.

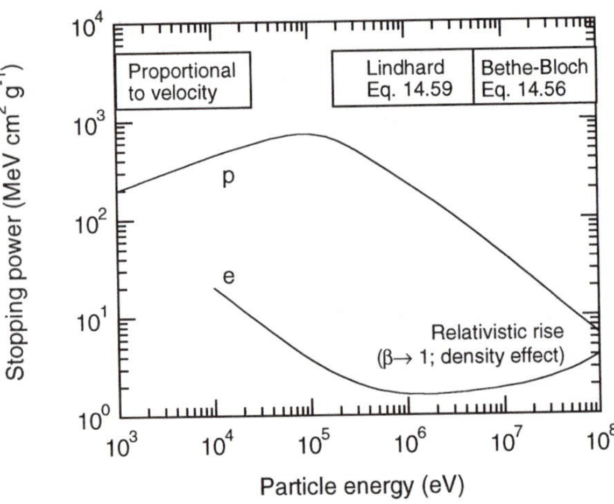

FIGURE 14.24. Proton and electron stopping power vs energy in carbon, showing the regions in which various models are valid.

experiments are quite difficult because of the thinness of the targets, contamination, etc. Figure 14.24 shows the regions where the various models apply for protons. For electrons, relativistic effects are important above about 500 keV. The rise in stopping power at high energies is due to the density effect (polarization of the electrons).

Another important effect at low energies is that the slowly moving ion can capture electrons, decreasing the value of z^2. Ziegler, Biersack, and Littmark (1985) discuss the scaling of data for different projectiles and the appropriate effective charge values. The average projectile charge follows a universal curve when plotted as a function of the appropriate combination of the speeds of the projectile and target electrons. They, and the ICRU Report 49 (1993), assume that for protons the effective charge is always unity. The theoretical justification is that the radius of the electron orbit in hydrogen is larger than the interatomic separation in solids.

14.11.2. Scattering from the Nucleus

The projectile can also scatter from the target atom as a whole. The recoil kinetic energy of the atom is lost by the projectile. Since the nucleus contains most of the mass, the kinematics are those of the bare projectile and the target nucleus, and this process is called *nuclear scattering*, with stopping power S_n. (Sometimes it is called *elastic scattering*, with a subscript that can cause it to be confused with electron interactions.)

Just as with Compton scattering, knowing the angle through which the projectile is scattered defines the amount of energy transferred to the target. The angle depends on the impact parameter. The problem can be solved for a given impact parameter if the force between the projectile and target is a function only of their separation and one knows the potential energy of their separation. The details are found in Ziegler, Biersack, and Littmark (1985). We will simply comment on the contributions to the potential energy. They are

1. The Coulomb force between the projectile and the target nucleus.
2. The Coulomb force between the projectile and the electron cloud of the target atom.
3. The Coulomb attraction between the target nucleus and any electrons surrounding the projectile.
4. The Coulomb repulsion between the electron clouds of the target and the projectile.
5. A term due to the Pauli exclusion principle if the

projectile is an ion with an electron cloud. To see how it arises, suppose that both the projectile and target have both of their possible K-shell electrons. If the nuclei get close enough, they effectively form a single nucleus that cannot have four K-shell electrons. Therefore two of the electrons have to move to unfilled shells. This requires energy that comes from the kinetic energy of the projectile. This is called *Pauli promotion*. Even though the electrons have time to return to their original orbits for a slow projectile, the effect changes the overall potential and hence the projectile orbit and the probability of a particular energy transfer.

6. An *exchange term* that also arises from the Pauli principle, related to whether the spins of the projectile and target electrons are parallel or antiparallel.

Because nuclear scattering is relatively unimportant for the charged particles we are considering and because it does not lead to ionization, we will not describe any details of the calculations.

14.11.3. Stopping of Electrons

Equations similar to Eq. (14.57) are obtained for electrons and positrons. Recall that energy loss in nuclear scattering is negligible for positrons and electrons because they are so light, and that bremsstrahlung transfers some of the electron kinetic energy to radiation. Electrons and positrons are assumed to collect no screening charge. Even at low energies, the electron velocities are high enough to that the Bethe–Bloch model is used. The collision stopping power for electrons is[10]

$$\frac{S_{\text{col}}}{\rho} = 4\pi N_A r_e^2 m_e c^2 \frac{1}{\beta^2} \frac{Z}{A} L_{\pm}. \tag{14.60}$$

The subscript $\pm$ indicates that stopping number per electron is slightly different for electrons and for positrons. The exact forms can be found in Attix (1986) or in ICRU Report 37 (1984). In both cases L depends on $I(Z)$ and the density effect. An accurate calculation of the shell correction for electrons has not been made; therefore ICRU Report 37 omits the shell correction from the tables for electrons and positrons. This omission makes the use of Eq. (14.59) less accurate for electrons below 10 keV. The best values of S_e/ρ for electrons and positrons are obtained from theoretical calculations using Eq. (14.60) and values of $I(Z)$ determined from proton data.

14.11.4. Compounds

In dealing with compounds, it is frequently assumed that each atom in the target interacts independently with the projectile, as we assumed for photons. The stopping power per molecule is then equal to the sum of the stopping powers for each atom in the molecule. This leads to a formula analogous to Eq. (14.32), known as the *Bragg rule*:

$$\frac{S}{\rho} = \sum_i w_i \left(\frac{S}{\rho}\right)_i. \tag{14.61}$$

This equation applies to the collision, radiative, nuclear, and total stopping powers. This approximation is quite inaccurate near the peak of the stopping power curve, where the errors can be greater than a factor of 2. This is not surprising, given the behavior of the scattering function I in Fig. 14.23(b). Most of the energy loss is to outer electrons—the conduction electrons if the substance is a metal. In a semiconductor there are gaps in the energy levels, and this precludes the low-energy transfers. As a result, the stopping power is lower in semiconductors. In crystals, channeling can occur. The stopping power depends on the orientation of the trajectory with the crystal symmetry axis.

Carbon poses a particular problem. It is an important element in the body, and it has chemical bonds that range from metallic to insulating in nature. Various investigators have shown variations in stopping power of 30% for ions in pure carbon, depending on how it was fabricated. Graphite can be made with different electrical conductivities, and there are associated differences in stopping power. Ziegler and Manoyan (1988) have applied charge-scaling techniques to several organic carbon compounds by considering separately the stopping due to closed atomic shells (''cores'') and the remaining ''bonds'' between different pairs of atoms.

ICRU Reports 37 (1984) 49 (1993) handle departures from the Bragg rule in the first approximation by using different values of I for electrons in compounds. The density effect is important for electrons and also does not follow the Bragg rule.

Some stopping-power values are found in Appendix O. ICRU Report 37 has more extensive tables for positrons and electrons. ICRU Report 49 has stopping powers for protons and α particles. A computer program for protons and ions, SRIM (Stopping and Range of Ions in Matter) is described by Ziegler, Biersack, and Littmark (1985).

14.12. LINEAR ENERGY TRANSFER AND RESTRICTED COLLISION STOPPING POWER

In modeling the effect of ionizing radiation on targets, whether they be radiation detectors, photographic emulsions, cells, or parts of cells, one often wants to know how much of a charged particle's energy is absorbed ''locally,'' that is, within some region around a particle's trajectory. An accurate calculation is difficult, since some of the electrons produced may leave the region of interest. Also, the energy

[10]The literature often replaces the 4π by 2π for electrons and makes L twice as large.

absorbed in some region of interest around a particle track comes both from energy lost by the particle while traversing that track segment and also from photons and charged particles produced elsewhere by the projectile. [This is discussed in detail in ICRU Report 16 (1970).]

An approximation to the desired quantity is the *linear energy transfer* (LET) or the *restricted linear collision stopping power* L_Δ. It is defined as the ratio dT/dx, where dx is the distance traveled by the particle and dT is the mean energy loss to electrons that result in energy transfers less than some specified Δ. This use of the symbol L should not be confused with the stopping number of Eqs. (14.57)–(14.60). The quantity L_Δ can be calculated by replacing W_{max} by Δ in the expression for the stopping power. The value of Δ is usually specified in electron volts.

The electron stopping power S_e is numerically equal to L_∞. However, S_e is defined in terms of the energy *lost by the particle*, while L is defined in terms of energy *imparted to the medium.*

Note that although the quantity actually of interest may be the energy imparted within some *region* around the trajectory, this definition is based on *energy transfers* less than Δ. A quantity based on the region of interest would be easier to measure; L_Δ is easier to calculate.

ICRU Report 37 calculates L_Δ for positrons and electrons for values of Δ down to 1 keV. The report points out that such calculations are inaccurate for smaller values of Δ, even in light elements. ICRU Report 16 provides values of L_Δ for protons and heavy ions.

14.13. RANGE

We can see in Fig. 14.27 that the α particles, entering from the bottom with the same energy, all travel about the same distance before coming to rest. This distance is called the *range* of the α particles. It will be defined more precisely below.

We can estimate the range in the following way. The stopping power represents an average energy loss per unit path length. The actual energy loss fluctuates about the mean values given by the stopping power. If these fluctuations are neglected and the projectiles are assumed to lose energy continuously along their tracks at a rate equal to the stopping power, then one is making the *continuous-slowing-down approximation* (CSDA). In this approximation one can calculate the range, the distance a particle with initial energy T_0 travels before coming to rest or reaching some final kinetic energy T_f. A factor ρ is introduced to express the range in mass per unit area:

$$R_{CSDA}(T_0,T_f)=\rho\int dx=\rho\int_{T_f}^{T_0}\frac{dT}{S_e+S_n+S_r}. \qquad (14.62)$$

ICRU Report 37 (1984) discusses the problem of carrying the integration to $T_f=0$.

The CSDA range is not directly measurable. Measurements of the fraction $F(R)$ of monoenergetic particles in a beam that passes through an absorber of thickness R gives a curve like that of Fig. 14.25. Various ranges can be defined using this curve. The most easily measured is the *median range* R_{50}, corresponding to an absorber thickness that transmits 50% of the incident particles. The *extrapolated range* R_{ex} is obtained by extrapolating the linear portion of the curve to the abscissa. The *maximum range* R_m is the thickness that just stops all of the particles; it is, of course, very difficult to measure. If $F(R)$ is known accurately one can define a *mean range* $\bar{R}=\int R(-dF/dR)dR/\int(-dF/dR)dR$. If the shape of the transmission curve is perfectly symmetrical about the mean, then R_{50} is equal to $\bar{R}$, even though they are conceptually quite different.

The fluctuations in the range are called *straggling*. The straggling distribution has also been calculated. The track of a heavy projectile such as an α particle is fairly straight, because the various scattering interactions result only in small angular deviations. The straggling results primarily from the fact that $S\,dx$ represents only an average energy loss in path length dx. The fluctuations can be integrated to give the spread in range; see Ahlen (1980) or ICRU Report 37 (1984) or ICRU Report 49 (1993) or the computer program SRIM [Ziegler, Biersack, and Littmark (1985)]. For heavy projectiles $\bar{R}$ (usually approximated by R_{50}) provides the best estimate of R_{CSDA}.

Electrons and positrons are so light that they undergo large-angle scattering (occasionally from an electron, more often from an atomic nucleus). The resulting electron trajectories are quite tortuous, as can be seen in Figs. 14.28 and 14.29. The median or mean range for an electron is considerably less than R_{CSDA}. For electrons and positrons the extrapolated range R_{ex} corresponds most closely to R_{CSDA}, at least in materials with atomic number up to silver [Tung *et al.* (1979)]. Figure 14.26 shows ranges in water. At medium energies they vary nearly as T^2.

Tables of ranges are found in Appendix O and in the stopping-power references cited above.

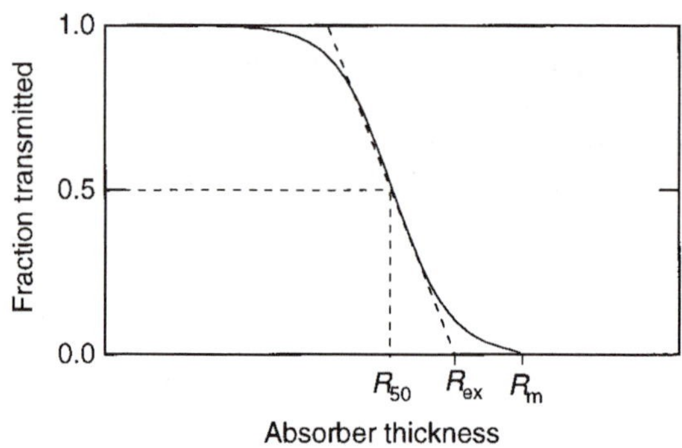

FIGURE 14.25. Plot of the number of particles passing through an absorber vs its thickness to show the definition of various ranges. R_{50} is the median range, R_{ex} is the extrapolated range, and R_m is the maximum range.

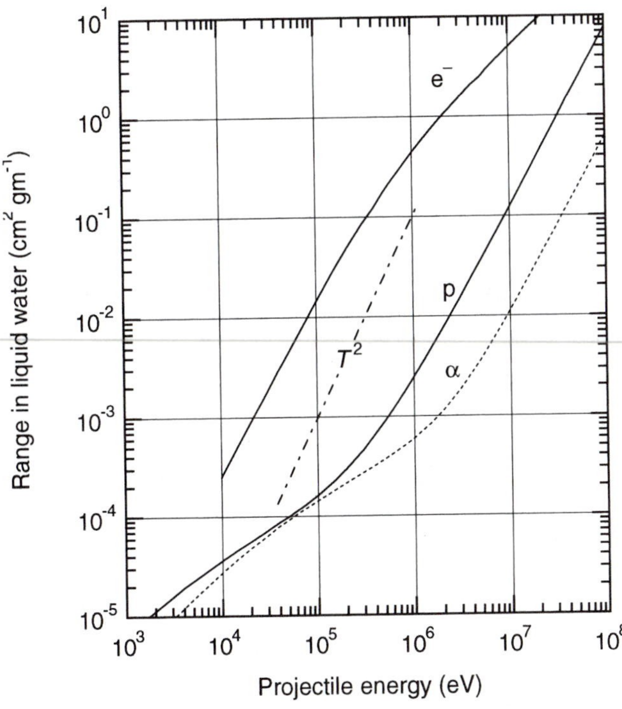

FIGURE 14.26. Range of electrons, protons, and α particles in liquid water. Data are from ICRU Reports 37 (1984) and 49 (1993). Note that for water the range in cm^2 gm^{-1} is the same as the range in cm.

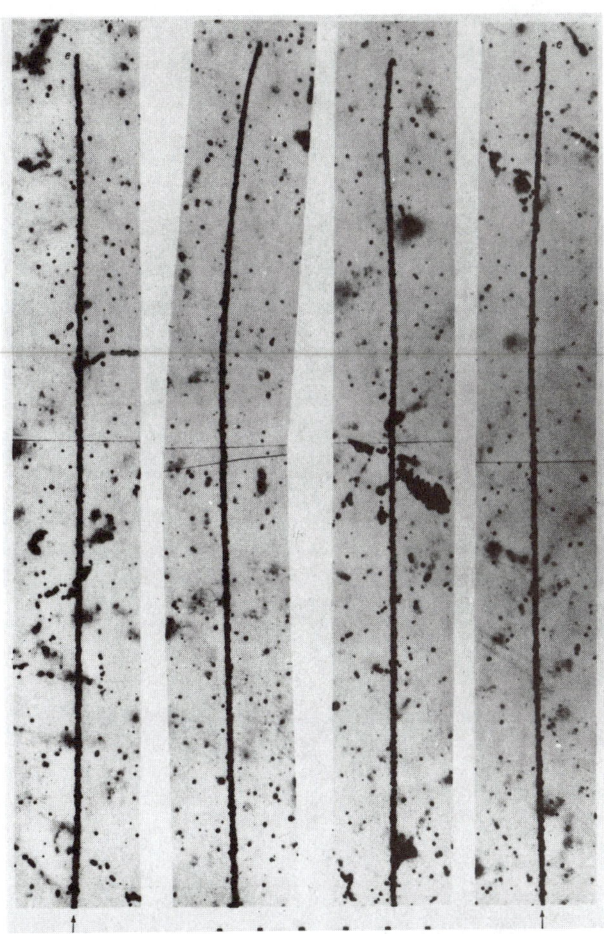

FIGURE 14.27. Tracks of 22-MeV α particles in photographic emulsion. The α particles enter at the bottom of the page and come to rest near the top. The features of the tracks are discussed in the text. From Powell *et al.* (1959). Reproduced by permission of D. H. Perkins.

The *radiation yield* is the fraction of the initial particle (usually electron) kinetic energy T_0 that is converted to bremsstrahlung photons as the electron comes to rest in the medium in question. The yield is calculated using the continuous-slowing-down approximation as (neglecting S_n)

$$Y(T_0)=\frac{1}{T_0}\int_0^{T_0}\frac{S_r(T)dT}{S_e(T)+S_r(T)}. \qquad (14.63)$$

14.14. TRACK STRUCTURE

One can gain insight into the interaction processes by examining tracks in nuclear emulsions or in cloud chambers. Figures 14.27 and 14.28 are taken from a classic atlas of tracks in nuclear emulsions [Powell *et al.* (1959)]. They show the difference between the interaction of heavy and light particles in matter. Figure 14.27 shows the tracks of four cosmic ray α particles, each of which entered the bottom of the figure and stopped near the top. The fiduciary marks along the bottom are 10 μm apart. Each track is about 195 μm long, corresponding to an initial α-particle energy of about 22 MeV. The emulsion has a density of 3.6×10^3 kg m^{-3}. Each black dot is a sensitive silver halide grain about 0.6 μm in diameter. At the beginning of the track, S is about 70 keV μm^{-1} or 42 keV per grain; 10 μm from the end of the track it is 200 keV μm^{-1} or 120 keV per grain. The energy that must be deposited in a grain to render it developable is about 2.8 keV. The amount of energy deposited in each grain is so much larger than this that the track density is uniform. Small bumps of 1–4 grains can be seen occasionally along each track. Some of these are due to δ *rays*, electrons that have received enough energy to travel a few micrometers in the emulsion. Others are artifacts due to the general background fog. Multiple small-angle scattering causes small deviations in each track, which become greater as the α particle slows down.

In Figure 14.28 an electron–positron pair has been produced in the lower left corner of the emulsion by a 1.5-MeV photon. Each particle has a kinetic energy of about 250 keV. One immediately notices the tortuous path of both particles due to large-angle scattering. The stopping power near the beginning of the track is about 0.8 keV μm^{-1}, so that about 0.5 keV is deposited in each grain. About 30 μm from the end, the stopping power and

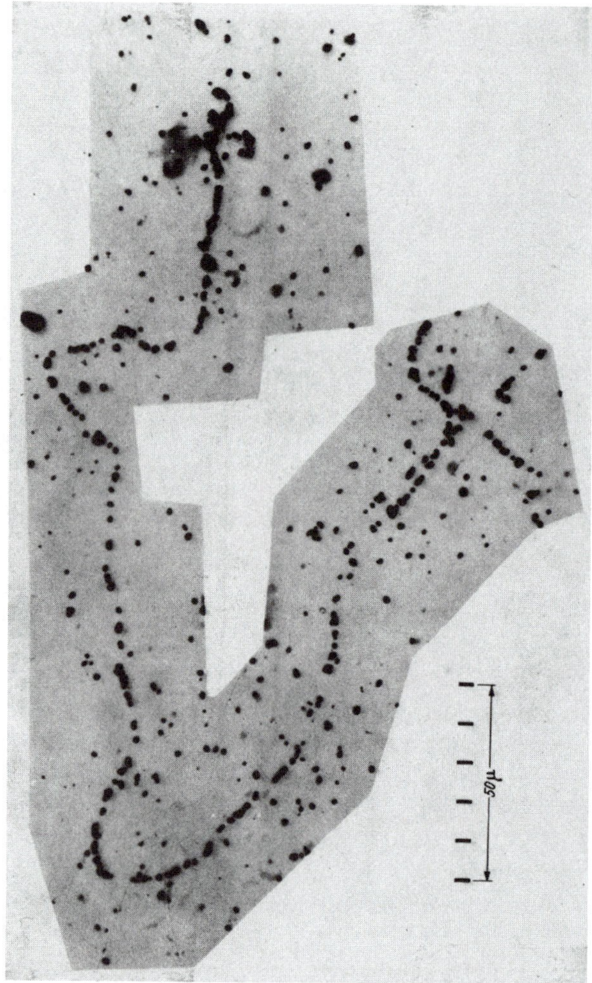

FIGURE 14.28. Tracks of electrons in emulsion. An electron–positron pair was produced in the lower left corner. Each particle has an energy of about 250 keV. The details are discussed in the text. From Powell *et al.* (1959). Reproduced by permission of D. H. Perkins.

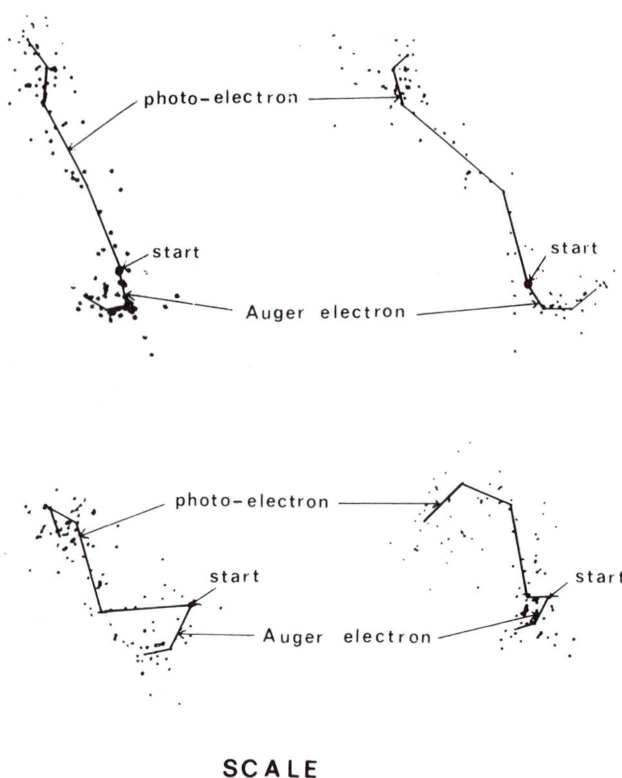

FIGURE 14.29. Tracks of electrons in a cloud chamber. An equivalent scale in water or tissue has been added. Photoelectrons and Auger electrons can be seen. From T. Budd and M. Marshall (1983). Reproduced by permission of Radiation Research Society.

the average amount of energy deposited in each grain are about 3 times larger. The upper track is considerably more dense near the end of its path. The failure of the other track to show this density increase could be due to annihilation of the positron in flight or to a large-angle scattering.

Figure 14.29 shows the ionization produced by an electron at a much different scale. It was produced from a cloud chamber photograph of an electron track in a low-density gas [Budd and Marshall (1983)]. The scale shows distances in liquid water or tissue that correspond to the same value of ρx, corrected for phase effects. Note that the scale shows 10 *nanometers*. An atomic diameter is 0.2–0.6 nm. In each case a photoelectron has been ejected from an atom. Auger electrons are also seen. One can see δ rays, energetic electrons knocked out approximately perpendicular to the electron's path and corresponding to high-W collisions.

14.15. ENERGY TRANSFERRED AND ENERGY IMPARTED; KERMA AND ABSORBED DOSE

The response of a substance to radiation, whether it is the darkening of a photographic film, an electrical pulse in an ionization chamber, or the response of a tumor to radiation therapy, is due, directly or indirectly, to the ionization produced by charged particles that lose their kinetic energy in the substance through the stopping mechanisms we have just discussed. We now define some quantities that are used to describe the transfer of energy from photons to charged particles and the energy lost by charged particles due to ionization.

Before discussing the formal definition of these quantities, let us consider some examples of energy transfer by photons. Figure 14.30 shows some schematic interactions of photons in a sample of water 10 cm thick. They are drawn to scale.[11] In Fig. 14.30(a) five photons of energy 100 keV

[11]These examples were constructed with a pedagogical simulation program called MacDose [see Hobbie (1992)]. At the time of this writing, it is available from Medical Physics Publishing Co., Madison, WI.

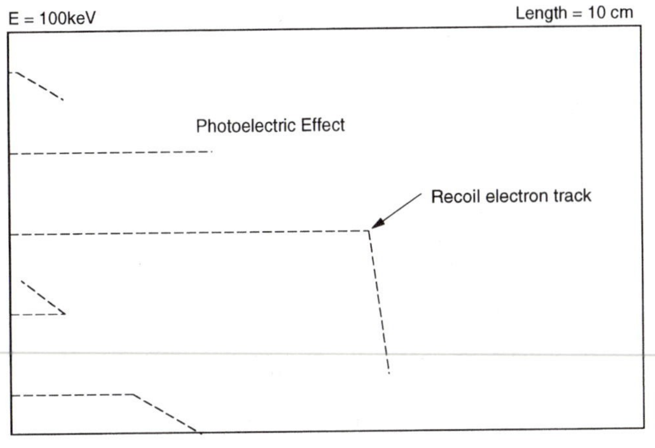

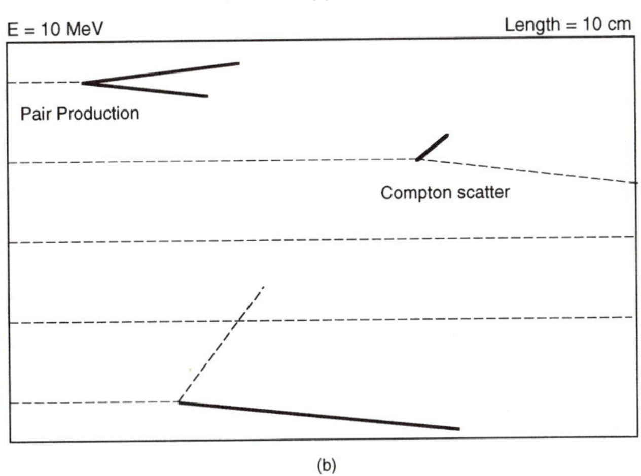

FIGURE 14.30. A simulation of photons passing through a layer of water 10 cm thick. (a) The photon energy is 100 keV. One photon has a photoelectric interaction. The other four are Compton scattered. (b) The photon energy is 10 MeV. Two photons do not interact, one produces an electron–positron pair, and two Compton scatter.

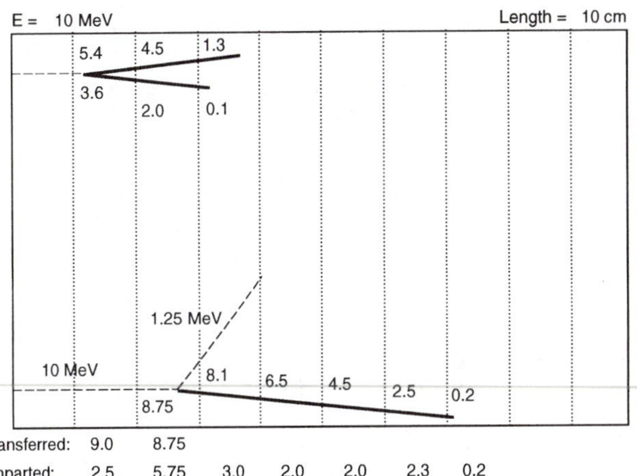

FIGURE 14.31. The difference between energy transferred and energy imparted. Two of the photons from Fig. 14.30(b) are shown. The water has been divided into ten 1-cm slices. The numbers on the drawing show the charged-particle energy at the entrance to each slice. The energy transferred and the energy imparted in each slice are shown at the bottom.

enter from the left. Photon tracks are dotted. One photon is absorbed by the photoelectric effect, and four are Compton scattered. The energy of the photoelectron and the Compton-scattered electrons is so low that the ranges are insignificant on this scale. In Fig. 14.30(b) the incident photons have 10 MeV energy. One has undergone pair production, two have Compton scattered, and two have passed through without interacting. The electron tracks are shown as thick solid lines. Their lengths are equal to the CSDA range of electrons or positrons of that energy. They are drawn as straight lines, even though the real tracks are tortuous.

One of the quantities of interest is the *energy transferred* to kinetic energy of charged particles in some mass of material. [We saw this briefly in the discussion surrounding Eq. (14.36)] Another is the *energy imparted* in some mass of material, which is the kinetic energy lost by charged particles as they come to rest. Figure 14.31 shows the distinction between the two quantities. It shows two photons from Fig. 14.30(b), one that underwent pair production, and one that was Compton scattered. The water has been divided into ten slices, each 1 cm thick. No energy is transferred in the first slice. Energy is transferred by pair production in the second slice and by Compton scattering in the third slice. In each case the electron (or positron) produced loses kinetic energy in that slice and also in other slices. There is energy imparted in slices 2–8, even though the energy is transferred only in slices 2 and 3.

Consider now the actual numbers. In keeping with the literature,[12] we will call the energy transferred $E_{\rm tr}$, even though we have been using T for kinetic energy. For pair production the energy transferred is

$$E_{\rm tr} = T_+ + T_- = h\nu_0 - 2m_e c^2 = 10 - 2\times 0.511$$
$$= 8.978 \approx 9.0 \text{ MeV}. \tag{14.64}$$

The partition of energy between the electron and positron is random or stochastic. We assume that about 60% (5.4 MeV) goes to one member of the positron–electron pair and 40% (3.6 MeV) to the other. These numbers are shown at the vertex of Fig. 14.31. From these energies the ranges can be determined. Measuring the distance from the end of the track to the boundary between each slice allows us to determine the energy of each charged particle as it enters the slice. For the Compton scattering, 8.75 MeV is transferred to the recoil electron and the scattered photon has 1.25

[12]See ICRU Report 33 (1980) or Attix (1986).

MeV. The energy imparted by the 5.4-MeV particle is $5.4-4.5=0.9$ MeV in slice 2, $4.5-1.3=3.2$ MeV in slice 3, and 1.3 MeV in slice 4. Similar calculations can be done for the other charged particles. The energy transferred and the energy imparted in each slice are shown at the bottom of Fig. 14.31. This ignores any interaction of the 1.25-MeV Compton-scattered photon and assumes it leaves the volume of interest. Because for the 100-keV photons the range of the charged particles is small compared to 1 cm, the energy transferred and the energy imparted in each slice are the same in Fig. 14.30(a).

Figure 14.32 shows a plot of the transferred and imparted energy for a uniform beam of 10-MeV photons all traveling to the right and striking a slab of water 20 cm thick. Both the energy transferred and the energy imparted are stochastic quantities. This simulation was done for 10^4 photons, and you can see the scatter in the points. The energy transferred falls exponentially as $\exp(-\mu_{\text{atten}}x)$.

We found the energy transferred by calculating the energy of each electron or positron produced. The standard definition uses slightly different bookkeeping. It subtracts the energy of the photons leaving the volume of interest from those entering, and adds a term Q for the energy going into the volume due to changes in rest mass. [This is the $2m_ec^2$ of Eq. (14.64).] The standard definition is

$$E_{\text{tr}}=(R_{\text{in}})_u-(R_{\text{out}})_u^{\text{nonr}}+\sum Q, \qquad (14.65)$$

The quantity R is radiant energy: the energy of particles (including photons) but not including rest energy. The subscript u means that it is the radiant energy of uncharged particles. The uncharged particles can be photons or neutrons.[13] Later we will use subscript c to denote the radiant energy of charged particles. The superscript nonr means that the quantity does not include radiant energy arising from bremsstrahlung or positron annihilation in flight from charged particles within the volume. The Q term is positive if mass is converted to energy (as in annihilation radiation) and negative if energy is converted to mass (as in pair production).

Using this method of calculating for Fig. 14.31 gives

$$E_{\text{tr}}=(R_{\text{in}})_u-(R_{\text{out}})_u^{\text{nonr}}+\sum Q=10-0-2\times0.511$$
$$=9.0 \text{ MeV}$$

for slice 2. For the third slice the equation gives

$$E_{\text{tr}}=(R_{\text{in}})_u-(R_{\text{out}})_u^{\text{nonr}}+\sum Q=10-1.25+0$$
$$=8.75 \text{ MeV}.$$

[13]Neutrinos, which we will discuss in Chap. 16, travel such long distances without interacting that they are not considered in the calculations. Energy carried by neutrinos, which come from nuclear β decay, is assumed to have left the body.

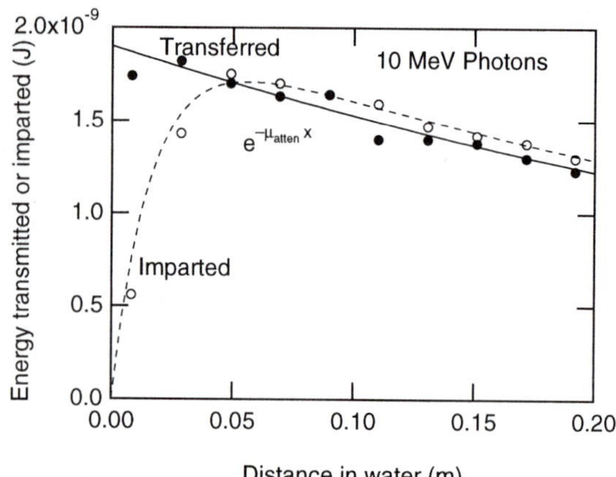

FIGURE 14.32. Plot of energy transferred and energy imparted for a simulation using 40 000 photons of energy 10 MeV. The filled circles are the energy transferred in each slice, and the open circles are the energy imparted in each slice.

For the fourth slice, the uncharged radiant energy in is equal to the uncharged radiant energy out. In the fifth slice, if the 1.25-MeV photon actually interacts as it appears to, we would have to include its energy transfer. In all the other slices the energy transferred is zero.

The energy transferred is a stochastic quantity, and so is the energy transferred per unit mass, dE_{tr}/dm. Its expectation value is the *kerma* (*k*inetic *e*nergy *r*eleased per unit *ma*ss):

$$K=\frac{d\overline{E}_{\text{tr}}}{dm}. \qquad (14.66)$$

If we consider monoenergetic photons of energy $h\nu$ and consider only the interaction of the primary photon beam (not any secondary photons, such as Compton-scattered photons or annihilation radiation), then the kerma is

$$K=\frac{\mu_{\text{tr}}}{\rho}\Psi, \qquad (14.67)$$

where Ψ is the energy fluence. To see why this is true, note that if the N photons are spread over area S, then $NE=\Psi S$ and $dm=\rho S\,dx$. The kerma is

$$K=\frac{d\overline{E}_{\text{tr}}}{dm}=\frac{\Psi S\mu_{\text{tr}}\,dx}{\rho S\,dx}=\frac{\mu_{\text{tr}}}{\rho}\Psi.$$

The *energy imparted E* is the net energy into the volume from all sources: uncharged particles, charged particles, and changes of rest mass:

$$E=(R_{\text{in}})_u-(R_{\text{out}})_u+(R_{\text{in}})_c-(R_{\text{out}})_c+\sum Q, \qquad (14.68)$$

The *absorbed dose* is the expectation value of the energy imparted per unit mass:

$$D=\frac{d\bar{E}}{dm}. \tag{14.69}$$

Another quantity used in the literature is the *net energy transferred*. It subtracts from the energy transferred the energy that is reradiated (bremsstrahlung and radiation from positron annihilation in flight), even if the reradiation takes place outside the volume of interest. It is

$$E_{\text{tr}}^{\text{net}}=(R_{\text{in}})_u-(R_{\text{out}})_u^{\text{nonr}}-R_u^r+\sum Q, \tag{14.70}$$

The *collision kerma* and *radiative kerma* are defined as expectation values per unit mass:

$$K_C=\frac{d\bar{E}_{\text{tr}}^{\text{net}}}{dm}=K-K_r=K-\frac{d\bar{R}_u^r}{dm}. \tag{14.71}$$

Considering only a primary beam of monoenergetic photons,

$$K_C=\frac{\mu_{\text{en}}}{\rho}\Psi. \tag{14.72}$$

14.16. CHARGED PARTICLE EQUILIBRIUM

There are three equilibrium conditions that sometimes exist or are assumed to exist, because they make it possible to calculate the relationship between energy transferred and energy imparted.

The first and most restrictive condition is *radiation equilibrium*. It is a useful model when considering an extended radioactive source that is distributed uniformly throughout some volume V (such as the body or a particular organ). The source is assumed to emit its radiation isotropically. The energy released to neutrinos is ignored. A point of interest within the large volume is surrounded by a smaller volume v. The volume v must be far enough from the edge of V so that any radiation emitted from v is absorbed before reaching the surface of V. The entire volume V is assumed to be of the same atomic composition and density. Because everything is isotropic, on average for every photon or neutron or charged particle entering volume v, another identical one leaves. This means that

$$(\bar{R}_{\text{in}})_c=(\bar{R}_{\text{out}})_c \tag{14.73a}$$

and

$$(\bar{R}_{\text{in}})_u=(\bar{R}_{\text{out}})_u. \tag{14.73b}$$

The average energy imparted is

$$\bar{E}=\sum \bar{Q}. \tag{14.74}$$

This means that when the conditions for radiation equilibrium are satisfied, the absorbed dose is the expectation value of the energy released by the radioactive material per unit mass. If there is no radioactive source, there is no energy imparted in radiation equilibrium.

A less restrictive assumption is *charged-particle equilibrium*, in which only Eq. (14.73a) is satisfied: the average amount of charged-particle radiant energy into the region is the same as the average amount leaving. The assumption of charged particle equilibrium is a useful model in several cases, but we will consider only the case of an external beam of photons striking volume V. Again we consider what happens in a smaller volume v, separated from the boundary of V by a distance larger than the maximum range of any secondary charged particles produced by the external radiation. We also assume that the medium is homogeneous and that only a small fraction of the primary radiation interacts within the volume so attenuation can be neglected. Then the average number of charged particles produced per unit volume and per unit solid angle in any given direction is the same everywhere in the volume. Though the charged particles need not be produced isotropically, on average for every particle that leaves volume v, a corresponding one will enter it, as shown in Fig. 14.33. For charged-particle equilibrium, the average energy imparted is

$$\bar{E}=(\bar{R}_{\text{in}})_u-(\bar{R}_{\text{out}})_u+\sum \bar{Q}.$$

Comparing this with the average of Eq. (14.70) shows that the average net energy transferred is

$$\bar{E}_{\text{tr}}^{\text{net}}=\bar{E}+(\bar{R}_{\text{out}})_u-(\bar{R}_{\text{out}})_u^{\text{nonr}}-\bar{R}_u^r.$$

Now recall that $(\bar{R}_{\text{out}})_u$ is the average value of all the uncharged radiation leaving volume v, $(\bar{R}_{\text{out}})_u^{\text{nonr}}$ is the average value of all uncharged radiation leaving excluding

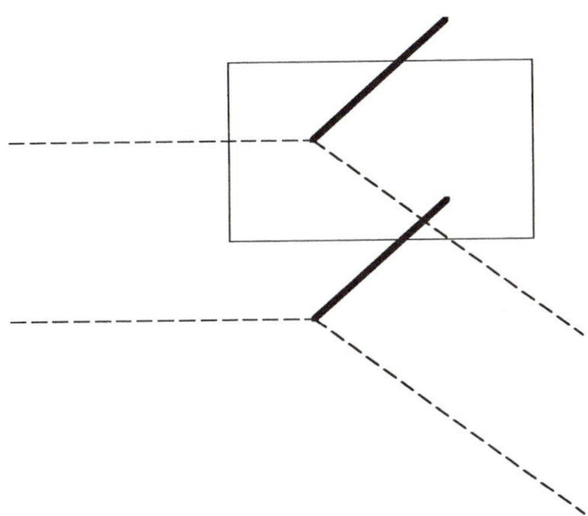

FIGURE 14.33. One of the conditions for charged-particle equilibrium is that on average, for every charged particle of a certain energy leaving volume v traveling in a certain direction, a corresponding particle enters the volume.

bremsstrahlung and photons from annihilation in flight that occur within the volume, and $\bar{R}_u^r$ is the bremsstrahlung and annihilation-in-flight radiation from charged particles originating in v regardless of where it occurs. If there is charged-particle equilibrium, any radiative interaction by a charged particle after it leaves the volume will on average be replaced by an identical interaction inside v. If the volume is small enough so that all radiative loss photons escape from the volume before undergoing any subsequent interactions, then

$$(\bar{R}_{\text{out}})_u = (\bar{R}_{\text{out}})_u^{\text{nonr}} + \bar{R}_u^r.$$

Therefore, for *charged-particle equilibrium,*

$$\bar{E} = \bar{E}_{\text{tr}}^{\text{net}},$$

and the dose is equal to the collision kerma:

$$D = K_C. \tag{14.75}$$

One situation where charged-particle equilibrium applies is for the thin slices in Fig. 14.30(a). The electron ranges are so short (10 μm for a 25-keV electron) that a slice can be thin compared to $1/\mu$ and yet all the electrons produced stay within the volume.

The conditions for charged-particle equilibrium fail if the source of photons is too close (Ψ is not uniform because of $1/r^2$), close to a boundary (as between air and tissue or muscle and bone), for high-energy radiation (as in Fig. 14.32), or if there is an applied electric or magnetic field that alters the paths of the charged particles (as in some radiation detectors).

In Fig. 14.32, if we look at the situation far enough to the right, the energy imparted is proportional to the energy transferred. This situation is called *transient charged-particle equilibrium.*

The dose for a monoenergetic parallel beam of charged particles with particle fluence Φ can be calculated by making three assumptions:

1. The volume of interest is thin enough so that S_e remains constant.
2. Scattering can be neglected, so every particle passes straight through the foil.
3. The net kinetic energy carried out of the foil by δ rays is negligible, either because the foil is thick compared to the range of the δ rays or because the foil is immersed in a material of the same atomic number so that charged-particle equilibrium exists.

Then the energy lost in collisions in a foil of thickness dz is

$$E = \Phi(\text{area})(S_e/\rho)\rho\, dz$$

and the mass is $\rho(\text{area})dz$, so the dose is

$$D = \Phi(S_e/\rho). \tag{14.76}$$

Attix (1986, pp. 188–195) discusses corrections for situations where these assumptions are not valid.

14.17. BUILDUP

We have been ignoring the interactions of secondary photons, primarily Compton-scattered photons and annihilation radiation. They can be quite significant. In fact, there can be a cascade of several generations of photons, though we will call them all "secondary photons." Figure 14.34 compares two simulations in which the secondary photons are allowed to interact. In Fig. 14.34(a) there is one secondary interaction before the scattered photon escapes from the water. In Fig. 14.34(b) there are a total of six Compton scatterings before the secondary photon escapes.

All of these secondary photons produce electrons that contribute to the energy transferred and energy imparted. Figure 14.35 compares two cases where 25 photons of energy 100 keV enter the water from the left. The primary interactions are the same in both cases. In Fig. 14.35(a) the small dots represent the electrons produced by the interaction of the primary photons. In Fig. 14.35(b) the electrons

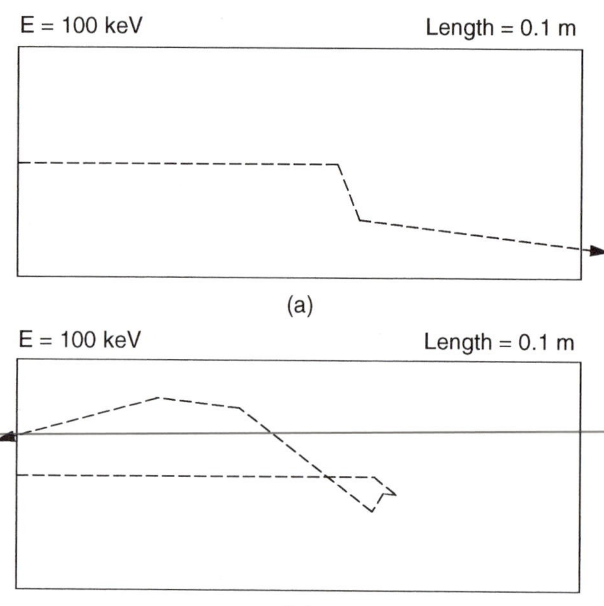

FIGURE 14.34. Secondary photons also interact. One 100-keV photon enters from the left in each panel. (a) The primary photon undergoes a Compton scattering. The Compton-scattered photon also undergoes a Compton scattering. The third photon escapes from the water. (b) The primary photon is Compton scattered. Each Compton-scattered photon undergoes another Compton scattering, until the sixth scattered photon leaves through the upstream surface of the water, traveling nearly in the direction from which the incident photon came.

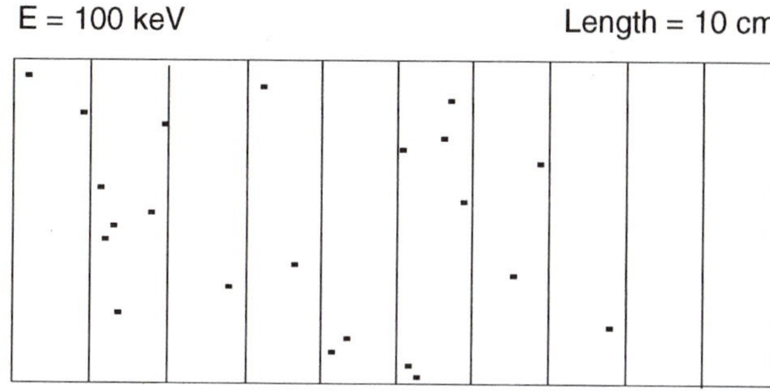

(a) Secondaries not included

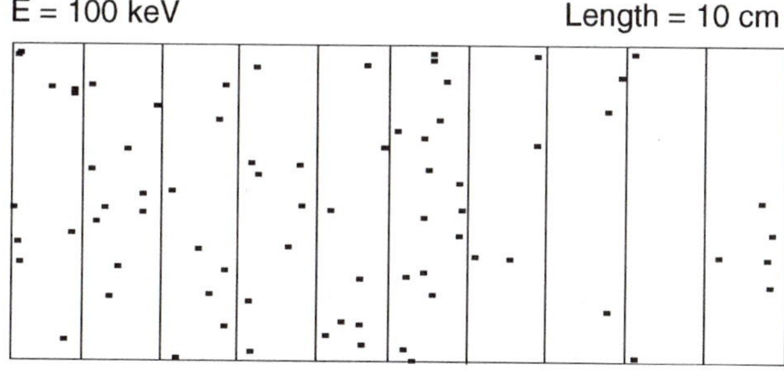

(b) Secondaries included

FIGURE 14.35. Twenty-five 100-keV photons entered the water from the left. The dots represent recoil electrons from Compton scattering or photoelectrons. (a) Only the first interaction of the primary photon is considered. (b) Subsequent interactions are also considered.

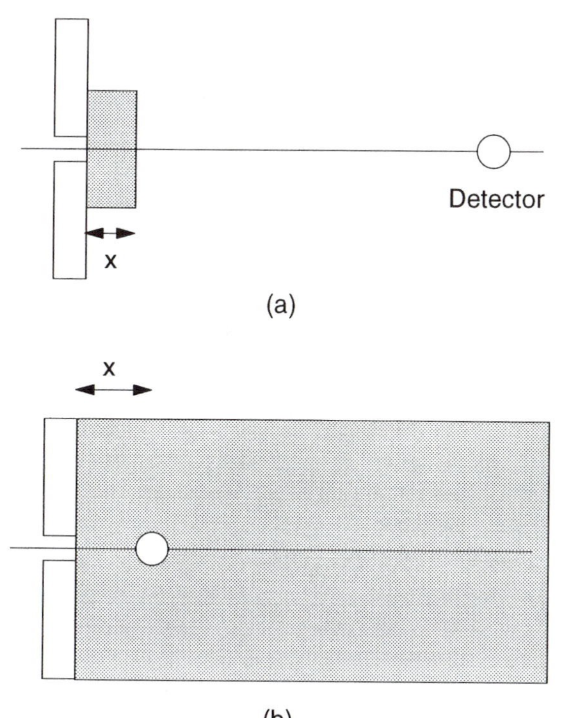

FIGURE 14.36. Two different detector geometries. (a) The detector is at a fixed location and the absorber thickness is increased. (b) The detector is at a varying distance from the source in a water bath.

produced by secondary and subsequent interactions are also shown. The energy transferred and energy imparted are much greater.

The *buildup factor* for any quantity is defined as the ratio of the quantity including secondary and scattered radiation to the quantity for primary radiation only. For example, if the primary beam has an energy fluence Ψ_0 at the surface, the energy fluence at depth x in the medium is

$$\Psi(x)=B(x)\Psi_0 e^{-\mu x}. \tag{14.77}$$

The buildup factor is quite sensitive to the geometry. Compare the two situations in Fig. 14.36. In Fig. 14.36(a) the detector is at a fixed location and the thickness of the absorber is increased. As the absorber thickness x approaches zero, the buildup factor approaches unity. In Fig. 14.36(b) the detector is at depth x in a water bath. Because of the backscattered radiation seen in Fig. 14.34(b), $B(x)>1$ as $x\rightarrow 0$. In this case, it is sometimes called the *backscatter factor*. For further discussion, see Attix (1986).

SYMBOLS USED IN CHAPTER 14

Symbol	Use	Units	First used on page
a	Acceleration	$m^2\ s^{-1}$	397
a_i	Number of atoms of constituent i		389
b	Impact parameter	m	399
c	Velocity of light	$m\ s^{-1}$	385
d	Diameter	m	397
$f, f_C, f_i, f_\kappa, f_\tau$	Fraction of photon energy transferred to charged particles		387
g	Fraction of energy of secondary electrons converted back into photons by bremsstrahlung		395
h	Planck's constant	J s	384
j	Total angular momentum quantum number		382
k	Spring constant	$N\ m^{-1}$	397
l	Orbital angular momentum quantum number		382
m	Mass	kg	397
m_e	Electron rest mass	kg	384

Symbol	Use	Units	First used on page
m_j	Quantum number for the component of the total angular momentum along the z axis		382
m_0	Rest mass	kg	385
n	Principal quantum number		383
p	Momentum	kg m s^{-1}	385
q	Charge	C	399
r_e	''Classical'' electron radius	m	386
s	Spin quantum number		383
t	Time	s	397
v	Velocity	m s^{-1}	397
w_i	Mass fraction of constituent i		391
x,y,z	Coordinate axes	m	389
x	Dimensionless energy ratio		386
z	Charge of projectile in multiples of e		396
A	Atomic mass number		389
A_i, A_{mol}	Atomic mass number of constituent i or molecule		391
B	Binding energy	J	384
B	Buildup factor or backscatter factor		411
C	Shell correction coefficient		401
D	Absorbed dose	J kg^{-1} or Gy (gray)	409
E	Energy	J	384
E	Electric field	N C^{-1}	400
F	Force	N	397
F	Fraction of charged particles passing through an absorber		404
I	Average ionization energy	J	401
I	Stopping interaction strength	J m^2	401
K, K_C	Kerma, collision kerma	J	408
L	Stopping number per atomic electron		401
L_Δ	Restricted linear collision stopping power	J m^{-1}	404
M	Mass	kg	389
N	Number of particles		389

Symbol	Use	Units	First used on page
N_A	Avogadro's number	(g mol)$^{-1}$, (kg mol)$^{-1}$	389
N_T	Number of target atoms per unit projected area	m^{-2}	391
N_{TV}	Number of target atoms per unit volume	m^{-3}	391
Q	Energy released from rest mass		408
R	Range	m	404
R_u, R_c	Radiant energy in the form of uncharged or charged particles	J	408
S	Area	m^2	408
S	Stopping power	J m^{-1}	396
S_{col}	Electron (collision) stopping power	J m^{-1}	398
S_n	Nuclear stopping power	J m^{-1}	398
S_r	Radiation stopping power	J m^{-1}	398
T	Kinetic energy	J	384
V	Velocity	m s^{-1}	399
$W_{K,L,M}$	Probability that hole in K, L, or M shell is filled by fluorescence		393
W	Energy lost in a single interaction	J	396
Y	Radiation yield		405
Z	Atomic number		382
β	v/c		396
δ	Average energy emitted as fluorescence radiation per photon absorbed	J	394
δ	Density-effect correction		401
θ, ϕ	Angles		386
κ	Pair production cross section	m^2	384
λ	Wavelength	m	386
$\mu, \mu_{\mathrm{atten}}$	Attenuation coefficient	m^{-1}	389
μ_{en}	Mass energy absorption coefficient	m^{-1}	395

Symbol	Use	Units	First used on page
μ_{tr}	Energy transfer coefficient	m^{-1}	394
ν	Frequency	Hz	384
ξ	Position	m	400
ρ	Density	kg m^{-3}	389
σ	Cross section	m^2	387
σ_C	Total Compton cross section for one electron	m^2	384
σ_{coh}	Coherent Compton cross section for one atom	m^2	388
σ_{incoh}	Incoherent Compton cross section for one atom	m^2	387
σ_{tot}	Total cross section	m^2	387
τ	Photoelectric cross section	m^2	384
Δ	Energy transfer	eV	404
Φ	Particle fluence	part m^{-2}	394
Ψ	Energy fluence	J m^{-2}	408
Ω	Solid angle	sr	386

PROBLEMS

Section 14.1

14.1. The quantum numbers $m_s = \pm\frac{1}{2}$ and $m_l = l, l-1, l-2, \ldots, -l$ are sometimes used instead of j and m_j to describe an electron energy level. Show that the total number of states for given values of n and l is the same when either set is used.

Section 14.3

14.2. The K-shell photoelectric cross section for 100-keV photons on lead ($Z=82$) is $\tau = 1.76 \times 10^{-25}$ m^2 $atom^{-1}$. Estimate the photoelectric cross section for 60-keV photons on calcium ($Z=20$).

14.3. Describe how you could use different materials to determine the energy of monoenergetic x rays of energy about 50 keV by using changes in the attenuation coefficient. What materials would you use?

Section 14.4

14.4. A 1-MeV photon undergoes Compton scattering from a carbon target. The scattered photon emerges at an angle of 30°.

(a) What is the energy of the scattered photon? What is the energy of the recoil electron?

(b) What is the differential scattering cross section for scattering at an angle of 30° from one electron? From the entire carbon atom ($Z=6$, $A=12$)?

14.5. What is the energy of a Compton-scattered photon at 180° when $h\nu_0 \gg m_e c^2$? At 90°?

14.6. Integrate Eq. (14.18) over all possible scattering angles to obtain Eq. (14.20). Use the solid angle in spherical coordinates (Appendix L).

14.7. Use the expansion $\ln(1+x) = x - x^2/2 + x^3/3 \ldots$ to show that Eq. (14.19) approaches Eq. (14.20) as $x \to 0$.

14.8. Suppose that attenuation is measured for 60-keV photons passing through water in such a way that photons scattered less than 5° still enter the detector. Estimate the incoherent Compton scattering cross section per electron for photons scattered through more than 5°.

Section 14.5

14.9. A beam of 59.5-keV photons from ^{241}Am scatters at 90° from some calcium atoms ($A=40$).

(a) What is the energy of a Compton-scattered photon?

(b) What is the energy of a coherently scattered photon?

(c) What is the recoil energy of the atom in coherent scattering?

Section 14.7

14.10. Use Fig. 14.07 to make the following estimates for 1-MeV photons. What is the mass attenuation coefficient for water? For Aluminum? For lead? What is the linear attenuation coefficient in each case?

14.11. Use Fig. 14.07 to estimate the attenuation coefficient for 0.1-MeV photons on carbon and lead. Compare your results to the tables in Appendix O. Repeat for 1-MeV photons.

14.12. Consider photons of three energies: 0.013, 0.02, and 0.03 MeV. What fraction of the photons at each energy will be unattenuated after they pass through 0.1 mm of lead (density = 11.35 g cm^{-3})? Comment on the differences in your results.

Section 14.8

14.13. Use Fig. 14.07 to find the mass attenuation coefficient for 0.2-MeV photons in a polyethylene absorber. The Compton effect predominates. Polyethylene has the formula $(CH_2)_n$.

14.14. What will be the attenuation of 40-keV photons in muscle 10 cm thick? Repeat for 200-keV photons.

14.15. Assume that a patient can be modeled by a slab of muscle 20 cm thick of density 1 g cm^{-3}. What fraction of an incident photon beam will emerge without any interaction if the photons have an energy of 10 keV? 100 keV? 1 MeV? 10 MeV?

14.16. Muscle and bone are arranged as shown. Assume the density of muscle is 1.0 g cm^{-3} and the density of bone is 1.8 g cm^{-3}. The attenuation coefficients are

E	$(\mu/\rho)_{\text{muscle}}$ (cm^2 g^{-1})	$(\mu/\rho)_{\text{bone}}$ (cm^2 g^{-1})
60 keV	0.200	0.274
1 MeV	0.070	0.068

Compare the intensity of the emerging beam that has passed through bone and muscle and just muscle at the two energies.

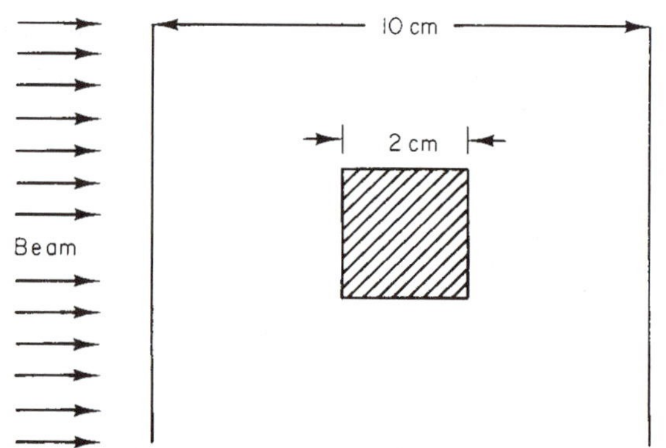

14.17. A beam of monoenergetic photons travels through a sample made up of two different materials, of unknown thickness x_1 and x_2, as shown below. The attenuation coefficients at two different energies, E_a and E_b, are accurately known. They are $\mu_1(a)$, $\mu_2(a)$, $\mu_1(b)$, and $\mu_2(b)$. One measures accurately the ratio of the number of photons emerging from the sample to the number entering, $R=\ln(N_0/N)$, at each energy so that R_a and R_b are known. Find an expression for x_2 in terms of R_a, R_b, and the attenuation coefficients.

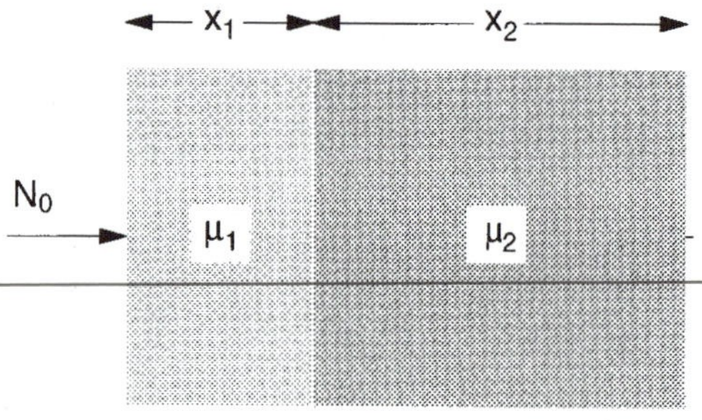

14.18. You wish to use x-ray fluorescence to detect lead that has been deposited in a patient's bone. You shine 100-keV photons on the patient's bone and want to detect the 73-keV fluorescence photons which are produced. The incident photon fluence is $\Phi_0=10^{12}$ photons m^{-2}. There are $X=10^{14}$ lead atoms (1 nanomole) in the region illuminated by the incident beam. The photoelectric cross section is 1.76×10^{-25} m^2 atom^{-1}. The fluorescence yield is $W=0.94$. Assume for simplicity that the fluorescence photons are emitted uniformly in all directions. The detector has a sensitive area 1×2 cm and is located 10 cm from the lead atoms. How many fluorescence photons are detected?

Section 14.10

14.19. A 5-keV photon strikes a calcium atom. The following events take place:

1. A K-shell photoelectron is ejected.

2. A K_α photon is emitted. This corresponds to the movement of a hole from the K shell to the L shell.

3. An electron in the M shell goes to the L shell and an M-shell photoelectron is emitted.

Give the excitation energy of the atom, the total energy in the form of photons, and the total energy in the form of electron kinetic energy at each stage. Use the following data for calcium: $Z=20$, $A=40$, $B_K=4000$ eV, $B_L=300$ eV (ignore differences in subshells), $B_M=40$ eV.

14.20. The following are the binding energies for hydrogen and oxygen.

	H	O
B_K	13.6 eV	532 eV
B_L		24 eV

(a) Determine f_τ for hydrogen from first principles.

(b) Use Eqs. 14.39 and 14.40 to estimate f_τ (K shell) for oxygen.

14.21. Use the Thomson scattering cross section, $d\sigma/d\Omega=(r_e^2/2)(1+\cos^2\theta)$, the total cross section $\sigma_C=8\pi r_e^2/3$, and the expression for the total energy of the recoil electron [Eq. (14.16)] to find an expression for f_C as $x\rightarrow0$. Plot $f_C\sigma_{C\text{ incoh}}$ on Fig. 14.07.

14.22. (a) For 50-keV photons on calcium, estimate f_τ.

(b) For 100-keV photons on calcium, the photoelectric cross section is $\tau=5.89\times10^{-28}$ m^2 atom^{-1}. Use $f_\tau=1.0$. Estimate μ_{tr}/ρ. Use the following data for calcium if you need them: $Z=20$, $A=40$, $B_K=4.000$ keV, $B_L=300$ eV, $B_M=40$ eV.

Section 14.11

14.23. Prove that if a particle of mass M_1 and kinetic energy T collides head on with a particle of mass M_2 which is at rest, the energy transferred to the second particle is $4TM_2/M_1$ or $2M_2V^2$ in the limit $M_2\ll M_1$. The maximum energy is transferred when the particles move apart along the line of motion of the incident particle.

14.24. The expression for $S_e=dT/dx$ has the SI units J m^{-1}. Therefore S_e/ρ in Eq. (14.57) has units J m^2 kg^{-1}.

(a) How must the coefficient in front of Eq. (14.57) be changed if T is in MeV? If x is in cm instead of m?

(b) What numerical factors must be introduced if N_A is in atoms per g mol and ρ is in g cm^{-3}?

(c) What is the average force on a 10-MeV proton in carbon? On a 100-keV proton? Use $I=78$ eV.

(d) What are the units of the stopping cross section [defined just below Eq. (14.47)]?

14.25. The peak in the stopping power occurs roughly where the projectile velocity equals the velocity of the atomic electrons in the target. Find an expression for the velocity of an electron in the $n=1$ Bohr orbit. Use Eq. (13.7), and the fact that the total energy is the sum of the kinetic and potential energies. Use the classical arguments and the fact that the electron is in a circular orbit to relate the kinetic and potential energies. The acceleration in a circular orbit is v^2/r.

14.26. A fishing lure is trolled behind a boat for a total distance D. Suppose that fish are distributed uniformly throughout the water at a concentration C fish m^{-3}, and that the probability of a fish striking the lure depends on b, the perpendicular distance from the path of the lure to the fish: $p=\exp(-b/b_0)$. Calculate the average number of fish caught.

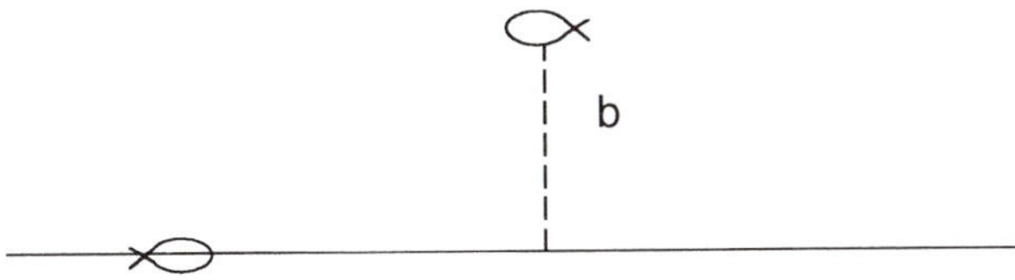

Section 14.13

14.27. What is the range energy relationship for high-speed non-relativistic particles if the variation of L with T is neglected and Eq. (14.57) is the dominant term?

14.28. Estimate the maximum electron range, and hence the radius of the δ-ray cloud surrounding the track of a 5-MeV α particle. (The rest energy Mc^2 of an α particle is about 4 times 938 MeV.) The range of a low energy electron in cm is about $10^{-2}\ \beta^2$.

Section 14.15

14.29. Suppose that a photon of energy $h\nu$ enters a volume of material and produces an electron–positron pair. Both particles come to rest in the volume, and the positron annihilates with an electron that was already in the volume. Both annihilation photons leave the volume. Show that the formal definition of energy transfer agrees with the common-sense answer that it is the kinetic energy of the electron and positron, which is $h\nu-2m_ec^2$. What is the energy imparted?

14.30. What are the energy transferred, the net energy transferred and the energy imparted in the volume shown?

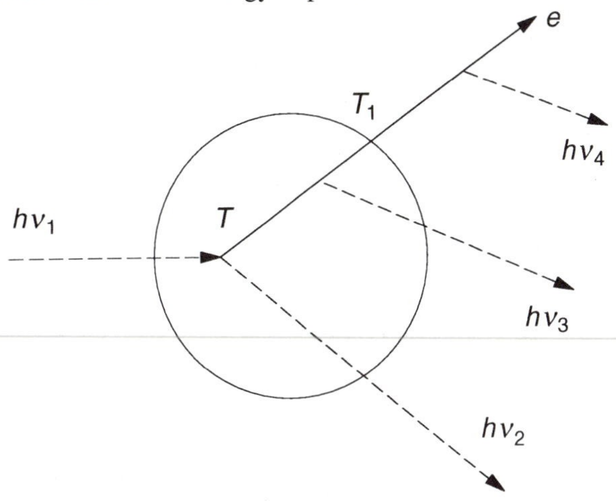

REFERENCES

Ahlen, S. P. (1980). Theoretical and experimental aspects of the energy loss of relativistic heavily ionizing particles. *Rev. Mod. Phys.* **52**: 121–173.

Attix, F. H. (1986). *Introduction to Radiological Physics and Radiation Dosimetry*. New York, Wiley.

Bambynek, W., B. Crasemann, R. W. Fink, H. U. Freund, H. Mark, C. D. Swift, R. E. Price, and P. Y. Rao (1972). X-ray fluorescence yields, Auger, and Coster-Kronig transition probabilities. *Rev. Mod. Phys.* **44**: 716–717.

Boone, J. M. and A. E. Chavez (1996). Comparison of x-ray cross sections for diagnostic and therapeutic medical physics. *Med. Phys.* **23**(12): 1997–2005.

Budd, T., and M. Marshall (1983). Microdosimetric properties of electron tracks measured in a low-pressure chamber. *Radiat. Res.* **93**: 19–32.

Carlsson, G. A., and C. A. Carlsson (1982). Calculation of scattering cross sections for increased accuracy in diagnostic radiology. 1. Energy broadening of Compton-scattered photons. *Med. Phys.* **9**: 868.

Hobbie, R. K. (1992). MacDose: A Simulation for Understanding Radiological Physics. *Comput. Phys.* **6**(4): 355–359.

Hubbell, J. H. (1969). *Photon Cross Section, Attenuation Coefficients, and Energy Absorption Coefficients from 10 keV to 100 GeV.* NBS 29. Washington, D.C. U.S. Gov't. Printing Office.

Hubbell J. H. (1982). Photon mass attenuation and energy absorption coefficients from 1 keV to 20 MeV *Int. J. Appl. Radiat. Inst.* **33**: 1269–1290.

Hubbell, J. H. and S. M. Seltzer (1996). *Tables of X-Ray Mass Attenuation Coefficients and Mass Energy-Absorption Coefficients 1 keV to 20 MeV for Elements Z=1 to 92 and 48 Additional Substances of Dosimetric Interest.* National Institute of Standards and Technology. Report No. NISTIR 5632—Web Version. Available at http://physics.nist.gov/PhysRefData/XrayMassCoef/cover.html

Hubbell, J. H., W. J. Biegle, E. A. Briggs *et al.* (1975). Atomic form factors, incoherent scattering functions and photon scattering cross sections. *J. Phys. Chem. Ref. Data* **4**: 471–538.

Hubbell, J. H., H. A. Gimm, and I. Øverbø (1980). Pair, triplet and total atomic cross sections (and mass attenuation coefficients) for 1 MeV–100 GeV photons in elements Z=1 to 100. *J. Phys. Chem. Ref. Data* **9**: 1023.

Hubbell, J. H., P. N. Trehan, N. Singh, B. Chand, M. L. Garg, R. R. Garg, S. Singh, and S. Puri (1994). A review, bibliography, and tabulation of *K*, *L*, and higher atomic shell x-ray fluorescence yields. *J. Phys. Chem. Ref. Data* **23**(2): 339–364.

ICRU Report 16 (1970). *Linear Energy Transfer.* Bethesda, MD, International Commission on Radiation Units and Measurements.

ICRU Report 33 (1980). *Radiation Quantities and Units.* Bethesda, MD, International Commission on Radiation Units and Measurements.

ICRU Report 37 (1984). *Stopping Powers for Electrons and Positrons.* Bethesda, MD, International Commission on Radiation Units and Measurements.

ICRU Report 49 (1993). *Stopping Powers and Ranges for Protons and Alpha Particles.* Bethesda, MD, International Commission on Radiation Units and Measurements.

Jackson, D. F., and D. J. Hawkes (1981). X-ray attenuation coefficients of elements and mixtures. *Phys. Rep.* **70**: 169–233.

Kissel, L. H., R. H. Pratt, and S. C. Roy (1980). Rayleigh scattering by neutral atoms, 100 eV to 10 MeV. *Phys. Rev. A.* **22**: 1970–2004.

Powell, C. F., P. H. Fowler, and D. H. Perkins (1959). *The Study of Elementary Particles by the Photographic Method.* New York, Pergamon.

Pratt, R. H. (1982). Theories of coherent scattering of x rays and γ rays by atoms, in *Proceedings of the Second Annual Conference of International Society of Radiation Physicists, Penang, Malaysia.*

Rossi, B. (1957). *Optics.* Reading, MA, Addison-Wesley, Chap. 8.

Seltzer, S. M. (1993). Calculation of photon mass energy-transfer and mass energy-absorption coefficients. *Radiat. Res.* **136**: 147–170.

Tung, C. J., J. C. Ashley, and R. H. Ritchie (1979). Range of low-energy electrons in solids. *IEEE Trans. Nucl. Sci.* **NS-26**: 4874–4878.

Ziegler, J. F., and J. M. Manoyan (1988). The stopping of ions in compounds. *Nucl. Instrum. Methods in Phys. Res.* **B35**: 215–228.

Ziegler, J. F., J. P. Biersack, and U. Littmark (1985). *The Stopping and Range of Ions in Solids.* New York, Pergamon.

CHAPTER 15

Medical Use of X Rays

X rays are used to obtain diagnostic information and for cancer therapy. They are photons of electromagnetic radiation with higher energy than photons of visible light. Gamma rays are photons emitted by radioactive nuclei; except for their origin, they are identical to x ray photons of the same energy. Section 15.1 describes the production of x rays. Section 15.2 introduces some new quantities that are important for measuring how the absorbed photon energy relates to the response of a detector, which might be a film, an ionization chamber, or a chemical detector. Several detectors are introduced in Section 15.3: film, fluorescent screens, scintillation detectors, semiconductor detectors, thermoluminescent dosimeters, and digital detectors. Section 15.4 describes the diagnostic radiograph, and the following section discusses image quality, particularly the importance of noise in determining image quality. Section 15.6 provides a brief mention of angiography, Sec 15.7 discusses some of the special problems of mammography, and fluoroscopy is described in Sec. 15.8. Computed tomography with x rays is discussed in Sec. 15.9. The final sections deal with the biological effects of x rays, cancer therapy, the relationship of dose to beam particle fluence, and the risk of radiation.

15.1. PRODUCTION OF X RAYS

When a beam of energetic electrons stops in a target, photons called x rays are emitted. *Characteristic x rays* have discrete photon energies and are produced after excitation of an atom by the electron beam. *Bremsstrahlung* (Sec. 14.11) is the continuous spectrum of photon energies produced when an electron is scattered by an atomic nucleus. Bremsstrahlung is responsible for most of the photons emitted by most x-ray tubes. The total bremsstrahlung radiation yield (Sec. 14.13) as a function of electron energy for various materials is shown in Fig. 15.1. High-Z materials are most efficient for producing x rays. Tungsten is often used as a target in x-ray tubes because it has a high radiation yield and can withstand high temperatures. For 100-keV electrons on tungsten, the radiation yield is about 1%, and most of the electron energy heats the target. We now consider these two processes in greater detail.

15.1.1. Atomic Electrons

Atomic energy levels are described in Sec. 14.1. The levels for tungsten are shown in Table 14.1 and Fig. 14.1. An electron bombarding a target can impart sufficient energy to a target electron to remove it from the atom, leaving an unoccupied energy level or *hole*. The deexcitation of the atom is described in Sec. 14.9. For a high-Z material with a hole in the K shell, the fluorescence yield is large (see Fig. 14.14). The hole is usually filled when an electron in a higher energy level drops down to the unoccupied level. As it does so, the atom emits a photon of characteristic energy equal to the difference in energies of the two levels. This leaves a new hole, which is then filled by an electron from a still higher level, with the emission of another x ray, or by an Auger cascade.

Because a hole moving to larger values of n corresponds to a decrease of the total energy of the atom, it is customary to draw the energy-level diagram for holes instead of electrons, which turns the graph upside down, as in Fig. 15.2. The zero of energy is the neutral atom in its ground state. Because this is a logarithmic scale, it cannot be shown. Creation of the hole requires energy to remove an electron. That energy is released when the hole is filled. Various possible transitions are indicated in Fig. 15.2. These transitions are consistent with these selection rules, which can be derived using quantum theory:

$$\Delta l = \pm 1, \quad \Delta j = 0, \pm 1. \tag{15.1}$$

The transitions are labeled by the letters K, L, M and so forth, depending on which shell the hole is in initially. Greek letter subscripts distinguish the x rays from transitions to different final states.

Analogous to the approximate formula of Eq. 14.3 is the

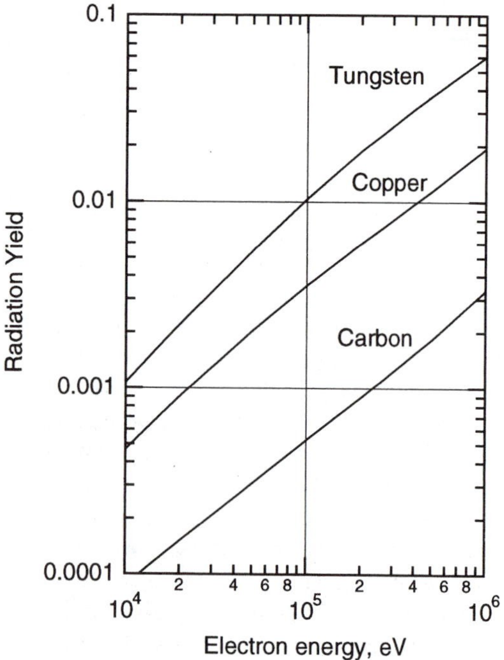

FIGURE 15.1. Radiation yield vs electron energy for carbon, copper, and tungsten. Plotted from data in ICRU Report 37 (1984).

following estimate of the energy of the K_α line (which depends on the screening for two values of n), which we have seen before as Eq. (14.34):

$$E_{K_\alpha} = \tfrac{3}{4}(13.6)(Z-1)^2. \tag{15.2}$$

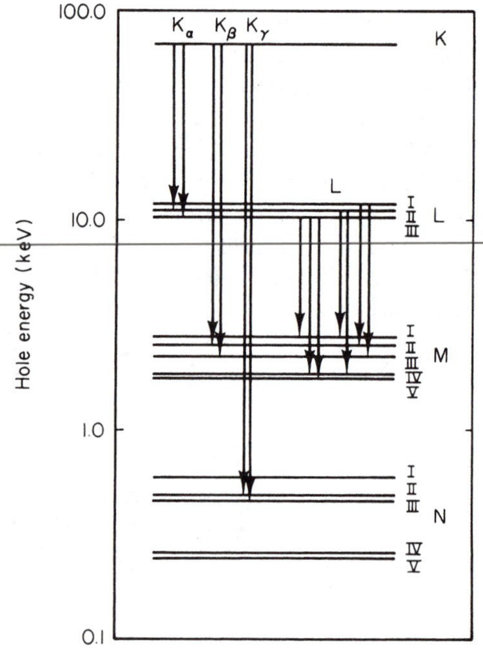

FIGURE 15.2. Energy-level diagram for holes in tungsten and some of the x-ray transitions.

The factor $\frac{3}{4}$ is what one would have for hydrogen if $n_i = 2$ and $n_f = 1$ are substituted in the Bohr formula, Eq. (13.8). The screening also depends strongly on l.

15.1.2. Nuclear Collisions

The other mechanism for x-ray production is the acceleration of electrons in the Coulomb field of the nucleus, described in Sec. 14.11. Classically, a charged particle at rest is surrounded by an electric field which is inversely proportional to the square of the distance from the charge. When in motion with a constant velocity it is surrounded by both an electric field and a magnetic field. When accelerated, additional electric and magnetic fields appear that fall off less rapidly—inversely with the first power of distance from the charge. This is classical electromagnetic radiation. Quantum mechanically, when a charged particle undergoes acceleration or deceleration, it emits photons. The radiation is called deceleration radiation, braking radiation, or bremsstrahlung. It has a continuous distribution of frequencies up to some maximum value.

A quantum-mechanical calculation shows that the photon *energy* fluence spectrum of bremsstrahlung radiation from monoenergetic electrons passing through a thin target is constant from a maximum energy $h\nu_0$ down to zero, as shown in Fig. 15.3. The maximum frequency is related to the kinetic energy of the electrons by $T = h\nu_0$, as one would expect from conservation of energy. (A photon of energy $h\nu_0$ is emitted when an electron loses all of its energy in a single collision.)

In a thick target, if we assume that all electrons at the same depth have the same energy (that is, we ignore straggling) and if we ignore attenuation of photons coming out of the target, then the spectrum is the integral of a number of spectra like that in Fig. 15.3. The thick-target spectrum is shown in Fig. 15.4. The spectral form is

$$\frac{d\Psi}{d(h\nu)} \equiv \frac{d\Psi}{dE} = CZ(h\nu_0 - h\nu) = CZ(T-E), \tag{15.3a}$$

where constant C depends on the target material and electron kinetic energy. For the photon particle fluence the spectrum is

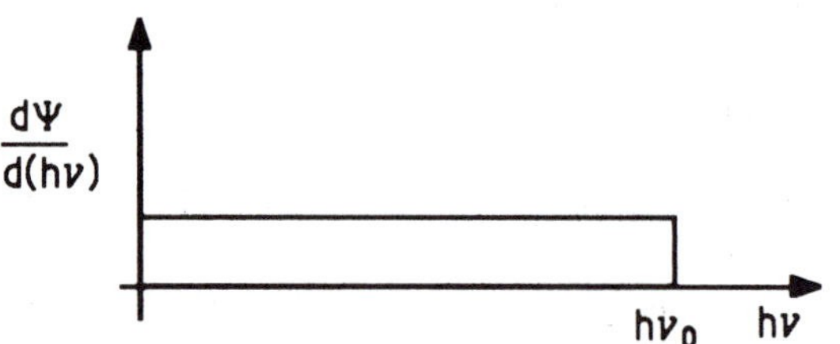

FIGURE 15.3. The energy fluence spectrum of bremsstrahlung x rays emitted when monoenergetic electrons strike a thin target.

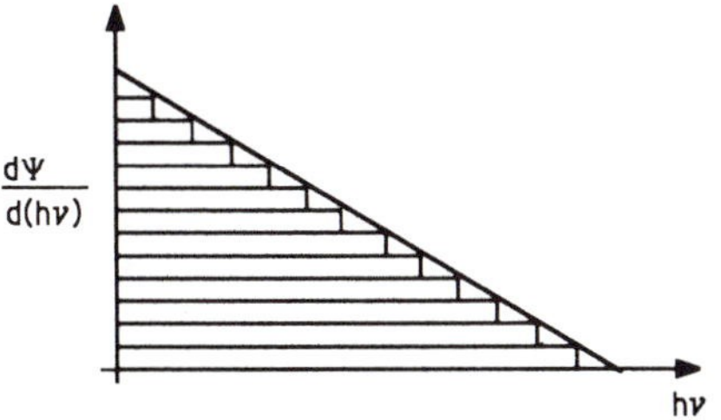

FIGURE 15.4. The energy fluence spectrum of bremsstrahlung x rays from a thick target, ignoring absorption of the photons in the target. The form is $d\Psi/d(h\nu)=CZ(h\nu_0-h\nu)$.

$$\frac{d\Phi}{dE}=\frac{1}{h\nu}\frac{d\Psi}{dE}=CZ\left(\frac{h\nu_0}{h\nu}-1\right). \tag{15.3b}$$

Low-energy photons that are generated within the target are attenuated as they escape. This cuts off the low-energy end of the spectrum. If the electron energy is high enough, the discrete spectrum due to characteristic fluorescence is superimposed on the continuous bremsstrahlung spectrum. Both of these effects are shown in Fig. 15.5, which compares calculations and measurements of the particle fluence spectrum $d\Phi/dE$. (Problem 15.3 shows that over the energy range shown, the shape of $d\Phi/dE$ differs only slightly from the shape of $d\Psi/dE$.)

Older works present the energy spectrum as a function of wavelength. The conversion is made using the same technique as in Sec. 13.5:

$$\frac{d\Psi}{d\lambda}=\frac{h^2c^2}{\lambda^3}\frac{d\Phi}{dE}.$$

Thin- and thick-target spectra are plotted vs wavelength in Fig. 15.6.

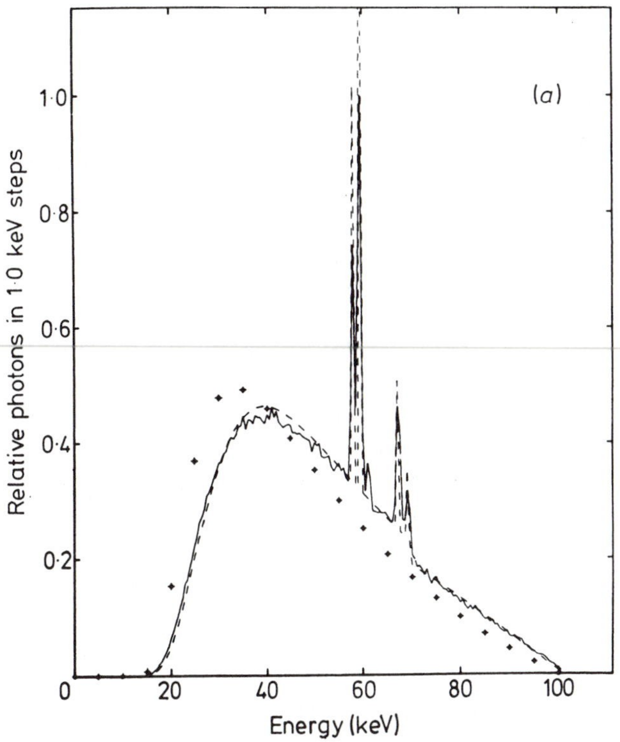

FIGURE 15.5. Plots of theoretical and measured photon particle spectra, $d\Phi/d(h\nu)$. The solid line represents measurements with a high-resolution semiconductor detector. The dashed line is the theory of Birch and Marshall (1979), which takes photon absorption into account, and the crosses show an earlier theoretical model. From R. Birch and M. Marshall. Used by permission of Institute of Physics Publishing.

15.2. QUANTITIES TO DESCRIBE RADIATION INTERACTIONS: RADIATION CHEMICAL YIELD, MEAN ENERGY PER ION PAIR, AND EXPOSURE

Section 14.15 introduced the quantities energy transferred, energy imparted, kerma, and absorbed dose, which are used to describe radiation and its effects. This section introduces some additional quantities [ICRU Report 33 (1980, 1992)] that are used to describe the interaction of the radiation with the detectors discussed in Sec. 15.3.

The *radiation chemical yield* G of substance x is the mean number of moles $\bar{n}$ of the substance produced, destroyed, or changed in some volume of matter, divided by the mean energy imparted to the matter:

$$G(x)=\frac{\overline{n(x)}}{\bar{E}}. \tag{15.4}$$

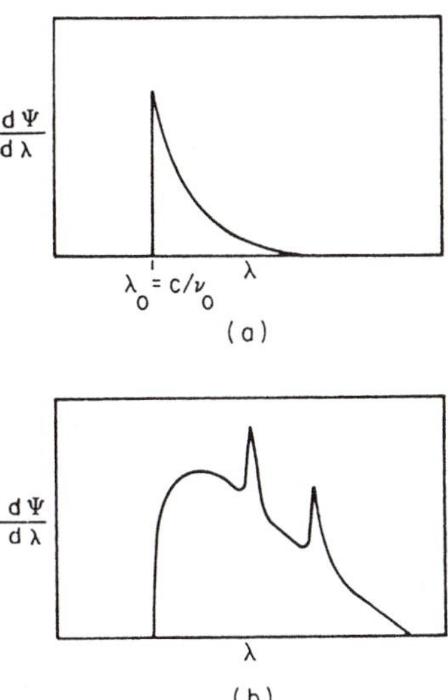

FIGURE 15.6. Intensity as a function of wavelength (a) for a thin target; (b) for a thick target with a line spectrum superimposed.

Its units are mol J^{-1}. (A related quantity expressed in non-SI units is the G *value*, expressed in molecules or moles per 100 eV of energy imparted.) The radiation chemical yield is particularly useful for describing chemical dosimeters. These are usually dilute aqueous solutions, so the radiation chemical yield of water is the important parameter.

Other detectors measure ionization in a gas produced by the radiation. The *mean energy expended in a gas per ion pair formed*, W, is the ratio

$$W=\frac{T_0}{\bar{N}}, \qquad (15.5)$$

where T_0 is the initial kinetic energy of a charged particle and $\bar{N}$ is the mean number of ion pairs formed when T_0 is completely dissipated in the gas. The units are joules or electron volts per ion pair. The mean energy expended per ion pair is not equal to the ionization potential. To see why, we must consider three processes that can take place. The first is ionization, with E_i being the average energy of an ionized atom. Second, the collision may promote an atomic electron to an excited state without ionization. The average excitation energy is E_{ex}. Finally, the charged particle may lose energy to impurities, without producing ionization, a process called subexcitation. The average subexcitation energy is defined to be the energy lost by this process divided by $\bar{N}_i$. Conservation of energy for this model leads to

$$T_0=\bar{N}_i\bar{E}_i+\bar{N}_{ex}\bar{E}_{ex}+\bar{N}_i\bar{E}_{se},$$

where T_0 is the initial projectile kinetic energy, $\bar{N}_i$ is the mean number of ion pairs produced, $\bar{E}_i$ is the mean energy of an ionized atom, $\bar{N}_{ex}$ is the mean number of atoms raised to an excited state but not ionized, $\bar{E}_{ex}$ is the average energy of an excited atom, and $\bar{E}_{se}$ is the "sub-excitation" energy.

Dividing each term by $\bar{N}_i$ leads to an expression for W. In general, W is determined experimentally, because the terms in this equation are quite difficult to calculate. However, they have been calculated for helium.[1] The mean energy of an ionized helium atom is only 62% of the value of W:

$$\underbrace{W}_{41.8\text{ eV}} = \underbrace{\bar{E}_i}_{\substack{25.9\text{ eV}\\62\%}} + \underbrace{(\bar{N}_{ex}/\bar{N}_i)\bar{E}_{ex}}_{\substack{0.4\times 20.8=8.3\text{ eV}\\20\%}} + \underbrace{\bar{E}_{se}}_{\substack{7.6\text{ eV}\\18\%}}.$$

Values of W are tabulated in ICRU Report 31 (1979). There are variations of a few percent depending on whether the charged particle is an electron or an α particle. Table 15.1 provides a few representative values. Though defined for a gas, W is also applied to semiconductors as the average energy per electron–hole pair produced. Values of W for semiconductors are much smaller than for a gas.

The *exposure* X is defined only for photons and measures the energy fluence of the photon beam. It is the amount of ionization (total charge of one sign) produced per unit mass of dry air when all of the electrons and positrons liberated in a small mass of air are completely stopped in air:

$$X=\frac{dq}{dm}. \qquad (15.6)$$

The units are coulomb per kilogram. Since the average amount of energy required to produce an ion pair is well defined, exposure is closely related to collision kerma in air. The definition of X does not include ionization arising from the absorption of bremsstrahlung emitted by the electrons, so there is a slight difference at high energies.[2] The relationship is

$$X=(K_c)_{air}\left(\frac{e}{W_{air}}\right)=\Psi\left(\frac{\mu_{en}}{\rho}\right)_{air}\left(\frac{e}{W_{air}}\right). \qquad (15.7)$$

The *roentgen* (R) is an old unit of exposure equivalent to the production of 2.58×10^{-4} C kg^{-1} in dry air; this corresponds to a dose of 8.69×10^{-3} Gy. (The relationship is developed in Problem 15.6.)

If charged-particle equilibrium exists, the dose in air is related to the beam energy fluence by Eqs. (14.72) and (14.75):

$$D_{air}=\left(\frac{\mu_{en}}{\rho}\right)_{air}\Psi.$$

The dose for the same energy fluence in some other medium is

$$D_{med}=\left(\frac{\mu_{en}}{\rho}\right)_{med}\Psi=\frac{(\mu_{en}/\rho)_{med}}{(\mu_{en}/\rho)_{air}}D_{air}. \qquad (15.8)$$

TABLE 15.1. *Some representative values of the average energy per ion pair, W.*

Gas	W (eV per ion pair for electrons[a])
He	41.3
Ar	26.4
Xe	22.1
Air	33.97[b]
Semiconductors	W (eV per electron–hole pair)
Si	3.68
Ge	2.97

[a]From ICRU Report 31 (1979).
[b]ICRU Report 31 (1979) recommends 33.85 J C^{-1}. Attix (1986) uses 33.97 J C^{-1}.

[1]See Platzman (1961); also Attix (1986, pp. 339–343).

[2]There is also a problem at high energies because the range of the electrons is large. If they are to come to rest within the chamber, the size of the chamber becomes comparable to the photon attenuation coefficient.

15.3. DETECTORS

Detectors are used for recording an image and also for measuring the quantity of radiation to which a patient is exposed. This section describes the most common kinds of detectors.

15.3.1. Film and Screens

Film is the classic x-ray detector, and it is still very widely used. A typical x-ray film has a transparent base about 200 μm thick, coated on one or both sides with a sensitive emulsion containing a silver halide (usually silver bromide). We will not discuss the rather complicated sequence of steps by which the absorption of photons or energy loss by charged particles leads to a latent or developable image in the film. When the film is developed, the emulsion grains that have absorbed energy are reduced to black specks of metallic silver. The film is then fixed, a process in which the silver halide that did not absorb energy is removed from the emulsion. The result is a film that absorbs visible light where it was struck by ionizing radiation.

The fraction of incident light passing through the film after development is called the transmittance, T. The *optical density* or "density" is defined to be

$$\mathrm{OD}=\log_{10}(1/T). \tag{15.9}$$

A film that transmits 1% of the incident viewing light has an optical density of 2.

The response of a film is described by plotting the optical density vs the log of the exposure in air immediately in front of the film (or equivalently, the absorbed dose in the film emulsion). Since the optical density is the logarithm of the transmittance, this is a log–log plot of the reciprocal of the fraction of the visible light transmitted when viewing the processed film vs the x-ray exposure before processing. A typical plot of film response is shown in Fig. 15.7. If the curve is linear, the transmittance is proportional to the exposure raised to some power:

$$T \propto X^{-\gamma}.$$

At very small exposures (the "toe") the transmission is that of the base and "clear" emulsion. At very high exposures (the "shoulder") all of the silver halide has been reduced to metallic silver, and the film has its maximum optical density. In between is a region which is almost linear. The ratio of maximum to minimum usable exposure is called the latitude of the film. The largest value of the exponent occurs at the inflection point and is called the "gamma" of the film. Both the exponent and the position of the curve along the log exposure axis depend on the development time, the temperature of the developing solution, and the energy of the x-ray beam. The *film speed* is the reciprocal of the exposure required for an optical density that is 1 greater than the base density.

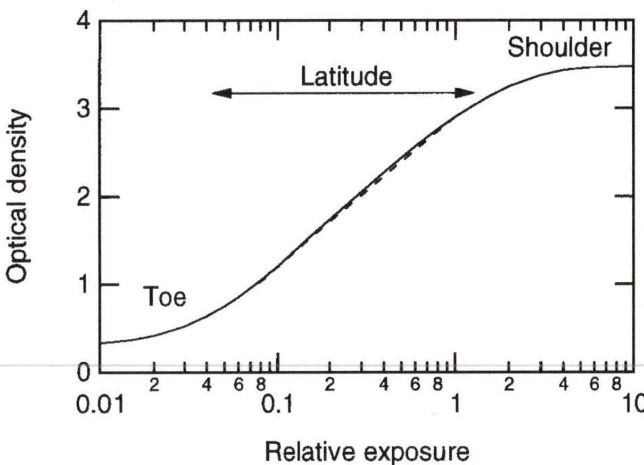

FIGURE 15.7. A plot of optical density vs the logarithm of the relative exposure for a hypothetical x-ray film.

A typical film has an emulsion containing AgBr. It requires a dose of 1.74×10^{-4} Gy in air just in front of the film to produce an optical density of 1. This might be where the body is not blocking the beam. The smaller dose to the film where there has been significant attenuation in the body gives a lighter region, as in the heart and bone shadows of Fig. 15.20.

The dose to the part of the body just in front of the film (the exit dose) can be written in several ways. For simplicity, we assume monoenergetic photons. In terms of the energy fluence of the photon beam, the exit dose is

$$D_{\mathrm{body}}=\left(\frac{\mu_{\mathrm{en}}}{\rho}\right)_{\mathrm{body}}\Psi. \tag{15.10a}$$

In terms of the dose in air just in front of the film it is

$$D_{\mathrm{body}}=\frac{(\mu_{\mathrm{en}}/\rho)_{\mathrm{body}}}{(\mu_{\mathrm{en}}/\rho)_{\mathrm{air}}}D_{\mathrm{air}}, \tag{15.10b}$$

and in terms of the dose in the film it is

$$D_{\mathrm{body}}=\frac{(\mu_{\mathrm{en}}/\rho)_{\mathrm{body}}}{(\mu_{\mathrm{en}}/\rho)_{\mathrm{film}}}D_{\mathrm{film}}. \tag{15.10c}$$

For 50-keV photons we find from Appendix O that $(\mu_{\mathrm{en}}/\rho)_{\mathrm{muscle}}/(\mu_{\mathrm{en}}/\rho)_{\mathrm{air}} = 0.004\,35/0.004\,10 = 1.061$. Therefore the exit dose would be $(1.74\times 10^{-4})\times(1.061)=1.85\times 10^{-4}$ Gy. Because of attenuation, the entrance dose can be much larger.

The dose can be reduced by a factor of 50 or more if the film is sandwiched between two fluorescent "intensifying" screens. The x-ray photons have a low probability of interacting in the film. The screens have a greater probability of absorbing the x-ray photons and converting them to visible light, to which the film is more sensitive. For 50-keV

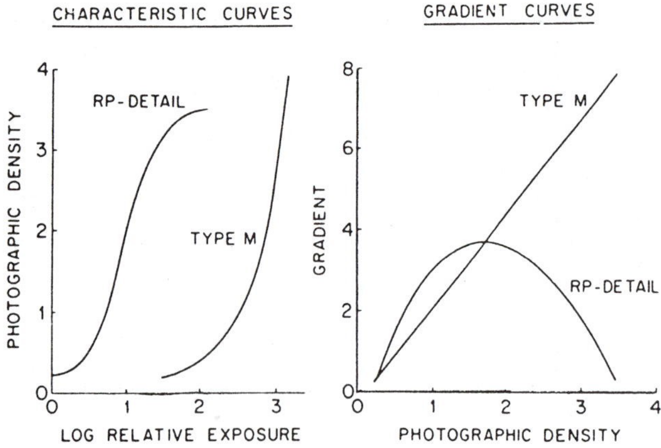

FIGURE 15.8. Plots of optical density vs logarithm exposure for a screen–film combination (RP) and a film (M). From K. Doi, H. K. Genant, and K. Rossman (1976). Comparison of image quality obtained with optical and radiographic magnification techniques in fine-detail skeletal radiography: Effect of object thickness. *Radiology* **118**: 189–195 (1976). Reproduced by permission of the Radiological Society of North America.

photons on typical emulsion, the value of μ_{en}/ρ is about $0.261\ \mathrm{m^2\,kg^{-1}}$. A typical value of $\rho\,\Delta x$ might be $0.02\ \mathrm{kg\,m^{-2}}$. Therefore $\mu_{en}\Delta x = 0.0052$. The fraction of incident energy absorbed in the emulsion is $1-e^{-0.0052} = 0.0052$.

A typical screen might consist of particles of gadolinium oxysulfide (Gd_2O_2S:Tb) suspended in a carrier about 150 μm thick (0.5–1.5 kg $\mathrm{m^{-2}}$). This layer is backed by a thin reflective layer. Two such screens (one on each side of the film) with a total thickness of 1.2 kg $\mathrm{m^{-2}}$ absorb 28% of the 50-keV photons that pass through them (see Problem 15.10). The overall effect is to produce the same optical density when the energy fluence in the x-ray beam is reduced by a factor of 54—the ratio of the incident radiation absorbed in the screen and in the film in each case.[3]

Figure 15.8 shows a plot of optical density vs the log of the exposure X for two typical radiographic systems. "RP-detail" is a screen–film combination. M is a film without a screen. At an optical density of 1 or 2, the difference in $\log_{10}X$ is about 1.9, corresponding to a factor of 80 in exposure difference, somewhat greater than the example given above. The curves are not linear. The slope at any point on the curve is[4]

$$\gamma = \frac{d\,\log_{10}(1/T)}{d\,\log_{10}X} = \frac{d\,\ln(1/T)}{d\,\ln X} = -\frac{dT/T}{dX/X}$$

$$= -\frac{X}{T}\frac{dT}{dX} = -\frac{1}{G}g, \qquad (15.11)$$

where $G = T/X$ is the *large signal transfer factor* and $g = dT/dx$ is the *incremental signal transfer factor*. This will be used in our discussion of detecting signals in noise in the next section.

15.3.2. Scintillation Detectors

When photons interact with matter, some of their energy is transferred to electrons. These electrons interact in turn, and some of their energy can become ultraviolet or visible photons. A *scintillator* is a substance that produces these photons with high efficiency, yet is transparent to them. The photons are then converted to an electrical signal by a *photomultiplier tube*. A representative setup is shown in Fig. 15.9. Light from the scintillator is transmitted to the photomultiplier tube either by a thin layer of coupling material or by a "light pipe." The light pipe is usually used when the shape or size of the scintillator does not match that of the photocathode. The entire assembly is enclosed in a light-tight case which allows entry of the radiation to be detected. Electrons are emitted from the photocathode into the vacuum of the photomultiplier tube by the photoelectric effect and are accelerated to the surface of a positively charged *dynode*. The kinetic energy they gain is sufficient to knock several electrons out of the surface of the dynode, a process called *secondary emission*. These electrons are then accelerated to a second dynode, where the process is repeated. If the average multiplication at each of n dynodes is m the overall charge multiplication is m^n. The pulse of current finally striking the anode is proportional to the number of photons striking the photocathode. Because the trajectories of the electrons are disturbed by even weak magnetic fields, some kind of magnetic shield must also be provided.

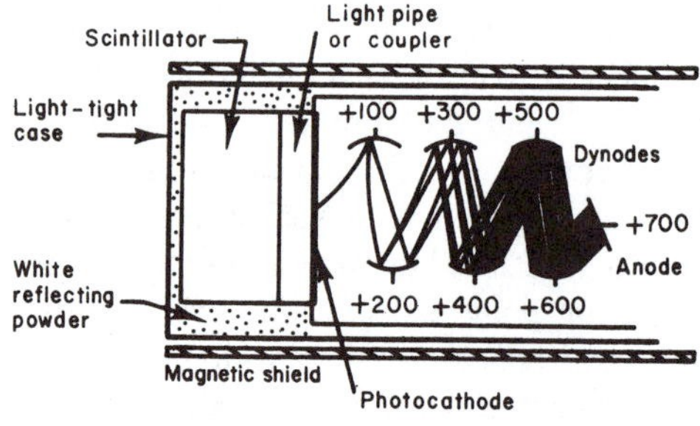

FIGURE 15.9. A representative scintillation detector, showing the scintillator, light pipe, and photomultiplier tube in a light-tight housing. The operation of the system is described in the text.

[3]The fluorescent radiation has a wavelength of about 545 nm, and each absorbed high-energy photon has sufficient energy to product about 14 000 fluorescence photons. However, the efficiency of production is only about 5% so 700 photons are produced. Some of these escape or are absorbed. Each x-ray photon produces about 150 photons of visible and ultraviolet light that strike the emulsion—more than enough to blacken the film in the region where the x-ray photon was absorbed by the screen.

[4]An argument based on Eq. (2.14) can be used to show that $\log_{10}x = (1/2.303)\ \ln x = 0.43\ \ln x$.

There are seven stages in the scintillation process:

1. Interaction of the photon, so that some fraction of its energy is converted to kinetic energy of the charged particles.
2. Slowing down of the charged particles, giving rise to ionization and excitation.
3. Immediate conversion of some of this energy to light. It can also become heat or be emitted later as light (phosphorescence).
4. Some of the resulting photons strike the photocathode.
5. Some of the photons that strike the cathode give rise to photoelectrons.
6. Electron multiplication takes place in the photomultiplier tube.
7. The pulse of electric current is analyzed.

Each of these steps involves an element of chance and gives rise to statistical fluctuations in the amount of current for a photon of a given energy. In addition, the efficiency of some of these processes depends on where in the scintillator the initial reaction takes place.

Each interacting photon produces an electrical current pulse at the output, called a *count*. When the number of counts is recorded vs the pulse height (total charge in the pulse), the result is a *pulse-height spectrum*. For monoenergetic photons, the ideal pulse height spectrum would consist of a single peak; all pulses would have the same height. This is not realized in practice for two reasons: statistical variations in the processes listed above cause the line to be broadened, and the entire energy of the incident photon is not converted into electrons.

An atom that has been excited by photoelectric absorption can decay by the emission of a fluorescence photon. If this photon is subsequently absorbed in the scintillator, all of the original photon energy is converted to electron energy so rapidly that the visible light is all part of one pulse. The pulse height then corresponds to the full energy of the original photon. However, if the initial photoelectric absorption takes place close to the edge of the detector, the fluorescence photon can escape, and the pulse has a lower height than those in the primary peak. This can be seen in Fig. 15.10. Photons *A* and *B* interact by photoelectric effect. All the energy for photon *A* is deposited in the scintillator, while the *K* fluorescence photon from *B* escapes. The effect on a pulse height spectrum is shown in Fig. 15.11 for a scintillator of sodium iodide. The photoelectric cross section for iodine is dominant.

In Compton scattering, only the energy of the recoil electron is transferred to the scintillator. The scattered photon may escape from the detector, as in *D* of Fig. 15.10. (If it is subsequently absorbed, as in mechanism C of Fig. 15.10, the pulse height will have the peak value.) The energy of

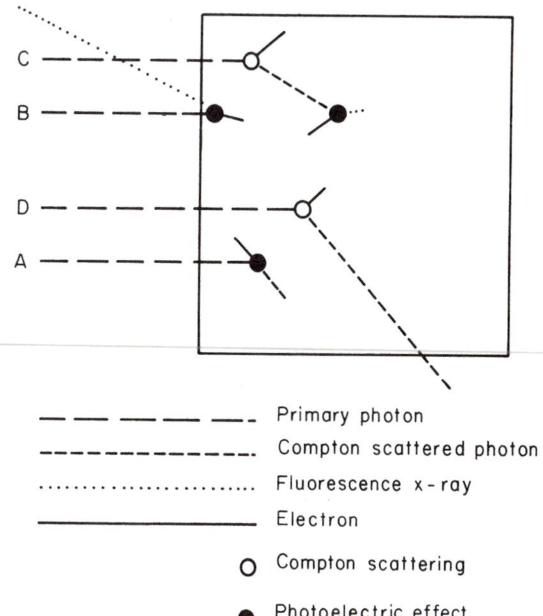

FIGURE 15.10. Mechanisms by which some of the energy of a primary photon can escape from a detector. Photons *A* and *B* undergo photoelectric absorption. All of the energy from *A* is absorbed in the detector, while the fluorescence x-ray from *B* escapes. Photons *C* and *D* are Compton scattered. The scattered photon from *C* undergoes subsequent absorption, while that from *D* escapes.

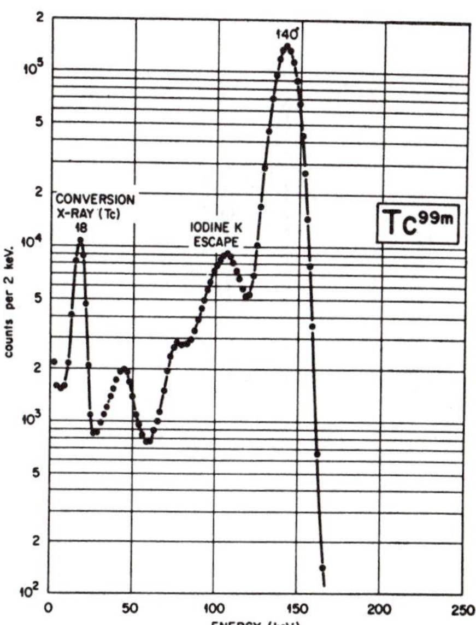

FIGURE 15.11. Spectrum of pulse heights for 140-keV photons from isotope ^{99m}Tc incident on a sodium iodide scintillator. The 140-keV total energy peak is prominent, as is the peak at 110 keV corresponding to the escape of the *K* fluorescence x ray from iodine. The Compton scatter continuum runs from 49 keV down to zero energy. The peak at 18 keV is from additional radiation from ^{99m}Tc (see Chap. 16). Reproduced by permission from H. N. Wagner Jr., ed. (1968) *Principles of Nuclear Medicine*, p. 162. Copyright by W. B. Saunders.

the recoil electron is given by Eq. (14.16). The maximum electron energy occurs when the photon is scattered through $\theta = 180°$. Then

$$T_{\max} = \frac{2h\nu_0 x}{1+2x} = \frac{(h\nu_0)^2}{h\nu_0 + m_e c^2/2}.$$

If the photon energy is in keV, this is

$$T_{\max} = \frac{(h\nu_0)^2}{h\nu_0 + 256}. \tag{15.12}$$

A spectrum for ''pure'' Compton scattering of 662-keV photons (as emitted by 137Cs) is shown in Fig. 15.12. The

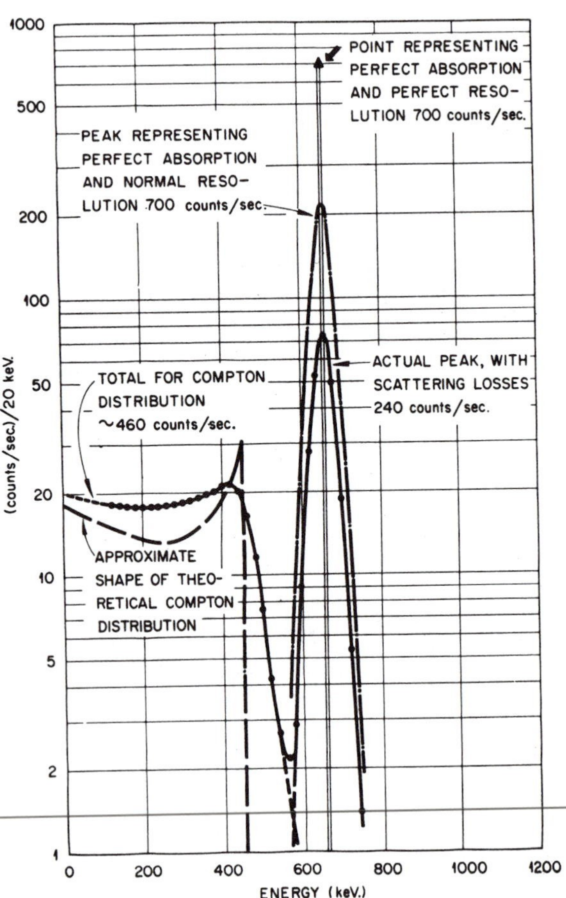

FIGURE 15.12. The response of a sodium iodide detector to 662-keV photons from ^{137}Cs. Theoretical responses are shown for a detector that absorbs the energy of all photons and has perfect resolution, for a detector with perfect absorption and finite resolution, and for a detector in which Compton-scattered photons can escape. Experimental data are for a 1½-in. by 1-in. NaI crystal. Redrawn from C. C. Harris, D. P. Hamblen, and J. E. Francis, Jr. (1969). *Basic Principles of Scintillation Counting for Medical Investigators.* ORNL-2808. Springfield, VA, Clearing House for Federal Scientific and Technical Information, National Bureau of Standards. Reproduced by permission from H. N. Wagner, Jr., ed. (1968). *Principles of Nuclear Medicine*, Philadelphia, Saunders, p. 153. Copyright by W. B. Saunders.

peak of the Compton continuum is at $T_{\max} = (662)^2/(662 + 256) = 477$ keV. The cases of perfect resolution with complete absorption and real resolution with complete absorption are shown, along with the Compton continuum with perfect resolution, and a real spectrum.

When the energy of the primary photons is so large that pair production is important, an additional escape mechanism must be considered. We know [(Eq. 14.23)] that

$$h\nu_0 = T_+ + T_- + (m_e c^2)_{e^-} + (m_e c^2)_{e^+}.$$

The positron will eventually combine with another electron to produce two annihilation radiation photons:

$$(m_e c^2)_{e^+} + (m_e c^2)_{\text{another } e^-} = 2E_\gamma.$$

The energy of each annihilation photon is 0.511 MeV. The initial photon energy is finally distributed as

$$h\nu_0 = T_+ + T_- + \gamma(0.511) + \gamma(0.511).$$

If all this energy is absorbed in the detector, the pulse height corresponds to the full energy of the incident photon. One or both of the annihilation photons can escape, giving the one-photon escape peak and the two-photon escape peak.

15.3.3. Gas Detectors

Ionization in gas can be used as the basis for an x-ray detector. A photon can produce photoelectric, Compton, or pair-production electrons. These then lose energy by electron collisions. Ion pairs are produced in the gas, the average number being proportional to the amount of energy lost in the gas. The average amount of energy required is W, as we saw in Sec. 15.2. Imagine that the ions are produced between the plates of a charged capacitor as shown in Fig. 15.13. The electrons are attracted to the positive plate and the positive ions travel to the negative plate. If all the electrons and ions are captured, the total charge collected on each plate has magnitude $q = Ne$, where e is the charge on the electron or ion and N is the number of ion pairs formed. If the capacitance is C, the change in voltage is $\delta v = q/C = Ne/C$. Such a device is called an *ionization chamber.* The cumulative discharge of the capacitor is

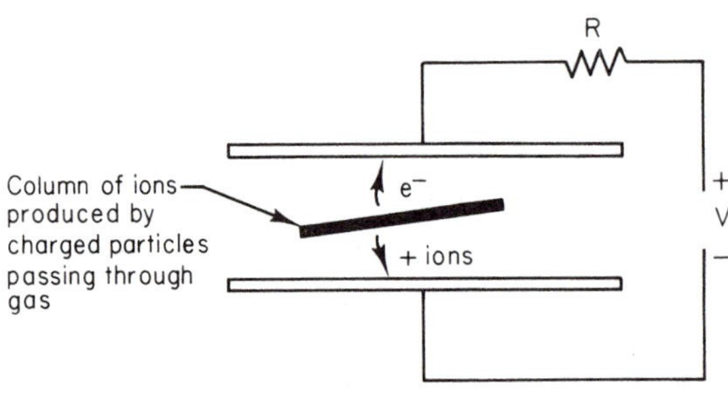

FIGURE 15.13. Schematic of an ionization chamber or proportional counter. The ions discharge the capacitor, which is recharged between counts through resistor *R*.

measured in some pocket dosimeters; in other cases, the capacitor is slowly recharged through a large resistance R so that each photon detected generates a voltage pulse of height δv.

A certain minimum voltage between the two plates is necessary to ensure that all the ions produced are collected, corresponding to the ion chamber region of Fig. 15.14. If the potential on the plates is raised further, the number of ions collected increases. Between collisions the electrons and ions are accelerated by the electric field, and they acquire enough kinetic energy to produce further ionizations when they collide with molecules of the gas, a process called *gas multiplication*. At moderate potentials, the multiplication factor is independent of the initial ionization, so the number of ions collected is larger than that in an ionization chamber but still is proportional to the initial number of ions. In this region of operation the device is called a *proportional counter*. Parallel-plate geometry is not used in a proportional counter. One electrode is a wire, and the other is a concentric cylinder. At still higher values of the applied voltage, pulse size is independent of the initial number of ion pairs. In this mode of operation, the device is called a *Geiger counter*.

Any gas detector used to detect high-energy photons suffers from the fact that the gas is not very dense. At low energies the photoelectric cross section is high and most photons interact. At higher energies, many photons pass through the gas and detector walls without interacting. A thin sheet of absorber in front of the gas detector can actually increase the counting rate, An example is shown in Fig. 15.15. The detector had an aluminum wall of thickness 0.3 kg m^{-2}. Electrons of 125 keV or more pass completely through the detector wall. The maximum energy of Compton electrons from the 1.1-MeV photons is 890 keV. Compton electrons produced in a thin layer of lead can pass

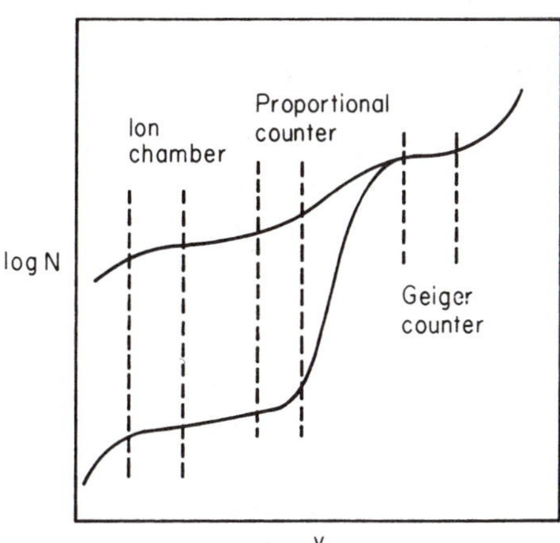

FIGURE 15.14. The number of ions collected vs collecting potential for two particles that deposit different amounts of energy in a gas detector. The voltage regions are indicated where the device operates as an ionization chamber, a proportional counter, and a Geiger counter.

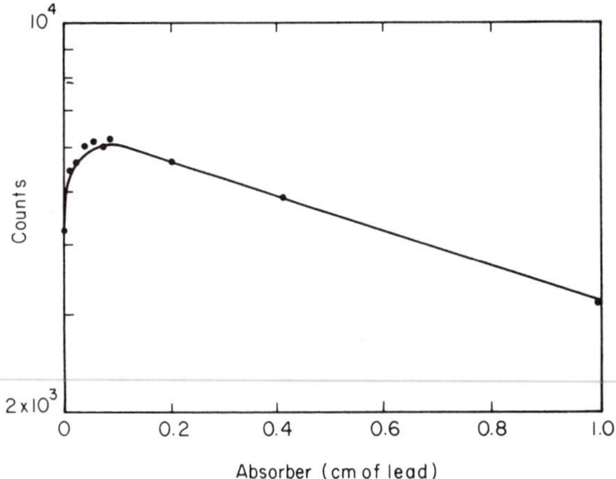

FIGURE 15.15. Counting rate of a Geiger counter vs the thickness of a lead absorber in front of the detector, showing the buildup of counting rate due to the conversion of photons to electrons in the lead by Compton scattering. These electrons pass through the thin wall of the counter and ionize the gas. The photons were from ^{60}Co and had an energy of 1.1 MeV.

through the aluminum and ionize the gas in the detector. Once the total thickness of lead and aluminum is sufficient to stop all the Compton electrons, the addition of more lead upstream does not increase the detector efficiency, and exponential attenuation is seen.

15.3.4. Semiconductor Detectors

A semiconductor detector is very much like an ionization chamber, except that a solid is used as the detecting medium. The "ion pair" is an electron that has received sufficient energy to be able to leave its atom and move freely within the semiconductor (but not enough energy to leave the semiconductor entirely) and the "hole" that the electron left behind. Electrons from neighboring atoms can fall into the hole, so the hole can move from atom to atom just like a positive charge.

A semiconductor detector has two principal advantages over a gas ionization chamber. First, the amount of energy required to create an electron–hole pair is only about 3 eV, one-tenth the value for a typical gas. This means that many more pairs are produced and the statistical accuracy is better. Second, the density of a solid is much greater than the density of a gas, so the probability that a photon interacts is larger. The cross section for interaction increases with high Z, so detectors made of germanium ($Z=32$) are better for photon detection than those made of silicon ($Z=14$). There have been technical problems making semiconductor detectors of large volume. Some of these have been overcome by using lithium-drifted germanium (Ge-Li) detectors. Their main drawback is that they have to be kept at liquid-nitrogen temperature to keep the lithium from drifting out once it has been implanted.

15.3.5. Thermoluminescent Dosimeters

Thermoluminescent phosphors consist of a small amount of dielectric material (0.1 g or less) that has been doped with impurities or has missing atoms in the crystal lattice that form metastable energy levels or traps. These impurities or defects are far from one another and are isolated in the lattice, so that electrons cannot move freely from one trap to another. When the phosphor is irradiated with ionizing radiation, some of the electrons are trapped in these metastable states. There are levels associated with the material at an energy E above the trap energy which allow electrons (or holes) to move throughout the phosphor. The probability that an electron escapes from the trap is proportional to a Boltzmann factor, $\exp(-E/k_BT)$. If E is large enough, the lifetime in the trapped state can be quite large—up to hundreds of years. Heating allows the electrons to escape to the higher levels, where they then fall back to the normal state with the emission of visible photons. Ordinary table salt (NaCl) exhibits this behavior. If it is irradiated and then sprinkled on an electric hot plate in a darkened room, one can see the flashes of light. The light emitted on heating is called *thermoluminescence*. In a thermoluminescent dosimeter (TLD), the light emitted is measured with a photomultiplier tube as the temperature is gradually increased. The total amount of light released is proportional to the energy imparted to the phosphor by the ionizing radiation.

Thermoluminescent dosimeters can measure an integrated dose from 10^{-5} to 10^3 Gy. Great care must be taken both in the preparation and reading of the phosphor. TLD chips are widely used to measure radiation doses because the chips are small and have the approximate atomic number and atomic weight of tissue. They are often made of LiF. A detailed description is found in Chap. 14 of Attix (1986).

15.3.6. Chemical Dosimetry

When radiation interacts with water, free radicals are produced. A free radical, such as H or OH, is electrically neutral but has an unpaired electron. Free radicals promote other chemical reactions. Typically, a dilute indicator of some sort is added to the water. A common dosimeter is the Fricke ferrous sulfate dosimeter. An 0.001M $FeSO_4$ solution is irradiated. The radiation changes the iron from the ferrous (Fe^{2+}) to the ferric (Fe^{3+}) state with a G of about 1.6×10^{-6} mol J^{-1}. The concentration of ferric ion can be measured by absorption spectroscopy. Details are found in Chap. 14 of Attix (1986).

15.3.7. Digital Detectors

A great deal of effort is under way to eliminate the use of film as a detector and storage medium. A digitally recorded image generally has a much greater dynamic range (latitude) than film. This is particularly important when radiographs are taken at the bedside in an intensive care unit, and the exposure may be slightly off. A factor-of-2 difference in exposure,[5] which renders a conventional radiographic image almost useless, is easily tolerated by digital recording. A digitally stored image allows easier retrieval, transmission, and creation of multiple copies. One digital detector uses a storage phosphor. Another uses an array of detectors fabricated in selenium. A third uses an array of detectors in amorphous silicon.

The storage phosphor technology is currently available commercially. The image is formed on a plate of phosphor crystals such as barium fluorobromide. Absorption of photons leaves the BaFBr crystals in a metastable state, like a TLD phosphor. Scanning by a thin laser beam in a horizontal and vertical raster pattern like a television image causes visible light to be emitted by the trapped electrons. The dynamic range of a storage phosphor can be as high as 10^4, compared to about 10^2 for radiographic film. Digital detectors have been reviewed recently by Yaffe and Rowlands (1997).

15.4. THE DIAGNOSTIC RADIOGRAPH

Figure 15.16 shows the typical elements for making a diagnostic x ray. An image recorded on film is called a *radiograph*. The x-ray tube ideally serves as a point source of photons. The photons are filtered and collimated to illuminate only the portion of the patient of interest. Typically, about 10% pass through the patient and strike the film–screen sandwich in a chest radiograph. In the abdomen the fraction is about 1%. There may be an optional collimator, as discussed below. We discuss each element below, and then discuss the quality of the image.

15.4.1. X-ray Tube and Filter

Most routine radiography is done with photons in the range from 35 to 85 keV. (Mammography uses lower voltage, and computed tomography is somewhat higher.) Figure 15.17 shows the loss of radiographic contrast as the energy of the incident photons increases and Compton scattering becomes more important.

The photons are typically produced by an x-ray tube running with a voltage between cathode and anode of about 100 kilovolts peak[6] (100 kVp). The anode is usually made of tungsten (which has a high radiation yield and withstands high temperatures) with a copper backing to conduct thermal energy away. The number of x rays produced for a given voltage difference depends on the total number of

[5]Even though the film may have a linear response over a larger range, doubling the exposure usually makes the film too dense to read.

[6]The word *peak* is included because the voltage from power supplies in older machines had considerable ''ripple'' caused by the alternating voltage from the power lines.

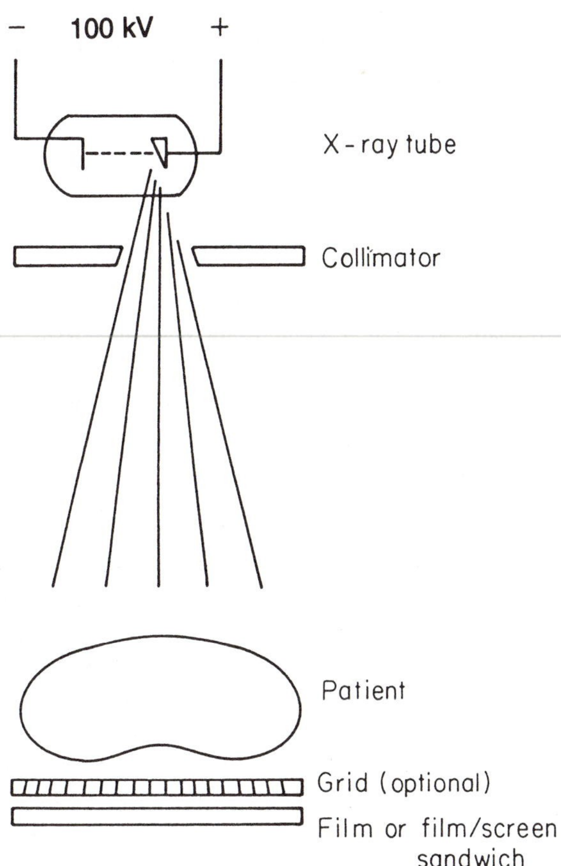

FIGURE 15.16. Overall scheme for making a radiograph. Photons are produced when electrons strike the tungsten anode. The beam is collimated to prevent unnecessary dose to the patient. The patient is placed directly in front of the grid (if any) and the film or a sandwich of film and screen.

electrons striking the anode, which is proportional to the product of the current and the duration of the exposure (mA s). The anode often rotates to help keep it cool. Additional filtration is usually used to remove low-energy photons that would not get through the body and would not contribute to the image. Figure 15.18 shows the effects of different thicknesses of aluminum on the particle fluence ($d\Phi/dE$) from a tube operating at 100 kVp. The average photon energy depends upon the filtration as well as the kVp, and is about 45 keV for 100 kVp and 2 mm of aluminum filtration (see Problem 15.16).

15.4.2. Collimation

The collimator is placed just after the x-ray tube. It has adjustable jaws, usually of lead, that limit the size of the beam striking the patient. Making the beam as small as possible has two important effects. It reduces the total energy absorbed by the patient. It also reduces the amount of tissue producing Compton-scattered photons that strike the film. Since the radiograph is based on shadows that assume the photons traveled in a straight line from the tube to the detector, scattered photons reduce the image quality.

15.4.3. Attenuation in the Patient: Contrast Material

The purpose of a radiograph is to measure features of the internal anatomy of a patient through differences in the attenuation of rays passing through different parts of the body. The photon fluence falls with distance from the x-ray tube as $1/r^2$. It also falls because of attenuation along the

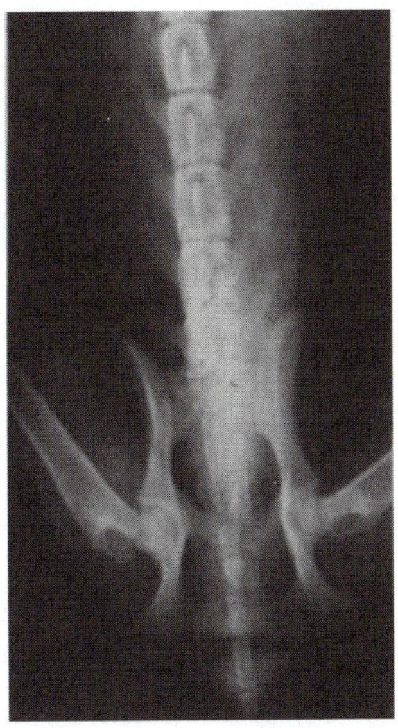

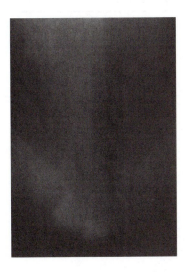

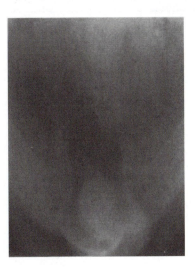

FIGURE 15.17. Radiographs taken at 70 kVp, 250 kVp, and 1.25 MeV (^{60}Co), illustrating the loss of contrast for higher energy photons. From W. R. Hendee and R. Ritenour (1992). *Medical Imaging Physics*. St. Louis, Mosby–Year Book. Used by permission of Mosby–Year Book, Inc.

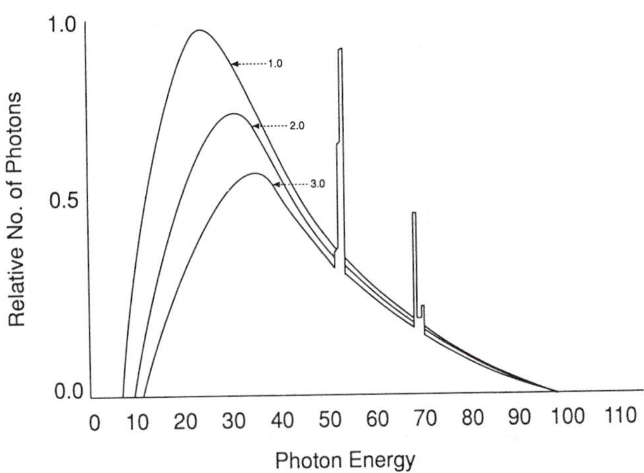

FIGURE 15.18. The particle energy spectrum ($d\Phi/dE$) from a tube operating at 100 kVp with 1, 2, and 3 mm of aluminum filtration. From W. R. Hendee and R. Ritenour (1992). *Medical Imaging Physics.* St. Louis, Mosby–Year Book. Used by permission of Mosby–Year Book, Inc.

path. (We ignore the fact that scattered photons may also strike the film.) We saw in Sec. 14.8 that the mass attenuation coefficient of a compound can be calculated as a weighted average of the elements in the compound:

$$\frac{\mu}{\rho}=\sum_i \left(\frac{\mu}{\rho}\right)_i w_i .$$

Table 15.2 lists various elements, their mass attenuation coefficients at 50 keV, and their composition in water, fat, muscle, and bone. Water and muscle are quite similar, fat has a somewhat smaller attenuation coefficient, and the attenuation of bone is significantly greater.

Figure 15.19 shows attenuation vs ρx for the beam in Fig. 15.18 with 2 mm of Al filtration in water and in bone. Bone contains calcium, which has a relatively high atomic number, and the attenuation coefficient rises rapidly as the energy decreases. Also shown as dashed lines are the corresponding values of $\exp(-\mu_{\text{atten}}x)$ for the average photon energy in the incident beam, which is 50 keV. In each case the transmitted fraction initially falls more steeply than the dashed line because there is more attenuation of the low-energy photons. For thicker bone the slope of the curve is less than the dashed line because only the high-energy photons remain. This shift of the beam energy and curvature of the attenuation curves is called *beam hardening*.

These differences in attenuation make it easy to distinguish bone from soft tissue. It is also easy to distinguish lungs from other tissues because they contain air and have much lower density. Air-filled lung has a density of about 180–320 kg m^{-3}, compared to about 1000 kg m^{-3} for water, muscle, or a solid tumor. Figure 15.20 shows a normal anterior–posterior (A-P) chest radiograph. You can see the exponential decay through layers of bone, the outline of the heart, the arch of the aorta, and the lacy network of blood vessels in the lungs. The patient in Fig. 15.21 has *pneumothorax*. Air has leaked into the pleural cavity and partially collapsed the lungs. You can see this collapse in the upper portion of each lung. Spontaneous pneumothorax can occur in any pulmonary disease that causes an alveolus (air sac) on the surface of the lung to

TABLE 15.2. *Relative composition of various tissues and the attenuation coefficient for 50-keV photons.*

		Fractional mass composition[b]			
Element	$\mu_{\text{atten}}/\rho^a$ (m^2 kg^{-1})	Adipose tissue, adult #1	Water	Skeletal muscle	Cortical bone, adult
H	0.0336	0.112	0.112	0.102	0.034
C	0.0187	0.517		0.147	0.155
N	0.0198	0.013		0.034	0.042
O	0.0213	0.355	0.888	0.71	0.435
Na	0.0280	0.001		0.001	0.001
Mg	0.0329				0.002
P	0.0492			0.002	0.103
S	0.0585	0.001		0.003	0.003
Cl	0.0648	0.001		0.001	
K	0.0868			0.004	
Ca	0.1020				0.225
μ_{atten}/ρ (m^2 kg^{-1})		0.0214	0.0227	0.0227	0.0424
ρ (kg m^{-3})		970	1000	1050	1920
μ_{atten} (m^{-1})		20.8	22.7	23.8	81.5

[a]Values of μ_{atten}/ρ are from Hubbell and Seltzer (1996).
[b]Fractional mass compositions are from ICRU Report 46 (1992).

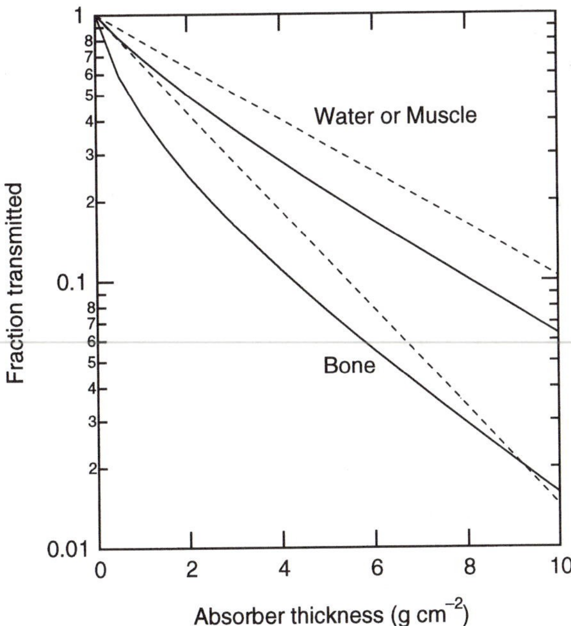

FIGURE 15.19. Attenuation of photons in water or muscle and in bone for the spectrum of Fig. 15.18 (100 kVp, 2 mm aluminum filtration). The dashed lines are for the attenuation coefficients at 50 keV.

rupture: most commonly emphysema, asthma, or tuberculosis. Pneumothorax can also be caused by perforating trauma to the chest wall. Spontaneous idiopathic (meaning cause unknown) pneumothorax occasionally occurs in relatively young people.

Abdominal structures are more difficult to visualize because except for gas in the intestine, everything has about the same density and atomic number. Contrast agents are introduced through the mouth, rectum, urethra, or bloodstream. One might think that the highest-Z materials would be best. However the energy of the K edge rises with increasing Z. If the K edge is above the energy of the x rays in the beam, then only L absorption with a much lower cross section takes place. The K edge for iodine is at 33 keV, while that for lead is at 88 keV. Between these two limits (and therefore in the range of x-ray energies usually used for diagnostic purposes), the mass attenuation coefficient of iodine is about twice that of lead. The two most popular contrast agents are barium ($Z=56$, K edge at 37.4 keV) and iodine ($Z=53$). Barium is swallowed or introduced into the colon. Iodine forms the basis for contrast agents used to study the cardiovascular system (angiography), gall bladder, brain, kidney, and urinary tract.

Some pathologic conditions can be identified by the deposition of calcium salts. Such ''dystrophic (defective) calcification'' occurs in any form of tissue injury, particularly if there has been tissue necrosis (cell death). It is found in necrotizing tumors (particularly carcinomas), atherosclerotic blood vessels, areas of old abscess formation, tuberculous foci, and damaged heart valves, among others.

15.4.4. Antiscatter Grid

Since the radiograph assumes that photons either travel in a straight line from the point source in the x-ray tube to the film or are absorbed, Compton-scattered photons that strike the film reduce the contrast and contribute an overall background darkening. This effect can be reduced by placing an antiscatter grid (or radiographic grid, or ''Bucky'' after its

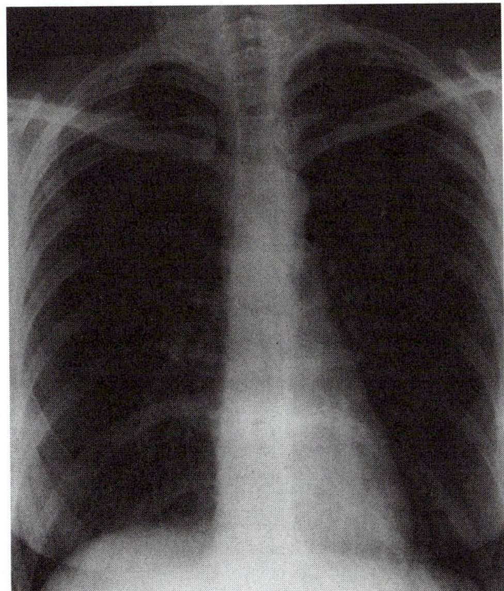

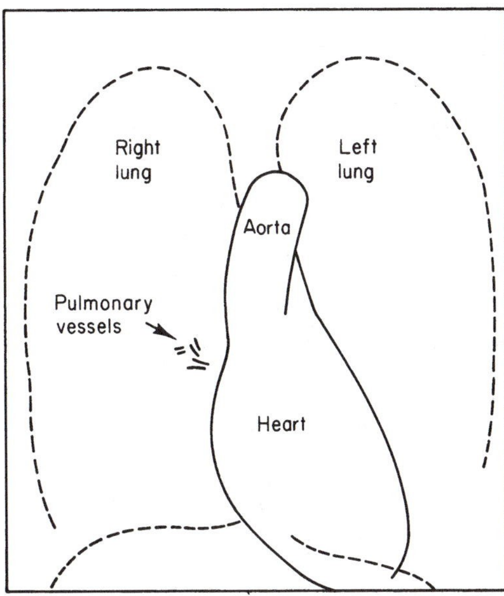

FIGURE 15.20. Radiograph of a normal chest. Some of the features are identified in the line drawing and are described in the text. Radiograph courtesy of D. Ketcham, M.D., Department of Diagnostic Radiology, University of Minnesota Medical School.

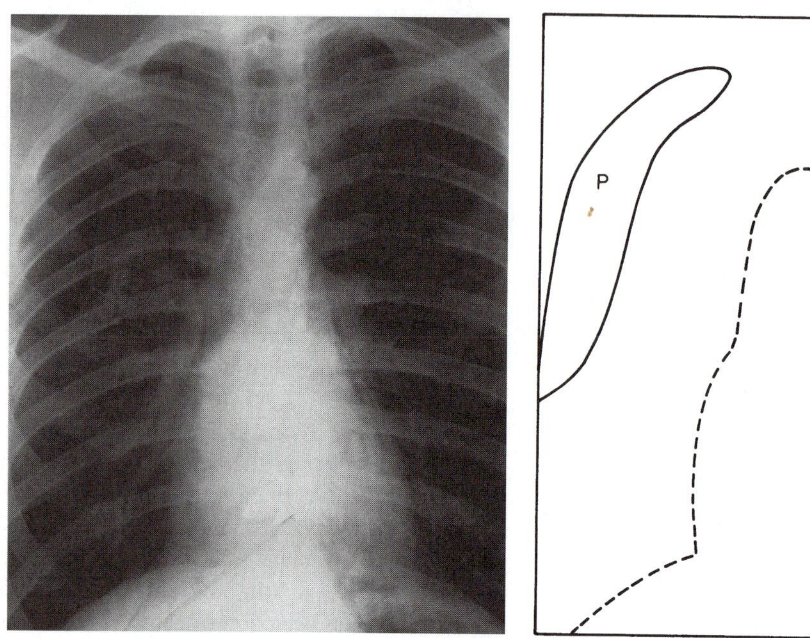

FIGURE 15.21. Radiograph of a patient with pneumothorax (*P*). Air has escaped from the lungs and caused them to collapse partially. Radiograph courtesy of D. Ketcham, M. D., Department of Diagnostic Radiology, University of Minnesota Medical School.

inventor) just in front of the film. Figure 15.22 shows how a grid works. The grid stops x rays that are not traveling parallel to the sides of the grid strips. A typical grid might have 10–50 strips of lead per centimeter that are 3 mm high and 0.05 mm thick, embedded in plastic or aluminum. The strips can be either parallel or ''focused,'' that is, slanted to aim at the point source on the anode of the x-ray tube. The grid can be either linear or crossed, with strips of lead running in both directions. A grid with a ratio of height to spacing of 10 improves the contrast by a factor of about 4, while increasing the exposure to the patient by a factor of about 3 (Hendee and Ritenour 1992, p. 360).

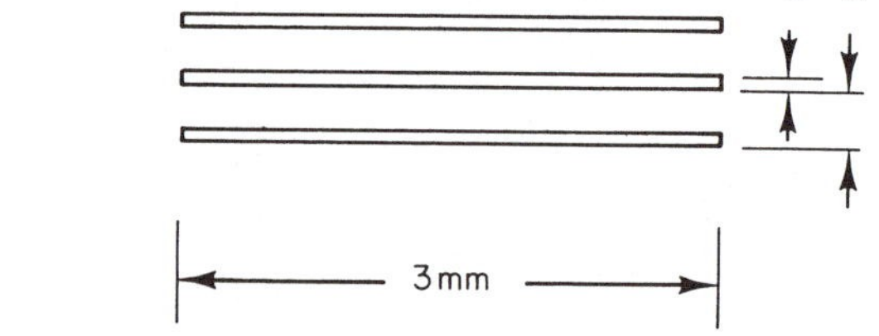

FIGURE 15.22. Scale drawing of the elements of a typical grid. The thin lead sheets absorb photons that have been scattered through more than a few degrees. As a result background fog due to scattering is reduced and the contrast is increased. Since the x rays come from a point source, the elements of the grid are usually tilted toward the source and are not parallel over the entire film surface.

15.4.5. Film–Screen Combination

The film–screen combination was described in Sec. 15.3. The fluorescent screens, by absorbing more of the photons, reduce the dose to the patient for the same optical density of the developed film. They also reduce the spatial resolution of the image, because fluorescence photons from the point where the x-ray photon interacts start out in various directions and then diffuse until they strike the film.[7]

15.5. IMAGE QUALITY

The quality of a radiographic image depends on three things: *resolution*, *contrast*, and *noise*. The resolution and contrast can be described by concepts introduced in Chap. 12 for a linear, shift-invariant system: the point-spread

[7]The point-spread function of a film or a screen–film system is easily measured. A point source is created by passing the x rays through a pinhole in a piece of lead placed directly on the screen. The resulting image in the film is the point-spread function. We saw in Chap. 12 how this is related to the modulation transfer function. Standard techniques have been developed for measuring the modulation transfer function (MTF) of the film–screen combination [ICRU Report 41 (1986); ICRU Report 54 (1996)].

function and its Fourier transform, the optical transfer function, whose magnitude is the modulation transfer function. The noise arises primarily from the fluctuations in the number of photons striking a given area of film—quantum noise—though granularity of the film is also important.

The transfer function for the entire system depends on many factors: the tube and spot size, filter, source–screen and source–patient distances, grid, film–screen combination, and scatter. If each of these subsystems operates in series, as in an audio system, one can successively convolve the point-spread functions or multiply together the (complex) optical transfer functions. It is also possible to have parallel[8] subsystems, each contributing to the final image, in which case the analysis is more complicated. An excellent review of the use of transfer-function analysis in radiographic imaging is the article by Metz and Doi (1979). The text by Macovski (1983) is at about the level of this book and presents many details of noise and convolution for radiographic, fluoroscopic, tomographic, nuclear medicine, and ultrasound images. The size of the spot where the electrons strike the anode of the x-ray tube is critical in determining the resolution of the final image, as discussed in detail by Wagner *et al.* (1974).

We define the *exposure contrast* as the change in exposure between two (usually adjacent) parts of the image divided by the average:

$$C_{\text{in}}=\frac{\Delta X}{X}. \tag{15.13}$$

This is similar to the modulation defined in Eq. (12.35). The *brightness contrast* is the analogous quantity for the light transmitted through the processed film when viewing the image:

$$C_{\text{out}}=\frac{\Delta T}{T}. \tag{15.14}$$

The exposure contrast and brightness contrast are related by

$$C_{\text{out}}=\gamma C_{\text{in}}. \tag{15.15}$$

The *radiographic signal* is a small change in optical transmission in adjacent areas of the image. Changes in transmission below a certain value are not detectable by the viewer. This is apparent in Fig. 15.23, which shows signals with different contrasts and different sizes on a uniform background. The smaller the diameter of the signal region, the more difficult the signal is to detect. We will develop a simple model to explain why.

Suppose first that there is no signal, but that the film (or screen–film combination) is illuminated with a uniform beam of x rays with a constant fluence. We make an exposure for a certain time and count the number of photons striking a ''sampling area'' of the film, S. Though the average fluence is constant across the film, the photons are randomly distributed. A somewhat different number of photons strike a nearby sampling area of the same size. This is a situation where the average number striking a sampling area of a given size is constant, the total number of photons is very large, and the probability that any one photon is absorbed in a given sampling area is small, so the situation is described by Poisson statistics (Appendix J). The mean number of photons striking a sampling area is ΦS and the standard deviation is $(\Phi S)^{1/2}$. Suppose that some fraction $f \leq 1$ of these photons actually interact with the silver halide grains in the emulsion. Then the mean number interacting is $m = f\Phi S$ and the standard deviation is[9] $(f\Phi S)^{1/2}$. Thus there are fluctuations in the brightness of the transmitted image across the uniformly exposed viewing region, just because of the Poisson statistics—quantum noise or shot noise—of the x-ray photons striking the film.

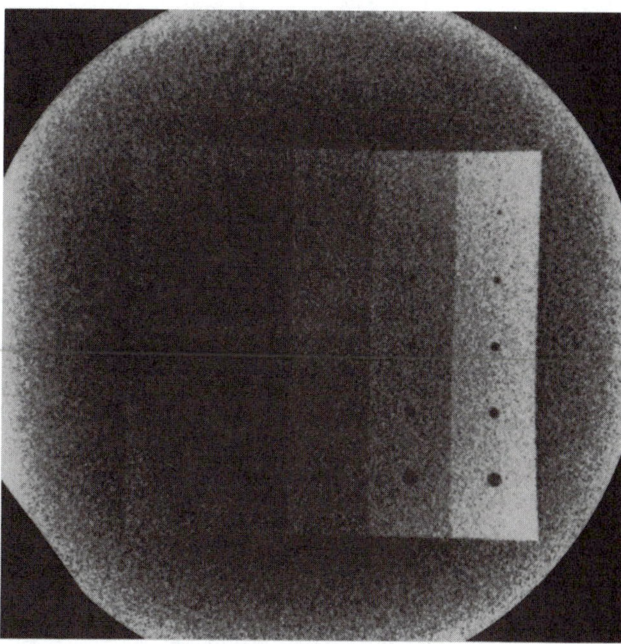

FIGURE 15.23. An example of the relationship among exposure, image size, and detectability. A type 1100 aluminum phantom was imaged with digital fluorography. It was exposed to an 80-kVp x-ray beam with 4.5-mm Al filter. As the image becomes lighter, the thickness of the aluminum changes in steps: 0.85, 1.3, 2.1, 3.2, and 5.2 mm. The holes are 1, 1.5, 2, 2.5, and 3 mm in diameter. The greater the attenuation in the aluminum, the lower the background, and the easier it is to detect the smaller holes. Photograph courtesy of Richard Geise, Ph.D., Department of Radiology, University of Minnesota.

The fluctuations in the number of photons striking area S can be related to fluctuations in the exposure of that area

[8]Examples of parallel subsystems are the two emulsion layers on double-coated film, and the effect of primary and scattered radiation on the formation of the image.

[9]This is very similar to the arguments about the fraction of photons absorbed by a visual pigment molecule in Eq. (13.57). Changes in the value of f in Fig. 13.34 shift the response curve along the axis. They also shift the film response curves in Fig. 15.8.

of the film, and hence to the transmission of visible light through the developed film. Since the exposure (measured in air just in front of the film) is proportional to the photon fluence, $X=A\Phi$, $\overline{(X-\bar{X})^2}=A^2\overline{(\Phi-\bar{\Phi})^2}$ and $(\Delta X)_{\text{rms}}=A(\Delta\Phi)_{\text{rms}}$. We define the *noise exposure contrast* to be the standard deviation of the number of photons affecting grains in area S divided by the average number affecting grains in an area that size:[10]

$$C_{\text{noise in}}\equiv\frac{(f\Phi S)^{1/2}}{f\Phi S}=(f\Phi S)^{-1/2} \tag{15.16}$$

The *noise brightness contrast* is then

$$C_{\text{noise out}}=\gamma(f\Phi S)^{-1/2}. \tag{15.17}$$

This says that the fluctuations in the noise, measured by noise contrast, are inversely proportional to the square root of the area of the lesion to be detected.[11] This is seen in Fig. 15.23. The noise in a screen–film system is measured by using a microdensitometer to measure the optical density across film that has received a uniform exposure. Variations with position can be described either in terms of its two-dimensional autocorrelation function or its Fourier transform, the *Wiener spectrum.* The radiographic noise consists of three components: *quantum mottle*, the statistical fluctuations in the number of photons absorbed in a small area (shot noise), *structure mottle* due to nonuniformities in the screen, and *film graininess*, variation in the size and distribution of the silver bromide grains in the emulsion. Here we discuss only quantum mottle.

Now introduce a signal, which is a small increase in the exposure or photon fluence: $\Delta X_{\text{signal}}=A\Delta\Phi_{\text{signal}}$. This gives a brightness contrast

$$C_{\text{signal out}}=\gamma\frac{(\Delta X)_{\text{signal}}}{X}=\gamma\frac{(\Delta\Phi)_{\text{signal}}}{\Phi}. \tag{15.18}$$

The ratio of the signal contrast to the noise contrast is called the *signal-to-noise ratio*:

$$\text{SNR}=\frac{C_{\text{signal out}}}{C_{\text{noise out}}}=\frac{\gamma(\Delta\Phi)_{\text{signal}}/\Phi}{\gamma(f\Phi S)^{-1/2}}=(fS)^{1/2}\frac{(\Delta\Phi)_{\text{signal}}}{\Phi^{1/2}}. \tag{15.19}$$

The signal will be detectable only if the signal brightness contrast exceeds the noise brightness contrast by a certain amount:[12]

$$\text{SNR}>k,\quad (fS)^{1/2}\frac{(\Delta\Phi)_{\text{signal}}}{\Phi^{1/2}}>k. \tag{15.20}$$

The larger the value of the signal-to-noise ratio, the greater the probability of detecting the signal. Many experiments on perception have been done and will not be discussed here.[13] Values of k that are used range from 2 to 5.

Let us apply the result in Eq. (15.20) to a simple model: a monoenergetic x-ray beam passing through a patient. The total thickness of the patient is L. The attenuation coefficient is μ. If an x-ray beam with fluence Φ_0 strikes the patient, the fluence of x-ray photons emerging is $\Phi_1=\Phi_0e^{-\mu L}$. Imagine a nearby region where for a distance x the attenuation coefficient is $\mu-\Delta\mu$. The x-ray fluence emerging along a line passing through this region is

$$\begin{aligned}\Phi_2&=\Phi_0e^{-\mu(L-x)-(\mu-\Delta\mu)x}=\Phi_0e^{-\mu L}e^{\Delta\mu x}\\&=\Phi_1e^{\Delta\mu x}\\&\approx\Phi_1(1+\Delta\mu x).\end{aligned} \tag{15.21}$$

The exposure contrast is therefore $C_{\text{in}}=(\Delta\Phi)_{\text{signal}}/\Phi_1\approx\Delta\mu\ x$. We combine this with Eq. (15.20) to obtain

$$(fS\Phi_1)^{1/2}(\Delta\mu\ x)>k, \tag{15.22}$$

where Φ_1 is the fluence leaving the patient or striking the film. (These are the same if variations in $1/r^2$ can be neglected, where r is the distance from the tube to the patient or the tube to the film.) The signal-to-noise ratio increases as the square root of the photon fluence or exposure, the square root of the area to be detected, and the square root of f, the fraction of the photons striking the film that are actually effective in rendering grains developable.

The fraction f in this Poisson model is equal to the *detective quantum efficiency* (DQE). It is easily visualized as the fraction of the photons striking the detector that actually affect it. The number of *noise equivalent quanta* (NEQ) in our model is $f\Phi_1$. [This simple equality exists only because we are using a model with Poisson statistics. The DQE is defined more generally as the square of the of the signal-to-noise ratio of the detector output divided by the square of the signal-to-noise ratio of the detector input. The more

[10]It is sometimes useful to write it as

$$\begin{aligned}C_{\text{noise in}}&\equiv\frac{(f\Phi S)^{1/2}}{f\Phi S}=\frac{1}{f^{1/2}S^{1/2}}\frac{(\Delta\Phi)_{\text{rms}}}{\Phi}=\frac{1}{f^{1/2}S^{1/2}}\frac{A(\Delta X)_{\text{rms}}}{AX}\\&=\frac{1}{f^{1/2}S^{1/2}}\frac{(\Delta X)_{\text{rms}}}{X}.\end{aligned}$$

[11]An analogous phenomenon is seen when counting individual photons with a radiation detector at a fixed average rate. The number counted in a given time interval fluctuates, with the fractional fluctuation inversely proportional to the square root of the counting time.

[12]There are statistical fluctuations in the signal as well as the noise. The variance of the difference between signal and noise will be the sum of the variances in the signal and in the noise. This has the effect of increasing the noise by a factor of $\sqrt{2}$, which can be absorbed in the value of k that is chosen. See Problem 15.17.

[13]The ability to detect the signal accurately is greater when the observer knows the nature of the signal and is only asked whether it is or is not present. That is, the ability of an observer to detect a signal is less in the more realistic situation where the observer does not know what the signal is or where it might be in the radiograph.

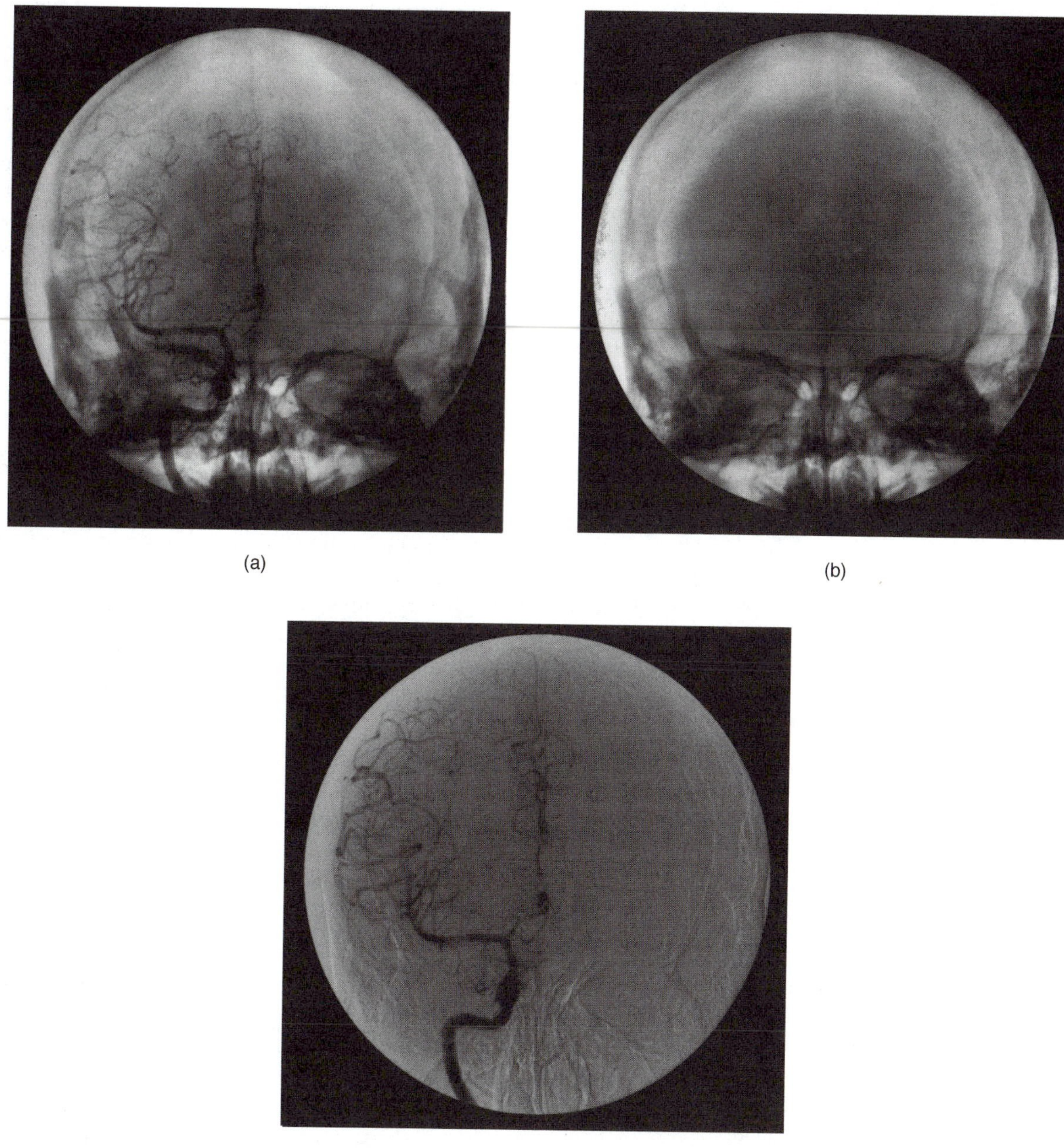

FIGURE 15.24. Digital subtraction angiography. (a) Image with contrast material. (b) Image without contrast material. (c) The difference image. Photograph courtesy of Richard Geise, Ph.D., Department of Radiology, University of Minnesota.

general definitions of DQE and NEQ are discussed in Wagner (1983), Wagner (1977), and Dainty and Shaw (1974).]

We can apply Eq. (15.22) to determine the number of photons that must be transmitted through the patient for a given image size and given signal-to-noise ratio. We assume that $f=1$. The required photon fluence emerging from the patient is (dropping the subscript on Φ_1)

$$\Phi S > \left(\frac{k}{\Delta \mu x} \right)^2 . \tag{15.23}$$

If the lesion thickness is $x=1$ cm and $\Delta\mu = 0.01\mu_{\text{water}} = (0.01)(22.7)\ m^{-1}$, then $\Delta\mu\ x = 0.002\,27$. For $k=4$, the number of photons in the image area must be greater than 3×10^6. The exit dose to the patient is (assuming monoenergetic photons)

$$D_{\text{body}} = \left(\frac{\mu_{\text{en}}}{\rho} \right)_{\text{body}} \Psi = \left(\frac{\mu_{\text{en}}}{\rho} \right)_{\text{body}} (h\nu)\Phi$$

$$= \left(\frac{\mu_{\text{en}}}{\rho} \right)_{\text{body}} \frac{(h\nu)(3\times 10^6)}{S} . \tag{15.24}$$

The dose increases as the area to be detected decreases. In order to detect an image 1 mm square using 50-keV photons, the exit dose in water would have to be at least 9.8×10^{-5} Gy.

15.6. ANGIOGRAPHY AND DIGITAL SUBTRACTION ANGIOGRAPHY

One important problem in diagnostic radiology is to image portions of the vascular tree. This can confirm the existence of and locate narrowing (stenosis), weakening and bulging of the vessel wall (aneurysm), congenital malformations of vessels, and the like. This is done by the injection of a contrast material containing iodine. If the images are recorded digitally, it is possible to subtract one without the contrast medium from one with contrast and see the vessels more clearly (Fig. 15.24).

In a typical angiographic study, 30–50 ml of contrast material is injected into a blood vessel. For a vessel with a diameter of 8 mm, ρx of the contrast material is about 4 mg cm^{-2}.

15.7. MAMMOGRAPHY

Mammography poses particular challenges for medical physicists. The resolution needed is extremely high (about 15 line pairs (lp) mm^{-1} compared to 5 lp mm^{-1} for a chest radiograph).[14] The radiologist may use a magnifying glass to inspect the image. The contrast in a breast image is inherently low. Fat and glandular tissue must be distinguished by the slight differences in their attenuation coefficients (see Problem 15.19). The dose must be made as small as possible.

These challenges have been met. Spatial frequencies of 14–16 lp mm^{-1} are routinely obtained. Noise limits the minimum size of a detectable object to >0.3 mm for microcalcifications or a few millimeters for soft tissue. The typical mammographic dose per view has been reduced from about 50 mGy in the 1960s and 4.1 mGy in the 1970s to 1.5 mGy in 1996.[15] One technique that has helped make these improvements is the molybdenum target x-ray tube, operating at 25–28 kVp. Figure 15.25 shows the photon fluence from such a tube, with the 17-keV K_α and 19-keV K_β lines being quite prominent. Filtration of the beam with the same material, molybdenum, further sharpens the spectrum. The dashed lines show the spectrum when a molybdenum filter is used. Meticulous attention to detail is required to obtain useful images [(AAPM Report 29 (1990)].

[14]Line pairs (abbreviated lp) are analogous to the period of a square wave.
[15]See NCRP Report 100 (1989) for early data; R. Geise, private communication for 1996 data.

15.8. FLUOROSCOPY

If a fluorescent screen replaces the screen–film cassette, one can observe an image that changes with time, as when the patient swallows a contrast agent. Although the x-ray tube current is several hundred times less than that for a radiograph, the dose absorbed by the patient would be quite high if the viewing time were very long. In order to reduce the dose, early fluoroscopic units were viewed by a radiologist using dark-adapted (scotopic) vision, which allows one to perceive a dimmer image but does not allow one to resolve great detail. Image intensifier tubes do not reduce the exposure very much, but they provide much brighter images so the radiologist can use photopic rather than scotopic vision and see more detail.

A cross section of a typical image intensifier tube is shown in Fig. 15.26. A fluorescent screen, the input phosphor, is sandwiched to a photocathode, similar to the one in a photomultiplier tube. An aluminum support reflects light from the fluorescent screen back into the tube. An accelerating voltage difference of about 30 kV exists between the photocathode and the anode. Electrons are focused by a series of electrodes to pass through a hole in the anode and strike another fluorescent screen, the output phosphor. The energy added to each electron by the accelerating field increases the brightness of the image, though some spatial resolution is lost. The brightness is also enhanced because the output phosphor is smaller than the input phosphor. The output phosphor is viewed with a television camera.

15.9. COMPUTED TOMOGRAPHY

A great drawback to radiographs is that they provide only an integrated value of the attenuation coefficient. That is, if $N_0(y,z)$ x-ray photons traverse the body along a line in the x direction after entering the body at coordinates (y,z), the number emerging without interaction is $N(y,z) = N_0(y,z)e^{-\alpha(y,z)}$, where

$$\alpha(y,z)=\int \mu(x,y,z)\ dx.$$

The radiograph measures $N(y,z)$ or $\alpha(y,z)$. The desired information is $\mu(x,y,z)$. The radiographic image is often difficult to interpret because of this integration along x. For example, it may be difficult to visualize the kidneys because of the overlying intestines.

Conventional tomography was used for many years to blur structures that are not in an image plane. (*Tomos* means slice.) The principle of the technique can be seen in Fig. 15.27. Both the x-ray source and the film are pivoted about a line in the plane ABC while the source is turned on. (As shown, the line is perpendicular to the page.) X rays passing

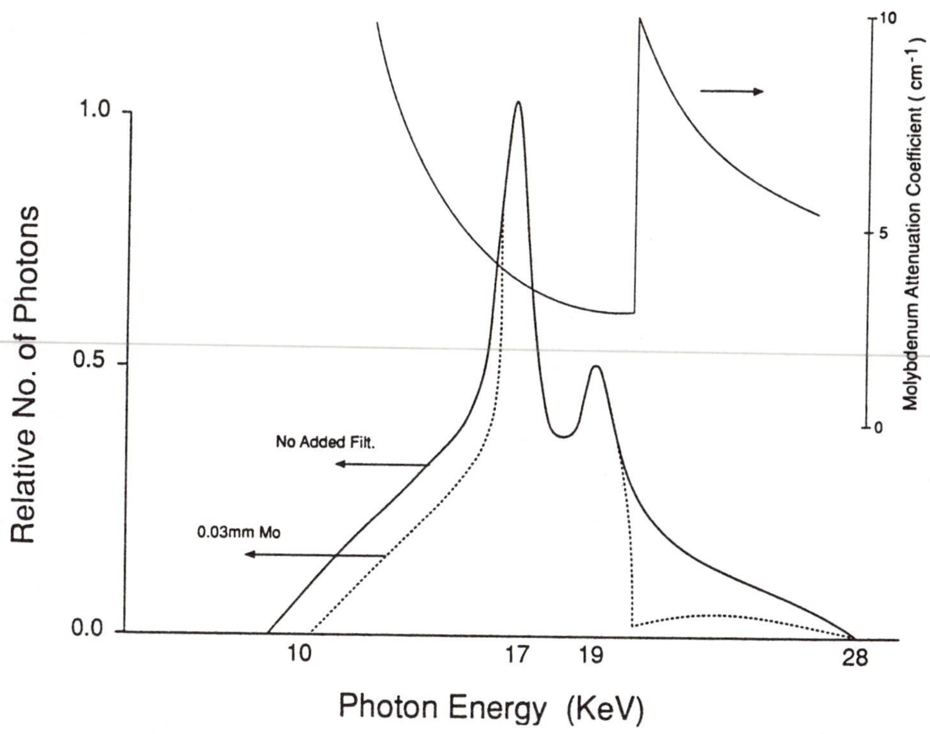

FIGURE 15.25. The x-ray spectrum from a molybdenum anode tube used for mammography, with and without filtration by a molybdenum foil. From W. R. Hendee and R. Ritenour (1992). *Medical Imaging Physics.* St. Louis, Mosby–Year Book. Used by permission of Mosby–Year Book, Inc.

through particular points in the plane always strike the same points on the film. X rays passing through point D, which is out of the plane, strike many different points on the film during the exposure, so any structure that is out of the plane is blurred. Pivoting may also be done about a point instead of an axis.

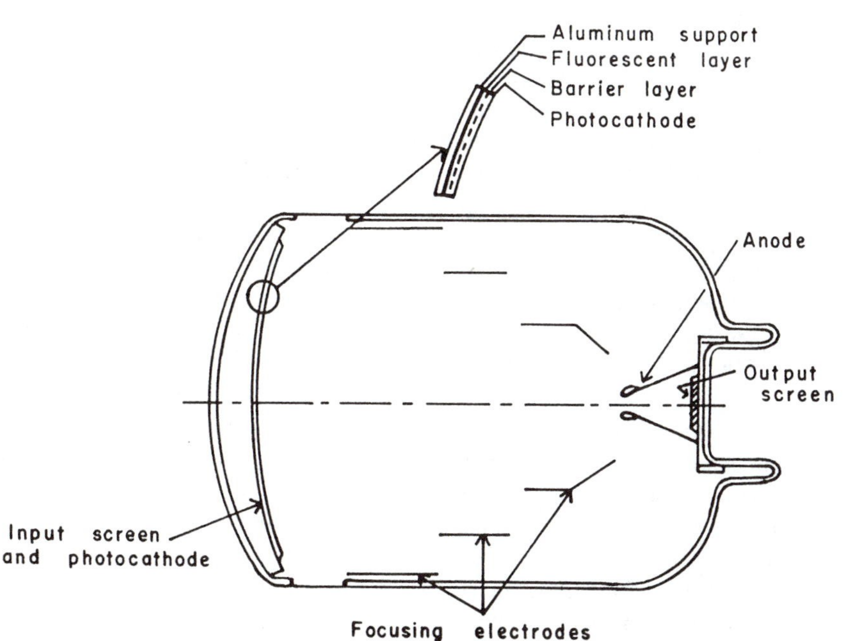

FIGURE 15.26. A schematic diagram of an image intensifier tube. From W. R. Hendee and R. Ritenour (1992). *Medical Imaging Physics.* St. Louis, Mosby–Year Book (1992). Used by permission of Mosby–Year Book, Inc.

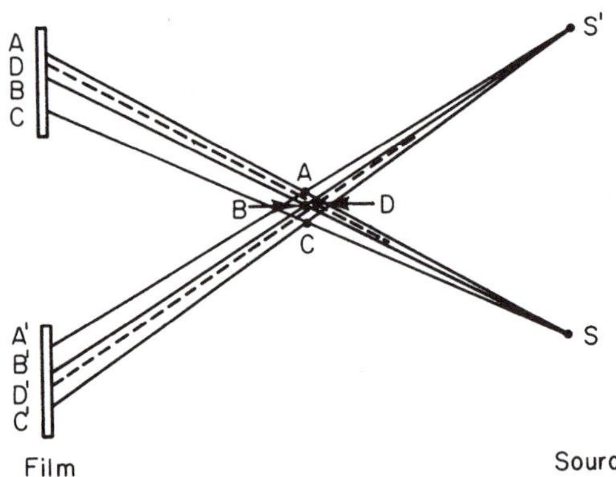

FIGURE 15.27. Conventional tomography. The x-ray source and the film are pivoted about a line in plane *ABC*. X rays passing through particular points in the plane always strike the same points in the film. X rays passing through points out of the plane such as *D* strike many points on the film, and the structure is blurred out.

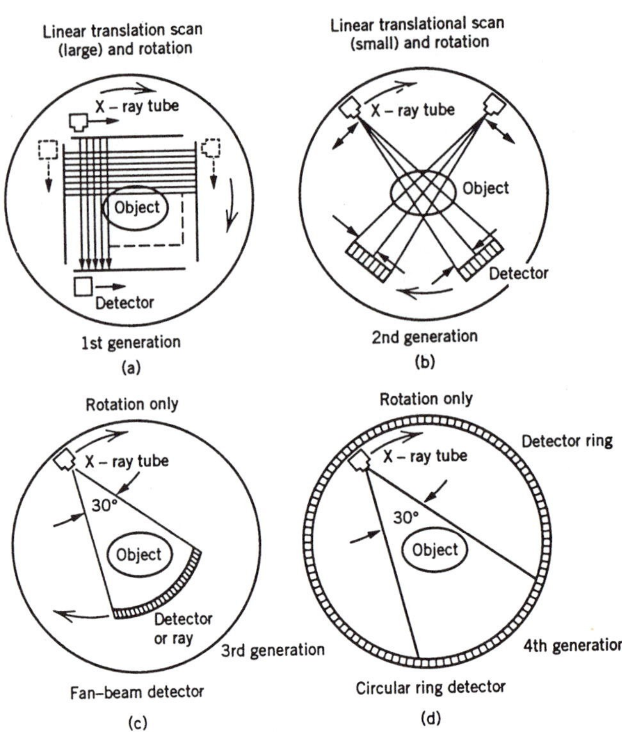

FIGURE 15.28. The scanning techniques used in the four generations of CT scanners. From Z-H. Cho, J. P. Jones, and M. Singh (1993). *Foundations of Medical Imaging.* Copyright © 1993 John Wiley-Interscience. Reprinted by permission of John Wiley & Sons, Inc.

Several types of computed tomography have been developed in the last few decades. They include *transmission computed tomography* (CT), *single-photon emission computed tomography* (SPECT), and *positron emission tomography* (PET). They all involve reconstructing, for fixed z, a map of some function $f(x,y)$ from a set of projections $F_\theta(x)$, as described in Secs. 12.7 and 12.8. For CT the function f is the attenuation coefficient $\mu(x,y)$. For SPECT and PET it is the concentration of a radioactive tracer within the body. They will be described in Chap. 16.

The history of the development of computed tomography is quite interesting [DiChiro and Brooks (1979)]. The Nobel Prize in Physiology or Medicine was shared in 1979 by a physicist, Allan Cormack, and an engineer, Godfrey Hounsfield. Cormack had developed a theory for reconstruction and done experiments with a cylindrically symmetric object that were described in two papers in the *Journal of Applied Physics* in 1963 and 1964. Hounsfield worked independently around 1970–71. The first clinical machine was installed in 1971. It was described in 1973 in the *British Journal of Radiology*. The Nobel Prize acceptance speeches [Cormack (1980); Hounsfield (1980)] are interesting to read. A neurologist, William Oldendorff, had been working independently on the problem but did not share in the Nobel Prize [See DiChiro and Brooks (1979), and Broad (1980)].

Early machines had an x-ray tube and detector that moved in precise alignment on opposite sides of the patient to make each pass. (The size of the machine allowed only heads to be scanned.) After one pass, the gantry containing the tube and detector was rotated 1° and the next pass was taken. After data for 180 passes were recorded, the image was reconstructed. A complete scan took about 4 minutes. Modern CT units use an array of detectors and a fan-shaped beam that covers the whole width of the patient. The scan time is reduced to a few seconds. Figure 15.28 shows the evolution of the scanning techniques. More recently all of the electrical connections have been made through slip rings. This allows continuous rotation of the gantry and scanning in a spiral as the patient moves through the machine. Interpolation in the direction of the axis of rotation (the z axis) is used to perform the reconstruction for a particular value of z. This is called *spiral CT* or *helical CT*. The pitch of the spiral is between 1 and 10 mm per rotation, and a typical machine makes 1–2 rotations per second. Many papers have been written about spiral CT. One by Kalender and Polacin (1991) discusses the physical performance of the machines.

Figure 15.29 shows a head scan with a line drawing showing some of the anatomical landmarks. Figure 15.30 is an abdominal scan showing a benign tumor in the liver.

There are some fundamental relationships between the dose to the patient and the resolution. It is often desirable to measure the attenuation coefficient with an accuracy of $\pm 0.5\%$. For water at 60 keV, $\mu = 20\ \mathrm{m}^{-1}$, so μ must be measured with an accuracy of $\delta\mu = 0.1\ \mathrm{m}^{-1}$. (A beam of 120 kVp with 2–3 mm of aluminum filtration has about this average photon energy.) It is customary to report the fractional difference between μ and μ_{water}. The *Hounsfield unit* is

$$H = 1000\,\frac{\mu_{\mathrm{tissue}} - \mu_{\mathrm{water}}}{\mu_{\mathrm{water}}}. \tag{15.25}$$

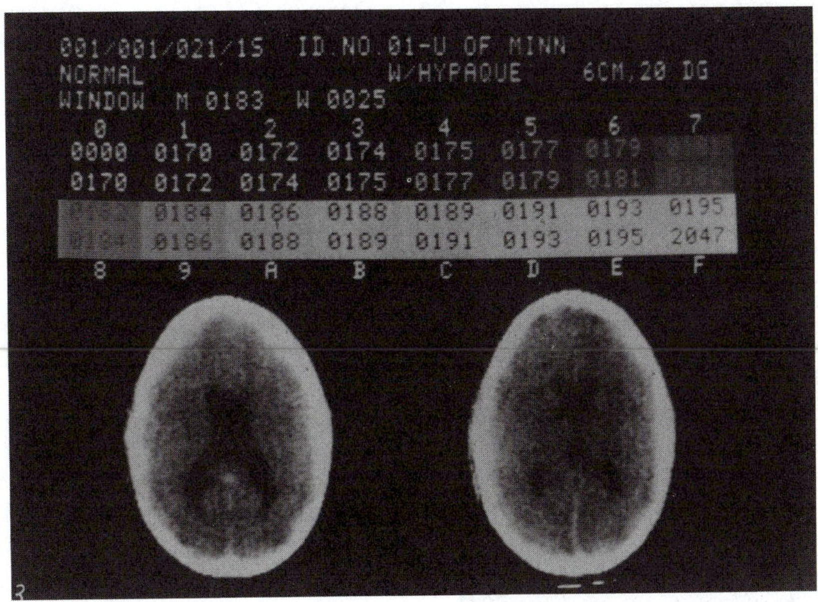

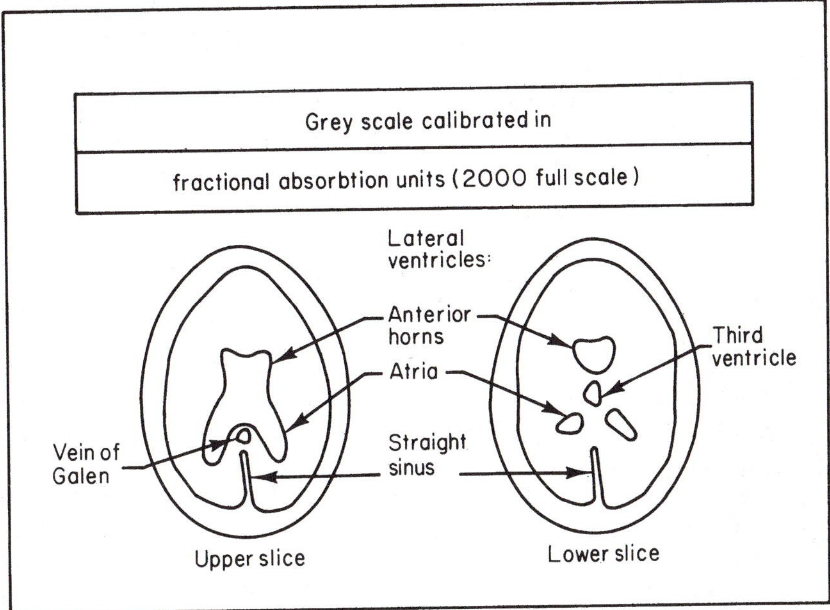

FIGURE 15.29. CT scan of the head, with a drawing showing some of the anatomical features. Scan courtesy of J. T. Payne, Ph. D., Department of Radiology, University of Minnesota Medical School.

The desired accuracy is ± 5 Hounsfield units.

Suppose we are reconstructing an object with a circular cross section as shown in Fig. 15.31. The object is to be resolved into volume elements or cells of length w on each side parallel to the x and y axes. The thickness of each cell along the z axis (perpendicular to the scan) is the slice thickness, b. The diameter of the object is L. For simplicity, we make the analysis assuming a first-generation machine, with rectilinear passes repeated at m different angles between 0° and 180°. The width of each sample in a pass is w. The number of samples in each pass is

$$n = \frac{L}{w}, \tag{15.26}$$

and the number of cells in the object is approximately the area of the circular object divided by the area of a cell: $\pi L^2/4w^2$ or $\pi n^2/4$. To determine $\pi n^2/4$ independent values of μ requires at least that many independent measurements. Since n measurements are made in each pass, we need m passes where $mn = \pi n^2/4$ or

$$m = \frac{\pi n}{4}. \tag{15.27}$$

With more passes the situation is overdetermined; with fewer it is underdetermined. If the values of μ are overdetermined, convoluted back projection (Chap. 12) or a similar procedure can be used to assign the values of μ.

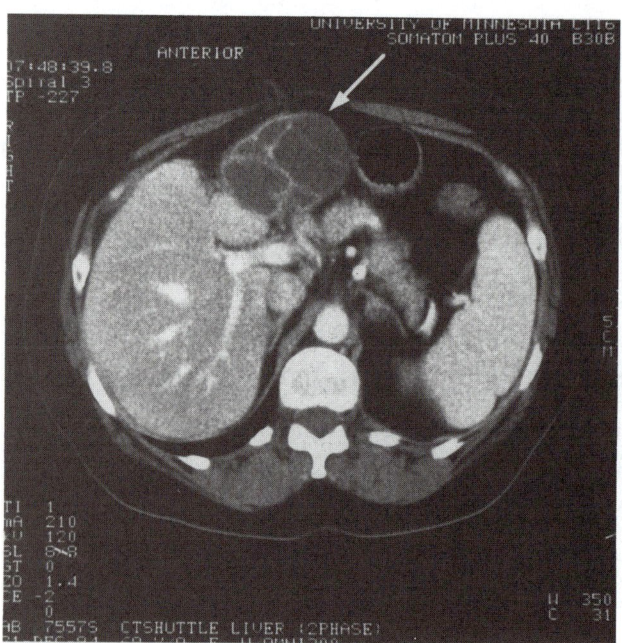

FIGURE 15.30. A spiral CT scan of the abdomen. The arrow points to a biliary cystadenoma, a benign tumor of the liver. Scan courtesy of E. Russell Ritenour, Ph.D., Department of Radiology, University of Minnesota Medical School.

Now consider the attenuation of photons in one sample along a diameter. In Sec. 15.5 we developed a relationship between the photon fluence in the beam needed to measure the attenuation with some desired accuracy [Eq. (15.21)]. The same arguments can be applied in this case:

$$\delta\Phi = \Phi_1 \delta\mu\ w = \Phi_0 e^{-\mu L} \delta\mu\ w.$$

The photons arrive at the detector, which we assume for simplicity to have 100% efficiency, at a constant average rate, so they are Poisson distributed. The standard deviation in the number of counts is $(\Phi_1 S)^{1/2} = (\Phi_1 wb)^{1/2}$. To detect the difference between the two samples, $w^2 \delta\Phi$ must exceed this by the minimum signal-to-noise ratio, k. This gives the number of minimum number of photons that must be detected:

$$N = \Phi_1 wb > \frac{k^2}{w^2(\delta\mu)^2}. \tag{15.28}$$

If $w = 0.8$ mm and $k = 4$, this means that $N > 2.5\times10^9$ photons. It can be shown [Brooks and DiChiro (1976)] that these counts can be divided among all m passes. Let N' be the number of photons detected in the sample at one angle:

$$N' = \frac{k^2}{mw^2(\delta\mu)^2} = \frac{k^2}{(\pi/4)Lw(\delta\mu)^2}. \tag{15.29}$$

The exit dose from one pass is

$$D' = \frac{\mu_{en}}{\rho} \Psi_1' = \frac{\mu_{en}}{\rho} h\nu \Phi_1',$$

where $\Phi_1' = \Phi_1 / m$, and the entrance dose is $e^{\mu L}$ times this. We multiply by m to get the total dose. With the help of Eq. (15.28) we obtain an expression for the total entrance dose for the entire scan from all m passes:

$$D = \frac{\mu_{en}}{\rho} h\nu e^{\mu L} \Phi_1 = \frac{\mu_{en}}{\rho} \frac{h\nu e^{\mu L} k^2}{w^3 b(\delta\mu)^2}.$$

Three additional factors must be added to this equation. A more careful analysis of the statistics gives a factor $\pi^2/12 = 0.82$. The second factor β is introduced for photons that miss the detector and a detector efficiency that is less than unity. Finally, the actual dose to the patient is less than the entrance dose. It must be averaged over all scans and must account for the attenuation of the beam as it goes through the patient. The result above is multiplied by the third factor γ (no relation to the film γ). The values of γ depend on the machine configuration and the photon energy. A typical value for an early machine described by Brooks and DiChiro (1976) is $\gamma = 0.34$ for 60-keV photons. With these added factors, the dose is

$$D = \frac{(\pi^2/12)\beta\gamma h\nu e^{\mu L}(\mu_{en}/\rho)k^2}{w^3 b(\delta\mu)^2}. \tag{15.30}$$

This equation shows a fundamental relationship between dose and resolution. Decreasing both w and b by a factor of 2 requires a 16-fold increase in dose, while improving $\delta\mu$ by a factor of 2 requires a dose that is 4 times as large.

We can insert typical numbers in Eq. (15.30) to calculate a minimum dose. Using $\beta = 2$, $\gamma = 0.34$, $k = 2$, $e^{\mu L} = 40$, $h\nu = 60$ keV, $\mu_{en}/\rho = 3.152\times10^{-3}$ m^2 kg^{-1} (from Appendix O for water), $w = 1.0$ mm, and $b = 1.0$ cm, we obtain

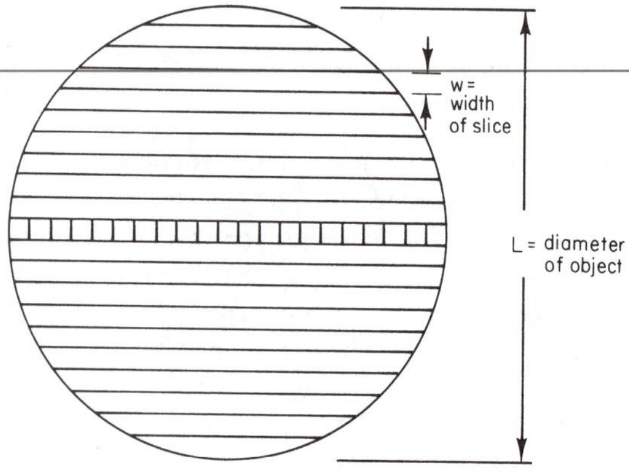

FIGURE 15.31. A circular object that is to be analyzed. The diameter of the object is L; the width of each slice in the scan is w. It is desired to resolve cells in each slice which have a length w, as shown for the center diameter.

$$D=\frac{(0.822)(2)(0.34)(40)(60\times 10^3\times 1.6\times 10^{-19}\ \text{J})(3.152\times 10^{-3}\ \text{m kg}^{-1})(2^2)}{(1.0\times 10^{-3}\ \text{m})^3(1.0\times 10^{-2}\ \text{m})(0.1\ \text{m}^{-1})}=2.7\times 10^{-2}\ \text{Gy}.$$

This number is somewhat larger than the dose found when small thermoluminescent dosimeters are placed in an artificial head phantom. Patient examinations usually involve several scans for different values of the distance along the z axis. Scattering from neighboring layers may double the dose.

Table 15.3 compares the doses due to various radiological procedures. The dose from a CT head scan is comparable to the doses from other procedures.

15.10 BIOLOGICAL EFFECTS OF RADIATION

Radiation at sufficiently high doses can kill cells, tumors, organs, or entire animals. Radiation, along with surgery and chemotherapy, is a mainstay of cancer treatment. Radiation can also cause mutations. *Radiobiology*, the study of how radiation affects cells and organs, has provided major improvements in our understanding of cell death and damage in recent years. This understanding has modified and improved our approach to radiation therapy. This section provides a brief introduction to radiobiology, but it ignores many important details, such as the relationship between sensitivity to radiation and when it is given during the cycle of cell division.[16] The discussion starts with some cell-culture (*in vitro*) results, presents the two most frequently used models for radiation damage, and then moves to *in vivo* tissue irradiation and the destruction of tumors.

15.10.1. Cell-Culture Experiments

Cell-culture studies are the simplest conceptually. A known number of cells are harvested from a stock culture and placed on nutrient medium in plastic dishes. The dishes are then irradiated with a variety of doses including zero. After a fixed incubation period the cells that survived have grown into visible colonies that are stained and counted. Measurements for many absorbed doses give *survival curves* such as those in Fig. 15.32. These curves are difficult to measure for very small surviving fractions, because of the small number of colonies that remain.

The shape of the survival curve depends on the linear energy transfer (LET) of the charged particles. For the α particles in Fig. 15.32 the LET is about 160 keV μm^{-1}, for neutrons it is about 12 keV μm^{-1}, and for the electrons from 250-kVp x rays it is about 2 keV μm^{-1}. The α particles and neutrons are called high-LET radiation; the electrons are low-LET radiation. Experiments with microscopic beams of radiation and short-range particles aimed at different parts of the cell have demonstrated that damage to the DNA is the most prominent cause of cell death or mutations.[17]

High-LET radiation produces so many ion pairs along its path that it exerts a direct action on the cellular DNA. Low-LET radiation can also ionize, but it usually acts indirectly. It ionizes water (primarily) according to the chemical reaction

$$H_2O \rightarrow H_2O^+ + e^-.$$

The H_2O^+ ion decays with a lifetime of about 10^{-10} s to the hydroxyl free radical:[18]

$$H_2O^+ + H_2O \rightarrow H_3O^+ + OH\cdot$$

This then produces hydrogen peroxide and other free radicals that cause the damage by disrupting chemical bonds in biological molecules. [Shabashon (1996), esp. pp. 64–69].

15.10.2. Chromosome Damage

Cellular DNA is organized into *chromosomes*. In order to understand radiation damage to DNA, we must recognize that there are four *phases* in the cycle of cell division:

M — Cell division. This stage includes both division of the nucleus (*mitosis*) and of the cytoplasm (*cytokinesis*). This phase may last one or two hours.

G_1 — The first "gap" phase. The cell is synthesizing many proteins. The duration of G_1 determines how frequently the cells divide. It varies widely by kind of tissue, from a few hours to 100 days or more.

S — Synthesis. A new copy of all the DNA is being made. This lasts about 8 hours.

G_2 — The second "gap" phase, lasting about 4 hours.

Figure 15.33 shows, at different magnifications, a strand of DNA, various intermediate structures which we will not discuss, and a chromosome as seen during the M phase of the cell cycle. The size goes from 2 nm for the DNA double helix to 1400 nm for the chromosome. In addition to cell survival curves one can directly measure chromosome damage. There is strong evidence that the radiation, directly

[16]For more information see the references by Hall (1994), by Moulder and Shadley (1996), by Orton (1997), and by Steel (1996).

[17]There are other mechanisms of cell damaged and cell death, such as *apoptosis* (programmed cell death), and there are also effects on cell cycle timing.

[18]*A free radical* has a single unpaired electron in its outer shell. Even though it is electrically neutral, it is highly reactive. See p. 426.

TABLE 15.3. *Typical skin doses for various radiological procedures.*

	Maximum skin dose (Gy)[a]
Chest (Anterior/Posterior)	1.3×10^{-4}
Skull (lateral)	10.1×10^{-4}
Abdomen (A/P)	31.8×10^{-4}
Lumbar-sacral spine	38.9×10^{-4}
Dental bitewing (posterior)	23.2×10^{-4}
CT head scan (single)	100.0×10^{-4}

[a]The first five entries are median doses at skin entrance, calculated from median exposures in a survey of 6606 sites in a federal survey [R. L. Burkart *et al.* (1985). *Recommendations for Evaluation of Radiation Exposure from Diagnostic Radiology Examinations.* Rockville, MD, U.S. Department of Health and Human Services, Center for Devices and Radiological Health, HHS Publication (FDA) 85-8247.].

or indirectly, breaks a DNA strand. If only one strand is broken, there are efficient mechanisms that repair it over the course of a few hours using the other strand as a template. If both strands are broken, permanent damage results, and the next cell division produces an abnormal chromosome. Several forms of abnormal chromosomes are known, depending on where along the strand the damage occurred and how the damaged pieces connected or failed to connect to other chromosome fragments. Many of these chromosomal abnormalities are lethal: the cell either fails to complete mitosis the next time it tries to divide, or it fails within the next few divisions. Other abnormalities allow the cell to continue to divide, but they may contribute to the multistep process that sometimes leads to cancer many cell generations later.

Even though radiation damage can occur at any time in the cell cycle (albeit with different sensitivity), one looks for chromosome damage during the next M phase, when the DNA is in the form of visible chromosomes as in the

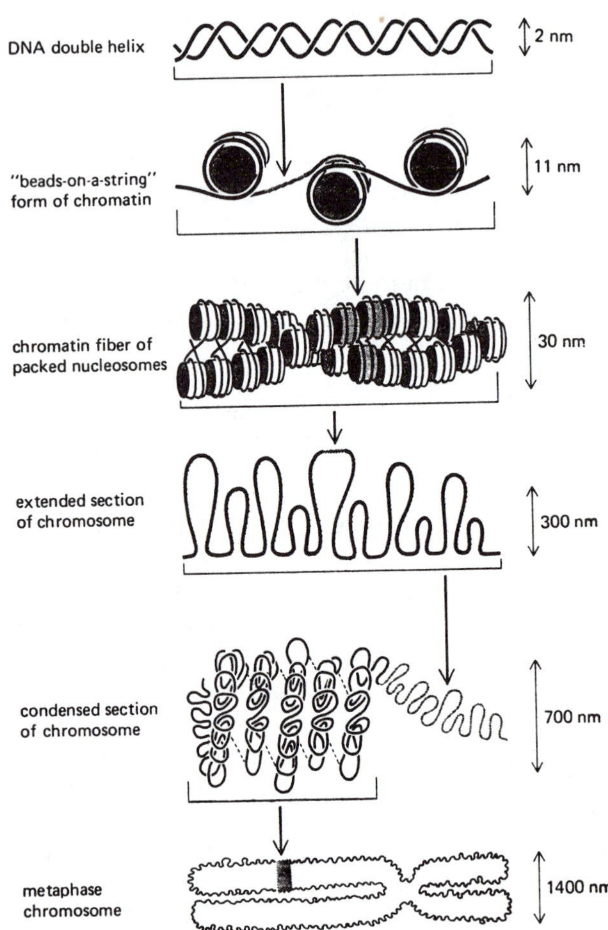

FIGURE 15.33. A schematic diagram of how the DNA is packed to give a chromosome, shown at metaphase of the cell cycle. From B. Alberts *et al. Molecular Biology of the Cell*, 1st. ed. New York, Garland (1983). p. 399. Used by permission of Professor Alberts. (A modified version of this illustration appears on p. 354 of the third edition.)

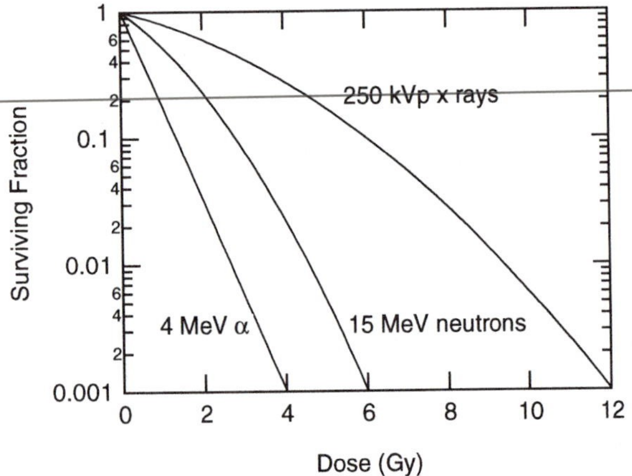

FIGURE 15.32. Typical survival curves for cell culture experiments, for 4 MeV α particles, 15-MeV neutrons, and 250-kVp x rays. These are representations of typical experimental data.

bottom example in Fig. 15.33. If the broken fragments have rejoined in the original configuration, no abnormality is seen when the chromosomes are examined. If the fragments fail to join, the chromosome has a "deletion." If the broken ends rejoin other broken ends, the chromosome will appear grossly distorted.

A sequence of processes leads to cellular inactivation. Ionization is followed by initial DNA damage. Most of this is repaired, but it can be repaired incorrectly. No repair or misrepair results in DNA lesions that are then manifest as chromosome aberrations, which may be nonlethal, may cause mutations, or may lead to cell death. The numbers quoted here are from the review by Steel (1996). A cell dose of 1 Gy leads to the production of about 2×10^5 ion pairs per cell nucleus, of which about 2000 are produced in the cell's DNA. It has been estimated that the amount of DNA damage immediately after radiation can be quite large: 1000 single-strand breaks and 40 double-strand breaks per Gy. Yet survival curves for different cell types show between 0.3 and 10 lethal lesions per gray of absorbed dose. Thus the amount of repair that takes place is quite large, and the model that is introduced below is an oversimplification.

A number of chemicals enhance or inhibit the radiation damage. Some chemical reactions can ''fix'' the DNA damage, making it irreparable; others can scavenge and deactivate free radicals. One of the most important chemicals is oxygen, which promotes the formation of free radicals and hence cell damage. Cells with a poor oxygen supply are more resistant to radiation than those with a normal supply.

The simplified model for DNA damage from ionizing radiation recognizes two types of damage, shown in Fig. 15.34. In type-A damage a single ionizing particle breaks both strands of the DNA, and the chromosome is broken into fragments. In type-B damage, a single particle breaks only one strand. If another particle breaks the other strand ''close enough'' to the first break before repair has taken place, then the chromosome suffers a complete break.

The probability of type-A damage is proportional to the dose. The average number of cells with type-A damage after dose D is $m=\alpha D=D/D_0$, and the probability of no damage is the Poisson probability $P(0;m)=e^{-m}=e^{-\alpha D}$. This is the dashed line in Fig. 15.35, which is redrawn from Fig. 15.32. For radiations with higher LET the proportionality constant α is greater, as seen in Fig. 15.32.

In type-B damage one strand is damaged by one ionizing particle and the other by another ionizing particle. The probability of fragmenting the DNA molecule is therefore proportional to the square of the dose. The average number of molecules with type-B damage is βD^2, and the survival curve for type-B damage alone is $e^{-\beta D^2}$. This is also shown in Fig. 15.35. This leads to the *linear-quadratic* model for cell survival:

$$P_{\text{survival}}=e^{-\alpha D-\beta D^2}. \qquad (15.31)$$

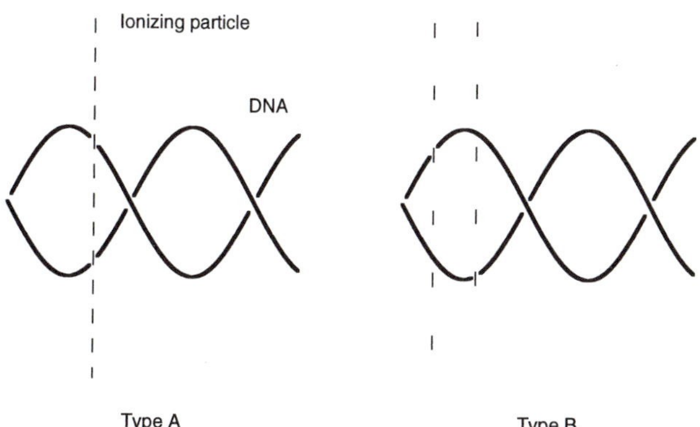

FIGURE 15.34. The two postulated types of DNA damage from ionizing radiation. In type-A damage a single ionizing particle breaks both strands. Two ionizing particles are required for type-B damage, one breaking each strand.

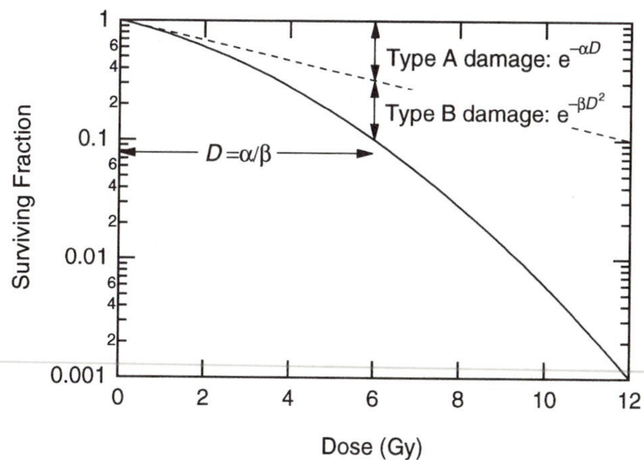

FIGURE 15.35. A survival curve, showing the linear exponent for type-A damage and the quadratic exponent for type-B damage.

The dose at which mortality from each mechanism is the same is α/β, as shown in Fig. 15.35.

An extension of the cell survival experiments is the *fractionation curve* shown in Fig. 15.36. After a given dose, cells from the culture were harvested and used to inoculate new cultures. After a few hours they were irradiated again. The survival curve plotted against total dose starts anew from the point corresponding to the first irradiation. The initial dose of 6 Gy caused both type-A and type-B damage. Before the second dose, all of the cells with single-strand damage had been repaired, and when the second dose was given, it acted on undamaged cells, so that only type-A damage occurred for small additional doses.

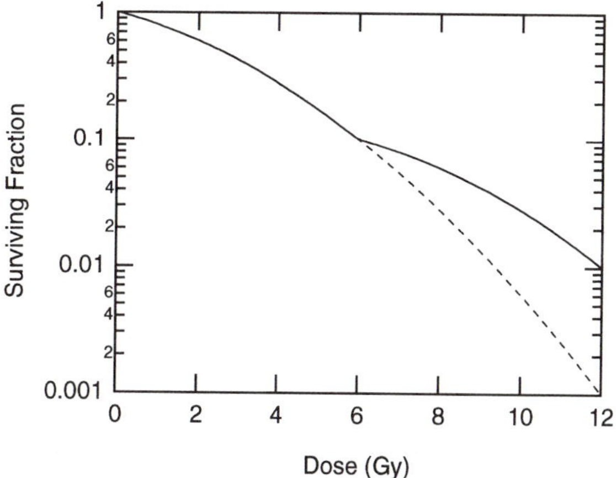

FIGURE 15.36. If the dose for low-LET radiations is divided into fractions, with a few hours between fractions, all of the single-strand breaks have been repaired, and survival follows the same curve as for the original fraction.

Another model, the *multitarget model* has been used in the past to describe cell-survival curves. Each cell contains r *target volumes* that are sensitive to the incident radiation. The interaction of radiation with a target is called a *hit*. A hit causes the target volume to change its state. When all r target volumes have been hit, the cell dies. The hits occur at random. The probability that one particular ionizing particle interacts with a given target volume is small. Therefore we can use the Poisson distribution to calculate the probability of x hits if the average number of hits per target is m:

$$P(x;m)=\frac{m^x}{x!}e^{-m}.$$

The probability that the target has received one or more hits and changed its state is $p'=1-P(0;m)=1-e^{-m}$. The probability that all r targets are hit is $(p')^r$. The probability of cell survival is

$$P_{\text{survival}}=1-(p')^r=1-(1-e^{-m})^r. \qquad (15.32)$$

This is plotted in Fig. 15.37 for a few values of r. You can see from the plots (and verify analytically) that the curves approach re^{-m} for large values of m (or dose). The value of r is the value of P extrapolated to zero dose. It is called the *extrapolation number*. The curve for x rays from Fig. 15.32 is also plotted in Fig. 15.37. The multitarget model has zero slope at small doses, which is not characteristic of these particular data. A "composite multitarget model" of the form

$$P_{\text{survival}}=e^{-\alpha D}[1-(1-e^{-\gamma D})^r] \qquad (15.33)$$

is sometimes used [see the review by Cohen (1993)]. The first factor represents survival due to type-A damage and gives an initial slope of $-\alpha$, while the second factor represents sublethal damage (analogous to type B) and dominates at high doses, approaching $re^{-\gamma D}$.

Survival curves for high doses are difficult to measure because of the small number of cells surviving in the culture. They seem to follow a straight exponential at large doses, which is characteristic of the multitarget model but not the linear quadratic model.

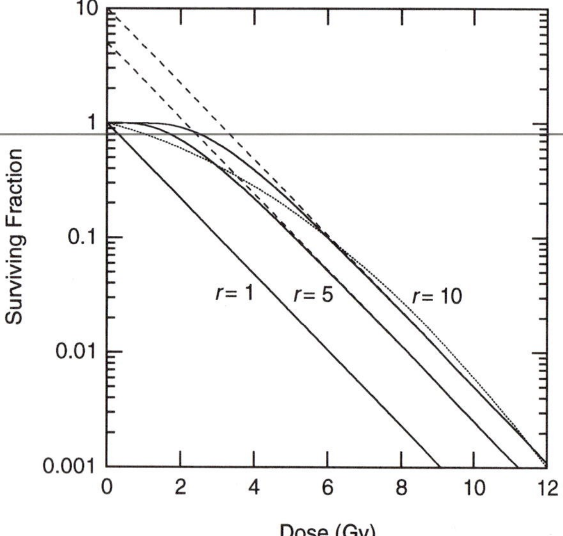

FIGURE 15.37. Calculated values of the surviving fraction using the multitarget model, for different values of *r*.

15.10.3. Tissue Irradiation

There is considerable variation in the shape of the survival curves for human cells (Fig. 15.38). The shaded area labeled "human A-T cells" is for cells from patients with a genetic disease, *ataxia-tangliectasia*, where repair mechanisms are lacking and breakage of a single strand of DNA leads to cell death.

The radiation damage to the DNA is not apparent until the cell tries to divide. At that point, the chromosomes are either so badly damaged that the cell fails to divide or the damage survives in later generations as a mutation. Some tissues respond to radiation quite quickly; others show no effect for a long time. This is due almost entirely to the duration of the G_1 phase or the overall time between cells divisions. Tissues are divided roughly into two groups: *early responding* and *late responding*. Early-responding tissues include most cancers, skin, the small and large intestine, and the testes. Late-responding tissues include spinal cord, the kidney, lung, and urinary bladder.

The central problem of radiation oncology is how much dose to give a patient, over what length of time, in order to

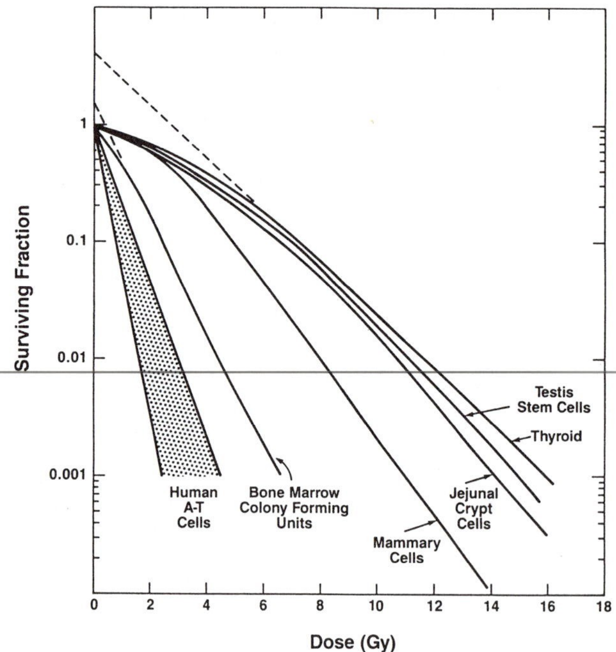

FIGURE 15.38. Survival curves for assays of human cells. There is a wide range in initial sensitivity, but not too much difference in final slope. The shaded area labeled "human A-T cells" is for cells from a disease, ataxia-tangliectasia, where repair mechanisms are lacking. Reproduced with permission from E. J. Hall (1994). *Radiobiology for the Radiologist*, 4th ed. Philadelphia, Lippincott, p. 60.

have the greatest probability of killing the tumor while doing the least possible damage to surrounding normal tissue. While the dose is sometimes given all at once (over several minutes), it is usually given in *fractions* five days a week for four to six weeks. Some recent treatment plans, primarily for brachytherapy (see Sec. 16.15), use fractions given every few hours.

What total dose (or dose per fraction) should be given in how many fractions, with what time between fractions? We try to answer these questions using the linear-quadratic model because the algebra is simpler, though the features of the argument depend on the shape of cell survival curve and any model with a shoulder could be used. Let the dose per fraction be D_f, the number of fractions be n, and the total dose be $D=nD_f$. We now need to plot survival vs total dose for different numbers of fractions. We assume that the time between fractions allows for full repair of sublethal damage (single-strand breaks). The fraction of cells surviving after n fractions have been delivered is

$$P_s=P_{\text{survival}}=S=(e^{-\alpha D_f-\beta D_f^2})^n=e^{-\alpha D-\beta D^2/n}. \tag{15.34}$$

As the number of fractions becomes very large for a given total dose, the survival curve approaches $e^{-\alpha D}$. This can be seen in Fig. 15.39, which plots survival vs total dose delivered in different numbers of fractions. With many fractions the dose per fraction is very small, all the single-strand breaks are repaired, and almost no type-B cell deaths take place.

Early-responding tissue and tumors have been found to have an α/β ratio of about 10 ($\beta/\alpha\approx0.1$). The survival curve is primarily due to type-A damage. Late-responding tissues have an α/β ratio of 2–3. There is considerable variation in these numbers.

Some of the problems of radiation therapy and the benefits of fractionation can be seen if we consider a strictly hypothetical example in which $\alpha=0.1$ for both the tumor and the surrounding tissue. The tumor is early responding with $\alpha/\beta=10$, and the surrounding tissue is late responding with $\alpha/\beta=2$. Figure 15.40 shows the cell-survival curves for 1 and 35 fractions. The tumor survival in each case is shown as a dotted line. The thicker curves correspond to delivering the dose in 35 fractions. In this example, both tissues have the same value of α and the surrounding tissue receives the same dose. (In real life, α for the tumor may be greater than that for the normal tissue, and the treatment will be more effective.) The surrounding tissue ($\alpha/\beta=2$) is actually more sensitive than the tumor ($\alpha/\beta=10$) and has a lower survival curve.

To see the benefit of fractionation, suppose that the patient can tolerate a dose at which only 10^{-6} of the cells of the surrounding tissue survive, represented by the horizontal line on the graph. (This is *not* realistic!) For a single fraction, this corresponds to a total dose of about 9 Gy, which, applied to the dotted line, shows that the surviving fraction of tumor cells is about 10^{-2}. For 35 fractions the normal tissue can tolerate about 32 Gy, yielding 5×10^{-5} as the fraction of tumor cells surviving.

Suppose next that it is possible to confine the radiation beam so that the dose to normal tissue is only about 0.6 times that to the tumor. This means that the tissue dose in Eq. (15.34) is multiplied by 0.6. The result is shown in Fig. 15.41 for 35 fractions. The dose can now be as high as 53 Gy for the same effect on surrounding tissues, leading to a tumor survival of only 10^{-8}. We will see how beam shaping is accomplished in the next section.

These calculations are solely to illustrate the basic principles. Clinically useful calculations must take several additional factors into account: the actual values of α for the

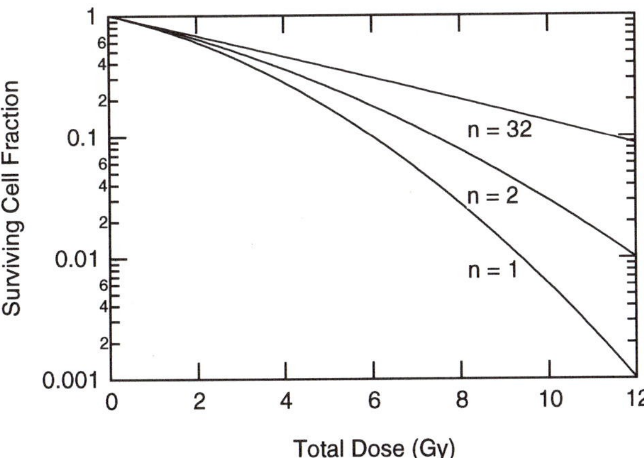

FIGURE 15.39. The fraction of cells surviving a total radiation dose when the dose is divided into 1, 2, and 32 fractions, showing how the curve approaches $e^{-\alpha D}$ as the number of fractions becomes large.

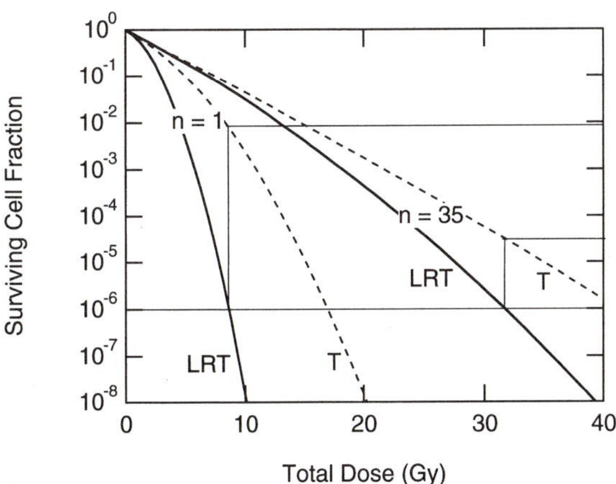

FIGURE 15.40. Cell survival curves for late-responding normal tissue (LRT) and for an hypothetical tumor (T), showing the improvement obtained by dividing the dose into fractions. With a single fraction, the tumor survives much better than the normal tissue. With 35 fractions, this discrepancy has been reduced. The details are discussed in the text.

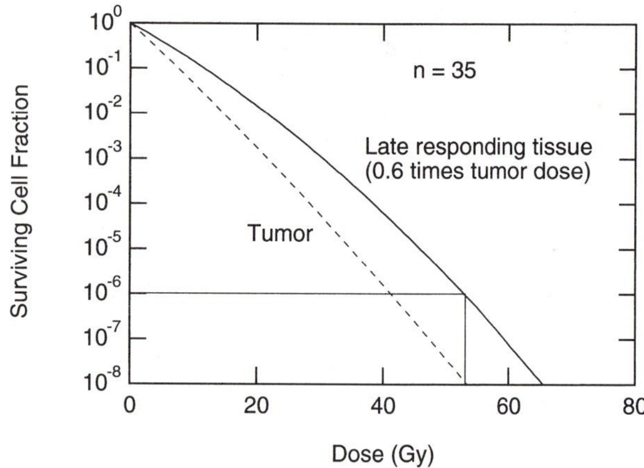

FIGURE 15.41. Survival curves for the same cells as in the previous figure, with the dose to the surrounding tissue reduced to 0.6 times that to the tumor. Now the probability of tumor survival at high doses is about 0.01 times that for the surrounding normal tissue. This shows the importance of confining the radiation to the tumor as much as possible.

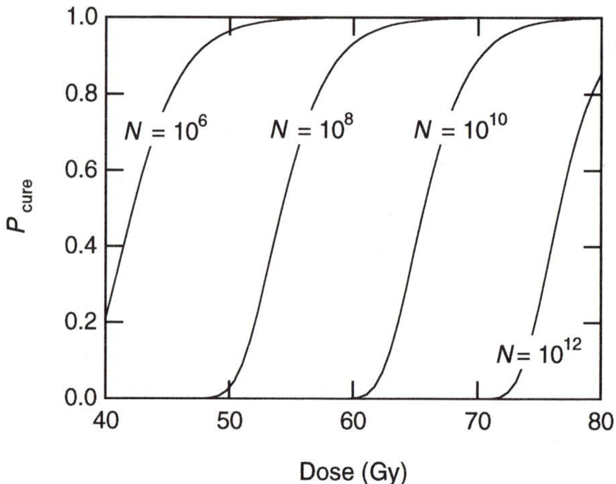

FIGURE 15.42. The probability of curing the tumor (no surviving tumor cells) as a function of dose for tumors containing different numbers of cells.

tissue and tumor under consideration, the effect of cell growth after irradiation, the effect of the first dose on synchronizing the cycles of the remaining cells, and the oxygen level in the tumor cells. (The greater the oxygen concentration the more sensitive the cells are, particularly for low-LET radiation. Rapidly growing tumors have often outstripped their blood supply and are less radio-sensitive.) A current review of fractionation is provided by Orton (1997). It is also necessary to take into account the fact that neither the tumor nor the surrounding normal tissue receives a uniform dose of radiation.

15.10.4. Tumor Sterilization

The target theory model can be applied to a collection of cells to give us insight into the central problem of radiation therapy: eradication of the tumor or *tumor sterilization.* Suppose that a tumor consists of N cells with identical properties. The cells are uniformly irradiated with dose D. If a collection of identical tumors were irradiated, the number of cells surviving in each tumor would fluctuate. The probability that a single cell survives is $p_s(D)$, which might be given by Eq. (15.32), (15.33), or (15.34). If this number is small and N is large, the number surviving follows a Poisson distribution. The average number surviving is $m = Np_s(D)$. The probability of a cure is the probability that no tumor cells survive:

$$P_{\text{cure}} = e^{-m} = e^{-Np_s(D)}. \tag{15.35}$$

This can be evaluated using your cell-survival model of choice. Figure 15.42 shows a tumor sterilization curve based on the 35-fraction curve in Fig. 15.40. The larger the tumor, the greater the dose required for cure. Figure 15.43 shows a plot of the probability of tumor cure and the probability of unacceptable damage to the surrounding tissue, both based on 10^8 cells and the cell-survival curves in Fig. 15.41 (with the normal tissue receiving 0.6 times the dose received by the tumor). For this model, at least 60 Gy are required in order to have a good probability of cure; once the dose is higher than 65 Gy, the damage to normal tissue is unacceptable.

15.11. RADIATION THERAPY

Doses for diagnostic radiology vary from about 10^{-2} to 10^{-4} Gy. Doses of 20–80 Gy are required to treat cancer. A great deal of physics is involved in the planning of treat-

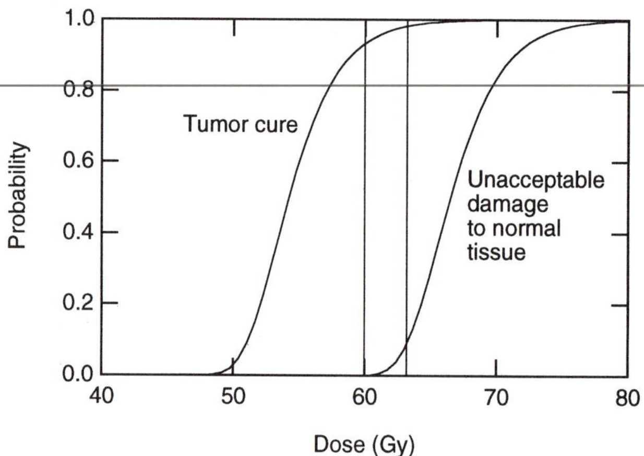

FIGURE 15.43. The probability of curing the tumor and the probability of unacceptable damage to normal tissue vs dose. Both the tumor and the normal tissue contain 10^6 cells. The 35-fraction doses shown in Fig. 15.41 have been used.

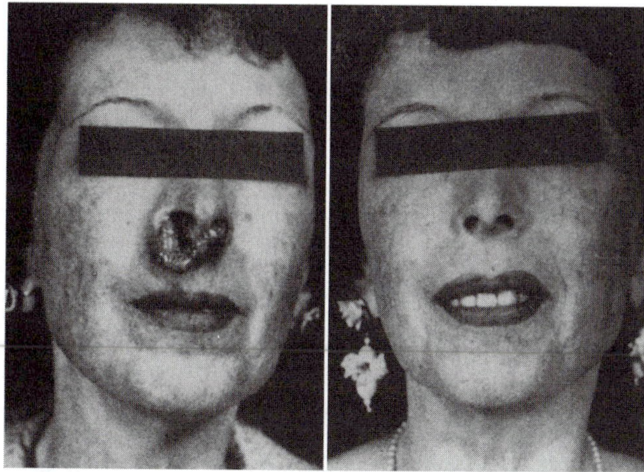

FIGURE 15.44. X-ray therapy was used to treat a carcinoma of the nose. *A* shows the original lesion; *B* is the result one year later. The patient remained asymptomatic five years after treatment. From William T. Moss and James D. Cox (1989). Radiation Oncology, 6th ed. St. Louis, Mosby.

ment for each patient.[19] There is a choice of radiation beams: photons of various energies, electrons, neutrons, protons, α particles, or π mesons. Photons and electrons are routinely available; the other sources require special facilities. Only a few of the beam issues will be raised here. Some of the dose measurement issues are discussed in the next section.

Two examples of the effectiveness of radiation therapy are shown in Figs. 15.44–15.46. In the first case, the patient developed a carcinoma of the nose and refused surgery. Radiation with a total dose of 50 Gy was used, and the results one year later are shown. It is ironic that the carcinoma probably developed because the patient was treated with x rays for acne many years earlier. In the second case, a basal-cell carcinoma involved the eyelids and the bridge of the nose. Figure 15.45 shows a lead shield placed under the eyelid to protect the lens of the eye during the irradiation. Fig. 15.46 shows the results six months later, after a total dose of 43 Gy.

Sometimes the results can be spectacular. *Orbital rhabdomyosarcoma* is a rare cancer of striated muscle that occasionally occurs in the muscles surrounding the eye (the orbit) of young children. It is now possible in many cases to treat it with a combination of radiation and chemotherapy without loss of eye function [Donaldson (1995)].

We have already seen the importance of reducing dose to tissue surrounding the tumor. Optimizing the dose determines the kind of radiation to be used and its energy, as well as the details of beam filtration and collimation and how it is aimed at the patient's body. For now, we discuss a photon beam. Attenuation and the $1/r^2$ decrease of photon fluence help spare tissue downstream from the beam. Since μ_{atten} decreases with increasing photon energy up to a few

[19]See Khan (1994).

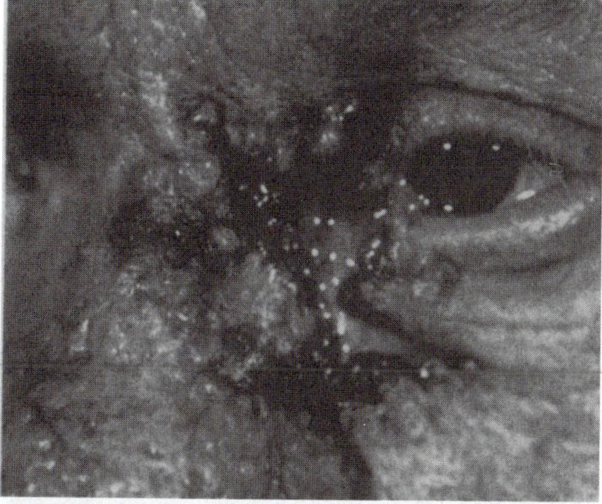

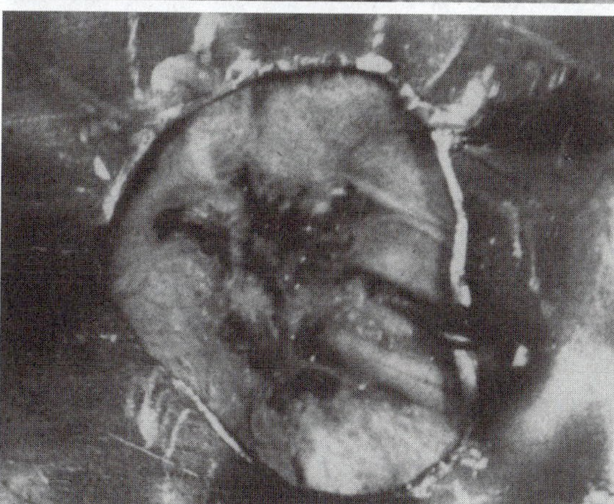

FIGURE 15.45. (A) An extensive basal-cell carcinoma of the skin involving the eyelids and the bridge of the nose. (B) Lead shields in place before irradiation. A shield between the lid and the eyeball protects the lens of the eye while allowing irradiation of the lids. An overlying shield limits the radiation to the suspected area of involvement. The total dose was 43 Gy. From William T. Moss and James D. Cox (1989). *Radiation Oncology*, 6th ed. St. Louis, Mosby.

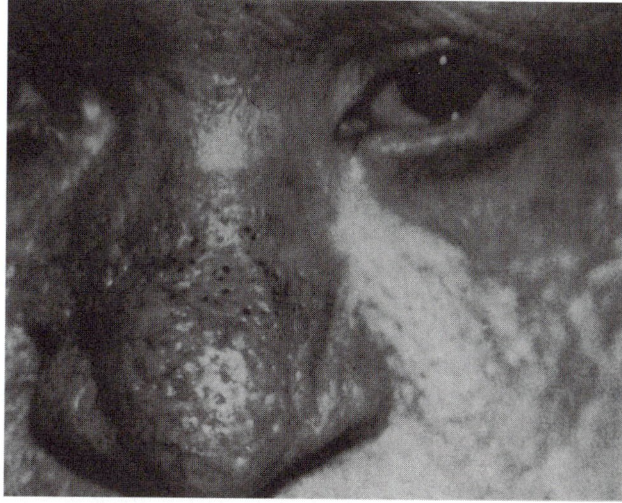

FIGURE 15.46. The patient shown in Figure 15.45, six months after treatment. From William T. Moss and James D. Cox (1989). *Radiation Oncology*, 6th ed. St. Louis, Mosby.

MeV, higher-energy photons penetrate more deeply and must be used for treating deeper lesions. There is also dose buildup with depth over distances comparable to the range of the Compton-scattered electrons. Both of these effects are shown in Fig. 15.47.

The beam is *collimated* with jaws of lead or with masks of a special lead alloy which are custom-made for each patient. Figure 15.48 shows *isodose contours* for various beams. In addition to the differences along the axis seen in Fig. 15.47, there are significant differences in the sharpness of the dose distribution across the beam. The extent of the lesion to be radiated must be carefully determined with radiographs or CT scans. This is sometimes a very difficult problem. Some new systems have computer-controlled collimator leaves to automatically change the field shape.

If the tumor is not on the surface, the ratio of tumor dose to normal tissue dose can be increased by irradiating the patient from several directions. Figure 15.49 shows how the relative dose at a deep tumor can be increased by irradiating with two "fields" on opposite sides of the patient. In Fig. 15.50 three and four fields are used. Care must be taken to reduce the dose to the patient due to scatter from the collimator. Special filters may also be used to shape the beam intensity across the radiation field.

An area of current research is *tomotherapy*. The goal is to use filtered back-projection techniques with continuously variable slits to shape the radiation field as the patient moves axially through the beam—the analog of a spiral CT study [Holmes, Mackie, and Reckwerdt (1995)]. It is, of course, impossible to make the filtered radiation field negative. This means that the dose outside the tumor is not strictly zero, but it can be made small. The limitation at the time of the reference cited is the long computation time required to do the back projection.

Electrons, typically between 6 and 20 MeV, are also used for therapy. Because of the range–energy relationship, the field falls nearly to zero in a centimeter or two. Electrons are used primarily for skin and lip cancer, head and neck cancer, and irradiation of lymph nodes near the surface. Figure 15.51 shows the dose vs depth as a percent of the maximum dose for electron beams of several different energies.

Protons and other heavy particles from nuclear accelerators have been used to treat tumors. Their advantage is the increase of stopping power at low energies. It is possible to make them come to rest in the tissue to be destroyed, with an enhanced dose relative to intervening tissue. Protons have been used to irradiate the pituitary gland of patients with acromegaly due to excessive activity of the pituitary from a (benign) pituitary adenoma. They have also been used to irradiate arteriovenous malformations in the brain. Proton therapy is also used for patients with malignant tumors of the retina (choroidal melanoma), with 98% local control of the tumor after five years and preservation of vision in about half the cases. Reviews are provided by Habrand *et al.* (1995) and by Suit and Urie (1992).

Negative π mesons (pions) have been tried as a treatment particle [Richman (1981)]. The pions come to rest in the tissue, giving greater ionization at the end of the path. However, there is an additional mode of energy deposition. Each pion goes into orbit around a nucleus because of its negative charge. Because they are 275 times as massive as an electron, they are closer to the nucleus. They interact with the nucleus, causing it to "blow apart" into many

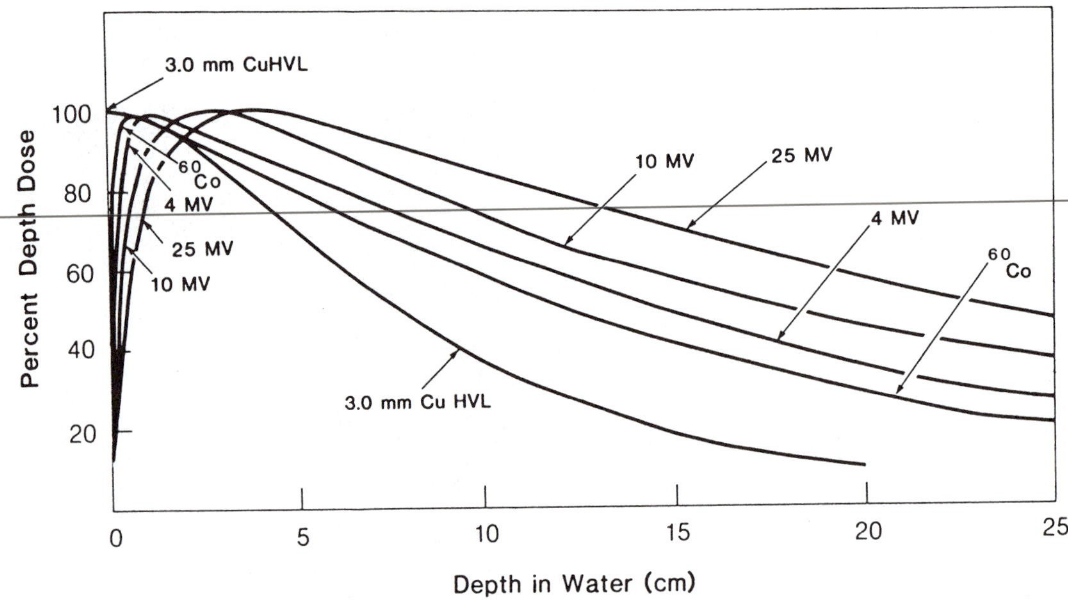

FIGURE 15.47. The dose vs depth for photon beams of different quality (energy) on the central axis of the beam. The source–surface distance (SSD) is 100 cm and the field size is 10 cm×10 cm. From F. Khan (1994). *The Physics of Radiation Therapy*, 2nd. ed. Baltimore, MD, Williams and Wilkins. © 1994 The Williams and Wilkins Co.

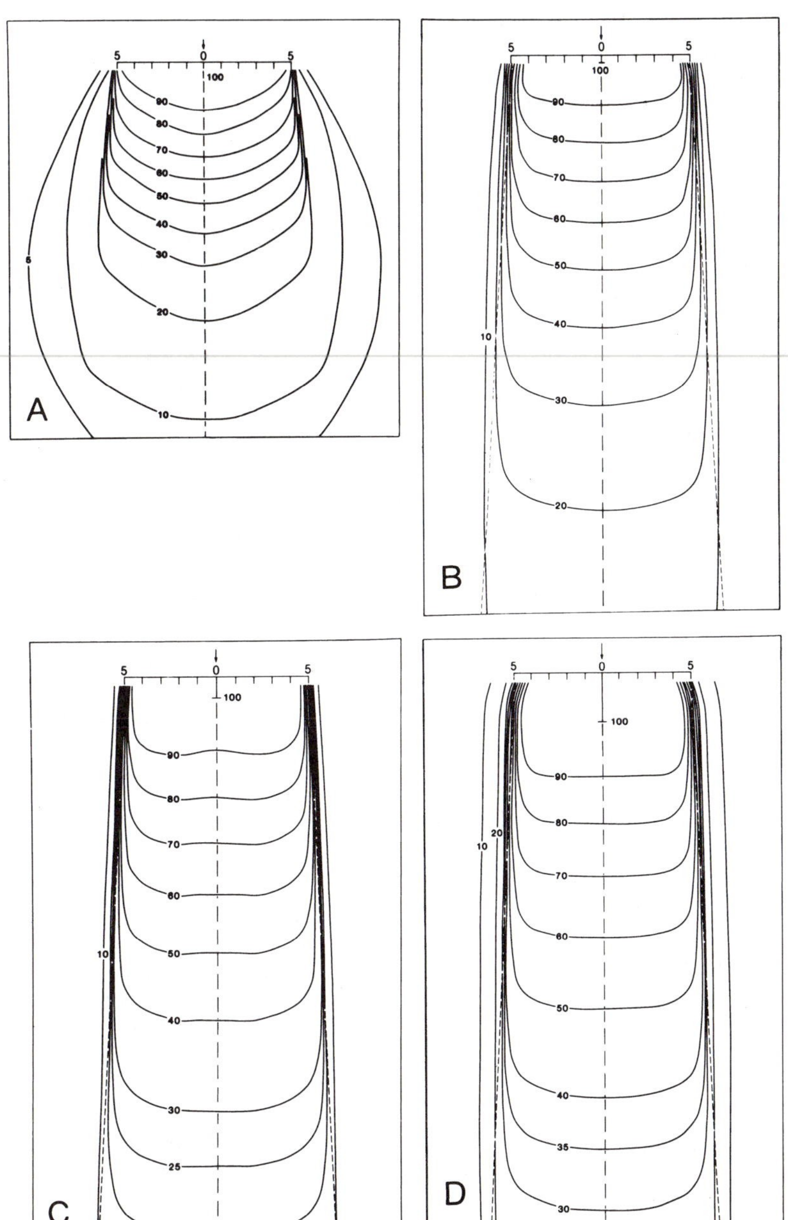

FIGURE 15.48. Isodose distributions for radiation under different conditions, all collimated to 10 cm×10 cm. (A) Radiation from an x-ray tube with 200 kVp, 0.5 m from the surface. (B) Photons from the radioactive isotope ^{60}Co, 0.8 m from the surface. (C) 4-MV photons, 1 m from the surface. (D) 10-MV photons, 1 m from the surface. From F. Khan (1994), *The Physics of Radiation Therapy*, 2nd. ed. Baltimore, MD, Williams and Wilkins. © 1994 The Williams and Wilkins Co.

short-range, high-z projectiles. Nearly all of the total pion rest energy, 140 MeV, is converted to ionizing radiation at the end of the path. Unfortunately, clinical trials did not show remarkable improvement over conventional therapy.

Fast neutrons have also been tried for therapy. The dose is due to charged particles: protons, α particles (^{4}He nuclei), or recoil nuclei of oxygen and carbon that result from interactions of the neutrons with the target tissue. All of these have high LET and the oxygen effect is less. So far they have not shown significant benefit over photon therapy [Duncan (1994)].

Brachytherapy (brachy means short) involves the implantation of radioactive isotopes in a tumor and will be discussed in Sec. 16.15.

15.12. DOSE MEASUREMENT

It is important to measure radiation doses accurately for radiation therapy in order to compare the effectiveness of different treatment protocols and to ensure that the desired protocol is indeed being followed. Accuracies of 2% are

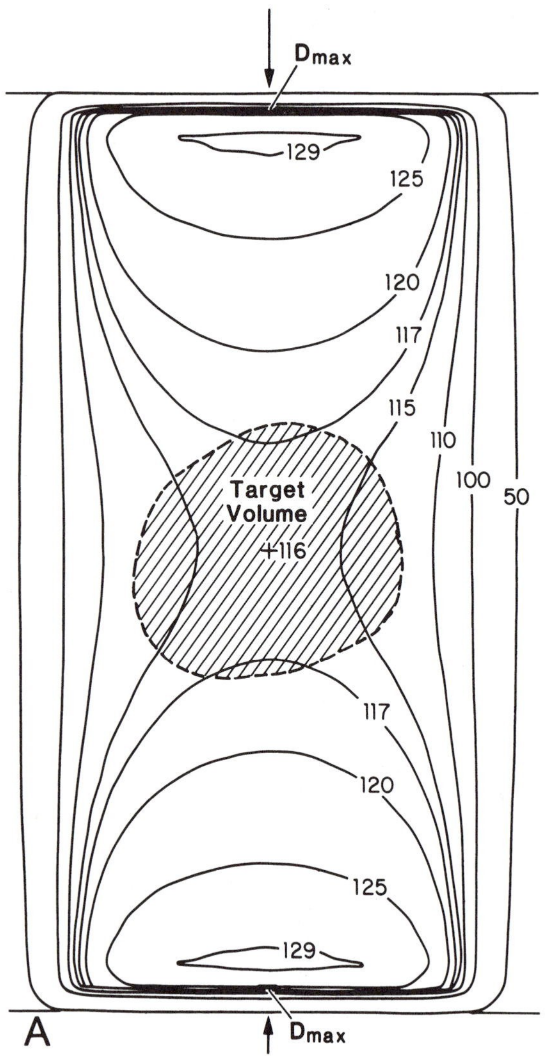

FIGURE 15.49. Isodose distribution when the patient is irradiated equally from opposite sides. From F. Khan (1994), *The Physics of Radiation Therapy*, 2nd. ed. Baltimore, MD, Williams and Wilkins. © 1994 The Williams and Wilkins Co.

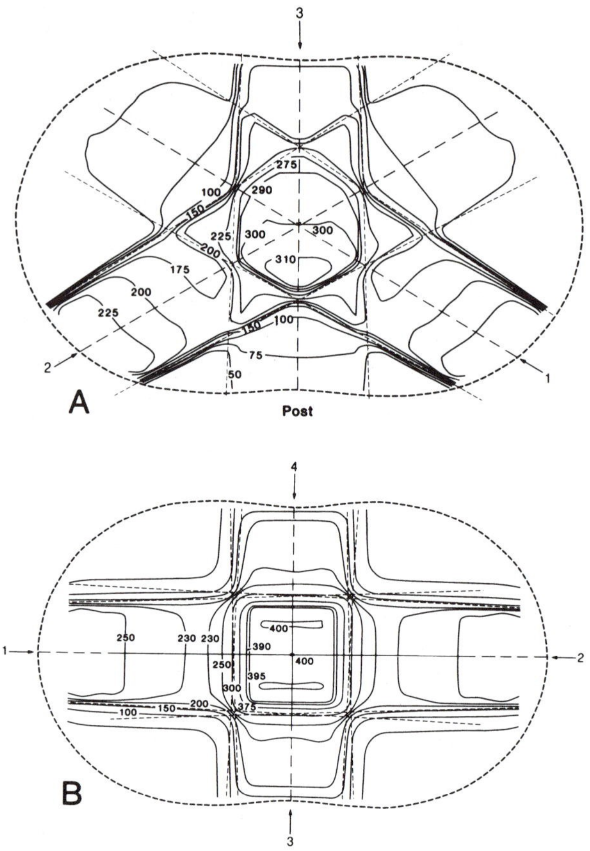

FIGURE 15.50. Isodose distribution for (A) three and (B) four radiation fields, each designed to give a relative dose of 100 at the center of the tumor. From F. Khan (1994). *The Physics of Radiation Therapy*, 2nd. ed. Baltimore, MD, Williams and Wilkins. © 1994 The Williams and Wilkins Co.

expected. An extensive literature about relating the dose in the measuring instrument to the dose in surrounding tissue exists.[20] Here we describe one of the techniques that is used.

A basic problem in dosimetry is that the measuring instrument has different properties than the medium in which it is immersed. Imagine, for example, that a gas-filled ionization chamber is placed in water. If the radiation field were very large and uniform, one could in principle use an ionization chamber whose dimensions are large compared to the range of secondary electrons, and the interaction of the radiation field with the chamber gas would be the dominant effect. This is not practical. At the other extreme, we imagine an ionization chamber that is so small that it does not alter the radiation field of the water. That is, its dimensions must be small compared to the range of the charged particles created in the water and passing through it.

We saw in Sec. 14.16 that the absorbed dose in a parallel beam of charged particles with particle fluence Φ is

$$D = \frac{S_c}{\rho} \Phi. \tag{14.76}$$

Usually the beam consists of particles with different kinetic energies T. Let Φ_T be the energy spectrum:

$$\Phi = \int_0^{T_{\max}} \Phi_T dT.$$

Then the dose is the integral of the number of particles with

[20] See Attix (1986), Chap. 10ff or Khan (1994), Chap. 8.

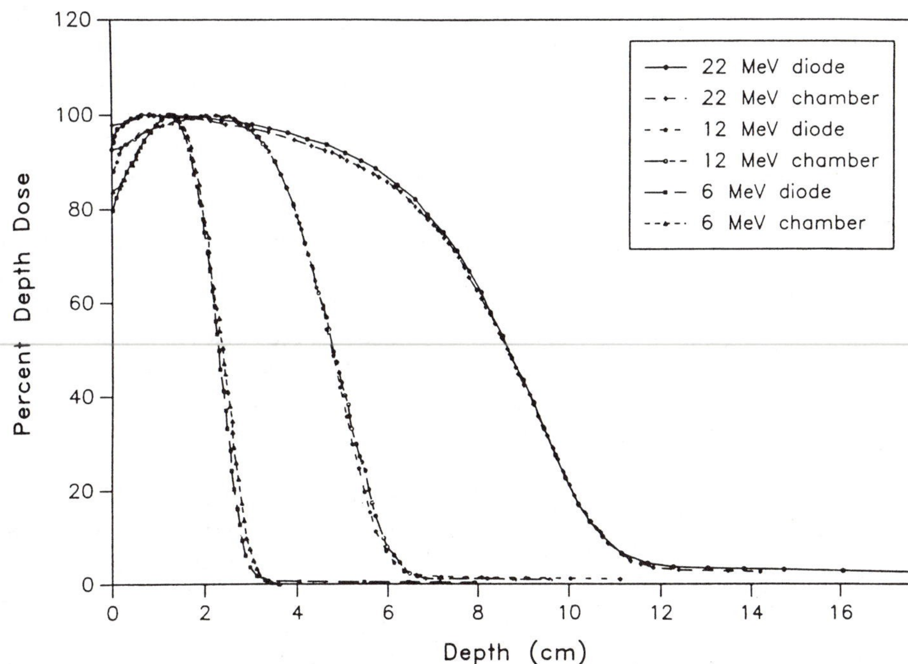

FIGURE 15.51. Depth–dose curves for electrons of different energies, measured with a solid-state detector and an ionization chamber. Both the range and the straggling increase with increasing energy. From F. Khan (1986). Clinical electron beam dosimetry. In J. G. Keriakes, H. R. Elson, and C. G. Born, eds. *Radiation Oncology Physics—1986.* College Park, MD, American Association of Physicists in Medicine. AAPM Monograph 15.

energy T times the mass stopping power for particles of that energy:

$$D = \int_0^{T_{\max}} \Phi_T \frac{S_c}{\rho} \, dT. \tag{15.36}$$

We can define an average mass collision stopping power:

$$\frac{\overline{S}_c}{\rho} = \frac{1}{\Phi} \int_0^{T_{\max}} \Phi_T \frac{S_c}{\rho} \, dT \tag{15.37}$$

so that

$$D = \Phi \frac{\overline{S}_c}{\rho}. \tag{15.38}$$

Let us apply this to the situation where a small detector (''gas'') is introduced in a medium (''water'') in which we want to know the dose. The charged particle fluence is not altered by the detector because it is small compared to the range of the charged particles. Applying Eq. (15.38) in both media, we obtain

$$\frac{D_w}{D_g} = \frac{(\overline{S}_c/\rho)_w}{(\overline{S}_c/\rho)_g} \equiv (\overline{S}_c/\rho)_g^w. \tag{15.39}$$

This is the Bragg–Gray relationship for the absorbed dose in the cavity. It is standard in the literature to denote the dimensionless ratio of the stopping powers in the two media by $(\overline{S}_c/\rho)_g^w$.

This equation is often used with ionization chambers. The charge created in an ionization chamber of mass m is the charge per ion pair e times the number of ion pairs formed in mass m. The number of ion pairs is the energy deposited, mD_g, divided by the average energy required to produce an ion pair, W:

$$q = e \frac{mD_g}{W}. \tag{15.40}$$

Combining this with Eq. (15.39) gives the dose in the medium in terms of the charge created:

$$D_w = \frac{q}{m} \left(\frac{W}{e} \right)_g (\overline{S}_c/\rho)_g^w. \tag{15.41}$$

The charge q created is usually greater than the charge collected in the ion chamber because of recombination of ions and electrons before collection. The collection efficiency and the chamber mass are deduced from calibration of the chamber. Once the chamber has been calibrated, the factor $(\bar{S}_c/\rho)_g^w$ accounts for placing the chamber in different media.

15.13. THE RISK OF RADIATION

What is the likelihood of an adverse effect on an individual who receives a dose of radiation for diagnosis or treatment? This question can probably never be answered, because of the variation between individuals. We will simply compare medical doses to natural background. There is no evidence that there is any risk from the natural background [Biological Effects of Ionizing Radiations Report V (BEIR V, 1990)]. We will also briefly address the epidemiologic question of radiation-induced disease in large populations.

We have already seen that the biological effect of radiation depends not only on the absorbed dose, but also on the LET and the nature of the tissue that is irradiated. In an attempt to take these differences into account, the International Commission on Radiation Protection (ICRP) introduced the *radiation weighting factor* W_R. It is multiplied by the dose to give the *equivalent dose*[21] H_T:

$$H_T = W_R D. \tag{15.42}$$

The unit of H_T is the *sievert* (Sv).[22] The radiation weighting factor depends on the *relative biological effectiveness* (RBE) of the radiation, which depends not only on the LET of the radiation but the biological endpoint being used. The variability in measured values for RBE led National Council of Radiation Protection and Measurements Report 104 to conclude that it is "a matter of judgment which Q value to use for radiation protection." [NCRP Report 104 (1990)].

In many cases the dose absorbed by the body is not uniformly distributed. For a particular configuration of absorbed dose, a *tissue weighting factor* W_T is assigned to each organ. It is the ratio of risk resulting from irradiation of that organ or tissue to the risk when the whole body is irradiated uniformly. The sum of W_T over all organs equals unity. The ICRP recommend W_T values in 1977 and slightly different values in 1991. It is added to Eq. (15.42) as another multiplicative factor, to define the *effective dose*[23] to an organ, E:

$$E = W_T W_R D.$$

We are continuously exposed to radiation from natural sources. These include cosmic radiation, which varies with altitude and latitude; rock, sand, brick, and concrete containing varying amounts of radioactive minerals; the naturally occurring radionuclides in our bodies such as ^{14}C and ^{40}K; and radioactive progeny from radon gas from the earth.[24] In a typical adult, there are about 4×10^7 radioactive disintegrations per hour from all internal sources. Table 15.4 and Fig. 15.52 summarize the various sources of radiation exposure. The radon entry in Table 15.4 is based on a W_R of 20 for α particles from radon progeny, the value used by NCRP.[25] There is considerable uncertainty in this determination: W_R could be as low as 3, in which case radon would contribute much less to the natural background.

Diagnostic procedures give doses that are in general comparable to the average annual background dose. One can explain to a patient that a chest x ray is equivalent to about one week of natural background, and a mammogram is equivalent to a month or two. The largest dose is for a conventional fluoroscopic study of the lower digestive system, which is equivalent to about a year of natural background.

We have considerably more information about human exposure to ionizing radiation than we have for any other known or suspected carcinogen [Boice (1996)]. Several studies at moderate doses show that radiation is a relatively weak carcinogen, though this is not the public view.

In dealing with radiation to the population at large, or to populations of radiation workers, the policy of the various regulatory agencies has been to adopt the *linear–no-threshold* (LNT) model to extrapolate from what is known about the risk of cancer at moderately high doses and quite high dose rates, to very low doses, including those below natural background. The LNT model ignores the repair mechanisms that were discussed above. There is currently disagreement over whether there is in fact a threshold for radiation-induced damage; if there is, the LNT model severely overestimates cancer risk. Anyone interested in this question should review the raw data of animal studies presented in NCRP Report 104 (1990).

Measurements at lower doses to determine if the response is linear are difficult to make, requiring large numbers of subjects, as the following simplified example shows. Suppose that we have two measurements of the probability of acquiring cancer: one at zero dose, which gives the "spontaneous" probability due to nonradiation causes, and one at a fairly large dose (say 1 Gy). At some low dose we want to distinguish between a linear increase

[21]The nomenclature here is quite confusing. One used to define the dose equivalent, also denoted by H, as QD, where Q was called the quality factor of the radiation. The radiation weighting factor is very similar, and essentially numerically equivalent, to the earlier quality factor, Q. Values of Q recommended by Nuclear Regulatory Commission (NRC) are 1 for photons and electrons, 10 for neutrons of unknown energy and high-energy protons, and 20 for α particles, multiply charged ions, fission fragments, and heavy particles of unknown charge. The ICRP also has recommendations that differ slightly for protons and neutrons.

[22]An older unit for H_T is the rem. 100 rem=1 Sv.

[23]An older, related unit is the effective dose equivalent, $H_E = W_T Q D$.

[24]Radon is chemically inert gas that escapes from the earth. Since it is chemically inert, we breathe it in and out. When it decays in the air (the decay scheme is described in Sec. 16.16), the decay products attach themselves to dust particles in the air. When we breathe these dust particles, some become attached to the lining of the lungs, irradiating adjacent cells as they undergo further decay.

[25]The dose to the lungs from radon progeny is about 1 mGy yr^{-1}. This is multiplied by $W_R=20$ and $W_T=0.12$ (lungs) to arrive at an effective dose of 2.4 mSv yr^{-1}.

TABLE 15.4. *Typical radiation doses from natural sources.*

Radiation source	Detail	Dose rate to target organ (Sv yr^{-1})	Average effective dose rate[a] (i.e., multiplied by w_T, Sv yr^{-1}) (NCRP 94)
Cosmic radiation	New York City	0.30×10^{-3}	0.27×10^{-3}
	Denver (1.6 km)	0.50×10^{-3}	
	La Paz, Bolivia (3.65 km)	1.8×10^{-3}	
	Flying at 40 000 ft	7×10^{-6} Sv h^{-1}	
Terrestrial (radioactive minerals)			0.28×10^{-3}
	Over fresh water	0	
	Over sea water	0.2×10^{-3}	
	Sandy soil	$(0.1-0.25)\times10^{-3}$	
	Granite	$(1.3-1.6)\times10^{-3}$	
Ingestion			0.4×10^{-3}
	Potassium	0.17×10^{-3}	
Inhalation of radon		25×10^{-3}	2.0×10^{-3}
Total			3.0×10^{-3}

[a]NCRP Report 94 (1987).

of probability with dose and a probability that remains at the ''spontaneous'' value because we are below some threshold dose for carcinogenicity. The probability p of acquiring cancer is small and the total population N is large. This means that if the experiment could be repeated several times on identical populations, the number of persons acquiring cancer, n, would be Poisson distributed with mean number $m=Np$ and standard deviation $\sigma=\sqrt{m}$. Figure 15.53 plots m vs dose for some value of N, with dashed lines to indicate $m\pm\sigma$. A measurement at the lower dose indicated by the dashed line will not distinguish between the two curves. The only way to reduce the width between the dashed lines would be to use a larger population N. To be quantitative, suppose that $p=e+\alpha D$, with $e=0.044$ and $\alpha=0.013$. At a dose of 0.25 Gy, $p=0.047$. For 10^5 persons, the constant curve (expected in the absence of radiation or below threshold) gives $m=4400\pm66$, while the linear curve gives $m=4730\pm69$. The two curves are distinguishable. At a dose of 0.01 Gy, m for the constant curve is still 4400 ± 66, while for the linear case it is 4410 ± 66. It is impossible to distinguish between the linear and constant curves. The problem of estimating cancer risks from low radiation doses is reviewed by Land (1980).

Whether the LNT model is correct or whether there is a threshold is a very important question. If the probability of acquiring a particular disease is p in a population N, the average number of persons with the disease is $m=Np$. If the LNT assumption is correct and the probability is proportional to dose, $p=\alpha D$, then the average number of people with the disease is

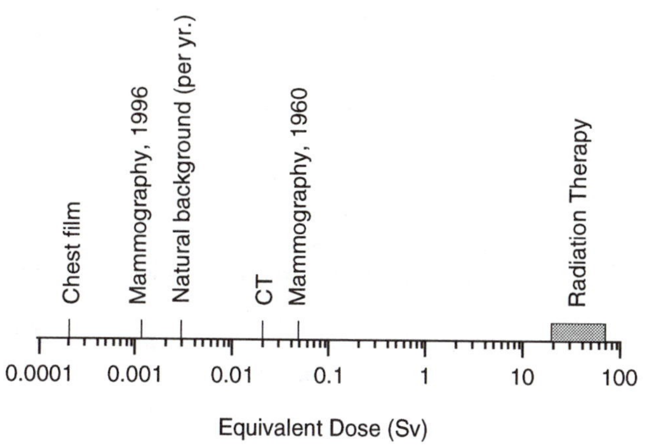

FIGURE 15.52. Various doses on a logarithmic scale. Natural background is per year; other doses are per exposure.

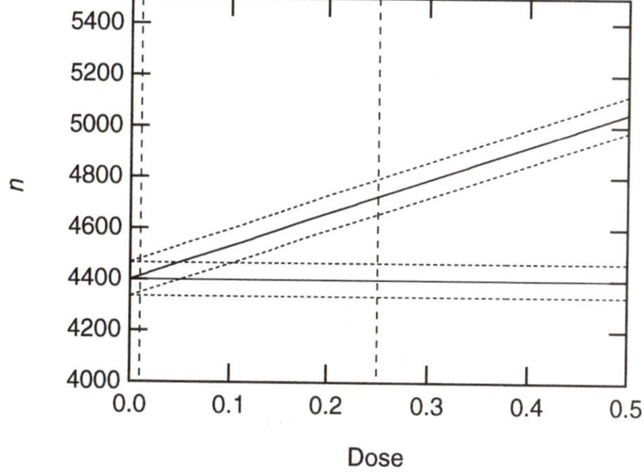

FIGURE 15.53. Plot of the average number of cases from a population N for a linear-no- threshold and a constant model. The dashed lines represent the mean ±1 standard deviation.

$$m = \alpha N D. \tag{15.43}$$

The product ND, expressed in person·Sv, is widely used in radiation protection, but it is meaningful only if the LNT assumption is correct. Small doses to very large populations can give fairly large values of m, assuming that the value of α determined at large doses is valid at small doses. If there is a threshold, then this can be a severe overestimate.

A number of investigators feel that there is evidence for a threshold dose, and that the linear–no-threshold model overestimates the risk [Kathren (1996); Kondo (1993)]. This may be particularly important for the case of radon, where remediation at fairly low radon levels has been proposed, but where studies of homes have shown a declining risk with increasing radon concentrations for small concentrations. Radon is produced naturally in many types of rock. It is a noble gas, but its radioactive decay products can become lodged in the lung. An excess of lung cancer has been well documented in uranium miners, who have been exposed to fairly high radon concentrations as well as high dust levels and tobacco smoke. Radon at much lower concentrations seeps from the soil into buildings and is claimed to contribute up to 55% of the exposure to the general population. Radon concentrations in the air are measured in the number of radioactive decays per second per cubic meter of air. One *becquerel* (Bq) is one decay per second. Figure 15.54 shows a study by B. L. Cohen (1995) that plots annual age-adjusted lung-cancer mortality rates in 1600 counties in the United States vs the average radon concentration measured in that county. The radon concentration is expressed as r/r_0, where r_0 is $37\ \mathrm{Bq\,m^{-3}}$ ($1.0\ \mathrm{pCi\,l^{-1}}$ in old units, which will be discussed in the next chapter). The upper two panels are for males, and the lower two are for females. The two panels on the right are corrected for the effects of smoking, using the radon-and-smoking model from BEIR Report V (1990). The dashed lines labeled *Theory* are based on the LNT model. The mortality rate falls with increasing radon concentration, though other studies have shown that it rises at radon concentrations higher than shown here. While epidemiological studies are difficult and can only be suggestive, Cohen has removed or addressed a number of "confounding" variables, and the effect remains. This is a possible example of *radiation hormesis*, a controversial hypothesis that low levels of radiation are actually beneficial. The proposed mechanism is that low radiation doses promote increased production of repair enzymes. Whether or not one considers this as evidence for radiation hormesis, it clearly suggests that the LNT model overestimates the risk at low doses.

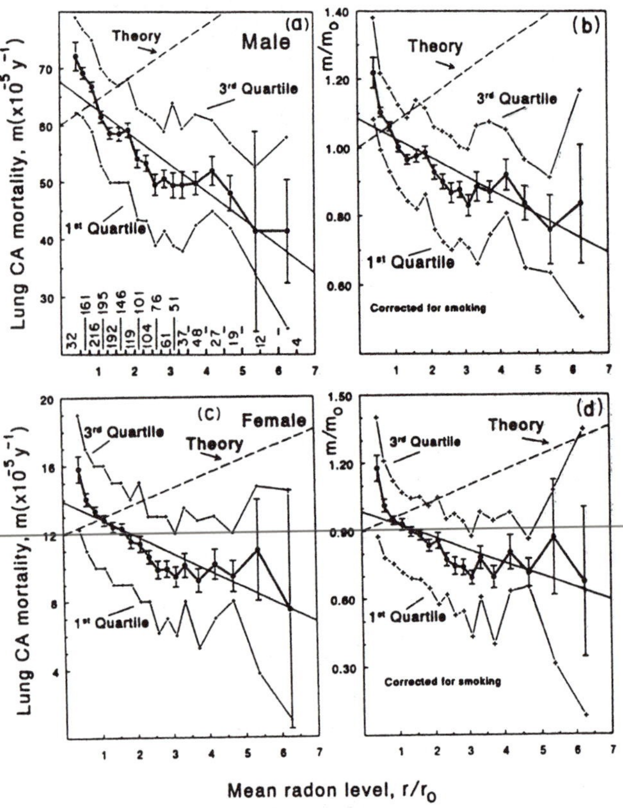

FIGURE 15.54. Lung-cancer mortality rates vs mean radon level in 1601 U.S. counties. Graphs (a) and (b) are for males; (c) and (d) are for females. Graphs (b) and (d) have been corrected for smoking levels. Error bars show the standard deviation of the mean. The meaning of radon level is discussed in the text. From Cohen (1995), Fig. 1. Used by permission of the Health Physics Society.

SYMBOLS USED IN CHAPTER 15

Symbol	Use	Units	First used on page
b	Thickness of slice being scanned	m	437
c	Velocity of light	$\mathrm{m\,s^{-1}}$	419
e	Electron charge	C	420
f	Fraction		431
f	General function to be reconstructed		436
f	Fraction of photons that interact or detective quantum efficiency		431
g	Incremental signal transfer function	$\mathrm{kg\,C^{-1}}$	422
h	Planck's constant	J s	418
j	Total angular momentum quantum number		417
k	Minimum signal-to-noise ratio		432
k_B	Boltzmann factor	$\mathrm{J\,K^{-1}}$	426

Symbol	Use	Units	First used on page
l	Orbital angular momentum quantum number		417
m	Mass	kg	420
m_e	Mass of electron	kg	424
m	Mean number		441
m	Number of scans		437
n	Principal quantum number		418
n	Number of slices in a scan		437
n	Number of moles of a substance		419
p	Probability		442
q	Charge	C	420
r	Number of targets		442
v	Voltage difference	V	424
w	Width of picture element		437
w_i	Mass fraction of constituent		428
x	Photon energy/rest mass energy		424
x	Number of hits		442
(x,y,z)	Coordinates	m	
A	Proportionality constant	$\text{C m}^2\text{ kg}^{-1}$	432
C	Constant	$\text{J}^{-1}\text{ m}^{-2}$	418
C	Capacitance	F	424
C_{in}	Exposure contrast		431
C_{out}	Brightness contrast		431
D	Absorbed dose	J kg^{-1} (Gy)	419
D'	Absorbed dose in one scan	Gy	438
D_0	Reciprocal of proportionality constant α	Gy	441
E	Energy	J or eV	418
E	Effective dose to an organ	Sv (J kg^{-1})	450
F	Projection (integral) of f along a path in some direction		436
G	Radiation chemical yield	mol J^{-1}	419
G	Large signal transfer factor	kg C^{-1}	422
H	Hounsfield (CT) unit		436
H	Dose equivalent	Sv (J kg^{-1})	450
H_T	Equivalent dose	Sv (J kg^{-1})	450
K_c	Collision kerma	J kg^{-1}	420
L	Length of object		432
N	Number		420
OD	Optical density		421
P	Probability		442
Q	Quality factor		450
S	Area	m^2	431

Symbol	Use	Units	First used on page
S	Surviving fraction		443
S_c	Collision stopping power		448
T	Kinetic energy	J or eV	418
T_0	Initial kinetic energy	J	420
T	Optical transmission		421
T	Temperature	K	426
W	Mean energy expended per ion pair formed	J or eV	420
W_R	Radiation weight factor		450
W_T	Target weight factor		450
X	Exposure	C kg^{-1}	420
Z	Atomic number		418
α	Integral of attenuation		434
α	Dose proportionality constant	Gy^{-1}	441
β	Coefficient		438
β	Squared dose proportionality constant	Gy^{-2}	441
γ	Fraction		438
γ	Film contrast		421
λ	Wavelength	m	419
μ, μ_{atten}	Attenuation coefficient	m^{-1}	428
μ_{en}	Energy absorption coefficient	m^{-1}	420
ν, ν_0	Frequency	s^{-1}	418
ρ	Density	kg m^{-3}	420
σ	Standard deviation		451
Φ	Particle fluence	m^{-2}	419
Φ_T	Particle fluence per unit energy interval	$\text{m}^{-2}\text{ J}^{-1}$	449
Ψ	Energy fluence	J m^{-2}	418

PROBLEMS

Section 15.1

15.1. Use Eqs. (14.3) and (15.2) to answer the following questions. Then compare your answers to values given in tables, such as those in the *Handbook of Chemistry and Physics*. What is the minimum energy of electrons striking a copper target that will cause the K x-ray lines to appear? What is the approximate energy of the K_α line? Repeat for iodine, molybdenum, and tungsten.

15.2. When tungsten is used for the anode of an x-ray tube, the characteristic tungsten K_α line has a wavelength of 2.1×10^{-11} m. Yet a voltage of 69 525 V must be applied to the tube before the line appears. Explain the discrepancy in terms of an energy-level diagram for tungsten.

15.3. Compare the shapes of the curves of Eqs. (15.3a) and (15.3b) for $h\nu_0 = 100$ keV, over the range 40–100 keV. Are the characteristics of Eq. (15.3b) reflected in the theoretical points of Fig. 15.5?

15.4. Express the formula for the thick-target x-ray energy fluence rate [Eq. (15.3a)] as $d\Psi/d\lambda$ and plot it.

Section 15.2

15.5. A beam of 0.08-MeV photons passes through a body of thickness L. Assume that the body is all muscle with $\rho = 1.0 \times 10^3$ kg m^{-3}. The energy fluence of the beam is Ψ J m^{-2}.

(a) What is the skin dose where the beam enters the body?

(b) Assume the beam is attenuated by an amount $e^{-\mu L}$ as it passes through the body. Calculate the average dose as a function of the fluence, the body thickness, and μ.

(c) What is the limiting value of the average dose as $\mu L \rightarrow 0$?

(d) What is the limiting value of the average dose as $\mu L \rightarrow \infty$? Does the result make sense?

15.6. The obsolete unit, the roentgen (R), is defined as 2.08×10^9 ion pairs produced in 0.001 293 g of dry air. (This is 1 cm^3 of dry air at standard temperature and pressure.) Show that if the average energy required to produce an ion pair in air is 33.7 eV (an old value), then 1 R corresponds to an absorbed dose of 8.69×10^{-3} Gy and that 1 R is equivalent to 2.58×10^{-4} C kg^{-1}.

15.7 During the 1930s and 1940s it was popular to have an x-ray fluoroscope unit in shoe stores to show children and their parents that shoes were properly fit. These marvelous units were operated by people who had no concept of radiation safety and aimed a beam of x rays upward through the feet and right at the reproductive organs of the children! A typical unit had an x-ray tube operating at 50 kVp with a current of 5 mA.

(a) What is the radiation yield for 50-keV electrons on tungsten? How much photon energy is produced with a 5-mA beam in a 30-s exposure?

(b) Assume that the x rays are radiated uniformly in all directions (this is not a good assumption) and that the x rays are all at an energy of 30 keV. (This is a very poor assumption.) Use the appropriate values for striated muscle from Appendix O to estimate the dose to the gonads if they are at a distance of 50 cm from the x-ray tube. [Your answer will be an overestimate. Actual doses to the feet were typically 0.014–0.16 Gy. Doses to the gonads would be less because of $1/r^2$. Two of the early articles pointing out the danger are Hempelmann, L.H. (1949), Potential dangers in the uncontrolled use of shoe fitting fluoroscopes, *New Engl. J. Med.* **241**: 335–337, and Williams, C.R. (1949), Radiation exposures from the use of shoe-fitting fluoroscopes, *New Engl. J. Med.* **241**: 333–335.]

Section 15.3

15.8. Rewrite Eq. (15.9) in terms of exponential decay of the viewing light and relate the optical density to the attenuation coefficient and thickness of the emulsion.

15.9. Derive the useful rule of thumb $\Delta(\text{OD}) = 0.43\gamma\, \Delta X/X$.

15.10. The atomic cross sections for the materials in a gadolinium oxysulfide screen for 50-keV photons are

Element	Cross section per atom (m^2)	A
Gd	1.00×10^{-25}	157
S	3.11×10^{-27}	32
O	5.66×10^{-28}	16

(a) What is the cross section per target molecule of GdO_2S?

(b) How many target molecules per unit area are there in a thickness ρdx of material?

(c) What is the probability that a photon interacts in traversing 1.2 kg m^{-2} of GdO_2S?

15.11. A dose of 1.74×10^{-4} Gy was estimated for part of the body just in front of an unscreened x-ray film. Suppose that a screen permits the dose to be reduced by a factor of 20. Calculate the skin dose on the other side of the body (the entrance skin dose) assuming 50-keV photons and a body thickness of 0.2 m. Ignore buildup, and assume that only unattenuated photons are detected.

15.12. Find an expression for photon fluence per unit absorbed dose in a beam of monoenergetic photons. Then find the photon fluence for 50-keV photons that causes a dose of 10^{-5} Gy in muscle.

15.13. A dose of 100 Gy might cause noticeable radiation damage in a sodium iodide crystal. How long would a beam of 100-keV photons have to continuously and uniformly strike a crystal of 1-cm^2 area at the rate of 10^4 photon s^{-1}, in order to produce this absorbed dose? For NaI $\mu_{en}/\rho = 0.1158$ m^2 kg^{-1}.

Section 15.4

15.14. Plot μ for lead, iodine, and barium from 10 to 200 keV.

15.15. Use a spreadsheet to make the following calculations. Consider a photon beam with 100 kVp.

(a) Use Eq. (15.3b) to calculate the photon fluence from a thick target at 1, 10, 20, 30, 40, 50, 60, 70, 80, 90, and 100 keV.

(b) The specific gravity of aluminum is 2.7. Make a table of the photon fluence at these energies emerging from 2 and 3 mm of aluminum. Compare the features of this table to Fig. 15.18.

(c) Use trapezoidal integration to show that the average photon energy is 44 keV after 2-mm filtration and 47 keV after 3-mm filtration.

(d) Repeat for 120 kVp and show that the average energies after the same filtrations are 52 and 55 keV.

Section 15.5

15.16. Suppose that two measurements are made: one of the combination of signal and noise, $y=s+n$, and one of just the noise n. One wishes to determine $s=y-n$.

(a) Find $s-\bar{s}$ in terms of y, $\bar{y}$, n, and $\bar{n}$.

(b) Show that if y and n are "uncorrelated," $\overline{(s-\bar{s})^2} = \overline{(y-\bar{y})^2} + \overline{(n-\bar{n})^2}$ and state the mathematical condition for being "uncorrelated."

(c) If y and n are Poisson distributed, under what conditions is the $\sqrt{2}$ factor of Ftn. 12 needed?

Section 15.7

15.17. A molybdenum target is used in special x-ray tubes for mammography. The electron energy levels in Mo are as follows:

K	20 000 eV	L_{I}:	2886 eV	M_{I}:	505 eV
		L_{II}:	2625 eV	M_{II}:	410 eV
		L_{III}:	2520 eV	M_{III}:	392 eV
				M_{IV}:	230 eV
				M_{V}:	227 eV

What is the energy of the K_α line(s)? The K_β line(s) (defined in Fig. 15.2)?

15.18. As a simple model for mammography, consider two different tissues: a mixture of 2/3 fat and 1/3 water, with a composition by weight of 12% hydrogen, 52% carbon and 36% oxygen; and glandular tissue, composed of 11% hydrogen, 33% carbon, and 56% oxygen. The density of the fat and water combination is 940 kg m^{-3}, and the density of glandular tissue is 1020 kg m^{-3}. What is the attenuation in 1 mm of fat and in 1 mm of glandular tissue for 50-keV photons? For 30-keV photons?

Section 15.9

15.19. For a CT scanner with $\mu=18$ m^{-1}, $\Delta\mu/\mu = 0.005$, $L=30$ cm, and $w=0.1$ cm, and $b=1$ cm, calculate, for a scan time of 15 s,

(a) N,

(b) m,

(c) N', and

(d) The dose.

15.20. It is often said that the number of photons that must be detected in order to measure a difference in fluence with a certain resolution can be calculated from $N=(\Delta\Phi/\Phi)^{-2}$. (For example, if we want to detect a change in Φ of 1% we would need to count 10^4 photons.) Use Eq. (15.20) to make this statement more quantitative. Discuss the accuracy of the statement.

15.21. Spiral CT uses interpolation to calculate the projections at a fixed value of z before reconstruction. This has an effect on the noise. Let σ_0 be the noise standard deviation in the raw projection data and σ be the noise in the interpolated data. The interpolated signal, α is the weighted sum of two values: $\alpha = w\alpha_1 + (1-w)\alpha_2$.

(a) Show that the variance in α is $\sigma^2 = w^2\sigma_0^2 + (1-w)^2\sigma_0^2$. Plot σ/σ_0 vs w.

(b) Averaging over a 360° scan involves integrating uniformly over all weights:

$$\sigma^2 = \int_0^1 [w^2\sigma_0^2 + (1-w)^2\sigma_0^2]\,dw.$$

Find the ratio σ/σ_0.

15.22. An experimental technique to measure cerebral blood perfusion is to have the patient inhale xenon, a noble gas with $Z=54$, $A=131$ [Suess *et al.* (1995)]. The solubility of xenon is different in red cells than in plasma. The equation used is

$$(\text{arterial enhancement}) = \frac{5.15\,\theta_{\mathrm{Xe}}}{(\mu/\rho)_w/(\mu/\rho)_{\mathrm{Xe}}} C_{\mathrm{Xe}}(t),$$

where the arterial enhancement is in Hounsfield units, C_{Xe} is the concentration of xenon in the lungs (end tidal volume), and

$$\theta_{\mathrm{Xe}} = (0.011)(\mathrm{Hct}) + 0.10,$$

and Hct is the *hematocrit*: the fraction of the blood volume occupied by red cells. Discuss why the equation has this form.

Section 15.10

15.23. Expand $(1-e^{-m})^r$ using the binomial theorem or a Taylor's series to show that the asymptotic form of cell survival in single-hit target theory for large m is $P \sim re^{-m}$.

15.24. Use Equations (15.32), (15.34) and (15.35) to obtain expressions for dose vs number of tumor cells for a probability of cure of 50%.

Section 15.11

15.25. The percent depth dose is defined to be the ratio of the dose in tissue (or a phantom) at depth d to the dose at a depth d_m where the dose is a maximum. The dose changes with depth because of three factors: the inverse square law, exponential attenuation, and scattering. Consider the two cases shown in the figure. Ignore the scattering factor, which depends on the size of the field. The figure is from Khan (1994, p. 187).

(a) What is the percent depth dose for each case as a function of f, d, and d_m?

(b) Show that if changes in scattering are ignored,

$$\frac{P(d,f_2)}{P(d,f_1)}=\left(\frac{f_2+d_m}{f_1+d_m}\right)^2\left(\frac{f_1+d}{f_2+d}\right)^2=F$$

F is called the *Mayneord F factor*.

(c) Show that if $f_2>f_1$ then $F>1$.

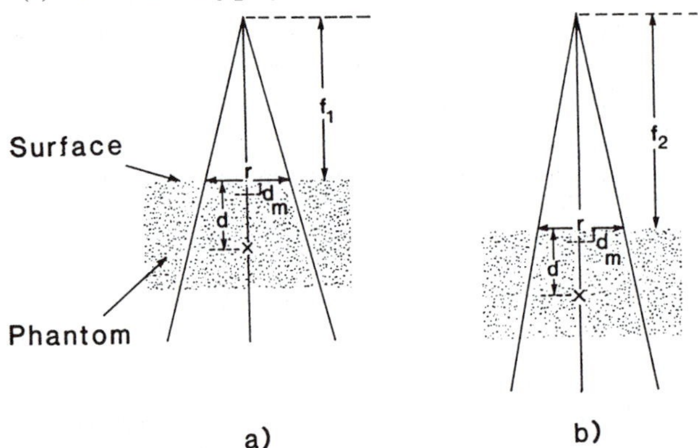

Figure from Khan (1994), p. 187. Used by permission.

Section 15.12

15.26. Calculate $(S_c/\rho)_g^w$ in argon for 0.1-, 1.0- and 10-MeV electrons. The values of S/rho for argon at these energies are 2.918, 1.376, and 1.678 cm g.

15.27. An ion chamber contains 10 cm^3 of air at standard temperature and pressure. Find q vs D for 0.5-MeV electrons.

Section 15-13

15.28. Suppose that the probability p per year of some event (death, mutations, cancer, etc.) consists of a spontaneous component S and a component proportional to the dose of something else, D:

$$p=S+AD.$$

The dose may be radiation, chemicals, sunlight, etc. Investigations of women given mammograms showed that if p is the probability of acquiring breast cancer, $S=1.91\times10^{-3}$ and $A=4\times10^{-4}$ per Gy. How many women had to be studied to distinguish between $A=0$ and the value above if $D=2$ Gy? If $D=10^{-2}$ Gy?

REFERENCES

AAPM Report 29 (1990). *Equipment Requirements and Quality Control for Mammography.* College Park, MD, American Association of Physicists in Medicine Report No. 29.

Alberts, B., D. Bray, J. Lewis, M. Raff, K. Roberts, and J. D. Watson (1983). *Molecular Biology of the Cell*, 1st ed. New York, Garland.

Attix, F. H. (1986). *Introduction to Radiological Physics and Radiation Dosimetry.* New York, Wiley.

BEIR Report V (1990). *Health Effects of Exposure to Low Levels of Ionizing Radiation.* Washington, D.C., National Academy Press. Committee on the Biological Effects of Ionizing Radiations Report V.

Birch, R., and M. Marshall (1979). Computation of bremsstrahlung x-ray spectra and comparison with spectra measured with a Ge(Li) detector. *Phys. Med. Biol.* **24**: 505–517.

Boice, J. D., Jr. (1996). Risk estimates for radiation exposures. In W. R. Hendee and F. M. Edwards, eds. *Health Effects of Exposure to Low-Level Ionizing Radiation.* Bristol, Institute of Physics.

Broad, W. J. (1980). Riddle of the Nobel debate. *Science* **207**: 37–38.

Brooks, R. A., and G. DiChiro (1976). Principles of computer-assisted tomography in radiographic and radioisotope imaging. *Phys. Med. Biol.* **21**: 689–732; Statistical limitations in x- ray reconstructive tomography. *Med. Phys.* **3**: 237–240.

Cho, Z.-H., J. P. Jones, and M. Singh (1993). *Foundations of Medical Imaging.* New York, Wiley-Interscience.

Cohen, B. L. (1995). Test of the linear–no threshold theory of radiation carcinogenesis for inhaled radon decay products. *Health Phys.* **68**(2): 157–174.

Cohen, L. (1993). History and future of empirical and cell kinetic models for risk assessment in radiation oncology. In B. Paliwal, *et al.*, eds. *Prediction of Response in Radiation Therapy: Radiosensitivity and Repopulation.* Woodbury, NY, AIP for the American Association of Physicists in Medicine.

Cormack, A. M. (1980). Nobel award address: Early two-dimensional reconstruction and recent topics stemming from it. *Med. Phys.* **7**(4): 277–282.

Dainty, J. C., and R. Shaw (1974). *Image Science.* New York, Academic.

DiChiro, G., and R. A. Brooks (1979). The 1979 Nobel prize in physiology and medicine. *Science.* **206**(30): 1060–1062.

Doi, K., H. K. Genant, and K. Rossman (1976). Comparison of image quality obtained with optical and radiographic magnification techniques in fine-detail skeletal radiography: Effect of object thickness. *Radiology.* **118**: 189–195.

Donaldson, S. S. (1995). Organ preservation in rhabdomyosarcoma: Fact or fiction? *Radiology* **197**(Suppl.): 35–36.

Duncan, W. (1994). An evaluation of the results of neutron therapy trials. *Acta Oncolog.* **33**(3): 299–306. This issue of the journal is devoted to fast-neutron therapy.

Habrand, J. L., P. Schlienger, L. Schwartz, D. Pontvert, C. Lenir-Cohen-Solal, S. Helfre, C. Haie, A. Mazal, and J. M. Cosset. (1995). Clinical applications of proton therapy. Experiences and ongoing studies. *Radiat. Environment. Biophys.* **34**(1): 41–44.

Hall, E. J. (1994). *Radiobiology for the Radiologist.* 4th ed. Philadelphia, Lippincott.

Hendee, W. R., and R. Ritenour (1992). *Medical Imaging Physics.* St. Louis, Mosby–Year Book.

Holmes, T. W., T. R. Mackie, and P. Reckwerdt (1995). An iterative filtered backprojection inverse treatment planning algorithm for tomotherapy. *Int. J. Radiation Oncolog. Biol. Phys.* **32**(4): 1215–1225.

Hounsfield, G. N. (1980). Nobel award address: Computed medical imaging. *Med. Phys.* **7**(4): 283–290.

Hubbell, J. H., and S. M. Seltzer (1996). *Tables of X-Ray Mass Attenuation Coefficients and Mass Energy-Absorption Coefficients 1 keV to 20 MeV for Elements Z=1 to 92 and 48 Additional Substances of Dosimetric Interest.* National Institute of Standards and Technology Report NISTIR 5632—Web Version. Available at http://physics.nist.gov/PhysRefData/XrayMassCoef/cover.html

ICRU Report 31 (1979). *Average Energy to Produce an Ion Pair.* Bethesda, MD, International Commission on Radiation Units and Measurements.

ICRU Report 33 (1980, 1992). *Radiation Quantities and Units.* Bethesda, MD, International Commission on Radiation Units and Measurements.

ICRU Report 37 (1984). *Stopping Powers for Electrons and Positrons.* Bethesda, MD, International Commission on Radiation Units and Measurements.

ICRU Report 41 (1986). *Modulation Transfer Function of Screen-Film Systems.* Bethesda, MD, International Commission on Radiation Units and Measurements.

ICRU Report 46 (1992). *Photon, Electron, Proton and Neutron Interaction Data for Body Tissues.* Bethesda, MD, International Commission on Radiation Units and Measurements.

ICRU Report 54 (1996). *Medical Imaging—The Assessment of Image Quality.* Bethesda, MD, International Commission on Radiation Units and Measurements.

Kalender, W. A., and A. Polacin (1991). Physical performance characteristics of spiral CT scanning. *Med. Phys.* **18**(5): 910–915.

Kathren, R. L. (1996). Pathway to a paradigm: The linear nonthreshold dose-response model in historical context: The American Academy of Health Physics 1995 Radiology Centennial Hartman Oration. *Health Phys.* **70**(5): 621–635.

Khan, F. M. (1994). *The Physics of Radiation Therapy.* 2nd ed. Baltimore, MD, Williams and Wilkins.

Kondo, S. (1993). *Health Effects of Low-Level Radiation.* Osaka, Japan, Kinki University Press. English translation: Madison, WI, Medical Physics.

Land, C. E. (1980). Estimating cancer risks from low doses of ionizing radiation. *Science* **209**: 1197–1203.

Macovski, A. (1983). *Medical Imaging Systems.* Englewood Cliffs, NJ, Prentice-Hall.

Metz, C. E., and K. Doi (1979). Transfer function analysis of radiographic imaging systems. *Phys. Med. Biol.* **24**(6): 1079–1106.

Moulder, J. E., and J. D. Shadley (1996). Radiation interactions at the cellular and tissue levels. In W. R. Hendee and F. M. Edwards, eds. *Health Effects of Exposure to Low-Level Ionizing Radiation.* Bristol, Institute of Physics.

NCRP Report 94 (1987). *Exposure of the Population in the United States and Canada from Natural Background Radiation.* Bethesda, MD, National Council of Radiation Protection and Measurements.

NCRP Report 100 (1989). *Exposure of the U.S. Population from Diagnostic Medical Radiation.* Bethesda, MD, National Council of Radiation Protection and Measurements.

NCRP Report 104 (1990). *The Relative Biological Effect of Radiations of Different Quality.* Bethesda, MD, National Council of Radiation Protection and Measurements.

Orton, C. (1997). Fractionation: Radiobiological principles and clinical practice. Chap. 11 in F. Khan and R. A. Potish, eds. *Treatment Planning in Radiation Oncology.* Baltimore, MD, Williams and Wilkins.

Platzman, R. L. (1961). Total ionization in gases by high-energy particles: An appraisal of our understanding. *Intl. J. Appl. Rad. Isotopes* **10**: 116–127.

Richman, C. (1981). The physics of cancer therapy with negative pions. *Med. Phys.* **8**: 273–291.

Shabashon, L. (1996). Radiation interactions: Physical and chemical effects. In W. R. Hendee and F. M. Edwards, eds. *Health Effects of Exposure to Low-Level Ionizing Radiation.* Bristol, Institute of Physics.

Steel, G. G. (1996). From targets to genes: a brief history of radiosensitivity. *Phys. Med. Biol.* **41**(2): 205–222.

Suess, C., A. Polacin, and W. A. Kalender (1995). Theory of xenon/computed tomography cerebral blood flow methodology. In M. Tomanaga, A. Tanaka, and H. Yonas, eds. *Quantitative Cerebral Blood Flow Measurements Using Stable Xenon/CT: Clinical Applications.* Armonk, NY, Futura.

Suit, H., and M. Urie (1992). Proton beams in radiation therapy. *J. Natl. Cancer Inst.* **84**(3): 155–164.

Wagner, R. F. (1977). Toward a unified view of radiological imaging systems. Part II: Noisy images. *Med. Phys.* **4**(4): 279–296.

Wagner, R. F. (1983). Low contrast sensitivity of radiologic, CT, nuclear medicine, and ultrasound medical imaging systems. *IEEE Trans. Med. Imaging* **MI-12**(3): 105–121.

Wagner, R. F., K. E. Weaver, E. W. Denny, and R. G. Bostrom (1974). Toward a unified view of radiological imaging systems. Part I: Noiseless images. *Med. Phys.* **1**(1): 11–24.

Yaffe, M. J. and J. A. Rowlands (1997). X-ray detectors for digital radiography, *Phys. Med. Biol.* **42**(1): 1–40.

CHAPTER 16

Nuclear Physics and Nuclear Medicine

Each atom contains a nucleus about 100 000 times smaller than the atom. The nuclear charge determines the number of electrons in the neutral atom and hence its chemical properties. The nuclear mass determines the mass of the atom. For a given nuclear charge there can be a number of nuclei with different masses or *isotopes*. If an isotope is unstable, it transforms into another nucleus through *radioactive decay*.

In this chapter we will consider some of the properties of radioactive nuclei and their use for medical imaging and for treatment, primarily of cancer.

Four kinds of radioactivity measurements have proven useful. The first involves no administration of a radioactive substance to a patient. Rather a sample from the patient (usually blood) is mixed with a radioactive substance in the laboratory, and the resulting chemical compounds are separated and counted. This is the basis of various *competitive binding assays*, such as those for measuring thyroid hormone and the availability of iron-binding sites. The most common competitive binding technique is called *radioimmunoassay*. A wide range of proteins are measured in this manner.

In the second kind of measurement, radioactive tracers are administered to the patient in a way that allows the volume of a compartment within the body to be measured. Examples of such compartments are total body water, plasma volume, and exchangeable sodium. Time-dependent measurements can be made, such as red-blood-cell survival and iron and calcium kinetics. Counting is of the whole body or of blood or urine samples drawn at different times after administration of the isotope.

For the third class of measurements, a *gamma camera* generates an image of an organ from radioactive decay of a drug that has been administered and taken up by the organ. These images are often made as a function of time.

The fourth class is an extension of these in which tomographic reconstructions of body slices are made. These include *single-photon emission computed tomography* and *positron emission tomography*.

Radioactive isotopes are also used for therapy. The patient is given a radiopharmaceutical that is selectively absorbed by a particular organ (e.g., radioactive iodine for certain thyroid diseases). The isotope emits charged particles that lose their energy within a short distance, thereby giving a high dose to the target organ. Isotopes are also used in self-contained implants for *brachytherapy*.

The first four sections introduce some of the nuclear properties that are important: size, mass, and the modes of radioactive decay and the amount of energy released. It is important to know the dose to the patient from a nuclear medicine procedure, and a standard technique for calculating it has been developed by the Medical Internal Radiation Dose (MIRD) Committee of the Society of Nuclear Medicine. Sections 16.5–16.8 show the steps in making these calculations. Section 16.9 describes some of the pharmacological considerations in selecting a suitable isotope, and Sec. 16.10 provides a sample dose calculation using this technique. Section 16.11 extends the dose discussion to the special considerations when an isotope emits Auger electrons and is bound to the DNA in the cell nucleus.

The next few sections describe various ways of forming images. Section 16.12 describes the gamma camera, and Sec. 16.13 extends this to single-photon emission tomography. Section 16.14 describes positron emission tomography, which can be done with special equipment when positron sources are available in the hospital.

Radiotherapy is described in Sec. 16.15, including both the relatively common brachytherapy and the less common injection of isotopes that target particular organs.

The final section describes the nuclear decay of radon and some of the considerations in calculating the dose and the risk to the general population. It supplements the material that was introduced in Sec. 15.13.

16.1. NUCLEAR SYSTEMATICS

An atomic nucleus is composed of Z protons and $N=A-Z$ neutrons. We call Z the *atomic number* and A the *mass number*. Neutrons and protons have very similar properties, as can be seen from Table 16.1. Therefore, they are classed as two different charge states of one particle, the *nucleon*.

The *rest energy* listed in Table 16.1 is the rest mass times the square of the speed of light. One can show using special relativity that the total energy E of an object with rest mass m_0 is related to its speed v and kinetic energy T by

$$E=\frac{m_0c^2}{(1-v^2/c^2)^{1/2}}=m_0c^2+T. \quad (16.1)$$

In this equation c is the velocity of light. Table 16.1 also lists the rest energy m_0c^2. The energy and mass of both the proton and neutral hydrogen atom are given; the distinction will be important later.

It is customary to specify a nucleus by a symbol such as the following for carbon ($Z=6$, $N=6$, $A=12$):

$${}^A_Z\mathrm{C} \text{ or } {}^{12}_{6}\mathrm{C}.$$

The mass number used to be written as a superscript on the right; however, this becomes confusing if the ionization state must also be specified. It is now customary to leave the right-hand side of the symbol for the chemists to adorn.

Different kinds of the same element with different numbers of neutrons are called *isotopes*. Another isotope of carbon is ^{11}C, which has five neutrons.

The sizes of atoms are roughly constant as one goes through the periodic table, with exceptions as shells are filled. On the other hand, the size of nuclei grows steadily through the periodic table. The nuclear radius R and atomic mass number are related by

$$R=R_0A^{1/3}. \quad (16.2)$$

The precise value for R_0 depends on how the nuclear radius is measured. If it is measured from the charge distribution, then

$$R_0=1.07\times10^{-15}\text{ m}. \quad (16.3)$$

Figure 16.1 shows how nuclear radii grow systemati-

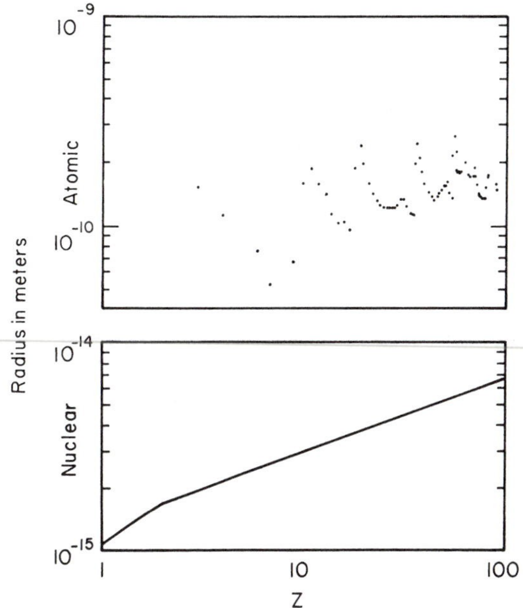

FIGURE 16.1. Plot of atomic radius and nuclear radius vs atomic number, showing the relative constancy of the atomic radius and the systematic increase of nuclear radius. Shell effects in atomic radii are quite pronounced; slight shell effects in the nuclear radius are not shown. [Atomic data are from Table 7b–3 of *The American Institute of Physics Handbook*, New York, McGraw-Hill, 1957. Nuclear radii are from Eq. (16.2), using the average atomic mass to estimate A from Z.]

cally, while atomic radii do not change appreciably with A (although they do change dramatically as shells close). The constancy of atomic size results from two competing effects: as Z increases the outer electrons have a larger value of the principal quantum number n. On the other hand, the larger charge means that Coulomb attraction makes the orbit radius smaller for a given n.

The $A^{1/3}$ dependence in the nuclear case means that the nuclear density is independent of A. To see this, note that the volume of a nucleus (assumed to be spherical) is $4\pi R^3/3=4\pi R_0^3A/3$. Since the mass and volume are both proportional to A, the density is constant. This implies that nucleons can get only so close to one another, and that as

TABLE 16.1. *Properties of nucleons, the electron, and the neutral hydrogen atom.*

Property	Neutron	Proton	Electron	H atom
Mass[a]	1.008 664 704	1.007 276 47	0.000 548 579 9	1.007 825 035
Charge[b]	0	$+e$	$-e$	0
Rest energy m_0c^2 (MeV)	939.565	938.272	0.5110	938.783
Half-life	≈12 min	Stable	Stable	Stable
Spin	$\frac{1}{2}$	$\frac{1}{2}$	$\frac{1}{2}$	...

[a]1 u is 1 mass unit; the mass of neutral ^{12}C is 12.000 000 0 u by definition; 1 u$=1.660\,540\times10^{-27}$ kg.
[b]$e=1.602\,177\times10^{-19}$ C.

more are added, the nuclear volume increases. This constant density is the same effect we see in the aggregation of atoms in a crystal or a drop of water.

Scattering experiments measure the force between two nucleons. At large distances, there is no force between two neutrons or between a neutron and a proton. (Between two protons, of course, there is Coulomb repulsion.) As two nucleons are brought close together, a strong attractive "nuclear" force exists; at still closer distances, the nuclear force becomes repulsive.

If we look at the nuclei that are stable against radioactive decay and are therefore found in nature, we find that for light elements, $Z=N$. As Z increases, the number of neutrons becomes greater than Z; this can be seen in Fig. 16.2.

From Eq. (16.1), it can be seen that when an object is at rest, its total energy (which is its internal energy) is related to its rest mass by

$$E=m_0c^2. \tag{16.4}$$

The measurement of nuclear masses has provided one way to determine nuclear energies. It is necessary to supply energy to a stable nucleus to break it up into its constituent nucleons (or else it would not be stable). The *binding energy* (BE) of the nucleus is the total energy of the constituent nucleons minus the energy of the nucleus:

$$\text{BE}=Zm_pc^2+(A-Z)m_nc^2-m_{\text{nucl}}c^2. \tag{16.5}$$

It represents the amount of energy that must be added to the nucleus to separate it into its constituent neutrons and protons.

Suppose we add Zm_ec^2 to the first term. Then we have the energy of Z protons plus the energy of Z electrons. Except for the binding energy of each electron, this is the same as the mass of Z neutral hydrogen atoms, which we call M_pc^2. Similarly, we can add the mass of Z electrons to $m_{\text{nucl}}c^2$ and neglect the electron binding energy to obtain $M_{\text{atom}}c^2$. Capital M represents the mass of a neutral atom, while m stands for the mass of a bare nucleus. For the neutron, $m=M$. In Eq. (16.5), we can add Zm_ec^2 to the first term and add $-Zm_ec^2$ to the last term, to obtain the binding energy in terms of the masses of the corresponding neutral particles:

$$\text{BE}=ZM_pc^2+(A-Z)M_nc^2-M_{\text{atom}}c^2. \tag{16.6}$$

This is fortunate, because neutral masses (or those for ions carrying one or two charges) are the quantities actually measured in mass spectroscopy.

Masses are measured in *unified mass units* u, defined so that the mass of neutral ^{12}C is exactly 12 u. Carbon is used for the standard because hydrocarbons can be made in combinations to give masses close to any desired mass. The carbon standard replaced one based on the naturally occurring mixture of oxygen isotopes in the early 1960s. (One of the troubles with the earlier standard was that the relative abundance of the various oxygen isotopes varies with time and with location on the earth.) The earlier unit was called the atomic mass unit, amu. One still finds confusion in the

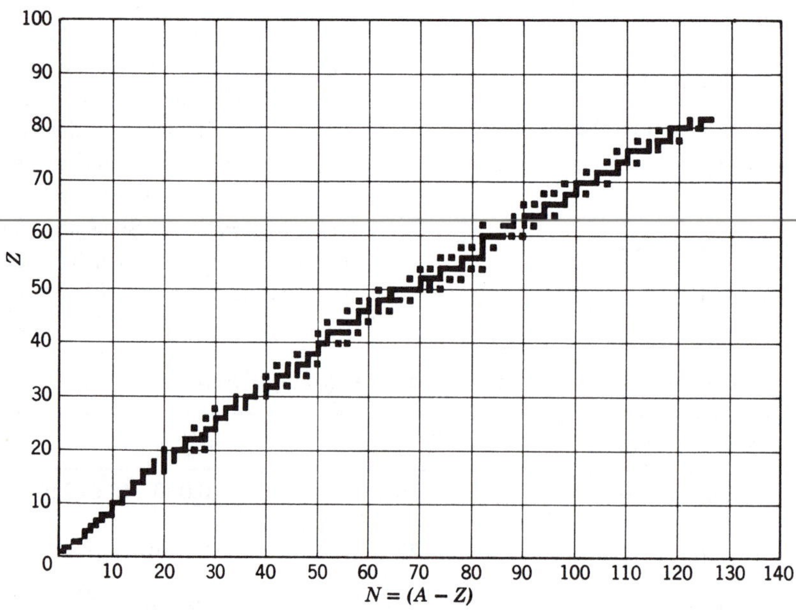

FIGURE 16.2. Stable nuclei. Solid squares represent nuclei which are stable and are found in nature. From Robert Eisberg and Robert Resnick, *Quantum Physics of Atoms, Molecules, Solids, Nuclei and Particles*, 2nd ed. New York, Wiley, 1985 p. 524. Reproduced by permission of John Wiley & Sons.

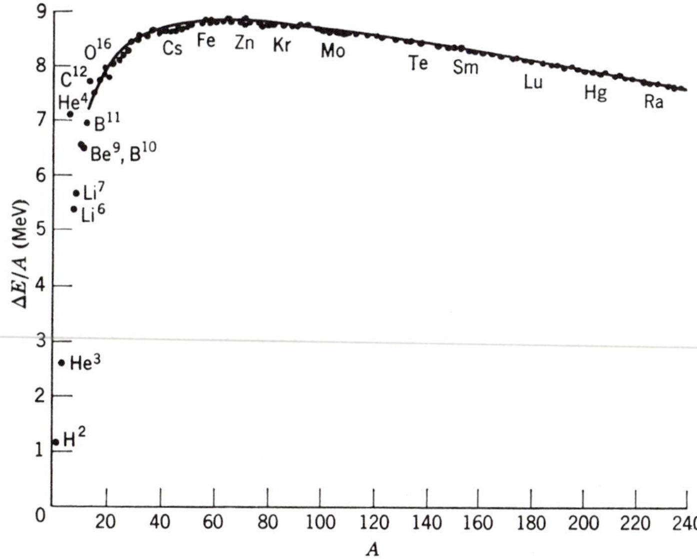

FIGURE 16.3. The average binding energy per nucleon for stable nuclei. From Robert Eisberg and Robert Resnick, *Quantum Physics of Atoms, Molecules, Solids, Nuclei and Particles*, 2nd ed. New York, Wiley, 1985, p. 524. Reproduced by permission of John Wiley & Sons.

literature about which standard is being used, and the carbon standard is sometimes called an amu.

One mass unit is related to the kilogram, the joule, and the electron volt by

$$1 \ \text{u} = 1.660\,54 \times 10^{-27} \ \text{kg},$$

$$(1 \ \text{u})(c^2) = \begin{cases} 1.492\,41 \times 10^{-10} \ \text{J} \\ 931.49 \ \text{MeV} \end{cases}. \tag{16.7}$$

If one plots the binding energy per nucleon versus mass number as in Fig. 16.3, one sees that the binding energy per nucleon has a maximum near $A=60$, and that the average binding energy (except for light elements) is about 8 MeV per nucleon. For less stable nuclei on either side of the stable line plotted in Fig. 16.2, the binding energy is less than that for the nuclei shown here.

The maximum near $A=60$ is what makes both fission and fusion possible sources of energy. A heavy nucleus with A near 240 can split roughly in half, giving two fission products. Since the nucleons in each of the products are more tightly bound on the average than in the original nucleus, energy is released. This energy difference comes almost entirely from the Z^2 dependence of the Coulomb repulsion of the protons in the nuclei. In fusion, two nuclei of very low A combine to give a nucleus of higher A, for which the binding energy per nucleon is greater.

16.2. NUCLEAR DECAY: DECAY RATE AND HALF-LIFE

If a nucleus has more energy than it would if it were in its ground state, it can decay. If the nucleus has sufficient energy, it can emit a proton, neutron, or cluster of nucleons (α particle, deuteron, etc.). When a nucleus has enough excitation energy to decay by nuclear emission it usually does so in such an extremely short time that the nuclei could never be introduced in the body after they were produced. An exception is the decay of a few elements near the upper end of the periodic table. They are found in nature, either because their lifetimes are very long or because they are formed as the result of some other decay process that has a long lifetime.

If a nucleus has just a small amount of excess energy, it emits a γ ray, a photon analogous to the x ray or visible photons emitted by an excited atom. Another process that can occur is the emission of a positive or negative electron, with the conversion of a proton to a neutron, or vice versa. This is called β decay. γ and β decay will be described in detail in the next two sections.

Each excited nucleus has a probability $\lambda \, dt$ of decaying in time dt. When there are N nuclei present, the average number decaying in time dt is[1]

$$-dN = N\lambda \ dt.$$

The rate of change of N is, therefore,

$$\frac{dN}{dt} = -\lambda N.$$

This leads to the familiar exponential survival of Chap. 2:

$$N = N_0 e^{-\lambda t}.$$

[1]The decay constant is called λ in this chapter to conform to the usage in nuclear medicine.

The half-life $T_{1/2}$ is related to λ by Eq. (2.10):

$$T_{1/2}=\frac{0.693}{\lambda}. \tag{16.8}$$

16.3. GAMMA DECAY AND INTERNAL CONVERSION

When a nucleus is in an excited state, it can lose energy by photon emission. The energy levels of the nucleus are characterized by certain quantum numbers, and γ emission is subject to selection rules analogous to those for x-ray emission by atoms. Half-lives for γ emission range from 10^{-20} to 10^{+8} s. Figure 16.4 shows an energy level diagram for ${}^{99}_{43}$Tc (technetium), an isotope widely used in nuclear medicine, along with some tabular material that we will need as we progress through this chapter. There are two important levels to consider in ${}^{99}_{43}$Tc. The ground state is not stable but decays by β decay, considered in the next section. However, its decay rate is so small (half-life of 2.12×10^5 years) that we can ignore its decay. There is a level at an excitation of 0.143 MeV above the ground state that has a half-life of 6 h for γ decay. This is an unusually long half-life; when the half-life is this long, we call it a *metastable state* and denote it by ^{99m}Tc. Looking at the table labeled "input data," we see that there are two modes of decay of the nucleus from this state. The first is the emission of a 0.0021-MeV γ ray followed by a 0.1405-MeV γ ray. It occurs 0.986 of the time. The other possibility is the emission of one γ ray, of energy 0.1426 MeV, which happens in 0.014 of the decays.

The notations $E3$, $M1$, and $M4$ in the last column of the input data refer to the selection rules and quantum number changes in the transitions. We will not need the details, although they can be used to estimate the decay rates. The labels $E1$ and $E2$ are called *electric dipole* and *electric quadrupole*, respectively, $M1$ means *magnetic dipole*, and so forth.

Whenever a nucleus loses energy by γ decay, there is a competing process called *internal conversion*. The energy to be lost in the transition, E_γ, is transferred directly to a bound electron (usually a K or L electron), which is then ejected with a kinetic energy

$$T=E_\gamma-B, \tag{16.9}$$

where B is the binding energy of the electron.

The *mean number per disintegration* in the table of *input data* is the mean number of times that the indicated *transi-*

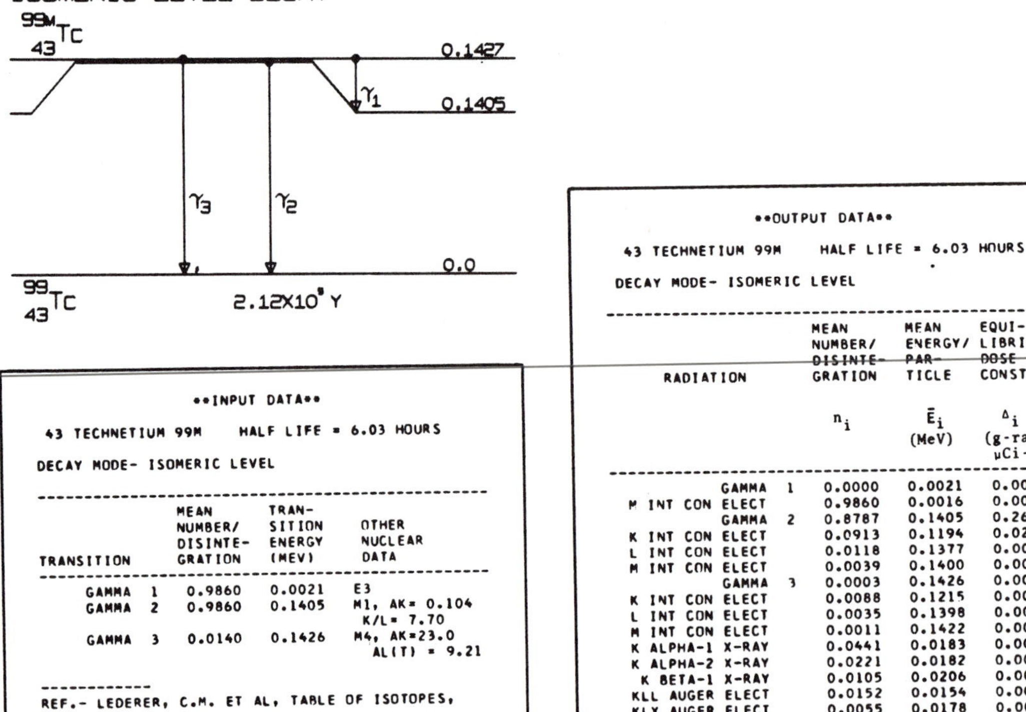

INPUT DATA

43 TECHNETIUM 99M   HALF LIFE = 6.03 HOURS

DECAY MODE- ISOMERIC LEVEL

TRANSITION	MEAN NUMBER/ DISINTE- GRATION	TRAN- SITION ENERGY (MEV)	OTHER NUCLEAR DATA
GAMMA 1	0.9860	0.0021	E3
GAMMA 2	0.9860	0.1405	M1, AK= 0.104 K/L= 7.70
GAMMA 3	0.0140	0.1426	M4, AK=23.0 AL(T) = 9.21

REF.- LEDERER, C.M. ET AL, TABLE OF ISOTOPES, 6TH ED.
LEGRAND, J. ET AL, INT. J. APPL. RAD. AND ISOTOPES 21, 139 (1970).

OUTPUT DATA

43 TECHNETIUM 99M   HALF LIFE = 6.03 HOURS

DECAY MODE- ISOMERIC LEVEL

RADIATION	MEAN NUMBER/ DISINTE- GRATION n_i	MEAN ENERGY/ PAR- TICLE $\bar{E}_i$ (MeV)	EQUI- LIBRIUM DOSE CONSTANT Δ_i (g-rad/μCi-h)
GAMMA 1	0.0000	0.0021	0.0000
M INT CON ELECT	0.9860	0.0016	0.0035
GAMMA 2	0.8787	0.1405	0.2630
K INT CON ELECT	0.0913	0.1194	0.0232
L INT CON ELECT	0.0118	0.1377	0.0034
M INT CON ELECT	0.0039	0.1400	0.0011
GAMMA 3	0.0003	0.1426	0.0001
K INT CON ELECT	0.0088	0.1215	0.0022
L INT CON ELECT	0.0035	0.1398	0.0010
M INT CON ELECT	0.0011	0.1422	0.0003
K ALPHA-1 X-RAY	0.0441	0.0183	0.0017
K ALPHA-2 X-RAY	0.0221	0.0182	0.0008
K BETA-1 X-RAY	0.0105	0.0206	0.0004
KLL AUGER ELECT	0.0152	0.0154	0.0005
KLX AUGER ELECT	0.0055	0.0178	0.0002
LMM AUGER ELECT	0.1093	0.0019	0.0004
MXY AUGER ELECT	1.2359	0.0004	0.0011

FIGURE 16.4. Energy levels and decay data for the isotope ^{99m}Tc. The various features are discussed in the text. Reprinted from Dillman and Von der Lage (1974, p. 62), with permission of the publisher.

tion between energy levels takes place. In the table of *output data*, on the other hand, it means the mean number of times that *radiation is emitted*. To see the distinction, compare the γ_2 entries in the input data and output data of Fig. 16.4. The γ_2 transition from the 0.1405-MeV level to the ground state occurs 0.9860 times per disintegration. From the output data, we see that this transition takes place by emission of a photon (γ_2) 0.8787 times, by K internal conversion 0.0913 times, by L internal conversion 0.0118 times, and by M internal conversion 0.0039 times (the sum is 0.9857, which agrees with the value 0.9860 times per disintegration).

To calculate the mean number of times a radiation is emitted from the mean number of transitions, we need numbers f that are the fraction of transitions involving a particular radiation. The fraction associated with γ decay is f_r, the fraction with K internal conversion is f_K, and so forth. The sum of all these fractions is unity:

$$f_r+f_K+f_L+f_M+\cdots=1. \tag{16.10}$$

Often these fractions are not listed in the literature. Instead, one finds the internal conversion coefficient α. For conversion of a K electron, it is $\alpha_K=f_K/f_r$. For the L shell it is $\alpha_L=f_L/f_r$, and so on. One also finds in the literature the ratio

$$\frac{K}{L}=\frac{f_K}{f_L}=\frac{\alpha_K}{\alpha_L}.$$

A useful empirical relationship is that $f_M \approx f_L/3$. In the input data of Fig. 16.4, α_K is called AK.

Once internal conversion has created a hole in the electronic structure of the atom, characteristic x rays and Auger electrons will be emitted. They must also be considered in calculating the total dose from the nuclear decay.

16.4. BETA DECAY AND ELECTRON CAPTURE

Nuclei that are not on the line of stability in Fig. 16.2 have greater internal energy and are susceptible to some kind of decay. They can lose energy by γ emission. In addition, nuclei above the line of stability have too many protons relative to the number of neutrons; nuclei below the line have relatively too many neutrons.

Two modes of decay allow a nucleus to approach the stable line. In beta (β^- or electron) decay, a neutron is converted into a proton. This keeps A constant, lowering N by one and raising Z by one. In positron (β^+) decay, a proton is converted into a neutron. Again A remains unchanged, Z decreases and N increases by 1. We find positron decay for nuclei above the line of stability and β decay for nuclei below the line. Figure 16.5 shows a portion of the line of stability, a line of constant A ($Z=A-N$), and the regions for β^+ and β^- decay.

We can plot the energy of the neutral atom for different nuclei along the line of constant A. Since there are one or two stable nuclei, there is some value of Z and N for which the energy is a minimum. The energy increases in either direction from this minimum. The first approximation to a curve with a minimum is a parabola, as shown in Fig. 16.6 for a nucleus of odd A.[2] When Z is too small, a neutron is converted to a proton by β^- decay. If Z is too large, a proton changes to a neutron by β^+ decay or electron capture (to be described below).

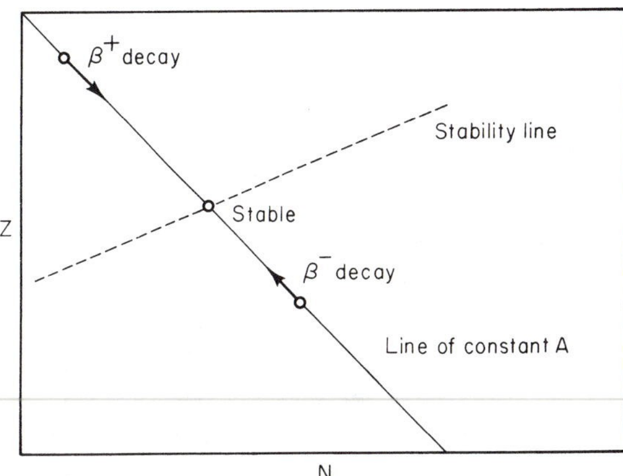

FIGURE 16.5. β decay and positron decay do not change A. They do change N and Z to bring the nucleus closer to the stability line.

When A is odd, there are an even number of protons and an odd number of neutrons (even–odd) or vice versa (odd–even). When we plot the energies of even-A nuclei, we find that the masses lie on two different parabolas (Fig. 16.7).

The one for which both Z and N are odd (odd–odd) has greater energy than the parabola for which both are even. The reason is that nucleons have lower energy when they are paired with one another in such a way that their spins are antiparallel. In the even–even case, the neutrons and the protons are all paired off and have this lower energy; in the odd–odd case there is an unpaired proton and an unpaired neutron and the energy is higher. As we change Z by one, we jump back and forth between the odd–odd and the even–even parabolas. For odd-A nuclei, either the neutrons are paired and one proton is not, or vice versa. There is always one unpaired nucleon as Z changes, so there is only one parabola.

The existence of the two parabolas means that there are usually (but not always) two stable nuclei with an odd–odd nucleus between them that can decay by either β^- or β^+ emission.

The emission of a β^- particle is accompanied by the emission of a neutrino (strictly speaking, an *antineutrino*):

$${}^{A}_{Z}X \rightarrow {}^{A}_{Z+1}Y+\beta^- + \bar{\nu}. \tag{16.11}$$

[2]This parabola and the general behavior of the binding energy with Z and A can be explained remarkably well by the semiempirical mass formula [Evans (1955, Chap. 11); Eisberg and Resnick, (1985, p. 528)].

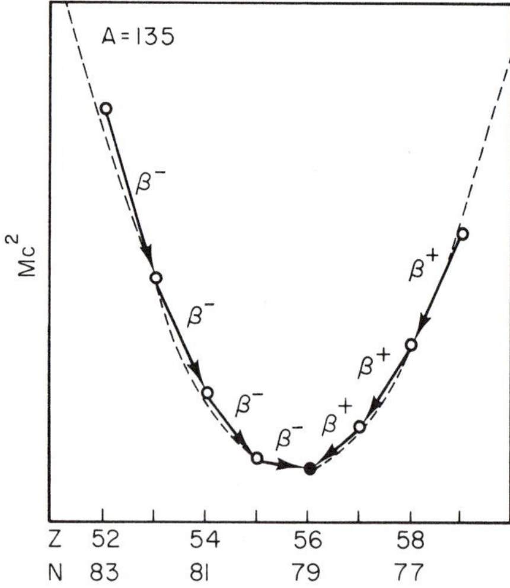

FIGURE 16.6. Energy of nuclei as a function of Z for an odd value of A ($A=135$). The only stable nucleus is $^{135}_{56}$Ba; nuclei of lower Z undergo β^- emission; those of higher Z undergo β^+ emission or electron capture.

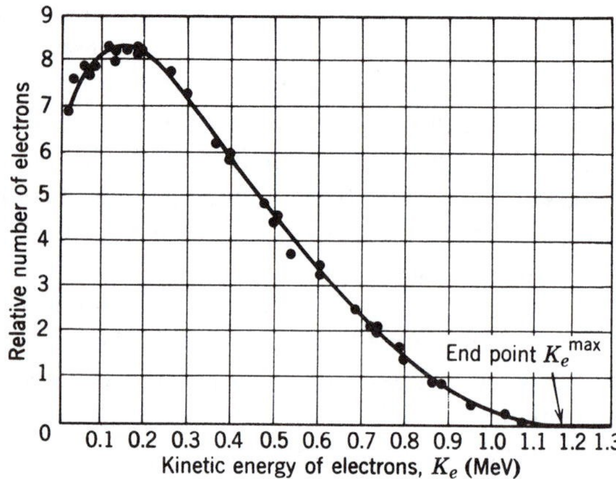

FIGURE 16.8. A typical spectrum of β particles. In this case it is for the β decay of $^{210}_{83}$Bi. From Robert Eisberg and Robert Resnick *Quantum Physics of Atoms, Molecules, Solids, Nuclei and Particles*. 2nd ed. New York, Wiley, 1985, p. 566. Reproduced by permission of John Wiley & Sons.

The antineutrino has no charge and no rest mass, so that like a photon, it travels with velocity c and its energy and momentum are related by $E=pc$. Yet it has spin $\frac{1}{2}$. Such a particle will not interact with matter very strongly, and neutrinos are quite difficult to detect. Nevertheless they have been detected through certain specific nuclear reactions that take place on the rare occasions when a neutrino does interact with a nucleus. What seemed originally to be an invention to conserve energy and angular momentum now has a strong experimental basis.

Suppose that β decay consisted of the ejection of only a β particle. If the original nucleus X were at rest,[3] then nucleus Y would recoil in the direction opposite the β particle to conserve momentum; the ratio of its velocity to that of the β particle would be given by their mass ratio. The recoil nucleus and the β particle would each have a definite fraction of the total energy available from the decay, and the β particles would all have the same energy. However, the β-particle energy spectrum is not a line spectrum but a continuum, as shown in Fig. 16.8. The different energies correspond to different angles of emission of the neutrino relative to the direction of the β particle. This kind of spectrum is characteristic of three bodies emerging from the reaction.

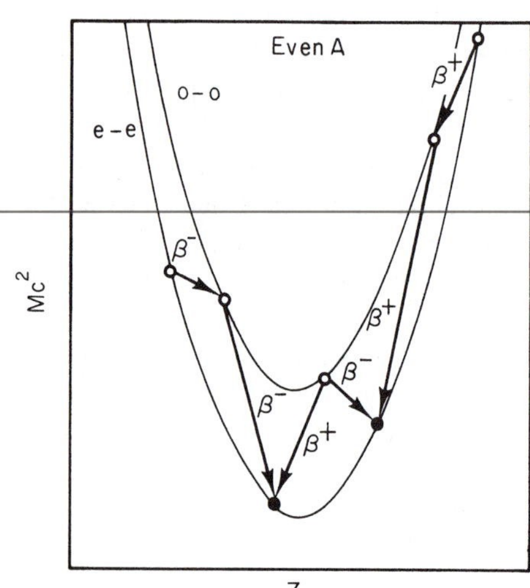

FIGURE 16.7. Energy of even-A nuclei as a function of Z. Nuclei with an odd number of protons and neutrons have higher energies than those with an even number of each. This makes it possible for the same nucleus to decay by either β^- or β^+ emission.

The total kinetic energy for the three emerging particle is

$$E_{\text{decay}}=m_{Z,A}c^2-m_{Z+1,A}c^2-m_ec^2.$$

If we add and subtract Zm_ec^2, the result is unchanged:

$$\begin{aligned}E_{\text{decay}}&=(m_{Z,A}c^2+Zm_ec^2)-(m_{Z+1,A}c^2+Zm_ec^2+m_ec^2)\\&=M_{Z,A}c^2-M_{Z+1,A}c^2.\end{aligned} \tag{16.12}$$

The energy released in the decay is given by the difference in rest energies of the initial and final neutral atoms. This energy is shared in different amounts by the three particles; it is shared mainly by the neutrino and electron, since the nucleus is so massive and its kinetic energy is $p^2/2m$. The maximum energy of the β spectrum in Fig. 16.8 corresponds to E_{decay}.

Figure 16.9 shows data for the decay of ^{24}Na, an isotope that has been used in nuclear medicine. The β transition labeled β_2 is overwhelmingly the most common. The β_2 emission is followed by two γ rays, and the end-point

[3]Its thermal energy of about $\frac{1}{40}$ eV is negligible compared to the energy released in β decay.

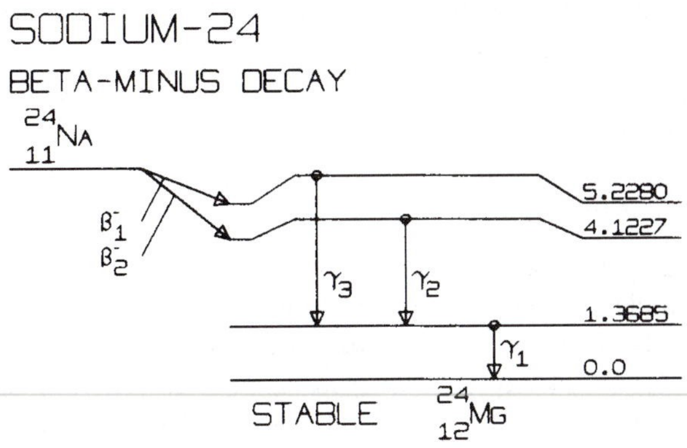

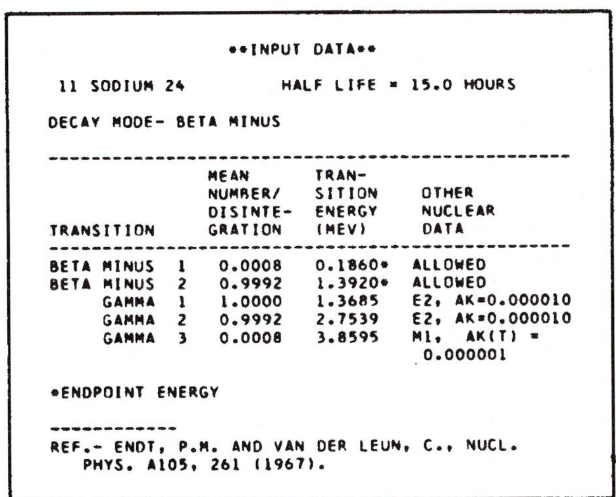

INPUT DATA

11 SODIUM 24 HALF LIFE = 15.0 HOURS

DECAY MODE- BETA MINUS

TRANSITION		MEAN NUMBER/ DISINTE- GRATION	TRAN- SITION ENERGY (MEV)	OTHER NUCLEAR DATA
BETA MINUS	1	0.0008	0.1860*	ALLOWED
BETA MINUS	2	0.9992	1.3920*	ALLOWED
GAMMA	1	1.0000	1.3685	E2, AK=0.000010
GAMMA	2	0.9992	2.7539	E2, AK=0.000010
GAMMA	3	0.0008	3.8595	M1, AK(T) = 0.000001

*ENDPOINT ENERGY

REF.- ENDT, P.M. AND VAN DER LEUN, C., NUCL. PHYS. A105, 261 (1967).

OUTPUT DATA

11 SODIUM 24 HALF LIFE = 15.0 HOURS

DECAY MODE- BETA MINUS

RADIATION		MEAN NUMBER/ DISINTE- GRATION n_i	MEAN ENERGY/ PAR- TICLE $\bar{E}_i$ (MeV)	EQUI- LIBRIUM DOSE CONSTANT Δ_i (g-rad/ μCi-h)
BETA MINUS	2	0.9992	0.5547	1.1805
GAMMA	1	0.9999	1.3685	2.9149
GAMMA	2	0.9991	2.7539	5.8610
GAMMA	3	0.0008	3.8595	0.0065

FIGURE 16.9. Energy levels and data for the β decay of ^{24}Na. Reprinted from Dillman and Von der Lage (1974, p. 19), with the permission of the publisher.

energy of the β decay is 1.392 MeV. On the other hand, the *average* energy of the β particle is only 0.5547 MeV, about 40% of the end-point energy.

Emission of a positron converts a proton into a neutron, and Z decreases by one. A neutrino is also emitted:

$$ {}^{A}_{Z}X \rightarrow {}^{A}_{Z-1}Y + \beta^{+} + \nu. \tag{16.13}$$

The decay energy is again given by

$$E_{\text{decay}} = m_{Z,A}c^2 - m_{Z-1,A}c^2 - m_e c^2.$$

However, this time, when we add Zm_ec^2 to the first term and subtract $(Z-1)m_ec^2$ from the second term to convert these to atomic masses, the electron masses do not cancel. Instead, we get

$$E_{\text{decay}} = M_{Z,A}c^2 - M_{Z-1,A}c^2 - 2m_e c^2. \tag{16.14}$$

Positron emission will not occur unless the initial neutral atomic mass exceeds the final neutral atomic mass by at least $2m_ec^2$. We remind ourselves of this by drawing a vertical line of length $2m_ec^2$ before drawing the slanting line for the β^+ decay, as in Fig. 16.10, the decay scheme for ^{18}F.

The first entry in the table of input data in Fig. 16.10 is for *electron capture*. The transition energy listed is $2m_ec^2$ more than for β^+ decay. Some of the inner electrons of the atom are close enough to the nucleus (quantum mechanically, the electron wave functions overlap the nucleus enough) so that the electron is captured by the nucleus, and a neutrino is emitted. In terms of nuclear masses, an electron rest energy is added to the parent nucleus (we ignore its kinetic energy).

$$E_{\text{e.c.}} = m_e c^2 + m_{Z,A}c^2 - m_{Z-1,A}c^2.$$

If we add and subtract $(Z-1)m_ec^2$, we have

$$E_{\text{e.c.}} = M_{Z,A}c^2 - M_{Z-1,A}c^2. \tag{16.15}$$

A K electron[4] is usually captured. The energy from the nuclear transition is given to a neutrino. No electron emerges, but there are K x rays and Auger electrons,[5] as there are any time a vacancy on the K shell occurs, and these contribute to the radiation dose. They are not listed in the output data of Fig. 16.10 because the K x rays only have an energy of about 530 eV.

There is also an entry under output data labeled *annihilation radiation*. Once a positron has been emitted, it slows down like any other electron. At some point it combines with an electron (since the positron and electron constitute a particle–antiparticle pair), and all of the rest energy of both particles goes into two photons.[6] The energy conservation equation is

$$2m_e c^2 = 2h\nu. \tag{16.16}$$

For each original positron emitted, two photons are produced, each of energy $m_ec^2 = 0.511$ MeV.

[4]See Chap. 14.
[5]See Chap. 14.
[6]Three photons are occasionally emitted.

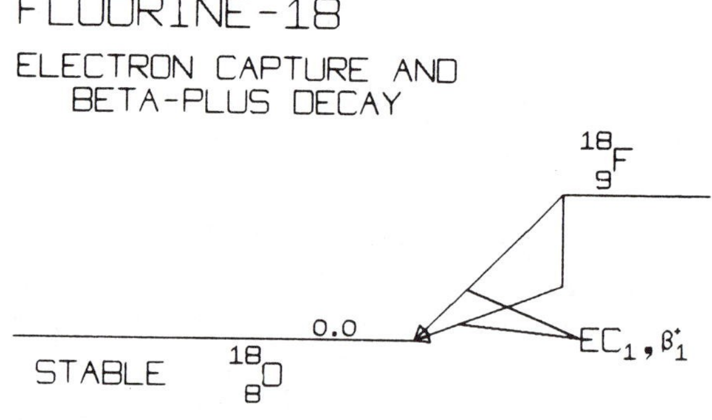

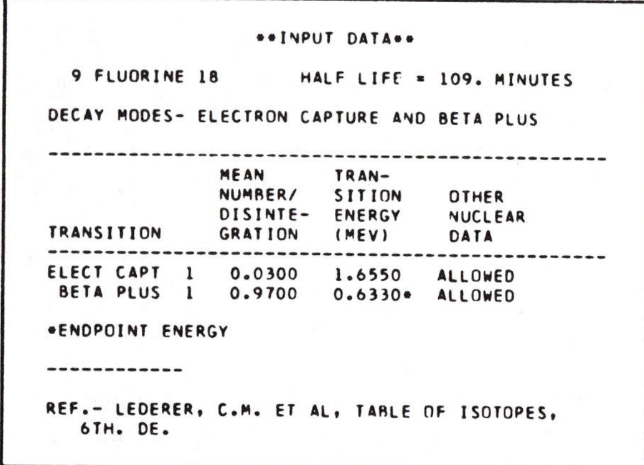

INPUT DATA

9 FLUORINE 18 HALF LIFE = 109. MINUTES

DECAY MODES- ELECTRON CAPTURE AND BETA PLUS

TRANSITION	MEAN NUMBER/ DISINTE- GRATION	TRAN- SITION ENERGY (MEV)	OTHER NUCLEAR DATA
ELECT CAPT 1	0.0300	1.6550	ALLOWED
BETA PLUS 1	0.9700	0.6330*	ALLOWED

*ENDPOINT ENERGY

REF.- LEDERER, C.M. ET AL, TABLE OF ISOTOPES, 6TH. DE.

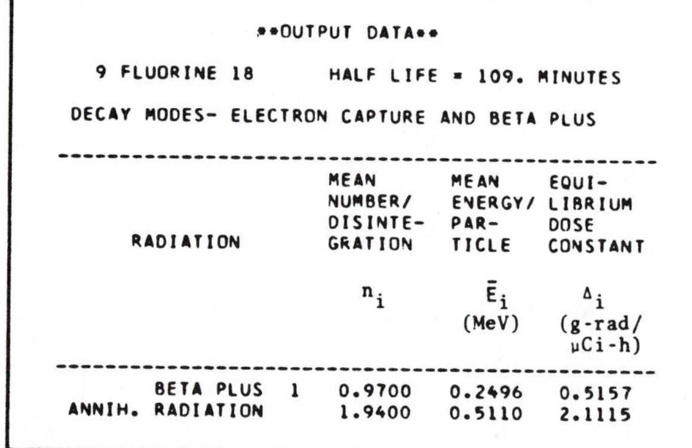

OUTPUT DATA

9 FLUORINE 18 HALF LIFE = 109. MINUTES

DECAY MODES- ELECTRON CAPTURE AND BETA PLUS

RADIATION	MEAN NUMBER/ DISINTE- GRATION n_i	MEAN ENERGY/ PAR- TICLE $\bar{E}_i$ (MeV)	EQUI- LIBRIUM DOSE CONSTANT Δ_i (g-rad/ µCi-h)
BETA PLUS 1	0.9700	0.2496	0.5157
ANNIH. RADIATION	1.9400	0.5110	2.1115

FIGURE 16.10. Energy levels and data for the β decay of ^{18}F. Reprinted from Dillman and Von der Lage (1974, p. 18), with the permission of the publisher.

16.5. CALCULATING THE ABSORBED DOSE FROM RADIOACTIVE NUCLEI WITHIN THE BODY

When a radiopharmaceutical is given a patient for either diagnosis or therapy, the nuclei end up in different organs in varying amounts, for example, ^{99m}Tc-labeled albumin microspheres injected intravenously lodge in the lungs. The problem is to calculate the whole-body-absorbed dose, the dose to the lungs, and the dose to other organs.

The dose calculation is carried out in the following way:[7]

1. Calculate the total number of nuclear transformations or disintegrations in organ h. It is called the *cumulated activity* $\tilde{A}_h$.

2. Calculate the mean energy emitted per unit cumulated activity for each type of photon or particle emitted.

a. If the radioactive nucleus can emit several types of particles or photons per transformation, call n_i the mean number of particles or photons of type i emitted per transformation. These include γ rays, electrons, x rays and Auger electrons. (In the "output data" of Dillman and Von der Lage (1974), n_i is called the *mean number per disintegration*. In the more recent report of Weber *et al.* (1989) it is called *particles per transition*. I have continued to show the older tables here because the newer ones contain only the output data.) The data are also available in electronic form [Eckerman *et al.* (1994)].

b. For each type determine E_i, the mean energy per particle or photon.

c. Calculate Δ_i, the mean energy emitted per unit cumulated activity, for each type of particle or photon emitted. (In earlier MIRD literature, this was called the equilibrium absorbed dose constant.)

3. Calculate $\phi_i(r_k \leftarrow r_h)$, the fraction of the radiation of type i emitted in source region r_h that is absorbed in target region r_k, and divide by the mass of the target region to get the *specific absorbed fraction*

$$\Phi_i(r_k \leftarrow r_h) = \frac{\phi_i(r_k \leftarrow r_h}{m_k}.$$

(Φ has the units of inverse mass.)

4. The mean absorbed dose in organ k due to activity in organ h, $\bar{D}$ (in J kg^{-1} or Gy) is

$$\bar{D}(r_k \leftarrow r_h) = \tilde{A}_h \sum_i n_i E_i \Phi_i(r_k \leftarrow r_h),$$

$$\bar{D}(r_k \leftarrow r_h) = \tilde{A}_h \sum_i \Delta_i \Phi_i(r_k \leftarrow r_h). \tag{16.17}$$

5. If several organs are radioactive, a sum must be taken over each organ:

$$\bar{D}(r_k) = \sum_h \tilde{A}_h \sum_i \Delta_i \Phi_i(r_k \leftarrow r_h). \tag{16.18}$$

The next three sections show how to determine $\tilde{A}$, Δ_i, and Φ_i. Some tables [Snyder *et al.* (1969, 1976)] give values of Φ_i for photons of various energies. It is necessary to multiply by Δ_i and sum for the isotope of interest. These

[7]Dose calculations in this chapter follow the technique and notation recommended by the Medical Internal Radiations Dose Committee of the Society of Nuclear medicine [Loevinger, Budinger, and Watson (1988)].

sums must be repeated over and over again for common radionuclides. The sum is called the *mean absorbed dose per unit cumulated activity*:

$$S(r_k \leftarrow r_h) = \sum_i \Delta_i \Phi_i(r_k \leftarrow r_h), \tag{16.19}$$

$$\bar{D}(r_k \leftarrow r_h) = \tilde{A}_h S(r_k \leftarrow r_h), \tag{16.20}$$

$$\bar{D}(r_k) = \sum_h \tilde{A}_h S(r_k \leftarrow r_h). \tag{16.21}$$

A table of S for many common radionuclides is available [Snyder *et al.* (1975)]. The tables cannot be summed over h because the values of A_h depend on how the isotope is administered. A computer program MIRDOSE is most commonly used for these calculations [Stabin (1996)].

We have used SI units. The dose for radiation type i is simply

$$D \text{ (Gy)} = [\tilde{A} \text{ (dimensionless)}][\Delta_i \text{ (J)}][\Phi_i \text{ (kg}^{-1})]. \tag{16.22a}$$

In an older system of units, where the dose is in rad and the total number of transitions is in microcurie-hour (see below), the equation is

$$D \text{ (rad)} = [\tilde{A}(\mu\text{Ci h})][\Delta_i \text{ (g rad } \mu\text{Ci}^{-1} \text{ h}^{-1})] \times [\Phi_i(\text{g}^{-1})]. \tag{16.22b}$$

The next three sections discuss cumulated activity, the mean energy emitted, and the absorbed fraction of the energy. Then all of these concepts are combined with examples of absorbed dose calculations.

16.6. ACTIVITY AND CUMULATED ACTIVITY

The *activity* $A(t)$ is the number of radioactive transitions (or transformations or disintegrations) per second. The SI unit of activity is the *becquerel* (Bq):

$$1 \text{ Bq} = 1 \text{ (transition) s}^{-1}. \tag{16.23}$$

The earlier unit of activity, which is still used occasionally, is the *curie* (Ci):

$$1 \text{ Ci} = 3.7 \times 10^{10} \text{ Bq},$$

$$1 \text{ } \mu\text{Ci} = 3.7 \times 10^{4} \text{ Bq}. \tag{16.24}$$

The *cumulated activity* $\tilde{A}$ is the total number of transitions that take place. The SI unit of cumulated activity is the transition or the Bq s. Both are dimensionless. The old unit of cumulated activity is the μCi h:

$$1 \text{ } \mu\text{Ci h} = 1.332 \times 10^{8} \text{ Bq s}. \tag{16.25}$$

Consider a sample of N_0 radioactive nuclei at time $t=0$. The total number of nuclei at time t is $N(t)=N_0 e^{-\lambda t}$, and the total activity is $A(t)=\lambda N(t)=A_0 e^{-\lambda t}$. The cumulated activity between times t_1 and t_2 is

$$\tilde{A}(t_1,t_2) = \int_{t_1}^{t_2} A(t)dt = \frac{A_0}{\lambda}(e^{-\lambda t_1} - e^{-\lambda t_2}). \tag{16.26}$$

If all times are considered, $t_1=0$ and $t_2=\infty$,

$$\tilde{A} = \tilde{A}(0,\infty) = A_0/\lambda = \frac{A_0 T_{1/2}}{0.693} = 1.443 A_0 T_{1/2}. \tag{16.27}$$

This is, as we would expect, N_0.

16.6.1. The General Distribution Problem: Residence Time

Suppose that a radioactive substance is introduced in the body by breathing, ingestion, or injection. It may move into and out of many organs before decaying, and it may even leave the body. The details of how it moves depend on the pharmaceutical to which it is attached.

The cumulated activity in organ h is

$$\tilde{A}_h = \int_0^\infty A_h(t)\, dt. \tag{16.28}$$

If initial activity A_0 is administered to the patient, the residence time in organ h is defined as the length of time that activity at rate A_0 would have to reside in the organ to give that cumulated activity:

$$\tau_h = \frac{\tilde{A}_h}{A_0} = \frac{\tilde{A}_h(0,\infty)}{A_0}. \tag{16.29}$$

The residence time for a given substance and organ must be determined by measurement, guided by the use of appropriate models. Many residence times have been determined and published. The presence of an abnormality in some organ can drastically alter the residence time. We now calculate the cumulated activity and residence time for some simple situations.

16.6.2. Immediate Uptake with No Biological Excretion

This is the simplest example. A certain fraction of the radiopharmaceutical is taken up very rapidly in some organ, and it stays there. This is a good model for ^{99m}Tc–sulfur colloid, which is used for liver imaging. About 85% is trapped in the liver; the remainder goes to the spleen and elsewhere [Loevinger, Budinger, and Watson (1988, p. 23)]. The activity in the organ is $A_h(t)=A_h e^{-\lambda t}$. [Note the difference between the activity in organ h as a function of time, $A_h(t)$, the initial activity in organ h, A_h, and the cumulated activity in organ h, $\tilde{A}_h$.] Let the fraction of the activity in the organ be F_h. The cumulated activity is

$$\tilde{A}_h = A_h \int_0^\infty e^{-\lambda t} dt = \frac{A_h}{\lambda} = \frac{F_h A_0}{\lambda}.$$

The residence time is

$$\tau_h = \frac{\tilde{A}_h}{A_0} = \frac{F_h}{\lambda} = 1.443 F_h T_{1/2}. \tag{16.30}$$

16.6.3 Immediate Uptake with Exponential Biological Excretion

Suppose that in addition to physical decay with decay constant λ, the pharmaceutical moves to another organ while it is still radioactive. Such a process can be complicated, involving storage in the gut or bladder, or it can approximate an exponential decay. From a particular organ, the disappearance may be close to exponential with a biological disappearance constant λ_j. (Assume for now that all nuclei can disappear biologically. If some are bound in different chemical forms, this might not be true.) If N is the number of radioactive nuclei in the organ (not the total number originally administered), then the rate of change of N is

$$\frac{dN}{dt} = -(\lambda + \lambda_j) N,$$

the solution to which is $N(t) = N_0 e^{-(\lambda+\lambda_j)t}$. The activity is λN, not dN/dt. Since it is proportional to N, we can again write

$$A_h(t) = A_h e^{-(\lambda+\lambda_j)t} = \lambda N_0 e^{-(\lambda+\lambda_j)t}. \tag{16.31}$$

Again, $N_0 = A_h/\lambda$. The decay constant $\lambda + \lambda_j$ is larger than the physical decay constant λ. The effective half-life is

$$(T_j)_{\text{eff}} = \frac{0.693}{\lambda + \lambda_j}. \tag{16.32}$$

In terms of the physical and biological half-lives T and T_j, this is

$$(T_j)_{\text{eff}}^{-1} = \mathrm{T}^{-1} + \mathrm{T}_j^{-1} \tag{16.33}$$

or

$$(T_j)_{\text{eff}} = \frac{T T_j}{T + T_j}. \tag{16.34}$$

The cumulated activity is

$$\tilde{A}_h(t_1, t_2) = A_h \int_{t_1}^{t_2} e^{-(\lambda+\lambda_j)t} dt$$
$$= \frac{A_h}{\lambda + \lambda_j} \left(e^{-(\lambda+\lambda_j)t_1} - e^{-(\lambda+\lambda_j)t_2}\right). \tag{16.35}$$

The cumulated activity for all time is

$$\tilde{A}_h = \frac{A_h}{\lambda + \lambda_j} = 1.443 (T_j)_{\text{eff}} A_h. \tag{16.36}$$

16.6.4. Immediate Uptake Moving through Two Compartments

Consider the simplest two-compartment model. A total of N_0 nuclei are administered that move immediately to the first compartment. They then move exponentially from the first compartment to the second but do not move back. The number in the first compartment is given by

$$\frac{dN_1}{dt} = -(\lambda_1 + \lambda) N_1. \tag{16.37}$$

The radioactive decay constant is λ and the biological disappearance rate is λ_1. In compartment 2, the substance enters from compartment 1 and is biologically removed with constant λ_2:

$$\frac{dN_2}{dt} = +\lambda_1 N_1 - (\lambda + \lambda_2) N_2. \tag{16.38}$$

If N_0 nuclei are injected in compartment 1 at t=0, then one can show (see Problem 16.8) that

$$N_1(t) = N_0 e^{-(\lambda+\lambda_1)t} \tag{16.39}$$

so

$$\frac{dN_2}{dt} = \lambda_1 N_0 e^{-(\lambda+\lambda_1)t} - (\lambda + \lambda_2) N_2, \tag{16.40}$$

the solution to which is

$$N_2(t) = N_0 \frac{\lambda_1}{\lambda_1 - \lambda_2} \left(e^{-(\lambda+\lambda_2)t} - e^{-(\lambda+\lambda_1)t}\right). \tag{16.41}$$

These solutions are worth examining. They are plotted in Fig. 16.11 for $\lambda=1$, $\lambda_1=2$, and $\lambda_2=1$. The number of nuclei in compartment 1 is $N_0 e^{-3t}$. At first, many of the particles leaving compartment 1 enter compartment 2, and N_2 rises. When there is no more of the substance entering the second compartment from the first, N_2 decays at a rate $\lambda+\lambda_2=2$. This corresponds to the vanishing of the second term in Eq. (16.41). The larger the value of λ_1, the faster the second term vanishes. For very large values of λ_1, the second term vanishes quickly, the factor $\lambda_1/(\lambda_1-\lambda_2)$

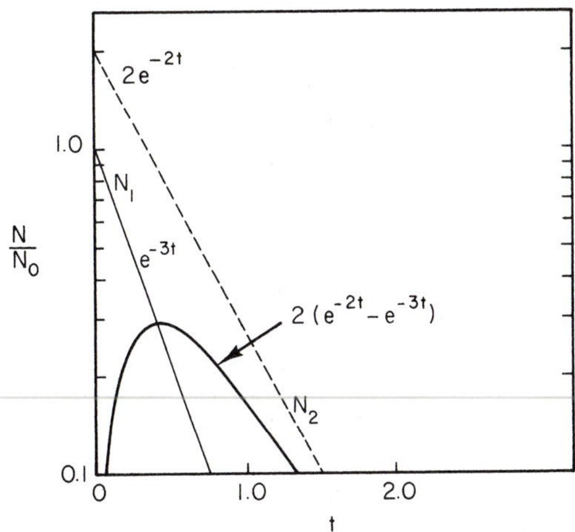

FIGURE 16.11. An example of two-compartment transfer when $\lambda=1$, $\lambda_1=2$, and $\lambda_2=1$.

approaches unity, and the decay is nearly $N_2(t) = N_0 e^{-(\lambda+\lambda_2)t}$. The case $\lambda_1=\lambda_2$ is discussed in Problem 16.10. The activities are

$$A_1(t)=\lambda N_1(t), \quad A_2(t)=\lambda N_2(t)$$

and the cumulated activities are obtained by integration:

$$\tilde{A}_1=\frac{A_0}{\lambda+\lambda_1}, \qquad \tilde{A}_2=\frac{A_0\lambda_1}{(\lambda+\lambda_1)(\lambda+\lambda_2)}. \tag{16.42}$$

The residence times are

$$\tau_1=\frac{1}{\lambda+\lambda_1}, \qquad \tau_2=\frac{\lambda_1}{(\lambda+\lambda_1)(\lambda+\lambda_2)}. \tag{16.43}$$

16.6.5. More Complicated Situations

A number of more complicated situations are solved by Loevinger, Budinger, and Watson (1988). These include situations where substances move between compartments in both directions, the experimental data for the activity have been fit with a series of exponentials, and convolution techniques are used. All of these cases are for isotopes and pharmaceuticals used in clinical practice.

16.6.6. Activity per Unit Mass

It is sometimes convenient to use the mean initial activity per unit mass

$$C_h=\frac{A_h}{m_h} \text{ Bq kg}^{-1} \tag{16.44}$$

and the cumulated mean activity per unit mass

$$\tilde{C}_h=\frac{\tilde{A}_h}{m_h}=\frac{\tau_h A_0}{m_h} \text{ kg}^{-1}. \tag{16.45}$$

Old units for these were μCi g^{-1} and μCi h g^{-1}.

16.7. MEAN ENERGY EMITTED PER UNIT CUMULATED ACTIVITY

The mean energy emitted per unit cumulated activity Δ_i is determined by knowing n_i and E_i for each particle or photon that is emitted. For a given nuclear transformation, the n_i and E_i must include all photons (whether γ rays or x rays) and all electrons (betas, internal conversion electrons, and Auger electrons). In SI units,

$$\Delta_i=n_i E_i \text{ J}. \tag{16.46a}$$

If E_i is expressed in MeV, we must use the conversion factor 1 MeV$=1.6\times10^{-13}$ J. In the old system of units, there is the conversion factor:

$$\begin{aligned}\Delta_i(\text{g rad }\mu\text{Ci}^{-1}\text{ h}^{-1}) &= n_i[E_i\ (\text{MeV})](3.7\times10^4\ \text{s}^{-1}\ \mu\text{Ci}^{-1})\\ &\quad\times(1.6\times10^{-13}\ \text{J MeV}^{-1})(10^7\ \text{erg J}^{-1})\\ &\quad\times(3.6\times10^3\ \text{s h}^{-1})(10^{-2}\ \text{rad g erg}^{-1})]\end{aligned}$$

$$\Delta_i=2.13 n_i E_i. \tag{16.46b}$$

Values of Δ_i in the old units are found in the "output tables" for each nucleus (see Figs. 16.4, 16.9, and 16.10). We will calculate some values in this section to show how it is done.

Although β particles have a complicated energy spectrum, they are the easiest radiation for which to determine Δ_i. Values of n_i are available in nuclear data tables that show the decay schemes for the nucleus under consideration, and average β energies have usually been measured.

For positron decay, there are also two 0.511-MeV annihilation photons to be considered.

Referring to ^{18}F, Fig. 16.10, we see that 0.97 of the decays were positron emission with an average energy of 0.2496 MeV$=4.00\times10^{-14}$ J. Therefore Δ_i is $0.97\times4.00\times10^{-14}=3.87\times10^{-14}$ J. Multiplying by $2.13/1.6\times10^{-13}=1.33\times10^{13}$, we get $\Delta_i=0.516$ g rad μCi^{-1} h^{-1}, which agrees with the entry under output data of Fig. 16.10. There are two annihilation photons for each nuclear transition, so $n_i=1.94$ and $E_i=0.511$ MeV for the annihilation radiation. This gives $\Delta_i=1.59\times10^{-13}$ J or 2.11 g rad μCi^{-1} h^{-1}.

Since electron capture competes with positron emission, it is also listed in the input data for ^{18}F as occurring 3% of the time. Most of the energy is taken away by the neutrino, and it does not contribute to the absorbed dose. The x rays emitted by the residual ^{18}O have such low energies that they are not listed in the output data. For a nucleus of higher Z, they would have been important, however.

We turn now to γ emission, which often occurs after β emission because the resulting nucleus is not in its ground state. Internal conversion competes with the γ emission. Let us use the specific example $^{99m}_{43}$Tc. Technetium-99 is an artificially produced radioisotope that is widely used as a radiopharmaceutical. It is the decay product of a longer-lived parent, so that although its half-life is only 6 h, it can be produced each day in a local hospital. The short half-life results in a reduced dose to the patient, since fewer of the radioactive transitions are "wasted" by occurring after the diagnostic procedure is completed. We will discuss more details of ^{99m}Tc in a subsequent section on radiopharmaceuticals.

The decay scheme of ^{99m}Tc is shown in Fig. 16.4. The 6-h half-life is associated with a metastable state, and the decay produces three possible γ rays. The "ground state" is actually not stable, but decays by β emission with a half-life of 2.1×10^5 yr. The problems will show that this can be neglected.

In addition to the data in the "input data" table of Fig. 16.4, we need some information about the x-ray characteristics of ^{99}Tc. These are listed in Table 16.2. The subscripts of n_i, E_i, and Δ_i in the following all refer to the line in the "output data" of Fig. 16.4.

Consider γ_1. Its energy (0.0021 MeV) is less than the binding energy of either a K or L electron, so there is no K or L internal conversion. However, α_M is very large, and we will assume that all of the γ_1 transitions are M-shell internal conversions. In the output table (Fig. 16.4), this means that Δ_1 for γ_1 is zero. The energy of the internal-conversion electron is the decay energy minus B_M: $E_2=0.0021-0.0005=0.0016$ MeV. Then

$$\Delta_2=n_2E_2=(0.9860)(0.0016)(1.6\times10^{-13})$$
$$=2.52\times10^{-16}\ \text{J}.$$

The remaining 0.0005 MeV is taken care of by M Auger electrons.

Now consider γ_2. There can be γ rays of this energy and internal conversion electrons from the K, L, or M shells. We are told (input data) that $\alpha_K=0.104$, and that $\alpha_K=7.7\,\alpha_L$. It is also known empirically that $\alpha_L\approx3\,\alpha_M$. Therefore, we can write

$$\alpha_K=0.104,$$
$$\alpha_L=0.0135,$$
$$\alpha_M=0.0045.$$

Now Eq. (16.10) combined with the definition of α gives

$$\begin{aligned} f_r+\alpha_Kf_r+\alpha_Lf_r+\alpha_Mf_r&=1,\\ f_r(1+\alpha_K+\alpha_L+\alpha_M)&=1,\\ f_r=\frac{1}{1.122}&=0.8913,\\ f_K=\alpha_Kf_r&=0.0927,\\ f_L=\alpha_Lf_r&=0.0120,\\ f_M=\alpha_Mf_r&=0.0040. \end{aligned} \tag{16.47}$$

Each of these must be multiplied by $n_{\gamma_2}=0.9860$ to get the overall number of photons or internal-conversion electrons per decay associated with γ_2:

$$\begin{aligned} n_3=n_{\gamma_2}f_r&=(0.9860)(0.8913)=0.8788,\\ n_4=n_{\gamma_2}f_K&=(0.9860)(0.0927)=0.0914,\\ n_5=n_{\gamma_2}f_L&=(0.9860)(0.0120)=0.0118,\\ n_6=n_{\gamma_2}f_M&=(0.9860)(0.0040)=0.0039. \end{aligned} \tag{16.48}$$

The energies are calculated as follows:

$$\begin{aligned} E_3&=E_{\gamma_2}=0.1405\ \text{MeV},\\ E_4&=E_{\gamma_2}-B_K=0.1405-0.0210=0.1195\ \text{MeV},\\ E_5&=E_{\gamma_2}-B_L=0.1405-0.0028=0.1377\ \text{MeV},\\ E_6&=E_{\gamma_2}-B_M=0.1405-0.0005=0.1400\ \text{MeV}. \end{aligned} \tag{16.49}$$

The values of Δ_i are the products of these:

$$\Delta_3=1.976\times10^{-14}\ \text{J}=0.2627\ \text{g rad}\ \mu\text{Ci}^{-1}\,\text{h}^{-1},$$
$$\Delta_4=1.748\times10^{-15}\ \text{J}=0.0232\ \text{g rad}\ \mu\text{Ci}^{-1}\,\text{h}^{-1},$$
$$\Delta_5=2.600\times10^{-16}\ \text{J}=0.0035\ \text{g rad}\ \mu\text{Ci}^{-1}\,\text{h}^{-1},$$
$$\Delta_6=8.736\times10^{-17}\ \text{J}=0.0012\ \text{g rad}\ \mu\text{Ci}^{-1}\,\text{h}^{-1}.$$

TABLE 16.2. *Additional properties of $^{99}_{43}$Tc for calculation of Δ_i.*

Fluorescence yield[a]	W_K	0.74
K-shell binding energy[b]	B_K	0.021 04 MeV
L-shell binding energy[b]	B_L	0.002 8 MeV
M-shell binding energy[b]	B_M	0.000 5 MeV

$$\frac{K_\alpha}{K_\alpha+K_\beta}=0.863\quad\text{(ratio of x rays emitted)}$$

$$\frac{K_\beta}{K_\alpha+K_\beta}=0.137\quad\text{(ratio of x rays emitted)}$$

[a]Estimated using Fig. 16.14.
[b]*Handbook of Chemistry and Physics*, 51st ed., p. E-186.

Similar calculations for the third γ ray are left to the problems.

We must next consider x-ray and Auger electron emission. A K-shell hole can be produced by a K internal-conversion electron from either γ_2 (n_4=0.0913) or γ_3 (n_8=0.0088). The total number of holes per nuclear transition is 0.1001. Of these, a fraction W_K=0.74 will give x rays. The total number per transition is 0.074. Of these, a fraction $K_\alpha/(K_\alpha+K_\beta)$=0.863 go to $K\alpha$ x rays. These are listed as n_{11} and n_{12} in Fig. 16.4. Our calculation gives $n_{11}+n_{12}$=0.0639. The table gives 0.0662 for this sum. Considering that our value of W_K was obtained independently of the table, this is good agreement. For $K\beta$ x rays, we get n_{13}=(0.1001)(0.74)(0.137)=0.0101. The energies of the x rays are obtained by subtracting the binding energies:

$$E_{11}(K\alpha_1)=B_K-B_L=0.0210-0.0028=0.0182 \text{ MeV},$$

$$E_{13}(K\beta)=B_K-B_M=0.0210-0.0005=0.0205 \text{ MeV}.$$

Of the holes in the K shell, a fraction $1-W_K$=0.26 decay by emission of Auger electrons. The total number of K holes is 0.1001; the total number of K Auger electrons should therefore be (0.1001)(0.26)=0.0260. The notation of the "output data" table is of the form.

KLM
- K — an electron falls to a hole in the K shell
- L — from the L shell
- M — with the emission of an Auger electron from shell M

The sum of n_i for KLL and KLX (X denoting shells above the L shell) is 0.0207, in good agreement with the 0.0260 we calculated.

16.8. CALCULATION OF THE ABSORBED FRACTION

The remaining part of the problem is the most difficult: the calculation of

$$\phi(r_k \leftarrow r_h),$$

the fraction of the radiation of a certain type emitted in region r_h that is absorbed in region r_k. A lot has been published on this problem; this section can provide only an introduction.

16.8.1. Nonpenetrating Radiation

The simplest case is for charged particles or photons of very low energy that lose all their energy after traveling a short distance. If the source volume is much larger than this distance, we can say that all the energy is absorbed in the source volume:

$$\phi(r_k \leftarrow r_h)=\begin{cases} 0, & r_k \neq r_h \\ 1 & r_k = r_h \end{cases}. \qquad (16.50)$$

16.8.2. Infinite Source in an Infinite Medium

Suppose that a radioactive source is distributed uniformly throughout a region that is so large that edge effects can be neglected. The activity per unit mass is C, so the total activity is $\tilde{A}=M\tilde{C}$, where M is the mass of the material. The energy released is $\tilde{A}\Delta$.

This is absorbed in mass M, so the fractional absorbed energy is 1, as in case 1. The dose is

$$D=\frac{M\tilde{C}\Delta}{M}=\Delta\tilde{C}. \qquad (16.51)$$

(This is why Δ used to be called the *equilibrium absorbed dose constant.*)

16.8.3. Point Source of Monoenergetic Photons in Empty Space

Another simple case is a point source of monoenergetic photons in empty space. The total amount of energy released by the source is $\tilde{A}\Delta_i$. If the energy of the radiation is E_i, the number of photons is $\tilde{A}\Delta_i/E_i$. At distance r the number per unit area is

$$\frac{\tilde{A}\Delta_i}{4\pi r^2 E_i}.$$

If a small amount of substance of area dS, density ρ, thickness dr, and energy absorption coefficient μ_{en} is introduced as in Fig. 16.12, the amount of energy absorbed in it is E_i times the number of photons absorbed:

$$\delta E=\frac{\tilde{A}\Delta_i dS\,\mu_{en} dr}{4\pi r^2}.$$

Therefore, the absorbed fraction is

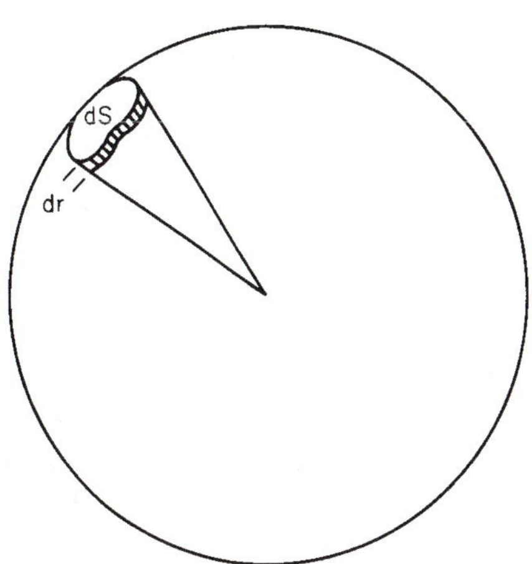

FIGURE 16.12. A small volume of absorbing material is introduced at distance r from a point source of γ rays.

$$\phi = \frac{\delta E}{\tilde{A}\Delta_i} = \frac{\mu_{en} dr\, dS}{4\pi r^2}. \tag{16.52}$$

This is exactly what we expect from the definition of ϕ. If the source radiates its energy isotropically, the fraction passing through dS is $dS/(4\pi r^2)$. The fraction of that energy absorbed in dr is $\mu_{en} dr$. The specific absorbed fraction is

$$\Phi = \frac{\phi}{m} = \frac{\phi}{\rho\, dS\, dr} = \frac{\mu_{en}}{4\pi r^2 \rho}. \tag{16.53}$$

16.8.4. Point Source of Monoenergetic Photons in an Infinite Isotropic Absorber

If the source is not in empty space but in an infinite, homogeneous, isotropic-absorbing medium, the number of photons at distance r from the source is modified by the factor

$$e^{-\mu_{atten} r} B(r),$$

where the *buildup factor* $B(r)$ accounts for secondary photons. Therefore,

$$\phi = \frac{\mu_{en} dr\, dS}{4\pi r^2} e^{-\mu_{atten} r} B(r) \tag{16.54}$$

and

$$\Phi = \frac{\mu_{en}}{\rho} \frac{e^{-\mu_{atten} r}}{4\pi r^2} B(r). \tag{16.55}$$

The buildup factor has been tabulated for photons of various energies in water [Berger (1968)].

16.8.5. More Complicated Cases—the MIRD Tables

For more realistic geometries, the calculation of ϕ is quite complicated. Tables for humans of average build have been prepared by the Medical Internal Radiation Dose Committee [Snyder *et al.* (1969, 1975, 1976)]. A ''Monte Carlo'' computer calculation was used. The description below shows how it works in principle; the actual calculations, though equivalent, are different in detail to save computer time. The radioactive nuclei are assumed to be distributed uniformly throughout the source organ. A point within the source organ is picked. The model emits a photon of energy E in some direction, picked at random from all possible directions. This photon is followed along its path; for every element ds of its path, the probability of its interacting, $\mu_{atten}\, ds$, is calculated. The computer program then ''flips a coin'' with this probability of having heads. If a head occurs, the photon is considered to interact at that point. If the interaction is Compton scattering, the angle is picked at random with a relative probability given by the differential cross section. The energy of a recoil electron for that scattering angle is calculated and deposited at the interaction site. Similar procedures are followed for the photoelectric effect and pair production. The scattered photon is then followed in the same way. If a tail occurred on the first flip, the photon is allowed to travel another distance ds and the probability of interaction is again calculated. This procedure is repeated until all the energy has been absorbed.

To determine what kind of material the photons are traveling through, a model of the body called a phantom is used. An example of a phantom is shown in Fig. 16.13. This entire procedure is repeated many times for each organ, until one has a map of the radiation deposited in all organs by γ rays leaving that point in the source organ.

The procedure is described in much greater detail by Snyder *et al.* (1969, 1976). Table 16.3 shows a portion of a table for ϕ.

A computer code (MIRDOSE) available from Oak Ridge National Laboratories is usually used to make the calculations [Stabin (1996)].

Often most of the isotope is taken up in one or two organs, and the rest of it distributes fairly uniformly through the rest of the body. Using the subscript h for the organs with the greatest activity, TB to mean total body, and RB to mean the rest of the body,

$$\tilde{A}_{RB} = \tilde{A}_{TB} - \sum_n \tilde{A}_h. \tag{16.56}$$

The dose is then

$$D_k = \sum_h \tilde{A}_h S(r_k \leftarrow r_h) + \tilde{A}_{RB} S(r_k \leftarrow \mathrm{RB}). \tag{16.57}$$

The quantity $S(r_k \leftarrow \mathrm{RB})$ cannot easily be tabulated, since it depends on what organs are included in the sum over h. Substituting the tabulated quantity $S(r_k \leftarrow \mathrm{TB})$ introduces errors because the ''hot'' organs that have significant activity are included a second time. One solution to this problem is to modify the cumulated activities [Coffey and Watson (1979)]. First, define a uniform total body cumulated activity that has the same cumulated activity per unit mass as the rest of the body:

$$\tilde{A}_u = \frac{m_{TB}}{m_{RB}} \tilde{A}_{RB}, \tag{16.58}$$

This activity in the total body would give a dose

$$D_k = \tilde{A}_u S(r_k \leftarrow \mathrm{TB}).$$

Then define for each organ of interest the quantity. $\tilde{A}_h^*$, which is the difference between the actual activity in organ h and that assuming the substance is uniformly distributed in the total body:

$$\tilde{A}_h^* = \tilde{A}_h - \frac{m_h}{m_{\mathrm{RB}}} \tilde{A}_{\mathrm{RB}}. \qquad (16.59)$$

Then the dose to organ k is

$$D_k = \sum_h \tilde{A}_h^* S(r_k \leftarrow r_h) + \tilde{A}_u S(r_k \leftarrow \mathrm{TB}). \qquad (16.60)$$

Problem 16.29 shows that Eqs. (16.58)–(16.60) are consistent with Eqs. (16.56) and (16.57) if

$$\frac{m_{\mathrm{RB}}}{m_{\mathrm{TB}}} S(r_k \leftarrow \mathrm{RB}) + \sum_h \frac{m_h}{m_{\mathrm{TB}}} S(r_k \leftarrow r_h) = S(r_k \leftarrow \mathrm{TB}), \qquad (16.61)$$

which is consistent with a uniform source $\tilde{A}_u$ distributed throughout the body. The dose can be determined either by calculating the modified activities and using the total body S in Eq. (16.60), or by calculating S for the rest of the body

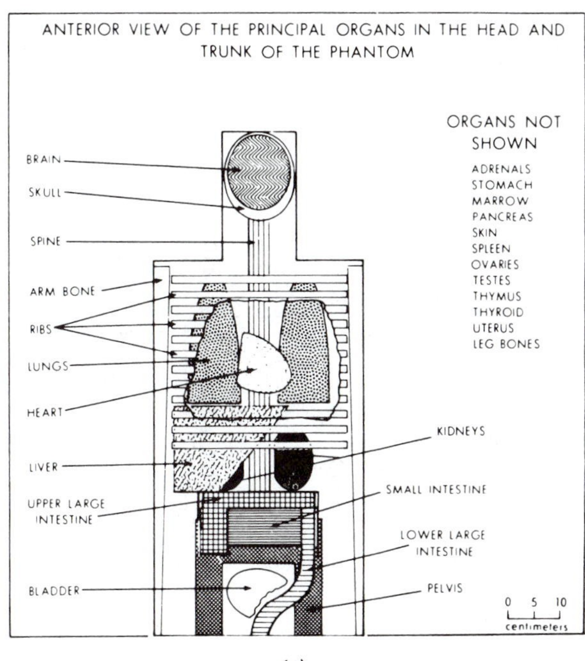

(a)

HEART

The heart is half an ellipsoid capped by a hemisphere which is cut by a plane. A rotation and translation are then effected. The heart (Fig. 4) is represented by

$$x_1 = 0.6943\,(x+1) - 0.3237\,(y+3) - 0.6428\,(z-51),$$
$$y_1 = 0.4226\,(x+1) + 0.9063\,(y+3),$$
$$z_1 = 0.5826\,(x+1) - 0.2717\,(y+3) + 0.7660\,(z-51),$$
$$\left(\frac{x_1}{8}\right)^2 + \left(\frac{y_1}{5}\right)^2 + \left(\frac{z_1}{5}\right)^2 \leqq 1,$$
$$x_1^2 + y_1^2 + z_1^2 \leqq (5)^2 \quad \text{if} \quad x_1 < 0,$$
$$\frac{x_1}{3} + \frac{z_1}{5} \geqq -1 \quad \text{if} \quad x_1 < 0$$

and has a volume of 603.1 cm³.

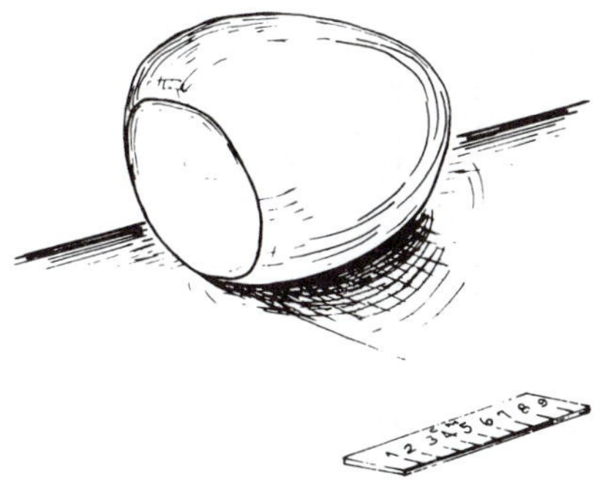

(b)

FIGURE 16.13. The phantom used by the MIRD Committee for calculations of the absorbed fraction. (a) A view of the whole body. (b) Details of the heart boundaries. Reprinted from Snyder *et al.* (1969, pp. 48, 49), with permission of the publisher.

TABLE 16.3. *Absorbed fractions for a uniform source in the lungs, calculated using the MIRD phantom, for photon energies from 10 to 200 keV, with percent errors.*

Target organ	Photon energy E (MeV)														Target organ
	0.010		0.015		0.020		0.030		0.050		0.100		0.200		
	ϕ	$\frac{100\sigma_\phi}{\phi}$	ϕ	$\frac{100\sigma_\phi}{\phi}$	ϕ	$\frac{100\sigma_\phi}{\phi}$	ϕ	$\frac{100\sigma_\phi}{\phi}$	ϕ	$\frac{100\sigma_\phi}{\phi}$	ϕ	$\frac{100\sigma_\phi}{\phi}$	ϕ	$\frac{100\sigma_\phi}{\phi}$	
Adrenals							0.159×10^{-3}	24	0.155×10^{-3}	19	0.164×10^{-3}	19	0.103×10^{-3}	24	Adrenals
Bladder											0.497×10^{-4}	41	0.145×10^{-3}	25	Bladder
GI (stomach)			0.124×10^{-3}	47	0.875×10^{-3}	16	0.263×10^{-2}	7.8	0.362×10^{-2}	5.6	0.263×10^{-2}	5.6	0.250×10^{-2}	6.4	GI (stomach)
GI (SI)							0.170×10^{-3}	34	0.879×10^{-3}	12	0.137×10^{-2}	8.2	0.160×10^{-2}	8.2	GI, (SI)
GI (ULI)									0.294×10^{-3}	19	0.446×10^{-3}	12	0.453×10^{-3}	12	GI, (ULI)
GI (LLI)											0.517×10^{-4}	31	0.973×10^{-4}	26	GI (LLI)
Heart			0.215×10^{-2}	11	0.983×10^{-2}	5	0.228×10^{-1}	2.9	0.203×10^{-1}	2.5	0.139×10^{-1}	2.6	0.130×10^{-1}	2.9	Heart
Kidneys							0.326×10^{-3}	20	0.923×10^{-3}	10	0.901×10^{-3}	8.7	0.934×10^{-3}	9.0	Kidneys
Liver	0.277×10^{-3}	34	0.420×10^{-2}	8.1	0.142×10^{-1}	4.3	0.258×10^{-1}	2.9	0.239×10^{-1}	2.5	0.173×10^{-1}	2.6	0.155×10^{-1}	2.8	Liver
Lungs	0.815	0.27	0.665	0.39	0.475	0.53	0.231	0.75	0.892×10^{-1}	0.97	0.493×10^{-1}	1.2	0.498×10^{-1}	1.4	Lungs
Marrow	0.989×10^{-4}	36	0.234×10^{-2}	6.9	0.133×10^{-1}	2.7	0.386×10^{-1}	1.4	0.466×10^{-1}	1.1	0.254×10^{-1}	1.4	0.161×10^{-1}	1.6	Marrow
Pancreas							0.379×10^{-3}	18	0.573×10^{-3}	10	0.607×10^{-3}	9.8	0.430×10^{-3}	13	Pancreas
Sk. (rib)	0.247×10^{-3}	36	0.576×10^{-2}	6.9	0.312×10^{-1}	2.8	0.710×10^{-1}	1.6	0.577×10^{-1}	1.5	0.243×10^{-1}	1.8	0.145×10^{-1}	2.3	Sk. (rib)
Sk. (pelvis)									0.196×10^{-3}	26	0.470×10^{-3}	15	0.348×10^{-3}	15	Sk. (pelvis)
Sk. (spine)					0.128×10^{-2}	13	0.148×10^{-1}	3.8	0.355×10^{-1}	2.3	0.235×10^{-1}	2.3	0.153×10^{-1}	2.7	Sk. (spine)
Sk. (skull)							0.264×10^{-3}	26	0.137×10^{-2}	11	0.120×10^{-2}	8.8	0.970×10^{-3}	9.7	Sk. (skull)
Skeleton (total)	0.247×10^{-3}	36	0.583×10^{-2}	6.9	0.331×10^{-1}	2.7	0.952×10^{-1}	1.4	0.113	1.1	0.611×10^{-1}	1.3	0.389×10^{-1}	1.6	Skeleton (total)
Skin			0.209×10^{-3}	34	0.176×10^{-2}	10	0.573×10^{-2}	4.3	0.599×10^{-2}	3.1	0.541×10^{-2}	3.1	0.547×10^{-2}	3.7	Skin
Spleen			0.128×10^{-3}	45	0.891×10^{-3}	16	0.182×10^{-2}	9.9	0.205×10^{-2}	7	0.146×10^{-2}	7.4	0.146×10^{-2}	8	Spleen
Thyroid									0.383×10^{-4}	35	0.329×10^{-4}	35	0.539×10^{-4}	38	Thyroid
Uterus											0.153×10^{-4}	47	0.157×10^{-4}	46	Uterus
Trunk	0.996	0.03	0.996	0.03	0.981	0.08	0.875	0.18	0.602	0.37	0.384	0.53	0.351	0.54	Trunk
Legs							0.475×10^{-4}	46	0.374×10^{-3}	27	0.390×10^{-3}	21	0.427×10^{-3}	20	Legs
Head					0.212×10^{-3}	35	0.310×10^{-2}	7.4	0.745×10^{-2}	4.4	0.793×10^{-2}	4.0	0.737×10^{-2}	4.0	Head
Total body	0.996	0.03	0.996	0.03	0.981	0.08	0.878	0.18	0.610	0.37	0.392	0.52	0.359	0.53	Total body

Source: MIRD Pamphlet No. 5, W. S. Snyder *et al.*, Society of Nuclear Medicine, 1969, pp. 30–31, with permission of the publisher.

from Eq. (16.61) and using the unmodified activities. Problem 16.30 shows how these reformulations work in a simple case.

16.9. RADIOPHARMACEUTICALS AND TRACERS

A radioactive nucleus by itself is not very useful. It must usually be attached to some substance that will give it the desired biological properties, for example, to be preferentially absorbed in the region of interest. It must also be prepared in a sterile form, free of toxins that produce a fever (pyrogens) so that it can be injected in the patient. This section surveys some of the properties of radiopharmaceuticals.

16.9.1. Physical Properties

The half-life must be short enough so that a reasonable fraction of the radioactive decays take place during the diagnostic procedure; any decays taking place later give the patient a dose that has no benefit. (This requirement can be relaxed if the biological excretion is rapid.) On the other hand, the lifetime must be long enough so that the radiopharmaceutical can be prepared and delivered to the patient.

For diagnostic work, the decay scheme should minimize the amount of nonpenetrating radiation. Such radiation provides a dose to the patient but never reaches the detector. This means that there should be as few charged particles (β particles) as possible. The ideal source then is a γ source, which means that the nucleus is in an excited state (is an isomer). Such states are usually very short-lived. Not only should the nucleus be a γ emitter, but the internal conversion coefficient should be small, since internal conversion produces nonpenetrating electrons. Positron emitters are more desirable than are β^- emitters because the positrons produce 0.5-MeV γ radiation that can reach an external detector. For therapy, on the other hand, nonpenetrating radiation is ideal.

It is also necessary that the decay product have no undesirable radiations. If the decay is a β^- or β^+ decay, the product has different chemical properties from the parent and may be taken up selectively by a different organ. If it is also radioactive, this can confuse a diagnosis and give an undesirable dose to the other organ.

Ease of chemical separation of the radioactive substance from whatever carrier it is produced with is also important. It is necessary to remove the radioactive isotope from stable isotopes of the same element, because the chemicals are usually toxic. This toxicity is avoided by giving the chemical in minute amounts, which can only be done if the specific activity is high.

16.9.2. Biological Properties

For diagnostic work, a pharmaceutical is needed that is taken up more by the diseased tissue to give a "hot spot" or taken up less to give a "cold spot." The former is easier to see with small amounts of radioactivity, but both techniques are used. For therapy one wishes to have selective absorption of the pharmaceutical so that the radiations will destroy the target organ but not the rest of the body. There are several mechanisms by which a pharmaceutical may be localized.

a. Active transport. The drug is concentrated by a specific organ against a concentration gradient. Examples are the selective concentration of iodine in the thyroid, salivary, and gastric glands. (It is rapidly excreted from the last two but is retained in the thyroid). This technique is also effective for certain drugs in the kidney.

b. Phagocytosis. Particles in the size range 1–1000 nm may be phagocytized—taken up by specialized cells of the reticuloendothelial system. This can take place in liver, bone marrow, and spleen. Particles of size 1 nm go to the Kupfer cells of the liver and to the marrow, while larger particles (100–1000 nm) are gathered by phagocytes in the liver and spleen.

c. Sequestration. Still larger particles, such as red blood cells that have been denatured by heat, are gathered in the spleen or liver by the process called sequestration. The particles are trapped as the blood percolates through the pulp of the spleen and are later phagocytized.

d. Capillary blockade. The capillaries have a diameter of 7–10 μm. Particles from 20 to 40 μm diameter injected into a vein will find progressively larger vessels as they work their way through the right heart and will be stopped in the capillaries of the lung.

e. Diffusion. It is also possible for a pharmaceutical to move through a membrane to a region of lower concentration. There is a blood–brain barrier between the blood and the central nervous system. It is relatively impermeable even to small ions. In a brain scan the chemical is not concentrated in normal brain tissue but leaks into tissue where the blood–brain barrier is compromised by a lesion.

f. Compartmental localization. A suitable pharmaceutical injected in the blood my remain there a long time, mixing well and allowing the blood volume to be determined.

The most widely used[8] isotope is ^{99m}Tc. As its name suggests, it does not occur naturally. We will consider it in some detail to show how an isotope is actually used. Its decay scheme has been discussed above. There is a nearly monoenergetic 140-keV γ ray. Only about 10% of the energy is in the form of nonpenetrating radiation. The isotope is produced in the hospital from the decay of its parent, ^{99}Mo, which is a fission product of ^{235}U and can be

[8]Other common isotopes are ^{201}Tl (thallium), ^{67}Ga (gadolinium), ^{123}I, and ^{111}In (indium).

separated from about 75 other fission products. The ^{99}Mo decays to ^{99m}Tc.

The technetium is made available to hospitals through a "generator" that was developed at Brookhaven National Laboratories in 1957 and is easily shipped. Isotope ^{99}Mo, which has a half-life of 67 h, is adsorbed on an alumina substrate in the form of molybdate (MoO_4^{2-}). From 8 to 100 GBq of ^{99}Mo can be provided. The heart of such a generator (without the lead shielding) is shown in Fig. 16.14. As the ^{99}Mo decays, it becomes pertechnetate (TcO_4^-). Sterile isotonic eluting solution is introduced under pressure above the alumina and passes through after filtration into an evacuated eluate container. After removal of the technetium, the continued decay of ^{99}Mo causes the ^{99m}Tc concentration to build up again. A generator lasts about a week.

Several steps must be taken to prepare the pertechnetate as a radiopharmaceutical. First, it must be checked for breakthrough of the ^{99}Mo. The Nuclear Regulatory Commission allows 1.5×10^{-4} Bq of ^{99}Mo per Bq of ^{99m}Tc. The purity is checked by placing the eluate in a lead sleeve that attenuates the ^{99m}Tc γ ray much more than the $\approx$750-keV γ rays from ^{99}Mo and measuring the activity. It is also checked with a colorimetric test for the presence of aluminum ion.

The eluate can be used directly for imaging brain, thyroid, salivary gland, urinary bladder, and blood pool, or it can be combined with phosphate, albumin or aggregated albumin, colloidal sulfur, or $FeCl_3$. Commercial kits are available for making these preparations.

For example, kits for labeling aggregated human albumin are commercially available. A vial containing 10 ml of saline solution is enough for ten doses. The aggregated albumin particles are 10–70 μm in diameter. Each milliliter of solution contains $(4–8)\times10^5$ particles. Tin is attached to the microspheres and serves to bind technetium. Up to 10^9 Bq of technetium pertechnetate is added to the vial by the user. A typical adult dose is 10–40 MBq (3.5×10^5 albumin particles).

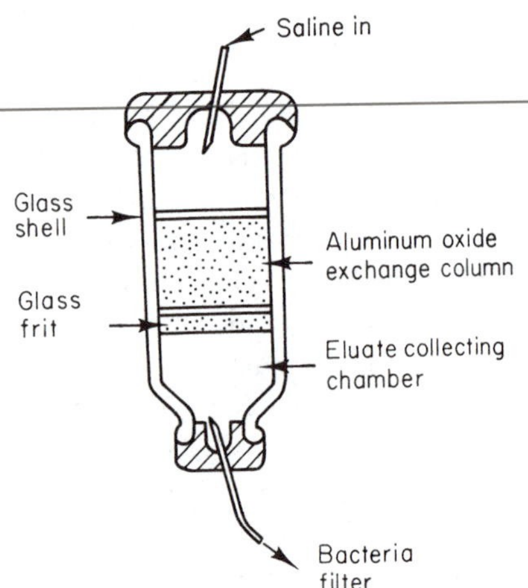

FIGURE 16.14. A ^{99}Mo–^{99m}Tc generator system. Molybdenum is trapped in the aluminum oxide layer. Eluant introduced at the top flows through and is collected at the bottom.

Other common isotopes are ^{201}Tl, ^{67}Ga, and ^{123}I. Thallium, produced in a cyclotron, is chemically similar to potassium and is used in heart studies, though it is being replaced by ^{99m}Tc–sestamibi and ^{99m}Tc–tetrofosmin. Gallium is used to image infections and tumors. Iodine-123 is also produced in a cyclotron and is used for thyroid studies.

16.10. SAMPLE DOSE CALCULATION

We pull this discussion together by making a simplified calculation of the dose to various organs from ^{99m}Tc-labeled microspheres used in a lung scan. We assume that 37 MBq of ^{99m}Tc is injected, that it all lodges in the capillaries of the lung, and that it remains there long enough so that the half-life is the physical half-life.[9] The residence time is then

$$\tau_h = 1.443 T_{1/2} = (1.443)(6) = 8.7 \text{ h},$$

so the cumulated activity is $\tilde{A}_{\text{lung}} = (3.7\times10^7)(8.7\times60\times60) = 1.16\times10^{12}$ Bq s.

The steps in calculating S and the dose are summarized in Table 16.4, which combines information on Δ_i with values of ϕ for lung, heart, liver, head, and the whole body. The values of Δ_i are from Fig. 16.15, which is a more recent version of the output table of Fig. 16.4. Table 16.4 refers to the line number for the entry in both Fig. 16.15 and Fig. 16.4. It is worth taking time to compare the entries with the lines in Fig. 16.4 to see the variations with more recent information. Values of $\phi_i(r_k \leftarrow r_h)$ are taken to be 1 in the lung and the whole body when the radiations are nonpenetrating, that is, photons with energy less than 25 keV or electrons. For the high-energy photons (line 3) the values of ϕ are obtained by interpolation to 140 keV from the values in Table 16.3 for 100 and 200 keV. The sum $S_k = \Sigma \Delta_i \phi_i$, the mass of each organ, and the dose for the cumulated activity of 1.16×10^{12} Bq s are shown on the bottom lines. The dose to the lungs is 4.3×10^{-3} Gy. The whole body dose is much less because the absorbed energy is divided by the mass of the entire body. The value of ϕ shows that about 38% of the photon energy is absorbed in the body.

The dose to the lungs is considerably greater than in a chest x ray; however, a chest x ray is almost useless for diagnosing a pulmonary embolus. The whole body dose is not unreasonable. The problems consider what fraction of the capillaries are blocked by this procedure.

[9]The last is not a good assumption. The ^{99m}Tc leaches from the microspheres into the general circulation. A more accurate calculation requires measurements and the use of a convolution integral. It is described in Loevinger, Budinger and Watson (1988, pp. 79–81). The principle residence times are 4.3 h in the lung, 1.8 h in the extravascular space, 0.83 h in the urine, 0.7 h in the kidney, and 0.6 h in the blood.

TABLE 16.4. *Values of Δ_i, E_i, and Φ_i for ^{99m}Tc in the lung.*

			ϕ_i						
i, Line in Fig. 16.15	Δ_i (Gy kg^{-1} Bq^{-1} s^{-1})	(keV)	Lung	Heart	Liver	Head	Whole body	Line in Fig. 16.4	Δ_i(g rad μCi^{-1}h^{-1}
	0.00							1	0.0000
1	2.56×10^{-16}	e	1	0	0	0	1	2	0.0034
2	2.64×10^{-17}	e	1	0	0	0	1		0.0004
3	2.00×10^{-14}	140.5	0.0495	0.0135	0.0166	0.0077	0.3785	3	0.2670
4	1.70×10^{-15}	e	1	0	0	0	1	4	0.0252
5–7	2.15×10^{-16}	e	1	0	0	0	1	5	0.0031
8	4.36×10^{-17}	e	1	0	0	0	1	6	0.0006
9	8.43×10^{-18}	e	1	0	0	0	1		0.0001
	0.00	142.6	0.0495	0.0135	0.0166	0.0077	0.3785	7	0.0000
10	1.08×10^{-16}	e	1	0	0	0	1	8	0.0014
11–13	3.84×10^{-17}	e	1	0	0	0	1	9	0.0005
14	7.64×10^{-18}	e	1	0	0	0	1	10	0.0001
15	1.17×10^{-16}	18.37	1	0	0	0	1	11	0.0016
16	6.14×10^{-17}	18.25	1	0	0	0	1	12	0.0008
17	2.26×10^{-17}	20.62	1	0	0	0	1	13	0.0003
18–22	3.42×10^{-17}	e	1	0	0	0	1	14	0.0005
23–25	1.62×10^{-17}	e	1	0	0	0	1	15	0.0002
26–28	3.37×10^{-17}	e	1	0	0	0	1	16	0.0045
29	7.27×10^{-17}	e	1	0	0	0	1	17	0.0010
$\Sigma\Delta_i\phi_i$			3.75×10^{-15}	2.70×10^{-16}	3.32×10^{-16}	1.54×10^{-16}	1.03×10^{-14}		
m (kg)			0.999	0.603	1.833	5.278	70.036		
$S=\Sigma\Delta_i\phi_i/m$			3.76×10^{-15}	4.48×10^{-16}	1.81×10^{-16}	2.92×10^{-17}	1.48×10^{-16}		
Dose (Gy)	A_0=37 MBq		4.32×10^{-3}	5.15×10^{-4}	2.08×10^{-4}	3.36×10^{-5}	1.70×10^{-4}		

Table 16.5 shows some typical doses from various nuclear medicine procedures.

16.11. AUGER ELECTRONS

In Sec. 14.9 we discussed the deexcitation of atoms, including the emission of Auger and Coster–Kronig electrons. The Auger cascade means that several of these electrons are emitted per transition. If a radionuclide is in a compound that is bound to DNA, the effect of several electrons released in the same place is to cause as much damage per unit dose as high-LET radiation. Linear energy transfer was defined in Chap. 14. A series of reports on this effect have been released by the American Association of Physicists in Medicine (AAPM) [Sastry (1992); Howell (1992); Humm *et al.* (1994)].

Many electrons (up to 25) can be emitted for one nuclear transformation, depending on the decay scheme [Howell (1992)]. The electron energies vary from a few eV to a few tens of keV. Corresponding electron ranges are from less than 1 nm to 15 μm. The diameter of the DNA double helix is about 2 nm. A number of experiments (reviewed in the AAPM reports) show that when the radioactive substance is in the cytoplasm the cell damage is like that for low-LET radiation in Fig. 15.32 with relative biological effectiveness (RBE)=1. When it is bound to the DNA, survival curves are much steeper, as with the α particles in Fig. 15.32 (RBE≈8). When it is in the nucleus but not bound to DNA the RBE is about 4. The fraction of the Auger emitter that binds to the DNA depends on the chemical agent to which the nuclide is attached.

A number of models have been constructed for both the electron spectrum and for the DNA strand breaks caused by radioactive nuclei at different distances from the DNA. Experiments, described in the reviews, have been done with cell cultures and *in vivo*, primarily with mouse testes.

If different chemicals containing ^{99m}Tc are injected in mouse testes, one can look for deformed sperm heads [Narra *et al.* (1994)]. Even though the chemicals bind to DNA in different amounts, the effective RBE values are all about 1, within experimental uncertainties. This seems to contradict what has been said above, but a careful analysis of the radiations explains the result. The expected RBE for a particular nuclide can be modeled by

$$\text{RBE}_{\text{expected}}=f_{\text{photon}}\text{RBE}_{\text{photon}}+f_{\text{ICE}}\text{RBE}_{\text{ICE}}$$
$$+f_{\text{Auger}}f_{\text{DNA}}\text{RBE}_{\text{Auger}}, \tag{16.62}$$

43-TECHNETIUM-99M

HALF-LIFE = 6.01 HOURS 11-AUG-86
DECAY MODE(S): β^-, IT

RADIATION	PARTICLES/ TRANSITION n(i)	ENERGY/ PARTICLE E(i) MeV	ENERGY/TRANSITION $\Delta(i)$ rad g/µCi h	$\Delta(i)$ Gy kg/Bq s
ce-M, γ 1	9.16E-01	1.748E-03†	3.41E-03	2.56E-16
ce-N⁺, γ 1	7.58E-02	2.173E-03†	3.51E-04	2.64E-17
γ 2	8.91E-01	1.405E-01	2.67E-01	2.00E-14
ce-K, γ 2	8.84E-02	1.195E-01	2.25E-02	1.70E-15
ce-L_1, γ 2	9.72E-03	1.375E-01	2.85E-03	2.15E-16
ce-L_2, γ 2	6.32E-04	1.377E-01	1.85E-04	1.40E-17
ce-L_3, γ 2	3.29E-04	1.378E-01	9.66E-05	7.26E-18
ce-M, γ 2	1.94E-03	1.401E-01†	5.79E-04	4.36E-17
ce-N⁺, γ 2	3.74E-04	1.405E-01†	1.12E-04	8.43E-18
ce-K, γ 3	5.53E-03	1.216E-01	1.43E-03	1.08E-16
ce-L_1, γ 3	9.34E-04	1.396E-01	2.78E-04	2.08E-17
ce-L_2, γ 3	1.94E-04	1.398E-01	5.78E-05	4.34E-18
ce-L_3, γ 3	5.92E-04	1.400E-01	1.77E-04	1.33E-17
ce-M, γ 3	3.35E-04	1.422E-01†	1.02E-04	7.64E-18
$K\alpha_1$ x-ray	3.99E-02	1.837E-02	1.56E-03	1.17E-16
$K\alpha_2$ x-ray	2.10E-02	1.825E-02	8.17E-04	6.14E-17
$K\beta_1$ x-ray	6.82E-03	2.062E-02	3.00E-04	2.26E-17
Auger-KL_1L_1	1.23E-03	1.487E-02	3.90E-05	2.95E-18
Auger-KL_1L_2	2.14E-03	1.512E-02	6.90E-05	5.17E-18
Auger-KL_1L_3	1.64E-03	1.524E-02	5.33E-05	4.00E-18
Auger-KL_2L_3	6.44E-03	1.547E-02	2.12E-04	1.60E-17
Auger-KL_3L_3	2.43E-03	1.559E-02	8.07E-05	6.07E-18
Auger-KL_1X	1.80E-03	1.755E-02†	6.73E-05	5.05E-18
Auger-KL_2X	1.42E-03	1.780E-02†	5.39E-05	4.05E-18
Auger-KL_3X	2.49E-03	1.792E-02†	9.51E-05	7.13E-18
Auger-L_2MM	1.98E-02	2.125E-03†	8.97E-05	6.74E-18
Auger-L_2MX	8.37E-03	2.538E-03†	4.53E-05	3.40E-18
Auger-L_3MM	4.81E-02	2.009E-03†	2.06E-04	1.55E-17
Auger-L_3MX	2.07E-02	2.422E-03†	1.07E-04	8.03E-18
Auger-MXY	1.11E+00	4.092E-04†	9.68E-04	7.27E-17
Listed x,γ and γ± radiations			2.69E-01	2.02E-14
Omitted x,γ and γ± radiations‡			3.22E-04	2.42E-17
Listed β,ce and Auger radiations			3.43E-02	2.58E-15
Omitted β,ce and Auger radiations‡			1.22E-04	9.20E-18
Listed radiations			3.03E-01	2.27E-14
Omitted radiations‡			4.45E-04	3.35E-17

† Average energy
‡ Each omitted transition contributes <0.100% to $\Sigma\Delta(i)$ in its category.
RUTHENIUM-99 daughter, yield 3.70E-05, is stable.
TECHNETIUM-99 daughter, yield 9.9996E-01, is radioactive.

FIGURE 16.15. The values of Δ_i for ^{99m}Tc used in Table 16.4 to calculate the dose from a source in the lungs. These are more recent values than those in Fig. 16.4. The reference at the bottom to ruthenium-99 is because ^{99m}Tc undergoes β decay in 3.7×10^{-5} of transitions. This does not affect any of the entries in the table. From Weber *et al.* (1989).

TABLE 16.5. *Some typical doses for nuclear medicine procedures.*

Study and agent	A_0 (MBq)	Organ and highest dose (mSv)		Total body dose (mSv)	Effective dose (mSv)
Bone ^{99m}Tc-pyrophosphate	555	Blader wall	51	2.0	4.4
Heart ^{201}Tl-chloride	55	Kidneys	20	3.6	13
Liver ^{99m}Tc-sulfur colloid	185	Bladder wall	17	0.9	2.6

Adapted from Table 9-3 in P. B. Zanzonico, A. B. Brill, and D. V. Becker, Radiation dosimetry, Chap. 9 in H. N. Wagner, Jr., Z. Szabo, and J. W. Buchanan, eds. *Principles of Nuclear Medicine*, 2nd. ed. Philadelphia, Saunders.

where f_{photon}, f_{ICE}, and f_{Auger} are the fractions of the absorbed dose to the organ (in this case the testes) resulting from photons, from internal-conversion electrons (ICE), and from Auger electrons. For ^{99m}Tc these three numbers are 0.058, 0.889, and 0.053. The internal conversion electrons have high energies and long ranges, so $RBE_{photon}=RBE_{ICE}=1$. Using a relatively large value $RBE_{Auger}=10$, we have

$$RBE_{expected}=0.058+0.889+(0.053)(10)f_{DNA}$$

$$=0.947+0.53f_{DNA}. \tag{16.63}$$

Thus, because such a small fraction of the ^{99m}Tc dose is in the form of Auger electrons, $RBE_{expected}$ can range from 0.95 to 1.5. The measured value of f_{DNA} for ^{99m}Tc–hydroxyethylene diphosphate is 0.13, leading to $RBE_{expected}=1.0$, in good agreement with an experimental value $RBE=1.1\pm0.16$.

16.12. DETECTORS; THE GAMMA CAMERA

Nuclear medicine images do not have the inherent spatial resolution of diagnostic x-ray images; however, they provide functional information: the increase and decrease of activity as the radiopharmaceutical passes through the organ being imaged.

Early measurements were done with single detectors such as the scintillation detector shown in Fig. 16.16. Directional sensitivity is provided by a collimator, which can be cylindrical or tapered. Single detectors are still used for *in vitro* measurements and for thyroid uptake studies.

Two-dimensional images can be taken with the *scintillation camera* or *gamma camera* shown in Fig. 16.17. The scintillator is 6–12 mm thick and 17–60 cm in diameter. Modern scintillators are rectangular. The scintillator is viewed by an array of photomultiplier tubes. Nineteen are shown in Fig. 16.17, but current cameras use 50–100 tubes. The tubes are arranged in a hexagonal array, as shown in Fig. 16.18. The tube nearest where the photon interacts receives the greatest signal. Signals from each tube are combined to give the total energy signal and also to give x and y position signals.

The collimator is a critical component of the gamma camera. The channels are usually hexagonal, with walls just thick enough to stop most of the photons which do not pass down the collimator opening. The collimator usually has parallel channels; single pinholes, diverging, and

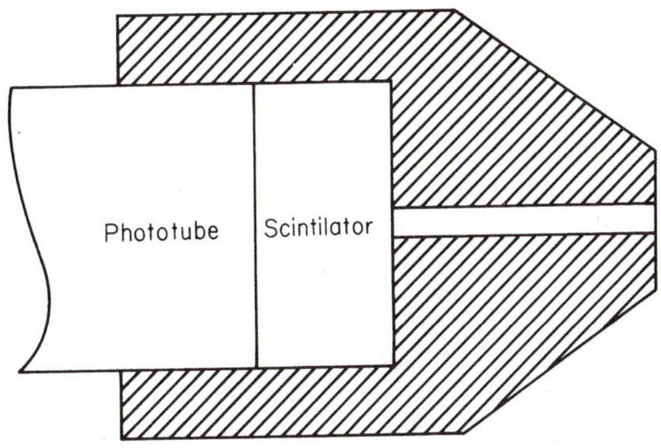

FIGURE 16.16. A scintillator with a lead collimator to give directional sensitivity.

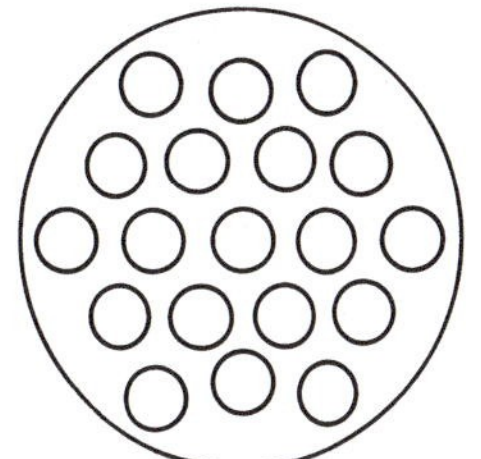

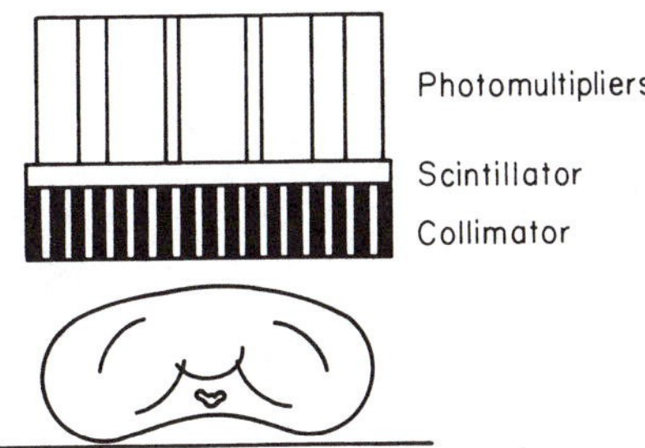

FIGURE 16.17. Schematic of a scintillation camera. A collimator allows photons from the patient to strike the scintillator directly above the source. An array of photomultiplier tubes records the position and energy of the detected photon.

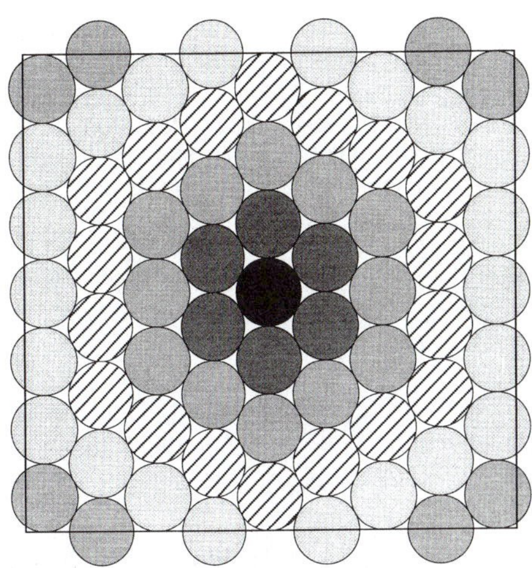

FIGURE 16.18. A rectangular scintillator viewed by an array of photomultiplier tubes. The hexagonal arrangement of the tubes above the scintillator in a gamma camera gives the closest spacing between tubes.

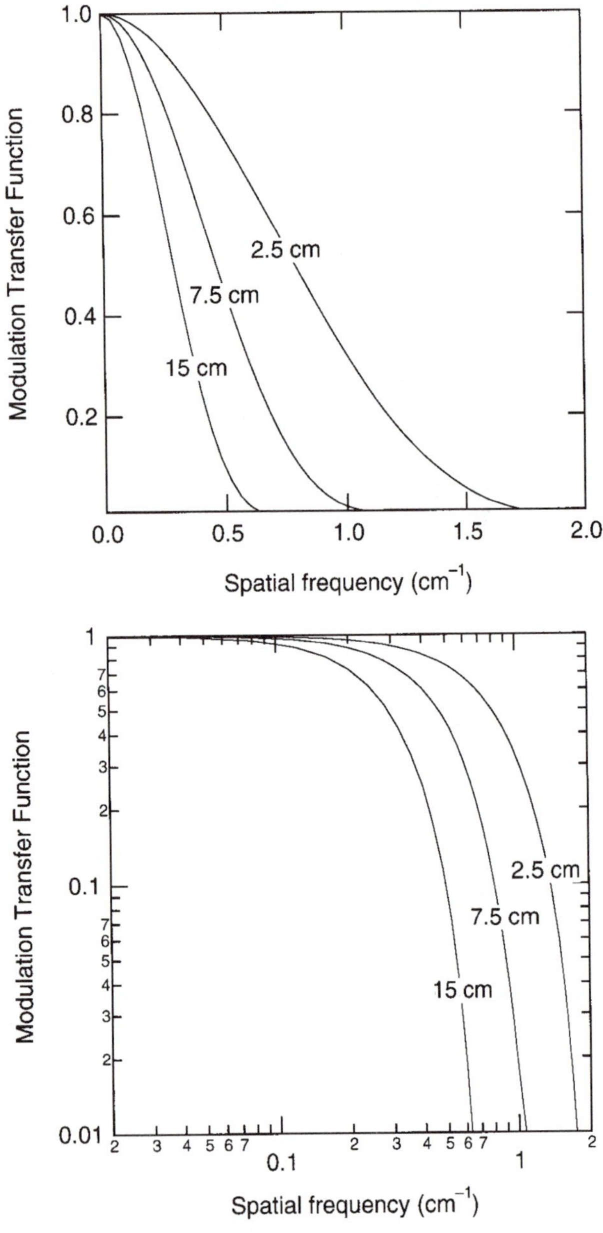

FIGURE 16.19. Modulation transfer function curves for a typical parallel hole collimator for different source-to-collimator distances. Both linear and log–log plots are shown. The source-to-collimator distances are 2.5, 7.5, and 15 cm. Data are from J. C. Erhardt, L. W. Oberly, and J. M. Cuevas, *Imaging Ability of Collimators in Nuclear Medicine*, Rockville, MD, U.S. Dept. HEW, Publ. No. (FDA)79-8077 (1978), p. 39.

converging channels are sometimes used and can lead to geometric distortions of the image [Sorenson and Phelps (1987, p. 311]. The spatial resolution depends on the distance from the source to the collimator, as shown for one collimator in Fig. 16.19. There are tradeoffs between sensitivity and resolution [Links and Engdahl (1995)]. Some of the aspects of collimator design are discussed in Problems 16.45–16.47.

Problem 16.47 shows that the detection efficiency,

$$g = K^2 \left(\frac{d}{l} \right)^2 \left(\frac{d}{d+t} \right)^2 \qquad (16.64)$$

is inversely proportional to the square of the collimator thickness l but is independent of the collimator-to-source distance b. The reason, discussed in the problems, is similar to what we saw for images formed by lenses in Chap. 13.

Figure 16.20 shows a bone scan of a young patient taken with a gamma camera. The ^{99m}Tc–diphosphonate is taken up in areas of rapid bone growth. This is a scan of a child, and the bone growth at the epiphyses at the end of each bone can be seen. There are also hot spots at the injection site, in one kidney, and in the bladder.

Nuclear medicine can show physiologic function. For example, if the isotope is uniformly distributed in the blood, viewing the heart and synchronizing the data accumulation with the electrocardiogram (*gating*) allows one to measure blood volume in the heart when it is full and contracted, and to calculate the *ejection fraction*, the fraction of blood in the

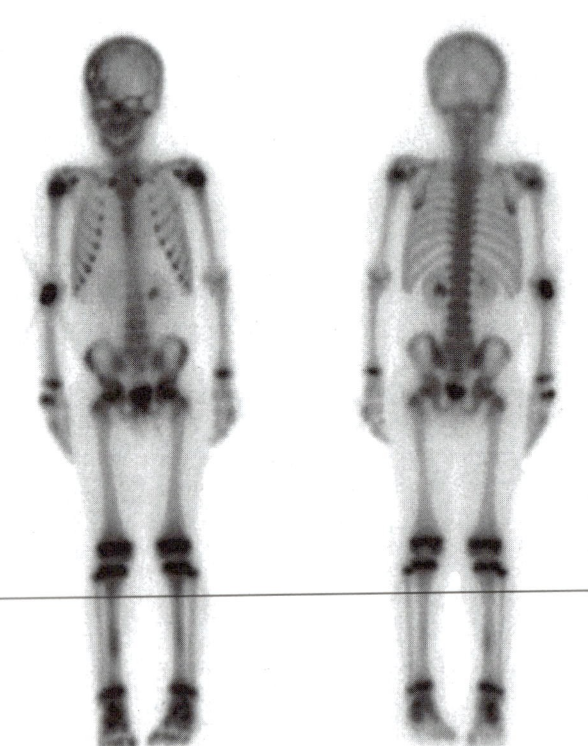

FIGURE 16.20. A scintillation camera "bone scan" of a 7-year-old male who received a ^{99m}Tc–diphosphonate injection. An anterior view is on the left, and a posterior view is on the right. The scan shows an area of decreased uptake in the right anterior skull consistent with an eosinophillic granuloma. Identifiable hot regions are the injection site in the right elbow, an attempted injection site in the right hand, the bladder, and the left kidney, which is probably not remarkable on this exam, along with the ends of the long bones. Photograph courtesy of B. Hasselquist, Ph.D., Department of Diagnostic Radiology, University of Minnesota.

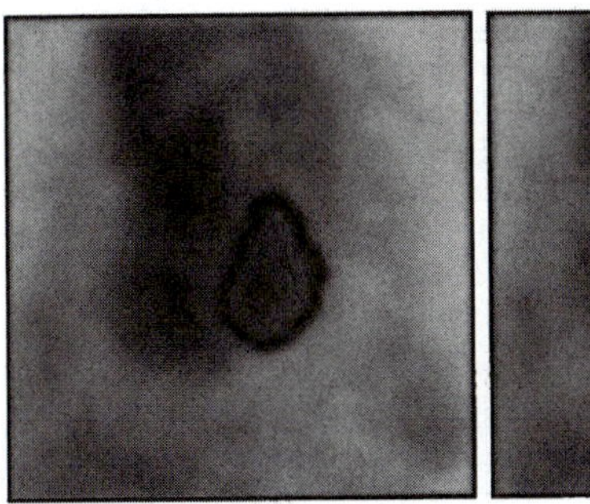
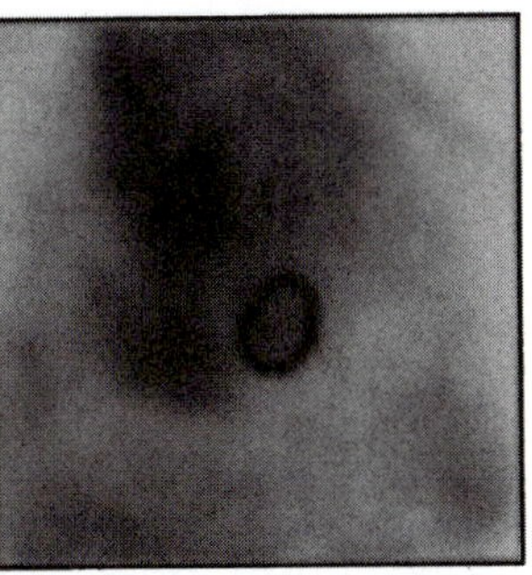

FIGURE 16.21. Two gated scintillation camera views of the heart, imaged with ^{99m}Tc-labeled red blood cells. The dots outline the left ventricle. On the left is end diastole (left ventricle filled with blood). On the right is end systole (left ventricle at smallest volume). The ejection fraction is 66%. Photograph courtesy of B. Hasselquist, Ph.D., Department of Diagnostic Radiology, University of Minnesota.

full left ventricle that is pumped out. This is shown in Fig. 16.21, which shows pictures and contours of the heart at end-systole and end-diastole. The imaging agent was ^{99m}Tc-labeled human red blood cells.

Figure 16.22 shows a series of images taken at six different angles around the patient. The patient has had a lung transplant. The left lung is new and shows considerably more activity than the diseased right lung.

16.13. SINGLE-PHOTON EMISSION COMPUTED TOMOGRAPHY

Still another detection scheme is analogous to computed tomography (CT). The detector is sensitive to all radioactivity along a line passing through the patient. The counting rate is thus proportional to a projection through the patient, and a cross-sectional slice can be reconstructed from a series of projections, just as was done with x-ray CT. This is *single-photon emission computed tomography* (SPECT). A series of images like those in Fig. 16.22, but at more angles, are used to reconstruct a three-dimensional image that can then be viewed from any direction, with slices at any desired depth. A SPECT scan is shown in Fig. 16.23. There are five reconstructed slices in planes parallel to the long axis of the heart. The left ventricle is prominent, and the right ventricle can be seen faintly in the last few slices.

One of the problems with SPECT is photon attenuation along the projection line. This is shown in Fig. 16.24 for a cylindrical source with uniform activity throughout. Let A be the activity per unit volume, and ignore variations in $1/r^2$. The projection $F(x)$ is

$$F(x)=\int_{-a}^{a} A(x,y)\Delta x\ \Delta z\ e^{-\mu(y+a)}dy,$$

where $dy\ \Delta x\ \Delta z$ is the volume detected. This can be integrated to give

$$F\propto \frac{1}{\mu}\,(1-e^{-2\mu(R^2-x^2)^{1/2}}). \tag{16.65}$$

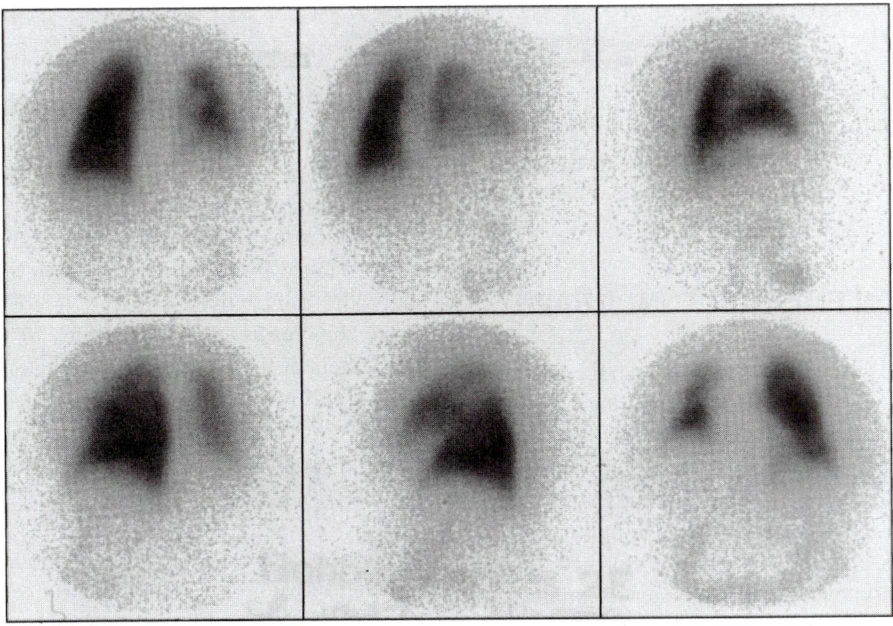

FIGURE 16.22. Lung scans of a patient who has received a lung transplant. The upper left is a posterior view; each successive view is rotated about the patient, ending with an anterior view on the lower right. The left lung is the transplant. It has much more activity than the diseased right lung. Photograph courtesy of B. Hasselquist, Ph.D., Department of Diagnostic Radiology, University of Minnesota.

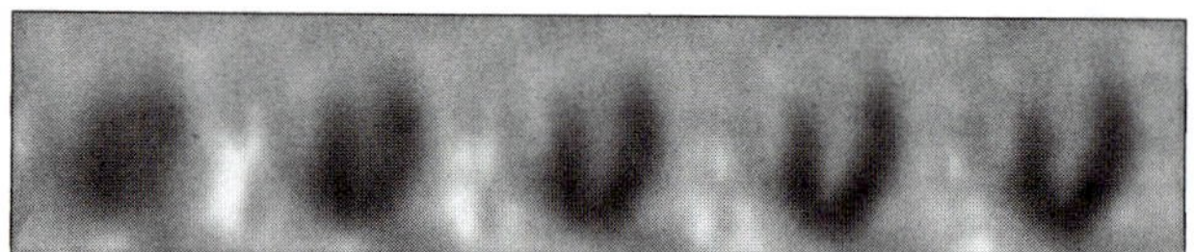

FIGURE 16.23. Single photon emission computed tomographic (SPECT) slices of the heart. The patient was injected with ^{99m}Tc–tetrofosmin, an agent that is taken up by myocardium. The images have been reconstructed in planes parallel to the axis of the heart. The obvious myocardium surrounds the left ventricle. In the last image on the right the right ventricle can also be seen. Photograph courtesy of B. Hasselquist, Ph.D., Department of Diagnostic Radiology, University of Minnesota.

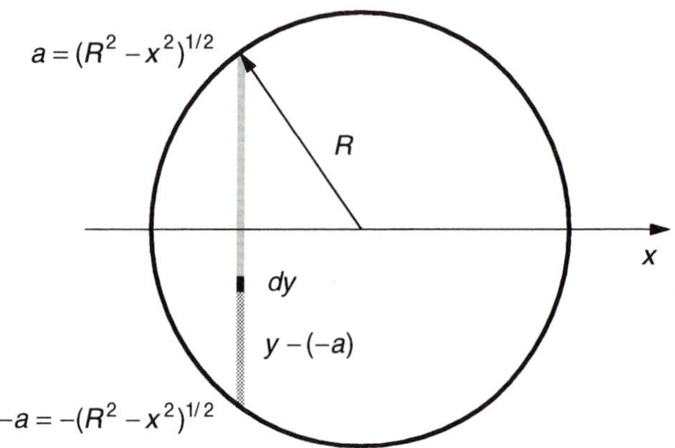

FIGURE 16.24. Projection perpendicular to the x axis for a radioactive source of uniform concentration, including the effect of photon attenuation.

For this uniform case the distortion is $1-e^{-2\mu a}$, where $2a$ is the thickness of the source. This is plotted in Fig. 16.25 for $\mu=0$, $\mu=10\ \mathrm{m}^{-1}$ (511-keV annihilation radiation) and $\mu=15\ \mathrm{m}^{-1}$ (140 keV ^{99m}Tc). Corrections are made in a number of ways.[10]

16.14. POSITRON EMISSION TOMOGRAPHY

If a positron emitter is used as the radionuclide, the positron comes to rest and annihilates an electron, emitting two annihilation photons back to back. These can be detected in coincidence. This simplifies the attenuation correction, because the total attenuation for both photons is the same for all points of emission along each ray through the body (see Problem 16.47). Positron emitters are short-lived, and for most it is necessary to have an accelerator for producing them in or near the hospital. Some of the lighter positron emitters have the advantage of being natural constituents of molecules in the body (Table 16.6).

[10]See Sorenson and Phelps (1987, pp. 409ff); Larsson (1980).

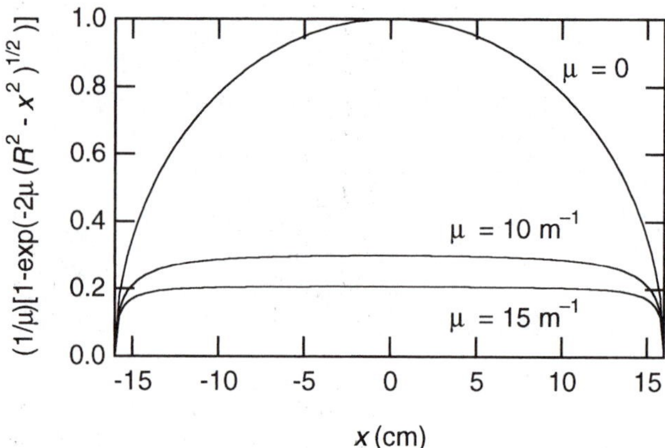

FIGURE 16.25. Plot of the projection including attenuation, Eq. (16.64), for $\mu=0$, $\mu=10\ \mathrm{m}^{-1}$ (corresponding to 511-keV annihilation radiation) and $\mu=15\ \mathrm{m}^{-1}$ (corresponding to the photons from ^{99m}Tc).

A positron emission tomography (PET) scan overlayed on an MRI image is shown in Fig. 16.26. On the bottom is a three-dimensional image viewed from above and to the right of the brain that has been sliced towards the back of the brain through the motor strip and cerebellum, with the positron image overlaid on the slice. The positron emitter is ^{15}O-labeled water. The subject is sequentially touching each finger of the left hand with the thumb. Activity can be seen in the right cerebral sensorimotor cortex the (slice, upper right) and in the left cerebellum (slice, lower left). The technique is described by Rehm *et al.* (1994) and Strother *et al.* (1995).

16.15. BRACHYTHERAPY AND INTERNAL RADIOTHERAPY

Brachytherapy (*brachy* means short) involves implanting in the tumor sources for which the radiation falls off rapidly with distance because of attenuation and also $1/r^2$. Originally the radioactive sources ("seeds") were implanted surgically, resulting in high doses to the operating room personnel. In the *afterloading* technique, developed in the 1960s, hollow catheters were implanted surgically and the sources were inserted after the surgery. Remote afterloading, developed in the 1980s, places the sources by remote control, so that only the patient receives a radiation dose.

We saw in Chap. 15 that fractionation of the dose results in better sparing of normal tissue for a given probability of killing the tumor. Afterloading allows the sources to be placed and removed, but it is often difficult for the patient to tolerate the catheters for long periods of time. This has led to the development of *high-dose-rate* brachytherapy (HDR), in which the dose is given in one or a few fractions

TABLE 16.6. *Positron emitters used in nuclear medicine.*

Nuclide	Half-life
$^{11}_{6}C$	20.3 min
$^{13}_{7}N$	10.0 min
$^{15}_{8}O$	124 s
$^{18}_{9}F$	109.7 min

over the course of a day or two [Nag (1994)]. Though this is much easier for the patient, tissue sparing is not as great as with a longer treatment. Current practice seeks to compensate for this by meticulous treatment planning based on an extended version of the linear-quadratic model, and by making sure that the tumor receives much higher doses than the surrounding normal tissue.

Radium was the first brachytherapy source, but it has been replaced by ^{60}Co, ^{192}Ir, and ^{137}Cs. Conventional low-dose-rate brachytherapy is delivered at 0.4–1.0 Gy hr^{-1}. High dose rates are about 1 Gy min^{-1}.

Internal radiotherapy treats the patient with a radionuclide in a chemical that is selectively taken up by the tumor.

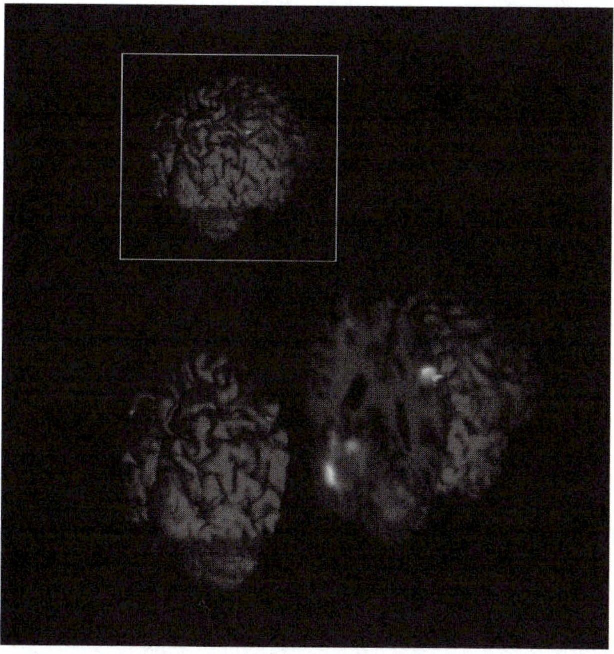

FIGURE 16.26. A positron emission tomography (PET) scan is overlayed on an MRI image. On the bottom is a three-dimensional image viewed from above and to the right of the brain, that has been sliced towards the back of the brain through the motor strip and cerebellum, with the positron image overlaid on the slice. The positron emitter is ^{15}O-labeled water. The subject is sequentially touching each finger of the left hand with the thumb. Activity can be seen in the right cerebral sensorimotor cortex (slice, upper right) and in the left cerebellum (slice, lower left). Image courtesy of the PET Imaging Service, Veterans Administration Medical Center, Minneapolis.

The classic example is the administration by mouth of capsules containing ^{131}I for treatment of hyperthyroidism and thyroid cancer. Other nuclides are being used experimentally for breast and neuroendocrine tumors and melanoma [Fritzberg and Wessels (1995)]. A radionuclide for this purpose should emit primarily nonpenetrating radiation, have a physical half-life long compared to the biological half-life, have a large activity per unit mass, and exhibit a high degree of specificity for the tumor. If the nuclide can be delivered within the cell nucleus, then the high RBE of Auger electrons can be exploited. One way to achieve high concentrations in the tumor (though not in the nucleus) is to tag monoclonal antibodies with the radionuclide [see the special issue of *Medical Physics* edited by Buchsbaum and Wessels (1993)]. The MIRD formulation can be adapted to the dose calculations [Watson, Stabin, and Siegel (1993)].

16.16. RADON

The naturally occurring radioactive nuclei are either produced continuously by cosmic ray bombardment or they are the products in a decay chain from a nucleus whose half-life is comparable to the age of the earth. Otherwise they would have already decayed. There are four naturally occurring radioactive decay chains near the top of the periodic table. One of these is the decay products from $^{238}_{92}$U, shown in Fig. 16.27. The half-life of uranium-238 is 4.5×10^9 yr. A series of α and β decays lead to radium-226, which undergoes α decay with a half-life of 1620 yr to radon-222.

Uranium, and therefore radium and radon, are present in most rocks and soil. Radon, a noble gas, percolates through grainy rocks and soil and enters the air and water in different concentrations. Although radon is a noble gas, its decay products have different chemical properties and attach to dust or aerosol droplets which can collect in the lungs. High levels of radon products in the lungs have been shown by both epidemiological studies of uranium miners and by animal studies to cause lung cancer [Committee on the Biological Effects of Ionizing Radiations, BEIR IV (1988)]. The deposition process is quite complicated. A certain fraction of the decay products attach to aerosol droplets. That fraction is an important parameter in estimating the dose, because the unattached particles are deposited in the airways and those that have attached to aerosols also are also deposited in the airways, the site depending on the droplet size. The rate at which natural mucus clearing from the lungs removes them is also variable.

The $^{222}_{86}$Rn decay scheme is shown in Fig. 16.28. (Alternate branches that occur very rarely are not shown.) The shaded nuclides have short lifetimes and are the greatest contributors to the dose, particularly the α particles from the decay of $^{218}_{84}$Po and $^{214}_{84}$Po. Radon dosimetry is described on pp. 137–158 of BEIR IV (1988). Typical uranium activi-

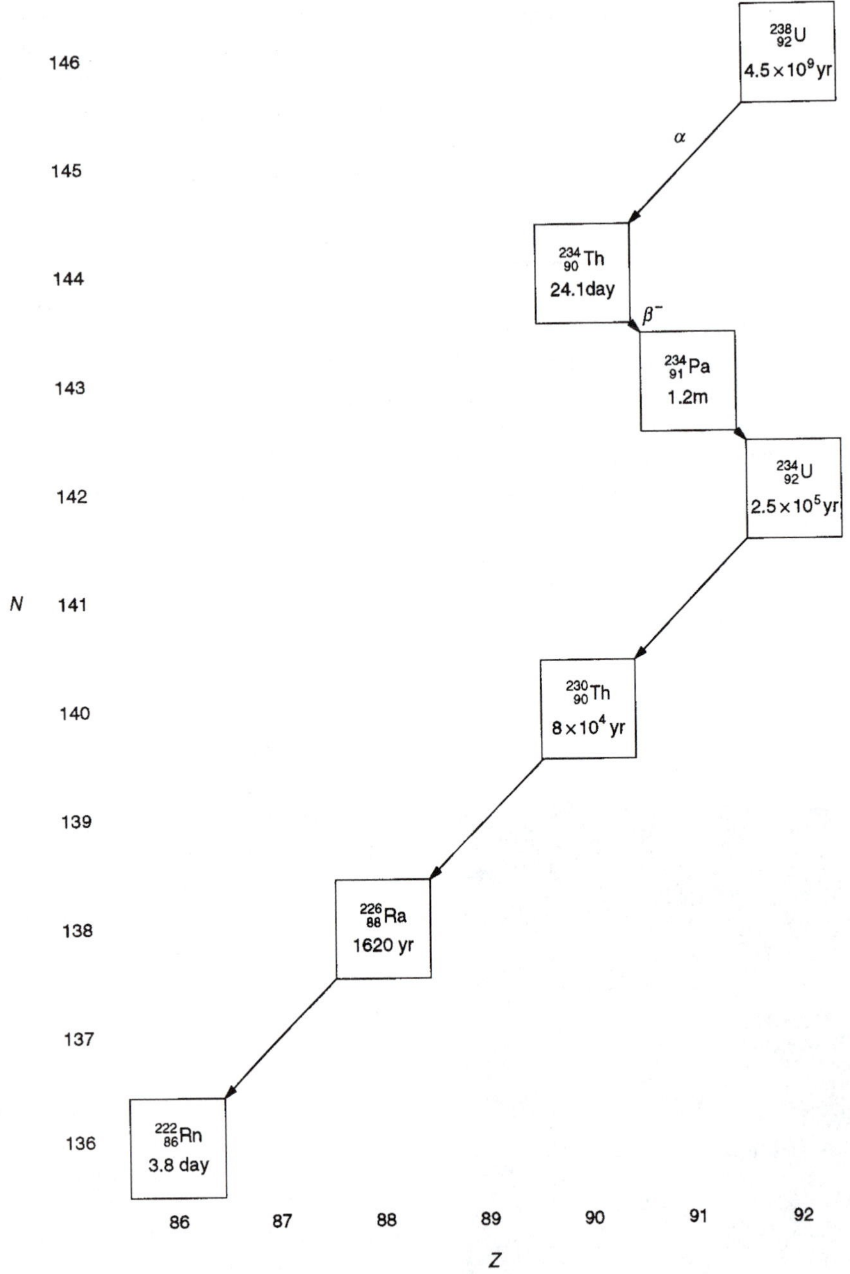

FIGURE 16.27. Decay of uranium-238 to radon.

ties in soil are 20 Bq kg^{-1} (range 7–40), leading to radon concentrations in the air over average soil of about 4 Bq m^{-3}.

The *working level* (WL) has been defined to be any combination of the shaded isotopes in Fig. 16.28 in 1 liter of air at ambient temperature and pressure that results in the ultimate emission of 1.3×10^5 MeV of α-particle energy. This is about the energy liberated by the decay products in equilibrium with 100 pCi (3.7 Bq) of radon. Thus 1 WL corresponds to 3.7 Bq l^{-1} or 3700 Bq m^{-3}. The *working-level month* (WLM) measures the duration of the exposure and is 1 WL for 170 h (1 month of 40-h work weeks).

Dose estimates for the miners and for the general population require models of aerosol size, unattached fraction, target cells, exercise level, and occupancy factors that are described in BEIR IV (1988). Averaging over all of these variables shows a dose in the lungs of about 6×10^{-3} Gy per WLM, with a factor-of-2 uncertainty because of these variables.

The report uses a time-since-exposure model to estimate the risk of lung cancer on the basis of four studies of groups of miners. The model predicts a relative risk ratio that is unity for no exposure and increases linearly to 3.5 for a

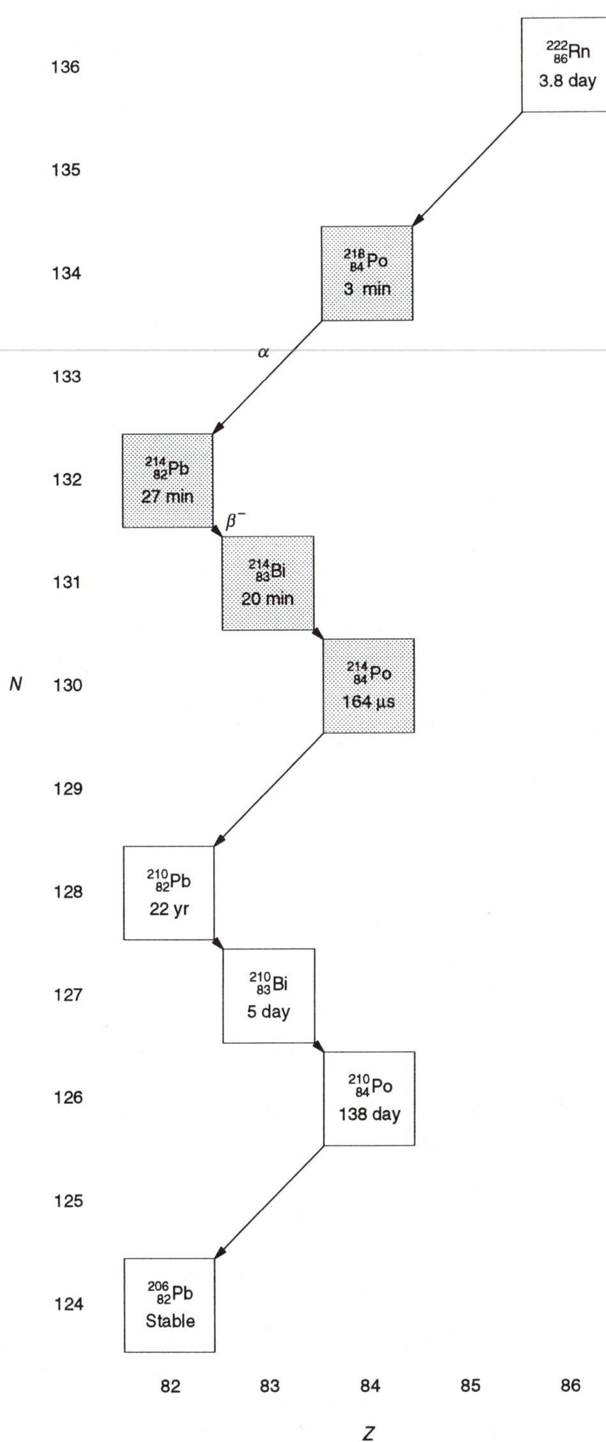

FIGURE 16.28. Decay of radon. The decay of the shaded nuclides is most significant in determining dose.

continuous exposure of 5 WLM per year over a lifetime.[11] The report uses the linear-no-threshold model to estimate risks to the general population at small exposures. Problems with application of the linear-no-threshold model were discussed in Sec. 15.13. See particularly the data from the Cohen study in Fig. 15.54. Typical radon concentrations in houses are usually less than $4r_0$ or 4 pCi l^{-1} (128 Bq m^{-3}) or 0.04 WL. Exposure to r_0 for 24 h per day for one year gives 0.5 WLM. The miners had exposures of 5–100 WLM per year, over periods of 3–20 years.

[11]BEIR IV (1988), Fig. 2.2. This is averaged by BEIR over smokers and nonsmokers and by me over sex.

SYMBOLS USED IN CHAPTER 16

Symbol	Use	Units	First used on page
b	Source to collimator distance	m	480
c	Speed of light	m s^{-1}	459
d	Width of collimator channel	m	480
f	Fraction		463
f_r	Fraction of transition energy released as photons		463
f_K, f_L	Fraction of transition energy released in K- or L-shell internal conversion		463
g	Detector efficiency		480
l	Collimator thickness	m	480
m_0	Rest mass	kg	459
m_X	Rest mass of particle type X	kg	459
n_i	Mean number (fraction) of emissions of type i per transition		466
r_n, r_k	Source and target regions		466
r	Distance	m	471
t	Time	s	461
t	Thickness of collimator wall	m	480
v	Velocity	ms^{-1}	459
w	Distance across collimator wall in the direction of photon travel	m	489
A	Mass number		459
A, A_0	Activity	Bq	467
$\widetilde{A}_h$	Cumulated activity in organ h		466
B	Buildup factor		472
B, B_K, B_L	Binding energy	eV	462
$C_h, \widetilde{C}_h$	Activity and cumulated activity per unit mass in organ h	Bq kg^{-1}; kg^{-1}	469

Symbol	Use	Units	First used on page
D	Dose	J kg^{-1}	466
E, E_γ	Energy	J, eV	459
F_h	Fraction of activity in organ h		468
F	Projection		481
K	Geometric factor		480
M, M_X	Mass	kg	460
N	Neutron number		459
N, N_0	Number of nuclei		461
N	Number of photons		459
R, R_0	Nuclear radius	m	459
R_t, R_0	True and observed counting rates	s^{-1}	489
S	Area	m^2	471
T	Kinetic energy	J, MeV	459
$T_{1/2}$	Half-life	s	462
T_j	Half-life for jth biological disappearance process	s	468
W_K	Fluorescence yield for K shell		470
Z	Atomic number		459
α_K, α_L	Internal conversion coefficient		463
α_h	Fraction of total activity in organ h		487
β^-, β^+	Electron and positron (in β decay)		463
λ	Physical decay constant	s^{-1}	461
λ_j	Decay constant for jth biological process	s^{-1}	468
μ_{atten}	Attenuation coefficient	m^{-1}	472
μ_{en}	Energy absorption coefficient	m^{-1}	471
$\nu, \bar{\nu}$	Neutrino, antineutrino		463
ρ	Density	kg m^{-3}	472
ϕ_i	Absorbed fraction		466
τ	Detector dead time	s	489
τ_h	Residence time in organ h	s	467
Δ_i	Mean energy emitted in radiation of type i per unit cumulated activity	J	466
Φ_j	Specific absorbed fraction	kg^{-1}	466

PROBLEMS

Section 16.1

16.1. The best current (1986) value for the mass of the proton is 1.007 276 470 u. the mass of the electron is $5.485\,799\,031 \times 10^{-4}$ u. The binding energy of the electron in the hydrogen atom is 13.6 eV. Calculate the mass of the neutral hydrogen atom.

16.2. The rest energy of the $^{184}_{74}$W nucleus is 171303 MeV. The average binding energies of the electrons in each shell are

Shell	Number of electrons	BE per electron (eV)
K	2	69 525
L	8	11 023
M	18	2 125
N	32	215
O	12	35
P	2	1

Calculate the atomic rest energy of tungsten.

Section 16.4

16.3. Refer to Figs. 16.2 and 16.5. Uranium splits roughly in half when it undergoes nuclear fission. Will the fission fragments decay by β^+ or β^- emission?

16.4. The following nuclei of mass 15 are known: $^{15}_{6}$C, $^{15}_{7}$N, and $^{15}_{8}$O. Of these, ^{15}N is stable. How do the others decay?

16.5. Look up the decay schemes of the following isotopes (for example, in the *Handbook of Chemistry and Physics*) and comment on their possible medical usefulness: ^{3}H, ^{15}O, ^{13}N, ^{18}F, ^{22}Na, ^{68}Ga, ^{64}Cu, ^{11}C, ^{123}I, and ^{56}Ni.

Section 16.6

16.6. Show that 1 μCi h$=1.332\times 10^8$ disintegrations or Bq s.

16.7. Obtain a numerical value for the residence time for ^{99m}Tc-sulfur colloid in the liver if 85% of the drug injected is trapped in the liver and remains there until it decays.

16.8. Derive Eqs. (16.39)–(16.41).

16.9. Calculate numerical solutions of Eqs. 16.39 and 16.41 and plot them on semilog paper. Use $\lambda=2$, $\lambda_1=0.5$, $\lambda_2=3$.

16.10. Eq. 16.41 is not valid if $\lambda_1=\lambda_2$. In that case, try a solution of the form $N_2=Bte^{-\lambda t}$ and obtain a solution.

16.11. Derive Eqs. (16.42) and (16.43).

16.12. The biological half-life of iodine in the thyroid is about 25 days. ^{125}I has a half-life of 60 days. ^{132}I has a half-life of 2.3 h. Find the effective half-life in each case.

16.13. For Sec. 16.6.4, with $\lambda=0.05$ h^{-1}, $\lambda_1=1$ h^{-1}, and $\lambda_2=0.1$ h^{-1}, find $\tilde{A}_1$ and $\tilde{A}_2$ in terms of the initial activity A_0 and in terms of the initial number of nuclei N_0.

16.14. N_0 radioactive nuclei with physical decay constant λ are injected in a patient at $t=0$. The nuclei move into the kidney at a rate λ_1, so that the number in the rest of the body falls exponentially: $N(t)=N_0e^{-(\lambda+\lambda_1)t}$. Suppose that the nuclei remain in the kidney for a time T before moving out in the urine. (This is a crude model for the radioactive nuclei being filtered into the glomerulus and then passing through the tubules before going to the bladder.)

(a) Calculate the cumulated activity and the residence time in the kidney by finding the total number of nuclei entering the kidney and multiplying by the probability that a nucleus decays during the time T that it is in the kidney.

(b) Calculate the cumulated activity and residence time in the bladder, assuming that the patient does not void.

16.15. Suppose that at $t=0$, ^{99m}Tc with an activity of 370 kBq enters a patient's bladder and stays there for 2 h, at which time the patient voids, eliminating all of it. What is the cumulated activity? What is the cumulated activity if the time is 4 h?

16.16. Suppose that the ^{99m}Tc of the previous problem does not enter the bladder abruptly at $t=0$, but that it accumulates linearly with time. At the end of 2 h the activity is 370 kBq and the patient voids, eliminating all of it. What is the cumulated activity?

16.17. A radioactive substance has half-life $T_{1/2}$. It is excreted from the body with biological half-life T_1. N_0 radioactive nuclei are introduced in the body at $t=0$. Find the total number that decay inside the body.

16.18. The *fractional distribution function* α_h is the fraction of the total activity which is in organ h: $\alpha_h(t)=A_h(t)/A(t)=A_h(t)/A_0e^{-\lambda t}$.

(a) Show that $\tau_h=\int_0^\infty \alpha_h(t)e^{-\lambda t}dt$.

(b) Calculate $\alpha_1(t)$ and $\alpha_2(t)$ for Eqs. (16.39) and (16.41) and show that integration of these expressions leads to Eqs. (16.43).

16.19. Suppose that the fractional distribution function (defined in the previous problem) is $\alpha(t)=1$, $t<T$; $\alpha(t)=b$; $t>T$ $(b<1)$. Find the residence time. [This is a simple model for the situation where a bolus (a fixed amount in a short time) of some substance passes through an organ once and is then distributed uniformly in the blood.]

16.20. The *distribution function* $q_h(t)$ is defined to be "the activity in organ h corrected for radioactive decay to a reference time." If the correction is from time t to time 0, find an expression for $q_h(t)$ in terms of $A_h(t)$.

16.21. The "official" definition of the fractional distribution function $\alpha_h(t)$ is "the ratio of the distribution function $q_h(t)$ produced by a bolus administration to the patient, divided by the activity A_0 in the bolus." Show that this is equivalent to the definition in Problem 16.18.

16.22. Show that if the uptake in a compartment is not instantaneous but exponential, with subsequent exponential decay, the cumulated activity is $\tilde{A}=1.443A_0(T_eT_{ue}/T_u)$, where T_e is the effective half-life for excretion, and $T_{ue}=T_uT_{1/2}/(T_u+T_{1/2})$. [Hint: see Eq. (16.42)].

Section 16.7

16.23. Use the output data of Fig. 16.4 to estimate values for W_K, $K_\alpha/(K_\alpha+K_\beta)$, and $K_\beta/(K_\alpha+K_\beta)$. Compare your values to those used in Table 16.2.

16.24. Rearrange the output data of Fig. 16.4. Find the total Δ for emission of photons below 30 keV and charged particles. Rank the radiations in the order they contribute to the dose.

16.25. Calculate the entries in the output data of Fig. 16.4 for γ_3, using the same techniques that were used in the text for γ_1 and γ_2.

16.26. The isotope $^{133}_{54}$Xe is used for studies of pulmonary function by inhalation. It decays by β emission with a half-life of 5.3 days to $^{133}_{55}$Cs. The maximum β energy is 0.346 MeV; the average is 0.100 MeV. The cesium then emits an 0.081-MeV γ ray. The Cs has the following properties:

$$\alpha_K=1.46, \quad \alpha_L=2.35, \quad \alpha_M=0.078$$

$$E_{K\alpha}=0.031 \text{ MeV}, \quad B_K=0.036 \text{ MeV},$$

$$B_L=0.006 \text{ MeV}.$$

Calculate Δ_i for the β particle, the γ ray, K and L internal conversion electrons, and $K\alpha$ x rays. Assume that all x rays are $K\alpha$.

16.27. Nitrogen-13 has a half-life of 10 min. All of the disintegrations emit a positron with end point energy 1.0 MeV (average energy 0.488 MeV). There is no electron capture. Make a table of radiations that must be considered for calculating the absorbed dose and determine E_i and Δ_i for each one.

16.28. A patient swallows 3.5×10^9 Bq of ^{131}I. The half-life of the iodine is 8 days. Ten minutes later the patient vomits all of it. If none had yet left the stomach and all was vomited, determine the cumulated activity and residence time in the stomach.

Section 16.8

16.29. Derive Eq. (16.60) by substituting Eqs. (16.58) and (16.59) in Eq. (16.57). You will also have to justify and use Eq. (16.61).

16.30. The body consists of two regions. Region 1 has mass m_1 and cumulated activity $\tilde{A}_1$. It is completely surrounded by region 2 of mass m_2 and cumulated activity $\tilde{A}_2=\tilde{A}_0-\tilde{A}_1$. We can say that the mass of the total body is $m_{TB}=m_1+m_2=m_1+m_{RB}$. A single radiation is emitted with disintegration energy Δ. The radiation is nonpenetrating so that

$$\phi(1\leftarrow1)=\phi(2\leftarrow2)=1,$$

$$\phi(1\leftarrow2)=\phi(2\leftarrow1)=0.$$

(a) What are $\phi(\text{TB}\leftarrow1)$ and $\phi(\text{TB}\leftarrow2)$?

(b) What are the corresponding values of Φ and S?

(c) Show that directly from the definition, Eq. (16.57)

$$D_1=\tilde{A}_1\Delta/m_1,$$

$$D_2=D_{RB}=\tilde{A}_2\Delta/m_2,$$

$$D_{TB}=\tilde{A}_0\Delta/(m_1+m_2)$$

(d) Calculate $\tilde{A}_u$ and $\tilde{A}_1^*$.

(e) What is $S(1\leftarrow TB)$? Remember that ϕ is calculated for activity uniformly distributed within the source region.

(f) Calculate the dose to region 1 using Eq. (16.60) and show that it agrees with (c).

(g) Evaluate $S(1\leftarrow RB)$ using Eq. (16.61) and show that it agrees with $S(1\leftarrow 2)$.

16.31. The body consists of two regions. Region 1 has mass m_1 and cumulated activity $\tilde{A}_1$. It is completely surrounded by region 2 of mass m_2 and cumulated activity $\tilde{A}_2$. A single radiation is emitted with disintegration energy Δ. The characteristics of the radiation are such that

$$\phi(1\leftarrow 1)+\phi(2\leftarrow 1)=1,$$

$$\phi(1\leftarrow 2)+\phi(2\leftarrow 2)+\phi(0\leftarrow 2)=1,$$

where $\phi(0\leftarrow 2)$ represents energy from region 2 that has escaped from the body. Obtain expressions for the dose to each region and the whole body dose.

Section 16.9

16.32. Consider the decay of a parent at rate λ_1 to an offspring which decays with rate λ_2.

(a) Write a differential equation for the amount of offspring present.

(b) Solve the equation.

(c) Discuss the solution when $\lambda_2>\lambda_1$.

(d) Discuss the solution when $\lambda_2<\lambda_1$.

(e) Plot the solution for a technetium generator that is eluted every 24 h.

16.33. N_0 nuclei of ^{99m}Tc are injected into the body. What is the maximum activity for the decay of the metastable state? When does the maximum activity for decay of the ground state occur if no Tc atoms are excreted? What is the ratio of the maximum metastable state activity to the maximum ground-state activity?

16.34. If 1 μCi of ^{99m}Tc is injected in the blood and stays there, relate the activity in a sample drawn time t later to the volume of the sample and the total blood volume. If the gamma rays are detected with 100% efficiency, what will be the counting rate for a 10-ml sample of blood if the blood volume is 5 liters? (Using non-SI units was intentional.)

16.35. Assume that aggregated human albumin is in the form of microspheres. A typical dose of albumin microspheres is 0.5 mg of microspheres containing 80 MBq of ^{99m}Tc and 15 μg of tin. There are 1.85×10^6 microspheres per mg.

(a) How many ^{99m}Tc atoms are there per microsphere?

(b) How many tin atoms per microsphere?

(c) How many technetium atoms per tin atom?

(d) What fraction of the surface of a microsphere is covered by tin? Assume the sphere has a density of 10^3 kg m^{-3}.

16.36. It is estimated that the total capillary surface area in the lung is 90 m^2. Assume each capillary has 50 segments, each 10 μm long, and a radius of 5 μm.

(a) How many capillaries are there in the lung?

(b) There are about 3×10^8 alveoli in both lungs. How many capillaries per alveolus are there?

(c) An alveolus is 150–300 μm in diameter. Are the above answers consistent?

(d) A typical dose of albumin microspheres is 0.5 mg with an average diameter of 25 μm. There are 1.85×10^6 spheres per mg. What fraction of the capillaries are blocked if there is good mixing?

Section 16.10

16.37. The half-life of ^{99m}Tc is 6.0 h. The half-life of ^{131}I is 8.07 day. Assume that the same initial activity of each is given to a patient and that all of the substance remains within the body.

(a) Find the ratio of the cumulated activity for the two isotopes.

(b) ^{99m}Tc emits 0.141-MeV photons. For each decay of ^{131}I the most important radiations are 0.89 β^- of energy 0.192 MeV and 0.81 photons of 0.365 MeV. If all of the decay energy were absorbed in the body, what would be the ratio of doses for the same initial activity?

16.38. A patient is given an isotope which spreads uniformly through the lungs. It emits a single radiation: a γ ray of energy 50 keV. There are no internal-conversion electrons. The cumulated activity is 40 GBq s. Find the absorbed dose in the liver ($m=1.83$ kg).

16.39. The decay of ^{99m}Tc can be approximated by lumping all of the decays into two categories:

Radiation	E_i (MeV)	Δ_i (Gy kg Bq^{-1} s^{-1})
γ	0.14	2×10^{-14}
Electrons and soft x rays		2.76×10^{-15}

Sulfur colloid labeled with 100 MBq of ^{99m}Tc is given to a patient and is taken up immediately by the liver. Assume it stays there. Find the dose to the liver, spleen, and whole body. Use the following information:

Target organ	Mass (kg)	Absorbed fraction for a source in the liver	
		$E(\gamma)$=0.1 MeV	$E(\gamma)$=0.2 MeV
Liver	1.833	0.165	0.158
Spleen	0.176	0.000 606	0.000 645
Whole body	70.0	0.454	0.415

16.40. An ionization type smoke detector contains 4.4 μCi of ^{241}Am. This isotope emits α particles (which we will ignore) and a 60-keV γ ray, for which $n=0.36$. The half-life is 458 yr.

(a) How many moles of ^{241}Am are in the source?

(b) Ignoring attenuation, backscatter, and buildup in any surrounding material (such as the cover of the smoke detector), what is the absorbed dose in a small sample of muscle located 2 m away, if the muscle is under the detector for 8 h per day for 1 year?

16.41. One mCi of a radioactive substance lodges permanently in a patient's lungs. The substance emits a single 80-keV γ ray. It has a half-life of 12 h. Find the cumulated activity and the dose to the liver (mass 1833 g).

16.42. The dose calculation for microspheres in the lung was an oversimplification because technetium leaches off the spheres. Footnote 9 lists some of the more realistic residence times. If none of the technetium is excreted from the body, the sum of all the residence times will still be 8.7 h. Assume that the residence time in the lungs is 4.3 h and the residence time in the rest of the body is 4.4 h.

(a) Show that $\tilde{A}_u=4.46\times3600\times A_0$ and $\tilde{A}^*_{\text{lung}}=4.24\times3600\times A_0$.

(b) For a source distributed uniformly throughout the total body, the absorbed fractions for 140-keV photons are $\phi(\text{lung}\leftarrow\text{TB})=0.0053$, $\phi(\text{TB}\leftarrow\text{TB})=0.3572$. Split the radiation into penetrating and nonpenetrating components:

$$S(\text{lung}\leftarrow\text{TB})=(\phi_{\text{nonpen}}\Delta_{\text{nonpen}}$$

$$+\phi_{\text{penetrating}}\Delta_{\text{penetrating}})/m_{\text{lung}}.$$

Remember that for activity uniformly distributed in the total body, $\phi(\text{lung}\leftarrow\text{TB})=m_{\text{lung}}/m_{\text{TB}}$ and use some of the information in Table 16.4 to show that

$$S(\text{lung}\leftarrow\text{TB})=1.463\times10^{-16}\ \text{J kg}^{-1}\,\text{Bq}^{-1}\,\text{s}^{-1},$$

$$S(\text{TB}\leftarrow\text{TB})=1.414\times10^{-16}\ \text{J kg}^{-1}\,\text{Bq}^{-1}\,\text{s}^{-1}.$$

(c) Calculate the dose to the lungs and the total body dose for an initial activity of 37 MBq. Compare the values to those in Table 16.4.

Section 16.12

16.43. Nuclear counting follows Poisson statistics. Show that for a fixed counting rate R (counts per second) the standard deviation of a sum of N measurements each of length T is the same as a single measurement of duration NT. (Hint: You will first have to consider the situation where one measures $y=x_1+x_2+\cdots$ and find the variance of y in terms of the variances of the x_i when there is no correlation between the x_i.)

16.44. The interaction of a photon in nuclear detector (an "event") initiates a process in the detector that lasts for a certain length of time. A second event occurring within a time τ of the first event is not recorded as a separate event. Suppose that the true counting rate is R_t. A counting rate R_o is observed.

(a) A nonparalyzable counting system is "dead" for a time τ after each recorded event. Additional events that occur during this dead time are not recorded but do not prolong the dead time. Show that $R_t=R_o(1-R_o\tau)$ and $R_o=R_t/(1+R_t\tau)$.

(b) A paralyzable counting system is unable to record a second event unless a time τ has passed since the last event. In other words, an event occurring during the dead time is not only not recorded, it prolongs the dead time. Show that in this case $R_o=R_te^{-R_t\tau}$. (Hint: Use the Poisson distribution of Appendix J to find the fraction of events separated by a time greater than τ. The probability that the next event occurs between t and $t+dt$ is the probability of no event during time t multiplied by the probability of an event during dt.)

(c) Plot R_o vs R_t for the two cases when τ is fixed. The easiest way to do this is to plot $R_o\tau$ vs $R_t\tau$.

16.45. Two channels of a collimator for a gamma camera are shown in cross section, along with the path of a photon that encounters the minimum thickness of collimator septum (wall).

(a) Show that if $(d+t)/l\ll1$, $w/t=l/(2d+t)$.

(b) If transmission through the septum is to be less than 5%, what is the relationship between t, d, l, and μ? Evaluate this for ^{99m}Tc and for a positron emitter.

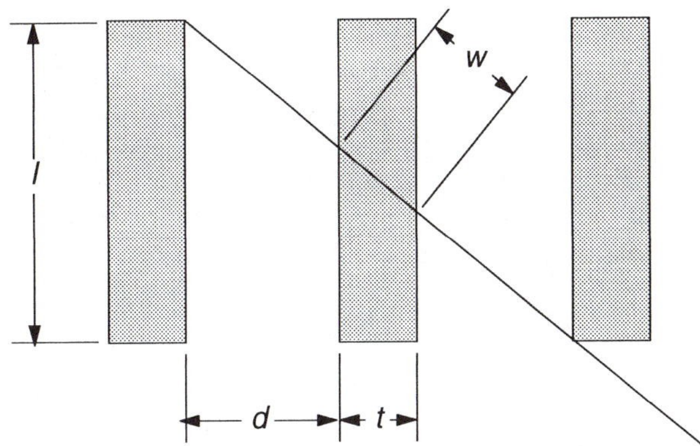

16.46. Photons from a point source a distance b below a collimator pass through channels out to a distance a from the perpendicular to the collimator passing through the source.

(a) Find an expression for a in terms of b, d, and l.

(b) Assume that a is related to the spatial frequency k for which the modulation transfer function (MTF)=0.5 in Fig.

16.19 by $a=K/k$, where K is a constant. Calculate the thickness l of the collimator.

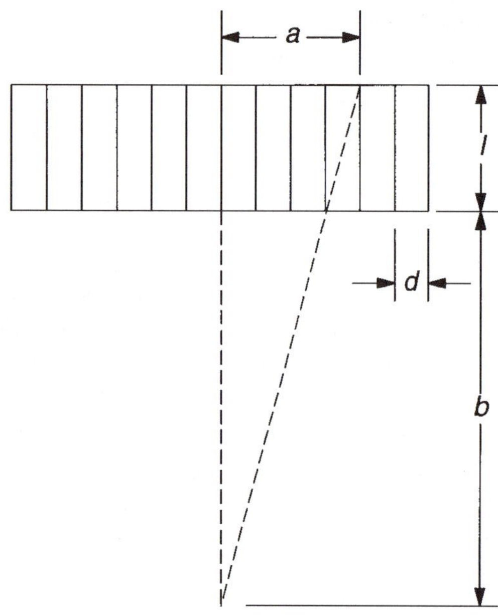

Point Source

16.47. The collimator efficiency of a gamma camera is defined to be the fraction of the γ rays emitted isotropically by a point source that pass through the collimator into the scintillator.

(a) Consider a circular channel of diameter d in the collimator directly over the source. Show that the fraction of the photons striking the scintillator after passing through that channel is $d^2/16(l+b)^2$. (Assume that any which strike the septum are lost).

(b) Use the result of the previous problem to estimate the number of channels through which at least some photons from the point source pass. Assume that the fraction of collimator area that is occupied by channels rather than lead is $[d/(d+t)]^2$.

(c) Calculate the geometric efficiency g assuming that all channels that pass any photons have the same efficiency as the one on the perpendicular from the source. Show that it is of the form

$$g=K^2\left(\frac{d}{l}\right)^2\left(\frac{d}{d+t}\right)^2.$$

and evaluate K. More detailed calculations show that K is about 0.24 for a hexagonal array of round holes and 0.26 for hexagonal holes.[12]

(d) How does the detector efficiency relate to the collimator resolution?

[12]Sorenson and Phelps (1987, p. 336); R. P. Grenier, M. A. Bender, and R. H. Jones (1974). A computerized multicrystal scintillation gamma camera, Chap. 3 in H. G. Hine and J. A. Sorenson, eds. *Instrumentation in Nuclear Medicine*. New York, Academic, Vol. 2.

Section 16.14

16.48. Suppose that A positrons are emitted at a point per second. They come to rest and annihilate within a short distance of their source. When a positron annihilates, two photons are emitted in opposite directions. Two photon detectors are set up on opposite sides of the source. The source is distance r_1 from the first detector, of area S_1, and r_2 from the second detector of area S_2. The area S_2 is large enough so that the second photon will definitely enter detector 2 if the first photon enters detector 1. Assume that both detectors count with 100% efficiency.

(a) Show that the number of counts in the first detector would be $2AS_1/4\pi r_1^2$ if there were no attenuation between source and detector, and that it is $(2AS_1/4\pi r_1^2)e^{-\mu a_1}$ if attenuation in a thickness a_1 of the body is considered.

(b) Detector 2 detects the second photon for every photon that strikes detector 1. Assuming a uniform attenuation coefficient and body thickness a_2, find an expression for the number of events in which both photons are detected.

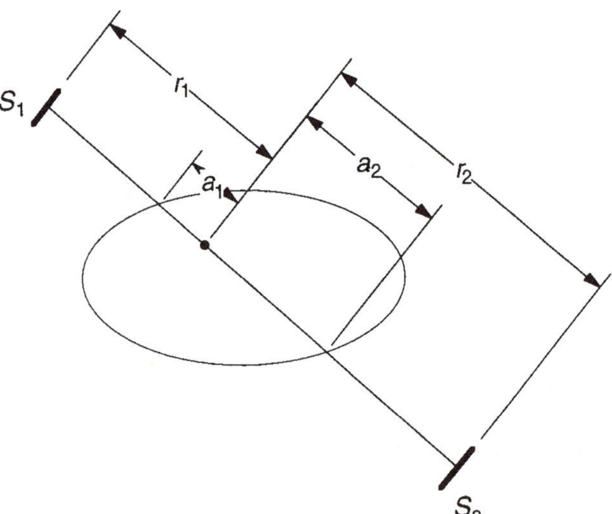

16.49. The attenuation distortion for SPECT can be reduced by making measurements on opposite sides of the patient and taking the geometric mean. The geometric mean of variables x_1 and x_2 is $(x_1x_2)^{1/2}$. Calculate the geometric mean of two SPECT measurements on opposite sides of the patient. Ignore possible $1/r^2$ effects.

REFERENCES

BEIR IV (1988). Committee on the Biological Effects of Ionizing Radiations. *Health Risks of Radon and Other Internally Deposited Alpha-Emitters*. Washington, D.C., National Academy Press.

Berger, M. J. (1968). Energy Deposition in Water by Photons from Point Isotropic Sources. NM/MIRD Pamphlet 2. New York, Society of Nuclear Medicine.

Buchsbaum, D., and B. W. Wessels (1993). Introduction: Radiolabeled antibody tumor dosimetry. *Med. Phys.* **20**(2, Pt. 2): 499–501.

Coffey, J. L., and E. E. Watson (1979). Calculating the dose from remaining body activity: A comparison of two methods. *Med. Phys.* **6**(4): 307–308.

Dillman, L. T., and F. C. Von der Lage (1974). *Radionuclide Decay Schemes and Nuclear Parameters for Use in Radiation Dose Estimation.* NM/MIRD Pamphlet 10. New York, Society of Nuclear Medicine.

Eckerman K. F., R. J. Westfall, J. C. Ryman, and M. Cristy (1994). Availability of nuclear decay data in electronic form, including beta spectra not previously published. *Health Phys.* **67**(4): 338–345.

Eisberg, R., and R. Resnick (1985). Quantum Physics of Atoms, Molecules, Solids, Nuclei and Particles, 2nd ed. New York, Wiley.

Evans, R. D. (1955). *The Atomic Nucleus.* New York, McGraw-Hill.

Fritzberg, A. R., and B. W. Wessels (1995). Therapeutic radionuclides. In H. N. Wagner, Jr., Z. Szabo, and J. W. Buchanan, eds. *Principles of Nuclear Medicine*, 2nd ed. Philadelphia, Saunders, pp. 229–234.

Howell, R. W. (1992). Radiation spectra for Auger-emitting radionuclides: Report No. 2 of the AAPM Nuclear Medicine Task Group No. 6. *Med. Phys.* **19**(6): 1371–1383.

Humm, J. L., R. W. Howell, and D. V. Rao (1994). Dosimetry of Auger-electron-emitting radionuclides: Report No. 3 of the AAPM Nuclear Medicine Task Group No. 6. *Med. Phys.* **12**(12): 1901–1915.

Larsson, S. A. (1980). Gamma camera emission tomography: Development and properties of a multi-sectional emission computed tomography system. *Acta Radiol.* (Suppl. 363).

Links, J. M. and J. C. Engdahl. (1995) Planar imaging. Chap. 17 in H. N. Wagner, Jr., Z. Szabo, and J. W. Buchanan, eds. *Principles of Nuclear Medicine*, 2nd ed. Philadelphia, Saunders.

Loevinger, R., T. F. Budinger, and E. E. Watson (1988). *MIRD Primer for Absorbed Dose Calculations.* New York, Society of Nuclear Medicine.

Nag, S. ed. (1994). *High Dose Rate Brachytherapy: A Textbook.* Armonk, NY Futura.

Narra, V. R., K. S. R. Sastry, S. M. Goddu, R. W. Howell, S.-E. Strand, and D. V. Rao (1994). Relative biological effectiveness of ^{99m}Tc radiopharmaceuticals. *Med. Phys.* **21**(12): 1921–1926.

Rehm, K., S. C. Strother, J. R. Anderson, K. A. Schaper, and D. A. Rottenberg (1994). Display of merged multimodality brain images using interleaved pixels with independent color scales. *J. Nucl. Med.* **35**: 1815–1821.

Sastry, K. S. R. (1992). Biological effects of the Auger emitter iodine-125: A review Report No. 1 of the AAPM Nuclear Medicine Task Group No. 6. *Med. Phys.* **19**(6): 1361–1370.

Snyder, W. S., M. R. Ford, and G. G. Warner (1976). *Specific Absorbed Fractions for Radiation Sources Uniformly Distributed in Various Organs of a Heterogeneous Phantom.* NM/MIRD Pamphlet 5, revised. New York, Society of Nuclear Medicine.

Snyder, W. S., M. R. Ford, G. G. Warner, and H. L. Fisher (1969). *Estimates of Absorbed Fractions for Monoenergetic Photon Sources Uniformly Distributed in Various Organs of a Heterogeneous Phantom.* NM/MIRD Pamphlet 5. New York, Society of Nuclear Medicine.

Snyder, W. S., M. R. Ford, G. G. Warner, and S. B. Watson (1975). *"S," Absorbed Dose per Unit Cumulated Activity for Selected Radionuclides and Organs*, NM/MIRD Pamphlet 11. New York, Society of Nuclear Medicine.

Sorenson, J. A., and M. E. Phelps (1987). *Physics in Nuclear Medicine*, 2nd ed. Philadelphia, Saunders.

Stabin, M. (1996). MIRDOSE—the personal computer software for use in internal dose assessment in nuclear medicine. *J. Nucl. Med.* **37**: 538–546.

Strother, S. C., J. R. Anderson, Jr., K. A. Schaper, J. J. Sidtis, J. S. Liow, R. P. Woods, and D. A. Rottenberg (1995). Principal component analysis and the scaled subprofile model compared to intersubject averaging and statistical parametric mapping: I. Functional connectivity of the human motor system studied with ^{15}O-water PET. *J. Cerebral Blood Flow Metab.* **15**(5): 738–753.

Watson, E. E., M. G. Stabin, and J. A. Siegel (1993). MIRD formulation. *Med. Phys.* **20**(2, Pt. 2): 511–514.

Weber, D. A., K. F. Eckerman, L. T. Dillman, and J. C. Ryman (1989). *MIRD: Radionuclide Data and Decay Schemes.* New York, Society of Nuclear Medicine.

CHAPTER 17

Magnetic Resonance Imaging

Between 1978 and 1985 magnetic resonance imaging (MRI) (formerly called nuclear magnetic resonance imaging) developed into a useful modality for medical diagnosis. It provides very-high-resolution images without ionizing radiation. There is also the potential for more elaborate imaging, including phase effects, flow, and the signature of particular atomic environments.

Magnetic resonance phenomena are more complicated than x-ray attenuation or photon emission by a radioactive nucleus. Magnetic resonance imaging depends upon the behavior of atomic nuclei in a magnetic field, in particular, the orientation and motion of the nuclear magnetic moment in the magnetic field. The patient is placed in a strong static magnetic field (typically 1–4 T). This is usually provided by a hollow cylindrical (solenoidal) magnet, though some machines are being made that use other configurations so that the physician can carry out procedures on the patient while viewing the MRI image. Other coils apply spatial gradients to the magnetic field, along with radio-frequency signals that cause the magnetization changes described below. Still other coils detect the very weak radio-frequency signals resulting from these changes.

First, we must understand the property that we are measuring. Section 17.1 describes the behavior of a magnetic moment in a static magnetic field, and Sec. 17.2 shows how the nuclear spin is related to the magnetic moment. Section 17.3 introduces the concept of the magnetization vector, which is the magnetic moment per unit volume, while Sec. 17.4 develops the equations of motion for the magnetic moment. In order to describe the motion of the magnetization, it is convenient—in fact, almost mandatory—to use the rotating coordinate system described in section 17.5.

To make a measurement, the nuclear magnetic moments originally aligned with the static magnetic field are made to rotate or precess in a plane perpendicular to the static field, after which the magnetization gradually returns to its original value. This relaxation phenomenon is described in Sec. 17.6. Sections 17.7 and 17.8 describe ways in which the magnetization can be manipulated for measurement or imaging.

Imaging techniques are finally introduced in Sec. 17.9. Sections 17.10 and 17.11 describe how chemical shifts and flow can affect the image or can themselves be imaged.

17.1. MAGNETIC MOMENTS IN AN EXTERNAL MAGNETIC FIELD

Magnetic resonance imaging detects the magnetic dipoles in the nuclei of atoms in the human body. We saw in Chap. 8 that isolated magnetic monopoles have never been observed [see Eq. (8.6)], and that magnetic fields are produced by moving charges or electric currents. In some cases, such as bar magnets, the external field is the same as if there were magnetic charges occurring in pairs or *dipoles*.[1] The strength of a dipole is measured by its *magnetic dipole moment* $\boldsymbol{\mu}$. (In Chap. 8 the magnetic dipole moment was called $\mathbf{m}$ to avoid confusion with μ_0. In this chapter we use $\boldsymbol{\mu}$ to avoid confusion with the quantum number m and to be consistent with the literature in the field.) The magnetic dipole moment is analogous to the electric dipole moment of Chap. 7; however, it is produced by a movement of charge, such as charge moving in a circular path. The units of $\boldsymbol{\mu}$ are J T^{-1} or A m^2. We saw that when a magnetic dipole is placed in a magnetic field as in Fig. 17.1, it is necessary to apply an external torque $\boldsymbol{\tau}_{ext}$ to keep it in equilibrium. This torque, which is required to cancel the torque exerted by the magnetic field, vanishes if the dipole is aligned with the magnetic field. The torque exerted on the dipole by the magnetic field is

$$\boldsymbol{\tau} = \boldsymbol{\mu} \times \mathbf{B} \tag{17.1}$$

[This is Eq. (8.4)].

[1]Dipoles can be arranged so that their fields nearly cancel, giving rise to still-higher-order moments such as the quadrupole moment or the octupole moment. (See Chap. 7). A configuration for which the quadrupole moment is important is two magnets in line arranged as N-S-S-N.

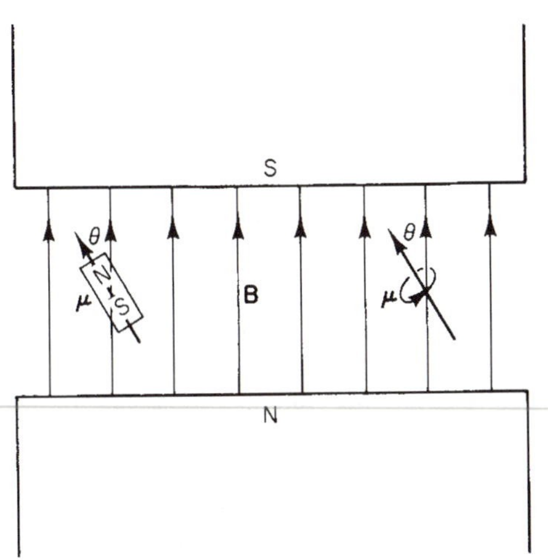

FIGURE 17.1. A dipole in a magnetic field. The dipole can be either a bar magnet or a current loop.

The potential energy of the dipole is the work that must be done by τ_{ext} to change the dipole's orientation in the magnetic field without changing any kinetic energy it might have. To increase angle θ by an amount $d\theta$ requires that work be done on the dipole–magnetic field system. This work is the increase in potential energy of the system:

$$dU = \mu B \sin\theta \, d\theta. \tag{17.2}$$

This can be integrated to give the change in potential energy when the angle changes from θ_1 to θ_2:

$$U(\theta_2) - U(\theta_1) = -\mu B(\cos\theta_2 - \cos\theta_1).$$

If the energy is considered to be zero when the dipole is at right angles to the z axis, then the potential energy is

$$U(\theta) = -\mu B \cos\theta = -\boldsymbol{\mu} \cdot \mathbf{B}. \tag{17.3}$$

In many cases the moving charges that give rise to the magnetic moment of an object possess angular momentum. Often the magnetic moment is parallel to and proportional to the angular momentum: $\boldsymbol{\mu} = \gamma \mathbf{L}$. The proportionality factor is called the *gyromagnetic ratio* (sometimes called the magnetogyric ratio). When such an object is placed in a uniform magnetic field, the resulting motion can be quite complicated. The torque on the object is $\boldsymbol{\tau} = \boldsymbol{\mu} \times \mathbf{B} = \gamma \mathbf{L} \times \mathbf{B}$. It is not difficult to show (Problem 17.1) that the torque is the rate of change of the angular momentum, $\boldsymbol{\tau} = d\mathbf{L}/dt$. Therefore the equation of motion is

$$\gamma(\mathbf{L} \times \mathbf{B}) = \frac{d\mathbf{L}}{dt} \tag{17.4a}$$

or

$$\gamma(\boldsymbol{\mu} \times \mathbf{B}) = \frac{d\boldsymbol{\mu}}{dt}. \tag{17.4b}$$

Solutions to these equations are discussed in Sec. 17.4.

17.2. THE SOURCE OF THE MAGNETIC MOMENT

Atomic electrons and the protons and neutrons in the atomic nucleus can possess both angular momentum and a magnetic moment. The magnetic moment of a particle is related to its angular momentum. We can derive this relationship for a charged particle moving in a circular orbit. We saw in Chap. 8 that the magnitude of the magnetic moment of a current loop is the product of the current i and the area of the loop S:

$$|\boldsymbol{\mu}| = \mu = iS. \tag{17.5}$$

The direction of the vector is perpendicular to the plane of the loop. Its direction is defined by a right-hand rule: curl the fingers of your right hand in the direction of current flow and your thumb will point in the direction of $\boldsymbol{\mu}$ (see the right-hand part of Fig. 17.1). This is the same right-hand rule that relates the circular motion of a particle to the direction of its angular momentum.

Suppose that a particle of charge q and mass m moves in a circular orbit as in Fig. 17.2. The speed is v and the magnitude of the angular momentum is $L = mvr$. The effective current is the charge q multiplied by the number of times it goes past a given point on the circumference of the orbit in one second:

$$i = \frac{qv}{2\pi r}.$$

The magnetic moment has magnitude

$$\mu = iS = i\pi r^2 = \frac{qvr}{2}.$$

Since the angular momentum is $L = mvr$ and $\boldsymbol{\mu}$ and $\mathbf{L}$ are both perpendicular to the plane of the orbit, we can write

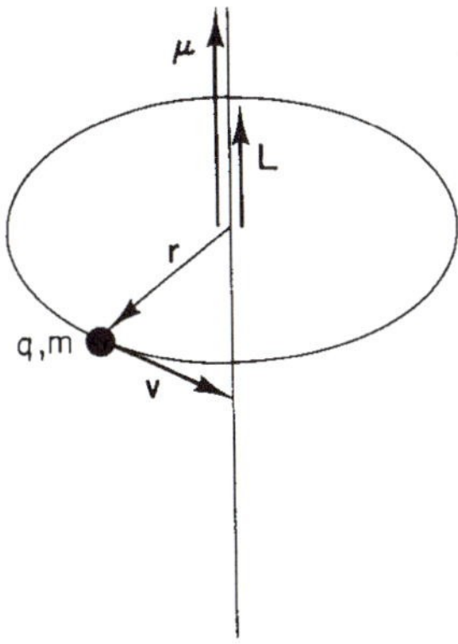

FIGURE 17.2. A particle of charge q and mass m travels in a circular orbit. It has a magnetic moment $\boldsymbol{\mu}$ and angular momentum $\mathbf{L}$. If the charge is positive, $\boldsymbol{\mu}$ and $\mathbf{L}$ are parallel; if it is negative they are in opposite directions.

$$\boldsymbol{\mu}=\left(\frac{q}{2m}\right)\mathbf{L}=\gamma\mathbf{L}. \qquad (17.6)$$

The quantity $\gamma=q/2m$ is the gyromagnetic ratio for this system. The units of γ are $T^{-1}\ s^{-1}$ (see Problem 17.2). The magnetic moment and the orbital angular momentum are parallel for a positive charge and antiparallel for a negative charge.

An electron or a proton also has an intrinsic magnetic moment quite separate from its orbital motion. It is associated with and proportional to the intrinsic or ''spin'' angular momentum **S** of the particle. We write

$$\boldsymbol{\mu}=\gamma\mathbf{S}. \qquad (17.7)$$

The value of γ for a spin is *not* equal to $q/2m$.

Two kinds of spin measurements have biological importance. One is associated with electron magnetic moments and the other with the magnetic moments of nuclei. Most neutral atoms in their ground state have no magnetic moment due to the electrons. Exceptions are the transition elements that exhibit paramagnetism. Free radicals, which are often of biological interest, have an unpaired electron and therefore have a magnetic moment. In most cases this magnetic moment is due almost entirely to the spin of the unpaired electron.

Magnetic resonance imaging is based on the magnetic moments of atomic nuclei in the patient. The total angular momentum and magnetic moment of an atomic nucleus are due to the spins of the protons and neutrons, as well as any orbital angular momentum they have inside the nucleus. Table 17.1 lists the spin and gyromagnetic ratio of the electron and some nuclei of biological interest.

If the nuclear angular momentum is **I** with quantum number I, the possible values of the z component of **I** are $I\hbar,(I-1)\hbar,\ldots,-I\hbar$. For $I=\frac{1}{2}$, the values are $\hbar/2$ and $-\hbar/2$, while for $I=\frac{3}{2}$ they are $3\hbar/2$, $\hbar/2$, $-\hbar/2$ and $-3\hbar/2$. The direction of the external magnetic field defines the z axis, and the energy of a spin is given by $-\boldsymbol{\mu}\cdot\mathbf{B}=-\gamma I\cdot\mathbf{B}=-\gamma m\hbar B$. The difference between adjacent energy levels is $\gamma B\hbar$, and the angular frequency of a photon corresponding to that difference is $\omega_{\text{photon}}=\gamma B$.

TABLE 17.1. *Values of the spin and gyromagnetic ratio for a free electron and various nuclei of interest.*

Particle	Spin	$\gamma=\omega_{\text{Larmor}}/B$ ($s^{-1}\ T^{-1}$)	ν/B (MHz T^{-1})
Electron	$\frac{1}{2}$	1.7608×10^{11}	2.8025×10^{4}
Proton	$\frac{1}{2}$	2.6753×10^{8}	42.5781
Neutron	$\frac{1}{2}$	1.8326×10^{8}	29.1667
^{23}Na	$\frac{3}{2}$	0.7076×10^{8}	11.2618
^{31}P	$\frac{1}{2}$	1.0829×10^{8}	17.2349

17.3. THE MAGNETIZATION

The MRI image depends on the *magnetization* of the tissue. The magnetization of a sample, **M**, is the average magnetic moment per unit volume. In the absence of an external magnetic field to align the nuclear spins, the magnetization is zero. As an external static magnetic field is applied, the spins tend to align in spite of their thermal motion, and the magnetization increases, proportional at first to the external field. If the external field is strong enough, all of the nuclear magnetic moments are aligned, and the magnetization reaches its saturation value.

We can calculate the magnetization. Consider a collection of spins of a single nuclear species in an external magnetic field. This might be the hydrogen nuclei (protons) in a sample. The spins do not interact with each other but are in thermal equilibrium with the surroundings, which are at temperature T. We do not consider the mechanism by which they reach thermal equilibrium. Since the magnetization is the average magnetic moment per unit volume, it is the number of spins per unit volume, N, times the average magnetic moment of each spin: $\mathbf{M}=N\langle\boldsymbol{\mu}\rangle$.

To obtain the average value of the z component of the magnetic moment, we must consider each possible value of quantum number m. We multiply the value of μ_z corresponding to each value of m by the probability that m has that value. Since the spins are in thermal equilibrium with the surroundings, the probability is proportional to the Boltzmann factor of Chap. 3, $\exp(-U/k_BT)=\exp(\gamma m\hbar B/k_BT)$. The denominator in Eq. (17.8) normalizes the probability:

$$\langle\mu_z\rangle=\frac{\gamma\hbar\sum_{m=-I}^{I}m\exp(\gamma m\hbar B/k_BT)}{\sum_{m=-I}^{I}\exp(\gamma m\hbar B/k_BT)}. \qquad (17.8)$$

At room temperature $\gamma I\hbar B/k_BT\ll1$ (see Problem 17.4), and it is possible to make the approximation $e^x\approx1+x$. The sum in the numerator then has two terms:

$$\sum_m m+\frac{\gamma\hbar B}{k_BT}\sum_m m^2.$$

The first sum vanishes. The second is $I(I+1)(2I+1)/3$. The denominator is

$$\sum_m 1+\frac{\gamma\hbar B}{k_BT}\sum_m m.$$

The first term is $2I+1$; the second vanishes. Therefore we obtain

$$\langle\mu_z\rangle=\frac{\gamma^2\hbar^2I(I+1)}{3k_BT}B. \qquad (17.9)$$

The z component of **M** is

$$M_z=N\langle\mu_z\rangle=\frac{N\gamma^2\hbar^2I(I+1)}{3k_BT}B, \qquad (17.10)$$

which is proportional to the applied field.

17.4. BEHAVIOR OF THE MAGNETIZATION VECTOR

A remarkable result of quantum mechanics is that the average or expectation value of a spin obeys the classical Equation (17.4b):

$$\frac{d\langle\boldsymbol{\mu}\rangle}{dt}=\gamma(\langle\boldsymbol{\mu}\rangle\times\mathbf{B}) \tag{17.11}$$

whether or not **B** is time dependent [Slichter (1978), p. 20]. Multiplying by the number of spins per unit volume we obtain

$$\frac{d\mathbf{M}}{dt}=\gamma(\mathbf{M}\times\mathbf{B}) \tag{17.12}$$

This equation can lead to many different behaviors of **M**, some of which are quite complicated.

The simplest motion occurs if **M** is parallel to **B**, in which case **M** does not change because there is no torque. Another relatively simple motion, called *precession*, is shown in Figure 17.3. With the proper initial conditions **M** (and $\langle\boldsymbol{\mu}\rangle$) precess about the direction of **B**. That is, they both rotate about the direction of **B** with a constant angular velocity and at a fixed angle θ with the direction of **B**. Since $\mathbf{M}\times\mathbf{B}$ is always at right angles to **M**, $d\mathbf{M}/dt$ is at right angles to **M**, and the angular momentum does not change magnitude. The analytic solution can be investigated by writing Eq. (17.12) in Cartesian coordinates when **B** is along the z axis:

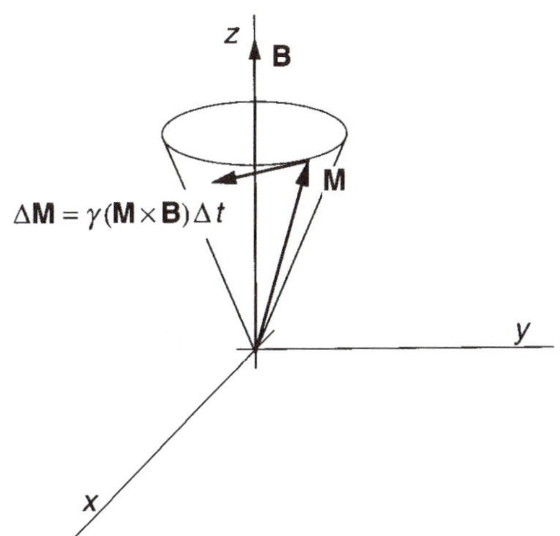

FIGURE 17.3. The system with initial magnetization **M** has been given just enough additional angular momentum to precess about the direction of the static magnetic field **B**. The rate of change of **M** is perpendicular to both **M** and **B**. For short time intervals, $\Delta\mathbf{M}=\gamma(\mathbf{M}\times\mathbf{B})\Delta t$.

$$\frac{dM_x}{dt}=\gamma M_y B_z,$$

$$\frac{dM_y}{dt}=-\gamma M_x B_z, \tag{17.13}$$

$$\frac{dM_z}{dt}=0.$$

One possible solution to these equations is

$$M_z=M_\parallel=\text{const},$$

$$M_x=M_\perp\cos(-\omega t), \tag{17.14}$$

$$M_y=M_\perp\sin(-\omega t).$$

You can verify that these are a solution for arbitrary values of $M_\parallel$ and $M_\perp$ as long as $\omega=\omega_0=\gamma B_z$. This is called the *Larmor precession frequency*. The minus sign means that for positive γ the rotation is clockwise in the xy plane. The classical Larmor frequency is equal to the frequency of photons corresponding to the energy difference given by successive values of $\boldsymbol{\mu}\cdot\mathbf{B}$. For this solution the initial values of **M** at $t=0$ are $M_x(0)=M_\perp$, $M_y(0)=0$, and $M_z(0)=M_\parallel$.

We need to modify the equation of motion, Eq. (17.12), to include changes in **M** that occur because of effects other than the magnetic field. Suppose that **M** has somehow been changed so that it no longer points along the z axis with the equilibrium value given by Eq. (17.10). Thermal agitation will change the populations of the levels so that M_z returns to the equilibrium value, which we call M_0. We *postulate* that the rate of exchange of energy with the reservoir is proportional to how far the value of M_z is from equilibrium:

$$\frac{dM_z}{dt}=\frac{1}{T_1}(M_0-M_z).$$

The quantity T_1, which is the inverse of the proportionality constant, is called the *longitudinal relaxation time* or *spin–lattice relaxation time*.

We also *postulate* an exponential disappearance of the x and y components of **M**. (This assumption is often not a good one. For example, the decay of M_x and M_y in ice is more nearly Gaussian than exponential.) The equations are

$$\frac{dM_x}{dt}=-\frac{M_x}{T_2},\quad \frac{dM_y}{dt}=-\frac{M_y}{T_2}.$$

The *transverse relaxation time* T_2 (sometimes called the spin–spin relaxation time) is always shorter than T_1. A change of M_z requires an exchange of energy with the reservoir. This is not necessary for changes confined to the xy plane, since the potential energy $(\boldsymbol{\mu}\cdot\mathbf{B})$ does not change in that case. M_x and M_y can change as M_z changes, but they can also change by other mechanisms, such as when individual spins precess at slightly different frequencies, a process known as *dephasing*. The angular velocity of precession of μ can be slightly different for different nuclear spins because of local variations in the static magnetic field; the angular velocity can also fluctuate as the field fluctuates with time. These variations and fluctuations are caused by neighboring atomic or nuclear magnetic moments or by inhomogeneities in the external magnetic field **B**. Figure 17.4 shows how dephasing occurs if several magnetic moments precess at different rates.

Combining these approximate equations for relaxation in the absence of an applied magnetic field with Eq. (17.12) for the effect of a magnetic field gives the *Bloch equations*:

$$\frac{dM_z}{dt}=\frac{1}{T_1}(M_0-M_z)+\gamma(\mathbf{M}\times\mathbf{B})_z,$$

$$\frac{dM_x}{dt}=-\frac{M_x}{T_2}+\gamma(\mathbf{M}\times\mathbf{B})_x, \tag{17.15}$$

$$\frac{dM_y}{dt}=-\frac{M_y}{T_2}+\gamma(\mathbf{M}\times\mathbf{B})_y.$$

While these equations are not rigorous and there is no reason for the relaxation to be strictly exponential, they have proven to be quite useful in explaining many facets of nuclear spin magnetic resonance.

One can demonstrate by direct substitution the following solution to Eqs. (17.15) for a static magnetic field **B** along the z axis:

$$M_x=M_0e^{-t/T_2}\cos(-\omega_0 t),$$

$$M_y=M_0e^{-t/T_2}\sin(-\omega_0 t), \tag{17.16}$$

$$M_z=M_0(1-e^{-t/T_1}),$$

where $\omega_0=\gamma B$. This solution corresponds to what happens if **M** is somehow made to precess in the xy plane. (We will see how to accomplish this in Sec. 17.5.) The magnetization in the xy plane is initially M_0, and the amplitude decays exponentially with time constant T_2. The initial value of M_z is zero, and it ''decays'' back to M_0 with time constant T_1. A perspective plot of the trajectory of the tip of vector **M** is shown in Fig. 17.5.

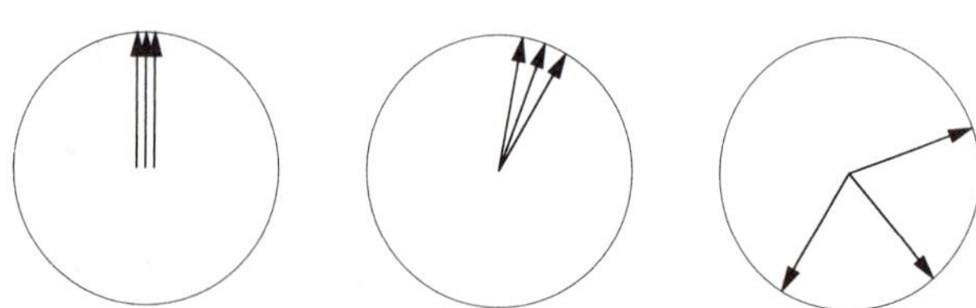

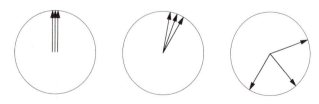

FIGURE 17.4. If two spins precess in the xy plane at slightly different rates, the total spin amplitude decreases due to dephasing.

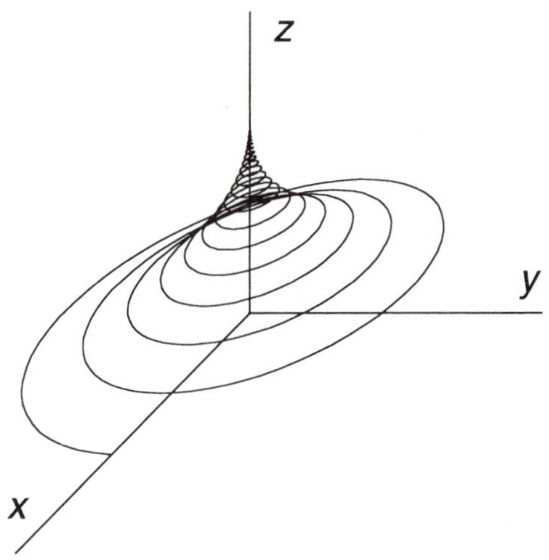

FIGURE 17.5. The locus of the tip of the magnetization **M** when it relaxes according to Eqs. (17.16).

17.5. A ROTATING COORDINATE SYSTEM

It is *much* easier to describe the motion of **M** in a coordinate system which is rotating at the Larmor frequency. Figure 17.6 shows a vector **M** and two coordinate systems, xy and $x'y'$. Components of **M** along each axis are also shown. By considering the components we see that

$$M_x=M_{x'}\cos\theta-M_{y'}\sin\theta,$$

$$M_y=M_{x'}\sin\theta+M_{y'}\cos\theta.$$

For a three-dimensional coordinate system rotating clockwise around the z axis, $\theta=-\omega t$, the z-component of **M** is unchanged, and the transformation equations are

$$M_x=M_{x'}\cos(-\omega t)-M_{y'}\sin(-\omega t),$$

$$M_y=M_{x'}\sin(-\omega t)+M_{y'}\cos(-\omega t), \tag{17.17}$$

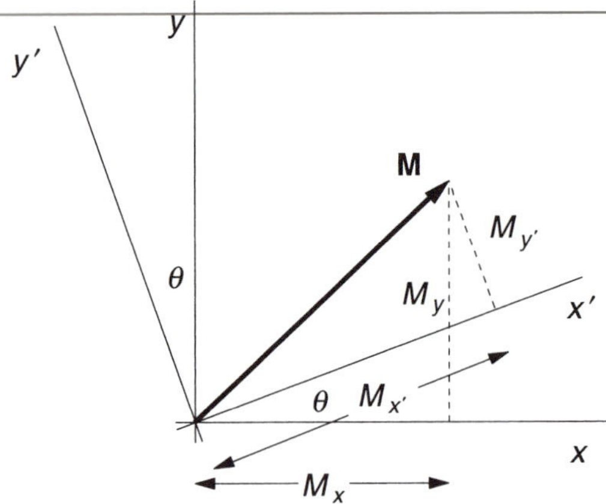

FIGURE 17.6. The vector **M** can be represented by components along x and y or along x' and y'.

$$M_z = M_{z'}.$$

The time derivative of **M** is obtained by differentiating each component and remembering that $\mathbf{M}'$ can also depend on t:

$$\frac{dM_x}{dt} = \frac{dM_{x'}}{dt}\cos(-\omega t) - \frac{dM_{y'}}{dt}\sin(-\omega t) + \omega M_{x'}\sin(-\omega t) + \omega M_{y'}\cos(-\omega t),$$

$$\frac{dM_y}{dt} = \frac{dM_{x'}}{dt}\sin(-\omega t) + \frac{dM_{y'}}{dt}\cos(-\omega t) - \omega M_{x'}\cos(-\omega t) + \omega M_{y'}\sin(-\omega t), \tag{17.18}$$

$$\frac{dM_z}{dt} = \frac{dM_{z'}}{dt}.$$

We can use these expressions to write the equations of motion in the rotating frame. First consider a system without relaxation effects and with a static field B_z along the z axis. We will show that the components of **M** in a system rotating at the Larmor frequency are constant. The equations of motion are given in Eqs. (17.13). In terms of variables in the rotating frame, the equation for dM_x/dt becomes

$$\frac{dM_{x'}}{dt}\cos(-\omega t) - \frac{dM_{y'}}{dt}\sin(-\omega t) + \omega M_{x'}\sin(-\omega t) + \omega M_{y'}\cos(-\omega t) = \gamma[M_{x'}\sin(-\omega t) + M_{y'}\cos(-\omega t)]B_z.$$

If the frame rotates at the Larmor frequency $\omega_0 = \gamma B_z$, the third and fourth terms on the left are equal to the right-hand side. The equation becomes

$$\frac{dM_{x'}}{dt}\cos(-\omega_0 t) - \frac{dM_{y'}}{dt}\sin(-\omega_0 t) = 0.$$

Under the same circumstances, the equation for dM_y/dt gives

$$\frac{dM_{x'}}{dt}\sin(-\omega_0 t) + \frac{dM_{y'}}{dt}\cos(-\omega_0 t) = 0.$$

Solving these simultaneously shows that $dM_{x'}/dt = 0$ and $dM_{y'}/dt = 0$. Therefore in the rotating system $M_{x'}$ and $M_{y'}$ are constant. Equation (17.13) showed that M_z is constant, so the components of **M** are constant in the frame rotating at the Larmor frequency. Using Eqs. (17.17) to transform back to the laboratory system gives the solution Eq. (17.14).[2]

The next problem we consider in the rotating coordinate system is the addition of an oscillating magnetic field $B_1\cos(\omega t)$ along the x axis, fixed in the laboratory system. We will show that if the applied field is at the Larmor frequency, the equations of motion in the rotating system, Eqs. (17.24), are quite simple but very important. They are given as Eqs. (17.24) below.

They are derived as follows. From the x component of Eq. (17.12),

$$\frac{dM_x}{dt} = \gamma(M_y B_z - M_z B_y),$$

we obtain (remembering that the $x'y'$ system is rotating at the Larmor frequency ω_0)

$$\frac{dM_{x'}}{dt}\cos(-\omega_0 t) - \frac{dM_{y'}}{dt}\sin(-\omega_0 t) + \omega_0 M_{x'}\sin(-\omega_0 t) + \omega_0 M_{y'}\cos(-\omega_0 t) = \gamma B_z[M_{x'}\sin(-\omega_0 t) + M_{y'}\cos(-\omega_0 t)].$$

Since $\omega_0 = \gamma B_z$, the last two terms on the left cancel the terms on the right, leaving

$$\frac{dM_{x'}}{dt}\cos(-\omega_0 t) - \frac{dM_{y'}}{dt}\sin(-\omega_0 t) = 0. \tag{17.19}$$

Similarly, the y-component of Eq. (17.12),

$$\frac{dM_y}{dt} = \gamma(M_z B_x - M_x B_z),$$

transforms to (remembering that $M_z = M_{z'}$)

$$\frac{dM_{x'}}{dt}\sin(-\omega_0 t) + \frac{dM_{y'}}{dt}\cos(-\omega_0 t) - \omega_0 M_{x'}\cos(-\omega_0 t) + \omega_0 M_{y'}\sin(-\omega_0 t) = \gamma M_{z'}B_1\cos(\omega t) - \gamma B_z[M_{x'}\cos(-\omega_0 t) - M_{y'}\sin(-\omega_0 t)],$$

which reduces to

$$\frac{dM_{x'}}{dt}\sin(-\omega_0 t) + \frac{dM_{y'}}{dt}\cos(-\omega_0 t) = \gamma B_1 M_{z'}\cos(\omega t). \tag{17.20}$$

The z-component of Eq. (17.12) is

$$\frac{dM_z}{dt} = \gamma(M_x B_y - M_y B_x), \tag{17.21}$$

which transforms to

$$\frac{dM_{z'}}{dt} = -\gamma B_1 M_{x'}\cos(\omega t)\sin(-\omega_0 t) - \gamma B_1 M_{y'}\cos(\omega t)\cos(-\omega_0 t).$$

It is possible to eliminate $M_{x'}$ from Eqs. (17.19) and (17.20) by multiplying Eq. (17.19) by $-\sin(-\omega_0 t)$, multiplying Eq. (17.20) by $\cos(-\omega_0 t)$, and adding. The result is

$$\frac{dM_{y'}}{dt} = \gamma B_1 M_{z'}\cos(\omega_1 t)\cos(-\omega_0 t). \tag{17.22}$$

[2]For those familiar with vector analysis, the general relationship between the time derivative of any vector **M** in the laboratory system and a system rotating with angular velocity $\mathbf{\Omega}$ is

$$\left(\frac{d\mathbf{M}}{dt}\right)_{\text{laboratory}} = \left(\frac{\partial \mathbf{M}}{\partial t}\right)_{\text{rotating}} + \mathbf{\Omega}\times\mathbf{M}.$$

This can be applied to the magnetization combined with Eq. 17.12 to give

$$\left(\frac{\partial \mathbf{M}}{\partial t}\right)_{\text{rot}} = \gamma(\mathbf{M}\times\mathbf{B}) - \mathbf{\Omega}\times\mathbf{M} = \gamma\mathbf{M}\times\left(\mathbf{B} + \frac{\mathbf{\Omega}}{\gamma}\right),$$

which vanishes if $\gamma\mathbf{B} = -\mathbf{\Omega}$.

A similar technique can be used to eliminate $M_{y'}$ from these two equations, giving

$$\frac{dM_{x'}}{dt}=\gamma B_1 M_{z'}\sin(\omega t)\cos(-\omega_0 t). \qquad (17.23)$$

Equations (17.21)–(17.23) are the equations of motion for the components of **M** in the rotating system. If $\omega \neq \omega_0$, the motion is complicated but averaged over many Larmor periods the right-hand side of each equation is zero. If the applied field oscillates at the Larmor frequency, $\omega=\omega_0$, then the $\cos^2(\omega_0 t)$ factors average to $\frac{1}{2}$ while factors like $\sin(\omega_0 t)\cos(-\omega_0 t)$ average to zero.

The averaged equations are our very important result:

$$\frac{dM_{x'}}{dt}=0 \qquad (17.24a)$$

$$\frac{dM_{y'}}{dt}=\frac{\gamma B_1}{2}M_{z'}, \qquad (17.24b)$$

$$\frac{dM_{z'}}{dt}=-\frac{\gamma B_1}{2}M_{y'}. \qquad (17.24c)$$

The first equation says that if $M_{x'}$ is initially zero, it remains zero. Let us define a new angular frequency

$$\omega_1=\frac{\gamma B_1}{2}. \qquad (17.25)$$

It is the frequency of nutation or rotation caused by B_1 oscillating at the Larmor frequency. It is much lower than the Larmor frequency because $B_1 \ll B_z$. In terms of it, Eqs. (17.24b) and (17.24c) become

$$\frac{dM_{z'}}{dt}=-\omega_1 M_{y'}, \qquad \frac{dM_{y'}}{dt}=\omega_1 M_{z'}.$$

These are a pair of coupled linear differential equations with constant coefficients. Differentiating one and substituting it in the other gives

$$\frac{d^2M_{z'}}{dt^2}=-\omega_1\frac{dM_{y'}}{dt}=-\omega_1^2 M_{z'}, \qquad (17.26)$$

which has a solution (a and b are constants of integration)

$$M_{z'}=a\sin(\omega_1 t)+b\cos(\omega_1 t). \qquad (17.27)$$

From Eq. (17.24c) we get

$$M_{y'}=-\frac{1}{\omega_1}\frac{dM_{z'}}{dt}=-a\cos(\omega_1 t)+b\sin(\omega_1 t). \qquad (17.28)$$

The values of a and b are determined from the initial conditions. For example, if **M** is initially along the z axis, $a=0$ and $b=M_0$. Then

$$M_{x'}=0,$$

$$M_{y'}=M_0\sin(\omega_1 t), \qquad (17.29)$$

$$M_{z'}=M_0\cos(\omega_1 t).$$

This kind of motion—precession about the z axis combined with a change of the projection of **M** on z—is called *nutation*. From Eqs. (17.29) it is easy to see that turning B_1 on for a quarter of a period of ω_1 (a 90° pulse or $\pi/2$ pulse, $t=T/4=\pi/2\omega_1$ nutates **M** into the $x'y'$ plane, while a 180° or π pulse nutates **M** to point along the $-z$ axis. **M** nutates about the rotating x' axis. Shifting the phase of B_1 changes the axis in the $x'y'$ plane about which **M** nutates. It may seem strange that an oscillating magnetic field pointing along an axis fixed in the laboratory frame causes rotation about an axis in the rotating frame. The reason is that B_1 is also oscillating at the Larmor frequency, so that its amplitude changes in just the right way to cause this behavior of **M**. Figures 17.7 and 17.8 show this nutation in both the rotating frame and the laboratory frame for a $\pi/2$ pulse and a π pulse.

Figure 17.7(c) emphasizes the difference between nutation and relaxation by plotting M_z vs the projection of **M** in the $x'y'$ plane. For nutation the components of **M** are given by Eqs. (17.29), the magnitude of **M** is unchanged, and the locus is a circle. For relaxation the components are given by Eqs. (17.16).

Another interesting solution is one for which the initial value of **M** is

$$M_{x'}(0)=M_0\cos\alpha,$$

$$M_{y'}(0)=M_0\sin\alpha,$$

$$M_{z'}(0)=0.$$

This corresponds to an **M** that has already been nutated into the $x'y'$ plane. Substituting these values in Eqs. (17.27) and (17.28) shows that $b=0$ and $a=-M_0\, sin\ \alpha$. Then the solution is

$$M_{x'}(t)=M_0\cos\alpha,$$

$$M_{y'}(t)=M_0\sin\alpha\cos(\omega_1 t),$$

$$M_{z'}(t)=-M_0\sin\alpha\sin(\omega_1 t) \qquad (17.30)$$

This solution is plotted in Fig. 17.9 in both the rotating frame and the laboratory frame for the case of a π pulse (a pulse of duration π/ω_1). The effect is to nutate **M** about the x' axis in the rotating coordinate system. We will see later that this is a very useful pulse.

17.6. RELAXATION TIMES

Since longitudinal relaxation changes the value of M_z and hence $\boldsymbol{\mu}\cdot\mathbf{B}$, it is associated with a change of energy of the nucleus. The principal force that can do work on the nuclear spin and change its energy arises from the fact that the nucleus is in a fluctuating magnetic field due to neighboring nuclei and the electrons in paramagnetic atoms.

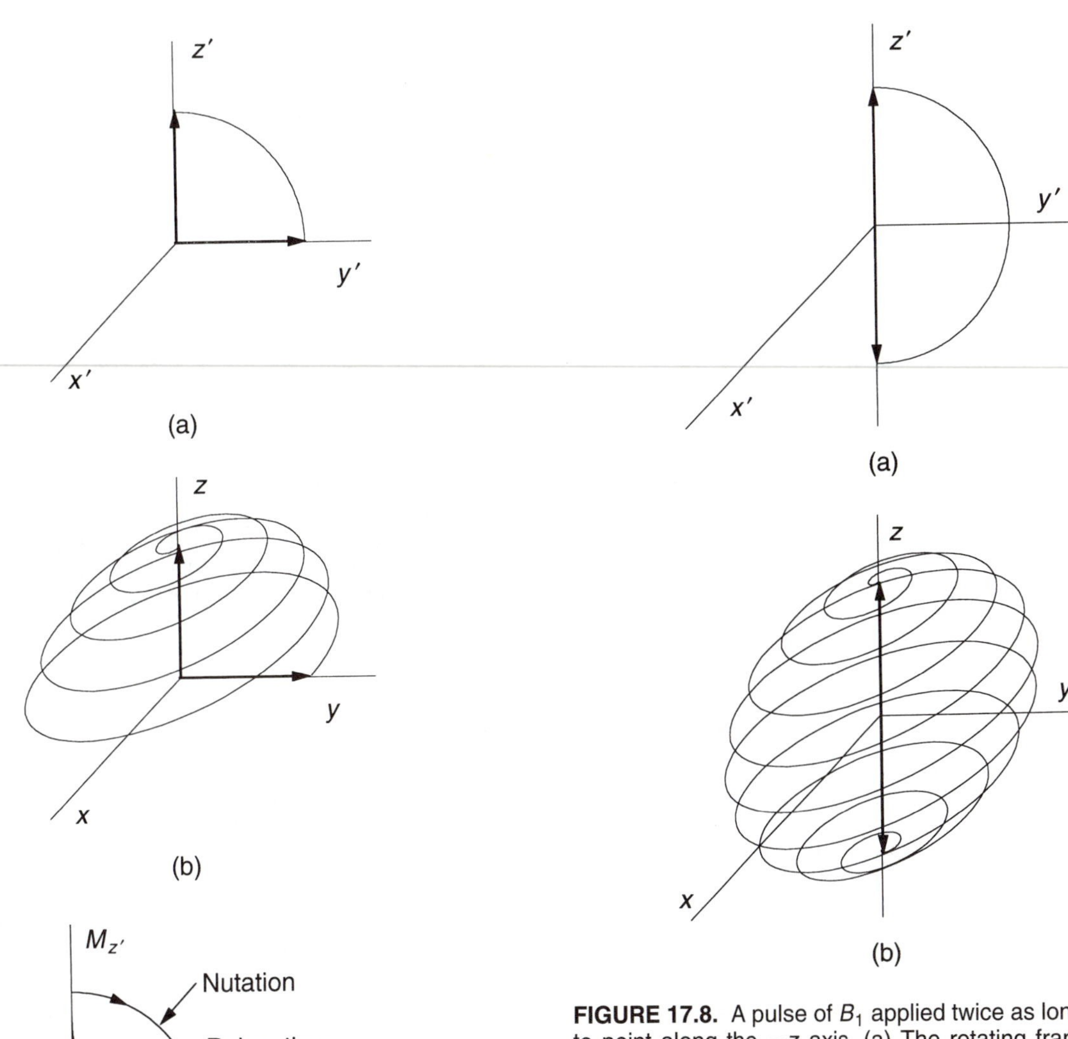

FIGURE 17.7. The locus of the tip of the magnetization **M** when an oscillating magnetic field B_1 is applied for a time t such that $\omega t = \pi/2$. This is often called a "$\pi/2$ pulse." (a) The rotating frame. (b) The laboratory frame. (c) Plots of M_z vs $(M_x^2+M_y^2)^{1/2}$ showing the difference between nutation and relaxation.

One way to analyze the effect of this magnetic field is to say that the change of spin energy ΔE is accompanied by the emission or absorption of a photon of frequency $\omega_{\text{photon}} = \Delta E/\hbar$, or $\omega_{\text{photon}} = \omega_0$. An increase of spin energy requires the absorption of a photon at the Larmor frequency. This will have a high probability if the fluctuating magnetic field has a large Fourier component at the Larmor frequency. A decrease of spin energy is accompanied by the emission of a photon. This can happen spontaneously in a vacuum (*spontaneous emission*), or it can be

FIGURE 17.8. A pulse of B_1 applied twice as long rotates **M** to point along the $-z$ axis. (a) The rotating frame. (b) The laboratory frame.

stimulated by the presence of other photons at the Larmor frequency (*stimulated emission*). These relative probabilities can be calculated using quantum mechanics. Stimulated emission or absorption is much more probable than is spontaneous emission. If the random magnetic field at the nucleus changes rapidly enough due to molecular motion, it will have Fourier components at the Larmor frequency that can induce transitions that cause M_z to change. To get an idea of the strength of the field involved, consider the field at one hydrogen nucleus in a water molecule due to the other hydrogen nucleus. The field due to a magnetic dipole is given by

$$B_r = \frac{\mu_0}{4\pi}\frac{2\mu}{r^3}\cos\theta,$$

$$B_\theta = \frac{\mu_0}{4\pi}\frac{\mu}{r^3}\sin\theta,$$

$$B_\phi = 0, \tag{17.31}$$

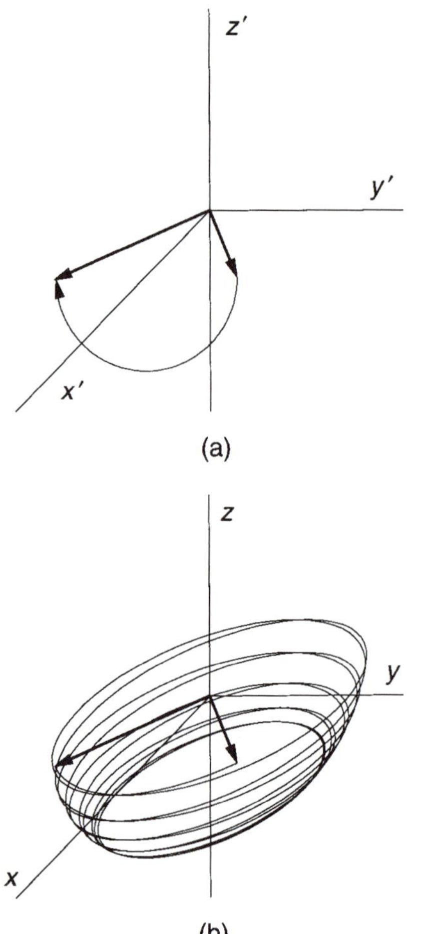

FIGURE 17.9. A magnetic field B_1 pointing along the laboratory x axis and oscillating at the Larmor frequency causes nutation of **M** around the rotating x' axis. In this case **M** was initially in the xy plane. The motion shown here is plotted from Eqs. (17.29) in (a) the rotating and (b) the laboratory frames.

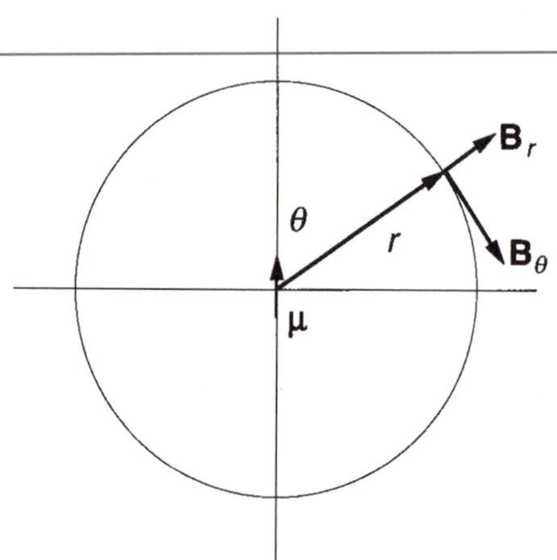

FIGURE 17.10. The magnetic field components of a dipole in spherical coordinates point in the directions shown.

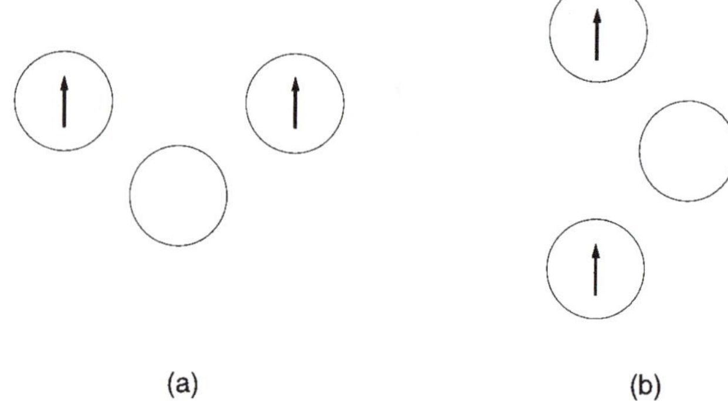

FIGURE 17.11. The z components of the magnetic moments of two protons in a water molecule are shown for two different molecular orientations, a and b. When the water molecule is fixed in space, as in ice, the magnetic field that one proton produces in the neighborhood of the other is static. When the water molecule tumbles, as in a liquid or gas, the field that one proton produces at the other changes with time.

where angle θ is defined in Fig. 17.10. (The factor $\mu_0/4\pi \equiv 10^{-7}$T m A^{-1} is required in SI units.) The magnetic field at one hydrogen nucleus in a water molecule due to the other hydrogen nucleus is about $(3-4) \times 10^{-4}$ T (see Problem 17.13). Consider the water molecule shown in Fig. 17.11. We refer to each hydrogen nucleus as a proton. The z components of the proton magnetic moments are shown. If the water molecule is oriented as in Fig. 17.11(a), the field at one proton due to the other has a certain value. If the water molecule remains fixed in space, as in ice, the field is constant with time. If the molecule is tumbling as in liquid water, the orientation changes as in Fig. 17.11(b), and the field changes with time.

When the molecules are moving randomly, the fluctuating magnetic field components are best described by their autocorrelation functions. The simplest assumption one can make[3] is that the autocorrelation function of each magnetic field component is exponential and that each field component has the same correlation time τ_C:

$$\phi_{11}(\tau) \propto \exp(-|\tau|/\tau_C). \tag{17.32}$$

The Fourier transform of the autocorrelation function gives the power at different frequencies. It has only cosine terms because the autocorrelation is even. Comparison with the Fourier transform pair of Eq. (11.100) shows that the power at frequency ω is proportional to $\tau_C/(1+\omega^2\tau_C^2)$. With the assumption that the transition rate, which is $1/T_1$, is proportional to the power at the Larmor frequency, we have [see also Slichter (1978), p. 167, or Dixon *et al.* (1985)]

[3] A more complete model recognizes that different atoms experience fluctuating fields with different correlation times and that frequency components at twice the Larmor frequency also contribute.

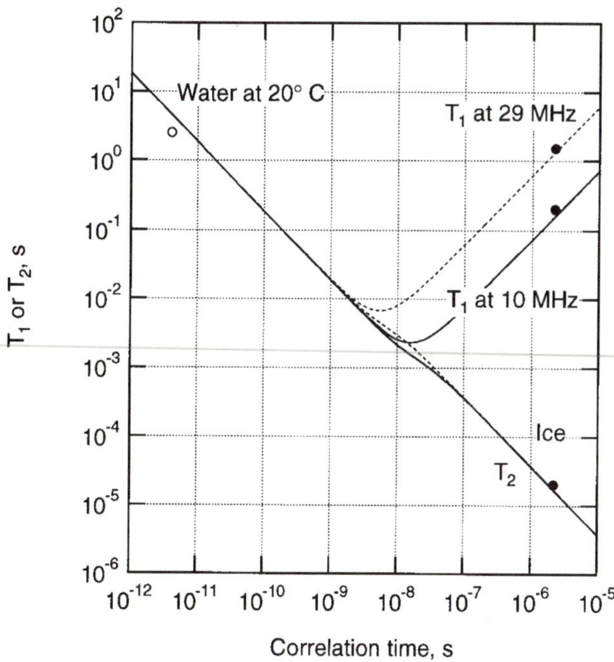

FIGURE 17.12. Plot of T_1 and T_2 vs correlation time of the fluctuating magnetic field at the nucleus. Experimental points are shown for water (open dot) and ice (solid dot).

$$\frac{1}{T_1}=\frac{C\tau_C}{1+\omega_0^2\tau_C^2}, \tag{17.33}$$

where C is the proportionality constant.

The correlation time in a solid is much longer than in a liquid. For example, in liquid water at 20 °C it is about 3.5×10^{-12} s; in ice it is about 2×10^{-6} s. Figure 17.12 shows the behavior of T_1 as a function of correlation time, plotted from Eq. (17.33) with $C=5.43\times10^{10}$ s^{-2}. For short correlation times T_1 does not depend on the Larmor frequency. At long correlation times T_1 is proportional to the Larmor frequency, as can be seen from Eq. (17.33). The minimum in T_1 occurs when $\omega_0=1/\tau_C$ in this model.

Table 17.2 shows some typical values of the relaxation times at 20 MHz. Neighboring paramagnetic atoms reduce the relaxation time by causing a fluctuating magnetic field. For example, adding 20 ppm of Fe^{3+} to water reduces T_1 to 20 ms.

Differences in relaxation time are easily detected in an image. Different tissues have different relaxation times. A contrast agent containing gadolinium is often used in magnetic resonance imaging. It is combined with many of the same pharmaceuticals used with ^{99m}Tc, and it reduces the relaxation time of nearby nuclei. The hemoglobin that carries oxygen in the blood exists in two forms: oxyhemoglobin and deoxyhemoglobin. The former is diamagnetic and the latter is paramagnetic, so the relaxation time in blood depends on the amount of oxygen in the hemoglobin. The technique that exploits this is called BOLD (blood oxygen level dependence).

TABLE 17.2. *Approximate relaxation times at 20 MHz.*

	T_1 (ms)	T_2 (ms)
Whole blood	900	200
Muscle	500	35
Fat	200	60
Water	3000	3000

It was pointed out in Sec. 17.4 that because of dephasing, T_2 is less than or equal to T_1. The same model for the fluctuating fields which led to Eq. (17.33) gives an expression for T_2:

$$\frac{1}{T_2}=\frac{C\tau_C}{2}+\frac{1}{2T_1}. \tag{17.34}$$

There is a slight frequency dependency to T_2 for values of the correlation time close to the reciprocal of the Larmor frequency.

Another effect that causes the magnetization to rapidly decrease is dephasing. Dephasing across the sample occurs because of inhomogeneities in the externally applied field. Suppose that the spread in Larmor frequency and the transverse relaxation time are related by $T_2\Delta\omega=K$. (Usually K is taken to be 2.) The spread in Larmor frequencies $\Delta\omega$ is due to a spread in magnetic field ΔB experienced by the nuclear spins in different atoms. The total variation in B is due to fluctuations caused by the magnetic field of neighbors and to variation in the applied magnetic field across the sample:

$$\Delta B_{\text{tot}}=\Delta B_{\text{internal}}+\Delta B_{\text{external}}.$$

Therefore

$$\Delta\omega_{\text{tot}}=\Delta\omega_{\text{internal}}+\Delta\omega_{\text{external}}.$$

The total spread is associated with the experimental relaxation time, $T_2^*=K/\Delta\omega_{\text{tot}}$. The "true" or "non-recoverable" relaxation time $T_2=K/\Delta\omega_{\text{internal}}$ is due to the fluctuations in the magnetic field intrinsic to the sample. Therefore

$$\frac{1}{T_2^*}=\frac{1}{T_2}+\frac{\gamma\Delta B_{\text{external}}}{K}. \tag{17.35}$$

T_2 is called the *nonrecoverable* relaxation time because various experimental techniques can be used to compensate for the external inhomogeneities, but not the atomic ones.

17.7. DETECTING THE SIGNAL

We have now seen that a sample of nuclear spins in a strong magnetic field has an induced magnetic moment, that it is possible to apply a sinusoidally varying magnetic field and nutate the magnetic moment to precess at any arbitrary angle with respect to the static field, and that the magnetization then relaxes or returns to its original state with two characteristic time constants, the longitudinal and transverse relaxation times. We next consider how a useful signal can be obtained from these spins. This is done by measuring the weak magnetic field generated by the magnetization as it precesses in the xy plane.

Suppose that one has a sample at the origin. The motions plotted in Fig. 17.7 suggest that one way to produce a magnetization rotating in the xy plane is to have a static field along the z axis, combined with a coil in the yz plane (perpendicular to the x axis) connected to a generator of alternating current at frequency ω_0. Turning on the generator for a time $\Delta t=\pi/2\omega_1=\pi/\gamma B_1$ rotates the magnetization into the xy plane. This is called a *90° pulse* or *$\pi/2$ pulse*. If the generator is then turned off, the same coil can be used to detect the changing magnetic flux due to the rotating magnetic moments. The resulting signal, an exponentially damped sine wave, is called the *free induction decay* (FID).

To estimate the size of the signal induced in the coil, imagine a magnetic moment $\boldsymbol{\mu}=\mathbf{M}\,\Delta V$ rotating in the xy plane as shown in Fig. 17.13. The voltage induced in a one-turn coil in the yz plane is the rate of change of the magnetic flux through the coil:

$$\Delta v=-\frac{\partial\Phi}{\partial t}=-\frac{\partial}{\partial t}\int \mathbf{B}\cdot d\mathbf{S}.$$

The magnetic field far from a magnetic dipole can be written most simply in spherical coordinates [Eqs. (17.31)]. We need the flux through the coil of radius a in the yz plane. However, Eqs. (17.31) are not valid close to the dipole. Since a fundamental property of the magnetic field is that for a closed surface $\iint \mathbf{B}\cdot dS=0$, the flux Φ through the coil in Fig. 17.13 is the negative of the flux through the hemispherical cap in Fig. 17.14:

$$\Phi=-\int B_r 2\pi a^2 \sin\theta\, d\theta=$$

$$-\frac{\mu_0}{4\pi}\frac{4\pi\mu_x}{a}\int_0^{\pi/2}\cos\theta\sin\theta\, d\theta$$

$$=-\frac{\mu_0}{4\pi}\frac{2\pi\mu_x}{a}. \tag{17.36}$$

At any instant $\boldsymbol{\mu}$ can be resolved into components along x and y. The component pointing along y contributes no net flux through the spherical cap of Fig. 17.14. Therefore, the flux for a magnetic moment $\boldsymbol{\mu}=\mathbf{M}\,\Delta V$, where $\mathbf{M}$ is given by Eqs. (17.16), is

$$\Phi=-\frac{\mu_0}{4\pi}\frac{2\pi M_0\Delta V}{a}e^{-t/T_2}\cos(-\omega_0 t).$$

The induced voltage is $-\partial\Phi/\partial t$:

$$v=\frac{\mu_0}{4\pi}\frac{2\pi M_0\Delta V}{a}e^{-t/T_2}\left(\frac{1}{T_2}\cos(-\omega_0 t)+\omega_0\sin(-\omega_0 t)\right).$$

Since $1/T_2\ll\omega_0$, this can be simplified to

$$v=\frac{\mu_0}{4\pi}\frac{\omega_0}{a}2\pi M_0\Delta V\, e^{-t/T_2}\sin(-\omega_0 t).$$

If the value of M_z which exists at thermal equilibrium has been nutated into the xy plane, then M_0 is given by the M_z of Eq. (17.10). For a spin-$\frac{1}{2}$ particle (and using the fact that $\omega_0=\gamma B_0$) we obtain

$$v=-\frac{\mu_0}{4\pi}\frac{\pi N\,\Delta V\,\gamma^3\hbar^2 B_0^2}{2k_B Ta}e^{-t/T_2}\sin(-\omega_0 t). \tag{17.37}$$

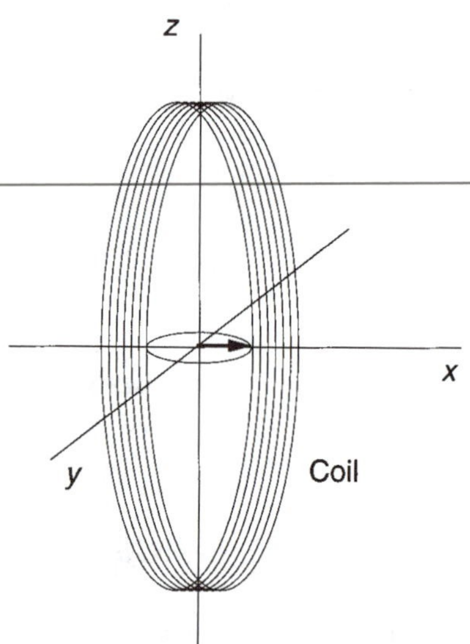

FIGURE 17.13. A magnetic moment rotating in the xy plane induces a voltage in a pickup coil in the yz plane.

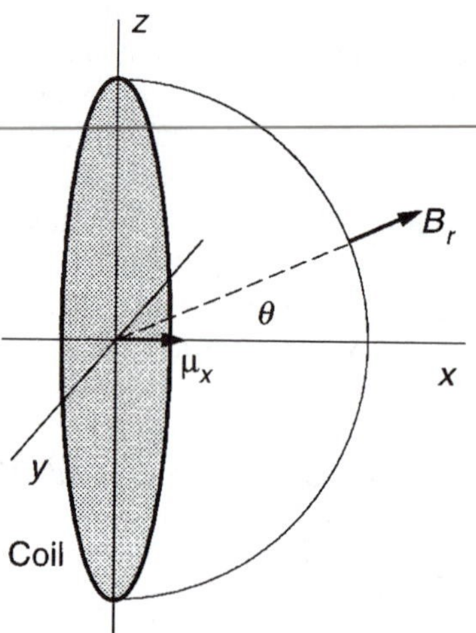

FIGURE 17.14. A dipole along the x axis generates a flux through the circle in the yz plane that is equal and opposite to that through the spherical cap.

Here $N\,\Delta V$ is the total number of nuclear spins involved, B_0 is the field along the z axis, and a is the radius of the coil that detects the free-induction-decay signal. For a volume element of fixed size, as in magnetic resonance imaging, the sensitivity is inversely proportional to the coil radius. If the sample fills the coil, as in most laboratory spectrometers, then the sensitivity is proportional to a.

17.8. SOME USEFUL PULSE SEQUENCES

Many different ways of applying radio-frequency pulses to generate B_1 have been developed by nuclear magnetic resonance spectroscopists for measuring relaxation times. There are five ''classic'' sequences, which also form the basis for magnetic resonance imaging.

17.8.1. Free-Induction-Decay (FID) Sequence

Free induction decay was described in Sec. 17.7. A $\pi/2$ pulse nutates **M** into the xy plane, where its precession induces a signal in a pickup coil. The signal is of the form $\exp(-t/T_2^*)\cos(-\omega_0 t)$, where T_2^* is the experimental transverse relaxation time, including magnetic field inhomogeneities due to the apparatus as well as those intrinsic to the sample. Figure 17.15 shows the pulse sequence, the value of M_x, and the value of M_z. The signal is proportional to M_x. The pulses can be repeated after time T_R for signal averaging. It is necessary for T_R to be greater than $5T_1$ in order for M_z to return nearly to its equilibrium value between pulses.

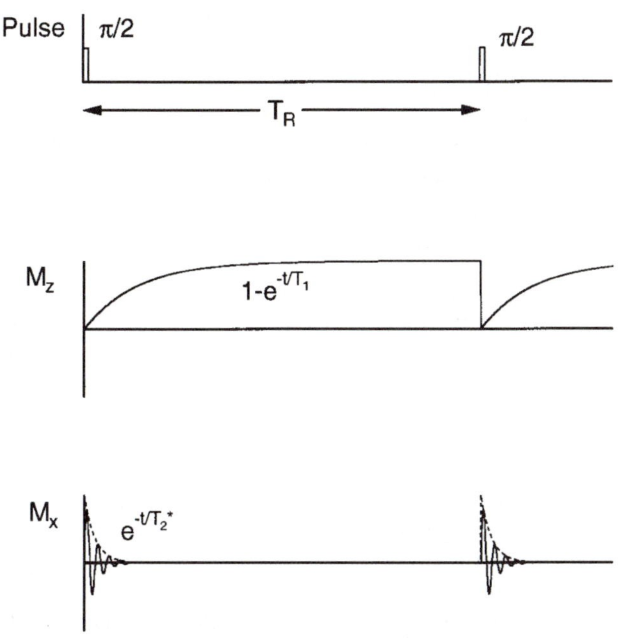

FIGURE 17.15. Pulse sequence and signal for a free induction-Decay measurement.

17.8.2. Inversion-Recovery (IR) Sequence

The inversion-recovery sequence allows measurement of T_1. A π pulse causes **M** to point along the $-z$ axis. There is not yet any signal at this point. M_z returns to equilibrium according to $M_z = M_0[1-2\exp(-t/T_1)]$. A $\pi/2$ interrogation pulse at time T_I rotates the instantaneous value of M_z into the xy plane, thereby giving a signal proportional to $M_0[1-2\exp(-T_I/T_1)]$, as shown in Fig. 17.16. The process can be repeated; again the repeat time must exceed $5T_1$.

You can see from Fig. 17.16 that there will be no signal at all if $T_I/T_1 = 0.693$. If T_I is less than this, the M_x signal will be inverted (negative). Unless special detector circuits are used which allow one to determine that M_x is negative, the results can be confusing.

Inversion recovery images take a long time to acquire and there is ambiguity in the sign of the signal. There are also problems with the use of a π pulse for slice selection [defined in Sec. 17.9; the details of the problems are found in Joseph and Axel (1984)].

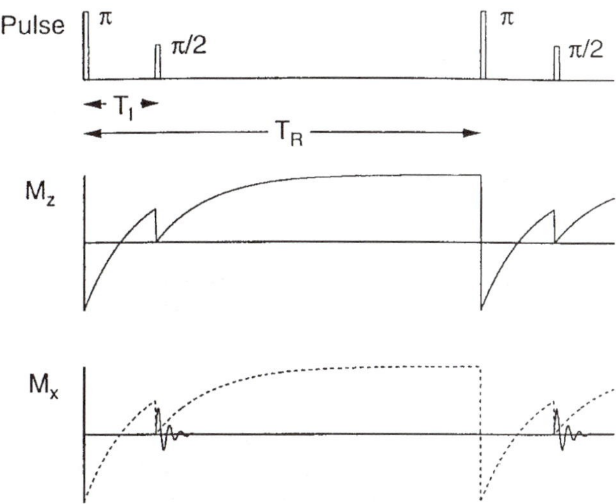

FIGURE 17.16. The inversion recovery sequence allows determination of T_1 by making successive measurements at various values of the interrogation time T_I.

17.8.3. Spin–Echo (SE) Sequence

The pulse sequence shown in Fig. 17.17 can be used to determine T_2 rather than T_2^*. Initially a $\pi/2$ pulse nutates **M** about the x' axis so that all spins lie along the rotating y' axis. Figure 17.17(a) shows two such spins. Spin a continues to precess at the same frequency as the rotating coordinate system; spin b is subject to a slightly smaller magnetic field and precesses at a slightly lower frequency, so that at time $T_E/2$ it has moved clockwise in the rotating frame by angle θ, as shown in Fig. 17.17(b). At this time a π pulse is applied that rotates all spins around the x' axis.

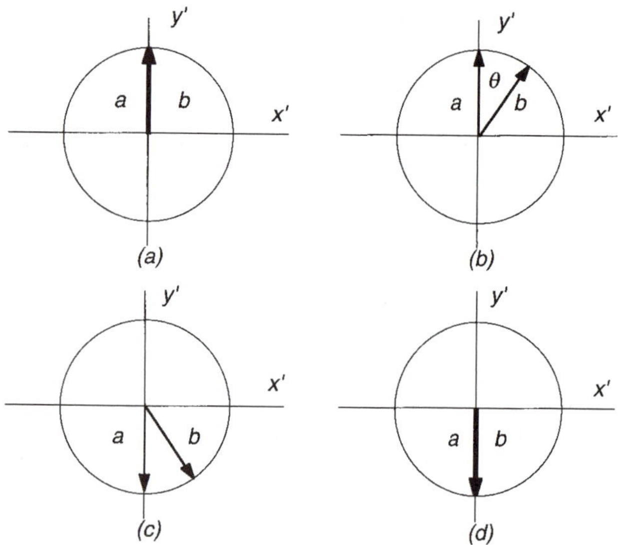

FIGURE 17.17. Two magnetic moments are shown in the $x'y'$ plane in the rotating coordinate system. Moment **a** rotates at the Larmor frequency and remains aligned along the y' axis. Moment **b** rotates clockwise with respect to moment **a**. (a) Both moments are initially in phase. (b) After time $T_E/2$ moment **b** is clockwise from moment **a**. (c) A π pulse nutates both moments about the x' axis. (d) At time T_E both moments are in phase again.

Spin a then points along the $-y'$ axis; spin b rotates to the angle shown in Fig. 17.17(c). If spin b still experiences the larger magnetic field, it continues to precess clockwise in the rotating frame. At time T_E both spins are in phase again, pointing along $-y'$ as shown in Fig. 17.17(d). This argument depends only on the fact that the magnetic field at the nucleus remained the same before and after the π pulse; it does not depend on the specific value of the dephasing angle. Therefore all of the spin dephasing that has been caused by a time-independent magnetic field is reversed in this process. There remains only the dephasing caused by fluctuating magnetic fields. Figure 17.18 shows the pulse sequence and signal.

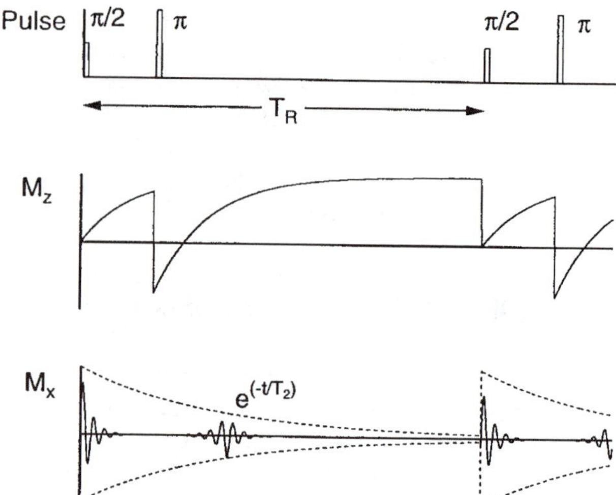

FIGURE 17.18. The pulse sequence and magnetization components for a spin–echo sequence.

17.8.4. Carr–Purcell (CP) Sequence

When a sequence of π pulses that nutate **M** about the x' axis are applied at $T_E/2$, $3T_E/2$, $5T_E/2$, etc., a sequence of echoes are formed, the amplitudes of which decay with relaxation time T_2. This is shown in Fig. 17.19. Referring to Fig. 17.17, one can see that the echoes are aligned alternately along the $-y'$ and $+y'$ axes. One advantage of the Carr–Purcell sequence is that it allows one to determine rapidly many points on the decay curve. Another advantage relates to diffusion. The molecules that contain the excited nuclei may diffuse. If the external magnetic field B_0 is not uniform, the molecules can diffuse to another region where the magnetic field is slightly different. As a result the rephasing after a π pulse does not completely cancel the initial dephasing. This effect is reduced by the Carr–Purcell sequence (see Problem 17.23).

17.8.5. Carr–Purcell–Meiboom–Gill (CPMG) Sequence

One disadvantage of the CP sequence is that the π pulse must be very accurate or a cumulative error builds up in successive pulses. The Carr–Purcell–Meiboom–Gill sequence overcomes this problem. The initial $\pi/2$ pulse nutates **M** about the x' axis as before, but the subsequent

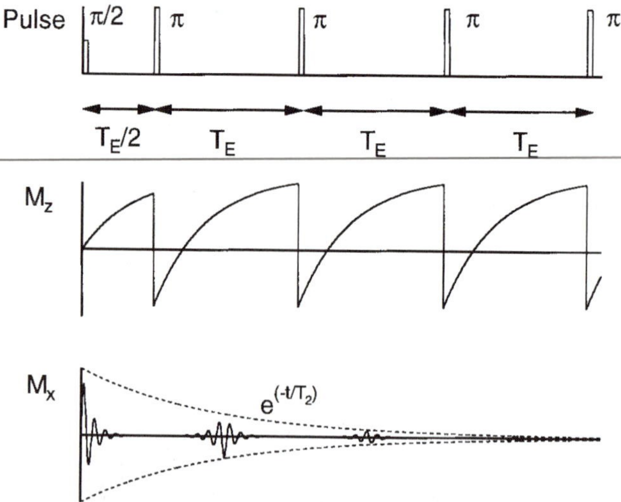

FIGURE 17.19. The Carr–Purcell pulse sequence. All pulses nutate about the x' axis. Echoes alternate sign. The envelope of echoes decays as $\exp(-t/T_2)$, where T_2 is the unrecoverable transverse relaxation time.

π pulses are shifted a quarter cycle in time and rotate about the y' axis. This is shown in Figs. 17.20 and 17.21.

17.9. IMAGING

There are many more techniques available for imaging with magnetic resonance than there are for computer tomography (CT). They are reviewed by Joseph (1985) and by Cho, Jones, and Singh (1993). An excellent new book that discusses pulse sequences in great detail along with signal-to-noise ratio and aspects of coil and electronics engineering has been written by scientists at Phillips Medical Systems [Vlaardingerbroek and den Boer (1996)].

We discuss two reconstruction methods here: projection reconstruction, which is similar to CT reconstruction, and a two-dimensional Fourier technique known as *spin warp* or *phase encoding*, which forms the basis of the techniques actually used in most machines. Our discussion is based on a spin–echo pulse sequence, repeated with a repetition time T_R as shown in Fig. 17.18.

17.9.1. Slice Selection

Suppose we were to apply a $\pi/2$ pulse in a 1.5-T machine ($\omega_0 = 401 \times 10^6\ \mathrm{s}^{-1}$; $f_0 = 63.9$ MHz). If the duration of this pulse is to be 5 ms, it will require a constant amplitude of the radiofrequency magnetic field

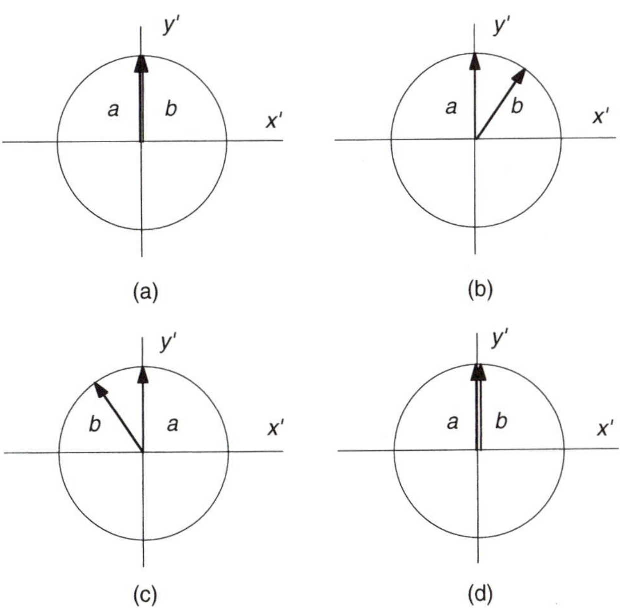

FIGURE 17.20. The effect of the Carr–Purcell–Meiboom–Gill pulse sequence on the magnetization. This is similar to Fig. 17.17 except that the π pulses rotate around the y' axis. Moment **b** rotates clockwise in the $x'y'$ plane. (a) Both moments are initially in phase. (b) After time $T_E/2$ moment **b** is clockwise from moment **a.** (c) A π pulse rotates both moments about the y' axis. (d) At time T_E both moments are in phase again.

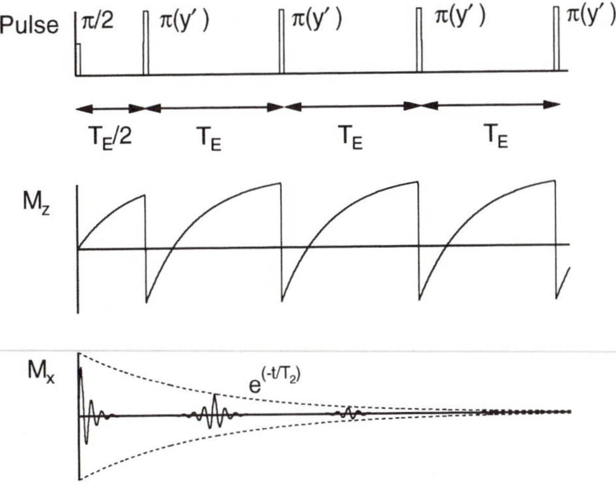

FIGURE 17.21. The CPMG pulse sequence.

$$B_1 = \pi/\gamma\Delta t = 2.35 \times 10^{-6}\ \mathrm{T}. \tag{17.38}$$

The pulse lasts for 3×10^5 cycles at the Larmor frequency. The frequency spread of the pulse is about 200 Hz. This would excite all the proton spins in the sample.

For MR imaging, we want to select a thin slice in the sample. In order to select a thin slice (say $\Delta z = 1$ cm) we apply a *magnetic field gradient* in the z direction while applying a specially shaped B_1 signal. In a static magnetic field B_0, the field lines are parallel. The field strength is proportional to the number of lines per unit area and does not change. With the gradient applied in the volume of interest, the field lines converge, and the field increases linearly with z as shown in Figs. 17.22(a) and 17.22(b):

$$B_z(z) = B_0 + G_z z. \tag{17.39}$$

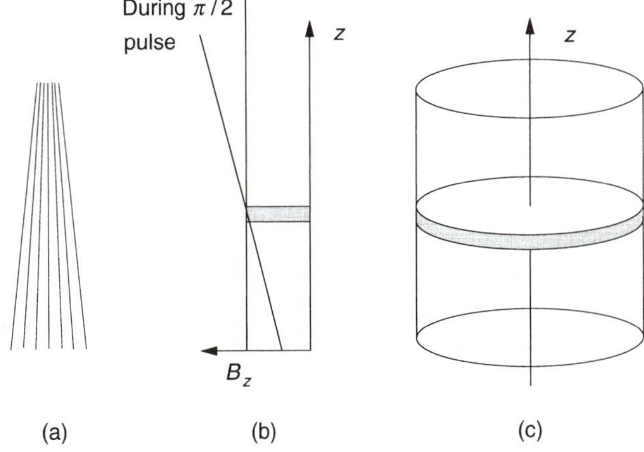

FIGURE 17.22. (a) Magnetic field lines for a magnetic field that increases in the z direction. (b) A plot of B_z vs z with and without a gradient. (c) After application of a field gradient in the z direction during the specially shaped rf pulse, all of the spins in the shaded slice are excited, that is, they are precessing in the xy plane.

We adopt a notation in which G represents a partial derivative of the z component of the magnetic field:

$$G_x \equiv \partial B_z / \partial x,$$
$$G_y \equiv \partial B_z / \partial y, \qquad (17.40)$$
$$G_z \equiv \partial B_z / \partial z.$$

In a typical machine, $G_z = 5\times 10^{-3}$ T m^{-1}. For a slice thickness $\Delta z = 0.01$ m, the Larmor frequency across the slice varies from $\omega_0 - \Delta\omega$ to $\omega_0 + \Delta\omega$, where $\Delta\omega = \gamma G_z \Delta z/2 = 6.68\times 10^3$ s^{-1} ($\Delta f = 1.064$ kHz).

It is possible to make the signal $B_x(t)$ consist of a uniform distribution of frequencies between $\omega_0 - \Delta\omega$ and $\omega_0 + \Delta\omega$, so that all protons are excited in a slice of thickness $\pm\Delta z/2$. Let the amplitude of B_x in the interval $(\omega, d\omega)$ be A. Using Eq. (11.55), $B_x(t)$ is given by

$$B_x(t) = \frac{A}{2\pi}\int_{\omega_0-\Delta\omega}^{\omega_0+\Delta\omega} \cos(\omega t)\, d\omega$$
$$= \frac{A\Delta\omega}{\pi}\frac{\sin(\Delta\omega t)}{\Delta\omega t}\cos(\omega_0 t). \qquad (17.41)$$

This has the form $B_1(t)\cos(\omega_0 t)$, where $B_1(t) = (A\Delta\omega/\pi)\sin(\Delta\omega t)/(\Delta\omega t)$. The function $\sin(x)/x$ has its maximum value of 1 at $x=0$. It is also called the sinc(x) function. The angle ϕ through which the spins are nutated is

$$\phi = \int_{-\infty}^{\infty}\omega_1(t)\,dt = \frac{\gamma}{2}\int_{-\infty}^{\infty} B_1(t)\,dt$$
$$= \frac{\gamma A\Delta\omega}{2\pi}\int_{-\infty}^{\infty}\frac{\sin(\Delta\omega t)}{\Delta\omega t}\,dt$$
$$= \frac{\gamma A}{2}.$$

For a $\pi/2$ pulse, $A = \pi/\gamma$. The maximum value of B_1 is therefore $\Delta\omega/\gamma = G_z\Delta z/2$, as shown in Fig. 17.23. The B_x pulse does not have an abrupt beginning; it grows and decays as shown. In practice, it is truncated at some distance from the peak where the lobes are small.

While the gradient is applied, the transverse components of spins at different values of z precess at different rates. Therefore it is necessary to apply a gradient G_z of opposite sign after the $\pi/2$ pulse is finished in order to bring the spins back to the phase they had at the peak of the slice selection signal. When the gradient is removed all of the spins in the slice shown in Fig. 17.22(a) precess at the Larmor frequency in the xy plane.

17.9.2. Readout in the *x* Direction

The voltage induced in the pickup coil surrounding the sample is proportional to the free induction decay of **M** in the entire slice. That is, the voltage signal induced in the pickup coil is proportional to $\int M(x,y,z)\cos(-\omega_0 t) f(t)\, dV$,

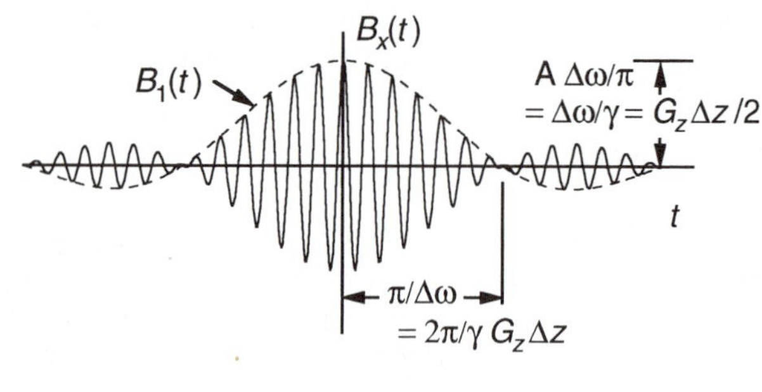

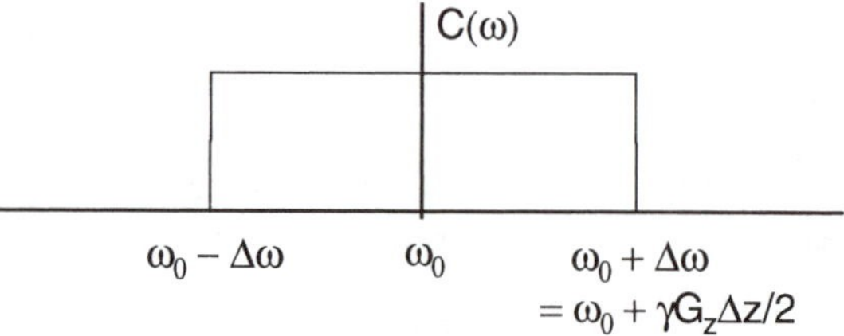

FIGURE 17.23. (a) The $B_x(t)$ signal shown is used to selectively excite a slice. It consists of $\cos(\omega_0 t)$ modulated by a "sinc(x)" or $\sin(x)/x$ pulse $B_1(t)$. (b) The frequency spectrum contains a uniform distribution of frequencies.

where $M(x,y,z)$ is the magnetization per unit volume that was nutated into the xy plane, $\cos(-\omega_0 t)$ represents the change in signal as **M** rotates in the xy plane at the Larmor frequency, and $f(t)$ represents relaxation, signal buildup during an echo, and so on. The initial free-induction-decay signal is ignored. Figure 17.24 shows the echo after a subsequent π pulse.

We assume that changes in $f(t)$ are slow compared to the Larmor frequency and neglect them here. Then the signal from an element $dx\, dy$ in the slice is

$$v(t) = A\, dx\, dy \Delta z\, M(x,y,z)\cos(-\omega_0 t). \qquad (17.42)$$

Constant A includes all the details of the detecting coils and receiver.

Suppose that B_z is given a gradient G_x in the x direction during the echo signal ("during readout"), as shown in Fig.

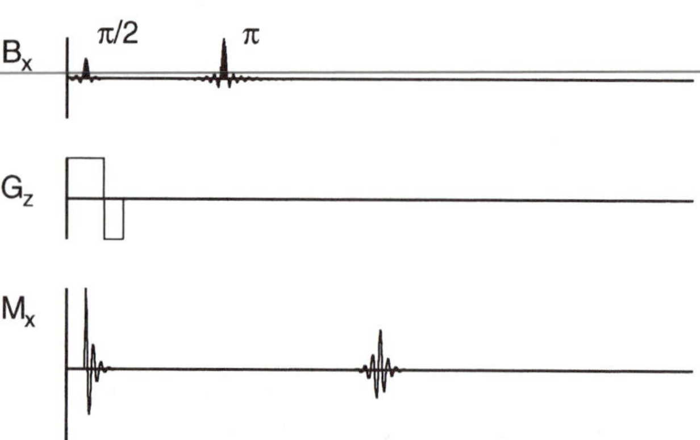

FIGURE 17.24. A slice selection pulse sequence. A $\pi/2$ B_x (rf) pulse while a gradient G_z is applied nutates the spins in a slice of thickness Δz into the xy plane. A negative G_z gradient restores the phase of the precessing spins. The echo after the π pulse is from the entire slice.

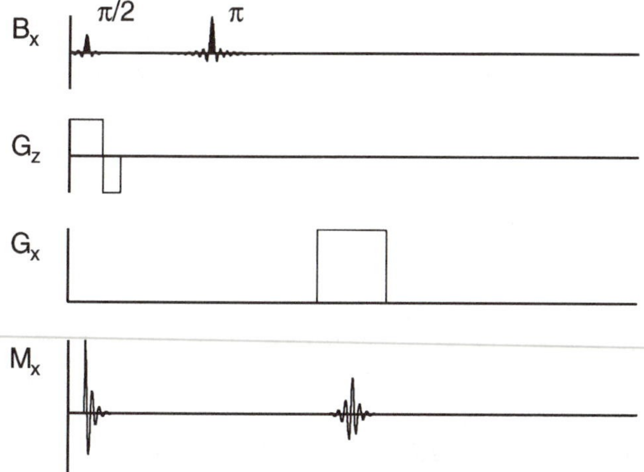

FIGURE 17.25. The pulse sequence for x readout. After slice selection, a gradient G_x is applied during readout. The echo signal between ω and $\omega+d\omega$ is proportional to the magnetization in a strip between x and $x+dx$, integrated over all values of y.

17.25. The spins that echo in the shaded slice between x and $x+dx$ in Fig. 17.26 will be precessing with a Larmor frequency between ω and $\omega+d\omega$, where $\omega=\omega_0+\gamma G_x x$. The signal from the entire slice is

$$v(t)=A\Delta z\int dx\left(\int dy\; M(x,y,z)\right)\cos[-\omega(x)t]. \tag{17.43}$$

We use the fact that $\omega(x)=\omega_0+\gamma G_x x$ to write the signal as

$$v(t)=A\Delta z\int dx\left(\int dy\; M(x,y,z)\right)\cos(\omega_0 t+\gamma G_x x t). \tag{17.44}$$

Since the z slice has already been selected, let us simplify the notation by dropping the z dependence of M. The electronics in the detector multiply $v(t)$ by $\cos(\omega_0 t)$ or

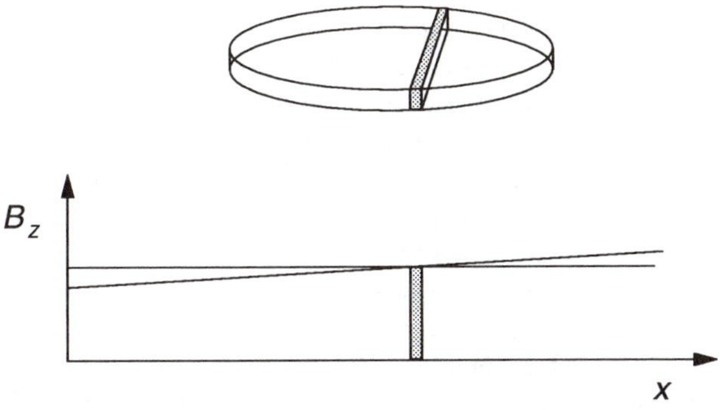

FIGURE 17.26. Because the gradient G_x is applied during readout, the Larmor frequency of all spins in the shaded slice is between ω and $\omega+d\omega$.

$\sin(\omega_0 t)$ and average over many cycles at the Larmor frequency. The results are two signals that form the basis for constructing the image:

$$s_c(t)=\overline{v(t)\cos(\omega_0 t)}\propto\int\int dx\; dy\; M(x,y)\cos(\gamma G_x x t),$$

$$s_s(t)=\overline{v(t)\sin(\omega_0 t)}\propto\int\int dx dy M(x,y)\sin(\gamma G_x x t). \tag{17.45}$$

The time average is over many cycles at the Larmor frequency but a time short compared to $2\pi/\gamma G_x x_{\max}$.

17.9.2.1. Projection Reconstruction

By inspection of Eq. (17.45) and remembering the relationship between ω and x, we see that the Fourier transforms of $s_c(t)$ and $s_s(t)$ are both proportional to $\int dy\; M(x,y)$. (Of course, the signals are digitized and one actually deals with discrete transforms.) This means that s_c or s_s can be Fourier analyzed to determine the amount of signal in the frequency interval $(\omega,d\omega)$ corresponding to (x,dx), which is proportional to the projection $\int M(x,y)\; dy$ along the shaded strip. In Sec. 12.8 we learned how to reconstruct an image from a set of projections. The entire readout process can be therefore be repeated with the gradient rotated slightly in the xy plane (that is, with a combination of G_x and G_y during readout). This is indicated in Fig. 17.27, which indicates many scans, with different values of G_x and G_y, related by

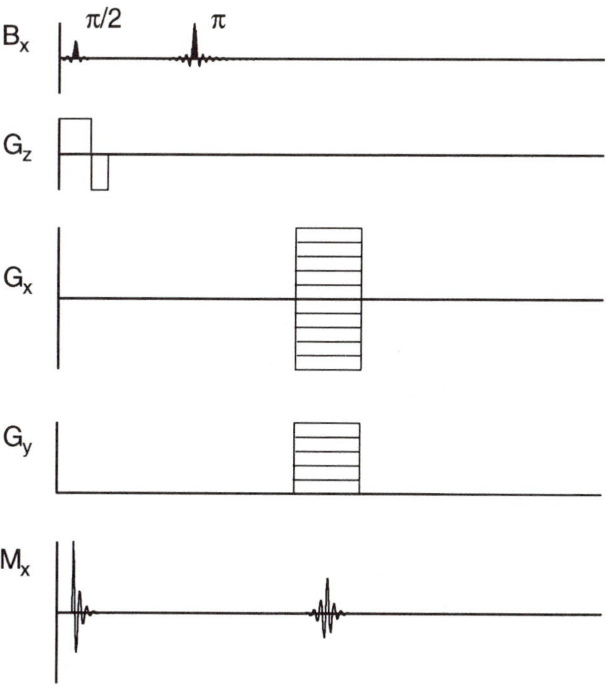

FIGURE 17.27. Projection reconstruction techniques can be used to form an image. A series of measurements are taken, each with simultaneous gradients G_x and G_y.

$G_x/G_y=\tan\theta$, where θ is the angle between the projection and the x axis. All of the techniques for reconstruction from projections that were developed for computed tomography can be used to reconstruct $M(x,y)$. Sending the proper combination of currents through the x and y gradient coils rotates the gradient; no rotating mechanical components are needed.

17.9.2.2. Phase Encoding

Techniques are available for magnetic resonance imaging that are not available for computed tomography. They are based on determining directly the Fourier coefficients in two or three dimensions. The basic technique is called *spin warp* or *phase encoding*.

We wish to construct an image of $M(x,y)$, modified by the function $f(t)$ that accounts for relaxation, etc. For simplicity of notation we again assume f is unity and suppress the z dependence, since slice selection has already been done. We will construct $M(x,y)$ from its Fourier transform. The Fourier transform of $M(x,y)$ is given by Eqs. (12.9):

$$M(x,y)=\left(\frac{1}{2\pi}\right)^2\int_{-\infty}^{\infty}dk_x\int_{-\infty}^{\infty}dk_y[C(k_x,k_y)\cos(k_xx+k_yy)+S(k_x,k_y)\sin(k_xx+k_yy)]. \quad (17.46a)$$

with the coefficients given by

$$C(k_x,k_y)=\int_{-\infty}^{\infty}dx\int_{-\infty}^{\infty}dy\ M(x,y)\cos(k_xx+k_yy), \quad (17.46b)$$

$$S(k_x,k_y)=\int_{-\infty}^{\infty}dx\int_{-\infty}^{\infty}dy\ M(x,y)\sin(k_xx+k_yy). \quad (17.46c)$$

Our problem is to determine C and S and from them construct the image.

The information from the x readout gives us $C(k_x,0)$ and $S(k_x,0)$ directly. We show this for the cosine transform. From Eq. (17.46b)

$$C(k_x,0)=\int dx\left(\int dyM(x,y)\right)\cos(k_xx). \quad (17.47)$$

Comparing this to the expression for $s_c(t)$ in Eq. (17.45), we see that

$$C(k_x,0)\propto s_c(k_x/\gamma G_x). \quad (17.48a)$$

Similarly,

$$S(k_x,0)\propto s_s(k_x/\gamma G_x). \quad (17.48b)$$

The times at which s_c and s_s are measured and therefore the values of k_x are, of course, discrete. The discussion in Sec. 12.6 shows that the values of k_x are multiples of the lowest spatial frequency:

$$k_x=m\ \Delta k=2\pi m/D.$$

The corresponding times to measure the signal are

$$t_m=\frac{2\pi m}{D\gamma G_x}.$$

The spatial extent of the image or "field of view" D determines the spacing Δk_x. The desired pixel size determines the maximum value of k_x or m:

$$\Delta x=\frac{\pi}{k_{\max}}=\frac{D}{2m_{\max}}.$$

The discrete values of k_x are shown in Fig. 17.28(a).

The next problem is to make a similar determination for nonzero values of k_y. To do so, a gradient of B_z in the y direction is applied at some time between slice selection and readout. This makes the Larmor frequency vary in the y direction. If the phase-encoding pulse is due to a uniform gradient that lasts for a time T_p, the total phase change is

$$\Delta\phi=\int\omega(t)\ dt=\gamma G_yT_py=k_yy. \quad (17.49)$$

The readout signal, Eq. (17.43), is replaced by

$$v(t)=A\ \Delta z\int dx\int dy\ M(x,y)\cos[-\omega(x)t+k_yy]. \quad (17.50)$$

Note that the added phase does not depend on t. However, the cosine term must now be included in both the x and y integrals. Carrying through the mathematics of the detection process shows that temporal Fourier transformation of the signals determines $C(k_x,k_y)$ and $S(k_x,k_y)$ for all values of k_x and for the particular of k_y determined by the phase selection pulse. Different values of the y gradient pulse give the coefficients for different values of k_y, as shown in Fig. 17.28. Both positive and negative gradients are used to give both positive and negative values of k_y. Application of a gradient G_x during the phase-encoding time (in addition to the readout gradient) changes the starting value of k_x. This allows one to determine the coefficients for negative values of k_x. This figure has been drawn without taking into account that the application of a π pulse changes k_x to $-k_x$ and k_y to $-k_y$. The gradients and signals for this spin–echo determination are shown in Fig. 17.29. The coefficients are substituted in Eq. (17.46a) to reconstruct $M(x,y,z)$ for the z slice in question.

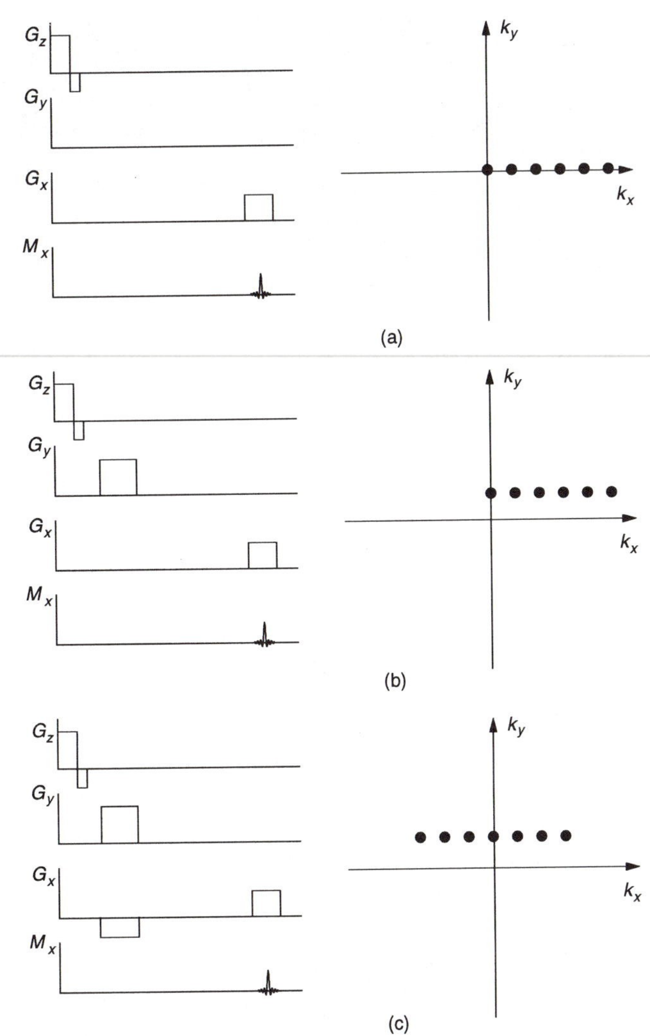

FIGURE 17.28. (a) The signal measured while the x gradient is applied gives the spatial Fourier transform of the image along the k_x axis. (b) The addition of a phase-encoding gradient sets a nonzero value for k_y so that the readout determines the spatial Fourier transform along a line parallel to the k_x axis. (c) Phase encoding along the x axis as well shifts the line along which the coefficients are determined.

17.9.3. Image Contrast and the Pulse Parameters

The appearance of an MR image can be changed drastically by adjusting the repetition time and the echo time. Problem 17.22 derives a general expression for the amplitude of the echo signal when a series of $\pi/2$ pulses are repeated every T_R seconds. The magnetic moment in the sample at the time of the measurement, considering both longitudinal and transverse relaxation, is

$$M(T_R,T_E)=M_0(1-2e^{-T_R/T_1+T_E/2T_1}+e^{-T_R/T_1})e^{-T_E/T_2}. \tag{17.51}$$

If $T_R \gg T_E$, this simplifies to

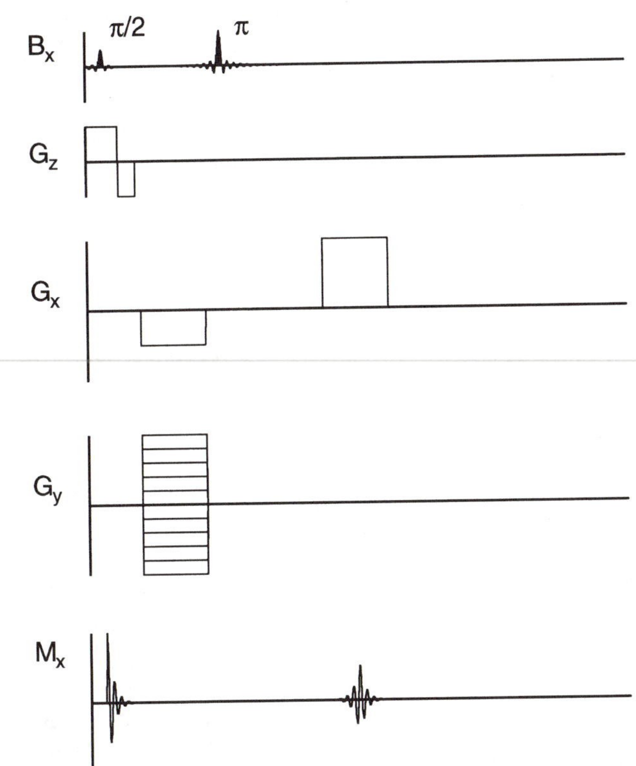

FIGURE 17.29. The signals in a standard phase encoding. The pulse sequence is repeated for each value of k_y.

$$M(T_R,T_E)=M_0(1-e^{-T_R/T_1})e^{-T_E/T_2}, \tag{17.52}$$

where M_0 is proportional to the number of proton spins per unit volume N, as shown in Eq. (17.10). We consider an example that compares muscle ($M_0=1.02$, $T_1=500$ ms, and $T_2=35$ ms) with fat ($M_0=1.24$, $T_1=200$ ms, and $T_2=60$ ms).

Figure 17.30 shows two examples where T_R is relatively long and M_0 returns nearly to its initial value between pulses. If the echo time is short, then the image is nearly independent of both T_1 and T_2 and it is called a *density-weighted image*. If T_E is longer, then the transverse decay term dominates and it is called a *T_2-weighted image*. The signal is often weak and therefore noisy because there has been so much decay.

Figure 17.31 shows what happens if the repetition time is made small compared to T_1. This is a *T_1-weighted image* because the differences in T_1 are responsible for most of the difference in signal intensity. Notice also that the very first pulse nutates the full M_0 into the transverse plane, so an echo after the first pulse would give an anomalous reading. Echoes are measured only for the second and later pulses. Suppose that the value of T_2 for fat had been shorter than the value for muscle. Then there would have been a value of T_E for which the two transverse magnetization curves crossed, and the two tissues would have been indistinguishable in the image. At larger values of T_E, their relative brightnesses would have been reversed. Figure 17.32 shows

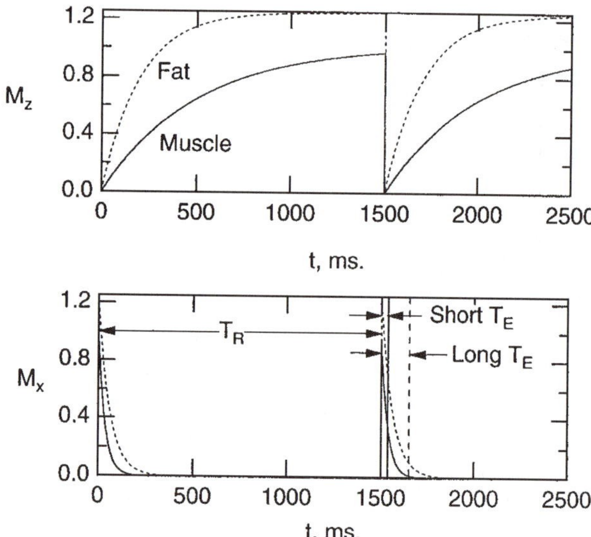

FIGURE 17.30. The intensity of the signal from different tissues depends on the relationship between the repetition time and echo times of the pulse sequence, and the relaxation times of the tissues being imaged. This figure and the next show the magnetization curves for two tissues: muscle (relative proton density 1.02, $T_1=500$ ms, $T_2=35$ ms) and fat (relative proton density 1.24, $T_1=200$ ms, $T_2=60$ ms). The repetition time is 1500 ms, which is long compared to the longitudinal relaxation times. A long echo time gives an image density that is very sensitive to T_2 values. A short echo time (even shorter than shown) gives an image that depends primarily on the spin density.

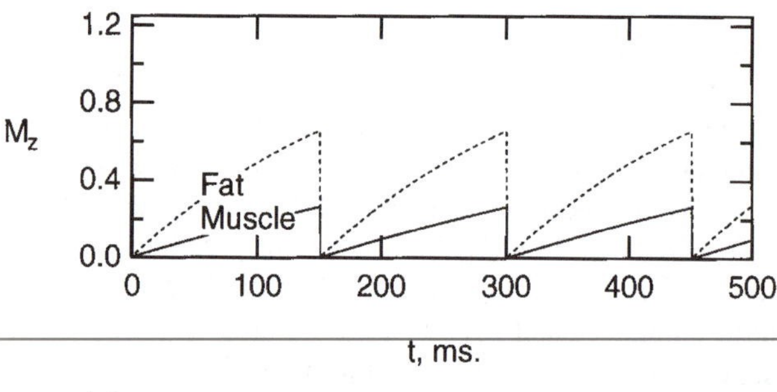

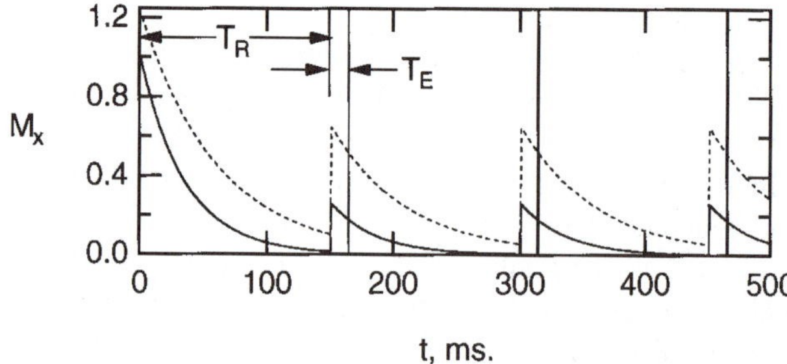

FIGURE 17.31. The tissue parameters are the same as in Fig. 17.30. The repetition time is short compared to the longitudinal relaxation time. As a result, the first echo must be ignored. With a short T_E, the image density depends strongly on the value of T_1.

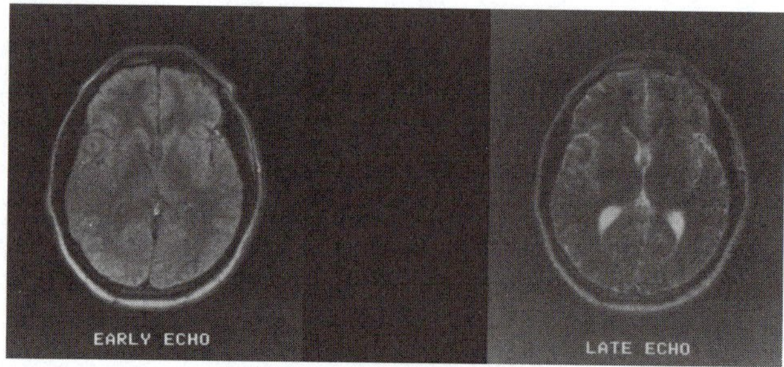

FIGURE 17.32. Spin–echo images taken with short and long values of T_E, showing the difference in T_2 values for different parts of the brain. Photograph courtesy of R. Morin, Ph.D., Department of Diagnostic Radiology, University of Minnesota.

spin–echo images taken with two different values of T_E, for which the relative brightnesses are quite different.

17.9.4. Other Pulse Sequences

There are a large number of other pulse sequences in use, all of which are based on the fundamentals presented here. We mention only a few.

One of the problems with conventional spin echo is that one must wait a time T_R between measurements for different values of k_y. One way to speed things up is to use the intervening time to make measurement in a slice at a different value of z.

Fast spin echo or *turbo spin echo* uses a single $\pi/2$ pulse, followed by a series of π pulses, as shown in Fig. 17.33. Each π pulse produces an echo, though the echo amplitudes decay and a correction for this must be made in the image reconstruction. Each G_y pulse increments or "winds" the phase by a fixed amount. A negative G_x pulse resets the positions of the k_x values. Faster image acquisition sequences not only save time, but they may allow the image to be obtained while the patient's breath is held, thereby eliminating motion artifacts.

A variation on this is *echo planar imaging* (EPI) which eliminates the π pulses. It requires a magnet with a very uniform magnetic field, so T_2 (in the absence of a gradient) is only slightly greater than T_2^*. A small constant G_y is applied, and G_x oscillates as shown in Fig. 17.34. One of several applications is interlaced EPI. The $\pi/2$ pulse is synchronized with the heart beat. Data for several k_y values spread across the entire range of k space are acquired during the first beat, with additional sets of interleaved values acquired during subsequent beats (Fig. 17.35).

Other fast sequences only partially flip the spins into the xy plane.

A 3-dimensional Fourier transform of the image can be obtained by selecting the entire sample and then phase encoding in both the y and z directions while doing frequency readout along x. One must step through all

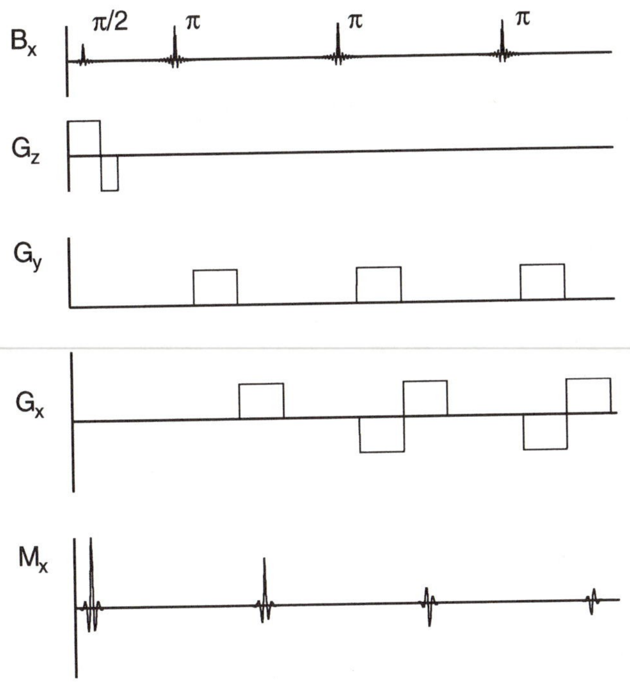

FIGURE 17.33. A fast spin echo sequence uses a single $\pi/2$ slice selection pulse followed by multiple echo rephasing pulses. A correction must be made for the transverse decay.

values of k_y for each value of k_z. This forms the basis for imaging very small samples with very high resolution (MRI microscopy).

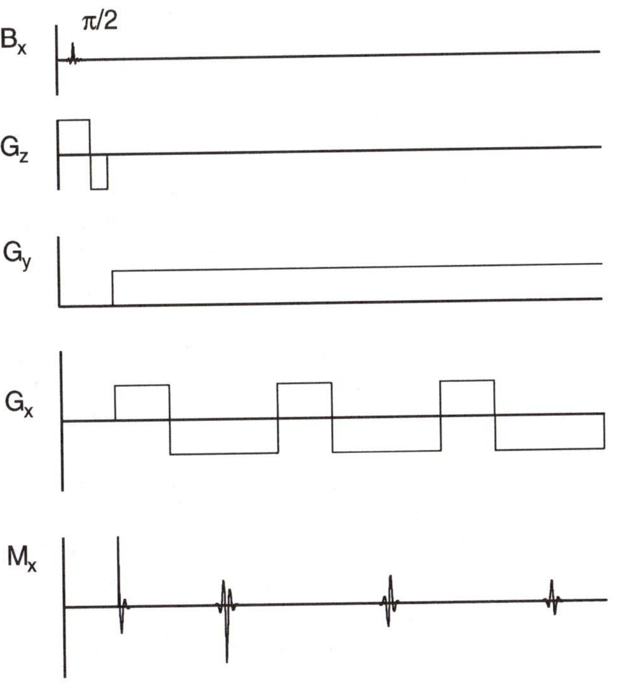

FIGURE 17.34. Echo planar imaging uses a very uniform magnet and eliminates the π rephasing pulses.

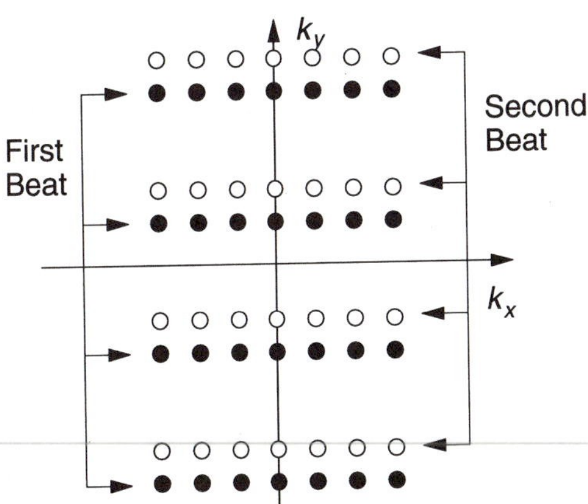

FIGURE 17.35. The excitation pulse for echo planar imaging (EPI) of the heart can be triggered by the electrocardiogram. Several values of k_y are sampled during the beat. Widely separated values of k_y are sampled on the first beat, with interleaved values on subsequent beats.

High spatial frequencies give the sharp edge detail in an image; the lowest spatial frequencies give the overall contrast. (We saw this in Figs. 12.16 and 12.17.) Changing the order of sampling points in k space can be useful. For example, when the image may be distorted by blood flow (see the Sec. 17.11), it is possible to change the gradients in such a way that the values of k near zero are measured right after the excitation. This gives the proper signal within the volume of the vessel. The higher spatial frequencies, which show vessel edges, are less sensitive to blood flow and are acquired later. Some acquisition sequences vary G_x and G_y as a function of time in such a way that a spiral path through k space is followed.

17.10. CHEMICAL SHIFT

If the external magnetic field is very homogeneous, it is possible to detect a shift of the Larmor frequency due to a reduction of the magnetic field at the nucleus because of diamagnetic shielding by the surrounding electron cloud. The modified Larmor frequency can be written as

$$\omega=\gamma B_0(1-\sigma). \tag{17.53}$$

Typical values of σ are in the range 10^{-5}–10^{-6}. They are independent of B_0, as expected for a diamagnetic effect proportional to B_0. Measurements are made by Fourier transformation of the free-induction-decay signal, averaged over many repetitions if necessary to provide the resolution required to detect the shift.

A great deal of work has been done with ^{31}P, because of its presence in adenosine triphosphate and adenosine diphosphate (ATP and ADP). Free energy is supplied for many processes in the body by the conversion of ATP to ADP. Figure 17.36 shows shifts in the ^{31}P peaks due to

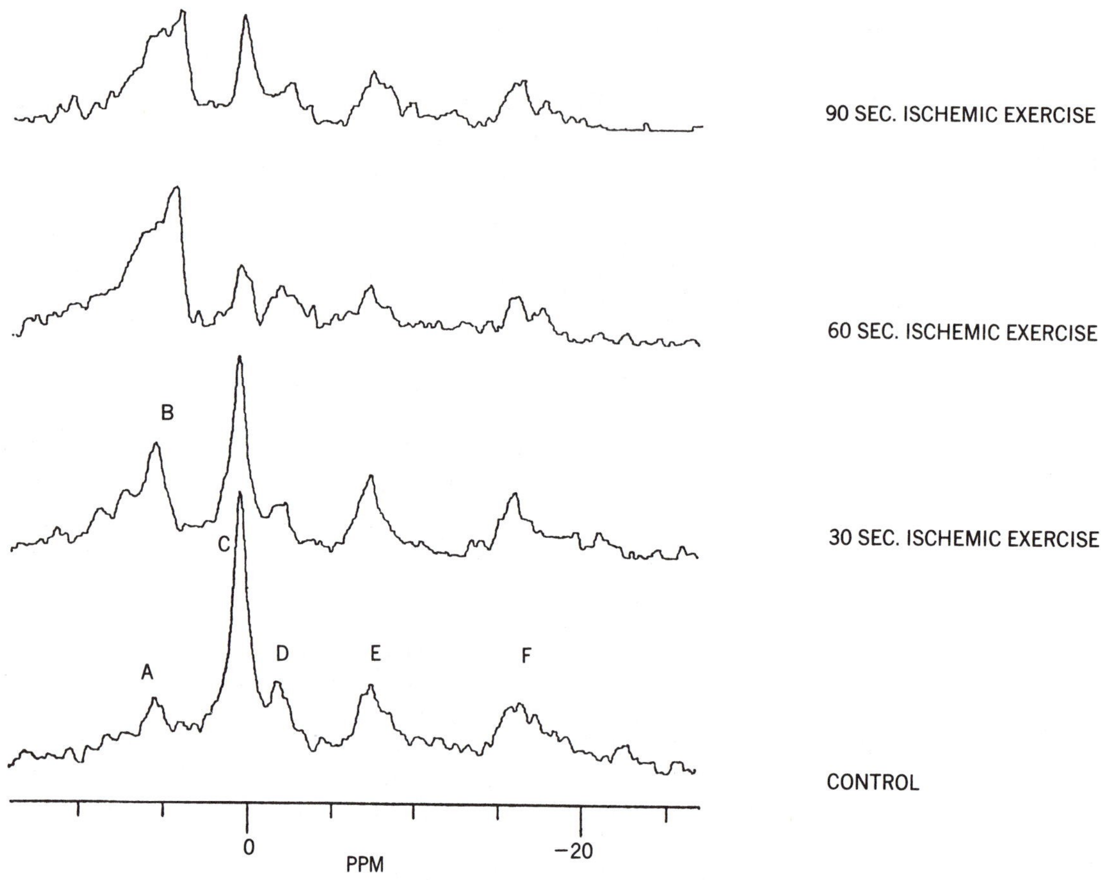

FIGURE 17.36. A series of ^{31}P NMR spectra from the forearm of a normal adult showing the change in various chemical-shift peaks with exercise. Peak A is inorganic phosphate; C is phosphocreatine; D, E, and F are from the three phosphates in ATP. One can see the disappearance of ATP and phosphocreatine with exercise, accompanied by the buildup of inorganic phosphate. From R. L. Nunally (1985). NMR spectroscopy for in vivo determination of metabolism; an overview, in S. R. Thomas and R. L. Dixon, eds. *NMR in Medicine: The Instrumentation and Clinical Applications,* College Park, MD, AAPM. Used by permission.

metabolic changes. With exercise the ATP and phosphocreatine peaks diminish and the inorganic phosphate peak increases.

It is also possible to make chemical shift images. An FID signal is measured for each volume element. Slice selection followed by phase encoding in two dimensions can be used (Fig. 17.37), or phase encoding can be used in all three directions. Because of the number of measurements required, spatial resolution is usually limited to 32×32 or 64×64. Figure 17.38 shows an ^{31}P image of the brain.

Metabolites containing hydrogen can also be measured, but special efforts are required to eliminate artefacts (distortions) due to the very strong signals from water and lipids [Hu *et al.* (1995)].

17.11. FLOW EFFECTS

Flow effects can distort a magnetic resonance image. Spins initially prepared with one value of **M** can flow out of a slice before the echo and be replaced by spins that had a different initial value of **M**. This is called *washout.* Spins that have been shifted in phase by a field gradient can flow to another location before the readout pulse is applied. Axel (1985) reviews the effect on images. This technique has also been used to measure blood flow [Battocletti *et al.* (1981)].

To understand the washout effect consider a simple model in which a blood vessel is perpendicular to the slice, as shown in Fig. 17.39. To simplify further, assume that all the blood flows with the same speed v, independent of where it is in the vessel. This is called *plug flow.*

First consider washout of the excited spins. Suppose that at time $T_E/2$ a π pulse is applied to the slice in Fig. 17.39 and that the echo is measured at time T_E. The shaded area in the vessel represents new blood that flows in during time t. If the flow velocity is zero, no new blood flows in, all of the blood in the slice was excited, and the signal has full strength. If the velocity is greater than $2\Delta z/T_E$, all of the spins that were flipped by the pulse will leave the sensitive region by the time of the echo, and there will be no signal.

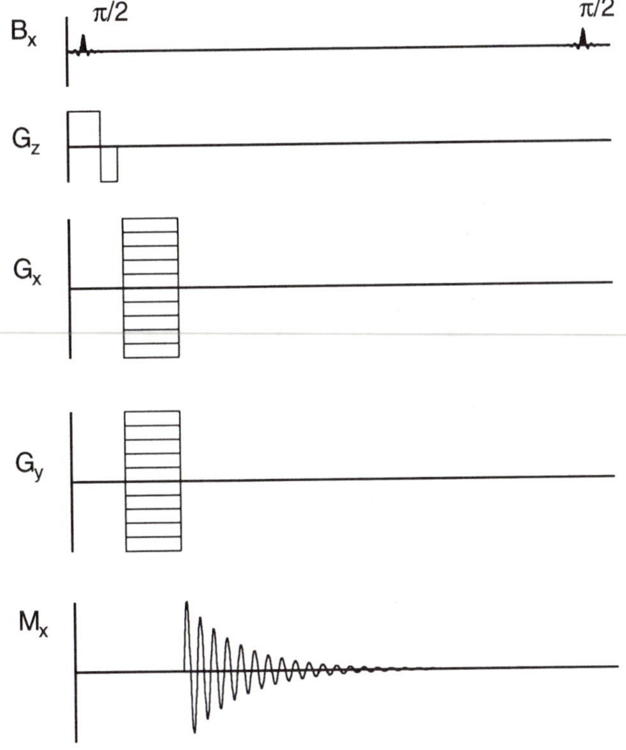

FIGURE 17.37. A possible pulse sequence to measure free induction decay (FID). Slice selection is used in the *z* direction, and phase encoding is applied in both the *x* and *y* directions before the FID signal is measured.

Because we assume plug flow, the fraction washed out is a linear function of velocity up to the critical value of v. The fraction of excited spins remaining at T_E is given by

$$f=\begin{cases} 1-\dfrac{vT_E}{2\Delta z}, & v<2\Delta z/T_E \\ 0, & v\geqslant 2\Delta z/T_E \end{cases}. \tag{17.54}$$

Now consider washout of spins between pulses. We saw that the effect of repetition and echo times on the MRI signal is given by Eq. (17.51), which, if $T_R \gg T_E$, simplifies to Eq. (17.52). For low velocities ($v<\Delta z/T_R$) there is an enhancement of the signal because blood with a larger value of M_z flows into the sensitive region. For $vT_R<\Delta z$, the factor in parentheses in Eq. (17.52) is replaced by

$$\frac{vT_R}{\Delta z}+\left(1-\frac{vT_R}{\Delta z}\right)(1-e^{-T_R/T_1}).$$

The first term represents spins that flow in and the second those that still remain and that are still affected by the previous pulse. This can be rearranged as

$$(1-e^{-T_R/T_1})+\frac{vT_R}{\Delta z}\,e^{-T_R/T_1}. \tag{17.55}$$

This factor has the value $1-\exp(-T_R/T_1)$ for small v and is replaced by unity for $v>\Delta z/T_R$. In this simple model, the effect on the signal is the product of Eqs. (17.54) and (17.55).

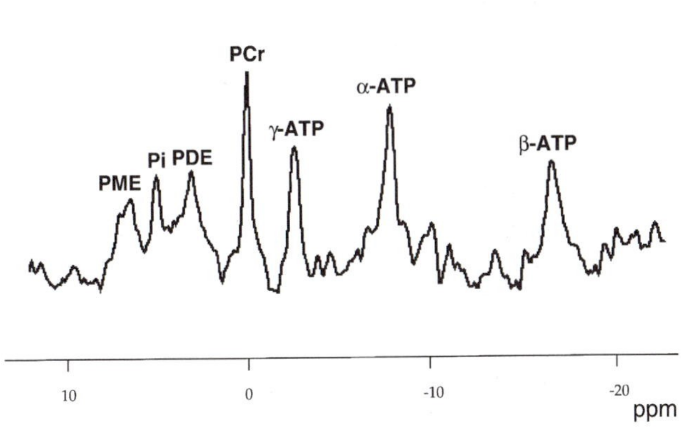

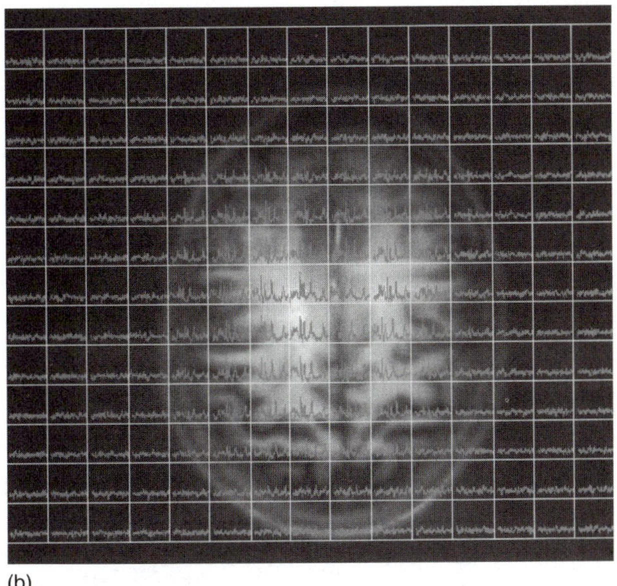

FIGURE 17.38. A ^{31}P chemical shift image obtained by making a Fourier transform of the FID signal to determine the various chemical peaks. (a) The chemical shift spectrum in a single volume element. (b) A map of the chemical shift image between $+10$ and -20 ppm. From X. Hu, W. Chen, M. Patel, and K. Ugurbil (1995). Chemical shift imaging: An introduction to its theory and practice. Chap. 65.4 in J. D. Bronzino, ed. *The Biomedical Engineering Handbook*. Copyright © CRC Press, 1995.

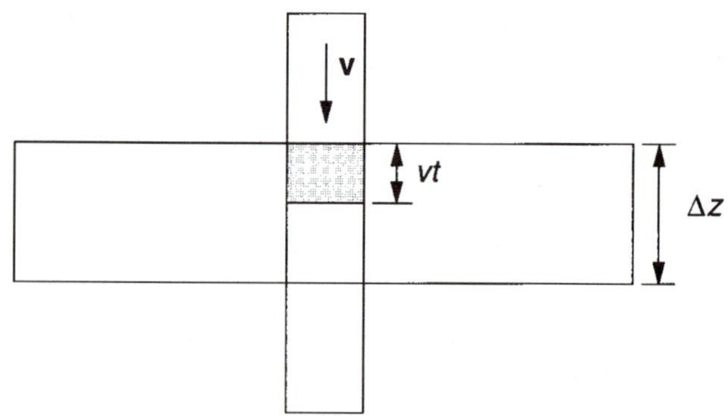

FIGURE 17.39. A blood vessel is perpendicular to the slice. The model developed in the text assumes plug flow, that is, all of the blood is flowing with the same speed v.

More complicated models can be developed, and phase changes because the blood flows through magnetic field gradients are also important.

SYMBOLS USED IN CHAPTER 17

Symbol	Use	Units	First used on page
a	Loop radius	m	502
a,b	Arbitrary constants	$\mathrm{J\,T^{-1}\,m^{-2}}$	498
f	Fraction		513
$\hbar$	Planck's constant (reduced)	J s	494
i	Current	A	493
k_B	Boltzmann factor	$\mathrm{J\,K^{-1}}$	494
k_x,k_y,k_z	Spatial frequency	$\mathrm{m^{-1}}$	508
m	Mass	kg	493
m	Azimuthal quantum number		494
q	Electric charge	C	493
r	Radius	m	493
t	Time	s	493
v	Velocity	$\mathrm{m\,s^{-1}}$	493
v	Voltage difference	V	502
x	Dimensionless variable		506
x',y',z'	Rotating axes	m	496
x,y,z	Axes	m	494
Δz	Slice thickness	m	506
$B,\mathbf{B}$	Magnetic field	T	492
C	Constant in expression for relaxation time	$\mathrm{s^2}$	501
$C(k),S(k)$	Fourier transforms	$\mathrm{T\,m^{-2}}$	508
E	Energy	J	499
G_x,G_y,G_z	Gradient of B_z in the x, y, or z direction	$\mathrm{T\,m^{-1}}$	506
$\mathbf{I}$	Nuclear angular momentum	$\mathrm{kg\,m^2\,s^{-1}}$	494
I	Nuclear angular momentum quantum number		494
K	Constant		501
$\mathbf{L}$	Orbital angular momentum	$\mathrm{kg\,m^2\,s^{-1}}$	493
$M,\mathbf{M}$	Magnetization	$\mathrm{J\,T^{-1}\,m^{-3}}$	494
N	Number of spins per unit volume	$\mathrm{m^{-3}}$	494
S	Area	$\mathrm{m^2}$	493
$\mathbf{S}$	Spin angular momentum	$\mathrm{kg\,m^2\,s^{-1}}$	494
S	Spin angular momentum quantum number		494
T	Temperature	K	494
T_E	Time of echo	s	503
T_I	Interrogation time	s	503
T_R	Repetition time between pulse sequences	s	503
T_1	Longitudinal relaxation time	s	495
T_2	Transverse relaxation time	s	495
T_2^*	Experimental transverse relaxation time	s	501
U	Potential energy	J	493
V	Volume	$\mathrm{m^3}$	502
α	Arbitrary angle		498
γ	Gyromagnetic ratio	$\mathrm{T^{-1}\,s^{-1}}$	493
$\mu,\boldsymbol{\mu}$	Magnetic moment	$\mathrm{J\,T^{-1}}$	492
μ_0		$\mathrm{T\,m\,A^{-1}}$	499
θ	Angle		493
σ	Chemical shift factor		511
$\tau,\boldsymbol{\tau}$	Torque	N m	492
τ	Shift time for autocorrelation	s	500
τ_C	Correlation time	s	500
ω_1	Angular frequency for B_1 rotation	$\mathrm{s^{-1}}$	498
ω	Angular frequency	$\mathrm{s^{-1}}$	494
ω_0	Larmor angular frequency	$\mathrm{s^{-1}}$	495
ϕ	Azimuthal angle		499
ϕ	Phase		508
Φ	Magnetic flux	weber ($\mathrm{T\,m^2}$)	502
$\boldsymbol{\Omega}$	Angular velocity vector	$\mathrm{s^{-1}}$	497

PROBLEMS

Section 17.1

17.1. Show that for a particle of mass m located at position $\mathbf{r}$ with respect to the origin, the torque about the origin is the rate of change of the angular momentum about the origin.

Section 17.2

17.2. Show that the units of γ are $\mathrm{T}^{-1}\ \mathrm{s}^{-1}$.

17.3. Find the ratio of the gyromagnetic ratio in Table 17.1 to the value $q/2m$ for the electron and proton.

Section 17.3

17.4. Evaluate the quantity $\gamma m\hbar B/k_B T$ and the Larmor frequency for electron spins and proton spins in a magnetic fields of 0.5 and 4.0 T at body temperature (310 K).

17.5. Verify that $\Sigma 1=2I+1$, $\Sigma m=0$, and $\Sigma m^2=I(I+1)(2I+1)/3$, when the sums are taken from $-I$ to I, in the cases that $I=\frac{1}{2}$, 1, and $\frac{3}{2}$.

17.6. Obtain an expression for the magnetization analogous to Eq. (17.10) in the case $I=\frac{1}{2}$ when one cannot make the assumption $\gamma\hbar B/k_B T \ll 1$. (This is called the *Langevin equation.*)

17.7. Calculate the coefficient of B in Eq. (17.10) for a collection of hydrogen nuclei at 310 K when the number of hydrogen nuclei per unit volume is the same as in water.

Section 17.4

17.8. Verify that Eqs. (17.16) are a solution of Eqs. (17.15).

17.9. Calculate the value of $M_x^2+M_y^2+M_z^2$ for relaxation [Eqs. (17.16)] when $T_1=T_2$.

17.10. Equations (17.16) correspond to a solution of the Bloch equations in the presence of a static field B. What would be the solutions if initially $M_x=0$, $M_y=0$, and $M_z=-M_0$?

Section 17.5

17.11. (a) Use Fig. 17.6 to derive Eq. (17.17). (b) Show that

$$M_{x'}=M_x \cos\theta+M_y \sin\theta,$$

$$M_{y'}=-M_x \sin\theta+M_y \cos\theta.$$

(c) Combine these equations with the equations for M_x and M_y to show that the application of both transformations brings one back to the starting point.

17.12. Calculate $M^2=M_x^2+M_y^2+M_z^2$ for the solution of Eqs. (17.29) and compare it to the results of Problem 17.9.

Section 17.6

17.13. Use Eqs. (17.31) to find the magnetic field at one proton due to the other proton in a water molecule when both proton spins are parallel to each other and perpendicular to the line between the protons. The two protons form an angle of 104.5° and are each 96.5×10^{-12} m from the oxygen.

17.14. The magnetic field at a distance of 0.15 nm from a proton is 4×10^{-4} T. What change in Larmor frequency does this ΔB cause? How long will it take for a phase difference of π radians to occur between a precessing spin feeling this extra field and one that is not?

17.15. Consider a collection of spins that are aligned along the x axis at $t=0$. They precess in the xy plane with different angular frequencies spread uniformly between $\omega-\Delta\omega/2$ and $\omega+\Delta\omega/2$. If the total magnetic moment per unit volume is M_0 at $t=0$, show that at time $T=4/\Delta\omega$ it is $M_0 \sin(2)/2=0.455M_0$.

17.16. What is the contribution to the transverse relaxation time for a magnetic field of 1.5 T with a uniformity of 1 part per million? The nonrecoverable relaxation time of brain is about 2.5 ms. What dominates the measured transverse relaxation in brain?

Section 17.7

17.17. In solving this problem, you will develop a simple model for estimating the radio-frequency energy absorption in a patient undergoing an MRI procedure.

(a) Consider a uniform conductor with electrical conductivity σ. If it is subject to a changing magnetic field $B_1(t)=B_1\cos(\omega_0 t)$, apply Eq. (8.18) to a circular path of radius R at right angles to the field to show that the electric field at radius R has amplitude $E_0=R\omega_0 B_1/2$. (Because this is proportional to R, the model gives the skin dose, along the path for which R is largest.)

(b) Use Ohm's law in the form $j=\sigma E$ to show that the time average power dissipated per unit volume of material is $p=\sigma E_0^2/2=\sigma R^2\omega_0^2 B_1^2/8$ and that if the mass density of the material is ρ, the specific absorbed rate (SAR) or dose rate is $\mathrm{SAR}=\sigma R^2\omega_0^2 B_1^2/8\rho$.

(c) If the radio-frequency signal is not continuous but is pulsed, show that this must be modified by the "duty cycle" factor $\Delta t/T_R$, where Δt is the pulse duration and T_R is the repetition period.

(d) Combine these results with the fact that rotation through an angle θ (usually π or $\pi/2$) in time Δt requires $B_1=2\theta/\gamma\,\Delta t$ and that $\omega_0=\gamma B_0$, to obtain $\mathrm{SAR}=(1/T_R\Delta t)(\sigma/2\rho)(R^2/4)B_0^2\theta^2$.

(e) Use typical values for the human body—$R=0.17$ m, $\sigma=0.3\ \mathrm{S\ m^{-1}}$—to evaluate this expression for a $\pi/2$ pulse.

(f) For $B_0=0.5$ T and $\mathrm{SAR}<0.4\ \mathrm{W\ kg^{-1}}$ determine the minimum value of Δt for $T_R=1$ s. Also find B_1.

(g) For 180 pulses, what is the dose in Gy? (This should not be compared to an x-ray dose because this is non-ionizing radiation.)

17.18. Use Eq. (17.37) to calculate the initial amplitude of a signal induced in a one-turn coil of radius 0.5 m for protons in an 1-mm cube of water at 310 K in a magnetic field of 1.0 T. (The answer will be too small a signal to be useful; multiple-turn coils must be used.)

Section 17.8

17.19. Plot the maximum amplitude of an inversion recovery signal vs the interrogation time if the detector is sensitive to the sign of the signal and if it is not.

17.20. (a) Obtain an analytic expression for the maximum value of the first and second echo amplitudes in a Carr–Purcell pulse sequence in terms of T_2 and T_E. (b) Repeat for a CPMG pulse sequence.

17.21. Consider the behavior of M_z in Figs. 17.19 and 17.21. The general equation for M_z is $M_z = M_0 + Ae^{-t/T_1}$. After several π pulses, the value of M_z is flipping from $-b$ to b. Find the value of b.

17.22. Consider a spin–echo pulse sequence (Fig. 17.18). Find

(a) M_z just before the π pulse at $T_E/2$,

(b) M_z just after the π pulse at $T_E/2$,

(c) M_z just before the $\pi/2$ pulse at T_R, and

(d) the first and second echo amplitudes as a function of T_E, T_R, T_1 and T_2. (The second amplitude is the same as all subsequent amplitudes.)

17.23. This problem shows how to extend the Bloch equations to include the effect of diffusion of the molecules containing the nuclear spins in an inhomogeneous external magnetic field. Since **M** is the magnetization per unit volume, it depends on the total number of particles per unit volume with average spin components $\langle \mu_x \rangle$, $\langle \mu_y \rangle$, and $\langle \mu_z \rangle$. In the rotating coordinate system there is no precession. In the absence of relaxation effects $\langle \boldsymbol{\mu} \rangle$ does not change. In that case changes in **M** depend on changes in the concentration of particles with particular components of $\langle \boldsymbol{\mu} \rangle$, so the rate of change of each component of $\langle \boldsymbol{\mu} \rangle$ is given by a diffusion equation. For example, for $M_{x'}$,

$$\frac{\partial M_{x'}}{\partial t} = D\nabla^2 M_{x'}.$$

If the processes are linear this diffusion term can be added to the other terms in the Bloch equations. Suppose that there is a uniform gradient in B_z, G_z, and that the coordinate system rotates with the Larmor frequency for $z=0$. When z is not zero, the rotation term does not quite cancel the $(\mathbf{M}\times\mathbf{B})_z$ term.

(a) Show that the x and y Bloch equations become

$$\frac{\partial M_{x'}}{\partial t} = +\gamma G_z z M_{y'} - \frac{M_{x'}}{T_2} + D\nabla^2 M_{x'},$$

$$\frac{\partial M_{y'}}{\partial t} = -\gamma G_z z M_{x'} - \frac{M_{y'}}{T_2} + D\nabla^2 M_{y'}.$$

(b) Show that in the absence of diffusion

$$M_{x'} = M_x(0)e^{-t/T_2}\cos(\gamma G_{zz} z t),$$

$$M_{y'} = M_y(0)e^{-t/T_2}\sin(\gamma G_{zz} z t).$$

(c) Suppose that **M** is uniform in all directions. At $t=0$ all spins are aligned. Spins that have been rotating faster in the plane at $z+\Delta z$ will diffuse into plane z. Equal numbers of slower spins will diffuse in from plane $z-\Delta z$. Show that this means that the phase of **M** will not change but the amplitude will.

(d) It is reasonable to assume that the amplitude of the diffusion-induced decay will not depend on z as long as we are far from boundaries. Therefore try a solution of the form

$$M_{x'} = M_x(0)e^{-t/T_2}\cos(\gamma G_z z t)A(t),$$

$$M_{y'} = M_y(0)e^{-t/T_2}\sin(\gamma G_z z t)A(t),$$

and show that A must obey the differential equation

$$\frac{1}{A}\frac{dA}{dt} = -D\gamma^2 G_z^2 t^2,$$

which has a solution $A(t) = \exp(-D\gamma^2 G_z^2 t^3/3)$.

(e) Show that if there is a rotation about y' at time $T_E/2$, then at time T_E M_x is given by

$$M_x(T_E) = -M_0 \exp(-T_E/T_2)\exp(-D\gamma^2 G_z^2 T_E^3/12).$$

Hint: This can be done formally from the differential equations. However it is much easier to think physically about what each factor in the expressions shown in (d) for M_x and $M_{y'}$ mean. This result means that a CPMG sequence with short T_E intervals can reduce the effect of diffusion when there is an external gradient.

17.24. A commercial MRI machine is operated with a magnetic gradient of 3 mT m^{-1} while a slice is being selected. What is the effect of diffusion? Use the diffusion constant for self-diffusion in water and the results of Problem 17.23. Compare the correction factor to $\exp(-T_E/T_2)$ when $T_2 = 75$ ms.

Section 17.9

17.25. Show that an alternative expression for the field amplitude required for a $\pi/2$ pulse is $B_1 = B_0\pi/\omega_0\,\Delta t = B_0/2\nu\Delta t$.

17.26. A certain MRI machine has a static magnetic field of 1.0 T. Spins are excited by applying a field gradient of 3×10^{-3} T m^{-1}. If the slice is to be 5 mm thick, what is the Larmor frequency and the spread in frequencies that is required?

17.27. Consider a pair of gradient coils of radius a perpendicular to the z axis and located at $z = \pm\sqrt{3}a/2$. The current flows in the opposite direction in each single-turn coil.

(a) Use the results of Problem 8.6 to obtain an expression for B_z along the z axis.

(b) For a gradient of 5×10^{-3} T m^{-1} at the origin and $a = 10$ cm, find the current required in a single-turn coil.

(c) Find the force on a coil in a field of 1.0 T.

17.28. Find a linear approximation for Eq. (17.52) for very small values of T_E and T_R, and discuss why it is called a T_1-weighted image.

17.29. How rapid is the transverse dephasing during a typical selection pulse if no compensating negative gradient is used?

17.30. Relate the resolution in the y direction to G_y and T_p.

17.31. Discuss the length of time required to obtain a 256×256 image in terms of T_R and T_E. The field of view is 15 cm square. Consider both projection reconstruction and spin warp images. Introduce any other parameters you need.

17.32. The limiting noise in a well-designed machine is due to thermal currents in the body. The noise is proportional to B_0 and the volume V_n sampled by the radio-frequency pickup coil. The noise is proportional to $T^{-1/2}$, where T is the time it takes to acquire the image. Show that the signal-to-noise ratio is proportional to $B_0 T^{1/2} V_v / V_n$, where V_v is the volume of the picture element.

Section 17.11

17.33. Use the model of Sec. 17.11 to plot the flow correction as a function of velocity for $T_E = 10$ ms, $T_1 = 900$ ms, and $T_2 = 400$ ms, when (a) $T_R = 50$ ms, (b) $T_R = 200$ ms.

REFERENCES

Axel, L. (1985). Flow effects in magnetic resonance imaging, in S. R. Thomas and R. L. Dixon, eds. *NMR in Medicine: The Instrumentation and Clinical Applications*. College Park, MD, AAPM.

Battocletti, J. H., R. E. Halbach, and S. X. Salles-Cunha (1981). The NMR blood flow meter—Theory and history. *Med. Phys.* **8**: 435–443.

Cho, Z.-H., J. P. Jones, and M. Singh (1993). *Foundations of Medical Imaging*. New York, Wiley-Interscience.

Dixon, R. L., K. E. Ekstrand, and P. R. Moran (1985). The physics of proton NMR: Part II—The microscopic description. In S. R. Thomas and R. L. Dixon, eds. *NMR in Medicine: The Instrumentation and Clinical Applications*. College Park, MD, AAPM.

Hu, X, W. Chen, M. Patel, and K. Ugurbil (1995). Chemical shift imaging: An introduction to its theory and practice. Chap. 65.4 in J. D. Bronzino, ed. *The Biomedical Engineering Handbook*. Boca Raton, CRC.

Joseph, P. M. (1985). Pulse sequences for magnetic resonance imaging. In S. R. Thomas and R. L. Dixon, eds. *NMR in Medicine: The Instrumentation and Clinical Applications*. College Park, MD, AAPM.

Joseph, P. M., and L. Axel (1984). Potential problems with selective pulses in NMR imaging systems. *Med. Phys.* **11**(6): 772–777.

Nunally, R. L. (1985). NMR spectroscopy for in vivo determination of metabolism: An overview. In S. R. Thomas and R. L. Dixon, eds. *NMR in Medicine: The Instrumentation and Clinical Applications*. College Park, MD, AAPM.

Slichter, C. P. (1978). *Principles of Magnetic Resonance*. New York, Springer.

Thomas, S. R., and R. L. Dixon, eds. (1985). *NMR in Medicine: The Instrumentation and Clinical Applications*. College Park, MD, AAPM.

Vlaardingerbroek, M. T., and J. A. den Boer (1996). *Magnetic Resonance Imaging: Theory and Practice*. Berlin, Springer.

APPENDIX A

Plane and Solid Angles

The angle between two intersecting lines is shown in Fig. A.1. It is measured by drawing a circle centered on the vertex or point of intersection. The arc length s on that part of the circle contained between the lines measures the angle. In daily work, the arc length is marked off in degrees.

In some cases, there are advantages to measuring the angle in *radians*. This is particularly true when trigonometric functions have to be differentiated or integrated. The angle in radians is defined by

$$\theta = \frac{s}{r}. \tag{A.1}$$

Since the circumference of a circle is $2\pi r$, the angle corresponding to a complete rotation of 360° is $2\pi r/r = 2\pi$. Other equivalences are

Degrees	Radians
360	2π
57.2958	1
1	0.01745

(A.2)

Since the angle in radians is the ratio of two distances, it is dimensionless. Nevertheless, it is sometimes useful to specify that something is measured in radians to avoid confusion.

One of the advantages of radian measure can be seen in Fig. A.2. The functions $\sin\theta$, $\tan\theta$, and θ in radians are plotted vs angle for angles less than 60°. For angles less than 15°, $y=\theta$ is a good approximation to both $y=\tan\theta$ (2.3% error at 15°) and $y=\sin\theta$ (1.2% error at 15°).

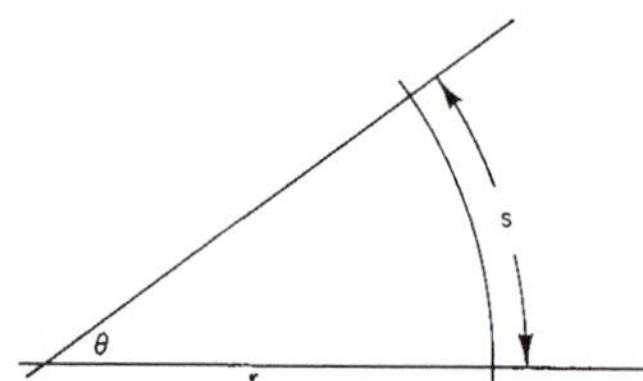

FIGURE A.1. A plane angle θ is measured by arc length s on a circle of radius r centered at the vertex of the lines defining the angle.

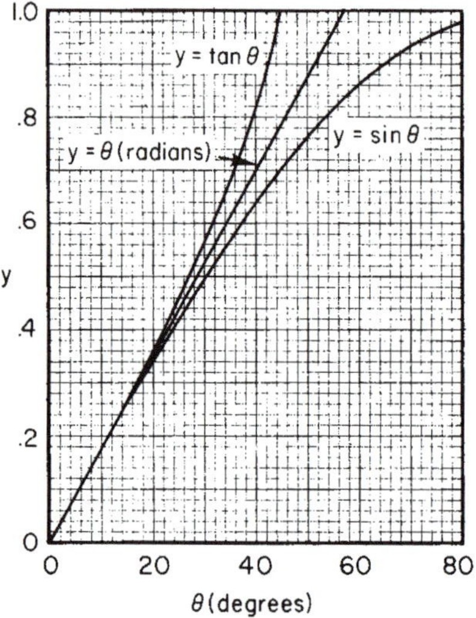

FIGURE A.2. Comparison of $y=\tan\theta$, $y=\theta$ (radians), and $y=\sin\theta$.

A plane angle measures the diverging of two lines in two dimensions. Solid angles measure the diverging of a cone of lines in three dimensions. Figure A.3 shows a series of rays diverging from a point and forming a cone. The solid angle is measured by constructing a sphere of radius r centered at the vertex and taking the ratio of the surface area S on the sphere enclosed by the cone to r^2:

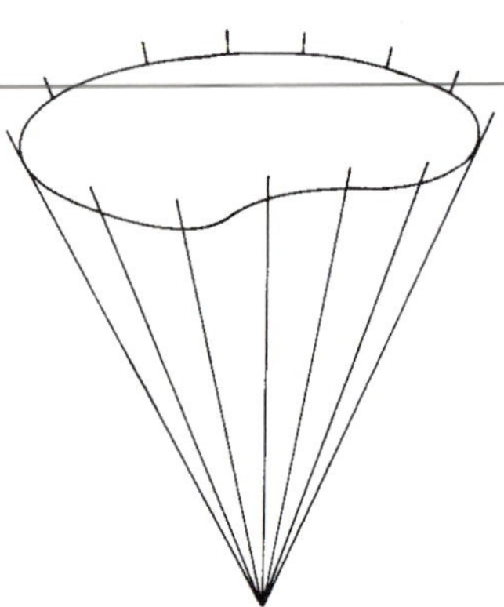

FIGURE A.3. A cone of rays in three dimensions.

$$\Omega = \frac{S}{r^2}. \qquad \text{(A.3)}$$

This shown in Fig. A.4 for a cone consisting of the planes defined by adjacent pairs of the four rays shown. The unit of solid angle is the *steradian* (sr). A complete sphere subtends a solid angle of 4π steradians, since the surface area of a sphere is $4\pi r^2$.

When the included angle in the cone is small, the difference between the surface area of a plane tangent to the sphere and the sphere itself is small. (This is difficult to draw in three dimensions. Imagine that Fig. A.5 represents a slice through a cone; the difference in length between the circular arc and the tangent to it is small.) This approximation is often useful. A 3×5-in. card at a distance of 6 ft (72 in.) subtends a solid angle which is approximately

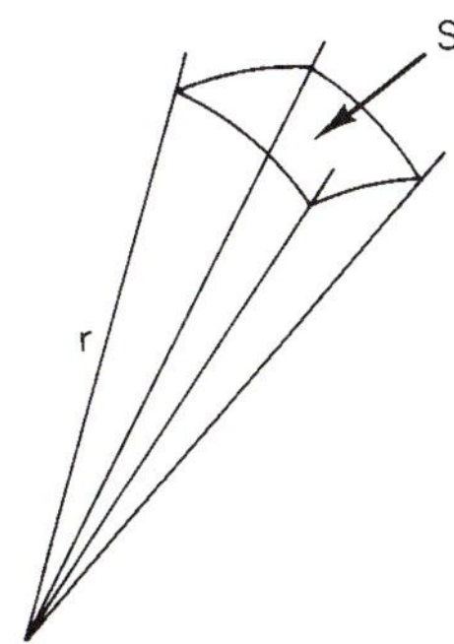

FIGURE A.4. The solid angle of this cone is S/r^2. S is the surface area on a sphere of radius r centered at the vertex.

FIGURE A.5. For small angles, the arc length is very nearly equal to the length of the tangent to the circle.

$$\frac{3\times 5}{72^2} = 2.9\times 10^{-3} \text{ sr.}$$

It is not necessary to calculate the surface area on a sphere of 72 in. radius.

Solid angles are occasionally measured in square degrees. To find the conversion between square degrees and steradians, consider a sphere of radius r. A plane angle of 1° with vertex at the center will define an arc length $s = r\theta = (r)(0.017\,45)$. Two such plane angles at right angles to each other will define a surface area which is approximately $s^2 = r^2(0.017\,45)^2$. Therefore,

$$(1°)^2 = 3.046\times 10^{-4} \text{ sr.} \qquad \text{(A.4)}$$

PROBLEMS

A.1. Convert 0.1 rad to degrees. Convert 7.5 deg to radians.

A.2. Use the fact that $\sin\theta \approx \theta \approx \tan\theta$ to estimate the sine and tangent of 3°. Look up the values in a table and see how accurate the approximation was.

A.3. What is the solid angle subtended by the pupil of the eye (radius=3 nm) at a source of light 30 m away?

APPENDIX **B**

Vectors; Displacement, Velocity, and Acceleration

B.1. VECTORS AND VECTOR ADDITION

A *displacement* describes how to get from one point to another. A displacement has a magnitude (how far point 2 is from point 1 in Fig. B.1) and a direction (the direction one has to go from point 1 to get to point 2). The displacement of point 2 from point 1 is labeled **A**. Displacements can be added: displacement **B** from point 2 puts an object at point 3. The displacement from point 1 to point 3 is **C** and is the sum of displacements **A** and **B**:

$$\mathbf{C}=\mathbf{A}+\mathbf{B}. \quad \text{(B.1)}$$

A displacement is a special example of a more general quantity called a *vector*. One often finds a vector defined as a quantity having a magnitude and a direction. However, the complete definition of a vector also includes the requirement that vectors add like displacements. The rule for adding two vectors is to place the tail of the second vector at the head of the first, the sum is the vector from the tail of the first to the head of the second.

A displacement is a change of positions so far in such a direction. It is independent of the starting point. To know where an object is, it is necessary to specify the starting point as well as its displacement from that point.

Because the displacement is independent of the starting point, displacements can be added in any order. In Fig. B.2, either of the vectors **A** represents the same displacement. Displacement **B** can first be made from point 1 to point 4, followed by displacement **A** from 4 to 3. The sum is still **C**:

$$\mathbf{C}=\mathbf{A}+\mathbf{B}=\mathbf{B}+\mathbf{A}. \quad \text{(B.2)}$$

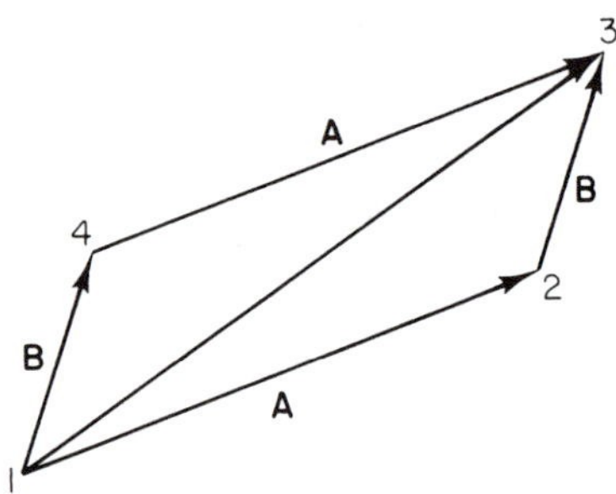

FIGURE B.2. Vectors **A** and **B** can be added in either order.

The sum of several vectors can be obtained by first adding two of them, then adding the third to that sum, and so forth. This is equivalent to placing the tail of each vector at the head of the previous one, as shown in Fig. B.3. The sum then goes from the tail of the first vector to the head of the last.

The negative of vector **A** is that vector which, added to **A**, yields zero:

$$\mathbf{A}+(-\mathbf{A})=\mathbf{0}. \quad \text{(B.3)}$$

It has the same magnitude as **A** and points in the opposite direction.

Multiplying a vector **A** by a scalar (a number with no associated direction) multiplies the magnitude of vector **A** by that number and leaves its direction unchanged.

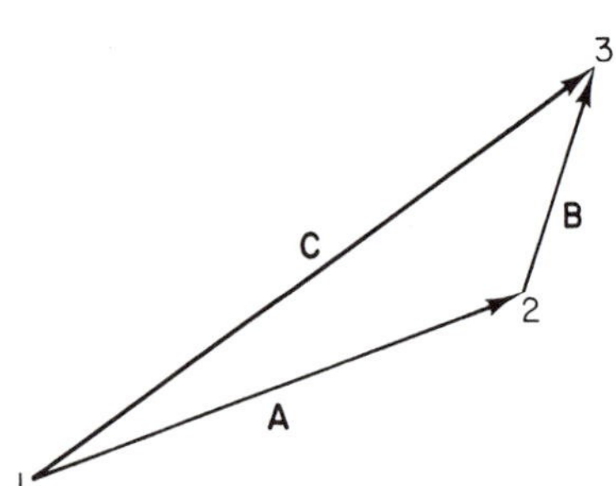

FIGURE B.1. Displacement **C** is equivalent to displacement **A** followed by displacement **B**: **C**=**A**+**B**.

B.2. COMPONENTS OF VECTORS

Consider a vector in a plane. If we set up two perpendicular axes, we can regard vector **A** as being the sum of vectors parallel to each of these axes. These vectors, $\mathbf{A}_x$ and $\mathbf{A}_y$ in

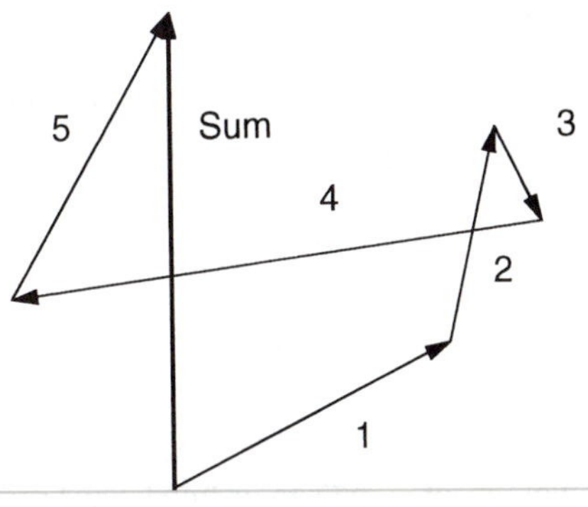

FIGURE B.3. Addition of several vectors.

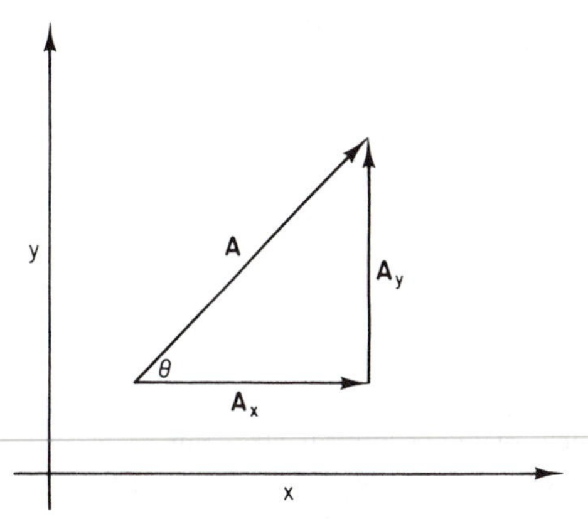

FIGURE B.4. Vector **A** has components $\mathbf{A}_x$ and $\mathbf{A}_y$.

Fig. B.4, are called the *components* of **A** along each axis[1]. If vector **A** makes an angle θ with the x axis and its magnitude is A, then the magnitudes of the components are

$$A_x = A\cos\theta,\qquad A_y = A\sin\theta. \tag{B.4}$$

The sum of the squares of the components is

$$A_x^2 + A_y^2 = A^2\cos^2\theta + A^2\sin^2\theta = A^2(\cos^2\theta + \sin^2\theta).$$

Since, by Pythagoras' theorem, this must be A^2, we obtain the trigonometric identity

$$\cos^2\theta + \sin^2\theta = 1. \tag{B.5}$$

In three dimensions, $\mathbf{A} = \mathbf{A}_x + \mathbf{A}_y + \mathbf{A}_z$. The magnitudes can again be related using Pythagoras' theorem, as shown in Fig. B.5. From triangle OPQ, $A_{xy}^2 = A_x^2 + A_y^2$. From triangle OQR,

$$A^2 = A_{xy}^2 + A_z^2 = A_x^2 + A_y^2 + A_z^2. \tag{B.6}$$

In our notation, $\mathbf{A}_x$ means a vector pointing in the x direction, while A_x is the magnitude of that vector. It can become difficult to keep the distinction straight. Therefore, it is customary to write $\hat{\mathbf{x}}$, $\hat{\mathbf{y}}$, and $\hat{\mathbf{z}}$ to mean vectors of unit length pointing in the x, y, and z directions. (In some books, the unit vectors are denoted by **i**, **j**, and **k** instead of $\hat{\mathbf{x}}$, $\hat{\mathbf{y}}$, and $\hat{\mathbf{z}}$.) With this notation, instead of $\mathbf{A}_x$, one would always write $A_x\hat{\mathbf{x}}$.

The addition of vectors is often made easier by using components. The sum hat $\mathbf{A}+\mathbf{B}=\mathbf{C}$ can be written as

$$A_x\hat{\mathbf{x}} + A_y\hat{\mathbf{y}} + A_z\hat{\mathbf{z}} + B_x\hat{\mathbf{x}} + B_y\hat{\mathbf{y}} + B_z\hat{\mathbf{z}}$$
$$= C_x\hat{\mathbf{x}} + C_y\hat{\mathbf{y}} + C_z\hat{\mathbf{z}}.$$

Like components can be grouped to give

$$(A_x+B_x)\hat{\mathbf{x}} + (A_y+B_y)\hat{\mathbf{y}} + (A_z+B_z)\hat{\mathbf{z}}$$
$$= C_x\hat{\mathbf{x}} + C_y\hat{\mathbf{y}} + C_z\hat{\mathbf{z}}.$$

[1]Some texts define the component to be a scalar. The magnitude of the component is defined here.

Therefore, the magnitudes of the components can be added separately:

$$C_x = A_x + B_x,\qquad C_y = A_y + B_y,\qquad C_z + A_z + B_z. \tag{B.7}$$

B.3. POSITION, VELOCITY, AND ACCELERATION

The position of an object is defined by specifying its displacement

$$\mathbf{R}(t) = x(t)\hat{\mathbf{x}} + y(t)\hat{\mathbf{y}} + z(t)\hat{\mathbf{z}}$$

from an agreed-upon origin. The *average velocity* $\mathbf{v}_{\mathrm{av}}(t_1,t_2)$ between times t_1 and t_2 is defined to be

$$\mathbf{v}_{\mathrm{av}}(t_1,t_2) = \frac{\mathbf{R}(t_2)-\mathbf{R}(t_1)}{t_2-t_1}.$$

This can be written in terms of the components as

$$\mathbf{v}_{\mathrm{av}} = \left(\frac{x(t_2)-x(t_1)}{t_2-t_1}\right)\hat{\mathbf{x}} + \left(\frac{y(t_2)-y(t_1)}{t_2-t_1}\right)\hat{\mathbf{y}} + \left(\frac{z(t_2)-z(t_1)}{t_2-t_1}\right)\hat{\mathbf{z}}.$$

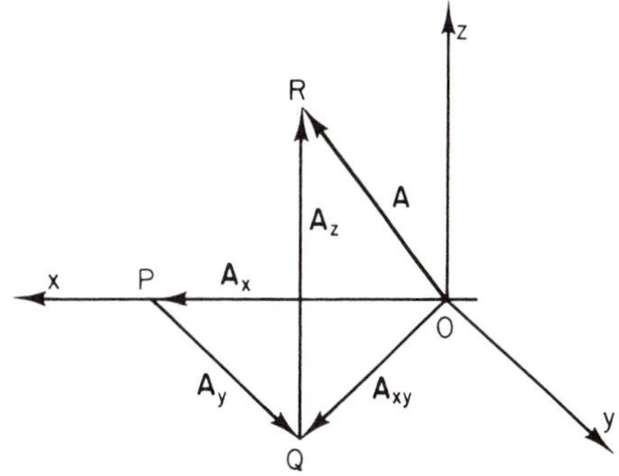

FIGURE B.5. Addition of components in three dimensions.

The *instantaneous velocity* is

$$\mathbf{v}(t)=\frac{d\mathbf{R}}{dt}=\frac{dx}{dt}\hat{\mathbf{x}}+\frac{dy}{dt}\hat{\mathbf{y}}+\frac{dz}{dt}\hat{\mathbf{z}}$$

$$=v_x(t)\hat{\mathbf{x}}+v_y(t)\hat{\mathbf{y}}+v_z(t)\hat{\mathbf{z}}. \qquad \text{(B.8)}$$

The x component of the velocity tells how rapidly the x component of the position is changing.

The *acceleration* is the rate of change of the velocity with time. The instantaneous acceleration is

$$\mathbf{a}(t)=\frac{d\mathbf{v}}{dt}=\frac{dv_x}{dt}\hat{\mathbf{x}}+\frac{dv_y}{dt}\hat{\mathbf{y}}+\frac{dv_z}{dt}\hat{\mathbf{z}}. \qquad \text{(B.9)}$$

PROBLEMS

B.1. At $t=0$, the position of an object is given by

$$\mathbf{R}=10\hat{\mathbf{x}}+5\hat{\mathbf{y}},$$

where $\mathbf{R}$ is in meters. At $t=3$ s, the position is

$$\mathbf{R}=16\hat{\mathbf{x}}-10\hat{\mathbf{y}}.$$

What was the average velocity between $t=0$ and 3 s?

B.2. If the leg of a person is modeled as a physical pendulum, the period is $T=K(L/g)^{1/2}$, where L is the length of the leg, g is the acceleration due to gravity, and K is a constant depending on the shape of the leg. Assume that L is proportional to a person's height, H. How does a person's natural walking speed depend on H? What other assumptions do you have to make?

APPENDIX C

Properties of Exponents and Logarithms

In the expression a^m, a is called the *base* and m is called the *exponent*. Since $a^2 = a\times a$, $a^3 = a\times a\times a$, and

$$a^m = \underbrace{(a\times a\times a\times \cdots \times a)}_{m \text{ times}},$$

it is easy to show that

$$a^m\times a^n = \underbrace{(a\times a\times a\times\cdots\times a)}_{m\text{ times}}\underbrace{(a\times a\times a\times\cdots\times a)}_{n\text{ times}},$$
$$a^m\times a^n = a^{m+n}. \tag{C.1}$$

If $m>n$, the same technique can be used to show that

$$\frac{a^m}{a^n} = a^{m-n}. \tag{C.2}$$

If $m=n$, this gives

$$1 = \frac{a^m}{a^m} = a^{m-m} = a^0,$$
$$a^0 = 1. \tag{C.3}$$

The rules also work for $m<n$ and for negative exponents. For example,

$$(a^{-n})(a^n) = 1$$

so

$$a^{-n} = \frac{1}{a^n}. \tag{C.4}$$

Finally,

$$(a^m)^n = \underbrace{(a^m\times a^m\times a^m\times\cdots\times a^m)}_{n\text{ times}},$$
$$(a^m)^n = a^{mn}. \tag{C.5}$$

If $y=a^x$, then by definition, x is the logarithm of y to the base a:

$$x = \log_a(y).$$

If the base is 10, since $100=10^2$, $2=\log_{10}(100)$. Similarly, $3=\log_{10}(1000)$, $4=\log_{10}(10\,000)$, and so forth.

The most useful property of logarithms can be derived by letting

$$y = a^m,$$
$$z = a^n,$$
$$w = a^{m+n},$$

so that

$$m = \log_a y,$$
$$n = \log_a z,$$
$$m+n = \log_a w.$$

Then, since $a^{m+n} = a^m a^n$,

$$w = yz,$$
$$\log_a(yz) = \log_a w = \log_a y + \log_a z. \tag{C.6}$$

This result can be used to show that

$$\log(y^m) = \log(y\times y\times y\times\cdots\times y)$$
$$= \log(y)+\log(y)+\log(y)+\cdots\log(y),$$
$$\log(y^m) = m\log(y). \tag{C.7}$$

All logarithms in this book, unless labeled with a specific base, are to base e (see Chap. 2). These are the so-called natural logarithms. We will denote the natural logarithm by ln, using $\log_{10}$ when we want logarithms to the base 10.

PROBLEMS

C.1. What is $\log_2(8)$?
C.2. If $\log_{10}(2)=0.3$, what is $\log_{10}(200)$? $\log_{10}(2\times10^{-5})$?
C.3. What is $\log_{10}(\sqrt{10})$?

APPENDIX D

Taylor's Series

Consider the function $y(x)$ shown in Fig. D.1. The value of the function at x_1, $y_1=y(x_1)$, is known. We wish to estimate $y(x_1+\Delta x)$.

The simplest estimate, labeled approximation 0 in Fig. D.1, is to assume that y does not change and say

$$y(x_1+\Delta x)\approx y(x_1).$$

A better estimate can be obtained if we assume that y changes everywhere at the same rate it does at x_1. Approximation 1 is

$$y(x_1+\Delta x)\approx y(x_1)+\left.\frac{dy}{dx}\right|_{x_1}\Delta x.$$

The derivative is evaluated at point x_1.

An even better estimate is shown in Fig. D.2. Instead of fitting the curve by the straight line that has the proper first derivative at x_1, we fit it by a parabola that matches both the first and second derivatives. The approximation is

$$y(x_1+\Delta x)\approx y(x_1)+\left.\frac{dy}{dx}\right|_{x_1}\Delta x+\frac{1}{2}\left.\frac{d^2y}{dx^2}\right|_{x_1}(\Delta x)^2.$$

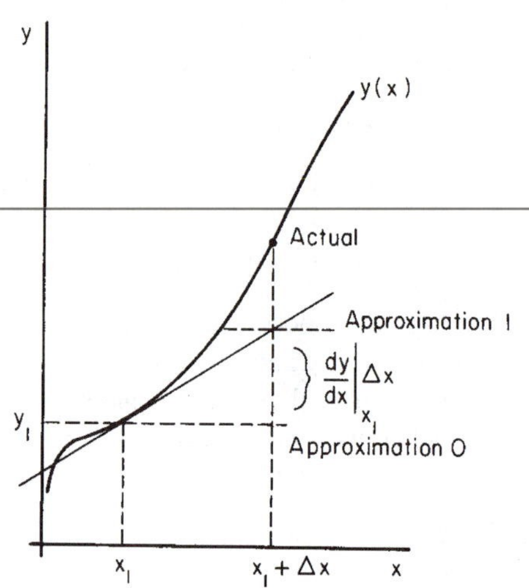

FIGURE D.1. The zeroth-order and first-order approximations to $y(x)$.

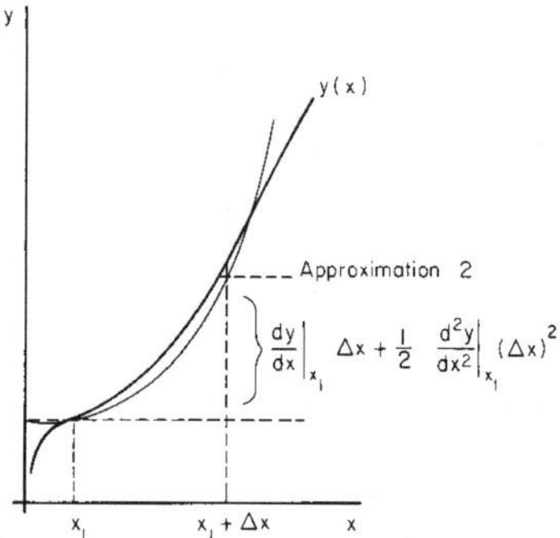

FIGURE D.2. The second-order approximation fits $y(x)$ by a parabola.

That this is the best approximation can be derived in the following way. Suppose the desired approximation is more general and uses terms up to $(\Delta x)^n=(x-x_1)^n$:

$$y_{\text{approx}}=A_0+A_1(x-x_1)+A_2(x-x_1)^2+\cdots+A_n(x-x_1)^n.$$

The constants $A_0,A_1,\ldots,A_n$ are determined by making the value of y_{approx} and its first n derivatives agree with the value of y and its first n derivatives at $x=x_1$. When $x=x_1$, all terms with $x-x_1$ in y_{approx} vanish, so that

$$y_{\text{approx}}(x_1)=A_0.$$

The first derivative of y_{approx} is

$$\frac{d}{dx}(y_{\text{approx}})=A_1+2A_2(x-x_1)+3A_3(x-x_1)^2+\cdots$$
$$+nA_n(x-x_1)^{n-1}.$$

The second derivative is

$$2A_2+3\times 2A_3(x-x_1)+\cdots+n(n-1)A_n(x-x_1)^{n-2},$$

and the nth derivative is

$$n(n-1)(n-2)\cdots 2A_n.$$

Evaluating these at $x=x_1$ gives

TABLE D.1. $y=e^{2x}$ and its derivatives.

Function or derivative	Value at $x_1=0$
$y=e^{2x}$	1
$\frac{dy}{dx}=2e^{2x}$	2
$\frac{d^2y}{dx^2}=4e^{2x}$	4
$\frac{d^3y}{dx^3}=8e^{2x}$	8

$$\left.\frac{d}{dx}(y_{\text{approx}})\right|_{x_1}=A_1,$$

$$\left.\frac{d^2}{dx^2}(y_{\text{approx}})\right|_{x_1}=2\times 1\times A_2,$$

$$\left.\frac{d^3}{dx^3}(y_{\text{approx}})\right|_{x_1}=3\times 2\times 1\times A_3,$$

$$\left.\frac{d^n}{dx^n}(y_{\text{approx}})\right|_{x_1}=n!A_n.$$

Equating these to the derivatives of $y(x)$ evaluated at x_1, we get

$$y(x_1+\Delta x)\approx y(x_1)+\sum_{n=1}^{N}\frac{1}{n!}\left.\frac{d^n y}{dx^n}\right|_{x_1}(\Delta x)^n. \quad \text{(D.1)}$$

Tables D.1 and D.2 and Figs. D.3 and D.4 show how the Taylor's series approximation gets better over a larger and larger region about x_1 as more terms are added. The function being approximated is $y=e^{2x}$. The derivatives are given in Table D.1. The expansion is made about $x_1=0$.

TABLE D.2. *Values of y and successive approximations.*

x	y	$1+2x$	$1+2x+2x^2$	$1+2x+2x^2+\frac{4}{3}x^3$
−2	0.0183	−3.0	5.0	−5.67
−1.5	0.0498	−2.0	2.5	−2.0
−1	0.1353	−1.0	1.0	−0.33
−0.4	0.4493	0.2000	0.5200	0.4347
−0.2	0.6703	0.6000	0.6800	0.6693
−0.1	0.8187	0.8000	0.8200	0.8187
0	1.0000	1.0000	1.0000	1.0000
0.1	1.2214	1.2000	1.2200	1.2213
0.2	1.4918	1.4000	1.4800	1.4907
0.4	2.2255	1.8000	2.1200	2.2053
1.0	7.389	3.0000	5.0000	6.33
2.0	54.60	5.0	13.0	23.67

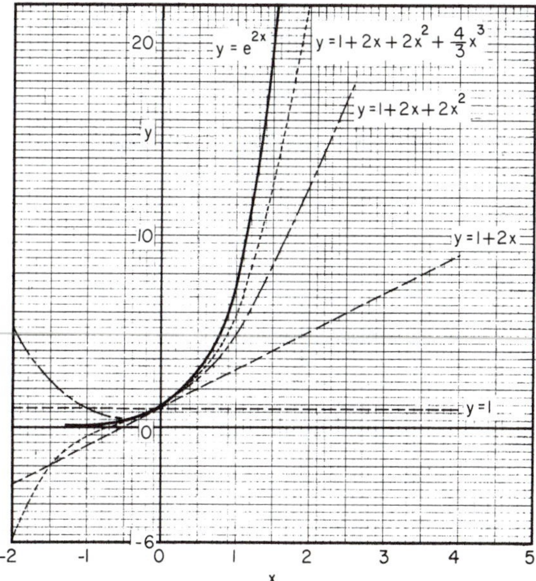

FIGURE D.3. The function $y=e^{2x}$ with Taylor's series expansions about $x=0$ of degree 0, 1, 2, and 3.

Finally, the Taylor's series expansion for $y=e^x$ about $x=0$ is often useful. Since all derivatives of e^x are e^x, the value of y and each derivative at $x=0$ is 1. The series is

$$e^x=1+x+\frac{1}{2!}x^2+\frac{1}{3!}x^3+\ldots=\sum_{m=0}^{\infty}\frac{x^m}{m!}. \quad \text{(D.2)}$$

(Note that $0!=1$ by definition.)

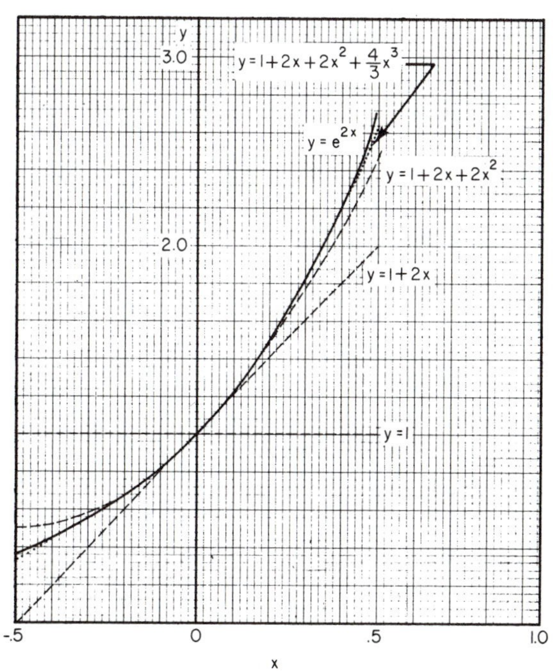

FIGURE D.4. An enlargement of Fig. D.3 near $x=0$.

PROBLEMS

D.1. Make a Taylor's series expansion of

$$y = a + bx + cx^2$$

about $x=0$. Show that the expansion exactly reproduces the function.

D.2. Repeat Problem D.1, making the expansion about $x=1$.

D.3. Make a Taylor's series expansion of the cosine function. Remember that $d(\sin x)/dx = \cos x$ and $d(\cos x)/dx = -\sin x$.

APPENDIX E

Some Integrals of Sines and Cosines

The average of a function of x with period T is defined to be

$$\langle f\rangle = \frac{1}{T}\int_{x'}^{x'+T} f(x)\,dx. \tag{E.1}$$

The sine function is plotted in Fig. E.1(a). The integral over a period is zero, and its average value is zero. The area above the axis is equal to the area below the axis. Figure E.1(b) shows a plot of $\sin^2 x$. Since $\sin x$ varies between -1 and $+1$, $\sin^2 x$ varies between 0 (when $\sin x=0$) and $+1$ (when $\sin x=\pm 1$). Its average value, from inspection of Fig. E.1(b) is $\frac{1}{2}$. If you do not want to trust the drawing to convince yourself of this, recall the identity

$$\sin^2\theta+\cos^2\theta=1.$$

Since the sine function and the cosine function look the same, but are just shifted along the axis, their squares must also look similar. Therefore, $\sin^2\theta$ and $\cos^2\theta$ must have the same average. But if their sum is always 1, the sum of their averages must be 1. If the two averages are the same, then each must be $\frac{1}{2}$.

These same results could have been obtained analytically by using the trigonometric identity

$$\sin^2 x=\tfrac{1}{2}-\tfrac{1}{2}\cos 2x. \tag{E.2}$$

The integrals of $\sin x$ and $\cos x$ are

$$\begin{aligned}\int \sin ax\,dx&=-\frac{1}{a}\cos ax,\\ \int \cos ax\,dx&=\frac{1}{a}\sin ax.\end{aligned} \tag{E.3}$$

These could be used to prove that the average value of $\sin x$ or $\cos x$ is zero. Then Eq. E.2 could be used to show that the average of $\sin^2 x$ is $\frac{1}{2}$. The integral of $\sin^2 x$ over a period is its average value times the length of the period:

$$\int_0^T \sin^2 x\,dx=\int_0^T \cos^2 x\,dx=\tfrac{1}{2}T. \tag{E.4}$$

We will also encounter integrals like

$$\begin{aligned}&\int_0^T \sin mx\,\sin nx\,dx, \quad m\neq n\\ &\int_0^T \cos mx\,\cos nx\,dx, \quad m\neq n\\ &\int_0^T \cos mx\,\sin nx\,dx, \quad n=m, \quad n\neq m.\end{aligned} \tag{E.5}$$

All these integrals are zero. Proofs may be found in calculus books or integral tables. Instead of proving these results, you can see why the integrals vanish by considering the specific examples plotted in Fig. E.2. Each integrand has equal positive and negative contributions to the total integral.

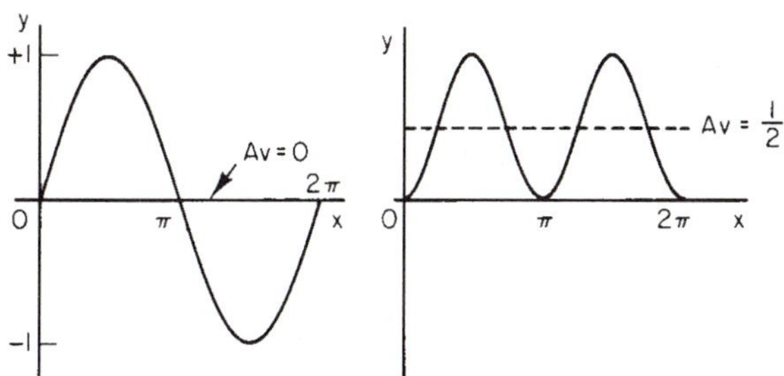

FIGURE E.1. (a) Plot of $y=\sin x$. (b) Plot of $y=\sin^2 x$.

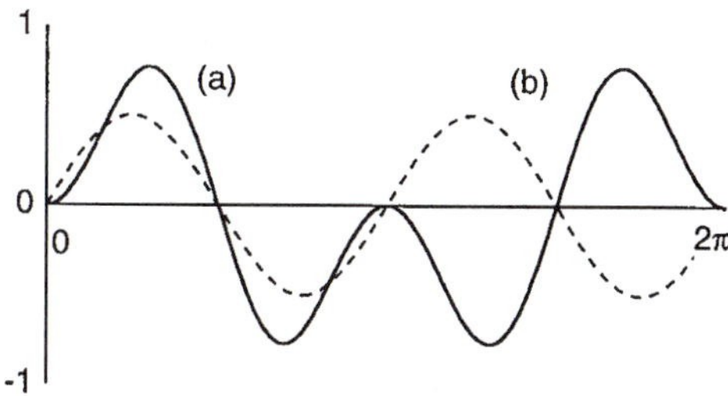

FIGURE E.2. Plot of one period of (a) $y=\sin x\sin 2x$; (b) $y=\sin x\cos x$.

APPENDIX F

Linear Differential Equations with Constant Coefficients

The equation

$$\frac{dy}{dt}+by=a \tag{F.1}$$

is called a linear differential equation because each term involves only y or its derivatives [not $y(dy/dt)$ or $(dy/dt)^2$, etc.]. A more general equation of this kind has the form

$$\frac{d^Ny}{dt^N}+b_{N-1}\frac{d^{N-1}y}{dt^{N-1}}+\cdots+b_1\frac{dy}{dt}+b_0y=f(t). \tag{F.2}$$

The highest derivative is the Nth derivative, so the equation is of order N. It has been written in standard form by dividing through by any b_N that was there originally, so that the coefficient of the highest term is one. If all the b's are constants, this is a linear differential equation with constant coefficients. The right-hand side may be a function of the independent variable t, but *not* of y. If $f(t)=0$, it is a homogeneous equation; if $f(t)$ is not zero, it is an inhomogeneous equation.

Consider first the homogeneous equation

$$\frac{d^Ny}{dt^N}+b_{N-1}\frac{d^{N-1}y}{dt^{N-1}}+\cdots+b_1\frac{dy}{dt}+b_0y=0. \tag{F.3}$$

The exponential

$$e^{st}$$

(where s is a constant) has the property that

$$\frac{d}{dt}(e^{st})=se^{st},$$

$$\frac{d^2}{dt^2}(e^{st})=s\frac{d}{dt}(e^{st})=s^2e^{st},$$

$$\frac{d^n}{dt^n}(e^{st})=s^ne^{st}.$$

If we let $y=Ae^{st}$, we will find that it satisfies Eq. (F.3) for any value of A and certain values of s. The equation becomes

$$A(s^Ne^{st}+b_{N-1}s^{N-1}e^{st}+\cdots+b_1se^{st}+b_0e^{st})=0,$$

$$A(s^N+b_{N-1}s^{N-1}+\cdots+b_1s+b_0)e^{st}=0.$$

This equation will be satisfied if the polynomial in parentheses is equal to zero. The equation

$$s^N+b_{N-1}s^{N-1}+\cdots+b_1s+b_0=0 \tag{F.4}$$

is called the *characteristic equation* of this differential equation. It can be written using summation notation in a much more compact form:

$$\sum_{n=0}^{N}b_ns^n=0. \tag{F.5}$$

with $b_N=1$.

For Eq. (F.1), the characteristic equation is

$$s+b=0 \quad \text{or} \quad s=-b,$$

and a solution to the homogeneous equation is

$$y=Ae^{-bt}.$$

If the characteristic equation is a polynomial, it can have up to N roots. For each distinct root, $y=A_ne^{s_nt}$ is a solution to the differential equation. (The question of solutions when there are not N distinct roots will be taken up below.) This is still not the solution to the equation we need to solve. However, one can prove[1] that the most general solution to the inhomogeneous equation consists of the homogeneous solution,

$$y=\sum_{n=1}^{N}A_ne^{s_nt},$$

[1] See, for example, G. B. Thomas. *Calculus and Analytic Geometry*, Reading, MA, Addison-Wesley (any edition).

plus *any* solution to the inhomogeneous equation. The values of the arbitrary constants A_n are picked to satisfy some other conditions that are imposed on the problem. If we can guess the solution to the inhomogeneous equation, that is fine. However we get it, we need only one such solution to the inhomogeneous equation. We will not prove this assertion, but we will apply it to the first- and second-order equations and see how it works.

The homogeneous equation has solution $y=Ae^{-bt}$. There is one solution to the inhomogeneous equation that is particularly easy to write down: when y is constant, with the value $y=a/b$, the time derivative vanishes and the equation is satisfied. The most general solution is of the form

$$y=Ae^{-bt}+\frac{a}{b}.$$

If the initial condition is $y(0)=0$, then A can be determined from

$$0=Ae^{-b0}+\frac{a}{b}.$$

Since $e^0=1$, this gives $A=-a/b$. Therefore

$$y=\frac{a}{b}(1-e^{-bt}). \tag{F.6}$$

A physical example of this is given in Sec. 2.7.

The second-order equation

$$\frac{d^2y}{dt^2}+b_1\frac{dy}{dt}+b_0y=0 \tag{F.7}$$

has a characteristic equation

$$s^2+b_1s+b_0=0$$

with roots

$$s=\frac{-b_1\pm\sqrt{b_1^2-4b_0}}{2}. \tag{F.8}$$

This equation may have no, one, or two solutions. If it has two solutions s_1 and s_2, then the general solution of the homogeneous equation is

$$y=A_1e^{s_1t}+A_2e^{s_2t}.$$

If $b_1^2-4b_0$ is negative, there is no solution to the equation for a real value of s. However, a solution of the form

$$y=Ae^{-\alpha t}\sin(\omega t+\phi)$$

will satisfy the equation. This can be seen by direct substitution. Differentiating this twice shows that

$$\frac{dy}{dt}=-\alpha Ae^{-\alpha t}\sin(\omega t+\phi)+\omega Ae^{-\alpha t}\cos(\omega t+\phi),$$

$$\frac{d^2y}{dt^2}=\alpha^2Ae^{-\alpha t}\sin(\omega t+\phi)-2\alpha\omega Ae^{-\alpha t}\cos(\omega t+\phi)$$
$$-\omega^2Ae^{-\alpha t}\sin(\omega t+\phi).$$

If these derivatives are substituted in Eq. (F.7), one gets the following results. The terms are written in two columns. One column contains the coefficients of terms with $\sin(\omega t+\phi)$, and the other column contains the coefficients of terms with $\cos(\omega t+\phi)$. The rows are labeled on the left by which term of the differential equation they came from.

Term	Coefficients	
	$\sin(\omega t+\phi)$	$\cos(\omega t+\phi)$
d^2y/dt^2	$\alpha^2-\omega^2$	$-2\alpha\omega$
$b_1(dy/dt)$	$-b_1\alpha$	$b_1\omega$
b_0y	b_0	0

The only way that the equation can be satisfied for all times is if the coefficient of the $\sin(\omega t+\phi)$ term and the coefficient of the $\cos(\omega t+\phi)$ term separately are equal to zero. This means that we have two equations that must be satisfied (call $b_0=\omega_0^2$):

$$2\alpha\omega=b_1\omega,$$

$$\alpha^2-\omega^2-b_1\alpha+\omega_0^2=0.$$

From the first equation

$$2\alpha=b_1,$$

while from this and the second,

$$\alpha^2-\omega^2-2\alpha^2+\omega_0^2=0$$

or

$$\omega^2=\omega_0^2-\alpha^2.$$

Thus, the solution to the equation

$$\frac{d^2y}{dt^2}+2\alpha\frac{dy}{dt}+\omega_0^2y=0 \tag{F.9}$$

is

$$y=Ae^{-\alpha t}\sin(\omega t+\phi) \tag{F.10a}$$

where

$$\omega^2=\omega_0^2-\alpha^2,\quad \alpha<\omega_0. \tag{F.10b}$$

Solution (F.10) is a decaying exponential multiplied by a sinusoidally varying term. The initial amplitude A and the phase angle ϕ are arbitrary and are determined by other conditions in the problem. The constant α is called the damping. Parameter ω_0 is the undamped frequency, the frequency of oscillation when $\alpha=0$. ω is the damped frequency.

When the damping becomes so large that $\alpha=\omega_0$, then the solution given above does not work. In that case, the solution is given by

$$y=(A+Bt)e^{-\alpha t}, \quad \alpha=\omega_0. \tag{F.11}$$

This case is called *critical damping* and represents the case in which y returns to zero most rapidly and with no overshoot. The solution can be verified by substitution.

If $\alpha>\omega_0$, then the solution is the sum of the two exponentials that satisfy (F.8).

$$y=Ae^{-at}+Be^{-bt},$$

where

$$a=\alpha+\sqrt{\alpha^2-\omega_0^2},$$
$$b=\alpha-\sqrt{\alpha^2-\omega_0^2}. \tag{F.12}$$

When $\alpha=0$, the equation is

$$\frac{d^2y}{dt^2}+\omega_0^2y=0. \tag{F.13}$$

The solution may be written either as

$$y=C\sin(\omega_0t+\phi) \tag{F.14a}$$

or as

$$y=A\cos(\omega_0t)+B\sin(\omega_0t). \tag{F.14b}$$

The simplest physical example of this equation is a mass on a spring. There will be an equilibrium position of the mass ($y=0$) at which there is no net force on the mass. If the mass is displaced toward either positive or negative y, a force back toward the origin results. The force is proportional to the displacement and is given by

$$F=-ky.$$

The proportionality constant k is called the spring constant. Newton's second law, $F=ma$, is

$$m\frac{d^2y}{dt^2}=-ky$$

or, defining $\omega_0^2=k/m$,

$$\frac{d^2y}{dt^2}+\omega_0^2y=0.$$

This (as well as the equation with $\alpha\neq0$) is a second-order differential equation. Integrating it twice introduces two constants of integration: C and ϕ or A and B. They are usually found from two initial conditions. For the mass on the spring, they are often the initial velocity and initial position of the mass.

The equivalence of the two solutions can be demonstrated by using Eqs. (F.14a) and (F.14b) and a trigonometric identity to write

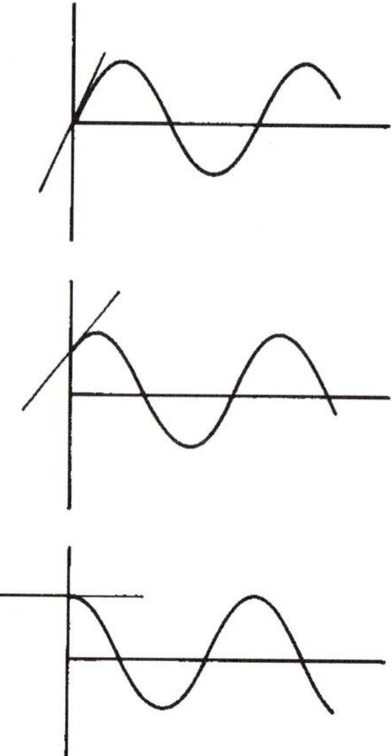

FIGURE F.1. Different starting points on the sine wave give different combinations of the initial position and the initial velocity.

$$C\sin(\omega_0t+\phi)=C[\sin(\omega_0t)\cos\phi+\cos(\omega_0t)\sin\phi].$$

Comparison with Eq. (F.14b) shows that

$$B=C\cos\phi, \quad A=C\sin\phi.$$

Squaring and adding these gives

$$C^2=A^2+B^2$$

while dividing one by the other shows that

$$\tan\phi=\frac{A}{B}.$$

Changing the initial phase angle ϕ changes the relative values of the initial position and velocity. This can be seen from the three plots of Fig. F.1, which show phase angles 0, $\pi/4$, and $\pi/2$. When $\phi=0$, the initial position is zero, while the initial velocity has its maximum value. When $\phi=\pi/4$, the initial position has a positive value, and so does the initial velocity. When $\phi=\pi/2$ the initial position has its maximum value and the initial velocity is zero. The values of A and B are determined from the initial position and velocity. At $t=0$, Eq. (F.14b) and its derivative give

$$y(0)=A, \quad \frac{dy}{dt}(0)=\omega_0B.$$

Increasing the damping α increases the rate at which the oscillatory behavior decays. Figure F.2 shows plots of y and dy/dt for different values of α.

The second-order equation we have just studied is called the harmonic oscillator equation. Its solution is needed in Chap. 8 and is summarized in Table F.1.

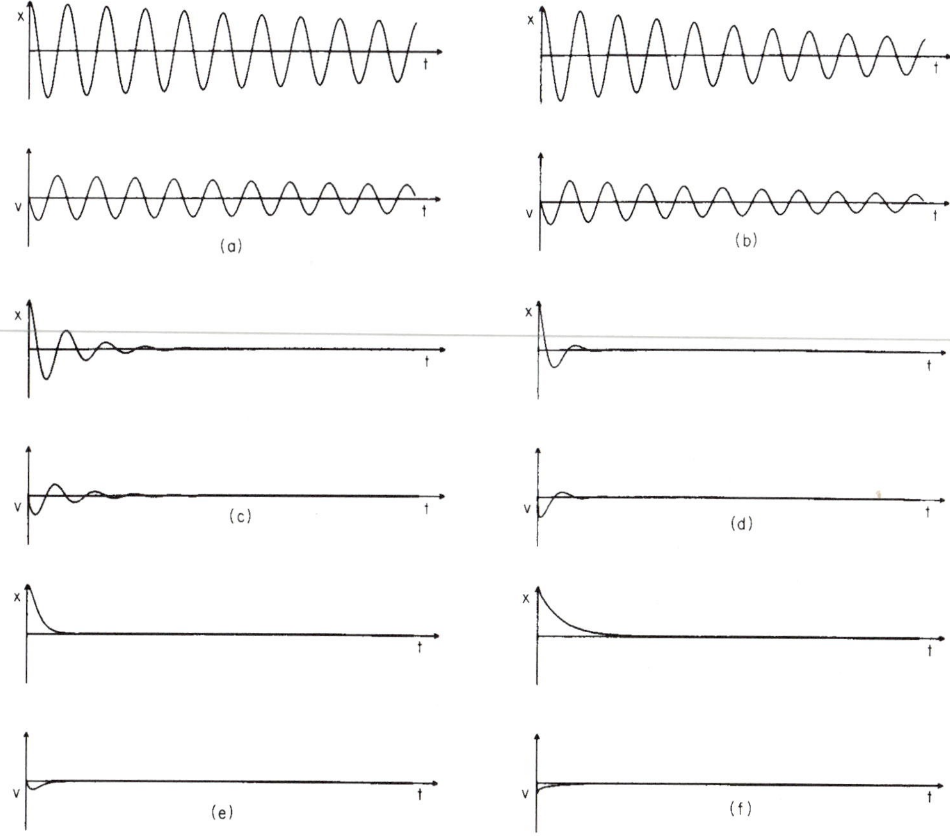

FIGURE F.2. Plot of $y(t)$ and dy/dt for different values of α. In (a)–(d) the system is underdamped. (e) Critical damping. (f) Overdamping.

PROBLEMS

F.1. If $\omega_0=10$, find A, B, C, and ϕ for the following cases:

(a) $y(0)=5$, $(dy/dt)(0)=0$.
(b) $y(0)=5$, $(dy/dt)(0)=5$.
(c) $y(0)=0$, $(dy/dt)(0)=50$.
(d) What values of A, B, and C would be needed to have the same ϕ as in case (b) and the same amplitude as in case (a)?

F.2. Verify Eq. (F.11) in the critically damped case.

F.3. Find the general solution of the equation

$$\frac{d^2y}{dt^2}+2\omega\frac{dy}{dt}+\omega_0^2y=\begin{cases}0, & t\leqslant 0\\ \omega_0^2y_0, & t\geqslant 0\end{cases}$$

subject to the initial conditions

$$y(0)=0,\quad \frac{dy}{dt}(0)=0$$

(a) for critical damping, $\alpha=\omega_0$,
(b) for no damping, and
(c) for overdamping, $\alpha=2\omega_0$.

TABLE F.1. *Solutions of the harmonic oscillator equation.*

$$\frac{d^2y}{dt^2}+2\alpha\frac{dy}{dt}+\omega_0^2y=0$$

Case	Criterion	Solution
Underdamped	$\alpha<\omega_0$	$y=Ae^{-\alpha t}\sin(\omega t+\phi)$ $\omega^2=\omega_0^2-\alpha^2$
Critically damped	$\alpha=\omega_0$	$y=(A+Bt)e^{-\alpha t}$
Overdamped	$\alpha>\omega_0$	$y=Ae^{-at}+Be^{-bt}$ $a=\alpha+(\alpha^2-\omega_0^2)^{1/2}$ $b=\alpha-(\alpha^2-\omega_0^2)^{1/2}$

APPENDIX G

The Mean and Standard Deviation

In many measurements in physics or biology there may be several possible outcomes to the measurement. Different values are obtained when the measurement is repeated. For example, the measurement might be the number of red cells in a certain small volume of blood, whether a person is right-handed or left-handed, the number of radioactive disintegrations of a certain sample during a 5-min interval, or the scores on a test.

Table G.1 gives the scores on an examination administered to 30 people. These results are also plotted as a histogram in Fig. G.1.

The table and the histogram give all the information that there is to know about the experiment unless the result depends on some variable that was not recorded, such as the age of the student or where the student was sitting during the test.

In many cases the frequency distribution gives too much information. It is convenient to invent some numbers that will answer the questions: Around what values do the results cluster? How wide is the distribution of results? Many different numbers have been invented for answering these questions. Some are easier to calculate or have more useful properties than others.

The *mean* or *average* shows where the distribution is centered. It is familiar to everyone: add up all the scores and divide by the number of students. For the data given above the mean is $\bar{x}=77.8$.

It is often convenient to group the data by the value obtained, along with the frequency of that value. The data of Table G.1, are grouped this way in Table G.2. The mean is calculated as

$$\bar{x}=\frac{1}{N}\sum_i f_i x_i=\frac{\sum_i f_i x_i}{\sum_i f_i},$$

where the sum is over the different values of the test scores which occur. For the example in Table G.2, the sums are

TABLE G.1. *Quiz scores.*

Student No.	Score	Student No.	Score
1	80	16	71
2	68	17	83
3	90	18	88
4	72	19	75
5	65	20	69
6	81	21	50
7	85	22	81
8	93	23	94
9	76	24	73
10	86	25	79
11	80	26	82
12	88	27	78
13	81	28	84
14	72	29	74
15	67	30	70

$$\sum_i f_i=30,$$

$$\sum_i f_i x_i=2335,$$

$$\bar{x}=\frac{2335}{30}=77.8.$$

If a large number of trials are made, f_i/N can be called the probability p_i of getting result x_i. Then

$$\bar{x}=\sum_i x_i p_i. \tag{G.1}$$

Note that $\Sigma p_i=1$. The average of some function of x is

$$\overline{g(x)}=\sum_i g(x_i)p_i. \tag{G.2}$$

For example,

$$\overline{x^2}=\sum_i (x_i)^2 p_i .$$

The width of the distribution is often characterized by the *dispersion* or *variance*:

$$\overline{(\Delta x)^2}=\overline{(x-\bar{x})^2}=\sum_i p_i(x_i-\bar{x})^2. \qquad \text{(G.3)}$$

This is also sometimes called the mean square variation: the mean of the square of the variation of x from the mean. The measure of the width is the square root of this which is called the *standard deviation* σ. The need for taking the square root is easy to see since x may have units associated with it. If x is in meters, then the variance has the units of square meters. The width of the distribution in x must be in meters.

A very useful result is

$$\overline{(x-\bar{x})^2}=\overline{x^2}-\bar{x}^2.$$

To prove this, note that

$$(x_i-\bar{x})^2=x_i^2-2x_i\bar{x}+\bar{x}^2.$$

The variance is then

TABLE G.2. *Quiz scores grouped by score.*

Score number i	Score x_i	Frequency of score, f_i	f_ix_i
1	50	1	50
2	65	1	65
3	67	1	67
4	68	1	68
5	69	1	69
6	70	1	70
7	71	1	71
8	72	2	144
9	73	1	73
10	74	1	74
11	75	1	75
12	76	1	76
13	78	1	78
14	79	1	79
15	80	2	160
16	81	3	243
17	82	1	82
18	83	1	83
19	84	1	84
20	85	1	85
21	86	1	86
22	88	2	176
23	90	1	90
24	93	1	93
25	94	1	94

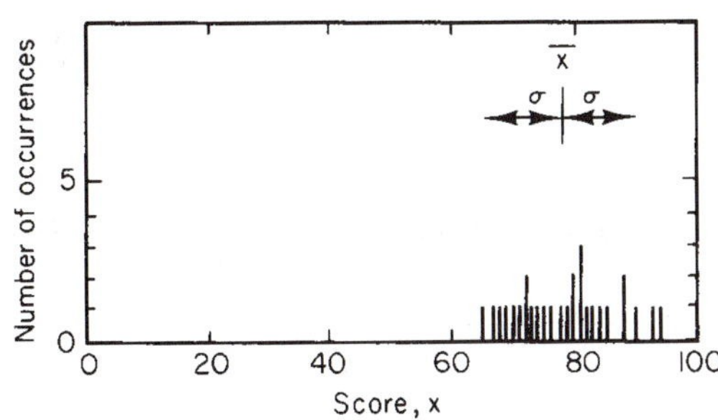

FIGURE G.1. Histogram of the quiz scores in Table G.1.

$$\overline{(\Delta x)^2}=\sum_i p_i x_i^2-2\sum_i x_i\bar{x}p_i+\sum p_i\bar{x}^{\,2}.$$

The first sum is the definition of $\overline{x^2}$. The second sum has a number $\bar{x}$ in every term. It can be factored in front of the sum, to make the second term $-2\bar{x}\,\Sigma x_ip_i$, which is just $-2(\bar{x})^2$. The last term is $(\bar{x})^2\Sigma p_i=(\bar{x})^2$. Combining all three sums gives Eq. (G.4). In summary,

$$\sigma=\sqrt{\overline{(\Delta x^2)}},$$

$$\sigma^2=\overline{(\Delta x)^2}=\overline{(x-\bar{x})^2}=\overline{x^{\,2}}-\bar{x}^{\,2}. \qquad \text{(G.4)}$$

This equation is true as long as the p_i's are accurately known. If the p_i's have only been estimated from N experimental observations, the best estimate of σ^2 is $N/(N-1)$ times the value calculated from Eq. (G.4).

For the data of Fig. G.1, σ=9.4. This width is shown along with the mean at the top of the figure.

PROBLEMS

G.1. Calculate the variance and standard deviation for the data in Table G.2.

G.2. In computing the standard deviation of a series of observations, you may find it easier to take the difference between x_i and x_0, where x_0 is an integer, than to calculate $x_i-\bar{x}$. Show

$$\sigma^2=\frac{1}{N}\left(\sum_{i=1}^{N}(x_i-x_0)^2-N(\bar{x}-x_0)^2\right).$$

This has been written for large N. For a sample of small size, it should be multiplied by $N/(N-1)$.

APPENDIX H

The Binomial Probability Distribution

Consider an experiment that can have two mutually exclusive outcomes and is repeated N times, with each repetition being independent of every other try. One of the outcomes is labeled "success"; the other is called "failure." The experiment could be throwing a die with success being a three, flipping a coin with success being a head, or placing a particle in a box with success being that the particle is located in a subvolume v.

In a single try, call the probability of success p and the probability of failure q. Since one outcome must occur and both cannot occur at the same time,

$$p+q=1. \tag{H.1}$$

Suppose that the experiment is repeated N times. The probability of n successes out of N tries is given by the binomial probability distribution, which is stated here without proof.[1] We can call the probability $P(n;N)$, since it is a function of n and depends on the parameter N. [Strictly speaking, it depends on two parameters, N and p. We should write $P(n;N,p)$.] It is[2]

$$P(n;N)=P(n;N,p)=\frac{N!}{n!(N-n)!}\,p^n(1-p)^{N-n}. \tag{H.2}$$

The factor $N!/[n!(N-n)!]$ counts the number of different ways that one can get n outcomes; the probability of each of these ways is $p^n(1-p)^{N-n}$. In the example of three particles in Sec. 3.1, there are three ways to have one particle in the left-hand side. The particle can be either particle a or particle b or particle c. The factor gives directly

$$\frac{N!}{n!(N-n)!}=\frac{3!}{1!2!}=\frac{3\times2\times1}{(1)(2\times1)}=\frac{6}{2}=3.$$

The remaining factor, $p^n(1-p)^{N-n}$, is the probability of taking n tries in a row and having success and taking $N-n$ tries in a row and having failure.

The binomial distribution applies if each "try" is independent of every other try. Such processes are called Bernoulli processes (and the binomial distribution is often called the Bernoulli distribution). In contrast, if the probability of an outcome depends on the results of previous tries, the random process is called a Markov process. Although such processes are important, they are more difficult to deal with and will not be discussed here.

Some examples of the use of the binomial distribution are given in Chap. 3. As another example, consider the problem of performing several laboratory tests on a patient. It has become fairly common to use automated machines for blood-chemistry evaluations of patients; such machines may automatically perform 6, 12, 20, or more tests on one small sample of a patient's blood serum, for less cost than doing just one or two of the tests. But this means that the physician gets a large number of results—many more than would have been asked for if the tests were done one at a time. When such test batteries were first done, physicians were surprised to find that patients had many more abnormal tests than they expected. This was in part because some tests were not known to be abnormal in certain diseases, because no one had ever looked at them in that disease. But there still was a problem that some tests were slightly abnormal in patients who appeared to be perfectly healthy. We can understand why by considering the following idealized situation. Suppose that we do N independent tests, and suppose that in healthy people, the probability that each test is abnormal is p. (In our vocabulary, having an abnormal test is "success"!). The probability of not having the test abnormal is $q=1-p$. In a perfect test, p would be 0 for healthy people and would be 1 in sick people; however, very few tests are that discriminating. The definition of "normal" vs "abnormal" involves a compromise between false positives (abnormal test results in healthy people) and false negatives (normal test results in sick people). Good reviews of this problem have been written by Murphy and Abbey[3] and by Feinstein.[4] In many

[1]A detailed proof can be found in many places. See, for example, F. Reif (1964). *Statistical Physics*. Berkely Physics Course, Vol. 5, New York, McGraw-Hill, p. 67.

[2]$N!$ is N factorial and is $N(N-1)(N-2)\cdots1$. By definition, $0!=1$.

[3]E. A. Murphy and H. Abbey (1967). The normal range—a common misuse. *J. Chronic Dis.* **20**: 79.

[4]A. R. Feinstein (1975). Clinical biostatistics XXVII. The derangements of the normal range. *Clin. Pharmacol. Therap.* **15**: 528.

cases, p is about 0.05. Now suppose that p is the same for all the tests and that the tests are independent. Neither of these assumptions is very good, but they will show what the basic problem is. Then, the probability for all of the N tests to be normal in a healthy patient is given by the binomial probability distribution:

$$P(0;N,p)=\frac{N!}{0!\,N!}\,p^0q^N=q^N.$$

If $p=0.05$, then $q=0.95$, and

$$P(0;N,p)=0.95^N.$$

Typical values are $P(0;\,12)=0.54$, and $P(0;\,20)=0.36$. If the assumptions about p and independence are right, then only 36% of healthy patients will have all their tests normal if 20 tests are done.

Figure H.1 shows a plot of the number of patients in a series who were clinically normal but who had abnormal tests. The data have the general features predicted by this simple model.

We can derive simple expressions to give the mean and standard deviation if the probability distribution is binomial. The mean value of n is defined to be

$$\bar{n}=\sum_{n=0}^{N} nP(n;N)=\sum_{n=0}^{N}\frac{N!\,n}{n!(N-n)!}\,p^n(1-p)^{N-n}.$$

The first term of the sum is for $n=0$. Since each term is multiplied by n, the first term vanishes, and the limits of the sum can be rewritten as

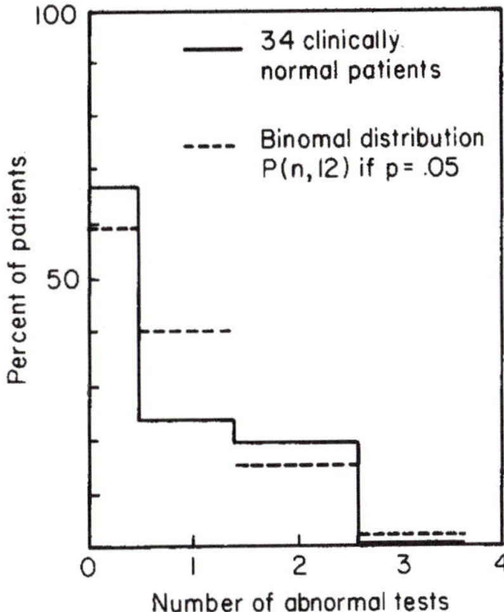

FIGURE H.1. Measurement of the probablity that a clinically normal patient having a battery of 12 tests done has n abnormal tests (solid line) and a calculation based on the binomial distribution (dashed line). The calculation assumes that $p=0.05$ and that all 12 tests are independent. Several of the tests in this battery are not independent, but the general features are reproduced.

$$\sum_{n=1}^{N}\frac{N!\,n}{n!(N-n)!}\,p^n(1-p)^{N-n}.$$

To evaluate this sum, we use a trick. Let $m=n-1$ and $M=N-1$. Then we can rewrite various parts of this expression as follows:

$$\frac{n}{n!}=\frac{1}{(n-1)!}=\frac{1}{m!},$$

$$p^n=pp^m,$$

$$N!=(N)(N-1)!,$$

$$(N-n)!=[N-1-(n-1)]!=(M-m)!.$$

The limits are

$$n=1,\quad m=0,$$

$$n=N,\quad m=M.$$

With these substitutions

$$\bar{n}=Np\sum_{m=0}^{M}\frac{M!}{m!(M-m)!}\,p^m(1-p)^{M-m}.$$

This sum is exactly the sum of a binomial distribution over all possible values of m and is equal to one. We have the result that, for a binomial distribution,

$$\bar{n}=Np. \tag{H.3}$$

This says that the average number of successes is the total number of tries times the probability of a success on each try. If 100 particles are placed in a box and we look at half the box so that $p=\frac{1}{2}$, the average number of particles in that half is $100\times\frac{1}{2}=50$. If we put 500 particles in the box and look at $\frac{1}{10}$ of the box, the average number of particles in the volume is also 50. If we have 100 000 particles and $v/V=p=1/2000$, the average number is still 50.

For the binomial distribution, the variance σ^2 can be expressed in terms of N and p using Eq. G.4. The average of n^2 is

$$\overline{(n^2)}=\sum_{n} P(n;N)n^2=\sum_{n=0}^{N}\frac{N!}{n!(N-n)!}\,n^2p^n(1-p)^{N-n}.$$

The trick to evaluate this is to write $n^2=n(n-1)+n$. With this substitution we get two sums:

$$\overline{(n^2)}=\sum_{n=0}^{N}\frac{N!}{n!(N-n)!}\,n(n-1)p^nq^{N-n}$$

$$+\sum_{n=0}^{N}\frac{N!n}{n!(N-n)!}\,p^nq^{N-n}.$$

The second sum is $\bar{n} = Np$. The first sum is rewritten by noticing that the terms for $n=0$ and $n=1$ both vanish. Therefore, $m=n-2$ and $M=N-2$:

$$\overline{(n^2)} = Np + N(N-1)\sum_{m=0}^{M} \frac{M!}{m!(M-m)!} p^2 p^m q^{M+2-m-2}$$

$$= Np + N(N-1)p^2$$

$$= Np + N^2p^2 - Np^2.$$

Therefore,

$$\overline{(\Delta n)^2} = \overline{n^2} - \bar{n}^2 = Np - Np^2 = Np(1-p) = Npq.$$

For the binomial distribution, then,

$$\sigma = \sqrt{Npq} = \sqrt{\bar{n}q}. \tag{H.4}$$

The standard deviation for the binomial distribution for fixed p goes as $N^{1/2}$. For fixed N, it is proportional to

$$\sqrt{p(1-p)}$$

which is plotted in Fig. H.2. For fixed N, the maximum value of σ occurs when $p=q=\frac{1}{2}$. If p is very small, the event happens rarely, if p is close to 1, the event nearly always happens. In either case, the variation is reduced. On the other hand, if N becomes large while p becomes small in such a way as to keep $\bar{n}$ fixed, then σ increases to a maximum value of $\sqrt{\bar{n}}$. This variation of σ with N and p is demonstrated in Fig. H.3. Figures H.3(a)–H.3(c) show how σ changes as N is held fixed and p is varied. For $N=100$, p is 0.05, 0.5, and 0.95. Both the mean and σ change. Comparing Fig. H.3(b) with H.3(d) shows two different cases where $\bar{n}=50$. When p is very small because N is very large in Fig. H.3(d), σ is much larger than in Fig. H.3(b).

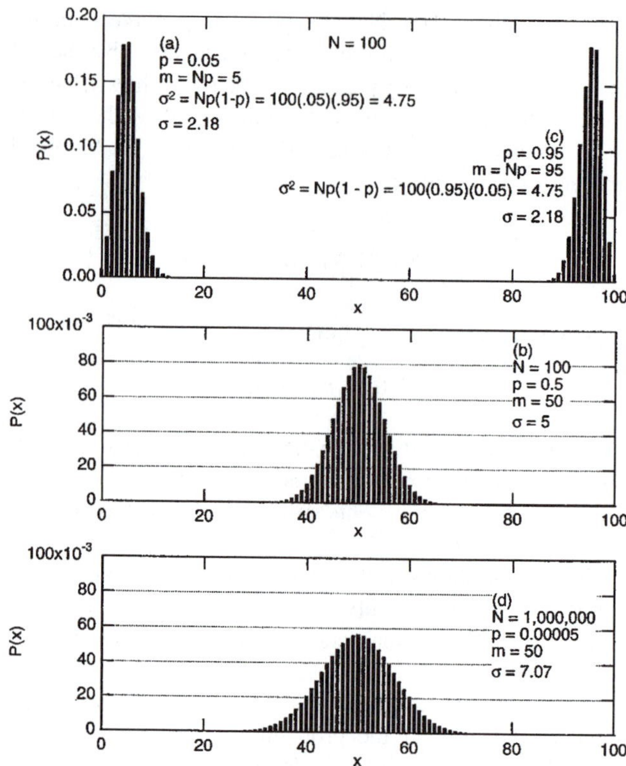

FIGURE H.3. Examples of the variation of σ with N and p. (a), (b), and (c) show variations of σ with p when N is held fixed. The maximum value of σ occurs when $p=0.5$. Comparison of (b) and (d) shows the variation of σ and p as p and N change together in such a way that $\bar{n}$ remains equal to 50.

PROBLEMS

H.1. Calculate the probability of throwing 0,1,...,9 heads out of a total of nine throws of a coin.

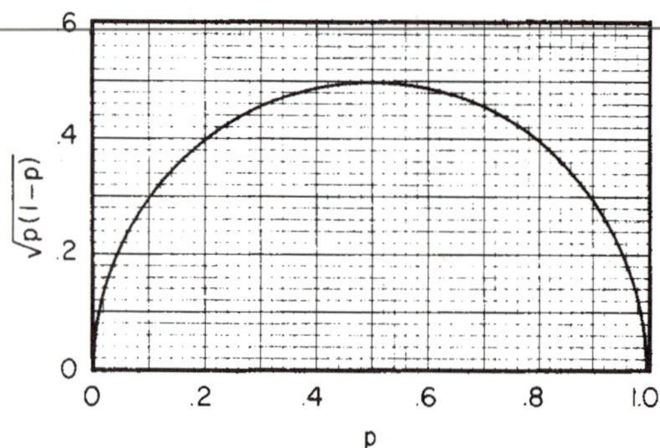

FIGURE H.2. Plot of $[p(1-p)]^{1/2}$.

H.2. Assume that males and females are born with equal probability. What is the probability that a couple will have four children, all of whom are girls? The couple has had three girls. What is the probability that they will have a fourth girl? Why are these probabilities so different?

H.3. The Mayo Clinic reported that a single stool specimen in a patient known to have an intestinal parasite yields positive results only about 90% of the time [R. B. Thomson, R. A. Haas, and J. H. Thompson, Jr. (1984). Intestinal parasites: The necessity of examining multiple stool specimens. *Mayo Clin. Proc.* **59**: 641–642]. What is the probability of a false negative if two specimens are examined? Three?

H.4. The Minneapolis *Tribune* on October 31, 1974 listed the following incidence rates for cancer in the Twin Cities' greater metropolitan area, which at that time had a total population of 1.4 million. These rates are compared to those in nine other areas of the country whose total population is 15 million. Assume that each study was for one year. Are the differences statistically significant? Show calculations to support your answer. How would your answer differ if the study were for several years?

APPENDIX I

The Gaussian Probability Distribution

Appendix H considered a process that had two mutually exclusive outcomes and was repeated N times, with the probability of "success" on one try being p. If each try is independent, then the probability of n occurrences of "success" in N tries is

$$P(n;N,p)=\frac{N!}{(n!)(N-n)!}p^n(1-p)^{N-n}. \tag{I.1}$$

This probability distribution depends on the two parameters N and p. It is possible to define two other parameters, the mean, which roughly locates the center of the distribution, and the standard deviation, which measures its width. These new parameters, $\bar{n}$ and σ, are related to N and p by the equations

$$\bar{n}=Np,$$

$$\sigma^2=Np(1-p).$$

It is then possible to write the binomial distribution formula in terms of the new parameters instead of N and p. At best, however, it is cumbersome, because of the need to evaluate so many factorial functions. We will now develop an approximation that is valid when N is large and which allows the probability to be calculated more easily.

The procedure is to take the log of the probability, $y=\ln(P)$ and expand it in a Taylor's series (Appendix C) about some point. Since there is a value of n for which P has a maximum and since the logarithmic function is monotonic, y has a maximum for the same value of n. We will expand about that point; call it n_0. Then the form of y is

$$y=y(n_0)+\left.\frac{dy}{dn}\right|_{n_0}(n-n_0)+\frac{1}{2}\left.\frac{d^2y}{dn^2}\right|_{n_0}(n-n_0)^2+\cdots .$$

Since y is a maximum at n_0, the first derivative vanishes and it is necessary to keep the quadratic term in the expansion.

To take the logarithm of Eq. (I.1), we need a way to handle the factorials. There is a very useful approximation to the factorial, called Stirling's approximation:

$$\ln(n!)\approx n\ln n-n. \tag{I.2}$$

To derive it, write $\ln(n!)$ as

$$\ln(n!)=\ln 1+\ln 2+\cdots+\ln n=\sum_{m=1}^{n}\ln m.$$

The sum is the same as the total area of the rectangles in Fig. I. 1, where the height of each rectangle is $\ln m$ and the width of the base is one. The area of all the rectangles is approximately the area under the smooth curve, which is $\ln m$. The area is approximately

$$\int_1^n \ln m\, dm=[m\ln m-m]_1^n=n\ln n-n+1.$$

This completes the proof of Eq. (I.2). Table I.1 shows values of $n!$ and Stirling's approximation for various values of n. The approximation is not too bad for $n\geqslant 100$.

We can now return to the task of deriving the binomial distribution. Taking logarithms of Eq. (I.1), we get

$$y=\ln P=\ln(N!)-\ln(n!)-\ln(N-n)!$$
$$+n\ln p+(N-n)\ln(1-p).$$

With Stirling's approximation, this becomes

$$y=N\ln N-n\ln n-N\ln(N-n)+n\ln(N-n)$$
$$+n\ln p+(N-n)\ln(1-p). \tag{I.3}$$

The derivative with respect to n is

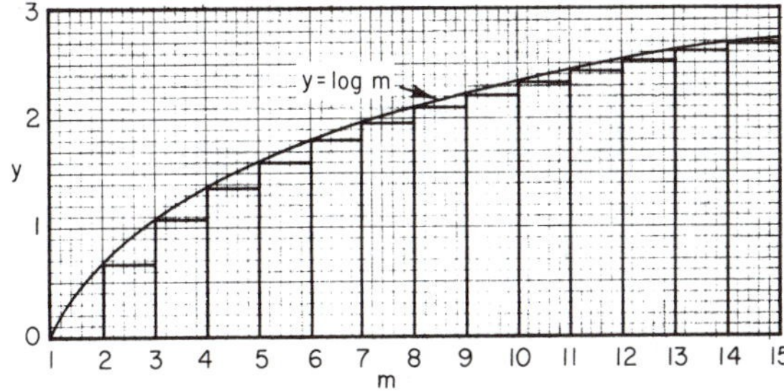

FIGURE I.1. Plot of $y=\ln m$ used to derive Stirling's approximation.

TABLE I.1. *Accuracy of Stirling's approximation.*

n	$n!$	$\ln(n!)$	$n \ln n - n$	Error	% Error
5	120	4.7875	3.047	1.74	36
10	3.6×10^{6}	15.104	13.026	2.08	14
20	2.4×10^{18}	42.336	39.915	2.42	6
100	9.3×10^{157}	363.74	360.51	3.23	0.8

$$\frac{dy}{dn} = -\ln n + \ln(N-n) + \ln p - \ln(1-p).$$

The second derivative is

$$\frac{d^2y}{dn^2} = -\frac{1}{n} - \frac{1}{N-n}.$$

The point of expansion n_0 is found by making the first derivative vanish:

$$0 = \ln \frac{(N-n)p}{n(1-p)}.$$

Since $\ln 1 = 0$, this is equivalent to

$$(N-n_0)p = n_0(1-p)$$

or

$$n_0 = Np.$$

The maximum of y occurs when n is equal to the mean. At $n = n_0$ the value of the second derivative is

$$\frac{d^2y}{dn^2} = -\frac{1}{Np} - \frac{1}{N(1-p)} = -\frac{1}{Np(1-p)}.$$

It is still necessary to evaluate $y_0 = y(n_0)$. If we try to do this by substitution of $n = n_0$ in Eq. (I.3), we get zero. The reason is that the Stirling approximation we used is too crude for this purpose. (There are additional terms in Stirling's approximation to make it more accurate.) The easiest way to find $y(n_0)$ is to call it y_0 for now and determine it from the requirement that the probability be normalized. Therefore, we have

$$y = y_0 - \frac{1}{2Np(1-p)}(n-Np)^2$$

so that, in this approximation,

$$P(n) = e^y = e^{y_0} e^{-(n-Np)^2/[2Np(1-p)]}.$$

With $Np = \bar{n}$, $e^{y_0} = C_0$, and $Np(1-p) = \sigma^2$, this is

$$P(n) = C_0 e^{-(n-\bar{n})^2/2\sigma^2}.$$

To evaluate C_0, note that the sum of $P(n)$ for all n is the area of all the rectangles in Fig. I.2. This area is approxi-

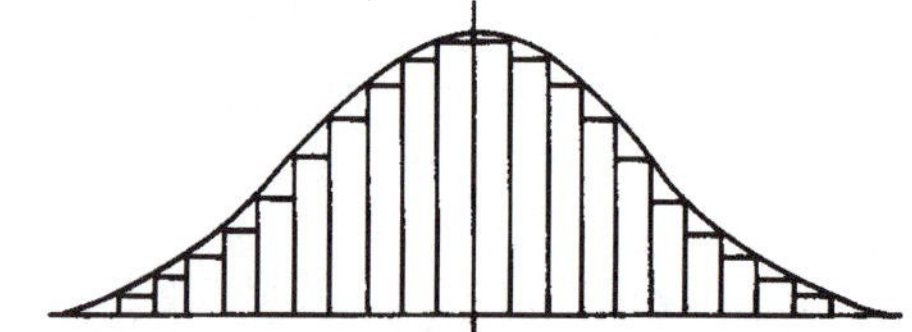

FIGURE I.2. Evaluating the normalization constant.

mately the area under the smooth curve, so that

$$1 = C_0 \int_{-x}^{x} e^{-(n-\bar{n})^2/2\sigma^2} dn.$$

It is shown in Appendix K that half of this integral is

$$\int_0^x dx\, e^{-bx^2} = \frac{1}{2}\sqrt{\frac{\pi}{b}}.$$

Therefore the normalization integral is (letting $x = n - \bar{n}$)

$$\int_{-\infty}^{\infty} e^{-(x^2)/2\sigma^2} dx = \sqrt{2\pi\sigma^2}.$$

The normalization constant is

$$C_0 = \frac{1}{\sqrt{2\pi\sigma^2}}$$

so that the probability is

$$P(n) = \frac{1}{\sqrt{2\pi\sigma^2}} e^{-(n-\bar{n})^2/2\sigma^2}. \tag{I.4}$$

It is possible, as in the case of the random-walk problem, that the measured quantity x is proportional to n with a very small proportionality constant, $x = kn$, so that the values of x appear to form a continuum. As shown in Fig. I.3, the number of different values of n [each with about the same value of $P(n)$] in the interval dx is proportional to dx. The easiest way to write down the Gaussian distribution in the continuous case is to recognize that the mean is $\bar{x} = k\bar{n}$, and the standard deviation is $\sigma_x^2 = \overline{(x-\bar{x})^2} = \overline{x^2} - (\bar{x})^2 = k^2\overline{n^2} - k^2(\bar{n})^2 = k^2\sigma^2$. The term $P(x)\,dx$ is given by $P(n)$ times the number of different values of n in dx. This number is dx/k. Therefore,

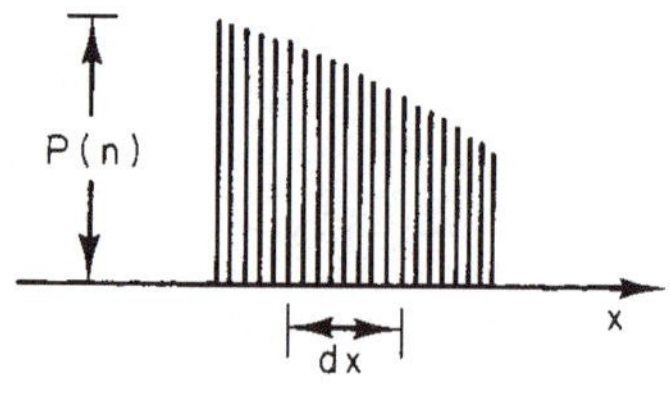

FIGURE I.3. The allowed values of x are closely spaced in this case.

$$P(x)dx = P(n)\,\frac{dx}{k} = \frac{1}{\sqrt{2\pi}\sigma}\,e^{-(x/k-\bar{x}/k)^2/2\sigma^2}$$

$$= dx\,\frac{1}{\sqrt{2\pi}\sigma_x}\,e^{-(x-\bar{x})^2/2\sigma_x^2}. \qquad \text{(I.5)}$$

To recapitulate, the binomial distribution in the case of large N can be approximated by Eq. (I.4), the Gaussian or normal distribution, for continuous variables. The original parameters N and p are replaced in these approximations by $\bar{n}$ (or $\bar{x}$) and σ.

APPENDIX J

The Poisson Distribution

Appendix H discussed the binomial probability distribution. If an experiment is repeated N times, and has two possible outcomes, with ''success'' occurring with probability p in each try, the probability of getting that outcome x times in N tries is

$$P(x;N,p)=\frac{N!}{x!(N-x)!}\,p^x(1-p)^{N-x}.$$

The distribution of possible values of x is characterized by a mean value

$$\bar{x}=Np$$

and a variance

$$\sigma^2=Np(1-p).$$

It is possible to specify $\bar{x}$ and σ^2 instead of N and p to define the distribution.

Appendix I showed that it is easier to work with the Gaussian or normal distribution when N is large. It is specified in terms of the parameters $\bar{x}$ and σ^2 instead of N and p:

$$P(x;\bar{x},\sigma^2)=\frac{1}{(2\pi\sigma^2)^{1/2}}\,e^{-(x-\bar{x})^2/2\sigma^2}.$$

The Poisson distribution is an approximation to the binomial distribution that is valid for large N and for small p (when N gets large and p gets small in such a way that their product remains finite). To derive it, rewrite the binomial probability in terms of $p=\bar{x}/N$:

$$P(x)=\frac{N!}{x!(N-x)!}\,(\bar{x}/N)^x(1-\bar{x}/N)^{N-x}$$

$$=\frac{N!}{x!(N-x)!}\,\bar{x}^x\left(1-\frac{\bar{x}}{N}\right)^N\left(1-\frac{\bar{x}}{N}\right)^{-x}. \quad \text{(J.1)}$$

It is necessary next to consider the behavior of some of these factors as N becomes very large. The factor

$$\left(1-\frac{\bar{x}}{N}\right)^N$$

approaches $e^{-\bar{x}}$ as $N\to\infty$, by definition (see p. 27). The factor

$$\frac{N!}{(N-x)!}$$

can be written out as

$$\frac{N(N-1)(N-2)\cdots 1}{(N-x)(N-x-1)\cdots 1}=N(N-1)(N-2)\cdots(N-x+1).$$

If these factors are multiplied out, the first term is N^x, followed by terms containing N^{x-1}, N^{x-2},..., down to N^1. But there is also a factor N^x in the *denominator* of the expression for P, which, combined with this gives

$$1+(\text{something})N^{-1}+(\text{something})N^{-2}+\cdots .$$

As long as N is very large, all terms but the first can be neglected. With these substitutions, Eq. (J.1) takes the form

$$P(x)=\frac{1}{x!}\,\bar{x}^x e^{-\bar{x}}\left(1-\frac{\bar{x}}{N}\right)^N. \quad \text{(J.2)}$$

The values of x for which $P(x)$ is not zero are near $\bar{x}$, which is much less than N. Therefore, the last term, which is really $[1/(1-p)]^x$, can be approximated by one, while such a term raised to the Nth power had to be approximated by e^{-x}. If this is difficult to understand, consider the following numerical example. Let $N=10\,000$ and $p=0.001$, so $\bar{x}=10$. The two terms we are considering are

$$\left(1-\frac{10}{10\,000}\right)^{10\,000}=4.517\times10^{-5},$$

which is approximated by $e^{-10}=4.54\times10^{-5}$, and terms like

$$\left(1-\frac{10}{1000}\right)^{-10}=1.001,$$

which is approximated by 1.

With these approximations, the probability is

$$P(x)=\frac{(\bar{x})^x}{x!}\,e^{-\bar{x}}$$

or, calling $\bar{x}=m$,

$$P(x)=\frac{m^x}{x!}\,e^{-m}. \tag{J.3}$$

This is the Poisson distribution and is an approximation to the binomial distribution for large N and small p, such that the mean $\bar{x}=m=Np$ is defined (that is, it does not go to infinity or zero as N gets large and p gets small).

This probability, when summed over all values of x, should be unity. This is easily verified. Write

$$\sum_{x=0}^{\infty} P(x)=e^{-m}\sum_{x=0}^{\infty}\frac{m^x}{x!}.$$

But the sum on the right is the series for e^m, and $e^{-m}e^m=1$ [see Eq. D.2]. The same trick can be used to verify that the mean is m:

$$\sum_{x=0}^{\infty} xP(x)=\sum_{x=0}^{\infty} x\,\frac{m^x}{x!}\,e^{-m}=\sum_{x=1}^{\infty} x\,\frac{m^x}{x!}\,e^{-m}.$$

The index of summation can be changed from x to $y=x-1$:

$$\sum_{x=0}^{\infty} xP(x)=\sum_{y=0}^{\infty}\frac{(y+1)}{(y+1)!}\,m^y m e^{-m}=m\sum_{y=0}^{\infty}\frac{m^y}{y!}\,e^{-m}=m.$$

One can show that the variance for the Poisson distribution is

$$\sigma^2=\overline{(x-m)^2}=m.$$

Table J.1 compares the binomial, Gaussian, and Poisson distributions. The principal difference between the binomial and Gaussian distributions is that the latter is valid for large N and is expressed in terms of the mean and standard deviation instead of N and p. Since the Poisson distribution is valid for very small p, there is only one parameter left, and $\sigma^2=m$ rather than $m(1-p)$.

The Poisson distribution can be used to answer questions like the following:

1. How many red cells are there in a small square in a hemocytometer? The number of cells N is large; the probability p of each cell falling in a particular square is small. The variable x is the number of cells per square.

2. How many gas molecules are found in a small volume of gas in a large container? The number of tries is each molecule in the larger box. The probability that an individual molecule is in the smaller volume is $p=V/V_0$, where V is the small volume and V_0 is the volume of the entire box.

3. How many radioactive nuclei (or excited atoms) decay (or emit light) during a time dt? The probability of decay during time dt is proportional to how long dt is $p=\lambda\ dt$. The number of tries is the N nuclei that might decay during that time.

TABLE J.1. *Comparison of the binomial, Gaussian, and Poisson distributions.*

Binomial	$P(x,N,p)=\dfrac{N!}{x!(N-x)!}\,p^x(1-p)^{N-x}$ $\bar{x}=m=Np$ $\sigma^2=Np(1-p)=m(1-p)$
Gaussian	$P(x;m,\sigma)=\dfrac{1}{(2\pi\sigma^2)^{1/2}}\,e^{-(x-m)^2/2\sigma^2}$
Poisson	$P(x;m)=\dfrac{m^x}{x!}\,e^{-m}$ $m=Np$ $\sigma^2=m$

The last example is worth considering in greater detail. The probability p that each nucleus decays in time dt, is proportional to the length of the time interval: $p=\lambda\ [dt]$. The average number of decays if many time intervals are examined is

$$m=Np=N\lambda\ dt.$$

The probability of x decays in time dt is

$$P(x)=\frac{(N\lambda\ dt)^x}{x!}\,e^{-N\lambda\ dt}.$$

As $dt\rightarrow 0$, the exponential approaches one, and

$$P(x)\rightarrow\frac{(N\lambda\ dt)^x}{x!}.$$

The overwhelming probability for $dt\rightarrow 0$ is for there to be no decays: $P(0)\approx(N\lambda\ dt)^0/0!=1$. The probability of a single decay is $P(1)=N\lambda\ dt$; the probability of two decays during dt is $(N\lambda\ dt)^2/2$, and so forth.

If time interval t is finite, it is still possible for the Poisson criterion to be satisfied, as long as $p=\lambda t$ is small. Then the probability of no decays is

$$P(0)=e^{-m}=e^{-N\lambda t}.$$

The probability of one decay is

$$P(1)=(N\lambda t)e^{-N\lambda t}.$$

This probability increases linearly with t at first and then decreases as the exponential term begins to decay. The reason for the lowered probability of one decay is that it is now more probable for *two or more* decays to take place in this longer time interval. As t increases, it is more probable that there are two decays than one or none; for still longer times, even more decays become more probable. The probability that n decays occur in time t is $P(n)$. Figure J.1 shows plots of $P(0)$, $P(1)$, $P(2)$, and $P(3)$, vs $m=N\lambda t$.

PROBLEMS

J.1. In the United States today, 400 000 people are killed or injured each year in automobile accidents. The total

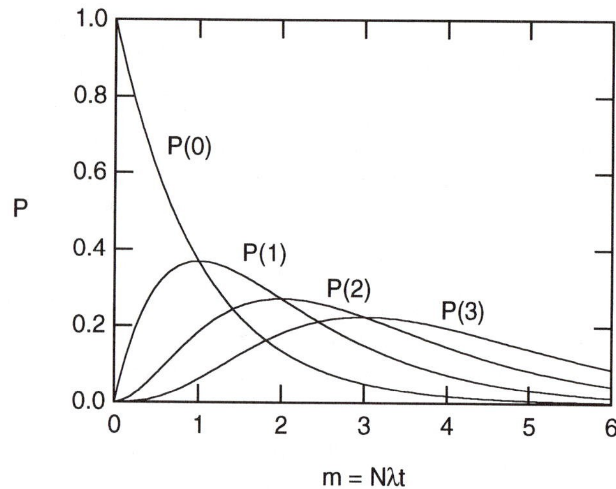

FIGURE J.1. Plot of $P(0)$ through $P(3)$ vs $N\lambda t$.

population is 200 000 000. If the probability of being killed or injured is independent of time, what is the probability that you will escape unharmed from 70 years of driving?

J.2. Assume that typographical errors are produced completely at random. A book of 600 pages contains 600 mistakes. What is the probability that a page

(a) has no misprints?

(b) has at least three misprints?

J.3. Large proteins consist of a number of smaller subunits that are stuck together. Suppose that an error is made in inserting an amino acid once in every 10^5 tries; $p=10^{-5}$. If a chain has length 1000, what is the probability of making a chain with no mistakes? If the chain length is 10^5?

J.4. The muscle end plate has an electrical response whenever the nerve connected to it is stimulated. Boyd and Martin [The end plate potential in mammalian muscle. *J. Physiol.* **132**: 74–91 (1956)] found that the electrical response could be interpreted as resulting from the release of packets of acetylcholine by the nerve. In terms of this model, they obtained the following data:

Number of packets reaching the end plate	Number of times observed
0	18
1	44
2	55
3	36
4	25
5	12
6	5
7	2
8	1
9	0

Analyze these data in terms of a Poisson distribution.

APPENDIX K

Integrals Involving e^{-ax^2}

Integrals involving e^{-ax^2} appear in the Gaussian and Maxwell–Boltzmann distributions. The integral

$$I=\int_{-\infty}^{\infty} e^{-ax^2}dx$$

can also be written with y as the dummy variable:

$$I=\int_{-\infty}^{\infty} e^{-ay^2}dy.$$

These can be multipled together to get

$$I^2=\int_{-\infty}^{\infty}\int_{-\infty}^{\infty} dx\ dy\ e^{-ax^2}e^{-ay^2}$$

$$=\int_{-\infty}^{\infty}\int_{-\infty}^{\infty} dx\ dy\ e^{-a(x^2+y^2)}.$$

A point in the xy plane can also be specified by the polar coordinates r and θ (Fig. K.1). The element of area $dx\ dy$ is replaced by the element $r\ dr\ d\theta$:

$$I^2=\int_0^{2\pi} d\theta \int_0^{\infty} r\ dr\ e^{-ar^2}=2\pi\int_0^{\infty} r\ dr\ e^{-ar^2}.$$

To continue, make the substitution $u=ar^2$, so that $du=2ar\ dr$. Then

$$I^2=2\pi\int_0^{\infty}\frac{1}{2a}e^{-u}du=\frac{\pi}{a}\int_0^{\infty}e^{-u}du$$

$$=\frac{\pi}{a}[-e^{-u}]_0^{\infty}=\frac{\pi}{a}[-0-(-1)].$$

The desired integral is, therefore,

$$I=\int_{-\infty}^{\infty} e^{-ax^2}dx=\sqrt{\frac{\pi}{a}}. \qquad \text{(K.1)}$$

This integral is one of a general sequence of integrals of the general form

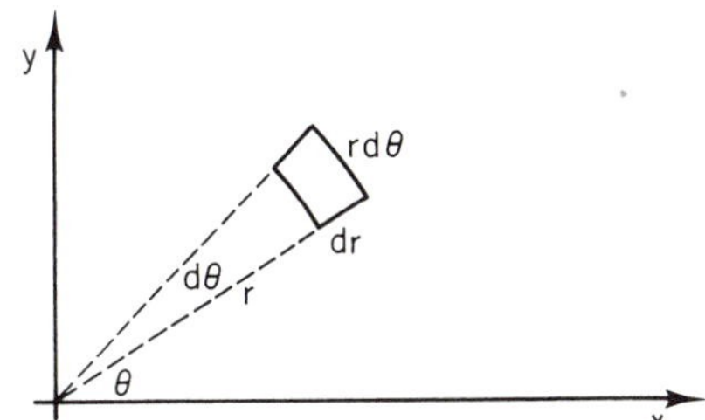

FIGURE K.1. An element of area in polar coordinates.

$$I_n=\int_0^{\infty} x^n e^{-ax^2}dx.$$

From Eq. (K.1), we see that

$$I_0=\frac{I}{2}=\frac{1}{2}\sqrt{\frac{\pi}{a}}. \qquad \text{(K.2)}$$

A value for I_2 can be obtained by integrating by parts:

$$I_2=\int_0^{\infty} x^2 e^{-ax^2}dx.$$

Let $u=x$ and $dv=xe^{-ax^2}dx=-(1/2a)d(e^{-ax^2})$. Since

$$\int u\ dv=uv-\int v\ du,$$

$$\int x^3 e^{-ax^2}dx=-\left(\frac{xe^{-ax^2}}{2a}\right)+\frac{1}{2a}\int e^{-ax^2}dx.$$

This expression is evaluated at the limits 0 and ∞. The next integral in the sequence can be integrated directly with the substitution $u=ax^2$:

$$I_1=\int_0^{\infty} xe^{-ax^2}dx=\frac{1}{2a}\int_0^{\infty}e^{-u}du=\frac{1}{2a}. \qquad \text{(K.3)}$$

The term xe^{-ax^2} vanishes at both limits. The second term is $I_0/2a$. Therefore,

$$I_2 = \frac{1}{2\times 2a}\sqrt{\frac{\pi}{a}}.$$

This process can be repeated to get other integrals in the sequence. The even members build on I_0; the odd members build on I_1. General expressions can be written. Note that $2n$ and $2n+1$ are used below to assure even and odd exponents:

$$\int_0^\infty x^{2n} e^{-ax^2} dx = \frac{1\times 3\times 5\times \cdots \times (2n-1)}{2^{n+1}a^n}\sqrt{\frac{\pi}{a}} \tag{K.4}$$

$$\int_0^\infty x^{2n+1} e^{-ax^2} dx' = \frac{n!}{2a^{n+1}} \quad (a>0). \tag{K.5}$$

The integrals in Sec. 3.10 and in Appendix I are of the form

$$\int_{-\infty}^\infty e^{-p^2/2mk_BT} dp.$$

This integral is $2I_0$ with $a=1/(2mk_BT)$. Therefore, the integral is $\sqrt{2\pi mk_BT}$.

Integrals like that in Eq. (3.50),

$$J = \int_0^\infty x^n e^{-ax} dx,$$

can be transformed to the forms above with the substitution $y=x^{1/2}$, $x=y^2$, $dx=2y\ dy$. Then

$$J = \int_0^\infty y^{2n} e^{-ay^2} 2y\ dy = 2\int_0^\infty y^{2n+1} e^{-ay^2} dy.$$

Therefore

$$\int_0^\infty x^n e^{-ax} dx = \frac{n!}{a^{n+1}} = \frac{\Gamma(n+1)}{a^{n+1}}. \tag{K.6}$$

The gamma function $\Gamma(n)=(n-1)!$ if n is an integer. Unlike $n!$, it is also defined for noninteger values. Although we have not shown it, Eq. (K.6) is correct for noninteger values of n as well, as long as $a>0$ and $n>-1$.

APPENDIX L

Spherical and Cylindrical Coordinates

It is possible to use coordinate systems other than the rectangular one (x,y,z): In spherical coordinates (Fig. L.1) the coordinates are radius r and angles θ and ϕ.

$$x = r \sin\theta \cos\phi,$$

$$y = r \sin\theta \sin\phi, \tag{L.1}$$

$$z = r \cos\theta.$$

In Cartesian coordinates a volume element is defined by surfaces on which x is constant (at x and $x+dx$), y is constant, and z is constant. The volume element is a cube with edges dx, dy, and dz. In spherical coordinates, the ''cube'' has faces defined by surfaces of constant r, constant θ, and constant ϕ (Fig. L.2). A volume element is then

$$dV = (dr)(r\,d\theta)(r \sin\theta\,d\phi) = r^2 \sin\theta\,d\theta\,d\phi\,dr. \tag{L.2}$$

To calculate the divergence of vector $\mathbf{J}$, it must be resolved into components $\mathbf{J}_r$, $\mathbf{J}_\theta$, and $\mathbf{J}_\phi$, as shown in Fig. L.2. These components are parallel to the vectors defined by small displacements in the r, θ, and ϕ directions. A detailed calculation shows that the divergence is

$$\nabla\cdot\mathbf{J} = \frac{1}{r^2}\frac{\partial}{\partial r}(r^2 J_r) + \frac{1}{r\sin\theta}\frac{\partial}{\partial\theta}[(\sin\theta)J_\theta] + \frac{1}{r\sin\theta}\frac{\partial}{\partial\phi}(J_\phi). \tag{L.3}$$

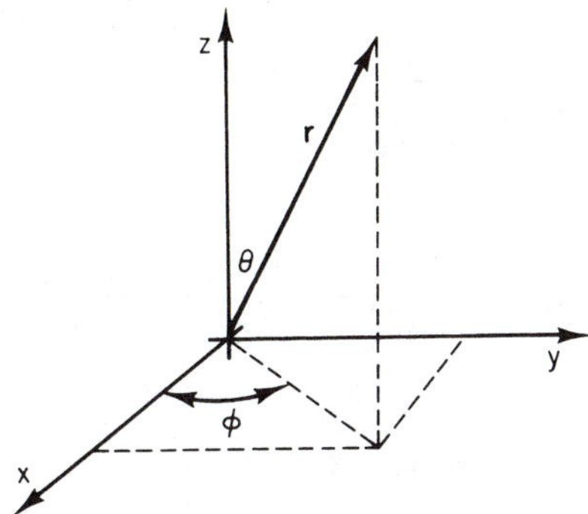

FIGURE L.1. Spherical coordinates.

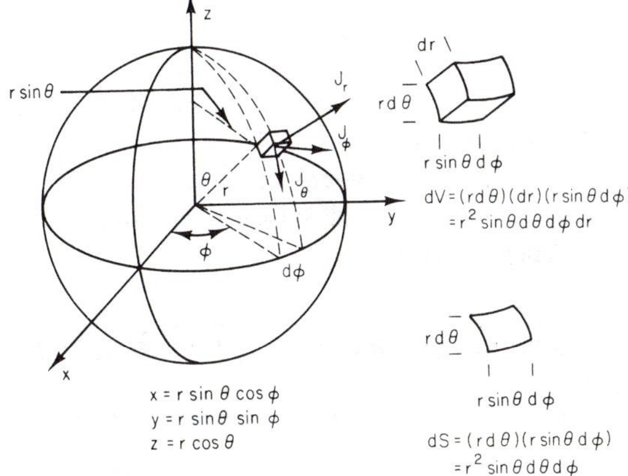

FIGURE L.2. The volume element and element of surface area in spherical coordinates.

The gradient, which appears in the three-dimensional diffusion equation (Fick's first law), can also be written in spherical coordinates. The components are

$$(\nabla C)_r = \frac{\partial C}{\partial r},$$

$$(\nabla C)_\theta = \frac{1}{r}\frac{\partial C}{\partial\theta}, \tag{L.4}$$

$$(\nabla C)_\phi = \frac{1}{r\sin\theta}\frac{\partial C}{\partial\phi}.$$

Figure L.2 also shows that the element of area on the surface of the sphere is $(r\,d\theta)$ $(r\sin\theta\,d\phi)$ $= r^2 \sin\theta\,d\theta\,d\phi$. The element of solid angle is therefore

$$d\Omega = \sin\theta\,d\theta\,d\phi.$$

This is easily integrated to show that the surface area of a sphere is $4\pi r^2$ or that the solid angle is 4π sr.

$$S=r^2\int_0^{\pi} d\theta \sin\theta \int_0^{2\pi} d\phi = 2\pi r^2 \int_0^{\pi} d\theta \sin\theta$$

$$=2\pi r^2[-\cos\theta]_0^{\pi}$$

$$=4\pi r^2.$$

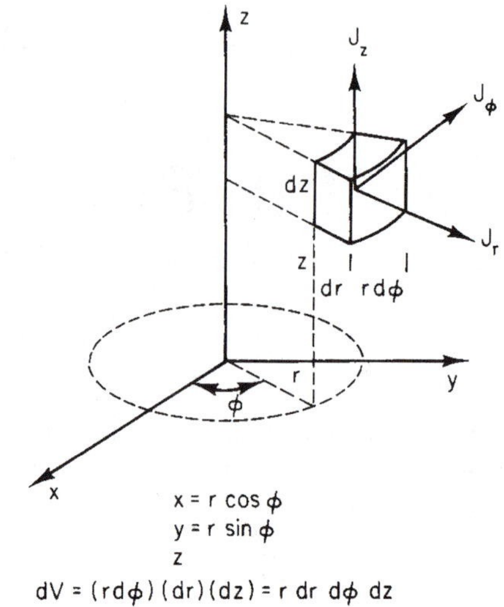

FIGURE L.3. A cylindrical coordinate system.

Similar results can be written down in cylindrical coordinates (r,ϕ,z), shown in Fig. L.3.

Table L.1 shows the divergence, gradient and curl in rectangular, cylindrical, and spherical coordinates, along with the Laplacian operator ∇^2.

Rectangular coordinates, x, y, z	Cylindrical coordinates, r, ϕ, z	Spherical coordinates, r, θ, ϕ
Gradient		
$(\nabla C)_x = \frac{\partial C}{\partial x}$	$(\nabla C)_r = \frac{\partial C}{\partial r}$	$(\nabla C)_r = \frac{\partial C}{\partial r}$
$(\nabla C)_y = \frac{\partial C}{\partial y}$	$(\nabla C)_\phi = \frac{1}{r}\frac{\partial C}{\partial \phi}$	$(\nabla C)_\theta = \frac{1}{r}\frac{\partial C}{\partial \theta}$
$(\nabla C)_z = \frac{\partial C}{\partial z}$	$(\nabla C)_z = \frac{\partial C}{\partial z}$	$(\nabla C)_\phi = \frac{1}{r\sin\theta}\frac{\partial C}{\partial \phi}$
Laplacian		
$\nabla^2 C = \frac{\partial^2 C}{\partial x^2} + \frac{\partial^2 C}{\partial y^2} + \frac{\partial^2 C}{\partial z^2}$	$\nabla^2 C = \frac{1}{r}\frac{\partial}{\partial r}\left(r\frac{\partial C}{\partial r}\right) + \frac{1}{r^2}\frac{\partial^2 C}{\partial \phi^2} + \frac{\partial^2 C}{\partial z^2}$	$\nabla^2 C = \frac{1}{r^2}\frac{\partial}{\partial r}\left(r^2\frac{\partial C}{\partial r}\right) + \frac{1}{r^2\sin\theta}\frac{\partial}{\partial\theta}\left(\sin\theta\frac{\partial C}{\partial\theta}\right) + \frac{1}{r^2\sin^2\theta}\frac{\partial^2 C}{\partial\phi^2}$
Divergence		
$\text{div}\,\mathbf{j} = \nabla\cdot\mathbf{j} = \frac{\partial j_x}{\partial x} + \frac{\partial j_y}{\partial y} + \frac{\partial j_z}{\partial z}$	$\text{div}\,\mathbf{j} = \nabla\cdot\mathbf{j} = \frac{1}{r}\frac{\partial(r j_r)}{\partial r} + \frac{1}{r}\frac{\partial j_\phi}{\partial\phi} + \frac{\partial j_z}{\partial z}$	$\text{div}\,\mathbf{j} = \nabla\cdot\mathbf{j} = \frac{1}{r^2}\frac{\partial(r^2 j_r)}{\partial r} + \frac{1}{r\sin\theta}\frac{\partial(\sin\theta\, j_\theta)}{\partial\theta} + \frac{1}{r\sin\theta}\frac{\partial j_\phi}{\partial\phi}$
Curl		
$(\nabla\times\mathbf{j})_x = \frac{\partial j_z}{\partial y} - \frac{\partial j_y}{\partial z}$	$(\nabla\times\mathbf{j})_r = \frac{1}{r}\frac{\partial j_z}{\partial\phi} - \frac{\partial j_\phi}{\partial z}$	$(\nabla\times\mathbf{j})_r = \frac{1}{r^2\sin\theta}\left[\frac{\partial(r\sin\theta\, j_\phi)}{\partial\theta} - \frac{\partial(r j_\theta)}{\partial\phi}\right]$
$(\nabla\times\mathbf{j})_y = \frac{\partial j_x}{\partial z} - \frac{\partial j_z}{\partial x}$	$(\nabla\times\mathbf{j})_\phi = \frac{\partial j_r}{\partial z} - \frac{\partial j_z}{\partial r}$	$(\nabla\times\mathbf{j})_\theta = \frac{1}{r\sin\theta}\left[\frac{\partial j_r}{\partial\phi} - \frac{\partial(r\sin\theta\, j_\phi)}{\partial r}\right]$
$(\nabla\times\mathbf{j})_z = \frac{\partial j_y}{\partial x} - \frac{\partial j_x}{\partial y}$	$(\nabla\times\mathbf{j})_z = \frac{1}{r}\frac{\partial(r j_\phi)}{\partial r} - \frac{1}{r}\frac{\partial j_r}{\partial\phi}$	$(\nabla\times\mathbf{j})_\phi = \frac{1}{r}\left[\frac{\partial(r j_\theta)}{\partial r} - \frac{\partial j_r}{\partial\theta}\right]$

APPENDIX M

Joint Probability Distributions

In both physics and medicine, the question often arises of what is the probability that x has a certain value x_i while y has the value y_j. This is called a *joint probability*. Joint probability can be extended to several variables. In Sec. 3.10 it was extended to six: we had the probability that a molecule was in the interval between x and $x+dx$ *and* between y and $y+dy$, *and* between z and $z+dx$, *and* had momentum components in the intervals (p_x, dp_x), (p_y, dp_y), and (p_z, dp_z). This appendix first derives some properties of joint probabilities for discrete variables, and then does the same for continuous variables.

M.1. DISCRETE VARIABLES

Consider two variables. For simplicity assume that each can assume only two values. The first might be the patient's health, with values *healthy* and *sick*; the other might by the results of some laboratory test, with results *normal* and *abnormal*. Figure M.1 shows the values of the two variables for a sample of 100 patients. The joint probability that a patient is healthy *and* has a normal test result is $P(x=0, y=0)=0.6$; the probability that a patient is sick and has an abnormal test is $P(1,1)=0.15$. The probability of a false positive test is $P(0,1)=0.20$; the probability of a false negative is $P(1,0)=0.05$.

		Health (x) healthy ($x=0$)	Health (x) sick ($x=1$)
Test Result (y)	normal ($y=0$)	60	5
	abnormal ($y=1$)	20	15

FIGURE M.1. The results of measurements on 100 patients showing whether they are healthy or sick, and whether a laboratory test was normal or abnormal.

The probability that a patient is healthy regardless of the test result is obtained by a summing over all possible test outcomes:

$$P(x=0)=P(0,0)+P(0,1)=0.6+0.2=0.8.$$

In a more general case, we can call the joint probability $P(x,y)$, the probability that x has a certain value independent of y $P_x(x)$, and so forth. Then

$$P_x(x)=\sum_y P(x,y)$$

$$P_y(y)=\sum_x P(x,y). \tag{M.1}$$

Since any measurement must give some value for x and y, we can write

$$1=\sum_x P_x(x)=\sum_x \sum_y P(x,y),$$

$$1=\sum_y P_y(y)=\sum_y \sum_x P(x,y). \tag{M.2}$$

M.2. CONTINUOUS VARIABLES

When a variable can take on a continuous range of values, it is quite unlikely that the variable will have *precisely* the value x. Instead, there is a probability that it is in the interval (x, dx), meaning that it is between x and $x+dx$. For small values of dx, the probability that the value is in the interval is proportional to the width of the interval. We will call it $p_x(x)dx$. The extension to joint probability in two dimensions is $p(x,y)dx\,dy$. This is the probability that x is in the interval (x, dx) and y is in the interval (y, dy). Figure M.2 shows each outcome of a joint measurement as a dot in the xy plane. The probability that x is in (x, dx) regardless of the value of y is

$$p_x(x)\,dx=\left(\int p(x,y)\,dy\right)dx. \tag{M.3}$$

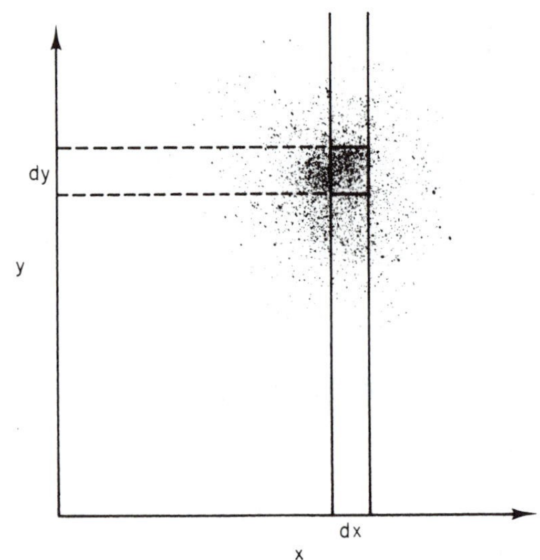

FIGURE M.2. The results of measuring two continuous variables simultaneously. Each experimental result is shown as a point.

It is proportional to the total number of dots in the vertical strip in Fig. M.2. Normalization requires that

$$1=\int p_x(x)\ dx=\int dx\int dy\ p(x,y). \qquad \text{(M.4)}$$

The first strip could be taken horizontally:

$$1=\int p_y(y)\ dy=\int dy\int dx\ p(x,y).$$

Figure M.3 shows a perspective drawing of $p(x,y)$. The shaded column is $p(x,y)\ dx\ dy$. The slice is $p_x(x)\ dx$. The entire volume under the surface is equal to one.

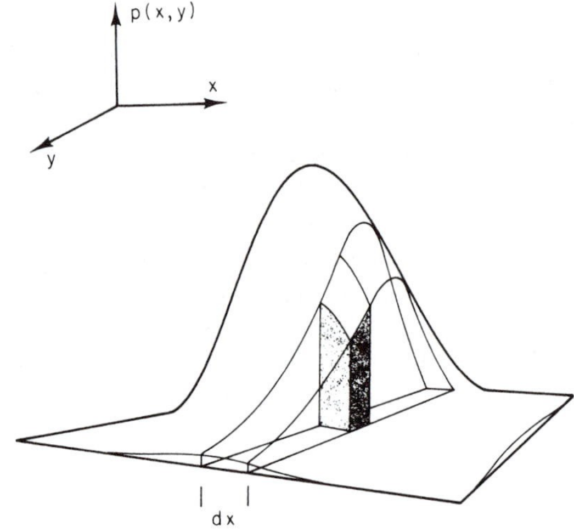

FIGURE M.3. Perspective drawing of $p(x,y)$.

APPENDIX N

Partial Derivatives

When a function depends on several variables, we may want to know how the value of the function changes when one or more of the variables is changed. For example, the volume of a cylinder is

$$V=\pi r^2 h.$$

How does V change when r is changed while the height of the cylinder is kept fixed?

$$V(r+\Delta r)=\pi(r+\Delta r)^2 h=\pi(r^2+2r\Delta r+\Delta r^2)h.$$

Subtracting the original volume, we have

$$\Delta V=\pi(2r\Delta r+\Delta r^2)h.$$

In the limit of small Δr, this is

$$dV=2\pi hr\,dr.$$

This is the same answer we would have gotten if h had been regarded as a constant. The *partial derivative* of V with respect to r is defined to be

$$\left(\frac{\partial V}{\partial r}\right)_h=\lim_{\Delta r\to 0}\left(\frac{V(r+\Delta r,h)-V(r,h)}{\Delta r}\right)=2\pi rh.$$

The subscript h in the partial derivative symbol means that h is held fixed during the differentiation. Sometimes it is omitted; when it is not there, it is understood that all variables except the one following the ∂ are held fixed.

If the cylinder radius is held fixed while the height is varied, we can write

$$V=V(r,h+\Delta h)-V(r,h)=\pi r^2\Delta h.$$

The partial derivative is

$$\left(\frac{\partial V}{\partial h}\right)_r=\lim_{\Delta h\to 0}\left(\frac{V(r,h+\Delta h)-V(r,h)}{\Delta h}\right)=\pi r^2.$$

Suppose now that we allow small changes in *both* r and h. The difference in volume is

$$\Delta V=V(r+\Delta r,h+\Delta h)-V(r,h).$$

We can add and subtract the term $V(r,h+\Delta h)$:

$$\begin{aligned}\Delta V&=V(r+\Delta r,h+\Delta h)-V(r,h+\Delta h)\\&\quad+V(r,h+\Delta h)-V(r,h)\\&=\frac{V(r+\Delta r,h+\Delta h)-V(r,h+\Delta h)}{\Delta r}\Delta r\\&\quad+\frac{V(r,h+\Delta h)-V(r,h)}{\Delta h}\Delta h.\end{aligned}$$

In the limit as Δr and $\Delta h\to 0$, the first term is

$$\left(\frac{\partial V}{\partial r}\right)_h\Delta r$$

evaluated at $h+\Delta h$. If the derivatives are continuous at (r,h), this is negligibly different from the derivative evaluated at (r,h). Therefore, we can write

$$dV=\left(\frac{\partial V}{\partial r}\right)_h dr+\left(\frac{\partial V}{\partial h}\right)_r dh.$$

This result is true for several variables. For a function $w(x,y,z)$

$$dw=\frac{\partial w}{\partial x}dx+\frac{\partial w}{\partial y}dy+\frac{\partial w}{\partial z}dz. \qquad \text{(N.1)}$$

The derivatives are evaluated as though the variables being held fixed were ordinary constants. If $w=3x^2yz^4$,

$$\frac{\partial w}{\partial x}=6xyz^4,$$

$$\frac{\partial w}{\partial y}=3x^2z^4,$$

$$\frac{\partial w}{\partial z}=12x^2yz^3.$$

It is also possible to take higher derivatives, such as $\partial^2 w/\partial x^2$ or $\partial^2 w/\partial x\partial y$. One important result is that the order of differentiation is unimportant, if the function, its first derivatives, and the derivatives in question are continuous at the point where they are evaluated. Without filling in all the details of a rigorous proof, we will simply note that

$$f=\frac{\partial w}{\partial x}=\lim_{\Delta x\to 0}\left(\frac{w(x+\Delta x,y)-w(x,y)}{\Delta x}\right)$$

$$g=\frac{\partial w}{\partial y}=\lim_{\Delta y\to 0}\left(\frac{w(x,y+\Delta y)-w(x,y)}{\Delta y}\right).$$

The mixed partials are

$$\frac{\partial f}{\partial y}=\lim_{\Delta y\to 0}\left(\frac{f(x,y+\Delta y)-f(x,y)}{\Delta y}\right)=\lim_{\substack{\Delta x\to 0\\ \Delta y\to 0}}\left(\frac{w(x+\Delta x,y+\Delta y)-w(x,y+\Delta y)-w(x+\Delta x,y)+w(x,y)}{\Delta x\ \Delta y}\right)$$

$$\frac{\partial g}{\partial x}=\lim_{\substack{\Delta x\to 0\\ \Delta y\to 0}}\left(\frac{w(x+\Delta x,y+\Delta y)-w(x+\Delta x,y)-w(x,y+\Delta y)+w(x,y)}{\Delta x\Delta y}\right).$$

The right side of each of these equations is the same, except for the order of the terms. Thus

$$\frac{\partial}{\partial x}\frac{\partial w}{\partial y}=\frac{\partial}{\partial y}\frac{\partial w}{\partial x}. \qquad \text{(N.2)}$$

PROBLEMS

N.1. If $w=12x^3y+z$, find the three partial derivatives $\partial w/\partial x$, $\partial w/\partial y$, and $\partial w/\partial z$.

N.2. If $V=xyz$ and $x=5$, $y=6$, $z=2$, find dV when $dx=0.01$, $dy=0.02$, and $dz=0.03$. Make a geometrical interpretation of each term.

APPENDIX O

Photon and Charged-Particle Tables

Table O.1 provides photon mass attenuation and mass energy absorption coefficients for a few elements and substances of biological interest. These have been taken with permission from J. M. Hubbell and S. M. Seltzer (1966). *Tables of X-Ray Mass Attenuation Coefficients and Mass Energy Absorption Coefficients 1 keV to 20 MeV for Elements Z=1 to 92 and 48, Additional Substances of Dosimetric Interest.* National Institute of Standards and Technology Report No. NISTIR 5632—Web Version. Available at http://physics.nist.gov/PhysRefData/XrayMassCoef/cover.html.

Another source of photon attenuation coefficients is J. M. Boone and A. E. Chavez (1996). Comparison of x-ray cross sections for diagnostic and therapeutic medical physics. *Med. Phys.* **23** (12): 1997–2005.

Stopping powers and ranges for electrons and positrons in water (Table O.2) are reproduced by permission from ICRU Report No. 37 (1984). *Stopping Powers for Electrons and Positrons.* Bethesda, MD, International Commission on Radiation Units and Measurements.

Table O-1. *Photon mass attenuation and energy absorption coefficients. To convert cm^2/g to $m^2 kg^{-1}$, divide by 10.*

	HYDROGEN		CARBON, GRAPHITE		NITROGEN		OXYGEN	
	Z=1		Z=6		Z=7		Z=8	
Energy (MeV)	μ/ρ (cm2/g)	μ_{en}/ρ (cm2/g)	μ/ρ (cm2/g)	μ_{en}/ρ (cm2/g)	μ/ρ (cm2/g)	μ_{en}/ρ (cm2/g)	μ/ρ (cm2/g)	μ_{en}/ρ (cm2/g)
1.00E−03	7.22E+00	6.82E+00	2.21E+03	2.21E+03	3.31E+03	3.31E+03	4.59E+03	4.58E+03
1.50E−03	2.15E+00	1.75E+00	7.00E+02	6.99E+02	1.08E+03	1.08E+03	1.55E+03	1.55E+03
2.00E−03	1.06E+00	6.64E−01	3.03E+02	3.02E+02	4.77E+02	4.76E+02	6.95E+02	6.93E+02
3.00E−03	5.61E−01	1.69E−01	9.03E+01	8.96E+01	1.46E+02	1.45E+02	2.17E+02	2.16E+02
4.00E−03	4.55E−01	6.55E−02	3.78E+01	3.72E+01	6.17E+01	6.09E+01	9.32E+01	9.22E+01
5.00E−03	4.19E−01	3.28E−02	1.91E+01	1.87E+01	3.14E+01	3.09E+01	4.79E+01	4.72E+01
6.00E−03	4.04E−01	2.00E−02	1.10E+01	1.05E+01	1.81E+01	1.76E+01	2.77E+01	2.71E+01
8.00E−03	3.91E−01	1.16E−02	4.58E+00	4.24E+00	7.56E+00	7.17E+00	1.16E+01	1.12E+01
1.00E−02	3.85E−01	9.85E−03	2.37E+00	2.08E+00	3.88E+00	3.55E+00	5.95E+00	5.57E+00
1.50E−02	3.76E−01	1.10E−02	8.07E−01	5.63E−01	1.24E+00	9.72E−01	1.84E+00	1.55E+00
2.00E−02	3.70E−01	1.36E−02	4.42E−01	2.24E−01	6.18E−01	3.87E−01	8.65E−01	6.18E−01
3.00E−02	3.57E−01	1.86E−02	2.56E−01	6.61E−02	3.07E−01	1.10E−01	3.78E−01	1.73E−01
4.00E−02	3.46E−01	2.32E−02	2.08E−01	3.34E−02	2.29E−01	5.05E−02	2.59E−01	7.53E−02
5.00E−02	3.36E−01	2.71E−02	1.87E−01	2.40E−02	1.98E−01	3.22E−02	2.13E−01	4.41E−02
6.00E−02	3.26E−01	3.05E−02	1.75E−01	2.10E−02	1.82E−01	2.55E−02	1.91E−01	3.21E−02
8.00E−02	3.09E−01	3.62E−02	1.61E−01	2.04E−02	1.64E−01	2.21E−02	1.68E−01	2.47E−02
1.00E−01	2.94E−01	4.06E−02	1.51E−01	2.15E−02	1.53E−01	2.23E−02	1.55E−01	2.36E−02
1.50E−01	2.65E−01	4.81E−02	1.35E−01	2.45E−02	1.35E−01	2.47E−02	1.36E−01	2.51E−02
2.00E−01	2.43E−01	5.25E−02	1.23E−01	2.66E−02	1.23E−01	2.67E−02	1.24E−01	2.68E−02
3.00E−01	2.11E−01	5.70E−02	1.07E−01	2.87E−02	1.07E−01	2.87E−02	1.07E−01	2.88E−02
4.00E−01	1.89E−01	5.86E−02	9.55E−02	2.95E−02	9.56E−02	2.95E−02	9.57E−02	2.95E−02
5.00E−01	1.73E−01	5.90E−02	8.72E−02	2.97E−02	8.72E−02	2.97E−02	8.73E−02	2.97E−02
6.00E−01	1.60E−01	5.88E−02	8.06E−02	2.96E−02	8.06E−02	2.96E−02	8.07E−02	2.96E−02
8.00E−01	1.41E−01	5.74E−02	7.08E−02	2.89E−02	7.08E−02	2.89E−02	7.09E−02	2.89E−02

Table O-1. (*Continued.*)

Energy (MeV)	HYDROGEN (Cont.) Z=1 μ/ρ (cm2/g)	μ_{en}/ρ (cm2/g)	CARBON, GRAPHITE (Cont.) Z=6 μ/ρ (cm2/g)	μ_{en}/ρ (cm2/g)	NITROGEN (Cont.) Z=7 μ/ρ (cm2/g)	μ_{en}/ρ (cm2/g)	OXYGEN (Cont.) Z=8 μ/ρ (cm2/g)	μ_{en}/ρ (cm2/g)
1.00E+00	1.26E−01	5.56E−02	6.36E−02	2.79E−02	6.36E−02	2.79E−02	6.37E−02	2.79E−02
1.25E+00	1.13E−01	5.31E−02	5.69E−02	2.67E−02	5.69E−02	2.67E−02	5.70E−02	2.67E−02
1.50E+00	1.03E−01	5.08E−02	5.18E−02	2.55E−02	5.18E−02	2.55E−02	5.19E−02	2.55E−02
2.00E+00	8.77E−02	4.65E−02	4.44E−02	2.35E−02	4.45E−02	2.35E−02	4.46E−02	2.35E−02
3.00E+00	6.92E−02	3.99E−02	3.56E−02	2.05E−02	3.58E−02	2.06E−02	3.60E−02	2.07E−02
4.00E+00	5.81E−02	3.52E−02	3.05E−02	1.85E−02	3.07E−02	1.87E−02	3.10E−02	1.88E−02
5.00E+00	5.05E−02	3.17E−02	2.71E−02	1.71E−02	2.74E−02	1.73E−02	2.78E−02	1.76E−02
6.00E+00	4.50E−02	2.91E−02	2.47E−02	1.61E−02	2.51E−02	1.64E−02	2.55E−02	1.67E−02
8.00E+00	3.75E−02	2.52E−02	2.15E−02	1.47E−02	2.21E−02	1.51E−02	2.26E−02	1.55E−02
1.00E+01	3.25E−02	2.25E−02	1.96E−02	1.38E−02	2.02E−02	1.43E−02	2.09E−02	1.48E−02
1.50E+01	2.54E−02	1.84E−02	1.70E−02	1.26E−02	1.78E−02	1.33E−02	1.87E−02	1.40E−02
2.00E+01	2.15E−02	1.61E−02	1.58E−02	1.20E−02	1.67E−02	1.29E−02	1.77E−02	1.36E−02

	ALUMINUM Z=13 Energy (MeV)	μ/ρ (cm2/g)	μ_{en}/ρ (cm2/g)
	1.00E−03	1.19E+03	1.18E+03
	1.50E−03	4.02E+02	4.00E+02
	1.56E−03	3.62E+02	3.60E+02
K	1.56E−03	3.96E+03	3.83E+03
	2.00E−03	2.26E+03	2.20E+03
	3.00E−03	7.88E+02	7.73E+02
	4.00E−03	3.61E+02	3.55E+02
	5.00E−03	1.93E+02	1.90E+02
	6.00E−03	1.15E+02	1.13E+02
	8.00E−03	5.03E+01	4.92E+01
	1.00E−02	2.62E+01	2.54E+01
	1.50E−02	7.96E+00	7.49E+00
	2.00E−02	3.44E+00	3.09E+00
	3.00E−02	1.13E+00	8.78E−01
	4.00E−02	5.69E−01	3.60E−01
	5.00E−02	3.68E−01	1.84E−01
	6.00E−02	2.78E−01	1.10E−01
	8.00E−02	2.02E−01	5.51E−02
	1.00E−01	1.70E−01	3.79E−02
	1.50E−01	1.38E−01	2.83E−02
	2.00E−01	1.22E−01	2.75E−02

	CALCIUM Z=20 Energy (MeV)	μ/ρ (cm2/g)	μ_{en}/ρ (cm2/g)
	1.00E−03	4.87E+03	4.86E+03
	1.50E−03	1.71E+03	1.71E+03
	2.00E−03	8.00E+02	7.97E+02
	3.00E−03	2.68E+02	2.65E+02
	4.00E−03	1.22E+02	1.20E+02
	4.04E−03	1.19E+02	1.17E+02
K	4.04E−03	1.02E+03	8.89E+02
	5.00E−03	6.03E+02	5.37E+02
	6.00E−03	3.73E+02	3.39E+02
	8.00E−03	1.73E+02	1.60E+02
	1.00E−02	9.34E+01	8.74E+01
	1.50E−02	2.98E+01	2.80E+01
	2.00E−02	1.31E+01	1.22E+01
	3.00E−02	4.08E+00	3.67E+00
	4.00E−02	1.83E+00	1.54E+00
	5.00E−02	1.02E+00	7.82E−01
	6.00E−02	6.58E−01	4.52E−01
	8.00E−02	3.66E−01	1.96E−01
	1.00E−01	2.57E−01	1.09E−01
	1.50E−01	1.67E−01	4.88E−02
	2.00E−01	1.38E−01	3.64E−02

	IODINE Z=53 Energy (MeV)	μ/ρ (cm2/g)	μ_{en}/ρ (cm2/g)
	1.00E−03	9.10E+03	9.08E+03
	1.04E−03	8.47E+03	8.45E+03
	1.07E−03	7.86E+03	7.85E+03
M1	1.07E−03	8.20E+03	8.18E+03
	1.50E−03	3.92E+03	3.91E+03
	2.00E−03	2.00E+02	1.99E+03
	3.00E−03	7.42E+02	7.35E+02
	4.00E−03	3.61E+02	3.55E+02
	4.56E−03	2.59E+02	2.54E+02
L3	4.56E−03	7.55E+02	7.12E+02
	4.70E−03	7.12E+02	6.72E+02
	4.85E−03	6.64E+02	6.27E+02
L2	4.85E−03	8.94E+02	8.38E+02
	5.00E−03	8.43E+02	7.90E+02
	5.19E−03	7.67E+02	7.20E+02
L1	5.19E−03	8.84E+02	8.28E+02
	6.00E−03	6.17E+02	5.82E+02
	8.00E−03	2.92E+02	2.78E+02
	1.00E−02	1.63E+02	1.55E+02
	1.50E−02	5.51E+01	5.21E+01
	2.00E−02	2.54E+01	2.36E+01
	3.00E−02	8.56E+00	7.62E+00
	3.32E−02	6.55E+00	5.74E+00
K	3.32E−02	3.58E+01	1.19E+01
	4.00E−02	2.21E+01	9.62E+00
	5.00E−02	1.23E+01	6.57E+00
	6.00E−02	7.58E+00	4.52E+00
	8.00E−02	3.51E+00	2.33E+00
	1.00E−01	1.94E+00	1.34E+00
	1.50E−01	6.98E−01	4.74E−01
	2.00E−01	3.66E−01	2.30E−01

Table O-1 (*Continued.*)

ALUMINUM (Cont.) Z=13			CALCIUM (Cont.) Z=20			IODINE (Cont.) Z=53		
Energy (MeV)	μ/ρ (cm2/g)	μ_{en}/ρ (cm2/g)	Energy (MeV)	μ/ρ (cm2/g)	μ_{en}/ρ (cm2/g)	Energy (MeV)	μ/ρ (cm2/g)	μ_{en}/ρ (cm2/g)
3.00E−01	1.04E−01	2.82E−02	3.00E−01	1.12E−01	3.15E−02	3.00E−01	1.77E−01	9.26E−02
4.00E−01	9.28E−02	2.86E−02	4.00E−01	9.78E−02	3.06E−02	4.00E−01	1.22E−01	5.65E−02
5.00E−01	8.45E−02	2.87E−02	5.00E−01	8.85E−02	3.02E−02	5.00E−01	9.70E−02	4.27E−02
6.00E−01	7.80E−02	2.85E−02	6.00E−01	8.15E−02	2.98E−02	6.00E−01	8.31E−02	3.60E−02
8.00E−01	6.84E−02	2.78E−02	8.00E−01	7.12E−02	2.88E−02	8.00E−01	6.75E−02	2.96E−02
1.00E+00	6.15E−02	2.69E−02	1.00E+00	6.39E−02	2.78E−02	1.00E+00	5.84E−02	2.65E−02
1.25E+00	5.50E−02	2.57E−02	1.25E+00	5.71E−02	2.65E−02	1.25E+00	5.11E−02	2.40E−02
1.50E+00	5.01E−02	2.45E−02	1.50E+00	5.21E−02	2.53E−02	1.50E+00	4.65E−02	2.24E−02
2.00E+00	4.32E−02	2.27E−02	2.00E+00	4.52E−02	2.35E−02	2.00E+00	4.12E−02	2.09E−02
3.00E+00	3.54E−02	2.02E−02	3.00E+00	3.78E−02	2.15E−02	3.00E+00	3.72E−02	2.06E−02
4.00E+00	3.11E−02	1.88E−02	4.00E+00	3.40E−02	2.05E−02	4.00E+00	3.61E−02	2.14E−02
5.00E+00	2.84E−02	1.80E−02	5.00E+00	3.17E−02	2.01E−02	5.00E+00	3.61E−02	2.25E−02
6.00E+00	2.66E−02	1.74E−02	6.00E+00	3.04E−02	1.99E−02	6.00E+00	3.66E−02	2.36E−02
8.00E+00	2.44E−02	1.68E−02	8.00E+00	2.89E−02	2.00E−02	8.00E+00	3.82E−02	2.55E−02
1.00E+01	2.32E−02	1.65E−02	1.00E+01	2.84E−02	2.02E−02	1.00E+01	4.00E−02	2.71E−02
1.50E+01	2.20E−02	1.63E−02	1.50E+01	2.84E−02	2.09E−02	1.50E+01	4.46E−02	2.98E−02
2.00E+01	2.17E−02	1.63E−02	2.00E+01	2.90E−02	2.14E−02	2.00E+01	4.82E−02	3.10E−02

BARIUM Z=56			LEAD Z=82		
Energy (MeV)	μ/ρ (cm2/g)	μ_{en}/ρ (cm2/g)	Energy (MeV)	μ/ρ (cm2/g)	μ_{en}/ρ (cm2/g)
1.00E−03	8.54E+03	8.52E+03	1.00E−03	5.21E+03	5.20E+03
1.03E−03	7.99E+03	7.97E+03	1.50E−03	2.36E+03	2.34E+03
1.06E−03	7.47E+03	7.45E+03	2.00E−03	1.29E+03	1.27E+03
M3 1.06E−03	8.55E+03	8.52E+03	2.48E−03	8.01E+02	7.90E+02
1.10E−03	7.96E+03	7.94E+03	M5 2.48E−03	1.40E+03	1.37E+03
1.14E−03	7.40E+03	7.38E+03	2.53E−03	1.73E+03	1.68E+03
M2 1.14E−03	7.84E+03	7.82E+03	2.59E−03	1.94E+03	1.90E+03
1.21E−03	6.86E+03	6.84E+03	M4 2.59E−03	2.46E+03	2.39E+03
1.29E−03	5.99E+03	5.97E+03	3.00E−03	1.97E+03	1.91E+03
M1 1.29E−03	6.26E+03	6.24E+03	3.07E−03	1.86E+03	1.81E+03
1.50E−03	4.50E+03	4.49E+03	M3 3.07E−03	2.15E+03	2.09E+03
2.00E−03	2.32E+03	2.31E+03	3.30E−03	1.80E+03	1.75E+03
3.00E−03	8.70E+02	8.62E+02	3.55E−03	1.50E+03	1.46E+03
4.00E−03	4.25E+02	4.18E+02	M2 3.55E−03	1.59E+02	1.55E+03
5.00E−03	2.41E+02	2.36E+02	3.70E−03	1.44E+03	1.41E+03
5.25E−03	2.14E+02	2.08E+02	3.85E−03	1.31E+03	1.28E+03
L3 5.25E−03	6.10E+02	5.69E+02	M1 3.85E−03	1.37E+03	1.34E+03
5.43E−03	5.63E+02	5.26E+02	4.00E−03	1.25E+03	1.22E+03
5.62E−03	5.17E+02	4.84E+02	5.00E−03	7.30E+02	7.12E+02
L2 5.62E−03	7.02E+02	6.49E+02			
5.80E−03	6.51E+02	6.03E+02			
5.99E−03	6.01E+02	5.58E+02			
L1 5.99E−03	6.93E+02	6.41E+02			
6.00E−03	6.90E+02	6.38E+02	6.00E−03	4.67E+02	4.55E+02
8.00E−03	3.33E+02	3.13E+02	8.00E−03	2.29E+02	2.21E+02
1.00E−02	1.86E+02	1.75E+02	1.00E−02	1.31E+02	1.25E+02

Table O-1 (*Continued.*)

	BARIUM (Cont.) Z=56			LEAD (Cont.) Z=82	
Energy (MeV)	μ/ρ (cm2/g)	μ_{en}/ρ (cm2/g)	Energy (MeV)	μ/ρ (cm2/g)	μ_{en}/ρ (cm2/g)
1.50E−02	6.35E+01	5.97E+01	1.30E−02	6.70E+01	6.27E+01
2.00E−02	2.94E+01	2.73E+01	L3 1.30E−02	1.62E+02	1.29E+02
3.00E−02	9.90E+00	8.88E+00	1.50E−02	1.12E+02	9.10E+01
3.74E−02	5.50E+00	4.77E+00	1.52E−02	1.08E+02	8.81E+01
K 3.74E−02	2.92E+01	9.35E+00	L2 1.52E−02	1.49E+02	1.13E+02
			1.55E−02	1.42E+02	1.08E+02
			1.59E−02	1.34E+02	1.03E+02
			L1 1.59E−02	1.55E+02	1.18E+02
			2.00E−02	8.64E+01	6.90E+01
			3.00E−02	3.03E+01	2.54E+01
4.00E−02	2.46E+01	8.84E+00	4.00E−02	1.44E+01	1.21E+01
5.00E−02	1.38E+01	6.53E+00	5.00E−02	8.04E+00	6.74E+00
6.00E−02	8.51E+00	4.66E+00	6.00E−02	5.02E+00	4.15E+00
8.00E−02	3.96E+00	2.50E+00	8.00E−02	2.42E+00	1.92E+00
			8.80E−02	1.91E+00	1.48E+00
			K 8.80E−02	7.68E+00	2.16E+00
1.00E−01	2.20E+00	1.47E+00	1.00E−01	5.55E+00	1.98E+00
1.50E−01	7.83E−01	5.31E−01	1.50E−01	2.01E+00	1.06E+00
2.00E−01	4.05E−01	2.58E−01	2.00E−01	9.99E−01	5.87E−01
3.00E−01	1.89E−01	1.03E−01	3.00E−01	4.03E−01	2.46E−01
4.00E−01	1.27E−01	6.13E−02	4.00E−01	2.32E−01	1.37E−01
5.00E−01	9.92E−02	4.52E−02	5.00E−01	1.61E−01	9.13E−02
6.00E−01	8.41E−02	3.74E−02	6.00E−01	1.25E−01	6.82E−02
8.00E−01	6.74E−02	3.01E−02	8.00E−01	8.87E−02	4.64E−02
1.00E+00	5.80E−02	2.66E−02	1.00E+00	7.10E−02	3.65E−02
1.25E+00	5.06E−02	2.39E−02	1.25E+00	5.88E−02	2.99E−02
1.50E+00	4.59E−02	2.22E−02	1.50E+00	5.22E−02	2.64E−02
2.00E+00	4.08E−02	2.07E−02	2.00E+00	4.61E−02	2.36E−02
3.00E+00	3.69E−02	2.04E−02	3.00E+00	4.23E−02	2.32E−02
4.00E+00	3.60E−02	2.14E−02	4.00E+00	4.20E−02	2.45E−02
5.00E+00	3.61E−02	2.25E−02	5.00E+00	4.27E−02	2.60E−02
6.00E+00	3.67E−02	2.36E−02	6.00E+00	4.39E−02	2.74E−02
8.00E+00	3.84E−02	2.57E−02	8.00E+00	4.68E−02	2.99E−02
1.00E+01	4.04E−02	2.73E−02	1.00E+01	4.97E−02	3.18E−02
1.50E+01	4.52E−02	3.00E−02	1.50E+01	5.66E−02	3.48E−02
2.00E+01	4.90E−02	3.13E−02	2.00E+01	6.21E−02	3.60E−02

ADIPOSE TISSUE (ICRU-44)			AIR (DRY, SEA LEVEL)			BONE CORTICAL (ICRU-44)		
Energy (MeV)	μ/ρ (cm2/g)	μ_{en}/ρ (cm2/g)	Energy (MeV)	μ/ρ (cm2/g)	μ_{en}/ρ (cm2/g)	Energy (MeV)	μ/ρ (cm2/g)	μ_{en}/ρ (cm2/g)
1.00E−03	2.63E+03	2.62E+03	1.00E−03	3.61E+03	3.60E+03	1.00E−03	3.78E+03	3.77E+03
1.04E−03	2.39E+03	2.39E+03	1.50E−03	1.19E+03	1.19E+03	1.04E−03	3.45E+03	3.44E+03
1.07E−03	2.18E+03	2.17E+03	2.00E−03	5.28E+02	5.26E+02	1.07E−03	3.15E+03	3.14E+03
K(Na) 1.07E−03	2.18E+03	2.18E+03	3.00E−03	1.63E+02	1.61E+02	K(Na) 1.07E−03	3.16E+03	3.15E+03
1.50E−03	8.62E+02	8.60E+02	3.20E−03	1.34E+02	1.33E+02	1.18E−03	2.43E+03	2.43E+03
2.00E−03	3.80E+02	3.79E+02	K(Ar) 3.20E−03	1.49E+02	1.46E+02	1.31E−03	1.87E+03	1.87E+03

Table O-1 (*Continued.*)

	ADIPOSE TISSUE (ICRU-44) (Cont.)			AIR (DRY, SEA LEVEL) (Cont.)				BONE CORTICAL (ICRU-44) (Cont.)		
	Energy (MeV)	μ/ρ (cm2/g)	μ_{en}/ρ (cm2/g)	Energy (MeV)	μ/ρ (cm2/g)	μ_{en}/ρ (cm2/g)		Energy (MeV)	μ/ρ (cm2/g)	μ_{en}/ρ (cm2/g)
	2.47E−03	2.05E+02	2.04E+02	4.00E−03	7.79E+01	7.64E+01	K(Mg)	1.31E−03	1.88E+03	1.88E+03
K(S)	2.47E−03	2.07E+02	2.06E+02	5.00E−03	4.03E+01	3.93E+01		1.50E−03	1.30E+03	1.29E+03
	2.64E−03	1.71E+02	1.70E+02	6.00E−03	2.34E+01	2.27E+01		2.00E−03	5.87E+02	5.85E+02
	2.82E−03	1.41E+02	1.40E+02	8.00E−03	9.92E+00	9.45E+00		2.15E−03	4.82E+02	4.80E+02
K(Cl)	2.82E−03	1.42E+02	1.41E+02				K(P)	2.15E−03	7.11E+02	6.96E+02
	3.00E−03	1.19E+02	1.18E+02					2.30E−03	5.92E+02	5.79E+02
	4.00E−03	5.05E+01	4.98E+01					2.47E−03	4.91E+02	4.81E+02
	5.00E−03	2.59E+01	2.53E+01				K(S)	2.47E−03	4.96E+02	4.86E+02
	6.00E−03	1.49E+01	1.45E+01					3.00E−03	2.96E+02	2.90E+02
	8.00E−03	6.30E+00	5.92E+00					4.00E−03	1.33E+02	1.30E+02
								4.04E−03	1.30E+02	1.27E+02
							K(Ca)	4.04E−03	3.33E+02	3.01E+02
								5.00E−03	1.92E+02	1.76E+02
								6.00E−03	1.17E+02	1.09E+02
								8.00E−03	5.32E+01	4.99E+01
	1.00E−02	3.27E+00	2.94E+00	1.00E−02	5.12E+00	4.74E+00		1.00E−02	2.85E+01	2.68E+01
	1.50E−02	1.08E+00	8.10E−01	1.50E−02	1.61E+00	1.33E+00		1.50E−02	9.03E+00	8.39E+00
	2.00E−02	5.68E−01	3.25E−01	2.00E−02	7.78E−01	5.39E−01		2.00E−02	4.00E+00	3.60E+00
	3.00E−02	3.06E−01	9.50E−02	3.00E−02	3.54E−01	1.54E−01		3.00E−02	1.33E+00	1.07E+00
	4.00E−02	2.40E−01	4.58E−02	4.00E−02	2.49E−01	6.83E−02		4.00E−02	6.66E−01	4.51E−01
	5.00E−02	2.12E−01	3.09E−02	5.00E−02	2.08E−01	4.10E−02		5.00E−02	4.24E−01	2.34E−01
	6.00E−02	1.97E−01	2.57E−02	6.00E−02	1.88E−01	3.04E−02		6.00E−02	3.15E−01	1.40E−01
	8.00E−02	1.80E−01	2.36E−02	8.00E−02	1.66E−01	2.41E−02		8.00E−02	2.23E−01	6.90E−02
	1.00E−01	1.69E−01	2.43E−02	1.00E−01	1.54E−01	2.33E−02		1.00E−01	1.86E−01	4.59E−02
	1.50E−01	1.50E−01	2.74E−02	1.50E−01	1.36E−01	2.50E−02		1.50E−01	1.48E−01	3.18E−02
	2.00E−01	1.37E−01	2.96E−02	2.00E−01	1.23E−01	2.67E−02		2.00E−01	1.31E−01	3.00E−02
	3.00E−01	1.19E−01	3.19E−02	3.00E−01	1.07E−01	2.87E−02		3.00E−01	1.11E−01	3.03E−02
	4.00E−01	1.06E−01	3.28E−02	4.00E−01	9.55E−02	2.95E−02		4.00E−01	9.91E−02	3.07E−02
	5.00E−01	9.70E−02	3.30E−02	5.00E−01	8.71E−02	2.97E−02		5.00E−01	9.02E−02	3.07E−02
	6.00E−01	8.97E−02	3.29E−02	6.00E−01	8.06E−02	2.95E−02		6.00E−01	8.33E−02	3.05E−02
	8.00E−01	7.87E−02	3.21E−02	8.00E−01	7.07E−02	2.88E−02		8.00E−01	7.31E−02	2.97E−02
	1.00E+00	7.08E−02	3.11E−02	1.00E+00	6.36E−02	2.79E−02		1.00E+00	6.57E−02	2.88E−02
	1.25E+00	6.33E−02	2.97E−02	1.25E+00	5.69E−02	2.67E−02		1.25E+00	5.87E−02	2.75E−02
	1.50E+00	5.76E−02	2.84E−02	1.50E+00	5.18E−02	2.55E−02		1.50E+00	5.35E−02	2.62E−02
	2.00E+00	4.94E−02	2.61E−02	2.00E+00	4.45E−02	2.35E−02		2.00E+00	4.61E−02	2.42E−02
	3.00E+00	3.96E−02	2.28E−02	3.00E+00	3.58E−02	2.06E−02		3.00E+00	3.75E−02	2.15E−02
	4.00E+00	3.38E−02	2.05E−02	4.00E+00	3.08E−02	1.87E−02		4.00E+00	3.26E−02	1.98E−02
	5.00E+00	3.00E−02	1.89E−02	5.00E+00	2.75E−02	1.74E−02		5.00E+00	2.95E−02	1.86E−02
	6.00E+00	2.73E−02	1.77E−02	6.00E+00	2.52E−02	1.65E−02		6.00E+00	2.73E−02	1.79E−02
	8.00E+00	2.37E−02	1.61E−02	8.00E+00	2.23E−02	1.53E−02		8.00E+00	2.47E−02	1.70E−02
	1.00E+01	2.15E−02	1.51E−02	1.00E+01	2.05E−02	1.45E−02		1.00E+01	2.31E−02	1.64E−02
	1.50E+01	1.84E−02	1.37E−02	1.50E+01	1.81E−02	1.35E−02		1.50E+01	2.13E−02	1.59E−02
	2.00E+01	1.70E−02	1.29E−02	2.00E+01	1.71E−02	1.31E−02		2.00E+01	2.07E−02	1.57E−02

Table O-1 (*Continued.*)

	MUSCLE, SKELETAL (ICRU-44)			WATER, LIQUID	
Energy (MeV)	μ/ρ (cm2/g)	μ_{en}/ρ (cm2/g)	Energy (MeV)	μ/ρ (cm2/g)	μ_{en}/ρ (cm2/g)
1.00E−03	3.72E+03	3.71E+03	1.00E−03	4.08E+03	4.07E+03
1.04E−03	3.39E+03	3.38E+03	1.50E−03	1.38E+03	1.37E+03
1.07E−03	3.09E+03	3.09E+03	2.00E−03	6.17E+02	6.15E+02
K(Na) 1.07E−03	3.10E+03	3.09E+03	3.00E−03	1.93E+02	1.92E+02
1.50E−03	1.25E+03	1.25E+03	4.00E−03	8.28E+01	8.19E+01
2.00E−03	5.59E+02	5.57E+02	5.00E−03	4.26E+01	4.19E+01
2.15E−03	4.58E+02	4.56E+02	6.00E−03	2.46E+01	2.41E+01
K(P) 2.15E−03	4.63E+02	4.61E+02	8.00E−03	1.04E+01	9.92E+00
2.30E−03	3.78E+02	3.76E+02			
2.47E−03	3.09E+02	3.07E+02			
K(S) 2.47E−03	3.14E+02	3.12E+02			
2.64E−03	2.60E+02	2.58E+02			
2.82E−03	2.15E+02	2.13E+02			
K(Cl) 2.82E−03	2.16E+02	2.14E+02			
3.00E−03	1.81E+02	1.80E+02			
3.61E−03	1.06E+02	1.05E+02			
K(K) 3.61E−03	1.10E+02	1.08E+02			
4.00E−03	8.13E+01	7.99E+01			
5.00E−03	4.21E+01	4.12E+01			
6.00E−03	2.45E+01	2.38E+01			
8.00E−03	1.04E+01	9.89E+00			
1.00E−02	5.36E+00	4.96E+00	1.00E−02	5.33E+00	4.94E+00
1.50E−02	1.69E+00	1.40E+00	1.50E−02	1.67E+00	1.37E+00
2.00E−02	8.21E−01	5.64E−01	2.00E−02	8.10E−01	5.50E−01
3.00E−02	3.78E−01	1.61E−01	3.00E−02	3.76E−01	1.56E−01
4.00E−02	2.69E−01	7.19E−02	4.00E−02	2.68E−01	6.95E−02
5.00E−02	2.26E−01	4.35E−02	5.00E−02	2.27E−01	4.22E−02
6.00E−02	2.05E−01	3.26E−02	6.00E−02	2.06E−01	3.19E−02
8.00E−02	1.82E−01	2.62E−02	8.00E−02	1.84E−01	2.60E−02
1.00E−01	1.69E−01	2.54E−02	1.00E−01	1.71E−01	2.55E−02
1.50E−01	1.49E−01	2.75E−02	1.51E−01	1.51E−01	2.76E−02
2.00E−01	1.36E−01	2.94E−02	2.00E−01	1.37E−01	2.97E−02
3.00E−01	1.18E−01	3.16E−02	3.00E−01	1.19E−01	3.19E−02
4.00E−01	1.05E−01	3.25E−02	4.00E−01	1.06E−01	3.28E−02
5.00E−01	9.60E−02	3.27E−02	5.00E−01	9.69E−02	3.30E−02
6.00E−01	8.87E−02	3.25E−02	6.00E−01	8.96E−02	3.28E−02
8.00E−01	7.79E−02	3.18E−02	8.00E−01	7.87E−02	3.21E−02
1.00E+00	7.01E−02	3.07E−02	1.00E+00	7.07E−02	3.10E−02
1.25E+00	6.27E−02	2.94E−02	1.25E+00	6.32E−02	2.97E−02
1.50E+00	5.70E−02	2.81E−02	1.50E+00	5.75E−02	2.83E−02
2.00E+00	4.90E−02	2.58E−02	2.00E+00	4.94E−02	2.61E−02
3.00E+00	3.93E−02	2.26E−02	3.00E+00	3.97E−02	2.28E−02
4.00E+00	3.37E−02	2.05E−02	4.00E+00	3.40E−02	2.07E−02
5.00E+00	3.00E−02	1.90E−02	5.00E+00	3.03E−02	1.92E−02
6.00E+00	2.74E−02	1.79E−02	6.00E+00	2.77E−02	1.81E−02
8.00E+00	2.40E−02	1.64E−02	8.00E+00	2.43E−02	1.66E−02
1.00E+01	2.19E−02	1.55E−02	1.00E+01	2.22E−02	1.57E−02
1.50E+01	1.92E−02	1.42E−02	1.50E+01	1.94E−02	1.44E−02
2.00E+01	1.79E−02	1.36E−02	2.00E+01	1.81E−02	1.38E−02

TABLE O-2. *Electron and positron stopping power and range in water.*

ELECTRONS IN WATER, LIQUID

I=75.0 eV — DENSITY=1.000E+00 g/cm³

ENERGY	STOPPING POWER COLLISION	STOPPING POWER RADIATIVE	STOPPING POWER TOTAL	CSDA RANGE	RADIATION YIELD
MEV	MeV cm²/g	MeV cm²/g	MeV cm²/g	g/cm²	
0.0100	2.256E+01	3.898E−03	2.257E+01	2.515E−04	9.408E−05
0.0125	1.897E+01	3.927E−03	1.898E+01	3.728E−04	1.133E−04
0.0150	1.647E+01	3.944E−03	1.647E+01	5.147E−04	1.316E−04
0.0175	1.461E+01	3.955E−03	1.461E+01	6.761E−04	1.492E−04
0.0200	1.317E+01	3.963E−03	1.318E+01	8.566E−04	1.663E−04
0.0250	1.109E+01	3.974E−03	1.110E+01	1.272E−03	1.990E−04
0.0300	9.653E+00	3.984E−03	9.657E+00	1.756E−03	2.301E−04
0.0350	8.592E+00	3.994E−03	8.596E+00	2.306E−03	2.599E−04
0.0400	7.777E+00	4.005E−03	7.781E+00	2.919E−03	2.886E−04
0.0450	7.130E+00	4.018E−03	7.134E+00	3.591E−03	3.165E−04
0.0500	6.603E+00	4.031E−03	6.607E+00	4.320E−03	3.435E−04
0.0550	6.166E+00	4.046E−03	6.170E+00	5.103E−03	3.698E−04
0.0600	5.797E+00	4.062E−03	5.801E+00	5.940E−03	3.955E−04
0.0700	5.207E+00	4.098E−03	5.211E+00	7.762E−03	4.452E−04
0.0800	4.757E+00	4.138E−03	4.762E+00	9.773E−03	4.931E−04
0.0900	4.402E+00	4.181E−03	4.407E+00	1.196E−02	5.393E−04
0.1000	4.115E+00	4.228E−03	4.120E+00	1.431E−02	5.841E−04
0.1250	3.591E+00	4.355E−03	3.596E+00	2.083E−02	6.912E−04
0.1500	3.238E+00	4.494E−03	3.242E+00	2.817E−02	7.926E−04
0.1750	2.984E+00	4.643E−03	2.988E+00	3.622E−02	8.894E−04
0.2000	2.793E+00	4.801E−03	2.798E+00	4.487E−02	9.826E−04
0.2500	2.528E+00	5.141E−03	2.533E+00	6.372E−02	1.161E−03
0.3000	2.355E+00	5.514E−03	2.360E+00	8.421E−02	1.331E−03
0.3500	2.235E+00	5.913E−03	2.241E+00	1.060E−01	1.496E−03
0.4000	2.148E+00	6.339E−03	2.154E+00	1.288E−01	1.658E−03
0.4500	2.083E+00	6.787E−03	2.090E+00	1.523E−01	1.818E−03
0.5000	2.034E+00	7.257E−03	2.041E+00	1.766E−01	1.967E−03
0.5500	1.995E+00	7.747E−03	2.003E+00	2.013E−01	2.134E−03
0.6000	1.963E+00	8.254E−03	1.972E+00	2.265E−01	2.292E−03
0.7000	1.917E+00	9.312E−03	1.926E+00	2.778E−01	2.608E−03
0.8000	1.886E+00	1.043E−02	1.896E+00	3.302E−01	2.928E−03
0.9000	1.864E+00	1.159E−02	1.876E+00	3.832E−01	3.251E−03
1.0000	1.849E+00	1.280E−02	1.862E+00	4.367E−01	3.579E−03
1.2500	1.829E+00	1.600E−02	1.845E+00	5.717E−01	4.416E−03
1.5000	1.822E+00	1.942E−02	1.841E+00	7.075E−01	5.281E−03
1.7500	1.821E+00	2.303E−02	1.844E+00	8.432E−01	6.171E−03
2.0000	1.824E+00	2.678E−02	1.850E+00	9.785E−01	7.085E−03
2.5000	1.834E+00	3.468E−02	1.868E+00	1.247E+00	8.969E−03
3.0000	1.846E+00	4.299E−02	1.889E+00	1.514E+00	1.092E−02
3.5000	1.858E+00	5.164E−02	1.910E+00	1.777E+00	1.291E−02
4.0000	1.870E+00	6.058E−02	1.931E+00	2.037E+00	1.495E−02
4.5000	1.882E+00	6.976E−02	1.951E+00	2.295E+00	1.702E−02
5.0000	1.892E+00	7.917E−02	1.971E+00	2.550E+00	1.911E−02
5.5000	1.902E+00	8.876E−02	1.991E+00	2.802E+00	2.123E−02

Table O-2 (*Continued.*)

ELECTRONS IN WATER, LIQUID
I=75.0 eV
DENSITY=1.000E+00 g/cm³

ENERGY	STOPPING POWER COLLISION	STOPPING POWER RADIATIVE	STOPPING POWER TOTAL	CSDA RANGE	RADIATION YIELD
MEV	MeV cm²/g	MeV cm²/g	MeV cm²/g	g/cm²	
6.0000	1.911E+00	9.854E−02	2.010E+00	3.052E+00	2.336E−02
7.0000	1.928E+00	1.185E−01	2.047E+00	3.545E+00	2.766E−02
8.0000	1.943E+00	1.391E−01	2.082E+00	4.030E+00	3.200E−02
9.0000	1.956E+00	1.601E−01	2.116E+00	4.506E+00	3.636E−02
10.0000	1.968E+00	1.814E−01	2.149E+00	4.975E+00	4.072E−02
12.5000	1.993E+00	2.362E−01	2.230E+00	6.117E+00	5.163E−02
15.0000	2.014E+00	2.926E−01	2.306E+00	7.219E+00	6.243E−02
17.5000	2.031E+00	3.501E−01	2.381E+00	8.286E+00	7.309E−02
20.0000	2.046E+00	4.086E−01	2.454E+00	9.320E+00	8.355E−02
25.0000	2.070E+00	5.277E−01	2.598E+00	1.130E+01	1.039E−01
30.0000	2.089E+00	6.489E−01	2.738E+00	1.317E+01	1.233E−01
35.0000	2.105E+00	7.716E−01	2.876E+00	1.496E+01	1.418E−01
40.0000	2.118E+00	8.955E−01	3.013E+00	1.665E+01	1.594E−01
45.0000	2.129E+00	1.021E+00	3.150E+00	1.828E+01	1.762E−01
50.0000	2.139E+00	1.146E+00	3.286E+00	1.983E+01	1.923E−01
55.0000	2.148E+00	1.273E+00	3.421E+00	2.132E+01	2.076E−01
60.0000	2.156E+00	1.400E+00	3.556E+00	2.276E+01	2.222E−01
70.0000	2.170E+00	1.656E+00	3.827E+00	2.547E+01	2.496E−01
80.0000	2.182E+00	1.914E+00	4.096E+00	2.799E+01	2.747E−01
90.0000	2.193E+00	2.173E+00	4.366E+00	3.035E+01	2.978E−01
100.0000	2.202E+00	2.434E+00	4.636E+00	3.258E+01	

POSITRONS IN WATER, LIQUID
I=75.0 eV
DENSITY=1.000E+00 g/cm³

ENERGY	STOPPING POWER COLLISION	STOPPING POWER RADIATIVE	STOPPING POWER TOTAL	CSDA RANGE	RADIATION YIELD
MEV	MeV cm²/g	MeV cm²/g	MeV cm²/g	g/cm²	
0.0100	2.483E+01	2.204E−03	2.483E+01	2.250E−04	4.226E−05
0.0125	2.077E+01	2.339E−03	2.073E+01	3.356E−04	5.394E−05
0.0150	1.795E+01	2.450E−03	1.796E+01	4.654E−04	6.570E−05
0.0175	1.587E+01	2.544E−03	1.583E+01	6.138E−04	7.750E−05
0.0200	1.427E+01	2.626E−03	1.427E+01	7.802E−04	8.933E−05
0.0250	1.196E+01	2.761E−03	1.196E+01	1.165E−03	1.130E−04
0.0300	1.036E+01	2.873E−03	1.036E+01	1.615E−03	1.365E−04
0.0350	9.135E+00	2.971E−03	9.188E+00	2.129E−03	1.599E−04
0.0400	8.236E+00	3.060E−03	8.289E+00	2.703E−03	1.832E−04
0.0450	7.574E+00	3.139E−03	7.577E+00	3.334E−03	2.064E−04
0.0500	6.995E+00	3.209E−03	6.999E+00	4.022E−03	2.294E−04
0.0550	6.516E+00	3.270E−03	6.519E+00	4.763E−03	2.522E−04
0.0600	6.111E+00	3.327E−03	6.114E+00	5.555E−03	2.747E−04
0.0700	5.466E+00	3.428E−03	5.469E+00	7.288E−03	3.192E−04
0.0800	4.974E+00	3.521E−03	4.978E+00	9.208E−03	3.627E−04
0.0900	4.587E+00	3.607E−03	4.591E+00	1.130E−02	4.053E−04

Table O-2 (*Continued.*)

POSITRONS IN WATER, LIQUID
I=75.0 eV DENSITY=1.000E+00 g/cm^3

ENERGY	STOPPING POWER			CSDA RANGE	RADIATION YIELD
	COLLISION	RADIATIVE	TOTAL		
MEV	MeV cm^2/g	MeV cm^2/g	MeV cm^2/g	g/cm^2	
0.1000	4.274E+00	3.690E−03	4.278E+00	1.356E−02	4.472E−04
0.1250	3.704E+00	3.883E−03	3.708E+00	1.986E−02	5.489E−04
0.1500	3.302E+00	4.065E−03	3.324E+00	2.700E−02	6.467E−04
0.1750	3.044E+00	4.243E−03	3.049E+00	3.487E−02	7.412E−04
0.2000	2.838E+00	4.423E−03	2.842E+00	4.338E−02	8.328E−04
0.2500	2.551E+00	4.794E−03	2.555E+00	6.200E−02	1.010E−03
0.3000	2.363E+00	5.183E−03	2.368E+00	8.237E−02	1.180E−03
0.3500	2.233E+00	5.588E−03	2.238E+00	1.041E−01	1.346E−03
0.4000	2.138E+00	6.016E−03	2.144E+00	1.270E−01	1.509E−03
0.4500	2.068E+00	6.468E−03	2.074E+00	1.507E−01	1.671E−03
0.5000	2.014E+00	6.939E−09	2.021E+00	1.751E−01	1.831E−03
0.5500	1.971E+00	7.426E−03	1.979E+00	2.001E−01	1.991E−03
0.6000	1.937E+00	7.928E−03	1.945E+00	2.256E−01	2.152E−03
0.7000	1.886E+00	8.979E−03	1.895E+00	2.778E−01	2.474E−03
0.8000	1.851E+00	1.009E−02	1.861E+00	3.310E−01	2.799E−03
0.9000	1.827E+00	1.125E−02	1.839E+00	3.851E−01	3.129E−03
1.0000	1.810E+00	1.245E−02	1.823E+00	4.398E−01	3.463E−03
1.2500	1.786E+00	1.563E−02	1.802E+00	5.778E−01	4.319E−03
1.5000	1.777E+00	1.904E−02	1.796E+00	7.169E−01	5.204E−03
1.7500	1.775E+00	2.264E−02	1.797E+00	8.560E−01	6.117E−03
2.0000	1.776E+00	2.639E−02	1.803E+00	9.949E−01	7.054E−03
2.5000	1.785E+00	3.429E−02	1.819E+00	1.271E+00	8.989E−03
3.0000	1.796E+00	4.265E−02	1.839E+00	1.545E+00	1.099E−02
3.5000	1.808E+00	5.135E−02	1.859E+00	1.815E+00	1.305E−02
4.0000	1.819E+00	6.032E−02	1.879E+00	2.082E+00	1.515E−02
4.5000	1.830E+00	6.955E−02	1.899E+00	2.347E+00	1.729E−02
5.0000	1.840E+00	7.901E−02	1.919E+00	2.609E+00	1.945E−02
5.5000	1.849E+00	8.867E−02	1.938E+00	2.868E+00	2.163E−02
6.0000	1.858E+00	9.850E−02	1.957E+00	3.125E+00	2.383E−02
7.0000	1.875E+00	1.185E−01	1.993E+00	3.631E+00	2.827E−02
8.0000	1.889E+00	1.391E−01	2.028E+00	4.129E+00	3.274E−02
9.0000	1.902E+00	1.601E−01	2.062E+00	4.617E+00	3.722E−02
10.0000	1.914E+00	1.814E−01	2.095E+00	5.098E+00	4.171E−02
12.5000	1.939E+00	2.362E−01	2.175E+00	6.269E+00	5.290E−02
15.0000	1.959E+00	2.926E−01	2.252E+00	7.399E+00	6.397E−02
17.5000	1.976E+00	3.501E−01	2.327E+00	8.491E+00	7.487E−02
20.0000	1.991E+00	4.086E−01	2.400E+00	9.549E+00	8.557E−02
25.0000	2.015E+00	5.277E−01	2.542E+00	1.157E+00	1.063E−01
30.0000	2.034E+00	6.489E−01	2.683E+00	1.349E+01	1.261E−01
35.0000	2.049E+00	7.716E−01	2.821E+00	1.530E+01	1.449E−01
40.0000	2.062E+00	8.955E−01	2.958E+00	1.703E+01	1.628E−01
45.0000	2.074E+00	1.021E+00	3.094E+00	1.869E+01	1.799E−01
50.0000	2.084E+00	1.146E+00	3.230E+00	2.027E+01	1.962E−01
55.0000	2.093E+00	1.273E+00	3.366E+00	2.179E+01	2.117E−01
60.0000	2.101E+00	1.400E+00	3.501E+00	2.324E+01	2.265E−01
70.0000	2.115E+00	1.656E+00	3.771E+00	2.599E+01	2.541E−01
80.0000	2.127E+00	1.914E+00	4.041E+00	2.855E+01	2.794E−01
90.0000	2.137E+00	2.173E+00	4.311E+00	3.095E+01	3.028E−01
100.0000	2.147E+00	2.434E+00	4.580E+00	3.320E+00	3.243E−01

APPENDIX P

Some Fundamental Constants

The values of the fundamental constants in Table P.1 are taken largely from E. R. Cohen and B. N. Taylor (1986). *The 1986 Adjustment of the Fundamental Physical Constants*, CODATA Bulletin No. 63, Nov. 1986.

TABLE P.1. *Some fundamental constants.*

Symbol	Constant	Value	ST units
c	Velocity of light in vacuum	$2.997\,925 \times 10^{8}$	$\mathrm{m\ s^{-1}}$
e	Elementary charge	$1.602\,177 \times 10^{-19}$	C
F	Faraday constant	$9.648\,53 \times 10^{4}$	$\mathrm{C\ mol^{-1}}$
g	Standard acceleration of free fall	9.806 65	$\mathrm{m\ s^{-2}}$
h	Planck's constant	$6.626\,076 \times 10^{-34}$	J s
$\hbar$	Planck's constant (reduced)	$1.054\,573 \times 10^{-34}$	J s
		$6.582\,122 \times 10^{-16}$	eV s
k_B	Boltzmann's constant	$1.380\,658 \times 10^{-23}$	$\mathrm{J\ K^{-1}}$
		$8.617\,385 \times 10^{-5}$	$\mathrm{eV\ K^{-1}}$
m_e	Electron rest mass	$9.109\,390 \times 10^{-31}$	kg
m_ec^2	Electron rest energy	$8.187\,114 \times 10^{-14}$	J
		$5.109\,99 \times 10^{5}$	eV
m_p	Proton rest mass	$1.672\,623 \times 10^{-27}$	kg
N_A	Avogadro's number	$6.022\,137 \times 10^{23}$	$\mathrm{mol^{-1}}$
r_e	Classical electron radius	$2.817\,941 \times 10^{-15}$	m
R	Gas constant	8.314 51	$\mathrm{J\ mol^{-1}\ K^{-1}}$
u	Mass unit (^{12}C standard)	$1.660\,540 \times 10^{-27}$	kg
uc^2	Mass unit (energy units)	$9.314\,94 \times 10^{8}$	eV
ϵ_0	Electrical permittivity of space	$8.854\,19 \times 10^{-12}$	$\mathrm{C^2\ N^{-1}\ m^{-2}}$
$1/4\pi\epsilon_0$		$8.987\,55 \times 10^{9}$	$\mathrm{N\ m^2\ C^{-2}}$
σ	Stefan–Boltzmann constant	$5.670\,51 \times 10^{-8}$	$\mathrm{W\ m^{-2}\ K^{-4}}$
λ_C	Compton wavelength of electron	$2.426\,31 \times 10^{-12}$	m
μ_B	Bohr magneton	$9.274\,015 \times 10^{-24}$	$\mathrm{J\ T^{-1}}$
μ_0	Magnetic permeability of space	$4\pi \times 10^{-7}$	$\mathrm{T\ m\ A^{-1}}$
μ_N	Nuclear magneton	$5.050\,787 \times 10^{-27}$	$\mathrm{J\ T^{-1}}$

APPENDIX Q

Conversion Factors

Some of the more useful conversion factors for converting from older units to SI units are listed in Table Q.1. They are taken from *Standard for Metric Practice*, ASTM E 380-76, Copyright 1976 by the American Society for Testing and Materials, Philadelphia.

TABLE Q.1. *Useful conversion factors.*

To convert from	To	Multiply by
angstrom	meter	$1.000\,000\times10^{-10}$
atmosphere (standard)	pascal	$1.013\,250\times10^{5}$
bar	pascal	$1.000\,000\times10^{5}$
barn	meter^2	$1.000\,000\times10^{-28}$
calorie (thermochemical)	joule	4.184 000
centimeter of mercury (0 °C)	pascal	$1.333\,22\times10^{3}$
centimeter of water (4 °C)	pascal	$9.806\,38\times10^{1}$
centipoise	pascal second	$1.000\,000\times10^{-3}$
curie	becquerel	$3.700\,000\times10^{10}$
dyne	newton	$1.000\,000\times10^{-5}$
electron volt	joule	$1.602\,18\times10^{-19}$
erg	joule	$1.000\,000\times10^{-7}$
fermi (femtometer)	meter	$1.000\,000\times10^{-15}$
gauss	tesla	$1.000\,000\times10^{-4}$
liter	meter^3	$1.000\,000\times10^{-3}$
mho	siemens	1.000 000
millimeter of mercury	pascal	$1.333\,22\times10^{2}$
poise	pascal second	$1.000\,000\times10^{-1}$
roentgen	coulomb per kilogram	2.58×10^{-4}
torr	pascal	$1.333\,22\times10^{2}$

Index

N

Volumes Published in This Series:

The Physics of Cerebrovascular Diseases, Hademenos, G.J., and Massoud, T.F., 1997

Lipid Bilayers: Structure and Interactions, Katsaras, J., 1999

Physics with Illustrative Examples from Medicine and Biology: Mechanics, Second Edition, Benedek, G.B., and Villars, F.M.H., 2000

Physics with Illustrative Examples from Medicine and Biology: Statistical Physics, Second Edition, Benedek, G.B., and Villars, F.M.H., 2000

Physics with Illustrative Examples from Medicine and Biology: Electricity and Magnetism, Second Edition, Benedek, G.B., and Villars, F.M.H., 2000

Physics of Pulsatile Flow, Zamir, M., 2000

Molecular Engineering of Nanosystems, Rietman, E.A., 2001

Biological Systems Under Extreme Conditions: Structure and Function, Taniguchi, Y., et al., 2001

Intermediate Physics for Medicine and Biology, Third Edition, Hobbie, R.K., 2001

Epilepsy as a Dynamic Disease, Milton, J., and Jung, P. (Eds), 2002

Photonics of Biopolymers, Vekshin, N.L., 2002

Photocatalysis: Science and Technology, Kaneko, M., and Okura, I., 2002

E. coli in Motion, Berg, H.C., 2004

Biochips: Technology and Applications, Xing, W.-L., and Cheng, J. (Eds.), 2003

Laser-Tissue Interactions: Fundamentals and Applications, Niemz, M., 2003

Medical Applications of Nuclear Physics, Bethge, K., 2004

Biological Imaging and Sensing, Furukawa, T. (Ed.), 2004

Biomaterials and Tissue Engineering, Shi, D., 2004

Biomedical Devices and Their Applications, Shi, D., 2004

Microarray Technology and Its Applications, Müller, U.R., and Nicolau, D.V. (Eds), 2004

Emergent Computation: Emphasizing Bioinformatics, Simon, M., 2005

Molecular and Cellular Signaling, Beckerman, M., 2005

The Physics of Coronary Blood Flow, Zamir, M., 2005